Advances in 21st Century Human Settlements

This Series focuses on the entire spectrum of human settlements—from rural to urban, in different regions of the world, with questions such as: What factors cause and guide the process of change in human settlements from rural to urban in character, from hamlets and villages to towns, cities and megacities? Is this process different across time and space, how and why? Is there a future for rural life? Is it possible or not to have industrial development in rural settlements, and how? Why does 'urban shrinkage' occur? Are the rural areas urbanizing or is that urban areas are undergoing 'ruralisation' (in form of underserviced slums)? What are the challenges faced by 'mega urban regions', and how they can be/are being addressed? What drives economic dynamism in human settlements? Is the urban-based economic growth paradigm the only answer to the quest for sustainable development, or is there an urgent need to balance between economic growth on one hand and ecosystem restoration and conservation on the other—for the future sustainability of human habitats? How and what new technology is helping to achieve sustainable development in human settlements? What sort of changes in the current planning, management and governance of human settlements are needed to face the changing environment including the climate and increasing disaster risks? What is the uniqueness of the new 'socio-cultural spaces' that emerge in human settlements, and how they change over time? As rural settlements become urban, are the new 'urban spaces' resulting in the loss of rural life and 'socio-cultural spaces'? What is leading the preservation of rural 'socio-cultural spaces' within the urbanizing world, and how? What is the emerging nature of the rural-urban interface, and what factors influence it? What are the emerging perspectives that help understand the human-environment-culture complex through the study of human settlements and the related ecosystems, and how do they transform our understanding of cultural landscapes and 'waterscapes' in the 21st Century? What else is and/or likely to be new vis-à-vis human settlements—now and in the future? The Series, therefore, welcomes contributions with fresh cognitive perspectives to understand the new and emerging realities of the 21st Century human settlements. Such perspectives will include a multidisciplinary analysis, constituting of the demographic, spatio-economic, environmental, technological, and planning, management and governance lenses.

More information about this series at http://www.springer.com/series/13196

T. M. Vinod Kumar
Editor

Smart Metropolitan Regional Development

Economic and Spatial Design Strategies

Springer

Editor
T. M. Vinod Kumar
Kozhikode, Kerala
India

ISSN 2198-2546 ISSN 2198-2554 (electronic)
Advances in 21st Century Human Settlements
ISBN 978-981-10-8587-1 ISBN 978-981-10-8588-8 (eBook)
https://doi.org/10.1007/978-981-10-8588-8

Library of Congress Control Number: 2018933464

Printed on acid-free paper

This Springer imprint is published by the registered company Springer Nature Singapore Pte Ltd.
part of Springer Nature
The registered company address is: 152 Beach Road, #21-01/04 Gateway East, Singapore 189721, Singapore

Foreword I

Smart Metropolitan Regional Development—A Western Point of View

In his book, 'Thank you for being late', Thomas Friedman (2016) describes the Western world as having both an epidemic of failing communities, as well as also having a plenty of thriving ones—because of strong leaders at the local level. He suggests a redesign of our geopolitics and communities, with special attention to the place of regional development.

The popular notion that America is a nation divided between two coastal metropolitans that are supposedly thriving, pluralizing and globalizing, comparing a vast interior where jobs have disappeared, drug addiction is endemic and everyone is hoping Trump can bring back the 1950s, is mistaken. The big divide in America is not between the coasts and the interior: it is between strong communities and weak communities. One can find weak ones along the coast and thriving ones in Appalachia. Yet, the model of urbanization over the past two decades has been far from 'smart'.

Cities, communities and metropolitans are at the forefront of today's economic changes and development. In an absurd way, globalization has resurrected urban governments in the world. While many old recipes to boost urban economy had their limits by benefiting only small groups, peer-to-peer markets, collectively known as the sharing economy, have shaken these established models and disrupted transportation, accommodation and an array of urban sectors. New bottom-up proposals, based on open participation and communal evaluation, are thriving with the implicit commitment to increase equity, sustainability, resilience and diversity. This dynamic urban economic landscape grows on cities. Committed to the UN's Sustainable Development Goals (SDGs), these cities must grow into sustainable metropolitans. Following a comprehensive review of smart metropolitans, this book examines whether this kind of regional development demands a more proactive urban policy, to grasp emerging opportunities.

According to former researches, smart and strong communities share a key attribute: they have created diverse adaptive coalitions, translating—in real time—the skills demanded by the global economy. Metropolitans can leverage these qualities by tapping local academia for talent and innovations that can diversify their economies and nurture unique local assets that would not go away. Local foundations and civic groups step in to fund supplemental learning opportunities, and local governments help to catalyze it all.

Interestingly, success stories of smart metropolitans appear to be all bottom-up. Examples from Pittsburgh, USA; Johannesburg, South Africa; Gujarat, India; and Stuttgart, Germany all portray the same pattern of a whole region that unites around one (or few) project(s), which create an adaptive coalition and attract investors based on the region's strengths. These innovative coalitions promote smart grids, transportation's solutions, green industrial strategies and more to draw attention to the power and necessity of regional development to maintain local solutions and stimulate the economy of the future.

Smart city definition, according to the new urban agenda, represents a shared vision for a better and more sustainable future, and has been formally adopted by national governments at Habitat III, on October 2016. Still, a smart city that promotes an energetic face-to-face communities and a vibrant urban economy is not enough. The city is not a computer. It must listen to all its people and communities and not only to those who have access to technology. The tactic must go hand in hand with strategy. While the strategy is the tool of the governance. The tactic is the tool of the citizens and their choice. Struggling with mass transit systems, education, sanitation and industrialization, metropolitans trying to leapfrog out of these challenges and catch up by engaging in a rapid digitization of their entire economy, transitioning energy to a decarbonized, digitized and decentralized system.

This ambitious project is a collection of inspiring stories about metropolitans rising to the challenges of growing population and incomplete infrastructures. It depicts the story of regions that developed extraordinary capabilities and are coming together to help their citizens to acquire the skills and opportunities to live their own futures. In a world of limited resources, regions can get away from the 'take, make, dispose of' model and capture the full potential of new economic paradigms, business models that based on circularity, such as circular supplies, resource recovery, product life extension, shared platforms and product as a service. Metropolitans can examine how this way of doing business can turn into the new normal in the twenty-first century. Western metropolitans are already embracing the newest technologies to increase efficiencies and bring down costs. Yet, they can only reach their full potential when they become a better place to live. Growing quality of life requires a mature reflection on the evolution of urban development and how to harness technology to provide better services, empower people and build a better future for all.

The metropolitans' stories presented here need to be shared. It is not anymore, the 'smart' meaning in which smart cities connected to efficiency in sort of a blind technological determinism, but huge urban projects with wide-ranging research conclusions that should be implemented in other urban clusters worldwide.

In today's world, different regions are rising together to leverage their unique assets to provide quality of life.

Employing unorthodox solutions to social challenges, instituting innovative social programmes, technical and managerial competence, accountability, transparency and citizens' participation is regarded as a key factor to ensure sustainable prosperity. Regional Development promotes the thinking of a bottom-up and top-down approach both. From a **city perspective**, it shall focus on building up a Smart Growth, Smart Environment, Smart Infrastructure, Smart Energy and Smart People.

From a **Regional Perspective**, it shall focus on Building Smart Connectivity and Smart Governance. In smart cities, citizens are taking an active part in the conception, design, steering and management of services. In the new smart metropolitans, it is the responsibility of urban managers to ease the transition to a regional economy and to provide citizens with the opportunity for a new source of wealth and well-being.

The space entrepreneur Elon Musk has forecasted that a city on Mars with a million inhabitants could be achieved within 50 years. 'Well, this is easy, building a new city from scratch to multi-planetary species without having to struggle with rapid population growth and unplanned development and numerous infrastructure challenges…; however, this book on Smart Metropolitan Regional Development: Economic and Spatial Design Strategies' required reading for a generation that is going to be asked to 'play science fiction'. Think of it as a guide to thriving and building resilience metropolitans in this age of global hurricanes, so spaceships to Mars will turn empty…

Tel Aviv, Israel

Dr. Yifat Reuveni
Collers School of Management
Tel Aviv University

Foreword II

Smart Metropolitan Regional Development—An African Perspective

Today, about four out of ten (40.4% in 2015) or nearly half a billion of the African population people lives in cities and towns. It is projected that the African urban population will reach about one billion in 2040. For over a century, a drastic change has, indeed, taken place in the distribution of African population, with an accelerated concentration in large urban agglomerations. Development of large urban agglomerations provides opportunities for economies of scale and agglomeration, but also calls for large investments in infrastructures to respond to the increased demand for basic social, health, environment and economic services such as water, sanitation, solid management, energy, streets and public spaces, and mobility, among others. They also require efficient institutions for the management of these services, as well as, for the protection of people against violence and insecurity. The just-concluded Millennium Development Goals (MDGs) show that, despite progress made towards sustainable urban development, two-thirds of African urban dwellers are still living in slums lacking either improved water, or improved sanitation, or durable housing or sufficient living area. Most of them also do not enjoy security of tenure.

Recognizing the unbalanced metropolitan regional development in Africa, and the multiple problems associated with lack of basic services, many African governments have taken steps towards policies that transform the urban landscape as echoed in the SDG11 adopted in 2015, the New Urban Agenda adopted in 2016 and the Africa Agenda 2063 adopted in 2014. The New Urban Agenda adopted in October 2016 in Quito encourages governments to commit themselves to adopting a **smart city approach** that makes use of opportunities from digitalization, clean energy and technologies, as well as innovative transport technologies, thus providing options for inhabitants to make more environmentally friendly choices and boost sustainable economic growth, and enabling cities to improve their service delivery (NUA, paragraph 66).[1] Recognizing the crucial role ICT can play in sustainable development, African Heads of States went further and created a holistic programme 'Smart Africa' in 2013. The programme aims to advance Africa's development through ICT by 2025. Five principles are the pillars of the programme: (1) ICT at the centre of national socio-economic development agenda; (2) development of broadband; (3) use of ICT to improve accountability, efficiency and openness; (4) private sector at the centre of economic development; and (5) use of ICT to promote sustainable development. African Heads of State are committed to put in place policies and regulatory environment that will foster partnerships, entrepreneurship, job creation and knowledge sharing. They recognize that 'ICTs have the ability to level the global playing field, unlock human capital and harness its full potential'.[2]

This publication 'Smart Metropolitan Regional Development: Economic and Spatial Design Strategies' featuring five African Cities: Conakry in Guinea, Nairobi in Kenya, Abuja in Nigeria, Dakar in Senegal and Johannesburg in South Africa, comes at an opportune time marked by political will to transform Africa urban development through the integration of ICT in planning, designing and managing cities. For the African cities included in this publication, a *Smart Metropolitan Regional Development* is viewed as a sustainable, inclusive, resilient and prosperous metropolitan regional development agenda that promotes a people-centric approach based on three core components—Smart Metropolitan Region Foundation, Information and Communications Technology (ICT) and Smart Institutions and Laws. Infrastructure Development complements the basic infrastructure services under each smart metropolitan foundation and extends to actual investment and advancement of services such as transport, ICT, industrial energy, education, health, etc. Environment Sustainability comprises elements of biodiversity, climate change, waste management, energy, transport, building and pollution. Social Inclusion includes aspects of participation in decision-making, as well as, equal opportunities for growth and prosperity. Social Development encompasses elements of education, health, public space, social inclusion and social capital. Disaster Exposure incorporates elements of mitigation and adaptation to various disasters such as flooding, droughts, storms and earthquakes. City Resilience is composed of elements of city foundation, environment, social capital and social development. Peace and security include the elimination of all forms of discrimination and violence and conflicts, including domestic violence, violence in public places, crime, armed conflicts, terrorism, etc. An insecure metropolitan region limits opportunities for investment and economic growth and cannot be a smart metropolitan region.

The publication key findings are that the digital dividends are real and the Africa metropolitan regional development must take the opportunity to efficiently integrate

[1]United Nations GA, October 2016. Draft outcome document of the United Nations Conference on Housing and Sustainable Urban Development (Habitat III).

[2]https://smartafrica.org/.

the use of ICT in all sectors of its development. From the beginning of the twenty-first century, a digital citizenship, particularly the 'Millennial' generation, has started to emerge in Africa as in other regions. The rapid adoption of digital technologies in the economy will have huge benefits both directly and indirectly. However, ICT alone will produce little in the smart metropolitan regional development. 'Smart' is not an end in itself; it is the way ICT is integrated into the metropolitan regional development that will determine the metropolitan region smartness. Maximizing the digital dividends requires better integration of ICTs with the other factors of smart metropolitan regional development such as city foundation, institutions and laws, infrastructure development, economic development, environment sustainability, social development, disaster prevention, resilience, peace and security. ICT can be seen as substitution and catalyst factors, but the other factors specific to the urban context are crucial in making smart metropolitan regional development.

Firenze, Italy

Priscilla Idele, Ph.D.
Deputy Director, UNICEF Office
of Research–Innocenti
Pidele@unicef.org

Contents

Author Biographies

Prof. T. M. Vinod Kumar had 47 years of experience in Urban Planning, as teacher, researcher and adviser/consultant and worked in India, China, Bhutan, Nepal, Malaysia, Indonesia and Hawaii, USA. He was Dean of Studies, Head of the Department of Urban Planning, Head Centre for Systems Studies and Analysis, Centre for GIS and Remote Sensing, and Centre for Urban Studies of School of Planning and Architecture, New Delhi; Visiting Professor National Institute of Technology, Calicut, and Institute of Technology Bandung, Indonesia and Professional Associates, East West Resources Systems Institute Honolulu, Hawaii, Fellow Centre for the Study of Developing Societies, Delhi, Project Manager in Council for Social Development, New Delhi, Regional Program Coordinator at the International Centre for Integrated Mountain Development (ICIMOD) and Planner-Engineer at the Ford Foundation. He is the author of many books and journal articles. He edited 'Geographic Information System for Smart Cities' (Copal:2014), 'E Governance for Smart Cities' (Springer:2015), 'Smart Economy in Smart Cities' (Springer:2016) and 'E-Democracy for Smart Cities' (Springer:2017). He now edits and coordinates this international project 'Smart Metropolitan Regional Development'. E-mail: tmvinod@gmail.com.

Dr. Femi Ola Aiyegbajeje is a Lecturer at the Department of Geography, Faculty of Social Sciences, University of Lagos. He was formerly a Teaching Assistant and Academic Advisor (Geography) at the Distance Learning Centre, University of Ibadan (2011–2016). Femi obtained a Bachelor of Science (B.Sc.), Master of Science (M.Sc.) and Doctor of Philosophy (Ph.D.) degrees in Geography from the University of Ibadan, Nigeria. He has a few publications in form of peer-reviewed journal articles, book chapters, monographs and conference proceedings. He recently co-authored 'Smart Economy in Smart Cities' (2017: Springer), 'E-Democracy for Smart Cities' (2017: Springer) and 'Readings in Human Geography' (2017: University of Port-Harcourt Press). His research interest is in Human Geography with specialization in Transport and Communication Geography, Medical Geography, Urban and Regional Development. E-mail: femidavid2002@yahoo.com.

Elina Alatalo graduated as an architect in 2009 by reacting to a blocked design task with a theoretical reinterpretation resulting in a book Open Source Architecture—Johannesburg, South Africa. This led to working and experimenting around the question of how to develop cities with bottom-up open processes. She is one of the initiators and facilitators of Finnish co-working network Mushrooming (www.mushrooming.fi), which has enabled both novel economical and spatial development in the cities. Mushrooming has taught the importance of real face-to-face encounters and local informal interaction beside smart technologies. Elina is currently working on her Ph.D. as part of a larger research consortium Dwellers in Agile Cities (www.agilecities.fi), in Environmental Policy of Tampere University, funded by Academy of Finland Strategic Research Council. Consortium sees that motivations, knowledge and activities of dwellers are central in co-creating diverse and sustainable cities. Elina enjoys teaching, and she is one of the editors of Versus—a forum of happenings and web publication aiming to sustain academic societal responsibility and discussions also in Finnish. E-mail: elina.alatalo@uta.fi.

Ibitayo Modupe Arije obtained a B.Sc. in Economics from the University of Ibadan, Nigeria and M.Sc. in Environmental Management from Coventry University, United Kingdom. With research and career interests centred on Sustainable Urban Development; she currently works as a research assistant to the lead consultant to the UN-Habitat in Nigeria. Ibitayo has actively contributed to the Global State of the Urban Youth Report 2015/2016: Urban Equity and Youth Development. She presented the Nigerian component of the report at the Asian Urban Youth Assembly, Melaka, Malaysia in March 2017. E-mail: arije_tayo@yahoo.com.

Cheikhou Balde is an Urban Planner graduated from the Institute of Urbanism of Paris (France) and University of Gaston Berger of Saint Louis (Senegal). Mr. Balde also holds a Post-Master Degree from the UNESCO CHAIR on Integrated Management and Sustainable Development. He has been carrying out many missions for 16 years as Urban Planner Expert. His main focus areas of interest are urban planning, spatial planning, land issues and resettlement procedures, environment, etc. For a decade, Mr. Balde contributes actively, in the framework of projects initiated by the government of Senegal such as: Program of Reinforcement and Equipment of Local Communities (PRECOL), Program of Actions for the Safeguarding and Urban Development of "Niayes" and green areas of Dakar (PASDUNE), Land Use Reconversion Program (PROSOL). As Senior Urban Planning Consultant of the Municipal Development Agency, Mr. Balde participates in the implementation of the multi-sectorial government project "the Stormwater Management and Climate Change Adaptation Project (PROGEP)", officially launched in November 2012 with a five year term (2013–2019). The PROGEP consists of four components that are: Component A "Flood Risk Mainstreaming in the Urban Sector"; Component B "Drainage Investment and Management"; Component C "Community Engagement in Urban Flood Risk Reduction and Adaptation to Climate Change"; and Component D "Project Coordination, Management, Monitoring and Evaluation". Under the PROGEP, Mr. Balde also oversees the Senegalese ambitious initiative "Sustainable City Initiative" supported by the Global Environment Facility (GEF), the World Bank and UNIDO; it consists of integrating climate change in urban planning, design and management. E-mail: cheikhou.balde@gmail.com.

Ms. Nene Mariama Balde holds a Master degree in Urban Planning and Territorial Management from the Ball State University (USA). She is also graduated in Geography from the University of Conakry (Guinea) and in Basic Infrastructures and Environmental Management from the Institute for Housing and Urban Development Studies (Netherlands). Since 2014, Ms. Balde is the Deputy Director of the Department of Territorial and Urban Planning at the City and Territorial Management Ministry of the Republic of Guinea. Prior to that, from 2006 to 2014, she was Sr. Consultant for the preparation of the National Housing Policy Vision 2021 of Guinea. From 2004 to 2006, she was the Deputy Director at the City Observatory. From 2002 to 2004, Ms. Balde was the National Director of Office for Studies and Strategies of the Ministry of Urban Development and Housing. She has been also consultant of several international organizations including the UNDP and UNFPA for the following projects: "Evaluation of the integration of climate change resilience into spatial planning instruments" (2013–2014); Support at the formulation of the National Housing Policy of Guinea (2008–2012); "Evaluation of National Population Policy—Component Urban Management"; and "Evaluation of the Goals of the Millennium Development Goals—Component Urbanization and Sanitation" (2004–2006). E-mail: baldette@yahoo.com.

Dr. Prabh Bedi is a Geographer, Urban and Regional Planner and Geospatial Technology expert. She completed her Post Baccalaureate Certificate in Geographic Information Systems from Pennsylvania State University, USA. She graduated from School of Planning and Architecture, New Delhi (India) with Master of Planning. Having over 20 years of experience in planning and geospatial technologies, she is the Founding Director of Nipun Planners and Infrastructure Pvt. Ltd., India. She serves as advisor on environment and micro-planning projects with Allianz Engineering Technology Corp, Cali Colombia. She has been visiting faculty since 2002 at School of Planning and Architecture, New Delhi and other architecture and planning institutes around Delhi. She has published articles on NUIS, Smart Cities, urban greens and hydrology in the context of sustainability. Her research interests include information systems, geospatial technologies, NUIS, smart cities, demography, hydrology and climate change. E-mail: prabhb@gmail.com.

Bhasker Vijaykumar Bhatt is an Engineer Town Planner and presently serving as a PG in charge (ME TCP) and Assistant Professor in Civil Engineering Department of Sarvajanik College of Engineering and Technology—SCET, Surat. He completed his masters in town planning with a research on Road Safety Audit at SVNIT, Surat. He possesses excellent academic records and awarded with several Gold Medals during Diploma, Degree and Postgraduate studies. He provided his services in the field of consultancy at national as well as international level. Research areas of him include Road Safety, Urban Planning, Infrastructure Planning, Statistical Analysis Techniques, GIS and Climate Change effects in urban areas. He at present is a research scholar and exploring climate change effects intervening urban planning and development. He is actively associated with several professional organizations and serving various prestigious posts for better interaction among educational, engineering and planning fraternity. He has more than 6 years' experience in technical consultancy and about 6 years of engineering teaching. He has guided about 21 PG Dissertations and 10+ UG projects as well as a few AMIE projects. Towards continuous upgradations of skills and technological advancements, he has attempted 50+ events consisting of workshops, training programmes, seminars, conventions and symposiums. He frequently organizes events for students and faculty members that ignite active academic upgradations and industrial interaction. He has published more than 45 technical research papers in conferences and journals. *All his works are available online for reference along with projects and dissertation reports.* Professor Bhasker is a recipient of '*Best Section Management Committee Member of ISTE Gujarat Section 2015*' with efforts going on. Along with ISTE Gujarat section, he is Hon. Treasurer for the Institute of Urban Transport (India) SVNIT Regional Chapter as well as serving as a Hon. Joint Secretary of the Institution of Engineers (India) South Gujarat Local Centre, a Managing Committee Member for the Institute of Civil Engineers & Architects (ICEA), Surat. He is chairing a number of committees at his institute such as GIC Committee Convener, SCETmate newsletter committee convener, Design Engineering Coordinator and PMMS Coordinator, to name a few. Also, he was honoured by '*Pedagogical Innovation Award 2015*' by the GTU for his efforts and contributions in engineering education. At the GTU, he was a '*Master Trainer for Faculty Development Programs*' conducted at the University wherein he trained 350+ teachers of diverse engineering discipline. 1500+ students have attended his sessions on Design Thinking approach that ignites the engineering insights and motivates the engineering brain to increase capabilities, drastically. He was also awarded as '*Best Club Secretary 2012–13 (Large clubs)*' for serving Rotary Club of Surat. He an active Member of the Rotary International since 2005. Recently, he has been active with Startup Weekend Surat (powered by Google Inc.) as a trainer. He has developed his own learning portal, www.bvbhatt.com. E-mail: er.bhasker@gmail.com.

Prof. Jignesh G. Bhatt received B.E. (Instrumentation and Control Engineering), Gujarat University, India, 1997 and M.Tech. (Electrical Engineering, Specialization: Measurement & Instrumentation) from IIT Roorkee, India, 2010. He acquired industrial experience in Managerial Positions during 1997–1999. Post industrial experience in managerial positions during 1997–1999, since 2000, he has been currently serving Dharmsinh Desai University (DDU), India as Sr. Assistant Professor (Instrumentation and Control Engineering Department). He served as Coordinator and Principal Investigator for Govt. Sponsored Research and Community Development Projects, PG Thesis Supervisor, Invited Expert-Session Chair in Conferences and Journal Reviewer-Editorial Board Member. His research interests include Automation, e-Governance, e-Learning, Instrumentation, Smart City, Smart Grid, Solar City, Solar Energy and Wireless Sensor Networks. E-mail: jigneshgbhatt@gmail.com.

Dr. Sweta Byahut is a teacher in the Community Planning Program at Auburn University. She earned a Ph.D. in Regional Development Planning from the University of Cincinnati. Her research focuses on improving urban land management, spatial planning practices and development regulation in Indian cities. In the US, she works on issues of urban sustainability, examining the influence of land use characteristics on travel behaviour, and the impact of environmental and open space amenities. Before joining academia, Dr. Byahut has a decade-long experience as an urban planner in India in consulting and applied research, spanning areas such as planning legislation and policy, development regulation, comprehensive/urban development plans, post disaster reconstruction planning and urban revitalization. E-mail: sweta.byahut@auburn.edu.

M. Ibrahima Camara is a Civil Engineer in Bridge Construction, Urban Planning and Urban Studies from the Polytechnic Institute of the University of Conakry in 1996. He also holds a post-graduate certificate on Integrated Urban Planning from Sweden and India where he focuses his research project on "Restructuring Informal Settlements in Guinea". Since 2011, M. Camara has been the National Director of Territorial and Urban Planning, and Roads and Infrastructures of the City and Territorial Management Ministry of the Republic of Guinea. From 2011 to 2014, he also was the Coordinator of the Project "Autonomous Sanitation of selected neighbourhoods in Conakry". Since 2011, he has been the Coordinator of the Project "Waste water Management of Moussoudougou" with financial support of the Government of Guinea and the Arab Bank for the Economic Development in Africa. Prior to that, from 2005 to 2011, M. Camara was the Chief of Section "Urban Restructuring" at General Mission for Urban Development (MGDU (**DATU**)/MUH). M. Camara has also been Sr. Consultant with several international firms: 2007–2011, Chief of Mission for the Monitoring of the civil engineering component of the Extension of the Network of the Telecommunication Company INTERCEL; 2010, Mission in Sierra Leone for the diagnostic and evaluation of the Freetown urban road network; 2006–2007, Chief of Mission for the Monitoring of the sanitation and road network project in Conakry implemented by the companies EGCEL, ECOTEL and SERICOM. E-mail: ibrahimacamara@gmail.com, ibrahimacamara@yahoo.fr.

Antonio Caperna Head of International Society of Biourbanism (ISB), Graduated from the Faculty of Architecture in Naples, Antonio has been awarded a Ph.D. in *Sustainability and Urban Environment* from Roma Tre University. He had previously taught '*Architectural Technology*' and '*Land and City Analysis*' at Faculty of Architecture, University of Rome '*La Sapienza*' and *Sustainable Urban Design* at Master course in *Sustainable Interactive Design and Multimedia* at Roma Tre University. His latest research is mainly oriented to the emergent field of Biourbanism and to the application of complexity theory, evolutionary biology, Biophilia and Morphogenesis to define procedure/process and tools for a human-oriented architecture and urbanism. E-mail: antonio.caperna@biourbanism.org.

Maria Cerreta graduated in Architecture, University of Naples Federico II, Ph.D. in 'Evaluation Methods for the Integrated Conservation of Architecture, Urban and Environmental Heritage'. She is Director of the Advanced Course in 'Real Estate Market and Urban Regeneration', Coordinator of the Second Level Master in 'Planning and Sustainable Design of the Port Areas', and member of the Ph.D. Program in Architecture, DiARC, University of Naples 'Federico II'. She has been senior researcher and coordinator of research projects and is actually involved in EU project Horizon 2020 'REsource Management in Peri-urban Areas: Going Beyond Urban Metabolism' (REPAiR). She is also the author of several papers and editor of a book with Springer. Her research work is mainly oriented to hybrid assessment of sustainability issues, Multi-Criteria Analysis and Multi-Group Analysis in collaborative and interactive decision-making processes for urban design and spatial planning. E-mail: maria.cerreta@unina.it.

Dr. Amit Chatterjee an Urban and Regional Planner wrote his doctoral dissertation on urban sustainability of Greater Mumbai and its satellite towns. Dr. Amit has a combined experience of more than a decade in teaching, research and industry and presently on the faculty of Planning, School of Planning and Architecture (SPA), Bhopal as Assistant Professor. Before joining academics, Dr. Chatterjee served the industry in various capacities. As principal investigator, Dr. Chatterjee was involved in different international collaborative research projects, namely, Co-benefits of Waste Management, Collaborating for Climate, UK-India Capacity Development project for Climate Change and Carbon Management, etc. Dr. Amit is currently involved in two of the research and consultancy projects: 'Shelter for All' and Seven Cluster Development Plans under National Rurban Mission. In academics, his research interest is primarily focused on future cities, urban sustainability models and climate change. Besides attending seminars at national and international level, Dr. Chatterjee contributes papers, book chapters to reputed international journals in planning. Dr. Chatterjee edited 11th Issue of International Journal of SPA, Bhopal 'SPANDREL' (Masking cities Smart and Competitive) and currently working on a book focused on Metropolitan Planning and Development. E-mail: amit.chatterjee@spabhopal.ac.in.

Dr. Krupesh A. Chauhan is serving as an Associate Professor and Section Head in Urban Planning Section of CED, SVNIT, Surat. He got his Master's in Water Resources Engineering and Town and Country Planning from South Gujarat University. He got his Ph.D. in Civil Engineering from SGU. He has field experience of 13+ years and teaching experience of 19+ years. His areas of interest area Traffic and Transportation planning, Housing, Infrastructure planning and management, etc. His special interests go along fusion of modern and ancient house planning techniques that include Vastu Shastra. He has published more than 20 International Journal papers and more than 40 conferences attended. He has guided several PG dissertation researches in urban planning. He also has co-authored book chapters and publications. He has guided two Ph.D. scholars and guiding six academic research scholars pursuing doctoral studies. As a guide, he is serving universities of SVNIT and GTU. He is a member of many professional organizations and has been serving on various posts. He has been involved in prestigious assignments by the State and Central Government such as PMGSY, Smart Village Development, Housing for all and Unnat Bharat Abhiyan, to name a few. E-mail: kac3srb@gmail.com.

Prof. Binayak Choudhary holds a Master's in Economics with specialization in Mathematical Economics and Econometrics. He graduated in Regional Planning from Indian Institute of Technology, Kharagpur and wrote his doctoral dissertation on Regional Economics. Currently, Binayak is Professor of Planning and Dean (Research and Development) at School of Planning and Architecture, Bhopal (SPAB). His research areas include urban and regional economics, urban governance and finance and quantitative techniques in planning. Professor Binayak is on different committees across various planning schools of India. He supervises/evaluates dissertation at undergraduate, graduate and doctoral level both at SPAB and other national planning schools. He contributes papers to international/national journals and conferences and is on the editorial board/ reviewer of national and international journals. He carries three-decade long experience in urban planning, development and management in various capacities across various institutions in different parts of India. E-mail: binayak@spabhopal.ac.in.

Dr. Ing. Manfred Dangelmaier studied mechanical engineering at the University of Stuttgart. After graduating as an engineer, he joined the Fraunhofer IAO in 1985 for research and development projects in the field of ergonomics. He received his doctorate in 2001 with the topic: "A method for the ergonomic evaluation of the driver place in passenger cars". From 2000 to 2001, he set up the Vehicle Interaction Lab at IAO to study human-vehicle interaction using driving simulation. From 2001 he headed the Virtual Environments Competence Center and coordinated national and international collaborative research projects focusing on the development of application areas for virtual environments in the engineering of products, production systems, buildings and services. Since 2009 he has been Director of the Engineering Systems business unit with the Virtual Environments, Visual Technologies, Digital Engineering and Human Factors Engineering teams. Manfred Dangelmaier publishes in the fields of ergonomics, human-machine interaction, digital product development and virtual and extended reality. He teaches Virtual Engineering at the University of Stuttgart and Virtual Product Design at the Technical University of Cluj-Napoca. www.iao.fraunhofer.de. E-mail: manfred.dangelmaier@iao.fraunhofer.de.

Dr. Mohammed Firoz is an architect and urban planner by profession. He holds a Ph.D. degree from Indian Institute of Technology Kharagpur, Post-graduation in Urban and Regional Planning from CEPT University and B.Arch. degree from NIT Calicut for which he was as a university rank holder. He has been involved in teaching, researching and consulting at NIT Calicut since July 2004. He was also engaged as a visiting teacher at the Architectural Association London (AA London) for the term May–June 2015. His field of interest includes rural-urban interface studies, sustainable design and planning, regional development and planning, etc. E-mail: firoz@nitc.ac.in.

Khalil Fofana is a partner with GORA Corp and currently working on the Conakry Smart City Project. He was Senior National Coordinator since January 2014 at Loblaw/Pharmaprix in the Design & Construction department in Toronto. Canada. Khalil is an accomplished senior executive who has accumulated multiple strategic, logistical and customer service achievements at the Toronto Operations Control Center over the course of more than 10 years at Bell Canada, Bell Satellite TV, IPTV and Fiber Optic TV and high-speed Internet in Canada. He began his professional career in the finance industry where he gained a solid experience in this field. First as a Consultant in Individual and Corporate Wealth Management at Investors Group in Montreal in 2000, then as Financial Analyst with the launch of the Tontine of the Astral Finance Mutual (AXA) in Strasbourg (France) and as Senior Analyst Personal Credit and Businesses at Citi Bank Masters Cards Canada in Toronto. He has always evolved as a team through the execution of tasks and initiatives to achieve business objectives, streamline processes, reduce costs in order to improve customer satisfaction Customers and partners. He holds a Bachelor's degree in Business Administration from the Québec university in Montréal (UQAM), a Prince 2 Project Management Certificate and an active member of the Project Management Institute (PMI). E-mail: khalilfof@gmail.com.

Dr. Leena Garg is an architect planner and a teacher. She completed her Graduation in Architecture from M.S.U. Baroda (1990) with First class, Post-Graduation (Master in Town and Regional Planning) batch topper from S.V.N.I.T. Surat (2004), and Ph.D. from S.V.N.I.T. Surat in Town Planning (2017). During her scholar studies, she explored housing settlement patterns in the expanding fringe area of a fast-growing metropolitan. She is having more than 26 years of industry and teaching experience. She is currently working as Principal at Mahavir Swami Institute of Architecture (BMEF Surat), besides actively working as volunteer in various social service activities and NGOs. She has written over 22 papers for various national and state level conference proceedings conducted all over India, mainly focused on town planning, transportation planning, gender issues related to urban governance, architecture and environmental issues. She has explored statistical techniques as well as spatial modelling tools for carrying out interventions based on researches in past. Around four publications in International Journals, notable being 'Development of Accessibility Index for Newly Expanded City Areas in Fuzzy Framework: A Case Study'—a paper presented and published in International Journal in the international conference (ICTLE 2012) held during 28–29 April 2012 at Chennai, India (pp. 173–176, Vol. 4, No. 2, April 2012, ISSN 1793-8236); and 'Temporal Residential Settlement Modeling for Recently Expanded Zones: A Case Study In Gujarat, India' (ITES Conference–2012) International Conference on Innovative Technologies in Engineering and Sciences, held during 22–23 December 2012 published in the Journal: IJSC, International Journal of Scientific Computing (pp. 75–79, Vol. 6, No. 2, July–December 2012, ISSN: 0973-578X) held at Rajkot, India. E-mail: leenagarg.1990@gmail.com.

Daniel Githira M.Sc. is an urban planner with interest in geographic information management, urban growth modelling and research. He holds a Master of Science degree in Geo-information Science and Earth observation from the University of Twente (ITC, Netherlands) and a Bachelor of Arts degree in Urban and Regional Planning from the University of Nairobi. For over 8 years, he has worked as a consultant spatial planner for the national government, private organizations, non-governmental organizations, particularly on refugee camp settlement planning. Currently, he is a consultant GIS trainer, a researcher, and a planning consultant working for the County Government of Taita Taveta in Kenya. He has membership in the two main professional bodies regulating spatial planning in Kenya: the Kenya Institute of Planners (KIP) and the Architectural Association of Kenya (AAK). E-mail: dangithira@gmail.com.

Dr. Sudeshna Ghosh is Assistant Professor of Regional Planning, in the Department of Geography and Regional Planning at Indiana University of Pennsylvania, Pennsylvania, USA. Dr. Ghosh is a scholar in the areas of urban and regional planning, community development planning, land use planning and planning in the developing world. She teaches planning history, global cities, planning methods and seminar courses. She also organizes study abroad courses in India, intended to explore urbanization issues in contemporary Indian cities. She got her training in architecture and urban planning in Kolkata and Kharagpur (India), and completed her doctoral studies from University of Cincinnati prior to joining IUP. She received several small grants for her research work in the areas of small town economic development in the US and issues of urban poor and slum settlements in the developing world. E-mail: Sudeshna.Ghosh@iup.edu.

Dr. Sujata S. Govada is an award-winning qualified urban designer, certified town planner and a registered architect in India, with over 35 years of diverse international experience in Hong Kong, PRC, India, Philippines and the United States working on design and planning projects. She is the Founding Director of the Institute of Sustainable Urbanisation and the Founding and Managing Director of UDP International, a boutique global practice. Her expertise and research interests include sustainable urbanization, smart city development, transit and people-oriented development, harbourfront planning, new town development, urban renewal, heritage conservation, affordable housing and community engagement. She is an Adjunct Associate Professor at the School of Architecture, Chinese University of Hong Kong and previously taught in the urban planning and urban design programmes at The University of Hong Kong. She is a Global Trustee of the Urban Land Institute (ULI) and the Founding Vice President of the Hong Kong Institute of Urban Design and Past President of the American Institute of Architects Hong Kong Chapter. E-mail: sujata@udpcltd.com.

Premjeet Das Gupta is an Architect and Urban Planner teaching as Assistant Professor in School of Planning & Architecture, Bhopal. He has around 12 years of combined professional and academic experience. He has been a part of the DFID-funded KUSP programme in West Bengal from 2005 to 2007. Between 2007 and 2009, he worked in a consulting role with the firms within the ICICI Bank and IL&FS groups. He was an Assistant Planner with Kolkata Metropolitan Development Authority from 2009 to 2013 during which he was involved with transportation and PPP projects. At present, he is involved in research on Non-Motorized Transport in urban peripheries. E-mail: premjeet@spabhopal.ac.in.

Suvankar Halder has completed his Master's degree in Remote Sensing and Geographic Information System (RS & GIS) from Indian Institute of Remote Sensing (IIRS) with specialization in Urban and Regional Studies and Bachelor's degree in Urban Planning (B.Planning) from School of Planning and Architecture, Vijayawada. Suvankar is currently working on Air Pollution Modelling to estimate the ambient concentration of air pollutants. The project focuses on estimating transport emissions using modelling approach and remote sensing techniques. His research interests lie in the field of Urban and Regional Studies, exploring human–environment interaction and its contribution to climate change, air pollution modelling, urban transportation planning and network analysis. E-mail: haldersuvankar02@gmail.com.

Michael Hertwig is working as a researcher at University of Stuttgart. His major research fields are development of suitable IT support of production processes and the procedure development for manufacturing in urban environments. Additional, he is doing research in the field of 'urban manufacturing' and 'holistic development of manufacturing in context of urban areas'. He is planning to do his Ph.D. in this research field. In 2008, he graduated in production technology at the University of Combined Studies, Thuringia. In 2013, he was awarded a Diploma in Mechanical Engineering at the University of Stuttgart. He is a member of the association of German engineers (VDI). Based on his expertise, he was invited to be on the committee of the initiative 'VDI STADT: DENKEN'. He is contributing to different conferences presenting the current state of research in the field of urban manufacturing. E-mail: michael.hertwig@iat.uni-stuttgart.de.

Dr. Priscilla Idele has more 25 years of research in the field of population, health and development in the Africa region and around the globe. After more than 12 years with UNICEF Headquarters (New York) in various senior positions, Dr. Idele was appointed in 2017 as the Deputy Director of the UNICEF Office of Research-Innocenti, Florence, Italy. Her responsibilities include the following: research, planning and monitoring; oversight of strategic partnerships, donor and external relations; resource mobilization; coordination and facilitation of selected research activities; and dissemination of research results. Prior to her current position, Dr. Idele had served UNICEF in various capacities as Chief, Data Analysis Unit, Data and Analytics Section; Chief, Planning and Monitoring, UNICEF Zimbabwe Country Office; and Senior Advisor for Statistics & Monitoring, Data & Analytics Section. Prior to joining UNICEF, she was a Senior M&E Expert with the USAID funded MEASURE Evaluation Project from 2000 to 2005. In 2003, she was a consultant with the UN-Habitat for the development of slum indicators use to monitor the Millennium Development Goals Slum Target as reported to the United Nations MDG report 2004. Dr. Idele has authored several books and articles in peer-reviewed journals on health, HIV/AIDS, population dynamics and other social and urban issues. She won various awards in the field of research on health and development. She is also a co-author in the forthcoming book **'Smart Economy in Smart African Cities'**. Dr. Priscilla Idele holds a Ph.D. in Social Statistics from Southampton University (United Kingdom), Master's degree in Population Studies from Nairobi University (Kenya), and B.Ed. in Business Education from Kenyatta University (Kenya). E-mail: Pidele@unicef.org.

Dr. Omkar K. Jani is currently the Principal Research Scientist (Solar Energy) at Gujarat Energy Research and Management Institute (GERMI). His primary goals include advisory for various solar-related government and private initiatives, developing and undertaking fundamental and applied research, and capacity building. Dr. Jani is a Member of the State Advisory Committee of the Gujarat Electricity Regulatory Commission. He is also an advisor to several clean-tech companies and NGOs. Dr. Jani received his Ph.D. in Electrical Engineering from Georgia Institute of Technology with specialization in Solar Photovoltaic Science and Engineering, and his post-doctoral fellowship from the Institute of Energy Conversion, Delaware. E-mail: omkar.j@germi.res.in.

Col K. K. Kakkar was born on 13 June 1965 in Jodhpur, Rajasthan. After completing his post-graduation in Mathematics from Punjab University Chandigarh, he joined the Indian Military Academy, Dehradun, and was commissioned into the Regiment of Artillery in December 1989. The officer received his second PG degree in M.Sc. Tech (Weapons Systems) in 2000 from Pune University. His thirst for newer areas of technology earned him his third Master's in Science (Information Technology) from Punjab Technical University, Jalandhar (Punjab) in March 2005. In 2011, he took 2 years study leave to complete M. Tech in Remote Sensing and GIS from IIRS Dehradun with specialization in Urban and Regional studies, where his thesis on Greater Chandigarh Region was recognized as a benchmark in geospatial technologies as applicable in the field of Urban and Regional Planning. He continues to be a strong votary of applying Geospatial technology for problem-solving even after his retirement in 2017 and to that effect, is presently pursuing a Ph.D. in Remote Sensing engineering from PEC (deemed to be University), Chandigarh. E-mail: kewalkakkar@gmail.com.

Prof. Ashok Kumar, Ph.D. is working as a Professor in the School of Planning and Architecture, New Delhi since 1993. His research interests include Inclusive City Planning, Spatial Justice and Planning Epistemologies. He has published nearly 100 papers in national and international refereed journals, written and edited eight books, and published 18 book chapters. He is the Editor of the *ITPI Journal* since 2002. A co-authored chapter has been published in the prestigious *The Routledge Handbook of Planning Theory* in 2017. Currently, Prof. Kumar is working on a co-authored book manuscript on 'City Planning in India' under contract with Routledge, which is expected to be published in mid-2018. E-mail: a.kumar@spa.ac.in.

Pramod Kumar is the Head, of the Urban and Regional Studies Department, at the Indian Institute of Remote Sensing, Dehradun, India. He is also the Group Head of Programme Planning and Evaluation Group of IIRS. He is an alumnus of IIT, Kharagpur, India and joined Indian Space Research Organization in 1991. Earlier, he has worked as Assistant Engineer at CES, New Delhi. He has been involved in more than 50 national level/technology demonstration and research projects using geospatial data and techniques to evolve solutions for natural resources management and brought out technical reports and research publications. He has published more than 40 papers in journals and conference proceedings and many technical reports. He is the recipient of ISRO Team Excellence Awards for two projects. At present, he has research interests in urban water utilities and urban hydrology. E-mail: pramod@iirs.gov.in.

Joachim Lentes is Head of Department and Member of the Management Board at the Fraunhofer Institute for Industrial Engineering IAO and Lecturer for Mathematical Methods for Production Planning at the University of Stuttgart. He acts as reviewer for international conferences and journals like International Journal for Production Research and International Journal for Production Economics as well as for governments. His main research areas are the integrated engineering of products and production systems as well as strategies for next-generation development and production as they can be summarized as Industry 4.0 and urban manufacturing, a field where he is active since 2011. E-mail: joachim.lentes@iao.fraunhofer.de.

Dr. Calvin Masilela is Professor of Geography and Regional Planning at Indiana University of Pennsylvania, USA. He also serves as Program Director, Bachelor of Science in Regional Planning, and Program Director of the Upward Bound Math and Science Program. His scholarly works on planning education, land reform, land use, development and urbanization issues have appeared in *African Geographical Review*, *Habitat International*, *Journal of Geography*, *Planning and Environment B*, *Small Town*, *The Third World Planning Review*, *The Pennsylvania Geographer* and *Regional Development Dialogue*. His most recent work, a book chapter on urban agriculture, has been accepted for publication by The Wiley Blackwell Encyclopedia of Urban and Regional Studies. E-mail: Calvin.Masilela@iup.edu.

Dr. Gora Mboup is the President and CEO of GORA Corp after more than 25 years of international development experience as a coordinator of the Demographic and Health Survey Programme of USAID for 10 years, and as a Senior Demographer and the Chief of the Global Urban Observatory, United Nations Human Settlements Programme (UN-Habitat) for another 10 years. In June 2014, UN-Habitat honoured Gora with a plaque "in recognition of ten years of distinguished and dedicated service to the United Nations". GORA Corp has signed several MoUs with international partners to advance sustainable, inclusive, resilient, and prosperous smart cities: the Arab Urban Development Institute (AUDI) in 2017, UN-Habitat in 2016, the Ministry of Urban Development of India in 2015, etc. Gora co-authored the book "Smart Economy in Smart Cities" published in September 2016 by Springer. He is presently coordinating the book project "Smart Economy in Smart African Cities" covering 30–50 African cities to be published by Springer in the first quarter of the year 2018. During his tenure with UN-Habitat, Gora authored a landmark urban planning and mobility book "Street as Public Spaces and Drivers of Urban Prosperity" in 2013 and four series of the State of World Cities: (1) SWCR 2012/13—Prosperity of Cities; (2) SWCR 2010/11—Bridging the Urban Divide; (3) SWCR 2008/09—Harmonious Cities; (4) SWCR 2006/30—Thirty of the Habitat Agenda. Since 2015, Gora collaborated with UN-Habitat in the monitoring of the Sustainable Development Goals (SDGs) 11 as well as in the Monitoring of the Security of Tenure. Since December 2016, Gora has also been an International Consultant of the Africa Transport Policy Program (SSATP) with the World Bank. He was also external member of the Technical Advisory Group of the Global Environment Facility (GEF)'s Sustainable Cities Programme (SCP). He was also panellist at the Inter-Ministerial panel at the African Union First meeting on Financing Infrastructure Programmes in Africa held in Togo in March 2017. Since September 2017, Gora is also assisting the United Nations Economic Commission for Africa (UNECA), UN-Habitat and the African Union on the Harmonized Monitoring Framework and the Regional Action Framework for the New Urban Agenda. Gora has also been an International consultant of the Senegal Sustainable Cities Initiative for the World Bank and the Government of Senegal. Gora is a Senior Fellow of the Global City Institute (GCI) of the University of Toronto and provides public lectures in several universities such as London School of Economics (UK), Columbia University, University (USA) of Twente (ITC, The Netherlands) and Royal Institute of Technology (Sweden). Gora holds a Ph.D. in Demography, a Master's degree in Demography, a Bachelor's Degree in Statistics, a Bachelor's Degree in Economics and a first degree in Mathematics and Physics. E-mail: gmboup@gora4people.org.

Dr. Paida Mhangara is a Remote Sensing and GIS professional by training and is currently employed as a Manager for Research and Applications Development at the South African National Space Agency (SANSA) in the Earth Observation Directorate. Paida has extensive experience and expertise in remote sensing, geographical information systems and environmental geography. He has been a principal scientist and project manager on a number of international earth observation science projects such as the FP7 projects, AMESD, TIGER, TIGER-Net, Garnet-e, EOPOWER and Geonetcab. Dr. Mhangara is currently a mission scientist responsible for co-leading the scientific process of defining technical specifications for a new South African earth observation satellite code-named EO-SAT1 which will be South Africa's contribution to the African Resource Management Constellation. Paida is a member of many international scientific committees in earth observation such as the PROBA-Vegetation satellite International Scientific User Committee, Landsat Ground Operators Technical Working Group, SANSA Earth Observation Mission Advisory Committee, and has participated in the Committee of Earth Observation Satellites Working Group on Capacity Building and Data Democracy and the Global Human Settlements Layer Project. In May 2014 was nominated as the representative for Africa to the Global Remote Sensing Water Initiative scientific review committee of the World Bank, Water Partnership Program (WWP) to review the Draft Remote Sensing for Water Resources Management Scoping Report. Within SANSA Paida is responsible for leading a team of scientists and plays a leading role in initiating and managing all earth observation research and application development projects. E-mail: pmangara@sansa.org.za.

Guglielmo Minervino is a Ph.D. student working on generative programmes for urban regeneration in historical settlements, a research part of the International Doctorate Program in Urban Regeneration & Economic Development (IDP URED)—Università Mediterranea of Reggio Calabria —Italy. He is also Member of the International Society of Biourbanism. He studied in Italy, UK and USA. Interest and experience in strategies and processes for a systemic urban development, participatory processes, community-based urban regeneration, Mediterranean urbanism, placemaking, self-building and social innovation. E-mail: guglielmo.minervino@unirc.it.

Naledzani Mudau is a Remote Sensing Scientist at the South African National Space Agency. She graduated at the University of Paris VI where she received her master's degree in Remote Sensing. Naledzani developed her interest in mapping and monitoring of human settlements during the master's scholarship application process where she was required to present a case study on the application of remote sensing. After her M.Sc. graduation she landed her first job at the private company where she worked as a Remote Sensing Technician responsible for satellite image processing and GIS data capturing. She then moved to a Council for Scientific and Industrial Research where she advanced her image processing and application development skills. Passionate about human settlements, Naledzani also worked at Eskom, as a Remote Sensing and GIS advisor where she introduced and managed development of remote sensing applications to support planning and maintenance of electricity infrastructure. She has since been focusing on the use of Earth observation technologies to support planning and monitoring of human settlement and infrastructure development. She is also focusing on the use of Earth observation data in the planning and monitoring of UN Sustainable Development Goals, African Agenda 2063 as well as National Development Plan. She currently co-chairs AfriGEOSS Sustainable Urban Development and South Africa-GEO Human Settlement working groups which promote the use of Earth observation to support sustainable development. E-mail: nmudau@sansa.org.za.

Dennis Mwaniki is a consulting urban and environmental planner and spatial data expert based in Nairobi. He is currently supporting methodological developments and data generation for monitoring of Sustainable Development Goal 11 "Sustainable Cities and Human Settlements". He has previously worked as a GIS and research consultant at the Global Urban Observatory (GUO) and the Best Practices units of UN-Habitat. In GUO, he developed and implemented the methodology for calculating the amount of land allocated to streets forover 100 cities. He contributed to the State of the World's Cities report: Prosperity of Cities in 2012–13; and to the publication "streets as public spaces and drivers of urban prosperity". He has also worked as a GIS and urban planning consultant with the Slum Dwellers International Kenya; the Center for Urban Research and Innovations at the University of Nairobi; and various private companies based in Nairobi. He holds a Masters in Environmental Planning and Management and a Bachelors degree in Urban and Regional Planning. His interests are in urban research, spatial modelling, smart city systems and sustainable development. Mr. Mwaniki was a co-author in the publications "Smart Economy in Smart Cities" and "E-Democracy for Smart Cities". E-mail: denmwa02@gmail.com, dangithira@gmail.com.

Mandiaye Ndiaye is a Sr. Urban Planner and Geographer (Geo-Spatial Analyst) and an Engineer in Territorial Management, Decentralization and Development. M. Ndiaye received an M.B.A. from the University of Paris Dauphine, a Postgraduate Degree in Decentralized Territorial Management and Territorial Development from the National School of Applied Economy of Senegal. He also holds a Postgraduate Degree in Geomatics (Development and Management of Geographical Information) from the University of Ile Ife (Nigeria), and Geography Master's degree from the University of Dakar (Senegal). He has also received certificates for professional training on 'Urban Planning and Development' from the National Centre of small towns (Beijing, China), and 'Urban Infrastructure Planning and Management' from the Human Settlement Management Institute (New Delhi, India). Since 2016, he has been the Chief of the Division Territorial Studies and

Urban Poles at the Department of Urbanism and Housing of the Ministry of Urban Renewal, Housing and Living Environment of Senegal. Prior to that, from 2013 to 2016, he was the Coordinator of the Coordination Committee of the Programme ZAC, and from 2004 to 2013 he was the Coordinator of the ZAC project at the Department of Urbanism and Architecture. He was the National Focal Point and Principal Coordinator of the National Committee of the Third Habitat Conference on Sustainable Urban Development and Housing. He also represented Senegal at the African Union Working Group in charge of the elaboration of the Africa Common Position Towards Habitat III. He also was Member of the Negotiation team of the Project Document of Habitat III. He was also Member of the Committee for Building Permits. Since 2006, he is the Representative of Ministry of Urban Development on Negotiation with Union Groups. From 2006 to 2014, he was the Focal Point of the National Statistical office at the Ministry of Urban Development. He also represented the Ministry at the Millennium Development Goals National Committee and several other national committees. E-mail: mnjaye@yahoo.fr.

Dr. Mame Cheikh Ngom is presently a Researcher and a Teacher at the University Cheikh Anta Diop (UCAD) of Dakar. He has an extended experience in management of local collectivities, territorial planning, local development and governance and decentralization. At the UCAD, his lecture at the Master degree covers the following: network configuration, territorial development and decentralized cooperation; at the Bachelor's degree, his lecture covers the geography of Senegal. Prior to joining UCAD, Dr. Ngom had 18 years of experience in local government management including 4 years on accounting system, finance and treasury with the city of Pikine and many years of administration with other municipalities of the department of Pikine. Dr. Ngom is the coordinator of GRED (Group for Research and Development Studies). He has also experience in rural community development as per his experience in the rural community of Diass in 2007 where he oversaw programmes on environment and natural resources management. He successfully implemented an urban development project commissioned by

ENDA GRAF, EVE and GRAIM ENDA in 2012. He is presently the coordinator of the Professional Association Cof Safy and President of the Association for the Development of Bandia. Dr. Ngom holds a Ph.D. in geography and regional planning, and municipal governance, a Master's degree in Urban Governance and Decentralization, a Master's degree in Marketing in Fisheries and a Bachelor's degree in Geography. E-mail: thieeko@hotmail.com.

Prof. Femi Olokesusi has a B.Sc. (Honours) Second Class Upper Grade from the University of Lagos, Akoka, a Master's degree in Environmental Studies from York University, Toronto, Canada, and a Ph.D. in Geography (with specialization in Environmental Planning and Management) from the University of Ibadan, Nigeria. A fellow of the Nigerian Institute of Town Planners (FNITP) and Hotel and Tourism Management Association of Nigeria (FHATMAN), he retired as a Research Professor from the Nigerian Institute of Social and Economic Research (NISER), Ibadan in 2016. At NISER, he was Director of Physical Development Department for about 6 years and among other professional and administrative duties, and also acted as the Director-General/Chief Executive Officer of the Institute from December 2010 to August 2011. A 1994/95 Fulbright African Senior Scholar at the University of Illinois, Urbana-Champaign, USA, Femi has conducted several studies for several national and international agencies. Examples the World Bank, United Nations Development Programme, UNCTAD, Geneva, African Technology Policy Studies Network, UK-DfID, London; International Development Research Centre, Ottawa, Canada; Federal Ministry of Environment and National Planning Commission, Abuja. Also, he served as a resource person on Nigeria Vision 20:2020 and the National Infrastructure Master Plan. Femi served as a consultant to the UN-Habitat in 2014/2015 on the City Structure Plan Projects for Lokoja and Dekina. He is currently a Professor in the Tourism and Events Management Programme, Afe Babalola University, Ado-Ekiti, Ekiti State as well as a member of the Independent Advisory Group of the World Bank-assisted Ibadan Urban Flood Management Project. He has over 100 peer-reviewed publications to his credit and is presently conducting research in the areas of Environmental Change and Tourism, Smart Cities and Smart Tourism Destinations. E-mail: femioloke@yahoo.com.

Romanus Opiyo Ph.D. is a lecturer in University of Nairobi, based in the department of Urban and Regional Planning. He has been part of a research team "African Centre of Excellence for Studies in Public and Non-motorised Transport (ACET) covering Nairobi, Dar es salaam and Cape Town". Currently he is also the coordinator for Centre of Urban Research and Innovations (CURI), University of Nairobi. He has over fifteen (15) years' experience in consultancy and research work with a number of them focussing on transportation policy and institutional framework, urban security and safety, preparation of master plans. Romanus has also been engaged in various Environmental and Social Impact Analysis, violence, security and safety, Institutional Governance and livelihoods projects and surveys. He has also done studies and consultancies for Government Departments and various international organizations in various areas including building capacities for Africa Planners through United Nations Centre for Regional Development (UNCRD) both in Kenya and Botswana as facilitator in data collection, analysis and projects evaluation techniques. He has Fourteen (14) years university lecturing and project supervision experience. He is a lead Environmental Impact Assessment/Auditor Lead Expert registered by Kenya National Environment Management Authority (NEMA). He has published in the area of urban and rural transport, smart cities, urban security and informal sector. Romanus has presented in various conferences both nationally and internationally including making presentation in World Urban Forum (WUF) which is one of the most respected forums for urban scholars and practitioners. E-mail: romanop2000@gmail.com.

Namratha Radhakrishnan has completed her Bachelor of Architecture from the University of Kerala and Master of Technology with specialization in Urban and Regional Planning from CEPT University. She has 9 years of international consulting experience across geographies such as the Middle East, North Africa, South East Asia, India and China, and sector experience includes industrial cities, SEZs and real estate. She is currently working as a faculty in the Department of Architecture and Planning, National Institute of Planning Calicut. E-mail: namratharadhakrishnan@gmail.com.

Dr. Yifat Reuveni directs the social finance innovation activity at the JDC Institute for Leadership and Governance, where she is responsible for the R&D and implementation of social finance models. She is also responsible for enhancing the entrepreneurial and social finance ecosystem building in Israel in order to sustain regulation and deal flows for impact investments. Dr. Reuveni teaches B.A. and M.B.A. courses in social finance, social and profitable entrepreneurship, and alternative economic models at the Coller Business School, Tel Aviv University and at the College of Management Business School, and is beneficiary of several EU grants such as Murie Curie (FAB-MOVE) and Erasmus + CLEVER. Until 2014, she was the Head of the Recanati Business School's Development and International Relations Unit. She completed a B.A. in Economics and Philosophy and an M.A. in Communication Technologies at the Hebrew University of Jerusalem, and a Ph.D. on the evolving of the new economy at McGill University, Montreal, Canada, focusing on the high-tech industries in North America and Europe. As a recipient of a Canadian International Development Agency (CIDA) scholarship, she participated in the McGill Program for Economic and Social Rights, working with the Mohawk people (first nation) in Kahnawake, Canada, and writing an assessment report of business opportunities in the Mohawk community (the KANATA 2000 Report). Prior to her time spent in Canada Yifat worked for 5 years with the Association for Human Rights in Israel as director of the consulting department, working with security forces and public service leaders. She has given courses on the new economy and the neo-liberal economy at the Recanati Business

School, Colman Business school and Ben-Gurion University's School of Business and Management. She mentors entrepreneurs from around the world for Innovation and International Development, is involved in numerous initiatives in Israel to promote social entrepreneurship and has acted as a judge for several enterprise pitching competitions. Her current academic activities consist of participation in an international research group on social entrepreneurship at the Center for Social Economy at the Liege HEC Business School, focusing on alternative finance models as the financial infrastructure for such entrepreneurship, acting as social and sustainable finance board member at OXFORD, beneficiary of Marie Skłodowska-Curie Research and Innovation Staff Exchange (RISE—FAB-MOVE: For a Better Tomorrow: Social Enterprises on the Move—2016–2018, beneficiary of CLEVER—Creative economy—EU Erasmus + 2016–2019 writing social finance case studies and leading two Israeli research groups in Tel Aviv University and in the College of Management Business School.

Timothy Rodgers is a Research Associate at the Institute for Sustainable Urbanisation, and Project Coordinator and Research Associate at UDP International. He holds a Master's degree in Transport Policy and Planning from the University of Hong Kong, and a Bachelor's degree in Economics and Geography (Urban Systems) from McGill University, Canada. His research experience at UDP International includes research and report writing on current issues such as Smart Cities and Affordable Housing, assisting in the development of a Smart City Framework with a focus on People, Place and Planet as core elements, as well as conducting research on public and private project proposals and reports. He was the team lead for the Hong Kong chapter of the collaborative international research publication 'Smart Economy in Smart Cities', and contributed to the Hong Kong chapter of the collaborative international research publication 'E-Democracy for Smart Cities'. E-mail: tim@udpcltd.com.

Mustafa Sangare is a partner with GORA Corp and currently working on the Conakry Smart City Project. He is also currently working at the IFC/World Bank Group over 10 years and as Meetings & Conference Services Coordinator. Mustafa was awarded a degree in Computer Service Specialist at TESST University of Alexandria Va. He has competencies in SQL Server, deploying applications, deploying database in different environments like test and production environments. Ability to write, tuning and optimizing query performance, troubleshooting connection related issues. Mustafa has also gained extensive professionalism and experience in investment management with a partnership with Diversified Energy Corp and ARCO Globus focusing on developing countries. E-mail: mustafasangare@msn.com.

Mfaniseni Sihlongonyane is an associate professor and Coordinator of Post-graduate Studies in the School of Architecture and Planning. He studied geography and Theology at the University of Swaziland in which he got a 2:1 in 1992. In 1993, he trained as a teacher to obtain a Post-graduate Certificate in Education (PGCE) with distinction. From 1994 to 1995, he did a Masters in Development Planning at Wits University and from 2004 to 2008 he obtained a doctorate (DPhil) at Oxford University. He has been the undergraduate coordinator in the School between 2008 and 2011. He was also director of the Planning Programme from 2011 to 2014. He also served as a member as well as the alternative chair of the South African Council of Planners in 2008-2012 and 2013 to 2017 respectively. He was chairperson of the organization committee for the 52nd Congress of ISOCARP which was held in Durban in 2016. He has been a hall coordinator at Esselen Residence between 199-2003 and warden for EOH Residence from 2007 to date. His research interests spread over a wide range of theoretical, applied and policy arenas in the global as well as African realms of development. The research encompasses principally the interface between development and urban studies largely within the context of the poignant dynamics of the political economy in Africa. Over the years, he has worked and published in areas of land reform, planning, gender, spatial development, urban politics, local economic development, community development, and African city. The analysis of these topics is drawn from a wide range of disciplines which include: economics, history, sociology, anthropology, political and geography. Moreso, his insights in the exploration of these areas are draw from his experience as an African. E-mail: Mfaniseni.Sihlongonyane@wits.ac.za.

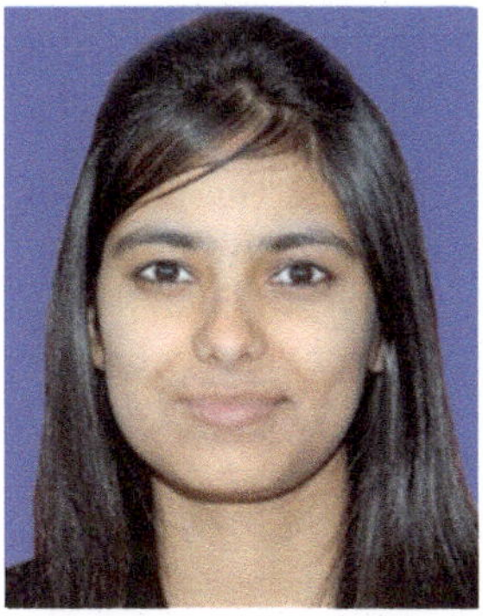

Asfa Siddiqui is currently working as a Scientist at Indian Institute of Remote Sensing (Indian Space Research Organization) since 2014. She did her Bachelor's in Architecture from Govt. College of Architecture, Lucknow in 2011 and Master's in Planning with specialization in Urban Planning from School of Planning and Architecture, New Delhi in 2013. She worked at NIT Kozhikode (Calicut) prior to joining ISRO. She joined Indian Institute of Remote Sensing (IIRS), ISRO in January 2014. Her work focuses on urban and regional areas with emphasis on energy and environment. E-mail: asfa.aas@gmail.com, asfa@iirs.gov.in.

H. B. Singh is a retired Professor of Urban and Regional Planning from School of Planning and Architecture, New Delhi. Apart from academics, he works as a development consultant/advisor to a few MNCs and has led their professional teams. An Architect and Town Planner by qualification from University of Roorkee and School of Planning and Architecture, New Delhi, he went on to do an advanced course in Urban Management Development from the University of Birmingham U.K. He was employed in responsible positions in the Town and Country Planning Department of the Government of Uttar Pradesh and Government of India, before his stint of 12 years as Chief Architect Planner in one of the largest universities—A.B.U. Zaria, Nigeria. After his return, he was selected to the position of the Principal, Government College of Architecture, Lucknow, through UP Public Service Commission. Finally, he joined School of Planning and Architecture New Delhi (An Institution of National Importance) and retired as Professor and

Head, Department of Regional Planning and Centre for Rural Development. He has written a large number of professional articles and co-authored three books, the latest being 'Urbanisation in India: Spatial Perspectives'. He has extensively travelled and delivered lectures in few prestigious American universities. His varied and global experience in academics and profession, in developing and developed countries, various tiers of Governments and MNCs, have provided him with deeper and contextual understanding of Urban and Regional Planning concerns. E-mail: hbsinghvats@gmail.com.

Dr. Eleni Tracada is Principal Tutor in the Built Environment (Architecture) at the University of Derby, UK; Chartered Member RIBA Part III; Senior Fellow of HEA. She graduated in architecture, Faculty of Architecture of Florence, Italy (1980) and worked as a self-employed architect in Florence (1983–1993). She was awarded an MA in Interior Design at Manchester Metropolitan University (1996) and a Ph.D. by Published Works at the University of Derby (2015). She taught Interior Design at Leeds College of Art & Design, Leeds (2001–2007). She is a Member of the Scientific Committee of the International Society of Biourbanism; she has been Editor in Chief of the Journal of Biourbanism (2011–2014). She participated in international research projects, such as BEST Leonardo Lifelong learning (2011–2013), Wor(l)ds which Exclude (2013–2014), Dance Architecture Spatiality (2012–2014). Her main research interests include architecture, human behaviours in urban spaces, architectural psychology and placemaking, ageing and vulnerable communities' needs in accommodation and urban design and smart cities and economies. Her outputs disseminate ideas and recommendations on planning frameworks and social policies. E-mail: E.Tracada@derby.ac.uk.

Dr. Neha Goel Tripathi is an environmental planner with over 15 years of experience in Environmental Management and Architecture. She is presently working as Assistant Professor in School of Planning and Architecture, New Delhi. She did her post-graduation from School of Planning and Architecture, New Delhi in Master of Planning, specialization in Environmental Planning (2003) and was awarded the Gold medal for overall performance. She is Associate Member, Institute of Town Planners, New Delhi, Member, Council of Architecture, New Delhi and Member, Indian Building Council, New Delhi. National Environmental Science Academy. Experience of more than 10 years in the field of environmental planning and architecture and worked on projects like Zonal Development Plan for Eco-sensitive Area, Mount Abu, Indian Institute of Management, Indore, Manali Redevelopment Plan, etc. She has published articles on field of Ecological footprint, Solar Zoning, Climate change, etc. Awarded the first prize from the Hon'ble Minister for Housing & Urban Poverty Alleviation, Kumari Selja for paper presentation, Climate Change and Indian Cities Perspectives presented for World Habitat Day 2011 organized by HUDCO, New Delhi. E-mail: nehagtripathi@gmail.com.

Gaurav Vaidya is an Infrastructure Planner and Asst. Professor in School of Planning and Architecture, Bhopal. He has around 11 years of professional experience in industry and academics as well. He had worked for more than 4 years (during 2009–2013) as Urban Planner in the Urban Development & Urban Housing Department, Govt. of Gujarat on various assignments, some of them are as follows: preparation of Gandhinagar Development Plan, Ahmedabad Gandhinagar Comprehensive Mobility Plan, Clean Green & Solar Gandhinagar Master Plan, 3 Town Planning Schemes, 12 City Development Plans, Detail Project Reports for Municipal Water Supply & Urban Poor Housing Projects, Urban Reforms Appraisal and Capacity Building Programs for Municipalities of Gujarat, etc. As civil engineer, he executed various small-scale civil engineering projects during 2004–2007 with the capacity of Asst. Consultant Engineer. At present, he is involved in the research of 'Management of urban sanitation services', 'Regional-urban infrastructure linkages' and 'Green infrastructure development'. E-mail: gaurav.vaidya@spabhopal.ac.in.

Nikolas Zimmermann is working as a researcher at Fraunhofer IAO. He has led several industrial and scientific projects and gained a lot of experience in the engineering domain. He has competencies in the field of additive manufacturing technologies, optimisation and digitalisation of development and production processes as well as semantic knowledge management. He is in charge of the Digital Engineering Lab, which is a demonstration centre in Stuttgart, Germany for research activities and demonstrators concerning Digitalisation and Industry 4.0 influences. Furthermore, he also participates in the Future Work Lab (FWL) initiative, which shows how industry work will look like in the future. Over 60 different Industry 4.0 solutions are presented in the FWL. He holds a Diploma in Mechanical Engineering with specialization in Production Technology from the University of Karlsruhe (KIT). E-mail: nikolas.zimmermann@iao.fraunhofer.de.

Part I
Introduction

Chapter 1
Smart Metropolitan Regional Development: Economic and Spatial Design Strategies

T. M. Vinod Kumar

Abstract An overview view of smart metropolitan regional development is presented as a backdrop for research studies of this book. First, metropolitan region is defined followed by its form and functions. Then, smart metropolitan regional development is defined. The chapter studies the global metropolitan cities development and shows how metro cities, megacities and meta-cities are emerging across the world in Asia, Africa, Europe and America. Metropolitan cities are continuously exposed to external economic stimuli and requires intermediate range strategies as responses to face it. These periodic challenges of cities call for differing and flexible spatial and economic strategies to intervene in emerging global situation. By and large such strategy making dynamics is rare to be found and it is an important gap which this book address. To face these emerging metropolitan challenges, there is a need to design economic and spatial strategies at the intermediate time horizon. Therefore, this chapter concludes with a critical analysis of economic and spatial design strategies of 17 metropolises in their official plans which of course is long range plans.

Keywords Metropolitan development form and functions · Smart metropolitan development · Metro cities · Megacities and meta cities evolution globally Global cities · World cities · Official metropolitan development plan and economic and spatial strategies

1.1 Introduction

In 2016 as per UNHABITAT statistics, an estimated 54.5% of the world's population lived in urban settlements. By 2030, urban areas are projected to house 60% of people globally and a one in every three people will live in cities with at least half a million inhabitants. This larger city shows more propensity for higher

T. M. Vinod Kumar (✉)
School of Planning and Architecture, New Delhi, India
e-mail: tmvinod@gmail.com

T. M. Vinod Kumar (ed.), *Smart Metropolitan Regional Development*,
Advances in 21st Century Human Settlements,
https://doi.org/10.1007/978-981-10-8588-8_1

employment opportunities and higher income and better quality of urban infrastructure that drives people to move there. When more households in the world for example in India decide to transform from rural to urban households, by the change to non-primary urban occupation (that is the village males adopts secondary and tertiary sectors of the economy instead of the primary sector as per Indian Census); migration to cities, urbanisation progress and then larger urban agglomerations are formed [1, 2]. Because of the secondary and tertiary sector occupation and its propensity for generating higher income than farm income, Metropolitan cities world over shares a larger percent of Gross National Products than other cities of lower population range and rural areas. There are also potential in cities to increase this share for National Domestic Product than rural areas. Yet there has been less of effort to work out an economic strategy suited to the strength and weakness of specific metropolis that directly benefits the whole population of the metropolis, even its hinterland or the Nation. Against this requirements today economic policies of Nations are formulated largely based on non-spatial and non-regional economic policies as if economy resides in national input output tables.

Often in urbanisation, existing inhabitants are replaced with new immigrants creating social tension due to lack of employment opportunities to the early settlers with legacy skill sets. There are many more vital and well documented issues in these large cities in developing countries such as that of Lagos in Nigeria, Delhi and Mumbai in India, Jakarta in Indonesia and many one million plus cities which remains un addressed due to absence strategic interventions in economic and spatial aspects. This exploratory research seeks some answer to some of this question.

There has been an attempt to convert many metropolitan cities to smart cities, largely using selective investment strategies using public funds and procedural administrative rules arbitrarily not sanctioned or related to constitutionally designed institutions in limited part of cities with no clear economic, social and spatial rationales that fits into the overall metropolis in a particular location as related to global system of cities. In many smart city projects such as 100 smart city programs of India where only a very small and insignificant and population share of the city has been taken up for the smart city program and in many cases, it is status and elitist area of cities like New Delhi Municipals Corporation area in Delhi that get selected under a set of criteria postulated and not where low-income people live, or centres of significant economic development potential in metropolises that can elevate people to a better way of life and give opportunities to accumulate more household income that pays for smart city infrastructure. A good example is the 100 smart cities project of India. This approach is adopted mainly because the smart city is conceived as merely a demonstrative smart urban infrastructure investment project with no thought of how it can be financed and governed by citizen with the existing constitutional institutions, relevant legislations and internal finances. It is known that these cities by themselves cannot be able to generate its own higher and additional income to convert the entire city into smart cities with expensive ICT and IOT enabled smart city infrastructure. Further existing Governance system is bypassed with the imposition another temporary administrative structure as special purpose vehicle to implement the project defeating the constitutionally established urban governance and city democracy. The existing urban plans as per the urban legislations of the respective states are not considered and largely ignored for smart city plan formulation weakening implementation of State level Town and Country Planning legislations. In short, the existing legitimate ecosystem of governance, democracy and planning system are bypassed in smart city programs. This book attempts to generate situation specific economic and spatial strategies that can convert the whole metropolitan region to smart metropolitan region. This will be defined subsequently in many chapters of this book which will not leave the earlier settlers with traditional occupations left out of the prosperity regime of smart metropolises. A societal and economic change will make these cities have their own smart infrastructure at their own expenditure and further strengthen the Governance capabilities as per constitution and existing legislations.

Spatial Plans are prepared for the metropolitan regional development to guide higher order infrastructure investment of the region. However spatial strategy that is integrated with the economic strategies which are inclusive of all sections of city population to transform them to a trajectory for household prosperity is not attempted in many of the Metropolitan Regional Plans. In short, metropolitan plans are to direct public investment, but no plan is prepared to charter people to prosperity, with profitable use the economic opportunities of the metropolis. Smart Community in a smart metropolis is fully capable of participating in a location and situation specific smart economy and build a smart city that is self-financing. Some of the studies will explore this possibility.

Designing economic and spatial strategies for Smart Metropolitan Regional Development shall be conducted by the spatially identifiable economic community at micro levels from time to time to suit the ever-changing scenarios of the global economic environment. However economic strategies may have a shorter lifespan than spatial strategies but both should fit in. Spatial strategies should be based on economic strategies each complementing the other. When this micro level communities cumulates to have a mega-community, they can make use of the vehicle of economic E-Democracy to formulate ever changing and dynamic strategies. E-Democracy for Smart Cities in this series of the book [3] discusses such issues, approaches and tools available for executing these designs of strategies.

1.2 Metropolitan Region

The term 'metropolitan region' is used to describe highly urbanised, city-regional areas that are characterised by a high population density as well as a concentration of economic, political and cultural activities [4, 5]. Furthermore, metropolitan regions form part of the interconnected city network and exhibit a specific governance structure that provides mechanisms of inter-jurisdictional cooperation between core cities and their hinterland. They constitute a link between the global network cities and locally embedded economic and social activities.

Brezzi, Monica; Piacentini, Mario; Sanchez-Serra, Daniel have studied the metropolitan form extensively [5]. While the urban area is the spatial form of the city, the metropolitan region, it has got a functional and economic form as studied by Burger, Meijers [6–8]. The spatial form is evident during the night when lights are on the metropolitan urban area and is visible from the sky while flying as continuous urbanised area with varying density of lighting. This area is unrelated to a municipal boundary, district boundary or even state and national boundary. The metropolitan region may be monocentric like Hyderabad or Bengaluru or polycentric like Mumbai or Delhi [7–9]. Urban area definition and terminology is developed by the national statistical authorities such as for example, the Census of India and the United Nations. They are at a liberty to use their own terminology and area definition. For example, Australia calls the urban area as urban centres, Canada Population centre, Denmark, Finland, the Netherlands, Norway, Sweden, and the United States calls it urban Area, United Kingdom the built urban area while India and the United Nations called it Urban Agglomerations in last few Indian censuses. A metropolitan area may include more than one urban area. For example, the Los Angeles metropolitan area includes several urban areas, such as Los Angeles, Riverside-San Bernardino, Mission Viejo, Santa Clarita, Simi Valley, Oxnard-Ventura, and Palm Springs. The United States designates combined statistical (metropolitan) areas, which are routinely used, as opposed to their smaller metropolitan statistical area (MSA) components. Some but not most nations formally designate metropolitan areas (such as the United States, France, Brazil, India, Argentina, and Canada). Caution is appropriate with respect to the term. Other countries have their own definition. A term "Greater" is often used to denote a metropolitan area, such as "Greater Los Angeles", "Greater Mumbai" or "Greater Chicago." Again, this term is imprecise, because it is also used in some situations to denote municipalities that are only a part of a metropolitan area as defined by Indian census or UNHABITAT. For example, the municipality of Mumbai is formally called the Municipal Corporation of Greater Mumbai and does not encompass the entire metropolitan area as defined by Indian Census. There is considerable confusion over the term "city" and urban terms, such as "urban area" and "metropolitan area". By its very nature, urbanisation in both the spatial sense and the economic sense is not defined by the borders of single municipalities, large or small. There is a requirement for standardisation of metropolitan areas definition jointly perhaps by a United Nations body to help the cause of scientific integrated metropolitan development.

Global Cities are not new but have been discussed by Saskia Sassan since the 1990s [10, 11]. Four functions that characterise metropolitan regions in the context of globalisation are,

1. innovation and competition
2. decision-making and control
3. gateway
4. symbol or branding.

World cities are identified as the control centres of the global capital flow, describing metropolitan regions based on a range of specialised metropolitan functions; and global cities or global city-regions as control centres and centres for the creation and marketing of business-orientated, knowledge-intensive services, and linkages between knowledge-intensive services as an indicator for analysing the rank and function a city occupies within the world city network [12, 13].

If a city also performs an important commercial, cultural and political function for its region or even the whole country can graduate to a global metropolis. Such a major importance can usually be assumed for cities boasting some 500,000 inhabitants or more as in the case of Europe and one million plus in India and other developing countries. On the other hand, Global cities are those selected few metropolises whose political, cultural and commercial influence extends across the entire globe (e.g. New York City, Tokyo or London). A metropolis is considered as a city which agglomerates major functions of coordination of complex activities and which fulfils these functions at a world scale [14]. Although the definitions highlight that the term 'metropolis' is to some extent like the concepts described above, it is both less clear and less focused on economic or social change. Furthermore, the metropolis concept fails to define the spatial extent of this type of city. Given the existence of functional relationships between these cities and their hinterlands or polycentric conurbations.

As social spaces, metropolitan areas or regions can be characterised by the following four dimensions: Metropolitan regions are defined as an accumulation of metropolitan facilities including public and private services. In terms of actors and actions, metropolitan regions constitute an arena for key regional stakeholders to exchange knowledge on joint regional objectives, strategies and projects, as well as on the necessary organisational structures. In the context of spatial development, metropolitan regions are a normative and guiding concept intended to contribute to innovation, creativity and economic growth. About the symbolic dimension of urban and regional development, metropolitan regions are the medium of symbols, norms and values which convey aspects associated with the metropolis and urbanity.

1.3 Smart Metropolitan City/Region

The term smart metropolitan region is rarely used. It calls for a definition. Authors of the book "Geographic Information System for Smart Cities" examined several definitions of smart cities and were not satisfied and came out with their own definition. I feel this definition can be used in this book [15].

> The smart metropolitan (centre/area/region) city is a knowledge-based city that develops extraordinary capabilities to be self-aware, functions 24 h and 7 days a week, communicate, selectively, knowledge in real time to citizen end users for a satisfactory way of life, with easy public delivery of services, comfortable mobility, conservation of energy, environment and other natural resources, and creates energetic virtual face to face communities and a vibrant urban economy even at time of national economic downturns.

All six components of smart cities such as Smart Economy, Smart People, Smart Governance, Smart Mobility, Smart Environment and Smart Living is integrated and implied in this definition which has been presented in detail in the four books of this series by the Author [3, 16–18]. The importance of E-Democracy and use of technology and innovative entrepreneurship will be the prime mover of the smart metropolitan regions which is the subject matter of one book in this series [3]. Smart Metropolitan (Regional) Development has not been tried extensively and this book will be an attempt to do so.

1.4 Functions of Metropolitan Regions in Global-Place-Competition

The concept global economy with partnership and division of labour of global network of cities is not new, but those who talk about a globalised economy insist that there have been distinct changes in its structure and modes of production [19]. Whereas earlier economic activities crossed national boundaries ("internationalisation"), globalisation includes a deeper integration, where transnational corporations orchestrate production from various locations. The term also includes other factors [20]. Global places indicate a rising of networked society globally [21] which can work effectively in a smart metropolis with high endowments of ICT and IOT.

One author boils globalisation and Global Urbanisation down to five basic elements [22]:

(1) "new innovative technology,
(2) the centrality of information made possible by instant communication,
(3) an increasing trend toward the standardisation of economic and social products,
(4) growing cross-national integration, and
(5) mutual vulnerability stemming from greater interdependence" [23].

As far as the decision-making and control function of metropolitan regions is concerned, the focus is to ensure the presence of those centres capable of steering international activities in business and politics [24]. In global place competition, these centres provide the region with greater influence and access to necessary networks. The current distribution of decision-making and control functions is not only the result of decisions made in the past, but also of specific conditions which have affected business locations over time. As soon as the number of control centres in a metropolitan region (e.g. corporate headquarters, ministries or international organisations) exceeds a critical mass, the importance of physical proximity induces a self-reinforcing process.

About the innovation and competition function, a similar picture is observable. The greater the importance of the knowledge economy, the greater the competitive edge of metropolitan regions as preferred locations for both national and international customers is. Among other things, increasing levels of efficiency enhance the attractiveness of locations for knowledge bearers, knowledge producers and creative individuals. In the knowledge sector, the risk of a 'brain drain' and the resulting loss of highly skilled labour is not to be underestimated. This can only be achieved by the instituting smart living which is one of six component of smart cities.

When it comes to integrating metropolitan regions into international and global flows, the gateway function is by far the most important. Alongside its focus on the efficiency of various infrastructure types (primarily of transport nodes), this function targets the ability of a metropolitan region to function as a 'gateway to the world' when it comes to the exchange of and access to, services, information, knowledge, ideas and opinions [25]. Given that metropolitan regions are the gateway for both in- and out-migration, they are the places where diverse cultures and lifestyles come together. Thus, the way in which these encounters can be harnessed for the region's development constitutes an additional indicator of the efficiency of metropolitan regions.

The efficiency of metropolitan regions can also be measured by the extent to which they succeed as key sources of sign and symbol production. Rather than focusing on the current trend towards essentially interchangeable festivals and events, this function focuses on a credible and unmistakable sense of uniqueness at the international level. This is particularly prevalent in milieus that are not only shaped by the nodal function of metropolitan regions in global networks, but also by the specific traditions, experiences and resources created by regional actors. The strength of these metropolitan images ('spatial brands') in global place competition increases the more they are based on an equal balance of economic, socio-cultural, spatial/physical and historical components, and the more they are 'lived' and borne by people in the region (creative individuals, 'ambassadors', etc.).

1.5 Metro Cities, Mega City and Meta-City

While many countries have their own definition related to the population of metro cities, metro cities are defined as 1 million plus agglomeration by the UN and the Indian Census. Considering increasing levels of urbanisation, the United Nations defined the 'mega-city' as a new population dependent category in 2012 with 10 million plus population. In 2011, 23 urban agglomerations qualified as megacities because they had at least 10 million inhabitants. Alongside the category 'megacity', UN-Habitat introduced the term [26] 'Meta city' which describes 'massive conurbations of more than 20 million people or above'. Termed by some as 'hyper cities', cities with more than 20 million inhabitants constitute a new type of settlement above and beyond the scale of megacities. Driven by economic development and rising population numbers, they gradually swallow rural areas, cities and towns, thus becoming single, yet multi-nuclear entities. Meta cities are conurbations of more than 20 million people, mega cities are 10 million population or above and metropolitan cities 1 million and above. These meta-cities speak of human vibrancy, an innovative and entrepreneurial spirit and commercial dynamism as discussed in earlier para. Side by side there is mounting issues such as environmental, transport, income equality, Governance and so on. These meta-cities comprise of people striving for more economic opportunities, better freedom and a bigger voice at the same time facing an increasing amount of urban issues and challenges. The area from Hong Kong to Shenzhen to Guangzhou (a region in South China that is 26 times larger than Greater London) is home to over 120 million people and a massive manufacturing base. A 932-mile Indian industrial corridor being developed between the cities of Mumbai and Delhi under construction may have a big population and bigger economic opportunities befitting a meta-city. Nigeria's Lagos, the fastest growing megacity in the world, is expanding at more than five percent a year with all its difficult urban issues heralding a beginning of a meta-city. Tokyo the largest meta city in the world is a good example of all attributes of a meta-city. Since ICT drives the mega city and meta city formation, these cities are smart cities in making.

1.6 Urban Agglomerations

These urban agglomerations called mega cities and meta cities, many of which accommodate populations larger than entire countries, are a truly unprecedented phenomenon and merit greater attention.

A map of countries by population range is shown in Fig. 1.1.

It can be seen that Population of Canada is 36,511,11 as on 2017 but that of meta city Tokyo is 38,241,000 in 2017. Similarly, there are many Nations smaller than meta-cities and mega cities in the world as shown above on the map.

Case studies from India, Africa, Europe, China and the United States are covered in this book. A comparative statement on India, China and the US for 2010 is presented below although 2011 census of Indian figure is available. This is meant to act as a background to case studies. Statistical data on size distribution, percent

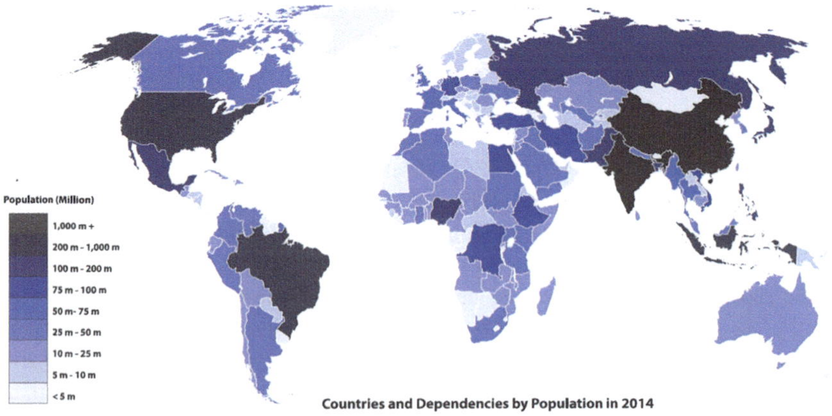

Fig. 1.1 Countries and population range 2014

urban growth rate of India, China and USA are as given below. Urban Population as a percentage of total Population by regions in 1970, 2000 and projection for 2015 is as given below. It can be seen that while North, Central and South America are reaching saturation points in urbanisation, there is much to catch up in Africa and Asia (Fig. 1.2).

In 2016, there were 512 cities with at least 1 million inhabitants (metropolitan centre/region/cities) globally. By 2030, a projected 662 cities will have at least 1 million residents. As hubs of trade, culture, information and industry, they will be vested with such power that at many levels they will act as city states that are independent of national and regional mediation. Today mega cities are home to less than 10% of the global urban population. In 2016, there were 31 megacities globally and their number is projected to rise to 41 by 2030. These 31 megacities by region are colour coded in the following Fig. 1.3 as per the region they belong to.

The list of megacities is given below for 2016 and projected 2030 by UNHABITAT in Table 1.1. There were eight meta cities above 20 million in 2016 in the universe, which is likely to be twice about 15–16 in number or little less in 2030 which may be considered as accelerated graduation of mega cities to meta cities. However, the number of mega cities increase during this period which includes meta cities were from 31 to 41 which is not as impressive as that of the meta-cities. Indian census defines urban agglomeration which UNHABITAT define as cities but there was no attempt to classify these urban agglomerations as

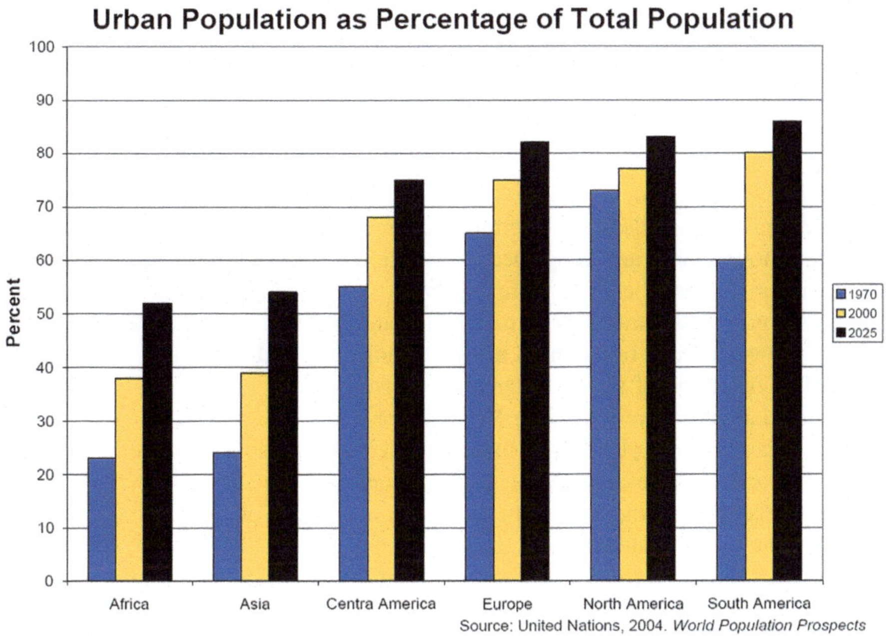

Fig. 1.2 Urban population growth 1970–2025 by geographic regions

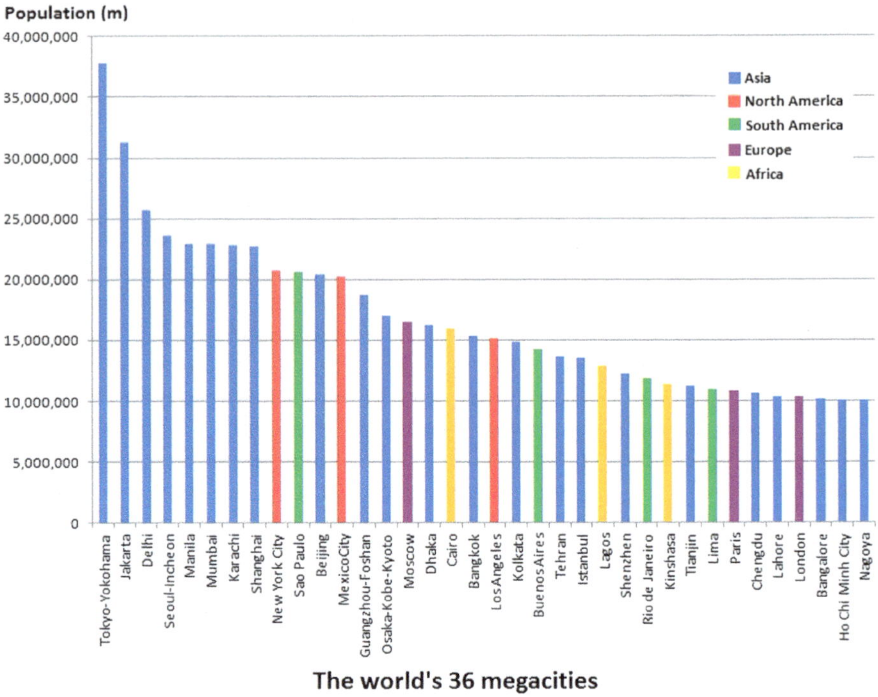

Fig. 1.3 Meta and megacities of the world by population and regions

meta-cities and megacities in Indian Census since it cuts across several districts as if there is a wall around districts for Indian Census. If you map using census town data of five metropolises namely Kannur–Kozhikode–Malappuram–Trissur–Kochi, there is already a megacity of 10 million plus population in 2016. In the northern part of Kerala, and Salem–Coimbatore another mega city in Tamil Nadu. Also, Indian census do not delineate metropolitan regions as against non-metropolitan districts although these are recognised constitutional areas as per 74th constitutional amendments. As can be seen there has been no special treatment in designing economic and spatial strategies of megacities and meta-cities which is a debatable point of view. No clear research on meta city functions and megacities functions as opposed to functions of metropolitan cities exist today.

As stated above, there is also the Metacity, or hyper city, an epithet that refers to massive sprawling conurbations of more than 20 million people. Tokyo became the first hyper city in the mid-1960s when it crossed the 20 million population thresholds. Today it is the largest meta-city in the world. Tokyoites—more than 35 million—outnumber Canadians. By 2020, Mumbai, Delhi, Mexico City, Sao Paulo, New York, Dhaka, Jakarta and Lagos all will have achieved meta-city status, someone dozen to fifteen meta-cities as per the listing above. These metropolises are so huge that they have changed the dynamics of urbanisation. People commute to work in meta/megacities from densely populated outlying villages or suburbs.

Table 1.1 Meta-cities and megacities of 2016 and 2030

Rank	City, Country	Population in 2010 (thousands)	Rank	City, Country	Population in 2030 (thousands)
1	Tokyo, Japan	36,140	1	Tokyo, Japan	37,190
2	Delhi, India	26,454	2	Delhi, India	36,060
3	Shanghai, China	24,484	3	Shanghai, China	30,774
4	Mumbai, India	21,357	4	Mumbai, India	27,797
5	Sao Paulo, Brazil	21,297	5	Beijing, China	27,706
6	Beijing, China	21,240	6	Dhaka, Bangladesh	27,374
7	Mexico City, Mexico	21,157	7	Karachi, Pakistan	24,838
8	Osaka, Japan	20,337	8	Cairo, Egypt	24,502
9	Cairo, Egypt	19,128	9	Lagos, Nigeria	24,239
10	New York-Newark, USA	18,604	10	Mexico City, Mexico	23,865
11	Dhaka, Bangladesh	18,237	11	Sao Paulo, Brazil	23,444
12	Karachi, Pakistan	17,121	12	Kinshasa, Democratic Republic of Congo	19,996
13	Buenos Aires, Argentina	15,334	13	Osaka, Japan	19,976
14	Calcutta, India	14,980	14	New York-Newark, USA	19,885
15	Istanbul, Turkey	14,365	15	Calcutta, India	19,092
16	Chongqing, China	13,744	16	Guangzhou, Guangdong, China	17,574
17	Lagos, Nigeria	13,661	17	Chongqing, China	17,380
18	Manila, Philippines	13,131	18	Buenos Aires, Argentina	16,956
19	Guangzhou, Guangdong, China	13,070	19	Manila, Philippines	16,756
20	Rio de Janeiro, Brazil	12,981	20	Istanbul, Turkey	16,694
21	Los Angeles-Long Beach-Santa Ana, USA	12,317	21	Bangalore, India	14,762
22	Moscow, Russian Federation	12,260	22	Tianjin, China	14,655
23	Kinshasa, Democratic Republic of Congo	12,071	23	Rio de Janeiro, Brazil	14,174
24	Tianjin, China	11,558	24	Chennai (Madras), India	13,921
25	Paris, France	10,925	25	Jakarta, Indonesia	13,812
26	Shenzhen, China	10,828	26	Los Angeles-Long Beach-Santa Ana, USA	13,257
27	Jakarta, Indonesia	10,483	27	Lahore, Pakistan	13,033
28	Bangalore, India	10,456	28	Hyderabad, India	12,774
29	London, UK	10,434	29	Bogota, Columbia	12,673

(continued)

Table 1.1 (continued)

Rank	City, Country	Population in 2010 (thousands)	Rank	City, Country	Population in 2030 (thousands)
30	Chennai (Madras), India	10,163	30	Paris, France	12,221
31	Lima, Peru	10,072	31	Moscow	12,200
			32	Bogota, Columbia	11,966
			33	Paris, France	11,803
			34	Jonesburg, South Africa	11,573
			35	Bangkok, Thailand	11,528
			36	London, UK	11,467
			37	Dar es Salam, Tanzania	10,760
			38	Ahmedabad, India	10,527
			39	Luanda, Angola	10,429
			40	Ho Chin Min city, Vietnam	10,200
			41	Chengdu, China	10,104

Source United Nations, Department of Economic and Social Affairs, Population Division (2016). The World's Cities in 2016—Data Booklet (ST/ESA/SER.A/392)

City centres stagnate as the economic base shifts outwards to peri-urban areas that are more attractive but less well regulated. Secondary cities and city systems become interconnected through manufacture and other business enterprises. Meta cities and Megacities are a key to globalisation, a state of interconnectedness around the globe that transcends and largely ignores national boundaries. Global urban economies rely on advanced producer services such as finance, banking, insurance, law, management consultancy, advertising and other services. The technology revolution has made it possible for business enterprises to hire these services anywhere in the world. There is no America First or Brexit impact on that and it has to continue if cities wants to retain the status as meta, mega and metro cities in the global network of cities in any part of the world.

Urban Growth the world's largest cities are given below. Thirty-six large mega and meta-cities by geography are also given in Fig. 1.3.

Asia leads in the number of megacities leaving far behind other regions such as Americas, Europe and African and there are 8 largest mega-cities are all in Asia. Further barring Moscow, all Europe mega cities are of smaller size. With urbanisation reaching almost saturation and lower population growth, it is unlikely these trends cannot change.

Urban growth in some largest cities in the world is given below. Growth rates of Asian large cities are much higher than that in the other regions which substantiate further the earlier statement (Fig. 1.4).

In 2016, 45 cities had populations between 5 and 10 million inhabitants. By 2030, 10 of these are projected to become megacities. UN Projections indicate that

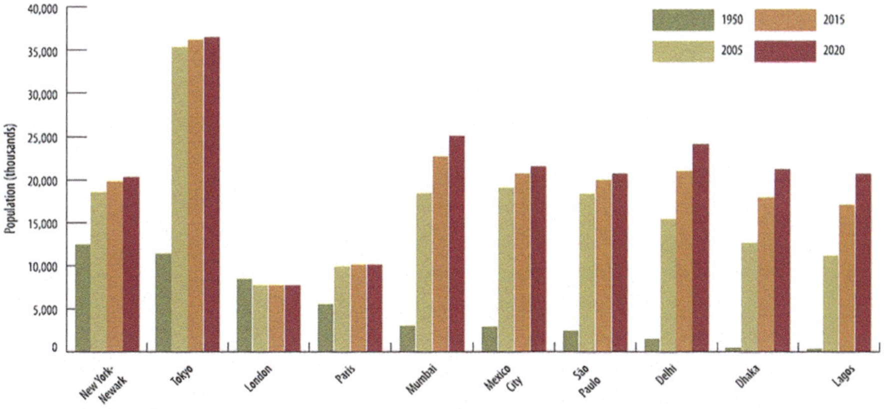

Fig. 1.4 Urban growth in the world's largest cities 1950–2020. *Source* United Nations, World Urbanization Prospects: The 2003 Revision. *Note* Population in 2020 was established in 2010 and 2015 assuming that trends for these years remain the same

29 additional cities will cross the 5 million marks between 2016 and 2030, of which 15 are in Asia and 10 in Africa. In 2030, 63 cities are projected to have between 5 and 10 million inhabitants. Megacities in 2016 and 2030 are given below.

An overwhelming majority of the world's cities have fewer than 5 million inhabitants. In 2016, there were 436 cities with between 1 and 5 million inhabitants and an additional 551 cities with between 500,000 and 1 million inhabitants. By 2030, the number of cities with 1–5 million inhabitants is projected to grow to 559 and 731 cities will have between 500,000 and 1 million inhabitants.

Mega-cities population were projected from 2016 to 2030 is given in Fig. 1.5. It can be seen again that Tokyo is slowing down while Delhi is growing faster and may overtake Jakarta approaching the second position. Many of the Asian cities and one African city Lagos are growing at much faster rate than European and American cities. If mega cities represent the economic hub of the future it is shifting towards Asia from America and Europe. Integrated spatial and economic strategies can help this transition. This is the subject matter of this book. As per the current trend GDP growth rate in Asia is much higher than Europe and Americas and it is likely to continue for few decades. It looks like an Asian and African era is emerging.

Growth rates of many mega cities in Asia and some in Africa are growing at higher rate than elsewhere.

In Table 1.2 Morphology of urbanisation of China, India and USA is compared. It is given here because this book carries case studies from these regions.

The largest urban population is in China followed by India and US. The total urban population in India is higher than the US. The percent of the urban population in mega cities in India are more than US and China in that order.

The population distribution of India and China are further studied in the following figures. China has all its metro and mega cities in the East facing the sea leaving a vast stretch in the western area without mega, meta and metro cities. China is fast moving towards a one billion urban population as shown in Fig. 1.6 by 2030.

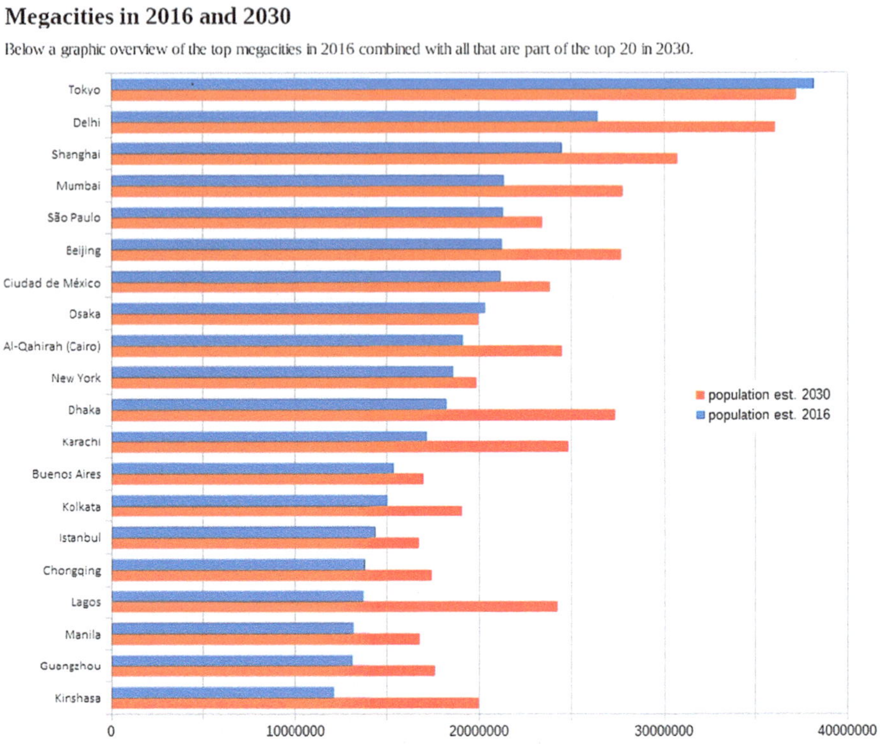

Fig. 1.5 Megacities population 2016 and 2030. *Source* United Nation, Department of Economic and Social Affairs, Population Division (2014). World Urbanization Prospects: The 2014 Revision

Table 1.2 Morphology of urbanisation India, China and USA 2010

Population size category of towns	% in urban population in 2010		
	India	China	United States
10 million or more	15	8	13
5–10 million	8	10	10
1–5 million	17	23	34
500,000–1 million	9	15	10
Fewer than 500,000	52	44	33
Total urban population (in 0000)	378,775	660,286	254,959
% urban population	30.9	49.2	82.1
Urban annual growth rate (%) 2005–2010	1.56	3.44	1.24

Source Population division of economic and social affairs of the United Nations secretariat, world population prospect. The 2010 revision and world urbanization prospects. The 2011 revisions 2014

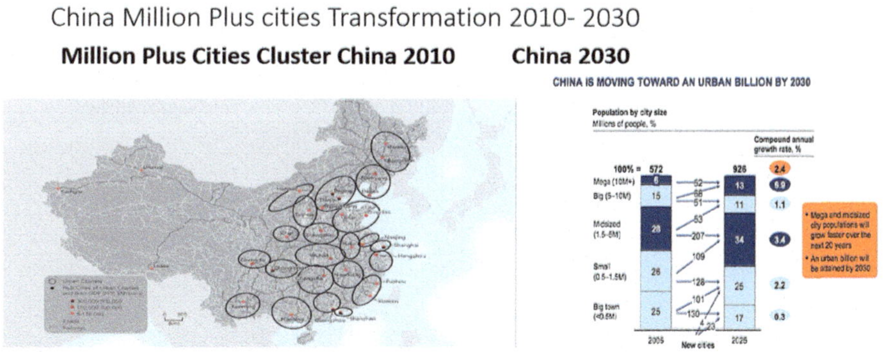

Fig. 1.6 Million plus city clusters in China 2010 and its transformation in 2030

In China, there are 655 cities, out of which 160 cities are 1 million and above which is more in number than India, 12 cities with a population of above one million to 5 million and 8 cities above 10 million population (megacities) by 2025 namely Beijing, Shanghai, Chengdu, Guangzhou, Shenzhen, Tianjin and Wuhan.

There is a large body of literature on Metropolitan development in India [27–34]. No attempt is made here to review of this literature but the data from these references is presented to showcase the momentum of metropolitan development in India. The growth of statutory towns and census towns in India for 2001 and 2011 is given below. It can be seen from the table below that the growth rate of census towns increased when compared with statutory towns (Table 1.3).

There were practically no immigrations to these census towns but showed a change in occupational pattern, (predominantly higher percent of non-agricultural in urban male occupation) and higher gross density, and several cases the population above 5000. The Urban situation of India from 1901 to 2011 is given below and several Urban Agglomeration and statutory towns in 2001 and 2011. The compound annual growth rate of urban growth has been fluctuating but India reached 31.2% urban in 2011 (Tables 1.4, 1.5 and Fig. 1.7).

Break up of cities in India which include 3 mega cities Delhi, Mumbai and Kolkata and remaining metro cities is given in Table 1.6. Further urban settlements and metro cities in 2011 are also given below. Among the urban population, the largest percent people in India live in megacities which are more than 10 million population (Fig. 1.8).

Table 1.3 Indian statutory and census towns in 2001 and 2011

Type of towns/UAs/OGs	Number of towns	
	2001	2011
Statutory towns	3799	4041
Census towns	1362	3894

Source Census of India 2001 and 2011 registrar general of census

Table 1.4 Urban situation in India 1901–2011

Census years	Number of towns/ UAs	Cities with population of 1 lakh and above	Urban population (in millions)	% urban population	Urban annual exponential growth rate
1901	1827	24	26	10.8	–
1911	1815	21	26	10.3	0.03
1921	1949	27	28	11.2	0.79
1931	2072	33	34	12.0	1.75
1941	2250	47	44	13.9	2.77
1951	2843	71	62	17.3	3.47
1961	2365	95	79	18.0	2.34
1971	2590	139	109	19.9	3.23
1981	3378	204	159	23.3	3.79
1991	4689	273	217	25.7	3.11
2001	5161	350	285	27.8	2.74
2011	7935	468	377	31.2	2.76

Source (i) Computed from Census of India 1991, Part-II A (ii) Towns and urban agglomerations classified by population in 1991 with variation since 1901 (iii) Census of India 2011, final population totals, registrar general of India. New Delhi

Table 1.5 Growth of metropolises in India 1901–2011

Census years	Number of metropolises	Population in millions	Decadal increase (%)	Population of metropolises as percentage of India's total population	Population of metropolises as percentage of India's total urban population
1901	1	1.5		0.6	5.8
1911	2	2.8	82.8	1.1	10.7
1921	2	3.1	13.4	1.3	11.1
1931	2	3.4	8.9	1.2	10.2
1941	2	5.3	5.7	1.7	12.0
1951	5	11.8	21.3	3.3	18.8
1961	7	18.1	54.0	4.1	22.9
1971	9	27.8	53.8	5.1	25.5
1981	12	42.1	51.3	6.2	26.4
1991	23	70.7	67.8	8.4	32.5
2001	35	107.8	52.8	10.5	37.0
2011	52	159.6	48.9	13.2	42.3

Source (i) Computed from Census of India 1991, Part-II A (ii) Towns and urban agglomerations classified by population in 1991 with variation since 1901 (iii) Census of India, 2011. Final population totals, registrar general of India, New Delhi

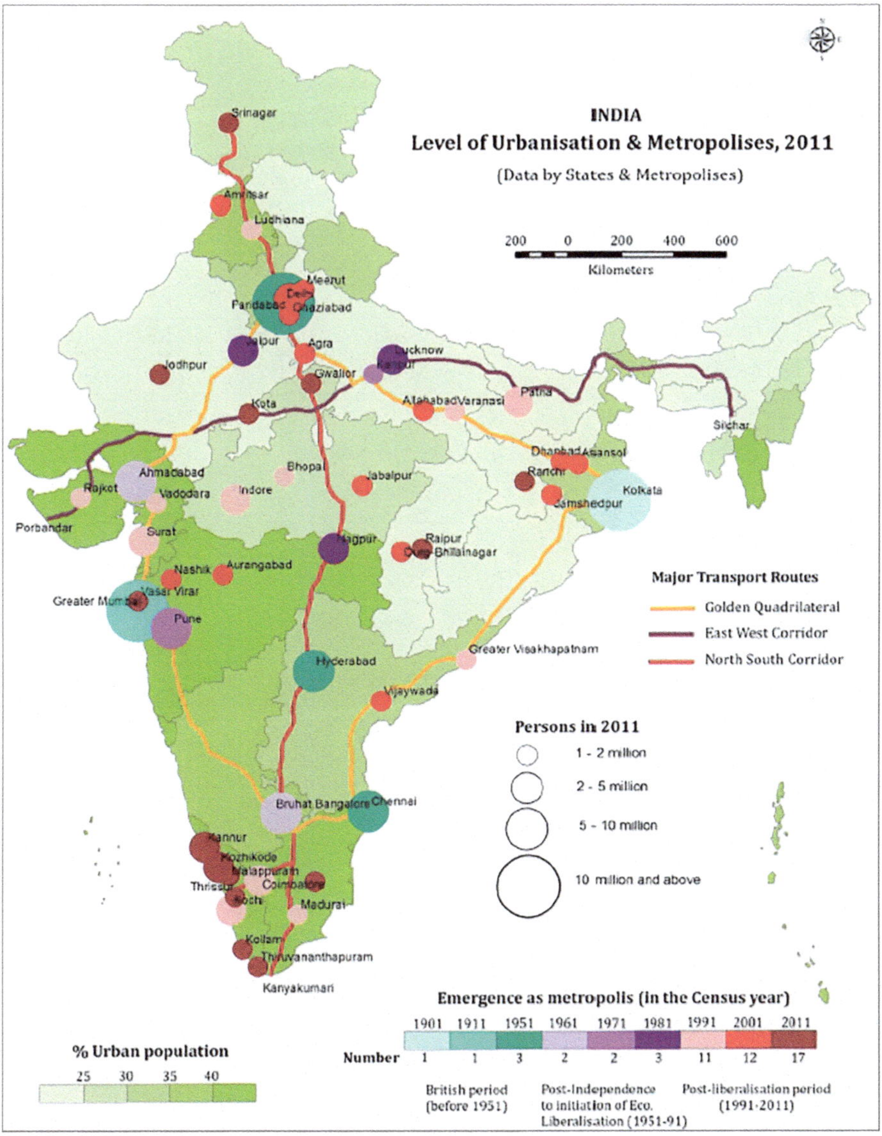

Fig. 1.7 Levels of urbanisation and metropolises in India 2011. *Source* Kumar [34]

Figure 1.8 represents the towns and cities in 1991 and number of metropolitan areas in 2011 in India. It can be seen unlike China the western part of India has more megacities and metropolitan cities than the eastern part. There is also near equitable distribution of mega and metro cities all around the Indian Union, unlike China.

Table 1.6 Configuration of metropolises in India (2011 Census)

State/Union Territory	Metropolises	No. of metropolises	Population of metropolises	State urban population	State % urban population	% share of metropolises
Kerala	Kochi UA, Kozhikode UA, Thrissur UA, Malappuram UA, Thiruvananthapuram UA, Kannur UA, Kollam UA	7	12,139,860	15,934,926	47.7	76.2
Uttar Pradesh	Kanpur UA, Lucknow UA, Ghaziabad UA, Agra UA, Varanasi UA, Meerut UA, Allahabad UA	7	14,025,098	44,495,063	22.3	31.5
Maharashtra	Greater Mumbai UA, Pune UA, Nagpur UA, Nashik UA, Vasai Virar (M.Corp.), Aurangabad UA	6	29,927,857	50,818,259	45.2	58.9
Gujarat	Ahmedabad UA, Surat UA, Vadodara UA, Rajkot UA	4	14,161,800	25,745,083	42.6	55.0
Madhya Pradesh	Indore UA, Bhopal UA, Jabalpur UA, Gwalior UA	4	6,428,127	20,069,405	27.6	32.0

(continued)

Table 1.6 (continued)

State/Union Territory	Metropolises	No. of metropolises	Population of metropolises	State urban population	State % urban population	% share of metropolises
Tamil Nadu	Chennai UA, Coimbatore UA, Madurai UA, Tiruchirappalli UA	4	13,278,580	34,917,440	48.4	38.0
Andhra Pradesh	Hyderabad UA, Vishakhapatnam (M.Corp.), Vijayawada UA	3	10,882,077	28,219,075	33.4	38.6
Jharkhand	Jamshedpur UA, Dhanbad UA, Ranchi UA	3	3,662,372	7,933,061	24.0	46.2
Rajasthan	Jaipur (M.Corp.), Jodhpur UA, Kota (M.Corp.)	3	5,186,157	17,048,085	24.9	30.4
Chhattisgarh	Raipur UA, Bhilainagar UA	2	2,187,780	5,937,237	23.2	36.8
Punjab	Ludhiana (M.Corp) Amritsar UA	2	2,802,428	103,991,346,375	37.5	26.9
West Bengal	Kolkata UA, Asansol UA	2	15,301,405	29,093,002	31.9	52.6
Bihar	Patna UA	1	2,049,156	11,758,016	11.3	17.4
Haryana	Faridabad (M.Corp.)	1	1,412,050	8,842,103	34.9	16.0

Source Census of India 2011, final population total, registrar general of India, New Delhi
Note M.Corp—Municipal Corporation

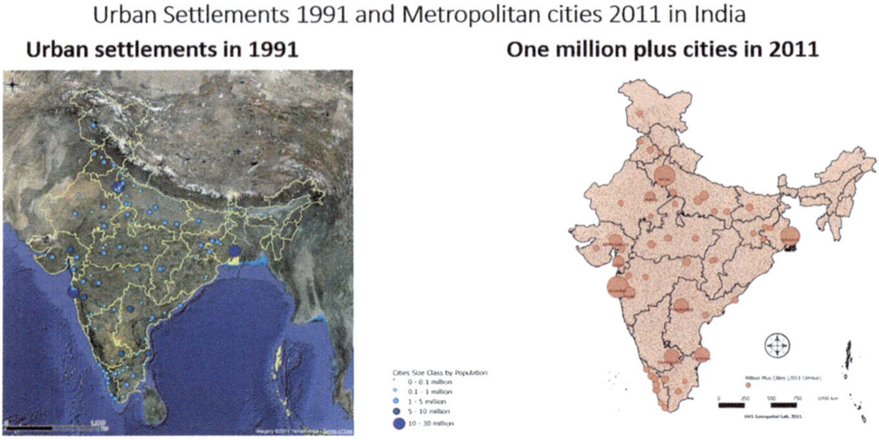

Fig. 1.8 The spatial distribution of urban settlement in 1991 and metropolitan and mega cities in 2011 in India

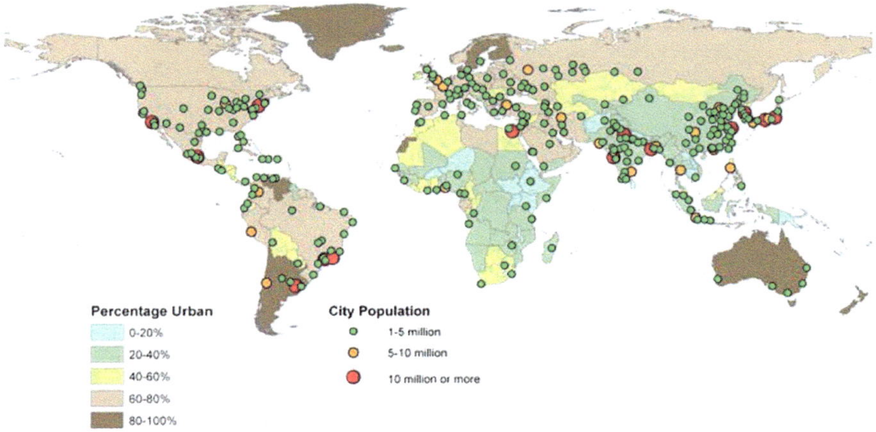

Fig. 1.9 Global pattern of urbanisation and large cities development in 1995. *Source* UNHABITAT (2016), World Cities Report (2016)

The global urbanisation and cities development is given in these two maps. It shows levels of urbanisation and settlement development 1995 and 2015 of metro cities and mega-cities as per UN statistics (Fig. 1.9).

Figure 1.10 gives Growth Rates of Urban Agglomerations in 2011–2025.

Figure 1.11 gives the spatial pattern of one million cities and above in 2015. Mega cities are more growing in Asia than Europe, Americas and Africa. These mega-cities have the capability to generate more per capita GDP. The rate of projected growth of Urban Agglomeration by size in the world is given below for the period 2011–2025 which substantiate the earlier statement.

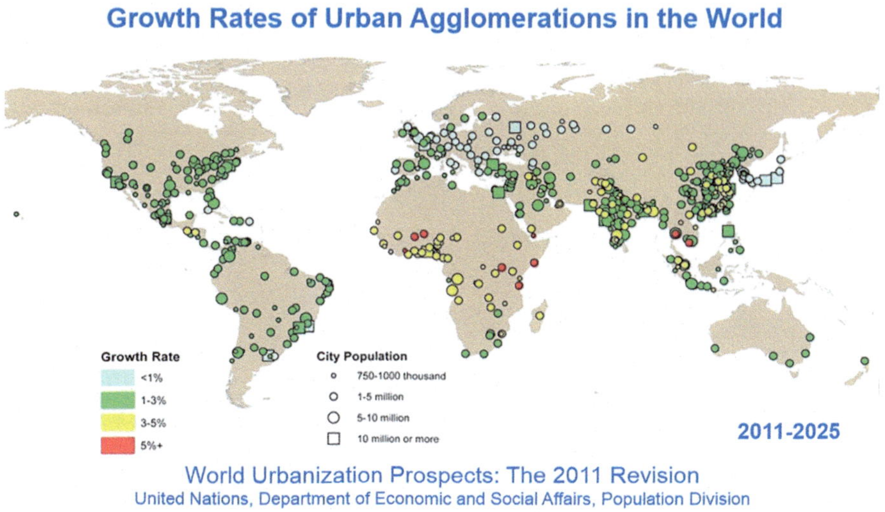

Fig. 1.10 Growth rates of urban agglomerations in 2011–2025

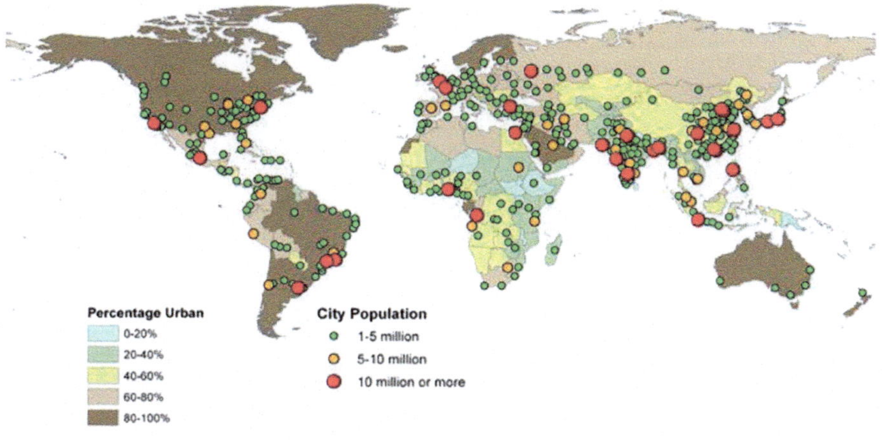

Fig. 1.11 Global patterns of urbanisation and large cities development in 2015. *Source* UNHABITAT (2016), World Cities Report (2016)

The following maps give how these growing urban agglomerations are connected there in the developed and developing the world. There are already networks and corridors cutting across the boundary of nations which has given rise to the formation of urban settlements in 2016 as given above in a globalised world. I am leaving to the readers to study these maps and establish the relationship of the emerging urban settlement morphology at the global level (Fig. 1.12).

Fig. 1.12 Urban networks

1.7 National GDP Share of Mega and Meta-Cities

Often the share of GDP produced in this mega and meta-city are considerable in comparison to total National GDP. The two graphs give how it varies in some selected cities. Figure 1.13 Share of GDP and National Population in Selected Developed Countries Cities. The graphs are self-explanatory (Fig. 1.14).

Table 1.7 arrange in the descending order, projected GDP 2025 of top 30 Urban Agglomeration and compute average real GDP/population growth rate projected during 2008–2025. This shows that GDP/population growth rates of largest 30 urban economies in the universe are generally low but there is few exceptions in the south and eastern urban agglomeration marked in Table 1.7.

Table 1.8 tabulates under descending order of high GDP growth rate of first 30 urban agglomerations by GDP growth rate 2008–25 and then cities are organised by

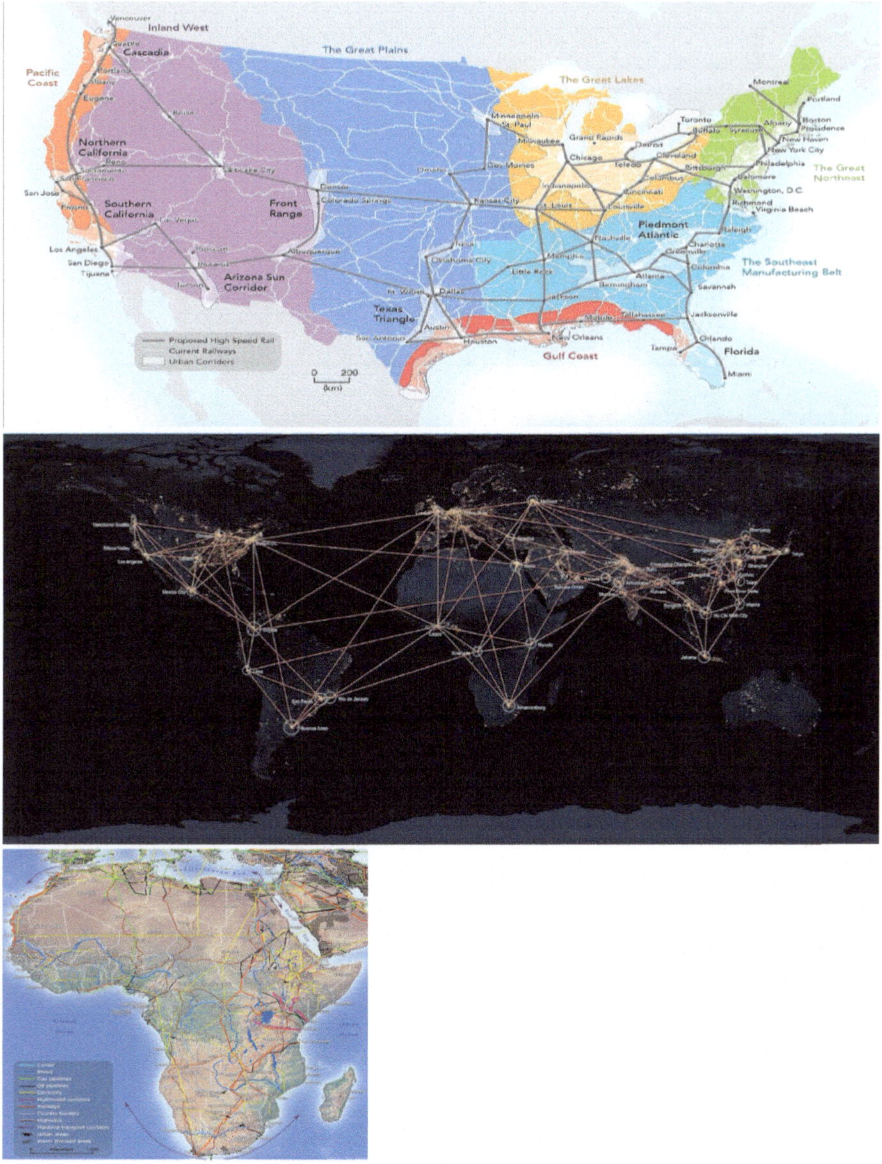

Fig. 1.12 (continued)

countries and found that these urban agglomerations are all in Asia and Africa. It can be seen the higher GDP performance cities are in large number in India followed by China and other countries. This justifies a maximum number of case studies in this book from Asia and Africa.

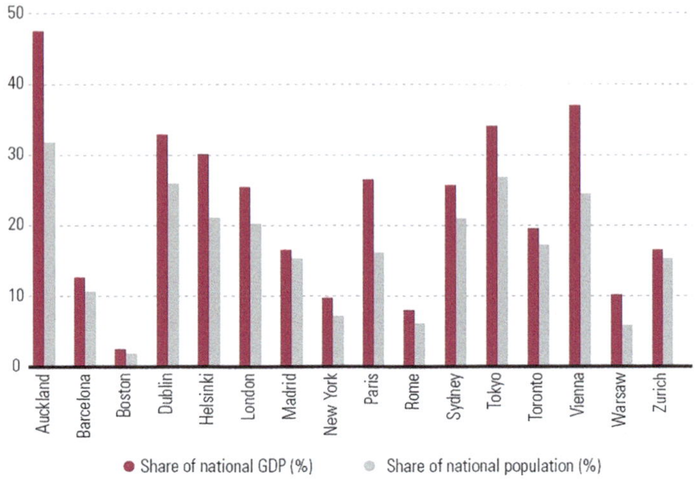

Fig. 1.13 Share of GDP of metropolitan cities in developed countries. *Source* UNHABITAT (2016), World Cities Report (2016)

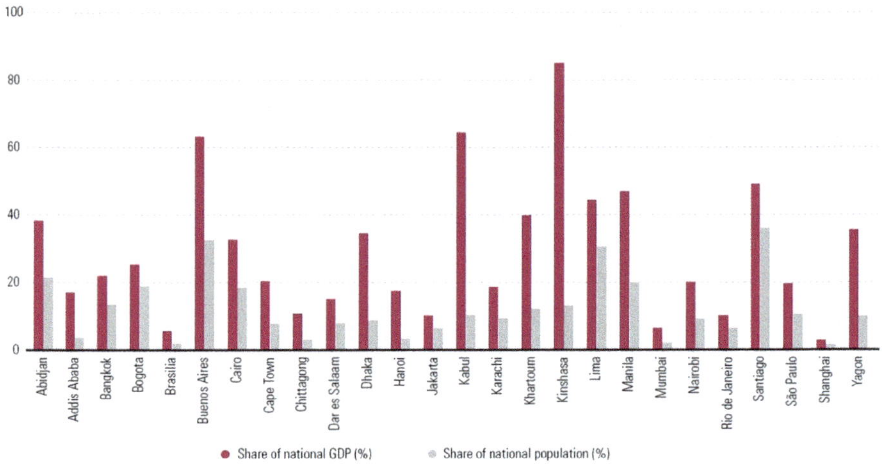

Fig. 1.14 Share of GDP and national population in selected cities in developing countries. *Source* UNHABITAT (2016), World Cities Report (2016)

1.8 The Metro/Mega/Meta City and Globalisation

Megacities, meta-cities and often metro cities are a key to globalisation, that brings about a state of interconnectedness around the globe that transcends and largely ignores national boundaries and slogans like America first, or Britain first. There is a division of labour across city regions transcending country boundaries sharing the

Table 1.7 Top 30 urban agglomerations by estimated GDP in 2025 (using UN population definitions and projections)

2025 GDP rank (2008 in brackets)	City		Estimated GDP in 2025 ($bn at 2008 PPPs)	Population in 2025 (millions)	Average real GDP growth (% pa: 2008–2025)
1(1)	Tokyo		1981	36.40	1.7
2(2)	New York		1915	20.63	1.8
3(3)	Los Angeles		1036	13.67	1.6
4(5)	London		821	8.62	2.2
5(4)	Chicago		817	9.93	2.1
6(10)	**Sao Paulo**	▲	**782**	**21.43**	**4.2**
7(8)	Mexico City		745	21.01	3.9
8(6)	Paris		741	10.04	1.6
9(25)	**Shanghai**	▲	**692**	**19.41**	**6.6**
10(13)	Buenos Aires		651	13.77	3.5
11(29)	**Mumbai (Bombay)**	▲	**594**	**26.39**	**6.3**
12(15)	Moscow		546	10.53	3.2
13(9)	Philadelphia		518	6.13	1.7
14(16)	Hong Kong		506	8.31	2.7
15(11)	Washington DC		504	4.89	1.8
16(7)	Osaka/Kobe		500	11.37	1.1
17(38)	**Beijing**	▲	**499**	**14.55**	**6.7**
18(12)	Boston		489	5.03	1.8
19(37)	**Delhi**	▲	**482**	**22.50**	**6.4**
20(14)	Dallas/Fort Worth		454	5.42	1.8
21(44)	**Guangzhou**	▲	**438**	**11.84**	**6.8**
22(21)	Seoul		431	9.74	2.3
23(17)	Atlanta		412	5.15	1.8
24(30)	**Rio de Janeiro**	▲	**407**	**13.41**	**4.2**
25(18)	San Francisco/ Oakland		406	3.90	1.8
26(19)	Houston		400	5.05	1.8
27 (20)	Miami		390	6.27	1.7
28(34)	**Istanbul**	▲	**367**	**12.10**	**4.2**
29(22)	Toronto		352	5.95	2.0
30(42)	**Cairo**	▲	**330**	**15.56**	**5.0**

▲ Rising by more than 3 places

Source PricewaterhouseCoopers projections

Table 1.8 Top 30 urban agglomerations by projected average real GDP growth 2008–25 (using UN population definitions and projections)

Growth rank	City	Country	Average real GDP growth in 2008–25 (% per annum)
1	Hanoi	Vietnam	7.0
2	Ho Chi Min City	Vietnam	7.0
3	Changchun	China	6.9
4	Guangzhou	China	6.8
5	Addis Ababa	Ethiopia	6.8
6	Xian	China	6.7
7	Surat	India	6.7
6	Beijing	China	6.7
9	Jaipur	India	6.7
10	Lucknow	India	6.6
11	Chengdu	China	6.6
12	Shenyang	China	6.6
13	Kanpur	India	6.6
14	Shanghai	China	6.6
15	Tianjin	China	6.6
16	Pune	India	6.6
17	Chongqing	China	6.6
18	Ahmedabad	India	6.5
19	Kabul	Afghanistan	6.5
20	Bangalore	India	6.5
21	Hyderabad	India	6.5
22	Dar es Salaam	Tanzania	6.5
21	Chennai (Madras)	India	6.5
24	Delhi	India	6.4
25	Lagos	Nigeria	6.4
26	Nairobi	Kenya	6.4
27	Kolkata (Calcutta)	India	6.4
28	Mumbai (Bombay)	India	6.3
29	Chittagong	Bangladesh	6.3
30	Kinshasha	Democratic Republic of Congo	6.3

Source PricewaterhouseCoopers projections using UN population definitions

global economic responsibilities as per capabilities. Global urban economies rely on advanced and standardises producer services such as finance, banking, insurance, logistics, law, management consultancy, advertising and other services. The technology revolution and uniformity of standard practices has made it possible for business enterprises to hire these services anywhere in the world. Intense use of ICT

in smart cities have an important role. Cities in the developing world have taken advantage of this to provide customer services at very competitive rates by drawing on a large labour force at low rates of pay. This trend, unfortunately, contributes to the creation of divided cities and the perpetuation of low income/slums. In recent time protectionism is emerging as a negative force against globalisation but only history can say how far protectionism became anti-globalisation and won the race, but it is more likely that Globalisation wins the race.
Examples of Cities that have become hubs of global economic activity are

1. Finance—Frankfurt, Hong Kong, Amsterdam, Singapore, Sao Paulo, Shanghai
2. Transport—Dubai, Rotterdam
3. Information technology—Bangaluru, Seattle, Silicon Valley.

Once the sole domain of Europe and North America, world cities are emerging in the developing world to compete with long-established capitals of commerce. Istanbul and Mumbai are culturally influential within their regions through film, literature, satellite TV networks and other forms of entertainment. Nairobi, Addis Ababa and Bangkok, amongst others, host regional headquarters for international agencies and development partners. Many of the metropolitan and megacities have become the seat of education and research bringing out innovations.

1.9 Emerging Global Functions of Large Cities

Large cities are interconnected and influence global and reinforce spatial interdependent functional structure with appropriate linkages. The connection is twofold within its city region and outside the city region transcending other national boundaries. The city is connected to hinterland and outside world simultaneously in a metropolitan region.

In 1950, there was 50% urban and 50% rural population in the world but thereafter urban population increased. Urban growth is often confronted with severe problems such as congestion, pollution, social segregation or high crime rates which threaten the achievement of a sustainable urban development. Consequently, many cities across the globe are engaging in an intense search for strategies which are suitable to address these issues. In several fields, such as public transportation, settlement and open space development as well as energy or water supply, solutions require a broader approach. Thus, both the city-region and metropolitan region are becoming increasingly important. The key role cities play in this context is two-fold: not only do they represent the centres in which economic and social change occurs, but also function as key 'actors' behind these changes. Globalisation and intense deployment of ICT in cities together result in the increasing importance of cities at the very top of the hierarchy, the so-called world cities or global cities.

1.10 Four Phenomena of Global Cities

1. Innovation capacity and the potential for economic development reside to a large part in metropolises and/or metropolitan regions. Research-intensive industries and knowledge-based services are becoming increasingly concentrated in metropolitan areas since there are reputed Universities and research institutions there. Innovation capacity and the potential for economic development reside to a large part in metropolises and/or metropolitan regions.
2. These areas not only enable innovative companies to exploit opportunities and to establish contacts but also allow them to access information and reduce risks. In addition, these areas also provide access to specialised resources and employees as well as to specific routines, traditions, values and other local institutions. Furthermore, urban agglomerations are characterised by a high degree of economic, social and cultural complexity functionally interconnected and embedded in the global division of labour, they, on the other hand, constitute the nodes of overlapping financial flows as well as trade, production, political, cultural and social networks.
3. The national hierarchy of cities and the spatial division of labour within the economy is superimposed by a global division of labour. Cities and metropolitan regions become part of an emerging international hierarchy based on a competitive division of labour at the global level by international connections that affect financial flows and the knowledge-intensive service sector. It leads to the emergence of a vertical hierarchy of globalised city-regions, the relationship between spatial decentralisation and territorial concentration alters the position of peripheral areas in this emerging spatial structure and new ones are created, unbalanced spatial development is reinforced and regions characterised by expansion, stagnation and negative growth co-exist. Highly-skilled employment, high-value infrastructure and investment increasingly concentrate in large metropolitan regions.
4. At the same time, these regions align their specialisation with the global division of labour and forge links with other dynamic metropolitan regions. In contrast, interconnections between metropolitan regions, their hinterlands and immediately surrounding (structurally weak) areas may well decrease. Thus, these areas are becoming increasingly disconnected from economic development. In view of the developments described above, the lack of suitable institutional structures that extend beyond the city-level becomes evident.

Alongside their effect on the role of metropolitan areas as the hubs of economic development, changing spatial patterns are also impacting the level of institutional action within the state hierarchy. State powers are not only being devolved to trans-national levels, but also to newly constituted tiers of metropolitan-regional governance. Considering the growing importance of regions, the need for adequate organisational structures ('for metropolitan governance') in metropolitan regions is greater than ever before.

1.11 World Cities and Global Cities

The term "world city" was coined by Patrick Geddes in his 1915 book, Cities in Evolution [35]. Geddes was "an unclassifiable polymath who officially taught biology (more probably, anything but biology) at the University of Dundee, gave India's rulers idiosyncratic advice on how to run their cities and tried to encapsulate the meaning of life on folded scraps of paper." Geddes's comments on world cities were mostly forgotten, however, in part because Geddes became so famous for his work on regional planning. Half a century later, however, Peter Hall catalysed a new generation of interest in the topic.

World cities are characterised by a sum of political power (both national and international) and organisations related to government; national and international trade, whereby cities function as gateway for their own and sometimes also neighbouring countries; providing superior banking, insurance and related financial services; advanced professional activities of all kinds; information gathering and diffusion. The form and extent of a city's integration with the world economy and the functions assigned to the city in the new spatial division of labour will be decisive for any structural changes occurring within it. Key cities throughout the world are used by global capital as 'basing points' in the spatial organisation and articulation of production and markets. The resulting linkages make it possible to arrange world cities into a complex spatial hierarchy. The global control functions of world cities are directly reflected in the structure and dynamics of their production sectors and employment. They are major sites for the concentration and accumulation of international capital. They are points of destination for large numbers of both domestic and/or international migrants. World city formation brings into focus the major contradictions of industrial capitalism, among them spatial and class polarisation. World city growth generates social costs at rates that tend to exceed the fiscal capacity of the state. Conspicuous consumption; arts, culture and entertainment and the ancillary activities that cater for them. Three main functions of world cities are headquarters function, financial centres function, and articulator' cities that link a national or regional economy to the global economy [28–30].

World Cities hypothesis of Freidman [36] is enumerated below.

1. The nature of a city's integration with the world economy is decisive for any structural changes occurring within it.
2. Key cities throughout the world are used by global capital as 'basing points' for the organisation of production and markets.
3. The global control functions of world cities are manifest in the structure of their industrial structure and job markets.
4. World cities are major sites for the concentration and accumulation of capital.
5. World cities are destinations for large numbers of migrants—both domestic and international.

6. World city formation exposes the major contradictions of industrial capitalism, particularly spatial class polarisation.
7. World city growth generates social costs that exceed the fiscal capacity of the state.

New technologies, telecommunications and information technology have led to both a decentralisation and agglomeration of economic activities. This combination of spatial dispersal and global integration 'has created a new strategic role for major cities', thus leading to a new type of city (essentially different from historical banking and trade centres). Global cities, for Sassan, form a 'virtual economic cycle' and function in four new ways: the demand for control creates cities as 'command points'; this leads to a demand for finance and business services, whereby cities become the 'key locations' for leading economic sectors; cities become sites of production and innovation for these leading economic sectors; cities constitute markets for main economic sector production [37].

1.12 Survey of Design of Economic and Spatial Strategies of Official Metropolitan Plans

There are two trends of thoughts about Meta cities, Megacities, and metro cities. One tries to attain a position of the global city by deliberately executing spatial and economic strategies to achieve that goal following what is discussed in the earlier paragraphs. This is in addition to solving many issues of metropolitan development for local people. The second approach is how to make a metropolitan area a smart metropolis by appropriate spatial and economic strategies. Smart is ICT and IOT implementation in all activities including economic activities. Economic strategies may involve converting the region with all its economic activities to the smart economy as discussed in "Smart Economy in Smart Cities" [18]. Since this book concentrates in a narrow area of design of economic and spatial strategies a survey, of such designs of several official plans is undertaken here. Cities are selected randomly with no sampling plans. It is also presented randomly with no regional clustering. This study is based on a classroom exercise conducted for Master's degree students of Planning in the National Institute of Technology Calicut by the author and the students involved are named in the acknowledgement before the References section. This survey will be conducted focussing on strategies will be briefly stated without any description of these metropolises and finally, these designs are analysed critically.

1.12.1 Case Study 1—Delhi NCR Metropolitan Region [38, 39]

The Delhi National Capital Region(NCR) comprises of regions from four different regions/states; National Capital Territory of Delhi, Haryana Sub-region comprising of Faridabad, Gurgaon, Rohtak, Sonepat, Rewari, Jhajjar, Mewat and Panipat districts, Rajasthan Sub-region comprising of Alwar district and Uttar Pradesh Sub-region comprising Meerut, Ghaziabad, Gautam Buddha Nagar, Bulandshahr and Baghpat. The central region is bustling with population leading to over-crowding and congestion due to its inherent capacity to generate more income and employment. The city limits are moving farther away from the city centre due to urban sprawl that creates spatial issues.

The main spatial strategy adopted was polycentrism by developing different nodes and the linkages via road and rail to arrest urban sprawl. This helps decentralise the core activities. NCR planning board has implemented zoning regulations to curb excessive urban sprawl and to preserve environmentally sensitive areas. Provision of ring roads to improve accessibility worsens the existing problem instead of the radial road system (Table 1.9).

The National Capital Region is rapidly emerging as a global economic hub and is among the fastest growing economic regions in India with a CAGR of GDP at 8.76% (2000–2008) to CAGR of GDP of India being 7.8% during the same period. There is a marked change in the economic structure of different Sub-regions. NCT of Delhi has become the main centre for tertiary sector activities. Uttar Pradesh and Rajasthan Sub-regions are still dominated by agriculture-led activities, while Haryana Sub-region is a mix of industrial and service sectors. Gurgaon district in Haryana sub-region has shown phenomenal growth in terms of per capita income, primarily due to the fast growth of IT and ITES sector, while Panipat district in Haryana Sub-region is evolving as a major trade centre. Alwar district in Rajasthan sub-region, on the other hand, remains an agriculture economy (Table 1.10).

The economic strategy adopted by the NCR planning board is not efficient enough for a city such as Delhi. Making trade and commerce barrier free which is the aim of Goods and Service Act 2017 nationwide within NCR will not suffice.

Table 1.9 Spatial concept and strategies of Delhi NCR

Objective	Strategy	Implementation
Polycentric model	Decentralisation	Infrastructure development is proposed at different nuclei to concentrate development at strategic locations
Socio-economic development	To provide access to education and health facilities, agricultural extension services and agro-industries based on local products	New townships are proposed along key transport corridors, expressways, orbital rail corridors
To contain urban sprawl	Controlled/development/ regulated zone	Urbanisable area, agricultural (rural) zone within controlled/development/ regulated area and green buffers
To preserve eco-sensitive areas	Zoning	Natural conservation zone, agriculture (rural) zone outside controlled/development/regulated areas

Table 1.10 Economic concept and strategies-Delhi NCR

Objective	Strategy	Implementation
To strengthen economic base functions	Counter magnet towns for decentralised economic development	To develop inter-regional linkages such as expressways, transport corridors
To remove barriers to trade with uniform tax	Single economic zone	To implement a uniform tax base and transparent system
To increase the ease of carrying out business	Single window facility	To set up single window system of registration, financing and industrial regulatory measures
To integrate and promote informal sector participation	Specialised streets/spaces	To provide parking and other open spaces To be located strategically near work centres, commercial areas, near schools, colleges, hospitals, transport nodes

Other specific issues mainly administrative issues are not addressed for further easing to do business and giving impetus for the development of industries. Interstate agreement on unified policies can create chaos and may reach a long time to reach consensus. This might be a hindrance to further development of NCR region. Increased ease of doing business, transparent laws and regulations for the same have not been given due consideration. Also, focus on people and skill development for economic development is not visible.

1.12.2 Case Study 2—Vancouver Metropolitan Region [40, 41]

The Metro Vancouver region includes twenty-one municipalities, one Treaty First Nation and one unincorporated area, each with its own unique physical, demographic and local economic characteristics. The region has great strengths on which to focus a prosperity initiative. It is recognised globally for its overall livability and has done especially well in international surveys of metropolitan areas on attributes such as tolerance, the presence of amenities and quality of place.

Metro Vancouver is a consortium of 21 municipalities which requires a very high level of agreement among the consortium municipalities on policy and actions. Regional Growth Strategy's ability to coordinate land use is governed by municipalities, and major transportation facilities are governed by the province. The future may require stronger Metro Vancouver political leadership through a part of the Board being directly elected rather than appointed by to provide equitable and efficient decision making and services (Table 1.11).

Table 1.11 Goals and spatial strategies Vancouver

Goals	Strategies
To create a compact urban area	– Contain urban development within the urban containment boundary – Focus growth in urban centres and frequent transit development areas – Protect rural areas from urban development
To support a sustainable economy	– Promote land development patterns that support a diverse regional economy and employment close to where people live – Protect the supply of industrial land – Protect the supply of agricultural land and promote agricultural viability with an emphasis on food production
To protect the environment and respond to climate change impacts	– Protect Conservation and Recreation lands – Protect and enhance natural features and their connectivity – Encourage land use and transportation infrastructure that reduce energy consumption and greenhouse gas emissions, and improve air quality – Encourage land use and transportation infrastructure that improve the ability to withstand climate change impacts and natural hazard risks
To develop complete communities	– Provide diverse and affordable housing choices – Develop healthy and complete communities with access to a range of services and amenities
To support sustainable transportation choices	– Coordinate land use and transportation to encourage transit, multiple-occupancy vehicles, cycling and walking – Coordinate land use and transportation to support the safe and efficient movement of vehicles for passengers, goods and services

Transit Oriented Development (TOD) and urban compaction have been deployed so that major employment and trip generating activities are concentrated in the centre to reduce the trip length and hence pollution. Mixed employment areas are provided while planning land use which is a preferred and unconventional approach. A mix of housing types suitable to an ageing population, changing family and household characteristics are provided which indicates inclusive development for all sects of the society. In city planning in Vancouver non-motorised mode and the barrier-free environment is given top most priority. Their strategies and actions also recognise the importance of providing connectivity throughout the region linking important natural features and enhance natural assets. The strategies adopted are smart, eco-friendly and sustainable and people inclusive in the long run.

Vancouver has a diversified economy dominated by the service sector which allows for greater responsiveness and resiliency in the face of changing economic trend. It supports the emergence of industry clusters important to competitive metropolitan economies. Metro Vancouver, situated on Canada's West Coast and at

the mouth of the Fraser River, became a major centre for trade. With the economic importance of the Asian Pacific economies, this role as a gateway will continue to grow. Business and labour migrate freely across municipal boundaries to perform activities. The number of businesses and jobs related to cargo handling, storage, distribution and transportation have grown along with the port. Tourists and conventions are attracted to the natural beauty as well as to the recreational and cultural opportunities within the region.
The main issues noted here are (Table 1.12):

- Deteriorating affordability
- Competition for land
- Investment in transportation and transit infrastructure
- Climate change.

In the case of Vancouver, there is a strong economic strategy in place. Local businesses and talents are nurtured which results in a booming diverse economy. Brain–drain is prevented by attracting foreign Canadians and immigrants with attractive and affordable housing and environment for families. This is a strategy to attract a skilled workforce. Integration of universities with business centres for productive R&D yet another important step in economic growth. They have thoroughly invested in clustered growth of industries with increased focus on green jobs to remain sustainable in the long run.

1.12.3 Case Study 3—Melbourne 2030—Planning for Sustainable Growth [42–44]

Melbourne 2030 Plan was based on inclusiveness, equity, leadership and partnership. The Goals laid out in the plan were a more compact city, better management of metropolitan growth, network with the regional cities, a more prosperous city, a great place to be, a fairer city, a greener city, better transport links, better planning decisions, and careful management.
A more compact city:

- Build up activity centres as a focus for high-quality development, activity and living for the whole community.
- Locate a substantial proportion of new housing in or close to activity centres and other strategic redevelopment sites that offer good access to services and transport.

Better management of metropolitan growth:

- Establish an urban growth boundary to set clear limits to metropolitan Melbourne's outward development.
- Concentrate urban expansion into growth areas that are served by high-capacity public transport.

Table 1.12 Economic concept and strategies-Vancouver

Focus areas	Goals	Strategy
Climate for growth	Promote innovation	– Innovation zones – Early adoption of technologies – Smart government practices – Pre-procurement strategies
	Protect and enhance job spaces	– Plan for job spaces of the future – Support speciality spaces for key sectors – Work/live neighbourhoods
	Leverage and support partnerships	– Building strategic global partnerships – Engage more effectively with all governments – Strengthen local partners and their work
	Communications	– Global communications campaign – Local outreach business and levering social media – Work with partners to deliver a stronger business message
Focus on people	Protect and enhance livability and affordability	– Engage businesses in affordability planning – More affordable day care – Address other livability/affordability barriers
	Attract, retain, retrain and repatriate talent	– Protect quality of living for talent – Retraining and skill development programs – Target Canadians abroad – Support employers with recruitment – Help strengthen enabling legislation
Supporting investment and trade	Levering event and programs	– Lever major events – Address gaps and establish Vancouver tech centre – Identify competitive advantages for cluster growth
	Global trade	– Launch global trade program – Support for mature and emerging sectors – Target new trade regions especially Asia
	Emerging sectors	– Execute greenest city jobs and economy – Launch creative city strategic planning – Support science and ICT clusters
	Business works	– Launch Business works targeting 450 companies a year – Business matching, incubation and coaching – Facilitate access to capital and markets – Build a database and research program

Networks with the regional cities:

- Promote the growth of regional cities and key towns on regional transport corridors as part of a networked cities model.
- Control development in rural areas to protect agriculture and avoid inappropriate rural residential development.

A more prosperous city:

- Maintain access to productive natural resources and an adequate supply of well-located land for energy generation, infrastructure and industry.
- Strengthen Central Melbourne's capital city functions and its role as the primary business, retail, sport and entertainment hub for the metropolitan area.
- Further, develop the key transport gateways and freight links and maintain Victoria's position as the nation's premier logistics centre.
- Create opportunities for innovation and the knowledge economy within existing and emerging industries, research and education.
- Encourage the continued deployment of broadband telecommunications services that are easily accessible.

A great place to be:

- Promote good urban design to make the environment more liveable and attractive. Improve community safety and encourage neighbourhood design that makes people feel safe.
- Protect heritage places and values: Promote excellent neighbourhood design to create attractive, walkable and diverse communities.
- Improve the equality and distribution of local open space and ensure long-term protection of public open space: Rectify gaps in the network of metropolitan open space by creating new parks and ensure major open space corridors are protected and enhanced.
- Maintain and develop metropolitan Melbourne as a desirable tourist destination.

A fairer city:

- Increase the supply of well-located affordable housing, plan for a more equitable distribution of social infrastructure, develop a strong cultural environment and increase access to arts, recreation and other cultural facilities.

A greener city:

- Protect habitat and areas of important biodiversity through appropriate land-use planning.
- Promote the concept of sustainability and develop benchmarks to measure progress.

Better transport links:

- Upgrade and develop the Principal Public Transport Network and local public transport services to connect activity centres and link Melbourne to the regional cities.
- Review transport practices, including design, construction and management, to reduce environmental impacts.
- Give more priority to cycling and walking in planning urban development and in managing our road system and neighbourhoods Promote the use of sustainable personal transport options.

Better planning decisions, and careful management:

- Achieve better planning decisions,
- Speed up resolution of appeals,
- Keep Melbourne 2030 up to date,
- Develop a strong partnership with local government, Implement Melbourne 2030 in an integrated way that involves the community.

Delivering a new integrated economic triangle consisting of Good Governance and strong partnerships, strengthening the competitiveness of Melbourne's employment and reduce the cost of living by increasing housing supply near services and public transport. Facilitate supply of affordable housing, expand the central city to retain competitive advantages and attract diverse value-added businesses. Prepare a plan for the central subregions to accommodate at least one million jobs and people. Facilitate the development of national employment clusters, support development of metropolitan activity centres, identify new development and investment opportunities.

Plan Melbourne does not identify how the government will respond to the impacts of climate change. The Metropolitan Planning Authority must work in close partnership with Local Government, in the future planning for Urban Renewal Areas. Initiatives to 'streamline' the planning system which limits community involvement in the planning process, or reduces Local Government's decision-making responsibilities, are not supported.

1.12.4 Case Study 4—Shanghai Metropolitan Plan 2001–2020 [45–51]

Shanghai is one of the four direct-controlled municipalities of the People's Republic of China. It is one of the most populous and dense cities in the world, with a population of more than 24 million as of 2014. It is a global financial centre and transport hub, with the world's busiest container port. The municipality borders the provinces of Jiangsu and Zhejiang to the north, south and west, and is bounded on the east by the East China Sea.

The issues encountered are a lack of carrying capacity of the land, the difficulty of land use structure, the difficulty of the renovation and expansion of municipal facilities, high-density city, environmental pollution, and energy saving pressure, rising sea level and other potential natural disasters.

Shanghai Metropolitan Master Plan (1999–2020) postulates "One dragon head, four centres"—dragon head of Yangtze River Region; international economic centre, financial centre, trade centre, and logistics centre in 2020 with Central Shanghai: financial and insurance; IT high tech, residents in 2020—16 million, and urbanisation rate of 85%. Infrastructure postulates "three harbours & two ways" (seaport, airport, cyber harbour; high-speed railway, expressway); "two networks" (rail transit network; expressway network) and green space in 2020—10 m^2 per capita.

Shanghai is China's most powerful economic centre, to drive the growth of the nation; with a leading portal for international business, to facilitate global cooperation and exchange; Global high-tech manufacturing industry centre; and a world-class metropolitan region.

"One Core, Six Development Zones": City development is based on "multi-axis, multi-layer, multi-core", space layout of central city "multi-nuclear, open" City evolved to Multi-centre network urban system trying to make Shanghai a Compact city.

The ecological strategy involves constructing an ecological network covering Shanghai region and connecting the Yangtze River Delta.

The main economic activities in core districts include commerce, tourism, professional services, and some traditional old industries. The main economic activities in the outlying districts are very diversified, ranging from manufacturing industries to commerce, agriculture, tourism, storage and transportation. Shanghai is now facing great opportunities in turning itself into an international economic, financial, and trade centre. China's continuous economic growth provides a solid base for Shanghai to move toward this goal. Shanghai has set its long-term strategic objectives for social and economic development. By 2010, Shanghai is planned to become one of the international economic, financial and trade centres of the world. Initially, to form the economic scale and comprehensive strength of a world metropolis; optimise urban spatial distribution, initially, modernise the city's physical infrastructure, participate in international labour division and the circular flow of the international economy, introduce the operational mechanism of a socialist market economy, and pursue the balanced social, economic and environmental development. The main goals were to build the economic scale and comprehensive strength of a world metropolis by optimise urban space distribution, modernise the city's infrastructure, participate in the international labour division and the circular flow of international economy, introduce the operational mechanism of a socialist market economy, and pursue balanced social, economic and environmental development. Economic Strategies plan is to optimise and upgrade the industrial structure, trying to improve the pivotal status of Shanghai in global city network and process to accelerate the technological innovation.

1.12.5 *Case Study 5—Mumbai Metropolitan Regional Development [52]*

Mumbai Metropolitan region consists an area of 4311.75 km^2 which includes two full districts of Mumbai City, Mumbai Suburban and three-part districts of Thane, Raigad and the newly created Palghar district. The boundary of the area includes river Tansa in North, administrative boundaries of Bhiwandi, Kalyan and Ambernath Tehsils in North-East, Sahyadri in South-East, river Patalganga in South and the Arabian Sea on West. Maharashtra is currently the third urbanised state in India. The urban population of Maharashtra constitutes 45.23% of the total population of the State and Urban population growth accounted for 62% of the total population growth in the State. Nearly 45% of this urban population lives in MMR (Mumbai Metropolitan Region). MMR includes 8 municipal corporations, 9 municipal councils, 35 census towns and 994 villages.

Rural-urban imbalance: There is an imbalance between the development of the urban and rural areas in the region in terms of access to social and physical infrastructure and livelihood opportunities. The urban areas of the region have better access to infrastructure while several pockets of inadequacy exist in the rural areas of the region, especially in the largely tribal northeastern belt.

Slum Proliferation: Lack of access to affordable formal housing and lack of access to housing finance because being employed in the informal sector are the two major reasons for the growth of slums.

Transport: The suburban rail transit network is the backbone of the region's economy. This suburban rail network is currently stretched beyond its maximum capacity. Peak hour commuters are subject to inhuman levels of crowding with each car carrying more than double the stated carrying capacity of 1800 passengers. Limited land availability is a major constraint to augmenting services which are currently run at maximum efficiencies. Not all existing rail lines are computerised. If services on such lines are augmented or if the suburban networks are extended, it would be beneficial.

Environment: Environmental pollution due to increasing urbanisation is resulting in poor quality of life and directly impacting the livelihoods of those engaged in fishing and agriculture in the region. Increased air and water pollution are observed across the region due to increased vehicular emissions, industrial pollutants, construction activity, quarrying, untreated or poorly treated effluent discharge from domestic as well as industrial sources as well as inadequate solid waste disposal capacities and measures.

Infrastructure: Dense living conditions in several metropolitan areas in the region, viz. Greater Mumbai, Bhiwandi, and Ulhasnagar have resulted in the inadequate provision of social infrastructure. Peri-urban areas have poor social and physical infrastructure given the lack of governance frameworks available in these areas. MMRDA was unable to provide infrastructure services where the land resource is not available to them to raise finances. A significant proportion of the urban areas

Table 1.13 Spatial concepts and strategies-Mumbai

Concept	Strategy
Increase public transit connectivity in the region	• Extend suburban rail connectivity to peripheries of MMR • Increase suburban stations • Create new transit corridors • Develop multi-modal corridor • Use transit to spur development in current under-developed areas
Framework for addressing the housing problem	• Policy to enable the market to meet the need • Creating land-banks for intervention by the govt. • Regulatory and institutional arrangement for housing in MMR
Create regional scale open space networks	• Create a blue-green regional network that links heritage sites and tourist sites • Create a greenway along the multi-modal corridor • Preserve all surface water bodies in the region • Maintain a buffer around all major rivers, water bodies and forests and create a network of open spaces throughout the region
Extend governance framework to peri-urban areas	• Extend current municipal boundaries to include peri-urban areas that are rapidly urbanising • Create decentralised sub-regional offices of MMRDA to realise local strategies
Creation of regional infrastructure	• Provide regional scale infrastructure like regional parks, sports complexes, large hospitals, educational institutions, fire services, landfills etc.
Zoning	• Simplified zoning
Regional information system	• Create a Regional Information System (RIS)

along with the rural areas in the region is underprovided in terms of sanitation systems and in terms of effective solid waste management systems (Table 1.13).

The Spatial strategies formulated tried best to cover every important aspect. The focus is given to transportation and housing.

The secondary sector in the region has been declining since the 1990s-following globalisation and liberalisation. This period has witnessed the flight of manufacturing from the region, starting with the closure of the textile mills in Greater Mumbai and the increasing non-availability of skilled jobs. The lack of jobs that match the skilled manpower available within the region has led to increasing formalisation of work. The CAGR of the contribution to the NGDP of MMR of the secondary sector between the periods 2000–2009 had decreased to 4.56 as against 6.27 during 1999–2003. Policies encouraging the conversion of industrial lands to residential use also aided this decline of the secondary sector and enabled the growth of the real estate market at the expense of manufacturing.

A concentration of employment opportunities in Greater Mumbai has resulted in imbalanced growth in the region. Currently, sixty percent of all the formal jobs in the region is in Greater Mumbai, though there is an increasing movement northward to the suburbs from the Island city. Since the Regional Plan 1996, new CBDs have emerged in Greater Mumbai at Bandra-Kurla, Lower Parel, Andheri and Powai, which is now indicating a polycentric growth pattern within Greater Mumbai. However, apart from Navi Mumbai and Bhiwandi, there are not many employment opportunities in the cities outside Greater Mumbai, resulting in long commutes from the cities outside Greater Mumbai that largely serve as dormitory cities (Table 1.14).

The Economic strategy given by the metropolitan plan is very broad. Mumbai Metropolitan Regional Plan 2016-36 mainly addressed issues of growing urbanisation, uneven distribution of jobs increasing commutes, lack of affordable housing and basic infrastructure in the region, environmental degradation and inadequate governance. The existing situation of Mumbai Metropolitan Region is analysed sectoral and cross-sectoral to arrive at issues that need to be addressed in the Regional Plan 2016-36.

1.12.6 Case Study 6—New York [53–55]

The New York metropolitan area, also referred to as the Tri-State Area, includes New York City; Long Island and the Mid- and Lower Hudson Valley in the state of New York. Lower Hudson Valley, consisting of Putnam, Rockland, and Westchester counties. New York City, consisting of Bronx, Queens, New York (Manhattan), Kings (Brooklyn), and Richmond (Staten Island) counties and Long Island, consisting of Nassau and Suffolk counties. The tristate region compares to other metropolitan areas with respect to economic disparities, segregation, access to quality primary schools, intergenerational mobility and quality of life indicators. The New York area fares worse than most comparable regions in terms of inequality and segregation. Intergenerational mobility in the region is high in

Table 1.14 Economic concepts and strategies-Mumbai

Concept	Strategy
Enable balanced regional development	• Create new growth centres/employment hubs dispersed across the region • Create Local Development Centres (LDCs) as rural hubs • Promote tourism and • Encourage primary sector livelihood opportunities
Encourage manufacturing in MMR	• Demarcate new industrial zone in the region • Provide logistic parks near the port • Encourage SMEs across the region
Increase public transit connectivity in the region	• Extend suburban rail connectivity to peripheries of MMR • Increase suburban stations • Create new transit corridors • Develop multi-modal corridor • Use transit to spur development in current under-developed areas

comparison to other parts of the country. The region is faced with a daunting set of challenges, yet has the potential to leverage strengths in promoting policies that afford equal opportunities to live full, healthy and productive lives regardless of starting point for all residents (Table 1.15).

The plan mainly focuses on the concept of Quality of Life. It tried to find a solution to the prevailing economic disparity and social segregation in New York. It gave less important to utility plan, but they have a separate transportation plan for New York Metropolitan area.

The tristate region has one of the largest, most dynamic and complex economies in the world. It produces nearly $1.5 trillion goods and services every year. Increasingly, this economy is driven by the region's ability to attract, educate and retain a workforce with diverse skills and creative talent.

Table 1.15 Spatial concepts and strategies-New York

Concept	Strategy
Revitalise urban and suburban downtowns	• Support transportation, water and sewer infrastructure investments that facilitate equitable and sustainable redevelopment • Promote tools such as generic environmental impact statements or tax increment financing that facilitate development
Increase supply of affordable housing	• Reform zoning and related regulations to permit higher densities, more multi-family development and accessory apartments • Revise permitting, approvals, labour regulations and other actions to reduce construction costs • Expand the use of inclusionary zoning to leverage market rate development for low- and moderate-income housing
Promote mixed-income, multi-racial neighbourhoods	• Create effective state fair share regulations to require and incentivize affordable housing in areas with high opportunity • Strengthen housing preservation in low-income and gentrifying neighbourhoods • Create regional institutions to develop metropolitan housing plan; set municipal targets and distribute housing vouchers
Create healthier and more resilient communities	• Promote policies and design guidelines, such as complete streets legislation, that improve safety, physical activity and public amenities • Advocate for reform of federal financing guidelines to promote mixed-use, walkable neighbourhoods • Improve access to quality parks and a network of regional greenways • Build in resiliency standards and guidelines to new development and building retrofits
Improve access to jobs and entrepreneurship	• Extend rail and bus service and improve service to underserved neighbourhoods • Expand and subsidise demand-response transportation service for "last mile" service or low-density areas • Implement reduced fares for low-income transit users • Build municipal and regional broadband services to expand coverage and lower cost
Reform property taxes and education finance	• Consolidate school districts to reduce costs and expand access to quality schools • Implement regional tax-sharing, building on models such as in the NJ Meadowlands • Replace local school taxes with statewide education taxes

(continued)

Table 1.15 (continued)

Concept	Strategy
Transportation strategy	
Enhance the regional environment	• Reduced traffic congestion and improved air quality; • Reduced greenhouse gas emissions; • Improved water quality; and • Preservation of open space, especially wetlands
Improve the regional quality of life	• Increased intra-regional mobility and accessibility for commuting, recreation and tourism • Increased ability to safely enjoy walking, bicycling and use of public space • Complete planning and/or environmental assessments for the following transit-oriented development • Transportation improvement vision projects linked to land use plans
Provide a convenient and flexible transportation system within the region	• A sufficient array of transportation choices; • Expanded connections, particularly across modes and between communities; • Increased reliability for passenger and freight trips; and • Implement congestion-related improvements and enhancements
Enhance the safety and security of the transportation system for both motorised and non-motorized users	• Reduced rate of annual injuries and fatalities on the region's transportation systems; • Promulgation of advanced safety and security measures throughout the region

Economic concepts and strategies

The Western New York has adopted an integrated approach in building economic strategies primarily to strengthen fundamentals the to build the target industries.

Strengthen Fundamentals
The strategy focuses on three fundamental issues shown here, that if addressed appropriately can create the right environment for job growth and wealth creation. With a stronger workforce, more efficient infrastructure, and a vibrant culture of entrepreneurship, New York can turn the economy around.

Build Target Industry Sectors
The strategy also puts forward ways to capitalise on opportunities in eight industry sectors. These are industries where New York already enjoy high concentrations of employment, potential growth in jobs and wages is greatest, and possess unique asset or advantage.

The economic strategy adopted clearly touches every aspect of the economy. The plan also emphasised on strengthening the fundamentals such as workforce which is very important in achieving economic development.

The New York Metropolitan Region have a separate spatial plan, transportation plan and economic development plan. The strategies adopted clearly ensure sustainable and long-term growth in jobs and income to contribute to the resurgence of the broader economy of New York State. Also, they have a detailed transportation development plan which has a shared vision for sustainability.

1.12.7 Case Study 7—Calgary Metropolitan Region [56]

The Calgary Region is an area centred on Calgary, Alberta, Canada. It consists of the City of Calgary, Rocky View County and the municipalities it encloses. The

Calgary Region is a major transportation hub for southern Alberta, Saskatchewan, eastern British Columbia, and parts of the northern United States. It is home to the Calgary International Airport, the third busiest airport in the country in terms of total aircraft movements. it the largest CMA in Alberta and the fifth largest in Canada. it had a population density of 237.9 people per square kilometre in 2011.

This region has plenty of natural landscapes, ecologies and water sources which are unidentified. There is a lot of potentials to develop and make the total region as harmonious. In the present situation, people are widely spread which implies less density and no continuous development throughout the region. Most of the land is not efficiently used. There is scope to develop scenic corridors. Presently percentage of mixed land uses is less.

Strategically adopted accommodating growth in more compact settlement patterns with higher density infill development across the region makes good sense for the environment, the economy, and for the quality of life. This type of managed growth and development has many benefits, including preserving agricultural land, reducing infrastructure expenses, promoting public transit and encouraging the effective use of green spaces. Development within priority growth areas should be built out in a contiguous and connected manner. New housing in priority growth areas will be near local and regional transit service. Unplanned lands (greenfield areas) in existing priority growth areas and lands identified for new priority growth areas will be designated and approved in local plans to achieve a minimum density of eight to 10 units per gross residential acre, and encourage higher densities where achievable. Member municipalities will strive to accommodate at least 25% of new population growth across the region through intensification of existing developed areas. Building sustainable and resilient communities require planners to consider all aspects of community life, including social, environmental and economic factors. The Calgary Metropolitan Plan helps by providing a framework for member municipalities to use as they decide how to create their own unique, sustainable communities. Member municipalities will identify and protect important historic and cultural resources that contribute to the unique character of existing communities. All new and existing priority growth areas will develop and maintain one or more viable mixed-use activity centres, integrated with local and regional transit stations, terminals and services, to promote local employment opportunities and transit accessibility. A minimum intensity threshold of 100 people or jobs/gross developable hectare, within walking distance of regional transit services, should be achieved in new mixed-use activity centres.

The Plan mention the intensification of density but did not mention what are the ways to increase density, in which all places need to be densified and what are the density values. Not much focus is on industry development activities which further leads to improvement of job opportunities which creates a quality of life. The Plan stated about infrastructure improvement but what's the present status and what is the desired level not clearly mentioned. Spatial representation is less clear.

In existing situation, there is need of retaining the regional workforce. Jobs distribution is uneven and there is need of making job distribution as uniform throughout the region. Finding of appropriate land for industrial development is essential.

The Calgary Regional Partnership (CRP) and member municipalities will work together to ensure a diversified and globally competitive region that continues to enjoy a high quality of life and is able to attract and retain a viable and adequate regional workforce and member municipalities will endeavour to achieve a distribution of jobs creation and economic activities throughout the region consistent with transit and complete mobility policies that encourage the location of jobs close to where people live. Member municipalities will strategically identify lands and corridors for future economic activities and work together with stakeholders to ensure identified lands successfully support economic activities. Member municipalities should encourage industrial and commercial sectors to establish in areas across the region that have been identified as compatible and strategic locations for specific types of activity and investment. And member municipalities will endeavour to maintain and enhance the region's strategic transportation networks (roadways, railways, airports, transportation hubs) to link regional industries to markets—locally, regionally, nationally and globally and to attract new industries to the region.

Recognising the connections and relationships that exist between communities, the Calgary Metropolitan Plan (CMP) acknowledges and respects the vital and historic importance of rural lands, industry and culture in our region. Calgary Regional Partnership (CRP) member municipalities will continue to support rural economic opportunities in the region. CRP and member municipalities will work proactively with the Province and the private sector to effectively manage resource development activities that impact the region's economy, natural environment and settlement pattern. The CRP and member municipalities will collaborate to provide regional servicing to identified commercial/industrial rural-regional priority growth areas for CRP members.

By coordinating efforts, instead of duplicating them, directly reduce the pressure on residential taxes to support constantly growing local needs. Through effective land use strategies, enhanced infrastructure provision and improved regional transit mobility options, the Calgary Metropolitan Plan (CMP) helps increase the economic competitiveness of the region and its municipalities. The Calgary Metropolitan Plan gives municipalities an opportunity to participate in region-wide solutions to common municipal issues while respecting local autonomy. The implementation of solutions is in the hands of the individual councils.

What are the industries and where those can be established to boost the economy and employment is not stated? What are the land management techniques suitable in this region are not mentioned? There is a lot of potentials to develop tourism which is not at all described. They mentioned that various studies need to be done to analyse the economic boost but what are those not stated. The Calgary Metropolitan Plan is the blueprint for accommodating growth in future. The plan giving the various strategies to make Calgary as a healthy environment in enriched communities, with sustainable infrastructure and a prosperous economy.

1.12.8 Case Study 8—Kuala Lumpur Metropolitan Regional Development Plan [57]

Kuala Lumpur is the largest city in Malaysia. Being rated as an Alpha world city, Kuala Lumpur is the only global city in Malaysia which covers an area of 243 km^2 (94 sq mi) and has an estimated population of 1.73 million as of 2016. Greater Kuala Lumpur, also known as the Klang Valley, is an urban agglomeration of 7.25 million people as of 2017. It is among the fastest growing metropolitan regions in South-East Asia, in terms of population and economy. Kuala Lumpur is one of three Federal Territories of Malaysia, enclaved within the state of Selangor, on the central west coast of Peninsular Malaysia. Kuala Lumpur has undergone rapid development in recent decades.

Kuala Lumpur and its surrounding urban areas form the most industrialised and economically, the fastest growing region in Malaysia. The city remains as the economic and business centre of the country. Kuala Lumpur is a centre for finance, insurance, real estate, media and the arts of Malaysia.

Residential land use increased from 3822 to 5490 ha between 1984 and 2000 and is the largest land use component in the City. However, residential land use in the City Centre has declined significantly between 1984 and 2000 and now accounts only for 26.4% of the total residential land use in 1984 and slow growth of residential land use in Bukit Jalil. Regarding commercial land use, the issues are a preponderance of commercial land use in the City Centre and Commercial growth outside the designated growth areas. Regarding industrial land use, the issue is many of the older industrial areas are in a dilapidated state. The problem for institutional land use is the future use of buildings and lands formerly occupied by federal government offices. However, open space, recreational and sports facilities only represent 6.5% of total land use, and the amount that is available as public open space is even less when private open spaces such as golf courses are excluded. There is a shortage of suitable sites for community facilities and under utilisation of utility reserves.

The strategies adopted to implement the spatial plan are as designate and develop International zones, designate and implement Comprehensive Development Areas (CDAs), encourage and facilitate the development of Malay Reservation Areas, traditional kampungs and new villages, initiate and implement the redevelopment of blighted areas, provide priority and incentives to development in areas around transit terminals, ensure the functional distribution of centres and facilities.

A prime concern of the City must be that its urban centres and facilities are distributed in such a way that they are easily accessible to most its population. To this end, a clear hierarchy of urban centres comprising the city Centre, district centres and neighbourhood centres shall be defined. Appropriate functions and facilities shall be determined for each genre of the centre according to their location, accessibility and catchment area or population Existing centres shall be consolidated and upgraded to meet the requirements of the local community. Larger and more sophisticated facilities, Mixed-use development incorporating high-density residential, high plot ratio commercial, as well as community and business facilities, shall be encouraged, thus greatly reducing reliance on private transportation by making accessibility flexible and convenient. Bus services shall be closely integrated with rail terminals and interchange facilities provided to facilitate fast, convenient and efficient transport. Pedestrian and traffic linkages, both within and from outside these zones, shall be improved to provide more convenient access to the transit terminals. Authorities shall prepare long-term comprehensive development plans and guidelines which will be implemented over the period. Improvement of basic infrastructures such as roads, utilities and drainage shall be the initial priority and, as opportunities present themselves, improvements to other amenities and community facilities shall also be implemented.

The Plan gave different strategies for improvement of the region and spatial representation also good but it did not make final land use map. The further detailing of different strategies is given in the local plan. The plan formulated is addressing most of the issues facing and the local plans made are clearly stating what are things to do to make **KUALA LUMPUR—A WORLD-CLASS CITY**.

The manufacturing component of employment has declined to 10.5% of total employment in 2000 from 16.8% in 1980, leading to a reduction in the range of employment opportunities in the manufacturing sector. The high rate of net out-migration and low population growth rate. Despite the growth of office and retail space outside the City and the City Centre, there is still an over-concentration of commercial floor space in the City Centre There is a significant quantity of older office buildings which are deficient in basic ICT facilities. There is a phenomenon of unsuccessful shopping complexes in Kuala Lumpur. The failure of some complexes is principally due to poor accessibility, insufficient catchment, unattractive design and the lack of proper market and financial studies. The traditional shopping areas in the City Centre have largely been superseded by the emergence of large-scale shopping malls in various parts of the City. Consequently, there is no longer a clearly defined major shopping area or spine within the City Centre. Increasing demand for service apartments.

A few new permanent hawker centres have been established intended to relocate hawkers. However unsuitable relocation premises for hawkers in terms of accessibility, catchment, comfort and inadequate support facilities are factors that have contributed to the failure of some hawker centres, resulting in hawkers re-establishing their businesses in their original location. There is an inadequate level of market promotion of the City compared to other major tourist cities. In some international tourist destinations, there are a city or regional based tourist boards heavily promoting the destination backed up by a mature tour and travel industry, hoteliers, national and regional airlines and business organisations. Tourist resources are not easily accessible, coherent, well linked to pedestrian movement or supported by services such as restaurants and related shopping. Although Kuala Lumpur has a diverse range of actual or potentially significant tourism resources, the present 'total tourism product', being the whole spectrum of tourism services, is inadequate and segmented. Inadequate promotion of traditional handicraft and souvenir industries. Lack of a clearly identifiable dining and entertainment area in the City Centre. There is under-utilization of major sporting venues.

To create an economic framework for the City which will enable it to achieve its vision to be a World-Class City, the plan aims to enhance the City's global and regional economic role as a leading centre of the knowledge-based Economy, attain a strong and well-diversified economic base integrate with and complement the activities within the Multimedia Super Corridor, attain an optimum population size and distribution.

To enhance Kuala Lumpur's role as an international commercial and financial centre, the Plan aims to:

- Promote Kuala Lumpur as a choice location for international organisations and business entities to establish their regional offices and headquarters.
- Create a technologically advanced city especially in the fields of building technology and design as well as information and communication technology.
- Enhance the City Centre as an international shopping and entertainment centre.

To enhance the role of Kuala Lumpur as an international commercial and financial centre, the Plan aims to:

- Develop Kuala Lumpur as an attractive international tourist destination.
- Increase the average length of stay (ALS) to 3.0 by the year 2010.
- Together with the Kuala Lumpur Tourism Action Council, market and promote tourism in Kuala Lumpur.

For Kuala Lumpur to become an International Commercial and Financial Centre, aims to:

- promote the development of industries related to the Knowledge-Economy;
- promote high-end industries employing highly skilled workers;
- provide to all residents a wide range of employment and business opportunities; and
- maintain a sectoral balance in industrial development.

For making the **KUALA LUMPUR—A WORLD-CLASS CITY** economy is more important. The strategies to achieve economic boost there are proposals for improving the industrial sector and the commerce sector. Very well stated tourism plan also made to attract global tourism by utilising tourism potential.

The vision and goals for Kuala Lumpur have been formulated with the aim of creating a sustainable city. City Hall Kuala Lumpur (CHKL) shall ensure that the planning of the City shall strike a balance between physical, economic, social and environmental development. Local Agenda 21 shall be adopted to encourage citizen participation towards creating a sustainable society. This is in line with government policies of implementing sustainable development strategies as stipulated in the Habitat Agenda of the Rio Declaration.

1.12.9 Case Study 9—London Metropolis [58–60]

Metropolitan profile of London: The 2000-year history of London has been one of constant change. It has grown from a port and river crossing point into a bustling centre of national Government and international commerce. It has been an imperial capital, and a city embracing villages and towns as it grew. It has been home for people in all walks of life, and from all parts of the world. London's population is likely to continue to grow. By the 2020s there are likely to be more Londoners than at any time in the city's history.

Although London's economy has been generally successful over the past twenty years, not everyone has benefited and the incidence of poverty has not fallen. London is an increasingly polarised city. On the one hand, it has seen a major growth in earnings, with significant rises both in the number of those earning high salaries and in the amount, they earn. This leaves those on low incomes or without employment further and further behind. A growing and ever-changing economy—London have always been at the forefront of enterprise and innovation. It already has a diverse range of economic specialisations extending beyond finance and business services. The next 20 years are likely to see continued changes to the London economy, with new sectors and enterprises emerging, building on the capital's rich resources of research and innovation and its world-class universities and specialist institutions.

A city that meets the challenges of economic and population growth in ways that ensure a sustainable, good and improving the quality of life and sufficiently high-quality homes and neighbourhoods for all Londoners and helps tackle the huge issue of deprivation and inequality among Londoners, including inequality in health outcomes.

A city that becomes a world leader in improving the environment locally and globally, taking the lead in tackling climate change, reducing pollution, developing a low carbon economy and consuming fewer resources and using them more effectively (Table 1.16).

The most efficient use will have to be made of London's limited reserves of land, identifying places with the potential for development on a strategic scale, and ensuring policies are in place to enable this to happen. In spatial terms, this will mean renewed attention to the large areas of unused land in east London where there are both the potential and need for development and regeneration (Table 1.17).

It makes clear the Mayor's overall economic development policy objectives to:

- promote London as the world capital of business, the world's top international visitor destination and the world's leading international centre of learning and creativity
- ensure London has the most competitive business environment in the world
- Support London to become one of the world's leading low carbon capitals by 2025

Table 1.16 Spatial strategies-London

Plan	Strategy
Vibrant, connected, and engaged neighbourhoods	
Support neighbourhood driven activities and decision making	• London strengthening neighbourhoods strategy 2015–2020 • Great near campus neighbourhood strategy
Fund and partner with the London public library to support people and neighbourhoods	London public library 2014–2017 strategic plan
Work with our partners in education to help keep neighbourhood schools open and use former school sites effectively	• The London plan (public facilities and services)—draft • Neighbourhood school strategy—new
Diverse, inclusive, and welcoming community	
Support immigrants and newcomers to be successful as they settle in our community	• London and Middlesex local immigration partnership strategic plan • Conference Board of Canada (resource for measurement)
Support all Londoners to feel engaged and involved in our community	• Community diversity and inclusion strategy —new • Consider a gender lens during the development and execution of new policies • Workplace diversity and inclusion
Work to always be a compassionate city to all	• Compassionate cities initiative
Amazing arts, culture, and recreation experiences	
Explore the potential for a multi-use performance venue(s) in London	• Cultural prosperity plan • London's Downtown plan
Invest in new parks and recreation facilities and pursue innovative models for programs and service delivery	• Parks and recreation strategic master plan 2009–2015—update • The London Plan (parks and recreation)—draft
Fund and partner with museum London, the London arts council, the London heritage council, Eldon house, the London public library, and others to strengthen culture in London	• Cultural prosperity plan • Museum London strategic plan • London public library 2014–2017 strategic plan

- give all Londoners the opportunity to take part in London's economic success, access sustainable employment and progress in their careers; and
- ensure prosperity is spread across the capital, addressing areas of deprivation across the city and fostering economic and employment growth in outer London, maintaining the global role of central London and maximising the benefits of investment.

In recent decades London's economy has been increasingly service-based, and this is likely to continue. As a result, ensuring there is enough office space of the right kind in the right places is a key task for the London planning system. Local plans and strategies should support the conversion of surplus offices to other uses and promote mixed use development in the light of integrated strategic and local studies of office demand.

Table 1.17 Economic strategies-London

Plan	Strategy
Diverse and resilient economy	
Work with partners to develop a community economic strategy	• Community economic roadmap—new
Buy and service industrial land to bring more jobs to London	• Industrial land development strategy • Various community improvement plans • The London Plan (City Structure Plan, growth management, and industrial place type)—draft
Support small businesses by improving city processes	• Streamlined approval processes • Service London business • Community economic roadmap—new
Promote culture as a key part of economic growth and quality of life	• Cultural prosperity plan • London music strategy • The London Plan (culturally rich and diverse city)—draft
Urban regeneration	
Create new partnerships to build, and support the building of, new affordable housing	• Housing development corporation strategy • Regenerating public housing plan
Use community improvement plans to coordinate city and private investment to meet both local and city-wide priorities	• Various community improvement plans • The London Plan (urban regeneration)—draft
Invest more in heritage restoration, brownfield remediation, urban regeneration, and community improvement projects	• Various community improvement plans • The London Plan (urban regeneration)—draft
Local, regional, and global innovation	
Use new and emerging technology to improve the quality of life and grow London's economy	• Smart cities strategy—The London Plan (smart city) • Fibre optic—last mile program —The London Plan (smart city) • Municipal best practices
Lead the development of new ways to resource recovery, energy recovery, and utility and resource optimisation with our local and regional partners to keep our operating costs low and assist businesses with commercialization to help grow London's economy	• Partnerships with Budweiser gardens, covent garden market, and London hydro • Community energy action plan • London waste to resources innovation centre • International water excellence centre
Diverse employment opportunities	
Improve workforce recruitment, development, and retention by working with local and regional partners	• London economic development corporation workforce development
Remove barriers to employment through the expansion of the City of London internship program	• City of London internship program (to include foreign trained professionals)
Attract and retain newcomers, including international students, foreign trained professionals	• Immigration strategy—new • London Middlesex local immigration partnership

Economic development strategy provides further detail on realising London's potential for economic growth

The London Plan sets out a new approach for planning in London. It emphasises growing inward and upward so that it can reduce the costs of growth, create walkable communities, revitalise our urban neighbourhoods and business areas, protect our farmlands, and reduce greenhouse gases and energy consumption. The plan sets out to conserve our cultural heritage and protect our environmental areas, hazard lands, and natural resources. Through the London Plan, the community is planning for vibrant, healthy, safe and fulfilling neighbourhoods, attractive and viable mobility alternatives and affordable housing that is accessible to those who need it. At the root of The London Plan is the goal of building a city that will be attractive as a place to live and invest in a highly competitive world and one that will offer the opportunity of prosperity to everyone—one their own terms and in their own way.

1.12.10 Case Study 10—Berlin Plan 2035 [61, 62]

The capital and the largest city of Germany, Berlin is a world city of culture, politics, media and science. With a population of approximately 3.5 million, Berlin is the second most populated city proper and seventh most populous urban area in the European Union. It has a population density of 4100/km^2. Its economy is based on high-tech firms and the service sector, encompassing a diverse range of creative industries, research facilities, media corporations and convention venues. Modern Berlin is home to world-renowned universities, orchestras, museums, entertainment venues and is host to many sporting events. The city is well known for its festivals, diverse architecture, nightlife, contemporary arts and a high quality of living.

Berlin possesses a special atmosphere that makes the city a desirable place to live, attracting people from all over the world. While many cities fail to achieve poly centrality, Berlin boasts of a polycentric spatial arrangement with short distances between centres and a wide array of local offerings in mixed-use centres in the inner and outer city. The city is continually changing and Berlin possesses a high degree of creativity to employ in changing it, to make it further development of the city of short distances. However, there separate centres with functional or aesthetic failings in the inner and outer city. With a high quality of life, the problems faced by Berlin are mostly environment related. These include strong air and noise pollution, low prevalence of renewable energies, large ecological footprint, climate change impact such as a rise in average annual temperature, a decrease in annual precipitation, more frequent extreme weather events.

This plan talks about continuing and improve the concept of neighbourhood diversity. Berlin mix and its special quality of life remain a unique feature of the growing city. These neighbourhoods are known for their unique social, multi-ethnic and multi-functional mix. The development proposes 25 major new residential development sites housing offering the potential for some 50,000 dwellings. Similarly, other development concepts are explained in Table 1.18.

Table 1.18 Spatial strategies-Berlin

Objective	Strategy
Supporting neighbourhood development	Forces are pooled to safeguard the social mix and to promote the distinct characters of different neighbourhoods, spaces for social interactions as well as safety and cleanliness
Creating living space	Housing and property development policy are designed to maintain reasonably priced living space and to create new, affordable housing
Safeguarding local shops and services	Tailored and resource efficient development of neighbourhood and centre friendly local retailers and social, cultural and transport infrastructure
Advancing high-quality inner development	Densification of the existing urban environment involving the re-use and conversion of existing structures
Connecting and enhancing free spaces	Improve connections between residential centres and open spaces in the city
Safeguarding and improving ecological qualities	Protection of natural resources such as soil, green spaces, water, climate and air
Preserving and developing green and open spaces	Green and open spaces are preserved and upgraded as recreation and climate change compensation areas
Making public transport more attractive	Expansion of the system to meet demand, to increase reliability, to maintain infrastructure and vehicles, and to ensure accessibility and affordability
Increasing bicycle and pedestrian traffic	Bicycle and pedestrian strategy and making eco-mobility more attractive
Developing an integrated commercial transport plan	Measures designed to reinforce e-city logistics, improve multi-modality and increase cooperation in commercial transport

Berlin's economic situation has seen positive progress in recent years, yet it continues to struggle in terms of job creation and job security as well as income levels. However, the city enjoys comparative advantages, particularly in its capacity as a location for knowledge and innovation. Berlin is a renowned hub of knowledge. Berlin plan proposes consists large-scale infrastructural investments, an efficient freight transportation hub, and highly effective networking within the surrounding region. Whilst enjoying a good number of outstanding universities and institutions of science, concentrations of scientific institutions in the inner and outer city, and private research institutions there is a grave problem of unemployment amongst the youth. The economic activity is predominantly in the inner city, the per capita income is low. A comparatively weak representation of companies and industries with high added value is observed.

Berlin 2030 is proposing to enhance as a business location by the influx of talented individuals from around the world, skilled workers and entrepreneurs who boost innovation and promote international networking. The plan proposes to further improve the innovative potential; every third company in the capital region is to be active in five innovative clusters. It also aims at upgrading the labour market, provides a comprehensive education and research landscape. The proposed strategies are discussed in Table 1.19.

The Strategy for Berlin, with its goals and fields of action, represents a citywide response to urban development and will affect the city. For its implementation, the development plan proposes ten transformation areas which provide focus attention on selected areas and offer responses to Berlin's major challenges and opportunities. They focus on the issues of population growth, economic structure, the capital city function, social cohesion, and climate change and energy transition as they affect Berlin. The transformation areas offer development potential relevant to the whole city in terms of both social issues and open spaces.

Table 1.19 Economic strategies-Berlin

Objective	Strategy
Intensifying knowledge and technology transfer	Networks and measurements used to consolidate science, research and business for exchange, impromptu meetings and transfer
Implementing the idea of Berlin as a smart city	Information and communication technologies used, post-fossil fuel society
Improving networking between learning institutions	Networks and partnerships are intensified across the state border
Developing multiple innovation hubs	Create space for innovative entrepreneurial activities, particularly those focused on future technologies
Safeguarding and developing important industrial and d commercial sites	Locations for the manufacturing sector are fortified and (potential) development sites are safeguarded and developed
Promoting start-ups	Conditions for start-ups (services, contacts, capital, and space) are improved to accelerate business development
Conditions for start-ups (services, contacts, capital, space) are improved to accelerate business development	Training and support provided to bolster employment, measures designed to improve the family work balance are promoted
Establishing a 'welcome culture'	An enlightened, non-bureaucratic attitude towards the integration of migrant workers and businesses, including the provision of foreign language capacity within local authorities
Safeguarding employment through education and skills	Providing a high-quality educational infrastructure Developing kindergartens and schools as the the foundation of the educational landscape Transforming libraries into centres of out-of-school learning and educational partners Turning educational establishments into inclusive establishments Safeguarding and strengthening out-of-school educational venues

Berlin development concept 2030 is the second part of the study, the first being the present situation study. With thorough analysis followed by extensive public participation in the planning process, the plan inculcates the opinions of the people. To safeguard the employment opportunities, the plan talks about enhancing and strengthening education right from the kindergarten level followed by various out of school-related activities in the higher level. The economic aspect of the city is planned efficiently, along with strong emphasis on the social and cultural aspects. It defines a progressive and innovative development plan.

The Berlin Strategy provides an inter-agency model for the long-term, sustainable development of the capital. With one-third of the city comprising of open spaces, a compact polycentric development, highly tolerant society, the development plan further proposes to enhance these key selling points of Berlin. The development plan builds its foundation on the strong points of Berlin and proposes strategies to tackle the challenges. Provisions of affordable housing, further increasing the short distances to amenities, enabling a start-up friendly environment are some of the key proposals of the plan. Cultural diversity and tolerance in the society are used as a selling point for the attraction of workforce and tourists. Community participation is given utmost importance. The economy focuses on the educational institutions, research centres, attracts skilled labour from all over the world. Using a range of strategies and goals, it sets out the areas and directions in which this growing city should develop and highlights the areas that will form the focus of its future development.

1.12.11 Case Study 11—Dhaka Structure Plan 2016–2035 [63–65]

Dhaka is the capital and largest city of Bangladesh. It is one of the most populous cities in the world with a population of 17 million in the Dhaka Metropolitan region. The area stretches over 1400 km^2. It is the third most densely populated cities in the world. The planning area includes four city corporations- Dhaka South, Dhaka North, Narayanganj and Gazipur City. Dhaka is home to thousands of businesses and international corporations and the biggest employment GDP generation of the country. Migration of people from rural areas is one of the reasons of the growing population. Dhaka will be home to 25 million people by the end of 2025 and will be a meta city as per UN estimate.

Like most of the mega-urban regions of the world today, there is a functional relationship between the Dhaka city centre and the smaller urban centres. More and more people are commuting to their workplace located within the core city of Dhaka from surrounding settlements like Narayanganj, Gazipur, Tongi, Savar, and Keraniganj. With the rapid increase in urbanisation, Dhaka experiences degradation of the urban environment, its air, water and soil have been polluted to dangerous levels. It has also caused heavy demands on urban utilities and services like electricity, gas, water, sanitation, sewerage, garbage disposal, transport, telephone, cables; and social services like health and education, etc. For the millions of rural poor in Bangladesh, Dhaka is still the most attractive destination for economic activities, leading to massive migration from rural to urban areas. About 63% of the total growth of Dhaka's population is due to migration and only 37% growth comes from natural increase. Most of these poor people are unable to afford habitable housing or other socioeconomic services. The consequence of this is the growth of unhygienic slums and squatter settlements. Wetlands encroachment is another issue because of excessive development pressure, any land would be lucrative for physical development and construction because there would be demand for it. Another serious issue is in the transportation sector, which includes poor public transport system, lack of coordination, the presence of rickshaw in major road corridors, inadequate pedestrian facility.

The planning area is divided into two strategic zones like Urban Promotion Area (UPA) and Urban Control Area (UCA), largely based on the possibility of future urbanisation. Urban Promotion Area has been further sub-divided into three strategic management areas like Central Urban Area, Outer Urban Area and Growth Management Area; while Urban Control Area has been sub-dived into two strategic management areas like, Agriculture and Conservation Area (Flood flow zones, Water Retention Area, National and Regional parks, Forest Area, large scale Heritage sites). The theme of the current structure plan is to diffuse the major functions performed by the central city and redistributing it to different urban centres (Polycentric approach). This will help create a hierarchical framework of different magnitudes of centres providing corresponding levels of services and employment opportunities. The main components of the concept are: Changing urban structure, reducing trip generation and traffic and enhancing local accessibility. Some of the strategies along with their implementation are listed in Table 1.20:

Table 1.20 Spatial strategies-Dhaka

Objective	Strategic action	Implementation
To enhance local accessibility	Decentralising urban functions and services, compact development	Division of planning area into 6 functional regions
To promote compact urban development	Utilize planned unit development (PUD) concept, promoting urban centres with appropriate densities, services	Make necessary amendments, development near high quality public transport
To mobilise the under-utilised lands within the city	Rigorous redevelopment or rehabilitation plan for these areas, Seek possibility of PPP with private sources of fund	Prepare revised land use plan of required areas, detailed plan for Dhanmondi to define and provide upgraded/rehabilitated utility services is required in view of densification
To enhance the linkage between land use and transport network	Coordinated approach to transport planning with the help of TOD	Promote proposed urban centre based development to reduce travel demand

The economy is the foundation of urban agglomeration and eventual progress of an urban community. Dhaka is basically a low-income economy. It is situated in a strategic location but currently not generating expected level of return for lack of infrastructure and services. With a huge population, there is no problem of labour. There is also a steady growth of export-oriented sectors for example garment industry. Other growing trends are the emergence of leather, information technology, finance infrastructure domestic financial resources including remittances and the service sector. The informal sector comprises of 65% of all employment in the city is a sector which needs to specialise. However, the poor infrastructure and services are inadequate for the economy as well as an institutional weakness to support business development. Other issues include traffic congestion causing loss of time and productivity, high interest on bank credit and a limited supply of urban land and infrastructure despite the high urban land price.

Specific policies have been provided for making Dhaka's economy increasingly productive and functional, which includes development of commercial hubs within different potential areas of DMR, promotion of compact and clustered industrial growth, promotion of exclusive industrial zones, relocating polluting industries, facilitate development of ICT sector in the Dhaka core area, promoting woven garment and knitwear in peri-urban areas of the metropolis and recommendation for limiting growth of garment factories in the central urban area, etc. Some of the strategies with their implementation have been discussed in Table 1.21:

A major drawback in the Dhaka Structure Plan 2016–2035, as pointed out by several concerned people, is non-involvement of the public's opinion or other concerned authorities. The local waste management is not addressed in the spatial planning. While the concept of discouraging plot based housing is a good concept, the option of building apartment housing and building individual housing can be made taken into consideration. The plan proposes major projects like Transit Oriented Development and five new ring roads. The process of acquiring land for such projects may not always be people friendly and is not efficient in a densely-populated brownfield settlement.

The Dhaka structure plan covers the basic issues such as effective land use management, transport for efficient connectivity, enhancing employment and productivity, public Facilities, protecting the natural and healthy environment. The spatial concept of dividing the planning area into two broad categories i.e. Urban Promotion Area (UPA) and Urban Control Area (UCA) is an efficient way to assure uniform distribution of development. Here, the basic requirements are made available at each functional region reducing the dependency on the core city. At the same time under the UCA flood flow zones, Water Retention Area, National and Regional parks, Forest Area, and large scale Heritage sites are preserved. Instead of focussing on individual plot based development, the concept of planned unit development is introduced. The plan talks about increasing the productivity of the informal sector, which is a good step towards mitigating urban poverty. The garment sector, leather manufacturing which boosts women employment is also planned to make it more efficient. While the report covers all the broad aspects

Table 1.21 Economic strategies-Dhaka

Objective	Strategic action	Implementation
Establish exclusive economic zone for leather industries in growth management area	Select strategic locations for establishment of new and relocation of existing leather industries and provide necessary infrastructure and services including ETP	Provide all services and facilities in the selected area-power, water, waste management, drainage, etc., arrange easy and reasonable interest credit facilities for investors
To elevate informal economic activities into higher productivity levels	Selection of site for providing informal business activities, formation of an appropriate tenure arrangement, establishment of a small scale replicable demonstration project	Space allocation to relocate informal enterprises with appropriate level of charges, skill development training to promote higher production and quality products, providing collateral free soft credit to upgrade business
Relocate and/or cluster polluting industries at suitable locations	Select strategic locations for establishment of noxious industries	Provide common ETP to treat pollutants at low cost before discharging into open water courses, provide incentives for relocation and export
Locate, declare and promote selected areas as exclusive industrial zones	Recognition of the as industrial clusters and priority facilitation, selected areas to be developed as a specialised centre where industrial agglomeration in planned way will be the major functions within those areas	Connect industrial areas by efficient public transport, Give priority in industrial plan approval, Provide incentive to industries in the form of tax holiday, duty-free import of machinery, credit facilities

required for the development of a region, the involvement of the public in the planning process would provide a more achievable and efficient plan.

1.12.12 Case Study 12—Master Plan for Patna Metropolitan Region 2031 [66]

Patna is the capital city of Bihar State in India situated 15 km along the confluence of the River Ganges. Patna is the important administrative and educational centre of the State. Out of 53 million plus cities (2011 Census) in India, Patna is 18th in rank with a total Urban Agglomeration population of 2046,652. Patna is the largest Corporation with 44.1% share of the population of 7 corporations, and 20% of the total urban population of Bihar State.

Patna being the capital of one of the economically backwards state of India with all kinds of disadvantages, it becomes necessary to develop Patna as a National

Level Competitive City so that it can reduce the impact of various kinds of disadvantages present in the Bihar economy. In last ten to fifteen years, Patna is developed as emerging trade and business hub in Eastern India. Therefore, the vision of The Master Plan of Patna should be to make it a National Level Competitive City and Regional Trade Centre in Eastern India by next two decades.

The Master Plan Vision for the year 2031 is "To develop Patna Planning Area as a modern economic region with locally competitive infrastructure and social amenities to address future requirements in harmony with its ecological resources".

The development strategies adopted in the Master Plan are as follows:

1. Creation of a compact city to achieve sustainable development.
2. Transit Oriented Development (TOD) is proposed along the major corridors, which will emphasise movement through public transport. The above-mentioned Centres are well-linked to the core city. Integration of the peripheral areas with the core and intermediate areas, so that the entire area acts as one entity (through TOD).

Creation of Growth Centres (GC) and Satellite Towns: Proposing multi-nuclei centres will help reduce pressure on the core area. Proposing Master Plan with the objectives of protecting rural livelihoods, catering to the needs of the population by providing agricultural zones and Provision of adequate infrastructure to influence overall development in the region are good proposals.

Spatial Strategy

A spatial strategy which is to be adopted in Patna rests on the basic concept of 'Multiple Nuclei Model'. Creating multiple nuclei centres/sub-centers will help to take the overload from the centre or core nuclei. These nodes/nuclei centres are identified as physical demarcation/accumulation of cluster of activities. Various urban nodes proposed in Master Plan are commercial, Transportation, and Industrial nodes. The different options for this development model are given.

1. Growth options for Multiple Nuclei Model.
2. Proposed Land use 2031.

The proposed land use shows the application of this spatial strategy. The Spatial Strategy focuses on achieving a single aim of decentralisation. The growth model proposed here is Multiple Nuclei model, which is achieved by land use. Though TOD is mentioned as a Strategy for development, it is not addressed in any of the sections and remains untouched. To promote industrial based economy 60% of secondary workers is accommodated in Industrial land use, while and rest 40% to be accommodated in Commercial/Institutional and Mixed land use. For secondary workers, the area per person required is 50 m^2/worker, therefore total area required for industrial land use is 966 ha.

The Economic approach in the Master Plan is only by addressing the land requirement for various employments generated. There is no mention about Economic Strategy.

The Metropolitan plan for Patna, though it takes into consideration, the Economic aspect of the city, doesn't provide a sufficient Economic Strategy for development. The plan has a concept for the spatial development of the city. The spatial strategy is fairly achieved. The Implementation plan simply explains a case study for TP schemes from Ahmedabad. There is no strategy adopted. Also, it doesn't consider each project in detail.

Patna being the only important city in the state has acquired the strong position in regional trade and business. It is necessary to have an economic strategy while planning for Patna. The Master plan lacks this. Instead, an overall development which also leads to economic development is attempted. Implementation plan doesn't consider every project in detail.

1.12.13 Case Study 13—Montreal Metropolitan Development Plan 2020 [67]

Montreal is the largest city in Canada's Quebec province. It's set on an island in the Saint Lawrence River and named after Mt. Royal, the triple-peaked hill at its heart. The planning area is a City of 500 km^2. There is an expected 60,000–75,000 new housing units over the next 10 years. There are 180 km of the waterside roadway to be enhanced and 11 km^2 of vacant land to be built throughout the City.

The Plan is the result of a planning and the cooperative process initiated at the Montreal Summit in June 2002. The Plan presents a planning and development vision for the City, as well as measures for implementing the goals and objectives resulting from that vision. The main issues raised were the following: Quality of life; Transportation problems; Presence of a healthy natural environment; Protection and enhancement of the built heritage; and Community facilities.

The City intends to consolidate the existing features of established areas, which cover most of Montreal's territory. Many areas are underused and ripe for transformation; The City intends to develop these areas by attracting new activities and buildings. The City favours high-quality urban design and architecture (Table 1.22).

Development strategies are categorised into seven sections. Under each section, strategies, their objectives and action plans are defined. Corresponding Projects and proposals are defined. The spatial strategy is based on three aspects. Under each aspect, the implementation program or the action plans are also given in detail.

The Economic centre of Quebec, the Montreal metropolitan area had 2,030,000 jobs in 2013 (Table 1.23).

There are three types of components in a Smart Metropolis: technological, human and institutional. Technological factors involve implementing physical infrastructure, creating applications and setting up platforms (ICTs). Human factors focus on individual interactions and spaces that foster such interaction. Institutional factors refer to policies; regulations and governance that will help make the metropolitan area smarter (Table 1.24).

Table 1.22 Spatial strategies-Montreal

Strategy	Implementation
Promoting an excellent living environment	• Developing tools to support a wide range of housing, including affordable accommodations • Promoting a full line of public facilities and ensuring easy access • Encouraging a commercial presence tailored to community needs
	• Promoting the use of public transit and active transportation through urban design and infrastructures • Adapting to climatic change by greening our spaces and building, along with effective rainwater management
Fuelling vitality of the urban agglomeration and its central core	• Supporting the central core's multiple functions, while boosting the sector's international renown • Expand and improve the public transit and active transportation network • Formulating an access plan for the different centres based on complementary forms of transportation • Building major road projects connected to intermodal platforms to optimise freight transportation
Enhancing areas of interest	• Stepping up regulatory measures to protect heritage and laying out rules for new construction • Expanding protected areas from 5.8 to 8% of the land • Producing conservation plans and adopting regulations to govern activities in ecological sectors • Protecting important scenic views and portions of the shoreline road • Including key heritage, ecology and scenery attractions on the green and blue nature network

Table 1.23 Economic strategies-Montreal

Advantages	Challenges
• Diversified economy with many industrial clusters of excellence • Mobilisation within industrial clusters • High quality of life • Low business operating costs • Attractive tax environment for businesses	• Low productivity • Low university graduation rate • Mismatch between workforce and business needs • Low private investment • Need for industrial space development • Deficient transportation infrastructure
• Abundance of research centres, colleges and excellent universities • Efficient freight transportation hub	• Untapped potential of the creative industries

Table 1.24 Smart economic strategies-Montreal

Focus	Strategy	Implementation
Focus on economy's strengths	Support the development of metropolitan clusters	›› Update the sectors of excellence profiles ›› Foster inter-cluster activities ›› Develop practices and tools that stimulate innovation ›› Develop an outreach strategy for metropolitan clusters
	Support consolidation among the industries of the creative economy	›› Evaluate the need for and feasibility of activating new metropolitan clusters to help structure the creative industries
	Promote the metropolis as a logistics hub	›› Develop a promotion strategy for logistics and transportation
Optimise production factors	Stimulate private investment	›› Implement the strategy for enhancing industrial spaces ›› Implement Montreal international's foreign direct investment strategy
	Provide the metropolis with a high-quality workforce	›› Implement Montreal international's talent attraction strategy ›› Support initiatives for retaining foreign students ›› Support initiatives for matching workforce to business needs ›› Support initiatives for improving student achievement, university graduation rates and job market access
Provide metropolitan coherence	Make Greater Montreal a leading smart city	›› Support and guide the smart city initiatives of municipalities in the region and encourage networking between them
	Mobilise metropolitan economic stakeholders	›› Hold a biennial meeting with the stakeholders of the Montreal model ›› Provide coherence between initiatives aimed at the economic positioning of Greater Montréal in partnership with the stakeholders of the Montréal model

Economic strengths are utilised and creative market and the smart economy are aimed to achieve through these strategies. A clear and comprehensive economic strategy is provided in the Master plan. The implementation strategy to carry out each strategy is also explained. The idea of Smart Metropolis is carefully dealt here which is considered as the most innovative and sustainable development concept in the current scenario.

Montreal being a large city with a diversified economy needed a spatial and economic strategy which is formulated and explained well in the Master plan. Issues and potentials were identified and addressed carefully.

1.12.14 *Case Study 14—Helsinki [68]*

Greater Helsinki is the metropolitan area including the smaller Capital Region urban kernel and commuter towns surrounding Helsinki, the capital city of Finland. It is in the south of Finland, on the coast of the Gulf of Finland, which is part of the Baltic Sea. The Population density is 2945.09/km^2 and the Population density for Greater Helsinki is 389.88/km^2.

The key strategic issues include:

- The growth of Helsinki and its region is beneficial to the whole country.
- The main city-centre will be expanded.
- Social unity is strengthened by satisfying people's needs now and in the future.
- The city-region grows urbaner and improves the environment.
- By building the region also towards the coast, the region will retain its vitality and the balance of the regional structure will be improved (Table 1.25).

A Lively Cityscape

- Creation of a high-quality urban environment together with the protection of the existing city landscape.
- Cities have created their own landscapes, employing the intersections between water, terrain and natural paths.
- Cultural and natural environments located in the city-region is to be made usable by the city dwellers for diverse outdoor and recreational activities Improving the quality of the green network. Special nature reserves will be preserved: Historical urban parks as well as modern parks of international quality, beach promenades and other public outdoor spaces.
- Recreational islands and their ferry connections will be improved.
- Conditions for boating, summer activities, hiking and tourism will be improved (Table 1.26).

The spatial strategy discusses are on the city landscape, green networks and traffic and transport. All these are linked together and the green city network enhances city life with fresh environment and aids in protecting the environment

Table 1.25 Spatial issues and potentials-Helsinki

European city-identity	– Tradition and history as in the typical European city in building and cultural landscapes
Climate change	– Create a low-carbon city-regional urban structure – Energy-efficient and ecological construction
Mitigating against urban sprawl	– Building more intensely within the city-region's development corridors – Communities will be made more compact
Radial and circular structure	– Cross-town and radial ring-road urban traffic structure

Table 1.26 Spatial strategies-Helsinki

Spatial strategy	Implementation
City landscape	– Continuous cityscapes extend beyond the administrative border and their characteristics will be emphasised and the areas will be developed
Green networks	– Accessibility to the green city network and recreational areas, preservation of natural diversity and landscape culture
Traffic and transport	– Access to green networks and recreational areas, choice of modes of transport but preference to walking, cycling and public transport

and heritage. The transportation is focusing on walking, cycling and public transport for a sustainable future. The inner city is connected to the residential areas by central park. Bay city landscape is very innovative and efficient—sustainable. Recreational island development promotes tourism and enhances the environmental quality of the place. Suburbs are placed such that it prevents urban sprawl. Extending the city centre and areas of specialised clusters:

- the main centre will be developed, strengthened and expanded as a regional and national centre.
- services related to finance, culture, leisure and tourism will be promoted.
- Busy passenger harbours connecting Helsinki to the other cities bring significantly tourist flows to the centre.
- business activities in the city become stronger—expanding airport logistical hub and improved international accessibility and fast connections within Finland.
- ring motorways and railway lines will act as a focus to improve business accessibility.

Income distribution is not evenly distributed, such that 20% of the highest income group uses 64% of the resources and 20% of the lowest income group uses 4% of the resources (based on 1994 statistics) (Tables 1.27 and 1.28).

The spatial structure for the economic/business development is mentioned. But the kind of economic activities is not clear even though the zones are marked. The spatial zoning of economic activities is done such a way that it is around the city centre and large-scale industries are placed away from the centre which has easy access from nearby places. Retail activities are evenly distributed in all area. The

Table 1.27 Basic strategies-Helsinki

European integration and Globalisation	– Helsinki city-region will be developed as a European city – Public transport projects will be developed
From monocultural to multicultural	– Need to take into account the increase in immigration as well as in the development of business activities
Governance	– Greater cooperation between the neighbouring municipalities a more integrated approach
Infrastructure	– Form a compact urban structure – Infrastructural changes as cost-effective and ecologically sustainable

Table 1.28 Economic concept and strategy

Economic strategy	Implementation
Business growth	– The increase in the number of inhabitants and jobs
Helsinki's city-centre	– The centre of Helsinki and the areas around it will offer competitive spaces for business activities
Workforce	– Different types of housing solutions for different built according to demand for business activity development
Connectivity	– Cross-traffic connections and goods traffic
Technological development	– Technological innovations
Workplaces	– Building residences and business premises side by side, particularly around public transport interchanges – Create mixed areas of residential and commercial – New logistical corridors of rail and Ring Road constitute the framework of the goods maintenance service hub – Business opportunities for growth near and around the city centre will be promoted

plan analyses the future of the city-region and its development needs and presents a Vision of the city region's future principles

- sets out the economic, social and environmental relationships and their impact physically upon metropolitan development for the next 30 years.
- integrated relationships are set out in a series of policies grouped around key issues of business activities (economic strategy), housing, city-landscape (spatial strategy) and regional structure.
- developed into a set of spatial strategies under each main heading which in turn form the physical demands of city-regional growth.

1.12.15 Case Study 15—Istanbul 2014–2023 [69]

The Istanbul Metropolitan Area is in the north-west of Turkey. It has an area of 5343.02 km^2 (2062.95 sq mi) and a population of 14,804,116 as of 2016. Istanbul Development Agency perceives development as a multidimensional holistic concept; including the development of social rights alongside economic development, environmental conscience, and protection of cultural values; and acts with an understanding of development which mobilises the potential of all segments of society. The Plan's vision for 2023 rests upon Istanbul's uniqueness. The vision of the 2014–2023 Istanbul Regional Plan is "Unique Istanbul; City of Innovation and Culture with Creative and Free Citizens". The milestones of the 2023 Vision consist of 23 Priorities and 57 strategies, which rise on 3 main axes determined for the city's economic, social and spatial development.

In the second half of the 20th century, the Asian side of Istanbul experienced major urban growth; the late development of this part of the city led to better infrastructure and tidier urban planning when compared with most other residential areas in the city and functions as a suburb of the economic and commercial centres in European Istanbul. Because of Istanbul's exponential growth in the 20th century, a significant portion of the city is composed of illegally constructed squatter buildings.

Large scale gentrification and urban renewal projects have been taking place. The Turkish government also has ambitious plans for an expansion of the city west and northwards on the European side. Istanbul does not have a primary urban park, but it has several green areas. Istanbul's unplanned urbanisation is to be transformed, and the city must be prepared for disasters, without neglecting urban arrangements; with full consideration for quality, contemporary design, aesthetics and architecture. With the intention of being an international logistics centre, the city's transportation system, logistical competitiveness, and quality of logistical infrastructure and services will be improved. Transportation capacity to urban zones, and to other regions and countries will be boosted. Natural resources under pressure from construction, such as water basins, forest areas, and green areas and parks in urban areas, will be protected, and their sustainable development will be ensured. The use of environment-friendly energy and sustainable waste management will be adopted. In 2023, creative and innovative people will freely lay out their potential in Istanbul with its enjoyable and authentic urban spaces and sustainable environment.

The spatial concept for Istanbul Development Plan is determined as "Joyful, Authentic Urban Spaces and Sustainable Environment". As the living standards of people in Istanbul to be enhanced, 9 priority areas and 21 strategies were determined for facilitation of their participation in social and economic processes and enabling them to live amongst nature simultaneously with urban life. In the plan period, smart and sustainable urban development of Istanbul is targeted, with an inclusive and holistic planning approach based on cooperation and participation. In this process, protection of environmental and natural heritage, and preservation of their sustainability, as well as Istanbul memory and the historical and cultural heritage of the city, which is the source of its authenticity, need to be protected (Table 1.29).

The spatial strategy considers urban planning, transportation, environment, architecture and heritage in a sustainable manner ensuring sustainable development in a public participatory approach. The implementation part is still lacking as if in the economic strategy which is very critical for the plans to become a reality.

Historically, Constantinople has been the centre of the country's economic life because of its location as an international junction of land and sea trade routes. In 2012, the City of Istanbul had a GDP of $332,4 billion. In 2008, companies based in Istanbul made exports worth $41,397,000,000 and imports worth $69,883,000,000; which corresponded to 56.6 and 60.2% of Turkey's exports and imports, respectively, in that year. In 2006 Turkey's exports grew a further + 16.1% while import grew +17.6% because of a rising demand for energy

Table 1.29 Spatial strategies-Istanbul

Domain	Spatial strategy
Sustainable urban development and participatory planning	• Ensuring sustainable urban development and smart growth, efficient use of space in the distribution of urban functions • Dissemination of participatory based collaborative, inclusive and holistic planning
Spatial quality, authentic design	Improving quality of urban function areas, their fair distribution, and the options available to urban inhabitants Ensuring high-quality urban design, spatial authenticity and diversity, in harmony with the urban identity
Holistic and inclusive urban transformation	• Ensuring urban transformation in Istanbul by integrative planning, taking into account regional needs and lifestyles and the balance of intra-regional development • Ensuring a spatial transformation which will facilitate the development of the industry, and that will reduce the adverse impacts of industry on the city
Protected Istanbul memory and cultural heritage	• Protecting the components forming the Istanbul memory, and tangible and intangible cultural heritage, with the historic urban landscape approach • Protection of historical areas and urban heritage as urban living areas, and ensuring acceptance by the citizens of Istanbul
Effective disaster management	Enabling the disaster management system to create a safe Istanbul with high quality of life and space
Sustainable transport and accessibility	Improvement of public transport infrastructure and services, and promotion of public transport Enhancement of transport facilities for and encouragement of walking and cycling • Improvement of accessibility to Istanbul • Effective management of transport demand and efficient use of existing transport infrastructure
Quality and sustainable environment	• Ensuring the sustainable management of basins and water sources • Protection and development of forests and agricultural areas • Protection of marine and coastal areas and improving their quality • Controlling and improving air quality • Reducing solid waste and wastewater, and ensuring their sustainable management • Reduction of resource use and waste generation in industrial activities, and ensuring sustainable waste management

resources and raw materials by the industrial manufacturers in the country. Income distribution is not evenly distributed in Istanbul, such that 20% of the highest income group uses 64% of the resources and 20% of the lowest income group uses 4% of the resources (based on 1994 statistics). In the late 1990s, the economy of Turkey, and Istanbul suffered several major depressions. The Asian financial crisis, as well as the crisis in Russia, had negative effects in all areas of the economy, particularly on exports. Following this setback, a slow reorganisation of the economy of Istanbul was observed in 1999.

The major earthquake which was epicentre in nearby Kocaeli on 17 August 1999, triggered one of the largest economic shocks for the city. Apart from the capital and human losses caused by the disaster, a decrease in GDP of approximately two percent occurred. Despite these downturns, Istanbul's economy has strongly improved and recovered in the recent years. The economy is designated to create a "Globally Decisive, High Value-Added, Innovative and Creative Economy" in Istanbul. 7 Priority areas and 19 strategies were specified to reach a globally decisive position in the economy beyond integrating into the global value chain. The aim is to transform Istanbul into a global centre of attraction with innovation, creativity and high value-added activities, and to accelerate this dynamism to gain a powerful position in the global value chain by attracting qualified labour and investments to the region (Table 1.30).

The economic strategy discusses all the sectors to make a globally decisive, high value added, innovative and creative economy. The focus has been given to entrepreneurship to encourage the work participation thereby eliminating unemployment that would lead to a better vibrant economy. The main idea of entrepreneurship is to be appreciated as the city will sustain its own economy rather than depending on external revenue as in trade. Even though the strategies and objectives are discussed elaborately the implementation is not detailed which makes it difficult for the region to put these concepts into reality. The Metropolitan Plan discusses a vision for Istanbul, spatial and economic strategy.

The very important part was the public participatory approach so that the people of the place make the place. All the strategical domains are made with the opinions of the citizens and thereby solve their issues and each project is for the development of the city and the citizens. The economic domains include the transformation of the industry, ascending R&D, encouraging entrepreneurship, increasing employment. All these directly or indirectly stand for the citizens. The spatial domains are increasing spatial quality, holistic urban transformation, protecting the culture and history, effective disaster management, sustainable transport.

1.12.16 Case Study 16—Limerick [70, 71]

Limerick city is in Ireland along the River Shannon. It is the third largest city in Ireland having a rich culture and history. The metropolitan area is newly formed and it is the principal urban centre of Mid-West. The city has a major role in

Table 1.30 Economic-strategies-Istanbul

Domain	Economic strategy
A strategic actor in the global economy	Becoming a global centre of attraction in the economy Strengthening Istanbul's foreign trade
Competitive position in the global value chain	Specialising in sectors with which Istanbul can demonstrate its global competitiveness and acquire high value-added functions in the international value chain, and developing these sectors
Transformation in the industry	Creating an industrial production structure which uses advanced technologies, produces high value-added, and employs skilled labour
Ascending R&D and innovation	Developing cooperation between stakeholders, initiatives that promote coordination and sharing Establishing social consciousness and awareness by expanding the R&D and innovation
Qualified entrepreneurship	Disseminating entrepreneurship culture, improving the quality of entrepreneurs Improving existing cooperation in the Istanbul entrepreneurship ecosystem and establishing new partnerships; providing fast and secure access to high-quality Information
Transforming the workforce, developing and increasing employment	Increasing employment, reducing unemployment Transforming the workforce in line with the changing economic structure Facilitating the matchup between labour supply and demand Improving the work environment to support creativity, innovation and inclusiveness
Urban image and effective publicity	Developing a corporate strategy, cooperation and institutional structuring for the publicity and urban image of Istanbul Expanding activities related to the publicity and image development of Istanbul, and conducting these activities effectively Preserving and using local resources and values to improve and publicise urban image; improving related infrastructure and services

defining the economic fortunes of the region as well as Ireland but now the city is having negative growth in terms of development.

Limerick is it is having strengths, but it is not being utilised. Limerick metropolitan area is surrounded by Irish economy and European economies which are performing very well and having good economic interactions with the world. So, the aim of Limerick Metropolitan area is to understand the strength potential opportunities and weaknesses of them and plan a better limerick with strong economic activity.

Limerick is a place having several educational institutions like the University of the Limerick, Limerick Institute of Technology, Research institutes, and ICT institutes. The academic institutes can act as nodes for pushing economy because these institutes are the places gifting new talents to the society if that assets are taken care of by providing job opportunities and facilities it can develop a rise to the current economy. Another positive of Limerick is its high-quality infrastructure facilities like Business Parks, Incubators and mentoring centres for giving training for start-ups, improved road links to Dublin and other Irish cities. These facilities may attract businesses to invest in the City. The well-established link with U.S is an asset for the city which has potential to up bring the economic activities in the Limerick. The presence of well-built and highly connected Airport and Docks which act as intermediate for economic interactions. Historic buildings can increase the tourism potential of the area if they are maintained in a good manner. the city is having some weak points also. Availability of resources is one of them. The major resources like manpower are missing because of degrading economy and brain drain is happening and the resources are going outside. The recession also impacted the economy of Limerick because various foreign investors stopped and went back to another country from Limerick and due to this the job opportunities got reduced and the people get unemployed. Considering the infrastructure facilities like business centres, roads etc. Limerick is showing very high quality and standard but coming to digital infrastructure the condition is very poor. In the coming years and the present condition broadband connections, internet facilities etc. are the game changers in economics sector due to the lack of quality digital infrastructure also makes a bad impression in the investors to invest in the Limerick city. There are so many educational institutions having high potential is there, but their underperformance is also affecting the economy.

The three main factors affecting the economy in Limerick is education, FDI investment and business growth. Education levels across the Limerick Metropolitan Area are average. At Metropolitan Area level 29.6% of the population qualified to a higher certificate or higher qualification and 25.2% have a degree or higher qualification. Even though many institutes are there less percentage of people is having a proper education. One of the main job providing sector is FDI but only 1% of FDI investments are there in Limerick when comparing Ireland economy. Considering the growth of business, the environment and facilities of future growth are less. One of the main things is internet facility. The lack of digital infrastructure makes the investors remove Limerick from their list.

The Economic Strategy sets the direction for the economy of Limerick and determines the ambition and the key objectives around scale, structure and value of the economy. So, the strategies are made considering the knowledge economy (educational institutions), outstanding business environment (providing adequate support to start-ups, incentives to businessmen which increases job opportunities) and long-term growth regarding digital infrastructure and free internet (Table 1.31).

The economic strategy of Limerick is concentrating the three weak sections in their economy and the strategies are well enough to up bring the positives and suppress the shortcomings. These strategies are making the educational institutions

Table 1.31 Spatial strategies-Limerick

Strategy	Rationale	Partnership and fund
Promote quality education through high-quality teachers Provide better and barrier-free facilities for education Promote lively and safe atmosphere	To promote high-quality education thereby economy	University of Limerick Limerick institute of technology Mary Immaculate College Limerick city and county council
Placements for students, research scholars	Educational institutions alone cannot promote business Generate new demands with available institutions	The university of Limerick, Economic development and Planning Directorate, fund-200k/annum
– Utilise existing opportunities and enterprise Ireland partnership with private, attract more businesses – Formation of innovation	– To expand the existing lapsing economy – Securing finance was risky in old days – For R&D fewer revenues are available – Banks are unable to support innovation due to risk factor	Ireland enterprise, Private sector and SVC fund
Provide quality spaces for businesses and making available for all leases Provide shared services Provide access to range of Business experts through Limerick Business Portal Accelerating route to market by raising fund and reduce risk	– Availability of vibrant spaces all over the city – Availability of facilities like ICT	Limerick city and county councils, university, private companies – in lease terms
– Raising the facilities by utilising the connectivity – provide high-quality business space and park to promote businesses	– Requirement of well-connected urban knowledge Or tech park – Close proximity to business and city centre	Limerick city County and councils Private sector – €25–30 m
– Clustering of medical facilities – Providing specialised treatment – Strengthen life science, healthcare and public health through clustering	Availability of medical facilities	Limerick city County and councils Private sector
– Raising fund for start-ups through business network and venture capital	To catalyst more investment where it is low	Limerick city and council €1 m
– Promote high potential start-up through Business portal	– Availability of support from stakeholders for start-ups – To support start-ups	Private sector, enterprises Ireland – Funding according to budget

(continued)

Table 1.31 (continued)

Strategy	Rationale	Partnership and fund
– Support on marketing and workforce development (through incentives)		
– Promote business by making funding available easily to all sized companies	– Lack of interconnection between companies	IDA, Limerick city and county council – Funding according to budget
– Improving Broadband facilities with high speed to attract Multiple service providers, business etc.	– Availability of high economic connections – High tech business location needs high-speed broadband	Internet service providers, Limerick city and County Councils €10–15 m
–Provide Skill development through centres to unemployed people by local authorities – Improving skill of workforce in ICT, tourism etc. by through experts – Continuous monitoring of skill through educational institutions – Collaborate school and colleges for training	Availability of educational institutes which can act as monitoring centres and skill development centres – Increase in no of residents in the area Increase in level of generic skills of young people and adults (communication, teamwork) – The increase in educational standards of people	Vocal education trainers Training providers, City and county councils
– Promote local workforce – Give proper training to labours and assign work	– Availability of historical monument – Regeneration of the historical buildings can provide employment	Community sector Construction firms Employment training boards Limerick city and county councils

well enough to bring new economic activities and it uses the R&D section for researchers and innovative technologies which may bring positive effect. Then various supporting programmes and incentives to start-ups may encourage start-ups providing better infrastructure facilities by clustering the medical facilities will attract people and the draining out population can be controlled. Quality spaces for businesses is a need for every business by providing high-quality business space with the impression of the place among the businessmen will increase and competition rise among people to get space in Limerick. This may cause an increase in quality of businesses leads to economic growth. For the economic growth to happen and continue Limerick must concentrate the future also. ICT enabled skill development centres to make the unemployed people to get trained and employed people to increase their skill so that the efficiency increases and future workforce efficiency increases. This is done through academic institutes which are the potential of Limerick County. Since the institutes are monitoring this will become effective. Improving the leadership among the stakeholders make efficient management of workforce, broadband connection to improve the technology and to know the future trends, improvement of airport creates more economic connections and interactions in future. Altogether the economic strategy is good enough to achieve the objectives or economic growth.

The city is having potentials like historic buildings at a nice site which can be turned into an economic generator when it is taken care off. The Gregorian quarter, opera site etc. are among them due to inefficient maintenance and management these buildings are depreciating and which are not improving the city. The second one is waterfront of the city, the Limerick city is having unique waterfront which is having the potential to become a tourist spot but the area is not well maintained. the economic strategy also is made spatially for economic development that is by providing high-quality sites for the economic generators like educational institutions and business centres (Table 1.32).

The spatial strategy is dealing with the position of educational institutions, tourism, recreational facilities, historical buildings, transportation and shopping destinations. The institutions are places in such a way that the facilities are available and transport are arranged in such a way that higher accessibility is there. Then only students will come and activities increases. Economic generators like parks, waterfronts, historic buildings are maintained and arranged in a good manner. Limerick implemented a connection between the parks and waterfront to attract tourists. The strategy is good enough to attract them and the potential is used very well. Similarly, the historic buildings are maintained. Limerick plan considered about the transport facility also they encouraged pedestrianisation of some streets connected to historic buildings for their conservation. These strategies are not ensuring that the City Centre fulfils its full economic potential by becoming a desirable place in which to 'do business'. Limerick only considering the sectors like tourism and recreation.

The Limerick city is a place where the economic activity is degrading due to various factors like recession, improper management of assets like tourist spots educational institutions etc. The people of the place are facing unemployment also. The investments from other countries decreased due to the recession were the reason. The strategies are made to overcome all these difficulties and to make the Limerick as a thriving economy better than any other surrounding Irish economies. The economic strategy they prepared had three divisions based on their weak points as well as potential but the spatial strategy is not a well prepared one. It does not take care of the start-ups and business people. It concentrated on the infrastructure like road and buildings with potential, also the quality of standard of living of people, the tourism and recreation will increase the economic activities but for long-term economic growth the facilities for the investors also to be taken care off.

1.12.17 Case Study-17—Tokyo [72–74]

The area under the Tokyo Metropolitan Government's (TMG) jurisdiction covers about 2000 km^2. Tokyo's population has been slowly but steadily growing. Tokyo

Table 1.32 Economic strategies-Limerick

Objectives	Spatial strategy	Implementation strategy
Improve economy through better education facilities	Attract young people to city centre by providing quality education and accommodation	High educational campus Limerick institute of technology and University of Limerick
Encourage and promote tourism	Utilise the heritage value and site value, convert that to mixed use by conserving some of the areas and others for commercial use	Through limerick culture centre and world-class waterfront
To increase recreational areas or improving existing recreational area	Improvement of park and connect waterfront to it	Interconnecting the park and waterfront
Increase economic activity in Gregorian centre	Attract people to Gregorian quarter through renovation of cultural heritage buildings	Utilise the heritage value keeping some portions as the same and converting others to mixed commercial use
Improve the transportation facilities and connectivity	– Improve Colbert station hub – Create integrated transport interchange with linked commercial development Including refurbished station building	Through community participation or using local workforce
To reposition the City Centre as the premier regional shopping destination	Enhance city centre retail offer	Convert the city centre to mixed use and increase the density Convert the informal markets to formal ones Open air street connecting the informal market and city centre
To make better quality living	Make some streets inaccessible to vehicles growth	Convert the highways to bicycle ways and pedestrian streets, reduce car parking Make streets more pedestrian friendly through lighting at junctions, signs etc.

is a Meta-City and ranks as the largest urban agglomeration in the world according to the UN. Tokyo contributes to about 40% of the national GDP. It stands as the 14th largest Gross Product in the ranking of national GDPs. Tokyo has been ranked as one of the most global cities in most international benchmarks on urban attractiveness for decades. In 2014 Tokyo is ranked as the fourth most global city, after New York, London and Paris. The market size, its economic dynamism and its highly qualified human capital make Tokyo a major business hub.

The TMG is one of the largest local governments in the world, as of 2014, it employed 165,425 members of staff, and the number is unlikely to decrease as the city prepares for the 2020 Olympic Games. Despite its size, the TMG is quite flexible and has initiated innovative measures to tackle environmental issues and promote sustainable development since the 1990s. The TMG is not the only authority when it comes to the governance of Tokyo in the field of sustainability. Especially when it comes to the Olympic and Paralympic Games-related planning, it shares competencies with other institutions from local, national and international levels, from the public and private sectors alike. Some research institutes and universities also influence the policy-making process related to the urban development in perspective of the Games. Tokyo's Energy and Environment Situation Reforms have been conducted since 1995 and allowed for the entrance of new utilities, often called Power Producers and Suppliers (PPS). There is a total of 195 power generation facilities, most of which are hydroelectric generating facilities, with a few thermal power plants, Renewable Energy power plants, and non-functioning nuclear power plants. Tokyo also ranks as the 33rd largest emitter of CO_2 in the world, between Finland and Singapore. The TMG contributed to the national effort by delocalizing factories from the metropolitan areas and by launching policies targeting the transportation sector especially. In 1997 Japan agreed to the Kyoto protocol and to achieve a 25% cut in CO_2 emissions by 2020.

The Smart Energy City strategy is based on measures taken following the Fukushima disaster and the resulting failure of generation plants to supply Tokyo to the same level as before. The Program included lowering the brightness standard, shifting to LED and keeping appliances on energy saving mode, and it targeted large facilities, small and medium facilities, households and TMG buildings.

If Tokyo, which is to remain the world's largest mature meta-city even in 2025, can show the world a model case of rebuilding as a sustainable city. The 2020 Tokyo Olympics will provide the best opportunity for "city sales." Three measures for rebuilding Tokyo into a sustainable city are:

i. Limiting vehicle use in city centre
ii. Promotion of city-centre residency
iii. Promotion of urban greening.

In the areas of business as well, Japan's slow rate of internationalisation stands out. Among major countries, Japan is one of the countries that receive very little foreign direct investment. Given this situation, the Japanese government set up the "Follow-Up Program for Promoting Japan as an Asian Business Centre and Direct Investment into Japan" in June 2012 and "Japan Revitalization Strategy Japan is Back" in June 2013 with the goal of doubling the inward foreign direct investment stock. As of 2012, the total debt of the Japanese national and local governments reached 224% of GDP. This is an overwhelmingly high rate. Given this financial situation, relying on the issuance of government bonds to finance infrastructure investment must be avoided. If funding from private sources proves difficult, it would be possible to procure funds from the private-public partnership infrastructure fund (Private Finance Initiative Promotion Corporation of Japan), which was established in October 2013.

Major construction works are planned by 2020. This is a non-exhaustive list. Some of these works had been planned regardless of the awarding of the Olympic and Paralympic Games to Tokyo.

- Airport facilities
- District redevelopment
- Metro network
- Road network.

A smart community is a community of a certain scale in which various consumers participate and one which has created a new social system. The new social system, while utilising a distributed energy system, including renewable energy and cogeneration, comprehensively manages energy supply and demand of the distributed energy system through an energy management system using IT, storage battery and other technologies to optimise use of energy and incorporates life support services, including the provision of care for elderly people. As smart community demonstration projects are cited in the Strategic Energy Plan that was decided by the Cabinet in 2014.

Major Smart Community Projects:

1. The Yokohama Smart City Project (YSCP).
2. The Toyota City Low-carbon Society Verification Project (Smart Meat).

Japanese-developed smart community systems will be disseminated abroad in the future through economic diplomacy by government leaders, the provision of appropriate platforms and enhancement of governmental support tools.

Most foreign companies that enter the Japanese market do so through a Local Partner. There are three main ways to go about this:

- Indirect business with clients via agent
- Indirect business with clients via non-exclusive distributor
- Indirect business with clients via sole-representative.

When Tokyo speak to potential partners, one should be aware that it will be a two-way assessment: it is not only you who will be trying to assess suitability; they will be assessing you. Japan is a very competitive, sophisticated and information-oriented market. The right partner can make all the difference. But Japanese partners can be discerning and there may be a host of issues which prevent them from entering into a partnership which on the surface may look obvious for them as well as you.

News about smart city-related development in Japan: Some websites are in Japanese only. It is possible to use an online translating application (such as Google automatic translation) to read a Basic English translation. While such translation is useful for an introduction, it is necessary to rely on more professional translations for accurate market information.

Most events at the Olympic and Paralympic Games Tokyo 2020 will take place in two thematic and operational zones, with the Athletes' Village located at the intersection of the two zones. The Heritage Zone houses several legacy venues used at the 1964 Games including the Yoyogi National Gymnasium and the Nippon Budokan. The Bay Zone serves as a model for futuristic urban development and includes the Dream Island Park and the Shiokaze Park venues—favourite places for

Tokyo residents to relax and play. The third pillar of Olympic and Paralympic Games: sustainability. The Tokyo Organising Committee of the Olympic and Paralympic Games is working to ensure the broad sustainability of the Tokyo 2020 Games, including in environmental, social, and economic terms. As part of that effort, the Committee has made a commitment to making sustainability a consideration in the procurement process during preparations for, as well as the hosting of, the Games.

The Committee's Sustainable Sourcing Working Group, which has been studying how to give shape to the Sourcing Code since that announcement, has compiled a draft Sustainable Sourcing Code. In addition to setting forth standards and methods that will apply to all goods and services procured by the Committee, the Code defines individual codes for agricultural, livestock and fishery products that take account of sustainability. The Working Group is now seeking public comment on the draft for its future study.

- Effective utilisation of Games-related facilities.
- Implementation of Urban Planning to Ensure Secured and Comfortable Living for Everyone.
- Communicate the Importance of Sustainability through Efforts Triggered by the Games.

Olympic Environmental Impact Assessment, a measure based on the current Environmental Impact Assessment (EIA) will be established. The EIA is initiated before the project is fully launched, and is over only after the project is completed, to consider both estimated and actual impacts on the environment. A similar EIA is planned to be applied to Olympic venues.

Challenges to sustainable 2020 planning and implementation: All construction, directly and indirectly, related to the mega-event will degrade the environment as CO_2 emissions and construction waste production is very likely to increase, considering the recent surge in carbon dioxide emission because of the shift back to fossil fuels, and the enduring problem of waste treatment and illegal dumping.

While Japan remains a major market for green technology and solutions, it is a fact that it has known better times in terms of economic performance. A three-part strategy relying on monetary regime change—with a 2% inflation goal, a fiscal stimulus—with JPY ¥ 10.3 trillion injected into the economy, and a series of structural reforms aimed at boosting long-term growth known as Abenomics is facing criticism since its proposal by the prime minister of Japan. Recently the whole Abenomics concept has been questioned as Japanese household's income have not kept pace with the inflation and taxation.

Despite the severe attacks on the results of the Abenomics and other shortcomings of Japan's public finance, the current Japanese economy presents attractive opportunities for business and cooperation between Japanese and European companies in the smart city sector. A fast-growing segment of the smart city market in Japan is the smart house service sector.

Research and innovation in the TMA: Additionally, many research institutes and R&D centres are in the TMA. Tokyo is ranked as the second metropolis submitting patent applications, right after San Francisco, with 5138 applications. The sectors where most patients are delivered are transported impact mitigation, potential or indirect contribution to emissions mitigation and general environmental management.

Olympic and Paralympic Games market and impact on Japanese economy: As mentioned in the introduction, hosting the Olympic and Paralympic Games has both tremendous advantages and drawbacks; the large costs to the host city are rarely compensated by the revenues or the legacy of Olympic venues. Indeed, hosting a mega event does not cause an increase in trade by itself, but sends a global signal of policy intentions. Thus, even candidates which do not host the Games are also likely to be more open and to experience an increase in exports. Although there is no specific mention of a direct effect of the Olympic and Paralympic Games on Japan's green economy in estimates so far, the market for smart technology has swelled since September 2013. The mega-event has significant potential in its educational value; in the case of Tokyo, the 2020 Olympics are the opportunity for increasing existing awareness about renewable energy, energy saving and waste management as is stated in the Games Foundation Plan.

While the real estate prices for sale and rental are already rising in Toyosu where the Olympic Village will be built, all wards where most of the Olympic venues are located can be expected to have higher land prices by 2020, making central Tokyo less and less attractive for renting, keeping or buying an office in Tokyo.

The electricity market reform was approved in 2013, and the first phase started from April 2015, with the establishment of the Organisation for Cross Regional Cooperation of Transmission Operators (OCTO). The OCTO has two functions: "to aggregate and analysed the supply-demand plans and grid plans, and order to change plans such as tie lines construction; and to order to reinforce generations and power interchanges under a tight supply-demand situation". In addition, they are not the most cooperative players on the market. They are very conservative companies with regional monopolies, which the reform intends to remove.

This section first highlights a few sectors which development is supported by Japan's Cabinet Office and looks at a few other sectors with market potential for European companies. Besides energy efficiency related business, a wide range of sectors in the smart city is considered, such as IT contributing to a more connected city. The opportunities may be directly linked to the Olympic and Paralympic Games operation, for example, the smart hospitality sector. Other opportunities may not be directly related to the Olympic and Paralympic Games, but are linked to the acceleration of green initiatives in Tokyo by 2020, as it is the case for the demand side management sector. The Cabinet Office defined this month 9 areas of Science and Technology Innovation to develop in perspective of the Games

• energy • next generation urban mobility • big data and sensors • weather forecast • smart hospitality • health monitoring • accessibility for handicapped individuals • audiovisual technology • green and flower arrangement.

Barriers and obstacles to business:
Complexity of networks and business practices in Tokyo is

- Difficulty in identifying the decision maker
- Lack of transparency from government agencies offering support
- Tendering process in Japanese
- Complexity of business practices.

Competitiveness of the market:

- Competition with Japanese and US companies
- Competition with "Olympic" companies.

The competition with companies and organisations from the UK involved in the 2012 Olympic and Paralympic Games is tough. The Tokyo 2020 Committee and the TMG are not only looking closely at the plans for the London Games but are very active in concluding agreements. Although there is no specific agreement between the UK and Japan about sustainability or smart urban development, it is possible that the Japanese Authorities would look more closely at the London case more than other cities.

The smart city strategy in Tokyo, the "Smart Energy City", was launched in May 2012 and builds on the emergency measures of energy savings which were enforced following the Fukushima accident in March 2011. Tokyo has further developed as a smart city through the promotion of renewable energy sources and smart meters in the central wards and the larger metropolitan area, along with several smart community initiatives. The Tokyo Metropolitan Government's (TMG) Bureau of Environment is very active in promoting energy efficiency and has achieved a 15% cut in CO_2 emissions in the industrial sector for instance. Despite such measures though, energy efficiency and carbon dioxide emission targets may be difficult to achieve by 2020.

The TMG is cooperating with the Tokyo Organising Committee of the Olympic and Paralympic Games (Tokyo 2020 Committee) to build on the existing green policies of the TMG and elaborate a concrete strategy for sustainable Games preparation and operation. Several obstacles can slow down enterprises' market access in Japan. The complexity of Japanese regulations and the lack of transparency of business practices can be challenging in some sectors for new companies which are unfamiliar with the language and the business culture. While 2020 is a catalyst and deadline for projects that are approved by the central and local government, it does not slow down the development of other smart community projects elsewhere in Japan.

1.13 Conclusion

This chapter is an introduction to international research studies of 16 smart metropolises of the book. The emerging scenario of metropolises, mega cities and meta-cities temporally and spatially across the world based on UN statistics and projection is discussed. GDP creation of nations is taking place very actively mostly in these three types of cities across the world which are projected for 205. Spatially and economically, higher population and GDP growth rate in these cities is now evident and happening mostly in the eastern and southern hemisphere than in the west. This book concentrates on more number of studies from the east and south than the west. City society through their legal framework of constitution brings about the creativity of design of an economic and spatial strategy to take these cities to the next level facing all local challenges through their official Master Plans. The design of economic and spatial strategies is presented for 17 cities based on official plans. Not all these cities give equal importance to economic strategies in contrast to spatial strategies. Often there are plans with less attention to strategies in general and economic strategies. Some of them do not mention economic strategies at all. Metropolitan Planning and implementation should move from geographical space to community and then to prosumer households both for planning and implementation, but this is happening only in a few cities presented in this chapter in a very limited way. These plans shall be considered as a design for societal change and should not be just for bringing out the infrastructure provision required for the bureaucracy to implement. What is happening is also not satisfactory. There is no possibility for a generalised design of spatial and economic strategies for metropolises since each city is unique and mobilising communities for the design is the only way through E-Democracy [18] with intensive use of ICT by society and continuing online training for metropolitan development by every household.

1. This first chapter explores large cities of one million and above with special references to the design of economic and spatial strategies of the smart metropolitan region. With such concentration of people in the limited area these cities have an important role in economic development of the nation. Further bigger the size of the city in population and area the spatial issues are more complex and challenging. Although Master Plan is expected to sort out these issues, the author would like to explore how far it is achieved in a study of 17 metropolises around the world.
2. Metropolitan cities are continuously exposed to external economic changes and require medium term strategies say 5 year to face it. These economic strategies of cities call for differing spatial strategies to intervene in emerging global situation.
3. There has been an attempt to convert many metropolitan cities, largely using selective investment strategies and procedural administrative rules arbitrarily not sanctioned constitution in a limited part of cities with no economic and spatial rationale to smart cities world over. In many smart metropolitan city projects, only a part of the city has been taken up for the smart metropolitan city program and in many cases, it is status and elitist area of cities like New Delhi Municipals Committee area in Delhi and not where low-income people live under 100 smart cities project of India. What is required is to take the whole metropolis to convert it to smart metropolis using smart economic and spatial strategies in many steps and in many years. Designing economic and spatial

strategies for Smart Metropolitan Regional Development shall be conducted by the spatially identifiable economic community at micro levels from time to time to suit the ever-changing scenarios of the economic environment.

4. If a city also performs an important commercial, cultural and political function for its region or even the whole country can be called a metropolis. Such a major importance can usually be assumed for cities boasting some 500,000 inhabitants in Europe and one million plus in India. On the other hand, Global cities are those selected few metropolises whose political, cultural and commercial influence extends across the entire globe (e.g. New York City, Tokyo or London).
5. As social spaces, metropolitan areas or regions can be characterised by the following four dimensions: Metropolitan regions are defined as an accumulation of specialised metropolitan facilities including public and private services. In terms of actors and actions, metropolitan regions constitute an arena for key regional stakeholders to exchange knowledge on joint regional objectives, strategies and projects, as well as on the necessary organisational structures. In the context of spatial development, metropolitan regions are a normative and guiding concept intended to contribute to innovation, creativity and economic growth. About the symbolic dimension of urban and regional development, metropolitan regions are the medium of symbols, norms and values which convey aspects associated with the specific brand image of metropolis and urbanity.
6. Authors of the book "Geographic Information System for Smart Cities" Edited by T. M. Vinod Kumar examined several definitions of smart cities and were not satisfied and came out with their own definition. "The smart metropolitan (area/region) city is a knowledge-based city that develops extraordinary capabilities to be self-aware, functions 24 h and 7 days a week, communicate, selectively, knowledge in real time to citizen end users for a satisfactory way of life, with easy public delivery of services, comfortable mobility, conservation of energy, environment and other natural resources, and creates energetic face to face communities and a vibrant urban economy even at time of national economic downturns". All six components of smart cities such as Smart Economy, Smart People, Smart Governance, Smart Mobility, Smart Environment and Smart Living is implied in this definition has been presented in detail in the four books of this series by the Author.
7. The concept global economy with partnership and division of labour of global network of cities is not new, but those who talk about a globalised economy insist that there have been distinct changes in its structure and modes of production. Whereas earlier economic activities crossed national boundaries, globalisation includes a deeper integration, where transnational corporations orchestrate production from various locations. Global places indicate a rising of networked society globally which can work effectively in a smart metropolis with high endowments of ICT and IOT.

8. The Global Urbanisation has five basic elements namely, new innovative technology, the centrality of information made possible by instant communication, an increasing trend toward the standardisation of economic and social products, growing cross-national integration, and mutual vulnerability stemming from greater interdependence.
9. UNHABITAT defined the metro cities as 1 million plus agglomeration 'mega-city' as with 10 million plus population and 'Meta city' which describes 'massive conurbations of more than 20 million people or above'. Since ICT drives the mega city and meta city formation, these cities are smart cities in making.
10. In 2016, there were 512 cities with at least 1 million inhabitants (metropolitan region/cities) globally. By 2030, a projected 662 cities will have at least 1 million residents. As hubs of trade, culture, information and industry, they will be vested with such power that at many levels they will act as city states that are independent of national and regional mediation. Today mega cities are home to less than 10% of the global urban population. In 2016, there were 31 megacities globally and their number is projected to rise to 41 by 2030.
11. There were eight meta cities above 20 million in 2016 in the universe, which is likely to be twice about 15–16 in number or little less in 2030 which may be considered as the accelerated graduation of mega cities to meta cities. However, the number of mega cities increase during this period which includes meta cities were from 31 to 41 which is not as impressive as that of the meta-cities.
12. Meta cities and Megacities are a key to globalisation, a state of interconnectedness around the globe that transcends and largely ignores national boundaries. Global urban economies rely on advanced producer services such as finance, banking, insurance, law, management consultancy, advertising and other services. The technology revolution has made it possible for business enterprises to hire these services anywhere in the world.
13. If mega cities including meta cities represent the economic hub of the future it is shifting towards Asia from America and Europe. Integrated spatial and economic strategies can help this transition. As per the current trend GDP growth rate in Asia is much higher than Europe and Americas and it is likely to continue for few decades. It looks like an Asian and African era is emerging. Growth rates of many mega cities in Asia and some in Africa are growing at higher rate than elsewhere.
14. When compared to China, India and the US; the largest urban population is in China followed by India and US. The total urban population in India is higher than the US. The percent of the urban population in mega cities in India are more than US and China in that order. China has all its metro and mega cities in the East facing the sea leaving a vast stretch in the western area without mega, meta and metro cities. China is fast moving towards a one billion urban population. Among the urban population, the largest percent people in India live in megacities which are more than 10 million population. In Kerala state, India 76% of the urban population lives in million plus cities, followed by Maharashtra state (59%), and Gujarat (55%). The lowest percentage of people

in India living in one million cities and above are Haryana (16%) and Bihar (17%). The largest number of million plus cities are in Kerala (7), and Uttar Pradesh (7) followed by Maharashtra (6) and Gujarat (4). The lowest number of one million plus cities are in Haryana and Bihar with one each.

15. There are already networks and corridors cutting across the boundary of nations which has given rise to the formation of urban settlements in 2016 in a globalised world.
16. Often the share of GDP produced in this mega and meta-city are considerable in comparison to total National GDP.
17. The projected GDP 2025 of top 30 Urban Agglomeration was used with projected population computed by UNHABITAT to derive average real GDP/population growth rate projected during 2008–2025. This shows that GDP/population growth rates of largest 30 urban economies in the universe are generally low but there are few exceptions in the south and eastern urban agglomeration.
18. Tabulating under descending order of high GDP growth rate of first 30 urban agglomerations by GDP growth rate 2008-25 and then cities are organised by countries and found that these urban agglomerations are all in Asia and Africa. It can be seen the higher GDP performance cities are in large number in India followed by China and other countries. This justifies a maximum number of case studies in this book from Asia and Africa.
19. There is a division of labour across city regions transcending country boundaries sharing the economic responsibilities as per capabilities. Global urban economies rely on advanced and standardised producer services such as finance, banking, insurance, logistics, law, management consultancy, advertising and other services. The technology revolution and uniformity of standard practices have made it possible for business enterprises to hire these services anywhere in the world. The national hierarchy of cities and the spatial division of labour within the economy is superimposed by a global division of labour. Cities and metropolitan regions become part of an emerging international hierarchy based on a competitive division of labour at the global level by international connections that affect financial flows and the knowledge-intensive service sector.
20. Large cities are interconnected and influence global and reinforce spatial interdependent functional structure with appropriate linkages. The connection is twofold within its city region and outside the city region transcending other national boundaries. The city is connected to hinterland and outside world simultaneously in a metropolitan region.
21. World cities are characterised by a sum of political power (both national and international) and organisations related to government; national and international trade, whereby cities function as gateway for their own and sometimes also neighbouring countries; providing superior banking, insurance and related financial services; advanced professional activities of all kinds; information gathering and diffusion. The form and extent of a city's integration with the world economy and the functions assigned to the city in the new spatial division of labour will be decisive for any structural changes occurring within it.

22. The world cities exhibit the following;

    i. The nature of a city's integration with the world economy is decisive for any structural changes occurring within it.
    ii. Key cities throughout the world are used by global capital as 'basing points' for the organisation of production and markets.
    iii. The global control functions of world cities are manifest in the structure of their industrial structure and job markets.
    iv. World cities are major sites for the concentration and accumulation of capital.
    v. World cities are destinations for large numbers of migrants—both domestic and international.
    vi. World city formation exposes the major contradictions of industrial capitalism, particularly spatial class polarisation.

23. There are two trends of thoughts about Meta cities, Megacities, and metro cities. One tries to attain a position of the global city by deliberately executing spatial and economic strategies to achieve that goal following what is discussed in the earlier paragraphs. This is in addition to solving many city specific issues of metropolitan development. The second approach is how to make a metropolitan area a smart metropolis by appropriate spatial and economic strategies. Smart is ICT implementation in all activities including economic activities. Economic strategies may involve converting the region with all its economic activities to the smart economy as discussed in the book in this book series "Smart Economy in Smart Cities".

    1. Since this book focuses on the area of design of economic and spatial strategies to achieve the overall goals of smart metropolitan development. A survey, of such designs of several official plans is undertaken here. Cities are selected randomly with no sampling plans. The following discusses only the appraisal of these economic and spatial strategies in several metropolises.

        i. **Delhi NCR Metropolitan Region**. Although Delhi will be the second largest meta city in the world in 2030 after Tokyo and therefore is a global city. There is no such consideration in the NCR Plan. Also, there is no deliberate promotion of the city to a smart metropolitan city with appropriate ICT-IOT design and related spatial strategies. The economic strategy adopted by the NCR planning board is not efficient enough for a city such as Delhi. The stated Delhi strategy of making trade and commerce barrier free which is the aim of Goods and Service Act 2017 nationwide within NCR will not suffice for Delhi. Other specific issues mainly administrative issues are not addressed for further easing to do business and giving impetus for the development of industries. Interstate agreement on unified policies can create chaos and may reach a long time to reach consensus. This might be a hindrance to further development of NCR region. Increased ease of doing business, transparent laws and regulations for the same have not been given due consideration. Also, focus on people and skill development for economic development is not visible. Smart mobility based spatial strategy is not stated.

ii. **Vancouver Metropolitan Region**. In the case of Vancouver, there is a strong economic strategy in place. Local businesses and talents are nurtured which results in a booming diverse economy. Brain–drain is prevented by attracting foreign Canadians and immigrants with attractive and affordable housing and environment for families. This is a strategy to attract a skilled workforce. Integration of universities with business centres for productive R&D yet another important step in economic growth. They have thoroughly invested in the clustered growth of industries with an increased focus on green jobs to remain sustainable in the long run.

iii. **Melbourne 2030**. Planning for Sustainable Growth. Plan Melbourne does not identify how the government will respond to the impacts of climate change. The Metropolitan Planning Authority must work in close partnership with Local Government, in the future planning for Urban Renewal Areas. Initiatives to 'streamline' the planning system which limits community involvement in the planning process, or reduces Local Government's decision-making responsibilities, are not supported.

iv. **Shanghai Metropolitan Plan**. Shanghai is now facing great opportunities in turning itself into an international economic, financial, and trade centre. China's continuous economic growth provides a solid base for Shanghai to move toward this goal. Shanghai has set its long-term strategic objectives for social and economic development. By 2010, Shanghai is planned to become one of the international economic, financial and trade centres of the world. Shanghai aims to form the economic scale and comprehensive strength of a world metropolis; optimise urban spatial distribution, modernise the city's physical infrastructure, participate in international labour division and the circular flow of the international economy, introduce the operational mechanism of a socialist market economy, and pursue the balanced social, economic and environmental development. Economic Strategies plan is to optimise and upgrade the industrial structure, trying to improve the pivotal status of Shanghai in global city network and process to accelerate the technological innovation.

v. **Mumbai Metropolitan Regional Development**. The Economic strategy given by the metropolitan plan is very broad. Mumbai Metropolitan Regional Plan 2016-36 mainly addressed issues of growing urbanisation, uneven distribution of jobs increasing commutes, lack of affordable housing and basic infrastructure in the region, environmental degradation and inadequate governance. The existing situation of Mumbai Metropolitan Region is analysed sectoral and cross-sectoral to arrive at issues that need to be addressed in the Regional Plan 2016-36.

vi. **New York**. The economic strategy adopted clearly touches every aspect of the economy. The plan also emphasised on strengthening the fundamentals such as workforce which is very important in achieving economic development. The New York Metropolitan Region has a separate spatial plan, transportation plan and economic development plan. The strategies adopted clearly ensure sustainable and long-term growth in jobs and income to contribute to the resurgence of the broader economy of New York State. Also, they have a detailed transportation development plan which has a shared vision for sustainability.

vii. **Calgary Metropolitan Region**. The Calgary Regional Partnership (CRP) and member municipalities will work together to ensure a diversified and globally competitive region that continues to enjoy a high quality of life and is able to attract and retain a viable and adequate regional workforce and member municipalities will endeavour to achieve a distribution of jobs creation and economic activities throughout the region consistent with transit and complete mobility policies that encourage the location of jobs close to where people live. Recognising the connections and relationships that exist between communities, the Calgary Metropolitan Plan (CMP) acknowledges and respects the vital and historic importance of rural lands, industry and culture in our region. What are the industries and where those can be established to boost the economy and employment is not stated? What are the land management techniques suitable in this region are not mentioned? There is a lot of potentials to develop tourism which is not at all described. They mentioned that various studies need to be done to analyse the economic boost but what are those not stated. The Calgary Metropolitan Plan is the blueprint for accommodating growth in future. The plan giving the various strategies to make Calgary as a healthy environment in enriched communities, with sustainable infrastructure and a prosperous economy.

viii. **Kuala Lumpur Metropolitan Regional Development Plan**. The vision and goals for Kuala Lumpur have been formulated with the aim of creating a sustainable city. City Hall Kuala Lumpur (CHKL) shall ensure that the planning of the City shall strike a balance between physical, economic, social and environmental development. Local Agenda 21 shall be adopted to encourage citizen participation towards creating a sustainable society. This is in line with government policies of implementing sustainable development strategies as stipulated in the Habitat Agenda of the Rio Declaration.

ix. **London Metropolis**. The London Plan sets out a new approach for planning in London. It emphasises growing inward and upward so that it can reduce the costs of growth, create walkable communities, revitalise our urban neighbourhoods and business areas, protect our farmlands, and reduce greenhouse gases and energy consumption. The plan sets out to conserve our cultural heritage and protect our environmental areas, hazard lands, and natural resources. Through the London Plan, the community is planning for vibrant, healthy, safe and fulfilling neighbourhoods, attractive and viable mobility alternatives and affordable housing that is accessible to those who need it. At the root of The London Plan is the goal of building a city that will be attractive as a place to live and invest in a highly competitive world and one that will offer the opportunity of prosperity to everyone—one their own terms and in their own way.

x. **Berlin Plan 2035**. The Berlin Strategy provides an inter-agency model for the long-term, sustainable development of the capital. With one-third of the city comprising of open spaces, a compact polycentric development, highly tolerant society, the development plan further proposes to enhance these key selling points of Berlin. The development plan builds its foundation on the strong points of Berlin and proposes strategies to tackle the challenges. Provisions of affordable housing, further increasing the short distances to amenities, enabling a start-up friendly environment are some of the key proposals of the plan. Cultural diversity and tolerance in the society are used as a selling point for the attraction of workforce and tourists. Community participation is given utmost importance. The economy focuses on the educational institutions, research centres, attracts skilled labour from all over the world. Using a range of strategies and goals, it sets out the areas and directions in which this growing city should develop and highlights the areas that will form the focus of its future development.

xi. **Dhaka Structure Plan 2016–2035**. The Dhaka structure plan covers the basic issues such as effective land use management, transport for efficient connectivity, enhancing employment and productivity, public Facilities, protecting the natural and healthy environment. The spatial concept of dividing the planning area into two broad categories i.e. Urban Promotion Area (UPA) and Urban Control Area (UCA) is an efficient way to assure uniform distribution of development. Here, the basic requirements are made available at each functional region reducing the dependency on the core city. At the same time under the UCA flood flow zones, Water Retention Area, National and Regional parks, Forest Area, and large-scale Heritage sites are preserved. Instead of focussing on individual plot based development, the concept of planned unit development is introduced. The plan talks about increasing the productivity of the informal sector, which is a good step towards mitigating urban poverty. The garment sector, leather manufacturing which boosts women employment is also planned to make it more efficient. While the report covers all the broad aspects required for the development of a region, the involvement of the public in the planning process would provide a more achievable and efficient plan.

xii. **Master Plan for Patna Metropolitan Region**. The Metropolitan plan for Patna, though it takes into consideration, the Economic aspect of the city, doesn't provide a sufficient Economic Strategy for development. The plan has a concept for the spatial development of the city. The spatial strategy is fairly achieved. The Implementation plan simply explains a case study for TP schemes from Ahmedabad. There is no strategy adopted. Also, it doesn't consider each project in detail. The Economic approach in the Master Plan is only by addressing the land requirement for various employments generated. There is no mention about Economic Strategy. Patna being the only important city in the state has acquired the strong position in regional trade and business. It is necessary to have an economic strategy while planning for Patna. The Master plan lacks this. Instead, an overall development which also leads to economic development is attempted. Implementation plan doesn't consider every project in detail.

xiii. **Montreal Metropolitan Development Plan 2020**. Economic strengths are utilised and creative market and the smart economy are aimed to achieve through these strategies. A clear and comprehensive economic strategy is provided in the Master plan. The implementation strategy to carry out each strategy is also explained. The idea of Smart Metropolis is carefully dealt here which is considered as the most innovative and sustainable development concept in the current scenario. Montreal is a large city with a diversified economy needed a spatial and economic strategy which is formulated and explained well in the Master plan. Issues and potentials were identified and addressed carefully.

xiv. **Helsinki**. The spatial structure for the economic/business development is mentioned. But the kind of economic activities is not clear even though the zones are marked. The spatial zoning of economic activities is done such a way that it is around the city centre and large-scale industries are placed away from the centre which has easy access from nearby places. Retail activities are evenly distributed in all area.

xv. **Istanbul 2014–2023**. The economic strategy discusses all the sectors to make a globally decisive, high value added, innovative and creative economy. The focus has been given to entrepreneurship to encourage the work participation thereby eliminating unemployment that would lead to a better vibrant economy. The main idea of entrepreneurship is to be appreciated as the city will sustain its own economy rather than depending on external revenue as in trade. Even though the strategies and objectives are discussed elaborately the implementation is not detailed which makes it difficult for the region to put these concepts into reality. The Metropolitan Plan discusses a vision for Istanbul, spatial and economic strategy. The very important part was the public participatory approach so that the people of the place make the place. All the strategical domains are made with the opinions of the citizens and thereby solve their issues and each project is for the development of the city and the citizens. The economic domains include the transformation of the industry, ascending R&D, encouraging entrepreneurship, and increasing employment. All these directly or indirectly stand for the citizens. The spatial domains are increasing spatial quality, holistic urban transformation, protecting the culture and history, effective disaster management, sustainable transport.

xvi. **Limerick**. The Limerick city is a place where the economic activity is degrading due to various factors like recession, improper management of assets like tourist spots educational institutions etc. The people of the place are facing unemployment also. The investments from other countries decreased due to the recession. The strategies are made to overcome all these difficulties and to make the Limerick as a thriving economy better than any other surrounding Irish economies. The economic strategy they prepared had three divisions based on their weak points as well as potential but the spatial strategy is not a well prepared one. It does not take care of the start-ups and business people. It concentrated on the infrastructure like road and buildings with potential, also the quality of standard of living of people, the tourism and recreation will increase the economic activities but for long-term economic growth the facilities for the investors also to be taken care of.

xvii. **Tokyo**. The smart metropolitan city strategy in Tokyo, the “Smart Energy City”, was launched in May 2012 and builds on the emergency measures of energy savings which were enforced following the Fukushima accident in March 2011. Tokyo has further developed as a smart metropolitan city through the promotion of renewable energy sources and smart meters in the central wards and the larger metropolitan area, along with several smart community initiatives. The Tokyo Metropolitan Government’s (TMG) Bureau of Environment is very active in promoting energy efficiency and has achieved a 15% cut in CO_2 emissions in the industrial sector for instance. Despite such measures though, energy efficiency and carbon dioxide emission targets may be difficult to achieve by 2020. The TMG is cooperating with the Tokyo Organising Committee of the Olympic and Paralympic Games (Tokyo 2020 Committee) to build on the existing green policies of the TMG and elaborate a concrete strategy for sustainable Games preparation and operation. Several obstacles can slow down enterprises' market access in Japan. The complexity of Japanese regulations and the lack of transparency of business practices can be challenging in some sectors for new companies which are unfamiliar with the language and the business culture. While 2020 is a catalyst and deadline for projects that are approved by the central and local government, it does not slow down the development of other smart community projects elsewhere in Japan.

Acknowledgements Author acknowledge the contribution of Arjun P., Bigly Sathyapal, Fathima Zehba M. P., Gopika C. B., Jose. B. Thomas, Kesowe-u Wetsah, Kuruva Manoj Kumar, L. V. Krishna Reddy, Mekhana S. S., R. S. Vishnu and Salimah Hasnah on an appraisal of official plans of 12 Asian and 12 Non-Asian metropolises and a smaller part of their work are summarised in this chapter.

References

1. Census of India (2011) Primary census abstract, UA/Towns. Office of the Registrar General & Census Commissioner, India, New Delhi (compact disk form)
2. Census (2006) "Population projections for India and states 2001–2026." Report of the technical group on population projections. Registrar General & Census Commissioner, India
3. http://www.oecd-ilibrary.org/docserver/download/0413091ec008.pdf
4. Boix R, Veneri P (2009) Metropolitan areas in Spain and Italy. IERMB Working Paper in Economics Nr. 09.01. Barcelona
5. Brezzi M, Piacentini M, Sanchez-Serra D (2012) Measuring metropolitan areas: a comparative approach in OECD countries. In: Fernández Vázquez E; Rubiera-Morollón F (eds) Defining the spatial scale in modern regional analysis. New challenges from data at local level. Advantages in spatial science—the regional science series, pp 71–89, Springer, Dordrecht
6. Burger MJ, Meijers EJ (2012) Form follows function? linking morphological and functional polycentric city. Urban Stud 49(5):1127–1149
7. Abrantes P, Pimentel D, Tenedorio JA (2010) Metropolitan dynamics typology of the portuguese urban system. Open Urban Stud J 3:68–77
8. Angel S, Parent J, Civco D, Blei A (2012) Atlas of urban expansion. Lincoln Institute of Land Policy, Cambridge, US
9. Champion AG (2001) A changing demographic regime and evolving polycentric urban regions: consequences for the size, composition and distribution of city populations. Urban Stud 38(4):657–677
10. Sassan S (1991) The global city: New York, London, Tokyo, Chapter 8. Princeton University Press, New Jersey
11. Sassan S (2000) Cities in a world economy, 2nd edn. Pine Forge Press, London, p 7
12. Beaverstock JV, Smith RG, Taylor PJ, Walker DRF, Lorimer H (2000) Globalization and world cities: some measurement methodologies. Appl Geogr 20:47
13. Beaverstock JV, Smith RG, Taylor PJ (2004) World-city network: a new metageography? Ann Assoc Am Geogr 90(1):124–125

14. Bourdeau-Lepage L, Huriot, J-M (2005) The metropolis in retrospect. From the trading metropolis to the global metropolis. Recherches Economiques de Louvain—Louvain Econ Rev 71(3): 257–284
15. Vinod Kumar TM and Associates (ed) (2015) Geographic information systems for smart cities. Copal Publishing Group, New Delhi
16. Vinod Kumar TM (ed) (2016) e governance for smart cities. Springer, Singapore
17. Vinod Kumar TM (ed) (2017) "Smart economy in smart cities" international collaborative research: Ottawa, St Louis, Stuttgart, Bologna, Cape Town, Nairobi, Dakar, Lagos, New Delhi, Varanasi, Vijayawada, Kozhikode, Hong Kong. Springer, Singapore
18. Vinod Kumar TM (ed) (2018) e-democracy for smart cities. Springer, Singapore
19. Davis DE (2005) Cities in global context: a brief intellectual history. Int J Urban Reg Res 29 (1):99
20. Davis, "Cities in Global Context," Child Hill R, Kim JW (2000), "Global Cities and the Developmental States: New York, Tokyo and Seoul," Urban Stud 37(12):2168
21. Castells M (2000) The rise of the network society, the information age: economy, society and culture, vol I. (2nd edn), Cambridge, MA. Blackwell, Oxford, UK
22. Savitch HV (2001) What's new about globalisation and what does it portend for cities? Int Soc Sci J 54(172):180–181
23. Child Hill R, Kim JW (2000) Global cities and the developmental states: New York, Tokyo and Seoul. Urban Stud 37(12):2168
24. Grant R, Nijman J (2002) Globalization and the corporate geography of cities in the less-developed world. Ann Assoc Am Geogr 92(2):320
25. Smith DA (2001) Global cities in East Asia: empirical and conceptual analysis. Int Soc Sci J 56(181):399
26. United Nations (2006) Urbanisation: Mega & Metacities, New City States; UN-HABITAT: State of the World's Cities 2006/7; United Nations: New York, NY, USA
27. Denis E, Mukhopadhyay P, Zérah M-H (2012) Subaltern urbanisation in India. Econ Polit Wkly 47(30):52–62
28. Bala R (1986) Trends in urbanisation in India. Rawat Publication, Jaipur
29. Bhagat RB (2011) Emerging pattern of urbanisation in India. Econ Polit Wkly 46(28):10–12
30. Bhagat RB (2011) "Migrants' (Denied) Right to the City". In: Zerah M-H, Dupont V, Tawa Lama-Rewal S (scientific eds), Faetanini, M (publication ed) Urban policies and the right to the city in India: rights, responsibilities and citizenship, pp 48–57. UNESCO and Centre de Sciences Humaines, New Delhi
31. Bhagat RB (2005) Rural urban classification and municipal governance in India. Singapore J Trop Geogr 26(1):61–74 (India urbanisation)
32. Davis K (1962) Urbanisation in India: past and future. In: Turner R (ed) India's urban future. University of California Press, Berkeley, CA, pp 3–26
33. Lahiri-Dutt K (2001) Growth and occupational structures of million cities of India. Asian Geographer 20(1–2):153–167
34. Kumar J (2015) Metropolises in Indian urban systems. Eur J Geogr 6-341-51
35. Birch E, Wachter S (eds) (2011) Global urbanisation. University of Pennsylvania Press, Philadelphia
36. Friedmann J, "The World City Hypothesis," at 317. Friedmann cites Manuel Castells' La question urbaine (1972, Maspero: Paris), and David Harvey's Social Justice and the City (1973, Edward Arnold: London)
37. Bazzanella L, Caneparo L, Corsico F, Roccasalva G (eds) (2012) The future of cities and regions: simulation, scenario and visioning, governance and scale. Springer, Dordrecht
38. Ministry of Urban Development, Government of India, National Capital Region—Regional Plan 2021 (2005), http://ncrpb.nic.in/regionalplan2021.php. Accessed 22 Jan 2017
39. Vancouver Economic Commission, Vancouver economic plan (Sept 2011), http://www.vancouvereconomic.com/. Accessed 28 Feb 2017
40. Greater Vancouver Regional District Board, Metro Vancouver 2040—Shaping Our Future (July 2011), http://www.metrovancouver.org/. Accessed 28 Feb 2017

41. Final Plan Melbourne (new Metropolitan Strategy) Released. http://www.portphillip.vic.gov.au/plan-melbourne.htm. Accessed 24 Feb 2017
42. Victoria State Government. October 2011. Policy and Strategy. https://www.planning.vic.gov.au/policy-and-strategy/planning-for-melbourne/melbourne-2030planning-for-sustainable-growth. Accessed 24 Feb 2017
43. Victoria State Government (2016) Plan Melbourne 2017–2050. http://www.planmelbourne.vic.gov.au/. Accessed 25 Feb 2017
44. Victoria State Government. 3rd April 2017. Melbourne Planning Schemes. http://planning-schemes.delwp.vic.gov.au/schemes/melbourne. Accessed 3 Apr 2017
45. AESOP 26th Annual Congress. July 2016. Implementation of the Shanghai Master Plan (2001–2020). Accessed 20 Jan 2017
46. Hu R (2008) China's metropolitan planning: Shanghai. http://www.globalisationandcities.com/uploads/1/5/7/5/15751464/chinas_metropolitan_planning__shanghai.pdf. Accessed 20 Jan 2017
47. Debin, Du. June 2015. Shanghai in 2050: Its Development Challenges and Strategy Choices
48. https://f.hypotheses.org/wp-content/blogs.dir/841/files/2015/01/Du-Debin-Shanghai-in2050.pdf. Accessed 20 Jan 2017
49. Na Z, Liu M (2012) Spatial transformation in Shanghai: the strategy, institutional arrangement and planning procedures—the case of EXPO 2010
50. http://www-sre.wu.ac.at/ersa/ersaconfs/ersa12/e120821aFinal00891.pdf. Accessed 21 Jan 2017
51. Yang G (2002) Shanghai's economic development: its opportunities and challenges in the 21st century. http://www.globalurban.org/GUD%20Shanghai%20MES%20Report.pdf. Accessed 21 Jan 2017
52. Draft Mumbai Metropolitan Regional Plan 2016–2036, September 2016 https://mmrda.maharashtra.gov.in/regional-plan. Accessed 8 Apr 2017
53. A Strategy for prosperity in Western New York, WNY Regional Economic Development Strategic Plan (2011) https://regionalcouncils.ny.gov/sites/default/files/regions/westernny/A_Strategy_for_Prosperty_WNY_Strategic_Plan.pdf. Accessed 8 Apr 2017
54. Spatial Planning and Inequality, Fourth Regional Plan Roundtable (2015) http://library.rpa.org/pdf/RPA-4RP-Whitepaper-Spatial-Planning-and-Inequality.pdf. Accessed 8 Apr 2017
55. New York Metropolitan Transportation Council, Regional Transportation Plan 2040 https://www.nymtc.org/portals/0/pdf/RTP/NYMTC_RTP_Complete.pdf. Accessed 8 Apr 2017
56. Calgary metropolitan plan—calgary regional partnership http://calgaryregion.ca/cmp/bin1/pdf/CMP.pdf
57. KL structural plan 2020 http://www.dbkl.gov.my/pskl2020/english/index.htm
58. The London plan the spatial development strategy for London consolidated with alterations since 2011 March 2016
59. Gordon IR (2004) 'Capital Needs, Capital Growth and Global City Rhetoric in Mayor Livingstone's London Plan', GaWC Research Bulletin 145, University of Loughborough
60. Hall PG, Pain K (2006) The polycentric metropolis: learning from mega-city regions in Europe. Earthscan, London
61. Senate Department for Urban Development and the Environment Berlin, Berlin Strategy Urban Development Concept 2030, 2013. Available: http://www.stadtentwicklung.berlin.de/planen/stadtentwicklungskonzept/download/strategie/BerlinStrategie_Broschuere_en.pdf. Accessed 23 Feb 2017
62. Senate Department for Urban Development and the Environment Berlin, Berlin Today, Status and Report, 2013. Available: http://www.stadtentwicklung.berlin.de/planen/stadtentwicklungskonzept/en/statusbericht/index.shtml. Accessed 23 Feb 2017
63. Rajdhani Unayon Kortipakhya—Dhaka, Dhaka Structure Plan 2016–2035, 2015. Available: http://www.rajukdhaka.gov.bd/rajuk/image/slideshow/1.%20Draft%20Dhaka%20Structure%20Plan%20Report%202016-2035(Full%20%20Volume).pdf. Accessed 17 Jan 2017

64. Qazi Azizul Mowla, Review of Dhaka Structure Plan 2016–2035, 2016. Available: https://www.researchgate.net/publication/291973110_Review_of_Dhaka_Structure_Plan_2016-2035. Accessed 20 Jan 2017
65. ADB-BBS (2012) The informal sector and informal employment in Bangladesh, Country Report, 2010 Available: https://www.adb.org/sites/default/files/publication/30084/informal-sector-informalemployment-bangladesh.pdf. Accessed 20 Jan 2017
66. Patna Master Plan 2031, Urban development department, Govt. of Bihar, 2016
67. Montréal Master Plan 2020, Montreal Metropolitan development board, Nov 2004
68. From City to City-Region—City of Helsinki Strategic Spatial Plan, City Planning Department 2009, City of Helsinki http://www.hel.fi/hel2/ksv/julkaisut/julk_2009-8.pdf. Accessed 24 Feb 2017
69. Istanbul Development Agency, Istanbul regional plan—2014–2023, Apr 2016. http://www.istka.org.tr/content/pdf/Istanbul-Regional-Plan-2014-2023.pdf. Accessed 9 Jan 2017
70. https://en.wikipedia.org/wiki/Limerick Limerick Metropolitan Plan
71. https://www.limerick.ie/limerick-2030-economic-and-spatial-plan. Accessed Mar 2017
72. Fujita K Hill RC (2007) The zero waste city: Tokyo's quest for a sustainable environment. J Comp Policy Anal Res Pract 9(4):405–425 https://www.msu.edu/~hillrr/The%20Zero%20Waste%20City%20whole%20text.pdf
73. IOC December (2014) Olympic agenda 2020, 20 + 20 Recommendations http://www.olympic.org/Documents/Olympic_Agenda_2020/Olympic_Agenda_2020-2020_Recommendations-ENG.pdf
74. June 2014 Olympic Games: legacies and impacts http://www.olympic.org/Assets/OSC%20Section/pdf/LRes_7E.pdf

Part II
China, Hong Kong

Chapter 2
Towards Smarter Regional Development of Hong Kong Within the Greater Bay Area

Sujata S. Govada and Timothy Rodgers

Abstract The Greater Pearl River Delta, more recently referred to as the Greater Bay Area (GBA), has and continues to rapidly develop into a significant and influential mega region in China, Southeast Asia and the world. The economic reforms of China and the Open Door Policy in 1978 transformed the primarily rural agricultural region into the highly populated, sprawling and urbanized high-tech value-added manufacturing region it is today. During this period, Hong Kong has played a significant role in the growth and development of the GBA with its more mature economy providing crucial support through its financial institutions, legal and professional services, and developed international trade and logistics hub, allowing businesses in the GBA to scale and reach global markets for their products. This in turn established economic, social, and political connections, collaboration and cooperation between Hong Kong and cities in the GBA such as Shenzhen, Guangzhou, Zhuhai and Macao among others, as well as improvements in infrastructure and transportation networks that have enhanced regional accessibility and integration. Going forward however, Hong Kong's role and competitive advantage in the region is being threatened as other cities in the GBA continue to grow larger than Hong Kong both demographically and economically. As such, Hong Kong is at a time where it must both reposition itself within the region, especially as Hong Kong is gradually becoming more integrated with the rest of the GBA in Mainland China. This chapter will present an overview of the GBA region, assess its current development and regional integration, and review the future spatial development plans and regional cooperation initiatives of the cities in the GBA. Finally, a discussion of how Hong Kong can more effectively integrate itself with the rest of the GBA from economic, social and political levels is presented, and recommendations of key areas to address are suggested in order for Hong Kong and the GBA to further develop into a smarter, more sustainable, and liveable integrated mega region.

S. S. Govada · T. Rodgers (✉)
Institute for Sustainable Urbanisation, UDP International, The Chinese University of Hong Kong, Suite 2207-9, Tower Two, Lippo Centre, 89 Queensway, Admiralty, Hong Kong
e-mail: tim@udpcltd.com

S. S. Govada
e-mail: sujata@udpcltd.com

T. M. Vinod Kumar (ed.), *Smart Metropolitan Regional Development*,
Advances in 21st Century Human Settlements,
https://doi.org/10.1007/978-981-10-8588-8_2

Keywords Hong Kong · Pearl River Delta · Greater Bay Area
Mega region · Meta city · Smart metropolitan region · Smart city smart region
Regional integration · Regional development

2.1 Hong Kong and the Pearl River Delta Region (Greater Bay Area)

2.1.1 Introduction and Background

The Pearl River Delta (PRD) region is a 55,000 km^2 area in Southern China comprised of the '9+2' cities from three jurisdictions as given below [1]. The nine cities comprise one jurisdiction and include the major cities of Dongguan, Foshan, Guangzhou, Huizhou, Jiangmen, Shenzhen, Zhaoqing, Zhongshan, and Zhuhai in the Pearl River Delta Region of Guangdong Province in Mainland China. The second and third jurisdictions comprise the two cities in the Special Administrative Regions; Hong Kong (HKSAR) and Macau (MSAR). Their location surrounding the Pearl River estuary on the coast of the South China Sea has benefitted this region during the past 50 years with a strategic geographic advantage by positioning itself as a gateway into Mainland China and hub for manufacturing and trade within South East Asia and the world.

More recently, the Pearl River Delta has been rebranded as the Greater Bay Area (GBA) to emphasize the updated regional development and integration plans of the Pearl River Delta. First mention of the GBA occurred in the English translation release of the "13th Five Year Plan for Economic and Social Development of the People's Republic of China" in December 2016, which introduced the concept of the "Guangdong-Hong Kong-Macau Greater Bay Area" [2]. In March 2017, Chinese Premier Li Keqiang announced the plan for a "development of a city cluster in the Guangdong-Hong Kong-Macau Greater Bay Area", and on 1st July 2017, the "Framework Agreement on Deepening Guangdong-Hong Kong-Macao Cooperation in the Development of the Bay Area", both further formalizing the term. Since then, all future discussion and mention of regional development and cooperation will be made with reference to the GBA [3, 4] (Figs. 2.1 and 2.2).

In 2015, the nine cities in the GBA region had a total population of around 58 million people [7], Hong Kong's population was at around 7.3 million people [8], and Macau had a population of about 646,000 people [9]. Therefore, the total population of the GBA was around 66 million people at the end of 2015. As discussed in Chap. 1, this classifies the GBA metropolitan region as a 'Meta City', which are "conurbations of more than 20 million people" [10]. The World Bank stated in 2015 that the GBA had become the largest urban area in the world in both size and population [11] (Table 2.1).

Although the GBA currently encompasses an expansive land area and numerous cities, over the years it has developed into a polycentric metropolis with three key

Fig. 2.1 Map of the GBA '9+2' and its location in Southern China. *Source* Li and Wong [5]

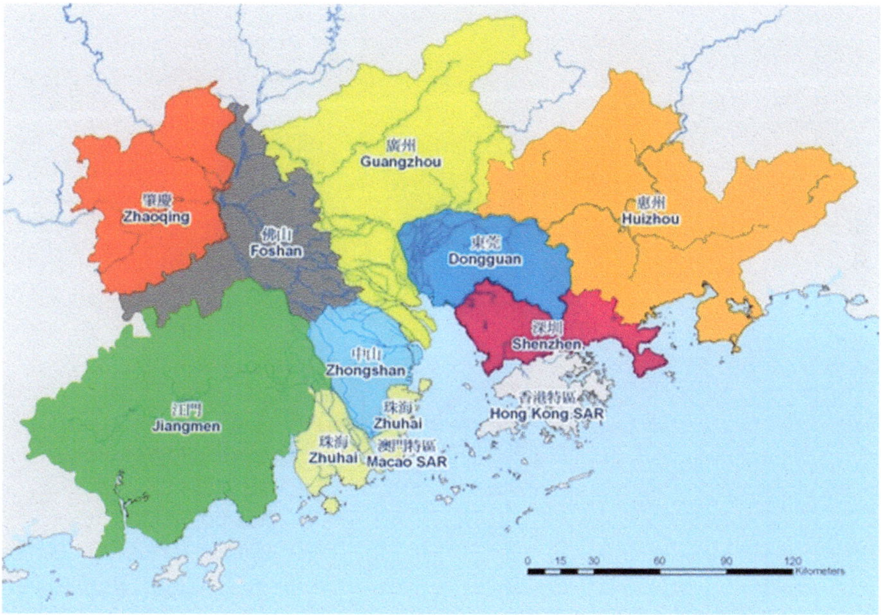

Fig. 2.2 More detailed map of the GBA's '9+2' cities. *Source* Daxue Consulting [6]

population centers, namely the Hong Kong and Shenzhen Metropolitan Area, the Guangzhou and Foshan Metropolitan Area, and the Macau and Zhuhai Metropolitan Area. The Hong Kong and Shenzhen, and Guangzhou and Foshan Metropolitan Areas make up the majority of the population in the GBA, with populations of 18.68

Table 2.1 Population and land area of major cities in the GBA region in 2015

Cities	Land area (km^2)	Population (millions)
Guangzhou	7434.0	13.50
Shenzhen	1953.0	11.38
Zhuhai	1688.0	1.63
Foshan	3848.0	7.43
Huizhou	11,158.0	4.76
Dongguan	2465.0	8.25
Zhongshan	1800.0	3.21
Jiangmen	9541.0	4.52
Zhaoqing	14,856.0	4.06
Hong Kong SAR	1105.0	7.30
Macau SAR	30.5	0.60
GBA total	55,878.5	66.64

Source Guangdong Statistical Yearbook 2016; Hong Kong Census and Statistics Department 2016; Macau Census and Statistics Service 2016

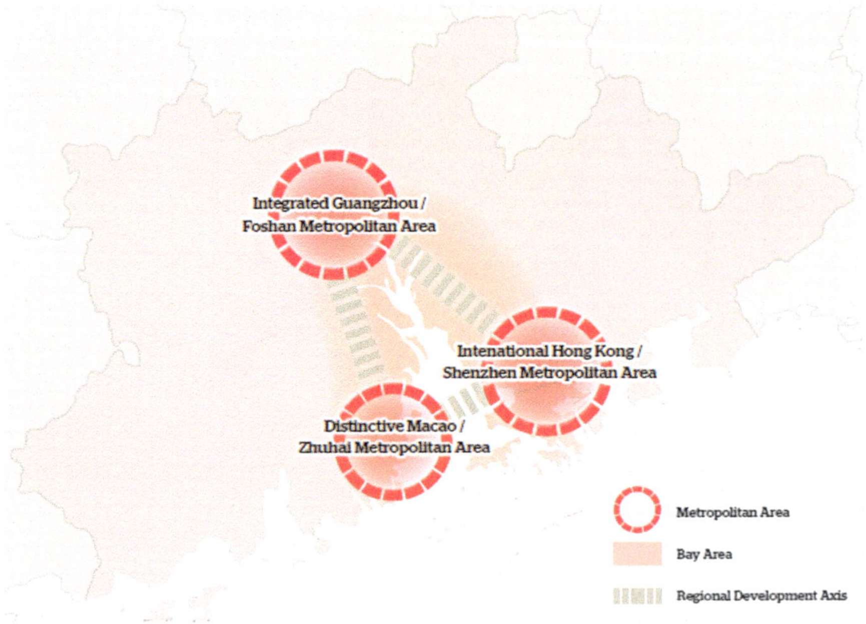

Fig. 2.3 Map of the 3 major metropolitan clusters within the GBA. *Source* Invest Hong Kong HKSAR [1]

and 20.93 million people respectively. These three population centres are where the majority of the economic growth and population migration have taken place, and now comprise the main nodes of development within the GBA (Fig. 2.3).

2.1.2 Comparison to Other Major Metropolitan (Bay) Areas

The Greater Bay Area became the largest metropolitan region in 2015 according to a World Bank Report, ahead of the Tokyo Bay Metropolitan Area, and the San Francisco Bay Area and New York City Metropolitan Regions [12]. A recent comparison of these major metropolitan regions shows that the GBA is largest in both area and population, and is catching up to New York City and the Greater Tokyo Area in terms of GDP contribution (Table 2.2).

2.1.3 Development of the GBA Throughout the 20th and 21st Centuries

While predominately known today as the low-cost manufacturing center of the world, the GBA was once drastically different and not nearly as large and economically significant as it is today.

The GBA region has evolved rapidly over time since the mid-20th century from a predominantly agricultural based primary economy, to an industrial manufacturing based secondary economy in the late 20th century, and in the 21st century it is now transitioning towards a more innovative high-technology and knowledge-based manufacturing and tertiary economy. The most significant catalyst in the growth of the GBA region was China's Open Door Policy initiated in 1979, which opened up China's economy to the global international marketplace and influenced a significant transformation of the region's industrial composition. At the time of initiation of the Open Door Policy, the GDP composition of the GBA was made up of 25.8% primary industries, 45.3% secondary industries, and 28.9% tertiary industries in 1980. Only 28 years later in 2008, primary industries made up only 2.8% of the GBA's GDP, secondary industries made up 49.9%, and tertiary industries made up 47.7%, signifying that a drastic economic transition, industrialization and urbanisation had taken place in the GBA during the late 20th century

Table 2.2 Comparison of major metropolitan regions

	Greater Bay Area	Greater Tokyo Area	San Francisco Bay Area	New York City
Area (km^2)	~56,000	~36,800	17,900	783.84
Population (million)	66.72	43.84	7.6	8.6
GDP (trillion USD)	1.3	1.8	0.8	1.7
Administrative units	9 cities + 2 SARs	1 capital + 7 counties	9 counties	5 counties

Source Fung Business Intelligence, 2015 [13]

following the opening up of the economy [7]. More recently in 2015, the GDP composition stood at 1.8% for primary industries, 43.6% for secondary industries, and 54.6% for tertiary industries, which signifies that the GBA's economy is further transitioning towards a more advanced secondary industry, and tertiary industry, focused on innovation, technology and knowledge-based manufacturing and services.

The main reasons this drastic demographic, economic and industrial transformation took place were due to policy changes and macro-economic reforms and adjustments by the local governments, and Chinese government's investment incentives for Hong Kong, Taiwan and foreign manufacturing enterprises at a national level. More specifically, the growth of the GBA was a result of the establishment of a Special Economic Zone (SEZ) in the GBA. As part of China's strategy to encourage investment and businesses, SEZs were developed at strategic locations in China. These SEZs were characterized by "special policies and flexible measures" that were less economically restrictive and enjoyed special financial, investment and trade privileges compared with the rest of the country, which attracted increased foreign investment into the region [14]. Shenzhen and Zhuhai were pioneered as two of the first Special Economic Zones (SEZs) in China and chosen for their geographic proximity to Hong Kong and Macau, as at the time Hong Kong and Macau were still under Colonial rule by the British and Portuguese respectively and thus had existing free-market economic trade connections to the rest of the world.

The success of these first SEZs led to their areas of influence being expanded within the GBA, as well as further SEZs developed in the GBA and other parts of China. Combined with the opening up of the Chinese economy, the SEZs propelled foreign direct investment (FDI) into the region and internal rural to urban migration, which rapidly transformed the GBA into a highly populated, highly urbanized, export oriented integrated manufacturing and logistics economy through the end of the 20th century, and now into a technology based industrial and service economy throughout the beginning of the 21st century.

The percentage of urban/built up land area in the GBA increased from 3.56% (1605 km^2) in 1990, to 15.18% (3839 km^2) in 2010 [15], as well as experienced accelerated urban growth as the "increase in built up land rose from 2234 km^2 during the 1990s to 2993 km^2 during the 2000s [15] (Fig. 2.4).

This economic, social and demographic transition has resulted in new challenges facing the GBA region, such as over-urbanisation, increasing densities, internal migration and overcrowding, and rising cost of living fuelled in part by speculative development and floating population due to migrant workers. As the GBA continues to grow economically, socially and demographically, authorities will need to think more carefully about the overall integrated regional planning of the GBA, the adoption of smart thinking, design and planning utilizing smart technologies and innovative approaches to ensure that the future growth of the GBA 'Meta-City' region will result in smart and sustainable urban development.

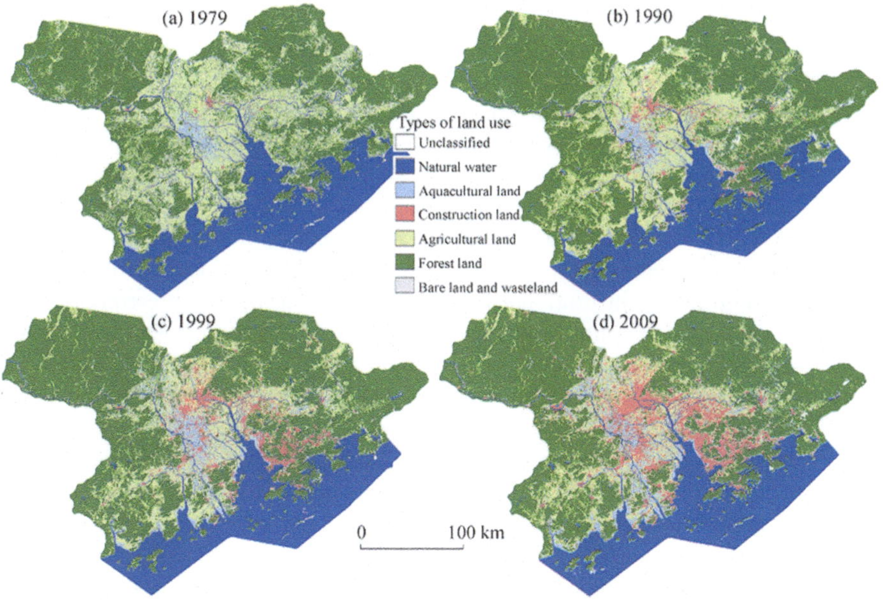

Fig. 2.4 Land use changes in the Pearl River Delta, 1979–2009. *Source* Zhijia et al. [16]

2.1.4 Hong Kong's Role in the Development of the GBA

Hong Kong, on the other hand, has followed a slightly different trajectory than the rest of the GBA. Its economic growth trajectory had been advanced by its history of colonial rule by Britain from the early 20th century to the end of the 20th century, when in 1997 Hong Kong was handed back to China, and transitioned to its current status of Special Administrative Region of the People's Republic of China under the One Country Two Systems arrangement for 50 years [17]. The effect of Hong Kong's colonial rule enabled the city to be governed by different laws and regulations than those cities in the GBA region of Guangdong Province on Mainland China, which enabled Hong Kong to grow and thrive economically during the 20th century in comparison to the rest of Mainland China. The British allowed Hong Kong to become an open market economy, which then grew its industrial manufacturing base and established itself as an important international trading hub between Europe and the emerging Asian markets throughout the 20th century. Hong Kong became the gateway to China, and even South East Asia through which goods, investments and human capital would flow, and where the majority of professional services such as financial, logistics, trading, investment and management services etc. would be located, transforming Hong Kong into the regional servicing hub for the GBA. Much of Hong Kong's industrial and manufacturing sector migrated to the SEZ, attracted by the incentives coupled with cheaper land and labour costs across the border. As a result of Hong Kong's significant investment, the manufacturing industry grew significantly in the GBA during the late 20th century. As a result, the focus of Hong Kong's economy shifted towards a knowledge-based service economy focused on exporting the low-cost goods manufactured in the GBA through its international trading networks, and providing high quality professional services for companies and businesses operating in Hong Kong and the GBA, and assistance to those wishing to do business internationally.

In the early 21st century, Hong Kong has been proactively repositioning itself within the region by looking to strengthen its economic ties with the GBA and Mainland China. Cooperation between Hong Kong and Mainland China began to develop through a new institutional framework in the form of a preferential trade agreement called the Mainland and Hong Kong Closer Economic Partnership Act (CEPA), as well as Hong Kong's inclusion into China's 11th and 12th Five Year Plans (2006–2015), which have enabled Hong Kong's growth to be more in tune with that of the GBA and Mainland China.

Currently, with a growing economy and established finance, commerce, logistics and professional services industries, the Government is beginning to invest more interest in improving other aspects of society and the environment through the adoption of technology and sustainability initiatives. This future direction has been laid out in the Government's most recent strategic planning study document titled "Hong Kong 2030+—Towards a Planning Vision and Strategy Transcending 2030" (HK2030+), which addresses the future territorial development strategy and spatial planning framework to "guide the future planning, land and infrastructure development and the shaping of the built environment of Hong Kong beyond 2030" [18]. This strategic planning effort is intended to lead the way for the sustainable future development and growth of Hong Kong, by addressing current issues and adopting innovative practices from around the world.

In addition to focusing on its own internal economic, social and political issues, Hong Kong must proactively plan and collaborate with the cities and authorities in the GBA to ensure that strategic regional growth and development is considered within its regional context, to retain Hong Kong's competitive advantages that earned the city its status as Asia's World City and as the gateway to Southern China. Moreover, involvement in the Belt and Road Initiative, promoted internationally by China to connect the trade routes of 60 countries from Asia to Europe, will be essential for Hong Kong and the GBA to further grow its market reach and economic development. The next few chapters will look at the existing regional development and the planned future development of the GBA. Then, assess the future development of Hong Kong to address some of the key issues and concerns, while exploring the potential opportunities presented by the GBA. Finally, some suggested recommendations are discussed for the future development of Hong Kong within the regional context of GBA to push towards a smart metropolitan regional development that benefits all.

2.2 Smart City Framework to Assess Smart Region

2.2.1 Smart Cities

The beginning of the 21st century has seen a rise of the 'Smart City' phenomenon, primarily spearheaded by technology companies such as IBM, Cisco and joined by numerous others with a focus on providing software and hardware solutions and expanding their businesses. Whereby new and existing cities have been developed and implemented with a number of smart technologies, both internet-connected hardware and software solutions by leveraging Internet of Things (IoT) and Big Data that aim to provide benefits in areas such as productivity, time savings, production efficiencies, cost savings, environmental sustainability and human health to improve the quality of life of local people and visitors alike. The initial growth of the Smart City phenomenon was focused primarily on technology with the idea using technology to evaluate and monitor the urban environment. Further the Smart City and sustainability lead to the development of some new master planned cities such as Songdo in South Korea, Tianjin Eco City in China and Masdar in the Middle East with big data and a focus on green technologies, although with some success offering valuable lessons. While many existing cities that adopted smart technologies still tend to suffer from the same critical urban challenges such as urbanization through migration, infrastructure development and mobility issues due to traffic congestion, car-oriented development, poor accessibility and walkability, lack of high quality public space with landscaped green spaces, higher densities with over-crowding, and poor quality of life. Technology alone cannot solve all of our current and future urban challenges, and bring about improved quality of life. While technology can clearly play a significant role to improve the efficiency and accessibility of certain city functions, such as e-governance, energy and water monitoring through the use of sensors, or real time public transit information with the help of GPS and mobile applications, these individual 'smart' initiatives divert attention away from the overarching characteristics that enable a city to be truly smart and sustainable. More recently, the definition of a Smart City addresses the city's ability to effectively solve or mitigate societal issues and the impact of urban and environmental challenges in cities by adopting more smart and sustainable development, through the creation of innovative and effective solutions that ensure the efficient use of resources and are enabled by embedded technologies, IoT and big data.

A more holistic approach to defining and assessing Smart Cities evolved over time, with one of the most prominent concepts and characterizations of Smart Cities was identified by Giffinger et al. [19] and developed further by Boyd Cohen in his Smart City Wheel [20]. Giffinger et al. identified six key elements that characterize a Smart City: Smart People; Smart Economy; Smart Environment; Smart Governance; Smart Living; and Smart Mobility. Boyd Cohen developed the Smart City Wheel in 2012, and by combining aspects and findings of Smart City research previously studied, the framework he constructed was one of the first to enable

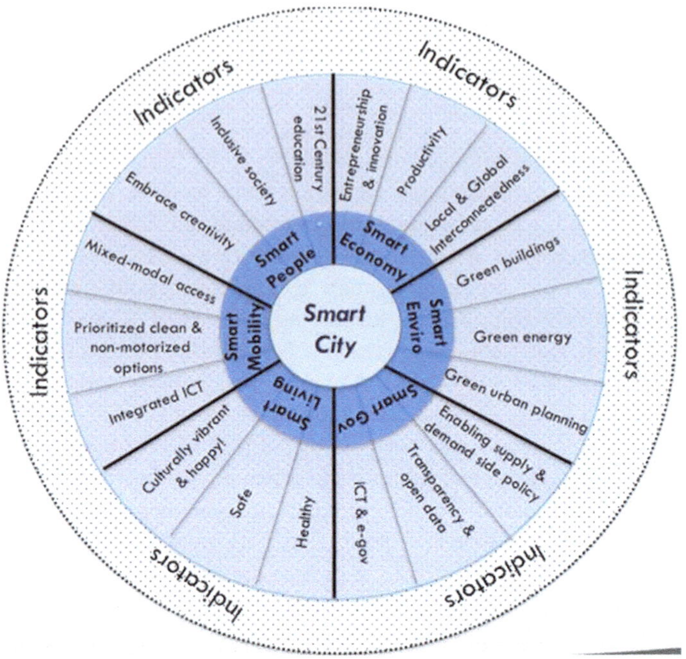

Fig. 2.5 Boyd Cohen's smart city wheel. *Source* Cohen, 2012 [20]

readers to easily understand what classifies as and comprises a Smart City. Different criteria and indicators define these Six Elements and facilitate comparisons between cities. Cohen's Smart City Wheel and other similar concepts are a good start in forming a holistic framework of city development, as Cohen includes and emphasizes the importance of all six Elements to the success of Smart Cities. It is considered, however, that a modified approach would be required to develop an enhanced framework to more closely define the purpose of a Smart City (Fig. 2.5).

2.2.2 Smart City Framework—People, Place and Planet

To address the gap in the existing smart city framework discussion surrounding urban planning, design and the built environment, an emphasis on People, Place and Planet with a focus on 'Smart Thinking, Design and Planning' as the core to Smart Cities has been developed as an enhancement to Boyd Cohen's Smart City Wheel [21]. By highlighting the importance of Smart Thinking, Design and Planning as central themes enabled with technology in developing successful and sustainable Smart Cities. This framework encourages governments, city leaders, businesses and the public to understand and approach the Smart City phenomenon in a more bottom-up people first perspective.

These central themes combine with three core values 'Smart People', 'Smart Place', and 'Smart Planet', to offer a greater holistic view of existing and future development. Like Cohen's model, there are six elements of city development, but "Smart People" is elevated to a higher level to signify its greater importance and contribution to developing truly smart cities. The values and indicators regarding Smart People remain the same, the only difference is its position and ranking within the Smart City Framework is heightened. An important new element, "Smart Infrastructure", is introduced following the elevation of People as a core value thereby maintaining the six smart city elements as represented in Cohen's model. Smart Infrastructure comprises of the both physical (example road network, public realm, electricity, water, drainage, sensors etc.) and non-physical infrastructure (ICT, internet etc.). Indicators that are able to evaluate and also encourage the planning, development, implementation of advanced both soft and hard infrastructure that support the vision of a smart city by ensuring a well-integrated and highly connected physical network and technologies for data collection, monitoring, analysis, evaluation and dissemination of information. This technological infrastructure must be well thought out, planned and designed, allowing for the efficient and ease of use while at the same time ensuring the safety and security of the data and information collected. Throughout the whole model, ever-advancing technology is viewed as an enabling factor to bring the potential of the elements to fruition, yet focusing on the People, Place and Planet as the core factors that define a Smart City, see Fig. 2.6.

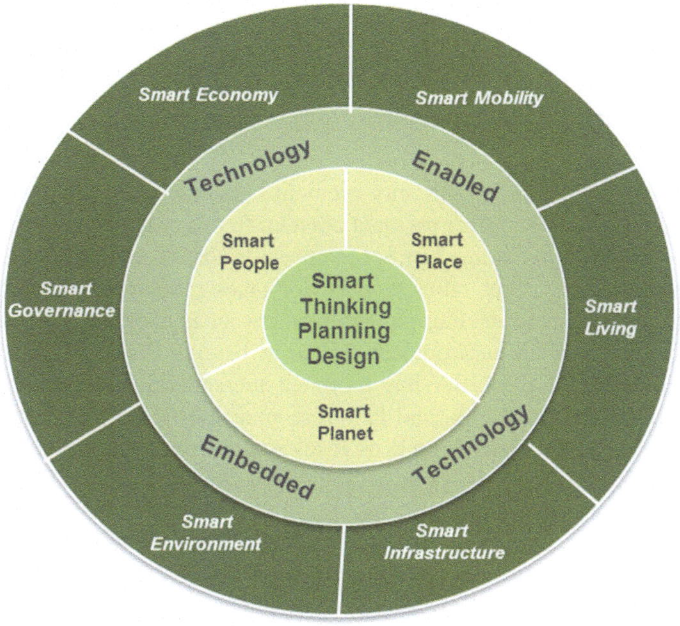

Fig. 2.6 Smart city framework wheel with people, place and planet as core values. *Source* UDP International/Institute for Sustainable Urbanisation [21]

This new Smart City Framework and model highlights that technology is a tool and should be effectively used but is not the primary driver to guide smart and sustainable city development. A poorly planned and designed city that does not focus on the people, place and planet, yet which is technologically advanced and internet-connected cannot become a truly smart and sustainable city as the fundamental premise of the city is lost. The current global attention to Smart Cities in its tech-driven connotation is in danger of overshadowing good city planning concepts and models. While the Smart City idea is being hailed as the modern way forward for new and existing urban areas both in the developed and the developing world, it is very important that the approach be based on smart thinking with a focus on people, place and planet, and technology enabled rather than being technology driven with little or no emphasis on sound planning and design principles. This new enhanced Smart City Framework attempts to point out that a truly smart and sustainable urban environment depends much more on the basic fundamental aspects of people, place and planet than on integrating the newest technological invention. Keeping the central focus on people, place and planet with Smart Thinking, Planning and Design, would enable a more comprehensive approach to our existing and future development of both urban and rural environments. A truly smart society is one where digital technology, thoughtfully deployed by the public and private sector, academia and civil society can significantly improve our existing and future cities and communities with five broad outcomes: the well-being of citizens, improving the quality of life, livability and affordability, the strength and diversity of the economy, opportunities for learning, employment, leisure and recreation with good governance and effective institutions.

2.2.3 A Smart GBA Region?

The same core values and smart city elements that define Smart Cities can be expanded and applied to a larger regional context. In this case, we are looking at the ever expanding and innovating metropolitan region that is the Greater Bay Area.

At an individual city level, cities such as Hong Kong, Shenzhen, Guangzhou and Macau have all begun to develop and implement smart city initiatives and technologies into their urban environment. Looking beyond the local city level, the urban development of the GBA has morphed the individual cities in the GBA region into one mega/meta city, and therefore smart city solutions should be considered at the regional scale as well as local scale. However it is not only technology that defines a smart city or smart region, but how the different cities interact and cooperate, the accessibility and mobility of citizens, quality of infrastructure and transportation network, environmental protection, education, healthcare, awareness and social cohesion, are all important issues that can also be addressed at a regional level that will have a positive and significant impact on the development and integration of the metropolitan region.

Smart Design and Planning

Central to Smart Design and Planning is recognition of the overall goal and impact of the city or regional area. Indicating a transition towards smarter thinking about regional planning and design is that the Chinese Central Government positioned the GBA as a 'National Optimized Development Zone', one of three under China's National Territory Development Plan released in December 2010, and the only region in southern China [22], which aimed to accelerate and support the economic and urban development of the region.

At a higher level, there have been studies, planning and policies implemented that address the need for an integrated development approach to the GBA region, with the recognition that the GBA has the potential to maximize its productivity and value by planning at a regional level as well as local city level within the GBA.

Smart People Place and Planet

When considering the GBA, the importance of smart thinking that goes into the planning and design within the GBA or in the various cities with an emphasis on the People, Place and Planet must be recognised.

An overview of some of the past and current projects, initiatives and cooperation agreements within the GBA between Hong Kong and Mainland China are outlined below, to provide an overview of the progress of the GBA into a smart, integrated and sustainable metropolitan region.

2.2.3.1 Smart Mobility

Smart Mobility concerns the efficient and sustainable movement of goods and people, and within a region as large as the GBA, sustainable mobility is a critical factor and one of the key drivers in the growth and sustainable urban development of the region.

For the movement of people, the most sustainable modes of transport are transit, biking and walking and these should form the basis for most movements, due to their resource, physical and mental health benefits, as well as minimal impact on pollution and congestion within the built environment. Given the large numbers of people moving about an urban environment, mass transit and other forms of public transportation are essential ingredients of smart mobility. Public transportation can readily be integrated with specific ICT services such as smart provision of information on real-time vehicle tracking and arrival times, scheduling, routing etc., that can greatly improve the user experience. A certain amount of private travel is inevitable in cities around the world, but smart mobility aims to discourage inefficient modes of transport, in particular motorized vehicles and single occupant rides. Zero-carbon forms of transportation, such as electric cars, commercial vehicles, taxis and buses, should also be encouraged to reduce roadside emissions and improve the pedestrian walking environment. Promoting ride sharing and driverless cars is another possible solution to help lower the dependence on automobiles and also reduce traffic congestion, the need for having parking spaces, and a resultant impact on the built environment ensuring more effective and efficient use of resources.

For the movement of goods, the use of technology to plan optimally efficient routes and tracking of truck fleet movements are examples of how freight travel can also be enhanced and regulated in smart ways to avoid excessive road side or maritime air and noise pollution, including promoting the use of more environmentally friendly fuels, engines or even electric commercial vehicles. All of the above is not possible without smart city planning that gives a central role to mass transit, walking and, where possible, biking. Public Transit and Pedestrian Oriented Developments (TPOD) is a promising way forward as development model in cities and metropolitan regions that are experiencing congestion challenges as a result of planning that was more car-oriented.

More importantly, with more integrated land use and transportation planning supported by more mixed use development ensures that smart mobility with high levels of accessibility on local, regional and international levels. Hong Kong should strive to retain its strategic competitiveness as the leading offshore investment and trading hub as other major Chinese cities, such as Shanghai, emerge and grow in national and international economic significance. In this regard, Hong Kong must look to continually differentiate itself and capitalize on its strengths by focusing on services that add value while enhancing its regional accessibility and connectivity. The city is working towards achieving this in a variety of ways, such as increasing transportation links and capacity, the number of border crossings, and technologies to assess optimal routes, monitor congestion, and improve the efficiency of travel throughout Hong Kong.

Border Control Points

The current border between Hong Kong and Mainland China, as a result of the colonization of Hong Kong by the British and effective till 2047, is controlled by several border control points. The number of border crossings, have increased over the years and more so since Hong Kong became a Special Administrative Region of Mainland China under the One Country Two Systems rule. Since 1997 the Hong Kong and Mainland governments have been coordinating on increasing the number of border crossing control points to enhance accessibility and mobility across the border for road, rail, air and marine transportation. As of 2017, there are a total of 12 border control points where immigration checks are conducted for individuals entering Hong Kong Special Administrative Region (HKSAR) and vice versa. The border control points include ground control points; comprising four road crossings (Lok Ma Chau, Shenzhen Bay, Man Kam To, and Sha Tau Kok) and three rail crossings (Lo Wu, Lok Ma Chau Spur Line, and Hung Hom), four marine port control points (Macau Ferry Terminal, Kai Tak Cruise Terminal, Tuen Mun Ferry Terminal, and China Ferry Terminal), as well as one aviation control point at the Hong Kong International Airport (HKIA) [23]. These control points provide entry and exit points to and from HKSAR and Mainland China, which are essential to

Hong Kong's economy and society for enabling the flows of people, goods and services across the border (Fig. 2.7).

With the Shenzhen Bay boundary control point, the Hong Kong and Shenzhen governments implemented the first 'co-location of boundary crossing facility' for pedestrians and vehicular traffic, meaning that vehicular passengers travelling by coaches that pass through the control point only need to get on and off the vehicle once rather than twice when passing through immigration authorities. This speeds up the border crossing time for passengers, making the process of travelling to and from Hong Kong more efficient and enjoyable.

Since 1999, the total average number of daily cross-boundary trips between Hong Kong and Mainland China, and Hong Kong and Macao has been steadily increasing, from 304,300 total average trips in 1999 to 701,600 in 2015 [25]. When considering passenger type of cross-boundary trips, there has also been a significant change. In 1999, 84.3% of passengers making cross-boundary trips were people living in Hong Kong, 6.6% were Hong Kong residents living in Mainland, 5.0% were Mainland visitors to Hong Kong, and 4.1% were people living in other places; however in 2015, the percentage of people living in Hong Kong making cross-boundary trips fell to 52.2% of total average trips, while the number of Hong Kong residents living in Mainland China increased to 15.5% and the number of Mainland visitors to Hong Kong increased to 30.2%. The number of cross-boundary trips by people living in other places fell slightly to 2.0%. This shift

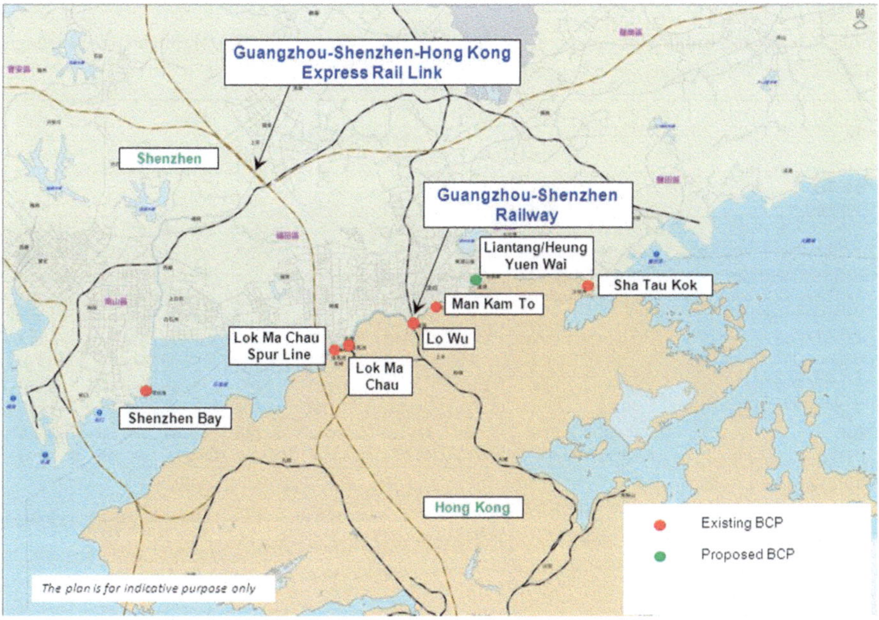

Fig. 2.7 Hong Kong and Shenzhen cross boundary infrastructure and border control points. *Source* Leung [24]

in the total average number and types of cross-border passenger trips indicates Hong Kong's growing attractiveness and importance to Mainland Chinese people, as well as Hong Kong people's closer connection and integration with living in Mainland China. Although the percentage of total average cross boundary trips made by people living in Hong Kong as decreased since 1999, the total number of passenger trips has continued to increase, from 239,900 in 1999 to 338,900 in 2015. In fact, the average number of cross boundary trips has increased for all categories of people since 1999, as shown in Table 2.3.

Moreover, between 1999 and 2015 the total number of vehicular average daily trips between Hong Kong and the Mainland has also increased from 30,000 to 43,200. While the total number of average daily vehicle trips has not increased substantially, the percentage of vehicle type has changed considerably. While in 1999, cargo vehicles (container trucks and goods vehicles) made up 85.6% of total average daily trips and passenger vehicles (private cars, coaches and shuttle buses) made up 14.4%, in 2015 cargo vehicles made up only 47.2% of trips and passenger vehicles made up 52.8% of trips, with private vehicles contributing 43.7% of total average trips [26]. This transition partly reflects the economic transition of Hong Kong towards a more knowledge-based service economy and the increase in cross-boundary passenger trips between the GBA and Hong Kong. However the decrease in cargo vehicles could also be a result of the opening of the Hong Kong International Airport in 1998, and a shift towards greater air cargo transportation, which increased from 1,974,300 tonnes in 1999 to 4,521,000 in 2016 [27, 28].

Table 2.3 Average daily passenger trips between Hong Kong and the Mainland by passenger type

Number of passenger trips

旅客類別 Passenger type	跨界旅運統計調查 Cross-boundary Travel Survey								
	1999	**2001**	**2003**	**2006**	**2007**	**2009**	**2011**	**2013/14**	**2015**
居於香港人士 People Living in Hong Kong	239 900 (84.3%)	275 400 (82.7%)	299 400 (78.2%)	329 300 (73.5%)	349 300 (71.2%)	342 600 (67.9%)	341 800 (60.8%)	314 200 (51.9%)	338 900 (52.2%)
居於內地的香港居民 Hong Kong Residents Living in the Mainland	18 900 (6.6%)	33 100 (9.9%)	36 200 (9.5%)	38 400 (8.6%)	49 500 (10.1%)	53 000 (10.5%)	70 800 (12.6%)	82 400 (13.6%)	100 800 (15.5%)
來自內地的旅客 Visitors from the Mainland	14 100 (5.0%)	17 600 (5.3%)	39 200 (10.2%)	65 200 (14.6%)	76 800 (15.6%)	94 400 (18.7%)	136 600 (24.3%)	195 800 (32.4%)	196 200 (30.2%)
居於其他地方人士 People Living in Other Places	11 700 (4.1%)	7 100 (2.1%)	8 100 (2.1%)	15 100 (3.4%)	15 200 (3.1%)	14 600 (2.9%)	13 200 (2.3%)	12 400 (2.1%)	12 800 (2.0%)
總計 **Total**	**284 600** **(100.0%)**	**333 200** **(100.0%)**	**382 800** **(100.0%)**	**448 100** **(100.0%)**	**490 900** **(100.0%)**	**504 600** **(100.0%)**	**562 400** **(100.0%)**	**604 900** **(100.0%)**	**648 800** **(100.0%)**

註釋： 數字已進位至最接近的百位數。

括號內的數字指以未經進位數字計算所佔往來香港及內地的旅客人次的百分比。

由於四捨五入關係，個別數字加起來可能與總數不符。

Notes : Figures are rounded to the nearest hundred.

Figures in brackets refer to the percentage shares of passenger trips between Hong Kong and the Mainland, which are derived from unrounded figures.

Figures may not add up to totals due to rounding.

Source Census and Statistics Department, Hong Kong SAR Government, 2017

To cope with the increase in border-crossings between Hong Kong and Mainland China, new additional transportation infrastructure and border-crossings are being planned and constructed. A new ground border crossing point is currently under construction on the Eastern side of the Hong Kong-Shenzhen boundary the new Liantang/Heung Yuen Wai Boundary Control Point (LT/HYW BCP) to alleviate the traffic congestion at the Lok Ma Chau border crossing control point, which is currently the busiest land boundary control point for cross boundary vehicles. When completed at the end of 2018, LT/HYW BCP will bring significant benefits to Hong Kong and eastern Shenzhen and Guangdong Province. The new LT/HYW BCP will reduce travel time between Hong Kong and eastern Guangdong region of the GBA and further towards southern Fujian and Jiangxi province, greatly enhancing and facilitating future regional cooperation and development along this eastern axis. Moreover, the existing border control points on the eastern side of the Hong Kong Shenzhen border at Man Kam To and Sha Tau Kok are served mainly by smaller busy local roads rather than high capacity expressways. The new LY/HYW BCP project also comprises a new 11 km dual-two lane trunk road connecting the BCP with Fanling Highway in Hong Kong, which will help redistribute the cross-boundary traffic from the busy Lok Ma Chau BCP, and the smaller Man Kam To and Sha Tau Kok BCPs, while ensuring smoother and more efficient pedestrian and cargo vehicular flows.

In addition to the LT/HYW BCP, two more new BCPs are under construction as part of new regional transportation infrastructure projects, the Hong Kong Express Rail Link (XRL) and the Hong Kong-Zhuhai-Macau Bridge.

Express Rail Link—High Speed Rail in China

With railways as the backbone of the GBA—for more smart and sustainable development, as railways are much more efficient in carrying people short and medium distances in terms of passenger capacity, energy consumption and pollution emissions, and land resource consumption needed to build tracks and stations. This sustainable development of transportation infrastructure is also reflected in China's plan to rapidly building out its High-Speed Rail Network throughout China to connect and enhance accessibility between major cities throughout the country by significantly reducing travel times.

In 2000, the HKSAR Government and the Governments of the GBA envisaged the need for a 'Regional Express Line' in the Rail Development Strategy 2000. The Express Rail Link (XRL) is the Hong Kong section of the Guangzhou-Shenzhen-Hong Kong Express Rail Link. Since 2010, Hong Kong began the construction of the XRL, which will be 26 km in length and run from the planned station in West Kowloon to the north to the Shenzhen/Hong Kong boundary, where it connects with the Mainland Section [29]. The new XRL will drastically reduce the travel time from Guangzhou to Hong Kong's West Kowloon station down from 1 h and 55 min by ordinary rail to only 48 min. This will significantly reduce the effective distance by improving the accessibility and mobility between Guangzhou and Shenzhen, and Hong Kong, enhancing regional

Fig. 2.8 Route map of the Guangzhou-Shenzhen-Hong Kong Express Rail Link

commuter capabilities for those living and working in different cities, as well as make other cities within the GBA and Hong Kong more easily reachable for tourists. Hence, fostering closer social and economic ties between Hong Kong and the GBA thereby supporting and creating new opportunities for development in the future (Fig. 2.8).

Hong Kong Zhuhai Macau Bridge

The GBA is situated around the Pearl River Estuary. This expansive water body separates the city of Hong Kong on the eastern banks of the GBA, and the cities of Macao and Zhuhai on the western banks of the GBA. Currently, there is a lack of direct road connectivity between the east bank and west banks, with vehicles currently having to travel to the north bank of the GBA and utilize the Humen Bridge to travel between the east and west banks. However, this is a considerable detour and increases both travel time and pollution emissions from private and cargo vehicles. The Hong Kong-Zhuhai-Macau Bridge (HZMB) was proposed in 2002 as an essential sea crossing to link the Hong Kong SAR, Macao SAR and Zhuhai in order to take advantages of the three cities through closer physical integration [30] (Fig. 2.9).

The HZMB is the first large scale infrastructure project jointly planned and developed by the three regional governments of Guangdong, Hong Kong SAR and Macao SAR under the "One Country, Two Systems" policy, which is a good sign

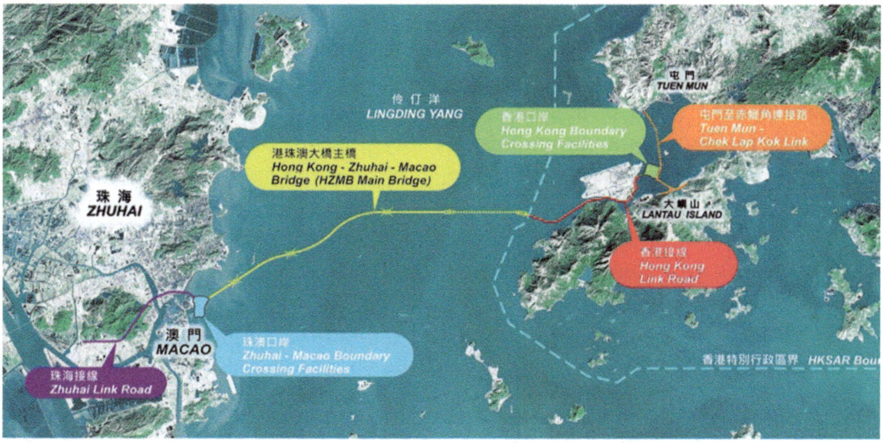

Fig. 2.9 HZMB connectivity map

for cooperation on future spatial and infrastructure development in the region. The bridge crossing has a total length of about 35.6 km from the Hong Kong Boundary Crossing Facilities at the eastern end to the Zhuhai/Macao Boundary Crossing Facilities at the western end of the bridge, of which 6 km is in Hong Kong territory and 29.6 km is in the territory of Guangdong. The HZMB is expected to reduce the road travel time between these two banks by 60–80%, from 3.5 to 4 h down to about 45–75 min depending on the final destination point [31]. The economic and social benefits of the HZMB include in addition to time saving, and also logistics cost savings for businesses, increased business between Hong Kong (East GBA) and Macau/Zhuhai (West GBA), as well as reduced vehicular emissions as a result of shorter journey times. In addition, the HZMB is likely to spur the development of the western part of the GBA including Zhuhai and Macau as well as Hong Kong.

Electric Vehicles

Pushing the envelope on sustainable low-carbon transportation is the growth of plug-in hybrid and electric vehicles in Hong Kong and cities in the GBA. Incentives such as tax rebates on newly registered electric vehicles and the piloting of electric vehicles, in particular from Chinese car manufacturer BYD, as taxis show that there is a vision for sustainable mobility [32]. While Mainland China is planning to extend its exemption of the 10% purchase tax on new electric vehicles through at least 2020, however, Hong Kong's adoption of electric vehicles has been stifled recently as on 31 March 2017, the Hong Kong government drastically reduced the tax rebate on newly registered private electric vehicles from a full waiver of first vehicle registration tax, which can range up to 115% of the vehicles retail value, down to a capped discount of only HK$97,500 [33, 34] giving the reason it was to curb the increase in road traffic congestion. This significantly affected the number of new electric private vehicle registrations in the months that followed, although it did not affect the full waiver on electric commercial vehicles (including goods vehicles, buses, light buses, and taxis) [35]. In this regard, Hong Kong and Shenzhen have been piloting the use of hybrid and fully electric buses on some bus routes, with Shenzhen looking to have the world's first all-electric public bus fleet, a positive sign for the transition to more environmentally transportation in the city [36].

Autonomous cars could be next in line to provide mobility between Hong Kong and the GBA, as well as within the various cities in the GBA. However, caution must be taken however as although the mass adoption of electric and autonomous vehicles could help reduce pollution and roadside emissions, it may not solve the problem of traffic congestion, or may even worsen it. As such, the spatial and infrastructure development should remain in line with sustainable development principles and not continue to expand so as to further encourage the use of other public transportation or active transportation such as walking and cycling.

Bike Sharing Systems

Dock-less Bike Sharing systems have sprung up across cities in Mainland China, whereby users can unlock and lock shared bikes at any location in the city, without being constrained to renting and returning bikes at docking stations at specific locations across the city, as the traditional bike sharing systems have done. Two of the most predominant across cities in the GBA are Oppo and Mobike. Currently, these bike sharing systems are most predominant in Mainland Chinese cities including Shenzhen and Guangzhou, whereas they have had less penetration into Hong Kong due to the city's slow adoption of cycling as a form of transportation. Mass cycling adoption is stifled in Hong Kong due to its compact high-density development and the government's stance that only recognizes cycling as a 'recreational activity' and mainly for transportation in the more spacious New Territories rather than a form of community in the dense urban areas of Kowloon and Hong Kong Island. It is promising to see one company, Gobee Bike that ventured into the Hong Kong market by starting operations in the New Territories where cycling infrastructure is more developed. However it is unlikely to translate into increased cycle commuting numbers in urban areas due to limited road space and lack of suitable cycling infrastructure. The spatial design of the cities has been transformed over time by the automobile as a result of rapid economic growth, urban development and the rising middle class in the GBA region. It should be noted that before the proliferation of private automobile usage, cycling was the norm for the majority of citizens in Hong Kong and cities in the GBA.

2.2.3.2 Smart Economy

Smart Economy is with regards to an open, transparent, diverse economy that adds value to the city and the region. Characteristics include a strong and diverse economy offering varied employment opportunities with labour market flexibility, promoting diversification, startup culture, entrepreneurship and innovation as well as more productivity through local, regional and global interconnectedness. These characteristics allow a smart economy to embody a high level of competitiveness globally as well as locally, and one that is well-connected to the global economy. A Smart Economy will facilitate an efficient and effective business environment, one that promotes and encourages innovation regardless of the outcome. It will also ensure a stable employment market with the resources and ability to adapt and transform if needed. A Smart Economy guarantees the economic success and growth of a city and the region, as well as the livelihood of its citizens. As such it should be a focus of any Government to be forward thinking, innovative and with a global outreach with regards to economic policies in order to remain attractive and competitive in the global economy. There should be availability of well-educated and skilled workforce to attract MNCs and large companies to set up within the city or the region.

In recent years, Hong Kong has been looking to expand and diversify its economy to supplement its strong financial, real estate, professional services, and trade and logistics sectors. It hopes to achieve this by opening up its economy in collaboration with the Mainland Chinese market, as well as invest and promote the high-tech, innovation and creative industries by forging economic partnerships and investments in institutions and infrastructure for better regional integration.

Mainland and Hong Kong Closer Economic Partnership Arrangement (CEPA)

A free trade agreement signed in June 2003 between the Mainland and Hong Kong Governments, which covers preferential provisions for trade in goods, services, investment, and economic and technical cooperation to facilitate increased business and economic integration within the GBA. CEPA has expanded the markets for Hong Kong goods and services, and will further strengthen the trade relationship, as well as foster trade and investment between Hong Kong and cites in the GBA, and further accelerate the close economic integration and trade development of the two areas. While Hong Kong business gain from greater access to the huge and growing GBA and Mainland markets, businesses in Guangdong and the rest of China will have easier access to the international and global markets through Hong Kong, allowing both sides to capitalize on the strengths of one another [37] (Fig. 2.10).

More recently in June 2017, two additional agreements to CEPA were signed: the Investment Agreement and Agreement on Economic and Technical Cooperation. The Investment Agreement aims to "promote and protect investments by investors of the Mainland and Hong Kong SAR; progressively reduce or eliminate substantially all discriminatory measures on investments between the two sides; protect the rights of investors; promote achieving progressive liberalization and facilitation of investments from both sides, and; enhance the level of bilateral economic and trade exchanges and cooperation" between both sides to strengthen investment opportunities [38]. The Agreement on Economic and Technical Cooperation aims to "promote trade and investment facilitation between the Mainland and Hong Kong, and to fully enhance the level of economic and technical exchanges and cooperation" [39].

Fig. 2.10 CEPA signing in 2003

With regards to technology and innovation, cooperation and collaboration has occurred in various forms over the years. In 2003, the "Guangdong-Hong Kong Expert Group on Cooperation in Innovation and Technology" was established, which was the first cooperation in technology between Hong Kong and Guangdong Province. In 2004, the "Guangdong-Hong Kong Technology Cooperation Funding Scheme" was launched which aimed to support and encourage research institutes and businesses to cooperate and collaborate on applied research and development projects [1]. In May 2007, the governments of Hong Kong and Shenzhen signed a cooperation agreement on the "Shenzhen-Hong Kong Innovation Circle", which aims to promote and enhance technology cooperation between the two cities by enhancing the exchange and sharing of innovative ideas and information, talent, equipment and resources through the launch of exchange programmes, cooperation between science and technology parks, academic and research institutions, and providing funding support for enterprises and R&D institutes to jointly conduct innovative projects and initiatives [40].

Science and Technology Parks

Hong Kong has invested in developing technology and innovation hubs to foster innovation and small business start-ups. Two main existing developments are Cyberport Business Park near Pok Fu Lam on Hong Kong Island and the Hong Kong Science and Technology Park Corporation (HKSTP) at Pak Shek Kok in the New Territories and the Inno Centre in Kowloon Tong. Cyberport was not as successful to begin with as it was considered more as a real estate development than a technology hub. More recently it has been successful in attracting more technology companies and starting new initiatives to promote hi-tech investment opportunities. The Inno Centre has been instrumental in promoting and supporting innovation and startup companies as an incubator within the urban area, while the Hong Kong Science and Technology Park located near the Chinese University of Hong Kong in the New Territories has been successful as well in promoting innovation and hi-tech industries.

Hong Kong Science Technology Park (HKSTP)

The HKSTP was set up as a statutory body in 2001 with the goal to transform Hong Kong into a regional hub for innovation and technology [41]. It aims to achieve this by raising Hong Kong's status as a global powerhouse through diversifying the city's economy, expanding employment opportunities across a number of industries, fostering talent and development in the Science, Technology, Engineering and Mathematics (STEM) sectors, and overall create a more sustainable future for the next generations. To facilitate its vision and mission, in 2002, the Hong Kong Science and Technology Park at Pak Shek Kok was constructed as a 330,000 m^2 knowledge-based campus environment for high technology startups and enterprises. The campus comprises a Science Park, Inno Centre and Industrial Estates which include R&D offices, modern infrastructure, laboratories, and technical centres with professional services to provide spaces and facilities for companies of all sizes and stages of development to nurture and accelerate the growth and innovation of companies, and promote interaction and innovation and the local, regional and global levels. As of June 2017, there are a total of 623 companies based in the park with a split of 72% local and 28% overseas technology companies, a working population of 12,791 of which about 9110 persons are involved in R&D related activities [42] (Fig. 2.11).

The HKSTP is smart and innovative solution for Hong Kong to promote and advance its technology and innovation sectors that delivers social and economic benefits to Hong Kong and the GBA region. By providing a collaborative space and

Fig. 2.11 Hong Kong Science and Technology Park

platform for connecting stakeholders, facilitating knowledge transfer and nurturing talent and enterprises, it will support the acceleration of technological innovation and commercialization towards a smarter society further up the value chain. Future plans for provision of housing within the site will bode well as it will make the HKSTP more convenient for the people working there.

Hong Kong—Shenzhen Innovation and IT Park

The Lok Ma Chau Loop is a future development site on the Hong Kong Shenzhen border that has been agreed to be jointly developed by the Hong Kong and Shenzhen governments into the Hong Kong Shenzhen Innovation and IT Park that fosters high-technology to support sustained economic growth of the region [43]. Announced in January 2017 with the singing of a "Memorandum of Understanding on Jointly Developing the Lok Ma Chau Loop of Hong Kong and Shenzhen", the site has a size of 87 hectares, four times the size of the current HKSTP at Pak Shek Kok, and is projected to provide an estimated total floor area of 1.2 million square meter [44]. Its strategic location close to the border in Hong Kong and Shenzhen makes it convenient for cooperation and collaboration between the top enterprises, research institutions and higher education institutes of Hong Kong and Shenzhen. With its location and size there is an opportunity it provide related higher education, cultural and creative, as well as other complementary facilities and create more interaction between Hong Kong, Shenzhen and the GBA. The joint development of the Innovation and IT Park aims to combine the competitive advantages of both cities, Shenzhen will contribute start-ups and capital, and Hong Kong will bring to the table its professional expertise in scaling, branding, intellectual property, and expanding and managing companies at an international scale (Fig. 2.12).

Nansha IT Park—Hong Kong/Guangdong Science Park Collaboration

Hong Kong is also involved in the investment and development of Science and IT Parks outside of Hong Kong in the GBA. For example, the Hong Kong University of Science and Technology (HKUST), the Fok Ying Tung Foundation and the Guangzhou Government have collaborated in building the 'Nansha IT Park', as a commitment to the advancement of technology and knowledge in the GBA region [45]. The Nansha IT Park is a 2.5 km^2. development built within Nansha City's Hi-Technology Zone, designated by the Guangzhou Government, for the purposes of "research, creativity, development and incubation of technology projects, and for training, education, cooperate/strategic planning meetings and retreats" [46]. The Nansha IT Park was completed in two phases in 2004 and 2007, and features modern facilities and infrastructure, such as campus wide fiber optic wireless intranet and internet connectivity, a central data center, building automation, smart card access, business centres, a fitness and health club and more, in order to foster a high quality, service oriented environment for technology and talent to converge, and innovative and creative ideas to flourish. The collaboration and development enabled by the Nansha IT Park, sets the vision to promote and support

Fig. 2.12 Lok Ma Chau Loop site for potential development of an IT and Innovation Park

the development of the GBA into a world class technology metropolis by nurturing and providing a platform for startups to incubate and grow, facilitating opportunities for Hong Kong based companies to expand into the region, and enhance the productivity and efficiency of companies within the GBA.

The construction, support and promotion of these high-tech, innovative Science, Technology and IT Parks in Hong Kong and cities within the GBA has facilitated the progress of research and innovation within the region, that will benefit educated and skilled workers, the economy, and society with new job opportunities and the advancement of technology to improve the quality of life of people.

Hong Kong as the Wealth Management Centre for Chinese Investors

Supporting the economic growth and integration of businesses and industries within the region are significant flows of investment and capital between Hong Kong and the GBA. Hong Kong's reputable and established financial institutions provide a secure and stable source of finance as well as investment opportunities for both businesses and individuals from Mainland China. Moreover, as China's economy grows, so has the size of the middle and upper-middle classes in China, and the number of high-net worth individuals. The success of the GBA has led to Guangdong Province having the highest number of high-net worth individuals in China, 17–18% of China's total high-net worth individual population, and is estimated to comprise approximately RMB28 trillion. As a result of this explosion in wealth, the flows of capital and investment into Hong Kong have been increasing, as it is the most popular destination for the Chinese to invest their wealth in financial products such as stocks and funds, due to Hong Kong's favourable tax policies, geographical location and open economic environment. This in turn, is leading to the growth in number of Mainland Chinese banks setting up branches and offices in Hong Kong to serve these Chinese high net worth individuals [47]. In addition to investing in financial products, many high net worth Chinese individuals are also heavily investing in Hong Kong real estate, as well as overseas in international markets such as North America, Europe and Australia. Combined with Hong Kong's limited land and new housing supply, this increased investment demand in the property market for both housing and commercial office space is causing the prices of real estate to rise [47]. On the one hand while this has led to healthy investment returns for property owners, it is causing housing and commercial space to become much less affordable for Hong Kong citizens.

2.2.3.3 Smart Living

Smart Living encompasses in addition to housing all aspects of human life that have an influence on the quality of life. Smart Living should aim to provide a high quality of life, with a priority placed on the people. A smart city that exhibits Smart Living characteristics should foster an inclusive society and social cohesion amongst all members of society. No one person or group of people should feel excluded or discriminated against in any way, as social equity and equality are important considerations. A Smart Living environment must also ensure the safety and security of its citizens and the urban environment. The public should not have to fear for their safety at any time of the day and night, and should feel safe with other people and protected by the authorities that are responsible for safeguarding the city. A healthy and liveable environment with high quality public space, educational, institutional and community facilities are a key indicator of Smart Living in a smart city, as these will influence the health and well being of the general population and future generations. Aside from accessible and affordable healthcare and education, there should be adequate provision of high quality public spaces such as parks, streets and plazas etc., with abundant green spaces and greening. Similarly, the urban life in the city should cultivate a civic culture between the public and Government, where citizens express characteristics of pride in their nation, freedom of speech, value in public participation in the decision-making process and to be treated fairly by authorities. Furthermore, a Smart Living environment implies there exists the provision of affordable quality housing, so everyone no matter of status or income has a minimum standard of living that is affordable. Like many other cities this is an area Hong Kong does not do too well and the Hong Kong Government under the leadership of the current Chief Executive Carrie Lam is focussing to provide more affordable and decent housing for the people of Hong Kong. Unless the affordable housing issue is resolved it will be difficult for Hong Kong people to be more welcoming and embrace regional integration.

To promote and highlight the growing integration and urbanisation of the GBA and its impact of quality of life, since 2007 every two years Hong Kong and Shenzhen have jointly held a Bi-City Biennale on Urbanism, Planning and Architecture, co-organised by the Hong Kong Institute of Architects, Hong Kong Institute of Planners, Hong Kong Designers Association, and in collaboration with the Shenzhen Planning Bureau. The aim of the exhibition is to share and showcase the work by leading professionals and academics in the fields of Urban Planning, Design, Architecture, and Arts, with a different common theme for each Biennale. This sharing of ideas and creative work facilities the interaction and integration of the two cities with regards to making smarter, more sustainable and liveable cities within the GBA and beyond [48].

Hong Kong and the GBA governments have cooperated on a number of educational initiatives, with Hong Kong public tertiary institutions providing support in establishing educational and even healthcare facilities in the GBA. In 1999, the Peking University (PKU) and the Hong Kong Science and Technology of Science and Technology (HKUST) established the PKU-HKUST Shenzhen-Hong Kong Institution as a joint venture between the Shenzhen Municipal Government, PKU and HKUST [49]. It acts as an incubator for high-tech professionals in Shenzhen-Hong Kong region and is located in the Shenzhen Hi-tech Industrial Park to capitalize on knowledge-transfer and proximity to related industries and services. In 2006, the Beijing Normal University-Hong Kong Baptist University United International College was founded in Zhuhai by the joint Mainland and Hong Kong University. Then in 2007, the HKUST established and set up the HKUST Fok Ying Tung Graduate School in Nansha, Guangzhou, to assist the university in "fulfilling its mission of technology innovation and advancement in the Pearl River Delta region through research and postgraduate education by fostering knowledge transfer and industry collaboration" [50]. In 2013, the Chinese University of Hong Kong signed an agreement with Shenzhen University and the Shenzhen Government to collaboratively develop the Chinese University of Hong Kong, Shenzhen, marking the establishment of another Hong Kong tertiary academic and research institution in the GBA to improve education offerings in Shenzhen and further promote the long term social and economic development of the region through collaboration in education, technology and culture [51] (Fig. 2.13).

Fig. 2.13 HKUST Fok Ying Tong Research Institute in Guangzhou

Schools in Hong Kong, especially international schools, are perceived to provide a higher quality standard of education and many Chinese parents believe that their children's chances of getting into prestigious universities overseas will be higher if their children attend international schools in Hong Kong at primary and secondary level institutions. Hong Kong universities are also typically ranked higher than the majority of universities on Mainland China, as for example in the 2013/14 QS World University Ranks, three Hong Kong universities (The University of Hong Kong, The Hong Kong University of Science and Technology, and The Chinese University of Hong Kong) ranked higher than the two most prestigious universities in Mainland China, Peking University and Tsinghua University [52].

Every year thousands of young students from Mainland China enroll in the primary, secondary and tertiary educational institutions in Hong Kong to learn but also to experience life in Hong Kong. Since 2011, the number of cross-border students from Shenzhen who come to attend school in Hong Kong has risen by 118% to 2016, from 12,865 students in the 2011–2012 school year to 28,106 students in the 2015–2016 school year [53]. This helps diversify the student bodies in Hong Kong but raises some challenges due to cultural differences and also tests the Hong Kong students further in their ability to accept the growing trend. Often, these students will find employment in Hong Kong after graduation, bringing knowledge and talent benefits to Hong Kong, as well as potential for Mainland China-Hong Kong future cooperation and collaboration, but also creates more competition for local graduates in finding suitable jobs. However, the number of Hong Kong students studying in China is not on the same scale as the GBA or the Mainland is less attractive for Hong Kong students and there are some issues such as cultural difference and also limited employment opportunities available for Hong Kong students after graduation in Mainland China.

Healthcare services are also perceived to be of better quality in Hong Kong than in many cities in the GBA and across China. As such, it is common for Mainland women to travel to Hong Kong to give birth so that they can receive better treatment, but also so that their children are eligible for Hong Kong citizenship, which is seen as an advantage to have. This also creates issues with local people as it is seen as a growing burden on Hong Kong and its resources to deliver services to people from the Mainland. There are also an increasing number of cross-border marriages between Hong Kong and Mainland Chinese citizens, thus indicating that there is more social interaction between the people of Hong Kong with those from the GBA or the Mainland. This indicates that there is a potential need for joint hospital, healthcare and/or urgent care collaborations and possible cooperation between Hong Kong and the GBA.

Recently, Hong Kong has fallen behind other major cities with regards to quality of life and smart city indicators. In 2012, a report by Boyd Cohen on the Top 10 Smart Cities on the Planet found Hong Kong to perform well in key areas of smart city development such as innovation (15) and digital governance (3), use of RFID in business and industry, and adoption of smart card technologies throughout society for everyday payment and access uses. However, its quality of life score was much lower than other areas as it ranked 70th, which is significantly lower compared to its other smart city rankings [54]. Released in November 2017, the EasyPark Smart City Index ranked Hong Kong 68th in the world, not an admirable achievement for one of the world's most globally connected financial, business, trade and logistics hub [55]. This demonstrates that much more needs to be done in terms of making Hong Kong a smart, sustainable and liveable city that will be able to capitalize on the rapid development of the GBA. Although, according to the recent EasyPark Smart City Index Hong Kong is the highest ranking city in China, above Beijing (81st) and Shanghai (85th), with no other city in the GBA making it in the top 100, this is far from satisfactory for Hong Kong's performance which looks at Singapore and other cities in the region as competition.

2.2.3.4 Smart Environment

Smart Environment encompasses concepts of green building and sustainable development concerning the natural and built environment in the city. A smart environment implements smart resource management for public open space, in which ecology and biodiversity should play an important role as a balance to the urban city and to provide a stimulating milieu for people to live, work and spend time in leisure and recreation. Open spaces should be plentiful and abundant in greenery, as these spaces provide a place for social interactions and leisure activities, facilitating an inclusive and cohesive society that will provide physical, psychological and social health benefits for individuals and communities. A smart environment also encompasses the built environment, promoting and encouraging Green Building designs and sustainable neighbourhoods that implement energy saving techniques, utilize sustainable materials, and manage waste, water and electricity usage efficiently can influence behavioural changes and significant energy savings.

The Government of a smart environment should also substantially advocate for environmental protection and play a leading role in educating the public of the dangers of unsustainable living, and also in introducing policies and regulations to safeguard the natural environment from excessive development or detrimental emissions, such as pollution control and management.

At a government and policy level, the governments of Hong Kong and Guangdong have been cooperating to improve air quality and reduce emissions across the border over the years. In January 2009, National Development and Reform Commission released the "Outline of the Plan for the Reform and Development of the Pearl River Delta (2008–2020)", in which transforming the GBA into a 'quality living area' was embraced. This plan aimed to enhance the environmental quality and ecology of the GBA, promote low-carbon development, cross-boundary cultural exchange, education, social welfare, food safety, and green transportation.

Two cooperative plans that directly address the issue of environmental emissions and air quality are the "Pearl River Delta Regional Air Quality Management Plan" enacted in 2003, and the "Emission Reduction Plan" introduced in 2012. These aim to collectively encourage the GBA region to have the same goals and ambitions for improving environmental air quality in the region, so as to enhance quality of life for people of Hong Kong and within the GBA.

Pollution emissions from marine vessels operating within the Hong Kong and the GBA has been a growing concern in the region, as emissions of harmful chemicals such as Sulfer Oxide (SOx), Nitrogen Oxide (NOx) and Respirable Suspended Particulates (RSPs), have been found to negatively affect both marine life and people living in coastal urban areas near shipping routes. Hong Kong took steps to reducing vessel emissions by introducing the first regulation in 2014 to cap the sulphur content of locally-supplied marine light diesel at 0.05%, and in 2015 mandated that ocean-going vessels switch to low-sulphur fuel while at berth, meaning having a sulphur content not exceeding 0.5%. This led to a roughly 50% reduction in Sulphur Dioxide levels [56]. In December 2015, a domestic emissions control area (DECA), was established in the waters of the GBA by the Ministry of Transport of the Mainland. Signifying that for the first time, the governments of Guangdong and Hong Kong will coordinate and jointly promote the establishment of a DECA in the GBA Waters to regulate and reduce emissions from marine vessels, to eventually achieve by 2019 enforcement that regulates all ships entering the DECA should use marine fuel with 0.5% m/m sulphur content or lower [57, 58] (Fig. 2.14).

Hong Kong is now focussing on the environment after a rather late start with the Hong Kong Green Building Council established in 2009, and the Zero Carbon Building to promote green and sustainable practices in building design, construction, and living. Also, it is good to note that the Hong Kong Environmental

Fig. 2.14 Zero carbon building in Hong Kong

Protection Department recently received the Urban Land Institute (ULI) award and American Institute of Architects, Hong Kong Chapter (AIAHK) Citation for its efforts in education and awareness among the public on waste reduction and recycling with the Hong Kong East Community Green Station [59].

However, Hong Kong falls far behind with regards to actual recycling efforts and public awareness and involvement in this regard. Although Hong Kong has done much to promote waste reduction and more sustainable living through the development of plans, frameworks and blueprints for future strategies, targets, policies and action plans for waste reduction, less progress has been made on the recycling front [60]. Other than promoting recycling efforts through the launch of a $1 billion Recycling Fund in October 2005, programmes such as the 2011 Community Recycling Network and 2005 Programme on Source Separation of Domestic Waste, development of a 20-ha Eco-Park aimed at facility, publicity and public education programmes, it has been found that still only a small percentage of waste in Hong Kong is actually recycled locally. In 2015, about 2.03 million tonnes of municipal solid waste was recovered in Hong Kong, of which only 2% was recycled locally and 98% exported to Mainland China and other countries for recycling [61]. Of the 2.03 million tonnes of solid municipal waste recovered in 2015, only 4.3% comprised recycling for plastics, with paper comprising 44%, ferrous metals comprising 42.4%, non-ferrous metals comprising 4.1%, and others comprising 4.6%. While both paper and ferrous metals were the most collected in 2015, none of it was recycled locally, with 100% being exported, and of the 4.3% of plastics collected, only 6% was recycled locally with the rest exported overseas [61]. This clearly indicates that although the policies, targets, programmes, and actions plans to promote sustainable measures, the local recycling infrastructure in Hong Kong is significantly lacking to make a significant difference. Moreover, it the government's efforts to encourage the public and private sector to recycle may not be as successful as the data may indicate, as figures released by the government for plastic recycling showed that of all the polyethylene terephthalate (PET) plastic bottles in Hong Kong, only 7.6% were actually recycled, and up to 96% of plastic bottles were in fact recycled by government bodies. This situation has in fact worsened since 2010, when the recycling rate of plastics as a whole in Hong Kong was 69%, it has since fallen to 11% in 2015 due to a combination of less incentives from the private sector to recycle due to falling crude oil prices making it less commercially attractive, as well as Mainland China cracking down on the imports of waste materials and dirty plastic waste such as electronics and white goods. In addition, in 2017 China enacted new restrictions on its national recycling policy, banning 24 types of polluting "foreign rubbish" imports. This has for example restricted the import of paper products for recycling to only paperboard, newspaper and office paper [62], meaning in the future both the public and government of Hong Kong must be more stringent on properly separating the types of plastic and paper in order for them to be accepted for recycling. Going forward, more public education and awareness including cooperation and collaboration between Hong Kong and the GBA should be considered regarding recycling efforts so as to facilitate a greater shift towards more environmentally friendly smart and sustainable region as a whole.

2.2.3.5 Smart Infrastructure

Smart Infrastructure comprises physical elements (road network, public realm, utilities, sensors, meters etc.) as well as non-physical infrastructure (ICT, internet). Both the physical and non-physical infrastructure are connected to one another and integrated with a software application that enables the communication and transfer of data to assist the efficient management of resources. Smart Infrastructure enables smart resource management for water, energy and waste through smart grids, sensors, and sustainable monitoring and usage of public utilities. Smart applications provide transparent data management for the public, business, organizations and government. By utilizing the advantages of the Internet of Things (IoT), the data and information acquired from the built environment and connected products should be able to be accessed by anyone through the Internet or smartphone applications. A smart city smart region should also have Resiliency and Disaster Management systems and procedures to react in the case of an emergency, and also provide rigorous security from cyber-attacks. As spatial and economic integration between Hong Kong and the GBA progresses in the coming years, planning and development of more integrated infrastructure systems and networks is also crucial to support the growth of both Hong Kong and the GBA and bring mutual benefits to cities within the GBA.

Existing integration of energy between Hong Kong and the GBA exists in some forms, as China Light and Power (CLP), one of the leading electric companies in Hong Kong, connected its power stations to the Chinese mainland grid and in 1979 began supplying power to Guangdong Province [63]. This energy transmission network allowed power to be supplied back and forth between Hong Kong and Guangdong. In 1985, CLP began building a 1968 MW pressurised water reactor nuclear power station in joint venture with Guangdong Nuclear Investment Company in Daya Bay in Guangdong Province, which was commissioned in 1994 and became the first nuclear power plant in China and remains the largest commercial nuclear installation today. Nuclear power from the plant was imported to Hong Kong, which enhanced the city's fuel supply reliability and enabled electric power to be supplied back to Guangdong Province from Hong Kong. Since then, CLP has invested in developing a number of power stations in Guangdong Province and Mainland China, including the Guangzhou Pumped Storage Power Station at Conghua in 1994, and the Huaiji hydro power project [63]. Additionally, since 1996 Hong Kong has been provided with natural gas from Guangdong province via an 800 km pipeline between Hainan and the Black Point Power Station in Tuen Mun District of Hong Kong [64].

Hong Kong and Guangdong province cooperate on clean energy provision, which was formalised in 2008 with the signing of a Memorandum of Understanding (MoU) on energy cooperation between Hong Kong and the China Central Government [65, 66]. This MoU promised the provision new sources of energy for Hong Kong, including new natural gas sources, Liquefied Petroleum Natural Gas and offshore gas supplies from the South China Sea that are to be developed.

Moreover, at a government and policy level the governments of Hong Kong and Guangdong have been cooperating to improve air quality and reduce emissions over the years. Two cooperative plans that address this issue are the "Pearl River Delta Regional Air Quality Management Plan" enacted in 2003, and the "Emission Reduction Plan" introduced in 2012. These aim to collectively encourage the GBA region to have the same goals and ambitions for improving environmental air quality in the region, so has to enhance quality of life for society.

2.2.3.6 Smart Governance

Smart Governance is arguably one of the most important elements of a Smart City's development since the content and vision of public policies, or the lack thereof, will dictate whether a particular issue is tackled 'smartly' or not. Smart Governance includes a promoting a common vision, public participation; public services, transparency; access to information; public-private and community partnerships; e-governance; pro-active public policy; and effective leadership. The smart and effective coordinated spatial and economic development of the GBA has begun to emerge between the three jurisdictions governments of Hong Kong, Macao and the Pearl River Delta through joint studies on regional planning and cross-boundary infrastructure links.

In October 2009, the regional governments of the three jurisdictions released "The Planning Study on the Coordinated Development of the Greater Pearl River Delta Township", which is the first planning exercise jointly commissioned by the governments of the three jurisdictions. It mainly comprises a high level strategic outline of recommendations on key planning issues for the region, and a blueprint for key cross-boundary infrastructure links. The overall development objectives included "to build: a world-class city-region of global competitiveness and influences; a world-class advanced manufacturing base with innovation abilities; a highly open world-class centre of modern services; a world-class domestic and international transportation hub; a cultural centre of global influences, and; a quality living area that is affluent, civilized, harmonious and livable" [67]. These objectives clearly indicate that the regional governments of the GBA are committed to ensuring the smart and sustainable development of the region in economic, social, and livability aspects. Moreover, in order to effectively plan this large-scale urbanization project, three major developments strategies for the region are recommended: (1) an optimized spatial structure that creates a "focus at the Bay Area and three Metropolitan Areas, development of three Axes and four Tiers, and the development of three Sub-regions in a poly-centric pattern"; (2) high accessibility to "establish a systematic transport network with the Bay Area as the hub of external links, an intercity one-hour commuting circle and seamless connections in cross-boundary traffic"; and (3) a quality living environment to "establish a comprehensive regional eco-security system, cooperate in regional environmental management and demarcate environmental protection duties of each city" [67]. Within these three major development strategies, a spatial structure of 'City

Regions', an intra-regional 'One Hour Intercity Commuting Circle', and a 'Multi-Tier Development Space' linked the GBA to its surrounding regions is devised. Since its publishing, most of the recommendations stated in the Planning Study have already been developed, or are in process.

The second joint study commissioned by the regional governments of the three jurisdictions, Guangdong Province, Hong Kong SAR, and Macao is the "Regional Cooperation Plan on Building a Quality Living Area", commissioned in 2012. This plan focuses more on sustainable development of the GBA at a regional level, and strives to create a balance between economic, social development, and environmental protection so as to transform the GBA into a "low-carbon, high -technology, and low-pollution city cluster for quality living" [68]. On 1st July 2017, the governments of the three jurisdictions signed the "Framework Agreement on Deepening Guangdong-Hong Kong-Macao Cooperation in the Development of the Bay Area", which formalizes the objective to cooperate in implementing the principle of "one country, two systems", improve and innovate the cooperation mechanism and establish a mutually beneficial cooperative relationship in order to effectively and jointly move forward with the development of the GBA [69]. This framework agreement is elaborated on in Sect. 2.3.4.3 (Fig. 2.15).

2.2.4 *Realizing Smart City Smart Region*

As outlined in the examples above, over the past 20–30 years Hong Kong and the Mainland have been working towards improving better interaction and promoting

Fig. 2.15 Signing ceremony for the framework agreement on deepening Guangdong-Hong Kong-Macau cooperation in the development of the Bay Area

more integration by developing both hard and soft connections between Hong Kong and the GBA. Investments in transportation and utilities infrastructure have supported the spatial, economic and social development of the cities and region, cooperation and collaboration through economic partnerships, environmental agreements, and investments in real estate, R&D, and technological innovation have fostered the growth of new and existing industries, and gradual integration of the Hong Kong and Mainland Chinese populations through education and corporate environments have begun to pave the way for future social integration and cooperation towards a more cohesive and mutually beneficial working and living environment for all people and businesses in the GBA metropolitan region. However, not too much of this is known to the general public especially in Hong Kong who are more focused on local issues such as wages, affordable housing, basic law etc. There seems to be a need to address the local issues but also communicate to the general public about these policies and initiatives and their impact in improving the quality of life of the people of Hong Kong and the GBA.

How this growth, development and integration will continue going forward will be crucial for the successful, smart and sustainable development of this mega region, and will require extensive collaboration and planning among the governments of Hong Kong, Shenzhen, Guangzhou and other major cities and stakeholders within the GBA.

2.3 Future Development Plans and Cooperation Initiatives of the Greater Bay Area Region

2.3.1 Introduction

This chapter will look at the published future plans and policy directions of the Greater Bay Area (GBA) region that will guide and influence its development in the coming years. These plans provide higher level cooperation and development initiatives between Hong Kong and cities in the GBA, which can be studied and discussed at the small, medium and large scales. The smaller scale focuses on cooperation between the two Special Administrative Regions (SAR)s, Hong Kong and Macau, and how the two SARs can cooperate for better economic and social integration; the medium scale focuses on the core GBA metropolis region of Guangdong-Hong Kong-Macau, looking at China's Thirteenth Five-year Plan, the China (Guangdong) Free Trade Zone, the future development of city clusters within the region, and the Framework Agreement on Deepening Guangdong-Hong Kong-Macao Cooperation in the Development of the Bay Area; and the larger scale looks beyond the GBA and towards the One-Belt-One Road vision, and what role Hong Kong will play in the bigger picture of the Belt and Road Initiative (BRI).

2.3.2 *Hong Kong*

2.3.2.1 Hong Kong 2030+ Towards a Planning Vision and Strategy Transcending 2030

"Hong Kong 2030+: Towards a Planning Vision and Strategy Transcending 2030" (Hong Kong 2030+)" provides spatial planning framework to plan and guide land and infrastructure development and shaping of built environment in Hong Kong beyond 2030 [70].

Positioning Hong Kong as "Asia's World City", it put forwards three building blocks.

Building Block 1: Planning for a Liveable High-density City

To create a liveable compact high-density city, Hong Kong 2030+ proposes four key strategic directions to enhance the quality of overall living environment: (i) Adopting the concepts of "age-friendly" planning, (ii) Leveraging green and blue spaces, (iii) Reinventing public space, and (iv) Rejuvenating urban fabric.

Building Block 2: Embracing New Economic Challenges and Opportunities

In recent years, the Gross Domestic Product growth in Hong Kong has been relatively modest while its neighbouring cities are experiencing rapid advancement, which poses economic challenges to Hong Kong. On the other hand, Hong Kong can capitalize on its geographical proximity, improved connectivity by regional transport infrastructure and new regional cooperative initiatives such as "Belt and Road Initiative", "China (Guangdong) Free Trade Zone", "Guangdong-Hong Kong-Macau Bay (the Plan)".

To embrace future challenges and opportunities, Hong Kong 2030+ proposes a few strategies highlighting the importance of providing adequate land for growth and supporting infrastructure, diversifying economic sectors, and promoting innovation and technology.

Building Block 3: Creating Capability for Sustainable Growth

Hong Kong 2030+ adopts an enhanced strategic planning approach to spatial development. It does not only aim to cater the predicted land use demand in the future, but also to enhance the quality of living and be responsive to any potential changes due to unforeseen circumstances. Few strategic directions are (1) Optimising the use of land; (2) Lifting transport and transport capacity; (3) Integrating biodiversity consideration into planning, and; (4) Adopting a smart, green and resilient (SGR) city strategy.

Ideas proposed under the three building blocks are then translated spatially to a conceptual spatial framework. It outlines the future development of Hong Kong with one metropolitan business core, two Strategic Growth Areas (SGAs) and three development axes.

One Metropolitan Business Core

The Metropolitan Business Core covers the mature central Business District (CBD) on Hong Kong Island, fast-growing CBD2 in Kowloon East and future CBD3 in the East Lantau Metropolis (ELM). Locating at a strategic location with its high proximity to Hong Kong International Airport and Hong Kong Island West, the proposed CBD3 at ELM plays a strong role in reinforcing the existing business core and also Hong Kong's connector function in the region.

Two Strategic Growth Areas (SGAs)

The two SGAs are East Lantau Metropolis (ELM) and New Territories North (NTN).

East Lantau Metropolis (ELM): The idea of ELM is creating a smart, liveable and low-carbon development cluster on reclaimed artificial islands near Kau Yi Chau and the Hei Ling Chau Typhoon Shelter and the underutilised land in Mui Wo. ELM enjoys the fortified regional transport infrastructure connecting both the main urban area and the GBA east and west.

New Territories North (NTN): The planning intention of NTN is to provide efficient use of land for new residential clusters, and modern industries, which can take advantage of its proximity to the border.

Three Development Axes

Three Development Axes are Western Economic Corridor, Eastern Knowledge and Technology Corridor and Northern Economic Belt.

Western Economic Corridor: The Western Corridor is designed to capture future economic opportunities in the GBA with various strategic transport infrastructures in place. It shapes the Western part of the territory become the international and regional gateway to Hong Kong. Coupled with strategic development projects, it can potentially give rise to substantial economic opportunities. These includes topside development at the Hong Kong Boundary Crossing Facilities Island of the Hong Kong-Zhuhai Macao Bridge (HZMB) and modern logistics development in Tuen Mun West,

Eastern Knowledge and Technology Corridor: The Eastern Corridor is comprised of six universities, industrial and service support centres, high- technology and knowledge industries in different districts on the Eastern part of Hong Kong. It goes all the way from northern Hong Kong (Tung North, the Lok ma Chau Loop and Liantang/Heung Yuen Wai) to the CBD2 in Kowloon East where growing number of innovation and technology related start-ups cluster around.

Northern Economic Belt: Situating close to Shenzhen, which is strong in R&D, the designated role of North Economic Belt is supporting the existing and future communities around the area with the development warehousing, R&D, modern logistics and other merging industries. The Belt also converges with the Eastern Knowledge and Technology Corridor at the proposed science park near future Liang Tang/Heung Yuen Wai Boundary Control Point to achieve greater synergy.

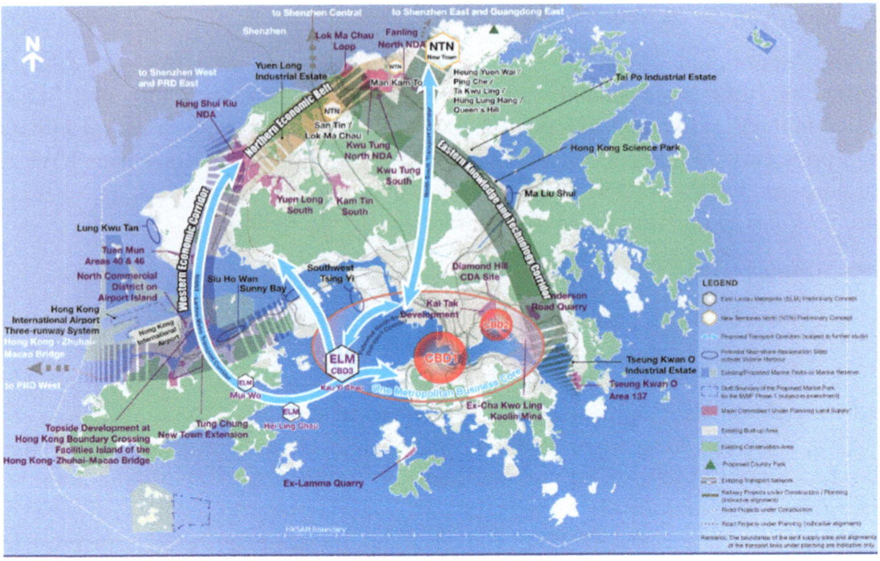

Fig. 2.16 HK2030+ spatial and territorial planning strategies. *Source* Development Bureau, Planning Department HKSAR [70]

These spatial and territorial planning strategies aim to guide and influence the future development of Hong Kong's infrastructure, transportation, businesses, and built environment to create more sustainable long-term development for Hong Kong and support its integration with the GBA region (Fig. 2.16).

2.3.3 *Shenzhen*

2.3.3.1 Comprehensive Plan of Shenzhen 2010–2020

The Comprehensive Plan for Shenzhen 2010–2020 stated that by 2020 Shenzhen aims to be a high-tech innovation base, international logistics hub, as well as a financial and cultural centre. Its current positioning in the GBA and China is that of a national special economic area, economic hub and international city. It plays a role as a service base for the prosperous and stable development of Hong Kong, as well as services in international finance, trade and shipping in co-development with Hong Kong under the "One Country, Two Systems" concept.

One of the major goals of Shenzhen's Comprehensive Plan is for greater regional coordination and cooperation in the GBA, and to strengthen cooperation with Hong Kong in the areas of regional infrastructure construction, industrial development, resource and energy utilization, and market cultivation regarding regional economic factors, social livelihood, technology innovation and environmental protection. To achieve greater cooperation between Shenzhen and Hong Kong, seven areas were identified which include to:

- Promote in-depth financial cooperation, such as free cash flow
- Encourage innovation, through cooperation between corporations, universities and research institutes in the field of R&D, technology innovation, intellectual property management, among others to develop the Shenzhen into an innovation centre and transformation base in the Asia Pacific region
- Focus on the pillar industries, such as hi-tech, financial services, modern logistics and the cultural sector
- Improve airport infrastructure by creating a railway connection between Hong Kong International Airport and Shenzhen Baoan International Airport
- Enhance the customs clearance and trans-boundary transportation process and procedure to facilitate more efficient cross-border movements
- Increased cooperation and development at territorial boundaries, such as construction of the Lok Ma Chau Loop, and Liantang Boundary Crossing
- Greater emphasis on ecological and environmental treatment and conservation.

The Comprehensive Plan also discussed the spatial structure for Shenzhen, where the government has strategically identified 3 north-south axis, 2 east-west belts, and 8 clusters each with a specialized focus to facilitate the future growth and development of Shenzhen, as shown in Fig. 2.17. The three north-south axes

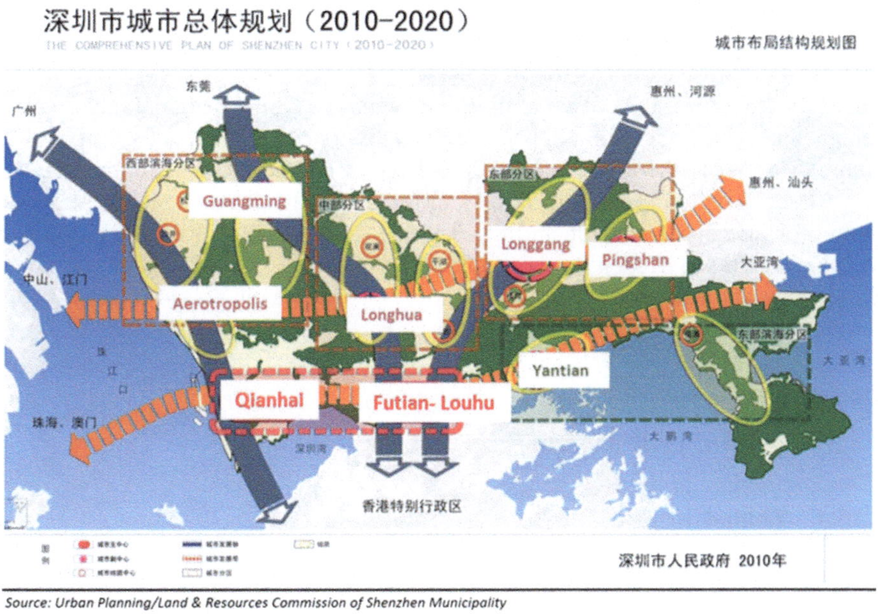

Fig. 2.17 Spatial master plan of Shenzhen (2010–2020). *Source* Fung Global Retail & Technology [71]

include the (1) western development axis, comprising mainly of service industry and high value added manufacturing, and key projects such as the Shenzhen airport expansion, and Shenzhen-Hong Kong airport connection railway line; (2) central development axis, comprising mainly service industry, hi-tech and advanced manufacturing, and key projects such as the North Railway Station, and Guangzhou-Shenzhen-Hong Kong Express Rail Link; and (3) the eastern development axis, comprising mainly hi-tech and advanced manufacturing services, and key projects such as the Liantang Boundary Control Point. The two east-west belts include the northern economic belt and the southern metropolitan belt, the latter of which will be the metropolitan functional belt connected and integrated with Hong Kong. The identified strategic clusters include:

(1) Qianhai: A Hong Kong-Shenzhen service center, focusing on high tech, education, art and logistics services.
(2) Futian: A political, financial and trading center.
(3) Luohu: A financial and trading services center.
(4) Longgang and Pingshan: A modern heavy industrial base.
(5) Longhua: A transportation and logistics hub.
(6) Guangming: A local high-tech and eco-agricultural base.
(7) Aerotropolis: An international logistics and supply-chain base.
(8) Yantian: A leisure tourism and logistics hub.

2.3.4 *Guangzhou*

Guangzhou has positioned itself as a national central city, international trade center, comprehensive transportation hub and shipping hub of south China. Its strategies to accomplish this at the Pan-Pearl River Delta level include strengthening regional cooperation, expanding the hinterland and for Guangzhou to serve as the regional centre for advanced manufacturing services. At the Pearl River Delta level, Guangzhou aims to deepen its cooperation between the rest of Guangdong Provice, Hong Kong and Macau. Within Guangzhou, Nansha Area is poised to become the new R&D Innovation area (Fig. 2.18).

2.3.5 *Zhuhai and Macau*

2.3.5.1 Zhuhai

The Master Plan of Zhuhai (2001–2020) stated the overall positioning of Zhuhai is as an international metropolitan area in collaboration with Hong Kong and Macao, and a core city in the west coast of the GBA. Functionally, it is positioned as a national special economic zone, and also as a business travel resort. Future

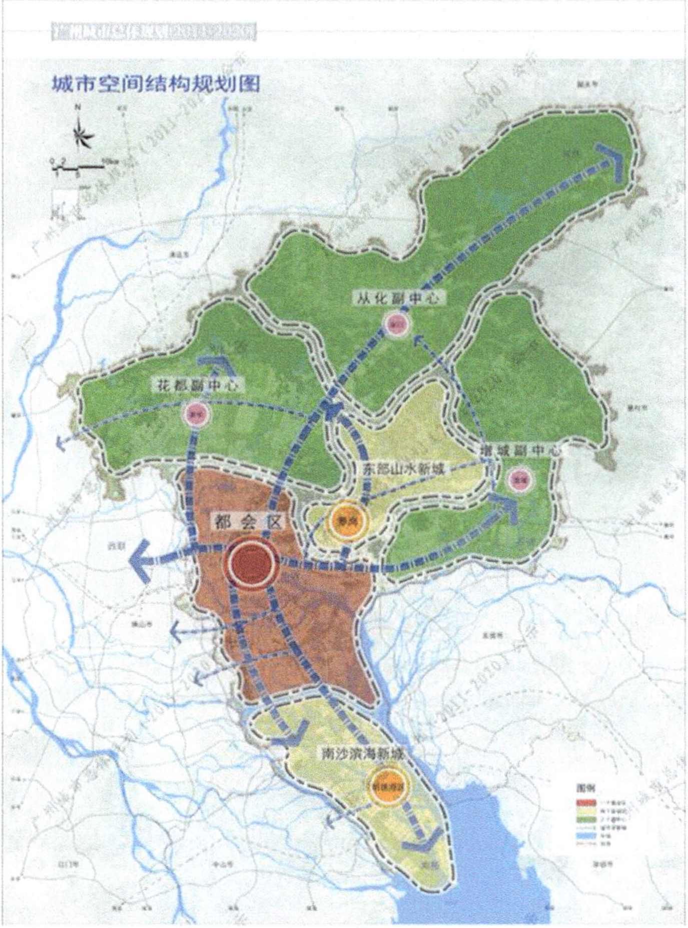

Fig. 2.18 Spatial plan for Guangzhou (2011–2020)

opportunities for Zhuhai's economic and spatial growth are coming in the form of the completion of the Hong Kong-Zhuhai-Macau Bridge (HZMB) in 2018, the Hengqin New Area economic zone, and the Guangzhou-Zhuhai intercity railway.

The regional development strategies to achieve its master plan vision include: greater collaboration with Hong Kong and Macau in marine ports and airports; increased infrastructure construction, enhanced industrial collaboration between the modern service industries, hi-tech and advanced manufacturing services; a high quality living environment; improved communication in legal affairs; more flexible and supportive business environment; and, greater regional resource utilization and environmental protection (Fig. 2.19).

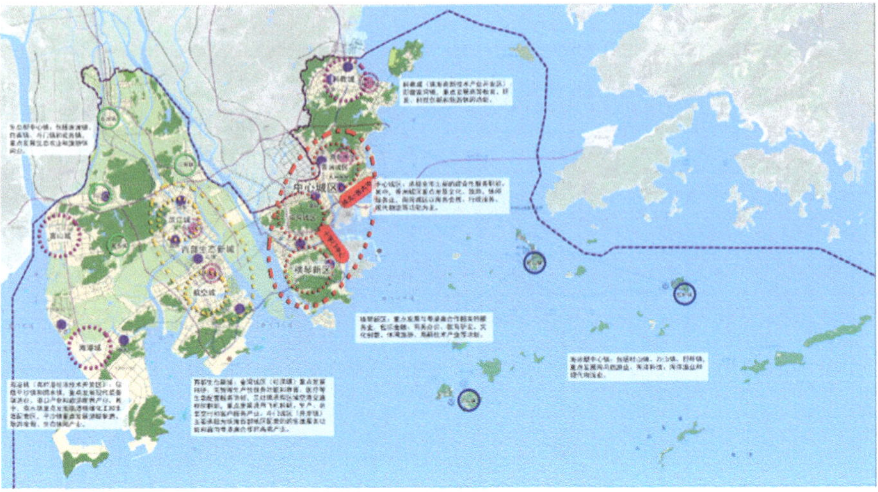

Fig. 2.19 Spatial plan for Zhuhai

2.3.5.2 Macau

Macau's Five Year Plan published in 2016, laying out its blueprint up to 2020, envisioned the city as a world class tourism hub around the year 2035 [72]. Its goals were to:

1. Ensure Stable development of the overall economy
2. Further optimization of industrial structure
3. Gradual formation of business environment for tourism and leisure industries
4. Continuous improvement in people's quality of life
5. Continuous development of culture and education
6. Efficient and effective environmental protection
7. Further improvement in administrative efficiency and legal governance.

Macau's development strategies to be adopted to achieve these goals are to:

1. Enhance the concept of innovative development and establish a cooperation network in this regard
2. Build a new image of "cultural Macao" and raise the city's competitiveness
3. Implement the strategy of "letting Macao thrive through education" and "building Macao with talent"
4. Improve soft and hard infrastructure and quality of tourism services
5. Expedite smart city development and facilitate integration industries of the internet
6. Optimise the public decision-making system and enhance the effectiveness of macro policies

7. Improve the collaborative governance system and coordinate the development of “One Centre, One Platform”
8. Deepen regional cooperation and platform-based economic strategies, promote moderate economic diversification and integrate with national development (Fig. 2.20).

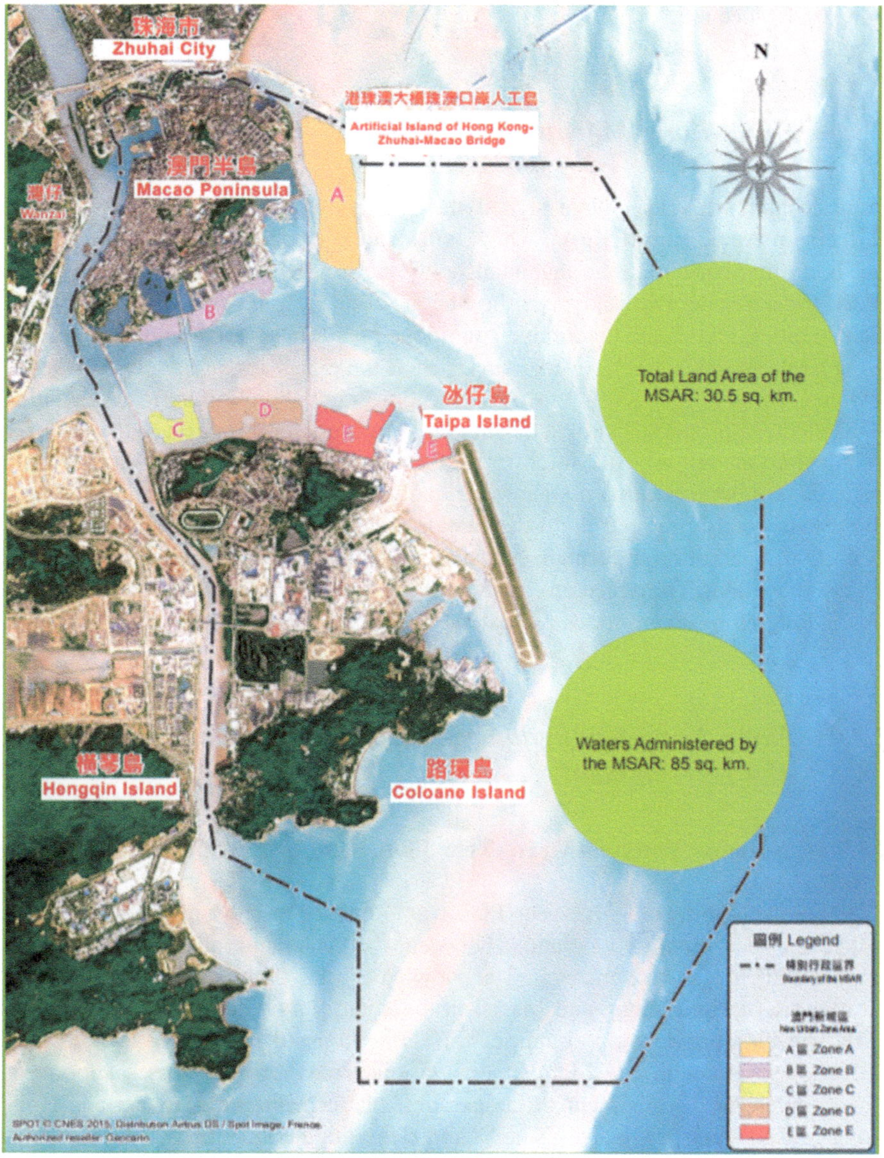

Fig. 2.20 Land area and waters administered by the Macau SAR government [72]

The 8th development strategy will be crucial for the future development of Macao, as Macao must leverage its unique complementary advantages in expanding and deepening its cooperation with the rest of Guangdong province and Hong Kong. This is in line with the development of a mutually beneficial relationship of the GBA, which is already beginning to be implemented with initiatives such as the Closer Economic Partnership Agreement (CEPA) with Hong Kong, and the Framework Agreement on Deepening Cooperation between Guangdong Hong Kong and Macau.

2.3.6 Hong Kong-Macau Cooperation

Hong Kong and Macau have signed the Hong Kong-Macau Closer Economic Partnership Agreement (CEPA), which will come into effect on January 1st 2018, and is the first free trade arrangement between Hong Kong and Macau. Once in effect, the agreement will advance the liberalization of trade and services between the two cities, and create favourable conditions for further regional integration. The Hong Kong-Macau CEPA covers items such as the expansion of trade in goods and services, the simplification of customs procedures, collaboration in matters relating to intellectual property, and establishment of a work program on economic and technical cooperation [73]. Moreover, the agreement states that the two cities will pledge to continue creating a business-friendly environment for companies in both cities, and jointly promote multi-destination itineraries, investment, and trade and exhibition. In addition, it formalizes that Hong Kong and Macau will work together with Guangdong province to press ahead with the development of the Guangdong-Hong Kong-Macau Greater Bay Area.

2.3.7 Hong Kong-Guangdong-Macao Cooperation and Development

2.3.7.1 National Thirteenth Five-Year Plan

The National Thirteenth Five-year Plan was officially enacted in March 2017, in which the deepening cooperation between the Mainland and Hong Kong is emphasized. It also recognizes and acknowledges the significant functioning and positioning of Hong Kong and Macao in the country's overall development plans [74]. The Plan provides opportunities for Hong Kong to strengthen its exchanges and cooperation with the Mainland in various areas, and expresses support in streamlining the development of cooperation platforms among Guangdong-Hong Kong-Macao, such as Qianhai, Nansha and Hengqin, the three Free Trade Zones of the GBA collectively known as the China (Guangdong) Free Trade Zone [75].

The Plan states much support for Hong Kong and Macau to play a greater important role in the Pan-Pearl River Delta Region cooperation, and the development of the Guangdong-Hong Kong-Macao Big Bay Area, in order to open up more opportunities for cooperation between Hong Kong and the Mainland. To maintain its competitive role in the region in this regard, Hong Kong has much to do in upgrading its financial services, business and commerce, logistics and professional services and associated industries into higher-end value added products and services, so as to enhance its position as the international financial, transport and trading centre in the GBA, as a strong and stable offshore Renminbi (RMB) business hub, international asset management and legal centre to both Hong Kong and Mainland Chinese businesses, and to foster greater innovation and technology advancement [76].

2.3.7.2 China (Guangdong) Pilot Free Trade Zone

The China (Guangdong) Pilot Free Trade Zone (GDFTZ) was established in December 2014 by the Chinese State Council, which covers a scope of 116.2 km^2 over three main areas: the Nansha Area of Guangzhou (60 km^2, including the 7.06 km^2 Guangzhou Nansha Bonded Port Area); the Qianhai-Shekou Area of Shenzhen (28.2 km^2, including the 3.71 km^2 Shenzhen Qianhai Bay Bonded Port Area), and the Hengqin Area of Zhuhai (a 28 km^2 district) [77]. Through smart planning and development, each of the three areas has its own industrial specialization: the Nansha Area will be established as a new modern manufacturing industrial base and comprehensive service hub prioritizing shipping logistics, special finance, international trade, high-end manufacturing and other industries; the Qianhai & Shekou Area will become an experimental financial and global service trade and port hub that prioritizes finance, modern logistics, information services, science and technology among other strategic energy innovative service industries, and; the Hengqin Area is being developed into a leading creative, cultural and educational cluster, and leisure and recreational base for international commercial services that prioritize tourism, leisure and health, business financial services culture, science, high-tech and other similar industries. By integrating and connecting the three significant areas of specialization, the GDFTZ will become a model to demonstrate the possibilities for deep meaningful collaboration within an expansive metropolitan region between Guangdong, Hong Kong and Macao, and an important hub of the 21st century's maritime silk road and pilot area for future reforms in China. This Free Trade Zone will also embrace Hong Kong and Macao to better serve the rest of the GBA and Mainland, and open itself up further to the rest of the world (Fig. 2.21).

Fig. 2.21 Map showing locations of the Free Trade Zones under the GDFTZ. *Source* HKTDC Research [78]

2.3.7.3 Development Plan for a City Cluster in the Guangdong-Hong Kong-Macau Bay Area

To facilitate the smart and effective development of the GBA under the National Thirteenth Five-year Plan, on 1st July 2017, a cooperation framework agreement on "Deepening Guangdong-Hong Kong-Macao Cooperation in the Development of the Bay Area" was signed by the governments of Guangdong, Hong Kong and Macau, and the National Development and Reform Commission (NDRC) of the People's Republic of China [69]. Since then, a draft development plan for a city cluster in the Bay Area was completed, which is aimed to provide a blueprint for building a world-class city cluster through complementary co-operation among the three places [79]. In doing so, the Government of Hong Kong is signifying that it will take an active participatory role in taking forward the development of the Bay

Area given the opportunities it can bring for Hong Kong, such as creating favourable conditions for diversifying its industries in particular the promotion of innovation and technology development, and seeking greater methods and support for the people of Hong Kong to study, work, start up and operate businesses, as well as live and retire in the Bay Area. Essential to this goal will be to further facilitate the flows of people, goods, capital and information between Hong Kong and cities in the GBA, and by creating a "quality living circle" for the people of Hong Kong.

In order to achieve the aims of the cooperation framework and Bay Area Plan, certain policy level changes have been made. To more effectively develop Hong Kong's integration with the GBA, the Steering Committee on Co-operation with the Mainland has been proposed to be renamed the "Steering Committee on Taking Forward Bay Area Development and Mainland Cooperation", which will help focus efforts on capitalizing on the unique advantages of Hong Kong with in the GBA, and formulate concrete work plans in a more focused manner. Furthermore, to improve cooperation and collaboration among key stakeholders and decision makers, such as relevant central authorities, the Guangdong Provincial Government, Macao SAR Government and relevant bureaux and departments of the HKSAR Government, the Constitutional and Mainland Affairs Bureau of Hong Kong will set up a "Guangdong-Hong Kong-Macao Bay Area Development Office" which will be specifically responsible for proactively engaging and working with trade associations, professional bodies and other relevant stakeholders to coordinate the effort towards implementation of the Bay Area Plan [79].

2.3.8 Hong Kong-Guangdong-Macao and the Belt and Road Initiative

China has grandiose plans to revive the historical Silk Road, as the Silk Road Economic Belt and the 21st Century Maritime Silk Road, the overall project is known as the Belt and Road Initiative (BRI). By connecting over 60 countries and regions from Asia to Europe via Southeast Asia, South Asia, Central Asia, West Africa and the Middle East, and offering a modern-day solution to the successful ancient trade routes, the BRI aims to foster inclusive growth and development in the 21st century [80]. For Hong Kong, this means even greater access to, and collaboration with, a much wider market for business, trade, investment and financial integration, talent, and cultural exchange, and policy coordination.

The integration, cooperation, and spatial and economic development of the GBA is bound to have an influence to inspire, fuel and in some cases even mold some of the BRI's key projects, and associated developments in the region. It will be essential for the progress and success of the GBA to give full scope to the dedicated roles of Qianhai (Shenzhen), Nansha (Guangzhou), Hengqin (Zhuhai), and Pingtan (Fujian) in opening up and cooperation, deepening their cooperation and relationships with Hong Kong, Macao and Taiwan, and to collective work towards building the Guangdong-Hong Kong-Macao Greater Bay Area.

2.3.9 Hong Kong Policy Address 2017

The 2017 Hong Kong Policy Address discussed a number of initiatives and policy directions for Hong Kong to enhance its integration and cooperation with the GBA and Mainland China. In addition to addressing various social and environmental issues the recent Policy Address by the current Chief Executive of Hong Kong SAR Government Mrs. Carrie Lam also highlights opportunities arising out of Hong Kong's regional context. As highlighted in the 2017 Policy Address Hong Kong faces grave challenges due to increased competition from other economies in the region as well as the rise of protectionism within in recent years. Therefore it is good to see the recognition for the need to develop high value added and a diversified economy with policy initiatives to explore development opportunities in the Mainland. These include (i) enhanced cooperation with the Mainland; (ii) actively participate in China's Belt and Road Initiative; and (iii) explore and capitalize on the opportunities in the development of the GBA.

Enhanced Cooperation with the Mainland

Enhanced cooperation with the Mainland is stated, including a more extended network and upgraded functions of the offices in Mainland China to better understand and capitalize on opportunities in the Mainland while strengthening economic and legal infrastructure.

Actively Participate in China's Belt and Road Initiative

China's Belt and Road Initiative is viewed as the new engine for Hong Kong's economic development. The Commerce and Economic Development Bureau (CEDB) is tasked to coordinate the work of the HKSAR on this initiative with the provision of additional resources and manpower. Building on the success of the second Belt and Road Summit recently held in Hong Kong, the HKSAR is keen to become the international commerce and trading platform for the BRI.

Capitalize on the Opportunities of the Greater Bay Area (GBA)

The GBA including the Guangdong-Hong Kong-Macau area has caught a lot of attention and interest since it was first announced in March 2017 by Premier Li Keqiang. The signing of the joint co-operation framework agreement by the governments of Guangdong, Hong Kong, Macau and the NDRC on July 1st 2017 in the presence of the President Xi Jinping is quite significant. The development plan of the city cluster within the GBA will provide the blueprint to guide the future development of the area. The HKSAR Government will work to create favourable conditions to diversify industries specifically promoting innovation and technology development.

The HKSAR Government will also facilitate and encourage more interaction within the GBA for Hong Kong people to explore opportunities to study, work, set up and operate businesses and also live and retire in the GBA. This will not only facilitate the flow of people, goods, capital and information between Hong Kong and the other cities within GBA but offer more choices for Hong Kong people for a better quality of life.

To capitalize on Hong Kong's strength in education and talent, further nurturing of talent in the city by enhancing training resources and policy initiatives to fully realize Hong Kong's soft power and strength in training capacity is needed. This would provide suitable talent for the high-value added and diversified development of Hong Kong in the long run, as well as contribute to training talents in the GBA region for greater social and economic benefits. Furthermore, increased government to government business, and political cooperation with the Mainland is crucial to effective coordinated growth and development. Hong Kong has upgraded the network and functions of its offices on the Mainland, with now 5 offices that cover 31 provinces, municipalities and autonomous regions. More dialogue between the governments is required, greater bilateral and multilateral ties will be explored with the Mainland and overseas. Doing so will better promote Hong Kong's advantages and attract more Mainland and international enterprises, investment and talent to Hong Kong.

Enhancing government to government connections at both the central and local levels, establishing platforms of direct communication between Hong Kong's business sector and relevant Mainland authorities, and strengthening the promotion of cooperation between Hong Kong and the Mainland would do much in supporting the Hong Kong people and businesses to fully grasp the development and social opportunities that the GBA will bring. Also needed to improve the efficiency and effectiveness of government communication and decision-making will be the greater facilitation of resolutions among cross-boundary civil and commercial disputes through a clear and user-friendly legal regime with the aim to further safeguard the rights of the parties from both Hong Kong and Mainland China. This will be achieved by expanding the scope of arrangements on mutual legal assistance between the two places in civil and commercial matters. Additionally, the 2017 Policy Address stated the need for a renewing of the "Co-operative Arrangement on Legal Matters" with the Shenzhen Municipal People's government to enhance legal collaboration and exchanges between Hong Kong and Shenzhen.

For Hong Kong to progress its economy and society in the right direction with the growth and development of the Greater Bay Area, a bold and innovative new policy direction must be adopted by the government. The government has stated it must enact policies that develop a high-value added and diversified economy, embrace the development of new economic sectors such as innovation, technology and creative industries, and to capitalise on the opportunities arising from the Guangdong-Hong Kong-Macao Bay Area development as this all will generate new incentives and support for the future economic development of Hong Kong.

2.3.9.1 Innovation and Technology

Although rather late in the game, finally in 2015 the Innovation and Technology Bureau was established in Hong Kong to promote the development of innovation and technology. Hong Kong is now beginning to realize the huge potential of focussing on Innovation and Technology can lead to new industries and create more employment opportunities for the young. However, recognizing there is a lot of catching up to do compared to other cities such as Shenzhen and Singapore, the HKSAR is stepping its efforts to develop innovation and technology in eight major areas.

- Increase resources in Research and Development: R&D resources have been increased to about HK$ 45 billion a year from 0.73 to 1.5% of Gross Domestic Expenditure. Encourage universities to increase research in innovation and technology and also provide tax incentives for the private sector;
- Pool together technology talent: A "Technology Talent Scheme" with a HK $500 million is started to recruit more postdoctoral students for scientific research and product development. The Government is also striving to attract overseas scientific research institutes to Hong Kong with MIT setting up an innovation node and Karolinska Institute from Sweden establishing a research centre in Hong Kong. There is a lack of local talent and it will take another 3–5 years to create a talent pool in Hong Kong;
- Provide investment funding: recently the Innovation and Technology Commission rolled out a HK$2 billion Innovation and Technology Venture Fund Scheme to co-invest with matching funds from venture capital funds to promote local technology startups;
- Provide technological research infrastructure: The joint development of the "Hong Kong-Shenzhen Innovation and Technology Park" at the Lok Ma Chau Loop to connect the upstream, midstream and downstream sectors of innovation and technology industries is aimed at developing an international innovation and technology hub in the GBA;
- Review existing legislation and regulations: Promote development of innovation and technology by removing outdated provisions in existing legislation and regulations to foster and nurture new industries creating embracing disruption through shared economies;
- Government to lead changes to procurement arrangements: Explore the inclusion of innovation and technology as a tender requirement and encourage local technological innovation and the ITB to promote the use of technology by government departments to enhance the quality of their services;
- Popularize science education: The Education Development Bureau (EDB) has updated the curricula of the Science, Technology and Mathematics Education Key Learning Areas to enhance students' knowledge, abilities and skills nurturing their creativity, collaborative problem-solving capabilities.

2.3.9.2 Smart City Initiatives

The HKSAR views Smart City development as a way to improve people's daily lives and make Hong Kong a more liveable city. With the consultation on the Smart

City Blueprint just completed and Government is targeting to release the Smart City Blueprint soon. An interdepartmental Steering Committee on Innovation and Technology will guide and oversee the implementation of innovation, technology and smart city initiatives.

The Hong Kong Government can be inefficient due to its many separate departments that oversee varying areas and functions of the city, this causes projects and simple decision making to drag on longer than necessary, or for departments to avoid action and responsibility if they feel it does not fall directly under their jurisdiction. Therefore, when developing the plan for Energizing East Kowloon, it was needed to break the silo view of individual departments in order to develop a effective and cohesive overall vision for the district. Kowloon East was selected as a smart city pilot area and since 2014 efforts have been made to improve Kowloon East.

In the case of Kowloon East, the vision and the master plan for Kowloon East was developed, and by starting with small changes that have localised positive impacts, quick wins were achieved. Then gradually worked up to larger changes with a broader impact, and the accumulation of small improvements leading to a big impact and influence on the area to improve the quality of life for residents and workers in the district. It seems that the Hong Kong government as a whole is too big to effectively handle and implement small beneficial changes at a local level, and decentralisation of decision-making at a local level as in EKEO would be more effective in getting things done. It is suggested to establish district level visions and responsibilities with executives at a higher level that make decisions at a geographical level, to break the barriers of traditional government process and thinking, and to do the right things, at the right place and at the right time.

In addition, the HKSAR is also investing HK$ 700 million in smart city development for the following key infrastructure projects.

- Provide an eID for all Hong Kong residents to facilitate the use of single digital identity and authentication to conduct government and commercial transactions online and provide key digital infrastructure for smart city development;
- Launch a pilot Multi-Functional Smart Lampposts scheme to provide real time data at selected urban locations to enhance city and traffic management and complement 5G mobile communications services for the future infrastructure development in Hong Kong;
- Reform the development technology of e-government systems and build a big data analytics platform to enhance operational efficiency and cyber security;
- Extend universal broadband coverage to cover the territory including the villages and remote areas of Hong Kong as well;
- Smart mobility initiatives and ITS including apps and smart parking meters to reduce traffic congestion. Provide an all-in-one platform to access information and real time traffic information for the public to better plan their trips and usage of public transit.

2.3.10 Hong Kong's Future Role in the Region

It is predicted that the demand from finance, business services and innovation and technology industries will push the GBA economy to reach USD 3.6 trillion by 2030 [81]. Demand for residential, office, and retail spaces across the GBA will rise as a result, but most significantly in Tier 2 cities in less developed areas of the GBA such as Futian, Zhuhai, Zhongshan, Dongguang and Huizhou, due to an influx of population and investment from more populated and expensive areas such as Shenzhen, Hong Kong and Guangzhou as a result of reduced travel times connecting all areas of the GBA, improving accessibility and mobility to areas that were previously less desirable. Moreover, the rise of e-commerce industries in the GBA will generate greater cross-border e-commerce services and transportation for Hong Kong, and cause shopping malls to reposition and rethink their strategies to stay competitive in the face of changing consumer behavior. Significant growth and change is imminent for Hong Kong and the GBA, and as such the successful implementation of the Belt and Road Initiative and further integration of the GBA will bring Hong Kong the opportunity and potential to continue to remain a top global city.

Based on the findings and discussions addressed in the preceding chapters of this section of the book, it is made clear that given the imminent growth and expansion of the GBA, the Central Government and regional governments in the GBA have a high-level spatial development plan in mind, as well as various cooperation agreements and initiatives to support and encourage the successful, smart and sustainable economic and social growth and expansion of the GBA region. The next and final chapter will present recommendations for the possible way forward for Hong Kong and its role within the GBA Regional Development.

2.4 Towards a Smart City Smart Region

2.4.1 Introduction

This chapter begins with a response to the existing development and future plans of the HK2030+ and further explores ideas for opportunities where Hong Kong, including the Government, businesses and citizens can improve their approach and mind-set towards smarter and more integrated growth and development within the Greater Bay Area (GBA). Views and opinions from the community, professional institutions, public and private sectors were obtained during a workshop that was organised to get multiple perspectives on developing a Smart City Smart Region. Key recommendations for Hong Kong's future development to reposition itself and integrate well within its regional context to make Hong Kong a more liveable, smart, sustainable city and improve the quality of life of Hong Kong people while capitalizing on the opportunities the GBA and BRI have to offer.

2.4.2 HK2030+ Within the Scope of Regional Plans

During the public consultation process for the "HK2030+ Towards a Planning Vision and Strategy Transcending 2030" (Hong Kong 2030+) Study and Public Engagement effort, the public were invited to comment and submit response papers to the study and plan. After reviewing the, we found the Study is shown to be a comprehensive study and guide for the public to gain awareness of the proposed planning, land and infrastructure development of Hong Kong towards and beyond 2030. After our review of the sections in Hong Kong 2030+ Public Engagement Digest, we at the Institute for Sustainable Urbanisation (ISU) and UDP International, as an Urban Design and Planning consultancy, have highlighted a few areas where we believe closer consideration should be taken to incorporate the inevitable integration with the region into the discussions on future visioning and planning for Hong Kong's repositioning within its regional context and to regain its competitive advantage within GBA.

As mentioned in Hong Kong 2030+, smart regional integration with Hong Kong's neighboring cities in the GBA (i.e. Shenzhen, Guangzhou, Zhuhai and Macau) will be increasingly important to the future positioning and competitiveness of Hong Kong as a global city, a leading financial centre and business hub, an attractive tourist destination, and a city with world-class infrastructure. It is critical to ensure that the future Development of Hong Kong will be Smart and Sustainable set within the regional context of the GBA and must encompass the Smart City Concept and characteristics of Smart Cities that will facilitate the effective and innovative integration and development of the major cities in the GBA. We have approached our response with regards to the Smart City Framework (Fig. 2.6), that was developed by ISU/UDP as part of a Smart Economy and Smart City publication where a preliminary assessment of Hong Kong as a Smart City was made while highlighting the recent smart city initiatives in Kowloon East. Our Smart City Framework emphasizes the core foundations of Smart Thinking, Design and Planning, with a key focus on Smart People, Place and Planet. It encompasses the six Smart City elements of Smart Economy, Smart Mobility, Smart Environment, Smart Living, Smart Infrastructure, and Smart Governance, in which technology is embedded to enable all aspects which makes a smart city, rather than the simply technology being a driver of smart cities as it is perceived to be.

Hong Kong 2030+ should include strategies for repositioning Hong Kong within the GBA, but also see how there could be more cooperation and better integration between Hong Kong and its regional context. This will mean not only building more physical infrastructure linkages such as the X-Rail and the Hong Kong Macau Zhuhai Bridge but there needs to be more interaction and understanding between the people from Hong Kong and the Mainland. As there are more people visiting Hong Kong from the Mainland compared to the number of Hong Kong people visiting the Mainland, perhaps more visits to the GBA could be encouraged. Also, there is an increasing number of students studying in Hong Kong universities such as the Chinese University of Hong Kong and the University of Hong Kong, opportunities to encourage more interaction, dialogue and communication between students from Hong Kong with students from the Mainland to foster better understanding and appreciation of both cultures bridging the gap between the people of Hong Kong and the people within the GBA.

Within the context of the above Smart City framework, the following points below are suggested be taken into consideration.

2.4.2.1 Smart Planning

Conceptual Spatial Strategy

With regards to the conceptual spatial strategy proposed, the Three Primary Axis comprising the Western Economic Corridor, Eastern Knowledge and Technology Corridor, and Northern Economic Belt is perhaps due to the existing land uses and infrastructures in place along these corridors. However it is noted that the proposed development of Qianhai Bay in Shenzhen as an innovative Modern Service Industry Cooperation Zone would be of great advantage for the knowledge and technology industries in Hong Kong, especially those located along the Eastern Knowledge and Technology Corridor. As such, connections between East of Hong Kong and Shenzhen West should be further enhanced through the development of a potential East-West Corridor to facilitate these direct flows of knowledge, people and information, and to take advantage of the Shenzhen Bay Border Control Point. Our view is that Hong Kong must ensure it can cooperate with the emerging innovation and technology industry in Shenzhen West, and seamless transport connections between Hong Kong's knowledge and technology industries and those emerging in Shenzhen West will be crucial to the competitiveness and growth of these enterprises and industries in Hong Kong. Alternatively, Hong Kong 2030+ could consider the potential for developing the Western Economic Corridor also as a 'Knowledge and Innovation Corridor" to capitalize on its proximity to Shenzhen West, in addition to it being an 'Economic Corridor'. Flexibility in this sense, with regards to planning, employment and industry, will be important to Hong Kong's ability to adapt and respond to rapid economic change and development in the region.

2.4.2.2 Smart Economy

Lok Ma Chau Loop Development

As integration with Shenzhen and the GBA increases over time, the economies, industries and human capital will inevitably also become integrated. This will allow start-ups, businesses and professionals to capitalize on the industry specializations and knowledge on both sides of the border. One key area where this premise can be tested is the Lok Ma Chau Loop development, which is proposed to be developed as an innovation and technology park. The Lok Ma Chau Loop has the opportunity to encompass characteristics of a Smart Economy—one that offers diverse employment opportunities, promotion of entrepreneurship and innovation, and encouraging productivity through local, regional and global interconnectedness—and it's planning, policies and development should reflect these features.

We feel that within Hong Kong 2030+, more attention and consideration should be paid to the unique situation of the Lok Ma Chau Loop, with administrative power belonging to Hong Kong and land ownership belonging to Shenzhen, and how it's positioning can be utilized to create a highly collaborative, mutually advantageous, green and sustainable community that fosters innovation and entrepreneurship in the hi-tech industries and technology sector. Our view is that there could be more discussion on the potential of the Lok Ma Chau Loop, the opportunities it presents for the knowledge and technology sector in Hong Kong, and how its proximity to Shenzhen will foster further integration while benefiting the industry as well as the people of Hong Kong.

Addressing Human Capital Needs

Furthermore, in preparing for our future economy and industry growth in finance and professional services, innovation and technology, and creative industries, Hong Kong must plan accordingly to willingly take a regional approach to the development of human capital. The Hong Kong 2030+ is now more important than ever before in addressing the needs for regional cooperation in entrepreneurial and technical education and training. It is clear that Hong Kong 2030+, under the section on "Sufficient and Suitable Human Capital" and "Innovation, Technology and Collaboration", understands the importance of establishing "world-class facilities to build up a pool of talent and to attract overseas investors to Hong Kong", "adequate supporting infrastructure, such as specialized incubation workspaces and suitable accommodation […] to nurture and retain talents", and "to adopt appropriate planning measures to promote and facilitate a tech-ecosystem, entrepreneurship and business start-ups and collaboration". Yet again, there should be more emphasis placed on how Hong Kong can utilize the proximity to the knowledge and talent in the GBA region to facilitate these goals and visions for Hong Kong. Both programs and facilities, including universities, can facilitate an environment where education, knowledge and training flows more freely across the border, allowing students and professionals, start-ups and established businesses, to mutually benefit from the opportunities both Hong Kong and the GBA have to offer.

2.4.2.3 Smart Living

Housing Affordability

In stating ballpark estimates of long-term land requirements, Hong Kong is actively promoting the need for greater land provisions to increase the housing supply for both public and private housing in response to the rising concerns regarding housing prices and affordability. While it is accepted that Hong Kong must plan for its future land provision towards land for more affordable housing development, it has been noted in recent years that housing prices aren't necessarily responsive to increases in supply or attempted suppressions of demand as normal market forces would expect. In this sense, the Hong Kong 2030+ should address the issue and importance of tackling the housing affordability issue through means other than simply providing more housing supply. One area that has been discussed as an alternative or additional approach is to tackle the housing affordability issue from a financial perspective, to make it easier for Hong Kong citizens to obtain sufficient funds for down-payments, mortgages etc. through reforms to the financial mechanisms that govern funding for property purchases, or even revisions to the existing mechanisms that govern the purchasing of land and development of new residential developments for private developers, such as land premiums etc. and the time it takes for development. Moreover, taking a regional approach to housing provisions and affordability in the future, combined with integrated and seamless transportation connections such as the Express Rail Link. Considering the availability of living options closer to Hong Kong along the border or within the GBA could be a viable solution to Hong Kong's housing affordability issue, and one that should be considered when planning and visioning beyond the year 2030 as the Hong Kong 2030+ is indicating. Providing choice and options alternative living accommodation within these areas are vital to the future development of Hong Kong and thus should be included in the Hong Kong 2030+ strategy to ensure these issues are addressed and considered in all future plans and visioning.

Besides the issue of housing affordability, the other key area that should be addressed is housing type and income mix in terms of both size of units and integration of public and private housing. In recent years, new housing from both the public and private sector has been focused on providing smaller and smaller units in response to rising housing prices. The view is that providing smaller units enables developers to offer lower prices and also a greater number of units to allow a greater number of the population to get on the housing ladder. The problem arises here where the choices of types of housing, regarding size, number of bedrooms etc. for Hong Kong citizens are diminishing as the focus is on housing supply and not quality. While a Key Action of Hong Kong 2030+ "An Inclusive and Supportive City" is "to encourage a variety of housing choices to be provided by the public and private sectors to enrich options available", but as an action, simply "encouraging" housing choice is not enough to ensure that this key action will be implemented in the near future. The Hong Kong 2030+ should state that it will "provide" or "ensure" a variety of housing choices, with decent accommodation and space standards rather than simply encourage, as this will ensure this is an accountable action and not one that can be forced on developers.

Provisions for Elderly Housing

It is well known that Hong Kong has an ageing population dilemma, as the proportion of elderly people is on the rise. While Hong Kong 2030+ promotes an inclusive and supportive society, and addresses the elderly population by promoting "concepts of universal design, age-friendliness, active ageing and "ageing in place" in the planning and design of the built environment", there needs to be greater considerations to the provision of quality living quarters for elderly people who may not be able to afford living in private housing or be eligible for public housing at a certain time. In this regard, opportunities across the border, outside of Hong Kong for elderly housing could be also considered to alleviate some of Hong Kong's ageing population issues as integration with the GBA progresses over time.

2.4.2.4 Smart Environment

Pollution Controls

As pollution levels, both air and water pollution, continue to be raised as issues in Hong Kong, these are areas that should be addressed in the Hong Kong 2030+ strategy as discussions for the future of Hong Kong. While the Hong Kong 2030+ promulgates sustainability through "Creating, Enhancing and Regenerating Environmental Capacity", by "planning for a low carbon city and better urban wind environment", and to "reduce air pollutants through environmentally friendly transport and green infrastructure", improvements to Hong Kong's environment regarding air and water pollution cannot come from Hong Kong alone. Cooperation and collaboration with the Mainland, specifically with cities in the GBA will be essential to ensuring mitigation and improvements to Hong Kong's air and water quality. The Hong Kong 2030+ should set out strategies and measurable actions in which cooperation can result in mechanisms such as integrated environmental pollution monitoring systems, common regional environmental goals and targets supported by regulations and standards, and a multi-disciplinary committee of sorts to guide, monitor, manage and ensure that environmental issues are being discussed and advanced on both sides of the border.

2.4.2.5 Smart Mobility

Integrated Transportation Data and ICT Systems

Although Hong Kong has adequate transport links to the border in terms of road, rail and ferry connections, and has smart transportation systems in place such as the Transport Department's ICT and mobile applications that provide real time traffic and routing applications. In the future, there will be a need to discuss ways to enhance the mobility and ease for people travelling across the border, more specifically with regards to cross-border transportation ICT data services such as real-time traffic, arrival and public transportation information, route-choice services etc. While these services exist to a certain extent in Hong Kong and the GBA, the data systems, applications and public access to information are segregated between the two cities. Moving forward, Hong Kong 2030+ will need to consider the development and implementation of an integrated transportation data-sharing platform and system will be mutually beneficial to regional integration of Hong Kong and the GBA. This will allow seamless visualization and awareness of traffic information and routing choices for those travelling cross-border for commuting, leisure or transportation logistics purposes.

Seamless Border Crossing Facilities

To enhance the mobility of those crossing the border on a regular basis, which will inevitably increase up to and beyond 2030, Hong Kong 2030+ should consider how cross-border controls could be simplified with cross-regional data sharing and expedited with biotechnology, such as facial or iris recognition systems, especially for the skilled labor frequently crossing Hong Kong and Shenzhen, who may facilitate the high-tech development in Hong Kong. Also, there could be series of border towns built with visa free access at border crossing points.

2.4.2.6 Smart Infrastructure

Providing Supporting Infrastructure Capacity—Water Supply

The Hong Kong 2030+ has sufficiently addressed the existing and future transportation infrastructure connections both within Hong Kong and between Shenzhen. However, given Hong Kong's increasing population up to and beyond 2030, although this in itself is questionable, more consideration could be given to the supporting infrastructure that is essential to the functioning of the city. One key area of concern is with Hong Kong's water supply, and our city's current reliance on importing water from China, specifically from the DongJiang River. Overtime the cost of importing water from China has increased, in addition to the growing water demands of the GBA due to its own rapid economic development. To foster Hong Kong's Smart Infrastructure capabilities—one that enables smart resource management for water, energy and waste through smart grids, sensors, sustainable monitoring and usage, as well as resiliency and disaster management systems and procedures—it is vital for Hong Kong to continually ensure the certainty of its own water supply in the future through progressive efforts to develop desalination plants within its territory. It is positive to see progress with the Tseung Kwan O Desalination Plant, in its First Stage of consultancy, yet we believe more consideration should be given in the Hong Kong 2030+ to alternative and innovative water supply measures such as desalination plants to guarantee these are important areas to be discussed and developed in the future of Hong Kong.

Utilities—Monitoring and Management

In addition, the Hong Kong 2030+ states that Hong Kong should "pursue an integrated smart, green and resilient infrastructure system" including things such as "district cooling systems, electric vehicle charging infrastructure, waste-to-energy conversion, effluent reuse and while life-cycle carbon assessment". These efforts for innovative and sustainable approach to developing Hong Kong's infrastructure systems are commendable, however the importance of ensuring that Smart Infrastructure is planned, designed and implemented appropriately and effectively needs to be stressed so as to gain the holistic benefits of an integrated smart green and resilient infrastructure system. This means ensuring that high-tech monitoring and measurement systems are in place to accurately assess the usage of utilities (i.e. water, waste, electricity etc.) in real-time and to address any issues or malfunctions in the most efficient and effective manner.

Enhanced ICT Infrastructure and Services

Building on the Hong Kong 2030+ provisions for "Adequate and Timely Provision of Supporting Infrastructure" in which a key action is to "better utilize ICT and free Wi-Fi services including those for helping businesses and tourists", and a "Smart, Green and Resilient City Strategy" that promotes a "common spatial data infrastructure and a robust network of ICT infrastructure". In addition to enhancing our infrastructure capabilities locally, the Hong Kong 2030+ should also significantly consider and address how the ICT, Big Data and data sharing platforms can be integrated and combined to utilize the systems and data of the cities in the GBA. Having a combined common data sharing platform and ICT infrastructure that expands its reach outside of Hong Kong would bring mutual benefits to both parties with regards to knowledge sharing, technological innovation and complementary economic, social and cultural benefits. Further details on how these ICT and data platforms can be implemented through industry and government cooperation must be discussed to ensure a holistic and common approach to the development of these infrastructure networks.

2.4.2.7 Concluding Remarks

As 2047 is fast approaching, it is good that Hong Kong is undertaking the Hong Kong 2030+ Study as a strategic plan to "guide Hong Kong's planning, land and infrastructure development, as well as shaping of built and natural environment, beyond 2030". However, there should be a further emphasis on regional integration and to use this as an opportunity to reposition Hong Kong within the GBA. There should be greater consideration and discussion of how Hong Kong and the GBA can collaborate and work together and integrate its economies, services, knowledge and information flows, quality of life and environmental aspects, in order Hong Kong to capitalize on and benefit from the regional proximity to the talent and resources that the GBA has to offer. Moreover, it is a crucial opportunity to address differences and gaps in things such as environmental regulations, data and information sharing, education and healthcare provisions etc., for the residents of both Hong Kong and cities in the GBA.

Regional integration issues that Hong Kong should prepare for now should be considered relevant to the Hong Kong 2030+, as these are issues that will affect Hong Kong's future development, planning and shaping of both its built and natural environment. Planning for Hong Kong's future without due consideration, cooperation and collaboration with the GBA, on not only the areas addressed above but on all areas of the Hong Kong 2030+, could result in a situation where Hong Kong is inadequately prepared or misguided on important planning and development issues. Although more visitors from Mainland China brings some strain to the existing infrastructure and resources of Hong Kong, but businesses are benefiting as a result. Creating more awareness among the local people that regional integration should be looked as an opportunity to reposition Hong Kong within GBA rather than viewed as a threat for Hong Kong and its people is needed. Our response to Hong Kong 2030+ is taken into account to assist in further refining the Hong Kong 2030+, for it to become a critical guide for the development and repositioning of Hong Kong not only within the global economy, but more importantly within the regional context of the GBA for the benefit to the citizens of Hong Kong.

2.4.3 Smart City Smart Region Workshop

2.4.3.1 Overview

The Smart City Smart Region Workshop was successfully held on October 25th 2017 by the Institute of Sustainable Urbanisation (ISU) in collaboration with the Research Institute for Sustainable Urban Development (RISUD) and UDP International. The aim was to promote and encourage the discussion of smarter regional integration of Hong Kong with Guangdong province within the GBA, and obtain a new perspective of views and opinions for the future direction of Hong Kong as a smarter more sustainable and liveable city.

Panel speakers included: Amy Cheung of the Hong Kong Planning Department, who spoke on the long term plan HK2030+ Towards a Planning Vision and Strategy Transcending 2030; Daniel Shih of Colliers International, who spoke on the GBA in a new development context by sharing new property investment opportunities from the surge of the Greater Bay Area; Donald Choi of Nan Fung Development, who spoke on the Urbanisation and Smart TOD approach of the GBA from a developers perspective; Professor Edwin Chan of RISUD and the Hong Kong Polytechnic University, who spoke on Green and Smart City regarding compact land use and smart living, and; Sujata Govada of ISU and UDP International, who spoke on Smart City Smart Region opportunities for Hong Kong, and how greater integration could take place.

Presentations were given by panel speakers on their unique perspective of Hong Kong's role in the Greater Bay Area (GBA), also known as the Pearl River Delta Region, and how the development of the GBA can be achieved in a smarter, more sustainable way. This was followed by breakout sessions of three groups focused on (1) Smart Living and Smart Environment, (2) Smart Mobility and Smart Infrastructure, and (3) Smart Economy and Smart Governance. The groups were then brought together for a group presentation and discussion session whereby a representative of each group shared the main points of their discussion with the rest of the workshop participants, and then the floor was opened up for an overall discussion on the ideas and topics raised.

2.4.3.2 Group Discussions

The breakout sessions and resulting group discussions provided a platform for workshop participants to share their views and experience regarding how Hong Kong can become a smarter city and improve integration at economic, social, environmental, quality of life, political and infrastructure levels with the GBA region, in order to enhance and maintain its regional competitiveness and attractiveness to local, Mainland Chinese and other people and businesses.

Group 1—Smart Living Smart Environment

Overall, two major issues were identified: (1) the perception and attitude of the people does not lead to a progressive smart city that can effectively integrate with the GBA region, and (2) a better bottom-up perspective for future opportunities is needed for Hong Kong to become a Smart City in a Smart Region. Hong Kong needs to be a more liveable and affordable city to then will be able to focus on regional integration and opportunities within the GBA.

To achieve a smarter living within a city or a region, data analysis is typically used to make decisions, whether they be good or bad. The amount of Big Data available with the advent and wide use of smart sensors and smartphones on a daily basis allows both public and private decision makers to harness the information and capitalise on it to better understand and justify decisions they make. However, as a city it is also crucial that we make smarter use of resources available. Cities must play on their own strengths, and not solely focus on what other cities are doing and try to replicate it. A better allocation of its resources among the local society should have effective benefits for people and businesses, and achieve a more comfortable and liveable lifestyle for local citizens given the city's strengths and constraints. Another area that can improve the living environment for citizens in a smarter way is for the definition of 'smart' to be defined in a better and more concise way. Smart City and Smart Living has generally implied that implementing technology solutions into the city automatically makes a city and its people 'smart', whereas it is felt that as a society we need to go beyond thinking about just the technology to develop smart cities, and that more thought should also be put towards focusing on people, place and planet that will both enable and facilitate smarter, sustainable and more liveable cities. Part of this could be achieved through education, by raising awareness of smarter, more sustainable city development through schooling and peer to peer educators. By teaching the younger generations at an early age, they would carry on these forward and innovative ways of thinking into their adult lives and also share their thoughts and ideas with their peers and elders to convey their needs to a higher level.

Essential to this approach is for governments and key decision makers in the public and private sector not to underestimate the power of 'regular' people in their views of what is needed and best for the future of their communities and cities. The power of normal citizens can lead to effective bottom-up approaches in search of smart city opportunities. Hong Kong, including cities within GBA and many other cities around the world are bound to bureaucratic procedures that can hamper the progress and development of new and innovative solutions and ideas that may have positive effects and benefits for people and society. A transparent and flexible process should be adopted by the government to allow for certain things to get done faster or more efficiently with available resources, while maintaining standards and regulations. Further, certain policies and regulations enacted in previous generations should be reviewed and revised, as many older regulations can be limiting in the current state of the market and economy. Also with bottom-up approaches, criticism of government procedure and decision making is inevitable, and being able to better absorb public criticism and turn it into positives and change should be the hallmark of an effective and 'smart' government and leadership, and would lead to addressing the precise needs and desires of smarter living and environment for all. In this regard, one of the most important aspects of cities that are able to adopt innovation and develop smartly are those with bold leadership who do not shy away from a different way of doing things. They take risks for the greater good of society, and take criticism as positive feedback when given.

Group 2—Smart Mobility Smart Infrastructure

Overall, for Hong Kong and its citizens to truly embrace itself as a smart city with a closer connection to the GBA, a change in mindset and mentality is needed for people to feel more open to embrace an integrated GBA that includes Hong Kong. One way to do this is address some of the needs of Hong Kong people with affordable housing with more open and inclusive public spaces for the public to integrate as well as enjoy the natural beauty of Hong Kong. Another suggestion to integration between the people in the GBA, including Hong Kong, is to organising sporting competitions between cities in the GBA. Although it may create a competitive atmosphere, it will also encourage the feeling of inclusion and integration between the various cities, and more specifically enable Hong Kong people to feel more a part of the GBA region. Engaging more young people as stakeholders is also critical, as they are the next generation and their forward thinking and innovative ideas will help contribute to the smart city development of Hong Kong and the GBA region as a whole.

To encourage greater smart mobility within Hong Kong and between the GBA, greater incentives should be provided by the government, and ideas to think outside of the box are also needed. For example, in addition to focusing on rail transport as the backbone of the transportation system, greater emphasis on the last mile journey should be considered. These could include park and ride facilities for those living on the outskirts of urban areas who do not have convenient access to public transportation. By enabling people to drive from home, and park their cars at public transport interchanges, it may reduce the desire for some to use their cars for the whole journey into downtown urban areas, and instead use public transportation to save time in traffic congestion and money on parking costs.

Regarding smart infrastructure in Hong Kong, in order to allow Hong Kong to truly benefit from technologies such as Open Data and Internet of Things (IoT), the Smart City platform needs to somehow be open to the public and to encourage more people to contribute their data and information for the greater good of society. However, this is where a change of mentality is needed. A greater level of trust needs to be built between the people and government of Hong Kong, one that shows that the government will do more for the people. One of the reasons cities in China develop so fast with regards to technological innovation and development is that mode of governance is different and there seems to be a greater level of trust from the people, and less of a worry about privacy of data. It is accepted to a greater extent that the government has access to data and will use it in ways they see fit, in particular to improving the lives of the citizens. In Hong Kong, more education is needed to enable people to understand more about what Open Data in the public and private sector can do for them and the advantages that Big Data and Analytics can bring. At the moment there is much opposition to the idea of handing over our data to the government, and as a result the government and people need to do more, in terms of sharing experiences and 'smart' ideas, and for citizens to give productive solutions to government and not simply criticise the actions that government takes.

Group 3—Smart Economy Smart Governance

To enhance itself as a smart city within the context of the GBA, Hong Kong should look to diversify its economy and industries beyond its traditional competitive advantages in financial and professional services, trade and logistics, and real estate. It cannot be argued that within the GBA, Hong Kong still plays a clear and essential role as the gateway to the rest of the world, in financial services, rule of law, intellectual property registering etc. However, a greater role needs to be played in promoting and supporting upcoming industries in technology and innovation so as to not fall behind cities in the GBA such as Shenzhen and Qianhai. Hong Kong has the talent and resources to facilitate development of innovative products and services, yet more support and recognition is needed for Hong Kong to capitalise on this strength to take the risks to venture into new innovative industries. Moreover, rather than competing with cities in the GBA on technology and innovation, Hong Kong should focus on the complementary qualities that can play up Hong Kong's strengths vis a vis other parts of the GBA, as well as build on its advantage as a gateway city, in particular to the rest of South East Asia and not only China.

In regards to Governance, Hong Kong and the GBA could improve on two main elements to further itself as a smart city and smart region: effective decision-making and input from stakeholders, in particular, citizens. The GBA clearly has the potential to become an economic powerhouse in the region and Asia, but a clearer delineation of who is in charge of strategic decision in the GBA is needed. While different municipalities and city regions in the GBA have their own spatial-functional plans, but to what extent the plans are overlapping or have been developed in coordination with each other and tie into one another is an important consideration. If there is an overarching commission looking over the overall development of the GBA, the amount of effective power and decision-making abilities it has will be a factor in the successful smart development of the GBA, as it will be necessary to carefully consider the complementary strengths and weaknesses of the different cities and nuclei in the polycentric GBA region. Within Hong Kong, there are numerous government departments that handle different functions and decision-making. Yet a layer of effective decision making is missing between the Chief Executive as Head of State, and these various departments. As in any large organisation, departments will dispute and disagree as a result of human nature; so therefore, the solution is not to try to avoid inter-departmental fighting, but to make sure they are adequately steered and controlled by a clear and effective leader above all departments who will minimise any inefficient bureaucracy. Hong Kong could benefit from a Mayor who could deal with city matters more effectively and support the Chief Executive who focuses on the next level towards regional integration and the Mainland.

Hong Kong does relatively well in terms of bottom-up input in decision-making, although not perfect. How this approach could be transferred or extended to the GBA region is another consideration to make. 'Smart' is not only use of technology but being clever and also having the ability to learn from oneself and progress in a productive and effective manner.

2.4.3.3 Summary of the Workshop Outcomes

In general, there is a consensus that much of the potential and opportunities for greater smart city integration with the GBA region will come from an approach that capitalises on Hong Kong's inherent strengths and unique resources, such as its strategic international positioning, established economy, industries and talent pool to further its dominant role as a hub in the region. Given its strengths, Hong Kong must not become complacent and to continue to strive to diversify its economy by embracing and supporting the creative, innovation and technology industries given its abundant resources in academia, talent, and research and development. At a social level, it is believed that a mind-set change is needed for the younger generations to accept and embrace the opportunities integration and cooperation with the GBA and the Mainland will bring. More exposure and better education including knowledge sharing among peers is seen as the best way to achieve this. Most importantly a common vision and strong leadership for effective guidance and direction of Hong Kong's key decision-makers and the community. At a political level, a bolder and less risk-averse Government is considered necessary to effectively make the transition to a truly smart, sustainable and more liveable city that will efficiently be able to integrate with the GBA region. Moreover, greater communication and understanding of what the younger generation need and want for the future of Hong Kong is critical in uniting its society for enhanced local and regional integration.

2.4.4 Recommendations for Hong Kong's Future Development Within the Context of GBA

In addition to the territorial development plans of HK2030+ and Hong Kong's stated initiatives to promote innovation, technology and smart city initiatives outlined in the 2017 Policy Address, further the recommendations to consider for Hong Kong's future development within the context of GBA are presented here to suggest how Hong Kong can better integrate not only spatially and economically, but also socially at a human level to improve quality of life for people of Hong Kong, and visitors as well.

i. The HK2030+ is the development strategy for Hong Kong going forward, and mainly addresses the issues of Hong Kong and focuses on improving the quality of life of the people of Hong Kong. However, in addition the future development of Hong Kong should be considered within the context of the Greater Bay Area offering a huge potential for Hong Kong and its people;
ii. Consider the greater harbor area as the CBD of Hong Kong with several nodes such as Central, Island East, Island West, Kowloon, Kowloon East, Kowloon West etc. Set up in 2012, the Energising Kowloon East Office (EKEO) was instrumental in not only developing the Kowloon East master plan but also engaged the various stakeholders throughout the process and successfully implementing improvements in street junctions, and the harbourfront as some of the preliminary place making solutions. Similar to the EKEO, there could be the Energising Kowloon Office (EKO) and the Energising Kowloon West Office (EKWO) on the Kowloon side. On the Hong Kong Island side there could be Energising Central Office (ECO), Energising Island East Office (EIEO) and Energising Island West Office (EIWO); These initiatives together could help develop Hong Kong into a livable, smart and sustainable city within the regional context of the GBA;
iii. Open space, public space enhancements with weekend retreats in country parks to create a means of leisure and recreation for the people of Hong Kong. Areas within Lantau can be further developed sensitively for leisure

and recreation. The East Lantau Metropolis currently proposed as CBD3 can be developed as a leisure, recreation and resort island sometime in the future;

iv. Continue the development of more public housing for rental as well as ownership by the Housing Authority and Hong Kong Housing Society (HKHS) but also ensure that more affordable housing developed by Urban Renewal Authority (URA), MTR, and the private sector as well. Encourage better quality housing with higher space standards with mixed income developments catering to the needs of the younger and elderly population promoting ageing in place;

v. Ensure that there is more exchange between Hong Kong universities and those in the Mainland with more exchange visits for faculty and students to encourage more teaching, research and learning opportunities between the people from either side. Ensure more exposure for Hong Kong students, graduates and faculty to experience living and working in the Mainland;

vi. Develop more internship and work opportunities for students and graduates between Hong Kong and the Mainland, so that Hong Kong students and graduates also get to experience life in China similar to the numerous students from the Mainland that go to universities in Hong Kong;

vii. Develop a platform by professional and other institutes to create more opportunities for exchange of ideas and experiences between the people of Hong Kong and Mainland China so there is better understanding and tolerance between people from either side;

viii. Create Visa free development zones where possible around border crossings such as Lo Wu, Lok Ma Chau, Lian Tang where a lot of activities, business transactions, social interactions can happen;

ix. Continue to promote and encourage innovation through education and creative industries, providing support for startups, incubation with proper infrastructure and technology and data sharing, and;

x. Explore and encourage disruption on different scales in various sectors, housing, retail, office, commercial real estate, transport, education, healthcare etc.

In summary, the spatial and economic strategies of the GBA aim to enable the complementary strengths of the different areas and cities within the GBA to thrive and achieve further growth and urban development, and is facilitated by investments in infrastructure and transportation. Given the existing spatial and economic development plans, policy initiatives and cooperation agreements, enhanced and more seamless social integration could be achieved given consideration of these recommendations. In general, much seems to be done at a physical, economic (hard) level that is visible to the public to further integrate Hong Kong with the GBA, however at a mental, social (soft) level, developments and progress are not as clear and noticeable to the general public. Hence, greater promotion and education of the benefits of increased regional integration is needed to further convince and encourage people and business to look beyond Hong Kong and to embrace the opportunities that the spatial and economic potential of the GBA has to offer.

References

1. Invest Hong Kong HKSAR (2014) The Greater Pearl River Delta, 7th edn. http://www.investhk.gov.hk/zh-hk/files/2014/05/InvestHK_GPRD-Book_Eng_Apr2014.pdf. Accessed 16 Mar 2017
2. Central Compilation & Translation Press (2016) The 13th Five-Year Plan For Economic and Social Development of the People's Republic of China (2016–2020). http://210.6.198.17/cache/en.ndrc.gov.cn/newsrelease/201612/P020161207645765233498.pdf?ich_args=0fcfc6c4af81270b7e3d674322521be9_1_0_0_12_b24c0892fe3aa3bbe6a64005ecb9d4d4f826777b874890c5748e9b207d343df0_dbe65f733b2e4e5ab3ba4a3dc5370e80_1_0&ich_ip=. Accessed 16 Mar 2017
3. Yeung P (2017) Greater Bay Area initiative a boost to Hong Kong's competitiveness. https://www.chinadailyasia.com/opinion/2017-04/18/content_15601299.html. Accessed 21 Mar 2017
4. HKSAR (2017) Framework agreement on deepening Guangdong-Hong Kong-Macao cooperation in the development of the Bay Area signed. http://www.info.gov.hk/gia/general/201707/01/P2017070100409.htm. Accessed 19 Mar 2017
5. Li JT, Wong D (2014) ECA continues its series of attraction market profiles with an overview and insights on the Greater Pearl River Delta (PRD) region in Southern China and theme park attraction development. ECA is currently working on multiple theme park destination developments in China including the PRD region. https://blooloop.com/features/eca-attraction-market-profile-overview-of-pearl-river-delta-prd-extended-theme-park-hub-cluster/. Accessed 19 July 2017
6. Daxue Consulting (2016) Pearl River Delta China. https://i.pinimg.com/originals/bf/f5/6f/bff56f428054e38892e0f12fca7faff9.jpg
7. HKTDC Research (2016) PRD economic profile. http://china-trade-research.hktdc.com/business-news/article/Facts-and-Figures/PRD-Economic-Profile/ff/en/1/1X000000/1X06BW84.htm. Accessed 28 July 2017
8. Census and Statistics Department HKSAR (2017) Hong Kong Annual Digest of Statistics 2016. http://www.statistics.gov.hk/pub/B10100032016AN16B0100.pdf. Accessed 28 July 2017
9. Statistics and Census Service Government of Macao Special Administrative Region (2017) Yearbook of statistics. http://www.dsec.gov.mo/Statistic.aspx?lang=en-US&NodeGuid=d45bf8ce-2b35-45d9-ab3a-ed645e8af4bb. Accessed 28 July 2017
10. UN-HABITAT (2006) Urbanization: Mega & Meta Cities, New City States? State of the World's Cities 2006/7. http://mirror.unhabitat.org/documents/media_centre/sowcr2006/SOWCR%202.pdf. Accessed 28 July 2017
11. The World Bank (2017) World Bank report provides new data to help ensure urban growth benefits the poor. http://www.worldbank.org/en/news/press-release/2015/01/26/world-bank-report-provides-new-data-to-help-ensure-urban-growth-benefits-the-poor. Accessed 28 July 2017
12. World Bank Group (2015) East Asia's changing urban landscape measuring a decade of spatial growth. http://www.worldbank.org/content/dam/Worldbank/Publications/Urban%20Development/EAP_Urban_Expansion_full_report_web.pdf. Accessed 4 Sept 2017
13. Fung Business Intelligence (2007) Fast facts of the Guangdong-Hong Kong-Macau Bay Area. https://www.fbicgroup.com/sites/default/files/IG_12_GDP.pdf. Accessed 4 Sept 2017
14. China Development Bank (2015) China's special economic zones. Available via World Bank. http://www.worldbank.org/content/dam/Worldbank/Event/Africa/Investing%20in%20Africa%20Forum/2015/investing-in-africa-forum-chinas-special-economic-zone.pdf. Accessed 4 Sept 2017
15. Du SQ, Shi PJ, Anton VR (2013) The relationship between urban sprawl and farmland displacement in the Pearl River Delta. http://www.mdpi.com/2073-445X/3/1/34/htm. Accessed 28 Sept 2017

16. Zhijia L, Heqing H, Saskia EW, Yan D (2016) Construction area expansion in relation to economic-demographic development and land resource in the Pearl River Delta of China. J Geogr Sci 26(2):188–20. Accessed 28 Sept 2017
17. Constitutional and Mainland Affairs Bureau HKSAR (2007) Joint declaration of the government of the United Kingdom of Great Britain and Northern Ireland and the government of the People's Republic of China on the question of Hong Kong. http://www.cmab.gov.hk/en/issues/jd2.htm. Accessed 2 Oct 2017
18. Development Bureau, Planning Department HKSAR (2016) Hong Kong 2030+ towards a planning vision and strategy transcending 2030 public engagement. http://www.hk2030plus.hk/document/2030+Booklet_Eng.pdf. Accessed 17 Oct 2017
19. Giffinger R, Fertner C, Kramar H, Kalasek R, Pichler-Milanovic N, Meijers E (2007) Smart cities—ranking of European medium-sized cities. Final Report, Vienna. http://www.smart-cities.eu/download/smart_cities_final_report.pdf. Accessed 26 Oct 2017
20. Cohen B (2012) What exactly is a smart city? Fast CoExist. Available via http://www.fastcoexist.com/1680538/what-exactly-is-a-smart-city. Accessed 13 Sept 2015
21. Govada S, Spruijt W, Rodgers T (2017) Chapter 7: smart city concept and framework. Smart economy in smart cities: international collaborative research: Ottawa, St. Louis, Stuttgart, Bologna, Cape Town, Nairobi, Dakar, Lagos, New Delhi, Varanasi, Vijayawada, Kozhikode, Hong Kong. In: Vinod Kumar TM (ed) Singapore: Springer Singapore, 2017, pp 189–197. Print. Advances in 21st century human settlements. Accessed 12 Oct 2017
22. Invest HongKong HKSAR (2014) The Greater Pearl River Delta, 7th edn. http://www.investhk.gov.hk/zh-hk/files/2014/05/InvestHK_GPRD-Book_Eng_Apr2014.pdf. Accessed 29 Oct 2017
23. Immigration Department HKSAR (2011) Control branch. http://www.immd.gov.hk/publications/a_report_2011/en/ch2/index.html. Accessed 29 Oct 2017
24. Leung J (2010) Hong Kong Shenzhen planning cooperation. Planning department, HKSARG. http://slideplayer.com/slide/4424848/. Accessed 14 June 2017
25. Census and Statistics Department HKSAR (2017) Hong Kong monthly digest of statistics cross-boundary travel survey 2015. http://www.statistics.gov.hk/pub/B71706FB2017XXXXB0100.pdf. Accessed 2 Nov 2017
26. Census and Statistics Department HKSAR (2017) Hong Kong monthly digest of statistics cross-boundary travel survey 2015. http://www.statistics.gov.hk/pub/B71706FB2017XXXXB0100.pdf. Accessed 7 Nov 2017
27. Hong Kong Airport (1999) Finalized Civil International Air Traffic Statistics at HKIA Year 1999. http://www.hongkongairport.com/eng/pdf/business/statistics/1999e.pdf. Accessed 9 Nov 2017
28. Hong Kong Airport (2016) Finalized Civil International Air Traffic Statistics at HKIA Year 2016. http://www.hongkongairport.com/eng/pdf/business/statistics/2016e.pdf. Accessed 11 Nov 2017
29. MTR Corporation Limited (2013) Key information Express Rail Link (XRL). http://www.expressraillink.hk/en/project-details/key-information.html. Accessed 1 Dec 2017
30. Hong Kong-Zhuhai-Macao Bridge Authority (2009) Introduction to Hong Kong-Zhuhai-Macao bridge. http://www.hzmb.org/en/bencandy.asp?id=2. Accessed 3 Dec 2017
31. Hong Kong-Zhuhai-Macao Bridge Authority (2009) Benefits of transportation. http://www.hzmb.hk/eng/benefits_transportation.html. Accessed 3 Dec 2017
32. Bloomberg (2017) China expected to extend electric vehicle tax rebate. Available via SCMP. http://www.scmp.com/tech/enterprises/article/2122887/china-expected-extend-electric-vehicle-tax-rebate. Accessed 15 Dec 2017
33. Transport Department (2017) Guidelines for importation and registration of motor vehicles. http://www.td.gov.hk/en/public_services/licences_and_permits/vehicle_first_registration/guidelines_for_importation_and_registration_of_mot/. Accessed 23 May 2017
34. Cheng R (2017) Decision to cut Hong Kong's electric vehicle tax waiver is "backwards" and sends wrong message, critics say. Available via http://www.scmp.com/news/hong-kong/

economy/article/2074016/decision-cut-hong-kongs-electric-vehicle-tax-waiver-backwards. Accessed 5 Mar 2017
35. Environmental Protection Department (2017) Promotion of electric vehicles in Hong Kong. http://www.epd.gov.hk/epd/english/environmentinhk/air/prob_solutions/promotion_ev.html. Accessed 15 Dec 2017
36. Cheung R (2017) Shenzhen set to have world's first all-electric public bus fleet. Available via Post Magazine. http://www.scmp.com/magazines/post-magazine/long-reads/article/2119175/why-shenzhen-set-have-first-all-electric-public. Accessed 17 Nov 2017
37. Trade and Industry Department HKSAR (2017) Mainland and Hong Kong Closer Economic Partnership Arrangement (CEPA). https://www.tid.gov.hk/english/cepa/cepa_overview.html. Accessed 29 June 2017
38. Trade and Industry Department HKSAR (2017) Mainland and Hong Kong closer economic partnership arrangement—investment agreement. https://www.tid.gov.hk/english/cepa/legaltext/files/cepa14_main.pdf. Accessed 29 June 2017
39. Trade and Industry Department HKSAR (2017) Mainland and Hong Kong Closer economic partnership arrangement—agreement on economic and technical cooperation. https://www.tid.gov.hk/english/cepa/legaltext/files/cepa15_main.pdf. Accessed 29 June 2017
40. HKSAR (2007) Agreement on Shenzhen/Hong Kong innovation circle signed. http://www.info.gov.hk/gia/general/200705/21/P200705210224.htm. Accessed Aug 2017
41. Hong Kong Science and Technology Park (2016) The corporation milestones. https://www.hkstp.org/en/about-hkstp/the-corporation/milestone/. Accessed 25 June 2017
42. Hong Kong Science and Technology Park (2016) About HKSTP. https://www.hkstp.org/en/about-hkstp/the-corporation/about-hkstp/. Accessed 25 June 2017
43. HKSAR (2017) Hong Kong and Shenzhen sign memorandum of understanding on jointly developing the Lok Ma Chau Loop. http://www.info.gov.hk/gia/general/201701/03/P2017010300609.htm. Accessed 25 June 2017
44. Siu P, Zhao S, Cheung G (2017) Hong Kong and Shenzhen settle border dispute as they join hands to develop technology park. Available via SCMP. http://www.scmp.com/news/hong-kong/economy/article/2058878/hong-kong-and-shenzhen-set-partner-innovation-and-technology. Accessed 27 June 2017
45. Nansha Information Technology Park (2006) Nansha IT Park partnership. http://www.nsitp.com/partnership.php. Accessed 8 July 2017. Accessed 27 June 2017
46. Nansha Information Technology Park (2006) An Oasis in the PRD. http://www.nsitp.com/eindex.php. Accessed 27 June 2017
47. Shih D, Geng K (2017) Greater Bay Area in new development context—targeting new property investment opportunities from the surge of South China's Greater Bay Area. Colliers Radar, Colliers International. Hong Kong. Retrieved from: http://www.colliers.com/en-gb/hongkong/insights/colliers-radar/greater-bay-area. Accessed 14 July 2017
48. Hong Kong and Shenzhen Bi-City Biennale of Urbanism\Architecture (2008) About the Biennale. http://hkszbiennale.asia/. Accessed 16 Sept 2017
49. Council on Foreign Relations (2014) Hong Kong Innovation Project Report No. 5 Hong Kong and the Pearl River Delta: Science and Technology Cooperation. http://demo.savantas.org/wp-content/uploads/2014/06/6_Segal.pdf. Accessed 14 July 2017
50. Hong Kong University of Science and Technology (2013) HKUST Fok Ying Tung Graduate School (FYTGS). http://www.vprg.ust.hk/rgs/eng/mainland_platforms/fytgs.html. Accessed 14 July 2017
51. Communications and Public Relations Office (2013) CUHK signs final agreements with Shenzhen University and Shenzhen Government inaugurating the development of the Chinese University of Hong Kong, Shenzhen. https://www.cpr.cuhk.edu.hk/en/press_detail.php?id=1509. Accessed 15 July 2017
52. Invest HongKong HKSAR (2014) The Greater Pearl River Delta, 7th edn. Available via http://www.investhk.gov.hk/zh-hk/files/2014/05/InvestHK_GPRD-Book_Eng_Apr2014.pdf. Accessed 18 July 2017

53. Harbour Times (2016) Cross-border students up 118% in five years. http://harbourtimes.com/2016/06/23/cross-border-students-up-118-in-five-years/. Accessed 18 July 2017
54. Boyd C (2012) The top 10 smart cities on the planet. Available via Co, Design. https://www.fastcodesign.com/1679127/the-top-10-smart-cities-on-the-planet. Accessed 22 Aug 2017
55. EasyPark (2017) 2017 Smart Cities Index. https://easyparkgroup.com/smart-cities-index/. Accessed 22 Aug 2017
56. Environmental Protection Department (2017) Data & statistics Hong Kong air pollutant emission inventory. http://www.epd.gov.hk/epd/english/environmentinhk/air/data/emission_inve.html. Accessed 14 June 2017
57. Ng KW (2016) Controlling ship emissions in the Pear River Delta. http://civic-exchange.org/cex_reports/201612AIR_PRDemission_brief_en.pdf. Accessed 14 June 2017
58. Huang H (2015) New regulation on ship emission control areas of the Pearl River Delta, the Yangtze River Delta and Bohai sea waters. https://www.ukpandi.com/knowledge-publications/article/new-regulation-on-ship-emission-control-areas-of-the-pearl-river-delta-the-yangtze-river-delta-and-bohai-sea-waters-134120/. Accessed 16 June 2017
59. Urban Land Institute (2017) Hong Kong East Community Green Station—2016 Global Awards for Excellence Winner. Awards & Competitions. https://americas.uli.org/awards/hong-kong-east-community-garden-2016-global-awards-excellence-finalist/. Accessed 14 Jun 2017
60. Environmental Protection Department (2017) Waste. An overview on challenges for waste reduction and management in Hong Kong. http://www.epd.gov.hk/epd/english/environmentinhk/waste/waste_maincontent.html. Accessed 14 June 2017
61. Environmental Protection Department (2017) Waste recycling statistics https://www.wastereduction.gov.hk/en/quickaccess/stat_recycle.htm
62. Tsang E (2017) Tear off the glossy covers or your magazines won't get recycled, Hongkongers told. Health & Environment. South China Morning Post. http://www.scmp.com/news/hong-kong/health-environment/article/2124680/tear-glossy-covers-or-your-magazines-wont-get
63. CLP Group (2014) Our history. https://www.clpgroup.com/en/about-clp/company-profile/our-history#tab3. Accessed 15 June 2017
64. CLP Group (2016) Towards a greener Pearl River Delta a roadmap for clean energy generation for Hong Kong. https://www.clp.com.hk/en/about-clp-site/media-site/resources-site/publications-site/Documents/CLPEnergyVisionPrimer_Eng.pdf. Accessed 17 June 2017
65. Loh C (2008) Hong Kong's energy future a cleaner energy future is Hong Kong's responsibility. Available via http://civic-exchange.org/cex_reports/200810EC_HKEnergyFuture1_en.pdf. Accessed 16 June 2017
66. HKSAR (2008) Memorandum of understanding on energy co-operation. http://www.info.gov.hk/gia/general/200808/28/P200808280188.htm. Accessed 18 June 2017
67. Planning Department HKSAR (2009) Planning study on the coordinated development of the Greater Pearl River Delta townships. http://www.pland.gov.hk/pland_en/misc/great_prd/news/pdf_en/summary.pdf. Accessed 18 June 2017
68. Invest HongKong HKSAR (2014) The Greater Pearl River Delta, 7th edn. Available via http://www.investhk.gov.hk/zh-hk/files/2014/05/InvestHK_GPRD-Book_Eng_Apr2014.pdf. Accessed 17 June 2017
69. HKSAR (2017) Framework agreement on deepening Guangdong-Hong Kong-Macao cooperation in the development of the Bay Area signed. Available via http://www.info.gov.hk/gia/general/201707/01/P2017070100409.htm. Accessed 27 Oct 2017
70. Development Bureau, Planning Department HKSAR (2016) Hong Kong 2030+ towards a planning vision and strategy transcending 2030 public engagement. Available via http://www.hk2030plus.hk/document/2030+Booklet_Eng.pdf. Accessed 3 July 2017
71. Fung Global Retail & Technology (2017) Deep Dive: Shenzhen—An International Hub of Hardware Innovation. https://www.fungglobalretailtech.com/research/deep-dive-shenzhen-international-hub-hardware-innovation/. Accessed 2 July 2017

72. Government of Macau Special Administrative Region (2016) The Five-Year Development Plan of the Macao Special Administrative Region (2016–2020). http://www.cccmtl.gov.mo/files/plano_quinquenal_en.pdf. Accessed 2 Nov 2017
73. Xinhua (2017). Macao, Hong Kong sign CEPA for closer cooperation, regional integration. http://news.xinhuanet.com/english/2017-10/27/c_136710636.htm. Accessed 2 Nov 2017
74. HKSAR (2017) A dedicated chapter on Hong Kong and Macao in the National 13th Five-Year Plan. http://www.info.gov.hk/gia/general/201603/17/P201603170899.htm. Accessed 7 Nov 2017
75. China (Guandong) Pilot Free Trade Zone (2016) News and laws. http://www.china-gdftz.gov.cn/en/. Accessed 7 Nov 2017
76. Central Compilation & Translation Press (2016) The 13th Five-Year Plan For Economic and Social Development of the People's Republic of China (2016–2020). http://210.6.198.17/cache/en.ndrc.gov.cn/newsrelease/201612/P020161207645765233498.pdf?ich_args=0fcfc6c4af81270b7e3d674322521be9_1_0_0_12_b24c0892fe3aa3bbe6a64005ecb9d4d4f826777b874890c5748e9b207d343df0_dbe65f733b2e4e5ab3ba4a3dc5370e80_1_0&ich_ip=. Accessed 9 Nov 2017
77. China (GuangDong) Pilot Free Trade Zone (2015) Introduction—China (Guangdong) Pilot Free Trade Zone. http://www.china-gdftz.gov.cn/en/ABOUT_GDFTZ/Introduction/201502/t20150214_587.html. Accessed 27 Oct 2017
78. HKTDC Research (2017) China (Guangdong) Pilot Free Trade Zone. http://china-trade-research.hktdc.com/business-news/article/Facts-and-Figures/China-Guangdong-Pilot-Free-Trade-Zone/ff/en/1/1X000000/1X0A2V1E.htm. Accessed 27 Oct 2017
79. HKSAR (2017) The chief executive's 2017 policy address—we connect for hope and happiness. https://www.policyaddress.gov.hk/2017/eng/pdf/PA2017.pdf. Accessed 29 Oct 2017
80. HKTDC (2017) Belt and road basics. https://beltandroad.hktdc.com/en/belt-and-road-basics. Accessed 24 Oct 2017
81. Kam KS (2017) Greater Bay Area economy to reach US$3.6 trillion by 2030, Colliers research says. http://www.scmp.com/property/hong-kong-china/article/2108083/greater-bay-area-economy-reach-us36-trillion-2030-colliers. Accessed 26 Oct 2017

Part III
Germany, Stuttgart

Chapter 3
Stuttgart Region—Sustainable Industrialization in Stuttgart Metropolitan Region

Michael Hertwig, Joachim Lentes, Nikolas Zimmermann and Manfred Dangelmaier

Abstract The Stuttgart region is a very prosperous area. The economical backbone of the region is the production of industrial goods. The globalization is a trigger for changing economical dependencies. Industrial processes are changing because of the new challenges. Digitalization is a current topic which has huge influence on economical procedures. A main topic in this context is the German initiative "Industry 4.0". New ways of thinking have a great impact on the way manufacturing is done and the digitalization opens up new possibilities. Current discussions on sustainability are influencing the economic thinking heavily. The importance of sustainable development with respect to environment and climate becomes more and more obvious to everybody. The climate conference in Paris was finished with a consensus of all present parties. Every important economic nation has a common sense on climate targets nowadays. Reasons for that are manifold, e.g. limited resources and huge amount of emissions. Nevertheless, society will only accept limitations, without a reduction of living quality. Based on that, it is necessary to implement a new procedures in existing structures. Additionally, it is required to implement changes without reducing economic potentials. The approach of symbiosis can support these developments. However, technology is a required extension to reach the target of sustainability. In this chapter, different views are discussed to facilitate long-term development. The first view is a holistic view. The idea of creating a symbiosis between enterprises in an industrial estate is easy to

M. Hertwig (✉)
Institute for Human-Factors and Technology Management IAT,
University of Stuttgart, Nobelstrasse 12, 70569 Stuttgart, Germany
e-mail: michael.hertwig@iat.uni-stuttgart.de

J. Lentes · N. Zimmermann · M. Dangelmaier
Fraunhofer Institute for Industrial Engineering IAO, Nobelstrasse 12,
70569 Stuttgart, Germany
e-mail: joachim.lentes@iao.fraunhofer.de

N. Zimmermann
e-mail: nikolas.zimmermann@iao.fraunhofer.de

M. Dangelmaier
e-mail: manfred.dangelmaier@iao.fraunhofer.de

T. M. Vinod Kumar (ed.), *Smart Metropolitan Regional Development*,
Advances in 21st Century Human Settlements,
https://doi.org/10.1007/978-981-10-8588-8_3

realize. On one hand the enterprise manager are able to shape the collaborations, and on the other hand the local administration is able to support the approach with small efforts. Besides the sustainable development of industrial estates the companies need a strategy for their own development. A development path towards urban manufacturing is helping the management of companies to create a strategy. A concrete step-by-step approach is the base. Additional technological drivers are analyzed in their capability of supporting sustainability.

Keywords Urban manufacturing · Industrial symbiosis · Sustainable industrial development · Additive technology · Long-term improvement

The Stuttgart Metropolitan Region is situated in the center of Baden-Wuerttemberg (see Fig. 3.1). This federal state is economically the most powerful area in Europe. It creates 3 percent of the economic value in the European Union [1]. The economic potential is mainly based on producing goods for industrial purposes, such as machines, process technology and automotive parts. Based on the generated turnover, companies in these fields are contributing around 80% of the total economic value in the production sector [2]. The companies in Baden-Wuerttemberg are mainly small to medium sized enterprises additionally a high number is still family-run.

Coming to Stuttgart Metropolitan Region a main focus lies on automotive products either cars or supply parts and products. In Stuttgart city and its surrounding which can be called Stuttgart Metropolitan Region live more than 2.7 million people [3]. The high density of people and the widely spread industrial areas of different sizes generate a high amount of commuters. The economic power with its potential for wealth and jobs let increase the value of all living areas in Stuttgart Metropolitan Region. Additionally, Stuttgart was rated number one for its richness in cultural possibilities of all Metropolitan Regions in Germany [3].

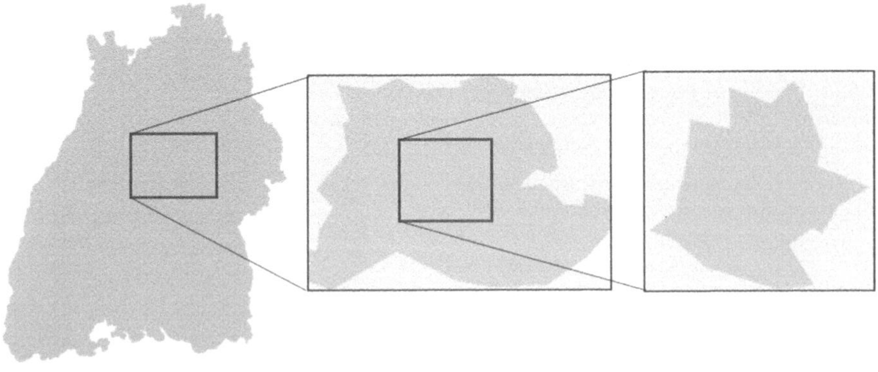

Fig. 3.1 Map of Baden-Wuerttemberg region (left), Stuttgart Metropolitan Region (middle), Stuttgart city (right)

The Stuttgart Metropolitan Region is also seen as High-Technology location. The reason is the high amount of headquarters and big production and research location of multi-national companies with technical products. Well-known are Daimler, Bosch, Porsche, IBM, HPE and Hewlett Packard. Characteristically is the mixture of global companies and highly innovative medium-sized companies. Innovation and research is well integrated in the Metropolitan Region. With a high number of universities and universities of applied science and world-known research institutions (5 Fraunhofer-institutes, 2 Max-Planck-institutes and 6 institutes of DLR, plus others) the region has a high potential for innovative research and development in technology, products and services [4].

Stuttgart Metropolitan Region is also well located in Europe. The east-west-connection is directly passing the region. Transportation connection to train and roadway is existing. A main connection is coming from London, via Paris, passing Stuttgart Region towards Istanbul via Munich, Vienna and Budapest. This is an additional advantage of the region which has an influence of the economic potential of the region.

Digitalization and Industry 4.0 is changing the way of performing business of German companies in the global economy. The distribution of information and data is increasing. The high competition of companies globally leads to new production methods and new products which can be customized faster and more efficient. The new technology has an impact on business processes and requirements of customers.

Locations of industrial production and living have short distances. Additionally it is common sense that sustainability is increasing in importance. Companies in Germany have already integrated the mindset for sustainability in their strategies because legal regulation force them to do so. However, the potential can even be extended when all entities in an industrial estate collaborate and use optimal technology. In this chapter it is presented how a framework for collaborative long-term development could be designed. It is presented how an implementation could be done on different levels of detail. This framework is currently tested in different industrial areas in the Stuttgart Metropolitan Region and Baden-Wuertemberg. As stated the distance between living and industrial manufacturing zones is decreasing which requires new ways of processing.

The presented work is structured in three sub-chapters. The work provides an insight on current ongoing changes. A transformation process is started and based on this, a study is presented which is currently in an evaluation process and its implementation in the Stuttgart Metropolitan region.

The first sub-chapter presents a way, how industrial estates can be transformed towards sustainability. It is presented which methods can be used to support the transformation process under respect of strategical guidance. Based on the size and capacities of the industrial estate, approaches of different detail and complexity are provided. The approaches guide not only the transforming companies itself but also their partners in the network. The approach will help to identify suitable partners and ease the matching for cooperation. The long-term orientated creation of networks is supporting sustainable links with a high potential to reduce negative effects without reduction of economic power.

In the second sub-chapter an approach for enterprise-internal transformation is explained. Starting as conventional production enterprise a way for sustainable manufacturing is presented whereby different areas have to change. The development path presented, is supporting the strategic change process. With simple steps in different fields, the companies are supposed to reach the target stage easier. The target stage is depending on the size and power of the company and can be defined freely without regarding boundary conditions.

Lastly in the third sub chapter, Additive Technology is presented with regard to a total change in production processes. Furthermore, the new technology supports sustainability aspects because there are huge potentials in reduction of used material, reduction of waste, energy efficient processes and less effort in adaption of processes even with big a variety of products. Nevertheless, the new technology requires a change in mindset and an adaption in processes. It is presented what areas in the value creation process are influenced by the changes. An insight on use-cases is completing the overview of the additive manufacturing.

3.1 Long-Term Oriented Holistic Development Towards Sustainability

Many influences enforce companies to rethink their doing in business. A main influence is the request for sustainability by customers. On the one hand all processes can be analyzed to find improvement potential. The focus of improvement is mainly reduction of resources and energy. Nevertheless, this can be broaden by a perspective on the employee. However, these improvement potential is mainly short-term driven. Therefore, an improvement is require which supports a long-term development towards sustainable processes, performance and thinking. The location and its surrounding are important to have supportive base.

3.1.1 Change in Location Selection and Company Settlement

The development of company locations has often been a decision based on individual interests of the company management and/or the owner. Society is changing, based on that processes for location identification and development have to be revised and adapted. Furthermore demands for sustainability are growing.

3.1.1.1 Sustainability as Core Element During Transformation

Sustainability and the associated long-term view are becoming increasingly important. In particular, the efforts in the area of climate protection, such as the Paris Conference [5], show the importance.

According to Chertow, the effective path to sustainability with value added in industrial eco-systems (see Fig. 3.2) is possible [6]. Within the industrial ecosystem, three main fields of activity—within the organization, between organizations and overarching relationships—are identifiable. In each of these fields of action, individual levels of action need to be placed which contribute to the sustainability of value creation.

The sustainability within an organization is decisively determined by the strategy of the company management. In the strategic orientation, goals and measures for the organization as well as employees are determined. This field is being put into focus by companies, in particular as a result of guidelines and requirements from regulatory bodies or customers. Within the scope of the concept study, only a subordinate focus is placed on this field of action.

The establishment of sustainable processes between organizations is more difficult since the internal strategy of the different parties is not always congruent or similar. Therefore, in addition to bilateral links between companies that can

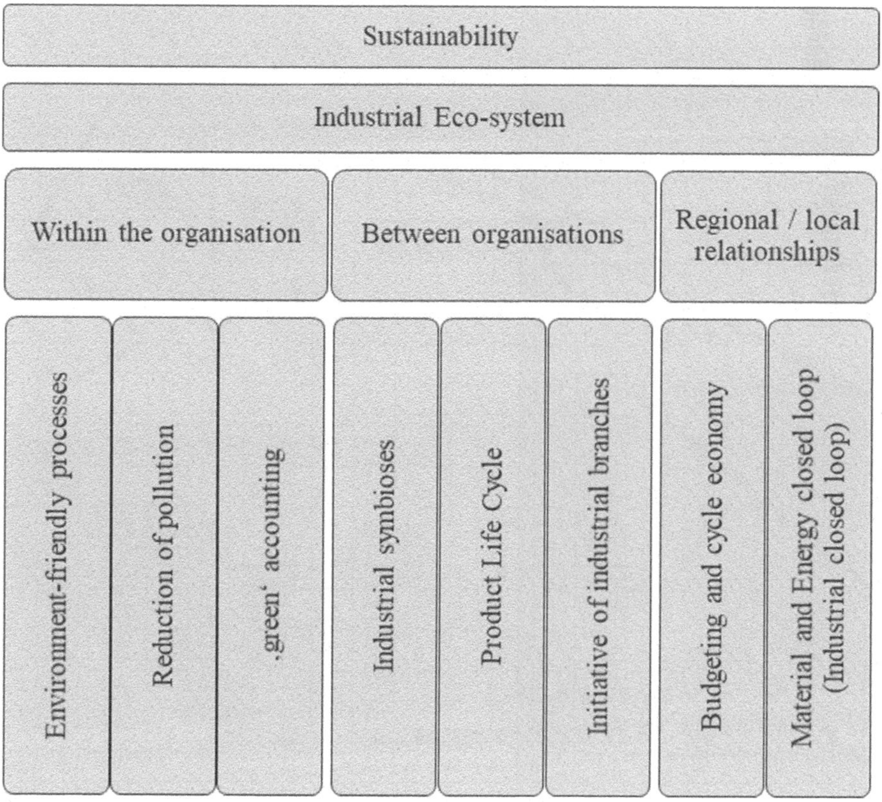

Fig. 3.2 Sustainability as a development driver [6]

successfully fulfill the symbiosis of their own interests, multilateral links are needed. Since the structures for the local compositions of manufacturing companies are diverse, methodological approaches for the creation and maintenance of multilateral symbioses are necessary. This area is the focus of the concept study.

More extensive relationships and links are addressed by approaches for circulatory systems. The complexity of this mechanism differs significantly from localized systems. The consideration of sustainability is necessary, but it will not be the focus of this study.

Taking the approach of Chertow as basis for the study, the concept study can be depicted as shown in Fig. 3.3. The focus is on the analysis of company internal processes and regional relations. As a system, the direct environment of companies is subjected to a detailed analysis. The focus is on the company settlements, which are classified according to BauNVO as commercial, industrial and mixed areas with a high commercial share [7].

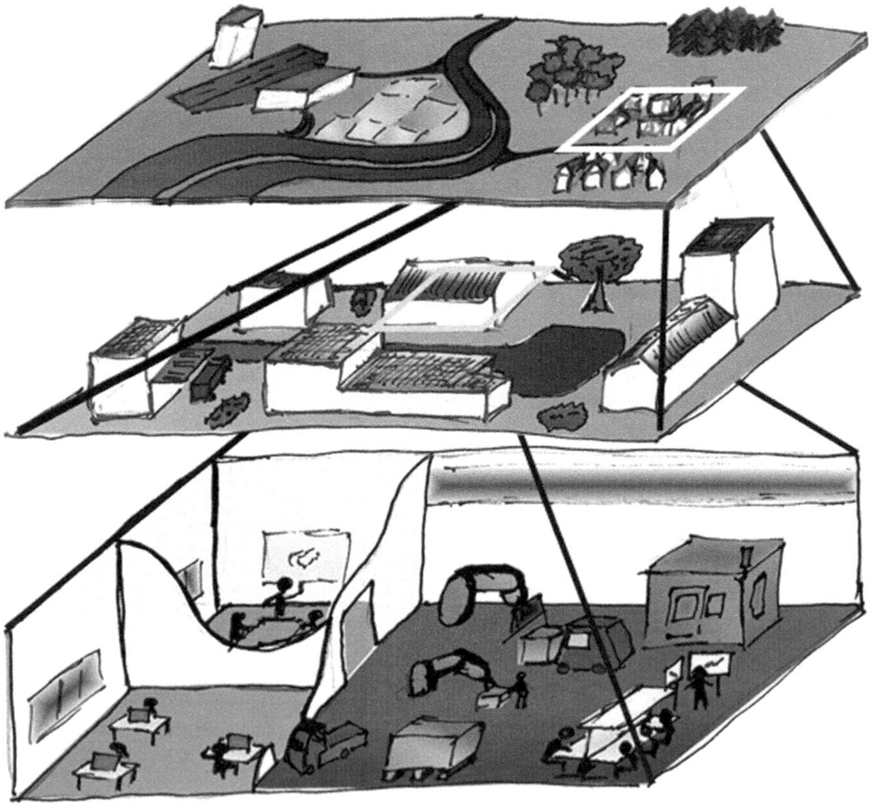

Fig. 3.3 Levels of analysis for implementation of sustainability (with local interdependencies, in direct surrounding of company, within the company—increasing degree of detail)

3.1.1.2 Manufacturing Companies in the Context of Sustainability

Industrialization is one of the triggers for climate change on the planet, which is largely accepted by government representatives. Based on this knowledge, an agreement [5] of 170 countries has been signed at the 21st World Climate Conference in Paris in 2015. The aim of this agreement is to limit CO_2 emissions in the long term to a level that is compatible with the environment [8]. Within the framework of further negotiations, such as the "Conference of Parties" in Marrakech in 2016, the partners of the Treaty will present national targets. These goals will locally lead to new requirements and regulatory frameworks, since the objectives are to be regularly checked for realization status. If companies and processes in manufacturing companies are considered, it becomes clear that approaches to resource and energy efficiency alone will not be sufficient. Due to the already existing environmental requirements, a large number of approaches are currently being implemented fully or at least partly. Collaboration between different stakeholders and a symbiosis of these partners will be necessary as they are already established in individual sectors in order to meet the requirements in the long term [9, 10].

For companies, it is therefore increasingly important to deal with the required production factors in the future and to use them optimally. Efficient use is not always possible. Therefore the understanding of one's own processes with the associated input and output becomes more important. A suitable modeling as an abstraction helps managers in companies to optimize the throughput (see Fig. 3.4). In order to carry out potential optimizations, it is necessary to know about these dimensions. In addition to the optimization of the throughput, potential users or interested parties can be identified for undesired output in order to establish possible links. These approaches allow the creation of small circulatory systems, which already offer great potential for resource efficiency. Through appropriate IT-support and tools, digitization can help create the required transparency and the complex control mechanism.

For the possible transformation processes, material, technology, energy, water, data, and human are central areas of interest in the context of resource and energy efficiency. The consideration of energy transformation already results from the efficiency efforts with regard to energy. Resources are to be understood as material, but also water and human beings are part of it. Water should be considered as a particularly important substance, as the global scarcity is to be taken into account in the context of sustainability efforts [11]. In the long term, this general scarcity could also create difficulties for companies [12]. In the context of value creation, people are often grouped together as a resource, which cannot longer be justified by the changes caused by digitization. Humans will less and less take up the task of production, instead takes agile decisions and exploits his creativity as a potential [13]. In the productive sector, new technology and data are also components which are becoming increasingly important for increasing efficiency and sustainability [14].

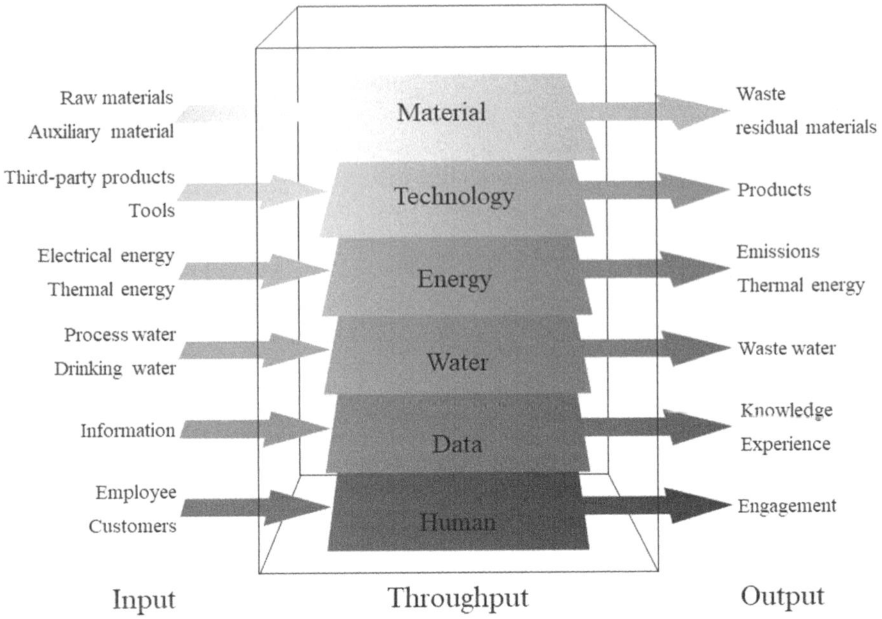

Fig. 3.4 White box model of a company with focus on sustainability

3.1.1.3 Transformation of Industrial Estates Under Respect of Sustainability

Through the publication of the Athens Charter, Le Corbusier created the basis for modern urban development, which envisaged a separation, from individual functional areas [15]. This is how to separate areas for work and living. Although the Charter of Athens is currently regarded as failed, some principles, especially sociological demands, are still anchored in current urban development, as is the distance from industry to housing. A further principle laid down in the Charter can now only be described as partial—subordination of private interests to Community interests [16]. The growing individualization of the individual's requirement for product, work and housing prevents it from addressing this principle in an appropriate form.

If business settlements are considered, a life cycle (see Fig. 3.5) becomes clear. The operator or owner starts with the planning. During this period, objectives, methods and approaches for the following periods are defined. After the planning, in part already during the planning, the development is started. In doing so, concepts are created that prepare the available surfaces based on this. After the development, the acquisition of business can be built up or filled with life by the acquisition of users. The start-up process is difficult to separate from the plant, especially in the initial phase, since the processes run parallel for some time. All

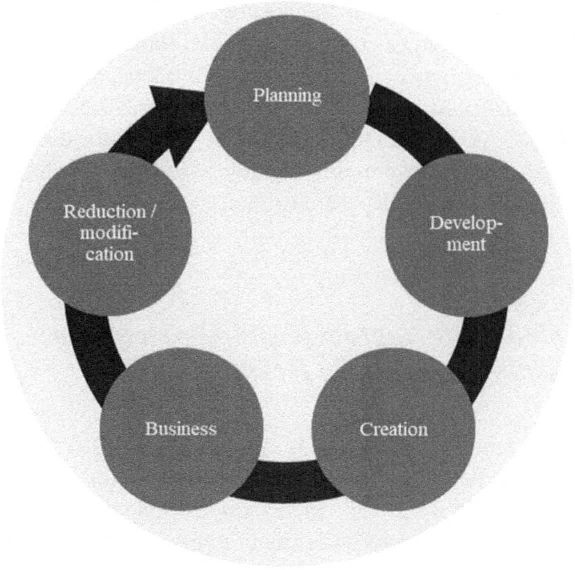

Fig. 3.5 Generic life cycle of an industrial estate

surfaces are used during operation. In operation, the optimal fulfillment of the user is the main focus. After the operational phase, which can be successful over many decades, the needs and requirements have to be re-analyzed. In this way, dismantling and conversion of existing infrastructure, utilization behavior and need are identified, which require a new planning or adaptation of the objectives. This starts the circulation [17] from the front.

The focus of the concept study is the period between high development activity and operation. For here, the settlement is always repeated, which is to be described with development and structure. This phase in the life cycle is so interesting, since there is a great potential for shaping the basic conditions. If the goals and methods for the appropriate design are established, expenses in the operating phase are less critical. The company stands, in particular, for the iterative continuous optimization, in order to realize goals and visions or simply to fulfill legal regulations.

In the context of resource-efficient and sustainable developments, company settlements are often referred to as eco-industrial parks. These company settlements can be subdivided into three types, which can be executed either as a new "Greenfield Development" or "Brownfield Development" portfolio development [18].

- Industrial Complexes—Industrial complexes are defined as spatially limited areas that contain industrial production processes. There is close chaining or coupling in the processes. Usually the use of material is low, but the processes are connected with enormous energy expenditures.

- Mixed industrial parks—Due to the locally limited settlement of companies, many SMEs as well as larger companies, mixed industrial parks can be defined and marked. The links between the companies are low, which is also reflected in the low chaining of production processes.
- Industrialization—An industrialized region is a spatially distributed enterprise group. Although the companies are usually assigned to different sectors, they have certain specializations, as are customary in clusters.

3.1.2 Methodological Approach and Development for Creating Synergistic Relationships

The complexity of creating synergies is not easy to master. Because the more complex relationships become, the more difficult it is for the user to draw the conclusions correctly. Because with increasing complexity and consideration of possible interactions, influences of individual factors in relation to other factors can no longer be described by simple causalities.

According to Romero and Ruiz an adaptable scaffold is needed for the modeling and operation of an eco-industrial park [19]. Only with appropriate modeling it is possible to assess changes and to assess their impact. The aim of the methodological approach is to establish and establish relationships and symbioses between the partners involved within commercial space. Existing loose relationships should be systematized and new relationships created. The aim of these relationships is to optimize the supply and the output in the long term. These are possible reductions resulting from the sharing and further utilization of the output of other parties within the commercial space (see Fig. 3.6).

3.1.2.1 Focus Area—Life Cycle of Industrial Estates

The selection of locations in the company usually takes place according to a clearly structured process. These location decisions are also becoming more and more important for companies in global competition. In particular, the optimal alignment and design of a production network or embedding in existing networks is crucial. Systematic decisions on site selection and development are described by Reuter and Prote and record the procedure within the selection process [20]. This is described with 5 steps (see Fig. 3.7). For the development of symbioses between participants within a business establishment, the period of observation is usually after step 4 or the settlement has already taken place.

These resulting states are significantly different. The analysis of these scenarios in consideration, selection of measures and long-term development will be analyzed and results are derived.

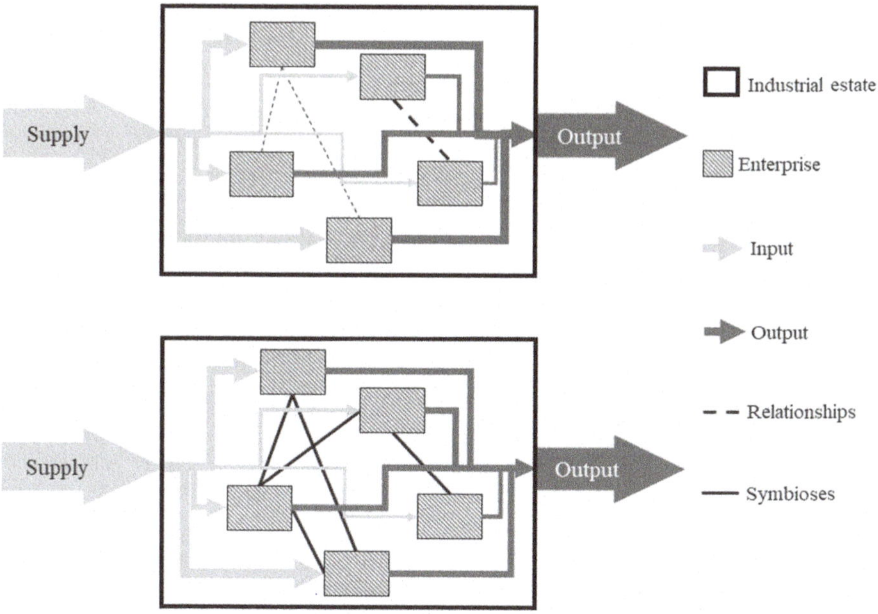

Fig. 3.6 Interactions to optimize the input and output of the overall system (based on 19)

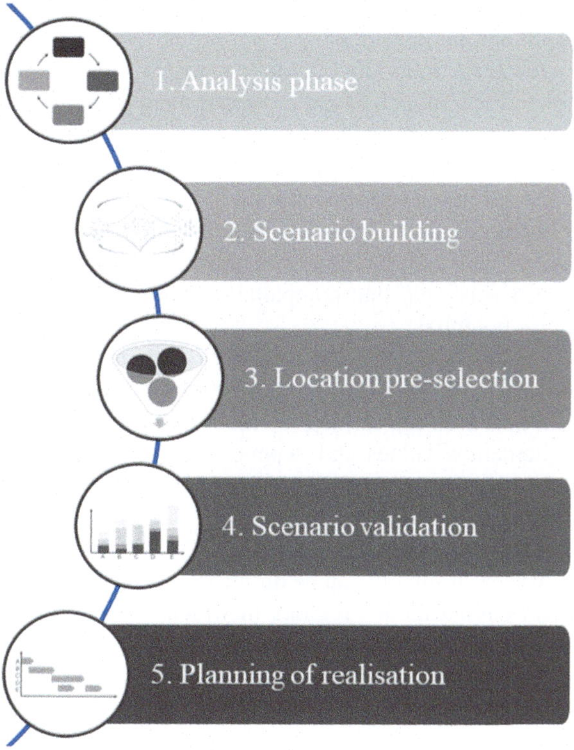

Fig. 3.7 Five phases of systematic site selection [16]

3.1.2.2 Model for the Creation of Complex Symbiosis

The high complexity of the modeling of interaction relationships is best realized through understanding. For this purpose, iterative pitfalls are used in the question, which allow even less experienced users to develop an understanding.

In order to be able to adapt the interaction models to the local specifications, a rough selection of potential stakeholders is necessary. This should be made possible by simply assessing the suitability of various potential stakeholders.

The approach provides for three levels with increasing level of information and increasing complexity with regard to networking and dependencies.

The first step is to match companies with the area. This is to distinguish unsuccessful or synergistically incompatible companies from potentially good symbiosis partners. Thus, an extensive reduction of possible combinations can be realized in the first step. To this end, the companies and the commercial space have to be structured by means of a transparent and objective classification approach in order to realize the allocation by means of a suitable assessment approach in order to address the next detail level.

In the next detail level, the company settlement is viewed as a collection of individual companies and thus a further selection for the optimal composition is supported. The desired selection of suitable partners should be supported by means of an abstraction model. Based on the Congruence model for the strategic management of organization is an adaptation [21]. This abstraction model is the basis for the assessment of matching. The congruence model consists of four main areas, which can be named as strategy, input, mechanisms and output. These main areas agglomerate individual factors or factor groups that influence the core in the Congruence model. For the development of interactions with symbiosis, the manufacturing company should be investigated, using the abstraction model. In addition, the company resettlement itself has to be regarded as a stakeholder with a specific congruence model approach.

In the third level of detailing, an interaction model is adapted from the determined data and findings and thus individualized for the application area. The objective is clear that companies are optimally represented in terms of their interaction potential.

On this basis, the assessment of matching as well as the derivation of possible measures for optimization are carried out. This assessment can be used in the long term based on the created interaction model. For this purpose, an adaptation of the implemented values and dependencies is necessary on a regular basis in order to adjust the actuality of the model to current circumstances.

Analysis and classification of commercial estates

The environment can be described by subsystems. According to Wiedemann, the urban environment—regions in which life and work are carried out with a small distance from each other—can be reduced by nine relevant subsystems [22]. The relevant subsystems are

- Production systems (production agglomerations)
- Culture
- Environment
- capital
- residents
- infrastructure
- Shops and services
- Knowledge/education and training
- Information and communication

Based on these relevant subsystems, a further reduction and adaptation is possible for the company settlement. There is a summary of the subsystems inhabitants and knowledge of employees. This is because only a part of the company can be used simultaneously. The integration of knowledge into the subset of employees is based on the assumption that knowledge is introduced by the employee into the company or is transferred to it from outside. Capital is generally important, but it is a subordinate part of the company settlement, since the fiscal-political links are to be considered only marginally. If the location is selected by site analysis, fiscal conditions have already been analyzed. For the commercial space, the transformation involves financial expenses, the source of the funds being dependent on the organizational form. Culture is a relevant commodity for society, but there is no direct link to company settlements. Therefore, culture is not analyzed and integrated as a separate subsystem (Fig. 3.8).

There are relationships between all remaining subsystems. Depending on the intensity of the characteristics of individual sub-systems, a classification of commercial space appears to be useful (Fig. 3.9).

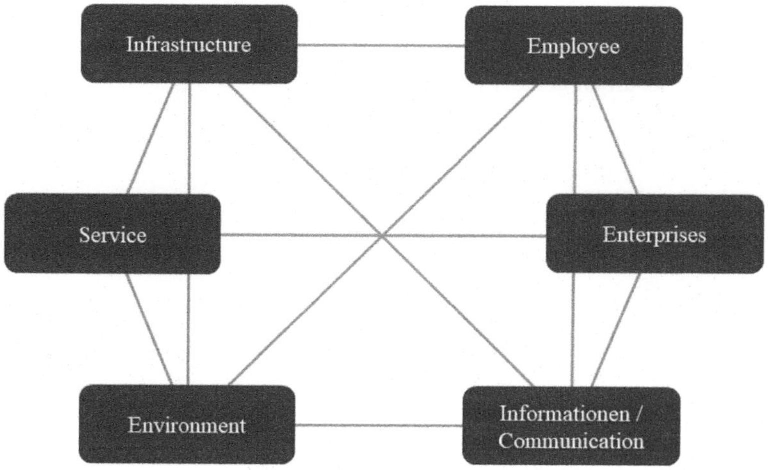

Fig. 3.8 Subsystems of relevance for industrial estates

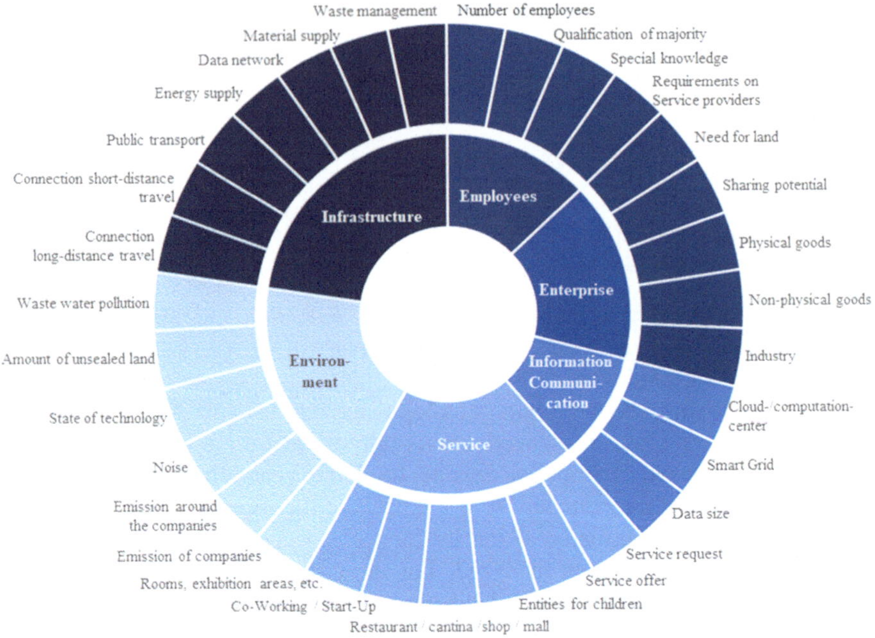

Fig. 3.9 Assessment factors for subsystems

The potential dimensions, which can have an influence on a classification, are manifold. This diversity, however, makes the consolidation and the related relationship difficult. A reduction of the dimensions to the essentials is therefore the goal. For the classification of the commercial space, five dimensions are identified that allow consolidation with sufficient diversity. The dimensions can be described as follows:

- Connection to urban centers (see Fig. 3.10),

Urban cities that are classified as central, suburban regions are the wider surrounding around these centers and rural are all other regions that have increased distances to the centers.

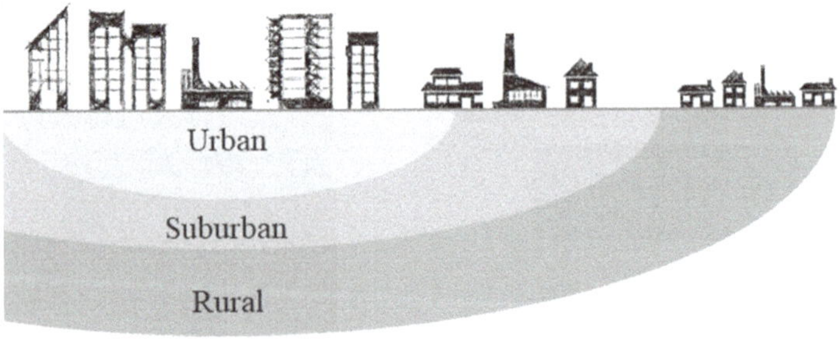

Fig. 3.10 Spatial connection and distance to centers

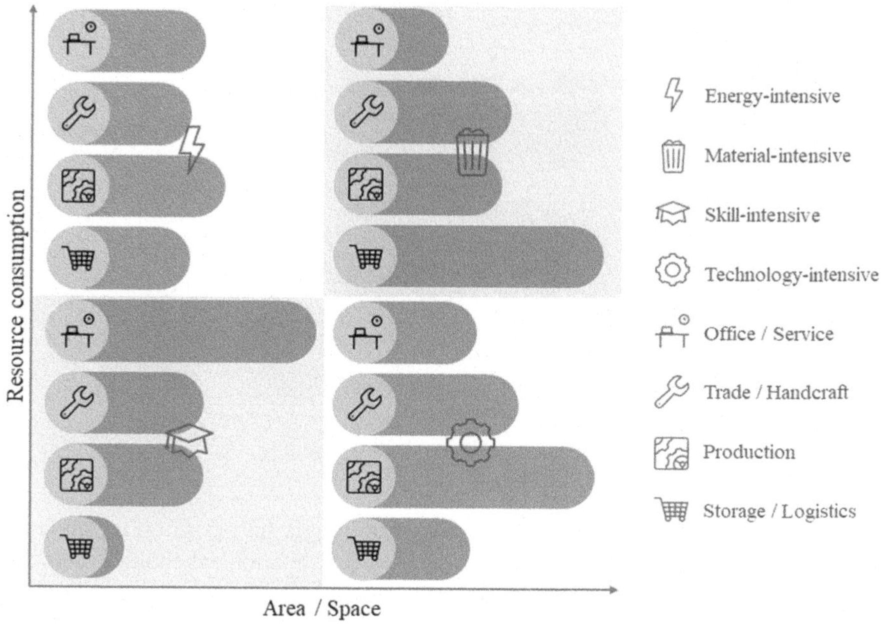

Fig. 3.11 Land use intensity related to the technical-economic classification

- Area utilization intensity (see Fig. 3.11),

The area utilization intensity describes how strongly a limited area is continually loaded by people. There can be many people on a very small area, then the area per employee is very small. If there are only a few people on a large area, the area per employee is large. Consideration is given to the total available net area for the work.

- company types according to technical-economic classification (see Fig. 3.11),

Companies can be classified technically and economically. The processes implemented in the company are considered. If the processes are associated with high energy consumption, the company is energy-intensive. If a large number of raw materials or large quantities of material are processed in the company, material intensive is the respective term. In processes with a high proportion of intellectual input and the need for highly qualified employees, these are knowledge-intensive and/or person-intensive. Processes that are characterized by high technology quality or a large number of systems, technology is intensively assigned.

- Control structure (see Fig. 3.12) and

The two extreme forms of the decentralized control and the central control system exist in the control system itself. In the case of decentralized control, all stakeholders are equally entitled and implement their own goals. One of the stakeholders can be in the de-central structure of the surface developers themselves.

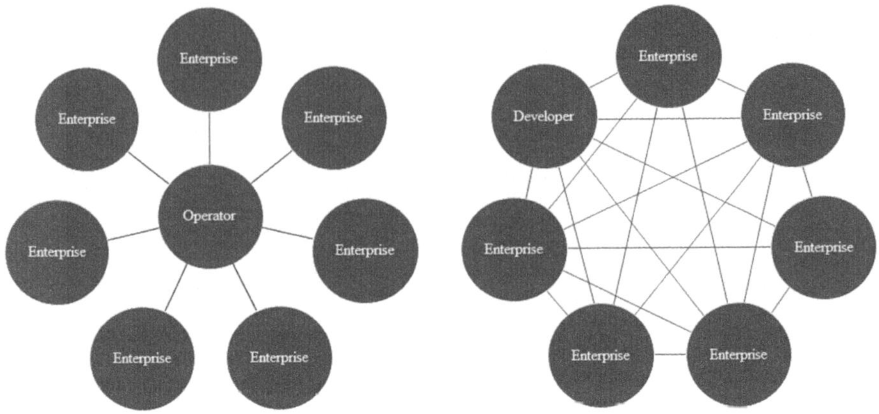

Fig. 3.12 Control structure (left: centralized, right: decentralized)

As a result of the ownership conditions, he is no longer in a position to influence other stakeholders. The central control structure has a superior stakeholder. This can be the owner of the whole area and the other parties have rented their land exclusively or an operator organizes the type of use of the areas in the settlement area. The central control system has the advantage of being able to influence the development stronger. On the other hand, the central organization does not always guarantee the optimal creation of symbiotic relationships, since not all dependencies are always clear and transparent.

- Special features of the supply.

In the simplest case, the commercial areas all have the same offer and offer similar facilities. Only the connection and the access to traffic infrastructure is specific to the location. Independently, there are many commercial settlements, which are distinguished by an additional offer. For example, in some locations, there are non-technical services, such as childcare, catering services or special counseling services. In many cases, however, technical services such as the provision of a computer center, building maintenance and facility management or energy from regenerative energies, infrastructure for electro mobility or energy storage are also highly developed. All these services, which are both centrally organized or provided by individual service providers, increase the supply on site.

Based on an analysis of the commercial settlements in Baden-Württemberg, a classification of settlements has been made. Five basic shapes are possible, which can be further specified with the corresponding design. Each of the basic forms has a certain appearance in the identified six subsystems (Fig. 3.13).

- Office

In the basic form of office, there is a great use of office and service facilities. These areas have a relatively high density of employees. Furthermore, services and

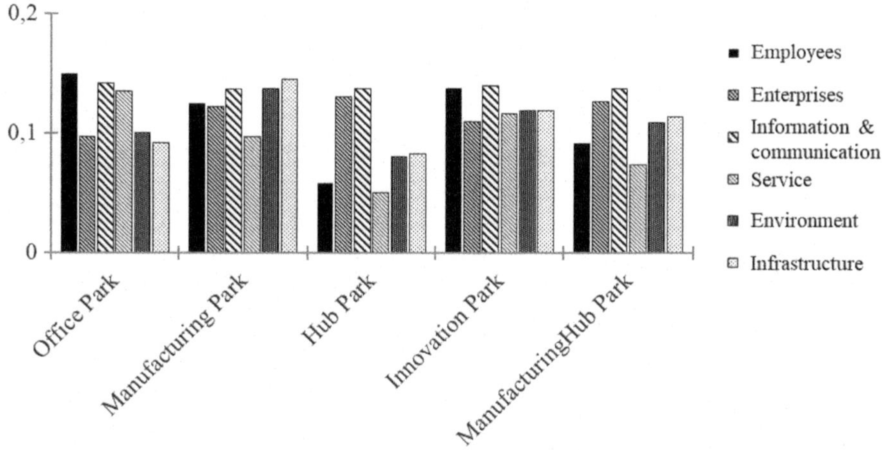

Fig. 3.13 Characteristics of the subsystems in comparison of basic form of industrial estate

services are more in demand and the requirements are higher. In infrastructure, particularly person-related criteria are paramount.

- Manufacturing

Manufacturing subsumes all areas that are increasingly characterized by manufacturing companies. The density of the employees is medium. Complex requirements are placed on the infrastructure, since, on the one hand, the movements of goods and materials must be reconciled with the additional demands on capacity and/or person-related requirements. Possible services are much more complex to identify, as the utilization of individual offers is significantly more demanding.

- Hub

Areas that are increasingly developed for goods handling or the storage of products and goods are defined as a hub in the basic form. The infrastructure of these areas is subject to requirements with regard to the connection and heavy load traffic. Furthermore, the density of employees is relatively low, which also has an impact on the services and services.

- Innovation

In the case of innovation, the use of office and service equipment with use for production and trade use is balanced. Due to the significant use of personnel, there is an increased density of employees. Furthermore, information and communication are a key element.

- ManufacturingHub

A balanced use of space through production and storage such as logistics is available in the ManufacturingHub. It is characterized by a reduced employee density. The requirements for services are therefore reduced. But the increasing amount of information and communication will be strongly identifiable as a result of the increasing digitalization in the areas.

The subdivision or specification of the description can be further expanded based on the basic forms. The view of the control approach allows the division into the park and area (Fig. 3.14). In a park, there is a central organizational structure, e.g. an operator, a private investor or a local authority. This organization provides targets for further development and also leads to developments with its own financial resources. Furthermore, these organizations can promote the attractiveness or the outside perception by creating an additional service offer.

Depending on the availability or availability of additional offers, extended parks or areas are possible. They are characterized by an increased range of services and increased information and communication. The strength of the infra-structure depends on the nature of the offer.

A more pronounced development of the environmental subsystem is characterized by the extension Eco. Corresponding to the nature of the subsystem environment, the sub-area infrastructure (Fig. 3.15) will also be identified.

In the US, the Eco Prelude has established itself as a collection of companies and organizations that have succeeded in improving their performance both environmentally and economically by jointly launching an initiative and co-operating in the management of environmental and resource requirements, They derive double benefits from this as efforts are reduced and the results are better than when individual companies carry out an optimization [23].

The link to local centers and the associated characteristics can also be examined. However, these are already integrated into the analysis of the subsystems infrastructure, employees and companies, which would lead to an additional reinforcement of effects in the sub-areas. This consideration seems to be of little use.

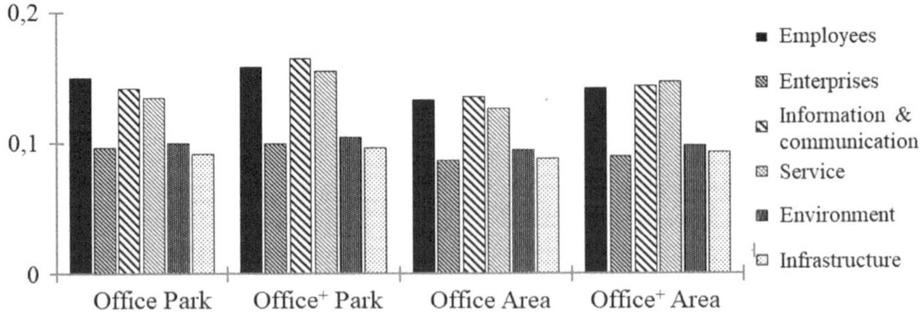

Fig. 3.14 Characteristics with extended viewing (park and area—with and without additional offers)

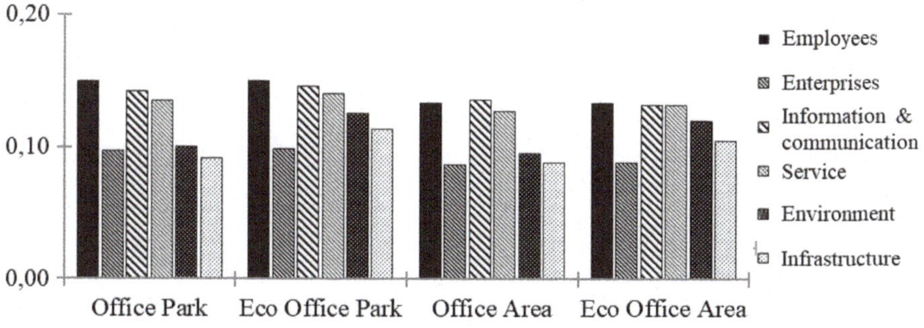

Fig. 3.15 Enhanced character of the environmental subsystem

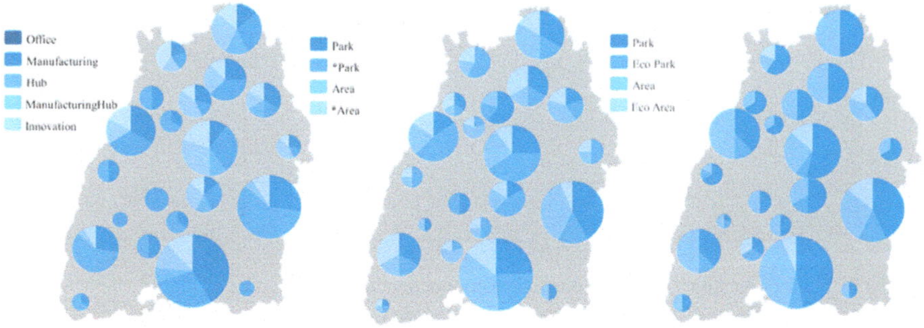

Fig. 3.16 Distribution of industrial estates according to basic form (left), characteristic of service (center) and focus on environment (right)

If the classification of the business areas is applied to Baden-Württemberg, an overview of the distribution in the state can be drawn up. When looking at more than 250 of the 1127 currently available industrial areas in Baden-Württemberg, the wider range of business areas is outstanding (Fig. 3.16). The business areas are consolidated to the upper centers and the counties of Main-Tauber, Schwäbisch Hall, Rottweil, Tuttlingen, Baden-Baden and Rastatt, in order to not be too small in the representation.

Classification on manufacturing enterprises for suitable combinations

Companies are very different, both in organization, products, technology and philosophy. Based on this, classification is difficult. The possible classifications are also different in relation to the viewing perspective.

For the further analysis and classification based on this, a reduction and compression was made on essential factors. The consolidation and summary result in 3 main business levels (see Fig. 3.17). These viewing plans result from the combination of grouped details. The aim of this summary is the reduction of degrees of freedom and the associated complexity in classification. The aim is to achieve a

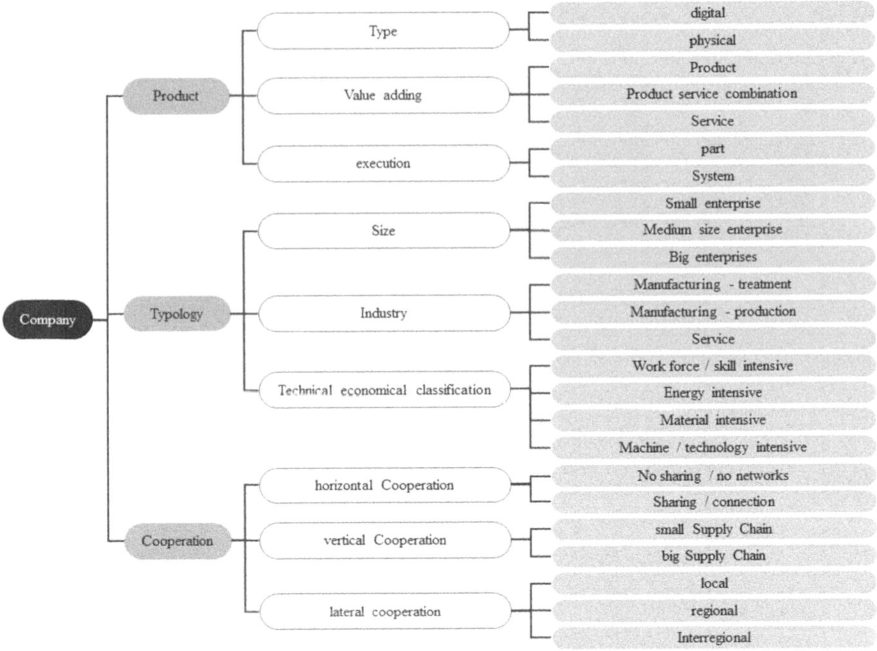

Fig. 3.17 Factors for the classification of companies in different degrees of detail

relatively simple classification of companies, which makes it possible to comprehend the breadth of the diversity of companies completely without simplifying them too much.

By analyzing and consolidating the factors for the classification of companies in the context of resource-efficient location development, the following company classification is conceivable:

- Tech Industries

Companies of the class Tech Industries produce complex products and product service combinations, so-called systems. The products can be digital as well as physical. The companies are more likely to be located in later stages of the supply chain and access the corresponding suppliers, which is why they can be found in the industry as processing companies. The technical-economic classification is characterized by system-intensive or technology-intensive processes.

Examples for companies in Baden-Württemberg: Agilent Technologies GmbH, Robert Bosch GmbH, [24].

- Heavy Industries

If individual parts or few complex components are produced by means of energy-intensive processes, these companies can be classified as heavy industries.

The products are mainly physical and are rarely combined with a service. Companies are more likely to be allocated to the processing companies.

Examples of companies in Baden-Württemberg: Albert Handtmann Metallguss GmbH, Hornschuh GmbH [24].

- Light Industries

If individual parts or few complex components are produced as material-intensive processes, the companies can be classified as light industries. The products are predominantly physical and in the case of services, these are rarely intended as an extension of the service offer.

Examples of companies in Baden-Württemberg: Amcor Flexibles Singen GmbH, Carl Edelmann GmbH, [24].

- Service Organizations

Service organizations are service companies that offer both digital and physical services. Services are classified as knowledge-intensive.

Examples of companies in Baden-Württemberg: Bertrandt AG, SCHOLPP AG [24].

- Transportation Organizations

Similar to service organizations, transportation organizations are service companies. They mainly offer physical services, which are usually classified as material intensive.

Examples for companies in Baden-Württemberg: Horst Mosolf GmbH & Co. KG, Müller - Die lila Logistik AG, Simon Hegele Company for Logistics and Service GmbH [24].

- Digital Organizations

If the company's product is digital and is often extended by a service, the company can be classified as a digital organization. The products are usually complex and the processes can be described as technology-intensive.

Examples of companies in Baden-Württemberg: Bechtle AG, Fiducia IT AG, RAFI GmbH & Co. KG, SAP AG [24].

- Innovative Organization

Innovative organizations produce products with a high percentage of service, mostly as a product service combination. The processes in the company are knowledge-intensive. Depending on whether the product is digital or physical, Innovative Organizations can be described as processing companies or service companies.

Examples for companies in Baden-Württemberg: Fraunhofer Institutes, Ensinger GmbH, HP Enterprise, IBM Germany Research & Development GmbH [24].

- Mixed Organizations

Companies which are not clearly assignable to one of the aforementioned categories are classified as mixed organizations.

Creation of congruence models

The Congruence model has been postulated for the first time by Nadler and Tushman [21] in the context of organizational modeling for management developments. In the congruence model, input and output parameters exist, as it is common in many currently used control models. Furthermore, the model is completed by control and disturbance variables. This approach is ideally suited for system thinking, since control and control systems are also used for systems.

As already mentioned before, two congruence model approaches are necessary since companies differ significantly from the settlement area. Therefore, the two possible approaches are discussed in detail below.

Congruence model for companies with focus on resources

The content of the four parameter fields must be defined by a manufacturing company (see Fig. 3.18).

A strategy (A) is important for a company because it serves as a guideline for controlling and controlling the company. The vision of long-term goals as well as the management of short-term steering play an important role in the company's strategy.

In order to enable the processes in the company, input values or an input (B) are required. For a manufacturing company, both measurable and non-measurable inputs are important. The measurable quantities include all raw materials, auxiliary

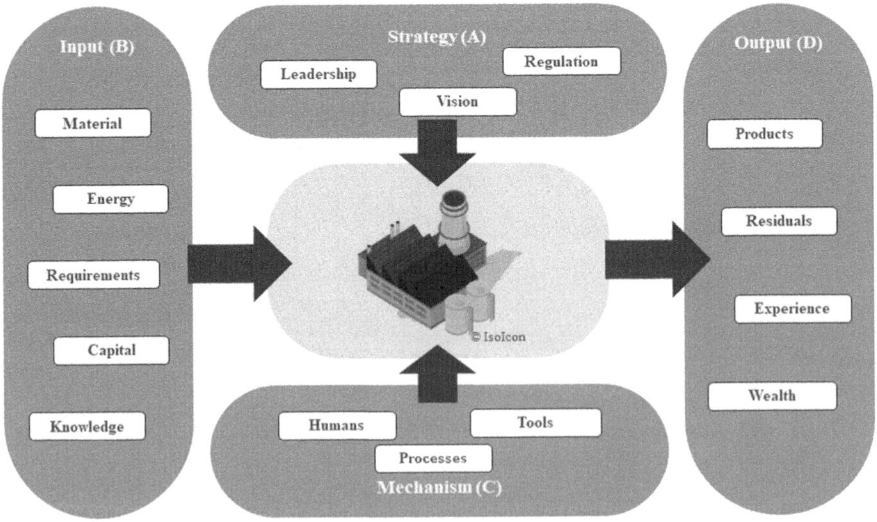

Fig. 3.18 Approach to a congruence model of a manufacturing company

materials, purchased parts—together the material. Energy is needed for all processes, especially in the production process. In a company, capital is always important in order to make investments or simply to bridge times with low capacity utilization. Non-measurable inputs are the knowledge of the organization, because without the appropriate knowledge it is not possible to produce products according to defined requirements.

There are mechanisms (C) in the company that influence the production process. The most important parts with influence on manufacture are the people, processes and available tools. If a mechanism falls behind a defined state, this always has negative effects on production and must be compensated by control.

According to classical understanding, an enterprise always has the goal of creating a capital increase through value creation [25]. The result of the added value is the product of the company as part of the company output and the output (D). In addition to the product, there are also residual substances, which are produced as a by-product in value creation. A less measurable output is the increase in experience in the organization, e.g. Employees, executive staff and corresponding units. In addition, prosperity is created for those involved in production.

In the course of resource efficiency, the input of the production company has to be optimized with appropriate approaches in the strategy and adapted use of mechanisms. Measurable input variables should be reduced and non-measurable quantities should be increased according to existing requirements.

The output should also be optimized. In doing so, the relationship between product and residual material is to be transformed in favor of the product.

Congruence models for commercial estates with respect of resource efficiency

Business settlements are very different from the established companies in objectives and systemic logic. For this reason, the Congruence model for an enterprise is not similar to the company. There are again the four parameter fields. However, the concrete content is different (see Fig. 3.19).

In the strategy (A) of the company settlement regulations, which are derived from laws, guidelines and guidelines are very rigid and strictly observable variables. In addition to the regulation, the motivation for the company's establishment is an important element. Similar to the company vision, long-term goals are to be achieved which are usually defined by the control authority. They have a great influence on the desired output.

The input (B) is a sum of the inputs of the companies located. Energy and materials, which are an important input for enterprises, are direct input of the settlement. In the case of a non-symbiotic business establishment, these variables can simply be combined as the sum of all the requirements of the companies. In addition, companies' knowledge is also a resource. This cannot be described by purely mathematical methods, whereby the exploitation of distributed knowledge offers enormous potential for the collective [26]. The situation is similar with the people who are employed in the company. They are a resource and a demand carrier and thus an important input in the system of company settlement.

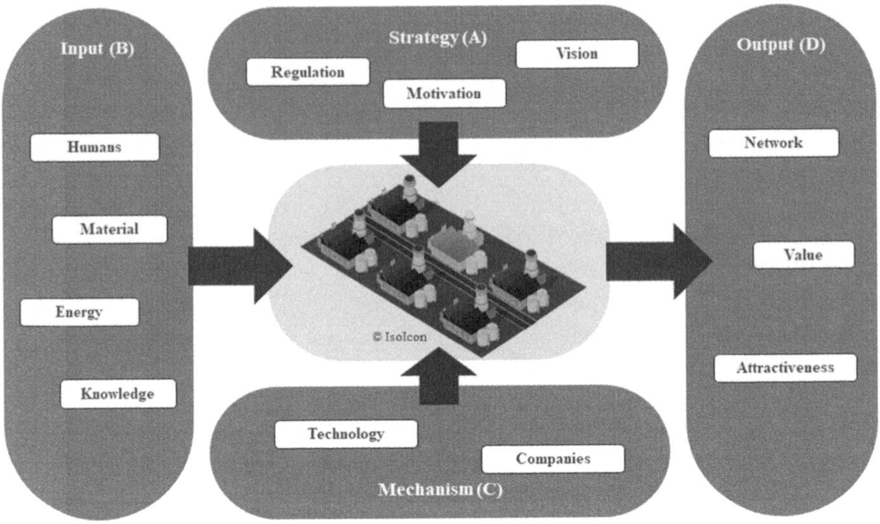

Fig. 3.19 Approach for a congruence model of industrial estates

The mechanisms HOW consist of the companies established and the technologies used. Because dependencies and interactions are established, the system is not balanced accordingly in the event of a company's failure. All relationships must then be analyzed again and appropriate adaptation must be made. Similarly, if a company replaces the established technology with another one, the resulting change in processes can significantly change the demands on energy and materials, which also entails a new analysis and adjustment e.g. usage of additive manufacturing.

The output (D) has only little to do with the congruence model of manufacturing companies. Because other resources are more important under the focus of resource efficiency. Thus an optimal symbiotic net value is striven for. The better the coordination of the network partners fits, the more successfully efficiency strategies can be implemented and sustained in the long term. The values that generates a settlement company as a result, on the one hand carries the result of the corporate value along the other hand, the perception/image as created value may be used. The increase in the attractiveness of the site is to be mentioned as a self-serving goal of the company's settlement. After all, a suitable successor can be quickly found if a network partner fails. Furthermore, the chances are increased to identify a suitable network partner and to settle in order to advance to a further increase of the other output factors by increasing the potential prospects.

Development of a holistic model for the system

Based on the developed congruence models, the development of a holistic interaction model for the system of company settlement is structured (Fig. 3.20). A combination of the various individual models is helpful as a preliminary step towards a holistic model. Because the congruence models according to the chapter

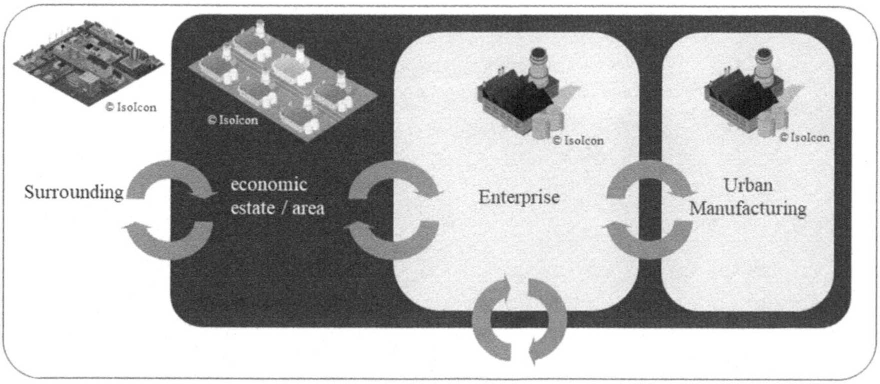

Fig. 3.20 Interaction links within the system view

before are created for all companies in the existing company resettlement or the potential network partners in new plans, input and output are all easily connected. In addition, the generic models are used to create a concept that is specific to each company. On this basis, a qualified evaluation can be carried out.

All relationships between the individual units of observation (participants) are documented and analyzed with regard to their active connection. The first step is a qualitative analysis. The qualitative analysis reveals whether the active relationship has a reinforcing or a diminishing effect. After the qualitative evaluation is complete, the quantification has to be carried out. The quantification is necessary in order to have a mathematical model for a simulation. The active relationships are described according to the qualitative evaluation by means of a mathematical relationship. The set of all mathematical descriptions of the active relationships yields a calculation approach for each variable in the system.

3.1.2.3 Data Acquisition and Information Gathering

As it is apparent from the foregoing, the data and information required are varied and complex. Irrespective of these hurdles, it is imperative to use high-quality data, as this is the only way to make long-term reliable statements. This is important in order to formulate and manage strategic developments. In addition, it is relevant to be up-to-date with the current information in order to always be able to work with an actual and thus realistic data base.

For the data and information as well as their origin, two viewing directions are possible. The "top-down" approach is based on the viewpoint of the business owner. This is almost exclusively to restrict access to publicly accessible data sources opened by licensing. But already from these data sources a large number of information can be aggregated. The "bottom-up" approach is taken by the company as a participant in the system. Here, the amount of potentially available data and

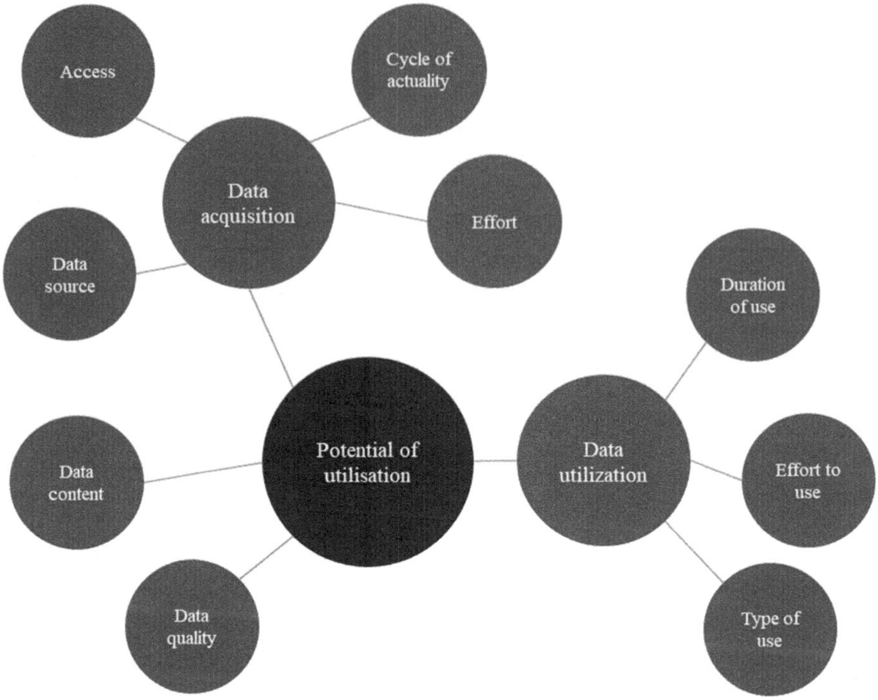

Fig. 3.21 Factors influencing the potential use of data and information

information is much more complex because there may already be close links and intermeshing. Furthermore, shares of the information are relevant to the competition and should not be directly visible to all. This increases the complexity of using and providing the available data.

There are various factors to be considered in the data to be used. These factors have a major influence on the usefulness of the data and information (see Fig. 3.21). For before the data can be used effectively, these must be collected or acquired, followed by a processing. If the resulting cost-benefit ratio is unfavorable, a check for reduction is useful [27].

3.1.2.4 Generic Approach for System Modelling

System modeling is carried out using system dynamics. Variable variables are defined for this purpose. For system dynamic modeling, the viewing space must be defined. According to the preceding considerations, consideration of the subsystems is suitable for this purpose. System dynamics are characterized by closed loop diagrams (see Fig. 3.22). The advantage of this representation is that all relationships between the viewing variables are directly visible. To create closed-loop

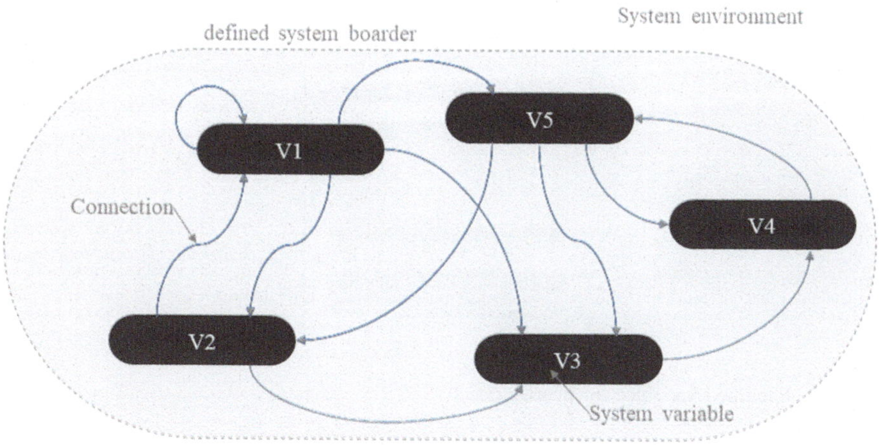

Fig. 3.22 Approach to system modeling [24]

diagrams, define the frame. Through the system boundary, the framework determines the viewing space, which is delineated by the system environment. The interaction between system and system environment takes place through a limited defined exchange. Regardless of these system input and system output variables, interactions are not considered in the system dynamics view. The analysis of the system variables as well as the established links is carried out in the system view.

The strengths and weaknesses of system dynamics are also associated with this. Strengths can be seen in the complexity of interactions. Large quantities of relationships can be described [27]. Furthermore, the system dynamics allow conclusions to be drawn about the backflow of information or the reaction. This allows dynamic dependencies to be assessed both in terms of time and importance. The weaknesses of system dynamics include the limited precision of the calculation results. This correlates with the exactness of the mathematical description of links. Since every relationship involves a number of unknowns, the results of system dynamics are difficult to make use of for a reliable statement of reality in absolute size. It is, however, very suitable for the estimation of trends and developments [28].

The analysis of the subsystems identifies the variables for the evaluation of interactions (see Fig. 3.23). This is the output information to derive the relationships. Each identified variable has been examined for connection to all other variables.

In interaction modeling, the effect is always analyzed, starting from a change in the factor. In the first approximation, the relationship is assessed only qualitatively. If factor A increases, the effect on factor B is positive, this results in an increase of factor B. If the influence is negative, an increase of factor A leads to a reduction of factor B. In order to obtain a complete system model, all factors are investigated

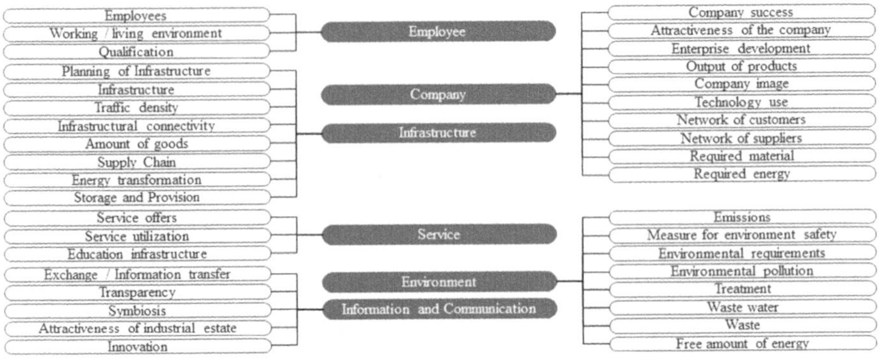

Fig. 3.23 Identified variables for interactions

with respect to each other. If all relationships are identified and evaluated, a visualization can be created. This presentation allows a quick understanding of the interactions and dependencies.

If the holistic model is created, the qualitative representation must be quantified. Only through the quantification simulation results can be calculated which allow to directly compare different scenarios.

An example of quantification can be found below. A simple example is used for the concrete representation. An increase in traffic density leads to an increase in emissions. This statement is comprehensible under current circumstances, that the multiplicity of transport emissions generate emissions, as more means of transport lead to more emissions [29]. Now the mathematical link must be established in order to create a calculation basis (Fig. 3.24).

An adaptation of the generic interaction approach (see Fig. 3.25) is made by specifying the interaction and linking to quantifiable dependencies. Furthermore, the company's image can be reproduced by replicating the company-specific variables in a generic approach. In particular, the networks of the various companies are focused on achieving improved networking and synergies.

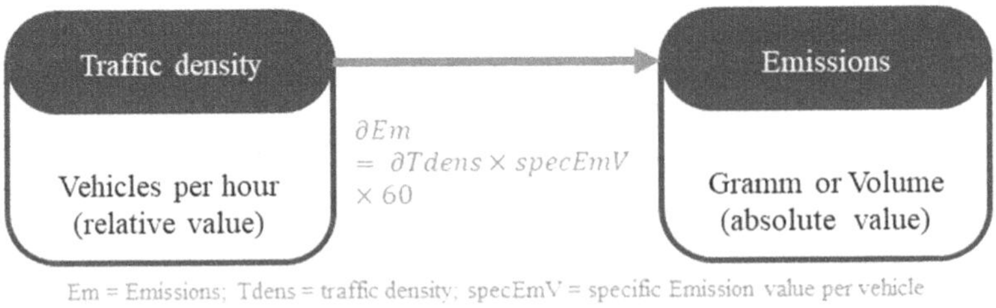

Fig. 3.24 Example for quantification of a relationship from the interaction model

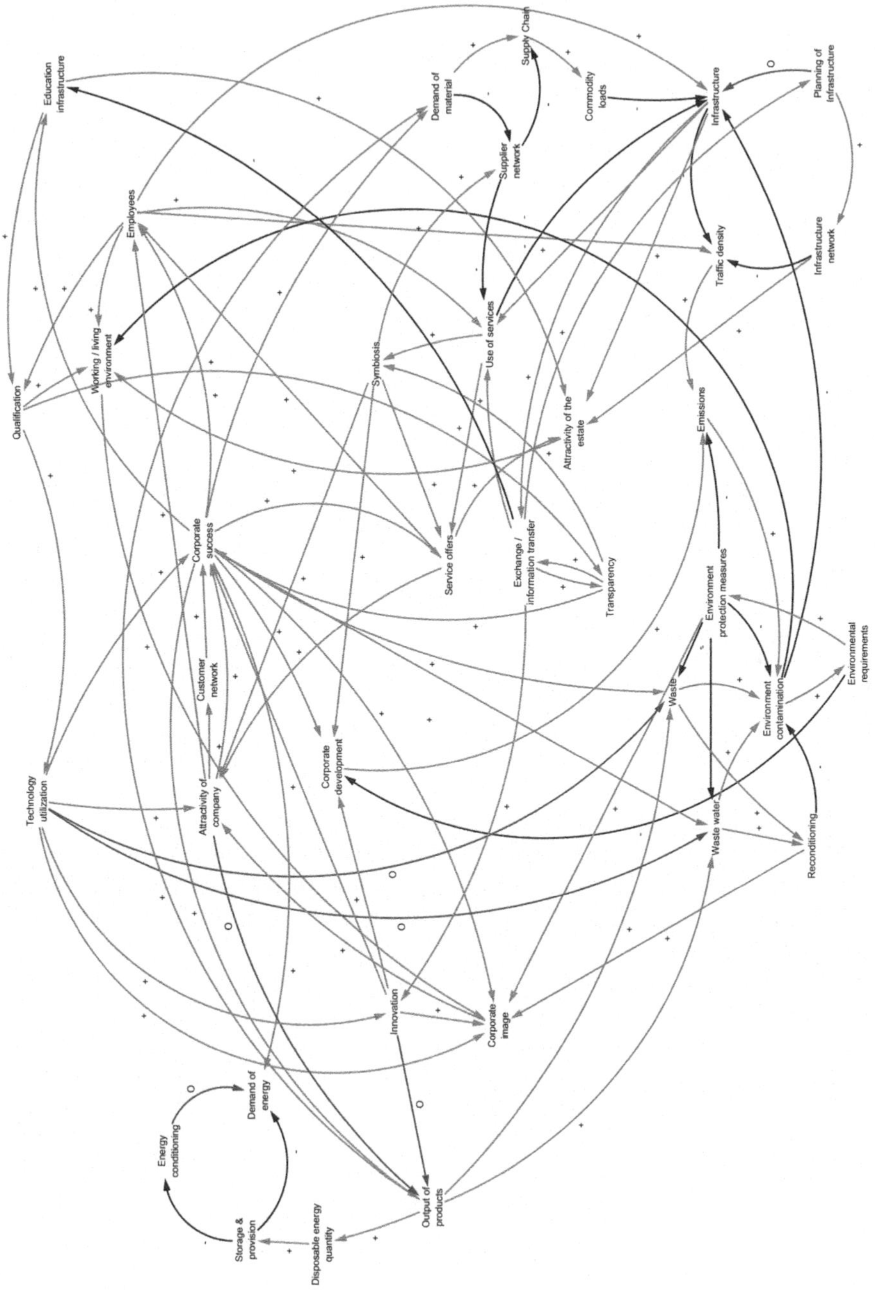

Fig. 3.25 System dynamic network for holistic sustainable development

3.1.3 Evaluation Approach for Long-Term Resource Efficiency

Using the activities described lastly, a status quo analysis needs to be carried out for the environment to be considered. This survey of the current situation is important in order to develop good-looking, meaningful scenarios. For a meaningful assessment, the distinction between existing development and the new settlement is necessary. If the new business development is planned again, the flexibility in the selection of potential interaction partners as well as the appropriate requirements are significantly higher than the development of existing areas.

Two objectives can be pursued, which are defined by the respective boundary conditions: stock development and the new settlement.

Inventory development is significantly characterized by grown structures. In particular, existing links and connections must be taken into account. In most cases, the development of the stock is about the resettlement of one or fewer companies on areas with a lot of space or a building complex. Rarely, the existing structures meet the requirements for the new use. This has to be considered, with all its advantages and disadvantages.

3.1.3.1 Evaluation of Classification Levels

Resettlement offers the developer or future operator all degrees of freedom during the development process. This also entails the requirements that the future operator will set up for collaboration and data provision for holistic control. In the case of resettlement, the development of the settlement can be divided into individual steps. The step-by-step construction always leads to situations that are similar to the development of the stock. Accordingly, the procedure for inventory development is significant for many situations.

In the case of a resettlement, but also in the case of integration into an existing commercial area, the extent to which the company is suited to the boundary conditions at the location and to the other companies must be examined. Here, statements early on in the decision-making process on the basis of classification data are useful and desirable.

If the classifications of the commercial area and the company can be carried out in a first step, a profile matching can be carried out.

By matching commercial space and companies on the basis of qualitative characteristics, it is determined how well a company fits into the commercial area. This helps the companies in the pre-selection as well as the municipal side and the management of the commercial area. Table 3.1 shows how the matching can be performed using a simple table as a tool. Such a profile can also be used for communication and negotiation between companies and local authorities or operators. Differences in supply and demand or demand become obvious and can be negotiated efficiently.

Table 3.1 Matching table between industrial estate and company

Part of settlement area	Evaluation factors, and other	Offer Industrial estate	Needs Companies	Matching
Infrastructure	Data infrastructure (Band width internet)	Glass fiber G5 test area	G5 Test area	1
	Waste management	Industrial hazardous waste treatment company locally	No special requirements	1
	Energy supply	District heating/ electricity	Gas	0
Employees	Number of employees	Rural area 4000 inhabitants at location, 80,000 inhabitants in radius of 10 km	1400 300 employees specialized in electronics	0
	Specialized expert knowledge	Vocational schools of all industries (5 km) Universities of applied science (15 km)	Experts electronics	1
Companies	Need of land	Parking spaces, trans-shipment center	Parking spaces	1
	Sharing potential	Vehicle fleet, Cantina, parking space	Own provision	0
	Non-material products	Yes/no	Yes	0
….	…			
Environment	Waste water pollution	Regulation nature protection area	Not applicable	0
	State of technology	Fully automated/ manually	Manually	0
	Noise	Regulation sleeping times	Day shift	1
0—bad matching 1—well matching				5

In addition to the coordination between companies and the infrastructure, the question arises of how well companies fit in a commercial area. This is true both in the sense of synergies through commonalities and symbiosis through differences. The matching between companies gives a first indication of the similarity of the companies. It is assumed that similar companies, through their similarities, also create prerequisites for synergies. A possible summarization of the factors in companies A and B can be found in Table 3.2.

Table 3.2 Matching of companies

	Factors	Enterprise A	Enterprise B	Matching
Product	Type	Digital/physical	Digital/physical	1
	Value adding	Part/system	Part/system	0
	Execution	Part/system	Part/system	1
Typology	Size	Small/medium/big	Small/medium/big	1
	Industry	Treatment/ production/service	Treatment/ production/service	1
	Technical economical classification	Personnel/energy/ material/machines	Personnel/energy/ material/machines	1
Cooperation	Horizontal	Sharing/no sharing	Sharing/no sharing	1
	Vertical	Small/big supply chain	Small/big supply chain	0
	Lateral	Local/regional/ interregional	Local/regional/ interregional	1
0—bad matching 1—well matching				7

Table 3.3 Similarity matrix of the companies A–E

Enterprise	A	B	C	D	E	Sum
A	–	7	6	3	5	21
B	7	–	9	2	6	24
C	6	9	–	8	5	28
D	3	2	8	–	7	20
E	5	6	5	7	–	23

The results of the matching of companies can be summarized and compared in tabular form. The sum of the matching values per company can be used to determine synergistic potentials. Table 3.3 shows an example of such a similarity matrix.

At this stage, companies that do not fit into the commercial zone can be excluded. However, the similarity between companies alone is not sufficient to derive conclusions about the quality and degree of possible symbioses. Often, different companies complement each other with regard to resource efficiency, since they are not competing for scarce resources. This can also be formally mapped in a matching matrix in which similarity or symbiotic matching is highly valued for certain criteria. Such algorithms are often dependent on the order of evaluation. Typically, most recently rated companies have disadvantages with limited resource criteria.

3.1.3.2 Assessment of Congruence Models

The data obtained is used in the development of the congruence models. To this end, the input variables and the output variables for companies and commercial settlements are firstly defined. In addition, the company strategies as well as the strategy of the business area developers are described and the mediating resources designated as "mechanisms", are characterized.

It is a good idea to create congruence models based on templates, and then join them together to get a good base for an integrated congruence model. Figure 3.26 shows such an integrated model. In doing so, the inputs and outputs of companies, their strategies and mechanisms/resources are brought together.

The integrated congruence model reveals whether the initial strategy of the business area is in harmony with the company's strategy and whether the understanding of mechanisms, inputs and outputs is covered. It is possible to identify contradictions in the system that go beyond the pairwise matching on classification level. In the present example, the sustainability strategies of the commercial region and of the three companies considered fit together well. This is also reflected at the operational level in the mechanisms.

Figure 3.26 also shows that the views of the companies can differ from the viewpoint of the operator of a commercial establishment or an economic developer. Thus, the increase in purchasing power in the region is an important argument for him, whereas this does not play an important role in the company's view.

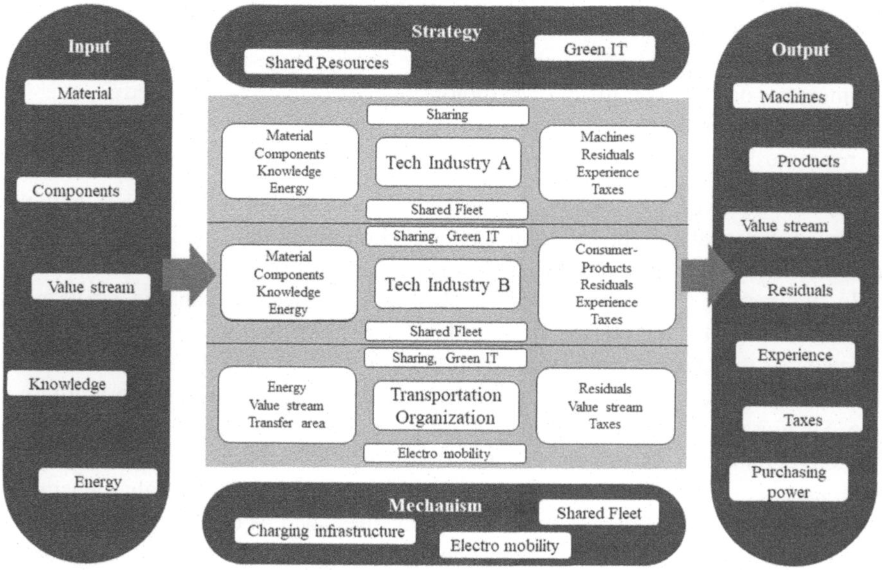

Fig. 3.26 Integrated Congruence model of an industrial estate

Important for the creation of congruence models is their specificity and meaningfulness for the respective situation and application. In this way, machines, consumer products and the flow of goods from the »Transportation Organization« can be combined again into the material output of the commercial area. The separate presentation, however, says more about the specific characteristics of the commercial settlement. The companies "Tech Industry A", "Tech Industry B" and "Transportation Organization", which are generically named here, should also be replaced by speaking names in the specific case. Here, company names can be found, or »mechanical engineering«, »manufacturers of consumer electronics« and »commodity distribution centers«.

3.1.3.3 Assessment of System Models

On the basis of the integrated congruence model, the input and output variables can also be operationalized and an estimation of magnitudes can be carried out. Thus, there are indications of bottlenecks. In the above example, a bottleneck could be assumed by the required load capacities for electric vehicles in the distribution network and the shared vehicle fleet. Are the transport capacities sufficient for the intensive inflow and outflow of the goods capacities in the distribution center? Does the environment of the commercial area provide enough technical personnel?

Typical questions for evaluation at the level of congruence models are:

- Comparison of input and output variables for network analysis (informational, physical). What input does a company need from other companies? What output does the company provide to other companies?
- Are company strategies compatible? Are there any conflicts of interest and interests, or can synergistic or symbiotic mechanisms be developed?
- Can the mechanisms in companies be reconciled? Are there potentials for shared resources?

On the basis of the understanding gained with the profile analysis and the congruence models qualitative and later quantitative system models of the commercial settlement can be created.

In contrast to Congruence models, this is not the case with the creation of individual models. As shown in Fig. 3.27, the modeling of key variables that characterize the commercialization (Start A) proceeds. Partially, they are extracted from the integrated congruence model. In the beginning, the most important target size, Such as the number of companies established or the number of employees in the commercial area. Further key variables can then be ecological balance variables such as energy consumption or traffic volumes. This can not only be simply quantifiable variables but also more abstract values such as company development or symbiosis potentials. However, in contrast to the representation of the congruence models, it is necessary that the variables for system modeling increase or decrease. System models are displayed as graphs. The key variables are written to a

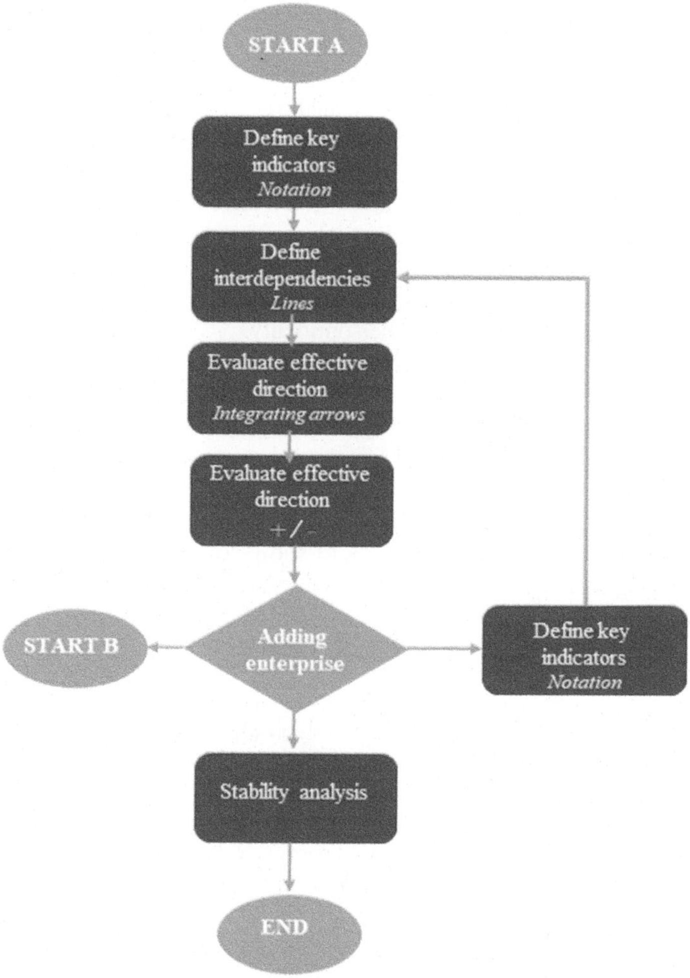

Fig. 3.27 Qualitative system modeling and evaluation

block diagram. Subsequently, the variables which influence each other are connected to one another. The direction of the interaction between the variables is also determined. For example, the number of established companies may have an impact on the attractiveness of the commercial area and not vice versa. Finally, the effect direction must be specified. Thus, the number of companies being established will have a positive effect, i.e. steadily, on the attractiveness of the industrial area.

It is recommended to start with the key variables from the conception of the commercial area and to network them. This model should include balance figures that also reflect company interests and needs. Subsequently, additional key sizes of the individual companies can be added to improve the specificity of the model and to incorporate the respective company interests into the model. In addition, all the companies involved are added to the modeling (Start B), the scheme being retained.

System models are typically no longer created on paper. There is relevant software for this. This also allows the analysis of qualitative system models. A dependency diagram shows how strongly variables can be influenced and whether they have a strong influence on the other. With this analysis, levers can be identified, in particular. These are variables which have a strong influence on others without being strongly influenced by themselves. These are the driving forces in the commercial settlement. Both the individual companies and the developers of the commercial area have their levers, which should be clear to all parties involved in order to drive the entire system together optimally.

In addition, the qualitative system models can be used to perform stability analyzes. Software-based systems for system modeling can identify amplifying circuits that promote desired growth (Fig. 3.28) but also show limitations of growth (Fig. 3.29).

However, the use of qualitative system models is limited. They do not allow the temporal development of the variables to be projected. This requires the functional relationships between the sizes. Figure 3.30 shows the procedure for quantitative system modeling. Non-quantifiable variables are either to be eliminated or replaced by quantifiable variables. The functional relationships between the variables are then to be defined as well as the time units used.

Fig. 3.28 Increasing feedback

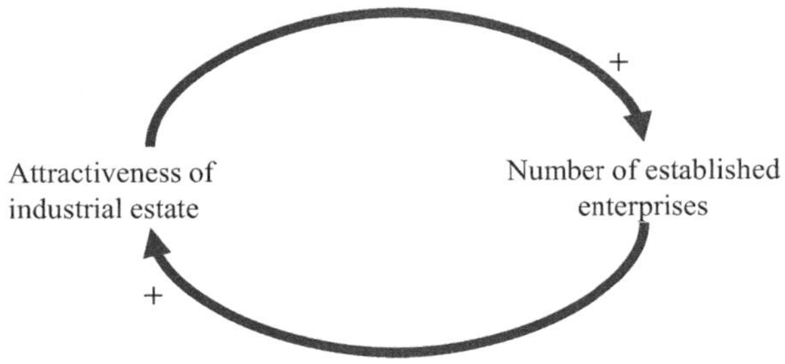

Fig. 3.29 Self-stabilizing feedback

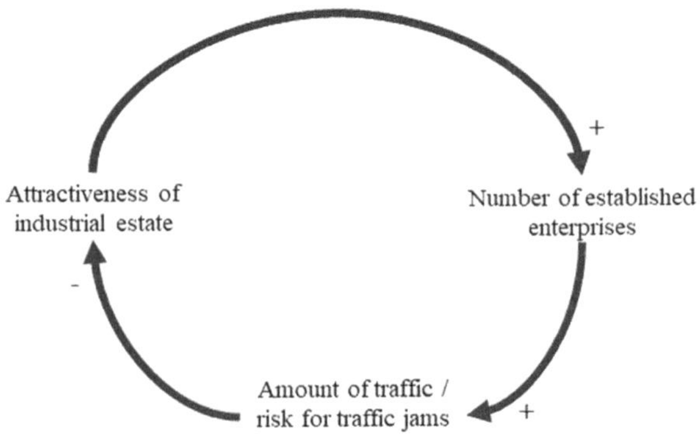

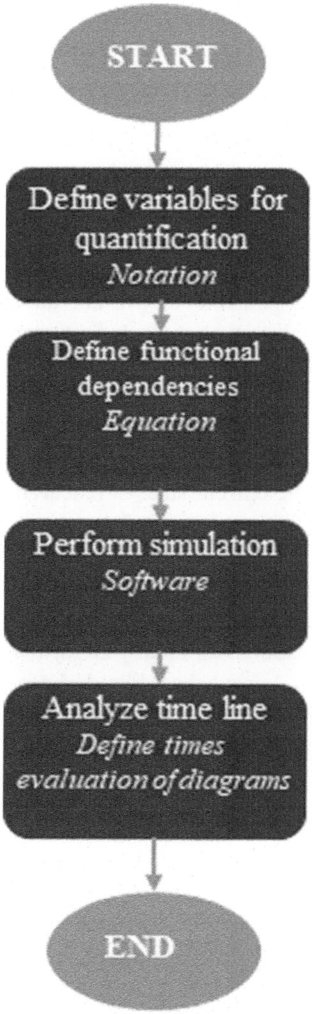

Fig. 3.30 Quantitative system modeling and evaluation

The system-dynamic evaluation can then be carried out with a corresponding software. The typical result of the analysis is time series of the key variables or target variables of the commercial settlement. They are interested in the development of the size of the companies, the number of workers employed in the settlement, and the tax revenue that can be achieved therewith. The same applies to the balance of sustainability, such as traffic volumes, energy consumption, water consumption, environmental pollution, etc. The individual sizes can also be broken down into the corresponding companies.

While qualitative models can also be created and analyzed at a higher level by a developer with basic knowledge in system modeling, dynamic system models require greater effort and are typically created by a team of specialist and specialist planners.

3.1.4 Summary and Outlook

Within the framework, it is possible to present an approach to the creation of resource-oriented commercial settlements. A first step towards establishing symbiosis with respect to resource efficiency can be presented through a correspondingly described procedure. With the method presented above, commercial settlements with narrow control logic should be able to establish symbiosis resulting in resource efficiency.

The concept is based on the current state of research and already implemented in several projects. The idea of the symbiosis is not new. The symbiosis has already been postulated in 2012 as the basis for an urban-focused production [30]. Even in these scientific treatises the symbiosis predicts a reduction of effort. Today, some exemplary approaches [31, 32] have been implemented, but the created connections are less for the creation of resource efficiency as for the compatibility of production in the context of urbanity. Approaches like Urban manufacturing can play an important role for the further development and implementation of sustainable procedures in manufacturing enterprises.

In the future, it is required to adapt the approach to different industries. With an industry specific model and approach the development will be even increased.

3.2 Strategic Approach for Enterprise Improvement

3.2.1 Introduction

Current trend is the individualization of products. This provides the challenge of smaller lot sizes and high numbers of variants in final assembly of companies. Parallel the rural areas where plenty of world leading manufacturing companies are situated in Germany, a lack of qualified personnel can be stated. Reasons for that are multiple.

Facing these challenges the following approach is promising to address the question occurring. A general approach is the differentiation between customer-neutral manufacturing and customer-specific individualization. Therefore, an approach could be to keep centralized manufacturing of costumer-neutral pre-products and to decentralize the customer-specific production steps, e.g. final assembly, into smaller units which are close to the customers and near to hubs of qualified personnel, like in urban surroundings. To support companies in the transition towards the resulting urban manufacturing, in this contribution, a development path is introduced. For this, at first urban manufacturing is characterized, then the concept of development paths is highlighted. Based on this, perspectives for a development path towards urban manufacturing are described and a generic development path to support the respective transition is presented. The contribution ends with a summary and an outlook to future work.

3.2.2 Urban Manufacturing

Urban manufacturing can be seen as an approach to tackle current challenges for industrial companies like the need for more flexibility and labor shortage [33]. Urban manufacturing is a manufacturing approach, which is so compatible with its surroundings that it may even take place in urban surroundings and furthermore results in advantages for all related parties [34]. Thereby, its application area is especially products with a high degree of customization, a high ratio of value to volume, and which require knowledge-based processes in production. So, branches like heavy industries are not seen in the focus of urban manufacturing. Obviously, the request to be city-compatible results in demands on a factory, based on potential disadvantages of manufacturing in urban surroundings (cf. [33]).

An essential disadvantage of urban locations is the low availability of areas, which tend to be expensive. Consequently, to be city-compatible, factories have to have a low land consumption. This results in the demand for products with low dimensions, which are produced in low quantities and which have short dwell times in the site. Furthermore, to be city-compatible, manufacturing has to produce a low amount of emissions. In addition, as traffic is mostly heavy in cities, logistics should be low to keep the impact of the factory on the traffic volume low, thereby not worsening the traffic situation. Thereby logistics not only concerns the transport of goods, like material and products but also of humans, i.e. employees and customers.

Essential advantages of urban manufacturing are based on the premise that urban manufacturing results in short distances to the markets. Proximity to the resourcing market supports the collaboration and cooperation with suppliers for goods and services like research and development as well as training. Short distances to the output market reduce efforts for distribution logistics and facilitate the integration and bonding of customers.

Furthermore, short distances between housing and workplaces of employees support flexibility in working times and therefore in production capacity and work-life-integration.

3.2.3 Concept of a Development Path

Typically, companies will not realize urban manufacturing from scratch, but a respective transition process is needed. This process has to lead from individual starting points to company-specific target states. To guide the transition process, we suggest the concept of a generic development path which can be applied to different kinds of companies. Thereby, the development path considers essential perspectives of the development of a company towards urban manufacturing and is based on a set of development steps for the different perspectives. Interpreted as a process, the development may be run through completely from initial state to target state or in a company-specific range. By this interpretation of the term development path we follow the work of Lopez et al. [35] but are in contrast to Perkins and Small [36]. Lopez introduced a Digital Business Development Path to support companies in assessing their position related to digitalization. In this development path, the state of a company can be characterized by 5 perspectives with 6 development steps. Here, it is not needed that the steps have to be passed one after the other. Perkins and Small presented a Product Development Path for Microbial Diagnostics consisting of 7 steps in 3 perspectives. For advancing on this development path, it is needed to successively go through the steps one after the other without skipping one or more steps. Furthermore, it is needed to run through the complete development path from the starting point to the final stage.

So, in the following, development paths and maturity models are seen as instruments to guide the transition of an object, e.g. a factory, from an initial state to a desired state along a series of steps or stages, respectively. Thereby the starting point of the transition is to be determined based on an as-is characterization of the object and is not necessarily at the initial stages of the development path, but the object may already have a specific maturity. During the advancement, no step or stage should be skipped completely, but all respective characteristics have to be fulfilled, which may be the case for several steps at once. Thereby we see the difference between a development path and a maturity model in the fact that in a maturity model along the steps more perspectives and more characteristics or criteria, respectively, have to be fulfilled, meanwhile the amount of perspectives and criteria along the transition stays constant along the transition for a development path.

3.2.4 Development Path for Urban Manufacturing

3.2.4.1 Structure

The development path for urban manufacturing is structured into perspectives, covering essential characteristics: the factory (representing the overall manufacturing distribution approach), the product, manufacturing and its organization, sales, the human side consisting of employees and citizens as well as of the cooperation perspective. Therefore, in the development path emphasis is given to leveraging the advantages of urban manufacturing based on short distances to the output and resourcing markets as well as for the employees. Along the 7 perspectives, the development path is divided into 5 steps including the initial state and the target state, thereby following the most popular way of evaluating maturity as stated by de Bruin [37]. Obviously, besides the considered perspectives, other elements of the company and its processes like logistics, low-emission equipment, and IT-infrastructure have to be shaped in accordance to the state of the company, too. These elements are not directly considered in the development path as they are not seen as drivers for urban manufacturing, but enablers which have to be shaped based on the occurrence of the driving perspectives which are represented in the development path. Nevertheless, elements as logistics are of high importance for the city-compatibility of factories and should be shaped in a low-impact manner, e.g. by using electrically driven vans instead of heavy trucks for the case of logistics.

3.2.4.2 Factory

With the perspective factory, we refer to the manufacturing distribution approach including decentralization and number of factories as well as the respective work distribution between the sites.

The first characteristic in the perspective factory is the number of sites. In the classical manufacturing approach, we assume that manufacturing takes place in one central site, thereby leveraging economies-of-scale. The location of this site is typically selected without considering advantages of urban manufacturing but with traditional locational factor analysis. Therefore, the sites are often placed in more rural areas, and may be surrounded by cities over time. In these central sites, all organizational functions are bundled and products are entirely manufactured, also based on supplied components. The central sites are controlled in a centralized manner.

In the final state of the development path, the urban manufacturing state, we assume a decentralized approach, consisting of one or more pre-factories which are dedicated to manufacture customer-neutral pre-products, thereby still leveraging economies-of-scale and city factories, dedicated to the final manufacturing of customer-specific products, e.g. by configuration and final assembly. The manufacturing sites also contain indirect organizational units, e.g. for internal control and administration, but also for product development. In the case of city factories, product development focusses on localization of products towards regional needs and costumer integration in engineering and innovation. The proximity of engineering and manufacturing also supports short feedback loops to advance products with regard to manufacturing, like manufacturability and quality issues. The distribution of organizational units for administration and control also enables a decentralized manufacturing planning and control approach, thereby reducing the complexity of managing the overall manufacturing network and its parts.

For the development path towards urban manufacturing, we suggest the following steps from classical factories to the decentralized urban manufacturing approach.

- Initial State:
  - Classical factory for complete manufacturing, which is often in an initially rural area and is operated in a centralized manufacturing approach.
- Step 1:
  - establishment of a first city factory for final manufacturing of customer-specific products, thereby emphasizing on a specific urban area,
  - Core factory remains unchanged but is complemented by first city factory.
- Step 2:
  - establishment of further city factories,
  - Core factory remains unchanged but gets focus on manufacturing of pre-products, and is complemented by city factory.
- Step 3:
  - Check if initial factory brings substantial market access, if not: question to transform it into pre-factory.
- Final State:
  - Urban manufacturing by a decentralized manufacturing approach of one or more pre-factories for the manufacturing of costumer-neutral pre-products and city factories for the final manufacturing of customer-specific final products.

Furthermore, it could be reasonable to decentralize also the innovation and engineering process as well as manufacturing control along the transition towards urban manufacturing. The product development-related area can thereby be seen as linked to customer-integration which is considered in the sales perspective and to its subject-matter the product, addressed in the following chapter. Manufacturing control is considered along with Industry 4.0 in the manufacturing.

3.2.4.3 Product

Another essential base for a transition of the manufacturing approach in a company is the product itself, delivering possibilities and boundary conditions for manufacturing.

As stated in Sect. 3.2, the development path for urban manufacturing especially addresses the manufacturing of discrete, complex series products with a high degree of customization, resulting in a high variety.

A significant property of the product with regard to the distribution of its production is its structure, i.e. in how far the product is composed from modules or single parts. Thereby, we assume for the initial state a classically developed product with no consideration of modularization, and for the final state a very high level of modularization of the products, which, in addition, are built on product platforms. Consequently, the following steps for the product perspective are suggested:

- Initial State:
  - Classical product without modularization or platform consideration.
- Step 1:
  - First modules in a part of the products.
- Step 2:
  - Cross-product platform/s.
- Step 3:
  - Modularization of the full product range.
- Final State:
  - Products, modules and platforms optimized for decentralized manufacturing with pre-factories and city factories, including late customization.

3.2.4.4 Manufacturing

To enable the envisaged decentralized manufacturing approach for customized products, classical production systems have to be transformed towards systems which enable the efficient manufacturing of very small lot sizes, i.e. down to lot size 1. This can be supported by applying lean approaches as well as by using technologies from the so-called Industry 4.0 [38], resulting in the following steps.

- Initial State:
  - Classical manufacturing based on push principle.
- Step 1:
  - Lean production with pull principle in sync with customer requests at city factory/factories.
- Step 2:
  - Decentralized production triggered by customer orders at city factories which order components at pre-factory.

- Step 3:
  - Manufacturing in lot size 1, enabled by Industry 4.0 to allow fine granular manufacturing control.
- Final State:
  - Self-organization by products which are cyber-physical systems (CPS) that control their own flow through the overall manufacturing network as well as the manufacturing parameters.

3.2.4.5 Sales

An essential advantage of a manufacturing location for customized goods which is situated in an urban surrounding is its proximity to potential clients with the respective possibilities for customer integration. To leverage these possibilities, measures besides typical integration aspects like internet-enabled open innovation should be taken. These measures can range from face-to-face open innovation approaches to exploit customer ideas for designs and product solutions to the incorporation of customers in the actual manufacturing processes. Therefore, steps for the advancement towards leveraging the potentials of urban manufacturing can be the following:

- Initial State:
  - Typical approaches for customer integration like internet-based open innovation.
- Step 1:
  - Face-to-face open innovation by incorporating customers in idea development and problem-solving.
- Step 2:
  - Integration of customers in indirect processes like product development, i.e. face-to-face co-creation.
- Step 3:
  - Advancing the transparency of the factory by providing customers access to the production of their personal products, potentially including manufacturing and quality data.
- Final State:
  - Opening the factory by integration of customers also in direct processes, e.g. by incorporating them in the final assembly of their personal products.

Reason for doing costumer involvement in early phases of product creation earlier than in later phases is that in these phases, a one-way information flow from customer to company can be established. In contrast, by incorporating externals like customers in later phases of the product creation, also insight in these value adding processes is provided—which is an approach that may be seen as critical by the company at first.

3.2.4.6 Employees

To leverage the potential to increase the flexibility of working hours and of production capacity based on the possibly short distances between housings and workplace, which may be accompanied with short transfer times, new working-time models can be applied. This may range to a very high short-term flexibility based on self-organization approaches like KapaFlexcy [39] which may result in new levels of work-life-integration, too. So, potential steps for the transition are:

- Initial State:
  - Rigid working-time model.
- Step 1:
  - Flex-time model, especially for indirect employees.
- Step 2:
  - (multi-)annual working time accounts.
- Step 3:
  - Participative working time planning.
- Final State:
  - Self-organization of working times.

Besides this flexibility-oriented perspective, depending on the company-case, an increase of the degree of employee participation may be reasonable but is not mandatory for urban manufacturing but desirable.

3.2.4.7 Citizens

Seeing a factory as integral part of a city, which in the best case acts as a "friendly neighbor", the relation between the factory and its neighbors, i.e. the citizens of the city, can be seen as crucial. Therefore, a perspective to present the relation between the factory and the citizens is suggested. Presentations for this feature may range from no active relation management up to an open factory, which is not only participative planned but which offers its infrastructure partly or as a whole to the public. Whereby the opening of the infrastructure as a whole may be difficult and even unachievable, the opening of parts could concern parking areas, canteens or day-care centers, so low critical areas, to externals. This results in the following steps for the perspective of the citizens of the development path:

Initial State:

- No specific relation-oriented measures towards citizens.

Step 1:

- Provision of information about the factory, its state or future, to citizens.

Step 2:

- Information exchange and citizen involvement by events like workshops.

Step 3:

- Participative planning of the factory and its further development by incorporating citizens.

Final State:

- Open factory which allows the usage of infrastructure, partly or fully, to externals, especially citizens.

3.2.4.8 Cooperation

The proximity to suppliers supports a close cooperation between the companies. This relates not only to suppliers of goods like material and components, but also of services. The mode of operation for this perspective may range from the classical supplier-client approach up to an integrated planning and operation of the supply chain including strategies, capacities and inventories in the sense of an Extended Enterprise as presented by Dold et al. [40] among others. Therefore, potential steps along the development path are the following:

Initial State:

- Classical supplier-client relations.

Step 1:

- dedicated identification of local suppliers nearby the factory and begin of respective relations for sourcing.

Step 2:

- sharing of forecast and inventory level, supported by linked IT-systems.

Step 3:

- Collaborative planning of the overall supply chain, thereby ensuring a global optimum in performance instead of local optimizations.

Final State:

- integrated management of the supply chain.

3.2.5 Application of the Development Path

The presented development path has the objective to support the transition of a manufacturing enterprise by guiding it towards the vision of an urban manufacturing through the identification of possible next steps for advancement. Thereby, the company may decide the actual target state, which is not necessarily the long-term oriented characterization of the perspectives of the final states, but company-specific.

The application of the development starts with an as-is analysis resulting in a classification of the company in the states of the development path. Each of the states identified shows the option of a next step for advancement of the company towards its goal by examining the difference between the current state and the next step or state in the development path.

Table 3.4 shows the example of a company which produces design-oriented, complex series products with a high amount of customization. Thereby, starting points for the advancement are marked in green, intended target states in red. The promising next steps for the advancement of the company along the perspectives of the development path are identified as follows.

Concerning the factory and overall manufacturing approach, the current state at the company is the manufacturing at one central, classical factory. The company intends to open city factories to get better access to the output markets but would continue to manufacture final products at the core factory in future, too, if this is reasonable—as shown by the intended target state. In accordance to the development path, the next step can be to create a first city factory to complement the existing factory and to gain first experiences in manufacturing in urban surroundings, while already addressing one dedicated market.

The products of the company are already composed out of first modules, whereby the intention is to introduce cross-product platforms in addition. At the same time, these gaps also represent the next step to be done for advancement with regard to the product.

For the perspective manufacturing, the as-is analysis resulted in a classification in the lean production-state. As objective the company identified to apply approaches and technologies of Industry 4.0 to support the manufacturing in lot size 1 with innovative information technology. Therefore, the next step can be to realize decentral manufacturing, triggered by customer orders, thereby controlling the

Table 3.4 Example of the application of the development path (red—current state, green—target state, yellow—same current and target state)

Perspective	Initial state	Step 1	Step 2	Step 3	Final state
Factory	Classical factory	First city-factory, core factory unchanged	Further city factories, core factory focus on per-products	Transform core factory into pre-factory	Pre-factory/ies and city factories
Product	Classical product	First modules	Cross-product platform/s	Modularization of full product range	Products, modules and platforms optimized
Manufacturing	Classical manufacturing, push principles	Lean production	Decentralized manufacturing triggered by orders	Manufacturing in lot size 1 , enabled by Industry 4.0	Self-organization by products
Sales	Typical approaches	Face-to-face open innovation	Face-to-face co-creation	Transparent factory	Open factory
Employees	Rigid working time model	Flex-time model	(Multi-)annual working time accounts	Participative working time planning	self-organization of working times
Citizens	No specific measures	One-way communication	Information exchange	Participative planning	Open factory
Cooperation	Classical supplier-client relations	Local suppliers	Sharing of forecast and inventory level	Collaborative planning	Integrated management of the supply chain

manufacturing in the system resulting from the establishment of a complementary city factory besides the original core factory.

Concerning sales, the company currently uses typical approaches and want to reach a state with high customer involvement including a transparent factory which enables customers to visit the factory and regard the manufacturing of its product, including the availability of manufacturing- and quality-related data. For this, the development path suggests as a next step, to start with customer involvement in local open innovation processes to address issues like design and the solution of customers' problems.

As shown in Table 3.4, the company currently applies a flex-time model for the working time of the employees where possible and wants to remain at this level. Therefore, no steps for advancement are needed for the perspective employees. Besides, it may be questioned if even if the working time model remains unchanged, if the degree of employee participation is advanced.

In the as-is state, the company has a one-way communication behavior towards the citizens, e.g. by press releases, information meetings or open days. For the perspective citizen, an information exchange was determined. So, the establishment of events for information exchange is a potential next step for advancement.

For the cooperation perspective, the as-is state of the company is based on classical supplier-client relations. The aim of the company is to source locally at suppliers but not to advance the supply chain by integration or joint activities. Therefore, a potential next step is to identify nearby to suppliers and to start the respective sourcing.

The application of the development path should not be once, but in the long-term on a regular base. Thereby, the actual achievements may be compared with the defined objectives and the objectives may be verified and adopted where needed.

3.2.6 Summary and Outlook

Companies face ever higher challenges based on trends like product individualization and talent shortage. A promising strategy to cope with the challenges is the implementation of an overall manufacturing approach, which is based on a decentralization of manufacturing into pre-factories and city factories. In this context, pre-factories are intended to manufacture customer-neutral components at economies-of-scale, and city factories are responsible for steps of product creation which are customer-specific—so an approach based on a partitioning by means of the customer decoupling point of order fulfilment. To support the transition of classical manufacturing companies into this decentralised "urban manufacturing" approach, a generic development path is suggested in this contribution.

The development path for the transition towards the urban manufacturing approach is divided into thematic perspectives and development steps. The perspectives of the development path are based on drivers of urban manufacturing. In the contribution, the following perspectives are suggested:

- factory, which also represents the overall manufacturing distribution approach,
- product,
- manufacturing and its organization,

- sales,
- employees,
- citizens, as well as
- Cooperation.

Besides these perspectives, other factors are important for the realization of an urban manufacturing approach, too. Among others, this refers to logistics, low-emission equipment and machinery as well as to the information technology. These factors have to be realized in a manner, which fits to the requirements of the company in its respective state and which supports the city-compatibility of the site.

As next steps, the development path will be advanced based on discussions and expert talks with the scientific community before first tests in company applications will be done. The tests will serve for the further development and detailing of the path.

3.3 Additive Manufacturing—A Technology Supporting Sustainable Urban Manufacturing

3.3.1 Introduction

As mentioned in the preceding statement, decoupling of highly individualized production steps will be a possible strategy. Therefore a transition to more flexible production processes will be needed. The ever faster developing technology of additive manufacturing (AM) is becoming a serious alternative for manufacturing small series of end customer products. Regarding a survey from KPMG [41] a majority of companies see impacts on four main fields - suppliers, production, logistics and customer relation. The results of the survey are summarized in Fig. 3.31.

The surveyed companies see the biggest impact in the forthright area of usage—the production area. Furthermore, the impact on Sales and Customer relation management is estimated quite high. The possibility for impacts in the field of Logistics and Suppliers is seen indifferent by the companies. Opposing to that, research describes AM as a "powerful complement to traditional manufacturing and the end-to-end supply chain" [42]. Even more drastically formulated by Gravier "It should be a warning sign for companies that if they don't innovate their supply chains, they may become irrelevant" [43]. These findings directly tie in with the hypothesis of distributed production networks, whereby additive manufacturing can be a main enabler to realize such networks.

In the following paragraph, additive manufacturing, the reasons for using it and the different levels of application will be described in more detail. For all applications of these technologies, Additive manufacturing is defined as the official industry standard term after ASTM F2792. It describes the process of joining materials layer upon layer to produce physical objects from 3D model data. In

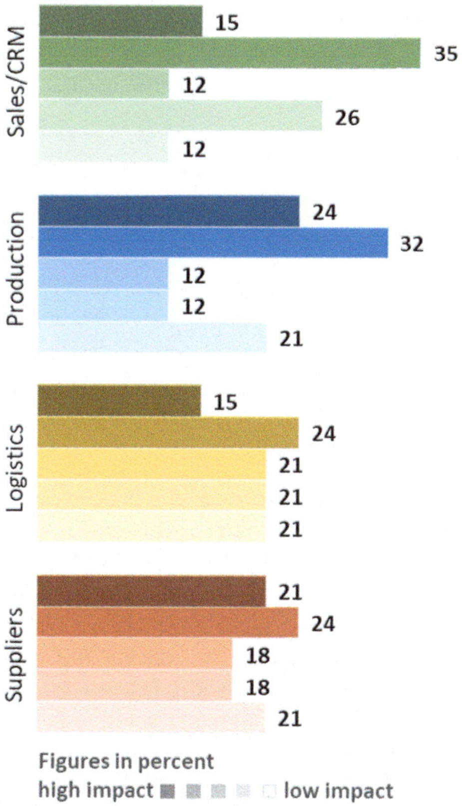

Fig. 3.31 Impacts on different areas [41]

contrast to subtractive manufacturing methodologies, where the "inverted" 3D model has to be removed from a blank of material. Widely used synonyms for AM are 3D-printing, additive fabrication, additive processes, additive technologies, additive layer manufacturing, layer manufacturing, and freeform fabrication [44].

The reasons for AM are manifold, it enables companies to flexibly react on volatile markets, produce spare parts on demand and realize individual customer requests directly. Therefore three main potentials of Implementing AM are:

- Realization of complex geometries, that were not possible with traditional methods
- Integration of functionalities through producing movable parts in one process without assembly
- Complete adoption of customer needs

A first differentiation of the field additive technology can be made by using a level approach. This differentiation is required, to have a definition of the wording 'rapid manufacturing' and is provided in Fig. 3.32. The field of additive technology can be examined either from a technological point of view or from an application point of view.

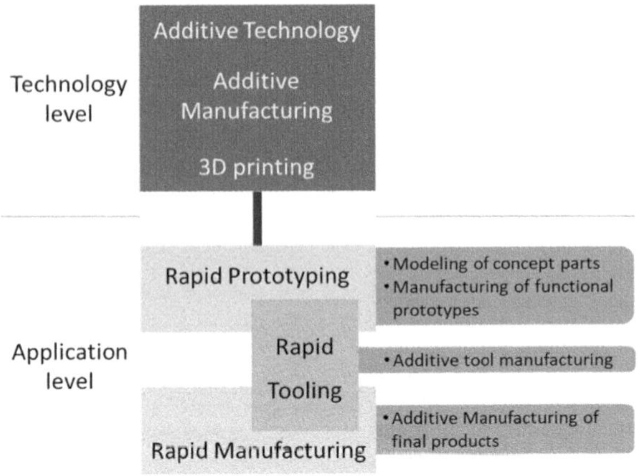

Fig. 3.32 Levels of AM application [41]

The technology level comprises technological aspects, like parameter setting, material specification etc. This will not be discussed in detail.

The application level determines the type of usage of AM. The required performance differs significantly according to the intended use. Lowest level of integration of AM is rapid prototyping. The target is fast manufacturing of prototypes for functional testing or concept models for decision support. The capacity utilization is not major priority in this case. Rapid Prototyping is already widely spread in the industry. Rapid manufacturing is used to produce parts for customers. Quality and cost efficiency requirements are high because the technology is measured against conventional manufacturing technologies. Rapid tooling is in-between those application fields. It is used to manufacture tools for conventional manufacturing procedures, e.g. moulding. For example, tools for plastics moulding have lower durability at lower costs. Changes and optimization of the tool can be performed fast and cost effective.

3.3.2 *Impacts on a Traditional Product Lifecycle Through AM*

The lifecycle of a product is heavily influenced by using AM technologies. The impact of potential changes is presented on a general value added chain (see Fig. 3.33). The value chain is divided in two threads, the product realization (grey) and the order processing (white). The scenario is based on the assumption all processes will be changed to additive manufacturing given a small lot size production.

In product realization, one factor is the impact on used material. The material is a key factor for additive manufacturing because the conditions of material required differ much from conventional manufacturing [44]. Depending on the used

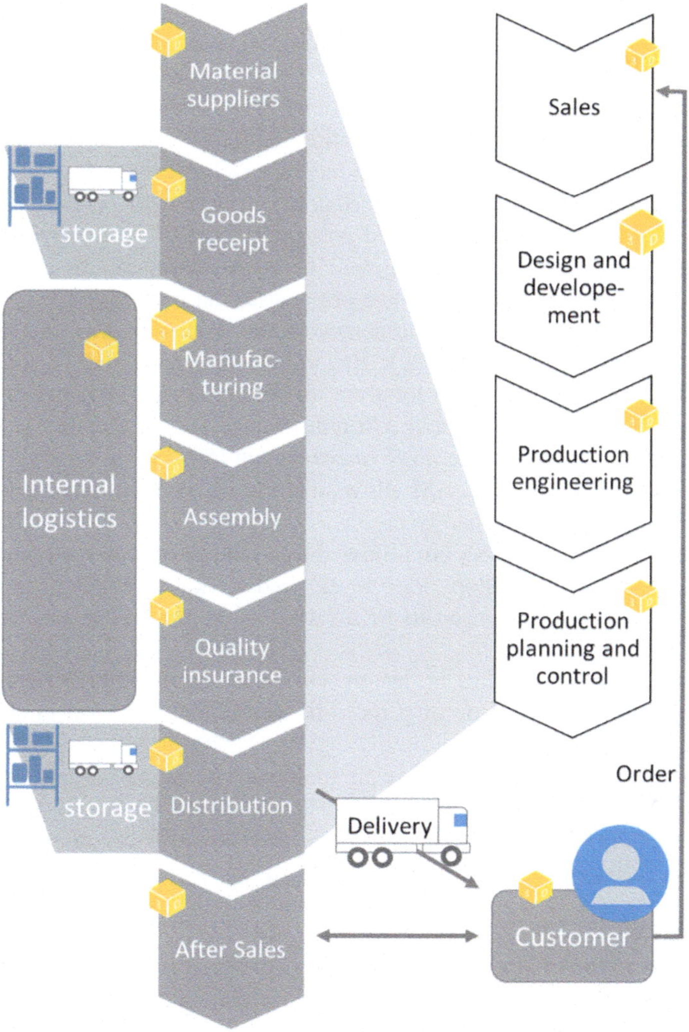

Fig. 3.33 Impacts on a traditional product lifecycle

technology, the material needs to be distributed in a different form and shape, ranging from solid blocks to powder or liquids. However, the number of different materials for AM is limited as of today, it will increase based on further application areas [45].

The required material specification does have an impact on the availability and distribution time of it [46]. Furthermore, the number of materials will have an impact on storage facilities. The different consistence of the raw material will change the way of storage as well as the storage strategy [47].

Internal logistics will change drastically in this scenario. Depending on the number of production machines, the material provision can be done automatically (many) or manually (some). Additionally, additive machine manufacturers are developing tools and approaches to automate part handling as well as material removal under respect of treatment regulations [48].

A change in technology has always a huge impact to the production processes, its KPIs, the factory layout and required tools. Efforts will be moving towards post-processing. Most of the parts need post-processing to fulfil the requirements. Additionally, the material processing will be more important for additive technologies to make sure a high percentage of material can be reused after processing. New feedback-systems are needed to achieve same process quality and stability as with conventional technologies [49]. Regarding cost calculation, economies of scale get less relevant with small lot sizes; therefore, competitiveness of AM will rise. This also means, that a general cost reduction can only be realized with better process efficiency through technology improvement [50, 51]. In addition, factories will change in layout and size. AM allows distributed production, which results in more but smaller facilities, mainly closer to the customer [51]. Changed requirements in tooling and machinery are followed by a change of required competencies. Currently, employees on shop floor level need to know about setting-up and adjustments, in future they need to be aware of post-processing and digital data handling.

Moreover, new possibilities in design will affect the assembly effort. By integrating functions and reduction of parts to be assembled, the effort in assembly is decreasing in general [52].

The part distribution will increase in frequency because the order-to-delivery time is decreasing. The distribution effort is minimized further, if the visionary business case "the customer produces its parts on his own" is becoming reality [53]. Physical part distribution is replaced by the distribution of 3D data sets to the customer. This will lead to a stronger inclusion of the customer in the product configuration process by adaption of relevant parts [45].

Regarding the order processing, additive technology is changing sales approaches extensively. The upcoming possibilities in design and customization require a basic knowledge of the sales personal. This expertise allows the right determination of price and delivery time. The service potential during consultancy in product design can widen the sales opportunities.

This has an influence on the philosophy to develop and shape products. The technology allows integration of functionality and high degree of design freedom. For that, the product data needs to be adaptable which requires consideration during the development and design phase to reduce efforts for the future [54]. This is even more important for complex products where assembly steps are eliminated [45]. On the other side, also restrictions and boundary conditions given by AM have to followed [55]. For optimal usage of the new degree of freedom, simulation within design procedures becomes more important. This is fundamental to secure the functionality of new complex structures.

Operations management is also changing. The planning of workflows is different. However, the creation of work plans is still required, but with different focus. The production planning will be more flexible. Due to smaller lot sizes, the adaption of changes needs to be done faster but with consistent quality [56].

Quality insurance is a field with high relevance for using AM technologies. Based on the process chain a higher effort is required to secure stable quality and a prediction of life expectancy for the parts [57]. Additionally, new processes for failure parts needs to be defined—e.g. repair or re-production. For hybrid production, the effort will be even higher [52].

Looking at the after sales area there was research done on the effect that additive manufacturing would have on the supply chain of spare parts that could possibly be manufactured using AM. Main result of this work is that only 7–23% of the parts have an economical benefit but the lead-time was lowered significantly [58]. Furthermore, the spare parts supply chain of the F-18 Super Hornet has been investigated in another work, resulting in: e.g. lower overall operation costs, lower down time, higher potential for customer satisfaction, lower capacity utilization and higher flexibility [59].

3.3.3 Flexible Spare Parts Production with AM by Deutsche Bahn AG

The application case from Deutsche Bahn AG (DB) is presented in this paragraph, it addresses the production of non-security-relevant spare parts. In the first phase of the project, the DB team has decided to work together with experienced additive manufacturing service providers. Therefore, a standardized process for admission of suppliers has to be put in place, when the demand for parts is rising. This is currently in research and development. Independently a process for qualification of spare parts is developed. A first version of this procedure is shown in Fig. 3.34.

The example of DB shows that a use-case of using AM in the sector of after-sales/maintenance has very high potential. The first impression of this use-case is that the realization is not very complex. However, along with the progress and the first realized prototypes the challenges of using AM became clear, even for this level of implementation. The demand for standardized processes and required knowledge is easily underestimated. Regarding a complete changeover towards AM technologies, the amount of influences and required changes increase dramatically. The adoption of production with its challenges concerning material, quality, process stability etc. is only a small step towards completeness. For successful overall implementation of AM, a well-defined framework is needed to tackle the extensive impacts on development, operational planning, organization and logistics. Furthermore, it becomes clear, that AM competencies for employees are needed across multiple divisions in the company.

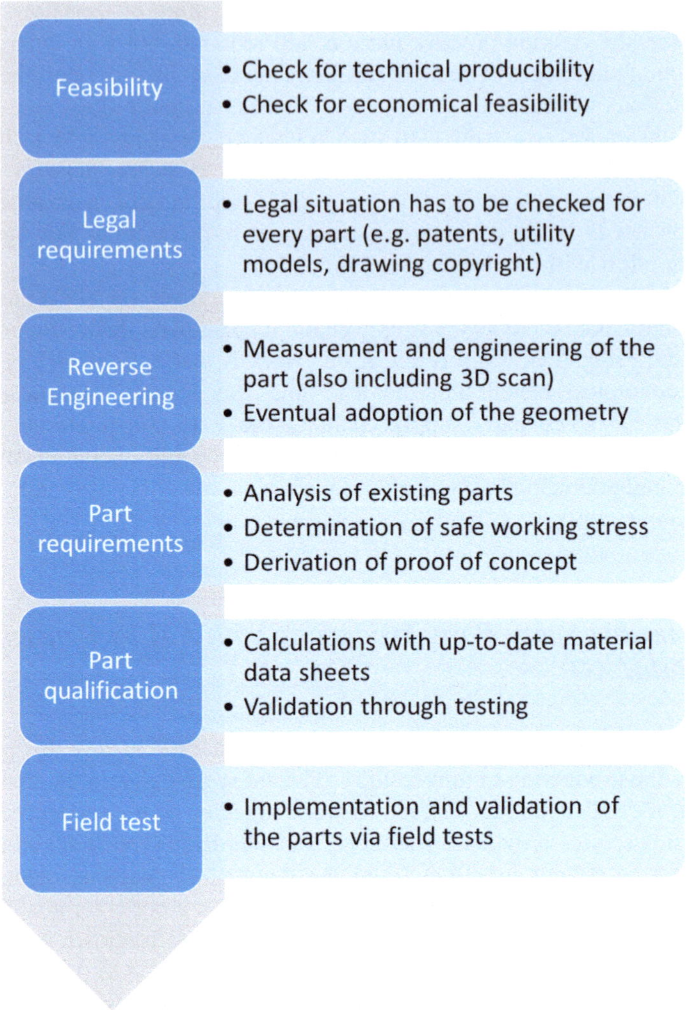

Fig. 3.34 Qualification process for spare parts

Furthermore the example shows the potential of AM for a sustainable, flexible and very fast way of producing spare parts for the public transport sector. Especially in this field idle times of trains have to be kept as short as possible to keep the transport operational.

Acknowledgements The colleagues of the team of authors are gratefully thanked for providing input of their topics as well as discussing different aspects. Furthermore, the number of assistant researchers has to be stated as well supporting the work by collecting data, adjust it and provide it during the research work. Prof. Dieter Spath, Head of the institute for human factors and technology management at University of Stuttgart, can be also named because his offer to discuss all details in context of a broader view helped the development as well.

Furthermore, the ministry of environment, climate and energy economy, of Baden-Württemberg is gratefully thanked for choosing the project and providing the funding. The funded project HoliPOrt—holistic development of manufacturing location was a very important base for the development of the content of the chapter.

Sven Gaede was doing his thesis under surveillance of Nikolas Zimmermann. The discussions about his work at Deutsche Bahn AG inspired to have a closer look on those aspects in additive technology.

Glossary

Area As an area, areas are described within the concept study, which are characterized by decentralized control. The reasons for this are the absence of an operating company or distributed ownership (each company owns the property with real estate). Each stakeholder in the system therefore has little or no dependency on others

Module Modules are composed of several components. The complexity and the range of functions are still very limited. The integration of external competencies in the manufacturing process is low. Individualization is still limited, since assemblies are also components of systems

Component/part A component is a product produced by processing from a raw material/ raw material. The complexity and the range of functions are limited. These products are usually required in larger quantities because they function as components of assemblies or systems

Bottom-Up Bottom-up approaches are given by the executive level. The impulse comes due to needs, needs or experiences that can be interesting from the perspective of the appropriate level for all. Usually, different impulses are agglomerated across the planes in the bottom-up approach in order to achieve a central image

Brownfield In the field of urban and land development, partial living areas are free from the current use through the life cycle. In the field of Brownfield development, the aim is to reuse these freed areas and properties. Since there are inventory areas, the prevailing framework conditions must also be taken into account or even be adapted. However, the flexibility is usually very small

Congruence English agreement or congruence—in the context of consistency, it is intended to serve as a representation of reality. It is intended to help shape certain contexts more structurally and also serve as a model through abstraction. It is especially used in management to describe complex processes and relationships in a simplified manner

Greenfield If new areas are indicated in urban and land development that is, also redefined in terms of their use one speaks of green-field development. The advantages of the Greenfield development are the high flexibility in design and usability

Industries The term is used in the study for companies and organizations that have a clear reference to production. Furthermore, the facilities are also companies of industrial production

Input The input can be any input variable or parameter. For the execution of processes, these are initial values that are necessary

Organizations In the case of organizations, the allocation is not clearly possible or there are suitable companies and facilities which are less suitable for industrial production

Output The output can be any result or part of the result. It is the result of a process

Park Within the framework of the concept study, a construct with centralized control and partly also ownership conditions is described with Park. This means that there is an owner or operator who manages the units in the settlement, provides guidelines and, if necessary, offers additional services and services. The operator may be either a private-law organization or a local authority

Rural Rural are rural areas, which are relatively remote from main and secondary centers. These areas are characterized by little flexible infrastructure connection. The size of people settlements is limited, which can correlate with the challenges faced by specialists and appropriate competencies in the immediate environment

Suburban Suburban are areas around of centers. Usually the cconnection and infrastructure are well developed and the access to suitable specialists and competences is moderate to good. The proximity to larger population settlements is moderate, so the conflict probability is lower than in the urban but higher than in the rural areas

System A system is a complex product that combines many different competences and disciplines. These products are customized on the basis of customer requirements and often lead to the integration of external and internal competencies throughout the entire production process. Also the support processes are characterized by the high complexity

System dynamic The system dynamics is a relatively young science, which in the 1950er years for the first time by J.W. Forrester [Forrester 1971] was used. The focus is on the viewing element, which is, however, viewed as a system with many dependencies and interactions with other system components. System dynamics is currently a recognized process for modeling complex contexts and simulations

Throughput The sum of the transformation processes can be seen as a throughput in the context of the study. An output is generated from input by transformation. In the study, the throughput is considered a white box. The transparency is preserved, but no more detailed consideration of the transformation processes takes place. However, the White Box offers the possibility to look more in detail and thus to analyze individual aspects

Top-Down Usually as a way of providing information from the lead to the leading organs. Visions and objectives are often formulated in this way and are then implemented further into the implementation via downstream levels. In the concept study, this is understood as a path from the holistic idea to the concrete implementation

Urban The term urban is referred to in the study on main and secondary centers and specifies areas that belong to the urban area or are directly connected. These areas are characterized by their proximity to large human settlements and residential areas. The available infrastructure and connection is excellent

References

1. Weber R, Hauser I (2008) Eine kleine politische Landeskunde (a little political applied geography), Landeszentrale für polische Bildung Baden-Württemberg, Version 6, Stuttgart 2008)
2. Wirtschaftsregion Stuttgart - Zahlen und Fakten (Region Stuttgart—data, information and figures), Industrie und Handelskammer Region Stuttgart, Stuttgart (2017)
3. Die Region Stuttgart – Hightechregion in Europa; Information paper, Wirtschaftsförderung Region Stuttgart GmbH (WRS), Stuttgart (2014)
4. Dispan J, Koch A, Krumm R, Seibold B (2015) Strukturbericht Region Stuttgart 2015 - Entwicklung von Wirtschaft und Beschäftigung (structurel report region Stuttgart 2015 – development of economy and employment, Verband Region Stuttgart, Handwerkskammer Region Stuttgart, Industrie und Handelskammer Region Stuttgart, IG Metall Region Stuttgart, Stuttgart, 2015
5. United Nations (2015) Adoption Of the Paris agreement—framework convention on climate change, Paris, 30 Nov–11 Dec 2015
6. Chertow MR (2000) Industrial symbiosis: literature and taxonomy. Ann Rev Energy Environ 25:Seite 313–337
7. Verordnung über die bauliche Nutzung der Grundstücke (Baunutzungsverordnung – BauNVO), Bundesministeriums der Justiz und für Verbraucherschutz, Letzte Änderung 11.06.2011, Berlin (2011)
8. https://www.bundesregierung.de/Content/DE/Artikel/2015/12/2015-12-12-klimaabkommen.html, veröffentlicht 12.12.2015, letzter Besuch 15.11.2016, 16:22
9. Wraneschitz H, Böckmann C, Eder SW (2016) Der CO2-Entzug, VDI nachrichten, Ausgabe 45, 10. Nov 2016, VDI Verlag, Düsseldorf
10. Tiilikainen K (2016) Finland addresses climate actions in agriculture, climate action, Marrakesch. http://www.climateactionprogramme.org/climate-action-publication

11. Gödecke M, Keller B, Munz J, Pflüger J (2015) „Wasser – ein knappes Gut? “, E-learning-Projekt, Lateinamerika-Institut, WS 2014/15, 2015. http://www.lai.fu-berlin.de/disziplinen/literaturen_und_kulturen_lateinamerikas/studiengaenge/elearning-projekte/brennpunkte_la/ressourcen/wasser.pdf
12. FM Global (2016) Wasser – das knappe Gut, TouchPoints – Resilienz als Wettbe-werbsvorteil, 27 April 2016
13. Spath D, Weisbecker A, Peissner M, Hipp C (2013) Potenziale der Mensch-Technik Interaktion für die effiziente und vernetzte Produktion von morgen. Fraunhofer IAO, Stuttgart
14. Bauer
15. Jeanneret-Gris CE (Le Corbusier) (1943) Entretien avec les étudiants des Ecoles d'Architecture. Paris, 1943
16. http://www.urban-is.de/Quellennachweis-Internet/StadtPlanung@CD/Charta_v_Athen.pdf, last viewed: 10 Nov 2016
17. Romero E, Ruiz MC, Álvarez R, Bayona E (2013) Eco-industrial areas modeling proposal. In: First international conference on resource efficiency in interor-ganizational networks—ResEff 2013—13 Nov–14 Nov 2013, Georg-August Universität Göttingen, Universitätsverlag Göttingen
18. Lambert AJD, Boons FA (2002) Eco-industrial parks: stimulating sustainable development in mixed industrial parks. Technovation 22 (2002) Seite:471–484
19. Romero E, Ruiz MC (2013) Framework for applying a complex adaptive system approach to model the operation of eco-industrial parks. J Ind Ecol. https://doi.org/10.1111/jiec.12032
20. Reuter C, Prote JP (2015) Standortentscheidungen systematisch herleiten, ZWF Online, Jahrg. 110 (2015) 10, Carl Hanser Verlag, München, 2015
21. Nadler DA, Tushman ML (1980) A model for diagnosing organizational behavior: applying a congruence perspective, Managing Organizations—readings and cases, reprint of Organizational Dynamics, Little, Brown and Company
22. Wiedemann C (2014) Identifikation und Modellierung von Wirkzusammenhängen zwischen Produktionssystemen und ihrem urbanen Umfeld. AKAD University of Applied Sciences, Stuttgart
23. Lowe EA, Moran SR, Holmes DB (1995) Field book of the development of eco-industrial parks. Report draft. Indigo Dev. Co., Oakland, CA
24. Očigrija A, Kreh O (2013) Die größten Unternehmen in Baden-Württemberg, In-dustrie- und Handelskammer Region Stuttgart, Stuttgart, 2013
25. http://wirtschaftslexikon.gabler.de/Definition/unternehmen.html, last viewed: 22 Dec 2016, 15:37
26. Bernet M (2010) Foren und Wikis: verteiltes Wissen, Social Media in der Medienar-beit, Seite 143-146, VS Verlag für Sozialwissenschaften, Wiesbaden
27. Vester F (2002) Die Kunst vernetzt zu denken – Ideen und Werkezuge für den neu-en Umgang mit Komplexität, Der neue Bericht an den Club of Rome, akt. Auf-lage, Deutscher Taschenbuch Verlag GmbH & Co. KG, München
28. Forrester J (1971) Counterintuitive behavior of social systems. Technol Rev 73(3):52–68
29. https://www.bundesregierung.de/Content/DE/Magazine/MagazinInfrastrukturNeueLaender/014/s1-verkehrssituation-auf-deutschlands-strassen.html, last viewed: 28 Dec 2016, 13:42
30. Spath D, Lentes J (2012) Flexibler produzieren in der Stadt, FtK 2012 – Fertigungstechnisches Kolloquium Stuttgart: schriftliche Fassung der Vorträge zum fertigungstechnischen Kolloquium on 25 and 26 September in Stuttgart. In: Proceeding, Gesellschaft für Fertigungstechnik, Stuttgart
31. N.N.: WITTENSTEIN AG – Urbane Produktion & Industrie 4.0. http://www.wittenstein.de/urbane-produktion-industrie-4.html, last viewed: 08 June 2014, 18:57
32. Standort GMD – Die Gläseren Manufaktur. http://www.glaeserenmanufaktur.de/de/idee/standort, last viewed: 06 June 2014, 19:27

33. Spath D, Lentes J (2013) Urban production to advance the competitiveness of industrial enterprises. In: Proceedings of ICPR22, Iguassu Falls, 5 pp
34. Spath D, Bauer W, Lentes J (2015) A quick-check to identify potentials by urban manufacturing for industrial enterprises. In: Proceedings of ICPR23, Manila, 6 pp
35. Lopez J, Meehan P, Prentice S, Raskino M, Howard C, Willis D (2014) Get ready for digital business with the digital business development path, Gartner Research. http://www.gartner.com/doc/2777417, last accessed 31 May 2017
36. Perkins M, Small P (2006) Partnering for better microbial diagnostics. Nat Biotechnol 24:919–921
37. De Bruin T, Freeze R, Kaulkarni U, Rosemann M (2005) Understanding the main phases of developing a maturity assessment model. In: Australasian conference on information systems, Sydney, 10 pp., authors version. http://eprints.qut.edu.au/25152/, last accessed 31 May 2017
38. Kagermann H, Wahlster W, Helbig J (2013) Securing the future of German manufacturing industry: recommendations for implementing the strategic initiative INDUSTRIE 4.0. Final report of the Industrie 4.0 working group. Forschungsunion im Stifterverband für die Deutsche Wirtschaft, Berlin
39. Bauer W, Hämmerle M, Gerlach S, Strölin T (2014) Planning flexible human resource capacity management. In: 19th IFAC world congress, Cape Town, pp 4459–4464
40. Dold C, Mauro P, Nøstdal R, Pierpaoli F (2003) Development and management of east-west European extended enterprises. In: Proceedings of PRO-VE 2003, Lugano, pp 467–474
41. KPMG (2015) Additive manufacturing—Strategy & Operations Fokusanalyse
42. Attaran M (2017) Additive manufacturing: the most promising technology to alter the supply chain and logistics. J Serv Sci Manage 10:189–205
43. Gravier M (2016) 3D printing: customers taking charge of the supply chain. Industry Week, April 12
44. Wohlers Associates Inc. (2016) Wohlers report 2016—rapid prototyping, tooling & manufacturing state of the industry, United States of America
45. Gebhardt A (2013) Generative Fertigungsverfahren: additive manufacturing und 3D Drucken für prototyping—tooling—Produktion. Carl Hanser Verlag, München
46. Deloitte University Press (2015) The future of manufacturing—making things in a changing world. Deloitte Development LLC
47. Jones Lang Lasalle GmbH (2013) Produktion im Wandel. Höhere Komplexität, größere Chancen
48. EOS (2017) Systeme und Zubehör zur additiven Fertigung von Metallteilen. URL: http://www.eos.info/systeme_loesungen/metall/systeme_und_zubehoer
49. National Institute of Standard and Technology (2013) Measurement science roadmap for metal-based additive manufacturing
50. Coenenberg AG, Fischer TM, Günther T (2012) Kostenrechnung und Kostenanalyse. Schäffer-Poeschel Verlag, Stuttgart
51. Seliger (2015) Additive manufacturing im Leichtbau - strategisches und betriebswirtschaftliche Herausforderungen und Perspektiven. Leichtbau BW GmbH
52. Breuninger J, Becker R, Wolf A, Rommel S, Verl A (2013) Generative Fertigung mit Kunststoffen, Konzeption und Konstruktion für Selektives Lasersintern, Heidelberg, Springer Verlag
53. Schmoll A, Ballestrem J, Hellenbrand J, Soppe M (2015) Dreidimensionales Drucken und die vier Dimensionen des Immaterialgüterrechts, in GRUR, S.1041-1050
54. x-Technik IT & Medien GmbH (2015) Additive und hybride Produktion mit 3D-Druck, Additive Fertigung
55. Deradjat D, Minshall T (2015) Implementation of additive manufacturing technologies for mass customization. In: International association for management of technology IAMOT 2015 conference proceedings

56. Werner H (2010) Supply chain management: Grundlagen, Strategien, Instrumente und Controlling. Gabler Verlag, Wiesbaden
57. Schmitt R, Pfeifer T, Beaujean P (2014) Qualitätsgerechte Organisationsstrukturen, Masing Handbuch Qualitätsmanagement, München. Carl Hanser Verlag
58. Augustsson R, Becevic D (2015) Implementing additive manufacturing for spare parts in the automotive industry. Chalmers University of Technology, Report no E2015:108
59. Khajavi S, Partanen J, Holmström J (2013) Additive manufacturing in the spare parts supply chain. Elsevier B.V

Part IV
Guinea, Conakry

Chapter 4
Towards a Smart Metropolitan Regional Development—Spatial and Economic Design Strategies: Conakry

Gora Mboup, M. Ibrahima Camara, Nene Mariama Balde, Mustafa Sangare and Khalil Fofana

Abstract The dynamic transformation of the Conakry Metropolitan Region from a population of 38,800 inhabitants in 1950 to a population 1.9 million inhabitants million in 2015 has occurred in different paces, times and places in terms of numbers, urban form and structure (density distribution, and compactness). When a city reaches one million or more of population size, it becomes more complex for its planning, design and management compared to a small city. For instance, when a city reaches one million inhabitants, its accessibility will be largely determined by elements of its urban form and structure as well as elements of transport infrastructures and, recently, elements of Information and Communication Technologies (ICT). This accessibility determines in large the interaction between economic and spatial design strategies. After the introduction, this study is divided in five sections. **The first section** presents the "Conceptual framework of the Smart Metropolitan Regional Development" of the study; **The second section** presents "Urbanization of the Conakry Peninsula in Time and Space"; **The third section** focuses on the "Urban Accessibility and Mobility"; **The fourth section** presents "Urbanization, Economic Growth and Metropolitan Regional Development" including economic growth, human development and sustainability, inclusion, resilience and prosperity; **The fifth section presents** "Towards a Smart Metropolitan Regional Development" including human development, environmental sustainability and city prosperity; **The sixth section**, based on the development of the previous sections, presents the "Spatial and Economic Design Strategies for a Smart Conakry Metropolitan Regional Development".

G. Mboup (✉)
Global Observatory Linking Research to Action (GORA) Corp, New York, USA
e-mail: gmboup@gora4people.org

M. I. Camara · N. M. Balde
Ministere de La Ville et de L'Aménagement Du Territoire, Conakry, Guinée

M. Sangare
Global Observatory Linking Research to Action (GORA) Corp, Washington DC, USA

K. Fofana
Global Observatory Linking Research to Action (GORA) Corp, Toronto, Canada

T. M. Vinod Kumar (ed.), *Smart Metropolitan Regional Development*,
Advances in 21st Century Human Settlements,
https://doi.org/10.1007/978-981-10-8588-8_4

Keywords Smart metropolitan region · Economic & spatial design strategies Urbanization · Urban system · Urban form and structure · Streets and other public spaces · Human development · Accessibility · Sustainable · Inclusive Resilience · Prosperity

4.1 Introduction

Large urban agglomerations such as Conakry provide opportunities for economies of scale and agglomeration, but it will also call for large investments in infrastructures to respond to the increased demand for water, sanitation, solid management, energy, streets and public spaces, and mobility, etc. They will also require efficient institutions for the management of social demand and equity such as on education and health as well as protection of people against violence and insecurity. Economies of scale and agglomeration economies are, for instance, greater in metropolitan areas where accessibility infrastructures are able to answer accessibility to services' needs with higher access to markets and resources than those where accessibility is impeded by deficient accessibility infrastructures. Efficient accessibility infrastructures will allow large-scale production of goods and services that can be distributed within the metropolitan regions and beyond with time, cost and reliability opportunities [1]. Without efficient accessibility systems, a metropolitan region losses its economic power and remains just clusters of disconnected settlements. The latter is the current situation of the Peninsula of Conakry as most African metropolitan region.

As a peninsula in the Atlantic Ocean, Conakry is located in a very low zone with an average elevation of 13 metres above sea level and with several watersheds mostly backfilled. Considering its geographical location, Conakry has the potential to be an agriculture, green city in addition to its huge potential of fish production. Wetlands areas are sources of income from agriculture, livestock, and crafts, among other activities. They play an important role in maintaining the water quality and the prevention of natural hazards. These are also ideal places for reception and reproduction of waterfowl which are indispensable elements for the ecological balance of aquatic environments and key links in the food chain, hence the importance the preservation of wetlands. If well planned and designed, Conakry can be a smart city where citizens enjoy high quality of life.

Conakry as most African cities has not been able so far to respond to growth in accessibility demand and several other needs such as access to water, sanitation, management of solid waste, and streets and other public spaces key elements of a city foundation. From a population of about 38,800 inhabitants in 1950, the Conakry urban agglomeration (the Conakry Peninsula and other connected urban centres) has a population of 1.9 million inhabitants million in 2015. For the Peninsula alone, its population is estimated at 1.7 million according the 2014 Guinea Population and Housing Census (Map 4.1).

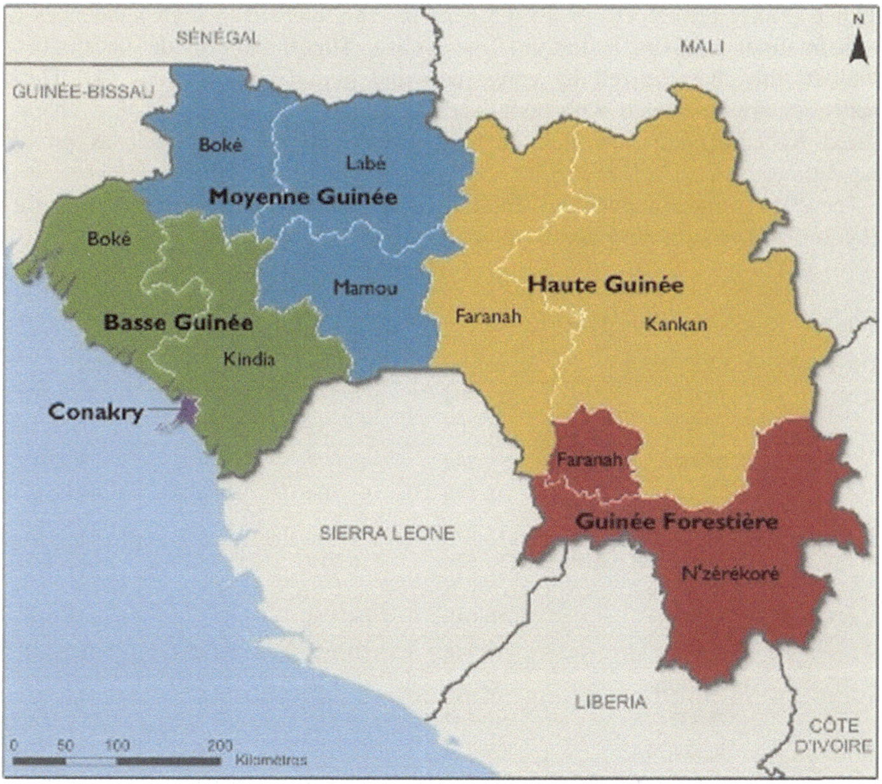

Map. 4.1 Map of the Republic of Guinea. *Source* National Institute of Statistics, Republique de Guinee, 2012, Demographic and Health Survey, and Multiple Indicators Cluster Survey

The existence of planning tools and instruments, particularly in the 1980s, has not prevented the proliferation of unplanned settlements along the urbanization of Conakry Peninsula. Due to lack of affordable housing and land, most people, particularly the urban poor, are forced to settle on the periphery of the Peninsula, particularly in the municipalities of Mototo and Ratoma. Since independence in 1958, Conakry extends considerably all along the peninsula limited only by its physical borders (the mangrove) and administrative boundaries (the prefectures bordering Dubréka and Coyah) [2]. The Conakry Peninsula has not been well planned with sufficient land allocated to streets and public spaces, and it lacks smart basic infrastructure and smart institutions and laws [3]. Many settlements in the Peninsula lack a sewerage system and rainwater drainage facilities, and adequate waste management sites are missing, which are key components of smart basic infrastructure along with connection to water and energy. Flooding during rainy seasons as well as uncollected garbage is frequent phenomena in all parts of the Peninsula, but particularly in poor settlements. Frequent energy shortages also affect the city's economy. In addition, infrastructure for non-motorized transport (e.g. pavements or sidewalks for walking and bicycle lanes for cycling) is often lacking, poorly developed, or on the decline. This has led to high incidences of traffic fatalities involving pedestrians and cyclists. Streets that provide space only to motorists are characterized by congestion and high CO_2 emissions [4]. These challenges are associated with poor land administration and governance, characterized by corruption and lack of transparency in public as well as private transactions.

The 2014 Guinea population (Table 4.1) and housing census provides the distribution of the population across region.

Table 4.1 Distribution of the Population of Guinea, 2014

REGIONS/PREFECURES		Ménages	Maseulin	Féminin	Total
BOKE	BOFFA	27 799	104 347	108 236	212 583
	BOKE	61 107	222 124	228 154	450 278
	FRIA	15 784	46 077	50 623	96 700
	GAOUAL	28 199	91 204	102 408	193 612
	KOUNDARA	17 635	63 167	66 807	129 974
	Sous-total région de Boké	**150 524**	**526 919**	**556 228**	**1 083 147**
CONAKRY	DIXINN	20 063	68 574	67 214	135 788
	KALOUM	9 345	30 810	31 697	62 507
	MATAM	20 133	71 536	71 719	143 255
	MATOTA	91 444	334 515	332 125	666 640
	RATOMA	95 786	328 320	324 463	652 783
	Sous-total Ville de Conakry	**236 771**	**833 755**	**827 218**	**1 660 973**
FARANAH	DABOLA	**25 297**	85 392	95 745	181 137
	DINGUIRAYE	26 827	94 387	102 082	196 469
	FARANAH	34 403	134 663	145 507	280 170
	KISSIDOUGOU	39 375	136 381	147 397	283 778
	Sous-total région de Faranah	**125 902**	**450 823**	**490 731**	**941 554**
KANKAN	KANKAN	46 867	233 617	239 742	473 359
	KEROUANE	**27 838**	102 126	105 421	207 547
	KOUROUSSA	29 148	130 55	138 072	268 630
	MANDIANA	22 760	167 873	168 126	335 999
	SIGUIRI	64 044	346 276	340 726	687 002
	Sous-total région de Kankan	190 657	980 450	992 087	1 972 537
KINDIA	COYAH	35 892	127 108	136 753	263 861
	DUBREKA	45 650	162 656	167 892	**330 548**
	FORECARIAH	32 716	116 500	126 442	242 942
	KINDIA	62 872	212 994	226 620	439 614

(continued)

Table 4.1 (continued)

	TELIMELE	49 153	129 954	154 455	284 405
	Sous-total région de Kindia	**226 283**	**749 212**	**812 162**	**1 561 374**
LABE	KOUBIA	16 633	45 533	54 637	100 170
	LABE	56 600	**143 583**	175 355	318 938
	LELOUMA	29 714	68 998	94 071	163 938
	MALI	42 168	134 957	153 044	288 001
	TOUGUE	21 472	55 788	68 492	124 280
	Sous-total région de Labé	**166 587**	**448 859**	**545 599**	**994 458**
MAMOU	DALABA	27 158	59 852	**73 825**	133 677
	MAMOU	58 139	148 157	170 824	318 981
	PITA	54 182	123 999	154 531	278 530
	Sous-total region de Mamou	**139 479**	**332 008**	**399 180**	**731 188**
NZEREKORE	**BEYLA**	38 089	**159 859**	166 223	326 082
	GUECKEDOU	45 455	138 407	**152 204**	290 611
	LOLA	28 232	81 654	89 907	278 561
	MACENTA	44 230	**133 045**	145 411	278 456
	N/ 1 RFKORI	60 970	**194 055**	202 894	396 949
	YOMOU	18 089	**55 262**	59 110	114 371
	Sous-total région de NZérékoré	**235 065**	762 281	**815 749**	1 578 030
Total Guinée		**1 471 268**	**5 084 307**	**5 438 954**	**10 523 261**

Source Republique de Guinee, 2015 (Decret D/2015/229/PRG/SGG), Portant Publication des Resultats Definitifs du Troisieme Recensement General de la Population et de l'Habitation realise du 1er mars au 2 avril 2014

Considering that the population of Grand Conakry Metropolitan Region will double the population of the Peninsula by 2040, the Government of Republic of Guinea commissioned a study for the "Grand Conakry Vision 2040". The study proposes various scenarios of territorial settlements. These scenarios are aimed at preventing disasters that can hamper ecological, social, economic and urban development. From these scenarios, a "synthesis, balanced and controlled scenario" is structured around three levels of intervention: (1) at the metropolitan level to develop and strengthen urban polarities; (2) at the agglomeration level to channel and structure the Conakry urban extensions; and (3) at the peninsula level to infill the city along a urban renewal approach. Recognizing that the Master Plan of the city is out-dated, the "Grand Conakry Vision 2040" will first focus in the development of a new generation of master plans: the master plan of the Grand Conakry and the national land use planning scheme (SNAT). The Grand Conakry Vision 2040 will consist of:

- At the metropolitan level, to develop and strengthen urban polarities: (1) Fixing and attracting populations in urban polarities; (2) Building a strong metropolitan region based on territorial solidarity and; (3) Preserving the natural and agricultural environment to enhance the productive force of the territory.
- At the agglomeration level, to channel and structure urban extensions: (1) Preventing the process of urban sprawl in order to preserve the natural environment through the establishment of perimeters for urban land use; (2) Structuring, reorganizing and controlling the urban extensions.
- At the peninsula level, to infill the city along the urban renewal approach: (1) Building an efficient, safe, healthy and pleasant capital and; Optimizing the use of the soil by densification of its structure to accommodate new urban areas with improved living conditions.

The Conakry case study will consist of analysing the economic and spatial strategies undertaken by the Government of the Republic of Guinea in the Grand Conakry Vision 2040 for a Smart Conakry Metropolitan Regional Development. It will be developed through the following sections:

The first section "Conceptual framework of the Smart Metropolitan Regional Development" is based on its City Foundation (urban planning, basic infrastructures and land tenure) and Institutions as well as on the Economic Development, the Infrastructure Development, Environmental Sustainability, the Social Development, the Social Inclusion, Disaster Risk Exposure and Resilience, and Peace and Security. One element of the City Foundation, which is the spatial design, and the element of Economic Development will play the role of interface vis-à-vis the other elements of the conceptual framework. Economic and Spatial Design Strategies along cannot make a regional metropolitan region smart, the ways they interplay with the other dimensions are crucial.

The second section is specifically dedicated to Urbanization of the Conakry Metropolitan Region in Time and Space covering: The Formation of the Conakry Metropolitan Region from 1950–2015; Guinea's Urban System; spatial planning and change in urban land use of the Conakry Metropolitan region, elements of the city foundation.

The third section focuses on the Urban Accessibility and Mobility: Streets as drivers of urban accessibility; and Transport modes and infrastructures, traditional focus of urban mobility studies.

The fourth section will present "Urbanization, Economic Growth and Metropolitan Regional Development" including economic growth, human development and sustainability, inclusion, resilience and prosperity. Though cities are potentially engine of prosperity with their power of economies of scale and agglomeration as well as of technology innovation and diffusion of ideas, the Relationship between urbanization and development had been uneven in the Conakry Metropolitan Region. This section will assess the main reasons, including the urban form and structure, and the urban accessible that have contributed to the relatively low economic performance of the Conakry metropolitan compared to many cities with similar size cities in other developed regions.

The fifth section presents "Towards a Smart Metropolitan Regional Development" including human development, environmental sustainability and city prosperity.

The sixth section, based on the development of the previous, presents the "Spatial and Economic Design Strategies for a Smart Conakry Metropolitan Regional Development". The main objective of the Grand Conakry Vision 2040 is to "improve the living conditions of the residents of Conakry and adapt land-use planning and planning policies to the rapid urbanization of the metropolitan region" [5]. The "Grand Conakry—Vision 2040" is conceived as the preliminary framework for the elaboration of a future Master Territorial and Urban Planning. Its objective is to define the main key points and themes to be considered in order to make Conakry a modern liveable metropolis safeguarding the environment. The study aims to be the reference framework for several ministerial departments, communities and local authorities to set a long-term urban development target by 2040 [6].

4.2 Conceptual Framework, Methodology and Data

The Grand Conakry Vision 2040 occurs in the era of the digital revolution in Conakry as it occurs everywhere in the world. For these past 15 years, the Guinean government has taken various initiatives favourable to the development and use of ICT at all levels. It has created legal institutional framework to support regulatory mechanisms on the development and use of ICT. We are taking it further with the development of a holistic approach in the making of Conakry a Smart Metropolitan region through a better integration of ICTs in the planning, designing and managing of metropolitan regions. The opportunities for ICT to support the overall urban challenges and opportunities are enormous, and the Conakry Metropolitan Region must integrate and use ICT solutions to facilitate the greater provisioning of urban services. Through increased efficiency and innovation, ICT increases economies of scale and agglomeration, and promote diffusion of knowledge, even at the smaller human settlements. Small settlements as small firms are benefiting from ICT in making available their goods and services beyond their territories. Virtual offices, virtual networks, teleconferences are increasing large public participation and inclusion. Digital firms can start and scale up quickly with relatively little staffing or capital investment. ICTs harness the benefits of agglomeration economies in easing circulation of goods and services and encouraging polycentric urban development and allowing synergies between centres and sub-centres. They intensify urban nodes and corridors to maximize the benefits of concentration.

Following the exigency of the city of the 21st century that calls for sustainability, inclusion, resilience and prosperity, the planning and management of human settlements must take into consideration the gain in knowledge on various conditions that make cities smart, green, ecological, liveable and healthy; and the progressive emergence of the ICT infrastructures and their correlates such as social media and in general big data. This is the context where we are introducing the smart metropolitan regional development to unlock the potential of Conakry metropolitan region to be sustainable, inclusive, resilient and prosperous.

4.2.1 *Smart Metropolitan Regional Development: Concepts and Components*

The New Urban Agenda adopted in October 2016 in Quito encourages governments, including the government of Guinea as a member of the United Nations, to commit themselves to adopting a smart-city approach that makes use of opportunities from digitalization, clean energy and technologies, as well as innovative transport technologies, thus providing options for inhabitants to make more environmentally friendly choices and boost sustainable economic growth, and enabling cities to improve their service delivery (NUA, paragraph 66) [7].

In our study, a smart metropolitan regional development is viewed as a sustainable, inclusive and prosperous metropolitan regional development that promotes a people-centric approach based on three core components—Smart Metropolitan Region Foundation, Information and Communications Technology (ICT) and Smart Institutions and Laws. Here The Metropolitan Region Foundation goes beyond the city foundation and includes elements of mobility across urban centres that form the metropolitan region. First each urban centre must have a smart city foundation and smartly connected to the other urban centres. Smart Metropolitan regional foundation, institutions and laws, and ICT are the pillars of the seven other dimensions of a smart metropolitan region: Infrastructure Development, Environmental Sustainability, Social Development, Social Inclusion, Disasters Exposure, Resilience, and Peace and Security. **Infrastructure Development** complements the basic infrastructure services under each smart metropolitan foundation and extends to actual investment and advancement of services such as transport, ICT, industrial energy, education, health, etc. **Environment Sustainability** is comprised of elements of Climate Change, Biodiversity, Waste Management, energy, transport, building and pollution. **Social Inclusion** includes aspects of participation in decision-makings as well as equal opportunities for growth and prosperity. **Social Development** encompasses elements of education, health, public space, social inclusion and social capital. **Disaster Exposure** incorporates elements of mitigation and adaptation to various disasters such as flooding, droughts, storms and earthquakes. **City Resilience** is composed of elements of city foundation, environment, social capital, and social development. **Peace & security** includes the elimination of all forms of discrimination and violence and conflicts, including domestic violence, violence in public places, crime, armed conflicts, terrorism, etc. An insecure metropolitan region limits opportunities for investment and economic growth and cannot be a smart metropolitan region (Fig. 4.1).

Numerous societal problems are explored and addressed in urban and regional planning agencies, including urban growth, unemployment and economic revitalization, transportation, environmental degradation and protection, neighbourhood decay and redevelopment, conservation of land and natural resources, provision of open space, parks and recreational facilities, etc. New urban planning instruments are becoming available with the worldwide spread of ICTs. They make it possible

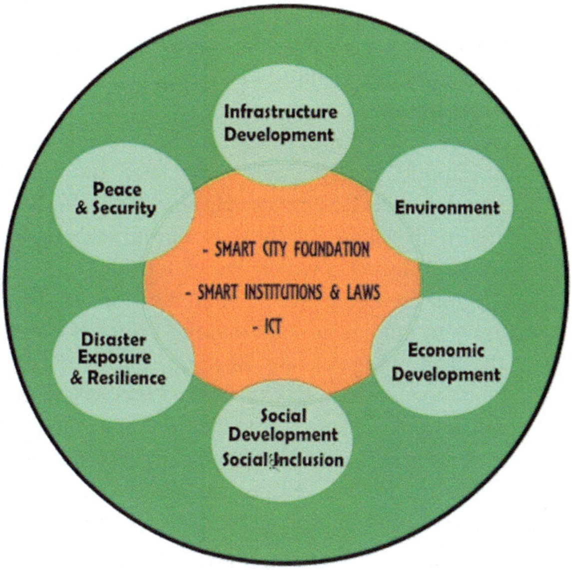

Fig. 4.1 Smart Conakry metropolitan regional development conceptual framework. *Source* Adapted from Mboup G et al. (2016). Smart city foundation—drivers of smart cities. In Vinod et al. (2016). Smart economy in smart cities, Springer

to adopt innovative e-planning approaches, strengthen communication between urban stakeholders, and make communication available at various stages of the planning process. Local governments can engage their citizens with real-time information to gain support for policy initiatives, identify unforeseen concerns, and recognize potential conflicts [8].

Cities are dynamic living organisms that are constantly evolving. ICT has begun to turn some places into real-time cities. This rapidly changing society makes the assessment and anticipation of future needs of city dwellers in terms of services, including transport, water, energy, employment, education and health, even more problematic. To address the complex problems of city planning it is not sufficient just to be concerned with the physical structure of the city; the interplay of intangible economic, social and environmental factors needs to be considered as well [9]. Planning procedures make use of models that show historic and present situations and communicate planned situations [10]. The introduction of ICTs allows planners and planning departments to carry out new actions or to implement conventional practices through new tools, such as GIS, virtual reality technologies, e-participation devices, including public participation GIS applications, among other tools, with the aim of improving conventional decision-making processes. The provision of better planning and urban management services, more efficient, with lower costs and, at the same time, a more collaborative and participative, transparent and accountable planning decision-making process are some of the basic objectives usually associated with the move from conventional urban planning to e-planning [11, 12].

ICT also can increase economic productivity and competitiveness through increased smart mobility that reduces traffic congestion and commuting costs. Efficient and fast transport, in turn, can increase labour productivity by reducing commuting times, and increasing worker productivity. ICT reduces the use of motorized means and transport and promotes walking and cycling. This will improve the overall health of city residents, thereby reducing the health and economic costs of workers who are absent due to illness. Another important benefit is on consumer expenditure, as less is spent on cars and fuel. Non-motorized means of mobility support specific industries, such bicycle shops, tourism, retail activity, construction and real estate development that highlight liveability. It can also lead to a drop in the share of the household budget devoted to motorized means of transport, such as cars, which are expensive.

The digital dividends are real and the Conakry metropolitan region must take the opportunity to efficiently integrate the use of ICT in all sectors of its development. From the beginning of the 21st century a digital citizenship, particularly the "Millennial" generation, has started to emerge in Conakry as in many African cities. The rapid adoption of digital technologies in the economy will have huge benefits directly as well as indirectly. However, ICT alone will produce little in the smart metropolitan regional development. 'Smart' is not an end in itself; It is the way ICT is integrated in the city development that will determine the city smartness. Maximizing the digital dividends requires better integration of ICTs with the other factors of smart cities such as: city foundation, infrastructure development, environment sustainability, social development, disaster prevention, resilience, peace and security. ICT can be seen as substitution and catalyst factors, but the other factors (or the analogue part) of the operation are crucial in making smart economy. Those factors include the city foundation, institutions and laws, infrastructure development, social development, social inclusion, environmental sustainability, disasters prevention and resilience (particularly in the context of climate change), peace and security. Most tasks have an aspect that cannot be automated by technologies and that requires human judgment, intuition, and discretion. This is where lies the crucial role of human capital associated with institutions and laws, efficient governance, management and administration. Without improvement on accountability at all levels, ICT alone cannot change the outcome of the economic productivity equation. African Countries that are able to swiftly adjust to this evolving digital economy will reap the greatest digital dividends, while the rest are likely to fall behind. This is the context where we are introducing economic and spatial design strategies for a smart metropolitan regional development to unlock the potential of Conakry metropolitan region to be smart, sustainable, inclusive, resilient and prosperous.

4.2.2 Urban Policies, Planning and Legislation

The Republic of Guinea has a number of spatial planning and management instruments at the territorial, regional and urban levels, produced from the second half of the 1980s along with the liberalization and decentralization of the national economy during the Second Republic. These instruments include, among others, the National Territorial Planning System in 1991 (Schéma National d'Aménagement du Territoire—SNAT), the Regional Planning and Development Schemes of Maritime Guinea, the Conakry Urban Development Plan, the Land and Estate Code, and the Urban Planning Code. To complement this system, other documents such as Urban Audits and Referential Urban Plans (Plans d'urbanisme de reference—PUR) were developed from 2000. Similarly, the Ministry of Urban Planning, Housing and Construction has also been in the process of adopting the Construction and Housing Code. The Guinean government also adopted the Conakry Urban Development Plan [Plan de Développement Urbain de Conakry (PDUC)]. It has also adopted four Regional Spatial Planning and Development Schemes (Schémas Régionaux d'Aménagement et de Développement—SRAD). At the local level, in order to guide major development actions, urban planning and development schemes (les Schémas d'Aménagement et d'Urbanisme—SDAU) were also conceived for 14 secondary cities out of 33 and four rural development communities [13].

From a situation analysis, the Conakry Urban Development Plan encompasses various land use planning for sustainable urban development including: (a) long-term urbanization scenarios; (b) urban form and structure; (c) residential, administrative, commercial and industrial zoning; (d) land tenure administration; (e) sectorial studies including on housing, equipment, transport, infrastructure and urban services; and (f) decennial programming. The National Territorial Planning System (SNAT) defines the structuring elements, which, among others, the territorial organization, the program or development areas and an urban framework of five levels of development centres, communication axes, large natural balancing areas, regional-level secondary structure and implementing institutions [14].

Along the development of urban and territorial planning guidelines and instruments, the Guinean government have adopted several codes land and planning codes, including: Urban Panning Code (Code de l'urbanisme), Land and Estate code (Code Foncier et Domanial), and Building and Construction Code (et du Code de la Construction et de l'Habitation) [15].

The land and Estate code was elaborated and adopted in 1992. It deals with the general rules applicable to the land development, the determination of land use plans concerning local authorities, parts of local authorities or groups of local authorities are determined by law or regulation. The code also prohibits the occupation of Maritime Public Land. Adopted by the law L98/N° 012/1998 the Urban Planning Code includes a regulatory part and a legislative part. The code deals with the foundations of town planning in the Republic of Guinea; it provides town planning rules that incorporate the national urban planning by-law, master plan of urban planning, detailed planning codes; building permit, alteration and demolition, etc. [15]. The Building and Construction code provide rules and regulations on, for instance: the distance to be respected in relation to the occupation of the DPM is fixed at 100 m. Due to global warming and rising sea levels, this distance may be reviewed on the basis of a preliminary study of this phenomenon (Article 26, Title II: Chapter 1); urban renewal operations, urban restructuring in accordance with the requirements of the local SDAU or POS (Article 58 Title III: Chapter 1).

Guinea has also made a special commitment at the international level to promote sustainable and resilient urbanization as promulgated in the Sustainable Development Goals (SDGs) adopted in September 2015, the New Urban Agenda adopted in October 2016, the COP 21 (Paris Declaration), the Sendai Framework on Disaster Risk Reduction, the Addis Ababa Declaration on Population and Development and the Africa Agenda 2063.

However, due to the low level of implementation of urban legislation, rules and regulations, as it is observed in most African cities, urban planning and management guidelines as well as the urban planning, land and building codes were not respected along the urbanization of Conakry. Most settlements in Conakry have occurred unplanned. Many households have been forced to settle in flood prone areas due to inaccessibility to planned land in the city of Conakry. Due to poor land administration and governance, there is no compliance with standards of occupancy of the space leading a high building density and irregularity of the urban fabric. Building in unplanned wetlands without adequate drainage systems expose the population of Conakry to flooding that cause various human, material and financial damages and losses. Industrial areas have also been transformed into human settlements exposing the population to air pollution. More than two-thirds of households in Conakry and other urban centres reside in informal settlements that lack most basic services such as connection to water and drainage facilities [13].

4.2.3 Methodology and Data

The Smart Metropolitan Regional Development is composed of several dimensions. Put them together to assess the degree of smartness will require the development of composite indices using sophisticated multi-level statistical analysis including Principal Component Analysis. The complexity of composite index is the fact that it is strongly influenced by several factors including: the normalization, the standardization and the degree of association between variables and the number of variables that compose the index. The analysis of the Conakry Smart Metropolitan Regional Development (SMRD) will consist first of the analysis of each of its components with the production of individual indicators such as the proportion of household with access to improved water, followed by a development a composite index, the SMRD index (SMRDI) with sub indices such as the Smart City Foundation Index (SCFI) built upon the spatial planning, the land tenure and the basic infrastructure indicators. However, an index can also be conceptualised differently using different variables. This calls for caution when interpreting a composite index. For instance the Quality of Life is conceptualised and measured in different ways across studies. The Human Development Index includes indicators of health and education, alongside an economic component (Gross Domestic Product or GDP); the Legatum Prosperity Index considers quality of life to be multidimensional, including both wealth and wellbeing; The Economist Intelligence Unit's Quality of Life Index links life satisfaction to health, family life and community life;

and the OECD's Well-being Initiative has two dimensions: 'material living conditions' and 'quality of Life' [16]. The following section will provide the different components of the SMRDI and the variables used for their development.

Putting the dimensions together to develop the Smart Metropolitan Regional Development Index (SMRDI)

The SMRDI, which is built here, is meant to serve research as well as policies. For the latter, the SMRDI will be built using existing mechanisms of the production of indicators globally as well as nationally. Building the SMRDI through existing mechanism of development of indicators such as the monitoring of the Africa Agenda 2063, the Sustainable Development Goals (SDGs), the New Urban Agenda, the Sendai Framework, the COP21, the Addis Ababa Declaration on Population and Development as well as the monitoring of national programmes and plans will ease its implementation at the local level for SMRD policies. As part of the international communities, Guinea has also endorsed all the above global agendas. Box 4.1 provides for each dimension of the SMDR variables, quantitative or qualitative, included for its measurement.

Box 4.1 Smart Metropolitan Regional Development Index (SMRDI)

Dimensions	Definitions/variables
City Foundation	**Spatial Planning** Planned settlements, Streets, Other Open Public Spaces **Land Tenure** Land: Documentation or perceived eviction Housing: Proportion of urban population living in slums, informal settlements or inadequate housing **Basic Infrastructures** Connection to piped water, Connection to sewerage system, Connection to electricity, management of solid waste
Institutions & Laws	Urban Planning Codes, Urban Policies, Urban Plans, Unplanned settlements, Resilient building codes, standards, development permits, land use by-laws and ordinances, and planning regulations
ICT	Connection to internet, Mobile phone, Computer, Broadband
Infrastructure development	Transport and road infrastructure, Energy, Knowledge infrastructure, Health infrastructure
Environmental sustainability	Air quality (PM10), CO_2 emissions, energy and indoor pollution
Economic Development	City product, Employment Primary, Secondary and Tertiary sectors

(continued)

(continued)

Dimensions	Definitions/variables
Social Development	Education: literacy, primary, secondary and tertiary enrolment Health: life expectancy, under-five mortality rates, HIV/AIDS, morbidity and nutrition: Social capital, Public space
Social inclusion	Income/consumption Inequalities: Gini coefficient, Health, Education, Access to basic services
Disaster Risk Reduction/Resilience Strengthen resilience and adaptive capacity to climate-related hazards and natural disasters in all countries	Number of deaths, missing persons and directly affected persons attributed to disasters per 100,000 population Adoption and Implementation of national disaster risk reduction strategies in line with the Sendai Framework for Disaster Risk Reduction 2015–2030 Proportion of local governments that adopt and implement local disaster risk reduction strategies in line with national disaster risk reduction strategies
Peace & Security	Proportion of population that feel safe walking alone around the area they live; Proportion of women and girls aged 15 years and older subjected to sexual violence by persons other than an intimate partner in the previous 12 months, by age and place of occurrence; Proportion of ever-partnered women and girls aged 15 years and older subjected to physical, sexual or psychological violence by a current or former intimate partner in the previous 12 months, by form of violence and by age; Average share of the built-up area of cities that is open space for public use for all, by sex, age and persons with disabilities; Proportion of population that feel safe walking alone around the area they live

Source Mboup, G. et al., 2018 (Forthcoming). The Relevance of Smart City in the African Context. Springer. All the variables of this table are not included in the Conakry SMRDI due to lack of data

Data sources and quality assurance

Data are generally obtained from various sources depending on the level of analysis. For the SMRDI, as an impact product, population-based data is most indicated since they provide information on people's quality of life. Since independence, Guinea has already conducted three population censuses (the last one was held in 2014) and several household surveys such as Demographic and Health Surveys, the

Multiple Indicators Cluster Surveys, the Living Standard Measurement Survey and several other household surveys. In addition to these population-based data, the analysis of spatial form such as density, compactness, streets call for the use of the Geographical Information System (GIS) for the analysis of remote sensing data with the integration of population-based data. However, these population-based as well as GIS sources feature limitations in providing information on urban planning, housing, land tenure, etc. For these variables, our study will heavily rely on official administrative reports including the Visions Habitat 2010, the Grand Conakry Vision 2040, the Urban Audits, the National Territorial Planning document (Schema National d'Amenagement du Territoire-SNAT), etc. The analysis of the component of the SMRDI such as Institutions and Laws call for a Policy Analysis of urban legislations, rules and regulations in Guinea, it is more a content analysis of administrative records rather than a spatial or a population-based analysis. Finally, the development of the SMRDI will be based on quantitative as well as qualitative data: population-based, spatial and content analysis, etc.

Challenges and emerging successes in the definition of urban components for the development of the SMRDI

The development of the SMRDI will face challenges such as data comparability, the delimitation of urban boundaries, etc. Critical issues and huge challenges the development of the SMRDI will face is the choice of data sources in the situation where it is noted significant discrepancies between national sources. Lack of data comparability and non-compliance with international standard make the development of SMRDI challenging. A lack of uniform definitions across data sources poses an even greater challenge for aggregating data and comparing levels of SMRDI [17]. Urban human settlements have various definitions: urban agglomeration, metropolitan region, city proper, municipalities, etc. These definitions also vary across countries [18]. The **urban agglomeration** is defined as the built-up or densely populated area containing the city proper; suburbs, and continuously settled commuter areas. This may be smaller or larger than the metropolitan area. Other similar UN definition: Comprises a city or town proper and the suburban fringe or thickly settled territory lying outside, but adjacent to, its boundaries. A single large urban agglomeration may comprise several cities or towns and their suburban fringes. (United Nations. 1998. Principles and Recommendations for Population and Housing Censuses). This has been regularly published by UNDESA and it forms the official reference of urban population as in the Urbanization Prospects publication. The **metropolitan area** is the set of formal local government areas, which are normally taken to comprise the urban area as a whole and its primary commuter areas. The **city proper** is the single political jurisdiction, which contains the historical city centre [19]. The Grand Conakry Vision 2040 considers the three components at three levels: the city of Conakry as the Peninsula, the urban agglomeration through the extension of the peninsula and the metropolitan region through the development or reinforcement of urban polarities.

Other important measurement, one can be interested is how the spatial growth occurs along the population growth; a crucial information for spatial planning. There is **emerging success in disaggregation of urban growth** made by UN-Habitat and partners using remote sensing and GIS to analyse changes in urban extent in different components such as infill, extension, inclusion and leapfrog. City growth within the same urban extent is considered as **city infill** that "consists of all built-up pixels added in the new period that occupy urbanized open space within the urban extent of the earlier period". City can also grow beyond its previous urban extent in a contiguous manner. This type of city growth is known as city **extension** that consists of all built-up pixels added in the new period that constitute **contiguous** urban clusters that are **attached** to the urban extent of the earlier period. The growth of city can also occur through **inclusion** that consists of all urban, rural, or suburban built-up pixels that were **outside** the urban extent in the earlier period and are now within the urban extent of the new period. There is also another trajectory of growth where city expand to **over rural open space** that were **not attached** to the urban extent of the earlier period or to new extension clusters. This is known as leapfrog [20]. All these categories of spatial growth have occurred along the urbanization of the Conakry Metropolitan Regional over a century and will be analysed here.

4.3 Conakry Peninsula in Time and Space

4.3.1 Urbanization of Conakry Peninsula

Cities form and grow in many different parts of the world for various different reasons such as rural-to-urban migration, natural population growth, economic opportunities, politics, environmental changes, social conflicts, etc. In a given country, the way these factors play are determinant for the size and the national distribution of cities that, in turn, form the urban system. Conakry became the capital of French Guinea in 1904 and prospered as an export port, particularly after a railway to Kankan ease accessibility to the interior of the country for the large-scale export of groundnut, for instance [21]. In decades after independence, the population of Conakry boomed, from 50,000 inhabitants in 1958 to 600,000 in 1980, year of economic liberalization and rural migration, to over 1.9 million in 2016. Since 1950, the city of Conakry was, indeed, marked by continuous increase of its share in the national urban population from 15 to 30% in 1960. This exponential growth made it to reach 58% in early year, 1970, and 62% in 1980. However From 1980, there is a decline of its share to less than 40% with the growth of other cities such as the second largest city in Guinea, NZerekore (343,000 inhabitants) [22].

Since the independence of the Republic of Guinea in 1958 to the advent of the Second Republic in 1984, urbanization of the Peninsula was mainly the product of the development of informal settlements. Public or "parapublic" production of housing was limited to the preparation of land and the construction of housing at a very high cost, without concern for profitability and even less respect for comfort standards. During that period, the Republic of Guinea received very limited financial assistance from the international communities [13]. For over a quarter century the Government of Guinea hardly benefited from certain forms of advantageous loans granted by international financial institutions. Private land development and housing production have been largely the responsibility of the informal sector. Faced with the weakness of the public offer and the private real estate development, the populations occupied space in an anarchic way. This resulted to the proliferation of informal settlements lacking most basic services such as connection to water, electricity and sewerage systems.

Today the Conakry Peninsula is composed of five municipalities: Kaloum, Dixin, Matam, Matoto and Ratoma. Along the urbanization process, the urban form and structure of Conakry is characterized by: (1) a linear geographical structure along the peninsula (five kilometres wide over forty kilometres); and (2) a historical centralized organization of the development of the city, with all port activities, decision-making functions and administrations, concentred in the municipality of Kaloum, in an area of just 3 km^2, less than 1.5% of the urbanized area [5]. The Population in Conakry Peninsula is estimated at 1.7 million (1,660,973) inhabitants according to the 2014 Population and Housing Census. While most of the administrative and commercial buildings are located in the municipality of Kaloum that hosts 4% of the Conakry population, the majority of the Conakry population are concentrated in two municipalities, Matoto and Ratoma, 40 and 39% respectively. The municipalities of Dixin and Matam have a share on the Conakry population of 8 and 9% respectively [23].

The five municipalities of Conakry were created like the other municipalities of the country by the ordinance N° 019/SGG/PRG/90 of April 21st, 1990 on the organization and the functioning of municipalities in the Republic of Guinea. The **municipality of Kaloum**, subdivided into 13 neighbourhoods ("quartiers") is a peninsula of 25 km^2 of surface that consists of an insular part and a continental part. It is surrounded on all sides by the Atlantic Ocean except the eastern side, which borders with the municipalities of Matam and Dixinn. **The municipality of Dixinn,** subdivided into 24 neighbourhoods, is a residential area where are located the University of Conakry and several embassies. It is bounded on the West by the municipality of Kaloum, on the East by the municipality of Ratoma, on the North by the Atlantic Ocean and on the South by the rails of Niger and the municipality of Matam. **The municipality of Matam**, subdivided into 24 neighbourhoods, is limited to the North by the rails of Niger, to the South by the Atlantic Ocean, to the West by the municipality of Kaloum, to the East by the arm of the sea separating the Bonfi market district from that of Dabondi. **The municipality of Ratoma**, subdivided into 34 neighbourhoods, is a semi-rural area located in the Northeastern periphery of the capital. It is bounded on the East by Dubréka prefecture, on the west by Dixinn commune, on the South by Matoto municipality and on the North by the Atlantic Ocean. **The municipality of Matoto**, subdivided into 38 neighbourhoods, is located in the coastal area southeast of the capital; it covers a territory, largely covered with mangroves; it extends from the bottom of Dabondy to Lansanayah. It is limited to the East by the Prefecture of Coyah, to the West by the Commune of Matam, to the North by the municipality of Ratoma through the railway Conakry-Niger and to the South by the Atlantic Ocean (Figs. 4.2 and 4.3).

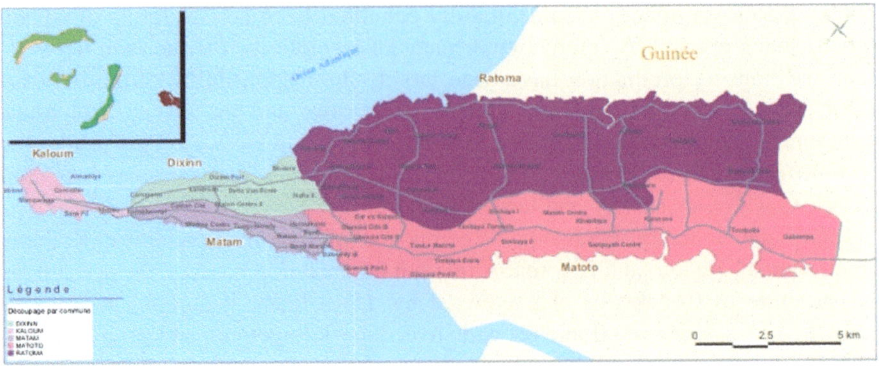

Fig. 4.2 Spatial illustrations of the municipalities of Conakry. *Source* Republic of Guinea, 2017. Administrative sub-divisions of the city of Conakry

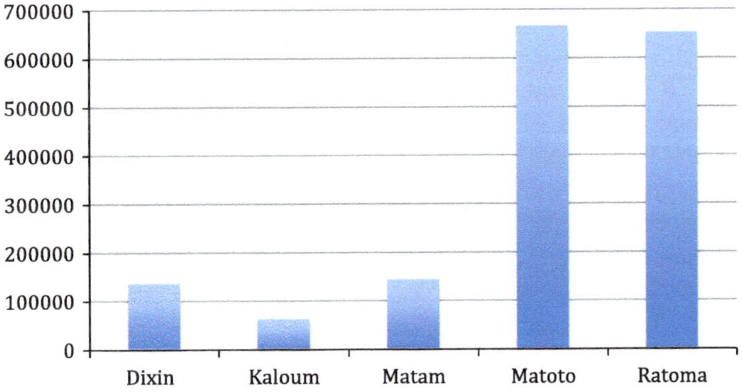

Fig. 4.3 Population of Conakry Peninsula, 2014. *Source* Institut National de la Statistique, Republique de Guinee, 2015. Census 2014

Data from the UNDESA's urbanization prospects 2014 revealed that during the period 1950–2015, the urban agglomeration of Conakry had increased from a population of 38,800 inhabitants in 1950 to 1.9 million inhabitants. The Conakry urban agglomeration population figure is slightly higher than the Conakry Peninsula figure as published in the 2014 Population and Housing Census due to the fact that it includes other urban settlements connected to the Peninsula (forming the whole urban agglomeration) (Fig. 4.4).

As shown in Fig. 4.5 (Growth rates of the urban population of Conakry, 1950–2015), the highest annual population growth rate occurred between 1950 and 1970 with a level of 12% or above. Since 1970, the population of Conakry urban agglomeration continues to increase but by a continuous smaller pace until it reach a growth of 3% in 1980. Since 1980, the annual growth rate of the urban agglomeration seems to stabilize around 3%.

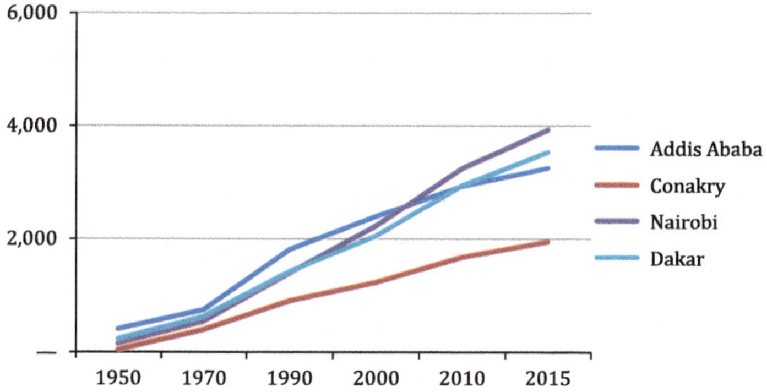

Fig. 4.4 Urban Population Growth of Conakry and other African cities of similar sizes in 1950. *Source* Data from United Nations (2015) World Urbanization Prospects: The 2014 Revision, UNDESA, Population Division, New York, NY, http://esa.un.org/unpd/wup/index.htm

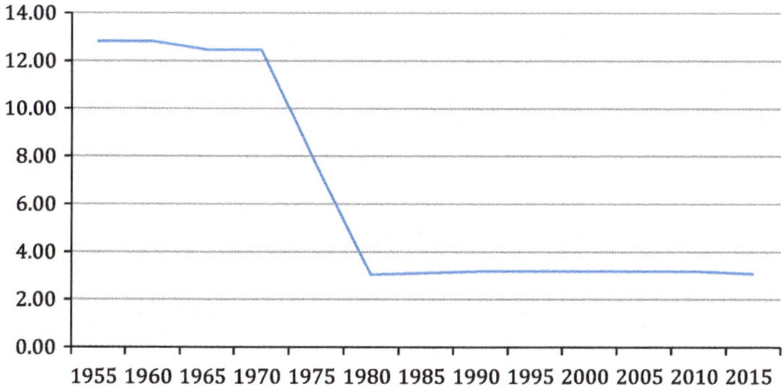

Fig. 4.5 Growth rates of the urban population of Conakry (1950–2015). *Source* Data from United Nations (2015) World Urbanization Prospects: The 2014 Revision, UNDESA, Population Division, New York, NY, http://esa.un.org/unpd/wup/index.htm

4.3.2 Guinea Urban System

The spatial distribution of cities of different sizes throughout the national space constitutes a country's urban system. The urban system is determined by the way cities are geographically distributed in the country according to their size. In the urbanization process, some countries were able to develop a balanced system of cities with no city does predominantly hold a large share of the urban population while others concentrate their population in one city that become a primate city. The conditions under which urban primacy emerges can vary greatly and evolve with country size, population density, and the stage in its urbanization process, among other factors [24]. Changes in primacy can be affected by all of the factors affecting urbanization more generally, including natural growth, migration and reclassification of settlements from rural to urban. A country's urban system can be linked to the organization of the government at the national, regional, and local levels [25]. The share of city to the total urban population can be a good indicator of a city performance, attractiveness or a simple political preference [26]. In turn, the urban system may determine the need for specific urban development policies, consistent with the size, growth and function of each city. For instance, when all the national investments, all health, education and commercial infrastructures are vested in a particular city, such as Conakry, people do not have other choices except moving to that city.

The Urban primacy as defined here characterizes the urbanization process in the Republic of Guinea during the last 60 years where Conakry has been a primate city with more than half of the national urban population living in the urban agglomeration. Since 1950, the city of Conakry was marked by continuous increase of its share in the national urban population from 15 to 30% in 1960. This exponential growth made it to reach 58% in early year, 1970, and 62% in 1980. However From 1980, there is a decline of its share with the growth of other cities such as the second largest city in Guinea, NZerekore (343,000 inhabitants). The regional capital of the regions of Kindia, Labé, Kankan and N'Zérékoré are only home to about one-sixth of the country's urban population; the third of the national urban population is distributed among the 29 other urban centres, including three regional capitals of the regions of Boké, Mamou and Faranah [13]. The permanent primacy status of Conakry is similar to the situation of Dakar in Senegal but opposite from most cities in East Africa where though they were primate cities at the onset had lost their primacy along the urbanization process. For instance in Kenya, the share of Nairobi in the national population was 49% in 1960, but in 2015 Nairobi represented only 33% of the national population. In Ethiopia there was a drastic reduction of the demographic power of Addis Ababa from 47 to 17%. In North Africa, expect in Egypt there is no single country that hold a share in the national urban population that reached 30%.

The permanent primacy status of Conakry can be associated with the fact that most administrative and commercial institutions are located in the city. According to the classification of the SNAT, in the hierarchy of city, only Conakry can be considered of higher order of development centre [13]. There is stagnation in many intermediate cities and the absence of a network of small urban centres offering basic services. There is need to empower other cities in other region in order to decentralize urban development in the country and create a balanced urban system. Conakry, alone, will not be able to distribute wealth to other regions (Fig. 4.6).

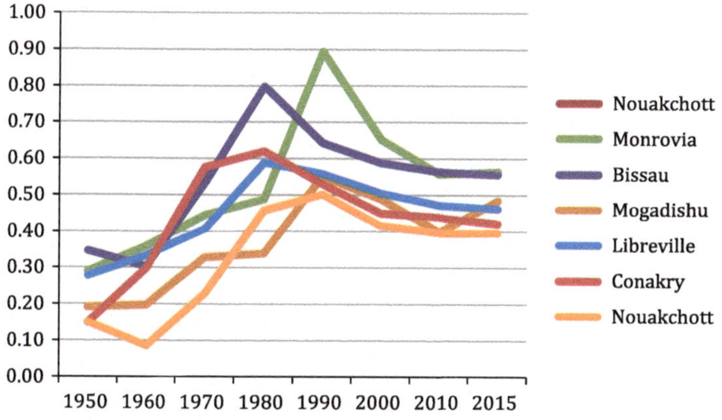

Fig. 4.6 Share of Conakry and other African cities in the national urban population (1950–2015). *Source* Computed using data from United Nations (2012) World Urbanization Prospects: The 2014 Revision, UNDESA, Population Division, New York, NY, http://esa.un.org/unpd/wup/index.htm

4.3.3 The Demographic Dividends of the Conakry Peninsula from 1950 to 2015

The monocentric form of the city of Conakry with most of the services located in the municipality of Kaloum has reached its limits leading to huge traffic congestion. The mobility in Conakry is heavily affected by the fact the city was conceived as a monocentric city where residential areas and workplaces are distinct, with the later concentrated in the centre of the city. Residential areas were also designed along economic class lines. This has led to social and economic fragmentation that disadvantage lower income in accessing basic services and prevent social interaction and integration. In addition to monocentric form of the city, most households are built in unplanned settlements that lack most basic services, particularly in Matoto and Ratoma, leading to frequent flooding and several other social and economic negative externalities.

Urban Densities as factor of economies of scale and agglomeration

The population size is an important component of urbanization, but to address various questions in urban economics, environment and infrastructure development, information on density of the population is valuable [27]. Positive production or residential externalities, as well as negative externalities such as congestion, are function of density among others [28]. Various policy responses as well as programmes and projects such as on transport and infrastructure are subject to how population is distributed across a city. Important factors of land use such as land prices should be also considered in this exercise as part of elements that can feed the land legislation and investment in infrastructure [29]. From an economic point of view, the higher the density of an urban agglomeration is, the lower are the costs of transactions, and the more prosperous is the economy. A deficient spatial structure fragments labour and consumer markets into smaller less efficient markets; it contributes also to higher transactions costs by unnecessarily increasing distances between people and places. It increases the length of the city infrastructure network and therefore increases its capital and operating costs. Lower densities and segregated land uses increase economic costs in terms of average commuting distances, public utility provision, and energy consumption. From an environmental point of view, a deficient spatial structure can decrease the quality of life by increasing the time spent on transport, by increasing air pollution, and contributing to the unnecessary expansion of urbanized areas in natural sites. Poor environmental quality can also lower a city's productivity.

Results from the 2014 Guinean Housing and Population Census show that Conakry is densely populated, across all its five municipalities (Dixin, Kaloum, Matam, Matoto and Ratoma) with an average of 3706 inhabitants per square kilometre on the total land area of the Conakry Peninsula of 450 km^2 (as 2009) [30].

With the right empowerment, Conakry's young population can lead the vision for the city of tomorrow and work for it. The ability of the youth, as the "Millennial", to adopt emerging technologies and transfer experiences from other parts of the world must be key drivers of Conakry's smartness in the long term. Various studies show that Africa's young people will be the driving force behind economic prosperity in future decades, but only if policies and programmes are in place to enhance their opportunities [31]. These policies and programmes include factors such as a strong rule of law, efficient bureaucracies, government stability, lack of corruption, and a stable business environment that encourages domestic and foreign investors. In order to benefit from its youthful population, the Conakry Metropolitan Region must embrace smart education driven by the availability of ICT and respond to the requirement of the job market of the 21st century. It is also important to encourage young people's participation in public life, and in policies, programmes, and services. However, youthful population could also present a significant risk and threat to social cohesion and political stability if Conakry fails to create sufficient economic and employment opportunities to support decent living conditions for them [32].

4.3.4 The Conakry City Foundation

4.3.4.1 Proliferation of Unplanned Settlements and Lack of Security of Tenure

The results of the survey carried out as part of the development of the "Vision Habitat 2021" strategic document reveal a "generalized precariousness" in terms of housing. Along unplanned rapid urbanization, human settlements occur in disaster-prone areas with lack of most basic services such as connection to water and sewerage systems. Sustainable urban development has been seriously hampered by proliferation of slums and informal settlements in the Conakry Peninsula. Over two thirds of human settlements in Conakry are in such conditions. This situation is

Fig. 4.7 Distribution of irregular settlements (lacking most basic infrastructures). *Source* Republique de Guinee, Ministere de l'Urbanisme de l'Habitat et de la construction, 2011. Politique nationale de l'habitat « Vision Habitat 2021 »

also observed in other Guinean cities. For instance out of 16 settlements in the city of Kindia, 11 are irregular and lack most basic services. In Labé, only a small portion (14%) of the total population lives in relatively well-equipped neighbourhoods. The city of Kankan, this figure is only 4% indicating the informality is general (96%). Similar situations are observed in the cities of Siguiri, Faranah and Kissidougou [13]. The lack of subdivision complicates the implementation of road networks, sanitation, water supply, electricity and other equipment reserves. This limits investment opportunities and plunges these areas into exclusion and marginality.

As illustrated in Fig. 4.7, irregular settlements (red colour) are predominant along the Peninsula, particularly in the municipalities of Matoto and Ratoma where reside 80% of the city's population. Only the municipalities of Kaloum, Matam and Dixin seem not to be affected by the proliferation of irregular settlements, but they represent only 20% of the Conakry population, 4, 9 and 8% respectively.

Information from the Vision Habitat 2021 document features a very dark picture of proliferation of slums and informal settlements in Conakry Peninsula. Towards the Northwest, along National 3, beyond Dubréka, the spontaneous extensions develop significantly in a very constrained environment, trapped between the reliefs and the mangrove. Heading South and South-east from Coyah and North-East along the railway line and from Nationale 1 to Kindia, the urban areas stretch along two axes: towards the commune of Wonkifong, on the one hand, where the constructions infiltrate on the dry mangrove and between the arms of rivers, and to the communes of Toguiron and Kendoumaya, on the other hand, where the urbanization unfolds in "fishbone" along the Nationale 4. Towards the Northeast, the urban extension develops mainly along the railway line, between PK45 and National 1 to Kouriah. On the other hand, the urban development is relatively limited along the National 1 on the portion located between Coyah and Kouriah because of a more accidental relief. This picture illustrates the fact that the proliferation of irregular settlements covers all corners of the Conakry moving towards the municipalities of Matoto and Ratoma as illustrated by the red colour of the map [13].

The Guinean government has put in place holistic urban development programmes for the "eradication of spontaneous habitat, rehabilitation of degraded housing in the central districts of Conakry" such as those in Vision Habitat 2021 and, recently, the Grand Conakry Vision 2040. These programmes aim at achieving, in two decades, "cities without slums" including rehabilitation, restructuring, and land regularization of irregular neighbourhoods as well as poverty reduction. It was expected that a pilot programme be launched in Conakry and in the fast-growing major cities (annual rate of urban growth above 5%, thus doubling the population every fifteen years) and very rapidly growing cities (annual rate of urban growth above 7%, thus doubling the population every ten years). The national program of cities without slums—diagnosis, inventory and adoption of an urban redevelopment law—was expected to be developed in 2012. The objective is to completely reduce the irregular habitat in Conakry, by 2021 and reduce by 50% the irregular neighbourhoods in large cities, while meeting the needs of other neighbourhoods [13].

4.3.4.2 Basic Infrastructures

In a smart city foundation, basic infrastructure such as piped water services, sewerage facilities, electricity sources and solid waste management are considered along the city planning. They are part of the city planning prior to settlements of households. In a sustainable city foundation, use of improved water from piped water services, sewerage facilities, solid waste management, energy for lightning is quasi universal. However, for the city of Conakry only connections to piped water facilities are quasi universal with a level of 90%, followed by connections to electricity (64%). The coverage of sewerage facilities is still very low, below 30% in 2012. In Conakry, the public collection of household waste concerns 55% of households. Industrial and toxic waste (contaminated waste from hospitals, laboratories, slaughterhouses and mining enclaves) is generally released in the nature, rivers or the sea, and it is treated in the same way as other wastes.

Due to lack of rainwater drainage system, the inadequacies and dysfunctions of the rainwater management network, and the obstruction of natural outlets by human settlements, flooding is frequent in Conakry. The annual rainfall in Conakry is 4000 mm, particularly in the months of July and August. Precipitation is reflected by very intense rainfall typical of the subtropical climate. Associated with this heavy rainfall, the particular topographical situation of the city of Conakry favours recurring floods during the raining season [5].

4.3.4.3 The City Foundation: Spatial Planning, Land Tenure and Basic Infrastructures

A smart city foundation is composed of three elements: Urban Planning & Design, Land Policies and Basic Infrastructure. For a city foundation to be smart, it must be

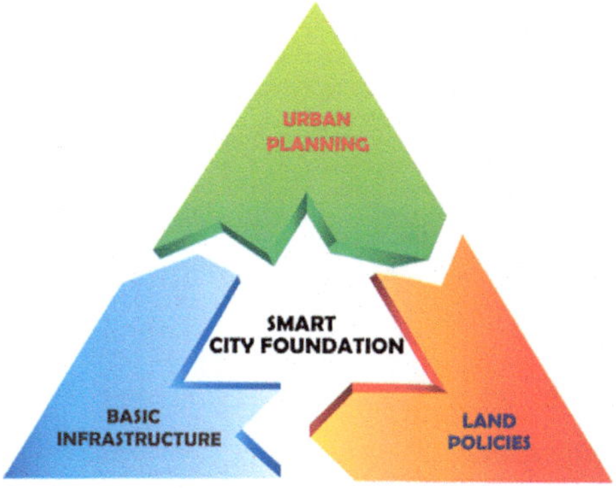

Fig. 4.8 City foundation conceptual framework. *Source* Mboup G et al. (2016). Smart city foundation—drivers of smart cities. In Vinod et al. (2016). Smart economy in smart cities, Springer

inclusive at the onset of the urban planning and promotes mixed neighbourhoods where social clustering is discouraged. Having all the poor living together creates slums and fuels instability and insecurity. Inclusive urban planning eases access to basic services (water, sanitation, housing, education & health) and to decent employment for all. A key element of smart urban planning is a smart street network that reduces travel time and encourages walking and social interactions. Smart urban planning enhances infrastructure development, environmental sustainability, economic and social development; makes cities resilient and prepared to overcome natural disasters; and promotes mixed neighbourhoods where services are walking distances from people's residences (Fig. 4.8).

Using the three main components of the city foundation, we have computed a composite index that represents the overall city foundation, the City Foundation Index (CFI) with values varying from 0 to 1. A value closes to 0 means that a city lacks most of the elements of the city foundation elements, and a value closes to 1 means that a city enjoys most of the elements of the city foundation. The Conakry Peninsula spatial configuration features the tale of a city with the municipality of Kaloum enjoying all key elements that make a city foundation smart: adequate urban planning, access to basic services (electricity, water, sewerage facilities) and a land tenure moderately well administered and governed. The municipality of Kaloum is followed by the municipalities of Dixin and Matam. The municipalities of Mototo and Ratoma lack most of the elements that make a city foundation smart.

While CFI of the municipality of Kaloun is estimated at 0.815, and the CFIs of Dixin and Matam at 0.644 and 0.643 respectively, the CFIs of Mototo and Ratoma are estimated at 0.312, 0.310 respectively. The low level of the CFI of Mototo and Ratoma is due to low coverage of a sewerage system, less than 30% compared to 64% in the municipality of Kaloum. The two municipalities also perform poorly in land governance and administration with a proliferation of informal settlements. While in Kaloum, the majority of households enjoy security of tenure, in the two municipalities less than 20% of households have security of tenure (Fig. 4.9).

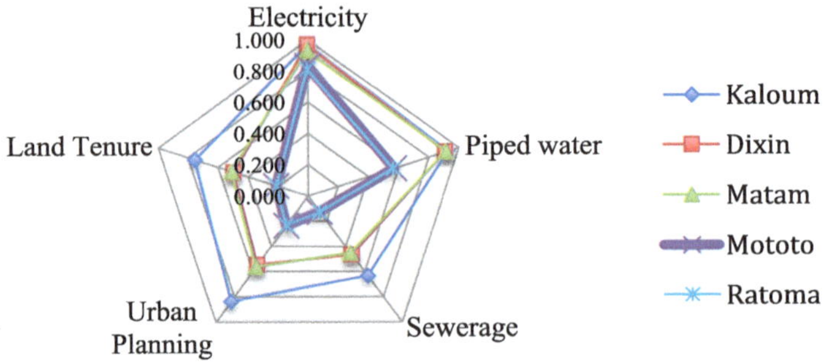

Fig. 4.9 Components of city foundation index. *Source* Computed by the authors

National and local authorities must improve the foundation of the city of Conakry, particularly of Mototo and Ratoma with smart planning, smart basic infrastructures and smart institutions and laws. Smart basic infrastructures, which are particularly of great urgency in these municipalities, include connection to sewerage facilities as well as development of efficient waste management systems. Effecting these changes will enhance the economic value of land, encourage investments, reduce risks from natural hazards, increase resilience and minimize the costs of infrastructure maintenance among various other positive impacts. This calls for use of ICT and GIS in spatial planning, land tenure governance and administration, and management and monitoring of the provision of basic infrastructures.

4.4 Urban Accessibility and Mobility

Development of large urban agglomerations like Conakry constitutes economic and social opportunities as well as challenges. It provides opportunities for economies of scale and agglomeration as well as for diffusion of ideas and innovations, but it will also call for large investments in infrastructures to respond to the increased demand for water, sanitation, solid management, energy and mobility, etc. It will particularly come with increased demand on mobility that must be satisfied with an efficient public transit accompanied with increased spaces for pedestrians and cyclists in order to safeguard the environment while creating economic growth. Economies of scale and agglomeration economies are greater in metropolitan areas where transportation infrastructures are able to answer mobility needs with higher access to markets and resources than those where people mobility is impeded by deficient transportation infrastructures. Efficient mobility will allow large-scale production of goods and services that can be distributed within the metropolitan regions and beyond with time, cost and reliability opportunities [1]. Without efficient mobility, a metropolitan region losses its economic power and remains just clusters of disconnected settlements. The latter is the situation in Conakry as in most African cities, marked by frequent traffic congestions, a major indication of the disjuncture between land-use planning and transport systems in the city. It not only exposes the limitation of a transport-oriented bias to mobility, but also reveals the inefficiency of land-use systems in Conakry. Urban accessibility is shaped by both spatial planning elements such as: density, compactness, type of city (monocentric or polycentric) and streets, and transport planning elements such as the streets infrastructures and the transport infrastructures. The first three elements of spatial planning—density, compactness, and type of city (monocentric or polycentric)—are presented in previous sections. This section will address one important element of spatial planning—streets—and elements of transport planning and system.

4.4.1 *Streets as Drivers of Urban Accessibility*

One key element of urban planning is the street, which defines the form and structure of city. In recent years streets have been recognized as an integral factor in the achievement of sustainable urban mobility. A connected street network reduces travel time and encourages walking and social interactions. One fundamental feature of sustainable streets is their connectivity in terms of planning as well as design. With regard to planning, sufficient land should be allocated to streets (at least 25% city's area) and the street network should be sufficiently long to cover all areas. There must be sufficient intersections available (at 100 intersection per square km) to facilitate shorter distances and reduce travel times, and encourage walking and social interactions [33]. A connected street network expands multimodal mobility systems with sidewalks and bicycle paths, ensures eco-efficiency of infrastructural systems, and supports density through integrated infrastructure development, thereby enhancing efficiency and access. In addition to accommodating all kinds of users (pedestrians, cyclists, motorists), sufficient land allocated to streets promotes connections to services that contribute to good health and productivity, such as clean water, sewerage facilities, drainage systems, power supply, and information and communication technologies. It enhances infrastructure development, environmental sustainability, and economic and social development. Streets that provide space only to motorists are characterized by congestion and high CO_2 emissions [34].

The Conakry metropolitan region is not benefiting from all the multiple advantages associated with well-connected streets. With only 10% of its land allocated to streets in its CBD, and less than 5% of land in its suburbs, Conakry suffers from traffic congestion along with the exclusion of pedestrians and other non-motorized means, and further marginalization of the most vulnerable segments of society who rely the most on foot and public transports if they are affordable. In most densely population in Conakry suburbs, there is no space left for mobility except very tiny streets where people have to squeeze through. The streets are also narrow, not paved, and lack sidewalks; this result in competition for space with motorists, often exposing pedestrians to accidents. While Conakry as in many African cities has high pedestrian flows, less than 20% have pedestrian footpaths [35]. In many of these settlements, public transport services are accessible in few arterial streets where people have to walk several kilometres to reach them. During the raining season, the already limited walking spaces are flooded, forcing people to

Table 4.2 Shares of various modes of transport in use in selected cities

Country	City	Walk
AFRICA		
Burkina Faso	Ouagadougou	–
Cameroon	Douala	60
Cote d'ivoire	Abidjan	22
Ethiopia	Addis Ababa	30
Ghana	Accra	12
Guinea	Conakry	78
Kenya	Nairobi	47
Mali	Bamako	–
Nigeria	Lagos	
Rwanda	Kigali	5
Senegal	Dakar	–
Uganda	Kampala	–
United republic of tanzania	Dar es Salaam	26

Source UN-Habitat, 2013. Streets as Public Spaces and Drivers of Urban Prosperity. [Author; Mboup, G.]. Nairobi, Kenya

stay at home, losing many productive hours. This in turn has serious social and economic impacts on the city's economy (Table 4.2).

Besides the low level of land allocated to streets, the street networks in Conakry as in most African cities are generally substandard. Streets lack service lanes, pavements and are poorly maintained, with limited street lighting the street planning and design do not anticipate the polycentric form of cites and the rapid increase in the use of private cars. In most African cities, few streets are paved and most lack sidewalks as shown in Fig. 4.10 [36], with Conakry having a percentage of paved road of 32%.

During the preparation of the Urban Development Plan of Conakry, the secondary and tertiary road network was evaluated at 590 km of roads distributed as

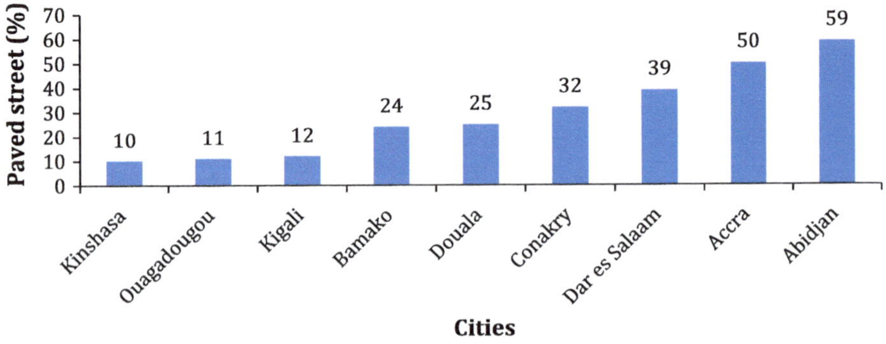

Fig. 4.10 Percentage of paved roads for selected African cities. *Source* Ajay Kumar and Fanny Barrett (2008) Stuck in traffic: urban transport in Africa, Africa infrastructure country diagnostic (AICD), obtained from city

follow: 120 km of paved tracks and 470 km of unpaved tracks. Only 119 km were covered, a proportion of 20% located mainly in the oldest sectors of the city of which the third was in an acceptable state. The rest of the network consisted of dirt tracks. During the past 15–20 years, the secondary network has, however, recorded some changes [14]:

- The total length of this network has increased from 590 to 970 km, an increase of 380 km (64%). This reflects a rapid expansion of urbanization by spreading despite the phenomenon of densification of some neighbourhoods. In the absence of support for urban extensions by the State or municipalities, the new roads created add to the inventory of unpaved roads and accentuate the degradation of the network;
- The coated line has been enriched by 60 km. It went from 120 to 180 km; an increase of 50% but its share in the total network remained the same 22% against 20% initially. The quality of the paved road has improved;
- The tracks in acceptable condition (good or average condition) represent more than 95% of the track lanes compared to only 33% initially, which indicates the effort undertaken for the development and the maintenance of the road network.

Composite Street Connectivity Index

Street connectivity is determined by the amount of land allocated to streets, the length of the street network and the number of intersections along the network. The proportion of land allocated to streets and the length of the street network is not sufficient to assess street connectivity. A city with wide streets within a very limited street network and low intersection density is considered a city with low street connectivity because the width of the streets is not complemented by a larger street network and higher street density. Similarly, a city with a lengthy street network and dense intersections may not qualify as a city with high connectivity if the streets are very narrow. The Composite Street Connectivity Index (CSCI) aims to assess the connectivity of a street considering its width, its length and the number of intersections, all in relation to the total land area of a city. Interestingly, some cities in this group have low levels of land allocated to streets, but higher intersection density (ID) increases the value of their CSCI. Conakry city core has a LAS index moderately low (10%). However, the Conakry suburbs suffer from low LAS, limited street density with few intersections. In these suburbs, which are mainly unplanned, provisions of basic services as well as means of transport remain challenges. Conakry belongs to the group of Cities with a CSCI below 0.500. Cities in this group have poor street connectivity due to low levels of land allocated to streets, low street density and low intersection density. Their CSCI is less than half the highest level of the CSCI, which is 1 (Figs. 4.11, 4.12).

Comprehensive city planning programmes are needed in these cities to improve the lives of urban dwellers. Streets where the needs of all users are considered and factored (ages, gender, economic status, modal means, etc.) are urgently needed in Conakry. Mobility must be multidimensional, easy, comfortable, and safe [37].

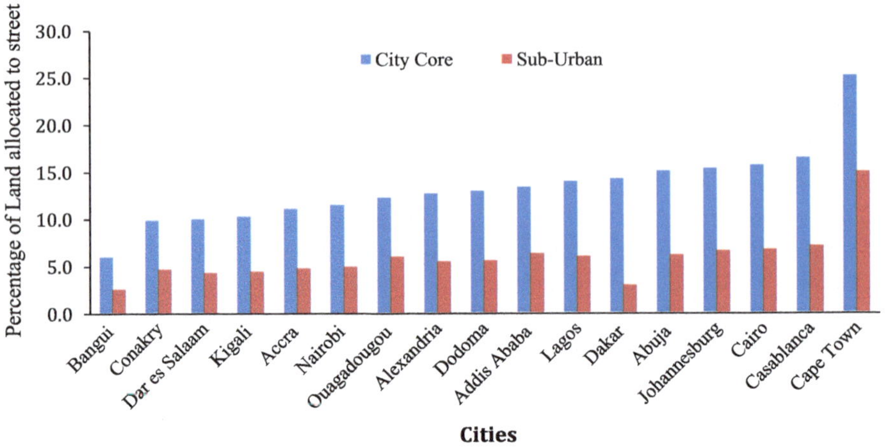

Fig. 4.11 Land allocated to street (LAS) in African cities. *Source* Figure prepared from Mboup G. et al. (2016). Smart city foundation—drivers of smart cities. In Vinod et al. (2016). Smart economy in smart cities, Springer

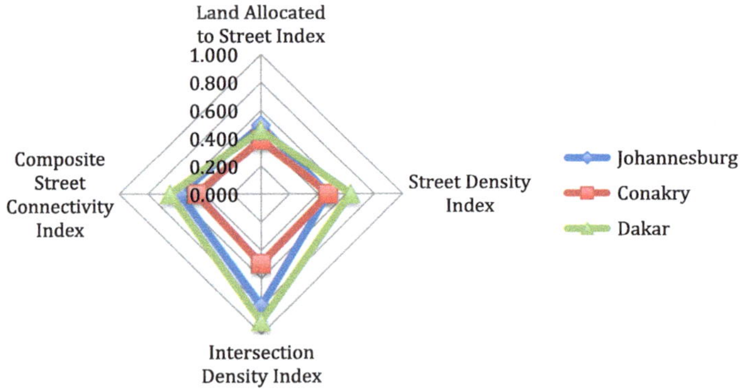

Fig. 4.12 Composite street connectivity index (CSCI)

4.4.2 State of Urban Mobility in Conakry

It is rightly recognized, in the Sustainable Development Goals (SDGs) adopted by United Nations' member states, that sustainable transport is crucial for urban economic development. One of the target of Goal 11 of the SDGs "*Make cities and human settlements inclusive, safe, resilient and sustainable*" is, indeed, to provide access to safe, affordable, accessible and *sustainable transport systems* for all, improving road safety, notably by expanding public transport, with special attention to the needs of those in vulnerable situations, women, children, persons with disabilities and older persons" [38]. In the SDGs and the COP 21, it is also recognized that transport is crucial for sustainable environment. It plays a key role in achieving commitments in the COP21 as a critical part of the climate change solution where it contributes to 23% of global CO_2 emissions. It can contribute to both reducing greenhouse gas (GHG) emissions and building economy wide resilience to the impacts of climate change [39]. Putting sustainable transport in global agendas is of importance for the African continent that has progressively been urbanized, and therefore will equally require efficient mobility of people, goods and services within and between localities.

"No aspect of urban form and travel has been more closely studied than the influences of urban densities on public transport ridership. It is widely accepted that high densities are essential for sustaining cost effective public transport services. The impact of densities on travel modes gained particular attention in the 1990s, in the wake of a global energy crisis and economic recession. Rail, with its high up-front capital costs and economies of scale, needs to attain a threshold density of trips, in order to cost less than accommodating the same trips by car or bus. The reliance of public transport on urban densities has prompted efforts to define the minimum density thresholds required to support successful public transport services. Cities need to average 3000 inhabitants per square km to support reasonably cost-effective public transport services" [40]. With low density, neither the investment required for BRT, nor that for rail, is likely to be viable because stations will lack sufficient proximate populations to generate demand. In spread-out cities, public transport has a difficult time competing with the private car for the middle and upper economic or walking for the urban poor. Public transport that is cost-effective can only be achieved through high urban densities and a large share of jobs and retail activities concentrated in the urban core, or in polycentric cities with multi-directional travel patterns [40]. In Conakry Peninsula, the population density exceeds this threshold indicating that Conakry transport problem is not associated to low density but to other urban form components such as a low coverage of street network, urban monocentrism, etc.

Infrastructure Development is one component of a smart metropolitan regional development. It is composed of several elements including transport, energy and ICT. Here, our focus is on the first element, which is transport. While transport data has been gathered for most cities in developed countries, information on transport is scarce in African cities, and when it exists it is poorly documented; metadata providing definition, method of measurement and sources of information is rarely available. The lack of reliable data on transport has held back the development of effective urban policies aimed to effectively tackle deficient transport systems in African cities. The most common available statistics are the distribution of trips by transport modes (large buses, minibuses, private cars, motorcycles, and walking). However, this information does not make distinction on the type of trips. Are they for going to work, to the health centres, to the school, to the market, or all together [41]. In the era of data revolution, this information must be processed, analysed in association with social, economic and demographic information and used to guide transport policies and planning in an open platform accessible to all stakeholders including the public. This information will help to overcome transportation challenges in African cities through informed planning, real-time monitoring and management that take into account multidimensional variables beyond the transport sector.

Though the statistics may be from different sources, there is a common finding which is that in most sub-Saharan African cities, people have limited mobility choice in getting to work: either they ride minibuses or they use their foot. Data collected in Conakry show that the share of minibuses in the public transport exceeds 90%. Despite various efforts taken by national and local authorities to boost the formal public transport, it remains with a very low share in Conakry, less than 5% in the total public transport, which is heavily dominated by informal means of mobility. Most of the formal public transport sector is composed with few medium-sized buses that must follow a specific itinerary, but they are barely seen in the streets submerged with the informal motorized means of transport.

The situation of transport and mobility is particularly critical and weighs heavily on the economic performance of the city and on the quality of life of Conakry residents. This inefficiency of the transport system is the result of the combination of two factors: the very peculiar shape of the territory, a long and narrow peninsula, associated with a very strong polarization of the administrative and commercial activities. The location of the port at Kaloum and the traffic it induces in the peninsula also contributes to the asphyxiation of the system and the deterioration of traffic conditions [5]. No efficient transport system can be implemented without a change in the distribution of activities throughout the peninsula through the development of centrality and a good articulation of functions around each polarity.

Informal public transport

Considering the lack of capacity by the formal public transport systems to meet the mobility demand of the residents of Conakry, dominance of the informal sector operators remains irreversible. The informal transport sector, composed of small-scale economic activities with unregulated employment, is largely dominated by minibuses operated by private entrepreneurs It represents 95% of the public transport. It supplies small-vehicles with low investment and minimal public support, low-performance services that fill the niche between formal taxis and formal public transport. However, the minibuses as means of public transport come with enormous externalities that authorities must not overlook in the transport planning process. In this sector, most vehicles do not fit to be in the road technically and with respect to traffic rules and regulations (technical control, license, insurance, etc.). This contributes significantly to congestion, air and noise pollution and traffic accidents. In addition, passengers, particularly women are exposed to harassment from the bus operators and other passengers. Violence against women is common in the public transport sector [42]. In a city like Conakry where the formal public transport sector is quasi absent, the role of informal transport in generating broader social and economic benefits must be assessed together with the costs entailed. Prohibiting informal transport is not a viable solution as it destroys jobs and blocks access to employment, but externalities must be assessed and reduced significantly. A combination of valuing the service and regulating them can be effective in managing related issues, such as traffic congestion, accidents and pollution.

Increased use of private cars hampers efforts to make Conakry a smart city

In absence of reliable public transport systems, households that can afford it, usually the upper middle class and the richest, possess private means of transportation or use taxis, which offer comfort compared to the public means of transport. The perceived advantages of convenience, privacy and status continue to make the private car an attractive means of transport in Conakry as in many African cities [43]. The growth of private motorized transport during the twentieth century had major impacts on the growth and development of the city of Conakry as many cities in developing countries. The rate of motor vehicle ownership in Conakry has significantly increased during these past 20 years, from 5% in 1995 to 13% 2014. The major factor behind the growth of car ownership in Conakry is at a cost that is becoming affordable for a growing number of people of the middle class that cannot tolerate the discomfort of the informal public transport and seek individual freedom. With the continuous urban growth, it is expected that Conakry will be an exclusive motorized city if the public transport is not reinforced and mixed neighbourhood promoted.

While private cars offer the convenience of individual choice, this advantage is traded for much more land being used for road space and parking. Cars' land consumption and infrastructure costs can be a significant part of a city's budget and this costs heavily subsidized by both drivers and passengers. Emissions of pollutants in Conakry seem to be higher than the thresholds defined by international standards. Estimates of safety costs from the World Bank show that in African cities these costs can exceed 2.7% of national GDP [44]. A city submerged with cars is prone to more accidents, which generate economic and human costs. Without adequate public transport systems in place, and good planning choices that increase connectivity and proximity, congestion, pollution and energy consumption will also increase exponentially with the growth in car ownership.

Two major impacts associated with low coverage of public transport and increased uses of private cars are traffic congestion and pollution. The monocentrism of the city of Conakry is a source of traffic congestion and an obstacle to smartness. The mobility in Conakry is heavily affected by the fact the city was conceived as a monocentric city where residential areas and workplaces are distinct, with the later concentrated in the centre of the city. Residential areas were also designed along economic class lines. This has led to social and economic fragmentation that disadvantage lower income group in accessing basic services and prevent social interaction and integration. This has impacted the traffic with every morning and evening people have to share a main street network to and from the downtown of Conakry.

Traffic congestion is a major indication of the disjuncture between land-use planning and transport systems. It not only exposes the limitation of a transport-oriented bias to mobility, but also reveals the inefficiency of land-use systems in a given city. Limited road capacity, in the face of growing demand for motorized mobility, partly explains deteriorating traffic conditions. Congestion has widespread impacts on the urban quality of life, consumption of fossil fuels, air pollution and economic growth and prosperity. Congestion accounts for significant percentages of the gross domestic product in many major cities as noted in World Bank studies from the 1990s. About 90% of the cost comes from the value of the time lost by residents, 7% from the fuel consumed and 3% from gas emissions. In addition to economic costs, congestion causes significant numbers of early deaths from respiratory illnesses, stress and physical and mental fatigue. It also degrades green areas, which, in turn, diminishes their carbon sequestration properties.

Walking—Key for Conakry smartness

In absence of affordable, reliable public transport, various options exist depending on level of economic conditions or purchasing power. Those that can afford it will own a car or ride a taxi, while those that cannot afford it, who constitute the majority, will use their foot. Those using their foot have limited opportunities compared to others that have access to public transport. They won't accept job where they have to walk very long distance, or if they accept job far away from their residence, they will encounter several days of absenteeism, and they end up by quitting the job. Despite lack of comparative data, it seems that the share of walking to access services remains the first option in Conakry due two main factors, namely: low coverage of public transport infrastructure and unaffordability of public transport for the poor where various needs compete. While in cities of developed countries, it is assumed that a walkable street is more attractive to people for various reasons and even defines the "liveability" of a city, in the city of Conakry walking is not a choice, but a necessity due to lack of other affordable transport alternatives.

Available information shows that 78% of residents of the city of Conakry walk to their place of work, to health centres, to schools, or to the market. However, the streets of Conakry are not designed for pedestrian use. First, the land allocated to the streets is low, only 10% in the city core and less than 5% in many sub-urban settlements. The streets are also narrow, not paved, and lack sidewalks; this result in competition for space with motorists, often exposing pedestrians to accidents. It is also noted that in many African cities, 95% of the city streets also have high pedestrian flows but only 20% have pedestrian footpaths [35]. In addition, streets in the suburbs are poorly connected with less than 40 intersections per square km. Lack of sufficient intersections make the street network of Conakry not friendly to pedestrians. It is demonstrated that for a street network to be well connected and walkable, at least 100 intersections per square km are required [45]. In areas such as Mototo and Ratoma, which are densely populated, there is no space left for mobility except very tiny streets where people have to squeeze through. In many settlements, public transport services are accessible in few arterial streets where people have to walk several kilometres to reach them. During the raining season, the already limited walking spaces are flooded, forcing people to stay at home, losing many productive hours. This in turn has serious social and economic impacts on the city's economy.

Despite challenges, walking has enormous economic, social and environmental advantages. Conakry of the 21st century must be planned as walkable with affordable means of public transport. Walking is an enabler of social cohesion and environmental sustainability with enormous social, Economic and Environmental benefits. It is now recognized and advocated that walk is the most efficient means of mobility for environmental sustainability as expressed in global agendas such as Sustainable Development Goals (SDGs) and the COP21 in 2015, and the New Urban Agenda in 2016. In addition to its social and economic benefits, walking has a major advantage in reducing energy consumption, greenhouse gas emissions and pollution (air, water and noise) substantially, as it does not rely on fossil fuels unlike other modes of transport in cities. Furthermore, as walking requires significantly less road space and parking, it enables the preservation of natural habitats and open spaces. Walking also provides the daily physical activity required for a healthy lifestyle. Based on this, many motorized cities of developed countries have been changing their street planning and designing, and promoting public transport in order to reduce private motorized use and boost walking and cycling. With the 60% of people that are already walking in city, Conakry is indeed in a good position for a healthy society in a sustainable environment. However, in order to sustain this advantage, it is high time to give to pedestrians their right share in the transport infrastructure. Providing adequate infrastructures to pedestrians is cost-effective considering the enormous benefits from walking.

Integrating public transport, walking and cycling along the urbanization of Conakry

Conakry does not benefit from the multiple advantages of high density due its unplanned urban growth and land expansion. While the advantage of a high density settlement is to ease accessibility, reduce cost of provision of basic infrastructures and other urban services, reduce erosion of natural resources, lower business costs and improve social equality, in Conakry none of these benefits are present or they are little. Considering the high population density in many unplanned settlements of Conakry where the urban growth still high (more than 2% annual growth), it is urgent to develop and implement city extension at the fringes of the built-up area. The extension of the city of Conakry will specifically host inhabitants of densely populated areas and settlements built in flood prone areas. Every year, flooding causes enormous economic, social and environmental damages and losses among people and communities. With the climate change associated with rise of sea level, there is urgent to operate a city extension that was supposed to be in the 20th century when the density is still manageable. Planning an extension requires vision and commitment [46]. National and local authorities have already identified areas where the extension of the city of Conakry will be implemented as the part of the Grand Conakry Vision 2040.

4.5 Urbanization, Economic Growth and Metropolitan Regional Development

Conakry is a hub of economic activities linking local, regional and global economies. Today, it has an added advantage associated with its high population density and its youthful population, two important drivers of economic productivity and growth. One of the components of a Smart Metropolitan Regional Development is Economic Development, which is analysed in this section.

4.5.1 Urbanization and Economic Growth

The Conakry Peninsula contributes 40 to 60% to the national Gross Domestic Product (GDP) [47]. Up to nine out of ten national civil servants, and from 70 to 90% employees in national's trade, transport, banking and industrial enterprises are in Conakry. Though the Conakry Peninsula had been administratively divided in five municipalities, in reality most of the formal economic activities are concentred in the municipality of Kaloum, other municipalities such as Mototo and Ratoma acting as suburbs. The municipalities of Mototo and Ratoma remain haunted by the early spatial and social division; they are not well planned, and they lack sufficient land allocated to streets, other public spaces, basic infrastructure and security of tenure.

Though Conakry Peninsula is engine of national economies, association between urbanization and GDP around the world shows that the Conakry Peninsula as many African cities perform much less than its counterparts in other regions. In the Republic of Guinea, urbanization has not led to expected economic growth as shown in Fig. 4.13, with Guinea below the curve of association between GDP per capita and urbanization.

While countries in Asia such as China, Indonesia, Malaysia, Thailand and Vietnam and in LAC such as Mexico, Costa Rica and Argentina display an increase in GDP per capita associated to high urbanization rates, a similar association between GDP per capita and urbanization was not observed in most sub-Saharan African countries among them Senegal, Guinea, Togo, Liberia and Burundi. In most of African countries, capital cities were not able to produce and distribute wealth nation-wide. Conakry has not been able to fully transform its large population in terms of economies of scale and agglomeration economies as many Europeans and American cities did more than 50 years ago, and LAC cities and Asian are recently doing. The low productivity of Conakry lies to several factors

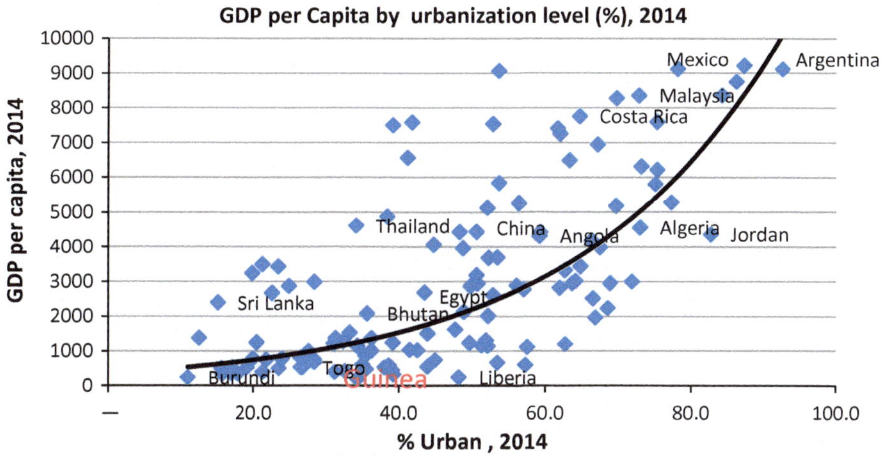

Fig. 4.13 GDP per capita by urbanization level (%), country with GDP below US$15,000. *Source* % urban (UNDESA), GDP per capita (World Bank)

including those spatial and mobility planning presented in previous sections as well as other factors associated to urban legislation and governance and financial capacities. Unbalanced urban system, unplanned dense settlements without basic services, poor street connectivity, and poor transport infrastructures impeded Conakry economies of scale and agglomeration potential, and create scale and agglomeration diseconomies.

Before the political normalization that took place in December 2011, Guinea experienced great political, economic and social fragility. From 1987 to 2007, the annual growth of GDP per capita was 0.9%. Moreover, GDP per capita has changed at a much slower pace since 2000. Between 2002 and 2005, the growth of the Guinean economy has average of 2.8% per year. Based on population growth of 2.2% per year, the resulting improvement in purchasing power is only 0.6% per year, which is a performance well below the target of 5% growth of the Poverty Reduction Strategy Programme (PRSP) [48]. In the national budget, the priority sectors are: Education (11.6%), Habitat and Social Affairs (11.2%), Health (4.9%), Road Infrastructures (8.7%) and rural development (24.7%) [14]. Thus, despite its considerable economic potential, Guinea is ranked among the least developed countries (LDCs) with a poverty rate that affects nearly 55% of the population in 2009 against 49% in 2002 [49]. The political, economic and social situation has particularly deteriorated from 2007 to 2010. But since December 2010, the country has embarked on the path of normalization, after a courageous and difficult democratic transition.

Guinean cities are essentially tertiary cities, based on trade and administration and where the informal sector occupies a prominent place. There is no correlation between the administrative hierarchy of cities and their economic role, nor between

Fig. 4.14 12 distribution of GDB by economic sectors, Guinea, 2012. *Source* Data from National Institute of Statistics, Guinea 2012

urban services and industrial production functions. The industrial employment rate remains below 5% everywhere, even in mining towns. Agriculture is important with 20 to 25% of jobs in regional cities but more than 30% in smaller cities [14]. When we consider the non-agriculture sector, trade, mostly informal, will represent quarter (26%) of the employment followed by mining (22%). Bank, assurance and other financial institutions represent 9.4%, similar to duties and important taxes (9.3%). Transport and Communication have a share of 7.0%, similar to public administration (6.6%). Manufacturing represents only 5.8%, followed by the sector of electricity, gas and water (1%) (Fig. 4.14).

Industry in Conakry is confronted with several challenges that include low production levels, inadequate competitiveness of the local market, lack of capacities of industrial firms to upgrade their production systems, geographical and structural weakness of the industrial fabric. In operational terms, initiatives concerning partnership between the State and the private sector are still minimal, particularly with regard to the promotion of entrepreneurial initiatives, the development of innovation through research application, the creation of integrated competitiveness poles and the training of future champions of the different sectors and the development of venture capital. The shortage and unreliability of power supply and the weak infrastructural platforms cripple the performance and competitiveness of industries, causing substantial additional costs. In addition to these drawbacks, industrial production suffers significantly from its strong concentration in the Conakry area as mentioned above, hampering the potential of provincial economic zones, as well as from lack of diversification and a system of product quality certification. Against this background, industry has not paid attention to the need to protect the environment as a general concept, but also as a strategy to improve their productivity and competitiveness [50].

Unemployment rates are particularly high in Conakry with a level 13.6% compared to less than 4% in all other regions. Conakry has not been able to satisfy the employment demand, particularly from young people (Fig. 4.15). The Conakry job market is predominantly informal. In addition to that, due to a deficient urban mobility, Conakry has not been able to create conditions for specialization. While urban agglomeration allows for job specialization, efficient market transactions and

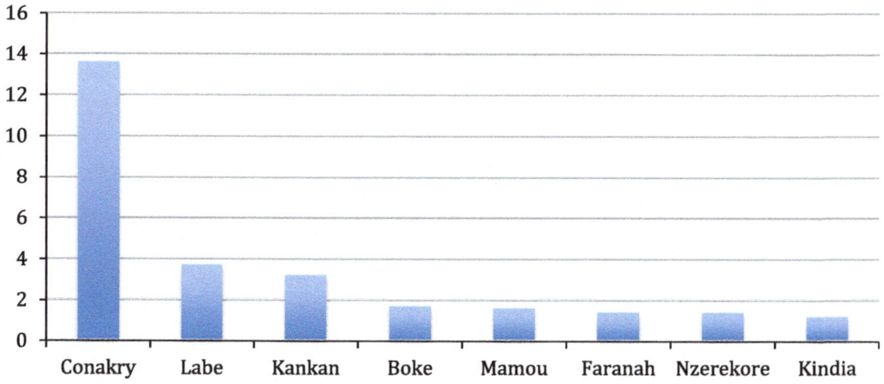

Fig. 4.15 Unemployment rates, Guinea, 2012. *Source* Data from National Institute of Statistics, Guinea 2012

knowledge diffusion, if concentrated growth is not well planned—such as the integration of urban growth with efficient public transit investments, the resulting economic benefits tend to erode. Agglomeration diseconomies—i.e. the inefficiency and loss resulting from poorly planned concentrations—are expressed in the form of lost labour productivity from extreme traffic congestion, increasing air pollution and an overall decline in the quality of urban living [51]. This is the case in Conakry as in many African cities with poor connectivity leading to disconnected settlements.

For cities to act as integrated labour markets and match jobs seekers and employers, they also need to make employment spatially accessible to all residents. The Conakry metropolitan region has failed to do so. Inaccessibility to public transport, indeed, limits the access to formal jobs for many people that rely on minibuses or their foot. For instance, it is noted that within 45 min, only 10% of minibus riders or pedestrians can access to formal job. Even with 60 min, the figures remain low with only 10% of pedestrians and 20% of minibus and pedestrians can do it. This means that people who count on their foot or on minibus are excluded from several formal jobs, unless they accept to be in the minibus for hours or to walk hours in order to reach their work place as illustrated. Though most people with cars (96%) can access their formal workplace within an hour, this figure drops to 77% when the traffic is congested, particularly during pick hours when everybody is rushing to reach their workplace [52]. This shows poor street connectivity associated to poor transport infrastructures limit job opportunities among the urban poor. They cannot afford neither the bus fare nor to have a private car. They cannot also relocated to where jobs are due to high cost of rental proprieties.

Economic inequalities are also very high in the Conakry Peninsula leaving municipalities with high unemployment and low access to basic services unable to afford many subsistence needs as illustrated in Table 4.3.

Table 4.3 Distribution of neighbourhoods of the City of Conakry according to degree of poverty

Degré de pauvreté	Communes	Quartiers
Degree of Poverty Very poor neighborhoods	Kaloum Dixinn Matam Matoto Ratoma	Coronthie, Fotoba, Koulewondy Dixinn port, Kenien, Carriére, Mafoco centre, Coleah cité Dabondy 1, Dabondy 2, Dabondy ecole, Dar-es-Salam 2, Gbessia cité, Matoto centre, Yimbaya école Kaporo centre, Symbaya gare, Sonfonia gare, Wanidara
Poor neighborhoods	Kaloum Dixinn Matam Matoto Ratoma	Boulbinet, Sandervalia, Tombo Bellevue école, Cameroun 1, Dixinn centre 1, Dixinn centre 2, Dixinn gare, Dixinn mosquée, Hafia 1, Hafia 2, Hafia minière, Hafia maosquée, Landreah, Minière cité Bonfi, Coleah centre, Domino, Imprimerie, Lanseboundji, Madina mosquée, Matam, Touguiwondi Cité de l'air, Dabondy 3, Dabondy rail, Gbessia centre, Gbessia cité 3, Gbessia école, Gbessia port 1, Gbessia port 2, Matoto Khabitaye, Matoto marché, Sangoyah marché, Sangoyah mosquée, Symbaya 2, Tanene marché Dar-es-Salam 1, Hamdallaye 1, Hamdallaye mosquée, Kipe, Kobayah, Lambandji, Nongo, Ratoma centre, Sonfonia centre, Yattayah
Neighborhoods with moderate living conditions	Kaloum Dixinn Matam Matoto Ratoma	Almamya, Kassa, Manquepas, Sans fll, Temenetaye Camayenne, Dixinn rail gare Bonfi marché, Boussoura, Hermakonon, Madina centre, Madina cité, Madina école, Madina marché, Mafanco, Matam lido Dabompa, Gbessia cité 1, Kissosso, Tanene, Yimbaya permanence, Yimbaya tannerie Hamdallaye 2, Kaporo rail, Kotoma 1, Koloma 2, Ratoma dispensaire, Taouyah
Wealthy neighborhoods	Kaloum Dixinn Matam Matoto Ratoma	Bellevue marché Behanzin, Symbaya 1, Tombolia

Source Republic of Guinea and Royaume du Maroc, 2017. Assistance technique pour la mise en oeuvre du projet d'assainissement de la ville de Conakry

The Population of Conakry was classified according to their degree of poverty (Table 4.3), from the highest living standard to the poorest living standard. Overall only one third of the population of Conakry has a decent living standard (32.7%); the remaining two-thirds are poor or very poor (67.3%). Extreme poverty, characterized by multiple deprivations, affects one fifth of the population (19.4%). Poverty is particularly high in Mototo, Ratoma and Dixin where the level exceeds the two-third level. In all municipalities, the majority of population are poor. A widening urban divide between social groups has also been underscored in the Grand Conakry Vision 2040 [5].

4.5.2 The Economics of Urban Land—The Hidden Assets [53]

4.5.2.1 The State of Land Tenure

The smartness of Conakry Peninsula has also been long time hampered by weak institutions and laws holding back various factors of the smart city foundation, making informality the norms in many sectors starting with the city foundation surrounded by informal settlements with irregular land tenure. Secure tenure goes beyond protection against eviction and includes economic and financial advantages. *Tapping in the Triangle of Economic Productivity—People, Land & Infrastructures* will create sustainable, inclusive prosperous and resilient cities. Cities are built by people; with their concentration offering economies of scale and agglomeration starting from their land and housing assets. With functioning institutions and laws, land and housing assets can contribute to the planning, management and provision of services in settlements. However, in absence of functioning institutions and laws providing legal propriety rights, as observed in Conakry, most of these assets remain dead investments sheltering only people. To tap into the potential of high densities, the Conakry Peninsula must formalize its land system, which will be the driver of many other components of its foundation such as streets and public spaces, provision of basic infrastructures such as water, sanitation and energy, and waste management. It is urgent that national and local authorities recognize the wealth of their citizens and involve them in the planning, the building and the management of their city. No city can claim to be sustainable, inclusive, prosperous and resilient, when the wealth of its citizens is not fully taken into consideration in the economy. Conakry as most African Capital cities are not, however, fully benefiting from its land assets due to the fact that most of its land is considered as irregularly acquired and lacks legal ownership document such as a title deed.

The land tenure system was not able to synchronize with the population growth of Conakry due to lack of implementation of urban plans. Finally people settle first where there is land, and then proceed later for regularization that will never happen in most cases. These settlements are no longer only in the suburbs but also in the city core of Conakry as observed in Kaloum and Dixin as well as in other municipalities. The indigenous people, regrouped in families, progressively abandoned agriculture and sold their land to whoever can pay. First they subdivided land in plots with the help of a geometer. In this case, the emphasis is more on the number of plots than in streets and other public spaces. However, they could only provide sale documents showing buyers and sellers. However, certain buyers can negotiate and have legal documents in an illegal area. This opens the door for corruption, lack of transparency, and poor recording. Since this situation seems comfort the officials, there was little or no interest to improve it.

Possession of ownership or tenancy document is low in Conakry where, less than half of households had a proof of legal ownership: title deed, sale certificate, power bill or other documents [54]. The problem of land is particularly complex in

Conakry (as in the rest of the country) because of two major obstacles: (a) the effects of the reforms, and in particular the content of the Land and Estate Code of 1992 with the reintroduction of customary law alongside modern law; (b) the absence of the Guinean State in the control of land-related regulations and the "laissez-faire" policy that has prevailed for years [5]. Measures to reduce the risk and stress associated with lack of documents and fear of eviction are based on recognizing and respecting a plurality of tenure systems, including intermediate forms of tenure arrangements and alternative forms of land administration and land records [55]. The legal institutional framework in a given country or city plays a key role on various elements of security of tenure such as acquisition or adjudication which is the process of final and authoritative determination of the existing rights and claims of people to land.

4.5.2.2 Financing Infrastructure Development and Economic Development Through Smart Land Tenure

Having legal land ownership documents has various social and economic advantages including access to the financial and economic market as illustrated by de Soto. De Soto argued that granting titles to the poor would liberate the plots they occupy and transform them into capital. This, in turn, could be used as collateral for loans to jumpstart their businesses, or improve their houses, among other gains that increase their quality of life. At the community level, regularization of land tenure will help the municipality to legally collect various taxes that can be used to improve basic infrastructures such as connection to water, sewerage facilities, energy sources and waste management facilities. This would also allow people to fully participate in the development of their communities at the policy as well as the implementation level instead of seeing proprieties as dead investments serving only for shelter [56]. Though having a title dead does not necessary lead to secure a bank loan; it may not be sufficient in itself to animate the dead capital interred in land and property, particularly in countries where banks lend only to workers with high wages and a stable job, as it seems happen in some countries covered in de Soto study. However, families with title dead may be likely to invest either to improve the quality of their homes or to increase their size. It is also important to note that land tenure goes hand and hand with urban planning [57]. Without an accurate mapping, legal title dead cannot prevent land disputes as it often occurred in African cities.

While there is argument against the de Soto theory in poor settlements, in Conakry regularization of land tenure will work for many middle and upper income households that have not been able to use their property as collateral due to lack of legal ownership document. Another important group land regularization will benefit is the Guinean from the diaspora who usually send money to their family and build houses in Conakry. Most of these houses are built in irregular settlements. Among the Guinean of the diaspora, certain have a high desire to return home, but they have not save enough to do so, and they cannot use their houses as collateral and start a business in Guinea. No choice, even those they reside in Conakry cannot get a title dead, what about those that just come to visit their family for a month or less. These are not the poor as described in the de Soto book, they have already the value of savings as underlined by the money they sent to their family in a monthly basis.

Urban land is a vital economic asset, and asset transactions are viable only where purchasers can rely on enduring extra-legal documentation of ownership. A formal market offers purchasers legal protection with transactions adequately recorded in land administration book; it also generates public good with accurate valuation [58]. Transparent property rights to urban land are a precondition for formal land markets. When these systems pose barriers to urban land access, they impede the consolidation of plots and the evolution of land use. Land transactions are long, costly, and complicated in Conakry as in most African cities [59]. Such market constraints reduce the collateral value of structures, giving developers little incentive to invest in residential height. In such a condition transactions are made through informal arrangements [60].

It is also noted that due its rapid growth, land prices are high in the city of Conakry but benefits less to the infrastructural development of the city. It does not create revenues for the cities as illustrated in the city income-expenditures sheet. In cities of developed countries, land-based financing has significantly contributed to urban investment. Taxes on land use will reduce the high level of inactive land in Conakry. Considering that they are paying taxes on the land, landowners will either develop their land in its most profitable use or they will rent/sell it. Improved valuation of land and properties closer to their market value, deepening the tax base; Improved enforcement of land and property taxes on a larger number of owners, broadening the tax base and; monetization of underused public land [61]. This will require functioning institutions in a transparent manner with inclusive, documented property rights using standardized and objective methods of land valuation. Good land governance and administration where corruption does not have a place will make trusted institutions and will encourage landowners to register their properties considering the high returns for them and for their community. Due to limited revenues of the city of Conakry, it is high time to tap to the potential capital of land through the regularization of land tenure.

Large infrastructure projects require huge investments. However, like any large structures, they depreciate very slowly over decades or even centuries [62]. The central government transfers on which Conakry often rely will not suffice to finance all infrastructures required to take advantage of the economies of scale and agglomeration that Conakry can offer its large population of about 2 million inhabitants with high density of over 4000 inhabitants per square km. Therefore national and local authorities should explore various financing options for infrastructure development. The costs of developing housing, infrastructure, and industrial premises must be coordinated with land markets and land use regulations in order to fully take advantage of the economies of scale and agglomeration of the city of Conakry. This calls for the formalization of the land tenure, which, in turn, will increase the land values that can be used to contribute to the development of basic infrastructures such connection to water, sanitation, drainage, solid waste management and streets. For instance, in Conakry due to lack of municipal finances most municipal street networks are not paved. Municipalities with formal land system have the large proportion of paved streets compared to others without formal land system.

Secure tenure goes beyond the legal character; it attracts investments as illustrated by the level of provision of infrastructure in municipalities considered as legal settlements comment to others considered as informal settlements. These municipalities also enjoy health centres and school facilities. They also attract the financial market because they have a legal urban plan with a sufficiently documented cadastre system. In most irregular settlements, there are few or no paved streets, and the few streets they have do not have light. Lack of documented urban plans affects the financial as well as the land market. It is noted that capacity and resource constraints are the main reason infrastructure in most municipalities have been lagging behind. In absence of formal land system, a settlement is trapped into poverty. With this, there is no doubt that wealth is associated to formal land system. Let make secure tenure work for people and communities. Conakry is expected to reach 5 million inhabitants in 2040 with the Grand Conakry 2040. This will propel new demand for infrastructure such as water, sanitation, sewerage, waste management and streets among others. To meet this demand, land transactions must be eased with transparent efficient land law, administration and governance with documented land and property rights; documented guidance of land valuation and prices and; coordinated land use and urban planning. Land tenure goes hand and hand with urban planning. Urban planning supported by accurate demarcation of public and private uses are of importance [57]. Without an accurate mapping, legal title dead cannot prevent land disputes as it often occurred in African cities. In Conakry, it may take several years before regularization, particularly when it is bought informally. Formal land Registration will ease transactions and boost economies of scale and agglomeration.

It is also necessary, in Conakry, to proceed with the renovation of the degraded habitat of municipalities. But this renovation can only give satisfactory results if it is part of an urban planning and concerted development project between the State, the city of Conakry and municipalities, on the one hand, and the private sector, on the other hand. In this system, the private sector will occupy a central place in the construction or rehabilitation of buildings. The public authority will be the first developer of equipment and regulation. The private sector will thus promote, and renovate buildings, for which it will provide financing through the channels it has itself identified, in consultation with the public authorities. An urban land-use law, adapting the rules of town planning and development, will put the land in the service of the renovation program. Sectors of social mix, with a minimum of 25 to

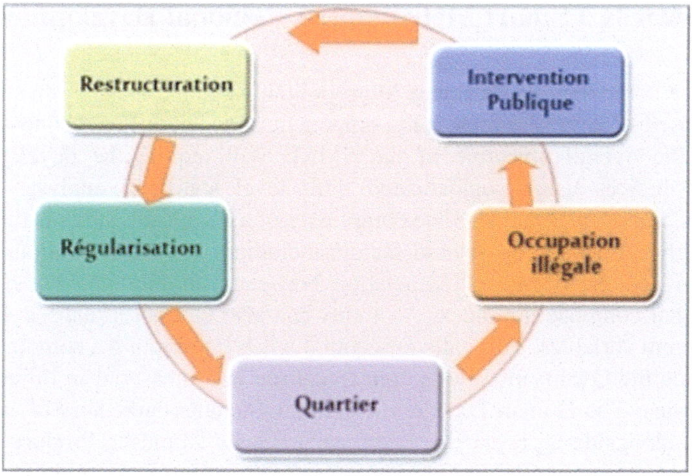

Fig. 4.16 Cycle of regularisation of irregular settlements. *Source* Vision Habitat 2021

30% of social housing, will be included in this renovation program. A functional mix (greater diversification of urban functions) will be encouraged as expressed in the Conakry Vision Habitat 2021 [49] (Fig. 4.16).

The State has a major responsibility in the implementation of land policies: (a) as a regulatory authority, it must facilitate access to an abundant land base, and a controlled cost, especially for the public operators in charge of the application of housing and city policies and major structuring projects; (b) as legislator, it determines the conditions of application of prerogatives of public authority (expropriation and declaration of public utility) and operational procedures, which must be justified by the notion of public interest, which must, in all circumstances, be indisputable; and (c) regarding public finances, it has the capacity to perceive the benefits of taxation on built and undeveloped land, traditionally oriented to the benefit of local communities [5]. The capacity of the State to conduct a land policy is decisive to enable the authorities to influence the areas of urban development at different scales: (a) at the level of national planning for the management of major infrastructures roads and railways, the programming of equipment of national interest, and the preservation of large natural reserves; (b) at the level of metropolitan land planning, with the location of new settlements, and structuring large agglomeration infrastructures; (c) at the level of local town planning, through the ability to conduct punctual development or urban restructuring operations based on clear and indisputable procedures.

4.6 Towards a Smart Metropolitan Regional Development

Most of the components of a Smart Metropolitan Regional Development (SMRD) are composed of several simple sub-components. Put them together to track progress on the overall objective of the SMRD will require the development of composite indices using sophisticated multi-level statistical analysis including Principal Component Analysis. The complexity of a composite index is the fact that it is strongly influenced by several factors including: the normalization, the standardization and the degree of association between variables and the number of variables that compose the index. An index can also be conceptualised differently using different variables. This calls for caution when interpreting a composite index. For instance the Quality of Life is conceptualised and measured in different ways across studies. The Human Development Index includes indicators of health and education, alongside an economic component (Gross Domestic Product or GDP); the Legatum Prosperity Index considers quality of life to be multidimensional, including both wealth and wellbeing; The Economist Intelligence Unit's Quality of Life Index links life satisfaction to health, family life and community life; and the OECD's Well-being Initiative has two dimensions: 'material living conditions' and 'quality of Life' [16].

In our study, a smart metropolitan regional development is viewed as a sustainable, inclusive and prosperous metropolitan regional development that promotes a people-centric approach based on three core components—Smart Metropolitan Region Foundation, Information and Communications Technology (ICT) and Smart Institutions and Laws, which are the pillars of the other dimensions of a smart metropolitan region: Infrastructure Development, Environmental Sustainability, Social Development, Social Inclusion, Disasters Exposure, Resilience, and Peace and Security. **Infrastructure Development** complements the basic infrastructure services under each smart metropolitan foundation and extends to actual investment and advancement of services such as transport, ICT, industrial energy, education, health, etc. **Environment Sustainability** is comprised of elements of energy, transport, building and pollution. **Social Inclusion** includes aspects of participation in decision-makings as well as equal opportunities for growth and prosperity. **Social Development** encompasses elements of education, health, public space, social inclusion and social capital. **Disaster Exposure** incorporates elements of mitigation and adaptation to various disasters such as flooding, droughts, storms and earthquakes. **City Resilience** is composed of elements of city foundation, environment, social capital, and social development. **Peace & security** included the elimination of all forms of discrimination and violence and conflicts, including domestic violence, violence in public places, crime, armed conflicts, terrorism, etc. An insecure metropolitan region limits opportunities for investment and economic growth and cannot be a smart metropolitan region.

The city of Conakry was included in the development for three indices published by the UN-Habitat and conceptualized and further developed by Mboup G. (2012, 2013 and 2014). Those indices include: the City Human Development Index, the

City Prosperity Index and the urban form-based City Prosperity Index integrating the Composite Street Connectivity Index. This section will take it further for the development of the Smart Metropolitan Regional Development Index.

4.6.1 Urbanization and Human Development

4.6.1.1 Urbanization and Education

A second priority axis of the Guinean Poverty Reduction Strategy (PRS) concerns education. Progress has been recorded during the most recent PRS evaluation period: between 2002 and 2005, the Gross enrolment rate has increased from 72 to 79%. This exceeds the PRS target (70%) but remains below that of PEPT 'Education for All Program' (81%); the girls' enrolment rate has reached 73%, which is 10 percentage points higher than of the PRSP, with the ratio of girls to boys going from 0.74 to 0.79. However the regional results remain very contrasted: the Conakry Gross enrolment rate is 128%, much higher than its levels in other regions such as Labé where it is only 60% [14]. The city of Conakry also has the highest literacy rate among the adult population 65.4% as recorded in 2012 against less than 35% in all other regions (Fig. 4.17).

Despite the overall progress recorded in the education sector, some structural problems persist, such as: (a) Insufficient coordination between educational administrative institutions; (b) low pro-poor educational policies towards the disadvantaged regions; (c) high repetition rate indicators estimated in 2005 as 23% in primary school, 27% in middle school and 19% in high school; (d) deficit in school infrastructures: annual needs in classrooms are over 1200 rooms while the average annual built classrooms is 345 rooms; and (e) lack of training of teachers [14].

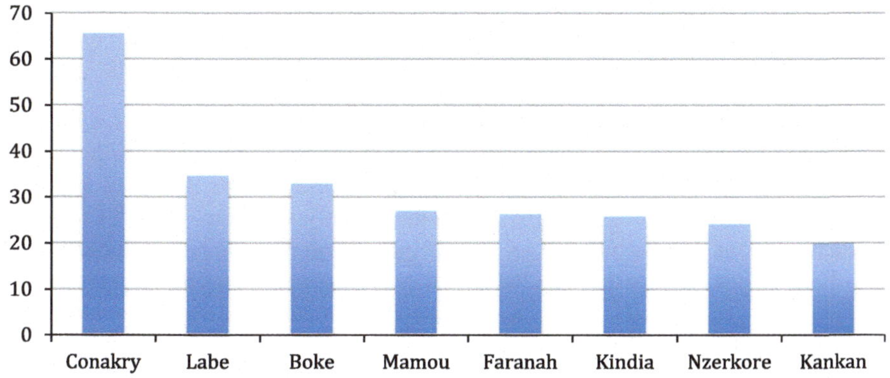

Fig. 4.17 Literacy rates, Guinea, 2012. *Source* Data from National Institute of Statistics, Guinea 2012

4.6.1.2 Urbanization and Health

A healthy population is critical to realizing economic growth through increased productivity [63]. Healthy workers are more productive, bringing greater income to families and higher levels of economic growth for nations. The Demographic and Health Surveys (DHS) held in Guinea in 1999, 2005 and 2012 show a constant progress in access to health services in Guinea during the past 15 years. Until 2000, access to health services was very limited and infant and child mortality rates were consequently high. With a steady improvement in health coverage, enhanced with the Millennium Development Goals (MDGs) with its Goal 4, 5 and 6, the improvement was accelerated during the 15 years of the MDGs. There is no doubt that improvement in health is significantly contributing to the Conakry smartness. Children born in Conakry have more access to health services such as antenatal (by their mothers), delivery cares and immunization than those born in other cities, town and villages of Guinea. While the child immunization coverage (with all vaccines) is estimated at 44% in Conakry in 2012, in some regions such as Faranah, Labe and Mamou, this level is less than 30% (29.4, 23.9 and 19.3% respectively. While in Conakry the quasi-totally of children had received at least a vaccine (99.7%), in some regions such as Labe and Mamou many children had not received any vaccine (23 and 19% respectively).

Significant decline of infant and child mortality rates along the urbanization of Guinea—In the context where access to health services has significantly improved, mortality has also significantly decreased, particularly among children under five years old. During the period of 1997–2012, the under-five mortality has decreased from 152 per 1000 to 123 per 1000. The decline of mortality rates has also been observed in other parts of Guinea, but to a lesser extent, making the figures of Conakry much better than the national figures. While the under-five mortality rate is estimated at 70 per 1000 in 2012 Conakry, in other regions this level is higher than 100 per 1000, with the highest recorded in Kankan (194 per 1000) followed by Faranah (163 per 1000) and Labe (141 per 1000) (Fig. 4.18).

4.6.1.3 The City Human Development Index

Despite its potential to be sustainable, inclusive, resilient and prosperous, Conakry has not be able to be the engine of the economic development of the Republic of Guinea as illustrated by its low City Human Development Index (CHDI) compared to other African cities. Among the cities analysed here, only Conakry, Antananarivo and Monrovia have a CHDI, which is equal or lower than the national HDI. This is partly due to the fact that Conakry urban agglomeration's development has been hampered by its weak city foundation characterised by proliferation of unplanned settlements, poorly served in basic services and without land tenure security (Fig. 4.19).

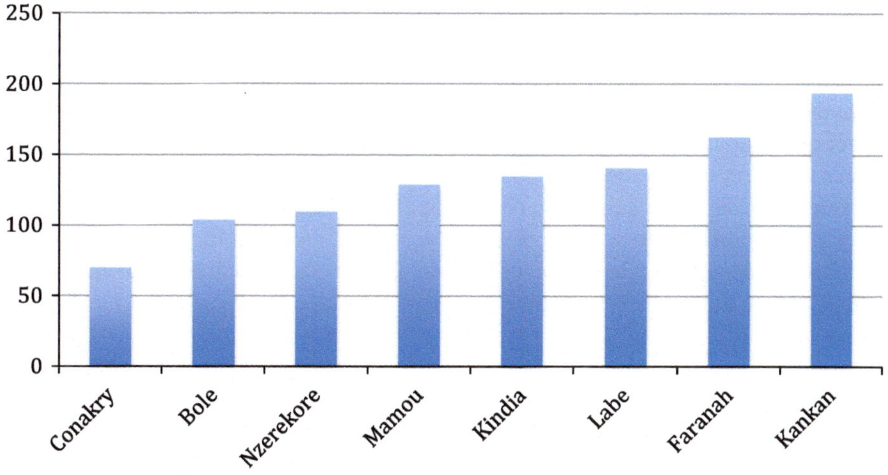

Fig. 4.18 Under-five mortality rates (per 1000). *Source* Computed from National Institute of Statistics, Republic of Guinea, 2013. Demographic and Health Survey and Multiple Indicators Cluster Survey (DHS-MICS, 2012)

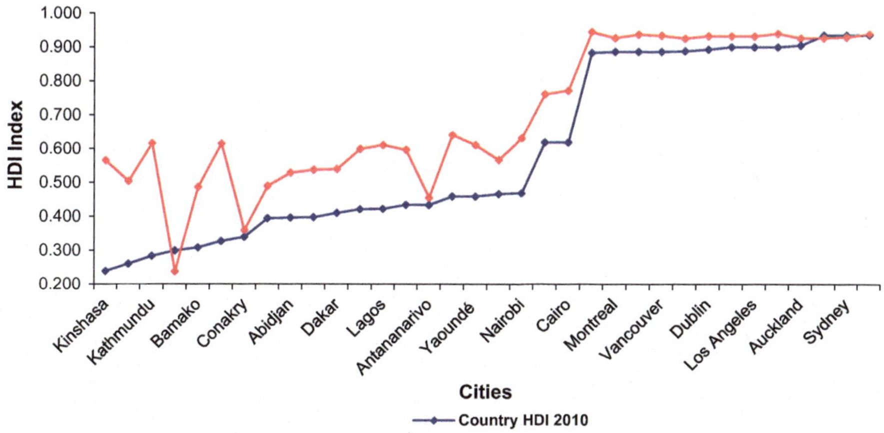

Fig. 4.19 City HDI and Country HDI, 2014. *Source* Mboup G (2016). Smart social development key for smart economy. In Vinod K. et al. (2016). Smart economy in smart cities, Springer

Though cities like Conakry are engine of economy growth, they need to be driven smoothly in an efficient administrative and governance environment. Thinking that having a large city and a high density is sufficient to produce economies and scale and agglomeration, is a simplistic view because they are many other factors that need to be present for urbanization to produce wealth and prosperity in general. In Conakry like in many African cities those ingredients are lacking leading to various negative externalities. Finally instead of enhancing economies of scale and agglomerations, Conakry as many African cities encourage diseconomies. Youth non-employment is high, and young people from impoverished urban areas can only find jobs in the informal sector with no social security coverage, paid and parental leave, retirement, and unemployment benefits. The high rate of non-employment of young people in Conakry as in many African large cities is reason for concern; the lack of decent, sustainable jobs promotes a sense of displacement in the general youth population and often leads to crime, under-development, and a cycle of poverty. Frustrations accompanying long-term unemployment among groups of urban young men may feed political and ideological unrest and provoke violence.

4.6.2 Urbanization and Environmental Sustainability

The Study for the Review of Coastal Town Planning Schemes of Guinea is a component of the Project "Strengthening Resilience and Adaptation to the Negative Impacts of Climate Change in Coastal Vulnerable Areas of Guinea-RAZC". The aim of the study is to establish that the urban planning management tools in Guinea integrate or not measures and actions on the adaptation of coastal cities to climate risks. Regarding the Conakry's development plan, the options selected include: (a) the creation of a large network of traffic corridors providing links between urbanized areas; the establishment of multi-functional centres playing the role of urban poles linked to the peninsula; (b) the organization of housing areas around small sector centres that will be served by cross-roads, etc. These measures can help protect people and their property from the effects of climate change as they can induce decongestion of the port centre located in Kaloum and better distribute urbanization throughout the city's territory and thus minimize the impacts of natural disasters such as floods, heat waves and possibly the consequences of extreme events such as hurricanes. The urban plan of Conakry also recommends land management actions that are likely to fight against the anarchic use of space and urban sprawl, that promote increase in energy consumption, and floods; Conakry being one of the twenty most exposed cities because of its geographical location, and high population density. In-depth investigations and highly targeted proposals and rigorous regulation are needed to deal with potential natural disasters and protect people, infrastructure and superstructures. The accelerated degradation of the environment in Maritime Guinea is due to this important population growth, anarchic urbanization and uncontrolled urban sprawl.

The coastal territory of the Greater Conakry metropolitan area is an extremely fragile environment that is also under extremely high anthropic pressure. In particular, the mangrove areas surrounding the peninsula are particularly fragile and are continuously deteriorating. These impacts are the result of the increase in urbanization in these areas. Marine environments are also extremely degraded due to the pollution generated by discharges of domestic and industrial wastes. The shoreline is often used as a rubbish dump for domestic and industrial waste and as a wastewater receptacle, causing biological and chemical pollution of the water and the parasitic and bacteriological infestation of living resources. These areas are fundamental to ecosystem equilibrium: the mangrove is a breeding ground, nursery and habitat for several species of fish, crustaceans as well as a place of refuge. Moreover, it constitutes a natural barrier that protects the coasts from marine erosion [5]. To these anthropogenic threats are added those related to climate change and rising water levels.

The proliferation of informal sectors also has led to the anarchic development of economic activities, industries and residential settlements. Growing industries located in or near the city, are key resource users as well as sources of pollution, waste, and greenhouse gases emissions. High use of informal modes of transport and private cars also contributes significantly to congestion, air and noise pollution.

4.6.3 Urbanization and City Prosperity

The City Prosperity Index (CPI) published by UN-Habitat in 2012 under the coordination of Mboup G includes various indices and indicators that are relevant to urban areas, and important for prosperity-oriented public policy-making. The first version of CPI published in 2012 was based on five dimensions: Productivity, Quality of Life, Infrastructure Development, Environmental Sustainability, and Equity and Social Inclusion [64]. In 2013, the CPI was revised to include the urban form dimension measured by street connectivity [65].

The resulting CPI values can be regrouped in six distinct brackets that range from cities with 'very solid' prosperity factors (CPI of 0.900 or above) to those where those factors are found to be 'very weak' (CPI below 0.500). The city of Conakry was classified among the cities with very weak prosperity factors. Cities in this bracket feature contrasted patterns among the sub-indices in the CPI. For some, the dispersion of index values across the 'spokes' reflects institutional and structural problems. For others, the five dimensions of prosperity do converge, only at very low values, a hallmark of dysfunctional systems, institutional failures, sluggish economic growth as well as widespread poverty and destitution [64]. In this group, much remains to be done there in terms of quality of life, infrastructure and environment in most of the cities in this bracket. Production of goods and services is still too low, a reflection of underdevelopment. Historic structural problems, chronic inequality of opportunities, widespread poverty, inadequate capital investment in public goods, and lack of pro-poor social programmes are critical factors behind such low degrees of prosperity. Various cities/countries from this group have a recent past marred by conflicts, political instability or economic crisis such as Monrovia (with a CPI of 0.285), Antananarivo (with a CPI of 0.446) or Conakry (with a CPI of 0.416). In all three cities, the city product is very low, 0.048, 0.133 and 0.171 respectively.

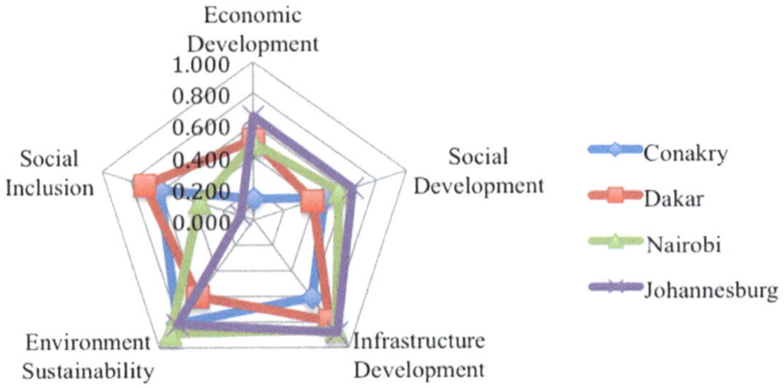

4.6.4 Towards a Smart Metropolitan Regional Development

Among the components of the Smart Metropolitan regional Development, five have been part of the City Prosperity Index (CPI): Productivity, Infrastructure development, Environmental Sustainability, Quality of Life and Social Inclusion. The SMRD index introduces other components relevant for sustainable urban development: Disaster Risk Reduction and Resilience, Peace and Security, Institutions and Laws; It also expands productivity to Economic Development, Quality of Life to Social Development, and make the distinction between Infrastructure Development and Basic Infrastructure. It also introduced the concept of city foundation built upon three dimensions: urban planning, land policies and basic infrastructure. It also introduces ICT at the centre of the SMRD along with the City Foundation and Institutions and Law. Considering that the CFI and the CPI was already measured in previous sections, the measurement of the SMRD is incremental and built upon those indices with the inclusion of Peace & Security as well as Disaster Risk Reduction and Resilience (Fig. 4.20).

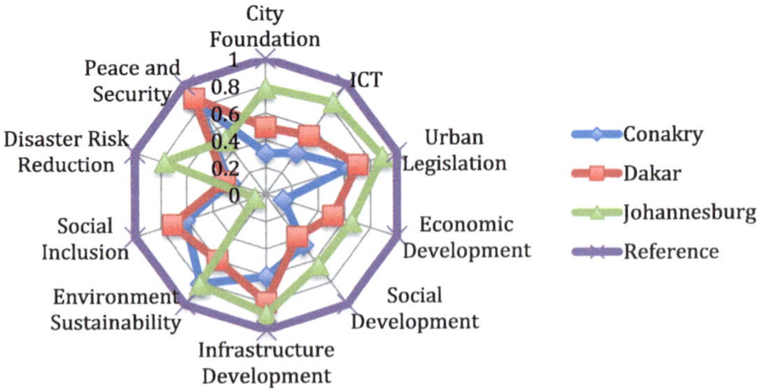

Fig. 4.20 Components of the SMRD. *Source* Provisional results from Mboup G (ed.) (2018) Forthcoming publication "Smart economy in smart African cities", Springer

In addition to the urban challenges analysed in previous sections, the population of Conakry Peninsula has been constantly exposed to disasters, particularly to flooding during raining seasons. Apart from habitat degradation, floods have caused considerable economic, social and environmental losses in Conakry. Most households in Conakry are built in unplanned settlements that lack most basic services, particularly in Matoto and Ratoma, leading to frequent flooding and several other social and economic negative externalities. Associated with this heavy rainfall, the particular topographical situation of the city of Conakry favours recurring floods during the raining season.[1] In the Conakry Vision 2040, to reduce flooding occurrence, strategies include: a) Protect and restore natural outlets; b) Initiate studies on flood zones; c) Deepen studies on rainy reference events; d) Put in place compensatory measures for the relocation of the affected populations; e) Initiate a reflection on new channel models; f) Continue rehabilitation of valleys and flood zones and; g) Continue to improve the drainage network. In addition to disaster exposure, Guinea was classified among the less peaceful countries based on the Global Peace Index (GPI) ranking in 2015. The GPI scores are the aggregates of 23 qualitative and quantitative indicators across three thematic domains: the level of Societal Safety and Security; the extent of Ongoing Domestic and International Conflict; and the degree of Militarisation. Out of 163 countries, Conakry was ranked 117 with a score of 2.214 compared to the lowest score, which is estimated at 1.148 for Iceland and the highest score, which is estimated at 3.645 for Syria. For South Africa, the GPI score is estimated at 2.376, ranked at 136 behind Guinea. In Africa, only four countries—Senegal, Mauritius, Botswana and Namibia—are among the top 50 peaceful countries globally.[2] With a weak City Foundation, a low Human Development Index and a weak City Prosperity Index, it is obvious that high exposure to disaster coupled with high insecurity will make it worse with a very low level of Smart Metropolitan Regional Development Index. Though this a preliminary assessment of the SMRDI since data are not able for some sub-indices, rapid assessment of available data point to the fact Conakry SMRDI is very low due do to historic structural failure on urban development, chronic inequality of opportunities, widespread poverty, inadequate capital investment in public goods, etc.

4.7 Spatial and Economic Design Strategies for a Smart Conakry Metropolitan Regional Development

4.7.1 Background, Justification, Objectives and Strategies

Considering that the population of Grand Conakry will double by 2040 to reach 5–6 million inhabitants, the Government of Republic of Guinea commissioned a

[1]Republique of Guinea, European Union and Louis Berger, 2016. Grand Conakry Vision 2040. The Original document Grand Conakry Vision 2040 is in French and remains the official document.

[2]https://reliefweb.int/sites/reliefweb.int/files/resources/Global%20Peace%20Index%20Results%20Map_0.pdf. Downloaded 8 March 2018.

study for the “Grand Conakry Vision 2040”. The study proposes various scenarios of territorial settlements. These scenarios are aimed at preventing disasters that can hamper ecological, social, economic and urban development. Based on a structured dialogue within a Technical Committee and a Steering Committee with the Representatives of various ministerial departments, a consensus was reached on “a synthesis, balanced and controlled scenario” which is structured around three levels of intervention: (1) at the metropolitan level to develop and strengthen urban polarities; (2) at the agglomeration level to channel and structure urban extensions and; (3) at the peninsula level to renew the city (urban renewal). Recognizing that the Master Plan of the city of Conakry has become obsolete to tackle the challenges the city is facing, the government has commissioned a study for the “Grand Conakry vision 2040”, which is the first step towards developing a new generation of master plans: the master plan of the great Conakry and the national land use planning scheme (SNAT).

The main objective of the Grand Conakry Vision 2040 is to “improve the living conditions of the residents of Conakry and adapt land-use planning and planning policies to the rapid urbanization of the metropolitan region” [5]. The “Grand Conakry—Vision 2040” is conceived as the preliminary framework for the elaboration of a future Master Territorial and Urban Planning. Its objective is to define the main key points and themes to be considered in order to make Conakry a modern liveable metropolis safeguarding the environment. The study aims to be the reference framework for several ministerial departments, communities and local authorities to set a long-term urban development target by 2040 (Figs. 4.21 and 4.22) [6].

This section is a prospective analysis of different ways economic and spatial transformations from these different programmes, reforms and plans will contribute to the smartness of the Conakry Metropolitan Regional Development. Most national and urban policies and programmes aim to spatially and economically transform the Conakry Metropolitan Region to be smart, sustainable, inclusive, resilient and prosperous. We will analyse the economic transformation expected in these programmes, reforms and plans. We will then proceed with a holistic analysis using the conceptual framework presented in the introduction in consideration of all dimensions of the smart metropolitan regional development such as: smart metropolitan regional foundation, ICT, smart Institutions and laws, smart economic development, smart infrastructure development, smart environment, smart social development, smart peace and security and smart disaster prevention and resilience.

The preferred scenario for the Grand Conakry Vision 2040 is an equilibrium scenario that pursues a strong ambition: the preservation of resources and the limitation of environmental damage through a polarized, oriented, controlled and optimized urbanization. This scenario is built on a triple strategy:

- The reinforcement of urban polarities relay to limit the demographic pressure on the peninsula
- The control and the structuring of the urban extensions to avoid a continuous development of the urbanization
- The intensification of the urbanized spaces of the peninsula

Fig. 4.21 Future National Master Plan. *Source* Republic of Guinea, European Union and Louis Berger, 2016. Grand Conakry vision 2040

Spatial Design and Planning Strategies will be at three levels: (1) Metropolitan region with the development and the reinforcement of urban poles connected to Conakry; (2) Urban agglomeration in increasing the urban extent of Conakry with the urbanization of neighbouring areas; and (3) City level with the densification of the city of Conakry (Fig. 4.23).

From the three scenarios, a resulting scenario is built with seven ambitious goals for a sustainable Conakry for all: (1) Strengthening territorial balances to create a network of solidarity-based cities in a preserved environment; (2) Controlling the city limits to prevent urban sprawl and preserve urbanization; (3) Optimizing the port system to decongest the city and protect the populations;

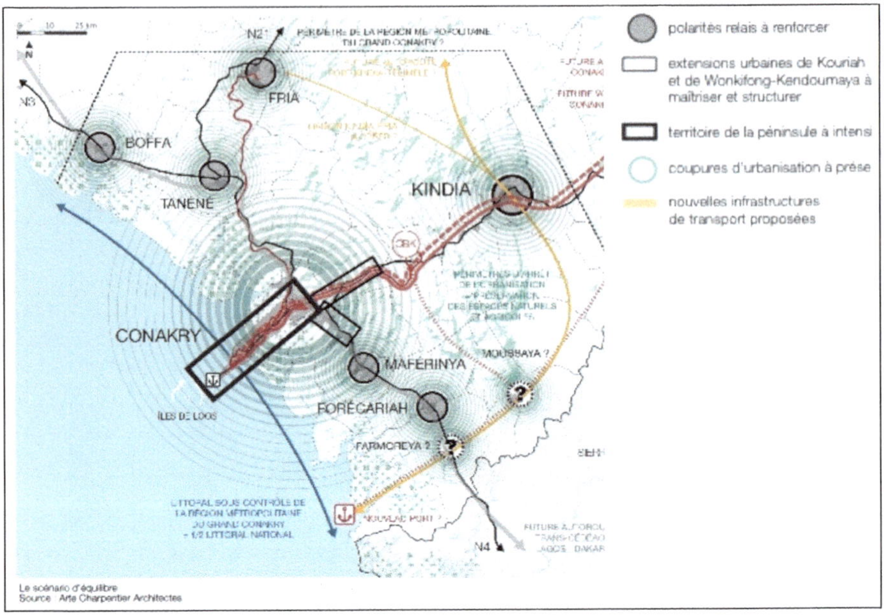

Fig. 4.22 A balanced scenario—Polarized and controlled urbanization to preserve the environment. *Source* Republic of Guinea, European Union and Louis Berger, 2016. Grand Conakry Vision 2040

Fig. 4.23 Spatial distribution of Conakry and surroundings. *Source* Figure prepared using data from Conakry vision 2040

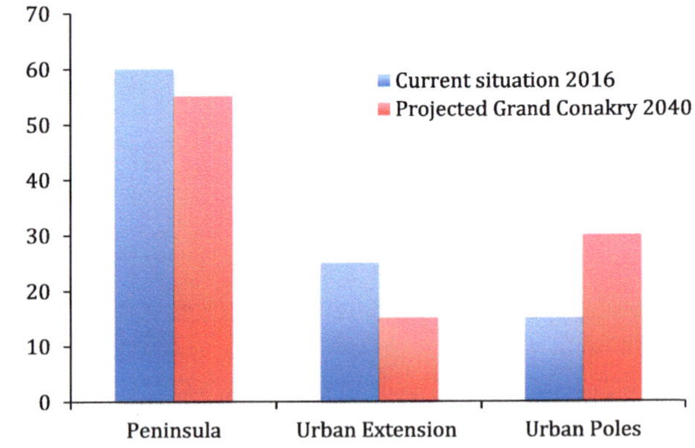

(4) Restructuring the centrality to rebuild an efficient urban system on the peninsula; (5) Restoring the landscapes to build a healthier, safer and more enjoyable city; (6) Making the city sustainable to provide housing for everyone in mixed and lively neighbourhoods and; (7) Thinking waste as a resource to protect people, the environment and generate wealth.

(1) Strengthening territorial balances will consist of: (a) Strengthening and organizing urban polarities in a modernized, hierarchical network connected to the regional network and; (b) Preserving fragile natural areas and agricultural economic zones, natural Park and protecting the islands and; (c) Develop the agro-industrial sector.
(2) Controlling the city limits will consist of: (a) Setting up new geographical boundaries according to the administrative, demographic and economic weight of the capital city; (b) Reviewing the administrative planning of the Grand Conakry and adapt the institutional and regulatory framework and; (c) Optimizing land use.
(3) Optimizing the port system will consist of: (a) Valuing the harbour tool in harmony with urban development (transfer of polluting and hazardous activities) and the creation of a dry port and; (b) Modernizing and streamlining existing facilities.
(4) Restructuring the centrality will consist of: (a) Reorganizing the urban fabric; (b) Redeploying the centralities along the peninsula and the metropolitan area around the structuring facilities; (c) De-locating tertiary activities from the tip of the peninsula (Kaloum) to appropriate and developed sites; (d) Implementing a genuine mobility strategy; (e) Restructuring the urban network; (f) Developing a Urban Travel Plan (UDP) at the metropolitan area scale; (g) Developing capacities systems (a network of heavy axes in its own right to alleviate traffic) and diversification of the transport supply by facilitating inter-modality by means of adapted structures and; (h) Creating an authority for the management and regulation of urban transport.
(5) Making the city sustainable will consist of: (a) Ensuring a balanced distribution of settlements, structuring proximity facilities as well as urban services in a network that promotes accessibility; (b) Initiating the opening up of equipment, and controlling densification of the peripheral sectors; (c) Proceeding with the creation of a public operator responsible for producing and managing the social housing stock; (d) Implementing an ambitious land tenure policy, financing mechanisms, operational procedures and legal arrangements, in order to allow the community to operate on urban development in the name of the general interest and to better frame the creation of public or mixed operators and; (e) Establishing mechanisms for regular participatory communication and dialogue with populations.
(6) Restoring urban landscapes will consist of: (a) Restoring water quality and natural flows of the peninsula; (b) Renovating public spaces for the comfort and living environment of the inhabitants; (c) Protecting and managing the coastline to guarantee the integrity of the maritime public land (DPM) by a “littoral law”; (d) Establishing safeguard perimeters to protect sensitive areas and create the Mangrove Natural Park, and protecting the islands from urban pressure; (e) Modernizing the agricultural sector and develop the potential of arable land and wetlands, including in urban areas.

(7) Restoring urban landscapes will consist of: (a) Establishing a coherent strategy and reorganize the sanitation network master plan; (b) Creating networks for the collection, sorting, treatment and recovery of urban waste and; (c) Enhancing communication and training.

4.7.2 *Conakry Metropolitan Region: Developing and Reinforcing Urban Polarities*

Spatially connected and economically inter-dependant

The extended territory of Grand Conakry is characterized by an exceptional and fragile natural environment, punctuated by small and medium-sized towns, distributed within a radius of 150 km around the capital. To limit the influx of more and more people in the direction of the peninsula (and more generally in the direction of the littoral territories) and to prevent major and potentially irreversible environmental degradation, it is essential to channel some of this demographic growth towards these secondary cities through the reinforcement of their attractiveness, in a logic of complementarity and balance with the capital city. In sum, the strategies will consist to develop and strengthen urban polarities by: (1) Fixing and attracting populations in urban polarities; (2) Building a strong metropolitan region based on territorial solidarity and; (3) Preserving the natural and agricultural environment to enhance the productive force of the territory.

Existing urban poles to be part of the Grand Conakry Metropolitan Regions are

(a) Kindia (140 km from PK36) that has some potential in terms of tourism development but also, and above all, a strong potential for agricultural and agro-industrial development. The realization of new transport infrastructures such as the future highway Conakry-Mamou and the future railway Conakry-Kankan will reinforce this potential;
(b) Fria (120 km from PK36), the old mining industrial city: the announced revival of its aluminium factory closed since 2012, combined with strategies for diversifying economic activity, could mark the rebirth of the city and stimulate a new development facilitated by the upgrading of existing transport infrastructure (railway Conakry-Fria, Nationale 21 and even Katourou airport);
(c) Tanéné (65 km from PK36): the city of Tanéné is a crossroads city located at the junction of the National 3 (Conakry-Boké-Guinea Bissau) and Nationale 21 (Tanéné-Fria); This particular cross-city situation and the dynamism it generates could serve as a basis for economic development in relation to the flow of goods and encourage the development of agricultural activities;

(d) Boffa (120 km from PK36): the town of Boffa has a diversified economy (fishing, agriculture, solar salt, tourism, mining, etc.), which continues to be dynamic and which could thus contribute to fixing some of the expected demographic growth as a result of these many employment sectors;
(e) Maférinya (40 km from PK36): the town of Maférinya has developed agricultural production and in particular its pineapple production. This rich natural environment and its proximity to the capital, could allow it to develop productions dedicated to both domestic consumption and exportation with the planned airport;
(f) Forécariah (65 km from PK36): the town of Forécariah also develop many agricultural activities, some of which are dedicated to export (banana, pineapple, mango, oil palm, etc.).

Creation of urban poles

Respectively located 80 and 100 km from PK36 by road, the towns of Farmoreya and Moussaya are today only small villages. In the hypothesis of the construction of this new port, they could be developed in connection with the planned new railway infrastructure (the line connecting the port to the Simandou mine eventually supplemented by an additional line connected to the mine of the CBK), with a view to economic development linked to the flows generated, particularly around mineral processing industries.

It is important to note the spatial design of the metropolitan is aligned with the needs for transport, most of the urban pole to strengthen or to develop are related through the road network PK36. A large metropolitan region provides opportunities for economies of scale and agglomeration, but it will also call for large investments in infrastructures to respond to the increased demand for water, sanitation, solid management, energy and mobility. It will particularly come with increased demand on mobility that must be satisfied with an efficient public transit and connection between urban centres, accompanied with increased spaces for pedestrians and cyclists in order to safeguard the environment while creating economic growth. Economies of scale and agglomeration are, indeed, greater in metropolitan areas where transportation infrastructures are able to respond mobility needs with higher access to markets and resources than those where people mobility is impeded by deficient transportation infrastructures. Efficient urban mobility systems increase accessibility to markets, employment and investments and therefore provide better access of people to economic and social opportunities. Deficient transport systems create negative externalities and are source of social inequalities in cities. Efficient mobility systems make transportation means accessible and affordable to all people, while deficient mobility systems exclude the urban poor from many urban advantages and opportunities. The mobility of people and freight reflects the level of accessibility of urban residents to the multiple economic opportunities that cities offer.

Another important aspect noted is that the spatial design of the metropolitan region is aligned with specialization: Kindia has a strong potential for agricultural and agro-industrial development as well as for tourism development; Fria is an old mining industrial city (including aluminium); Tanéné, as a crossroads city, will ease the circulation of good in the metropolitan region and beyond; Boffa has a diversified economy (fishing, agriculture, solar salt, tourism, mining, etc.); Maférinya has developed agricultural production (particularly in pineapple); and Forécariah has various agricultural activities (banana, pineapple, mango, oil palm, etc.), particularly for export. However, this specialization calls for efficient transport infrastructures that must be assessed, planned and implemented. Efficient mobility allows localities of urban agglomerations to specialize in the production of goods and services for which they have comparative advantages and ease inter-localities cooperation. This will also allow large-scale production of goods and services that can be distributed within the metropolitan regions and beyond with time, cost and reliability opportunities [1]. This is the way the economies of scale and agglomeration associated to large urban agglomerations and high densities can be materialized.

An important forgotten component in the transport sector in Conakry as in most African cities is the movement of freights. Today urban logistics, as a strategy ensuring efficient freight movements and innovative responses to urban customer and business demands, is an emerging field of investigation making the movement of freight efficient while mitigating congestion and environmental externalities [66]. The development of the Grand Conakry, the Conakry metropolitan region with specialized urban poles will, indeed, result in growing quantities of freight moving between cities. There will be an impressive variety of supply chains servicing a wide array of economic activities such as agro-industrial from Kindia, agriculture products from Forécariah, industrial products from Fria, etc. As freight traffic commonly shares infrastructures with the circulation of passengers, the mobility of freight in between and within urban areas will be problematic along the development of the metropolitan region. It has been noted that congestion associated to transportation of freight starts to be a serious issue once a threshold of about one million inhabitants is reached. Considering the growing level of production, distribution and consumption of goods, the metropolitan region should have planning and circulation management schemes where urban freight distribution is preeminent [66]. Though transport infrastructures are factored in the development of the metropolitan region, it must be further underscored.

4.7.3 Extension of the Conakry Urban Agglomeration

Extension of the Conakry Urban Agglomeration to channel and structure urban extensions will consist of: (1) Preventing the process of urban sprawl in order to preserve the natural environment through the establishment of perimeters for urban land use; (2) Structuring, reorganizing and controlling the urban extensions.

In order to protect the natural environment and preserve urbanization for territorial ecological balances, it is urgent to stop the rampant urbanization and to engage in strategies for structuring territories that have already been impacted. It must define precise and stabilized limits and engage the optimization of the use of land by connecting together these "pockets of urbanization" disconnected from each other, and planning in these new urban spaces to accommodate a significant portion of the population expected by 2040.

It is along the National 4, which is much affected by the unplanned urbanization, to limit the urban development to the limit of the present sub-prefecture of Coyah and thus impose a limit of urbanization between the commune of Kendoumaya and that of Maferenya (which, along with that of Friguiagbé near Kindia, is one of the main pineapple producing areas in the region and of its surrounding agricultural areas must be protected) as well as protection limits of the northeastern reliefs and mangrove protection on the southeast side.

Beyond this limit, it is a question of favouring multi-polarity (rather than continuous linear extension) and a controlled development of the existing polarities (Maférinya and Forécariah) or in the future, guaranteeing the preservation of the natural and agricultural environment in which they operate and from which they derive their wealth and potential.

Along National 1, around Kouriah, it is again to limit the urban development on the edge of the current sub-prefecture of Coyah and to impose an urbanization limit before Kolibaya, in limit of the plateau rocky, which limits the valley. Transversally, it is a question of setting limits adapted to the preservation of the slopes of the reliefs and the agricultural zones.

In these peri-urban areas, there are many different spaces (natural spaces, agricultural spaces, built-up areas, contemporary habitats and traditional habitats) and many uses that are as much linked to the urban context as to the rural context: this spatial mix land use must be preserved and serve as a basis for structuring the territory biodiversity. This will particularly include:

- Reintegrate hierarchical centralities for the programming of equipment and services to the population
- Structure a hierarchical road service to the heart of neighbourhoods;
- Structure a network of public spaces and green spaces articulated with the spaces of nature which can still be preserved on the still free spaces of any urbanization
- Develop pilot operations of habitat and typological classification;
- Ensure the management of rainwater through the preservation of valleys; and
- Implement pilot projects such as Farms of Guinea and the "Filter counter" for the management of solid and liquid waste.

These areas of fringes could in particular enable the implementation of particular strategies for managing the city/nature interface through pilot operations of peri-urban agriculture, agro-forestry, development of gardening in the lowlands, maintaining traditional agriculture practices, etc.

- At the peninsula level, to renew the city will consist of: (1) Building an efficient, safe, healthy and pleasant capital and; Optimizing the use of the soil by densification of its structure to accommodate new urban areas with improved living conditions.

4.7.4 (Re) Build the City on the City: A Dense, Restructured, Efficient, Safe, Healthy and Liveable City

The protection of the environment, and in particular the natural and agricultural areas of the greater territory of the metropolis, today requires a strong ambition capable of controlling urban sprawl while allowing the reception of new populations on territories already urbanized. This need for a more rational use of land must take place through the (re-) construction of the city on itself, necessarily integrating a infill effort, and must be understood as the opportunity to improve the living conditions of today's Conakry residents by overcoming the dysfunctions of the constituted urban space.

Thus, through this effort of densification and intensification of already urbanized spaces, the aim is to implement a more efficient use of resources in the broad sense: land resources, energy resources, water resources, resources in materials, etc. The compact city, the dense city, is the one that allows efficient use of resources through better management of networks and opportunities for pooling and implementation of short circuits for recovery, reuse, recycling, etc.

Combined with the implementation of a larger functional diversity makes it possible in particular to act on a functional mix, in particular the mobility needs over distances of displacement, and modal shares, and consequently on energy consumption and gas emissions energy consumption and greenhouse gas emissions. But, it's not just about acting quantitatively on the urbanized area of the peninsula to maximize on the peninsula's urbanized space to maximize its use by aiming at the "technical performance" the urban system, but to engage in actions of the urban system, but to urban renewal coherent with the urban renewal set consistent with all the urban strategies to be implemented at scale territorial guarantee of territorial improvement, the only guarantee of an improvement in the quality of life as a whole and sustainable urban development (see § NOTES OF FRAMING).

Therefore, these urban renewal actions must not simply consist of an increase of the volume built on the surface considered but they must be oriented towards the implementation of a densification "qualified", inscribed within a global strategy of improvement. It is an uncompromising development to establish and/or restore urban development, capable of establishing and/or restoring a number of balances that have now been broken.

This (re-) construction of the city on the city must thus contribute to:

- The redeployment and rebalancing of central and urban functions at the metropolitan level—the integration of greater diversity Functional on all scales of the urban up to all the scales of the urban, to the heart of neighbourhoods;

- The mobility network, the restoration of continuity between neighbourhoods and their connection to continuity between neighbourhoods and their connection to the metropolitan structure;
- Support for housing policies and in particular the elimination of precarious housing;
- Upgrading neighbourhoods access to basic urban services (access roads to basic urban services (highways servicing, water, electricity, sanitation, and public waste facilities, ICTs, etc.) and public facilities nearby (school equipment, health, etc.);
- The preservation and reintroduction of spaces nature in the city (nature of nature ornamentation nature in the city (nature of ornament, ecological nature, productive nature, etc.)—the restoration of public spaces—etc.

This third scale of the recommended development strategy is resolutely a long-term task, necessarily transversal (the question of urban renewal deals with all the problems: urban, social, economic, cultural, environmental issues, etc.) and inscribed in a global vision. In this perspective, the role of the National Agency for Urban Renovation (ANRU) will be determining and will require adapted means.

4.7.5 Sectorial Programmes for the Metropolitan Region, the Urban Agglomeration and the Peninsula

The spatial design of the metropolitan region, the urban agglomeration as well as the re-structuration of the Peninsula will be supported by various sectorial programmes such as on: Housing, Land, Transport and Mobility, Port and Port Infrastructure (Fig. 4.24).

At the Housing sector, strategies include: development of a city and housing policy; creation of a public social housing agency; supplement and strengthen other instruments of the implementation of the housing policy; promote the use of local building materials; re-structure urban governance and; put in place sustainable financing mechanisms.

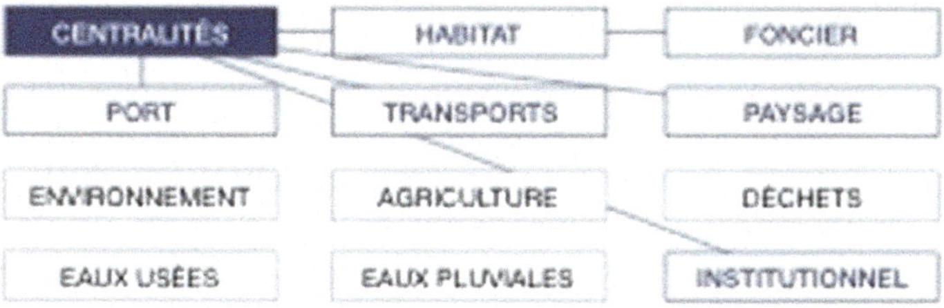

Fig. 4.24 Necessary conditions of the urban re-structuration. *Source* Republic of Guinea, European Union and Louis Berger, 2016. Grand Conakry Vision 2040

At the Land sector, strategies include: (a) Clarify the scope and conditions of application of customary law alongside modern law; (b) Complete, consolidate and modernize the cadastral system; (c) Assess the state's land assets; (d) Reclaiming the state's land assets; (e) Develop land planning documents; and (f) Engage a policy of "production" of land.

At the Transport and Mobility sector, strategies include: (a) Establishment of traffic restriction; (b) Development of a fleet of taxi boats; (c) Structuring a road network at the scale of the Greater Conakry metropolitan area; (d) Create a bus station at PK36; (e) Reinforcing the railway; (f) Organize maritime connections; (g) Develop a BRT system on its own site and improve the existing urban network; (h) Make accessible the districts of the peninsula; (i) Implement strategies for stationing; (h) Gather typologies of structures and modes of maintenance of the road network.

At Port Infrastructure sector, strategies include: (a) Engage and optimize the port extension project; and (b) Improve conditions for pre- and post-shipment of goods.

At the Urban Space level—including Public Spaces, coastal areas (link between the sea and the city), urban nature spaces (such as the lowlands, urban forests and woodlands as well as urban and peri-urban areas of agriculture), and in general green spaces. The strategies include: (a) Reintroducing the question of public space in the structuring of the metropolis and neighbourhoods; (b) Initiate a coastal reclamation project; (c) Strengthen the notion of green city; (d) Preserve and develop alignment trees; (e) Protect classified urban forests and develop urban woodlands.

At the Environment Sector, the strategies include: (a) Set up urbanization perimeters; (b) Preserving and restoring mangroves; (c) Continue to support the development of solar "culture"; (d) Modify practices related to mangrove rice cultivation; (e) Implement accompanying measures and protection against climate change; (f) Promote reforestation actions; and (g) Create a large National Nature Park on the mangrove territories.

At the Agriculture Sector, strategies include: (a) Promote the agricultural potential of relay polarities; (b) Rethinking the agricultural organization; c) Better channel the development of palm groves; (d) Promote endogenous positive and sustainable knowledge and practices; and (e) Promote urban and peri-urban agriculture.

Sanitation—solid waste (An ecological, social and economic emergency), strategies include: (a) Start a characterization campaign; (b) Restart the collection service; (c) Set up one (or more) new landfill centre (s); (d) rehabilitate the site of the "Mining area"; (e) Develop and promote local waste recovery methods; (f) Promote a centralized treatment of biomedical waste; (g) Optimize the collection service "Filter Counter"; (h) Develop waste management units oriented towards the circular economy;

Sanitation—wastewater, strategies include: (a) Rehabilitate and optimize the Petit Bateau boat station; (b) Manage and valorise waste materials through pilot projects of multifunctional platforms; (c) Establish the sanitation scheme of the individual network of the city; and (d) Extend the collective network.

Sanitation—rainwater, strategies include: (a) Protect and restore natural outlets; (b) Initiate studies on flood zones; (c) Deepen studies on rainy reference events; (d) Put in place compensatory measures for the relocation of the affected populations; (e) Initiate a reflection on new channel models; (f) Continue rehabilitation of valleys and flood zones; (g) Continue to improve the drainage network.

Institutional Setting for the creation of a metropolitan region, strategies include:

(a) **Create a solid legal base favouring inter-communality**: Promulgate and implement the law on the code of local authorities; immediately transfer the 32 competencies listed in Article 29 of the Local Government Code to local authorities, as well as the financial means involved; clearly define the status and competences of the urban and rural communes mentioned in articles 18 and 19 of the Code of Local Authorities; and develop a public policy of the city.

(b) **Create the EPCI of the urban community of Grand Conakry**
An EPIC is an administrative structure governed by the provisions of the Code of Local Authorities, which includes municipalities that have chosen to develop several common skills, such as spatial planning, public transport or the management of public buildings. The Grand Conakry Vision 2040, to be viable, must be based on a solid juridical framework, privileging the inter-communality, as: —policies aiming to unite the territories, to pool skills, fiscal resources, projects in a more integrated way—a real legal tool in the service of municipalities. A clear definition of inter-communality must therefore be inserted in the Community Code, recalling that this is for urban or rural communes to constitute a Public Institution of Inter-communal Co-operation (EPCI). The EPCI receives by its statutes competences that it exercises in place of the communes. It is therefore essential that the state transfer the 32 competencies to the communes so that they can transfer them to the EPIC, if necessary, etc.

(c) **Create the Metropolitan Pole of Grand Conakry**
Once launched, the Grand Conakry could favour the Metropolitan Pole, connecting the Metropolis (very integrated inter-municipal structure) and the network of cities. The Metropolitan Pole is a flexible tool with a few legislative specificities: —a minimum population is required—its territory is in one piece and without enclave. Its actions of metropolitan interests are: —economic development—development transport infrastructure and services—the promotion of innovation, research, higher education and culture.

Centrality and Programming

The creation of an efficient and sustainable metropolis imposes a deep restructuring of the urban system through the redeployment of balanced centralities spread over the entire territory of the peninsula and associated with the implementation of greater functional diversity at all scales. The implementation of an efficient urban system, and through it an efficient transport system requires a profound transformation of the existing urban structure and in particular a redeployment of the centrality at the scale of the new limits of the urban area (Framework: Transport and

Mobility). A first step in this necessary rebalancing of urban functions is being completed through the announced realization of the new Koloma administrative centre, which is expected to accommodate a portion of the administrations and ministries now located in Kaloum. However, if this first step marks a necessary prerequisite for the construction of a more efficient metropolitan structure, to create a balanced and efficient urban system, it is imperative to structure other points of attractiveness on the linear peninsula. In particular, the new territorial scale of the metropolis, considered not within the limits of the peninsula but in a wider area, extended to Dubréka, Coyah, Kouriah, Wonkifong and Kendou-maya, but also to relay polarities within a defined centre of gravity.

Use commercial functions and markets to structure the city and its neighbourhoods

Trade is a formidable driver of economic activity and urban animation. Markets in particular are one of the most polarizing elements of urban life. They are the central point of complex networks of supply, storage, distribution: a sum of functions which generates a large number of activities but which constitutes today a major obstacle to the fluidity of the traffic and transport, which regularly leads to network congestion. The district of Madina is the perfect illustration: the concentration of commercial activities generates daily significant traffic jams on the highway. Because the commercial function invades everywhere the urban space, beyond the limits of the markets, on the axes of transport and the intersections. This dynamism of the commercial function must be used to accompany, control, guide and channel new urban developments and the renewal of the city on itself. The structuring of the commercial offer must thus be hierarchical at different scales of the metropolis and be articulated to the structuring of the mobility. These different market centres will have to be associated with exchange hubs integrated with urban transport and the public transport network, to organize intra-city and intra-neighbourhood flows.

Consideration of the concept of centrality at all scales of urban space

In addition to the commercial function, all urban functions must be broken down at different scales of the city: metropolitan scale, municipal scale and neighbourhood scale, but also at the sector level. At these smaller scales, there are "neighbourhood" centralities that could be built around a public area nearby: primary school, mosque, small shops, neighbourhood house dedicated to associations, etc. (Fig. 4.25).

ICT can ease this process with innovative e-planning along with design, management and monitoring. From the beginning of the 21st century a digital citizenship, particularly the "Millennial" generation, has started to emerge in Conakry as in many African cities [67]. Maximizing the digital dividends requires better integration of ICTs with the other factors of smart cities such as: city foundation, infrastructure development, environment sustainability, social development, disaster prevention, resilience, peace and security. ICT can be seen as substitution and catalyst factors, but the other factors (or the analogue part) of the operation are crucial in making smart economy. ICT alone will produce little in the smart

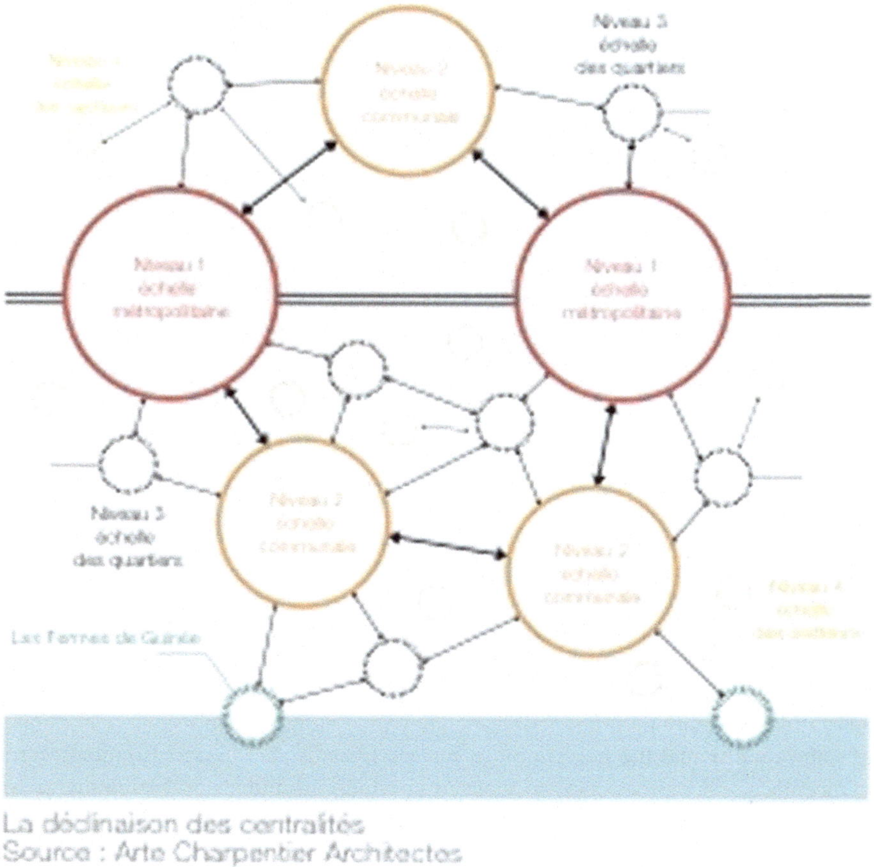

Fig. 4.25 Centralities diagram

metropolitan regional development. ‘Smart’ is not an end in itself [68]. It is the way ICT is integrated in the city development that will determine the city smartness.

ICT can in particular impact the transport sector in many ways among them we can enumerate two which will be the focus of this paper: (a) the digitalization of the transport sector as it happens in all sectors and; (b) substitution (partial or full) of mobility to perform or access services [69]. The ICT revolution with the rapid development and use of Internet, digital mobile communication, and “big data” analysis enable to create a less costly and more powerful “intelligent transport systems” (ITS) [70]. The ITS have a greater potential to more efficiently manage transportation assets, improve road safety, reduce traffic congestion and travel time. This will boost productivity and reduce greenhouse gas (GHG) emissions [71] (Fig. 4.26).

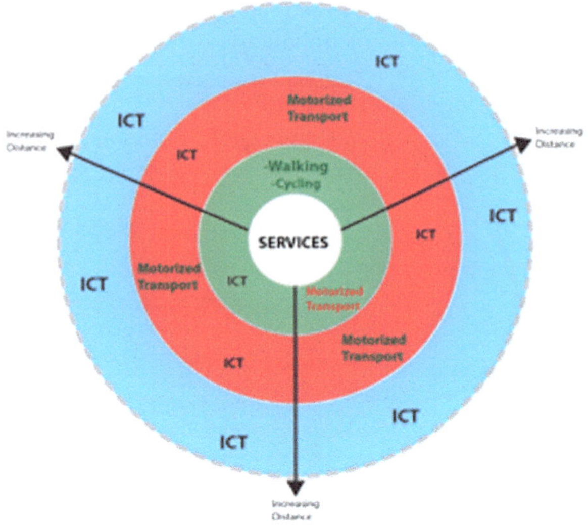

Fig. 4.26 Multiple choices to access services. *Source* Mboup G (2016) in Smart economy in smart cities (ed. V. Kumar et al.), Springer

Conclusion

Considering its geographical location as a peninsula in the Atlantic Ocean, Conakry has the potential to be an agro-industrial, green city in addition to its huge potential of fish production. Wetlands areas are sources of income from agriculture, livestock, and crafts, among other activities. They play an important role in maintaining the water quality and the prevention of natural hazards. These are also ideal places for reception and reproduction of waterfowl which are indispensable elements for the ecological balance of aquatic environments and key links in the food chain, hence the importance of the preservation of wetlands. If well planned and designed, Conakry can be a smart metropolitan region where citizens enjoy high quality of life. In addition to its geographical advantages, the Conakry Peninsula has a population of nearly of 2 million that, as large urban agglomerations, constitutes economic and social opportunities as well as challenges depending on how the urban growth is planned and managed. However, Conakry as most African cities has not been able so far to respond to growth in accessibility demand and several other needs such as access to water, sanitation, management of solid waste, and streets and other public spaces, key elements of a city foundation.

Considering the proliferation of informal settlements along the rapid urbanization of the Conakry Peninsula and surroundings, the Government of the Republic of Guinea is developing a holistic approach, the "Grand Conakry Vision 2040", for a smart metropolitan regional development through economic as well as spatial design strategies. The Grand Conakry Vision 2040 considers three spatial design strategies: (1) at the metropolitan level to develop and strengthen urban polarities; (2) at the agglomeration level to channel and structure urban extensions and; (3) at the peninsula level to develop urban renewal approaches. The main objective of the Grand Conakry Vision 2040 is, indeed, to "improve the living conditions of the residents of Conakry and adapt land-use planning and policies to the rapid urbanization of the metropolitan region”. This is built on seven ambitious goals for a sustainable Conakry for all: (1)Strengthening territorial balances to create a network of solidarity-based cities in a preserved environment; (2) Controlling the city limits to prevent urban sprawl and preserve urbanization; (3) Optimizing the port system to decongest the city and protect the populations; (4) Restructuring the centrality to rebuild an efficient urban system on the peninsula; (5) Restoring the landscapes to build a healthier, safer and more enjoyable city; (6) Making the city sustainable to provide housing for everyone in mixed and lively neighbourhoods; and (7) Thinking waste as a resource to protect people, the environment and generate wealth.

An important aspect noted in the Grand Conakry Vision 2040 is that the spatial design of the metropolitan region is aligned with specialization: (1) Kindia has a strong potential for agricultural and agro-industrial development as well as for tourism development; Fria is an old mining industrial city (including aluminium); Tanéné, as a crossroads city, will ease the circulation of good in the metropolitan region and beyond; Boffa has a diversified economy (fishing, agriculture, solar salt, tourism, mining, etc.); Maférinya has developed agricultural production (particularly in pineapple); and Forécariah has various agricultural activities (banana, pineapple, mango, oil palm, etc.), particularly for export.

However, this specialization calls for efficient transport infrastructures that must be assessed, planned and implemented. Efficient mobility allows localities of urban agglomerations to specialize in the production of goods and services for which they have comparative advantages and ease inter-localities cooperation. This will also allow large-scale production of goods and services that can be distributed within the metropolitan region and beyond with time, cost and reliability opportunities. The spatial design of the Grand Conakry Vision 2040 is, indeed, aligned with the needs for transport with most of the urban centres of the metropolitan region connected to the main road of the city (PK36). In addition to that, transport strategies for the Grand Conakry Vision 2040 include: (a) Establishment of traffic restriction; (b) Structuring a road network at the scale of the Greater Conakry metropolitan area; (c) Reinforcing the railway; (d) Organize maritime connections; (e) Develop a BRT system on its own site and improve the existing urban network; (f) Implement strategies for stationing; and (g) Gather typologies of structures and modes of maintenance of the road network.

References and Notes

1. See Rodrigue et al (2013) https://people.hofstra.edu/geotrans/eng/ch1en/ch1menu.html
2. Julie Kébé-Gangneux (2017) Op. Cit.
3. Mboup G (2015) Sustainable city foundation—key for sustainable, inclusive and prosperous city. In: World statistics congress, Rio de Janeiro (Brazil)
4. Mboup (2015) Sustainable city foundation—key for sustainable, inclusive and prosperous city. In: World statistics congress, Rio de Janeiro (Brazil), UN-Habitat, 2013, Streets as public spaces and drivers of urban prosperity, Nairobi, Kenya
5. Republique of Guinea, European Union and Louis Berger (2016) Grand Conakry Vision 2040
6. Under the Ministry of the City and Territorial Planning in collaboration with: the Ministry of Economy and Finance, the Ministry of Territorial and Decentralization Administration, the Ministry of Planning and International Cooperation, The Ministry of Transport, the Ministry of Public Works and the Ministry of Environment, Water and Forests. The Study was funder by the European Commission
7. United Nations GA (October 2016) Draft outcome document of the United Nations Conference on Housing and Sustainable Urban Development (Habitat III)
8. Conroy & Berke, 2004; Conroy & Gordon, 2004; Wild & Marshall, 1999
9. Chen Y, Hamilton A, Borning A (2010) Modeling & matching and value sensitive design: two methodologies for E-planning systems development
10. Marcinkowski (2007)
11. Silva CN (2010) The E-planning paradigm—theory, methods and tools: an overview. In: Silva CN (ed) 2010 Handbook of research on E-planning: ICTs for urban development and monitoring (http://www.igi-global.com/book/handbook-research-planning/41793) (http://www.igi-global.com/viewtitlesample.aspx?id=43177&ptid=41793&t=the+e-planning+paradigm+%E2%80%93+theory%2c+methods+and+tools%3a+an+overview), downloaded, 9 April 2017, The use of social media for urban planning: virtual urban landscapes created using twitter data
12. Yigitcanlar T (2013) Planning online: a community-based interactive decision-making model (http://www.igi-global.com/viewtitlesample.aspx?id=43178&ptid=41793&t=planning+online%3a+a+community-based+interactive+decision-making+model) downloaded 9 April 2017
13. Republique de Guinee, Ministere de l'Urbanisme de l'Habitat et de la construction (2011) Politique nationale de l'habitat « Vision Habitat 2021 »
14. Republique de Guinee (2007) Audits urbain (Urban Audits), organisationnel et financier de la ville et des communes de Conakry
15. Republique de Guinee, Direction Nationale de l'Aménagement du Territoire et de l'Urbanisme (2011) Resultat des analyses des outils de planification spatiale et de gestion urbaine en vigueur en Guinee
16. Mboup G (2014) Urban quality of life—concepts and measurements. http://deliver2030.org/?wp_blog=urban-quality-of-life-concepts-and-measurements (Accessed 9 Nov 2017)
17. Kiregyera (2007); Mboup G (2003)
18. UNDESA (2014) Urbanization prospects The 2014 revision
19. UN-Habitat (2009) Urban indicators guidelines
20. Angel et al (2016) Atlas of urban expansion—the 2016 edition. Volume 1: Areas and densities. New York: New York University, Nairobi: UN-Habitat, and Cambridge, MA: Lincoln Institute of Land Policy. Further analysis of these data has been done by Mboup G. (2017) "African Cities—Urban Form & Structure" in the forthcoming book "Smart Economy in Smart African Cities" by Springer, Winter 2018
21. Julie Kébé-Gangneux (2016) Whose right to the city? urban environment, Vol 10/2016
22. Mboup G African cities in time and space—a historical perspective and emerging trends, in forthcoming publication Smart economy in smart African cities, 2018, Springer

23. Republique de Guinee (2015) (Decret D/2015/229/PRG/SGG), Portant Publication des Resultats Definitits du Troisieme Recensement General de la Population et de l'Habitation realize du 1er mars au 2 avril 2014
24. OECD (2012) Short and Pinet-Peralta 2009
25. Kim and Law (2012)
26. This definition is brought from UNDESA "Urbanization Prospects 2014"
27. See also Kevin A. Bryan, Brian D. Minton, and Pierre-Daniel G. Sarte
28. Kevin A. Bryan, Brian D. Minton, and Pierre-Daniel G. Sarte, The evolution of city population density in the United States. Economic Quarterly—Volume 93, Number 4—Fall 2007—Pages 341–360; (Chatterjee and Carlino 2001, Lucas and Rossi-Hansberg 2002)
29. Betraud et al.
30. http://guinea.opendataforafrica.org/apps/atlas/Conakry
31. Ashfor S. Lori (2007) Africa's youthful population: risk or opportunity? Population Reference Bureau, Washington DC., USA
32. http://www.africaneconomicoutlook.org/en/theme/youth_employment/, downloaded on 21 Oct 2015. Promoting Youth Employment in Africa
33. See Mboup G (2013) Streets as public spaces and drivers of urban prosperity, UN-Habitat's publication, November 2013, Nairobi; Mboup, G. 2015. Streets as Public Spaces and Drivers of Sustainable, Inclusive and Prosperous Cities in Africa, World Bank's Land and Poverty Conference, March 2015, Washington DC; Mboup, G. 2015, City Foundation—Key for Sustainable, Inclusive and Prosperous Cities, World Statistics Congress, July 2015 Rio de Janeiro, Brazil
34. Mboup G (2016) Smart infrastructure development will make smart cities. Promoting Smart Transport and ICT in Dakar. In: Smart economy in smart cities, Springer, Sept 2016
35. UN-Habitat (2013a) Planning and design for sustainable urban mobility global human settlements 2013. Earthscan from Routledge
36. Sources: Kumar, A., and Barrett, F. (2008 Stuck in Traffic, Urban Transport in Africa, Africa Infrastructure Country Diagnostic (AICD); Vivien Foster and Cecilia Briceño-Garmendia (Editors), 2010. Africa's Infrastructure—A time for Transformation; UN-Habitat (2013) Planning and Design for Sustainable Urban Mobility: Global Report on Human Settlement 2013; UN Millennium Cities database
37. Gehl's early work in Copenhagen suggested the need to promote non-motorized means of mobility in order to create livable streets. Based on Gehl's findings, in 1962, Copenhagen made a shift towards increased use of bicycles as an alternative to cars. Durin
38. United Nations (2015a) Transforming our world. The 2030 agenda for sustainable development, 70th United Nations general assembly, New York Sept 2015
39. United Nations (2015b) Adoption of the Paris agreement. In: Conference of the Parties, twenty first session, Paris, 13 Nov to 11 Dec 2015
40. UN-Habitat (2013a) Op. Cit.
41. Sources of information used here are from different sources such as the Street report of UN-Habitat in 2013, the International Association of Public Transport, the Global Human Settlements Report of UN-Habitat (2013), Kumar A. and F. Barret, 2008 (Stuck in Traffic; Urban Transport in Africa) and Lall, S.V., et al. 2017, Africa's Cities: Open Door to the World, World Bank)
42. www.wds.worldbank.org/external/default/WDSContentServer/WDSP/IB/2010/10/07/000356161_20101007005655/Rendered/PDF/569540NWP0Tran10Box353751B01PUBLIC1.pdf/access on 15 January 2016
43. UN-Habitat (2013a) Planning and design for sustainable urban mobility global human settlements 2013. Earthscan from Routledge on 15 Jan 2016
44. World Bank (2005) Sub-Saharan Africa transport policy program. In: Working paper 80, Washington DC, USA
45. UN-Habitat (2013b) Streets as public spaces, drivers of urban prosperity. Nairobi, Kenya
46. UN-Habitat (2013) Guide for city planners. New York's Manhattan Commissioners' Plan of 1811 in the United States is one far-sighted extension plan

47. UN-Habitat (2012) Prosperity of cities. State of world's cities, 2011/2012. Earthscan
48. Republic of Guinea (2007) Audits urbain, organisationnel et financier de la ville et des communes de Conakry, 3eme Projet de Developpement Urbain (PDU3). Rapport proviso ire (Volet diagnostic)
49. Ministere de l'Urbanisme, de l'Habitat, et de la Construction Republic of Guinea (2011) Vision Habitat 2021
50. UNIDO (2016)
51. Rodrigue et al (2013) Op. Cit.
52. Statistics re-calculated by the author from Lall, S.V., et al (2017) Africa's cities: open door to the world, World Bank
53. Mboup, G., (2017). Land the Hidden Assets in African Cities, World Bank Land and Poverty Conference, Washington DC, March 2017. This is an adaptation from Mboup G., 2017.
54. Mboup G (2015) Op. Cit.
55. GLTN (2008)
56. De Soto (2000)
57. Toulmin (2005)
58. Lall SV, Henderson JV, Venables AJ (2017) Africa's cities: opening doors to the world. World Bank, Washington, DC. License: Creative Commons Attribution CC BY 3.0
59. World Bank (2015) cited by Lall et al 2017
60. Collier (2016) cited by Lall et al 2017
61. Lall et al (2017) Op. Cit.
62. Philibert (2007)
63. African Union and ECA, COM2013 (2013) Toward a demographic dividend: invest in health and education. In: Conference industrialization for an emerging Africa, Abidjan, March 2013
64. UN-Habitat (2012) Prosperity of cities (Author: Mboup et al.)
65. Mboup G (2013) Streets as public spaces and drivers of urban prosperity, Publisher: UN-Habitat, 2013
66. See Rodrigue et al (2013) Op. Cit.
67. According to Karen Mossberg (2008, pp. 1–2)
68. Vinod K, Dahiya B (2016) Op. Cit.
69. World Bank (2016) Digital dividends. World development report 2016; http://www.ist-africa.org/home/: Accessed on 15 Jan 2016; http://www.ist-africa.org/home/default.asp?page=doc-by-id&docid=5557
70. http://www.worldbank.org/en/topic/transport/brief/connections-note-26. By Winnie Wang, Raman Krishnan, and Adam Diehl."
71. Krambeck H (January 2015) Mapping Manila Transit A New Approach to Solving Old Challenges. "Haddad, Ryan; Kelly, Tim; Leinonen, Teemu; Saarinen, Vesa. 2014. Using Locational Data from Mobile Phones to Enhance the Science of Delivery. World Bank, Washington, DC. © World Bank. https://openknowledge.worldbank.org/handle/10986/19316 . License: CC BY 3.0 IGO."

Part V
India, Ahmedabad-Gandhinagar

Chapter 5
Smart Development of Ahmedabad-Gandhinagar Twin City Metropolitan Region, Gujarat, India

Jignesh G. Bhatt and Omkar K. Jani

Abstract With India transforming as matured democracy, the government is focusing upon improving quality of life of citizens by urban renewal and infrastructure development vide ambitious smart cities project. Energy, the electrical power in particular, has been the most crucial and the resource always in scarcity in India and proving itself as a major bottleneck. Therefore, India has been transforming legacy conventional non-smart non-intelligent unidirectional electrical power grids into modern smart grids which are bidirectional and intelligent in nature by leveraging ICT, IoTs, e-Governance and e-Democracy. Smart grids are likely to serve as energy backbones of smart cities and involve high interactive participation of citizens in energy management, based on humanitarian and customer centric approach. Different types of Prosumers (Producers + consumers), their different energy requirements at different timings, different types of energy resources and their switching feasibilities considering different aspects have been integrated. The Ahmedabad-Gandhinagar twin city metropolitan region (Naroda area) has been considered as region of interest and study. The region is surrounded by reputed industrial, commercial, educational-research organizations, heritage monuments and demonstrates extremely encouraging potential for creative research and technological developments with variety of Prosumers in particular. To study existing economic and spatial strategies and recommend suggestions for smart metropolitan region development have been main objectives of the work presented. Useful approach for smart metropolitan region development has been presented by effective energy management, active citizen participation and e-governance by proposing deployments of smart grid and smart buildings with integration of

J. G. Bhatt (✉)
Department of Instrumentation and Control Engineering,
Faculty of Technology, Dharmsinh Desai University (DDU),
Nadiad 387 001, Gujarat, India
e-mail: jigneshgbhatt@gmail.com

O. K. Jani
Solar Energy Division, Gujarat Energy Research and Management
Institute (GERMI), Pandit Deendayal Petroleum University (PDPU) Campus,
Gandhinagar 382 007, Gujarat, India
e-mail: omkar.j@germi.res.in

T. M. Vinod Kumar (ed.), *Smart Metropolitan Regional Development*,
Advances in 21st Century Human Settlements,
https://doi.org/10.1007/978-981-10-8588-8_5

renewables, ICT and IoTs. Ensuring 24 × 7 electricity with limiting carbon footprint has been the major challenge.

Keywords Electricity · Smart buildings · Smart city · Smart grid
Smart metropolitan region development · Information and communication technology (ICT) · Internet of things (IoT)

List of Abbreviations

AMI	Advanced Metering Infrastructure
AT&C	Aggregate Technical and Commercial
BAS	Building Automation System
BPO	Business Process Outsourcing
CCTV	Closed Circuit Tele Vision
CO_2	Carbon Dioxide
CT	Communication Technology
DDU	Dharmsinh Desai University
DISCOM/DisCom	DIStribution COMpany/Distribution Company
DR	Demand Response
DSM	Demand Side Management
GAV	Gandhinagar-Ahmedabad-Vadodara
GDP	Gross Domestic Product
GEDA	Gujarat Energy Development Authority
GERMI	Gujarat Energy Research and Management Institute
GHG	Green House Gas
GIDC	Gujarat Industrial Development Corporation
GIFT	Gujarat International Finance-Tec city
GoG	Government of Gujarat
GoI	Government of India
GUI	Graphical User Interface
HAN	Home Area Network/Home Automation Network
ICT	Information and Communication Technology
IMRB	Indian Market Research Bureau
IOT/IoT	Internet of Things/Internet of Things
ISGF	India Smart Grid Forum
IT	Information Technology
ITES	Information Technology Enabled Services
M2M	Machine to Machine
MRD	Metropolitan Region Development
NAN	Neighborhood Area Network
NASSCOM	National Association of Software and Services Companies
NCR	National Capital Region
NCRB	National Crime Record Bureau
NITC/NIT-C	National Institute of Technology Calicut
NRI	Non Resident Indian

O&M	Operations and Maintenance
PDPU	Pandit Deendayal Petroleum University
PHEV	Plug-in Hybrid Electric Vehicle
PLCC	Power Line Carrier Communication
Prosumer	Producer + Consumer
PV	Photo Voltaic
R&D	Research and Development
RAPDRP/R-APDRP	Restructured Accelerated Power Development and Reforms Programme
RBI	Reserve Bank of India
RCP	Representative Concentration Pathway
RF	Radio Frequency
RTP	Real Time Pricing
SCADA	Supervisory Control And Data Acquisition
SEZ	Special Economic Zone
SG	Smart Grid
SGCT	Smart Grid Communication Technology
SPV	Special Purpose Vehicle
STPI	Software Technology Parks of India
SWOC	Strengths, Weaknesses, Opportunities, Challenges
T&D	Transmission and Distribution
TCS	Tata Consultancy Services
TOU	Time Of Use
UGVCL	Uttar Gujarat Vij Company Limited
UNESCO	United Nations Educational, Scientific and Cultural Organization
V2G	Vehicle to Grid
WAN	Wide Area Network

5.1 Introduction [1–8]

Transforming lifestyles and rising population eventually causing rising expectations for better quality life and steadily driving people towards urban areas, resulting in the growth of megacities across the globe. Against problems such as rapid rise in population, fast urbanization, sustainable livelihood, employment creations, and rising migration from rural to urban areas, smart cities are emerging as novel solutions [1, 2].

Cities are engines of growth for the economy of every nation, including India. Nearly 31% of India's current population lives in urban areas and contributes 63% of India's GDP (Census 2011). With increasing urbanization, urban areas are expected to house 40% of India's population and contribute 75% of India's GDP by 2030. This requires comprehensive development of physical, institutional, social and economic infrastructure. All are important in improving the quality of life and attracting people and investments to the city, setting in motion a virtuous cycle of growth and development. Development of Smart Cities is a step in that direction [1, 2].

Around 90% of the world's urban population growth will take place in developing countries, with India taking a significant share of that. Between 2015 and 2030, India's GDP is expected to multiply five times, with over 70% of new employment generated in cities. Close to 800 million m^2 of commercial and residential space needs to be built to serve this population. Urban areas of India contributed and are expected to continue contributing a higher share of the GDP. While the urban population is currently around 31% of the total population, it contributes over 60% of India's GDP. It is projected that urban India will contribute nearly 75% of the national GDP in the next 12–15 years. Therefore, cities are referred to as the '**engines of economic growth**', and ensuring that they function as efficient engines is critical to our economic development. This trend of urbanization that is seen in India over the last few decades will continue for some more time, in fact it is likely to pick up further. In this context, the government of India has decided to develop 100 smart cities in the country [1, 2].

In such context, reliable electricity supply is essential and vital for development of any metropolitan region or a smart city. Therefore, as suggested by [2], smart grids—as energy backbones, are observed being essential and mandatory at the core of smart city developments globally.

Referring to our earlier works, in [3], detailed technical review of smart grid along with identification of critical applications and parameters, while in [4], e-governance of rooftop based solar photo voltaic rooftop system has been covered with special focus on Gandhinagar solar city project. Next, in [2], smart grid pilots along with interesting applications have been discussed including various initiatives of UGVCL and GERMI. This work presents interesting details of smart metropolitan region development of Ahmedabad-Gandhinagar twin city metropolitan region by smart grid installation with Naroda area at the focus.

5.1.1 The 'Smart City' Concept [1–7]

People migrate to cities primarily for employment. To support their happy and comfortable living, people also need good quality housing; cost-effective physical and social infrastructure, such as water, sanitation, electricity, clean air, education, healthcare, security, entertainment, etc.

The first and most basic need people have is the need for survival: their physiological requirements for food, water, and shelter. If any of these physiological necessities is missing, people are not motivated enough to meet the growth needs [1]. Seven categories of basic needs common to all people have been identified and represented as a hierarchy in the shape of a pyramid in [1] as shown in Fig. 5.1. This hierarchy is an arrangement that ranks people or concepts from lowest to highest. According to [1], individuals must meet the needs at the lower levels of the pyramid before they can successfully be motivated to tackle the next levels. The lowest four levels represent deficiency needs, and the upper three levels represent growth needs [1].

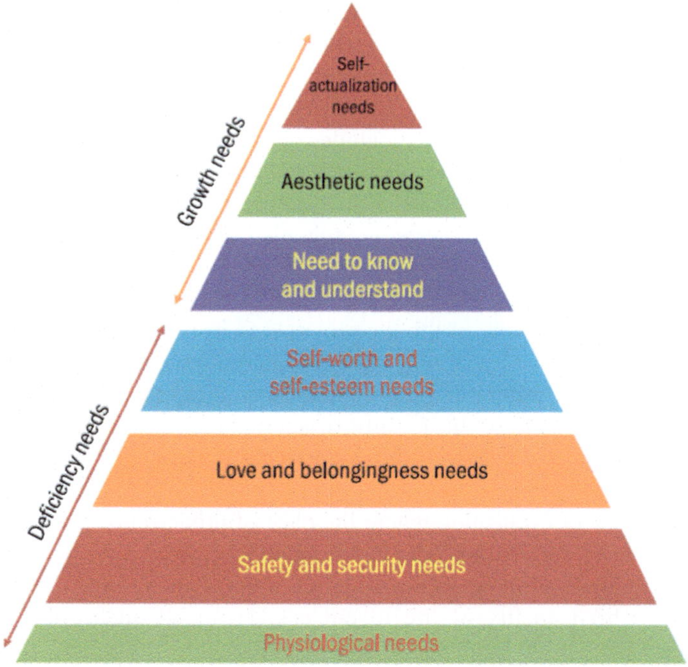

Fig. 5.1 Maslow's hierarchy of needs [1]

One will need to fulfill growth needs once their deficiency needs are fulfilled. In India, many cities have the infrastructure where deficiency needs are fulfilled and Smart City concept will fit in, while in other cities citizens are even struggling for their deficiency needs to be fulfilled. Due to dense population and lack of streamlined civic facilities and processes, such deficiencies remain unattended leading to complex problems in cities. To overcome this difference, the Government needs proper strategy that helps in successful implementation of smart city concept. India is at a point of transition where the pace of urbanization will speed up. The relatively low base allows us to plan our urbanization strategy in the right direction by taking advantage of the latest developments in technology especially in ICT [1].

There is no universally accepted definition of smart city [5]. The conceptualisation of smart city, therefore, varies from city to city and country to country, depending on the level of development, willingness to change and reform, resources and aspirations of the city residents.

According to the Smart Cities India Foundation [6], a 'Smart City' is defined as a developed urban area that creates sustainable economic development and high quality of life by excelling or becoming "smart" in multiple key areas; economy, mobility, environment, people, living, and government. Excelling in these key areas can be done through strong human capital, social capital, and/or Information Communication Technology (ICT) infrastructure.

As described by Wikipedia [7], a smart city is an urban area that uses different types of electronic data collection sensors to supply information used to manage assets and resources efficiently. This includes data collected from citizens, devices, and assets that is processed and analyzed to monitor and manage traffic and transportation systems, power plants, water supply networks, waste management, law enforcement, information systems, schools, libraries, hospitals, and other community services.

It is a well-planned and well-developed urban area where sustainable economic development and high quality of life are provided to its citizens by enhancement in different key areas; such as economy, mobility, environment, people, living, government, etc. through strong human capital, social capital, and/or Information Communication Technology (ICT) infrastructure. The new internet and smart technologies have opened new ways for collective action and collaborative problem solving.

India is at a point of transition where the pace of urbanization will speed up. In this context, the Government of India has launched the Smart City Mission in 2014 and has decided on developing 100 smart cities in the country, which are planned to provide a very high quality of life to citizens, i.e. good quality but affordable housing, cost efficient physical, social and institutional infrastructure (water, sanitation, 24/7 electricity), clean air, quality education, cost effective health care, security, entertainment, high speed connectivity and efficient mobility; it must also attract investments, experts and professionals. The 100 smart cities are to be developed in different population ranges of up to one million, one to four million and over four million.

The smart city concept integrates ICT and various physical devices connected to the network (popularly known as IoT) to optimize the efficiency of city operations and services and connect to citizens. Smart city technology allows city officials to interact directly with both community and city infrastructure and to monitor what is happening in the city and how the city is evolving. ICT is used to enhance quality, performance and interactivity of urban services, to reduce costs and resource consumption and to increase contact between citizens and government. Smart city applications are developed to manage urban flows and allow for real-time responses. A smart city may therefore be more prepared to respond to challenges than one with a simple “transactional” relationship with its citizens.

Recent interest in smart cities is motivated by major issues and challenges, including climate change, economic restructuring, the move to online retail and entertainment, ageing populations, and pressures on public finances.

As mentioned by [1], growth and deficiency needs can be classified under smart city characteristics as shown in Table 5.1.

Obviously as mentioned in Table 5.1, the energy grid (smart grid) forms not only the most primary and basic smart city component under the classification of ‘Growth Needs’, but it also serves as mandatory and fundamental requirement as to

Table 5.1 Classification of smart city components according to Maslow's hierarchy [1]

	Domain	Objective
Growth needs	Energy grids	Automated grids that employ ICT to deliver energy and enable information exchange about consumption between providers and users, with the aim of reducing costs and increasing reliability and transparency of energy supply systems
	Public lighting, natural resources and water management	Managing public lighting and natural resources. Exploiting renewable resources, such as heat, solar, cooling, water, and wind power
	Waste management	Applying innovations in order to effectively manage the waste generated by people, businesses, and city services. It includes waste collection, disposal, recycling, and recovery
	Environment	Using technology to protect and better manage environmental resources and related infrastructure, with the ultimate goal of increasing sustainability. It includes pollution control
	Transport, mobility, and logistics	Optimizing logistics and transportation in urban areas by taking into account traffic conditions and energy consumption. Providing users with dynamic and multi-model information for traffic and transport efficiency. Assuring sustainable public transportation by means of environmental friendly fuels and innovative propulsion systems
	Office and residential buildings	Adopting sustainable building technologies to create living and working environments with reduced resources. Adapting or retrofitting existing structures to gain energy and water efficiency
	Healthcare	Using ICT and remote assistance to prevent and diagnose diseases and deliver the healthcare service. Providing all citizens with access to an efficient healthcare system characterized by adequate facilities and services
	Public security	Helping public organizations to protect citizens' integrity and their goods. It includes the use of ICTs to feed real-time information to fire and police departments
Deficiency needs	Education and culture	Capitalizing system education policy, creating more opportunities for students and teachers using ICT tools. Promoting cultural events and motivating people participation. Managing entertainment, tourism, and hospitality
	Social inclusion and welfare	Making tools available to reduce barriers in social learning and participation, improving the quality of life, especially for the elder and disabled. Implementing social policies to attract and retain talented people

(continued)

Table 5.1 (continued)

	Domain	Objective
	Public administration and (e-) government	Promoting digitized public administration, e-ballots, and ICT-based transparency of government activities in order to enhance citizens' empowerment and involvement in public management
	Economy	Facilitating innovation, entrepreneurship, and integrating the city in national and global markets

provide sustainability of other components in both 'Deficiency Needs' as well as 'Growth Needs' [1].

5.1.2 The Smart Cities Framework

An integrated smart city component interconnection presented in Fig. 5.2 comprising the key enablers like **Smart Governance, Smart Living, Smart People, Smart Mobility, Smart Environment, and Smart Economy** to serve conceptual smart city objectives and to facilitate implementation [1].

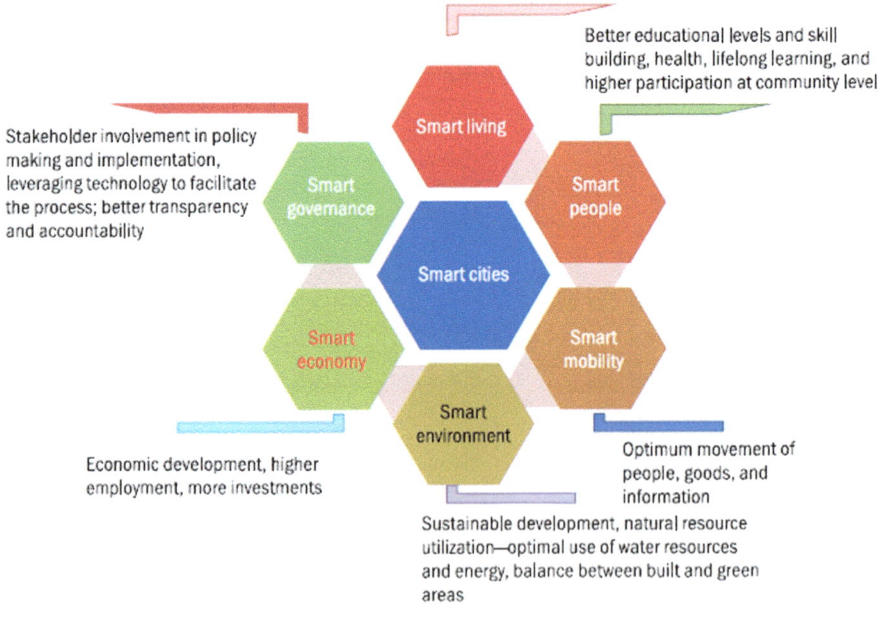

Fig. 5.2 Smart city components [1]

A key enabler is **Smart Environment**, which includes optimal use of water resources and energy, to maintain balance between built and green areas. Smart grid could be proved a game changing mechanism here [1].

5.1.3 Conversion of Existing City to Smart City [8–10]

The genesis of smart cities in India is linked to the post economic liberalization and reform since 1991. Indian economy is going through a process of significant structural change between 1991 and 2011, with the declining share of agriculture in the GDP, stagnating manufacturing sector and leap-frog growth of service sector with its contribution to GDP rising from 44.1 to 56.4%. The improved economic performance is powered by the growth of urban based knowledge intensive services (e.g. IT-ITES-BPO, education and healthcare), automobiles, pharmaceuticals, biotechnology, garments, hotels and recreation sector activities. Initially seven big urban agglomerations in the cities of Bengaluru, Hyderabad, Chennai, Mumbai, Pune, Kolkata and Delhi NCR started due to growth of ICT sector, and subsequently, more grandiose vision of smart cities has become integral to the idea of industrial corridor being launched for economic turnaround of the nation through development of green-field industrial cities, where high-end infrastructure, sensors, smart grids, big data and analytics considered as the elementary tools for urban governance [8].

As depicted an illustration in Fig. 5.3 [1], for successful conversion of an existing city into smart city, fulfillment of certain features have been expected and desired. It could be observed that while 'Energy' forms the core feature along with 'Sustainable transport', 'Water management' and 'Spatial programming', ICT remains as vital connecting link to all the four core features.

The smart city infrastructure is mentioned as the introductory step for establishing the overall smart city framework and architecture. A smart city is a city that uses digital technologies to improve the quality of life and standard of living of its citizens. Smart cities anticipate and mitigate current and future challenges by using the power of the communication network, distributed wireless technology and intelligent business management system [9].

A rapid urbanization increases day by day because of which there has been a steady increase in migration from rural to urban areas. It is expected that around 70% of the global population will be living in cities by the year 2050. This needs about 500 new cities to accommodate the inundation.

The announcement of '100 Smart Cities' by government of India falls in line with the vision of providing ample living space in the urban regions. This also allows for investment opportunities in the infrastructure sector in India. In a smart city, the information technology plays a pivotal role in providing the essential services to its residents. The information technology is the major infrastructure of these cities. Automated sensor networks and data centers are the examples of technological platforms involved in such smart cities. These smart cities have

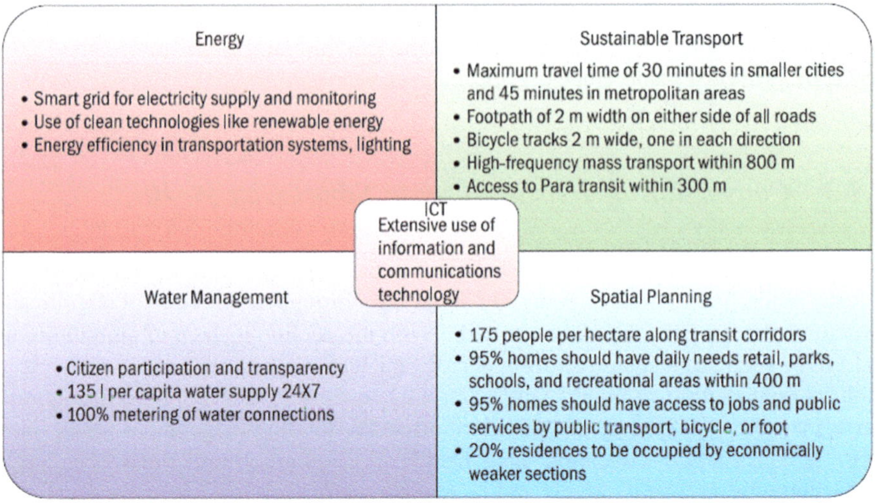

Fig. 5.3 Features required for smart cities [1]

sustainable economic development, which benefits everyone, including citizens, businesses, the government and the environment.

Challenges remain, especially in relation to access to financing of large scale reforms like smart grid in the electricity sector [10].

Also, due to issues such as electricity theft, etc. energy management is emerging as one of the major challenges in the growth of smart cities in India. Government of India has started various initiatives to provide smart energy solutions via smart grid and finalized following initiatives:

(i) Establish smart grid test bed and smart grid knowledge centre
(ii) Implementation of eight smart grid pilot projects in India with investment of USD 10 million
(iii) Addition of 88,000 MW of power generation capacity in the 12th Five Year Plan (2012–17)
(iv) The Power Grid Corporation of India Limited has planned to invest USD 26 billion in the next five years
(v) Installation of 130 million smart meters by 2021

5.2 Smart Regional Development [9]

There are under mentioned six dimensions of development of a smart city or smart metropolitan region [9]:

5.2.1 Smart Governance

City is a private and civil organization that works at its best as one organism fuelled by infrastructure, hardware, software and data mining. Smart governance is about transparency and can be achieved from government enabled applications in terms of citizen decision making and e-public services. It includes involvement of the public in decision-making, public and social service and government transparency.

5.2.2 Smart People

It includes a culture of life-long learning, social and ethnic diversity, flexibility, creativity and community participation. People fed since childhood by e-skill that promotes creativity, critical thinking, and independence fostering innovation by all means.

5.2.3 Smart Mobility

Smart mobility means integrated transport and logistics system supported by wise infrastructure, hardware, software, and data mining. It includes local and national accessibility, safe and sustainable transportation systems, and access to ICT infrastructure.

5.2.4 Smart Economy

E-business process and E-commerce are to boost sustainable growth and productivity. It also includes entrepreneurship and productivity, economic progression, flexibility in the labor market, and an overall culture of innovation.

5.2.5 Smart Living

Smart living means healthy and safe living through smart technologies. It also includes cultural and educational facilities, quality health conditions and public safety, accessibility to quality housing, tourists' attractions and social integration.

5.2.6 Smart Environment

It means renewable and clean energy managed by ICT. The main objective is a clean environment with pollution and trash disposal under control. It also includes

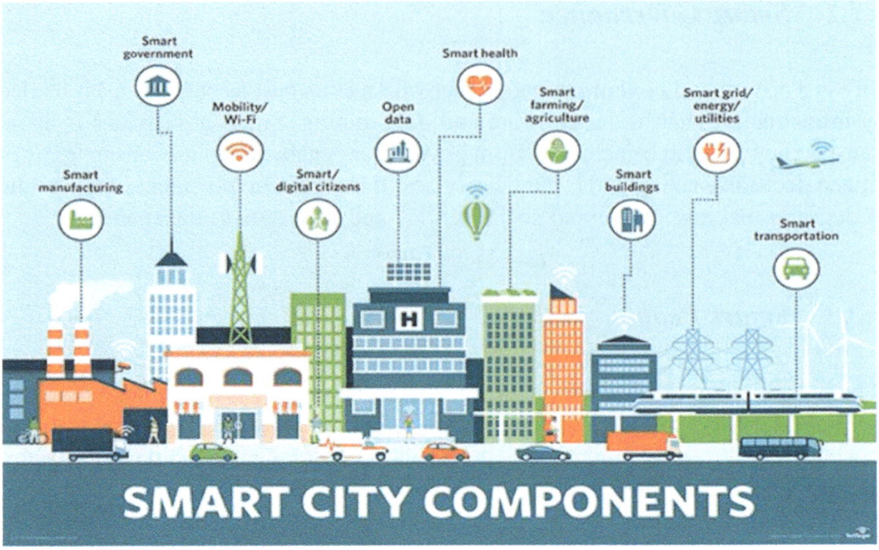

Fig. 5.4 Smart city components [9]

attracting natural conditions, reduction in pollution, and increase in environmental protection and sustainable resource management.

ICT is serving as vital interconnecting link between all the smart city components mentioned above and shown in Fig. 5.4 [9].

5.3 Drivers of Growth [10–13]

According to [10], India's smart grid top market rankings have been improving by a fast-growing economy and electricity sector. Growing electricity consumption and ambitious government policies for energy access, renewable resources deployment and "development of smart cities" are among major strengths for the smart grid developments.

Comparing the smart city initiatives by China and India, [11] pointed out various differences in methodologies adopted by both countries, and indicated that to meet huge demands of electricity, China has already started smart grid development in 2012, with targets of installing smart meters by 2017 to enable time based electricity pricing which complements smart cities mission, on the other hand, India is still working with pilot projects in experimental stages.

The challenging physical, economic and technological environment across the globe necessitates smart cities which help to enhance livability, workability and sustainability. Such powerful drivers which are responsible for growth of smart cities are identified by authors in [12] as presented below:

5.3.1 Growing Urbanization

Cities deliver many benefits—greater employment opportunities, easier access to healthcare, education, entertainment, culture and the arts. As a result, people are moving to cities at an unprecedented rate. In India, over 200 million people will be added to urban populations over the next 15 years.

5.3.2 Rapidly Improving Technology

Improvement in technology has made things much easier than before, and will go a long way in making cities smart. In many developed countries, the installation of millions of smart meters and smart sensors will produce data of value to a smart city. The adoption of smart thermostats and building management systems in many cities is resulting in smart buildings. Other smart technologies such as intelligent transportation management software, roadway sensors, smart parking apps, navigation apps and equipment displaying real-time traffic are effectively planning and managing traffic. The use of electric vehicles helps reduce pollution levels. High-bandwidth networks worldwide connect one billion computers and four billion cell phones approximately.

5.3.3 Growing Environmental Challenges

Cities house half of the world's population but use two-thirds of the world's energy and generate three-fourths of the world's CO_2 emissions. Every city has to learn to proactively mitigate the effects of such environmental concerns and other climate changes. Smart technologies can help to do this, and to spread awareness.

5.3.4 Rising Expectations

Accustomed to instant, anywhere, anytime, personalized access to information and services via mobile devices and computers, citizens are now expecting the same kind of access to city services. In developed countries, ICT plays an important role in all type of city services, from governance to smart management of buildings. This helps increase transparency between the government and citizens. Easy access to all city services causes citizens to interact more and participate in the long term development and maintenance of their city.

5.3.5 Increasing Economic Competition

Cities across the world are competing to secure investments, jobs, businesses and talent. This may lead to a transition in the workforce and migration of people from one city to another. Cities will increasingly need to focus more on physical infrastructure and social infrastructure to meet the needs of a growing population.

5.3.6 Inadequate Infrastructure

Urbanization is putting a significant strain on city infrastructure that was, in most cases, built for populations a fraction of their current size. Currently, most cities, towns, districts, and states in India lack adequate base infrastructure such as roads, power, water supply, sewerage, and sanitation. The need for base infrastructure to be addressed to prepare the foundation for 'Make in India' a major new national program, designed to facilitate investment, foster innovation, enhanced skill development, protect intellectual property. and build best-in-class manufacturing infrastructure.

5.3.7 Rising Stress

The challenges and competitions from increasing population, increased costs, increase in travel distances, unemployment, unavailability of quality time with family, increase in crime rates have led to many physical and psychological problems. Hence, the city should have good transportation, public amenities and leisure facilities, public security, smart technologies etc. to mitigate the growing stress.

In the Indian context, following are identified as key drivers of growth of smart grids in [13]:

(i) Reduction in AT&C losses
(ii) Demand side management (TOU Tariff—Dynamic pricing based for peak demand and peak supply)
(iii) 24 × 7 Power for all
(iv) Outage reduction
(v) Renewable energy integration
(vi) Improved energy efficiency
(vii) Reliable grid stability
(viii) Faster restoration of electricity after fault or disturbances
(ix) Grid flexibility (backup power)

(x) Reduction in peak demand
(xi) Reduction in power purchase cost

5.4 Needs and Impacts of the Grid [14, 15]

5.4.1 Needs [14]

Following needs and requirements of the smart grid have been outlined [14]:

(i) *Primary energy shortage*: Energy should be transmitted from external place to city in a distant way in the form of coal, gas or electric power. Relying on long distance transmission channels is indeed expensive and non-reliable.
(ii) *Imperfect network structure*: The capacity to accept power from external generations needs to promote. Substations are over loaded and in low automation level. The distribution system has a high density of lines, a high fault rates, low efficiency and operates inflexibly.
(iii) *More requires*: Consumers in city, residents and industries, require more in service quality and varieties.
(iv) *Great damage risk*: Once blackout occurs, it will cause a great damage to people's daily life and economics, even to the national security.
(v) *Limited land*: With the processes of urbanization and modernization, the capacity is still expanding, but the land needed to build new substations and lines is getting limited.

5.4.2 Impacts [15]

In [15], following impacts of the smart grids have been presented:

5.4.2.1 Environmental Impact

By enabling distributed generation, smart grid can work with fewer generating plants, fewer transmission and distribution assets in order to cater the growing demand of electricity. With the possible expectation of wind farm sprawl, landscape preservation is one of the evident benefits. Since maximum generation today results in emission of greenhouse gas, a smart grid reduce air pollution and plays a significant role in combating global climate change issue.

A smart grid has the capability to accommodate technical difficulties of integrating renewable resources like wind and solar to the grid, providing further reduction in greenhouse gas emissions.

5.4.2.2 Cost

The ability to bypass the cost of the plant and grid development is a major advantage to both the utilities and customers. Further with the aid of smart grids less generating units would be required in order to fulfill the energy demand of the growing population and cost of setting up more and more plants can be deferred. At that point of time, more emphasis will be on overall development of T&D efficiency based on demand response, load control, and many other technologies.

Energy efficiency would be the second priority in order to save cost with reference to the customers. With timely and detailed information provided by Smart Grids, customers would be encouraged to limit waste, adopt energy-efficient building standards, and invest more and more in energy efficient appliances.

5.4.2.3 Utility Operations

Smart Grids can assist the utilities, as the principal focus of the utilities is to improve business processes. Many utilities have an extensive list of projects that they would like to fund in order to improve the customer service or to ease workforce's burden of repetitive work. Calculating smart grid benefits by the cost/benefit analysis it puts emphasis in favor of the change and can also significantly decrease settlement/payback periods.

Mobile workforce group and asset management group work collectively to organize assets and then maintain, renovate, and replace them. This can result in increased productivity and fuel saving from superior methods.

Similarly, smart grid provides customers with real time information and encourages them to do online payments, thus lowering billing costs.

5.4.2.4 Theft Control

This is not an issue in developed countries like US, but in developing countries like India as well, where people have a little insight of the grid and higher poverty rate, power theft is quite common. With development of smart grid, power theft can be controlled to a greater extent, thereby improving the efficiency of distribution system. Thus, grids will provide higher quality and reliable power supply, and there will be fewer blackouts.

5.5 Benefits of Smart Grids [6, 16–20]

According to [16], various anticipated benefits of smart grid are as follows:

(i) Better power quality as well as reliability by better matching of demand-supply and improvement in grid congestion

(ii) Reduction in spinning reserves and improved ancillary service
(iii) Increased efficiency, security and capacity utilization of grid
(iv) Power disruption management and analysis
(v) Integration of renewable energy resources, ultimately leading to reduction of greenhouse gases in the anticipated climate change
(vi) Opportunity for new markets
(vii) Energy management savings in the monthly electricity bills, improved efficiency of appliances
(viii) Self-healing nature of the grid, which will improve the maintenance of the grid and will also allow the islanded operation whenever required
(ix) Demand response will enable various choices of utility generated power and green power with differential pricing for consumers
(x) Transparency in operation of the utility grid and optimization of its asset utilization
(xi) Reduction in expenditure spent on the maintenance and operation of the utility grid
(xii) Innovative decision assistance and business intelligence will be available at both automatic and manual monitoring level through optimization tools
(xiii) PHEV and V2G integration would help in peak shaving and valley filling
(xiv) Improved electric load forecasting through data mining of the information collected from two-way M2M communication of utility grid
(xv) Encourages energy independence
(xvi) Enables distributed generation
(xvii) Allows customers to manage their consumption level and to take benefit of pricing and supply options
(xviii) Better catering of environmental issues

As presented in [17], Real Time Pricing (RTP) based tariff is more advantageous to the flat rate tariff as it encourages customers to use their appliances during off-peak period. By applying this type of tariff, peak load on the power plant reduces and amount of monthly bill of customers also reduces.

5.5.1 Economic Benefits [6, 18, 19]

References [6] and [18] mentioned that no standardized method currently exists for assessing the economic and environmental impacts of SG systems. Therefore, the context, boundaries, and ICT technologies included should be made very clear so that comparison and extrapolations can be made. Significant variation exists among studies in their estimates of SG systems, so the precise costs, benefits, and GHG emission reductions are uncertain. Standardizing some methodologies and key assumptions (time horizon, discount rates for costs), as well as scrutinizing some key input data (e.g. data related to electricity losses), can result in more consistent estimates of costs, benefits, GHG emission reductions, and energy savings

estimates. Despite these variations, the analysis shows that SG systems may not results in cost-savings but contribute to energy and GHG savings due to the large deployment of renewable energies.
Reference [19] identified following economic benefits of smart grids:

(i) Improved asset utilization
(ii) T&D capital savings
(iii) T&D O&M savings
(iv) Theft reduction
(v) Energy efficiency
(vi) Electricity cost savings
(vii) Power quality

5.5.2 Employment Generation Capabilities [20]

In [20], nine different types of renewable power generation technologies have been compared for their employment generation capabilities, particularly comparing installation and O&M works, summarized in Table 5.2 [20].

From the data presented in Table 5.2, it is evident that after biogas, solar PV (residential + large scale PV) has been second largest employment generator [20].

From charts in Fig. 5.5 [20], it is interesting to note that as compared to biomass-the highest employment generator, the second highest employment generator-solar PV provides more employment opportunities in installation and relatively lesser employment opportunities in O&M.

5.5.3 Social Benefits [19]

Reference [19] identified following social benefits of smart grids:

(i) Offering flexibility to reschedule operation of consumer appliances via DSM
(ii) Detailed feedback of energy consumption patterns to consumers via smart meters
(iii) Enhancement in quality of lifestyle
(iv) Simplicity, transparency and convenience
(v) Improved user awareness and sharing of resources

5.5.4 Environmental Benefits [19, 20]

Reference [19] suggested that due to under-mentioned environmental benefits of smart grids, the living environment and lifestyle quality enhances:

Table 5.2 Breakdown of lifecycle employment factors (persons-years per GWh) for the nine different renewable power generation technologies [20]

	Construction stage			Operation & maintenance stage			Total
	Direct	Indirect	Subtotal	Direct	Indirect	Subtotal	
Residential PV	0.67	1.43	2.10	0.33	0.30	0.63	2.73
	(24.6%)	(52.4%)	(77.0%)	(12.0%)	(11.0%)	(23.0%)	(100.0%)
Large-scale PV	0.59	1.12	1.71	0.89	0.23	1.13	2.84
	(20.8%)	(39.5%)	(60.3%)	(31.5%)	(8.2%)	(39.7%)	(100.0%)
Wind	0.24	0.70	0.94	0.50	0.45	0.95	1.89
	(12.7%)	(36.9%)	(49.7%)	(26.5%)	(23.8%)	(50.3%)	(100.0%)
Large-scale geothermal	0.27	0.40	0.67	0.15	0.20	0.35	1.01
	(28.1%)	(36.0%)	(64.1%)	(23.2%)	(12.7%)	(35.9%)	(100.0%)
Small-scale hydro	0.56	0.71	1.27	0.46	0.12	0.58	1.85
	(30.0%)	(38.6%)	(68.6%)	(25.0%)	(6.4%)	(31.4%)	(100.0%)
Wood biomass	0.22	0.17	0.39	0.65	2.85	3.50	3.89
	(5.7%)	(4.4%)	(10.1%)	(16.6%)	(73.3%)	(89.9%)	(100.0%)
Sewage sludge biogas	0.41	0.88	1.30	0.91	0.80	1.72	3.01
	(13.7%)	(29.3%)	(43.0%)	(30.3%)	(26.7%)	(57.0%)	(100.0%)
Animal waste biogas	1.01	0.95	1.96	1.14	1.78	2.92	4.88
	(20.7%)	(19.4%)	(40.1%)	(23.4%)	(36.4%)	(59.9%)	(100.0%)
Food waste biogas	0.63	0.98	1.61	0.95	2.48	3.43	5.04
	(12.5%)	(19.4%)	(31.9%)	(18.9%)	(49.2%)	(68.1%)	(100.0%)

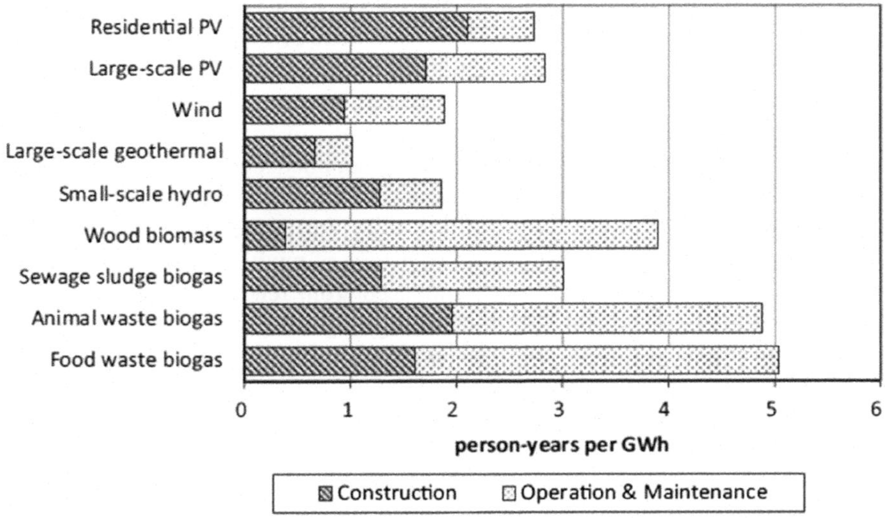

Fig. 5.5 Life cycle employment factors (person-years per GWh) for the nine different renewable power generation technologies [20]

(i) Cleaner air, reduced CO_2 emissions
(ii) Reduction in Green House Gas (GHG) emissions by reducing fossil fuels and encouraging renewables
(iii) Reduction in COx, NOx, and PM-10 emissions

5.6 Challenges and Solutions [9, 15, 21]

5.6.1 Challenges

As described by [21], there are several challenges during the execution & post execution of the smart grid pilot project like transition of legacy equipment/system, lack of standard & interoperability, policy & regulatory framework, lack of awareness, cyber security & data privacy etc., the government should encourage financial and technical development of smart grid standards including pilot models, for the expected key benefits from these projects like lowering of peak demand, differed capital investment by avoiding network capacity cost, reduction in outage duration & outage restoration costs, increased efficiency of network & empowerment of consumers. There are various inherent risks involved in restructuring the utility grid, which are required to be addressed for the modernization of utility grid: Information integration across all units of the smart grid is a necessity.

Non-conventional sources of energy may have critical impact on the reliability of smart grid such as faulty forecasting of load, distribution and transmission congestion, little correlation with requisite load profiles and so on, making it more vulnerable to failures. Further, absence of policies, regulations and standards for smart grid, cyber-attacks on smart grid, such as theft of important information or manipulation of the whole or part of system through smart meters, AMI, supervisory control and data acquisition control, energy storage subsystem, V2G infrastructure, etc. may cause impairment of the vital units of smart grid.

Integration and information security (interoperability, encryption, data protection, cyber terrorism) pose a major challenge for the implementation of smart grid and extensive requirement of capital for ICT, smart meters, communication infrastructure, etc. Energy markets have a constraint of minimum production, which makes it impossible for the electricity produced from renewable sources to participate in the energy market. Lack of knowledge regarding smart meters, its usage and literacy related to power rating of appliances, peak load, etc. make this technology less lucrative. Large information will be generated on a real-time basis. If any wrong decision is made by the consumer due to some incorrect information or due to lack of needed information, it may lead to power outages.

One of the major challenges involved in the implementation of smart grid is investment. Generation of electricity through non-conventional sources of energy needs extensive capital and large lead times. A huge cost would be incurred for the automation of whole grid. The investment required would need contribution from government, utilities and also through credits and issuance of bonds.

Some other ill effects are inherent, such as the communication infrastructure for smart grid would lead to the formation of electro smog; exposing individuals to continuous RF radiation; emitted vampire energy, which would ultimately lead to various health hazards. These RF radiations may interfere with the ecosystem, thus disrupting the movement of birds, butterflies, marine animals and so on who use electromagnetic radiations as a guide for their migrations.
Mentioned below are some more challenges identified in [9]:

(i) Retrofitting existing legacy city infrastructure for making it smart

There are a number of latent issues to consider when reviewing a smart city strategy. The most important is to determine the existing city's weak area that needs utmost consideration i.e. 100% distribution of water supply and sanitation. The integration of formerly isolated legacy system to achieve citywide efficiencies can be significant challenge.

(ii) City development plan

Most of our cities do not have good master plan or development plan, which is important to smart city planning and implementation.

(iii) Governance

There should be successful implementation of smart city solution needs effective horizontal and vertical coordination between various institution providing various municipal amenities as well as effective coordination between central govt., state government and local government agencies.

(iv) Building program

To build 100 smart cities is not an easy task and most ambitious project is delayed owing to lack of quality manpower, both at the state and central level. In terms of funds only around 5% of the central allocation may be allocated for capacity building program that focus on training. Investment in capacity building program have a multiplier effect as they help in time bound completion of project and in designing program.

(v) Reliability of utility services

For any smart city in the world, the focus is on reliability of utility services, whether it is electricity, water, telephone or broadband services. Smart city should have universal access to electricity 24 × 7, this is not possible with the existing system. Cities need to shift towards renewable sources and focus on green building and green transport to reduce the need for electricity

As identified by [15], the challenges for smart grid development are:

(i) Policy and regulation

The current policy and regulatory frameworks were typically designed to deal with the existing networks and utilities. To some extent the existing model has encouraged competition in generation and supply of power but is unable to promote clean energy supplies. With the move towards smart grids, the prevailing policy and regulatory frameworks must evolve in order to encourage incentives for investment. The new frameworks will need to match the interests of the consumers with the utilities and suppliers to ensure that the societal goals are achieved at the lowest cost to the consumers. Generally, governments set policy whereas regulators monitor the implementation in order to protect the consumers and seeks to avoid market exploitation. Over the last two decades, the trend of liberalized market structure in various parts of the world has focused the attention of policy-makers on empowering competition and consumer choice. The regulatory models have evolved to become more and more effective to avoid market abuse and to regulate the rates of return. Moving forward, the regulatory model will have to adopt the policy which focuses much on long term carbon reduction and security of supply in the defined outcomes and they need to rebalance the regulatory incentives to encourage privately financed utilities to invest at rates of return that are commensurate to the risk. This may mean creating frameworks that allow risk to be shared between customers and shareholders, so that risks and rewards are balanced providing least aggregate cost to the customer.

(ii) Business scenario

The majority of examples results in negative business cases, undermined by two fundamental challenges:

a. High capital and operating costs
   Capital and operating costs include large fixed costs linked to the chronic communications network. Hardware costs do not cause in significant growths in economies of scale and software integration possess a significant delivery and integration risks.
b. Benefits are constrained by regulatory framework
   When calculating the benefits, organizations tend to be conservative in what they can gather as cash benefits to the shareholders. For example, in many cases, line losses are considered to be put on to the customer and as a result any drop in losses would have no net impact on the utility shareholder. The smart grid benefits case may begin on a positive note but, as misaligned policy and regulatory incentives are factored in, the investment becomes less attractive. Therefore regulators are required to place such policies and regulations in place which could provide benefits both to the utilities and the consumers. Therefore the first factor to be considered is to provide incentives to the utilities in order to remove inefficiencies from the system. They should be aptly remunerated for the line losses on their networks. On the budget side of the calculation, there is no avoiding the fact that smart technologies are expensive to implement, and at the

present level it is right to factor in the risk associated with delivery. But the policy makers and regulators can mitigate that risk by seeking economies of scale and implementing advanced digital technologies.

(iii) Technology maturity and delivery risk

Technology is one of the essential constituents of Smart Grid which include a broad range of hardware, software, and communication technologies. In some cases, the technology is well developed; however, in many areas the technologies are still at a very initial stage of development and are yet to be developed to a significant level. As the technologies advances, it will reduce the delivery risk; but till then risk factor have to be included in the business situation.

On the hardware side, speedy evolution of technology is seen from vendors all over the world. Many recently evolved companies have become more skeptical to the communications solutions and have focused on operating within a suite of hardware and software solutions. Moreover the policy makers, regulators, and utilities look upon well-established hardware providers for Smart Grid implementation. And this trend is expected to continue with increasing competition from Asian manufacturers and, as a consequence, standards will naturally form and equipment costs will drop as economies of scale arises and competition increases.

On the software and data management side, the major challenge is to overcome the integration of the entire hardware system and to manage high volume of data. With multiple software providers come multiple data formats and the need for complex data models. In addition, the proliferation of data puts stresses on the data management architecture that are much similar to the telecommunications industry than the utilities industry. Many of these issues are currently being addressed in pilots such as smart grid task force and, as a consequence, the delivery risk will reduce as standards will be set up.

(iv) Lack of awareness

Consumer's level of understanding about how power is delivered to their homes is often low. So before going forward and implementing Smart Grid concepts, they should be made aware about what Smart Grids are? How Smart Grids can contribute to low carbon economy? What benefits they can drive from Smart Grids? Therefore:

a. Consumers should be made aware about their energy consumption pattern at home, offices … etc.
b. Policy makers and regulators must be very clear about the future prospects of Smart Grids.
c. Utilities need to focus on the overall capabilities of Smart Grids rather than mere implementation of smart meters. They need to consider a more holistic view.

(v) Access to affordable capital

Funds are one of the major roadblocks in implementation of Smart Grid. Policy makers and regulators have to make more conducive rules and regulations in order to attract more and more private players. Furthermore the risk associated with Smart Grid is more; but in long run it is expected that risk-return profile will be closer to the current situation as new policy framework will be in place and risk will be optimally shared across the value chain.

In addition to this, the hardware manufacturers are expected to invest more and more on mass production and R&D activities so that technology obsolescence risk can be minimized and access to the capital required for this transition is at reasonable cost.

(vi) Skills and knowledge

As the utilities will move towards Smart Grid, there will be a demand for a new skill sets to bridge the gap and to have to develop new skills in analytics, data management and decision support. To address this issue, a cadre of engineers and managers will need to be trained to manage the transition. This transition will require investment of both time and money from both government and private players to support education programs that will help in building managers and engineers for tomorrow. To bring such a change utilities have to think hard about how they can manage the transition in order to avoid over burdening of staff with change.

(vii) Cyber security and data privacy

With the transition from analogous to digital electricity infrastructure comes the challenge of communication security and data management; as digital networks are more prone to malicious attacks from software hackers, security becomes the key issue to be addressed.

In addition to this; concerns on invasion of privacy and security of personal consumption data arises. The data collected from the consumption information could provide a significant insight of consumer's behavior and preferences. This valuable information could be abused if correct protocols and security measures are not adhered to.

If above two issues are not addressed in a transparent manner, it may create a negative impact on customer's perception and will prove to be a barrier for adoption.

5.6.2 Solutions [15]

In [15], following solutions have been proposed to overcome challenges identified in earlier sections:

5.6.2.1 Forming Political and Economic Frameworks

Policy makers and regulators have to implement a framework which optimally spread the risk over the whole value chain i.e. to guard the investors from risk and to yield the result at lower cost to the customers. They have to form a robust incentive model in order to attract more and more private investment. Also rate of return should be based on the output generated. Rewards and penalty mechanism should be considered in order to monitor the performance of the utilities and to encourage them to deliver the outcomes in the most efficient manner.

Technological and delivery risk associated with Smart Grid are significant. And this can be overcome over a due course of time as more issues arise and are addressed. Risks associated with Smart Grid have to be shared by every member across the value chain. While making the framework regulators must consider how much of that risk a utility can pass on to the contractors, suppliers and consumers. By maintaining the proper balance, there will be an improved alignment of the incentives. And further they have to tackle numerous policy disputes and recommend potential solutions.

5.6.2.2 Moving Towards a Societal Value System

The major challenge for the transition from analogous to digital infrastructure will be to move from utility-centric investment decision to societal-level decisions which determine wider scopes of the Smart Grid. This would help in the accelerated adoption of Smart Grid Technology by the society.

5.6.2.3 Achieving Greater Efficiency in Energy Delivery

Smart Grid Technology should consider building greater efficiency into the energy system which would result in reduction of losses, peak load demand and thereby decreasing generation as well as consumption of energy. New regulatory framework which incentivizes utilities for reducing the technical losses would help utilities to perform more efficiently.

5.6.2.4 Enabling Distributed Generation and Storage

Smart grids will change where, when and how energy is produced. Each household and business will be empowered to become a micro-generator. Onsite photovoltaic panels and small-scale wind turbines are the predominant examples; developing resources consist of geothermal, biomass, hydrogen fuel cells, plug-in hybrid electric vehicles and batteries. As the cost of traditional energy sources continues to rise and the cost of distributed generation technologies falls, the economic situation for this evolution will build.

5.6.2.5 Increasing Awareness on Smart Grids

There is an imperative need to make the society and the policy makers aware about the capabilities of a Smart Grid. The main step is to form a perfect, universal description on the common principles of a smart grid. Beyond agreement on a characterization, the matter also needs to be debated more holistically as a true enabler to the low-carbon economy, rather than as an investment decision to be taken within the meeting room of distinct utilities. The importance of consumer education is not to be under estimated. The formation of user-friendly and state-of-the-art products and services will play a significant role in convincing the society about Smart Grids.

Also the utilities are required to scrutinize the major challenges in implementation of Smart Grid and their impact on their business model and operations.

5.6.2.6 Creating a Fresh Pool of Skills and Knowledge

Successful implementation of the smart grid will require a large number of highly skilled engineers and managers mainly those who are trained to work on transmission and distribution networks. As a result to on-job training and employees development will be vital across the industry. Simultaneously, there is a requirement for investment in the development of relevant undergraduate, postgraduate and vocational training to make sure the availability of a suitable workforce for the future. The investment in T&D should not be limited and neither in research and knowledge development, which would be essential for the development of this sector.

5.6.2.7 Addressing Cyber Security Risks and Data Privacy Issues

Smart Grid success depends on the successful handling of two major IT issues: Security and Integration and data handling.

With increase in computers and communication networks comes the increased threat of cyber-attack. The Government should look into this matter because consumer's consumption data can be misused by the utilities and the third party. Utilities have to give assurance to the consumers that their valuable information is handled by authorized party in ethical manner. The government has to adopt high standard level in order to withstand cyber-attacks.

5.7 Study Area Information [22–30]

5.7.1 Historical Perspectives

Ahmedabad [22–25]:

The historic city of "Ahmedabad" or "Amdavad" or "Ahmadabad", the largest city of Gujarat, has been declared as *India's first UNESCO World Heritage City* in July, 2017.

Ahmedabad has been the fifth most populous city and seventh most populous urban agglomeration in India. Currently, with a population of more than 6.3 million and an extended population of 7.8 million, it is the sixth largest city and seventh largest metropolitan area of India. According to the 2011 census, the urban agglomeration of Ahmedabad was 6,361,084.

Ahmedabad is located on the banks of the Sabarmati River, 30 km (19 mi) from the state capital Gandhinagar, which is twin city of Ahmedabad forming Ahmedabad-Gandhinagar twin city metropolitan region, focused in this chapter.

The Ahmedabad city, founded by Sultan Ahmad Shah in 1411 AD, lies on the banks of the Sabarmati river, is amongst the major metropolitan cities in India. The history of Ahmedabad stretches as far back as in the 11th century and linking itself with old towns of Ashaval and Karnavati about 1000 years ago. In the year 1411 AD, Sultan Ahmed Shah built citadel and encouraged development of trade and commerce. In 1456 AD, an enclosing wall was constructed defining a periphery to the city-limits. The city within this wall got structured into wards, organized by 12 main roads each terminating at a gate in the wall.

The Ahmedabad city is the administrative headquarter of Ahmedabad district and is the judicial capital of Gujarat as the Gujarat High Court is located here.

With a population of more than 5.8 million and an extended population of 6.3 million, it is the fifth largest city and seventh largest metropolitan area of India. With the increasing opportunities for trade and commerce and as a center for higher education, this heavy growth continues.

In 2003, Ahmedabad has the *lowest crime rate* of the 35 Indian cities with a population of more than one million according to the National Crime Records Bureau (NCRB) report. In 2010, Forbes magazine rated Ahmedabad as *one of the fastest-growing city in India and the world*. In 2011, Ahmedabad was rated *India's best megacity to live in* by leading market research firm IMRB. Again in 2012, The Times of India chose Ahmedabad as *India's best city to live in* (Fig. 5.6).

Gandhinagar [26, 27]:

Gandhinagar is the capital of the state of Gujarat in Western India. Gandhinagar is located approximately 23 km north of Ahmedabad, on the west central point of the Industrial corridor between Delhi, the political capital of India, and Mumbai, the financial capital of India.

Gandhinagar, Gujarat's new capital city, lies on the west bank of the Sabarmati River, about 545 km (338 miles) north of Mumbai, the financial capital

Fig. 5.6 Panoramic view of Ahmedabad city [24]

Fig. 5.7 Panoramic view of Gandhinagar city [26]

of India and 901 km (560 miles) southeast of Delhi, the political capital. Built with parks, extensive plantation and recreational areas along the river, Gandhinagar has a green garden-city atmosphere (Fig. 5.7).

Naroda [28, 29]:

Naroda is fast growing area in northeast side in Ahmedabad city, situated on the emerging Gandhinagar-Ahmedabad-Vadodara (GAV) corridor. With the establishment of the Naroda industrial area in the 1980s, it progressed well as separate town and later incorporated into Ahmedabad in 1996. Being eight km from Ahmedabad international airport and located on the SP Ring Road, Naroda has over the last two years transformed from a neglected industrial area to desirable location for homes. The Naroda GIDC industrial park hosts national and multinational corporations. Several major township projects are being developed along the Ahmedabad-Vadodara expressway with Naroda being at the center of the development. In the recent Vibrant Gujarat summit, 24 projects worth more than Rs. 1000 Crore were slated to be developed in this part of city and along the Ahmedabad-Vadodara expressway including hospitals, hotels and clubs, educational institutions, NRI residential colonies, business parks, etc. In the satellite map of Fig. 5.8, Naroda's strategic presence surrounded by diversified types of locations has been depicted which is region of our research study interest in the Ahmedabad-Gandhinagar smart twin city metropolitan region (Table 5.3).

5.7.2 *Regional Contexts [23, 29]*

5.7.2.1 Demographic Attributes [23]

Ahmadabad District is the central region district of Gujarat with its administrative headquarters located at Ahmadabad city. Naroda area is located in Ahmadabad city on the junction of Ahmedabad-Gandhinagar cities, in Ahmedabad district of Gujarat state of India.

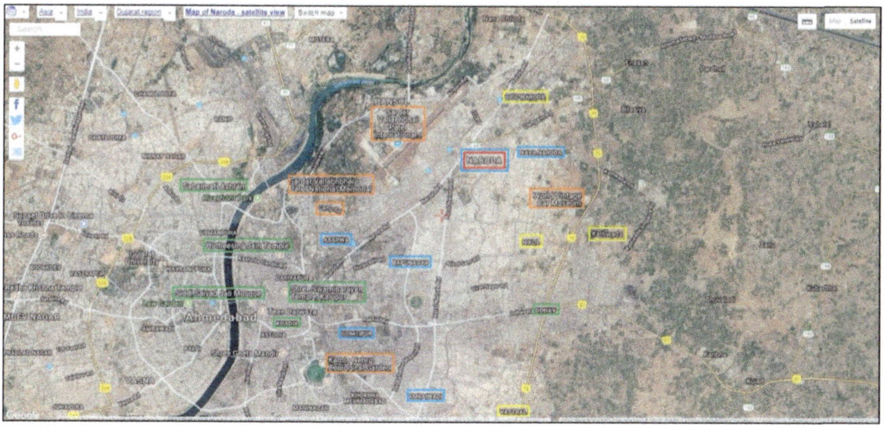

Fig. 5.8 Satellite map of Naroda area [29]

Table 5.3 Details of diversified locations surrounding Naroda in satellite map [29]

Sr. No.	Highlighting colour	Location classification	Examples in the map
1		Main location	Naroda
2		Residential area	Naroda, Nava Naroda, Asarwa, Bapunagar, Gomtipur, Amraiwadi
3		Heritage monuments	Sabarmati Ashram (Gandhi Ashram), Hathising Jain Temple, Shree Swaminarayan Temple, Siddi Saiyad Jali Mosque, Khadia
4		Industrial area	GIDC Naroda, Nikol, Kathwada, Vastral
5		Other important location	Sardar Vallabhbhai international airport, Sardar Vallabhbhai National Memorial, World Vintage Car Museum, Kamla Nehru Zoological Garden

Geographical area: 8107 km^2 (total cropped area 6410 km^2, forest area 149 in sq km (as per 2015 data)
Population: total 72,14,225, 37,88,051 males and 34,26,174 females (according to 2011 census)
Sex ratio: 904 females for every 1000 males
Literacy rate: 85.31% persons, 90.74% males and 79.35% females
Spoken languages: Gujarati (77.05%), Hindi (12.41%) and Urdu (5.14%)
Labour force participation rate: 36.25%
Year of Establishment: 1411 A.D.
Commencement of the municipal works by 'Town wall fund committee': 1834.
Establishment of Municipality—1858
Establishment of Municipal Corporation—1950

Area of the city: 464.16 km^2.
Languages spoken Gujarati, Hindi, English
Urban Population of the State is 42.6%, which used to be at 37.4% in 2001. Rural population in the state in 2011 fell to 57.4% from 62.6% in 2001.

Ahmedabad is the most populated District in the State, with 7.20 million people, up 11.94% from 2001, followed by Surat with 6.07 million people, up 10.07%, as per Gujarat's Directorate of census operations.

5.7.3 Physical Aspects (Climatic and Weather Conditions Including Rain, Etc.) [23, 30]

Latitude—22° 58N, Longitude—72° 35EN, Altitude—49 Mts. above MSL
Average annual rainfall—750 mm (July to September)
Climate: Summer—24–39 °C (May exceed 42 °C or above.)
Winter—10–24 °C (It may dip to 5 °C)

From Figs. 5.9 and 5.10 of [30], it could be observed that average temperature in the region is continuously rising every year causing rapid rise in the electrical energy being consumed every summer as compared to monsoon in both residential as well as commercial areas.

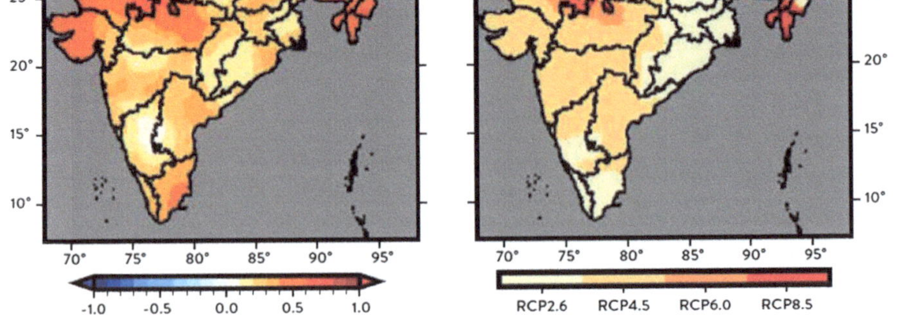

Fig. 5.9 **a** Change in mean (2006–13) annual temperature as compared to historical (1951–2005) period and **b** Representative Concentration Pathway (RCP) [30]

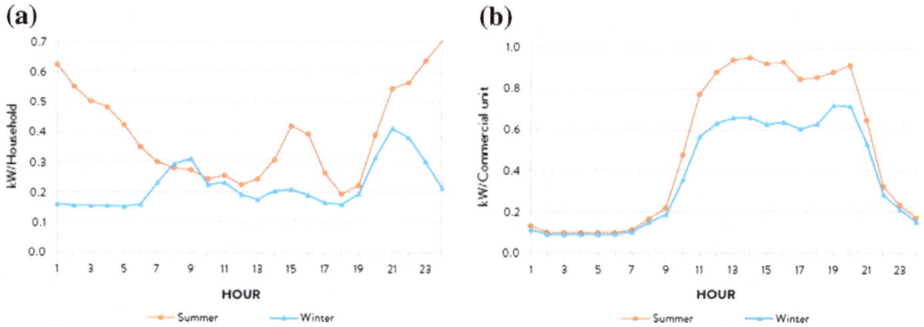

Fig. 5.10 **a** Load analysis curve for Gujarat—residential area and **b** Load analysis curve for Gujarat—commercial area [30]

5.7.4 Rise in the New Economy and Post-liberalization Growth [22–24]

The GDP of Ahmedabad was estimated at US$64 billion in 2014. The RBI ranked Ahmedabad as the seventh largest deposit centre and seventh largest credit centre nationwide as of June 2012. In the 19th century, the textile and garments industry received strong capital investment.

On 30 May 1861, Mr. Ranchhodlal Chhotalal founded the first Indian textile mill, the Ahmedabad Spinning and Weaving Company Limited, followed by the establishment of a series of textile mills. The textile industry further expanded rapidly during the First World War, and benefited from the influence of Mahatma Gandhi's Swadeshi movement, which promoted the purchase of Indian-made goods. Ahmedabad was known as the "Manchester of the East" for its textile industry. The city is one of the largest suppliers of denim and exporters of gemstones and jewellery in India. In 1966, Mr. Dhirubhai Ambani incorporated Reliance Textiles Industries Private Limited in Maharashtra and established a synthetic fabrics mill at Naroda in Gujarat. In 1975, the company expanded its business into textiles, with "Vimal" becoming its major brand in later years.

Post independence, Ahmedabad has emerged as an important commercial, enterprising and industrial hub in India. It has been one of the largest producers of cotton in the country, and with the country's second oldest stock exchange. Impacts of liberalization of the Indian economy have nurtured economy of this city towards various diversified sectors finance, trade, communication, housing, construction, etc.

The automobile industry also flourished in the city with Tata, Ford and Maruti Suzuki established their plants near Ahmedabad. IT industry significantly developed in Ahmedabad, with presence of giant IT companies such as Tata Consultancy Services (TCS). Ahmedabad was ranked fifth among the top nine most competitive cities in the country in 2002 NASSCOM survey on the "Super Nine Indian Destinations" for IT-enabled services. Indian as well as foreign students and young skilled workers have been attracted by educational and industrial institutions of the city.

Software Technology Parks of India (STPI) was established at Infocity, Gandhinagar. Gujarat International Finance Tec-City (GIFT-City) is an under-construction central business district between Ahmedabad and Gandhinagar. It will be built on 359 ha (886 acres) of land with core objective is to provide high quality physical infrastructure (electricity, water, gas, district cooling, roads, telecoms and broadband), so that finance and tech firms can relocate their operations there from Mumbai, Bangalore, Gurgaon etc. where infrastructure is either inadequate or very expensive. It will have a special economic zone (SEZ), international education zone, integrated townships, an entertainment zone, hotels, a convention centre, an international techno park, units, shopping malls, stock exchanges and service units.

5.7.5 Importance and the Role of the Region

Ahmedabad-Gandhinagar twin city has been already selected as potential smart city in phase-I under Government of India's flagship Smart Cities Mission. As the smart grid is the energy backbone of this upcoming smart city, well established and successful smart grid pilot has been conducted at Naroda area at the junction of the twin city by utility UGVCL. This project has been funded by Government of India funding under R-APDRP scheme.

5.8 Scope Analysis [30, 31, 32]

As the cities are becoming smart, citizen and governance support applications are rising in quantities and complexities. This has direct impact on overall electrical energy requirements. Production, transmission and distribution of electricity are getting expensive. Higher dependencies on fossil fuel based thermal power stations as well as low penetration of renewables has been resulting in rising costs. Bad habits, lack of discipline of citizens, heavy wastages, etc. are also important factors.

Therefore, it is an urgent need of today to transform all smart city homes into smart homes with smart HANs and relevant solutions, which would result into effective and optimal utilization of electricity.

5.8.1 Potential

In India, home automation sector is largely unexplored and rarely any full-scale installations could be found.

In general, popular understanding of home automation is believed as security solution with CCTV and detectors for smoke-fire-gas leak only.

Integration of technologies like IoT, HAN, BAS, Green buildings, Solar PVs and Smart Grid has strong potential to build up 'Sustainable and Energy Efficient Smart City', which result in effective and convincing justification for investment.

5.8.2 Feasible Applications

(i) Advanced Metering Infrastructure (AMI)
(ii) Monitoring and automation of substations
(iii) Home Automation Network (HAN)
(iv) Power network monitoring
(v) Demand Response (DR)
(vi) Integration of renewables
(vii) Supervisory Control And Data Acquisition (SCADA) system
(viii) Plug-in Hybrid Electric Vehicles (PHEV)

5.8.3 Technological Advancements

5.8.3.1 IoT Based Smart Buildings [30]

Smart buildings are realized by technologies such as HAN and/or BAS. Home Area Network (HAN) is the network within the premises of a home enabling devices and electrical loads to communicate with each other and dynamically respond to externally sent signals (e.g. price, etc.) Building Automation System (BAS) is a data acquisition and control system that incorporates various functionalities provided by central control system of a building.

Figure 5.11 presents conceptual architecture of AMI containing integration of different electricity and communication networks and sub-systems. Figure 5.12 provides modified version of AMI architecture of Fig. 5.11 after deployment of IoT components with wireless connectivity at necessary points.

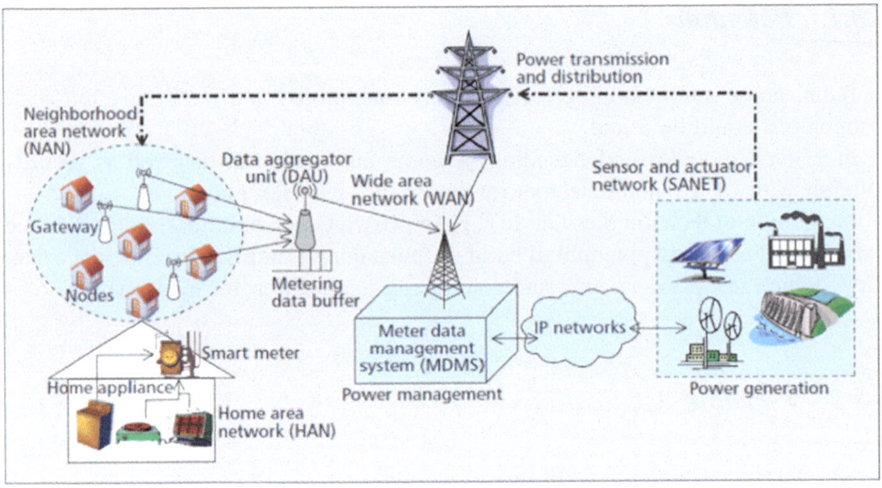

Fig. 5.11 Automated Metering Infrastructure (AMI) [31]

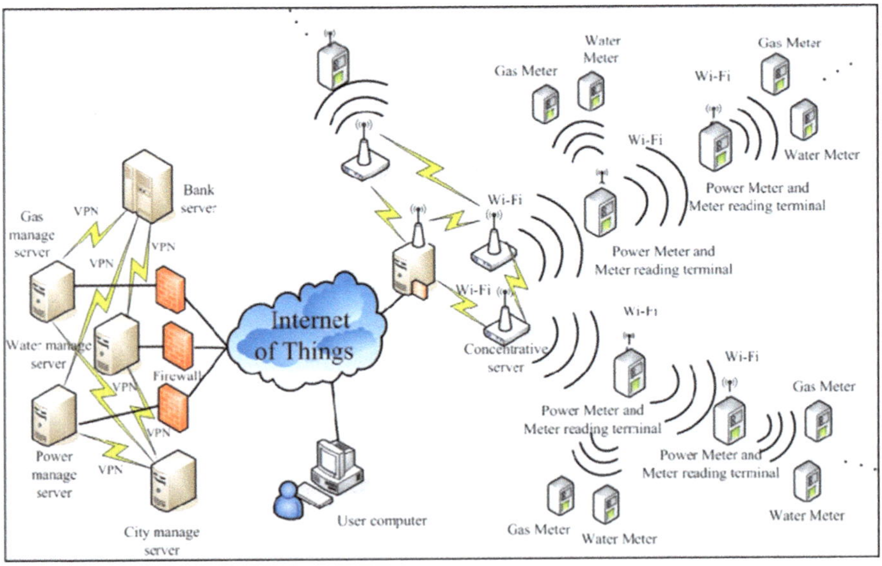

Fig. 5.12 IoTs and AMI [32]

5.9 Present Situation [8]

Referring [8], data related to Naroda smart grid pilot are as mentioned below:

Consumers: 39,422 (Residential, Commercial and industrial)

Input energy: around 1700 GWh approx.
Total Project Cost: Rs. 48.78 Cr approx.

Benefits/objectives/Expected outcomes:

(i) Effective and efficient utilization of energy
(ii) Reduction in energy wastage
(iii) Reduction in AT&C losses
(iv) Savings in peak power purchase cost by reduction of peak load
(v) Reduction in meter reading cost, cost of payment collection, etc.
(vi) Reduction in transformer failures
(vii) Reduction in number of outages
(viii) Reduction in overall carbon footprint

Features of smart grid proposed to be installed:

(i) AMI R: AMI for residential consumers focusing on Demand Response (DR)
(ii) AMI I: AMI for industrial consumers focusing on Demand Side Management (DSM)
(iii) OM: Outage Management
(iv) PLM: Peak Load Management

Unmetered consumers were accounting for input energy of around 1700 MU approx. The functionalities of Peak load management, Outage Management, Power Quality Management are proposed by implementing Automated Metering Infrastructure (AMI) for Industrial, Commercial and Residential Consumers. Some additional functionalities like Load forecasting and Asset Management are also proposed and functionalities of load forecasting, peak power management and outage management are also considered at utility level which will impact all consumers of utility (i.e. 27 Lac consumers) indirectly. Renewable energy integration has also been proposed to be carried out at Patan Solar Park and few roof top installations at some of the universities. The funding scheme for the programme is RAPDRP, Part-C.

Naroda is ready with successful pilot of smart grid with smart meters already installed and NAN-WAN connectivity already tested. Meters are fetching sampled data and proper interpretation and analysis are efficiently being carried out at UGVCL head end. Citizens in the area have been found satisfactory with the setup and ready to participate at next level of development.

5.9.1 SWOC Analysis

5.9.1.1 Strengths

(i) Preliminary level of foundation work completed
(ii) Pilot has been successful
(iii) Participant citizens are cooperative

5.9.1.2 Weaknesses

(i) Connectivity and ICT infrastructure

5.9.1.3 Opportunities

(i) The work itself
(ii) Motivation and support from central and state governments
(iii) Training of citizens for upcoming changes
(iv) Possible employment and commercial opportunities for everyone

5.9.1.4 Challenges

(i) Costing and revenue sharing, subsidies, etc.
(ii) Security of data and access authentication

5.10 Proposed Workplan [3, 5, 33–35]

Naroda area of smart grid pilot is ready for the next stage in which data received from consumer is being analyzed and post-AMI applications shall be implemented-integrated.

Plan of [33] shown in Fig. 5.13 could serve as valuable reference for preparation of the way ahead. While writing this, the region at the start of 2018 is exactly in the middle of the plan, wherein step 1 seems almost completed, work for steps 2 and 3 is undergoing and likely to continue till end 2020. Next and final phase of the work plan for 2021–2024 shall begin.

5.10.1 Solution Approach

First, existing economic and spatial strategies and recommend suggestions for smart development of the metropolitan region have been studied. Then after, a solution approach has been prepared for metropolitan region development by effective energy management, active citizen participation and e-governance. This is carried out by recommending deployments of smart grid and smart buildings with integration of renewables, ICT and IoTs. To ensure 24×7 electricity supply along with limiting carbon footprint has been observed as the major challenge and connectivity has been found major bottleneck in the process.

A five step methodology has been suggested for the implementation of recommended approach:

TABLE III.
PLAN TO IMPEMENT SG

Main steps	Up to 2015		2016 to 2020		2021-2024	
	Case	Action/ implementation	Case	Action/ implementation	Case	Action/ Implementation
1	Deploy base Technology	• Smart meter installed • OMS/DMS system • Micro grid pilot	Automated outage detection, restoration customer notification	• Expanded SCADA& line devices • Both way information • Self-healing grid technology in place	Customer supply side and storage decisions	• Significant DER penetration • Additional micro grids where cost effective • Customer as a resource
2	Customer program offered by utilities	• Dynamic pricing • EE demand response • HAN, energy management	Mature new services for customers	• Load control with DR • Bundled services • DER aggregation	PHEV adoption rises	• PHEV adoption emerges as a critical component of DER • Charging in frame structure in place • PHEV rates in place
3	Many smart grid components are initially deployed	• Self-healing grid technologies • Micro grid technology& self-sustaining community concept • PHEV infrastructure pilots	Major regulatory issues	• Data ownership and access • Cross jurisdictional conflicts • T&D renewable strategy	Advanced grid technology	• CBM cable Diagnostics • Advanced energy storage to support RPS goals • Self-healing grid is a reality

Fig. 5.13 Tabular plan to install smart grid [33]

(i) Integration of renewables

This has been covered in [5] with interesting case of Gandhinagar Solar Photo Voltaic Rooftop Program in by including case-study of application of hybrid communication technologies deployed to serve need based data along with *Development of Remote Energy Parameter Monitoring System*. The proposed and implemented system has distinct features such as an affordable cost, scalability and *anytime-anywhere* monitoring, to encourage inclusion of more sensors for enhanced data acquisition, improved spatial resolution for more fine-grained measurements and better monitoring of critical regions.

(ii) Smart energy (Deployment of Smart Grid)

In [3] presented democratic and citizen-centric approach of design-implementation of architectural details along with presenting a useful framework in order to make Smart Grid more inclusive, effective and comprehensive. Descriptions on communication technologies in form of instrumentation telemetry deployed to timely serve need based bidirectional information between utility and end users have also been included.

(iii) Smart People (Prosumers) and their participation

In [3], smart grid has been presented as an energy backbone of Smart City is immensely vital and serving at the core of Smart City realization. Evolving e-Democracy, smart grid includes highly interactive participation of citizens in energy consumption domain, based on humanitarian and customer centric approach. Different types of customers, their different energy requirements at different timings, different types of energy resources and their switching feasibilities considering different aspects have been integrated. Critical smart grid subsystems (such as BAS, HAN, AMI, DR, etc.), ICT integration and GUIs have been identified as some of the major design considerations. 'Transformation of Conventional Consumer into Smart Prosumer' has been major outcome of the work presented, since the customer has been now enabled as producer of electricity, thereby contributing to the grid and getting credits which is adjustable against consumption.

(iv) Smart buildings with energy efficiency

In [34] and [35], indicating the transformation of legacy stand-alone security systems into intelligent computerized-network based building automation systems, authors presented design-development of IoT based working models for security and HAN-BAN. Development of web-based virtual instrument to run-time couple local/remote monitoring and control of the building has been major outcome. The presented proof-of-concepts of IoT systems could be employed with suitable ICT for converting existing buildings into smart buildings with improved energy efficiency.

(v) Implementation with minimal financial implication

Massive national level awareness campaigns should be organized for citizens' awareness and encouragement for enthusiastic participation. Funding could be arranged from The World Bank, IMF, United Nations and developed countries to government of India, which could be sent to state governments via Special Purpose Vehicles (SPV) such as RAPDRP. State governments should encourage the Prosumers via various state level programs in form of subsidies and other techno commercial support mechanisms. Under 'Make in India', indigenous manufacturers (SME and entrepreneurs in particular) should be encouraged for local manufacturing of devices, systems, software and engineering integration-maintenance supports. Entire mechanism should be operated as single integrated system with single window clearance and e-governance.

5.11 Institutional Collaboration Setup

5.11.1 Gujarat Energy Research and Management Institute (GERMI), Gandhinagar

Gujarat Energy Research and Management Institute (GERMI), Gandhinagar, is a centre for excellence in industry learning and is set up to develop human resource assets to cater to the renewable as well as non-renewable energy sectors, improve knowledge base of policy makers and technologists and provide a competitive edge to leaders to compete in the global arena. GERMI is a Scientific and Industrial Research Organization recognized by the Department of Scientific and Industrial Research (DSIR), Govt. of India; and an ISO 9001:2008 certified by Bureau Veritas, France.

5.11.2 Dharmsinh Desai University (DDU), Nadiad

Dharmsinh Desai University (DDU), Nadiad, is a progressive and one of the leading universities of India, excelling in research and education in the fields of Engineering-Technology, Management, Dental Science, Pharmacy, Information Science, and Commerce. DDU is a Scientific and Industrial Research Organization recognized by the Department of Scientific and Industrial Research (DSIR), Govt. of India; DDU is approved institution under 2(f) and 12(B) of the UGC Act, 1956; DDU is western India's Premier Education Institution established in 1968, ISO 9001:2008 certified by ISOQAR, Manchester, UK.

5.11.3 Association for Collaborative Research

Under Vibrant Gujarat 2013, GERMI and DDU signed a Memorandum of Understanding (MoU) for a collaborative research project 'Solar City-Smart Grid Project'. During 2013–2016, Phase-I of the said research project with detailed title "Development of a City-Level Smart Communication System for Monitoring, Recording, Assessment and Forecasting of Power from Distributed Source of Generation" has been already executed. The MoU has been extended for 2016–2019, wherein phase-II of the research project with detailed title "Undertaking joint research project in Smart Grids and Microgrids for Smart communication system for monitoring, recording, assessment and forecasting of power from distributed sources of Generation" is currently in progress.

5.12 Excerpts from Research Group Activities

5.12.1 International Consultative Meeting, 17–18 December, 2016 at NITC

International Consultative Meeting on 'Design of Spatial and Economic Strategy for Smart Metropolitan Region Development', 17–18 December, 2016, was organized at Department of Architecture and Planning, National Institute of Technology Calicut (NITC), Calicut, Kerala, India. The main objective of this meeting was to follow up the ideas expressed earlier and share-encourage discussions regarding smart metropolitan regional development as well as to exchange-evolve novel perspectives of views and opinions for the future research works.

Considering convenience of participant teams representing different countries across the world, two different modes of presentation have been offered: Physical and Virtual. The virtual mode of presentation was by A-VIEW Video conferencing software platform developed and supported by Amrita University, while entire arrangements for this 2-days workshop were provided by NITC. On day-1 post inauguration, Indian teams representing Calicut, Surat, Bhopal and the team from Naples, Italy made their presentations. Similarly, on day-2, teams representing Hong Kong-China, Dakar-Senegal, Johannesburg-Africa, Ahmedabad-Gandhinagar-India, Abuja-Nigeria, Nairobi-Kenya, Conakry-Guinea, Jaipur-India made their presentations. Each day ended with concluding sessions with valuable comments and suggestions for enhancement of future works.

On day-2, 18/12/2016, 10.30 to 11.15 a.m., authors of this chapter made presentation as research team representing Ahmedabad-Gandhinagar twin city metropolitan region development. Our presentation was made using virtual mode and based upon smart grid and its contribution to spatial-economic development of Ahmedabad-Gandhinagar twin city smart metropolitan region with a special focus Naroda pilot project as case study. After presenting details of need and scope, region identification and its present state, SWOC analysis, etc., a methodology along with workplan and envisaged outcomes has been proposed for smart MRD of region. The presentation ended with review comments, useful guidelines and suggestions for enhancement of the work in progress.

5.12.2 Bulletin Contribution

Out of total three bulletins published, teams representing Pittsburg, Stuttgart and Naples contributed articles to first bulletin; teams representing Dakar, Conakry, Abuja, Johannesburg and Nairobi contributed articles to second bulletin and Ahmedabad-Gandhinagar, Bangalore, Chandigarh, Hong Kong, Kozhikode, New Delhi, Jaipur and Surat contributed articles to the third bulletin.

Our article 'Ahmedabad-Gandhinagar, Gujarat, India: Towards sustainable growth by energy reliability' was contributed to Bulletin-3. The article has been in line with earlier contributions as well as post-presentation suggestions in the international consultative meeting.

5.12.3 Summary of Outcomes

Proposed approach for metropolitan region development by effective energy management, active citizen participation and e-governance has been major outcome this book chapter. This is achieved by proposing deployments of smart grid and smart buildings with integration of renewables, ICT and IoTs. The approach suggested is useful, adaptive and simple, but effective for implementation, if applied with suitable customization as necessary.

5.13 Recommendations

5.13.1 For Regulators

(i) Create a regulatory framework which aligns incentives of each member in the value chain.
(ii) Allocation of risk and reward efficiently, considering both utilities and customer while making policies.
(iii) Adopt output based regulatory system (Reward/Penalties) which stresses on utilities to perform better.

5.13.2 For Utilities

(i) Adopt more holistic approach about Smart Grids, so that they can convey its future benefits to the customers.
(ii) Reduce the risk of technology obsolescence by R&D activities.
(iii) Undertake large scale pilot projects and analyze the benefits.
(iv) Transformation from utility-centric investment decision to societal-level decisions.

5.13.3 For Vendors

(i) Required to play important role in policy making process.
(ii) To help utilities to adopt flexible design and compatibility of Smart Grid fast.
(iii) To convince customers about the acceptance of changing trend by product and service offering.

5.13.4 For Customers

(i) Plays critical role by demanding for more flexible service.
(ii) To encourage more players to enter in this field and in order to make the market competitive.
(iii) To help utilities and regulators to set goals and make conducive policies.
(iv) To increase the awareness in society.

5.14 Summary and Conclusions

5.14.1 Summary

(i) Proposal regarding HAN-BAS development in the region presented for smart grid development
(ii) Feasible applications suggested
(iii) Earlier works cited
(iv) Ubiquitous and effective usage of different ICT technologies integrating IoT and WSN
(v) Significant optimization in overall energy consumption and efficient utilization
(vi) Advancements like demand response could be made possible

5.14.2 Conclusions

Smart grids are energy backbones of smart cities. Without smart grids implemented, smart cities cannot be realized. However, in case where existing city like Ahmedabad-Gandhinagar twin city are being transformed into smart cities, existing grids are operational and they are to be converted into smart grid while their operation is continued, this is a challenging task considering technical complexities involved as well as cost implications and many humanistic-socialistic issues involved. Therefore, rather than converting complete grid, area wise pilot installations such as Naroda could be planned out and phase wise conversion can be carried out. Continuously feedback from all the stakeholders, Prosumers in particular, should be acquired and required customizations should be made in the solution approach for easy adaptability of the change.

Proposed approach for metropolitan region development should be utilized for effective energy management and could be realized by deployments of smart grid, smart buildings with integration of renewables, ICT and IoTs.

Acknowledgements The authors take this opportunity to express their sincere thanks to Editor Prof. TM Vinod Kumar and his entire team, bulletin editors and Springer staff members for their valuable guidance, excellent cooperation and timely help extended.

Cooperation received in development of this book chapter from the faculty members, management and office bearers of GERMI, PDPU and DDU, India is acknowledged with thanks. Useful contributions and cooperation received from ISGF, UGVCL, GEDA, Govt. of Gujarat, Govt. of India and all the cited sources of references are also gratefully acknowledged.

References

1. Shukla P (2015) Smart cities in India
2. Smart cities: mission statement & guidelines. New Delhi, India (2015)
3. Bhatt J, Jani O (2017) Smart grid: energy backbone of smart city and e-democracy. In: Kumar TMV (ed) E-democracy for smart cities, vol 3. Springer, pp 319–366
4. Bhatt J, Shah V, Jani O (2014) An instrumentation engineer's review on smart grid: critical applications and parameters. Renew Sustain Energy Rev 40:1217–1239
5. Bhatt J, Jani O (2015) E-governance for photo voltaic powergrid: solar city Gandhinagar, Gujarat, India. In: Kumar TMV (ed) E-governance of smart cities. Springer, pp 177–230
6. What is 'a smart city' online article dtd. 11 Aug 2015, Smart Cities India Foundation. https://www.thescif.in/single-post/2015/08/11/What-is-a-%E2%80%98Smart-City%E2%80%99. Accessed 26 Dec 2017
7. Smart city, Wikipedia (2017) https://en.wikipedia.org/wiki/Smart_city. Accessed 26 Dec 2017
8. Roy S, Chatterji T (2017) Smart city initiative in India : a policy review. In: 3rd international conference on public policy (ICPP3) 28–30 June 2017—Singapore, pp 1–29
9. Waghmare M, Datar M, Hire R (2017) Democratic interdisciplinary approach to transform existing city pattern into smart city. Int J Recent Trends Eng Res 3(11):68–73
10. U. International Trade Administration (2016) 2016 top markets report smart grid: country case study—India
11. Chandrasekar KS, Bajracharya B, O'Hare D (2016) A comparative analysis of smart city initiatives by China and India-Lessons for India. In: 9th international urban design conference: smart cities for 21st century Australia: how urban design innovation can change our cities, Hyatt, Canberra, pp 339–357
12. Shankar A, Patra P (2010) Smart cities: an opportunity to transform Indian cities into global destinations
13. India Smart Grid Forum (ISGF) (2017) Smart grid handbook for regulators and policy makers (draft for comments)
14. Zhang Q, Gang S, Li P (2010) Smart city grid: the start to develop smart grid. In: 2010 International Conference on E-product E-service E-entertainment, ICEEE2010, pp 0–3
15. Kaushal R (2011) Challenges of implementing smart grids in India
16. Thakur J, Chakraborty B (2015) Intelli-grid: moving towards automation of electric grid in India. Renew Sustain Energy Rev 42(2015):16–25
17. Joshi H, Pandya V (2015) Development of smart plug and energy price forecasting technique to reduce peak demand and cost of energy in smart grid, vol 3, no 3, pp 4–8
18. Moretti M, Djomo SN, Azadi H, et al (2016) A systematic review of environmental and economic impacts of smart grids. Renew Sustain Energy Rev http://dx.doi.org/10.1016/j.rser.2016.03.039
19. Mikalauskas I (2015) Economic, social and environmental benefits of smart grids. Eur J Interdiscip Stud Young Res Sect 7(2):19–28
20. Hondo H, Moriizumi Y (2017) Employment creation potential of renewable power generation technologies: a life cycle approach. Renew Sustain Energy Rev 79(August 2016):128–136
21. Chakraborty S, Chowdhury A, Chakraborty S (2017) Smart grids & smart grid technologies in India. Int Res J Eng Technol 4(1):1536–1541
22. Ahmedabad, Wikipedia (2017) https://en.wikipedia.org/wiki/Ahmedabad. Accessed on 26 Dec 2017
23. Ahmedabad Municipal Corporation, "Ahmedabad-Presentation," pp 1–122
24. https://etravelsnow.blogspot.in/2012/11/ahmedabad-india-tourist-attractions.html. Accessed on 03 March 2018
25. Ahmedabad city, http://ahmedabadcity.gov.in/portal/index.jsp. Accessed on 26 Dec 2017
26. http://www.ddcti.ac.in/aboutgandhinagar.php

27. Gandhinagar, Wikipedia (2017) https://en.wikipedia.org/wiki/Gandhinagar. Accessed on 26 Dec 2017
28. Naroda, Wikipedia (2017), https://en.wikipedia.org/wiki/Naroda. Accessed on 26 Dec 2017
29. Naroda, Google Maps (2017) https://www.google.co.in/maps/@22.9908557, 72.7632085, 10. 5z?hl = en&authuser = 0. Accessed on 26 Dec 2017
30. Garg A, Mohan P, Shukla S, Kankal B, Vishwanathan SS (2017) High impact opportunities for energy efficiency in india
31. Desai B, Lebow M (2010) Needed: ASAP approach. IEEE Power Energy Mag 8(6):53–60
32. Gamroth C, Wu K, Marinakis D (2012) A smart meter based approach to power reliability index for enterprise-level power grid. In: Smart grid communications (SmartGridComm), 2012 IEEE third international conference on 2012, pp 534–539 (IEEE)
33. Acharjee P, Gunda J (2010) Development Prospect of Smart Grid in India. 2010 IEEE Int Conf Power Energy (PECon2010), Nov 29–Dec 1, 2010, Kuala Lumpur, Malaysia. pp 953–957
34. Bhatt J, Verma H (2015) Design and development of wired building automation systems. Energy Build (Mar 2015)
35. Bhatt J, Verma H (2010) RS-485/MODBUS based intelligent building automation system using LabVIEW. In: 4th international conference on computer applications in electrical engineering-recent advances (CERA-09), p 5
36. Smart grid when energy meets the internet of things, http://ictpost.com/smart-grid-when-energy-meets-the-internet-of-things-2. Accessed on 26 Dec 2017

Part VI
India, Bangalore

Chapter 6
Towards a Smart Metropolitan Region: A Roadmap for Transforming Bangalore Metropolitan Region

Amit Chatterjee, Binayak Choudhary, Premjeet Das Gupta and Gaurav Vaidya

Abstract Smart Metropolitan Region, as this chapter traces, is the one that adopts new approaches of the concept of economic growth compatible with space within a minimum possible time. Bangalore Metropolitan Region (BMR) within an area of 8005 km^2, and a population of 11.69 million is one of the fast growing metropolitan region and becoming large global conglomerates. This chapter argues about an alternative approach to smartness which inculcates the principals of innovation, efficiency and inclusion. This study also focused on moves beyond the constructed ontology of mega city-centric concentration and urbanism and looked through the lens of dynamic city-region perspective.

Keywords BMR · Economy · Smart · Spatial · Transforming

6.1 Introduction

The relationship between urbanization and economic growth has been perplexing and vital for policy concern especially in developing world. We are living in the century of the metropolis with more than half of the population living in the metropolitan region [1]. Metropolitan regions are flourished in every region of the world and they present multifarious challenges and opportunities. The metropolitan context in which contemporary metropolitan regional planning is formulated differs

A. Chatterjee (✉) · B. Choudhary · P. Das Gupta · G. Vaidya
Department of Planning, School of Planning and Architecture, Bhopal, Neelbad Road, Bhauri, Bhopal 462030, Madhya Pradesh, India
e-mail: amit.chatterjee@spabhopal.ac.in

B. Choudhary
e-mail: binayak@spabhopal.ac.in

P. Das Gupta
e-mail: premjeet@spabhopal.ac.in

G. Vaidya
e-mail: gaurav.vaidya@spabhopal.ac.in

T. M. Vinod Kumar (ed.), *Smart Metropolitan Regional Development*,
Advances in 21st Century Human Settlements,
https://doi.org/10.1007/978-981-10-8588-8_6

significantly from one region to another. It requires understanding of region specific problems and localized solutions. Urban India accommodated 377 million people (31.2% of total population), the second largest urban population in the world as per Census 2011 [2]. In the decade 2001–2011, the pattern of India's urbanization underwent an important shift, which was characterized by an increase in the number of metropolitan cities/Urban Agglomerations (UAs) from 35 to 52 and increase in the percentage share of population from 37.8 to 42.3%. According to United Nation's estimate in 2014, by the year 2050, half of India's populations are expected to live in urban areas [3]. Many large Indian cities, especially the large metropolises and metropolitan regions are facing problems with respect to their growth, composition, spatial spread, congestion, environmental factors, housing, infrastructure availability as well as accessibility. Bangalore (also known as Bengaluru), the 'Silicon valley of India' is no exception to these trends. In the year 2015, Government of India launched Smart City Mission the major thrust of which is to "develop sustainable and inclusive development and create compact and replicable model" [4]. The Mission will cover 100 cities over the five year time period, including Bangalore. It is but obvious that urban economy not only dictates the urban fabric of a city, it also influences the city's socio economic profile.

Bangalore Metropolitan Region (BMR) with an area of 8005 km^2, and a population of 11.69 million is divided into 3 districts namely—Bangalore Urban, Bangalore Rural and Ramanagra. BMR has only one Corporation namely the Bruhat Bangalore Mahanagar Palika i.e. BBMP and ten Urban Local Bodies (ULBs). Bangalore is a rapidly developing metropolis located in Bangalore district of the Karnataka state. Indian Census 2011 has ranked it as the fifth largest urban agglomeration of India. Bangalore city covers a space of 10% of the geographical area and has a population share of almost 72% of Bangalore Metropolitan Region (BMR). Currently within BMR, majority population and employment are concentrated in Bangalore city which makes it a primate city. The mistake that we often commit is to regard the growth of BMR as being endogenous when the fact remains that BMR is nothing but conurbations. BMR not only engulfs its degenerated peripheries, but also the small towns around, thereby leading to the total obliteration of the latter. The present study will analyze of the economic and spatial strategies undertaken by national and local authorities for a Smart Bangalore Metropolitan Regional Development.

This chapter is divided into nine sections and following this section of introduction, Sect. 6.2 presents the regional settings of BMR. Section 6.3 discusses the population dynamics of BMR. Section 6.4 explains the methodology to analyze spatial and economic transformations for a smart BMR. Section 6.5 presents spatial embeddedness of the region, followed by Sect. 6.6 on current economic status of Bangalore and its region. Section 6.7 discusses about the current challenges faced by region. Section 6.8 examines the smart ways to move ahead by linking the economy with space followed by conclusions drawn from the research findings and the consequent policy implications.

6.2 Regional Setting

Bangalore city is located at 12° 58′N and 77° 34′E in the southern part of India on Deccan Plateau. The city is the capital of the state of Karnataka state since 1956, when the state was formed. The Bangalore Metropolitan Region (BMR) is a planning region recognized under the Bangalore Metropolitan Region Act, 1985. Bangalore Metropolitan Region with an area of 8005 km^2 constitutes 4.17% of total area of Karnataka. 19.13% of the population of Karnataka lives in Bangalore Metropolitan Region. Bangalore region is located in the South Karnataka Region (SKR) of the state. The districts of Karnataka are divided into 5 regions namely, South, West, Central, North-East and North-West Region. These are conceptual entities envisioned by BMR Structure Plan 2011 and do not possess any statutory status. The SKR contains of 10 districts. The BMR is consisted of 3 districts namely, Bangalore Urban, Bangalore Rural and Ramanagaram districts. The rest of the districts in SKR surround the BMR. National Highway (NH) 7 and NH4 (part of North South Corridor and Golden Quadrilateral, respectively) and NH 209 pass through Bangalore forming five important radial roads within the Bangalore Metropolitan Area. State Highways linking Bangalore with Mysore, Bannerghatta and Magadi form other major radial corridors. Developed as a radial town, Bangalore does not have a strong circumferential road system, except for the Outer Ring Road, despite the intervening space between the corridors having been developed. With respect to urbanisation at state level, Bangalore urban comes out to be the most urbanised district with 90.94% of urban population, whereas the second most urbanised district, Dharwadis has just 56.82% urban population (See Fig. 6.1). The urbanization level however is much lower at 27.12 and 24.73% for Bangalore Rural and Ramanagram respectively, indicating the concentration of density at regional level.

As regards Gross District Domestic Product (GDDP) too, Bangalore urban has highest annual GDDP with INR26,851.77 million (2013–14) in contrast to INR 4697.01 million for Dakshin Kannada district. The GDDP of other two districts, Bangalore Rural and Ramnagaram is much lower than Bangalore Urban at INR1124.49 million and INR1122.50 million respectively [5].

6.3 Population Dynamics of BMR

The BMR is comprised of three districts, namely Bangalore Urban District, Bangalore Rural District and Ramanagaram District (the latter being a newly created district carved out of Bangalore Rural District with Ramanagaram, Chennapatna, Magadi and Kanakpura taluks or sub districts) covering a total of 8005 km^2. The three districts of BMR are further divided into 11 taluks. The 11 taluks in all have 11 Urban Local Bodies (ULBs) and 284 g panchayats (village council) covering 2511 villages. The details administrative unit of BMR is shown in Fig. 6.2.

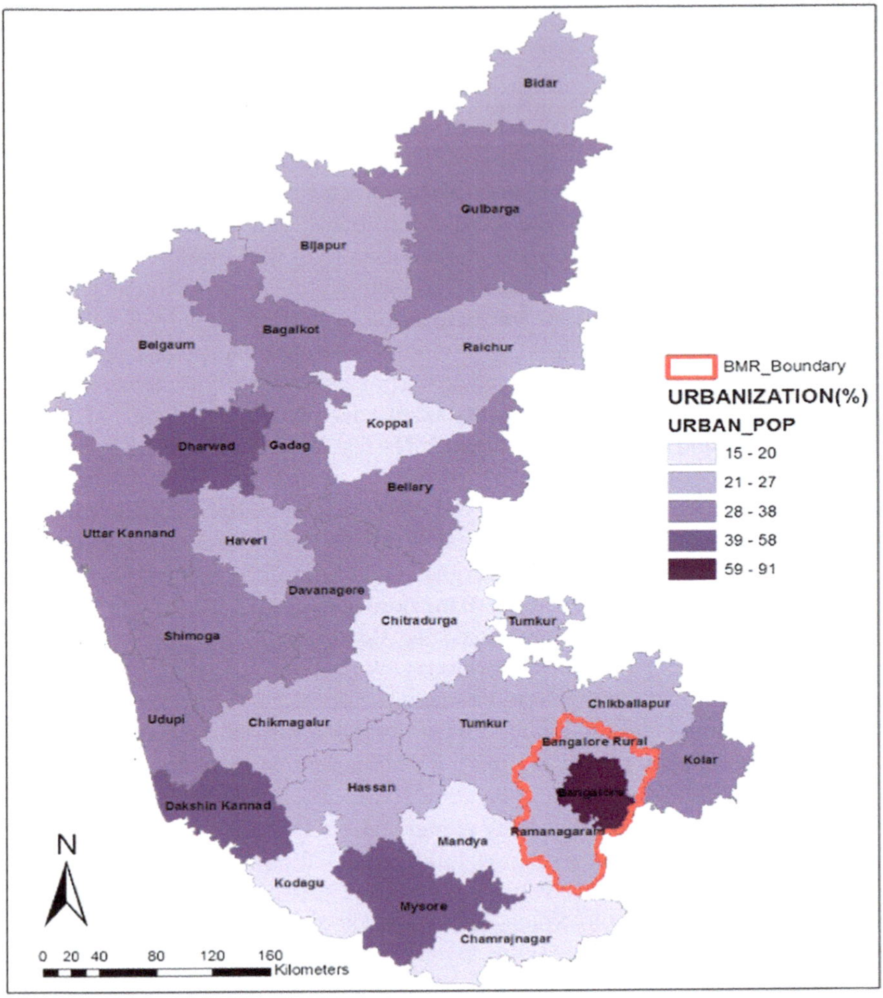

Fig. 6.1 District wise percentage urban population in Karnataka state—2011. *Source* Prepared by Authors

6.3.1 Population Distribution: Districts and Taluks of BMR

Although all the three districts have witnessed an increase in populations from 1991 to 2011, Bangalore Urban shows a steep increasing trend when compared to other two districts. Bangalore Urban witnesses this huge overload due to the presence of Bangalore city [6, 7]. The population of Bangalore Urban accounts for 82% of the total BMR population, with a population density of 4393 persons/km^2 which is maximum when compared to Bangalore Rural and Ramanagra. The urban population density of Bangalore Urban district is 11,400 persons/km^2, while that of

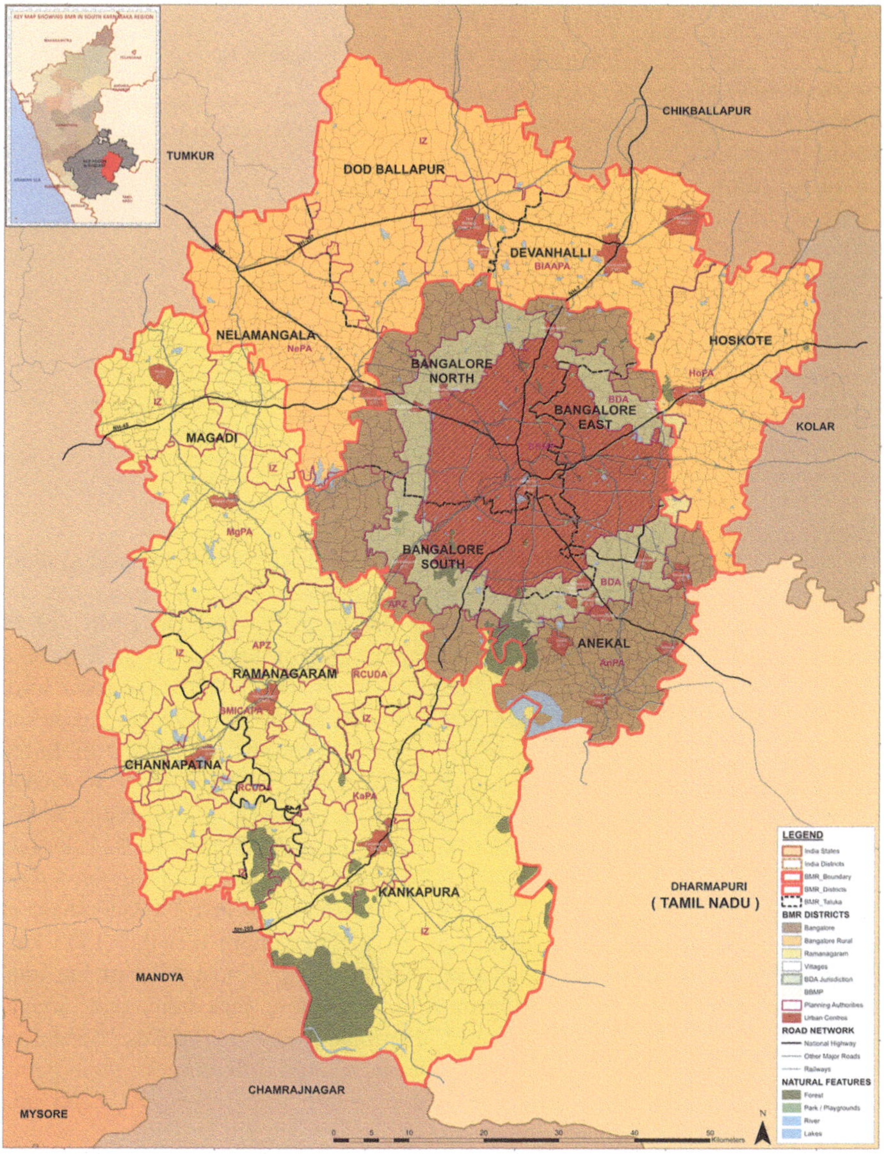

Fig. 6.2 Administrative jurisdictions within BMR. *Source* Prepared by Authors

Bangalore Rural district is 3923 persons/km^2 and that of Ramanagra is 6315 persons/km^2 which is almost half of Bangalore Urban district. Table 6.1 shows the comparison amongst the three districts with respect to various demographic features.

Out Growth (OG) should be a viable unit such as village or a part of the a village contiguous to a statutory town and possess the urban features in terms of

Table 6.1 District wise comparisons of demographic attributes

	Bangalore Urban	Bangalore Rural	Ramanagra
Population (million)	9.62	0.99	1.08
Area (km^2)	2190	2298	3517
Density (person/km^2)	4393	431	308
Growth rate (2001–2011)	4.72	1.64	0.51
Compound annual growth rate (2001–2011)	3.94	1.53	0.49
No. of villages	562	957	820
No. of towns	16	7	6
No. of out growths (OGs)	12	2	0
Percentage urban population	90.9	27.12	24.73
Urban area (km^2)	767.5	68.5	42.4
Percentage urban area	35	3	1.2

Source Compiled by Authors from various sources

infrastructure and amenities such as pucca roads, electricity, taps, drainage system, education institutions, post offices, medical facilities, banks, etc. Examples are railway colonies, University Campuses etc. There are 14 OGs in BMR, 12 of which are of BBMP and 2 of Dod Ballapur. Out of the 14 OGs, 7 have a population below 5000, 4 have a population between 5000 and 15,000 and 2 have a population of more than 50,000. BMR comprises of 12 talukas. Figure 6.3 shows the population distribution in the talukas according to Census of India, 2011. It is observed that, 16% of the total population of BMR lives in Anekal taluk of Bangalore urban district which covers about 7% of the total area of BMR, 3% of the population of BMR lives in Bangalore East Taluk. Considering the fact that Bangalore East has been formed in 2011, it is the taluk with the least population covering an area of 129.37 km^2 which accounts for just 2% land share of BMR. On the other hand, Kanakpura covers the maximum area i.e. 1553 km^2 (20%) of BMR and accommodates a population which is just 11% of the total BMR population. On the whole, apart from Bangalore East taluk and Anekal Taluk, the population in 2011, is evenly distributed over the talukas. Bangalore South and Bangalore North talukas experience a huge difference in their population from 2001 to 2011 because Bangalore South in 2011 was divided into Bangalore South and Bangalore East talukas.

6.3.2 Rural Urban Population Distribution

Karnataka has a rural to urban population ratio of 8:5, whereas, the rural to urban population ratio of BMR is 1:4 and the land ratio is 9:1 respectively as per 2011 census. This shows the gap between the demand and supply of land which is an important aspect that needs to be looked in the BMR regional plan. BMR has 2339

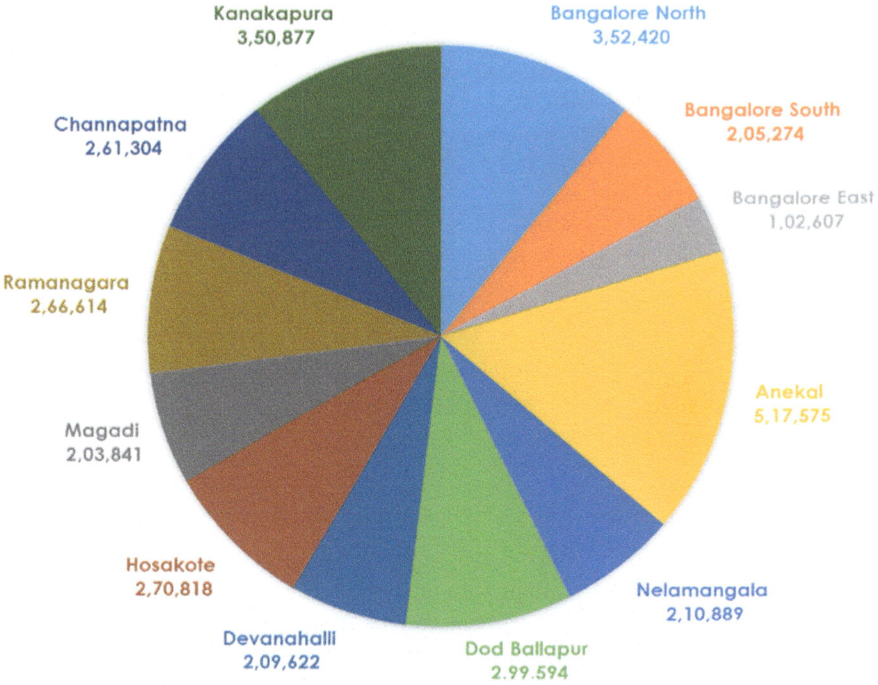

Fig. 6.3 Taluk wise population distribution within BMR—2011. *Source* Census of India [41]

villages and 29 urban centres, out of which Bangalore Urban has 562 villages, Bangalore Rural has 957 villages and Ramanagra has 820 villages. The urban population increased from 69% in 1991 to 73% in 2001 and thereafter to 79% in 2011, the obvious reasons behind being rural to urban migration and job opportunities in urban areas. The urban population over the past three decades, in the BMR has shown a steady increase, whereas the rural population, though increased over the last three decades, does not show a steep variation. Rural and urban population share is shown in Fig. 6.4. In rural area out of the total inhabited villages in BMR, 66% (1538) of the villages are having a population less than or equal to 1000. 23% (542 villages) have a population between 1000 and 2000 and also 9% (221 villages) have a population between 2000 and 5000. The remaining 1% (38) villages have a population greater than 5000.

The urban population of BMR, has increased from 68% in 1991 to 73% in 2001 and to 80% in 2011. The decadal growth rate from 2001 to 2011 was 50.6%. The urban population is distributed across 29 towns, out of which 5 have a population below 10,000 and 10 have a population between 10,000 and 20,000. There are 8 towns that have a population between 20,000–50,000 and 6 have population of more than 50,000. According to Census 2011, the 29 towns are classified into five classes of towns, namely—Class I (more than 100,000 population), Class II (Population from 50,000 to 99,999), Class III (population from 20,000 to 49,999),

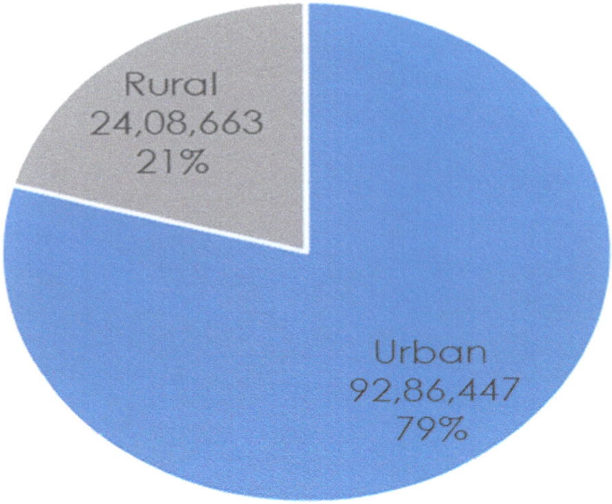

Fig. 6.4 Rural urban population share within BMR—2011. *Source* Census of India [41]

Class IV (population from 10,000 to 19,999) and Class V (population from 5000 to 9999). There is only one class I town, that of BBMP Municipal Corporation (with a population of 8.4 million), 5 class II towns, 8 class III towns, 10 class IV towns and 5 class V towns. Population distribution of various classes of towns is shown in Fig. 6.5 and decadal growth rate (2001–2011) of various classes of town is shown in Table 6.2.

Both the BMR and BBMP, over the past decades have shown a steady growth in terms of population size. As per Census 2001, the BMR and BBMP had 8.42 million and 6.17 million population respectively. As per 2011 census, the BMR and BBMP jumped to 11.6 million and 8.4 million population respectively. However, in terms of rate of growth of population, BBMP shows a higher rate over BMR as BBMP constitutes the majority of the BMA. During 1991–2001, the BBMP had grown at a rate of 43.5% and BMR at a rate of 29.3%. During 2001–2011, the BBMP has grown at a rate of 92% and BMR at a rate of 50.6%. The density profile of BMR shows a dense density in the core with BBMP having a density of approximately 12,006 persons/km^2. However, the average urban population density of BMR is 10,503 persons/km^2.

6.3.3 *Population Growth in Bangalore City*

Bangalore city population grew at a high rate, and the population of Bangalore increased from 5.7 million to 8.4 million during the decade 2001–2011 (Fig. 6.6). A significant rise in the growth rate (4.95%) had taken place during 2001–2011 and it is presumed that the same trend may possibly continue for the coming decades as well.

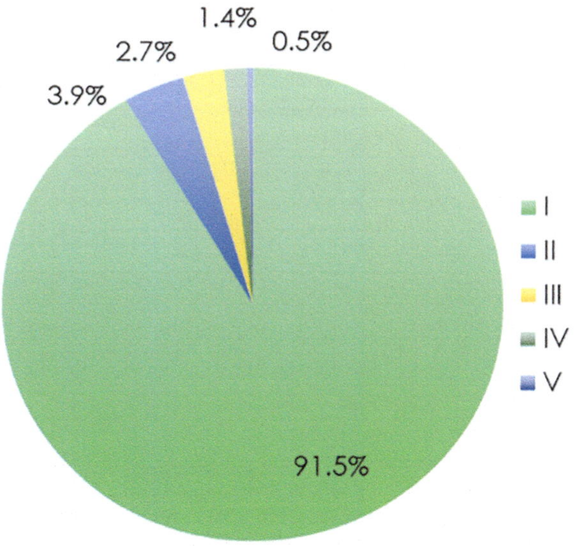

Fig. 6.5 Population distribution of various classes of towns within BMR—2011. *Source* Census of India [41]

Table 6.2 Decadal growth rate (2001–2011) of various classes of towns within BMR

Town classification	Decadal growth rate in percentage (2001–2011)
Class-I	96.3
Class-II	21.7
Class-III	45.6
Class-IV	91.3
Class-V	54.1

Source Census of India 2001 [42] and 2011 [2, 41]

6.4 Methodology to Analyze Spatial and Economic Transformation

Smart region can be defined as 'a cluster of small urban areas which work together in order to frame a diffused smart city, sharing visions, ideas, common goals and resources'. Transition from 'the urban level of smart city to the territorial level of the smart region' is the problem that needs to be addressed. The important issue in transitioning from smart city to smart city or from a cluster of smart cities and villages to smart region is not the technology but the capability to collaborate. It is about transcending administrative boundaries in a technological sense and in other ways too. Experimentation to this effect has been carried out in Italy. In the case of Bangalore, there is a strong need to think beyond the city of Bangalore.

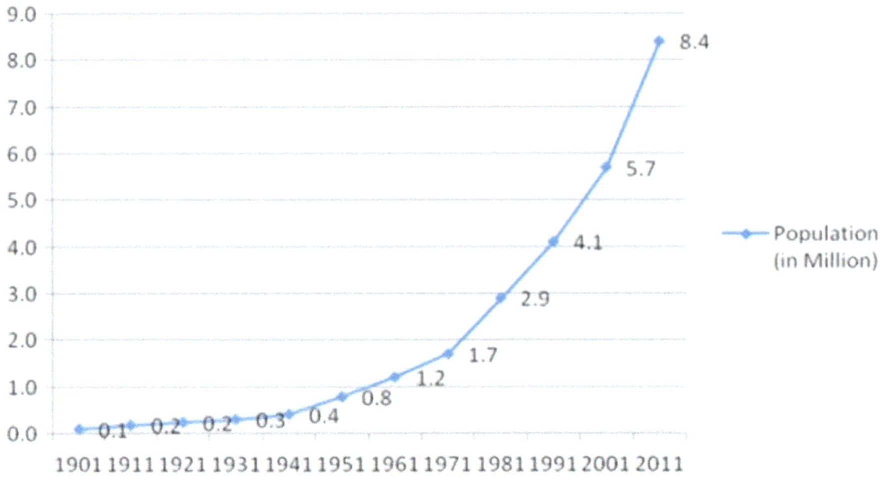

Fig. 6.6 Population growth of Bangalore City (1901–2011). *Source* Census of India (1901–2011)

The businesses have to transform into digitally oriented enterprises. Bangalore is already the technology capital of India, driven largely by IT. While the IT industry in India is likely to emerge as much stronger from the current downturn riding on the back of technology transformation, it is the need of the hour to support them, and also other industries, by providing facilities for digital innovation and incubation. The role of industry bodies will be of prime importance here, but again, the need is to connect them with the smart network of local self governments, so that a synergy is achieved in terms of spatial decision-support and parameter-based prioritization without requiring complex interfacing among industry and multiple layers of government agencies. To achieve this, each of the regional cities around BMC has to be developed as smart cities first. Recently Bangalore city has broken into the smart-city club. But it needs to be emphasized that merely promoting Bangalore City as a smart city will be a self-defeating exercise as the irreversible process and the tremendous rate of urbanization means that the region itself is transforming beyond limits of imagination and has to be integrated in the context of a 'smart region'. After all, a smart Bangalore city which gets promoted in isolation will only distort the market further and lead to lopsided urbanization, costing society dear in terms of crowding, quality of public spaces, travel times, etc. If we look at the underlying concepts of smart city, we see themes like smart energy, smart health, smart living, smart mobility, and smart economy. However, if we have to graduate to smart region paradigm we need to look beyond making cities smart. Thus transformation of BMR into a smart region would mean more than location specific actions within the area. While smart energy interventions could be a thrust in the BMC area, piloting of such interventions could be easier in the smaller City Municipal Councils or Town Municipal Councils in the region. In fact, the population explosion happening in these regional centres outside BMC demands that they should also be considered for smart energy solutions. Smart health also needs to be scaled up to regional level through linkage between health institutions spread across the region, thus benefiting the rural settlements also. The first steps toward smart living at a regional scale would involve promotion of a dispersed settlement strategy dovetailed with transit oriented development and backed by incentives for affordable housing development. Smart mobility would entail smart demand-responsive integrated regional connectivity catering to diverse commuter segments. Smart mobility strategies, with fast and high-capacity regional networks supported by efficient intermodal options, would complement smart living strategies as discussed. According to BMR Revised draft Structure plan, 2031, it is projected that, if the population growth patterns continue to remain the same, by 2031, 80% of the population will be residing in the core of BMR and just 20% at the periphery [6].

6.5 Spatial Embeddedness

The BMR revised structure plan (draft)-2031, had divided the entire BMR (excluding the planning area of Bangalore Development Authority) into five Area Planning Zones (APZs) and six Interstitial Zones (IZs). While APZs were identified as the development zones, the IZs were classified as conservation/preservation zone. As per the Draft Regional Plan 2016–2036, built-up area increased by 197 km^2 (2.46%) whereas area under water bodies reduced by 53 km^2 and wastelands reduced by 91 km^2 in the eight year time period (See Table 6.3) and land utilization pattern of BMR in 2008 is shown in Fig. 6.7.

BMR has approximately 11,000 km of road length for an area of 8005 km^2. BMR has two Expressways, Five National Highways and Major District roads connecting major cities and towns within BMR and beyond:

Table 6.3 Changing land utilization pattern (in %) of BMR (2000–2008)

Sl. No	Land utilization categories	2008	2000
1	Built up	11.49	9.03
2	Agricultural land	66.06	66.49
3	Forest	10.38	10.36
4	Grass land, grazing land	0.07	0.27
5	Waste land	7.48	8.61
6	Water bodies and wetland	3.58	4.25
7	Others	0.94	0.99
	Total	100	100

Source BMRDA [7]

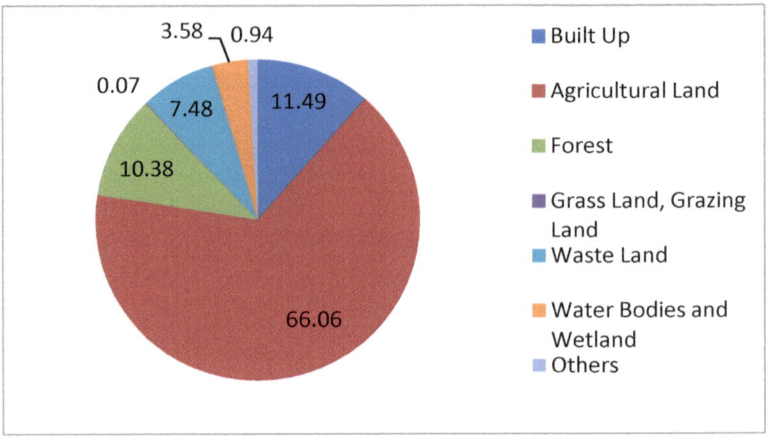

Fig. 6.7 Land utilization of BMR (in %)—2008. *Source* BMRDA [7]

Expressways:

- Nandi Infrastructure Corridor Enterprises road, commonly known as NICE Road and officially called Bengaluru–Mysuru Infrastructure Corridor (BMIC), is a proposed 4–6 lane private tolled expressway that intends to connect the two important cities of Bengaluru and Mysuru.
- Bangalore Elevated Tollway or Hosur Road Elevated Expressway is a 9.985 km long elevated highway from Madiwala to Electronics City.

National Highways (NH) passing through BMR.

- NH-7 connecting Varanasi to Kanyakumari passes through BMR and links Hyderabad in the North and Mysore to the South.
- NH-4 part of Golden Quadrilateral links Mumbai and Chennai with BMR.
- NH-48 links Mangalore with NH-4 through Hassan.
- NH-209 connects Kanakpura to Kerala and
- NH-207 connects NH-4 and NH-7 through SH-74 and 12 state highways (SH).

Figure 6.7 shows the influence zones of bus depots/stations. Following are the important facts related to Bangalore Metropolitan Transport Corporation (BMTC) service in Bangalore Metropolitan Region [6].

- No. of Buses: 6316
- No. of Schedules: 6095
- Service Km: 12.11 Lakhs
- Avg. Vehicle utilization: 224.6 km per vehicle per day
- No. of Bus trips: 73,726
- Daily Passengers: 5.02 million
- No. of Bus Depots: 43
- No. of Bus Stations: 53

Further, the existing rail network of BMR runs through three districts, namely Bangalore, Bangalore Rural and Ramanagara connecting to Kolar on North-East, Mysore on South-West, and Chennai on South-East and Mangalore on North-West. It enjoys a ridership of 150,000 with a projected ridership of 2.2 million in 2031 [8]. Figure 6.8 shows the passenger travel pattern in BMR with highest travel of 76% taking place from Baiyyappanahalli to Whitefield, followed by 75% from Yesawanthpur to Tumkur, 64% from Tumkur to City Railway Station and 60% from Hebbal to Hosur. This gives an insight to intra-sectoral movement where location of terminals do not influence as short journeys would be higher in comparison to long outside sector journeys and inter-sectoral movement towards Bangalore sector is higher showing terminal preference sector in that sector. Termination of traffic at suburban terminals will force transshipment.

An attempt has been made to calculate functionality Index of rural BMR. Functionality is defined by the number of facilities available in a village by the total facilities available in the taluka. Functionality index tells us how functional a location is or in other words, where are the major functions located. Considering the 8 facilities in all the 2339 villages, functionality index was calculated for each village. The villages with maximum index or more facilities from each taluka were then looked upon. The 8 facilities are as follows:

1. Primary School
2. Primary Health Centre
3. Secondary School
4. Community Health Centre
5. Dispensary
6. Family Welfare Centre
7. Primary Health Sub-Centre
8. Maternity and child welfare centre

The top five villages from each taluk is represented in Fig. 6.9 and the overall top five villages of BMR with maximum facilities is listed in the Table 6.4.

Through this study, we can infer that the top five villages with maximum functionalities are located in the periphery of Bangalore Urban district. The other top five villages from each taluk—with maximum facilities are located in Bangalore Rural and Ramanagra Districts. Lesser number of villages are seen in Bangalore Urban district.

Further, land suitability analysis has been performed to identify the best suited land for an identified land use. Future developable land of BMR in 2016 is shown in Fig. 6.10.

6.6 Economy of Bangalore and BMR

The city of Bangalore or 'Bengalooru', as has been recently rechristened, is a city to reckon with in India's urban trajectory. The name of the city is found to be first recorded in a ninth century Ganga inscription (hero-stone) from a place called,

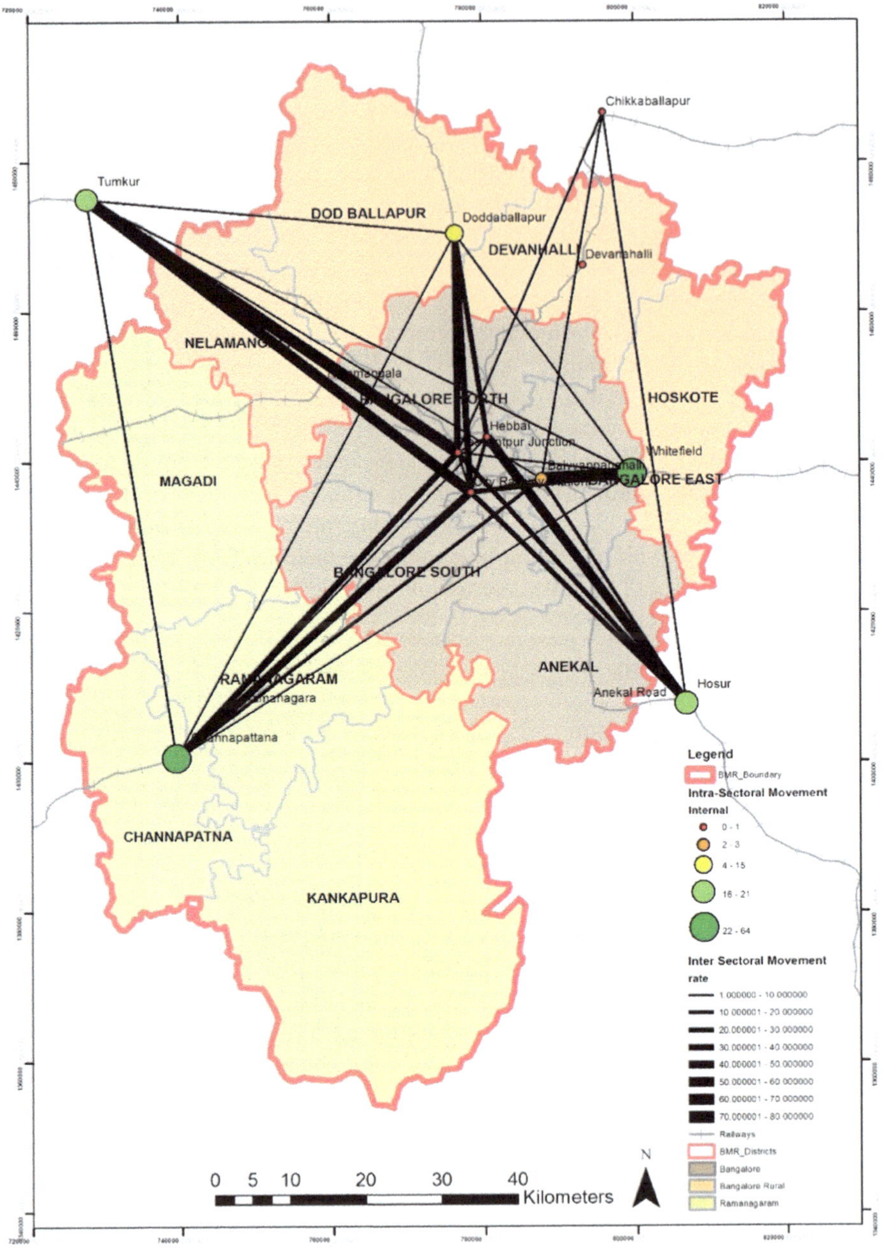

Fig. 6.8 Travel volume and pattern by commuter rail within BMR

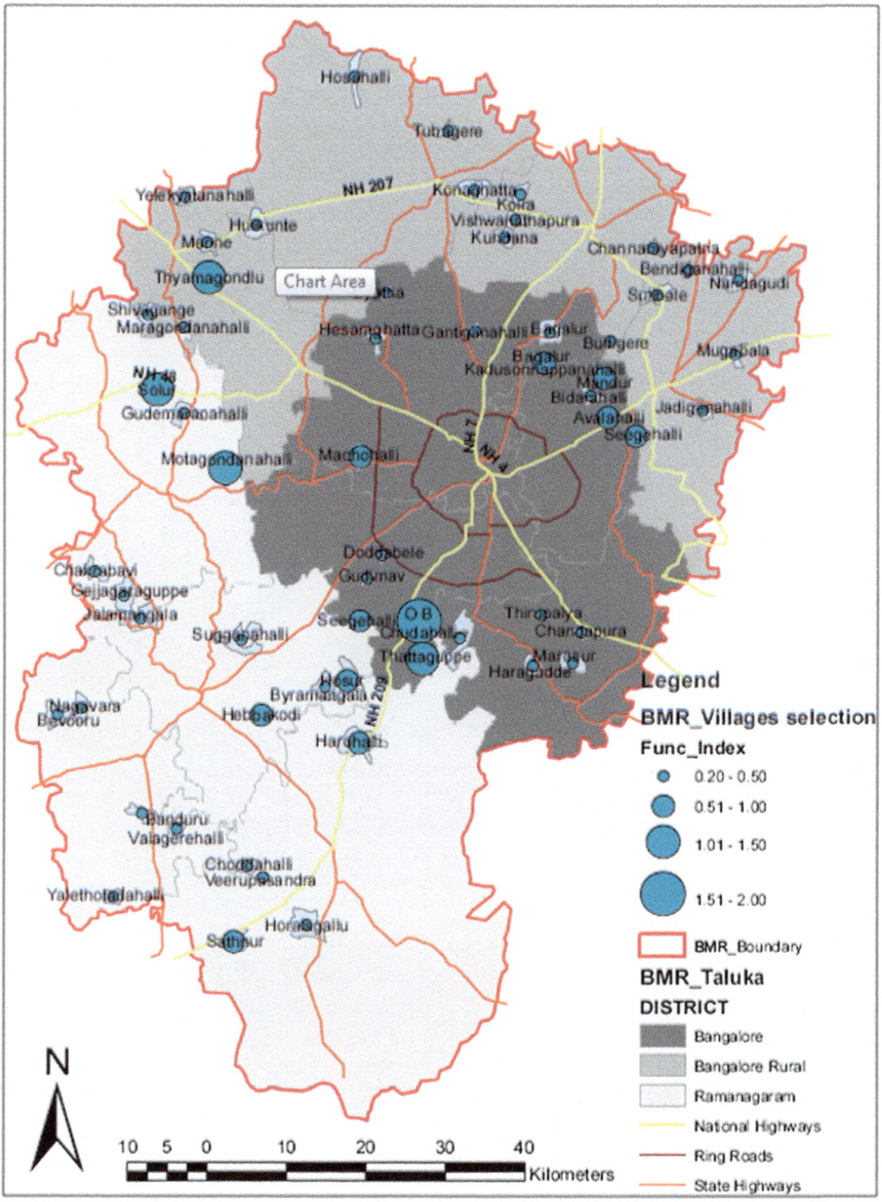

Fig. 6.9 Functionality index of rural BMR

Begur, where a battle was fought. Bangalore is also said to have got its name from 'Benga' and 'ooru', meaning 'pterocarpus marsupium' (a species of dry and moist deciduous tree) and 'town' respectively in Kannanda language. However, the founding of modern Bangalore is attributed to Kempe Gowda, a scion of the

Table 6.4 BMR villages with maximum functionality index value

Sub-district	Village	Functionality	Population[a] (2011)
Bangalore South	O.B.Chudahalli	1.68	1255
Magadi	Solur	1.24	3919
Nelamangala	Thyamagondlu	1.20	8160
Magadi	Motagondanahalli	1.04	2330
Bangalore South	Thattaguppe	1.02	2503

[a]*Source* Census of India [2, 41]

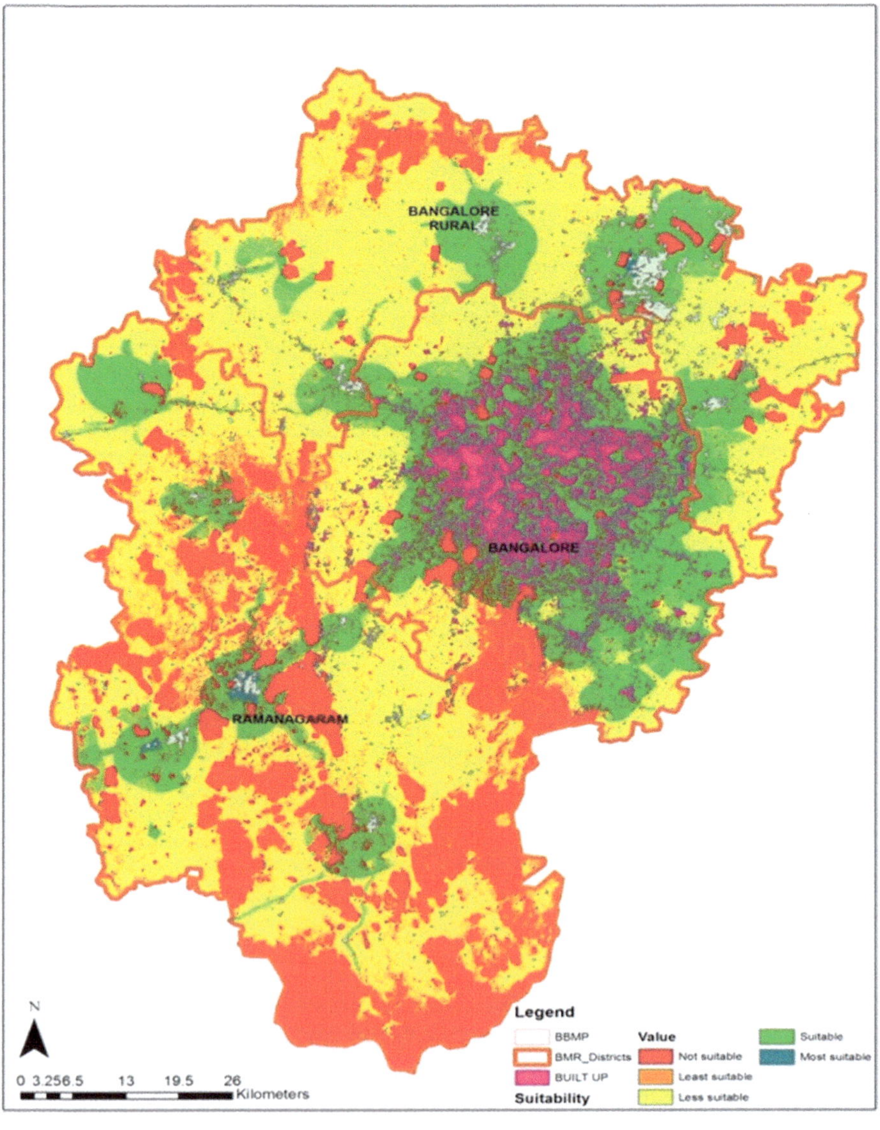

Fig. 6.10 Land suitability analysis for BMR

Yelahanka line of chiefs, in 1537 [8]. As one looks into the historical foundation of the city, the fort near the present City Market or Krishna Rajendra Market and the barracks near the Halsoor tank were the main foundations of the city built in 1537. Later on, the city was administered by the Wodeyars, rulers of Mysore, until it was given as Jagir (with rights for general administration and collection of taxes) to Hyder Ali during late 18th century. Hyder Ali and later, his son, Tippu Sultan, were responsible for growth and development of Bangalore in a significant way with the construction of summer palace and Lalbagh. Indeed, Bangalore was already the commercial capital during Tippu's time and the second important city after Srirangapatna, Tippu's capital. During the early 19th century, the city was known to have almost all coins in circulation from different places and kingdoms, thus evidencing a flourishing trade and commerce [9].

This historical foundation was gradually developed with unique specificity. The Fort neighbourhood was developed on the basis of the philosophy of a city by temple builders, agriculturist turned warriors who patronised wholesale and retail traders, highly skilled artisans etc. Hence, the Fort neighbourhood was developed as a typical native town with traditional characteristics of bazaar, temples and residential neighbourhoods etc. The barracks neighbourhood, on the other hand, was developed to cater to the needs of British troops and officers who were relocated from Srirangapatnam near Mysore. Bangalore was developed with dual characteristics since its foundation as a result of which the city has been attracting various categories of people from different regions of the country. While the attraction to the city was previously based on its importance as a specialized centre for trade, commerce and industry, it has now been shifted as the city has become the global destination for Information Technology (IT) and Bio-technology (BT) [10].

The state of Karnataka was carved out in 1956 based on linguistic boundaries, with regions dominated by Kannada speaking people. Bangalore was retained as the capital of the state, with Kannada as the official language while accommodating other languages like Tamil, Telugu, Malayalam, English and Hindi, true to its cosmopolitan status. Population growth during the 1970s could be ascribed to numerous public sector industries and other defence establishments that came up during the period and fuelled significant immigration. By this time, incidentally, Bangalore had lost its tag of 'Pensioners Paradise', gained before Independence. Although the advent of IT is attributed to the late 1980s, major growth and expansion of this industry happened only during the late 1990s [11]. Existing industrial estates of BMR is shown in Fig. 6.11.

Long before Bangalore was called the Silicon Valley of India, the city made its name as headquarters to some of the largest national heavy industries of India. The Hindustan Aeronautics Limited (HAL) headquarters was set up in Bangalore. Although HAL was set up to promote R&D activities for indigenous fighter aircraft for the Indian Air Force, it now maintains an impressive fleet of fighter aircraft for the Indian Air Force including Sukhoi 30, Flankers and Jaguars. Air shows showcasing inventories from HAL and international corporations such as Sukhoi, Lockheed Martin, Mirage and BAE Systems are held at the Yelahanka Airforce base near Bangalore once every two years. The National Aerospace Laboratories

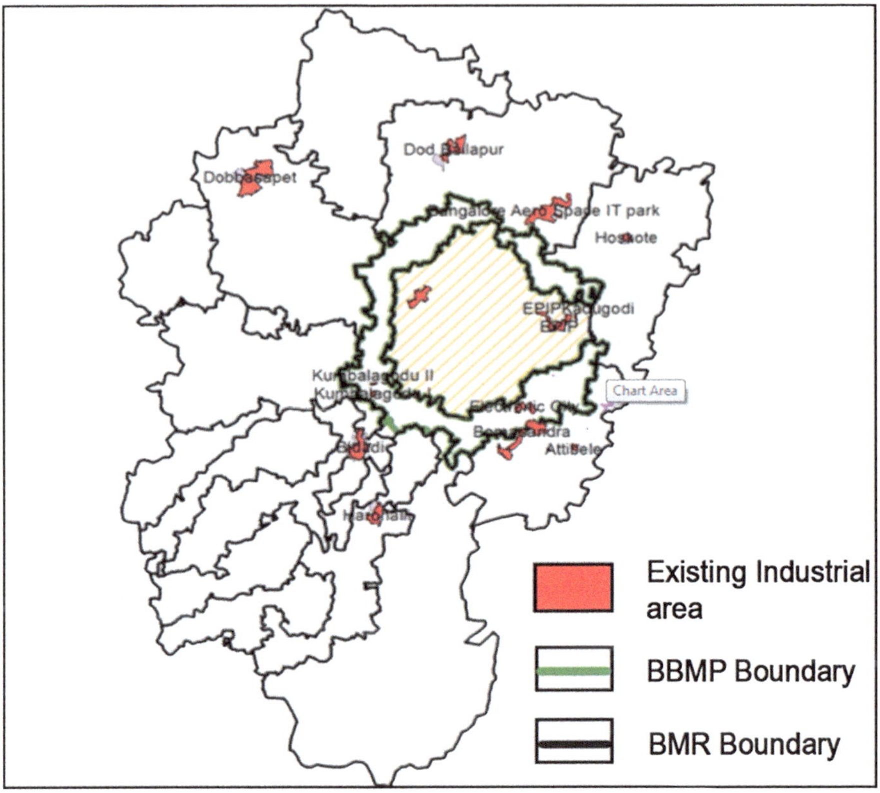

Fig. 6.11 Existing industrial estates of BMR

(NAL) is also headquartered in Bangalore and is dedicated to the development of aerospace technologies. NAL has a staff strength of over 1300 employees and often works in conjunction with HAL [12].

In June 1972, the Government of India set up the Space Commission and Department of Space (DOS). India's premier space research organization, the ISRO was created under the DOS and headquartered in Bangalore. The main objective of ISRO includes development of satellites and launch vehicles. Aryabhatta, India's first satellite, was developed and successfully launched by ISRO. Since then, the organization has successfully launched numerous other satellites such as Bhaskara, Rohini, APPLE and the INSAT series, and successfully deployed PSLVs and GSLVs. ISRO also heads India's ambitious moon program. Bangalore is also a major manufacturing base and houses such public sector manufacturing giants as Bharat Heavy Electronics Limited (BHEL), Bharat Earth Movers Limited (BEML), Hindustan Machine Tools Ltd. (HMT), Indian Telephone Industries Ltd. (ITI), and Bharath Electronics Limited (BEL) etc. An important feature of the economic activities of Bangalore is the huge concentration of Small and Medium Enterprises

(SMEs) in diversified sectors across the city. Bangalore has more than 20 industrial estates/areas comprising large, medium and small enterprises. Of these, Peenya Industrial Estate located in the northern part of the city houses about 4000 SMEs and is considered to be the largest industrial estate in South and South East Asia [13]. A majority of the SMEs function as ancillaries/subcontractors to large enterprises in the field of engineering and electronics industries, among others.

One of the important factors spurring Bangalore's growth was that the Central Government invested heavily in public sector industries in Bangalore, partially due to the fact that it is geographically disconnected from India's rivals, Pakistan and China. This led to the concentration of technical and scientific manpower in Bangalore which facilitated the 'IT revolution' in the city.

Bangalore is home to numerous high-tech knowledge hubs evident from the establishment of premier centres like Indian Institute of Science (IISc), National Institute of Advanced Studies (NIAS), Tata Institute for Fundamental Research (TIFR), Jawaharlal Nehru Centre for Advanced Scientific Research (JNCASR), Indian Space Research Organisation (ISRO), National Aerospace Laboratories (NAL), Defence Research and Development Organisation (DRDO), Indian Institute of Management (IIM), Institute for Social and Economic Change (ISEC), Indian Institute of Information Technology (IIIT) and several professional engineering and medical colleges at undergraduate and graduate levels. Existence of such high-tech education and knowledge hubs and several professional engineering and medical colleges at undergraduate and graduate levels in Bangalore are itself offering an additional thrust to regional economy [6].

The best multi-speciality hospitals in Bangalore enable the city to become popular destination in medical tourism which provides advanced treatment to international patients at an affordable rate. There are also affordable provisions for stay-back facilities for family members who are accompanying the patients. Most of the top hospitals are located within the Bangalore city limits like; Fortis Hospital, Bannerghatta Road, Manipal Hospital, HAL Airport Road, Columbia Asia Referral Hospital, Yeshwanthpur, Aster CMI, Hebbal, Apollo Hospital, Jayanagar, Fortis Hospital, Cunningham Road, Columbia Asia, Whitefield and Columbia Asia, Hebbal, Narayan Multi Speciality Hospital, HSR Layout, Manipal Hospital, Whitefield, Hosmat Hospital, Magrath Road, which are offering the core specializations like Cardiology, Neurology, Dentistry, Orthopaedics, Nephrology, Neurosurgery, Oncology, Gynaecology and Obstetrics, Infertility, Paediatrics and Neonatology and many more.

Another reason why Bangalore has become a major health care center is because most of the hospitals in Bangalore also provide their patients with post-treatment care and follow-up facilities. The treatment ranges from the ordinary treatment to the critical ones. All around the world people prefer to visit Bangalore for its expertise in liver transplantation. The city hospitals also have world class infrastructures in cancer treatment, knee replacement, heart bypass surgery etc. All these make the city a blooming medical tourism destination because of the presence of proficient medical experts, the adaption of updated technologies and an overall health care system that is at par with the best clinical services in the world. Most of the hospitals in Bangalore have the required accreditation from national and international associations to ensure the best standards of medical treatments [14].

Bangalore is called the 'Silicon Valley of India' due to the large number of information technology companies having been located there. Many multinational corporations, especially computer hardware and software giants, have operations in Bangalore. As per India SEZ, Bangalore has fourteen operational IT/ITES based SEZs which are majorly clustered around White field and Electronic City area. There are two SEZ at Ramnagara on Research and Development and textile respectively. Total area under all the operational SEZs are around 745 ha which engages approximately 287,445 employees in 2014–15. The total investment is estimated as INR 107.40 billion. In addition to the operational ones, seven more SEZs have been proposed in Bangalore.

Electronics City, located in the southern outskirts of Bangalore, is an industrial park spread over 330 acres (1.3 km^2). Whitefield, located in the north-eastern outskirts of the city is another technology hot spot. The government has plans to develop an Information technology corridor linking Whitefield and Electronics City. Over 200 Information Technology corporations have facilities in Bangalore. At the peak of the dot-com boom in the late 1990s, Koramangala—a suburb of Bangalore, was believed to have had the highest density of telecom software companies per square mile in the world. Infosys and Wipro, India's 2nd and 3rd largest software companies are headquartered here and are now billion dollar companies, looking to reach 2 billion in 2005. The IT industry of Bangalore itself accounts for about 30% of total IT work-force of the country, which subsequently becomes the reason of Bangalore having higher personal disposable income than the other Indian cities. This has also resulted in a trickling down effect within the urban economy. Further, investments in other industries (non-IT), infrastructure and other services, have significantly increased the purchasing power among the people and have nurtured real estate with consequent land market dynamics, apart from creating numerous secondary employments in services [10].

Bangalore's information technology (IT) industry is divided into three main 'clusters'—Software Technology Parks of India, Bangalore (STPI); International Technology Park Ltd. (ITPL); and Electronics City. Infosys and Wipro, India's 2nd and 3rd largest software companies are headquartered in Electronics City. Today, Bangalore is home to 66 Fortune 500 companies, 682 multi-national corporations (MNCs), 1685 IT/ITES and 131 Biotech companies. Bangalore accounts for half of the approximately 260 biotechnology companies in India. Biocon, headquartered in Bangalore, is the nation's leading biotechnology company and ranks 16th in the world in revenues. Jack F. Welch Technology Research Center in Whitefield, Bangalore, is the second largest research facility of General Electric (GE) in the world. Founded in 2000, The Research Center employs around 1600 people, including more than 1000 doctorates [15].

Biotechnology is a growing field in the city. Bangalore houses at least 97 of the approximately 240 biotechnology companies in India. Interest in Bangalore as a base for biotechnology companies stems from Karnataka's comprehensive biotechnology policy, described by the Karnataka Vision Group on Biotechnology.

In 2003–2004, Karnataka attracted the maximum venture capital funding for biotechnology in the country—$8 million. Biocon, headquartered in Bangalore, is the nation's leading biotechnology company and ranks 16th in the world in revenues. Institute of Bioinformatics and Applied Biotechnology (IBAB) initiated by Biotechnology Vision group, Biocon, and located in ITPL is nurturing scientists in this field. Like the software industry which initially drew most of its talent from the local public sector engineering industries, the biotechnology industry had access to talent from the National Center of Biological Sciences (NCBS) and the Indian Institute of Science (IISc).

Bangalore has been spearheading the growth of Indian industry, particularly in terms of high-technology industries in the areas of electrical and electronics, information & communication technology (ICT), biotechnology and, more recently, nanotechnology. The industrial structure presents a blend of modern high-tech capital goods and knowledge intensive industries on the one hand and traditional consumer goods industries on the other. The modern economic history of Bangalore city has registered a shift from it being a centre of textile hub to the current global software technology hub. The intervening period between 1900 and the present are divided in four distinct economic phases [16].

(i) The Phase of Textile Production 1900–1947 (Pre-colonial to Colonial Phase): Initially, the city was dependent on a trade of agro-produce. After, The Great Plague epidemic, by 1900 Bangalore worked on expansion of the health and sanitation facility and on her modernization—the only city commissioned with telephone services and electricity by 1905. The textile production started in the late nineteenth century, continued producing cotton, silk and wool products till the end of the 1940s. The period also formed the foundation for the role of the City as a centre for high technology research and development.

(ii) Phase of Public Sector Production 1947–1960 (National Phase): In view of the already established tradition for Bangalore as a centre of textile and electronics-electrical related production, the city was chosen for the establishment of public sector institutions for research and production. In the first phase after independence, the large public sector giants such as Indian Telephones Industries (1950), Hindustan Machine Tools (1953), Bharat Electronics Ltd. (1954) were set up. Large dedicated townships were set up on the periphery of the city, which encompassed the industry as well as housing and services for the employees. While the Indian Institute of Science (1909), Bangalore Stock Exchange (1957), Defense Research and Development Organization (1958), the Central Power Research Institute (1960) and Indian Space Research Organization (1972) were among the other pioneer establishments in the city.

(iii) Phase of Bureaucratic Growth (1960–1980): Bangalore became the capital of the unified princely state of Greater Mysore in 1956, later renamed as state of Karnataka. This period saw the establishment of offices and institutions that were responsible for running the government and organizing the State. The major transformations during this period were development of institutions and civic agencies at the public level and the expansion of the business services at the market and private level. The growth of the research and educational institutions also continued along with an increase in channeling of public funds for defense research in the institutions already established.

(iv) Phase of Software Production (1980–present): Until recently, innovation oriented high technology production was strictly under the control of the Central Government or other public agencies. The current regime of an export oriented Information Technology sector was initiated in the early 1980s. Texas Instruments was the first MNC to establish in Bangalore in 1984, but there already existed some local firms which were trying to make a thrust into Software production. Post 1984, there was an influx of MNC's and Domestic firms in Bangalore to capitalize on cheap skilled labour which was easily available. Till the 1990s the communication infrastructure of the country was very poor and had no direct satellite link up. However, after 1991 with the onset of the process of liberalization of the economy of India and the setting up of the Software Technology Parks (STP) Scheme, an earth station was established in Bangalore to provide efficient communication and the controls to entry of foreign firms made easier. A large number of MNCs established their back-end operations in India where labour intensive work can be done at cheaper costs. In the last two decades the standing of the software programmers from India has been firmly established in the global market.

As per Bangalore Metropolitan Region (BMR) Revised Structure Plan, 2031, BMR contributes to 36.8% in State's Domestic Product and Bangalore Urban has the major share of domestic product which is especially from tertiary sector. Workforce classification is shown in Fig. 6.12.

The gross district domestic product (GDDP) of BMR at current price for 2013–14 is estimated to be INR. 2943.38 billion while the net district domestic product (NDDP) is estimated to be INR. 2691.98 billion. In case of Bangalore Urban, the sectoral contribution to GDDP is dominated by the tertiary sector which is 76.7%. Primary activities show very less contribution which is mainly from Ramanagaram District and Secondary activities are majorly spread across Bangalore Rural and Ramnagra districts. Primary sector has only 2.2% GDDP contribution with 15.2% employment generation in BMR as evident from the declining trend from 2004–05 to 2013–14. Location Quotient in primary sector is 0.18, which signifies the less concentration of primary activities within BMR (see Fig. 6.13).

The workforce participation rate (WFPR) for the BMR is observed to be 44.7% with 52, 38,277 workers which is higher than that of the district (28.7%). While the rural WFPR is observed to be 22.7% with a workforce of 11, 87,102, the Urban WFPR is 77.3% indicating that a large volume of urban population contributes to economic activity as compared to the rural areas. According to Census of 2011, the growth in the work force population has been 185.8% from 1991 to 2011 against the total population growth at 79.5% and a decline in marginal workers by

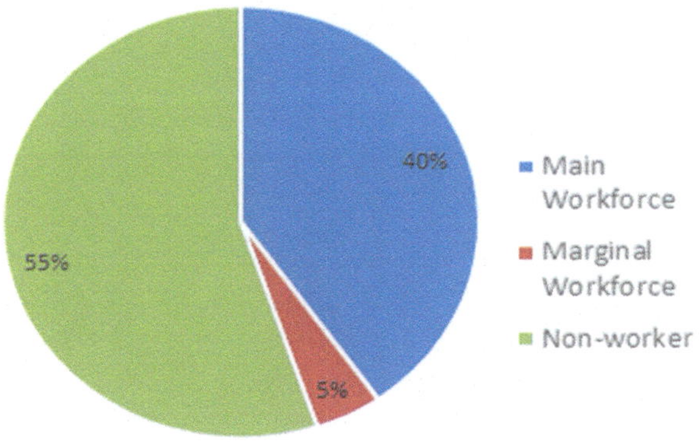

Fig. 6.12 Work force classification of BMR—2011. *Source* Census of India [2, 41]

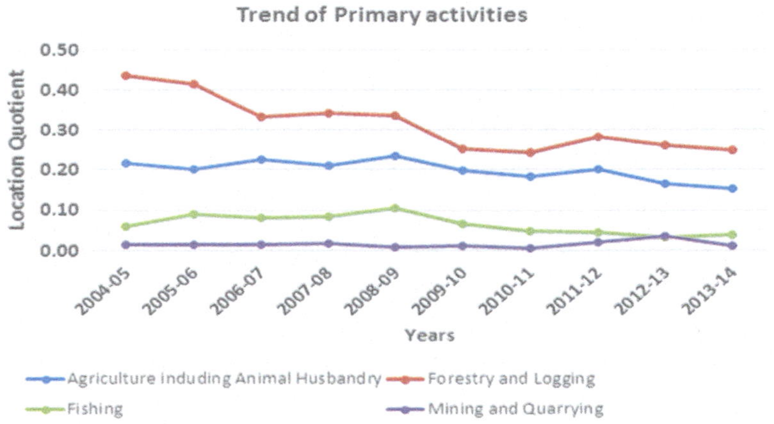

Fig. 6.13 Location quotient in primary sector (2004–2014) in BMR *Source* BMRDA [7]

−0.5% from 1991 to 2011 [7]. District wise workforce distribution is shown in Fig. 6.14.

In case of secondary sector, readymade textiles, chemicals and engineering industries are leading in the region, whereas automobile, leather, electrical, wood, paper printing, rubber and plastic are other major contributors. At the same time, fast moving consumer goods (FMCG) and first order agro processing industries are also growing with time, which ensure the growth prospect for primary sector produce of this region.

Bangalore city is a key growth centre in the whole BMR with 81% of the working population residing in the city. The diversification of economic activity is quite extensive in Bangalore and a majority of the population is engaged in non-agricultural economic activities. The city development plan for Bangalore, 2006 reported that less than 1% of the workforce was engaged in agriculture. The

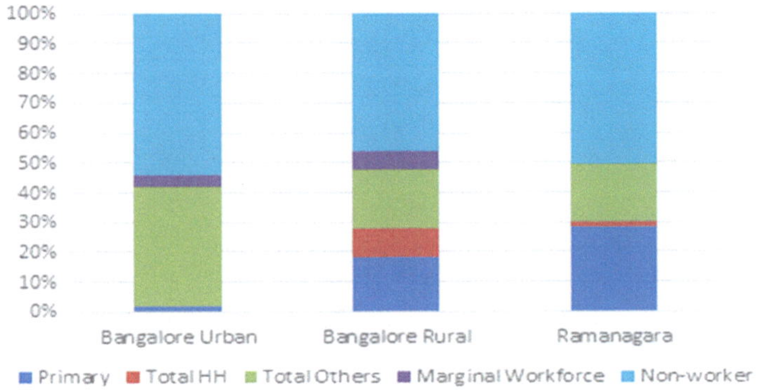

Fig. 6.14 District wise workforce distribution within BMR—2011. *Source* Census of India [41]

biggest employment generators in the city are the manufacturing accounts for 43.36% and Service sector accounts for 31.51% respectively. Approximately 11% of the total workforce is also engaged in Banking and finance related activities. Transportation related activities engage approximately 7.2% of the total workforce. Thereafter, tertiary sector appears to be the most promising sector for generation of employment in coming decade as it is expected to create work opportunities for huge population in formal as well as informal service activities. Besides IT and ITES services, financial services, real estate, communication, storage and trade and hospitality industry etc. are shaping the city's economic landscape (See Fig. 6.15).

In tune with the recent trends, Bangalore now has numerous malls and multiplexes that are swarmed during weekends. With an active night life and fondness for fast-food, a large number of restaurants, pubs and 'eat-outs' throng the city.

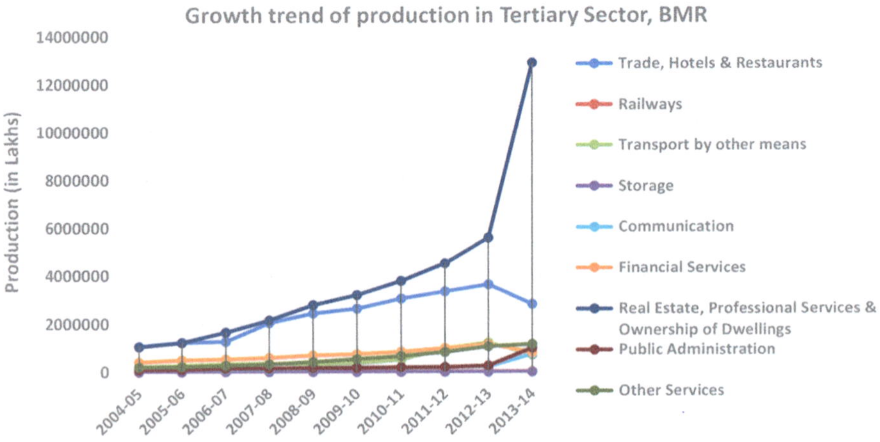

Fig. 6.15 Growth trend of production in tertiary sector within BMR (2004–2014) *Source* BMRDA [7]

Thus, the thriving economy of the city has resulted in a per capita income of INR 39,420, a little more than twice the State's average per capita income of INR 18,360 [16]. Based on the economy based spatial analysis, potential economic activities with proposal of specific industries are proposed at different sub-regional growth centers (Fig. 6.16).

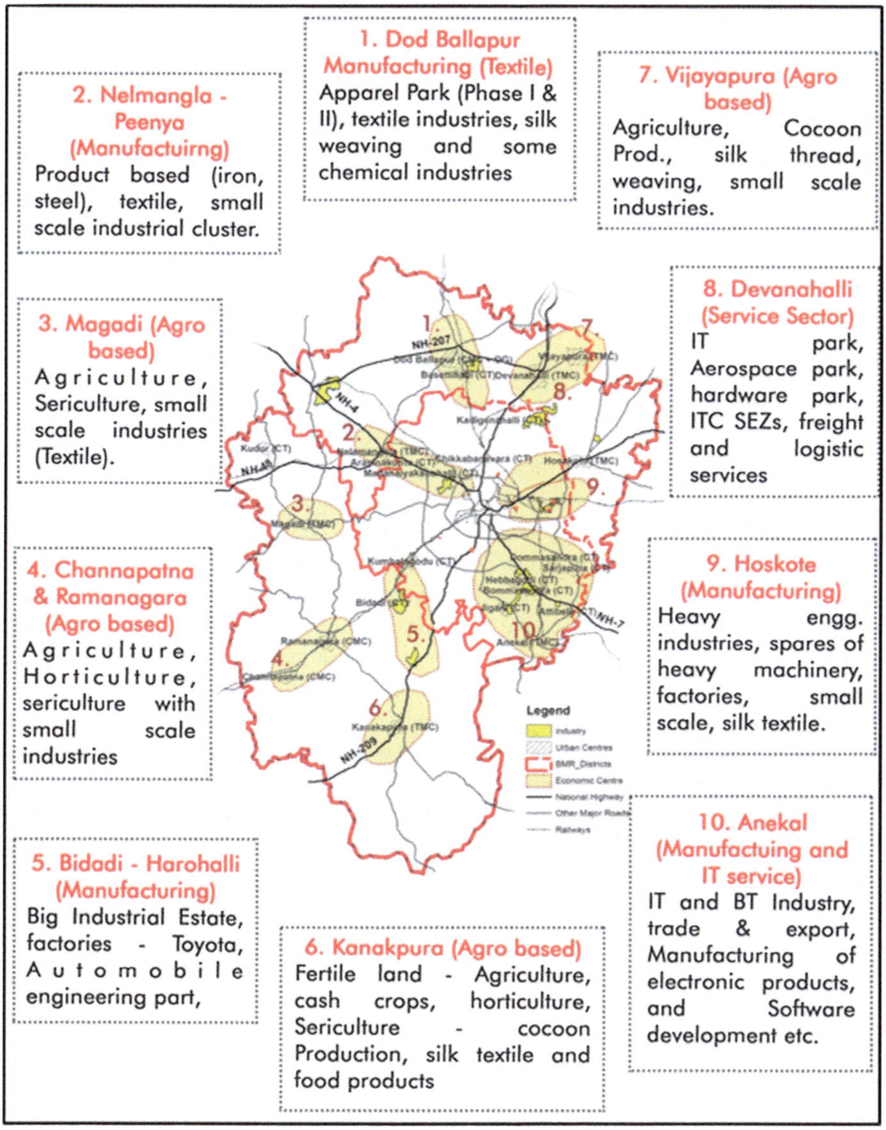

Fig. 6.16 Potential economic activities at different sub-regions of BMR

From last decade, economic stagnation is seen in the region, which is expected to withhold the remarkable and dynamic growth story of Bangalore. So, to maintain the momentum of growth and to further explore the potential of this region, it is essentially required to rework on the current weaknesses and start afresh with a new approach of smart economy to achieve sustainable development. Now to revive the identity of BMR, it is essentially required to boost regional economy by addressing the basic regional governance issues and weak links of local urban transformation with the help of smart innovations to develop the global competitive capacities. One of the recent popular approaches for the same is known as smart specialization. Smart specialisation is an industrial and innovation framework for regional economies that aims to illustrate how public policies, framework conditions, and especially R&D and innovation investment policies can influence economic, scientific and technological specialisation of a region and consequently its productivity, competitiveness and economic growth path [17].

6.7 Existing Challenges

Bangalore is transitioning from a formerly mononucleated growth pattern to a polycentric pattern, with the fastest growth taking place around multiple peripheral areas [18]. This pattern of growth, with the city core becoming increasingly saturated, and new urbanization centers developing at the periphery, can be traced to multiple factors. The high land prices and lack of large spaces in the city center have led to the location of many public sectors, local and multinational companies, and prominent educational institutions at the city periphery [19]. As a result, in periphery areas of Bengaluru, loss of agricultural land, water body, and green spaces are faster and it is imperative that urban planning efforts concentrate to city periphery where unplanned growth is taking place at an extremely rapid rate. The partial implementation of decentralization agenda has further added to the current problem resulting in functional, political and economic gaps [7].

Even though previous regional plans have focused towards maintaining the primacy of Bengaluru over the region, yet the externalities of a primate city cannot be ignored (Fig. 6.18). Primacy in any region always hints the existence of regional disparity and the primate city always face problems of high immigration, high demand for urban infrastructure and deteriorating quality of life. All these problems are much evident in Bangalore. A study conducted by the Indian Institute of Science (IISc) presents an alarming picture. It shows 525% growth in the built-up area in the last four decades, 78% decline in vegetation 79% were decline in water bodies. These are not just figures, the lakes and trees that dotted across the city have quietly disappeared as a result of the urban sprawl.

Further, according to Urban and Regional Development Plan Formulation Guidelines (URDPFI) 2015, the minimum per capita open space at 10–12 m^2 [20], the per capita open space in Bangalore city was found to be just 2.01 m^2 [21]. This may be attributed to the high concentration of people in Bangalore and subsequent uncontrolled real estate development. Increasing urbanization has an adverse effect on the Bangalore city. The most alarming impact has been on primary health. According to Ravichandar, urban catalyst, air pollution is a major contributor to health hazards in Bangalore. About 15 years back, only 2% of the kids had asthma, but as on 2008, the percentage has risen to 28.5% [21]. Bangalore has currently one perennial water source i.e. river Cauvery which is situated at a distance of 108 km from the proper city. Apart from that, it is dependent on lakes and ground water. However, due to over-exploitation and pollution both the sources are getting deteriorated [22]. The biggest problem ailing the Silicon Valley of India, is the traffic of the city. Ask any Bangaloreans, what is that they don't like about the city, and the answer will be traffic. Unlike all other metropolis around the world, Bangalore never grew along a proper trunk transport network. Due to an inefficient public transit system, the motorisation grew at an unprecedented rate and now Bangalore is regarded as the sixth-most congested city worldwide for commuters and second-worst for parking after New Delhi, according to a 2011 survey of 20 cities by International Business Machines Corporation [23].

Also in Bangalore another set of issues concerned with large and rapidly growing domestic and regional economies. There are questions about their methods of integration into the global economy and the cost to the grassroots communities. Perhaps the challenge facing urban researchers is to better understand local conflicts surrounding integration into the global economy in order to inform the operation of more socially just or equitable urban management. In Bangalore the growth of the IT industry has led to large-scale encroachment on peripheral land in the shape of technology parks and related development supported with public resources. Such developments often preclude the delivery of land to other economic activities. At the same time, however, changes in the local economy have created opportunities for the growth of other IT-related support facilities as well as non-IT small-scale production and service functions. These include construction, catering and food provision, furniture and interior facilities, garments and textiles. Urban researchers and policymakers in Bangalore failed to identify mechanisms that optimized the results for both the overall urban economy and local economic development [24].

All the above-mentioned issues clearly depict that the city is struggling to keep up with the level of demand. According to Prof. T.V. Ramachandra of CES at IISC, Bangalore is more likely to be a dead city in next five years due to unplanned urbanization [21]. This actually shows the need to regulate the volume of influx of population into Bangalore city. Except for the regional centre i.e. parts of Bangalore urban, entire region is characterised with low economic potential and very low densities. Hence, the Bangalore regional plan must focus towards a balanced growth of the region by developing regional centres all across the Metropolitan area. These centres must act as centrifugal forces to deconcentrate the development from Bangalore city to the entire region. Hence the approach of the Draft Bengaluru Structure Plan-2036 will be dispersed growth across the Metropolitan region to achieve a sustainable balanced growth.

6.8 Smart Ways to Move Ahead: Linking Economy and Space

It has been seen that many times policymakers are awfully uninformed of the effect of macro-economic policies on urban advancement [25]. Free market economy, liberalization of trade and hands-off approach of State Governments towards national economies significantly affect the physical aspects of urban settlements. Emergence of mega-cities is strongly related to the increasing concentration of Multi National Company (MNC) led manufacturing and service industries in urban areas [26] and Bangalore is no exception. It has been observed that MNCs in collaboration with local industry effectively controlled the shape, frame and character of urbanization in emerging mega-cities like Bangkok [27]. Therefore, in metropolises like Bangalore which are experiencing runaway population growth, one has to explore whether changing the pattern of industrialization to alter the urban form of the city would at all be effective. Would there be takers for government-offered incentives to relocate outside the metropolitan core? It is a very critical question and holds the key to transformation of the city to a smart economy with commensurating quality of life.

The above observation has to be now seen in a different context: rapid urban transition within the BMR metropolitan region, comprising UAs and their outgrowths. The rapid transformation of the areas around the metropolitan core of BMR has been taking place since the 1990s, which is typical of metropolitan areas in India, driven by increasing demand for housing the middle class leading to population movement towards peripheries of the city where land is cheaper [28]. It has been observed that rapid transformation of the peri-urban spaces in large cities is offshoot of neo-liberal policies which encourage private capital, promote MNCs, and induce the creation of special economic zones [24, 28, 29]. Conscious efforts by government at industrial decentralization along with the need to reduce pollution in major cities also contribute to this rapid transformation. As enterprises start operating in the peripheral zones, land use patterns change and land prices are driven upwards, leading to gradual monetization of land in the periurban areas [30]. This process of periurbanization is witnessed conspicuously in major Indian metropolises including Bangalore where a real estate boom has transformed the pace of development. Land has been acquired by the State and private corporations for several industrial, residential and recreation purposes by changing the land use away from agriculture and allied activities. The question to be asked is: what is the long-term cost to BMR of this emerging pattern of development along the peripheries of the main city within the envelope of the broader region? To usher in a smooth transition to a smart economy, there is a need to first balance the changing settlement pattern within the region with spatially-integrated location allocation for industry and service sector drawing real time information and predicting scenarios in a dynamic environment. The BMRDA needs to be equipped with such tools and capabilities. Systems should be in place to constantly access, capture and process information on land rates, residential choices, natural resources, flows of people and goods along with proposed interventions in transport, trunk infrastructure, etc.

Like other large metro cities, the peri-urban areas within BMR can be broadly classified into two categories following the pattern of evolution and the performance of the primary and secondary sectors therein. Thus, theoretically we can have 'neo peri-urban' and the 'old peri-urban' with differing scale and quantum of channelization of agricultural land into real estate, industries and/or infrastructure. In the newly urbanizing pockets within BMR, we can experience transformation driven mostly by neo-liberal policies and new infrastructure. The older peri-urban pockets within BMR started developing in the pre-liberalization era (pre-1990s) in response to the demand for housing from the growing number of secondary sector workers employed across public and private sector enterprises located in the metropolitan core. Thus, in the context of smart economic growth, it is the 'neo peri-urban' that requires special attention.

The presence of large tracts of rural settlements within BMR, besides peri-urban areas of various characters, brings into sharp focus the need to analyze the rural-urban interactions in the context of smart economy development. It has been noted by Tacoli [31] that the nature of economic, social and cultural transformations impacts the scale and strength of rural-urban interactions. Already, reform-oriented macroeconomic policies at the national level have impacted rural-urban interactions. Within BMR, several factors impact the nature and extent of rural-urban interactions. Besides geographical and demographic characteristics, these include agricultural systems, availability of transport infrastructure and services linking local settlements to a number of urban centres where markets and services are located. The latter set of factors need to be specially addressed in the context of smart economy, keeping in mind the fact that the rural—peri-urban continuum within BMR itself is dynamic in nature and responds independently to economic realities on the one hand and to infrastructure thrusts on the other.

The rural-urban divide that exists in BMR has largely stayed out of focus for policy-makers and calls for interventions with reference to local contexts while avoiding generalizations. Taking guidance from the observations of Tacoli [32], in the light of wide diversity in the occupational structure and in economic activity in the peripheries of BMR, policies should be framed for the periurban areas aimed at protecting livelihoods and facilitating occupational diversification. Policy interventions should focus on enabling communities to access diverse range of assets which would create more non-farm based livelihood opportunities, as it is observed that periurban livelihoods span both rural and urban resources and spaces.

Globalization has brought in its wake economies of scale in production and distribution, which has further led to spawning of land-intensive enterprises which seek out large tracts of land often found on the outskirts of mega cities like Bangalore. Tacoli [33] talks about spill-over growth on the outskirts driven by the need for accommodating growing workplaces and workforces. Availability of cheaper infrastructure, land and labour in the peripheries further encourages periurbanization. Policies favourable to SEZs and also IT and ITES industries have been very influential in shaping the periurbanization processes within BMR. Now the time has come to put in place a metropolitan decision support system to balance the skewed growth patterns within BMR.

Keivani et al. [24] noted that globalization had a profound economic impact on peripheral land and the local economies of major cities of South India. In many cases economic activities expanded or encroached into the peripheral lands around the core cities, leading to significant alteration of the spatial development pattern. There has been a restructuring of the institutional set up too and the government has shown eagerness to put in place a planning framework and regulatory structures like single window clearances to enable the MNCs to accelerate the process of setting up their units. In the case of Bangalore, these aspects cause not only conflicts of interest between new demands of international capital and existing demands of local firms and populations, but also conflicts between different tiers of administrative and governance circuits as they vie for influence and areas of competence.

While the IT led growth in Bangalore yielded positive results in terms of creating large-scale job opportunities in IT support services and greater demand for housing and commercial developments, Keivani et al. [24] have observed that urban researchers and policymakers in Bangalore were not pro-active in identifying mechanisms that could have optimized the results for both the overall urban economy and local economic development. They suggest that the institutional decision making system being driven by parastatal agencies instead of elected councils was subject to little accountability and this insulation from popular politics led to a disconnect. Overemphasizing the needs of MNCs and tech companies by and large may have led to a neglect of such local considerations. Similarly, based on a study of Hyderabad in South India, Kennedy [34] noted that Hyderabad was successful as a region in attracting both domestic and international private firms and creating wealth and employment. But certain lacunae was pointed out with respect to governance. It was pointed out that large enterprises have dominated the decision making platforms bypassing the locally elected representatives who have largely left out from the policy process, along with civil society groups. Large tracts of peri urban space are being transformed into special purpose enclaves, governed by specific regulatory frameworks. By subdividing peri-urban spaces and increasing spatial differentiation, such policies appear to weaken prospects for governance institutions at the metropolitan scale. One key question that Kennedy [34] raises here is the extent to which such top-down decision-making, in conjunction with regulatory tools that carve out selected areas for development within largely undeveloped peri-urban spaces, are effectively weakening prospects for participatory urban governance at the metropolitan scale. Local actors are conspicuously absent from the policy process. Governance continues to be characterised by centralised political institutions. There is thus an issue of governance which needs to be addressed in the context of smart economy development of Bangalore.

The process of urbanization in the peripheries of BMR is susceptible to threat from unplanned, informal and illegal developments, with frequent struggles over land use since the peri-urban areas are generally between legal and administrative boundaries of Bangalore city (BBMP and BDA jurisdictions) and the other urban entities (governed by respective CMCs and TMCs). Typically in such situations the capacity of government authorities to regulate economic activity is particularly weak [35]. IT enclaves have been a new major claimant of land and water, as have been gated communities, up-market medical facilities as well as new and upcoming educational institutions. In the move towards smart economy, policy makers in Bangalore need to smoothen out such confusions which have the potential for creating negative externalities of growth.

For the reasons mentioned above, policy making for smart Bangalore should focus on governance reforms, to the decentralization of resources and responsibilities, and to the strengthening of local authorities. The role of local authorities in supporting economic development and poverty alleviation has to be recognized besides their traditional role as infrastructure and service providers. It has to be ensured at the same time that social infrastructure such as education and health enjoy the same attention as physical infrastructure which is believed to produce tangible effects on economic growth.

Expansion and extension of metropolitan core of BMR, brought about by concentration of trade and services and resulting population pressure, has manifested itself in spatial sprawl. However, the governance structures have not responded adequately to this process of change. While Urban Local Bodies govern over entities which are defined as the Statutory Towns by Census of India, the rural spaces which host land uses and population densities close to that of urban, are yet to be declared urban by the Census and are still being governed by rural administrative governance institutions called panchayats. The reciprocal relationships between the entities are not smooth and there are conflicts over the administrative jurisdiction between the urban and the urbanizing rural in the peripheries. Typically this results in ambiguity over service delivery obligations, especially in relation to solid waste and sanitation, which are essential to the support ecosystem of smart economy of Bangalore. While BDA is performing the function of land use planning at the level of the statutory planning area defined as per the relevant Act, there are other defined statutory planning areas within BMR corresponding to the areas under smaller urban entities. BMRDA is preparing strategic plans at the level of BMR as defined under the given statutes. While there is integration between the different levels in theory, real time information sharing systems need to be in place involving these three entities namely BBMP and the rural governance entities need to be integrated into this network to make available the infrastructure conditions and service delivery status on a real time basis.

The criteria used by Census of India to define ‘urban’ is as: (a) population of five thousand (b) population density of four hundred persons per hectare and (c) at least seventy-five percent of the male working population engaged in non-agricultural pursuits. Any settlement meeting this threshold qualifies as a ‘Census Towns’. States enjoy the authority to frame their own criteria to further classify these urban areas or to confer the status of urban statutory bodies on settlements other than census towns. The 74th Constitutional Amendment Act enacted in 1992 has paved the way for constitutional classification of settlements in transition from rural to urban as ‘Nagar Panchayat’. However, the capacity of these intermediate tiers of governance to fit into the transition framework of smart economy does remain questionable. This holds special significance in the case of BMR where the rural entities are experiencing exponential population growth.

Many authors have commented that the peri-urban areas of large metropolitan cities act as a source and sink for human resources for activities in the metropolitan core. While framing their vision for a smart economy for Bangalore, policymakers in BMR should avoid the pitfall of utilizing the peri-urban areas primarily towards an exponentially growing urban economy without any appreciation for the negative social and ecological effects befalling these areas. If this concern is not addressed, the smart economy dream will rest thinly on external drivers in the form of inaccessible markets and imported human resource, making it exclusionary. Within the scenario of capital-driven economic growth and growth-driven urbanisation, there is tremendous potential for diminishing the divide between the rural and the urban, as well as between rich and the poor, provided the linkages shared by peri-urban areas with rural and urban entities are nurtured through governance structures which are based in the realm of information sharing rather than that of jurisdiction protection.

The undercurrent of tensions between peri-urban dwellers and municipalities in BMR has the potential to complicate the formulation and implementation of policy interventions for smart economy. In the march towards smart economy, conflicts may arise among the constituent units of BMR regarding components of the support ecosystem like solid waste management and water resources and also regarding changes in land use. Since these issues are likely to become increasingly central, mechanisms for inter-local government negotiation need to be developed [32]. Residents of periurban Gurgaon, in National Capital Region (NCR) of India, for instance, were found to harbour feelings of resentment against urban planning and development authorities for forced acquisitions of land, delayed compensation and their inability to participate in the opportunities thrown up by the rapidly growing and expanding Gurgaon city [29].

Creating mediations for the periurban interface requires both urban and provincial experts to work together, for which there might be no point of reference. A case is sewage streaming into periurban territories, which is mostly the duty of the urban government and somewhat of the rustic government. Tacoli [32] noticed that ensuring synergies amongst different tiers of governments, NGOs, neighborhood common society and the private area can assume a critical role in supporting the positive provincial urban collaborations while decreasing their negative effects. Specifically, governments at the suburban level can assume a vital part in supporting positive country urban linkages. While neighborhood leadership, upheld by sufficient assets, can encourage this, more extensive issues like arrive residency frameworks, institutional structures of business sectors and various programmes and policies are probably going to influence neighborhood activity [36].

The series of ring and radial road proposals of the BMR draft Structure Plan 2011 would improve the connectivity of towns, cities and villages within the BMR region and would also provide for smooth transit of non BMR-destined traffic, therefore helping decongest Bangalore city. These include Peripheral Ring Road, Intermediate Ring Road, Town Ring Roads, Radial Roads and Satellite Town Ring Road (Fig. 6.17).

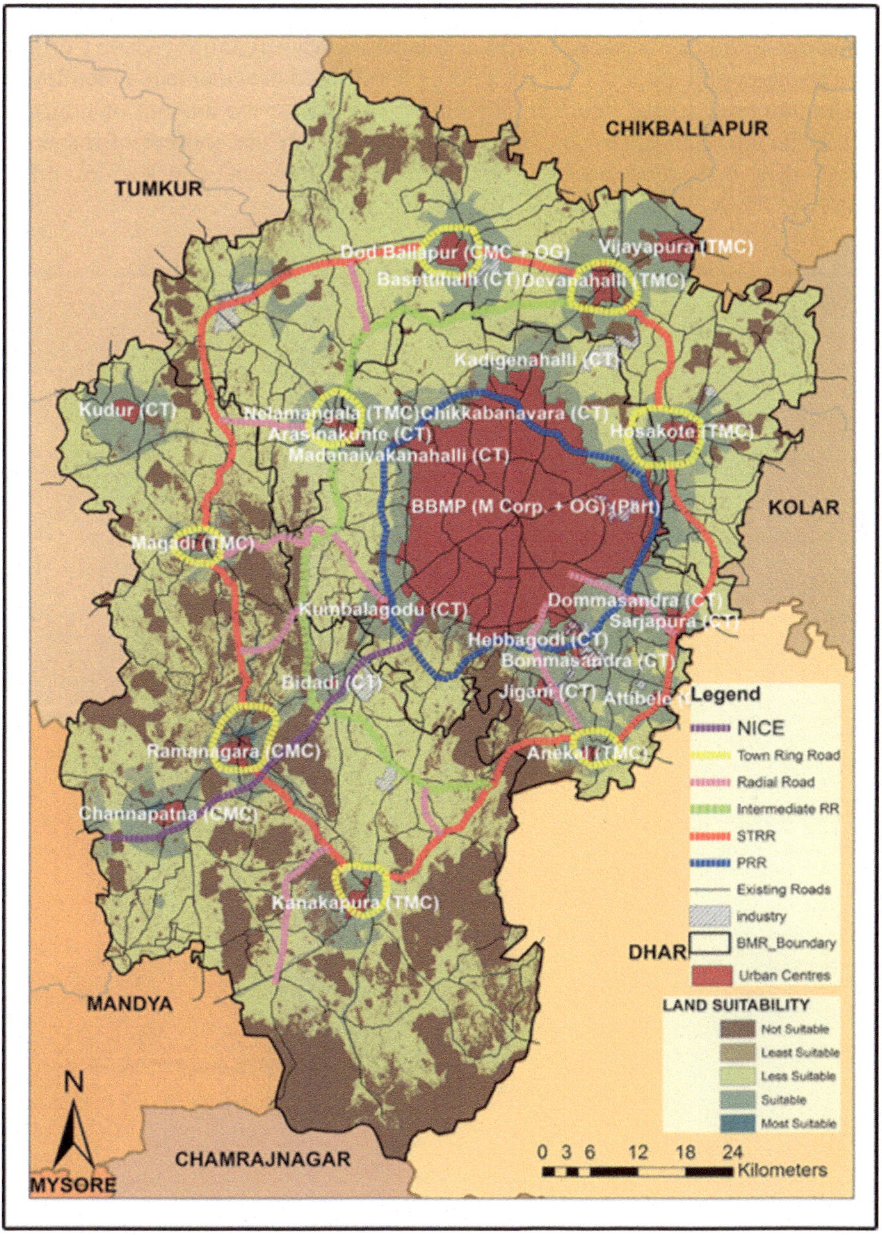

Fig. 6.17 Proposed BMR road network-2031

Further, the buffer taken was five hundred meters of both existing and proposed roads (Fig. 6.18). The analysis depicts that the proposed roads have a good coverage within the region and would be increasing the accessibility of the remote areas. Further there are mega infrastructure projects particularly road connectivity proposed by Government. These include Chennai–Bangalore Industrial Corridor, Bangalore–Mumbai Economic Corridor, Bangalore–Mysore Infrastructure Corridor etc. The target of these projects is to achieve accelerated development within BMR.

The proposed spatial structure is required to regulate the amount of influx of population into Bangalore city. This can be done with the concept of dispersed growth (Fig. 6.19) and potential developable land available in various urban Centers of BMR.

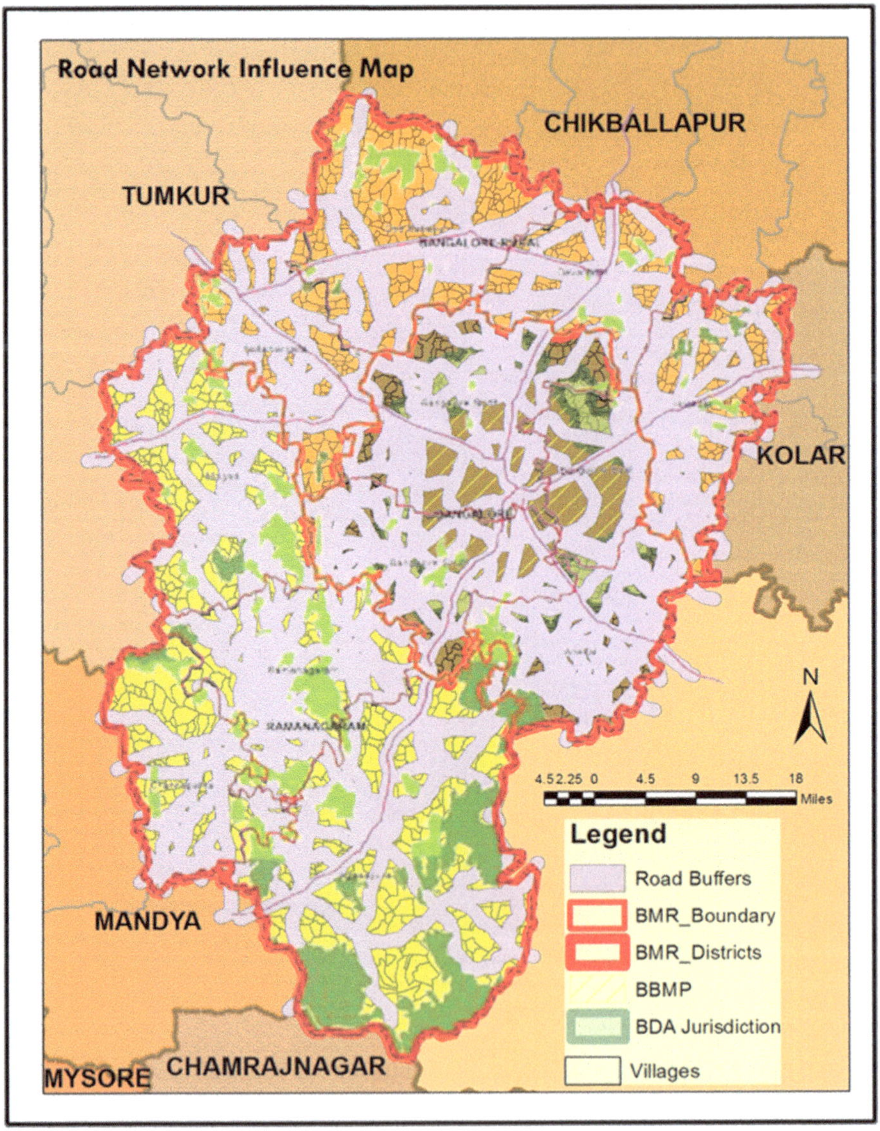

Fig. 6.18 Road network influence map of BMR

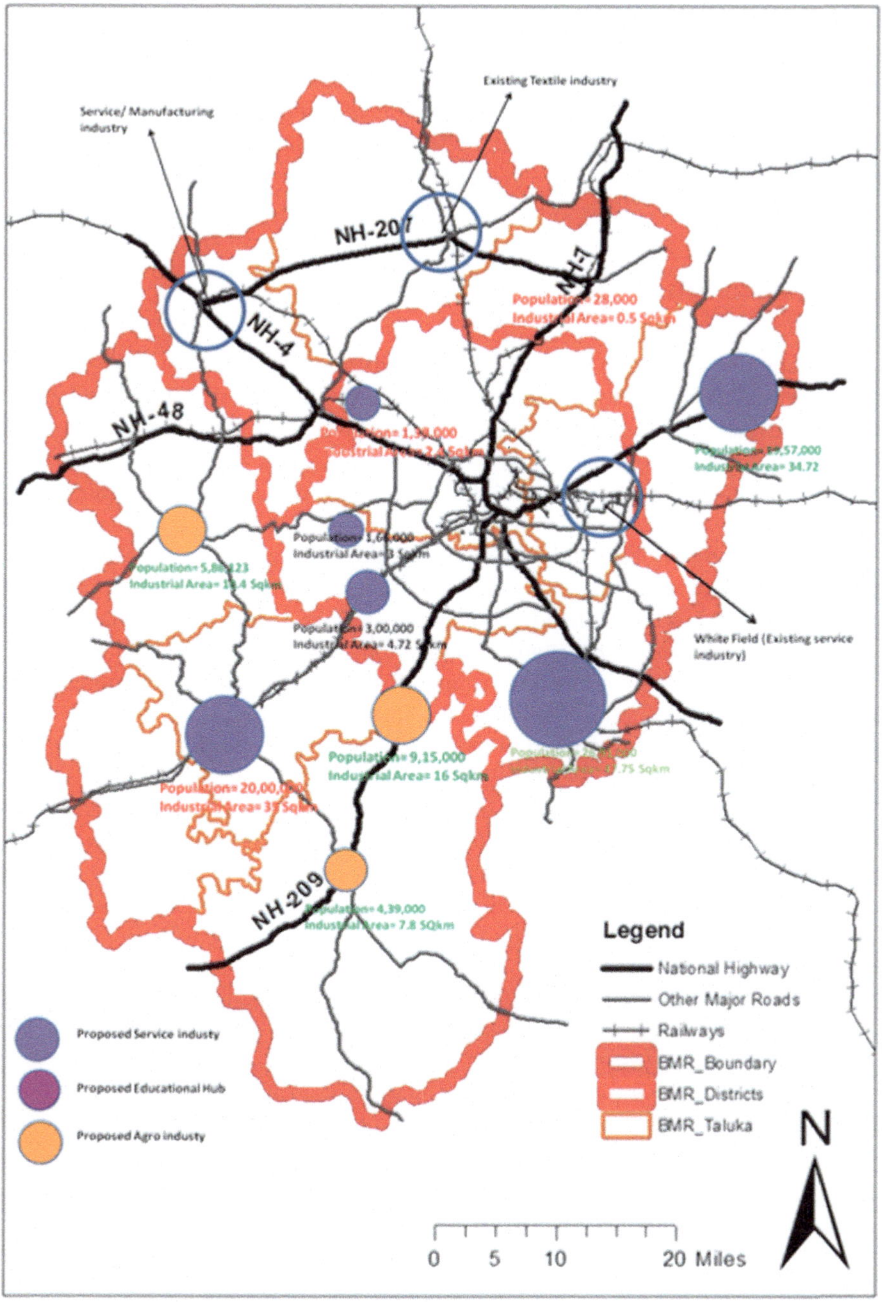

Fig. 6.19 Proposed population dispersal plan of BMR

The 74th Amendment to the Constitution of India accommodates the making of District Planning Committees (DPC) and Metropolitan Planning Committees (MPC) for spatial and sectoral improvement for the whole region [30]. There is a crying need to constitute DPC and MPC and delegate the necessary functional and fiscal autonomy to enable them script the roadmap for a smart economy and a smart region.

6.9 Conclusion

It is but obvious that urban economy not only dictates the urban fabric of a city, it also influences the city's socio economic profile. The mistake that we often commit is to regard the growth of BMR as being endogenous when the fact remains that BMR is nothing but conurbations. BMR not only engulfs its degenerated peripheries, but also the small towns around, thereby leading to the total obliteration of the latter.

It is now being widely held that alongside superseding the national economy, globalisation has resurrected urban governments in India. Riding on this globalisation juggernaut, BMR has long entered an era of rejuvenation by inheriting its rich cultural and economic heritage with sustained innovation in its spatio-economic evolution. There is perhaps no ambiguity in terming BMR urbanisation as 'entrepreneurial urbanisation' [37] and in this enterprise, the smart city narratives of BMR has been synonymous with both brownfield as well as green field development with the latter having a relatively larger share.

The BMR economy although largely depends on its IT based knowledge economy, a significant share is also held by other services and manufacturing. BMR should not be carried away by its IT sector since a significant part of its economy is attributed to textile, automobile, machine tool, aviation, space, defence, and biotechnology based industries. The region is replete with huge concentration of small and medium industries [10]. 89% of BMR areas falling under rural areas signify the contribution of rural economy through agro based products and handlooms. Besides all these, BMR is bustling with services ranging from education, health care, to hospitality, banking and transport logistics services.

Launching a new business can be a daunting experience in the face of uncertainty of securing funding and sufficient skills. It is here where BMR should not only secure and promote its stature as India's IT hub, it must also incubate its intellectual capital in order to sustain and upgrade its being the investment destination for entrepreneurial capital. BMR must draft its regulatory frameworks to host an ecosystem that does not blindly replicate the popular incubator and accelerator models but customises it to develop local versions. Apart from incentivising the potential investors, BMR may encourage and create platforms for civil society organizations to take resort to crowdfunding to supplement its initiatives. Urban finance shall be a formidable task for BMR administration. The municipal fiscal health should be strong enough to build the credit worthiness of BMC to access international funding for effecting transformational change and towards achieving this, BMC should explore non conventional fiscal tools to replenish its coffer. Alongside BMC, BMRDA should also harness the potential of all sources of infrastructure finance, namely: external borrowing, FDI in infrastructure, long term financing (including take out finance/mezzanine finance, if need be), service or management contracts, different variants of concessionaire.

BMR appears desperate to go more digital to improve quality of life of the citizenry. While doing so, BMR banks on all the stakeholders in order to co-create tomorrow's urban services with corporate sectors and the civil society (Janagraha). Although BMR strives to steer its economy on technology-driven urbanism, it must take cognizance of the fact that today's decision-making is compounded by an information overload. BMR also must not downplay the offshoots of pollution, traffic congestion, and mushrooming of informal settlements. Let alone informal settlements, informal economic agents (hawkers, vendors, rovers etc.) should also be adequately addressed to. BMR may go more and more wired, the everything-on-demand Bangaloreans may go for higher speed connectivity, tech gizmos may satiate their appetite, but BMR must ensure that its labour markets and education systems mutually support each other to address the skill gap in order to accommodate 'youth bulge' (millennials) and prevent a 'lost generation'. With technological obsolescence exponentially picking up, BMR will find newer occupations forming a new segment of the labour market with many occupations getting disappeared. BMR must address the concern of this surplus labour force through newer skill formation and through product market and labour market scan. BMR must rope in the corporate sector in this endeavour.

The concept of 'smart economy' requires new qualitative approaches of the concept of economic growth, focused on the quality of life and on the standard of living. It is a dynamic process [38] adaptable to contextual situations in which the individual and the community act in time and space for which the foundation is represented by a set of moral, cultural, traditional, political, democratic, leisure precepts [39].

Being a global city, BMR's strategic vision must become complementary and compatible with internal realities on the one hand and the development trends of the global economy on the other. BMR must promote entrepreneurship both ways: mimetic (technology transfer through FDI and innovative innovations suitable for local economies). Certainty is a key word, diversity is a key reality and clarity is a key prerequisite. BMR must be able to tie a knot and build a trinity. In this context, the intensity of primary energy (and not secondary energy) is of utmost importance to green the economy and for this, BMR needs collaboration between ecologists and economists, between the development of well-informed environmental policies and economic decisions.

Keeping in view the universality of public goods such as education, culture, health and research, BMR should adopt a different approach to the concept of 'smart economy' through an indissoluble bond with the science of marketing,—cibermarketing, neuromarketing, virtual marketing. BMR must promote public private partnership: promote up-skilling and work based training, set up networked economy and ensure liaison and information sharing between business and economic development agencies.

Given the fact that we are fast moving to Industry 4 and the *third age* (retirement) industry is seen as having more development potential (BMR once being the Pensioners' Paradise), given the fact that rural BMR contributes considerably towards its economy, it is high time that BMR readies itself accordingly. BMR should also take necessary steps to multi skill the rural workforce, skilled and unskilled, to absorb the uncertain structural shocks of BMR economy.

BMC must acknowledge that everything that can be digital will be digital. The economy is shifting towards subscription based business models, product life cycles are getting shorter, users are migrating from 'possession' to 'use' and look for ultra personalization. Information is becoming more and more transparent and businesses are experiencing increasing disintermediation and automation. Processes become data centric and mobile. Businesses are preferring transition to a pull rather than a push approach. Either concentration or fragmentation is being seen to be adopted by service industries. Data monetization, especially of big data analytics and its trading is going to emerge as a big employer. BMC and BMRDA should put in place the enabling platforms to smoothen these inevitable transitions [40].

But we should never forget what urban sociologist, Robert Park, once remarks, "If the city is the world which man created, it is the world in which he is henceforth condemned to live." But unfortunately the creators and the condemned are not necessarily the same, nor do they stand on equal ground.

Smart economy requires an adequate supportive ecosystem, an important component of which is transport infrastructure and services. In the context of BMR, as we talk of smart spatial allocation processes for enterprises and balancing the growth at the peripheries of the metropolitan core, our focus has to be on enhancing the quality of regional transport networks. Although widening of existing highways through land acquisition and development of ring roads, bypasses and expressways catering to the metropolitan core have been the priority of decision makers in the transport sector in BMR, connecting roads to periurban settlements have received comparatively lesser attention. This calls for a focus on improving and meeting local transportation needs more effectively. Often this entails the creation of some intermediary form of transport that connects the village to the city centre, or to the main highways. An important dimension of periurban transport is the reliance on semi-public transport such as auto-rickshaws that have gained popularity due to the poor reliability of public buses. However, this form of transport remains unorganized, and is associated with the common problems of crowding, unreliable service and overcharging by the operators. Tacoli [32] identifies high transport cost as a bottleneck in shaping access to markets. Improved roads greatly enhance rural-urban linkages by increasing the value-added and marketability of cash crops and other commercial produce.

Bangalore Metropolitan Planning Committee, a constitutionally recognized institution to look after the spatio- sectoral planning of BMRDA area has been almost defunct as is evident from having only two meetings till now since its constitution in 2015. It is regretted to note that Government of Karnataka had to enact Bangalore Metropolitan Planning Committee Act in 2014 under duress following direction from the Karnataka High Court which gave it a 48 h deadline to enact the Act.

Given that globalisation has resurrected the urban governments, it is high time that BMC assumes its role assertively since governance responsibilities are shifting from the national to the subnational and supranational levels. BMC need not however be hyperglobalist. Any programme or policy adopted by BMC is not only bound to impact areas beyond its administrative boundary, in turn, BMC also stands impacted by the decisions made by urban and rural local bodies situated within the BMR. Thus, it is only obvious that the newly created Bangalore Metropolitan Planning Committee (BMPC) strives to forge the relationship between BMC and other local self governments around its (BMC's) frontier more intensely and ensure the maximum possible convergence of different programmes. BMPC should appreciate the problem arising out of the absence of a hierarchically integrated government structure where different local governments and parastatal agencies compete over the development of infrastructure, fragment the implementation process and make metropolitan planning agencies ineffective.

BMPC must not only embrace both 'deterritorialisation', the process that expands and accelerates the movement of commodities, capital, money, people and information through space, but also 'reterritorialisation' that fixes spatial infrastructures and reconfigures and/or transforms spaces to allow these flows to happen.

Although the intensely competitive nature of the market within the global system is forcing specialisation upon cities, Bangalore should not try mono functionality in IT and ITES, it should rather promote and sustain its multi functionality nature in providing different other services besides manufacturing industries. Bangalore must retain its identity as the entrepôt centres for IT, ITES, Medical and Health Care, Higher Education Centres in the regional city system and act as the international hub with spokes spread across the State of Karnataka, the rest of India, and the Asia-Pacific region.

Since environmental issues are likely to be of preeminence, BMC, BMRDA, BMPC should together ensure that permission to use land should be granted based both on environmental planning and economic planning objectives. BMPC must constitute an Urban Design Review Committee allowing it to go in for flexible land use planning with predictable pattern of mixed land uses with adequate concern for public health issues.

The spatio economic expansion of Bangalore should not lead to a 'Shrinkage' in other cities in the region. Bangalore becoming 'global' should not relegate others as 'local' experiencing the outflow of capital and human resources, lack of entrepreneurship and low levels of innovation and intellectual engagement. We must widen the scope of the current urban planning paradigm focused on 'growth' to one that integrates 'growth and decline' as simultaneous and interrelated urban processes.

Both BMC and BMPC should assume special roles in this globalization juggernaut since the restructuring of space involves a restructuring of governance. Pressures on cities are seen to be evident from two angles,—(i) socio-physical structures, such as major development projects, social polarisation or gentrification and (ii) influences that impact the decision-making processes. Since both these influences interact in a political context, decisions are made on whether to promote certain projects and undertake certain policies which will affect the degree to which the forces of globalisation are accentuated, modified or prevented.

We must admit the onset of a 'transnational' urban system and accordingly attune our urban agenda at the national and provincial level. In this transnational journey, we should not however make our cities become disconnected from their regional and national contexts. Bangalore must be allowed to serve the 'command functions' within the framework of provincial, national, and transnational framework keeping in mind the international treaties and environmental internationalism. Bangalore must facilitate the marriage between neo-liberal economics and environmental sustainability.

Acknowledgements We acknowledge the support received from BBMP and BMRDA officials regarding sharing of data, information and documents. We also thank SPA Bhopal B.Plan 2013–2017 batch students, who put their best efforts to make this research possible.

References

1. Angotti T (2013) The new century of the metropolis. Routledge, New York
2. Census of India (2011) Provisional population totals paper 2 of 2011 India series 1. Office of the Registrar General & Census Commissioner, India
3. United Nations (2014) World urbanization prospects. Available at: http://esa.un.org/unpd/wup/Highlights/WUP2014-Highlights.pdf. Accessed 30 July 2015
4. Ministry of Urban Development (2015) Smart city mission statement and guidelines, Ministry of Urban Development, Government of India, New Delhi
5. Directorate of Economics and Statistics (2014) Economic survey of Karnataka, Government of Karnataka
6. Bangalore Metropolitan Regional Development Authority (2016) Comprehensive traffic and transportation study for Bangalore Metropolitan Region, Government of Karnataka
7. Bangalore Metropolitan Regional Development Authority (2016) Bangalore Metropolitan Region revised structure plan-2031, Government of Karnataka
8. Kamath S (1990) Places of interest, Karnataka State Gazetteer: Bangalore District. Government of Karnataka, Bangalore
9. Buchanan F (1870) A journey from Madras through the countries of Mysore, Canara and Malabar, 2nd edn., vols 1–3. Higginbotham and Co., Madras
10. Sudhira HS, Ramachandra TV, Bala Subrahmanya MH (2007) City profile: Bangalore, Elsevier Publication. http://wgbis.ces.iisc.ernet.in/energy/water/paper/bangalore/TVR24_p11_Bangalore.pdf
11. Shastry GS (2008) Emerging development issue of greater Bangalore, published as Working Paper No. 194 by Institute for Social & Economic Change, Bangalore
12. Nair J (2005) The promise of the metropolis: Bangalore's twentieth century. Oxford University Press, New Delhi
13. PIA (2003) PIA Technical Directory. Peenya Industries Association, Bangalore
14. Upadhaya S (2014) Prospects in medical and wellness tourism in India. J Tourism: A Contemp Perspect 1(1):18–24

15. Sareen D (2005) Innovation and IT in India: case study of Bangalore. In: Proceedings of UNIDO 2nd International Conference on 'Process of Innovation & Learning in Dynamic City Regions of India', Bangalore
16. Heitzman J (1999) Corporate strategy and planning in the science city: Bangalore as 'Silicon Valley'?. Econ Polit Wkly 34(5):2–11 January 30, 1999
17. Organisation For Economic Co-Operation and Development (2013) Innovation-driven growth in regions: the role of smart specialisation. OECD Press, Paris
18. Taubenbock H, Wegmann M, Roth A, Mehl H, Dech S (2009) Urbanization in India—spatio temporal analysis using remote sensing data. Comput Environ Urban Syst 3(33):179–188
19. Shaw A, Satish MK (2006) Metropolitan restructuring in post liberalized India: separating the global and the local. Cities 24(2):148–163
20. MoUD (2014) URDPFI guidelines, Ministry of Urban Development, Govt. of India
21. Ravichandar V (2014) Citizens gasp as Bangalore's lung space reduces. The Times of India, Bangalore city, July 30, 2014. https://timesofindia.indiatimes.com/city/bengaluru/Citizens-gasp-as-Bangalores-lung-space-reduces/articleshow/39255217.cms. Accessed 15 Sept 2017
22. Central Ground Water Board (2011) Ground water—Bangalore urban. Bangalore Rural and Ramanagra, Government of Karnataka
23. Bloomberg (2012) Bangalore 'sixth-most painful' in world for traffic congestion. The National, June 28, 2012, https://www.thenational.ae/world/asia/bangalore-sixth-most-painful-in-world-for-traffic-congestion-1.361748. Accessed 15 Sept 2017
24. Keivani R, Mattingly M (2007) The interface of globalization and peripheral land in the cities of the south: implications for urban governance and local economic development. Int J Urban Reg Res 31(2):459–474
25. Becker CM, Morrison AR (1996) Public policy and rural-urban migration. In: Gugler J (ed) Cities in the developing world: issues, theory and policy. Oxford University Press, Oxford, pp 88–107
26. Potter R, Unwin T (1995) Urban-rural interaction: physical form and political process in the third world. Cities 12(1):67–74
27. Parnwell M, Wongsuphasawat L (1997) Between the global and the local: extended metropolitanisation and industrial location decision making in Thailand. Third World Plann Rev 19(2):119–138
28. Shaw A (2005) Peri-urban interface of indian cities—growth, governance and local initiatives. Econ Polit Wkly 40(2):129–136
29. Narain V (2009) Expanding city, shrinking hinterland, land acquisition, transition and conflict in periurban Gurgaon. Environ Urbanization 21(2):501–512
30. Brook R, Purushothoman S, Hunshal C (eds) (2003) Changing frontiers: the peri-urban interface Hubli–Dharwad, India, Books for Change, Bangalore, 146 p
31. Tacoli C (1998) Rural-urban interactions: a guide to literature. Environ Urbanization 10:147–168
32. Tacoli C (2002) Changing rural-urban interactions in sub-Saharan Africa and their impact on livelihoods: a summary. Working Paper 7. London: International Institute for Environment and Development. 40 pp
33. Tacoli C (1999) Understanding the opportunities and constraints for low-income groups in the peri-urban interface: the contribution of livelihood frameworks. p 7. Draft for discussion. London: Peri-urban interface project, development planning unit, University College London
34. Kennedy L (2005) Regional industrial policies driving peri-urban dynamics in Hyderabad, India. CNRS, Centre d'E´tudes de l'Inde et de l'Asie du Sud (CNRS-EHESS), 24(2):95–109
35. Parkinson J, Tayler K (2003) Decentralised wastewater management in periurban areas in low-income countries. Environ Urbanization 15(1):75–90
36. Tacoli C (2003) The links between urban and rural development. Environ Urbanization 15 (1):3–12
37. Datta A (2015) New urban utopias of post colonial India: 'entrepreneurial urbanisation' in Dholera smart city, Gujarat. Dialogues in Hum Geogr 15(1):3–22

38. President's Council of Advisors on Science and Technology (PCAST). (2016) Technology and the future of cities Available from: https://www.whitehouse.gov/sites/default/files/microsites/ostp/PCAST/pcast_cities_report_final_3_2016.pdf. Accessed 9 Aug 2017
39. Diana A, Cristina B, Eleonora MC (2010) Smart-economy concept—facts and perspectives. J Econ Lit 48(1):264–277
40. Deloitte (2015), Smart cities, Version, 1.0, Available from: https://www2.deloitte.com/content/dam/Deloitte/tr/Documents/public-sector/deloitte-nl-ps-smart-cities-report.pdf. Accessed 5 Aug 2017
41. Census of India (2011) Primary census abstract, UA/Towns. Office of the Registrar General & Census Commissioner, New Delhi, India (compact disk form)
42. Census of India (2001) Primary census abstract, UA/Towns. Office of the Registrar General & Census Commissioner, New Delhi, India (compact disk form)

Part VII
India, Chandigarh

Chapter 7
Smart Chandigarh Tri-City Region: Spatial Strategies of Transformation

Asfa Siddiqui, K. K. Kakkar, Suvankar Halder and Pramod Kumar

Abstract The inception of the "*City Beautiful*" dates back in the early 1950s, right after India attained independence. The planned visionary modernist settlement designed by famous architect Le Corbusier started gaining a 'magnet' status. The regional phenomenon observed around Chandigarh gave way to the establishment of counter magnets Panchkula and Mohali, together referred to as "*Tri-city*". Soon, the green buffer zones around the city got into urbanisation processes and started showing signs of uncoordinated growth. The demographic profile suggested a slowing pace of growth in Chandigarh Union Territory (UT) while other settlements in the Chandigarh region (the 16 km periphery Control Area) kept expanding. A need for regional development was felt in 1970s resulting in various regional plans severely lacking legal status and hence had no prominent effect. A region can attain a planned development through careful understanding in incorporating sub-regional objectives through *Smart Information, Smart Decision-making, Smart Connections* and ultimately leading to *Smart Development*. Smart tools and techniques offered by the geospatial technology can play a pivotal role in analysing the existing scenarios of an urban purlieu and aid significantly in the decision making process for better development. The study asseses the spatial growth of Chandigarh region using Neural Network based modelling growth scenarios and suggests that nearly 400 sq. km. of area will be urbanised by 2048. The study effectively demonstrates the importance of geospatial techniques and recommends spatial strategies in order to have a holistic development in the region.

A. Siddiqui (✉) · K. K. Kakkar · S. Halder · P. Kumar
Indian Institute of Remote Sensing, 4, Kalidas Road, Dehradun 248001, India
e-mail: asfa@iirs.gov.in

K. K. Kakkar
e-mail: kewalkakkar@gmail.com

S. Halder
e-mail: haldersuvankar02@gmail.com

P. Kumar
e-mail: pramod@iirs.gov.in

T. M. Vinod Kumar (ed.), *Smart Metropolitan Regional Development*,
Advances in 21st Century Human Settlements,
https://doi.org/10.1007/978-981-10-8588-8_7

Smart strategies incorporating inclusive planning can for transform settlements into Smart Metropolitan Regions.

Keywords Smart city · Tri-city · Geospatial technology · Chandigarh region

7.1 Introduction

After India attained Independence in 1947, a new capital for undivided Punjab was approved in March, 1948 called Chandigarh-'The City Beautiful'. With a vision of modernist thinking conceived by Pandit Jawahar Lal Nehru, the then Prime Minister of India, a 70 sq. km. space was earmarked between rivulets Sukhna Choe (in the east) and Patiala-ki-Rao Choe (in the west) for developing Chandigarh by Dr. M.S. Randhawa, the then Deputy Commissioner of Ambala. The Master Plan of the Chandigarh city was designed by the famous and renowned Architect- Le Corbusier and was assisted by Pierre Jeanneret, Jane B Drew and Maxwell Fry. In 1966, after the reorganization of the state of Punjab, Chandigarh was announced the capital of Punjab and Haryana, bearing an area of approx. 114 sq. km. (70 sq. km. urban area with 44 sq. km. covering an area of 26 adjoining villages). Le Corbusier designed the iconic city with a vision of 'modernism' and thus translated his geometric modular thought into reality. The work is still considered an iconic piece of work shaped through deep thinking explicitly demonstrating freedom, wisdom and perseverance. It is one of the live examples of a symbiotic relationship between the built and the un-built. The foundation stone of the beautiful city was laid in 1952.

Earlier in Punjab, Chandigarh was declared a Union Territory directly under the Central Government after reorganization of Punjab state on November 1, 1966 into three states, namely Punjab, Haryana and Himachal Pradesh. The first planned city of India, Chandigarh, and a Union Territory (UT) is the hub of political and bureaucratic activities of adjoining states. Its existence as a planned capital designed with a view of urban containment theory, as a result of Garden city concept, makes it a design to cherish and emulate. Although, the city was conceived to have a regulated planned growth, unfortunately the subsequent peripheral growth (areas like SAS Nagar, Zirakpur, Panchkula, etc.) suggests that this vision of the designers was not fulfilled. Some defend sprawl as an inescapable phenomenon, while others advocate its essentiality in preserving the character of the city core, the reality can be understood as the details of the Chandigarh and its environs are studied in detail.

The need of understanding the intra-city growth and the inter-city growth in the outskirts (called the Periphery) like Mohali in Punjab and Panchkula in Haryana (regarded as satellite towns) becomes inevitable in order to understand the regional dynamics of the area. There may be certain drivers of growth that also need to be studied. It is pertinent to mention that Chandigarh faces a legal framework restricting a regional plan that can be adhered to while designing the city. The laws guiding the development scenario is studied in detail in this chapter. This study attempts to understand the sprawl in time line from 1990 onwards with the help of remote sensing, GIS and online information available and ends in providing recommendations and strategies for Smart Growth leading to a Smart Metropolitan Regional Development.

7.2 The Shaping of Chandigarh: The Inception Phase

The land designated at the foothills of the Himalayas was selected for developing the 'dream' city of Pandit Nehru's vision. His notions of having a futuristic city catering to the need of tomorrow, bounded by a gambit of organized planning was to be achieved by a collaborative Indian and team of European architects and planners. While the location of the site was a constraint, first site specific Master Plan was prepared by the American lead planner Albert Mayer and Mathew Nowicki. Although, not much credit has been given to Albert Mayer for his rigorous efforts and hard work in bringing out the neighbourhood block, curvilinear fan shaped plan in merely six months' time from the approval of the city to be constructed. The fan shaped module was inspired by the Garden City concept (proposed by Ebenezer Howard in 1902) keeping enough circulation space and green belts. Emphasis in Mayer's plan was laid on placing the neighbourhood block in the centre [1]. The rest was built upwards from it. Mayer's aim was to build a simplistic urban centre possible [2]. The planner duo were replaced later due to administrative hitches and other unforeseen circumstances. The big project fell into the laps of the then lead, Le Corbusier (see Fig. 7.1).

In light of understanding the significance of self-contained neighbourhoods and essence of green spaces in an urban setting, Chandigarh was planned with focus on preservation of the unbuilt, urban design principles and aesthetic appeal. Urban containment theory was inherent in Garden City concept proposed by Ebenezer Howard adopted by the planners of Chandigarh.

The design of Chandigarh was more inspired from the original fan-shaped design proposed by Mayer. The improvisations in the plan restricting a grid-iron movement approach and phasing of city functionalities resembling intrinsic body parts into work, living and leisure, were worth admiring. The city was decided to be built in two phases consisting of Sectors 1–30 in phase 1 efficient to accommodate 1.5 lakh people and a rather dense development accommodating 3.5 lakh people in sectors 31–47 making it a total of 5 lakh people.

It was a vision of Le-Corbusier where his emphasis not only dwelled on the urban core development but also on the rural-urban transitional zones in the fringes. His work clearly depicts his passion for preserving agricultural activities and open spaces for a serene site like Chandigarh amidst Shivaliks. The perfect assimilation of town and country was addressed through the design keeping in view the problems faced by an urban setting due to urban sprawl. This decentralized solution was also helpful in driving planned growth and at the same time posed limitations on the ad hoc growth pattern resulting due to poor management of land resources and Indian way of organic city development. According to Howard, green and open

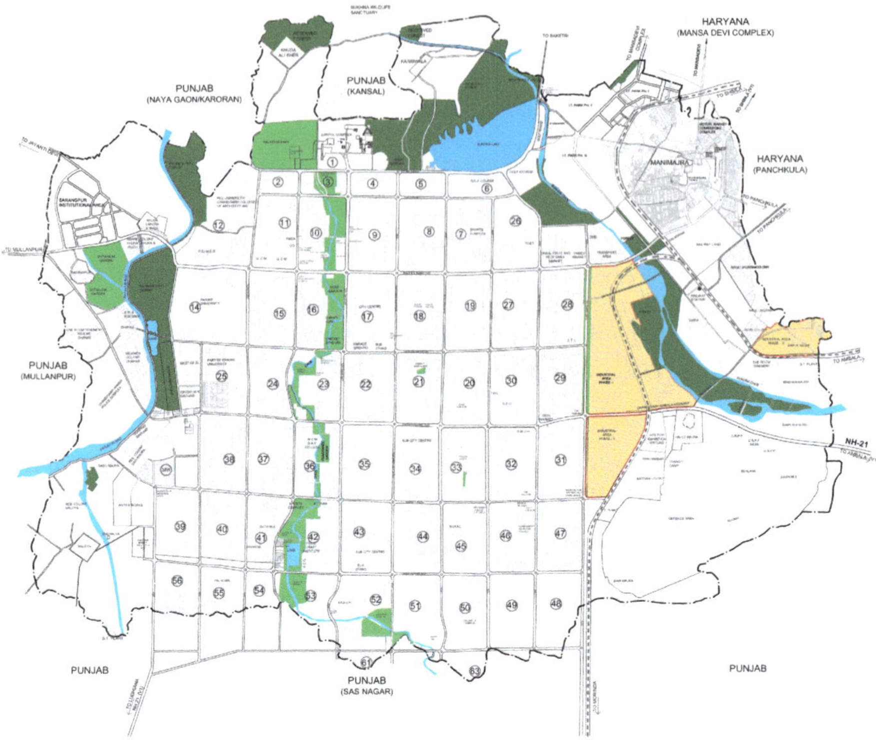

Fig. 7.1 Chandigarh and its sectors. *Source* Chandigarh Master Plan-2031

spaces around the cities paves way for a rationalistic self-contained developments scenario where the satellite towns shall grow and help cores to grow economically and socially, subsequently. But as is said, land is a resource and it is scarce, greenbelts were gradually sacrificed in to the process of urbanization.

7.3 Chandigarh: Planning of the City

The planning of Chandigarh does not follow the typical planning principles and is seen as a departure from conventional planning approaches in Indian framework [3]. With aspirations of Nehru in looking forward overlooking the past, modernism in planning and architecture found its way in the so called 'Nehru era' soon after Independence. The idea was to build modern but connecting with the character inherent in an Indian city. Had it been draping an Indian city into western town planning, the job would have been easier. The intermingled localism approach suffering the architect's design need the vision of parties/government involved in the project was a challenging task [4, 5]. Prior to the development of cities like Chandigarh and Bhubaneswar [6], the town planning principles were followed by trained Architects and city planners, mostly abroad. We heavily relied on the west and their expertise in this subject. But later in 1930s, emphasis was laid on training young minds in this direction. It was also observed that India was fascinated by the idea to 'modernise' but lacked indigenous professionals who could delve into city planning and design without guidance and help. It was also noticed that city planning was more seen as a piecemeal approach to parts of the city and lacked holistic approach [5].

After partition in 1947, the capital of Punjab was given to West Pakistan (now called Pakistan) and there was a dual need to accommodate refugee resultant of partition and to have an administrative and political capital for the state of Punjab [1]. Bureaucrats A.L. Hether and P.L. Verma were the first to examine the need for a new city and later together with P.N. Thapar, member of Indian Civil Service and Administrative Head of the Capital Project in 1949, guided the development of the city [4, 5].

Le Corbusier made important amendments on the plan developed by Albert Mayer and Mathew Nowicki on the 8500 acres of fertile land site [1]. Le Corbusier built the city cantering monumental government buildings [7]. Le Corbusier was an Architect who utilised his experience from the America and translated them in India in the form of hierarchical roads (V1–V7), roundabouts and big self-sustained sectors which he regarded as 'essential function' of his plan [5, 8]. His grid-iron pattern (replacing the curved streets in Mayer's plan) also revealed Le-Corbusier's imposing and dominant personality. He incorporated philosophies picked from Ebenezer Howard called Garden City concept introduced in 1898 (followed in planning of cities like Radburn) and changed it to principles based on CIAM (Congress Internationaux d'Architecture Modern) [1, 5, 6, 9–11]. The Architect invested in picking concepts like posting network of green spaces in the urban core functioning as 'lungs of the city' as per the CIAM principles [12].

Chandigarh, around 240 km North of New Delhi was planned on a mildly sloping terrain amidst Himalayan Shivalik range foothills. Standing currently at 30° 50′ N and 76° 48′ E longitude and 304.8–365.76 m (above sea level) altitude, is the legendary example of a planned city. Pandit Jawahar Lal Nehru mentioned the city in a way "Let this be a new town, symbolic of freedom of India unfettered by the traditions of the past…an expression of nation's faith in the future".

Symbolic to human body, the planned city had a head (the Sector 1 Capitol Complex), the heart (Sector 17 City Centre), the Lungs (the open spaces), the intellect (institutions), the circulatory system (the hierarchy of roads, the 7 V's) and "*the viscera*" (Industrial Area). The city caters to four main functions: ***living, working, leisure and circulation.*** Capitol Complex houses three architectural marvels-Secretariat, high Court and Legislative Assembly. Large piazzas separate the three monument/buildings, where the symbol of Chandigarh-The open hand (demonstrating the philosophy of open to give, open to receive) stands upright. Enough circulation space, green spaces within the sectors, roundabouts, market places, shopping complex, gives a holistic look to the city. Out of the 20,000 acres (80.93 sq. km.) required for Chandigarh in the first phase, 2000 acres (8.09 sq. km.) was meant for recreation spaces. In a grid of horizontal (northwest/southwest) and vertical (northeast/southwest) roads, the city depicted a decorum of circulation and movement. The entire logic behind the designing of the city Chandigarh was to ensure a sustainable neighbourhood, hierarchy of circulation pattern and ample amount of breathing space in the form of green/open spaces.

Initially, the city had Phase 1 and Phase II due to various constraints, administratively and financially. Phase I comprised of 30 sectors which were developed with keeping an agenda of low density spread over 9000 acres (Sectors 1–30). The sectors were capable of accommodating 1.5 lakh population. The Phase-II consisted of 17 sectors having relatively higher density (Sectors 31–47) capable of accommodating 3.5 lakh population in 6000 acres of unbuilt land. The sector was the smallest unit bearing a population varying between 3000 and 20,000 persons. The circulation within the city was forecasted to be planned in a way that it is able to cater to the motorised vehicular requirement of the future. Emphasis was laid on systematically designing wide enough roads for easy movement so that the city escapes from the burden of higher traffic volume and does not lead to congestion in the future [13].

Apart from the inner city planning regulations, certain architectural controls were also imposed on the city. To preserve the natural environment and to prevent deterioration of the cityscape, it was made inadvertent to ensure a dignified living for all classes and proceed with sustainability in one of the top agendas to cater. The vernacular touch to the architectural built-ups including local building materials, fenestrations, shading mechanism, pilotis, courtyards, louvers and brick jalis were just made in conjunction with the micro-climatic conditions. The stability as evident in stringent planning of the grid-iron city can also be seen in wise utilization of the local materials and using them for sustainable built forms. The high rise complexes were avoided in the city's planning, as a result of a written contract between P.L. Thapar and Le Corbusier, the cityscape witnesses a low skyline [13].

7.4 Acts for Chandigarh's Development: Legal Framework

While we turn the leaves of historical documents, the growth pace of the urban areas was negligible enough to have foreseen the scenario where Le-Corbusier's spirit of a greenbelt can be defeated.

a. **Punjab New Capital (Periphery) Control Act of 1952**: Going back to history and remembering notes on the effectiveness of having a green belt around the main city core, a fringe area termed as 'Periphery' was introduced. This philosophy was adopted in various states in the United States and the United Kingdom. Same philosophy reflects in the designing of Letchworth and Welwyn. The Periphery Zone consorted to having a large chunk of agrarian land, 8 km (5 miles) around the urban core. The greenbelt was legally restricted to have any developments under the Punjab New Capital (Periphery) Control

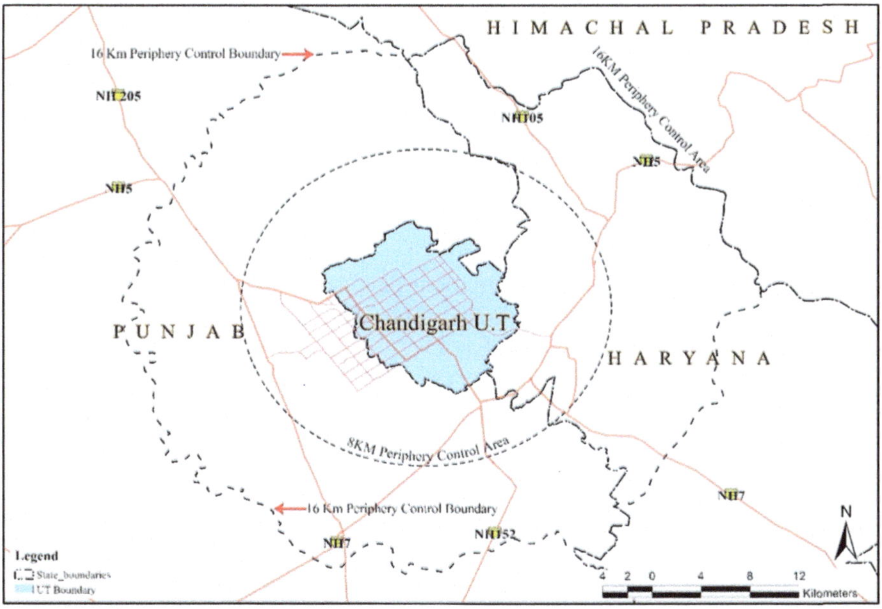

Fig. 7.2 8 and 16 km Punjab New capital (periphery) control boundary. *Source* Author and Chandigarh Master Plan Report-2031

Act of 1952 (see Fig. 7.2) [14]. The essence of 8 km buffer space can only be realized if we understand the basics of regional planning and the laws governing the same. The need of the Act is the resultant of concerns raised due to development in and around the city in a planned manner.

b. **Revised Punjab New Capital (Periphery) Control Act of 1962**: Punjab New Capital (Periphery) Control Act of 1952 was revised in 1962 suggesting a 16 km (10 miles) buffer as a result of establishment of an Army Cantonment, coming up of an Air Force and new township (see Fig. 7.2). However, in 1966 when Punjab was reorganized as Punjab and Haryana State, out of the total peripheral area called Periphery Control Area 1315 sq. km., 1021 sq. km. went to Punjab and 295 sq. km. went to Haryana and 114 sq. km. was reserved for capital city Union Territory (UT) Zone.

The Punjab New capital (Periphery) Control Area (5 mile or 8 km periphery) was a zone of fertile land comprising agricultural lands of the Indo Gangetic plans and forests of the Shivaliks. This periphery called "controlled area" had development restrictions beyond the urban context. This area consists of small agricultural settlements, villages like Manimajra (east of Chandigarh urban) and Kansal (north of Chandigarh urban) and Naya gaon (north of Chandigarh urban). The Act preserved the chances of development potential in the land for future development/ expansion (indicated by Lal Dora) for residential activities. The 8 km periphery was falling under Punjab's jurisdiction (Ambala District) [12, 15]. It was clear that development rights were reserved with the state's government as per the public acquisition clauses for ***"public purpose"*** [16]. The Govt. of Punjab justified the clauses by ensuring *"future extension of the capital" according* to a healthy and coordinated development and saving it from sprawl [12].

The prime concern behind demarcating the "Control Area" called "Periphery" was:

(a) Conserving the rural area comprising of village settlements and fertile agricultural land.
(b) To refrain from unplanned development outside the urban core.
(c) To discourage sprawl and development of suburbs.
(d) To preserve land prices in the urban core.
(e) To adhere to concepts proposed for Garden city.
(f) To maintain the importance of the built—Le Corbusier's built creation.

Le Corbusier's confidence in this dream creation, away from haphazard development (called sprawl evident in the cities of UK and UK) made him develop Chandigarh. His linkage with the city and the country made him propose the ideal situation, difficult to be maintained in the authoritative land of Punjab and Haryana [5, 17]. Le Corbusier was fascinated by the agricultural area and said that the "rural site must remain intact" [5, 17]. He wished to conserve the rural way of living where the urban areas could be dependent on rural for various agricultural produce and poultry requirements. He conceived that the rural periphery be maintained without any much infrastructure development possibility. Hence, for his dream, the sets of urban amenities were stitched into the preexisting rural fabric through minor or "light roads" restricting too much development. This harmonious-rural-urban agenda for an Indian scenario was more of a camouflage concept unaware of the Indian social way of living and the organic pattern of growth that it has witnessed in the past.

7.5 Growing Chandigarh: Development of Squatter

The Periphery Control Act of 1952 could not control the development in and around Chandigarh in an unplanned manner. The development took place in many phases.

The Le-Corbusier plan is considered to have overlooked the social fabric of the country and the essence of the economic strata was deeply neglected. The planning of Chandigarh catered to the need of all the sections of Indian society. It did not have housing provisions for many other non-governmental workers. Indian cities are hugely dependent on sweepers, maids, rickshaw pullers and drivers. The land allocated for this section was too costly and hence was beyond affordable limits for the daily wagers with limited income source. The city accommodated temporary shelters for this section, at the same time, areas in the adjoining villages within the periphery were urbanised due to the needs of the informal sector and slums.

The need was enhanced due to the in-migration the city witnessed due to the construction activities that was flourishing during 1955 and later years. The vernacular architecture approach in utilising the skill and material availability locally, made Pierre Jeanneret built nearly 50 styles of housing overall including housing, service buildings, schools, health centres, etc. [5]. The sectors were planned for setting a new style of urban living in the country, the needs of migrants made them settle in the periphery. Moreover, it was observed that Chandigarh's housing plan was restricted to public servants and had no consideration for floating population. Temporary migrants involved in construction activities were considered to leave once the task allotted was completed. But, the economic viability and Chandigarh's development (in phases) made way for their permanent settlement in the periphery abutting the urban area [5, 18].

In earlier designs developed by the principle, architects, sector 22 was earmarked for the lowest-level government workers. In one of the literature by Annapurna Shaw it is mentioned that "*Like all other sectors, it is fairly self-contained and crossed north-south by a land of greenery and east-west by a bazaar or Commercial Street lived with shops. The green area contains the educational and recreational features of daily life and lies within walking distance of the houses. The houses looked inward toward the green and away from the perimeter road that carries faster moving traffic. The sector is fed by a figure-eight slow road that connects all the housing areas. Some of the groups of houses possess their own internal green space. The cheapest housing provided two rooms, a veranda, kitchen courtyard and water supplied sanitation and washing facilities. Many devices were used to provide modern innovations, maximise space and still provide protection from hot sun*" [5].

When the need in 1959 was felt, then less developed sector like 15, 9, 20 & 24 accommodated low cost houses for low income worker known as 'labour colonies'. Ownership was transferred on nominal payment basis. However, as the other temporary sites were developed during 1950's and 60's, labour colony residents were moved to the periphery. Few households found place in resettlement colonies while others were accommodated in "transit camps" near the Periphery around existing villages. The initial settlements (called unauthorized settlements) were located near the capital complex abutting construction sites. Later in mid-1970's allotment of serviced plots with one room tenants were designated on the land acquired in villages surrounding master plan area. As per a survey by Estate office in 2006, mostly unauthorised settlements are located near the southern and eastern edge of the Union territory. At present there are 18 unauthorised settlements and 23,974 households. The unauthorised slums as per Chandigarh Master Plan-2031 [14] are:

(1) Kalyan Colony
(2) Kumhar Colony
(3) Shahpur Colony
(4) Rajiv Colony
(5) Guru Sagar Colony

(6) L.B.S Colony
(7) Nehru Colony
(8) Pandit Colony
(9) Kuldip Colony
(10) Mazdoor Colony
(11) Colony no.5 Labour
(12) Ambedkar Colony
(13) Kabari Colony
(14) Sanjay Colony
(15) Labour Colony
(16) S.B.S Colony
(17) Madrasi Colony
(18) Janta Colony

Within the lal dora boundary, currently unauthorised colonies bear a population of approx. 70,000 persons occupying 20,911 units in 2006 as per Census 2011 [19], the population residing in slums was 94,950 persons.

7.6 Chandigarh: The Demographic Profile

The earliest Master Plan prepared by Le Corbusier suggest phase-wise development. Phase 1, 2 & 3. Initially planned for 5 lakhs population in Phase-I, 36 sq. km. of land was acquired by the administration of the city for construction of 30 sectors. In the second phase, land was acquired for 3.5 lakhs population (Table 7.1).

The development in the third phase was not planned initially, but has now started in certain sectors beyond sector 47. The statistics of the population reveal that between 1951 and 1961, the decadal growth rate was 394.13%. The high growth rate was due to post independent time of housing requirement. Between 1951 and 1971, the population growth rate was 114.59%, one of the highest within any urban area at that time period. Subsequently, as the holding capacity of the two phase development ceased, the growth rate started declining from 75.55% between 1971 and 1981 to 40.28% during 1991–2001 (see Fig. 7.3). The current population within Chandigarh Union Territory as per census 2011 is 10.54 lakh [20], way beyond the actual carrying capacity limit estimated by Le Corbusier and team. It was during 1981 and 1991, the growth of the population rose beyond the estimated

Table 7.1 Composition of growth in Chandigarh UT

Composition of growth	1981–1991	% of total	1991–2001	% of total
Natural increase	63,505	1.41	86,110	1.34
In-migration	126,900	2.81	172,510	2.69
Total increase	190,405	4.22	258,620	4.03

Source Chandigarh Master Plan report–2031

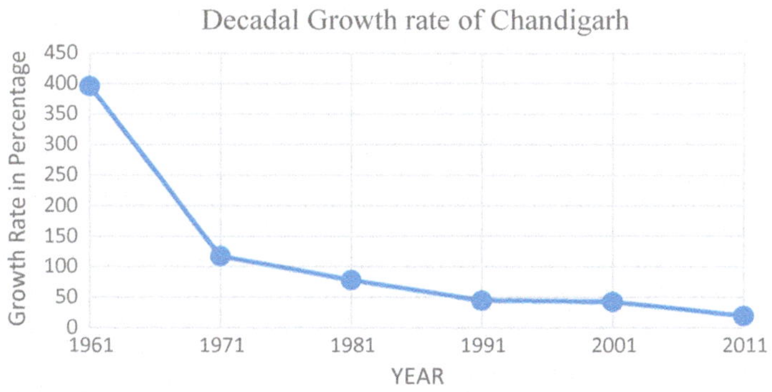

Fig. 7.3 Decadal growth rate of Chandigarh (1961–2011). *Source* District Handbook Census of India, 1951–2011

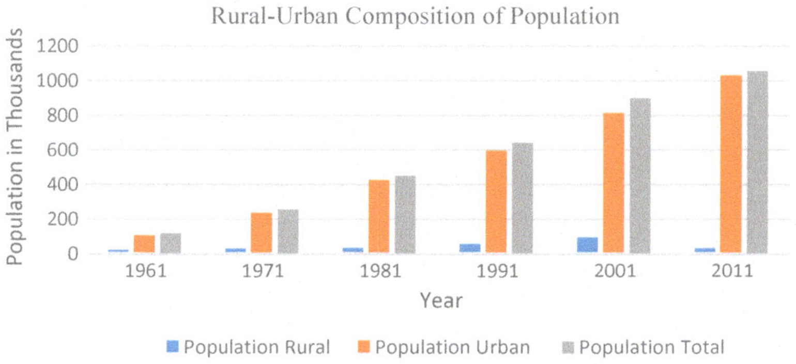

Fig. 7.4 Rural-Urban composition of Chandigarh UT. *Source* District Handbook Census of India, 1951–2011

5 lakh population limit for Chandigarh. It is also noteworthy that, as per the census 2011, 97.25% (10.25 lakh) of its population was urban and only 2.75% (29,004) was rural as compared to 82.80% urban and 17.20% rural in the year 1961 (see Fig. 7.4) [20].

The growth rate of the Chandigarh if observed and projected for the future, proposes an increase to 19.5 lakh (if the current rate of growth is considered)—almost four time to which it was originally built. It can also be observed the rural composition has declined over the period. This can be attributed to the fact that Chandigarh UT (Chandigarh and Manimajra settlement) has witnessed the need of outmigration. Also, the fact that the population residing in rural are resettled in the rehabilitation colonies falling under the gambit of urban area [14].

Another important issue while understanding the demographic dynamics of the city is to understand the population density pattern. The density of population has witnessed a nine fold increase from 1051 person per sq. km. to 2257 person per sq. km.. Subsequent decades also saw increase in the population density with 3961 person per sq. km. in 1981 to 5632 person per sq. km. in 1991, 7900 person per sq. km. in 2001 to 9252 person per sq. km. in 2011, recently. The population and population density increase from 1961 to 2011 shows the requirement of the settlement within Chandigarh UT or to develop counter magnets around the UT so as to absorb the pressure of in-migration. In the context of planning, it is a well-known phenomenon, wherever cities do not expand horizontally due to constraint of the land or the natural barrier like in the case of Chandigarh, where land for UT is limited and abutted by lake and Shivalik ranges, it either tends to grow vertically with minor FAR modifications (under governance agenda) or the nearby settlement start developing and result in conurbations. Once the city planned for having a density of 16 persons per acre has grown to 26 person per acre in phase I and from design density of the 59 person per acre in phase II to 60 person per acre in 2001 [14]. The population worked on the basis of holding capacity, the density of phase II and 100 person/acre (highest density) for phase III sectors. It is also observed that sector in each phase have varying density of population. As per Chandigarh Master Plan-2031 [14], the sector III is expected to accommodate the rising need of population and settlement in the form of group housing and large scale rehabilitation of unauthorised settlements in these sectors. Population projection made by census through various methods for Chandigarh reveal that the population may remain between 14 and 33.5 lakhs approximately in 2031 (Refer Table 7.2)

It is inadvertent to understand that the land resource within the Chandigarh UT is limited and is exploited at high pace. In order to preserve the image of city Chandigarh as an administrative city, the influx of the population in the form of in-migration should be avoided to a greater extent. Moreover, since the land resource is limited within the UT boundary, the uses pertaining to administration sector and the governance related centres shall get priority during the allocation of available land. Any additional population beyond the holding capacity of the city

Table 7.2 Sector-wise density pattern of Chandigarh

Phase	Planned density (persons/acre)	Density as per census 2001 (persons/acre)	Density as per holding capacity (persons/acre)	Area planned (in acres)	Planned population
I—(Sector 1–30)	16	26	34 (33.37%)	9398.83	1.5 lakhs
II—(Sector 31–47)	59	60	83 (18.31%)	5158.76	3.5 lakhs
III—(Sector 48–56, 61, 63)	~	Under process of development	100 (6.64%)	1870.54	~

Source Chandigarh Master Plan report–2031

i.e. 15.5 lakhs, need to be diverted to settlements adjoining the city, keeping in the view of the regional context of planning.

About 60 acre of land was acquired by the Chandigarh administration in 1989 for building more dwelling units called modern housing complex and understanding the need of housing due to population increase. Chandigarh UT houses 201,878 total households as per 2001 and 235,061 households in 2011 [21, 22].

7.7 Chandigarh: Land Uses/Land Cover

The existing land use of Chandigarh comprises of planned developments of Phase-I & II and the proposed subsequent development of the phase-III (the extended sector grid). Predominantly, in Phase-I & II, 39 out of 46 sectors are Residential, while the remaining are governed by Institutional and commercial land use. Some section of the light industrial and Sukhna lake also fall under the same category. (Refer Fig. 7.5 and Table 7.3)

Out of the total development area i.e. 16427.73 acres, 9398.83 acres (nearly 33.37% of the UT area) is developed under Phase-I. The area Under Phase-II is 5158.76 acres (nearly 18.31% of UT area). Phase-III has designated 1870.54 acres of land (nearly 6.64% of UT area) for development purpose. Out of total 28,170 acres, 41.68% falls within the remaining periphery area within UT category.

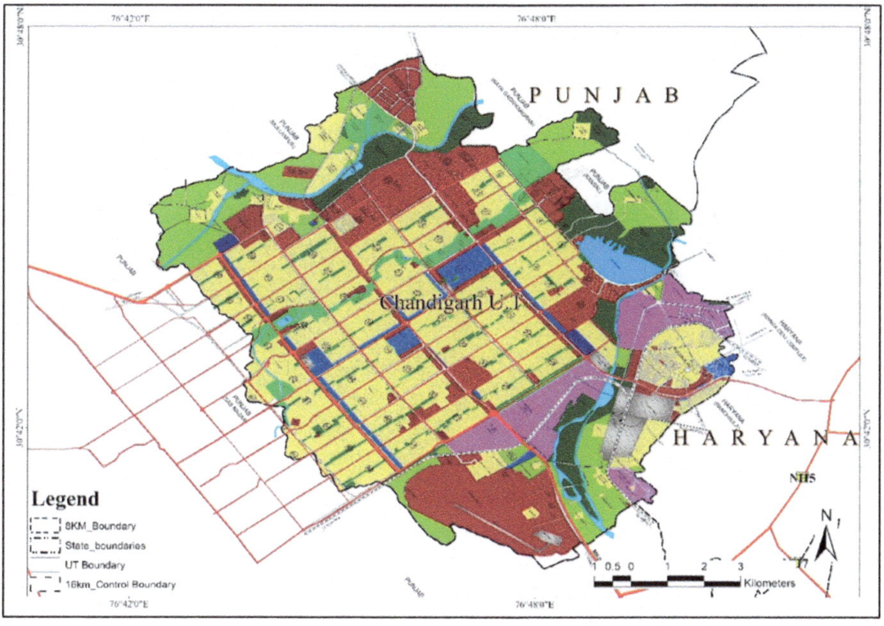

Fig. 7.5 Chandigarh existing land-use map 2008. *Source* Chandigarh Master Plan Report-2031

Table 7.3 Chandigarh existing and proposed land-use distribution—2008

Chandigarh UT land use	Existing area (ha)	Percentage (%)	Proposed Area (ha)	Percentage (%)
Residential	4319	37.9	4398	38.6
Commercial	542	4.8	599	5.3
Transportation	828	7.3	849	7.4
Indus/IT Park	537	4.7	652	5.7
Pub./Semi-Public	1201	10.5	1386	12.2
Recreational	983	8.6	1141	10
Agricultural	0	0	272	2.4
Public Utilities	122	1.1	138	1.2
Railway Land	128	1.1	128	1.1
Defence	637	5.6	637	5.6
Forest	855	7.5	1038	9.1
Reserved	0	0	125	1.1
Vacant land	1248	10.9	46	0.4
Total	11,400	100	11,400	100

Source Chandigarh Master Plan report–2031

Phase-III has the lowest area possible (1/4th of the Master Plan Area). The peripheral area within the Chandigarh UT boundary comprises of village abadis (population), agriculture land and unplanned need based development (accounting to 11741.88 acres) (Refer Fig. 7.6).

7.7.1 Phase-I, II & III

Largest proportion of land within the residential area is allocated to residential (42.69%) in Phase-I. In Phase-II, the residential component is the highest i.e. 67.08% of the total area. Residential area within Phase-III also occupies nearly 3/4th of the total area.

7.7.2 UT Periphery

Large chunk of land in the periphery have been acquired for accommodating new land use. The road networks were accordingly enhanced for catering to the need of development. 1/6th of the area has been allotted for residential use. This area caters 4.5% industrial area or industries. Total area under forest and agriculture covers 21.67% within the periphery. Le Corbusier also envisioned Chandigarh as a radio-concentric urban area surrounded by agriculture land. He thought that the city should not experience the dirt of industrial establishment. According to him,

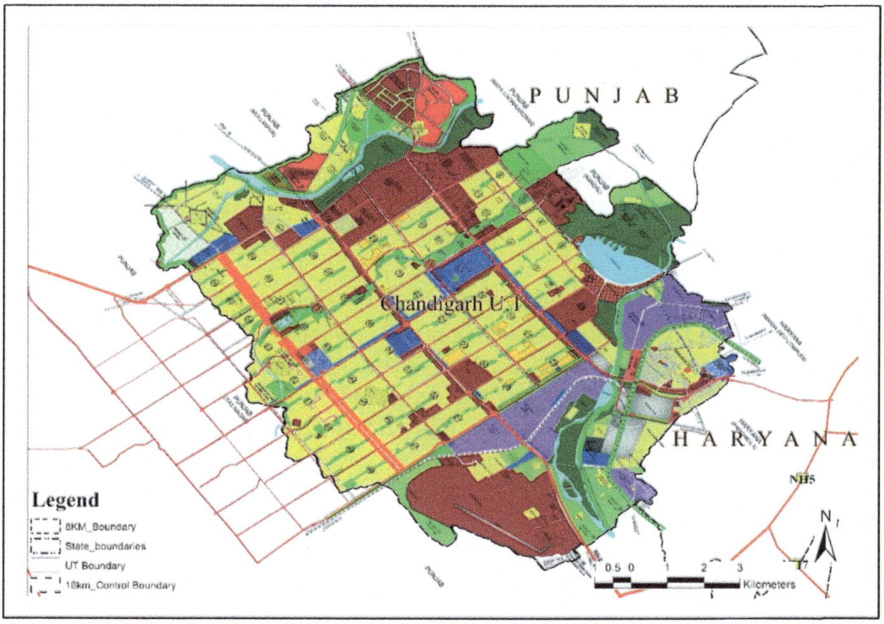

Fig. 7.6 Chandigarh proposed land-use map 2031. *Source* Chandigarh Master Plan Report-2031.

Chandigarh had to be developed as an administrative city. He writes that "Chandigarh is an administration-city and in consequence is a 'RADIAL-CONCENTRIC' city. It has never to become 'industrial-city". Regardless of his aspirations of an industrial free administrative centre, under administrative and official pressure, Le Corbusier reserved 580 acres at the south-eastern edges of the urban core for industrial setup. Out of 580 acre, initially for Phase-I, 338 acres were made available for development [12]. Today, the total industrial area within the UT has risen to 1326.50 acres, almost thrice the allocated land during the 1960s.

7.7.3 *Development Within the 16 Km Periphery Control Area*

Although, no development was initially planned for periphery, but as per the requirement and need of national defence by the government, Chandigarh's location prompted its choice for the government defence settlement. In spite of Le Corbusier's vehement opposition to all proposed military projects on the periphery, the Cantonment (called Chandigarh Cantonment, headquarter of western command of Indian Army) was proposed near lake Sukhna and capitol complex. Apart from Cantonment, Air force Station and township of state-owned Hindustan Machine Tools (HMT) as also planned. All the project called 'Special project' were identified in the 'Periphery Control Plan of Chandigarh' after 1962. It was in 1962 that 5 miles (8 km) radius boundary was extended to 16 km radius boundary [23]. It was later after Le-Corbusier's death in 1965, the Army Cantonment was built which was otherwise a non-appreciated construction requirement as per the principle Architect & Planner, Le Corbusier and his team. Today, the Chandigarh cantonment is serving as an economic base to Chandigarh. It is spread across 385 acres and contain all essential establishments like family quarters, stadium, hospitals, schools etc. The terminal Ballistve Research Laboratory also sprawls within the periphery.

Subsequently, in 1966, the state Punjab was further divided into the state of Haryana and Punjab based language speaking capability. After two decades of partition between India and Pakistan, this partition left Chandigarh serving as the capital to both the state. This proved the threefold richer administrative position of Chandigarh. With the creation of the Union Territory of Chandigarh having 70 sq. km. as the city and a very meagre part of the Periphery i.e. 44 sq. km. as a part it. The remaining 1021 sq. km. of the Periphery Control Area went to Punjab (75%) and 295 sq. km. (22%) fell into Haryana's jurisdiction. The sanctity of the Periphery Control Act was maintained but also paved ways for complex decision making related to development scenario [24]. An unwanted but natural competition now existed pertaining to the development of the two states of Punjab and Haryana, covering the majority of the Periphery Area of the UT of Chandigarh.

7.7.4 Establishment of Mohali and Panchkula

Mohali, also called Sahibzada Ajit Singh Nagar (SAS Nagar) was conceived in late 1966. The site of Mohali, immediate south of Chandigarh was a home to Harappa's around 8000 years ago. In 1967, around the existing Mohali Village was conceived to be developed as an industrial hub. On 1st November 1975, foundation stone for Mohali was laid. The land allocated for Mohali, industrial Estate was 5500 acres in Punjab. The sectorial grid of Mohali which were known as Phases here, few sectors fall under Chandigarh and Mohali both, like sector 48, 51, 52, 54 and 56 onwards. The development of Mohali was a landmark event in the history of development potential into an industrial area in Punjab. Impressed by the Chandigarh's grid-iron design, not much emphasis was laid on the designing of sectors and the extension in contiguity was picked up.

Simultaneously, plans for Panchkula in Haryana were approved in 1970, 4 km east of Chandigarh bearing 500 acre land. The Mohali made a contiguous development leaving no buffer and following same design, Panchkula maintained a slight buffer of 4 km and used a fan shaped grid plan. It seems, plan of Panchkula was inspired by the original plan of Chandigarh proposed by Albert Mayer. As a matter of fact, Chandigarh was more set built as a residential area, Panchkula and Mohali both were setup as industrial establishment in their urban development strategy.

It is notable to mention, that despite 'No Construction' in the Periphery, the two major counter-magnets, Panchkula in Haryana and Mohali in Punjab came up to accommodate population influx. It is also observed that the growth of the city of Chandigarh was slower as expected by the planner. This involved the need of satellite town for attracting the population to Panchkula and Mohali in search of jobs.

Meanwhile in Chandigarh, the village of the Manimajra, south-east of Sukhna lake and after the Phase-I industrial establishment, was given the urban status. Immediate construction with apartments and malls are passed. Unlike the actual agenda, Information Technology (IT) Parks like Rajiv Gandhi Chandigarh Technological Park (was developed south of Lake Sukhna) in 123 acres transforming agriculture land into the IT setup. Desires to setup IT industries due to demand and additional 272 acre of land was acquired for the same. Chandigarh was never meant to delve into competitive endeavours relating to IT sector and SEZ (Special Economic Zone).

Now, the tri-city—Chandigarh–Mohali–Panchkula can be seen to developing with visions of development within the periphery paving ways for a regional development with each passing decade. Contradictory to the agriculture based periphery the 16 km periphery control area has seen even more small settlements behaving as counter magnat to the tri-city. All attempts of planned development are compromised and the regional development plans hold no legal status in proving its eminence.

7.8 Regional Planning Initiatives

Township in the periphery, now emerged due to the pressures of development and resulted in regulated and unregulated growth. In 1975, a coordination committee was set up chaired by the Secretary, Ministry of Urban Development, GoI and chief secretaries of the state of Punjab and Haryana with chief commissioner of UT Chandigarh to resolve the matters related to development scenario in and around the UT without compromising the original intent of the original Chandigarh master plan. However, the development of Panchkula and Mohali in close proximity abutting the UT boundary within the 16 km Periphery Control Area had already compromised the original sanctioned and approved plan. It was only in 1977, a Regional Plan for Chandigarh known as **Chandigarh Urban Complex (CUC)** was delineated as the immediate region covering an area of 330 sq. km. (see Fig. 7.7). It consisted of the Chandigarh UT, part of Mohali with adjoining 27 villages and parts of Panchkula with 23 villages. The CUC covered the region only within the 8 km (5 mile) periphery region. It was felt that region conceptualised is too small and may need revision later.

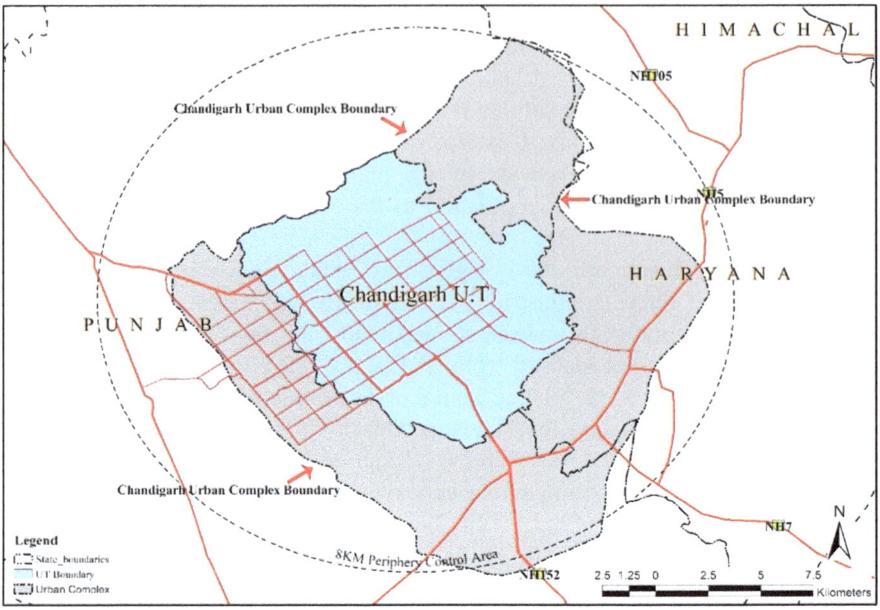

Fig. 7.7 Chandigarh Urban Complex (CUC) Boundary. *Source* Author and Chandigarh Master Plan Report-2031

Within the CUC plan, north of the capitol complex was declared as '*No development Zone*'. In 1984, the CUC plan was reviewed and need for a regional plan was felt. Therefore, an 'Interstate Regional Plan for Chandigarh region' [25] was prepared in 2001 by Town and Country Planning Office for 25 lakh population distributed in 7 categories of settlements (see Fig. 7.8). This plan was also known as the structure plan. This plan was an outcome of the forth coordination committee meeting held on 22nd May 1997. The committee figured a need to give statutory shape to the regional plan. It was also decided to follow up the preparation of Regional Plan by the respective state/UT Chandigarh and areas around it considering Ambala district (Haryana), Solan district (Himachal Pradesh) and Rupnagar and Patiala district (Punjab) covering an area of 2421 sq. km., averaging to a radius of 35 km from the core of Chandigarh city.

ISCR-2001: The ISCR-2001 plan (Refer Fig. 7.8) was a comprehended document indicating the delineation criteria for the region catering to the administrative boundaries already existing within the identified area. It laid emphasis on the Regional Development policy control and made recommendation for strategies Regional Development. The ISCR was more of a policy-cum-advisory document focusing on integrated, coordinated and wholesome development for the region. The plan had its own policy agendas to cater to following reasons:

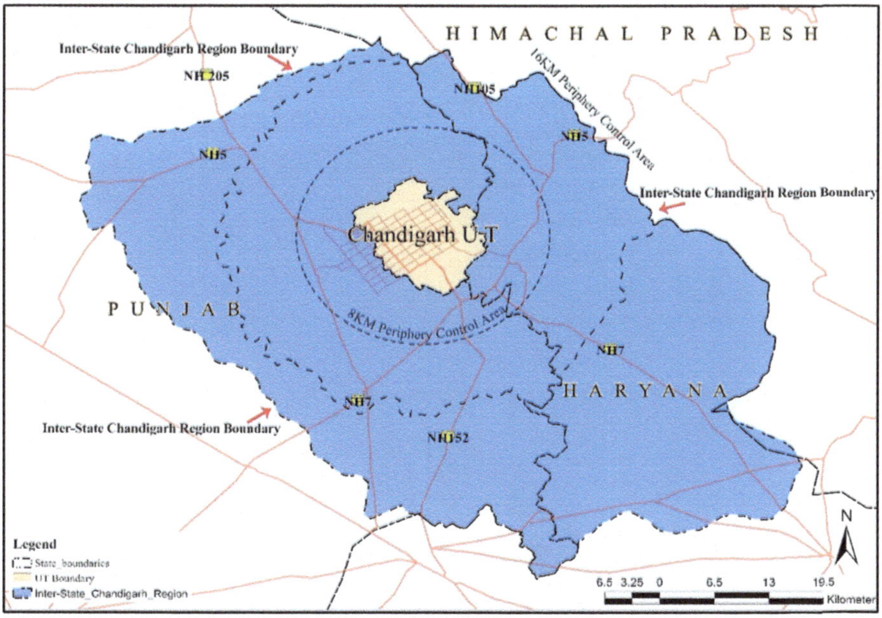

Fig. 7.8 Inter-State Regional Boundary for Chandigarh-2001. *Source* Author and 'Concept note on Chandigarh and its region', 2011

(a) Hierarchy of settlement: First order to Seventh order
(b) Development of rural areas
(c) Providing housing
(d) Promoting trade and commerce
(e) Development of Industries
(f) Conservation of environment and ecology
(g) De-centralisation of Government (quasi and semi-govt.) offices from Chandigarh.

The hierarchy of settlement identified by ISCR plan is as under (Table 7.4). The ISCR was a holistic regional development plan suitable for a planned and entangled relationship with the surrounding regions. However, the plan could not be implemented due to lack of legal stature and no enabling provision and notification. It can inferred that in 1984, effect were made for planned regional development in and around Chandigarh laying emphasis on two important aspect of development, trade & commerce and industrial development as a part of globalization era in India.

Chandigarh Inter-state Metropolitan Regional Plan 2021 (***CISMeR***): Professor E.F.N Riberio emphasised on the need of having a comprehensive framework for metropolitan regional development (Refer Fig. 7.9). The CISMeR Plan was proposed for a total of 5702 sq. km. (more than double the area identified for the ISCR-2001 plan. The Metropolitan plan proposed development in six layers [26]. Layer-I & II consist of Chandigarh's Phase-I & II catering to an area of

Table 7.4 Hierarchy and number of settlements within ISCR zone

Order	Settlement units	Number
1st	CUC	1
2nd	Regional Town	10
3rd	Sub-Regional Town	4
4th	Growth Centres	10
5th	Sub-Growth Centres	10
6th	Focal Villages	–
7th	Basic Villages	–

Source Concept note on Chandigarh and its region, 2011

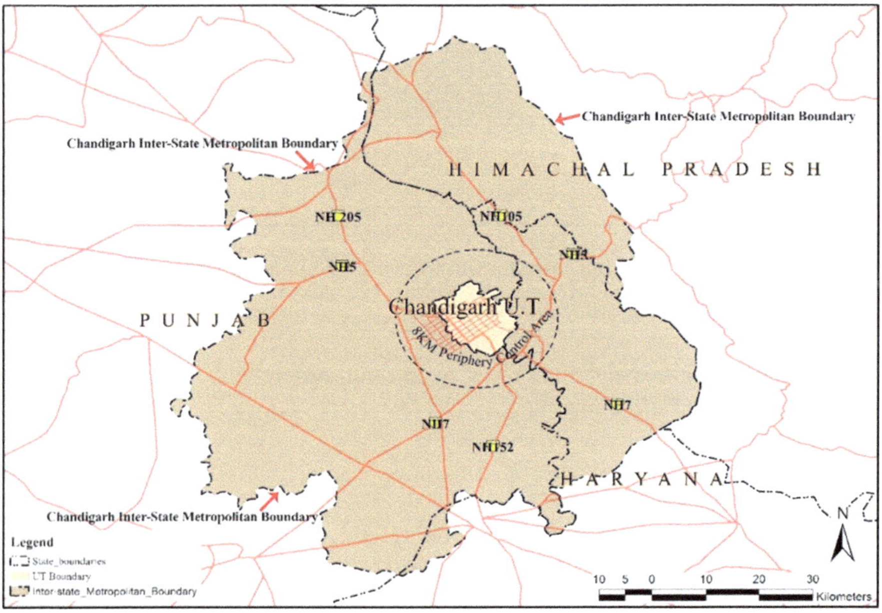

Fig. 7.9 Chandigarh Inter-State Metropolitan Regional Boundary-2021. *Source* Author, Concept note on Chandigarh and its region, 2011

70 sq. km. Layer-III is the old 8 km peripheral belt (1952), layer-IV deals with the Chandigarh Metropolitan Complex i.e. the CUC outside Chandigarh UT in Punjab and Haryana possessing an area of 435 sq. km. Subsequently, layer-V with 613 sq. km., a 16 km peripheral control belt in Punjab and Haryana and finally layer-VI, possessing 4158 sq. km. with a radius of 50 km area beyond layer-I & II totalling to an area of 5702 sq. km. (refer Tables 7.5 and 7.6)

Basically, the development scenarios proposed by CISMeR is a layering concept spread to approximately a fifty km (31 mile radius) for a project population of 65 lakhs. Considering Chandigarh to be the major metropolitan centre of growth, the CISMeR-2021 focused on developing all classes of town with villages. The primary

Table 7.5 Layers of Chandigarh Inter-State Metropolitan Region Plan

Layer	Area (sq. km.)	Constituents	Remarks
Layer I	43	Phase I: Sector 1–30	Original planned and designed part of Chandigarh
Layer II	27	Phase II: Sector 31–47	Original planned but not designed part of Chandigarh
Layer III	44	Sector 48–56	Part of old 8 km peripheral belt of 1952
Layer IV	435	Chandigarh Metropolitan Complex	Outside Layer I–III in lieu of the 1975 Chandigarh Urban Complex outside Chandigarh UT in Punjab and Haryana.
Layer V	613	16 km Peripheral Control belt in Haryana and Punjab	Outside Layer I–IV
Layer VI	4158	Radius of 50 km beyond layer I & II	10 tehsils (4 in Punjab, 3 in Haryana & 3 in Himachal Pradesh
Total Area	5702		Layer I–VI

Source Concept note on Chandigarh and its region, 2011

Table 7.6 Development proposal in six layers of CISMeR

Layer	Designed for population	Planning and development proposal
Layer I	3 lakh	Layer I to be preserved as per its original character
Layer II	5 lakh	Layer II to be preserved as per its original character, however 4 urbanised villages are to be included
Layer III	13 lakh	Layer I + II + III planned development to accommodate 13 lakh population by 2021
Layer IV	12 lakh	Punchkula + Mansadevi + Mohali + Kharar
Layer V		16 km Peripheral Control belt in Haryana and Punjab
Layer VI	Around 38 lakh by 2021	Outside Periphery Control Area of 16 km

Source Concept note on Chandigarh and its region, 2011

objective was to create a regulated development framework for residential and industrial zone which shall accommodate a uniformly distributed population influx. This was essential for refraining the region from experiencing uncoordinated growth in ecologically sensitive areas, agriculture land and already populated area like Chandigarh [3, 27]. CISMeR-2021 was more so an effort for developing the three state Haryana, Punjab, Himachal Pradesh and one UT Chandigarh in a decentralised manner [3].

Again due to lack of enabling powers, legal status and lack of support from states from Punjab and Haryana, the CISMeR-2021 plan could not be implemented [28]. Lack of carrying capacity analysis, incorporation of environmental perspective and lack of inclusion of financial implication has put the plan in question.

7.8.1 Chandigarh Master Plan-2031

In 2008, the requirement to control inconsistence prevailing due to land acquisition surrounding the Rajiv Gandhi Chandigarh Technological Park RGCTP (built on agriculture land, followed by protest from farmers and environment & ecology community), Chandigarh Master Plan-2031 Chandigarh Master Plan [14] was released in July 2015 [29]. The plan focuses on planning control within the UT portion of the Periphery covering the urban core.

7.8.2 Greater Mohali Region: Regional Plan-2058

The Regional Plan for Mohali discusses the proposals and strategies for Greater Mohali Region (GMR) covering an area of 1190 sq. km. located westwards to Chandigarh. It is divided in six planning zones viz. SAS Nagar, Zirakpur, Kharar, Mullanpur, Banur and Derabassi. The report discusses that the 60% population of the 0.7 million is rural and the other 40% is concentrated in SAS Nagar and Zirakpur coming out as satellite establishments acting as counter magnets to Chandigarh. The comprehensive document deals with the planning strategies and recommendations for almost all sections of the development sector. Recently, the Local Planning Area of Mullanpur is renamed as 'New Chandigarh' and a Master Plan for its development was proposed to be prepared in 2007. The primary objective was to ease out the pressure in Chandigarh and Mohali due to economic developments proposed to boom in the next 20 years.

7.9 Defining Chandigarh Region

The entire region of Chandigarh bearing a legal status can be regarded to be that of the 16 km periphery as per the Periphery Act, 1984 and consists of three major region Chandigarh, Punjab, Haryana (see Table 7.7). The 16 km periphery comprises of the many major settlements.

The growth and development in and around Chandigarh has finally resulted in exerting pressure on the services and utilities within Chandigarh (refer Figs. 7.10, 7.11 and 7.12). This has eventually given birth to adjoining towns within the Greater Mohali Region and Panchkula Region in Punjab and Haryana, respectively. The three major urban centres called the Tri-city viz. Chandigarh, Mohali and Panchkula form a triangle of activities and the other towns adjacent to the three towns support its development in the form of satellite towns. Due to legal constraints, no metropolitan regional plan could be devised for Chandigarh, but the 16 km Periphery has legal sanctity, hence the Periphery Region is considered further for recommendations and strategies helping in translating the Tri-city Region into a Smart Metropolitan Region.

Table 7.7 Major settlements within 16 km Periphery region and its functions

State/UT	Settlement	Population			Function(s)
		1991	2001	2011	
Chandigarh	Manimajra	44,710	5521	15,489	IT Hub
Punjab	Mohali/SAS Nagar	78,547	123,484	166,864	Industrial City
	Kharar	26,109	42,289	74,460	Education hub
	Zirakpur		25,022	95,553	Business hub
	Banur	10,014	15,013	187,755	Institutional hub (Knowledge-Business-Technological)
	Derabassi	9602	15,841	26,295	Industrial City
	Mallanpur	419	6147	6165	Resort centre and Regional Playground
Haryana	Panchkula	70,375	140,925	2,11,355	Industrial City
	Pinjore	13,529	29,609	35,192	Industrial City
	Kalka	27,281	30,830	34,134	Army Base
Himachal Pradesh	Parwanoo		8609	8758	Industrial City
	Baddi		22,601	25,639	Industrial City
	Barotiwala		2460	1743	~
	Kasauli		4990	3885	Cantonment Town

Source Census of India-1991–2011; Chandigarh Master Plan-2031; Greater Mohali Regional Plan; Report on Formulating a Master Plan for Chandigarh, 2009

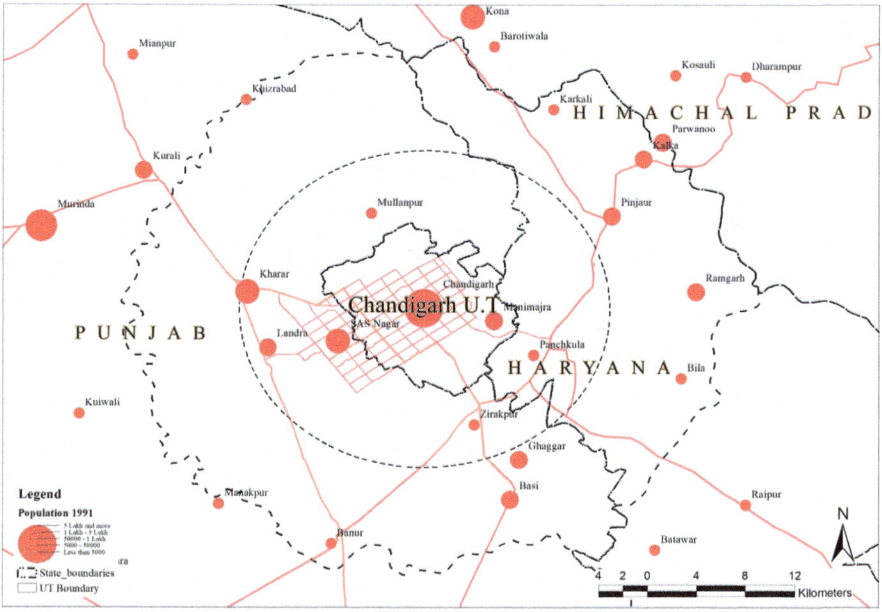

Fig. 7.10 Chandigarh and its region including major growth points-1991. *Source* Author and Concept note on Chandigarh and its region, 2011

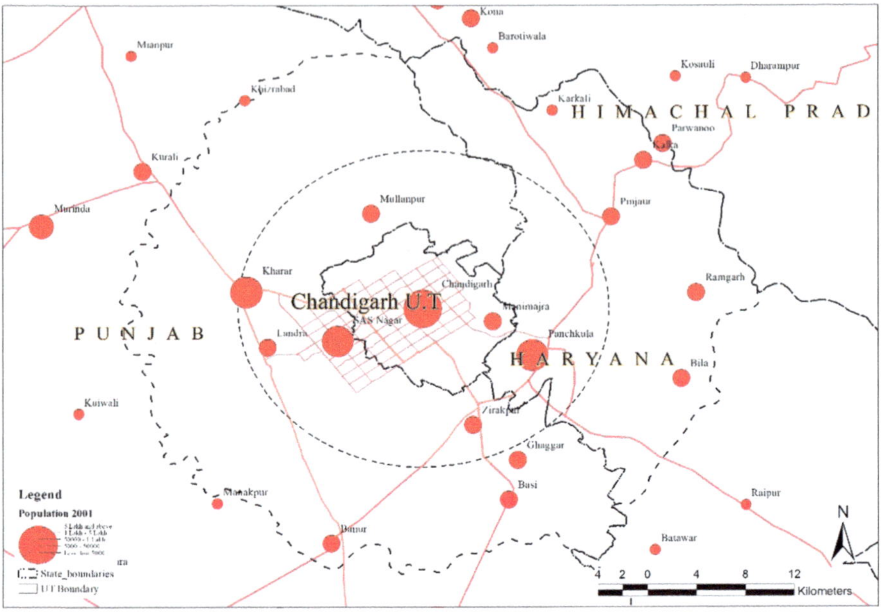

Fig. 7.11 Chandigarh and its region including major growth points-2001. *Source* Author and Concept note on Chandigarh and its region, 2011

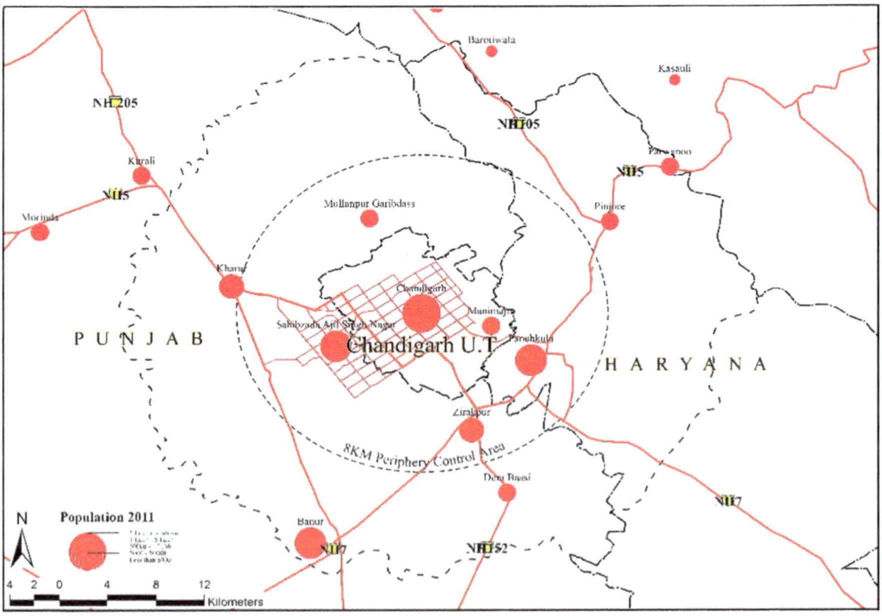

Fig. 7.12 Chandigarh and its region including major growth points-2011. *Source* Author and Concept note on Chandigarh and its region, 2011

The 16 km Periphery Region currently consists of the following areas:

1. Chandigarh
2. HMT Township
3. Chandimandir Cantonment
4. Manimajra
5. 22 villages
    a. Sectoral Villages (Burail, Badheri, Buterla, Attawa, Kajheri, Palsora)
    b. Non-sectoral villages (Kaimba, Khuda Ali Sher, Khuda Lahora, Khuda Jassu, Sarangpur, Dhanas, Dadumajra, Maloya, Kishangarh, Hallomajra, Behlana, Daria, Mauli Jagran, Makhan Majra, Raipur Kalan, Raipur Khurd)
6. Sukhna Wildlife Sanctuary

[1–6: Chandigarh UT (114 sq. km.)]

7. Panchkula
8. Panchkula Extension-1
9. Panchkula Extension-2
10. Shree Mata Mansa Devi Complex
11. Kalka Pinjore Urban Complex

[7–11: Haryana Part of Periphery (235 sq. km.)]

12. Mohali (SAS Nagar)
13. Mullanpur
14. Derabassi
15. Banur
16. Kharar
17. Zirakpur
18. Proposed New Chandigarh

[12–18: Greater Mohali Region (GMR)]

19. Kansal
20. Naya Gaon

[12–20: Punjab Part of Periphery (1021 sq. km.)]

Primarily, Chandigarh UT is developed and further refinements in the name of densification is the process being followed to accommodate the increasing population in Phase 3 construction sectors. Chandigarh UT comprising of 114 sq. km. of area comprises of fully developed 70 and 44 sq. km. periphery area within UT. The 44 sq. km. of the area within Chandigarh UT is regulated by the Punjab New Periphery Control Act, 1952 with the exception of abadi areas of the villages falling within it like Dadumajra, Palsora, Kajheri, Sarangpur, Khudda Ali Sher, Khudda Jassu, Khudda Lahora, Maloya, Raipur Kalan, Raipur Khurd, Behlana, Hallomajra, Makhanmajra, Kishangarh and Manimajra. It also comprises of two river channels called Sukhna Choe and Patiala ki Choe [14, 30].

As per Le Corbusier, the primary intent of the plan was to make a CITY as container and the PERIPHERY as provider. But recently, it was observed that the over spilling of abadi of the peripheral villages led to the changed character of PERIPHERY region within the 16 km radius. To understand the unregulated, unprecedented and haphazard growth has to be systematically monitored from time to time, functional relationships has to be understood leading to growth and its

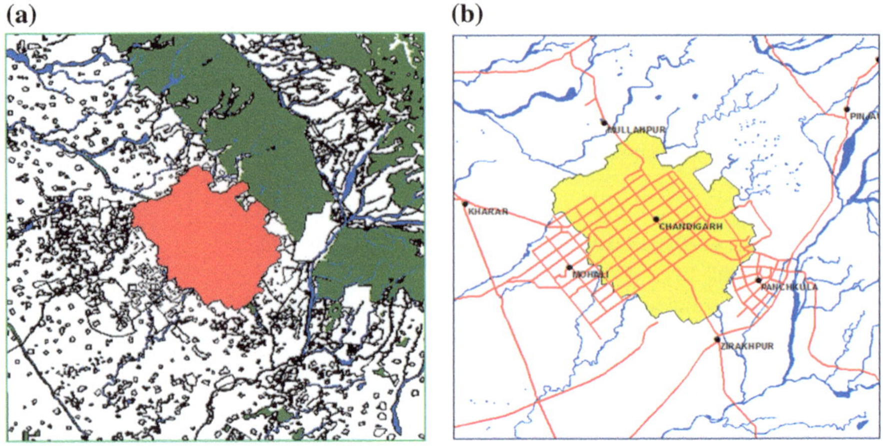

Fig. 7.13 Shivalik foothills running NW to SE, NE of UT influencing the local climate and Drainage NE to SW with Chandigarh's grid-iron pattern

directions. Finally, a suitable framework of spatial design strategies in terms of demand, present condition and future demand scenario can be assessed keeping in mind the legal stature of the available land.

7.9.1 Regional Setting: The Periphery Zone of 16 Km

The total area within the periphery zone (approx. 2431 sq. km.) falls under three jurisdiction of Chandigarh UT, Punjab and Haryana. The area around Chandigarh is surrounded by fragile and environmentally sensitive Himalaya ecosystem zone. This area comprises of protected reserved forests. The entire region forms the foothills of the Shivalik hills in Himalaya and has mainly a flat topography with a gentle slope towards southern-western part of the area. Ghaggar River & its tributaries form the major surface hydrological feature in the area. The area experiences an average annual rainfall of 111.4 cm/year. The soil is extremely fertile with annual deposition of river silt suitable for raising multiple crops. The subsurface formation comprises of beds of boulders, pebbles, gravel, sand, silt, clay and Kankar. The average elevation of the region ranges from 400 m above MSL to about 200 m MSL in the plains (refer Fig. 7.13).

Climate: this area has humid subtropical climate characterised by very hot summer, mild winter with great variation in the temperature pattern. Areas closer to Chandigarh receives cold wind from the north near Shimla and Jammu & Kashmir. The area can be regarded to have a dry weather where evaporation exceeds precipitation.

Linkages Transport: located at the northern part of the country, four national highway intercepts the region, namely NH21 to Ropar, NH95 to Ludhiana, NH22 to Shimla and NH1 to New Delhi via Ambala.

The entire area has an excellent road network link. Apart from highways, railways provide a means of intercity transport for the majority population. Twin railway tracks meet at Ambala. Single railway track joins at Chandigarh & forward to Kalak. The north-east and north-west part of the region is served by the newly build single track from Chandigarh to Morinda. An airport at Chandigarh (domestic) serves the connections with most part of the nation, but no international airport is built in this region.

The region is fertile, nearly flat and observes a view of Shivaliks with healthy environment. The prevalent ecosystem is an added advantage which prevents unscrupulous builders from building in the sensitive sites. Apart from Sukhna Lake as a major tourist destination within Chandigarh, the periphery houses Siswan Lake (about 13 km NW of Chandigarh) and Satluj Yamuna link canal (about 17 km SW of Chandigarh).

7.9.2 Factors Affecting Growth

The growth of any region is governed by some stimulating factors that control the pace and direction of growth in an area. The Periphery Zone is governed by few such major urban centres/settlements that have their unique reasons depicting scenarios of growth. In order to understand the spatio-temporal analysis of urban growth scenario in cities, it is important to identify and analyse the factors that drive urban development. There are various factors responsible for the process of sprawl like demographic, social, economic, cultural, physical/spatial, environmental, governance, proximate causes, biophysical, existing conditions, etc. The most indicating factor for urban growth to take place is demographic characteristics of the city which can be due to population growth or the migration scenarios. Economic factors are such as the job and business opportunities, trade and commerce, industrialization, land prices, market failures, etc. which have huge impact on the changes in the growth pattern of cities. Upcoming new industries or infrastructure projects also have huge impacts on the same. These indicators build competency between cities which is an invisible economic and social factor of urban growth. The planning and governance policies, also the regulatory framework play a major role in dealing with urban sprawl which are formulated to regulate and channelize the growth of cities. Land use zoning, investment plans, transportation policy, Govt. schemes, land management, FAR/FSI restrictions in the city are examples of this.

A hypothesis believes that nations having stringent land use policy controls, considerate and less dispersed governance system, very well manage the land development available in the area. It is also inadvertent that urban sprawl is also considered an intrinsic phenomenon of cities, sometimes promoted by local governance as well. Some local authorities like municipalities may facilitate growth directions to attract new inhabitants and increase the population and some restrict the growth. The evolution, spatial patterns of growth since the city existed, land use, density, regional setting and infrastructure are some of the spatial factors responsible for growth. The biophysical factors refer to the characteristics and processes of the natural environment like climate change, topography, soil types, availability of natural resources, etc. Proximity causes are one of the most talked about in researches where the areas located in the proximity to infrastructure facilities, roads, rivers, central business districts, etc. are studied which generally tend to grow in future due to potential benefits like ease of access and social services. Existing built-up is also one of the game changing factors around which the future urban growth is indicated to take place. Political situations also affect the growth pattern.

The growth scenario in Chandigarh can be attributed to several factors listed below:

- **Regional Centre**: Chandigarh serves as the capital for two prime states: Punjab and Haryana. Being a regional centre, it is a centre of administrative and political activities from the adjoining states. As a result, the UT attracts population from various parts of the adjoining states. Chandigarh has seen an in-migration of 1.269 lakh to 1.725 lakh in 1981–91 to 1991–2001, respectively.
- **Location**: The picturesque location of Chandigarh is an added advantage for the residents residing in the foothills of Himalayas.
- **Socio-Economic Status**: There is a close co-relation between social infrastructure, social performance and economic development in any setting that drives settlement in those areas. The UT has gained good economic status when compared with other cities of the nation. With an average per capita income of Rs. 67,350 (Indian average Rs. 23,241 as per 2001, three times the national average) [31], Chandigarh is rated as the '*Wealthiest Town*' in India. Also, it was rated sixth most prosperous city in terms of family wealth. The city is regarded to have first rank in overall Human Development Index with a relative positioning of Chandigarh UT based on health (life expectancy), education (literacy) and standard of living (per-capita income adjusted for inequality) as 0.784 in 2007. Chandigarh holds 3rd position in Health Index and Education Index, and 1st position in Income Index and Human Development Index [32]. Chandigarh has one of the least percentage of population below poverty line amongst north Indian states with only 7.1% of population below poverty line. The income inequality is not prevalent as per the Planning Commission report, Reverse Bank of India [33]. The Infant Mortality Rate (IMR) in Chandigarh is 20 deaths per 1000 live births in 2012 [34]. As per the Educational Development Index (EDI) of Indian States, with a score of 0.690, Chandigarh obtains 5th positions all over India [35].
- **Connectivity**: The UT has good road connectivity intra-territory and inter-territory. The internal roads are fully functional and in good shape designated as V1 to V7. The total road network within Chandigarh Municipal Corporation jurisdiction is approximately 1250 km. Moreover, the good connectivity with states like Haryana, Chandigarh, Delhi, Himachal Pradesh and Punjab paves way for effective in-migration.
- **Economic Base**: The transformation in economy of Chandigarh is witnessed in the form of knowledge based economy. Information Technology (IT) and IT

enabled services (ITES) due to policies of the Government is gaining momentum. Industrial technology and Bio-technology is flourishing in the area. This has two major implications: educational front in-migration and employment centric in-migration. With the promotion of tertiary sector over primary and secondary sector, a significant proportion of jobs is reserved paving way for employment. Recent advancements depict Chandigarh as 'large and thriving mega IT hub' through initiatives like Rajiv Gandhi IT Park, development of hi-Tech city, encouragement to private sector, Software Technology Park in Chandigarh, e-governance initiatives.

It can be observed that Chandigarh's population is growing and the state has the power to flourish dependent on economy, commerce, tourism, infrastructure, communication, quality of life and other service and tertiary sectors.

Within the Tri-city Region, there are other factors of growth in various settlements which not only propels in-migration in Chandigarh but diverts population in surrounding regions explained below:

- **Mohali (SAS Nagar)**: Mohali is seen as a financial district with a medical hub and is an administrative capital. In the North-eastern end of the city; airport and manufacturing hub can be found. Most of the developments in the Mohali region is basically an extension to the Southern peripheral area of Chandigarh. It is the most urbanized area within the Greater Mohali Region. The Aerotropolis-airport city planned in the Greater Mohali Region (GMADA) is one of the major triggers of growth in the area and acts as a catalyst for growth and a major economic boost to the area. Its close connectivity with Chandigarh makes it one of the most conducive places to reside in where residents can enjoy educational and social facilities from Chandigarh. Mohali, with super-speciality hospitals is also being regarded as a medical hub.
- **Panchkula**: Panchkula, the Haryana state's initiative is a fast growing residential and financial hub.
- **Kharar**: Kharar is primarily an agricultural hub but is slowly finding residential development due to its close connectivity with Chandigarh and Mohali and also due to district change from Rupnagar district to Mohali in 2006.
- **Zirakpur**: Zirakpur, south-east of Mohali has recently been recognised as a fast growing warehousing hub with manufacturing and wholesale trade. Its close proximity with Mohali and Panchkula also paves way for its development.
- **Derabassi and Banur**: The vision for Derabassi is identified as a core industrial town for Punjab.
- **Mullanpur**: Mullanpur is a low density country living centric settlement. Its locational proximity to proposed film city and educational city developments of Chandigarh offers potential to develop north of Chandigarh through a regional leisure space and adding tourist value to the place.

It is quite evident that the Regional scenario near, in and around the Tri-city area is Chandigarh development centric. Number of developments in the form of film city, educational city, new international airport, IT hub, etc. is paving ways for regional development within the 16 km Periphery region of Chandigarh. Proximity to Chandigarh is regarded as one of the major reasons for the development of settlements near the Chandigarh Delhi Highway and the advent of New Chandigarh being proposed North-West of Chandigarh.

7.10 Components of Smart Regional Planning

The Regional Plan of any area focuses on the physical development of the region keeping the future demand into consideration in terms of population, economic drivers, connectivity and governance. The location and the historic importance of Chandigarh makes it one of the most important locations for living offering the masses with a good quality of life. Hence, drivers of growth in Chandigarh suggests that the area is suitable for living and planning lives ahead. The region is slowly reaching a stage where it is unable to accommodate population any further. Ever since Chandigarh's inception, it has acted as a growth driver to the entire region and not alone to Chandigarh city. People have migrated and settled in Chandigarh enjoying better housing, infrastructure, and employment opportunity in various sectors of the economy acting as a magnet. Limited geographical area of the UT posed restrictions and exerted an outward pressure on the expansion on the periphery areas. It was obvious that the existing scenario would infuse changes in functional relationships amongst the regions surrounding Chandigarh UT per say.

In the case of Chandigarh, the primary components to be considered while assessing Regional Planning agenda can be:

(a) Self-sufficiency in terms of employment, basic services, facilities, amenities and infrastructure
(b) Area (s) acting as magnet (givers) while others as supporting suppliers and satellite towns
(c) Fruitful functional relationships around and with Chandigarh
(d) Supporting economic activities driving the wheels of progress, growth and prosperity
(e) Wholesome growth in terms of identified Magnets suitable for settling down, making investments, industrial growth, service sector growth, production or transient stay
(f) Incorporating the governance issues and within the legal framework.

For all kinds of planning, the major resource available is Land. Land is a scarce resource and has to be managed so as to utilise its potential for the benefit of the area it is falling in. There can be many ways in which an area can be smartly planned with smart agendas pertaining to planning. One of the primary objectives of Smart Planning that can be applied to Chandigarh and could focus on the following:

- **Smart Information**: Tools and Techniques can be used for basic data assimilation through platforms like Remote Sensing, Geographic Information System (GIS), and Global Navigation Satellite System (GNSS).
- **Smart Decision Making**: Decisions related to monitoring and allocation of land through suitability can be easily performed with the help of remote sensing techniques and specialised software for specific requirements like Urban Growth Modelling Software, Air Pollution Modelling Software, etc. catering to city and region issues both.
- **Smart Connections**: Spatial, Economic, Social, Governance and Functional relationships should be considered before designing for a balanced development scenario.
- **Smart Development**: Development scenario can be achieved only through careful inter-city and intra-city or inter-city and intra-region strategies having a long-term vision and a holistic approach incorporating innovative models of development like Compact City, Green City, Solar City, Transit Oriented Development, Smart Regional Transportation Plans, Comprehensive Mobility Plans, Inter-city Special Economic Zones, etc.

7.10.1 Smart Information-Tools and Techniques for Smart Regional Planning

Every planning exercise involves a comprehensive framework of data base generated in various layers helpful in decision making for planning of a city or region. This information extraction is extremely essential for proper planning and helps reduce the possibility of misguided and erroneous data. Smart Regional Planning can also be aided through proper monitoring of an urban area. The regular monitoring of an urban area helps in understanding the development scenario of a region and helps in careful land use mapping. Mapping of information and data is a basic component of data driven Smart Planning. Creation of suitable base maps delineating natural features from built forms, natural boundaries, and reserved forests can be done through Remote Sensing.

The recent developments in the field of geospatial technology has facilitated the availability of new tools and services for data management. The evolution of Remote Sensing (RS), Geographic Information System (GIS) and Global Navigation Satellite System(GNSS) has enabled easy collection and analysis of data in the most easiest and convenient way. Multi-dimensional and multi-faceted data and its immense capability has several useful dimensions.

A. **Global Navigation Satellite System (GNSS)**

The term GNSS stands for Global Navigation Satellite System(s). A GNSS typically consists of three major segments—(a) Space segment: which typically includes the satellites orbiting the Earth, (b) Control segment: includes the stations on the ground to track and monitor the satellites, (c) User segment: comprises of the users who rely on the satellites to compute their position and motion. Hence, GNSS is considered as one of the paramount technological interventions for provision of geo-enabled location data. There are several independent GNSS systems in operation around the world which are as follows: GPS, GLONASS, Galileo, BeiDou, IRNSS, etc.

GNSS system caters to high-precision positioning services with such accuracy that could be envied by conventional surveyors. Presently, the GNSS technology has been adopted by the consumer market; in an ever increasing range of products, Transportation, Port automation, Parking automation, Machine control, Precision Agriculture, Surface mining, Mapping and Geodetic data capture, Aerial Photogrammetry and Marine navigation. Its use has seen an increment in day to day life with increasing use of navigational applications and services.

B. Geographic Information System (GIS)

Geographic Information System (GIS) or Geo-Spatial Information System is an amalgamation of technologies that helps us to visualize, question, analyse and interpret data to understand relationship patterns and trends. In crude terms, GIS is a combination of cartography, statistical analysis and technology. The graphical displays that can be further used for analysis are prepared using GIS enabled software platforms. The inherent capability for statistical analysis of data and establishing new data base through spatial analysis is an added advantage to most extraordinarily working platform. The many benefits/advantages of using GIS in Regional planning and development include;

(a) Thematic Mapping
(b) Land use management
(c) Land availability and development monitoring
(d) Management of land use records
(e) Land use/Transport strategic planning
(f) Environmental Impact Assessment
(g) Public facilities and shops catchment areas/accessibility analysis
(h) Facility Planning
(i) Management of Land Registries
(j) Social and Economic analysis

C. Remote Sensing (RS)

The technology of Remote Sensing (RS) is very useful for accessing datasets and information of the area of interest. Its efficiency can be enhanced by using it in combination with GIS permitting multiple analytical domains. RS can widely be used for regional planning and development using a sequential framework of resource inventory and its collection/preparation, analysis of existing scenarios along with time-series data analysis. RS and GIS together can be considered as the Smart means of data collection and presentation techniques.

7.10.2 Smart Decision-Making

For sustainable urban development and smart regional development, geospatial technology can provide a useful platform for assimilation and formulation of methodological spatial analysis. Examining the various phases of strategic regional development, database inventory is the pre-requisite. However, geospatial technology can aid in all phases of strategic planning starting from objective framing to providing alternate scenarios of planning, evaluation of proposed plan and final plan monitoring [36]. For following the above mentioned methodology, geospatial technology could be of great help for operationalising the proposed strategic framework. GIS can provide tools for display and analysis of datasets using location based specific information systems [37]. Hence, it can be concluded that mapping, monitoring, planning and modelling of land cover can all be performed using the geospatial technology [38].

The pattern of urban growth can be analysed critically using satellite data products in the form of imageries and other ground based measurements backed by secondary data. Linear and radial growth patterns can be easily depicted using image processing tools available in any standard software. The spatial data in the form of topographical maps, geological maps, forest types, vegetation type, soil type, and land cover type, etc. can easily be generated and several layers can be extracted representing data in the form of thematic maps. The underlying spatial pattern of the urban growth phenomena over a different periods of time can be not only systematically mapped and monitored but also accurately assessed from satellite data. Land use/Land Cover mapping, spatial metrics, buffer analysis, urban growth driving factors' analysis, network analysis, urban growth modelling and many other terms are the ways to study the spatial and temporal changes of any place. This can help decision makers to find out the future growth directions or previously wrong practiced planning to efficient plan for a better future. Researches have applied different techniques to understand and model the land dynamics at various levels of planning i.e. micro, meso or micro scale [38].

One such information system portal developed by Indian Space Research Organization potentially for urban and regional planners is called Bhuvan-NUIS. Thematic maps, Land Cover maps, administrative boundaries, etc. are different layers available for browsing in Bhuvan-NUIS (refer Fig. 7.14).

A. **Classification and Modelling**

Remote sensing and GIS techniques using information sources from satellite imagery and other information from open source information portals can be used for identifying land use/land cover changes. Classification of satellite imageries of different time periods can be achieved through standard techniques like pixel based classification or object based classification with the help of maximum likelihood classifier (MLC). Later, change detection can be done for understanding the process of urbanization in a region. The change detection methods can divided into two categories viz. pre-classification and post-classification techniques like image

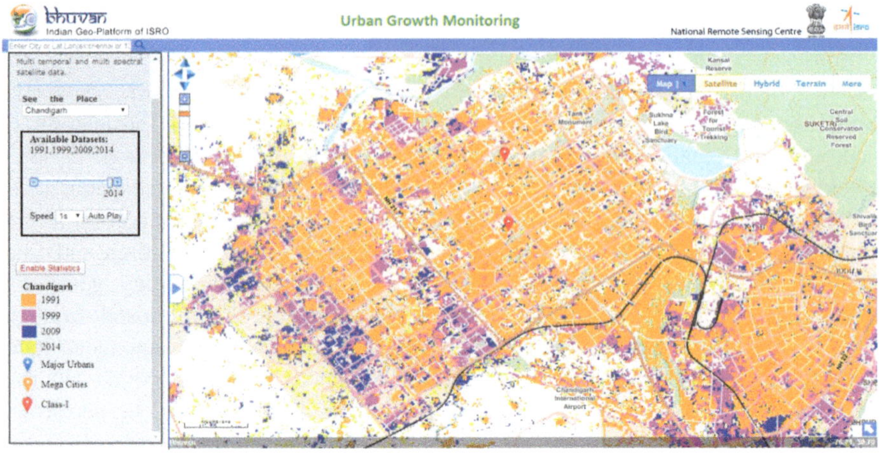

Fig. 7.14 Urban growth monitoring using Bhuvan Portal. *Source* http://bhuvan.nrsc.gov.in/urban/sprawl/urbangrowth.php

difference, image ratioing to single/multiple bands, etc. [39–41]. It is important to map changes as it helps in providing quantitative analysis for the spatial distribution and spatial pattern. This helps in decision making to the planners for further allocation of services and infrastructure corresponding to magnets of growth.

It is important to include drivers of growth for the definite understanding of urban growth scenario. There are various factors responsible for the process of sprawl like demographic, social, economic, cultural, physical/spatial, environmental, governance, proximate causes, biophysical, existing conditions, etc. The most indicating factor for urban growth to take place is demographic characteristics of the city which can be due to population growth or the migration scenarios. Economic factors such as the job and business opportunities, trade and commerce, industrialization. Land prices, market failures, etc. which have huge impact on the changes in the growth pattern of cities. Upcoming new industries or infrastructure projects also have huge impacts on the same. These factors build the competency between cities which is an invisible economic and social factor of urban growth. The planning and governance policies and also the regulatory framework play a major role in dealing with urban sprawl which are formulated to regulate and channelize the growth of cities. Land use zoning, investment plans, transportation policy, Govt. schemes, land management, FAR/FSI restrictions in the city are typical examples of this process. The evolution, spatial patterns of growth since the city existed, land use, density, regional setting and infrastructure are some of the spatial factors responsible for sprawl. The degree to which these driving forces can contribute can also be assessed using modelling software (refer Table 7.8).

Towards spatial growth modelling, various modelling techniques are available for building scenarios that support in the process of land use planning and policy making. There are many land use models available which based on mathematical expressions. In a nutshell, there are three types of growth modelling softwares:

Table 7.8 Overview of driving factors of urban growth

Factors	Types
Bio-physical	Slope, forest, wetland, waterfront, degree of steepness/slope percentage/elevation, rivers, lakes, soil
Demographic	Population (natural and migrated), Population density, Annual population growth, Rate of population growth
Plans, policies and governance	Evolution, Change in boundaries, city development policy, protected lands, cantonment policy, farmland protection policy, zoning, land reforms, investments, Master Plans, Government interventions like
Proximity causes	Distance to major roads, Distance to minor roads, Access to water supply, Access to sewer line, Distance to CBD, Proximity to State Highway, Access to electricity, Distance to educational institutes, Distance to medical facility, Proximity to open spaces/green spaces/ recreational facility
Existing scenario	Distance to existing urban area/cluster, Number of cells within 5 × 5 window of cells
Economic	Per capita income, employment rate, poverty rate, land value, high rental value

empirical estimation based models, dynamic simulation based models and rule based simulation based models [42, 43]. The models developed for growth modelling are Cellular Automata (CA), spatial statistical models, Agent Based Models (ABM), Artificial Neutral Network (ANN), fractal based modelling and Chaotic and Catastrophe modelling concerned with simulating future urban growth scenarios [44, 45]. There are many other softwares that not only support in simulations but also help in future predictions and aid planners in decision making for future cities (refer Table 7.9).

Tools and techniques constantly aiding in growth monitoring and modelling can be useful in decision making to a very large extent. In a similar attempt to understand the growth dynamics of Chandigarh 16 km Periphery region, a growth modeling using "*CA Markov*" and "*Multi-Layer Perceptron* (*MLP*)" was performed for the year 2024 and 2048 using datasets for the years 2000, 2006 and 2012 [11].

7.10.3 Smart Connections

A region shall cater to all major spatial, Economic, Social, Governance and Functional relationships before planning for a region. In case of Chandigarh, the spatial connectivity is facilitated with the help of a well-defined 7-level road network. Pedestrian facilities are available to 90% of the municipal roads. With more than 85% households owning vehicles and approximately 227 four-wheelers per thousand population (highest in the country), traffic congestion is now bubbling as a

Table 7.9 Land use models and tools of modelling spatial growth

Application	Models	Source
Land use models	Cellular Automata Models (SLEUTH, Clarke Urban Growth Model (UGM) and Deltatron Land Use/Land Cover Model (DLM), Fuzzy Cellular Automata Urban Growth Model (FCAUGM), MOLAND), Agent Based Models (ABM), Agent Based Cellular Automata (ABCA), Logistic Regression (LR), Artificial Neural Network (ANN), California Urban Futures Model (CUF), CUF-2, UPlan, FUTure Urban-Regional Environment Simulation (FUTURES) in R, UrbanSim,	http://www.urbansim.org/Main/WebHome
Tools for growth modelling	STARLOGO	http://www.media.mit.edu/starlogo/
	SWARM	http://www.santefe.edu, http://www.swarm.org/
	REcursive Porous Agent Simulation Toolkit (REPAST)	http://repast.sourceforge.net/
	ASCAPE	http://www.brook.edu/es/dynamics/models/ascape/default.htm
	NetLogo	http://ccl.northwestern.edu/netlogo/
	RAISE	
	Agent Building and Learning Environment (ABLE)	http://www.alphaworks.ibm.com/tech/able
	ZEUS	
	Java Agent DEvelopment Framework (JADE)	
	TerraMe	http://www.terrame.org/doku.php

key concern for the area [14]. The densification of areas around and within Chandigarh increases the demand of a by-pass and city to city connectivity in Chandigarh. Chandigarh, Panchkula, Mohali and all the other satellite settlements are connected through well-defined National and State Highways. The presence of an airport provides an exposure to the other parts of the nation. The airport in Chandigarh catering to the Tri-city Region in the 16 km Periphery is a domestic airport at present. All spatial connections in the form of built up area, corridor development, road connectivity, rail connectivity, air connectivity and natural barriers need to be carefully understood. Also, the social status of the population residing in the area should be carefully examined to design and develop for all sections of the society preventing unauthorised constructions in and outside the city.

Sometimes, the vision envisaged for development are not materialised due to restricting policies in the regions and differing jurisdictional limits. Hence, acts and policies play a pivotal role in guiding a region's development and shall be understood before arriving at spatial planning decisions. Also, the Government envisions a role for all cities in the jurisdictional control of the state or UT. The city is expected to grow in a particular way with the drivers of growth either modelled, superimposed or growing in a particular way. These drivers are to be understood and connections between drivers of the region has to be established resulting in a functional relationship. This aids in deciding factors of growth for the newer settlements.

7.10.3.1 Smart Developments

In India, cities are either developed naturally or as a result of demand in living due to natural growth and in-migration in search of facilities and employment opportunity. Chandigarh is planned city and subsequently, Mohali and Panchkula got developed as planned establishments following a planned town planning approach. Smart approaches of further developing a planned city could be intra-city or intra-region approaches. Some of the approaches that could be considered to develop according to the requirement and potential of the region like Smart ICT, Smart Mobility, Smart Energy, Smart Environment initiatives, Compact City development, Transit Oriented Development, Green City Development, etc. These visions for the city if understood and applied, can result in fruitful regional development scenario where the individual nodes are developed keeping in view the future.

7.11 Spatial Planning Strategies and Recommendations

The planning strategies and recommendations of an area can only be recommended with the help of proper analysis of the growth pattern of the study area into consideration regarded as Chandigarh Region. An attempt was made to study the same using geospatial technology.

7.11.1 Land Cover/Land Use (LCLU)

Keeping into consideration the need for the study at a regional scale, the study of transition of land use in terms of growing urbanization in the area, the classification scheme was devised. A total of ten classes relating to urban areas were chosen and the LULC map was prepared for the years 2000, 2006 and 2012. The classes defined were: (a) Built-up Urban, (b) Built-up Rural, (c) Built-up mixed, (d) Recreational or Urban Green, (e) Agricultural Land, (f) Forest Land, (g) water bodies, (h) Open/Barren land and, (i) Restricted Areas, (j) Land Under Plotting [11]. Both supervised and unsupervised classification schemes were adopted. Due to unsatisfactory results, Object Based Classification system was tested. Object Based Image Analysis (OBIA) is a fuzzy technology based classification unlike the regular pixel based classification approaches. Later, it was realised that pixel based mage classification and OBIA shall nor render the desired results and hence, visual interpretation was considered to be the most viable method of image classification producing an accuracy of 85% and even more [11].

A fused LISS IV and IRS 1C PAN product was felt suitable for classification purpose. Considering WGS 84 datum, UTM projection and map boundaries as defined and made available through Survey of India, Land cover maps were prepared. An approximation based area of 20 times the core area (UT) was used covering an approx. area of 2300 sq. km. for the purpose of digitization. This formula was decided by understanding the National Capital Region (NCR) where the influence area was found to be approximately. 23 times the National Capital Territory (NCT) area. The existing development status was also taken into account while preparing the land cover maps. The Master Plans for Panchkula, Pinjore-Kalka urban complex, SAS Nagar, Banur, Dera Bassi, Zirakpur, Kharar, Mullanpur were included for easy understanding of the study. Under the study, the entire 16 km periphery area has been divided into six rings or buffers of 4 km each (refer Fig. 7.15).

Fig. 7.15 Land cover/land use maps for the year 2000, 2006 and 2012 (clockwise starting from above-left). *Source* Author [11] LULC maps 2006, 2010, and 2016

7.11.2 Chandigarh Region: Possible Urban Forms

In 1966, Prof. LR Vagale of School of Planning and Architecture, presented a paper on "*A Case Study of Chandigarh and its Environs in the Regional Setting*" at UN sponsored seminar at Japan. It was later published also. From his understanding, it can be derived that in Punjab, the settlement pattern was characterised by presence of large towns developed along intersections of highways and like other small developments, settlements were either along the river banks and river courses or ribbon developments along corridors. Conventionally, urban areas are surrounded by smaller settlements forming a hierarchy of settlement pattern. In order to understand the growth pattern in and around Chandigarh, few types of growth patterns can be considered. The possible six conceptual patterns could be:

- Expansion of Chandigarh Urban area
- 'Linear township' along the transport corridors
- New township around existing smaller township
- New town on new site functioning as satellite/Counter magnet towns
- Combination of new satellite towns surrounding Chandigarh and its vicinity
- Expansion of Chandigarh in the form of corridor developments and wedges.

7.11.3 Growth Pattern: An Analysis

The analysis of the existing urban form is actually a combination of the proposed urban form on 1966. The development of adjacent areas in mid 1990s can be considered a beginning of 'Regionalization' of the area. Settlements like Kharar, Zirakpur, Dera Bassi are acting as satellite settlements to Chandigarh. No wedge and corridor development could be noticed in Chandigarh.

7.11.4 Prediction of Growth

For estimating the future growth scenarios of Chandigarh Region, few drivers of growth were identified like (a) Proximity to existing settlements, (b) Proximity to roads, (c) Proximity to railways, (d) Proximity to UT boundary. It was of utmost importance to assess the mathematical relationship between drivers of growth. If the relationship is not strong, it may be rejected as a potential driver contributing towards assessing the future growth areas. For evaluation, a seven step procedure was adopted where buffers of fixed distances at 200 m intervals up-to 1400 m were generated. The growth area within each buffer zone was estimated using overlay analysis. Using a reclassification approach, growth areas in each buffer was extracted. Cross tabulation was used for extraction of data within the buffer area in comparison to the total growth in terms of percentage against each buffer distance. The relationship between each year's growth with each growth driver is observed showing high degree of correlation except for Proximity to railways showed insignificant relationship with the growth area (refer Fig. 7.16)

7.11.5 Modelling Future Growth Scenarios in Chandigarh Region

The drivers were used to predict the growth scenarios of the future. First, the dataset for 2000 and 2006 were used for predicting the growth scenarios of 2012. The 2012 projected growth pattern was compared with the actual growth pattern observed in 2012. Towards a better and simplified understanding, the LULC classes were reclassified to obtain binary data consisting of two classes- Built-up (BU) and Non Built-up (BBU) outside the UT boundary and within the 16 km boundary.

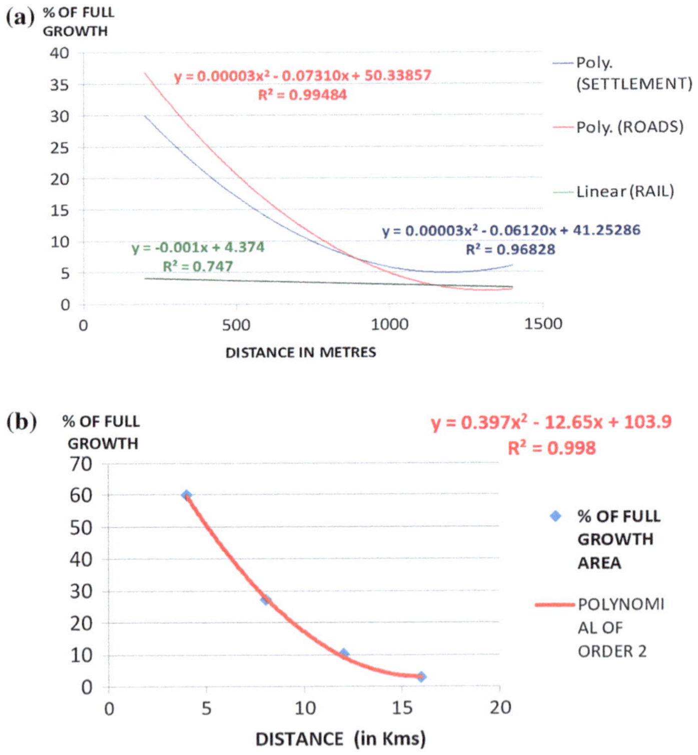

Fig. 7.16 Relationship between growth and growth drivers **a** the relationship between drivers and growth, **b** the relationship between the driver and the effect of distance of the UT boundary. *Source* Author [11] (observed 60% growth within the radius of 4 km)

Literature and site visits indicate that many planed interventions were observed east and south of Mohali around Chandigarh at a very rapid rate. Hence, the BU category consisted of amalgamating five classes together viz. BU Urban, BU Rural, BU Mixed, Urban Green or Recreational Area, Land under Plotting (LUP). The NBU category consisted of Agricultural and Barren land classes. Other three major classes like Forest Land, Restricted Area and Water Body were merged with the area under the UT boundary as masked area. These classes are static in nature and nearly show negligible signs of changes over due course of time at the regional level. It can be a criticism to the rationale on inclusion of plotted land in the built-up area leading to over classification and ultimately over prediction of results. This argument can be counter argued by the fact that prediction models are dependent upon transition probability following linear transformation procedures, whereas actual growth is generally considered to be non-linear in nature.

(a) **Built-up Area (BUA) Outside UT Boundary**

The maps for the year 2000, 2006 and 2012 suggest that there was a 31.13% increase in BUA outside the UT boundary for 2006–2012 as compared to 8.54% in 2000 to 2006. This clearly indicates the presence of unprecedented growth outside the UT boundary adjoining the Chandigarh UT.

(b) **Validation for Predicted LULC-2012**

The projected LCLU map for 2012 was validated with the actual 2012 map. Within the Chandigarh boundary, the UT Administration is effectively managing the growth with the help of building regulations such as "Floor Area Ratio". Using three binary BU images of 2000, 2006 and 2012 were prepared. Using BU 2000 and 2006, two models were tested for accuracy viz. "*CA Markov*" and "*Multi-Layer Perceptron* (*MLP*)".

The process of predicting the future scenario using 'CA Markov' is as follows:

- **Generation of transition suitability matrix**: The transition suitability matrix is generated for six years and the 'transition area' files are used to locate the future changes in different iterations.
- **Suitability Images**: Suitability maps is created using the various drivers of growth. Using min-max linear transformation, the values were normalised between 0 and 1.
- **Grouping of suitability images**
- **Running the model for output indicating future scenarios**

The process of predicting the future scenario using 'Multi-Layer Perceptron (MLP) Neural Network' Model was a three stage process as follows:

- Input as the BUA and NBU area maps for 2000 and 2006
- Change analysis using transition sub-model followed by evaluation of the growth driver's explanatory power using Cramer's V
- Drivers used for running the model for output.

The prediction accuracy evaluated for 2012 year revealed an accuracy of 86.12% using MLP and 81.34% using CA Markov. Using a cell size of 50 m 50 m, MLP showed superior results.

(c) **Predictions for the Year 2024 and 2048**

Using the BU map for 2000 and 2012, the Land Cover for 2024, 2036 and 2048 were estimated using the validated, tested and accurate model, MLP neural network based procedure. A total of six drivers were incorporated whose validity and affect tested with the help of Cramer's V where a value of more than 0.4 is considered 'Good' and values greater than 0.15 is considered useful or acceptable. The drivers were transformed using 'Evidence Likelihood Utility' for values between 0 and 1. The driver called 'Proximity to Railways' was also included as a growth driver for predicting the future scenarios of 2024 and 2048. The reason was to incorporate the actual scenarios and planning interventions subject to real world urbanization. If the urbanization pattern outside the Chandigarh UT is to be observed after 2006 and the release of GMADA plan for Mohali, most of the 'Agricultural Lands' were transformed into 'Plotted Land' subsequently in southern parts of Mohali, near the railway lies indicating the presence/proximity of railways as a potential indicator of urbanization. Railways indicated an extended boundary of built-up land near Mohali.

Using the different proximity drivers, The Cramer's V value for all the six drivers were assessed indicating the potential in decreasing order of influence as: Gains and losses (Changes between 2000 and 2012) (0.4569), Proximity of Chandigarh UT (0.4050), Proximity to CBD (0.3711), Proximity to existing settlement of 2012 (0.3668), Proximity to Major roads (0.3334) and Proximity to Railways (0.2427). The built-up scenario for 2024 suggested an area of 267.58 sq. km. and its further increase to 371.64 sq. km. in 2048 subject to existing drivers of growth. The built up area for 2024 indicate an area under built up category to be 267.58 sq. km. covering an area of 17.63% urban while in 2048 it would rise to 371.64 sq. km. with the total built-up percentage as 24.76% (refer Figs. 7.17 and 7.18).

7.11.6 Results

Keeping in view the Planning Zones proposed by "*Chandigarh Inter State Region (CISR)*" briefed in "*Concept Note on Chandigarh and its Region*" described by TCPO, a new planning zone system was analysed for Chandigarh region. If the NCR of Delhi is studied deeply, the core area of NCT is acting as a magnet to all the other 14 districts adjoining the city (8 districts from Haryana, 5 districts from Uttar Pradesh and 1 district from Rajasthan) making a ratio of 1:22 of the geographical area occupied by the UT to the surrounding region. It was observed that Chandigarh has a very less influence on the neighbourhood but indeed its influence cannot be ignored. The UT bearing an area of 114 sq. km. and the surrounding 16 km periphery, the ratio of UT area to influence zone is 1:14. The analysis of growth scenarios reveal the 4 km buffer from the UT boundary as the most potential

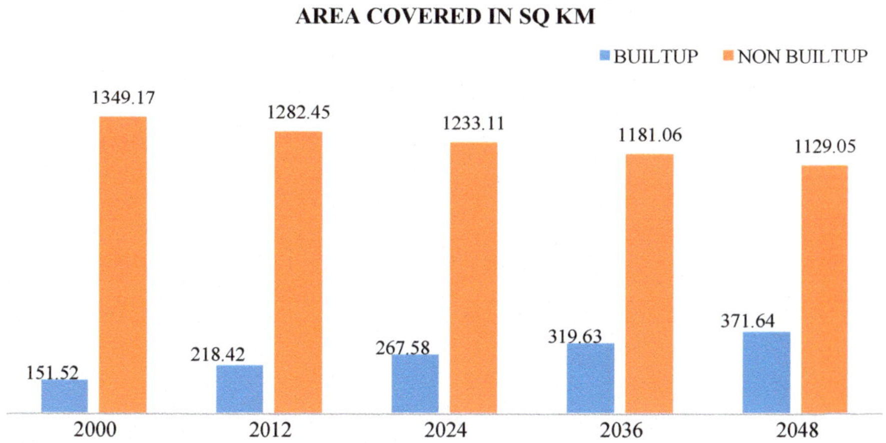

Fig. 7.17 Built-up and non built-up: 2000–2048 using MLP prediction

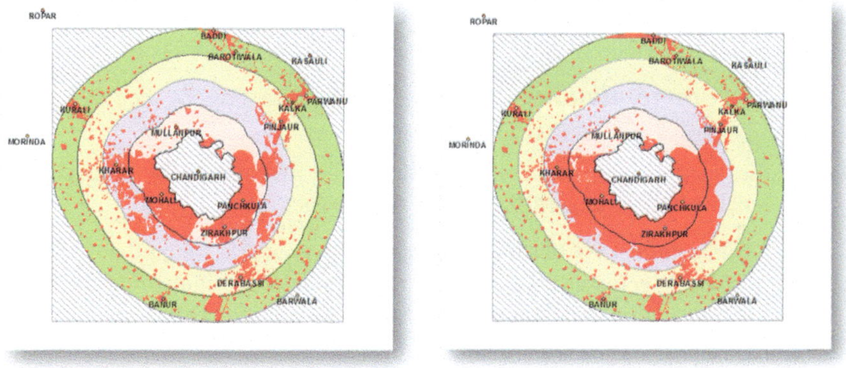

Fig. 7.18 Built-up results within the buffers for 2024 and 2048

area for future development scenarios which shall undergo rapid pace of urbanization till 2025. It is also analysed that further in 8 and 16 km buffers, the growth may be fragmented into small pockets in the future.

Hence, keeping in view the above discussion, the areas/zones mentioned in CISR can be further re-examined and classified based on the study. Zone 1 of CISR has been divided into two layers (Sector 1–56: Layer 1, rest of the UT: Layer 2), Zone 2 is divided into four layers (4 km periphery of UT: Layer 3, 4 to 8 km periphery: Layer 4, 8 to 12 km periphery: Layer 5 and further till 16 km as Layer 6) (refer Table 7.10).

Table 7.10 The description of planning zones proposed for the different layers of Chandigarh Region

Area	Chandigarh region-proposed layers	Planning zones: CISR	Remarks
UT sectors (Sectors 1–56)	Layer 1	Zone 1	Predominantly residential
Rest of UT	Layer 2		Airport, Railways, Industrial area, Forest area
4 km periphery of UT	Layer 3	Zone 2	Mohali, Panchkula, Zirakhpur, Mullanpur
4–8 km periphery	Layer 4		Kharar
8–12 km periphery	Layer 5		Dera Bassi, Pinjore
12–16 km periphery	Layer 6		Banur, Kurali, Kalka, Parwanoo, Baddi (partially), Barotiwala
Areas beyond periphery control belt	Layer 7	Zone 3	Kasauli, Barwala, Baddi (partially). Areas to be included in this layer not yet defined

7.11.7 Recommendations

Some of the spatial recommendations that can be understood by analysing the existing scenario in Chandigarh Region is as follows:

(a) ***Eco-sensitive zones***: As observed, certain regions within the buffer areas shall be earmarked as eco-sensitive zones. This shall help in maintaining the natural beauty of the region intact and aid in acting as a buffer between natural surroundings and the urban area.

(b) ***Water body***: The area within water body suggest that the region adjoining water bodies shall be earmarked as recreational spaces and potential tourist hotspots.

(c) ***Coordinated growth***: The spatial growth should be in sync with the existing social, economic and physical infrastructure.

(d) ***Holistic development***: The Chandigarh Region should develop holistically keeping in view the GMADA and Panchkula area being two major sub-regions of the Region.

(e) ***Smart Planning***: Growth scenario indicate the potential areas of growth. Subsequently, spatial planners may be aided to link the spatial growth pattern with the infrastructure development scenarios. It may be essential for incorporation in the planning and plan making process. It ca also help in decision making for better planning and placement of facilities and utilities. The datasets are one of the most inevitable inputs for regional planer as a tool for Smart Planning.

Table 7.11 The description of planning zones and the area proposed under Chandigarh Region

	Total area (sq. km.)	Agricultural land	Existing settlement	Forest land	Afforestation (proposed)	Urbanisation (proposed)
LAYER 1	55.80	–	55.80	–	–	–
LAYER 2	59.11	9.9	29.57	9.81	–	9.83
LAYER 3	235.17	–	86.69	70.31	–	78.17
LAYER 4	322.11	53.14	11.55	107.17	–	150.25
LAYER 5	421.57	172.61	8.83	102.71	43.38	94.04
LAYER 6	521.73	252.28	15.29	129.55	56.13	68.48
	1615.49	487.93	207.73	419.55	99.51	400.77

(f) ***Agriculture***: Analysis reveal decrease of agricultural area in Layer 3 and 4. Agricultural area should be separately reserved and earmarked for restricted use in the regional plan for Chandigarh.

(g) ***Forest***: Existing forest land is approx. 419.5 sq. km. A buffer of 200 m around the forest is proposed for enforcing no-construction activity

(h) ***Environment***: In order to cater to the concerns of environmental deterioration, afforestation should be proposed. Total afforestation can be proposed near seasonal rivers that shall not hinder the privately owned land in and out of the UT boundary.

(i) ***Growth***: A growth area of around 400 sq. km. can be proposed within the Chandigarh Periphery catering to the future growth requirement. Micro-level planning may need detailed planning and townships like New Chandigarh, Aero-city, education-knowledge health corridors, transportation zones, industrial townships and industrial belts as well as extension of Panchkula and Kalka-Pinjore area (refer Table 7.11).

(j) ***Traffic and Transportation***: Medium and long term traffic and transportation plans should be made for the Region. The recommendations on mass rapid transit system, bus rapid transit system, etc. can be made available for relieving the pressure of densification, in and around Chandigarh.

7.12 Conclusion

Chandigarh's development scenario is a realization of accommodating population that has seen a rise from designed limit of 5 lakhs to over 10 lakhs as per Census 2011 [22]. It is needless to mention that migration phenomenon cannot be avoided in any settlement especially in a city like Chandigarh offering good quality of life, green spaces, regularized legal framework and density control with a beautiful Shivalik backdrop and Sukhna Lake. It is regarded as one of the cleanest and greenest cites in India. However, statistics pertaining to demographics reveal that the city is unsafe from the effect of population explosion unless roads are paved for accommodating the additional need. This is the sole reason where public and private developments have started eating up the peripheral lands in the form of unplanned settlements. This is the rising demand that is forcing to act against Le-Corbusier's vision of development scenario aimed for the City Beautiful.

Regional Plan evolution is a major agenda for any important growing urban centre like Chandigarh. The existing deficiencies within the Master Plans need to be understood and further fruitful deliberations are required for evolving a wholesome solution towards planning. Legal status and sanctity to Regional plans for Chandigarh region is lacking due to lack of legislative provisions under which regional plans may be forwarded. This is one of the major reason leading to haphazard and unplanned growth in sensitive areas around the Chandigarh UT. There is a grave need for enablement of legal provisions as a cohesive approach for a 'Region' unlike piecemeal sub-regional approaches. While these sub-regions help in shaping the growth direction ad magnitude, the entire area is not bound by legal document hindering a holistic planned development scenario. Thus, it can be stated that a comprehensive and balanced development is yet to be executed. There is a need felt for Chandigarh Regional Planning Act for Chandigarh and its Periphery for provision of a comprehensive Master Plan covering all major and minor sub-regions governed through a Planning Board, in general. Chandigarh requires a Regional Plan extending to the three states keeping in view to restore the visions of the initial city planners.

The development of a region focuses on understanding the fragments of a region and then realizing the region as a whole. Regional Development promotes the thinking of a bottom-up and top-down approach both. Chandigarh requires a legal framework backed by efficient tools and techniques of Smart Development scenarios for prosperous futuristic scenarios of growth. From a city perspective it shall focus on building up a Smart Growth, Smart Environment, Smart Infrastructure, Smart Energy and Smart People. From a Regional Perspective it shall focus on Building Smart Connectivity and Smart Governance. The need of Transit Oriented Development, Smart Grids, catering the needs of informal settlements through inclusive planning, building a good transport network and provision of expressways, setting of new townships like Aerocity, New Chandigarh and other new planned industrial and residential townships could be some of the regulatory measures at the planning level in transforming a region into a Smart Metropolitan Region. This chapter dealt with providing recommendations on strategies that can be adopted for translating an urban setup like Chandigarh into a Smart Metropolitan Regional Plan. The drivers of growth shall help us in understanding the growth phenomenon supported by online web information systems like Bhuvan-NUIS Thematic Mapping. The efficacy of Geospatial Technology suggests its usefulness in the analytical and decision making phase for better planning as a component of Smart City Planning.

References

1. Kalia R (1999) Chandigarh: the making of an Indian City. Oxford University Press, New York
2. Mayer A (1950) The new capital of the Punjab. J Am Inst Archit 14(4):166–175
3. Town and Country Planning Organization (2011) Concept note on Chandigarh and its region, M.o. Development, Editor, Town and Country Planning Organization, Chandigarh, p 1–30
4. Prakash V (2002) Chandigarh's Le Corbusier: the struggle for modernity in Postcolonial India. Studies in modernity and national identity. University of Washington Press, Seattle
5. Shaw A (2009) Town planning in Postcolonial India, 1947–1965: Chandigarh re-examined. Urban Geogr 30(8):857–878
6. Lang JT, Desai M, Desai M (1997) Architecture and independence: the search for identity—India, 1880–1980, Lang JT (ed), Oxford University Press, Delhi, New York, p 347
7. Nilsson S (1973) The New Capitals of India, Pakistan, and Bangladesh. Scandinavian Institute of Asian Studies. Monograph series, vol 12. Routledge/Curzon, New Delhi, London, p 230
8. Marg (1961) Special issue on Chandigarh. In: Corbusier L et al (eds) A new planned city. Marg, Bombay
9. Perera N (2006) Chandigarh: India's modernist experiment. In: Gordon D (ed) Planning twentieth century capital cities. Routledge, London and New York, pp 226–236
10. Scott JC (1998) Seeing like a state: how certain schemes to improve the human condition have failed. In: Scott JC (ed) The Yale ISPS series, vol 3. Yale University Press, New Haven and London, p 432
11. Kakkar K (2013) Evolving a regional perspective on 'Greater Chandigarh Region' (GCR) using RS & GIS, in Urban and Regional Studies Department. Indian Institute of Remote Sensing, Dehradun, p 57
12. Chalana M (2014) Chandigarh: city and periphery. J Plann Hist 14(1):62–84
13. Evenson N (1966) Chandigarh. Environmental design and development series, vol 1. University of California Press, New Delhi and Berkeley, p 116
14. Chandigarh Master Plan, Master Plan of Chandigarh 2031, C. Department of Urban Planning, Editor. 2031, Chandigarh Administration, Chandigarh
15. Administrative Atlas of India (1971) Census of India 1971 Administrative Atlas of India. M.o. H. Affairs (ed). Office of the Registrar General & Census Commissioner, India
16. Punjab Government Gazette (1952) The Punjab New Capital (Periphery) Control Act, 1952. P.G.G. Extraordinary (ed). Government of Punjab, Punjab, p 677
17. Corbusier L (1979) The three human establishments. Punjab Government, Department of Town & Country Planning, Chandigarh
18. Sarin M (1975) Planning and the urban poor: The Chandigarh experience (1955–1975), London
19. Census of India (2011) District census handbook Chandigarh: Village and Town Directory. In: District census handbook Chandigarh. Office of the Registrar General & Census Commissioner, Delhi, India
20. Census of India (2011) Primary census abstract report 2011, A-1: Series. In: Primary census abstract. Office of the Registrar General & Census Commissioner, Chandigarh, India
21. Census of India (2001) Primary census report 2001, H-H: Series. In: Primary census abstract. Office of the Registrar General & Census Commissioner, Chandigarh, India
22. Census of India (2011) Primary Census Report 2011, H-H: Series. In: Primary census abstract. Office of the Registrar General & Census Commissioner, Chandigarh, India
23. Corbusier L (1964) The periphery control plan of Chandigarh. Collection of Foundation Le Corbusier, Paris
24. Saini SS, Kaushik SP (2011) Land use changes in Haryana Sub-region of Chandigarh periphery controlled area: a spatio-temporal study. Inst Town Planners, India J 8(4):96–106

25. Inter-State Regional Plan (2001) Inter-State regional plan for Chandigarh Region. Town and Country Planning Organization, Chandigarh
26. CISMeR (1999) Chandigarh Interstate Metropolitan Region Plan (CISMeR), 2021. In: Riberio EFN (ed). Chandigarh UT Administration, Chandigarh
27. Ribeiro EFN, Ansari JH (2002) A proposed concept plan for the Chandigarh Inter-state Metropolitan Region-2021. In: Takhar J (ed) Celebrating Chandigarh. Grantha Corporation, Chandigarh, pp 431–435
28. Upadhyay R (2012) Centre to revive Ribeiro's development plan. Daily Post India, Chandigarh
29. Kumar V (2012) UT administration fails to finalise master plan yet again. In: Hindustan Times. Hindustan Times, Chandigarh
30. New Chandigarh Master Plan (2015) Detailing of Master Plan New Chandigarh, P. Department of Town & Country Planning (ed). Department of Town & Country Planning, Punjab, SAS Nagar, p 117
31. Economic Statistical Organisation Punjab (2017) Economic statistical Chandigarh. Central Statistical Organisation, New Delhi
32. Ministry of Women & Child Development (2009) Gendering human development indices: recasting the gender development index and gender empowerment measure for India. Ministry of Women & Child Development, New Delhi
33. Confederation of Indian Industry (2011) Annual Report 2010–11. In: Confederation of Indian Industry Annual Report, New Delhi
34. India, C.o. (2013) Sample registration system statistical report 2013. In: SRS statistical report, New Delhi
35. National University of Educational Planning and Administration (2006) Annual report 2005–06. In: National University of Educational Planning and Administration Annual Report, New Delhi
36. Yeh AG, Li X (1996) An integrated remote sensing and GIS approach in the monitoring and evaluation of rapid urban growth for sustainable development in the Pearl River Delta, China. Int Plann Stud 2(2):193–210
37. Nicholas M (2008) Remote sensing tutorial. National Aeronautics and Space Administration
38. Yesserie AG (2009) Spatio-temporal land use/land cover changes analysis and monitoring in the Valencia Municipality, Spain, March 2009. Masters Program in Geospatial Technologies, Dept. Lenguajes y Sistemas Informaticos Universitat Jaume I, Castellon, Spain
39. Ridd MK, Liu J (1998) A comparison of four algorithms for change detection in an urban environment. Remote Sens Environ 63(2):95–100
40. Singh A (1989) Review article digital change detection techniques using remotely-sensed data. Int J Remote Sens 10(6):989–1003
41. Yuan D, Elvidge CD, Lunetta RS (1998) Remote sensing change detection: environmental monitoring methods and applications. Chapter: Survey of multispectral methods for land cover change analysis. Ann Arbor Press
42. Hu Z, Lo CP (2007) Modeling urban growth in Atlanta using logistic regression. Comput Environ Urban Syst 31(6):667–688
43. Zeng YN et al (2008) Modeling spatial land use pattern using auto-logistic regression. Int Arch Photogrammetry Remote Sens Spatial Inf Sci 37(B2):115–118
44. Guiping W et al (2009) Simulation and analysis on the land-use patterns of Nanjing city based on AutoLogistic method, p 1–6
45. Cheng J, Masser I, Ottens H (2003) Understanding urban growth system: theories and methods. In: Conference proceedings

Part VIII
India, Delhi

Chapter 8
Actualizing Smart Regional Aspirations: A Case of the National Capital Region, India

Ashok Kumar

Abstract National Capital Region (NCR) is one of the best known and economically fastest growing planning regions of India. Innovative real estate developments in the cities of Gurugram, Noida, and Greater Noida have caught the attention of policy makers and academic scholars alike. Most scholars however have focused on challenges presented by this region and critiqued inequitable developments. This chapter focuses on challenges as well as opportunities presented by this metropolitan planning region. This chapter speculatively seeks to answer the question of what holds for the people living in the NCR for the next 20 years. After brief introduction, the second part of the chapter used to set the context by examining planning and governance of the NCR. The third part of the chapter focuses on five smart spatial challenges and opportunities. Emanating from these challenges and opportunities are policy proposals that are discussed in the fourth part of the chapter that policy proposals would convert challenges into realizable opportunities, making regional aspirations a real possibility. The chapter ends with some conclusions.

Keywords Smart aspirations · National Capital Region · Smart economy
Environment and mobility · Delhi-Mumbai Industrial Development Corridor
Smart Meta City Delhi

8.1 Introduction

Although, it took nearly 40 years after independence to set up the Planning Board under the National Capital Region Planning Board Act, 1985, the first draft regional plan for the NCR was published in 1971 by town and Country Planning Organization Delhi [2: 224–225]. Prior to this, a number of regional and sub-regional studies were undertaken by the Town and Country Planning

A. Kumar (✉)
Department of Physical Planning, School of Planning and Architecture, New Delhi, India
e-mail: a.kumar@spa.ac.in

T. M. Vinod Kumar (ed.), *Smart Metropolitan Regional Development*,
Advances in 21st Century Human Settlements,
https://doi.org/10.1007/978-981-10-8588-8_8

Organization. Even if we take stock of regional planning attempts made after independence, a number of regional planning studies were carried out by the Town and Country Planning Organization, Government of India, which include Growth Pattern of Public Sector Offices in the NCR (1969); Cropping Pattern for the National Capital Region (1973); NCR Plan Implementation: A Programme for Action (1977); Industrial Profile of Rajasthan Sub-Region of NCR (1985); Industrial Profile of Uttar Pradesh Sub-Region of NCR (1985); and Industrial Profile of Haryana Sub-Region of NCR of 1986 [17: 12–49]. Above inventory of research reports shows that before formal commencement of regional planning in the National Capital Region under an Act of Indian Parliament, not only there was awareness and conversation within government about regional planning among planners, number of significant studies was carried out by the TCPO on behalf of Government of India. As Tridib Banerjee underscores: "Although talks about preparing a plan for the National Capital Region began as early as the late fifties, serious efforts were under way only in the late sixties, when the Town and Country Planning Organization of the Ministry of Works, Housing and Urban Development began its preliminary research and studies" [2: 223].

Established in 1985, the NCR Planning Board was made responsible "for the preparation of a plan for the development of the National Capital Region and for coordinating and monitoring the implementation of such plans and for evolving harmonized policies for the control of land-uses and development of infrastructure in the National Capital Region so as to avoid any haphazard development of that region …" [7: 3]. So largely based on studies undertaken by the TCPO between 1960 and 1980, the NCRPB prepared and enforced the National Capital Region Plan, 2001 in January 1989. The National Capital Region Plan, 2021 was notified on 17 September 2005. The NCRPB took nearly four year to prepare each of the two regional plans, which is not a long time to prepare and get approved plan for a large and complex region like the NCR. Since the current regional plan of the NCR will come to an end in 2021, preparations are being made to prepare and enforce the third regional plan for the NCR. This time however, the NCR has engaged the National Institute of Urban Affairs (NIUA) Delhi to support the efforts of making the regional plan for NCR, whose mandate is largely urban planning rather than regional planning as previous works of NIUA show that it has primarily focussed on urban studies and not on urban planning.

This brief overview about the NCR of India focuses on four elements of the smart regional planning and development. These are smart regional economy, smart regional accessibility, smart regional environment, and smart regional settlement pattern, which is also largely in line with the objectives of the Regional Plan 2021 (see [11: 18]. But before I do this, a brief discussion about the planning history of the NCR is imperative. Rapid urbanization and growth of the National Capital of Delhi was a cause of concern for the state and central government from the start. During the formulation of the first master plan of Delhi in 1959, the need of a regional plan was recognized to regulate and control growth of Delhi in a planned manner. National capital region was perceived as a vehicle to regulate population of Delhi with little emphasis on the intrinsic value of the region and regional planning in the NCR. Thus Master Plan for Delhi in 1962 made a recommendation for the formation of the National Capital Regional Planning Board to strategize the growth of Delhi and the National Capital Region. National Capital Region Plan is consistent in its focus on two issues: "(a) that it is possible to define the carrying capacity of the regional environment and of its settlement system, and (b) that the growth of the core city must be limited by decentralizing future growth to other urban centers in the region" [2: 225]. This utopian policy of containment, control and regulation continues even today in the currently enforced regional plan for the NCR.

So it is in this context that the National Capital Region Planning Board was formed under the National Capital Region Planning Board Act, 1985. NCR is the largest metropolitan region in India and the third largest metropolitan region in the world covering approximately 33,600 km^2. The constituents of National Capital Region are the National Capital Territory of Delhi (going to the smart meta city with population of 20 million or more in 2021) spanning over 1483 km^2; Haryana Sub-region comprising of Faridabad, Gurgaon, Rohtak, and Sonepat districts in addition to Rewari, Bawal, and Panipat tehsils covering an area of 13,413 km^2; Rajasthan Sub-region constituting of Alwar, Ramgarh, Behror, Mandawar, Kishangarh and Tijara tehsils with a total area of 4493 km^2; and Uttar Pradesh Sub-region constituting Meerut, Ghaziabad and Bulandshahr districts with an area of 10,853 km^2. The National Capital Region is also the largest populated metropolitan region in the world with a population of approximately 46.07 million (Census of India; for comparison see [6: 18]; also see Fig. 8.1). Present total area of the NCR is 45,888 km^2 after the Ministry of Urban Development vide gazette notification of 1 October 2013 included Bhiwani and Mahendragarh districts of Haryana state, and Bharatpur district of Rajasthan state in the NCR. This unfathomable and indecipherable geographical area of the NCR itself poses a great challenge of governance for regional planners for the realization of regional aspirations, hopes that are pinned on regional planning of such a large region. This also raises a concern about the delineation of the NCR, which appears to be made on the basis of political considerations (inclusion of three districts Bhiwani, Mahendragarh, and Bharatpur) rather than on the basis of any technical parameters (also see [9: 669]).

8.2 Planning and Governance of the Region

Planning process of the National Capital Regional Plan involves preparation of a draft regional plan by the NCRPB followed up with an exercise in public participation for inviting comments on the draft of the regional plan. The process of public participation is mandated under section 10 of the National Capital Region Planning Board Act, 1985. After obtaining objections and suggestions from the public, central government, state governments, local bodies and other agencies, regional plan is accordingly modified and submitted to Government of India for final

Fig. 8.1 National Capital Region of India. *Source* NCRPB [11]

approval. After approval by the government, regional plan is ready for implementation, which could be periodically reviewed. As could be imagined, planning process in the NCR is integrative and hierarchical. Since planning is considered to be integrative by nature, it is hoped that various parts of the NCR Plan would also reinforce each other. It is also hoped that the regional plan would continually speak to and guide sub-regional plans prepared by different state governments for their sub-regions. Integration among regional and sub-regional plans becomes critically important because plans are not prepared in a top down order, regional plan first and sub-regional plans later, but in any random order. For example, this time sub-regional plans for Haryana and Uttar Pradesh were prepared after the preparation of the NCR plan. But this order could reverse in future. Hierarchy of plans for National Capital Region is given in Fig. 8.2).

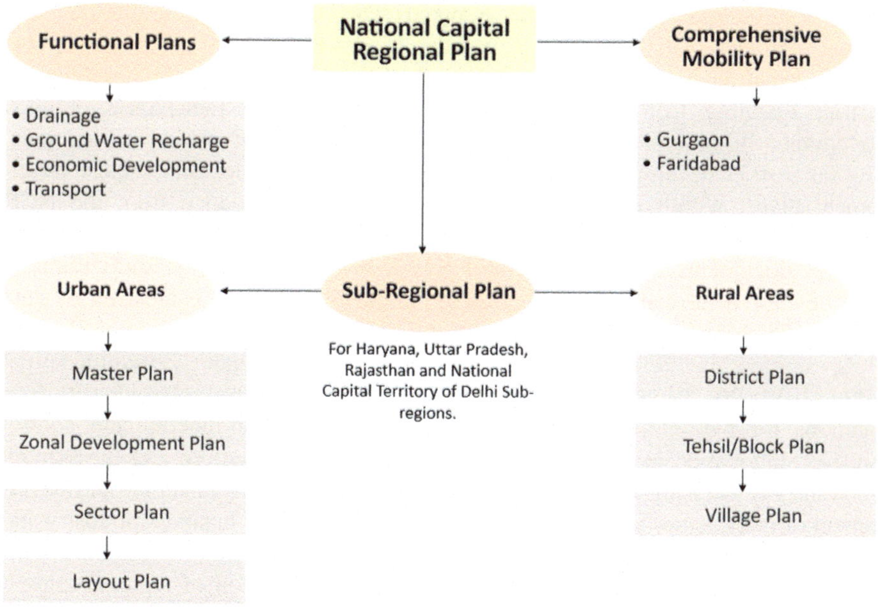

Fig. 8.2 Hierarchy of plans in the NCR of India

Two regional plans have been formulated till date for the National Capital Region with perspective years of 2001 and 2011. Regional plan for the National Capital Region is followed by Sub-regional Plans for the sub-regions of Uttar Pradesh, Haryana, Rajasthan and Delhi. However, from the start sub-regional plan for Delhi was never prepared, arguing that the area of Master Plan for Delhi and proposed Sub-regional Plan for Delhi is coterminous i.e. area of master plan and sub-regional plan is same. This argument is flawed because even if area is coterminous, a sub-regional plan would throw up different planning issues than the Master Plan for Delhi. For example, a regional plan would focus on forests while a master plan would concern itself with green spaces in a city, both implying different emphasis on environmental issues. Nonetheless, in order to implement the regional and sub-regional plans and elaborate a particular element of the regional plan, functional plans are prepared for drainage, ground water recharge, economic development of the region as whole and seamless transport connectivity. Finally, for the urban areas of Faridabad and Gurgaon, comprehensive mobility plans have been also prepared to ease traffic problems in both these areas.

Apart from the regional plan and sub-regional plans, the urban areas in the metropolitan region formulate their own master plans (example Master Plan of Delhi, 2021). Rural areas in the region make their own plans (largely economic plans) at three levels including district development plans, block development plans, and village development plans, mandated under local government acts of the relevant states since 1992 when 73rd and 74th constitutional amendments were made. Why economic plans are not prepared? The simple answer is that collector centered project planning remains popular in government as a British legacy. Dominance of the collector culture makes all other administrative structures such as District Planning Committee, Block Samiti and Village Panchayats defunct. Governance becomes abstruse in this maze of development plans. For example, how far district development plans, block development plans, and village development plans, within the NCR conform to and seek guidance from the NCR Regional Plan 2021. Since development plans and the NCR Regional Plan are statutory documents, which one should prevail if there is a policy conflict among these development plans? Additionally, each development plan is a carrier of state politics, which may or may not gel with the NCR Regional Plan.

As quoted in Deutsche Gesellschaft für Internationale Zusammenarbeit (GIZ) (2014: 26), Blotevogel and Schulze [4] and Taylor et al. [16] identify four functions for metropolitan regions. These include decision-making and control function; innovation and competition function; gateway function; and the symbol function. The decision making and control function involves planning and development of place nodes which are capable of attracting global business, industry and politics, for example headquarters of corporate houses, and global organisations. The innovation and competition function involves production of regional knowledge economy where efficiency is the key to attracting highly skilled and creative people. The gateway function presents the region as a gateway to the world for access and exchange of information, knowledge, ideas and services. If a region is able to perform the gateway function effectively it should be able to benefit regional development from a diversity of people and cultures. The symbol function involves production of authentic cultural signs and symbols, which represent uniqueness about the region. Authenticity would depend on a mixture of cultural, historical, economic and spatial elements, and also if these elements are experienced on a daily basis.

More in line with ground conditions of the present day economic and political environment, Tridib Banerjee [2] argues that the chief purpose of regional planning and development is "to develop a dynamic approach to allocating and distributing future population growth. Growth management, rather than growth control, should be the operative concept. That requires effective tools for continually monitoring the stresses and strains within the regional settlements system, and for guiding the policy process and investment decisions according to that intelligence" [2: 225]. The NCR plans have followed the rigid and static planning strategy of growth control rather than growth management.

The NCR governance is coherent to the extent that three state governments excluding NCT Delhi have established dedicated administrative offices at sub-regional level and tasked these with matters of regional planning including preparation of sub-regional plans. However, Government of the NCT of Delhi does not have any dedicated office for regional planning because Delhi Development Authority is responsible for planning and development of the city of Delhi and also land is controlled by DDA not Delhi Government. Second aspect of the NCR governance is that it is highly fragmented because there are a large number of districts, tehsils, blocks, villages, towns and cities, making coordination among these entities impossible and adversely affecting efficiency. When coordination at a decentralized level is not possible, this situation causes decisions to be made at state level, again affecting efficiency. Furthermore, District Planning Committees being largely dysfunctional adversely affects regional planning by creating inefficient administrative and planning framework. These elected development authorities should be given adequate funds to start with after which they could be allowed to raise funds from the markets because raising funds in a buoyant region is not difficult. Lastly, a number of private developers colonizing different parts of the region under different statutes give rise to fragmented spatial development in the region, which is least, connected to larger utility networks. Developments around Gurugram provide a relevant example.

Due to peculiar nature of administration of the national capital Delhi, little scope is left for the constitution of the Metropolitan Planning Committee in Delhi. Delhi state government does not even have enough powers to enact a law for setting up of a Metropolitan Planning Committee. Even if it is made, it will add to further confusion with the DDA. Metropolitan Committee becomes useful only if it is set up for the Central NCR including Delhi and DDA is abolished altogether and its powers are handed over to Delhi state government.

8.3 Smart Regional Planning and Development

8.3.1 Smart Regional Settlement Pattern

A region where all rural and urban settlements are planned and developed on the basis of the principle of 'balanced growth' would be able to obtain smart regional settlement pattern. In such regions, all settlements would be allowed to develop according to their different potentials in terms of natural and human resources. Regional planning would not make deliberate attempts to seek to equally develop all parts of the region. Regional inequalities, particularly, economic inequalities, would remain because of diverse potentialities of each area in the region. Therefore, to obtain a smart regional settlement pattern, no deliberate economic equality is emphasised. Certain parts of a region would grow faster than others; equality being addressed through free movement of people and goods, and also through welfare policies of government. However, deliberate attempts (read planning) would be made to create decent access to housing and basic services to all people of the region.

Having defined the idea of a smart regional settlement pattern, let us now examine national capital Regional Plan 2021. Regional Plan 2021 proposes six-tier hierarchy of settlements including Metro Centre (population 10 lakh and above), Regional Centre (3–10 lakh), Sub-regional Centre (0.5–3 lakh), Service Centre (10,000–50,000), Central Village (5000–10,000), and Basic Village (below 5000). A robust system of mass public transport is proposed to be developed to make these settlements accessible for movement of people and goods [11: 29]. This excludes

Central National Capital Region or CNCR. Regional Plan 2021 continues to maintain the primacy of Delhi, which is further reinforced by the development of proposed seven Metro Centres with combined population of 127.69 lakh excluding population of Delhi. Metro Centres are expected to "act as powerful growth nodes to attract capital functions and activities and help in population dispersal from the national Capital. Because of their special functional status and size, a very high level of physical, social and economic infrastructure better than that in the Capital is required to be developed within these towns/complexes. This would include efficient intra-urban mass transportation systems as well as strong transport and communication linkages with Delhi, other Metro Centres and NCR towns" ([11: 30]; also see Table 8.1).

Proposed settlement pattern partially continues to perpetuate the existing settlement pattern dictated by the existing road and rail networks (see Fig. 8.3). Accessibility continues to play crucial role in developing regional settlement pattern for 2021. However, with proposals like RRTS, circular movement of people and goods is expected to increase manifold in future.

Another crucial proposal made by central government, which would have significant consequences for the region is setting up of mega infrastructure project of Delhi-Mumbai Industrial Corridor (DMIC) worth nearly US$ 100 billion by central government, a part of which is located in the NCR. This corridor project significantly involves building of Multi-Modal High Axle Load Dedicated Freight Corridor between Delhi and Mumbai covering the length of 1483 km. This freight corridor will pass through the six states of U.P., Delhi, Haryana, Rajasthan, Gujarat and Maharashtra. Starting from Dadri in the National Capital Region, the freight corridor will terminate at the Jawaharlal Nehru Port near Mumbai. This corridor is to be used for long haul freight movement with additional purpose of sparing the existing track space for short haul freight movement. Former Economic Adviser to the Prime Minister of India in his Brief Conceptual Note: Development of the Delhi-Mumbai Industrial Corridor explains that the DMIC will have nine junctions and three terminal stations along the route, and the ports of Mumbai, Navi Mumbai,

Table 8.1 Metro Centres, 2021

S. No.	City and complex	Proposed population	
		2011	2021
1.	Faridabad-Ballabgarh	16.00	25.00
2.	Gurgaon-Manesar	4.50	16.50
3.	Ghaziabad-Loni	19.00	30.19
4.	Noida	6.00	12.00
5.	Sonepat-Kundli	3.50	10.00
6.	Greater Noida	7.00	12.00
7.	Meerut	15.00	22.00
Total		71.00	127.69

Source NCRPB [11]

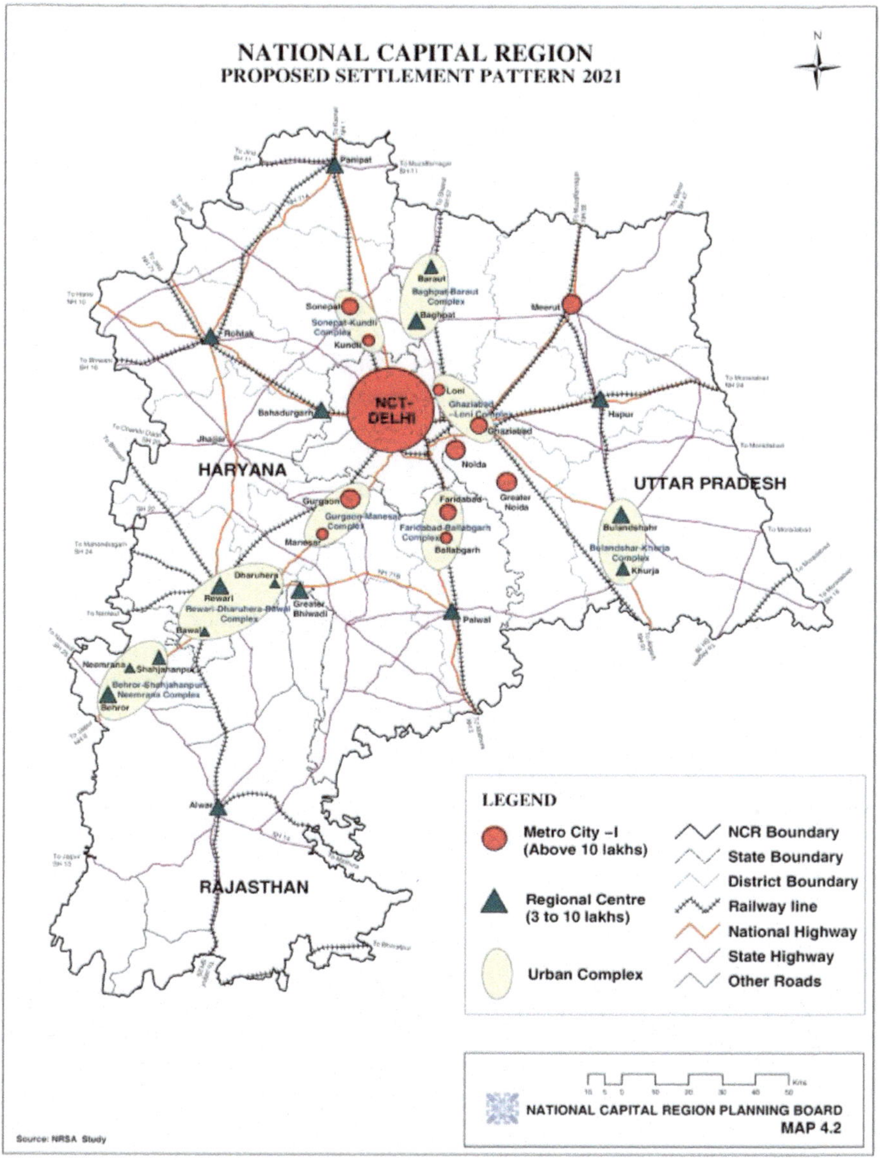

Fig. 8.3 Settlement pattern, 2021. *Source* NCRPB [11]

Pipavav, Kandla, Dholera, Navlakhi, Dahej, Mundra, Mandvi, Mahuva, Rewas and Dighi would serve this corridor (Narayan, undated: 2). Dadri-Noida-Ghaziabad Investment Region would take direct positive economic impact of the DMIC. Evidently, the DMIC is expected to speed up economic growth of the NCR, which is already one of the fast growing regions of the country. As government's own concept paper also notes: "The objective of DMIC, supported by world-class infrastructure, would be to optimize on the present potential, enhance investment climate and promote the economic development of the region through creation of a long term enabling environment [8: 1]. However, there is also a real possibility of other consequences such as increased regional 'inequality and environmental degradation' [14: 15].

8.3.2 Smart Regional Economy

Smart regional economy refers to sustainable economic growth that simultaneously protects physical environment and enhances general quality of life. National Capital Region of India is a buoyant region, its economy being one of the fastest growing regional economies in India. Economic growth is sustained over the last two and half decades, fulfilling the first condition of smart regional economy. But it is evident that the NCR planning policies and the kind of development it gives rise to lag much behind on protection of physical environment. A degraded environment such as Delhi Ridge in the southern parts of Delhi is just one notable example. Rampant mining activity continues unabated in spite of regular interventions by various courts.

Nevertheless, economy of the National Capital Region is surging ahead in the wake of sustained growth of the service sector and the real estate sector. Measured in terms of per capita income, the NCR stands among the top five regions of India. National Capital Region is counted among the top economic performers of the country and Delhi and the Haryana Sub-Region play an important role in making NCR an economic powerhouse. NCR contributes 6.7% to the GDP of India [13: 42]. As noted below:

The NCR occupies an important place in terms of share of GDP of India. India's GDP at constant (2004–05) prices in 2009–10 was Rs. 45160.7 billion whereas the GDP of NCR for the same period was Rs. 3193.4 billion accounting for 7.1%. In the year 2009–10, the GDP of India has seen a growth rate of 8.6% as per CSO and total GDP in year 2013–14 was recorded at Rs. 57417.91 billion. The average growth rate of GDP in the NCR (11.2%) is greater than that of India (8.7%) during 2004–05 to 2009–10 [13: 52].

Accordingly the 'Study of Economic Profile of NCR' conducted by Apex Cluster Development Services Private Limited in 2015, points out that the tertiary sector is the main contributor to the GDP of the NCR. For instance, contribution of the territory sector to NCR's GDP is 67% while secondary sector contributes to the tune of 25% and primary sector only contributes 8% to the GDP of NCR. The same study concludes that "Delhi is the largest economic contributor to the GDP of the NCR. Growth rate of GDP is the fastest in Haryana sub-region amongst all states. Uttar Pradesh sub-region's share in GDP has been declining consistently. On a yearly basis, Uttar Pradesh Sub-Region has touched the highest growth rate at 21.39% and lowest by Rajasthan Sub-Region at 0.01%. 2006–07 is the year with highest growth rate at 13.64% followed by 2009–10 at 10.60%. 2008–09, 2007–08 and 2005–06 followed at 10.45, 9.99 and 9.51% respectively in NCR economy" [13: 18–19]. While Delhi and the Central NCR are fast emerging as the mainstay in respect of service sector, the other three sub-regions are gradually shifting from agriculture to manufacturing. More precisely and in comparative sense, Uttar Pradesh and Rajasthan remain dominated by agriculture while the Haryana Sub-Region is restructuring itself into an area predominated by services and manufacturing. Districts of Gurugram and Panipat contribute greatly to the Haryana Sub-Region's economic growth. Gurugram district in the Haryana Sub-Region has been experiencing faster economic growth due to IT and ITES and Panipat has distinguished itself as a prime trade centre. As a whole, per capita income of the NCR remains higher than the per capita income of the country for the periods between 2004–05 and 2009–10. Restructuring of the regional economy is fast transforming the NCR. For example, over 71% of the total workers would be engaged in trade, commerce and services by 2021. This is an increase from nearly 64% in 1991. Overall worker participation rate has been consistently increasing and is expected to touch 40% by 2021 [11: 40]. Worker participation rate of NCR is slightly higher than India's worker participation rate, which is 39.8% [5].

Hastened process of industrialization presents a number of challenges. Environment of the region, for example, remains one of the major concerns. The ridge, wetlands and sanctuaries need to be protected, particularly from the process of industrialization, and unauthorized urbanization. Another important aspect is that productive agriculture land should also be protected from industrialization and urbanization. Government is steadfast in dealing with these issues as it has already published regulations for setting up SEZ in areas, which are not agriculturally productive. Regional Plan 2021 intends to enforce Environment Protection Act, 1986. However, mechanisms for enforcement are largely ineffective, leaving authorities responding to violations after they have already caused major adverse environmental impacts.

Another less known aspect of the regional economy of the NCR is the function of the NCRPB to provide interest bearing loans to state governments, parastatals of state governments and urban local bodies under section 7 of the NCRPB Act 1985. Section 7 (ii) of the NCRPB Act 1985 provides that the Board would "arrange for, and oversee, the financing of selected development projects in the NCR through Central and State Plan funds and other sources of revenue". Further section 7(e) of the NCRPB Act, makes the NCRPB responsible for arranging and overseeing financing of selected development projects in the NCR and in CMAs. The report titled 'National Capital Region Urban Infrastructure Financing Facility (RRP IND 41598-01) notes: "In fulfilment of its mandate, the NCRPB has been playing a significant role in the development of infrastructure in the NCR by financing infrastructure development projects selected under powers vested in it by section 8 (e) of the NCRPB Act. These projects cover a wide spectrum of basic infrastructure development, sewerage, drainage and solid waste management, water supply, power, transport, hospitals, abattoir and civic centre" [1: 1]; also see Table 8.2). Traditional sources of funds have been shrinking, central government has almost

Table 8.2 National Capital Region: budget allocation and expenditure

Five year/ annual plans	Plan funds released by Central Government	GNCTD contribution	Loans released	Total expenditure including loan released (plan)
Eleventh five year plan				
2007–08	100	50	705.37	741.97
2008–09	50	50	723.6	777.85
2009–10	50	00	814.56	894.88
2010–11	50	00	598.77	948.24
2011–12	50	00	613.65	790.14
Twelfth five year plan				
2012–13	55	00	418.51	637.05
2013–14	60	00	355.44	643.76
2014–15	80	00	237.91	523.56
2015–16	80	00	165.15	514.40

INR in crores
Source http://ncrpb.nic.in/ncrconstituent.php

halved its contribution and Government of the National Capital Territory of Delhi (GNCTD) has stopped funding since 2009–10; Delhi has not made any contribution to NCRPB since 2009–2010. Accordingly loans released by the NCR have also been drastically reduced from INR 705 crore in 2007–08 to merely INR 165 crore in 2015–16; a fall of over four times.

One of the crucial questions is: who is benefitting from the NCR funding? Of the total loan sanction by the NCRPB for all sub-regions, nearly 60% of that loan went to Haryana, and 29% of the sanction loan was given to Uttar Pradesh. Of the NCR sub-regions, the National Capital Territory of Delhi got least amount of loan from the NCR. Some amount of loan was also sanctioned to Punjab and Madhya Pradesh, which is not part of the region but both, has been identified as counter magnets to the NCR. Loans are rightly sanctioned based on transparently laid down criteria rather than population size (see Table 8.3).

Table 8.3 National capital region project summary as on 30 September 2016

S. No.	States	Status	Number of projects	Estimated cost	Loan sanctioned	Loan released by NCRPB
1.	Rajasthan including CMA Kota	Ongoing	6	285	214	68
		Completed	30	1679	631	594
		Sub-total	36	1964	845	662
2.	Uttar Pradesh including CMA Bareilly	Ongoing	9	7237	2654	329
		Completed	49	1949	834	609
		Sub-total	58	9186	3488	938

(continued)

Table 8.3 (continued)

S. No.	States	Status	Number of projects	Estimated cost	Loan sanctioned	Loan released by NCRPB
3.	Haryana including CMA Hisar	Ongoing	30	2971	2050	921
		Completed	160	12,249	5178	4615
		Sub-total	190	15,220	7228	5536
4.	NCT Delhi	Ongoing	1	102	76	20
		Completed	2	521	310	310
		Sub-total	3	623	386	330
5.	CMA Patiala, Punjab	Ongoing	0	0	0	0
		Completed	2	79	46	46
		Sub-total	2	79	46	46
6.	CMA Gwalior, Madhya Pradesh	Ongoing	2	104	63	32
		Completed	4	134	101	101
		Sub-total	6	238	164	133
Total			48	10,669	5057	1370
			247	16,610	7100	6275
Grand total			295	27,309	12,157	7645

INR in crore
Source http://ncrpb.nic.in/ncrconstituent.php

Total amount of loan sanctioned to the sub-regions is too little when compared with the costs of regional level projects. Even a part of the RRTS project would cost several times more than the total sanctioned funding of INR 12,157 by the NCR. For example, the National Highways Authority of India, which granted in principle approval for the construction of the Delhi-Panipat and Delhi-Alwar Regional Rapid Transit Systems corridors of the RRTS. Both corridors would cost INR 53,500 for 291 km high speed rail track [10].

8.3.3 Smart Regional Accessibility

Smart regional accessibility is crucial to the realization of regional planning and development. Without smart regional accessibility—fast, safe and affordable regional commuting, and economically viable movement of freight—regional planning and development would remain a pipedream. Movement for work, leisure, trade, education and other purposes remains at the heart of regional planning and development. Moreover, accessibility is pivotal to real estate development. Accessibility is crucially critical for inter-settlement linkages in order enhance interactions among small as well as large settlements. Overall accessibility is one of the primary aspects that determine overall prosperity of a region by directly influencing economic, social and cultural relations. Transport facilities in the

National Capital Region play a very important role by providing connectivity to and from the National Capital Territory of Delhi with respect to its neighbouring areas as well as within the region. A discussion below briefly examines policies of the Regional Plan of NCR of 2001 and 2021 pertaining to the transport sector.

National Capital Region is spread across approximately 33,578 km^2 with portions from three states other than the Union Territory of Delhi. In view of this large geographical expanse, it is important to plan and develop an extensive intra-regional as well as inter-regional transport networks in order to meet the needs of people in the region. Here it should be stressed that Delhi acts as the primary collection and distribution centre for freight goods that makes the transport sector so critical for the functioning of the NCR as well as the areas surrounding the region.

Currently existing transport network has developed in the form of a well laid out roadway system as well as rail network that extends out radially. Major policies and strategies as part of Transport Plan 2021 aim to decongest road and rail terminals by diversion of through traffic, and by providing improved connectivity to regional and sub-regional centres, even outlying areas of the region. In fact the road network was proposed to be developed in the form of a three tier hierarchical network consisting of primary roads (expressways, national and state highways), secondary (major district roads) and tertiary (single, intermediate or double lane metalled roads) networks.

The primary road network in the region consists of ten national highways and ten state highways. Five of the national highways meet in Delhi forming a part of the Ring Road, i.e. NH1 (Delhi to Panipat), NH2 (Delhi to Palwal), NH8 (Delhi to Behror), NH10 (Delhi to Rohtak) and NH24 (Delhi to Hapur). The proposal was made to extend these national highways to CNCR towns i.e. NH1 (Delhi to Kundli), NH2 (Delhi to Ballabgarh), NH8 (Delhi to Gurgaon), NH10 (Delhi to Bahadurgarh) and NH24 (Delhi to Ghaziabad). Above mentioned national highways along with NH58 (Ghaziabad-Meerut-Muzaffarnagar) and NH91 (Ghaziabad-Khurja-Aligarh) extend radially out from NCT Delhi connecting it to regional or priority towns and are in the form of four lane divided highways on most stretches, proposed to be developed as expressways. The other national highways in the region include NH71 (Rewari to Rohtak), NH71A (Rohtak to Panipat) and NH71B (Rewari–Sohna–Palwal).

Another major policy decision with reference to the primary road network was taken to connect all CNCR towns through the Eastern and Western Peripheral Expressway (WPE and EPE). Kundli-Manesar-Palwal (KMP) Western Peripheral Expressway and Kundli-Ghaziabad-Palwal (KGP) Eastern Peripheral Expressway were proposed to be implemented on priority in the first phase (2001–2011), both about 135 km long. In addition all metro and regional centres are proposed to be connected with perimeter roads of expressway standard to serve the purpose of bypass for enabling through traffic.

The EPE and WPE constitute a combined project forming a ring road around Delhi, connecting NH-1 and NH-2 from western and eastern side of Delhi with a total length of 270 km; about 183 km length passing through Haryana and about 87 km length passing through Uttar Pradesh.

The Western Peripheral Expressway, being implemented by HSIIDC, Haryana Government, is proposed to be completed by November 2017, the stretch from Palwal to Manesar is ready for use while the section between Manesar and Kundli is under construction, and is targeted to be completed in August 2018. There were several delays in the project including termination of the initial contract with KMP Expressways Limited. However, the project was revived in 2015 with fresh bids. The work on EPE also began in 2015 and was given about 900 days for completion. However, it faced several issues along the way. Mostly the project was stalled due to unsatisfactory compensation with respect to land being acquired. Its implementation is being done by the National Highways Authority of India.

The two expressways together would serve as a bypass for traffic not meant for Delhi and headed towards the neighbouring states of Uttar Pradesh, Uttarakhand, Haryana, the Punjab, and Jammu and Kashmir. This will not only decongest Delhi but also cut down pollution by almost 50%. EPE is estimated to provide a network for almost 80,000 freight vehicles every night, which would have passed through Delhi in its absence. The EPE is also being promoted as the 'first smart and green highway' with an intelligent highway traffic management system (HTMS), and video incident detection system (VIDS). Bringing in the concept of 'green and sustainable', it is proposed to be electrified by solar panels and faster electronic toll collection system at toll booths.

On the whole enhanced accessibility through these two expressways is expected to boost the economy of the region by providing better connectivity among settlements and also by managing traffic within the region. It would give a major boost to transport infrastructure of the region, making it speedier, efficient and less polluting, which in turn would make the economy buoyant as a whole. Moreover looking at the economic impact from the point of view of employment generation, these projects have created both direct and indirect jobs. The construction work of EPE has generated 4,000,000 man days of employment with 2100 engineers and 5200 workers employed on a basis. Besides this, local population would be involved in creating and maintaining green belt and roadside amenities such as petrol pumps, shops, etc. Furthermore these projects would add markets for local produce.

Other than road network, connectivity is provided by rail network. Existing rail network consists of five broad gauge railway lines converging at Delhi and also two special lines known as the Goods Avoiding Lines (GAL) and Delhi Avoiding Lines (DAL). The GAL provides a direct entry from Ghaziabad to New Delhi by bypassing the congested Delhi Railway Station Complex. The DAL provides a direct passage from the major yards such as Tughlakabad and Ghaziabad directly into Delhi-Ambala-Kalka section and through Lajpat Nagar, Patel Nagar, Daya Basti and Azadpur link.

Looking at these proposals about enhancing rail network in the Regional Plan 2021, one of the major policies is to build a primary regional rapid transit system connecting regional towns to each other and with Delhi along the following corridors with feeder links at sub regional centres (Table 8.4).

Table 8.4 Transport corridors in the NCR

Order of priority	Name of the corridor	Length in km
1.	Delhi-Ghaziabad-Meerut	90
2.	Delhi-Gurgaon-Rewari-Alwar	180
3.	Delhi-Faridabad-Ballabgarh-Palwal	111
4.	Delhi-Faridabad-Ballabgarh-Palwal	60
5.	Ghaziabad-Khurja	83
6.	Delhi-Bahadurgarh-Rohtak	70
7.	Ghaziabad-Hapur	57
8.	Delhi-Shahdra-Baraut	56

Source Revised as per feasibility report

Feasibility studies of the first three prioritized corridors have been completed. Draft Detailed Project Reports are at an advanced stage of finalization. It is important to note that Metro Rail would be extended to the CNCR towns and integrated with the Regional Rapid Transit System (RRTS) with feeder services. In order to provide connectivity to the CNCR towns of Gurgaon, Faridabad and Bahadurgarh in Haryana Sub-region, and Noida and Ghaziabad including Vaishali in U.P. Sub-region through Delhi Metro, work is continually being pushed on a fast track with Delhi Metro Rail Corporation and metro links has already been extended to Gurgaon, Noida and Ghaziabad. In addition the Rapid Metro has been commissioned in Gurgaon. Work on Badarpur-Faridabad-Ballabgarh and Mundka-Bahadurgarh Metro line is in progress, which is likely to be completed by December 2017.

The orbital rail corridor, parallel to proposed Peripheral Expressways connecting radial rail corridor of the Indian Railway in CNCR towns and directional terminals was required to be made. No progress has been seen with regard to this proposal. Looking at airways connectivity, at present two airports are located in the NCR where Delhi-Indira Gandhi Airport is used for international flights and Palam Airport for domestic flights. A proposal is under active consideration for the development of another international airport in Greater Noida and also a suitable domestic air terminal. In June 2017, the second international airport for this region has been approved in Greater Noida's Jewar Region in Uttar Pradesh. The new international airport would cater to 30–50 million passengers per year over the next 10–15 years. The only difficulty with this project is that it is a Greenfield project, which would permanently encroach upon a large piece of fertile agriculture land in western U.P.

International airport project to be built in the Greater Noida's Jewar Region in Uttar Pradesh is proposed to be completed by 2021. However, looking at the speed of progress being made, it appears to be hardly satisfactory. Some major projects scheduled for the first phase are yet to be completed, which implies postponement of projects to be completed in the second phase. The Peripheral Expressways and rail transit systems including metro rail link with CNCR towns, once completed would ease traffic burden on the roads of the region and provide improved connectivity and also pave the way for further economic development.

From the above review of the transport related major proposals in the region, it appears that the National Capital Region Planning Board clearly recognised regional accessibility as central to the development of the NCR. In this line of thinking, the National Capital Region Planning Board has prepared a Functional Plan on Transportation for NCR. This is an integrated multimodal transport plan with planning perspective of 2032. According to the National Capital Region Planning Board [11: 91], the major objectives of transportation in NCR Regional Plan, 2021 are:

- To provide transport system that tends people to shift towards use of public transport more.
- To provide efficient and economical rail and road based transportation.
- To discourage the movement of passengers and goods transit through core city by planning peripheral road expressway and bypasses.

As is obvious, the objectives of the NCR Plan place greater emphasis on promotion of public transport in opposition to travel through private automobile. Although air transport is directly mentioned in the transport objective of the Plan, all three modes—road, rail and air—when integrated provide the most efficient system of movement of people and goods. Hence the network of road, rail and air together form the transportation system in the NCR Delhi. The NCR Plan also intends to discourage through traffic from NCR Delhi. In order to achieve the regional plan's objectives, one of the actions taken by the NCRPB is that it has proposed Regional Rapid Transit System (RRTS) in the National Capital Region Plan 2021 notified on 17 September 2005. Functional Plan on Transport for National Capital Region, 2032 was prepared by a private consulting firm under the close supervision of the NCRPB in order to detail out implementation of the transport policies entailed in the National Capital Region Plan, 2021. The Functional Plan for Transportation in NCR, 2032 was approved by the NCRPB on 11 November 2009.

Even before the proposed RRTS is implemented, other major transport projects that would considerably enhance accessibility are at various stages of implementation. Notable among them include ([12: 84–88]; also see Fig. 8.4):

- Upgradation of national highway within NCR
- Peripheral Expressways around Delhi
- Upgradation of roads through NCR financed projects
- Delhi-Meerut Expressway
- New rail links on Rewari-Jhajjar-Rohtak, Sonepat-Gohana-Jind, Tuglakabad-Palwal Fourth Line and Sahibabad-Anand Vihar third and fourth line
- Extension of Delhi Metro to NCR Towns
- Dedicated Freight Corridors (DFCs)

Fig. 8.4 Integrated transport plan for the NCR: road networks and airports. *Source* www.ncrpb.nic.in

  - Western DFC: Mumbai-Delhi (Dadri)—1515 km
  - Eastern DFC: Ludhiana–Dadri–Son Nagar—1278 km

- Delhi-Mumbai Industrial Corridor
- Yamuna (Taj) Expressway
- Ganga Expressway

Nonetheless, the Functional Plan for Transportation in NCR, 2032 makes some radical transport planning proposals. Extensive network of 1107 km long regional expressways has been proposed connecting major urban nodes in the region with lane length of 9398 km. to be developed through public private partnership mode, the regional expressways would be developed with the right of way of 100 m and design speed of 120 km [12: 93–94]. Following are expressway links:

- Delhi–Sonipat–Panipat
- Delhi–Bahadurgarh–Sampla–Rohtak
- Delhi–Gurgaon–Manesar–Rewari
- Gurgaon–Faridabad
- Delhi–Faridabad–Ballabgarh–Palwal
- Delhi–Ghaziabad–Hapur
- Dadri–Ghaziabad–Meerut
- Loni–Baghpat (with potential to extend to Baraut and beyond)
- Sonipat–Baghpat–Meerut

- The entire outer grid from Panipat–Gohana–Rohtak–Jhajjar–Rewari–Palwal–Jewar–Bulandshahr–Hapur–Meerut

Marked as high speed, high capacity, safe and affordable mode of public transport, the Regional Rapid Transit System is expected to connect major regional nodes in the region with Delhi and with Sub-regional Centres; orbital rail and Mass Rapid Transit System (MRTS) for Delhi and CNCR towns. RRTS is proposed to run at an average speed three times that of the Delhi Metro with higher acceleration and deceleration. Average speed of Delhi Metro ranges from 30 to 40 km per hour. In order to cover the NCR, the RRTS stations would be placed at an average distance of 5–10 km.

Website of the National Capital Region Transport Corporation (NCRTC) shows that it is a Joint Sector company of Government of India and States of Delhi, Haryana, Rajasthan and U.P. and is mandated for implementing the Regional Rapid Transport System (RRTS) project across the National Capital Region, ensuring a balanced and sustainable urban development through better connectivity and access (also see Figs. 8.5 and 8.6).

At this stage it is worthwhile to take an example of a sub-regional plan to understand how far regional planning policies of the NCRPB are being translated

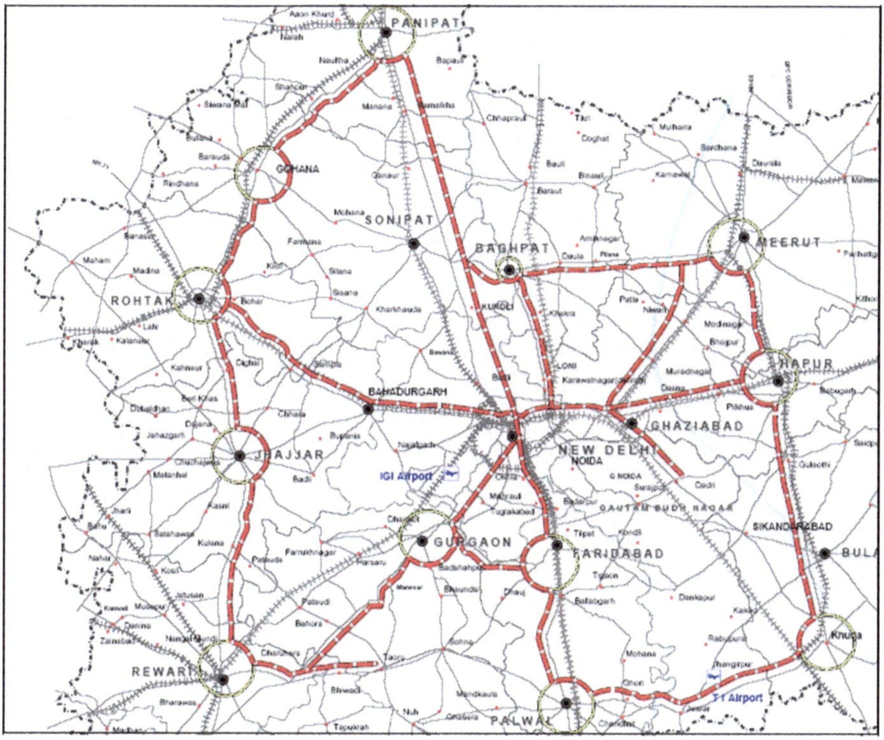

Fig. 8.5 Network of regional expressways in the NCR. *Source* www.ncrpb.nic.in

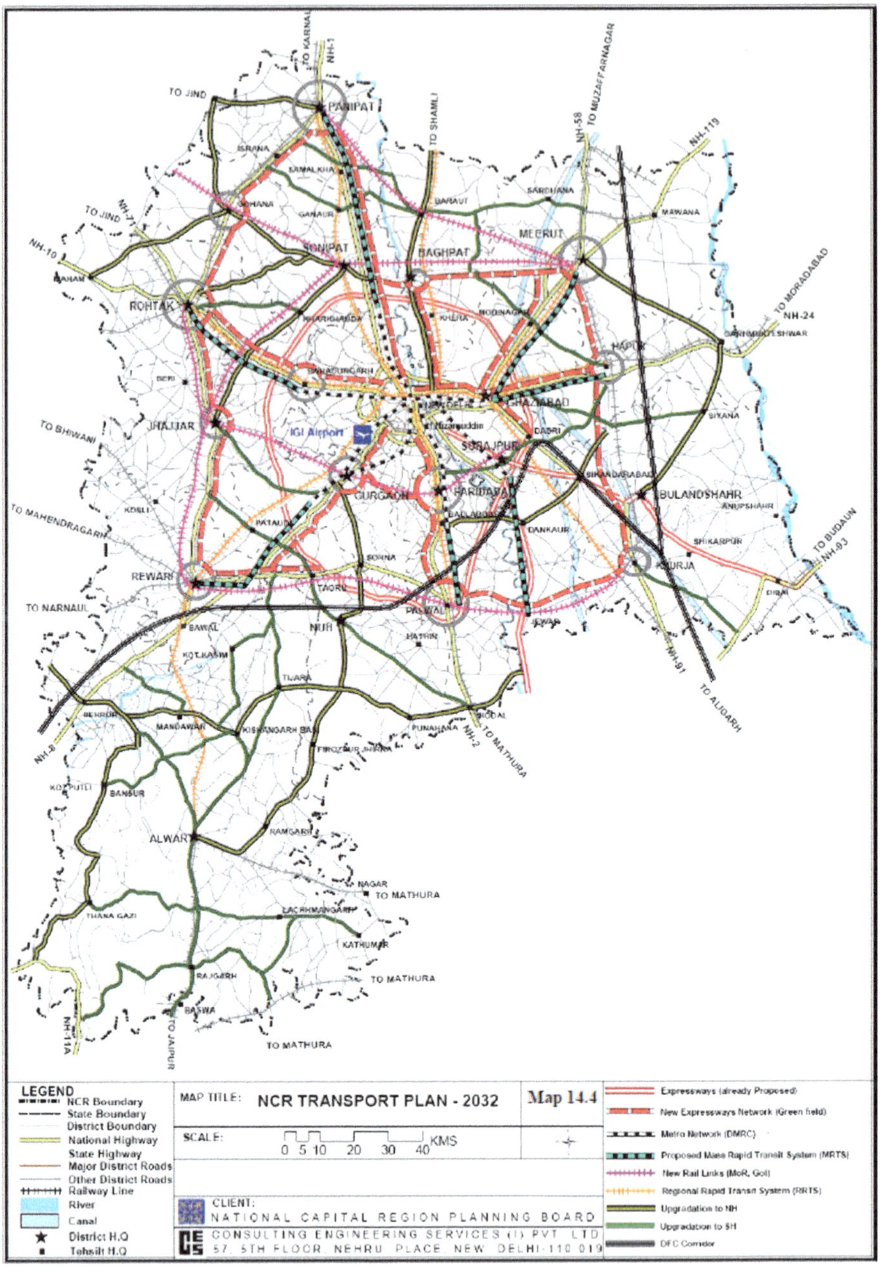

Fig. 8.6 NCR transport plan, 2032. *Source* www.ncrpb.nic.in

through lower order plans. For these purposes I will take the Sub Regional Plan of Haryana, which details out major transport proposals of the NCR Plan 2021. The Sub-Regional plan rightly points out that transportation is a system that promotes mobility of people and goods from one location to another. The Sub Regional Plan of Haryana, 2021 makes the following policies and strategies in order to achieve its objectives:

- To provide linkages among all six tier hierarchy of settlements and urban settlements.
- Connect Metro/Regional Centres with the Capital by an efficient and effective transport network.
- Unrestricted movement of public transport within NCR.
- Develop activities keeping in view rail and road linkages in the Central NCR.
- Transit Oriented Development (TOD) need to be promoted.
- Proposals of Integrated Mass Rapid Transport System (MRTS), Regional Rapid Transport System (RRTS), Expressways and Bypasses shall be implemented.
- Provide multi-modal stations wherever required.
- Promote unified fare collection system.
- Promote non-motorised transport in the Sub-region.
- Use of Intelligent Transport System for traffic management.
- Integration of land use and transport planning.
- Roads shall be made more pedestrian and disabled friendly.
- Efforts shall be made to provide cycle tracks.
- Regular training shall be conducted for the capacity building of road users and pedestrians (Sub Regional Plan—Haryana 2021: 122).

There are several proposals, which translate Regional Plan policies such as RRTS, MRTS, and enhancing accessibility of metro and regional centres. However, there are other proposals such as ‘integration of land use and transport planning’, which are generic in nature. There are other proposals including ‘the capacity building of road users’ that are completely incomprehensible.

In Haryana Sub-Region, there is only one expressway i.e. Delhi–Gurugram, 6 National highways, 12 state highways and 12 major district roads. Comparative analysis of road density shows that Delhi has the highest road density with 2103 km per 100 km^2 followed by Haryana Sub-Region with 59.30 km per 100 km^2, Rajasthan with 50.79 km per 100 km^2 and UP with 50.71 km per 100 km^2. The state highways cover 60% of the sub-Region of Haryana. National Highway 44 is the busiest route of the sub-region with 60,000 PCU traffic volume. Level of services describes operational conditions of road networks. These include parameters such as speed and travel time, freedom of manoeuvre, traffic interruptions, comfort, convenience and safety. Based on that the services has been designated from level A to F, whereas Level A represents best operating conditions, Level F represents the worst. In this sub-region many lane configurations operate at level of service F and only few roads and lanes operate at level of service A. Fewer vehicles on the roads and more people moving to public transport could also improve the level of service in the Haryana Sub-Region. Transit oriented development is such a planning concept that is designed to overcome traffic congestion. It is a designed to build urban area consisting of mixed use residential and commercial areas in order to promote and maximise access to public transport. In the Haryana Sub-Regional Plan, high density corridors are proposed to encourage people to settle in these areas. Master plans are required to identify several potential TOD zones (Sub Regional Plan—Haryana 2021: 91–102).

Delhi Metro Rail Corporation is continually expanding its network within Delhi and the NCR. Regional Plan 2021 for the NCR has proposed Delhi metro rail system for the NCR region. As far as Haryana Sub-region is concerned, metro is further expanding to Gurugram, Faridabad and Bahadurgarh. Delhi Metro Rail Corporation's future proposals cover extension of metro rail network further into the Haryana Sub-Region in order to improve connectivity with the NCT, Delhi and among settlements within the sub-region. Delhi Metro Rail Corporation's upcoming metro extension projects would cover Sonipat, Panipat, Rohtak, Palwal, Manesar and Rewari, making a large part of the Haryana Sub-Region accessible through metro. Accessibility within the Haryana Sub-Region will get big propulsion through the construction of Regional Rapid Transit System as the following links will get connected through the RRTS.

- Delhi–Sonipat–Panipat
- Delhi–Bahadurgarh–Sampla–Rohtak
- Delhi–Gurgaon–Manesar–Rewari
- Gurgaon–Faridabad
- Delhi–Faridabad–Ballabgarh–Palwal
- Sonipat–Baghpat–Meerut
- The entire outer grid from Panipat–Gohana–Rohtak–Jhajjar–Rewari–Palwal–Jewar–Bulandshahr–Hapur–Meerut

Except the last two links, other links promote accessibility between Delhi and other urban areas located in the Haryana Sub-Region. This goes against the grain of the National Capital Region Plan 2021, which aspires to enhance connectivity between urban settlements other than Delhi. But it must be emphasised that the Integrated Transportation Plan for NCR 2032 has identified eight regional rail based rapid transit corridors. Out of the eight corridors, and most these corridors have some link with the Haryana Sub-Region.

Apart from the RRTS, regional Bus based Public Transport System (BPTS) has been conceptualized for the NCR. For the Haryana Sub-Region 9283 buses are required for the base year, which could increase in future. In order to deal with the situation, the State has made provisions for bus terminals in the respective Development Plans and transport department acted actively to provide buses for intra city movement across cities like Gurugram and Faridabad. All buses have sophisticated features such as GPS systems, Intelligent Transport systems, passenger feeding systems and electronic ticket verification machines. All buses run on clean fuel Like CNG. For air travel, the Haryana Sub-Region depends on the Indira

Gandhi International Airport. However, in the Integrated Transportation Plan for NCR, small airports are proposed in Rewari, Jhajjar and Sonipat district. However these airports come under the influence zone of IGI, Delhi. Haryana state government also intends to set up an international cargo terminal on the confluence of Hisar, Rohtak and Bhiwani. Similarly a small airport is proposed in Panipat. Both projects will be undertaken after approval from the Ministry of Aviation, Government of India (Sub Regional Plan—Haryana 2021: 102–103).

Road development programme in the NCR assumes equal significance. The NCRPB proposes to develop road network as indicated below:

- NCR regional plan proposes to develop existing Ring Road, Outer Ring Road and Five radial roads as expressway upto NCR towns.
- The Plan proposes fast development of Eastern Peripheral Expressway (Kundli–Ghaziabad–Palwal) and West Peripheral Expressway (Kundli–Manesar–Palwal corridor).
- The Plan also proposes Metro and Regional Centres to be connected with perimeter roads of expressway standard to act as a bypass for through traffic.

In a modern and globalized world, it is necessary to develop a world class infrastructure, high speed connectivity and favourable environment in order to attract inward investment in the region. Hence, the Haryana Sub-Regional Plan envisages the Sub-Region as an arena where it is possible to create a strong economic base through world class infrastructure. Thus, the Sub-Regional Plan focuses on connectivity, economic activities, logistics hubs, and infrastructure. Presently Haryana has developed a good road network infrastructure. It supports the socio economic growth of the state. This may be the reason behind the high density of vehicles plying on these available roads. But to cater to future needs of traffic volume on roads, present roads and lanes need to be widened, new roads to be constructed and proposed bypasses and parallel expressways to be built.

On the basis Primary surveys and secondary sources information, it has been found necessary to the augmentation of existing roads as well as additional corridors will have to be implemented in a phased manner. The additional corridor in the form of expressway can cater to large volumes of traffic as it will provide better mobility and safety. The Sub-Regional Plan of Haryana has advantage of accessibility provided by KMP expressways, proximity to national capital and its geological situation. Government of Haryana proposes to develop 12 new hubs along expressway, designated as 'Global Corridor'. There are also plans to upgrade roads from two lanes to four lanes and from four lanes to six lanes (Sub Regional Plan—Haryana 2021: 104–106).

In the Haryana Sub-Regional Plan, transit interchanges—a place where passengers and cargo are exchanged between vehicles or between transport modes—can be designed along the Kundli–Manesar–Palwal Expressway where many national highways meet across the corridor. Orbital Rail Corridor is also planned along this Expressway. For promoting public transport to facilitate faster travel within the Sub-Region, the existing and proposed metro network and the proposed

RRTS is being developed in an integrated manner. Further in order to improve user facilities on roads stretches, it needs to facilitate roads with 'highway facilities centres' on comprehensive basis including common services like parking, fuel stations, restaurants, etc.

Haryana currently has about 1700 km of rail network. About 625 km i.e. 35% of rail routes in the state serve the sub-region. Presently all rail lines are used both by freight and passenger trains. The Regional Orbital Rail Corridor (RORC) was referred to enable the bypassing of a number of trains, which presently pass through Delhi. The proposed RORC would aim at integrating the existing and proposed radial rail lines originating and passing through the NCT Delhi and also to provide commuter services to the proposed urbanization along the RORC. In addition to the RORC, three other rail lines within the Haryana Sub-Region are proposed to strengthen connectivity through the rail system which will form the 'Inner Regional Orbital Rail Corridor' (IRORC).

Ministry of Railways, Government of India, is developing DMIC, a dedicated freight corridor between Delhi and Mumbai. A band of 150 km on both sides is being developed as Delhi Mumbai Industrial Corridor. DMIC Sub-Region in Haryana covers about 29,362 km^2 area, which covers about 66.4% of the total area of the state. According to the Perspective Plan Report, the value of output in the DMIC Sub-Region of Haryana is estimated to grow by 15 times over the next 30 year period (2010–2040). Development of proposed road network, which includes Expressways, NH, SH, MDRs, bypasses, flyovers and ROBs in the first phase would cost INR 17,442 crore and in the second phase it would cost INR 12,015 crores by the year 2021 (Sub Regional Plan—Haryana 2021: 113–121).

8.3.4 Smart Environment

NCR Regional Plan, 2021 encourages a balanced approach for incorporation of environmental features in the planning process for sustainable development of its sub-regions. Good quality environment is essential in order to achieve sustainable development. It is mandatory that development takes place with environmental considerations. It is also important to point out the factors, which are crucial for environment depletion and degradation, which might cause irreversible damage to regional water resources, forest cover, flora and fauna, ecological sensitive areas, good quality agricultural lands, and heritage areas, etc. It is in this context that we may to have focus on environment factors i.e. forests, air, water, and other environmentally sensitive zones in the region and sub-regions.

Importance of the forest cover could not be overstated under any circumstances because the forest cover has great bearing on ground water occurrence, soil erosion, reduction of floods, and protection of the environment. National Forest Policy, 1988 aims to have a minimum of one third of the total land area of the country under forests and tree cover. But the ever increasing demand for food, fuel, fodder and growing demand of land for urbanization has put tremendous pressure on the scarce forest resources of the Haryana Sub-Region. At an all India level, forest cover forms 21% of the total geographical area. In the National Capital Region, forests are spread over an area of 2131.2 km^2, which is 6.2% of the total land area of the NCR. Haryana Sub-Region of NCR has 3.94% area i.e. 529.41 km^2 under forests.

National Capital Delhi, which host a large percent of Delhi's population has been recently suffering from 'very poor' air quality (see Fig. 8.7). Several times in November 2017, Air Quality Index nearly touched the 500 level whereas the safe limit is only 40.

Present status of forest cover in the Haryana Sub-Region is that this sub-region is primarily an agriculture area with almost 80% of its land under cultivation. As per India State of Forest Report (2011) the forest cover in the state is 1608 km^2, which is 3.64% of the state's geographical area and the tree cover in the state is 1395 km^2, which is 3.16% of the geographical area. Thus the forest and tree cover of the state is 6.80% of its geographical area. The forests in Haryana are classified under Reserved, Protected, Unclassified and areas closed under section 38 of Indian Forest Act and under section 4 and 5 of the Punjab Land Preservation Act, 1900. The forest cover is unevenly distributed in the Haryana Sub-Region. The southern districts of Rewari, Gurugram and Faridabad have large areas under the reserve forest. Total area under reserve forest is 10.59 km^2 in Haryana Sub-Region and protected forest covers 290.96 km^2 followed by other forest, which covers 227.85 km^2 geographical area of the region. It is interesting to understand the temporal change in the forest cover in Haryana Sub-Region during 2001–11. From various studies, it is quite clear that forest cover in the Haryana Sub-Region has been continuously increasing since 2001. During the year 2011, total forest cover of the sub-region increased to 456 km^2. Gurgaon district has the highest forest cover i.e. 231 km^2. In the state, there are 9 wildlife sanctuaries and 2 national parks out of which 3 wildlife sanctuaries and 1 national park exist in the Sub-Region. A Zonal Master Plan for the Eco-sensitive Zones has to be prepared by the state with due involvement of all concerned state departments such as environment, forests, urban local bodies, tourism, municipal and revenue department and the Haryana State Pollution Control Board for integrating environmental and ecological

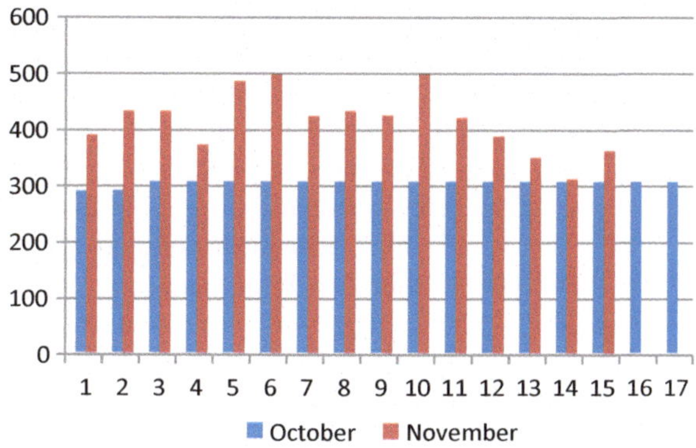

Fig. 8.7 Air quality index from 15 October to 15 November 2017. *Source* Central Pollution Control Board (2017) as quoted in Sunday Times [15]

considerations. This will include restoration of denuded areas, conservation of existing water bodies, management of catchment areas, watershed management, groundwater management, soil and moisture conservation, needs of local community and such other aspects of the ecology and environment that need attention (Sub Regional Plan—Haryana, 2021: 281–285).

Hills of the Aravalli Ranges constitute a critically significant part of the environment of the Haryana Sub-Region. Government of India has devised several statutory mechanisms for the protection of the Aravalli Ranges. In order to protect these fragile areas falling under Aravalli Ranges and restrict activities, which are likely to cause environmental degradation in the Aravalli Region, Ministry of Environment and Forests, Government of India prohibits carrying out of any of the following operations:

- Any kind of mining activity to be stopped with immediate effect
- Cutting of trees
- Electrification (laying of new transmission lines)
- Any kind of construction work e.g. dwelling units, roads, etc.

This appears to be a list of reasonable restrictions for the protection of the Aravalli Range. However, due to lack strict enforcement of laws and rules by concerned government agencies, depletion of the physical environment in the Aravalli Ranges continues unabated in spite of the fact that various courts have adjudged against environmental violations. For example, vested commercial interests continue to carry out mining activities in this area. Environmental arguments are continually defeated in the face of market centered arguments premised on profitable commercial exploitation of natural resources.

Recently air pollution has grabbed headlines in the media and all attention of political leaders and administrators in the wake of very poor air quality in northern India. Chief Ministers are holding parleys to combat air pollution. Central government has been sending advisories to state governments. People are asking questions with the hope of making politicians accountable. Air quality meanwhile continues to deteriorate. Air pollution occurs when harmful substances including particulate matter and biological molecules are introduced into Earth's atmosphere. Continuous mixing, transformation, and trans-boundary transportation of air pollutants make air quality very poor. Since the last one month people in the NCR are suffering from extremely poor air quality. Responding to this situation, government has declared this condition as a health emergency. Stubble (crop residual) burning has been considered as the chief source of air pollution. Some newspapers have reported quoting studies that claim that nearly 90% pollution in the NCR Delhi is caused by stubble (crop residual) burning in the neighbouring states of Punjab and Haryana. Historically high growth of population, polluting industry, and the most crucial factor being rapid growth of vehicles has been regarded as the main causes of air pollution. For example, it is said that vehicular pollution contributes more than 70% of total air pollution in towns of Haryana Sub-Region. This does not add up and perhaps long term research could answer these complex questions.

There are at present 3 online air monitoring stations at Gurugram, Faridabad and Rohtak and 9 other manual air monitoring stations within the Haryana Sub-Region. Ambient air quality status provides for permissible limits of 12 air pollutants, out of these 4 pollutants are regularly being monitored. These are PM 10, PM 2.5, SO_2, and NO_2. After analysing the air quality of the sub-region, it could be concluded that the concentration of PM 10 and PM 2.5 has been high; in fact both the PM levels have been higher in the entire NCR. However the SO_2 levels have been recorded within the limits in all the three monitoring stations. However, NO_2 levels were recorded beyond the permissible limits in Faridabad and Rohtak.

Maintenance of water quality standards and monitoring is another important challenge facing the Haryana Sub-Region. Haryana has two major river basins namely, the Ganga and the Indus. Yamuna is the only perennial river in the Sub-Region. The Yamuna sub-basin covers parts of Panipat, Sonipat, Rohtak, Gurgaon and Faridabad districts. The Aravallis bring a number of small rainy streams in Gurugram, Rewari and Jhajjar. In the Sub-Region a number of small ponds or lakes also exist. Canals are the most popular and important source of irrigation in the Sub-Region due to easy supply and regular flow of water to agriculture fields (Sub-Regional Plan—Haryana 2021: 285–288).

Water (Prevention and Control of Pollution) Act, 1974 was enacted with a basic objective to maintain and restore surface aquatic resources. CPCB's classification of surface water in India notifies primary water quality criteria for the following five uses.

- A—Drinking water source without conventional treatment but after disinfection
- B—Outdoor bathing
- C—Drinking water source after conventional treatment and disinfection
- D—Propagation of wildlife and fisheries
- E—Irrigation, industrial cooling, controlled waste disposal

National Water Quality Monitoring Program is meant for water quality monitoring and it collaborates with the State Pollution Control Boards (SPCBs) and Pollution Control Committees (PCCs) in order to implement various water related policies and regulations governing water quality management and also monitors water quality of various water bodies. In the Haryana Sub-Region of NCR, HSPCB regulates control of pollution and implements National Water Quality Monitoring Program. For water quality monitoring, state government has established eight monitoring stations in the Haryana Sub-Region.

Water quality in rivers is one particular challenge. The maximum and minimum values of core parameters in the River Yamuna have been observed at two monitoring stations. Biochemical Oxygen Demand and pathogenic bacteria indicate that there has been degradation in water quality. While Nitrates and Nitrites are in the range of 0–13.9 mg/l, pH level is within limit for all uses in the River Yamuna. As per CPCB, the contribution of pollution load from NCR is 80% and non-NCR states contribute only 20%. Thus River Yamuna water is not suitable for class A, B and C uses in almost the entire length of the NCR due to high total coliform, high BOD and low DO levels. Significantly water quality of canals in the Sub-Region is also deteriorating. Major canals in the Sub-Region are Yamuna Canal, Gurugram Canal and Agra Canal. It is found that water in these canals is not suitable for Class A, B and C uses. High BOD and high coliform levels make water quality below permissible limits. Environmentally Sensitive Conservation Zone is another cause for concern. Some of the natural features in the Sub-Region have been identified as environmentally sensitive areas. Extension of Aravalli ridge, sanctuaries and other ecologically sensitive areas are to be conserved with utmost care and afforested with suitable species. An area measuring approximately 340 ha has been reserved as 'Water Recharge Zone' to check run-off and replenish ground water (Sub Regional Plan—Haryana 2021: 289–291).

Depleting underground water table around Sohna town in district Gurugram has become a major concern for the state government planners. Further a complete sector namely 72A has been carved out in Gurugram for developing artificial water body. The City Forest in an area of 33 acres has been set up in Sector 52-A, Gurugram. A Biodiversity park is also proposed to be established on the Gurugram-Mehrauli Road near Delhi-Haryana boundary. Aravalli ranges have been designated as Natural Conservation Zone in the Plan in accordance with Regional Plan 2021. Some of the important proposals for reducing water pollution include:

- Industries should treat their effluents and conform the specified requirements.
- Disposal of garbage, solid, semi-solid wastes into river, its tributaries and drains should be restricted.
- Existing water bodies, lakes and ponds in the NCR should be revived by filling with water by connecting them to existing tail-clusters.
- Rain water harvesting structures may be constructed along with drains under the MNERGA program.
- Community participation in various Yamuna water quality restoration programs should be encouraged.

For the Haryana Sub-Region, a number of environmental issues and challenges have emerged. First, State of Forests Report, 2011 published by Forest Survey of India shows that forest and tree cover area is only 6.8% as against the national average of 21%, leading to further degradation of forests in the Aravalli ranges. To promote sustainable development of forest resources in the state, Haryana Government has fixed the goal of achieving 20% forest and tree cover in the state in a phased manner.

Air pollution has emerged as a major challenge in the NCR. Haryana state government has setup two continuous online stations at Gurugram and Rohtak along with one fully operational in Faridabad. The number of ambient air quality monitoring stations is inadequate in the Haryana Sub-Region. As per IS codes Haryana Sub-Region requires well over 50 monitoring stations. In order to address this situation, Sub Regional Plan—Haryana 2021 makes a number of important proposals:

- Along with the existing monitoring station, efforts would be made to set up monitoring stations in all district headquarters of the sub-region.
- CNG stations are being set up in Faridabad and Gurgaon. Industries at Faridabad and Sonepat have switched over to PNG and LNG, which will further reduce air pollution.
- Environmental clearances have to be obtained by the applicant for construction projects.
- There is need to operationalize the proposed monitoring stations in Haryana under NAMP on priority basis (Sub Regional Plan—Haryana 2021: 292–293).

Further proposals could be made to reduce air pollution. For example, governments could make efforts to reduce air pollution from transport activities by making available clean fuels such as CNG, PNG, etc. in whole of NCR for public and private vehicles. Land use planning and zoning regulations could be enforced to control dust from construction activities.

Extension of Aravalli ridge, sanctuaries and other ecologically sensitive areas are to be conserved with utmost care and afforested with suitable species. Development in this area should be made in compliance with the Environment Act, 1986. Aravalli ranges in the Haryana Sub-Region have been designated as Natural Conservation Zone in the Plan in accordance with Regional Plan 2021 except those areas which are coming in the urbanizable area (Sub Regional Plan—Haryana 2021: 294).

- The goal of achieving 20% forest and tree cover in the state in a phased manner should be pursued. This goal can be achieved by adopting agro forestry practices.
- Requirements of prior environmental clearances for projects mentioned in EIA notification, 2006 should be strictly enforced by the state.
- To increase green cover in the Sub-Region, it is necessary to encourage planting of trees alongside roads, railway lines, canals, rivers, rivulets and streams, and other unutilised lands.
- Since the Sub-Region is highly urbanized and densely populated, village panchayats should take up community lands for cultivation of tree crops and fodder resources.

8.4 Smart Regional Planning Strategies for the NCR

As shown above governance appears to the most important challenge of regional planning and development. Apart from regional governance, I have focussed on the smart economy, smart transport, smart settlement pattern, and smart environment. I will make recommendations for these five sectors only. This however does not imply that other sectors in the NCR are less important.

8.4.1 Smart Governance Strategy

I propose that smart governance strategy should locate regional level coordination at its core. Implementation of 74th amendment to the Indian Constitution is critical for carrying out effective coordination, and for this to happen all district planners must proactively prepare District Development Plans as mandated under the Constitutional provisions. Further, all districts should be viewed by the NCR as sub-divisions of the three sub-regions and urban settlements as urban nodes within these districts. Thus a hierarchy of plans starting at the NCR, moving onto four sub-regions, a number of districts, and ending with equally good number of urban and rural settlements plans. In order to plan and develop the NCR, preparation and implementation of regional, sub-regional and district development plans should be made the central concern of the NCR, Sub-regional Planning and Development Authorities, and constitutionally mandated District Development Authorities. A constitutional amendment to create District Development Authorities is required. Similarly, an amendment in the NCRPB Act is also needed to establish Sub-regional Planning and Development Authorities. Delhi Government should also become full partner in NCR's regional planning efforts by setting up exclusive regional planning authority by amending the Government of National Capital Territory of Delhi Act, 1991 as has become evidently clear after severe air pollution in Delhi and the NCR. Planning and development authorities at all three levels must be duly elected with a small percentage of nominated members from the concerned state governments and central government.

8.4.2 Smart Spatial Planning Strategy Including Meta City Strategy

National Capital Region Planning Board is crucially important organization for planning and development of the NCR, its urban and rural settlements, agriculture and industry among other sectors. Current practice is that the NCRPB extends loans for various projects to state governments, urban local bodies and parastatals of state governments for development in the NCR. These projects are too small and insignificant and do not make any regional level impact. The smart economic strategy would be that the NCRPB undertakes and implements projects of 'regional significance'. For example, projects like the Regional Rapid Transport System (RRTS) should be implemented by a regional agency. Appropriately, the recently established National Capital Region Transport Corporation, which is a joint sector company of Government of India and States of Delhi, Haryana, Rajasthan and U.P., is mandated to implement the Regional Rapid Transport System (RRTS) projects across the National Capital Region with funding coming from a diversity of sources including its own funds and central state and governments funds. Such joint sector companies could be created for other important sectors like water, sewerage, drainage, etc. under the overall administrative control of the NCR. However, these companies must be accountable to respective elected regional planning and development authorities.

The NCRPB has prepared four functional plans on drainage, economy, transport and water. All four aspects covered by the four functional plans are hugely important to the economy and population of the region and the country. The NCRPB is also mandated to bring about balanced development of the region. So preparation of functional plans is insufficient; more needs to be done by the Board. I recommend that proposals of 'regional significance' should be implemented by the NCRPB because no other agency has an integrative and hierarchical overview of the region better than the Board. Regional drainage system, regional economic programme, regional mobility and accessibility, and regional water management would be better served if these major proposals are implemented by the NCRPB. Implementation of these proposals of regional significance is like setting up scaffolding within which district development plans, master plans and village plans could be fit appropriately depending on varied local contexts. Planning and implementation of regionally significant planning proposals could not be left to the private sector or any form of partnerships involving the private sector. Project centered developments in Gurugram led by the private sector has promoted disintegrated development causing splintering of physical infrastructure. This could have been avoided if regional infrastructure was in place in advance.

On the conclusion of Census 2021, Delhi will have become a meta city alongside other Indian cities like Mumbai. With population exceeding 20 million, a different kind of planning is required. I believe that more rather than less devolution of power is required for governance of a meta city and at the same time more rather than less centralization is also required. We need to know what this really means. For certain functions like planning both centralized as well as decentralized governing institutions and plans are required because we cannot do away with a hierarchy of plans before plans get implemented. For sectors like transport, water, sewerage, drainage and power, a mix of centralization and decentralization is required. However, for services like basic education and health, decentralized governance is the best option. For meta cities like Delhi, largeness warrants transparency and technologies based on ICT and IoT would greatly help in making decision making regimes transparent. Day to day services such as grant of building permissions could be easily governed through online technologies if property ownership records are clearly recorded. So modern ICT and IoT will play critical part in managing the meta city Delhi, an economic hub of the NCR.

8.4.3 Smart Transport Strategy

Another significant proposal is about the speed of implementation of proposals pertaining mobility and accessibility. Proposals like Regional Rapid Transport System, metro links with all the Central NCR urban settlements should be implemented by the end of the Regional Plan 2021. We still have about four years. While progress about linking all the Central NCR urban settlements with metro is on course although speed of implementation could be accelerated, the RRTS is yet to take off and year 2032 is too far to completion of the RRTS. As it is truism that there is no life without mobility, it is equally true that there is no economic buoyancy without accessibility, without unhindered movement of people and freight. Delhi-Mumbai Industrial Corridor will promote efficient movement of long haul freight in Uttar Pradesh, Haryana and Delhi. It is time that all concerned agencies should make plans for the use of spared track space for short haul freight movement.

In order to comprehend the actual import of the DMIC for the NCR, the NCRPB should prepare another functional plan focussed on this issue. At present it appears that economic activities in the DMIC would accentuate regional inequality proactively, going in a direction that is completely in opposition to balanced development of the region.

8.4.4 Smart Environment Strategy

Air pollution is a classless phenomenon because all of us breathe the same air. Unlike sewerage system, water distribution system, solid wastes management, education and health, middle classes in India could not possibly create their own exclusive citadels and fortify themselves with exclusive enclaves leaving behind the common man and woman. Since air pollution affects all of us and also because all of us have caused this problem, all of us need to solve this problem. Based on Indicus Foundation's latest study, I make the following six recommendations. First, Crop stubble burning should be managed by changing the cropping pattern from paddy to other crops such as pulses, cotton, fruits and vegetables in Kharif season. Simultaneously regional planners could consider delineating some area around Delhi as the green belt. Second, all roads should be made dust free with construction and repair of roads by using world class construction materials and technologies so that these roads and areas around them could be easily cleaned with the help of machines. Innovative planting alongside roads could further reduce the impact of vehicular pollution. Third, solid wastes burning should be largely replaced with waste to energy situation by use of better technologies. All plants processing solid wastes should be kept at 100% capacity at all times. New landfill sites should be identified much in advance before the existing landfill sites could be no longer used. There is no place for solid waste dumping sites in modern city or region. In order to make a coherent planning statement on environment, the NCR should prepare a functional plan for energy like "The Tokyo Initiative on Smart Energy Saving toward a Smart Energy City" prepared by the Bureau of Environment of Tokyo Metropolitan Government in May 2012 (also se Chap. 1). An editorial in the Guardian published from London argued that wind energy makes economic sense too. "The precipitous drop in the price of electricity from offshore wind turbines should be a tipping point for green technology. In 2014 the current generated by a forest of giant whirling fans out at sea was priced at around £150 per MWh. In the latest auction this week the comparable cost dropped as low as £57.50/MWh. Even when the cost of providing back-up capacity for still days is added, the cost of producing energy from offshore wind is little more than £70/ MWh. Compared to the new Hinkley C nuclear plant which produces electricity at a cost of £92.50/MWh, one has to wonder whether as a nation we should persist with nuclear energy as an option to reduce our greenhouse gas output". Similar arguments hold true for India not only for wind energy but also for energy harvested from the sun. Britain's Climate Change Act stipulates that by 2050 the country's net emissions of greenhouse gases should be 80% lower than in 1990. The British Government is steadfast in producing enough wind energy that the country would no longer need nuclear energy by 2050.

Fourth, although government is planning to introduce Bharat VI fuel in 2018, all vehicles need to be made Bharat VI fuel compliant. Enhancing capacity of metro, speedy construction of metro in the Central NCR, and faster construction and completion of projects like RRTS would reduce pollution. Fifth, it has now become imperative that we move renewable energy to non-renewable energy as harnessing energy from sun has become affordable with declining prices per unit of energy and availability of modern technologies. With this use of coal would become unattractive with banning its use. Sixth, construction industry has to be reformed in various ways including covering of construction materials when they are transported to the site; covering of the construction sites, etc. (also [3]).

8.4.5 Smart Economic Strategy

Flexible governance, high mobility and accessibility, round the clock availability of smart energy (largely based on wind and solar power) and better physical environment free of pollutants is the mainstay of smart economy of the NCR. Free flow of information based on high speed internet; high speed and affordable movement of people, goods and services; a large number of housing options both for individual ownership as well as for rental purposes, etc. are critical for NCR's economy to continue to economically grow.

Four policy zones in the NCR include NCT Delhi, Central National Capital Region (CNCR), Highway Corridor Zone and Rest of the NCR. The 2021 NCR Plan proposes no further industrialization, wholesale trade and commerce in NCT Delhi. This is untenable policy because some industrialization, wholesale trade and commerce is required for a mega polis to sustain in the long run. Kolkata is one recent illustration. However, history has also shown that cities devoid of diverse economic activities soon begin to decline. On the other hand, there is not much mention of Information and communications technology like automotive safety applications, smart fleet management, and intelligent transportation systems, etc. in the NCR Plan 2021. A Delhi specific statement about e-governance, e-citizens and e-services is made. Similarly there is no mention of Internet of Things. Regional plan may or may not make a mention about ITC or IoT. But on ground the region has developed a niche in providing services to the global companies as shown by cities like Gurugram and Noida. The first thing the NCR should consider when making a smart strategy for economic growth is that it should include plans about how it could use advanced technologies for regional development. No doubt automotive safety applications, smart fleet management, and intelligent transportation systems are some of the well-known applications. There are other areas where these technologies should be considered for use such as easily available and affordable broadband internet networks are hugely important for producers as well as consumers of various services and goods. Internet of things could offer better trade opportunities if affordable broadband internet networks become available all times at all places. Most importantly, Internet of Things should be embedded in the development of physical infrastructure. The next NCR plan should consider making governance and economic policies for the development of ITC or IoT through massive public private partnerships and alliances.

In the Central National Capital Region including the towns of Ghaziabad-Loni, Noida, Greater Noida, Gurgaon-Manesar, Faridabad-Ballabgarh and Sonipat-Kundli, the regional plan prohibits hazardous and polluting industries and allows only high-tech industries. Existing industries are also required to conform to the provisions of the Environment Protection Act, 1986. In the Highway Corridor Zone, seven national highways are identified where 500 m on both sides of the right of way of the national highways is proposed for development inclusive of green belt outside the controlled, developed or regulated areas [11: 20]. It is left to the state governments to determine the kind of activities to be permitted in this zone and the state governments will also delineate this zone. It is unclear how much area would be left for the green belt. Purpose of the green belt appears to be to separate highway traffic from activities proposed to be carried out beyond the green belt. Service roads and controlled access to highways are other proposed strategies for the segregation of highway traffic. This appears to be a smart strategy for economic growth as it uses the existing high accessibility for economic growth. With very little investment, greater economic dividends could be harvested with the implementation of this strategy.

Rest of the NCR occupying nearly 30,000 km^2 area is proposed for accelerated development particularly identified 16 Metro Centers and Regional Centers. The regional plan proposes workers participation rate at about 30%, which is low [11: 45]. By 2021 two third of population of the NCR will be living in urban areas. All these people will need jobs and with 30% worker participation rate, it is not possible to absorb all job seekers. A major shift from primary sector (that still predominates at least in Rajasthan and Uttar Pradesh) to territory and secondary sectors needs to be made to make the region economically sustainable. Unfortunately, even the functional plan proposes that the highest number of workers will continue to be employed in agriculture [13: 233]. Economically speaking, the Regional Plan 2021—NCR is not sufficiently ambitious. It appears to be written in the idiom of slow economic growth before the end of 1970.

8.5 Conclusions

Governing the NCR is a complex task as it involves a large number of urban and rural settlements spread out in a large geographical area, large population, spatial and non-spatial plans, private sector, multicity of planning, and industrial policies, uneven development but most importantly differential governance systems. Smart governance strategy, as outlined above, involves creation of elected planning and development authorities at regional, sub-regional, and district levels. These authorities should be armed with technical staff of planners, engineers, etc. and equipped with substantial funding (government as well as market funding) for effectively carrying out regional level infrastructure projects apart from implementation of other planning policies.

This contribution also advances the smart region approach, which seeks to address environmental and quality of life issues simultaneously. While economic growth is necessary for providing jobs and creating wealth for everyone living in the NCR of India, it is equally crucial to protect the environment for current and future generations. A preliminary sketch is presented here by constructing four dimensions of the smart region—smart regional economy, smart regional accessibility, smart regional settlement pattern, and a smart regional settlement pattern. Smart regional settlement pattern could not be planned and developed without smart regional economy, and smart regional accessibility. Major objectives of the Regional Plan 2021 appear to remain same as that of the Draft Regional Plan of 1971—contain and control population growth of the central core of the region, and direct population and economic growth to the secondary cities of the region. Market forces rather than regional planners are now able reverse the trend of population growth of the core in comparison to other parts of the region. However, in order to realize regional aspirations of citizens and planners in the NCR, all the above proposals need to be implemented.

8.6 Material for the Conclusions of the Book

National Capital Region centered on Delhi is a massive but complex region. With area exceeding 45,000 km^2, the NCR is counted among the largest regions of the world. A multiplicity of plans (hierarchy plans with implications for the built environment), planned (citizens), and planners (planning organizations), the NCR is also one of the most complex regions in the country. Complexity increases and transparency is reduced because of less use of ICT and IoT in the governance of the NCR.

References

1. Asian Development Bank (undated) National Capital Region Urban Infrastructure Financing Facility (RRP IND 41598-01). https://www.adb.org/sites/default/files/linked-documents/41598-01-ind-oth-01.pdf
2. Banerjee T (1996) Role of indicators in monitoring growing urban regions—the case of planning in India's National Capital Region. J Am Plann Assoc 62(2):222–235
3. Bhandari L, Mehra CS (2017) Fixing Delhi's pollution. Indicus Foundation, New Delhi
4. Blotevogel HH, Schulze K (2009) Zum Problem der Quantifizierung der Metropolfunktionen deutscher Metropolregionen. In: Knieling J (ed) Metropolregionen. Innovation, Wettbewerb, Handlungsfähigkeit, Forschungs- und Sitzungsberichte der ARL, vol 231. Hanover, pp 30–58
5. Census of India (2011) Population of India. Registrar General of India, New Delhi
6. Demographia (2017) Demographia World Urban Areas, Retrieved November 5 Aug 2017. demographia.com/db-worldua-index.htm
7. Government of India (1985) The National Capital Region Planning Board Act, 1985. Ministry of Urban Development, Government of India, New Delhi
8. Government of India (2007) Delhi-Mumbai industrial corridor: concept paper. Department of Industrial Policy and Promotion, Ministry of Commerce and Industry, Government of India, New Delhi
9. Harris N (2005) Towards new theories of regional and urban development. Econ Polit Weekly 40(7):669–674
10. Mint (2017) NHAI approves Delhi RRTS projects worth Rs. 53,500 crore, Wednesday, 19 Jul 2017
11. NCRPB (2005) Regional Plan-2021, National Capital Region. NCRPB, Ministry of Urban Development, Government of India, New Delhi
12. NCRPB (2009) Functional plan on transport for National Capital Region-2032. NCRPB, Ministry of Urban Development, Government of India, New Delhi
13. NCRPB (2015) Economic Profile of NCR. NCRPB, Ministry of Urban Development, Government of India, New Delhi
14. Schindler S, Sharma S (2017) Delhi-Mumbai industrial corridor, economic and environmental consequences. Econ Polit Weekly LII(44):12–16
15. Sunday Times (2017) Choked for a while. Sunday Times, Delhi Edition, 19 Nov 2017, p 3
16. Taylor P, Ni P, Derudder B, Hoyler M, Huang J, Witlox F (eds) (2011) Global urban analysis: a survey of cities in globalization. Earthscan, London
17. TCPO (2001) Retrospect and prospect. TCPO, Ministry of Urban Development, Government of India, New Delhi

Part IX
India, Jaipur

Chapter 9
Smart Tourism Innovations for Smart Region, Case of Jaipur Metropolitan Region, India

Prabh Bedi, Neha Goel Tripathi and H. B. Singh

Abstract Jaipur city since its inception has flourished as an administrative, economic and cultural center. Jaipur gained the status of a Million Plus City in 1991 and was recognized as a metropolitan region in 2010. During the post-independence planning era, there have been marked deviations in the growth and development in Jaipur, be it related to the spatial spread of the urban area or the growth of population. With the spatial spread and population growth, the city and its region are not bereft of problems. Understanding of Jaipur region has highlighted that secondary and tertiary sector play a dominant. The dynamics and interdependence between the sectors has led to identification of the prime potential segment that may lead the region out of the environmental imbalance into which the region is fast plunging. The problems that have come to be associated with these very large regions need to be address with the use of technology and innovation. The answers to these may lie in the smart context which has been explored in this chapter in the context of Jaipur Region. Authors have suggested a vision based on the issues and potential of the region within current policies of central and state government. Strategies are proposed for increasing the regional economic resilience through locally-based, regionally-driven development initiatives based on technology and innovation focused on smart tourism.

Keywords Jaipur · Smart city · Smart region · Smart eco-tourism
Industry innovation · Smart specialization and · Regional competitiveness

P. Bedi (✉)
Spatial Planner and Geospatial Technology Consultant,
School of Planning and Architecture, 4, Block-B, I. P. Estate, New Delhi, India
e-mail: prabhb@gmail.com

N. G. Tripathi · H. B. Singh
School of Planning and Architecture, 4, Block-B, I. P. Estate, New Delhi, India
e-mail: nehagtripathi@gmail.com

H. B. Singh
e-mail: hbsinghvats@gmail.com

T. M. Vinod Kumar (ed.), *Smart Metropolitan Regional Development*,
Advances in 21st Century Human Settlements,
https://doi.org/10.1007/978-981-10-8588-8_9

9.1 Introduction

City-region is a functional entity that extends across urban administrative areas [1]. The focus, especially in the Indian context has been mainly on their internal relations and the demarcation of regional boundaries. As the pace of urbanization of the city-regions has been increasing due to globalization and technological advancements, there is need focus on innovation, local expertise, and governance so as to emerge as smart regions.

Census of India defines Million Plus Cities as those having a population of over one million. Census of India considers these as Urban Agglomeration defined as a continuous urban spread constituting a city and its adjoining outgrowths, or two or more physically contiguous towns together with or without outgrowths of such towns. An Urban Agglomeration must consist of at least a statutory town and must satisfying the basic condition of contiguity [2]. The Million Plus Cities are considered synonymous to the term metropolis internationally, wherein, a metropolis is a large city or conurbation. It is a significant economic, political, and cultural center for a country or region, and an important hub for regional or international connections, commerce, and communications. The criteria of population for defining the metropolis varies in countries.

Definition of metropolitan areas adopted by the planners in India is that of a very large area including villages, whether urbanizing or otherwise, which are at the periphery or intervening in an urban agglomeration but is essentially contiguous. Kulshrestha [3] defines a metropolis as a city having population above one million, a city governed by one or more municipal bodies and functionally serving its influence region as the dominating center of trade, commerce, art, culture, health care, recreation, education, research, administration and political activity. A metropolitan region is the area under the influence of the development impulse of a metropolis and comprising the metropolitan core and the metropolitan periphery [3].

Jaipur, the capital of the state of Rajasthan crossed one million population in the decade of 1981–1991 and was recognized as a Million Plus city by the Census of India in 1991 [4]. The earliest Master Plan for the city recognized the role of the city and its influence on the surrounding settlements and planned for the city and its satellite towns, giving it a regional flavor.

Jaipur like other metropolitan regions of the country is based on modern economies which attracts the migrants from the rural areas and other urban centers. Manufacturing industries and services sector constitute a major activity in these regions. The rate of migration from rural to urban areas is very high with majority of the migrant population is below the poverty line, which leads to overcrowding and environmental deterioration. The growth and distribution of population within the metropolitan regions is not uniform. There is a trend of corridor development in these regions along transport networks [5].

Jaipur has been selected as one of the hundred smart cities under the Smart City Mission of Ministry of Urban Development (MoUD), Government of India launched in June 2015. Jaipur Smart City aspires to leverage its heritage and tourism, and through innovative and inclusive solutions, enhance the quality of life of its citizens.

9.2 Introducing Jaipur

Jaipur, the capital of Rajasthan state of India is located at 26° 54’ North latitude and 75° 49’ east longitude. The city is wholly a part of Jaipur district situated in north eastern part of the state surrounded by the districts of Alwar in the north, Sikar in the north-west and Bharatpur and Dausa in the east. Towns and cities of Ajmer, Sikar, Alwar, Kotputli, Bandikui and Tonk located within 150 km of Jaipur have been playing an important role in Jaipur’s growth. The immediate influence zone of Jaipur city extends to towns up to a distance of 70 km to towns like Dausa, Lalsot, Niwai, Phagi, Dudu, Phulera, Renwal, Reengus and Shahpura. The metropolitan region delineated by Jaipur Development Authority includes the contiguous area between Jaipur Municipal Corporation, Amber, Sanganer and the towns of Bassi, Chandlai, Shivdaspura, Bagru, Chomu, Achrol, Jamwa Ramgarh (Refer Map 9.1).

Climatically, Jaipur falls under the semi-arid region [6] having mean maximum temperature of 41 °C and mean minimum temperature of 6.5 °C. The average annual rainfall is 62 cm and the pre-dominant wind direction is North west and west. The Region is drained by a number of seasonal rivers of which Banganga, Dhundh and Bandi are prominent. Major part of the Region is covered with thick mantle of soil, wind-blown sand and alluvium. The eastern and northern parts are formed of Aravalli hill ranges [6].

As per the Master Development Plan 2025 of Jaipur Region [7], the population of the region was 3,044,946 in 2001 which had increased by nearly 40% from 1991. The region has been experiencing a declining growth rate since 1981 (Refer Fig. 9.1).

The decade 2001–2011 saw a further decline in the growth rate to 33% (Refer Table 9.1). The density of population has been doubling in a two-decade period, despite increase in the areas of the region over time.

The density of the region increased from 1036 persons per km^2 to nearly 1400 persons per km^2 in 2011. The density in the rural part of the Region was much lower at 383 persons per km^2 compared to 4686 persons per km^2 in urban parts of the Region resulting due to in-migration from rural to urban areas as well as migration from smaller towns to the capital city [8].

The sex-ratio in the Region is lower than the state and national figures with the urban core having a lower sex ratio than the rural counterpart [8, 2], which is indicative of the male dominant migration into the region, especially to the urban core (Refer Table 9.2).

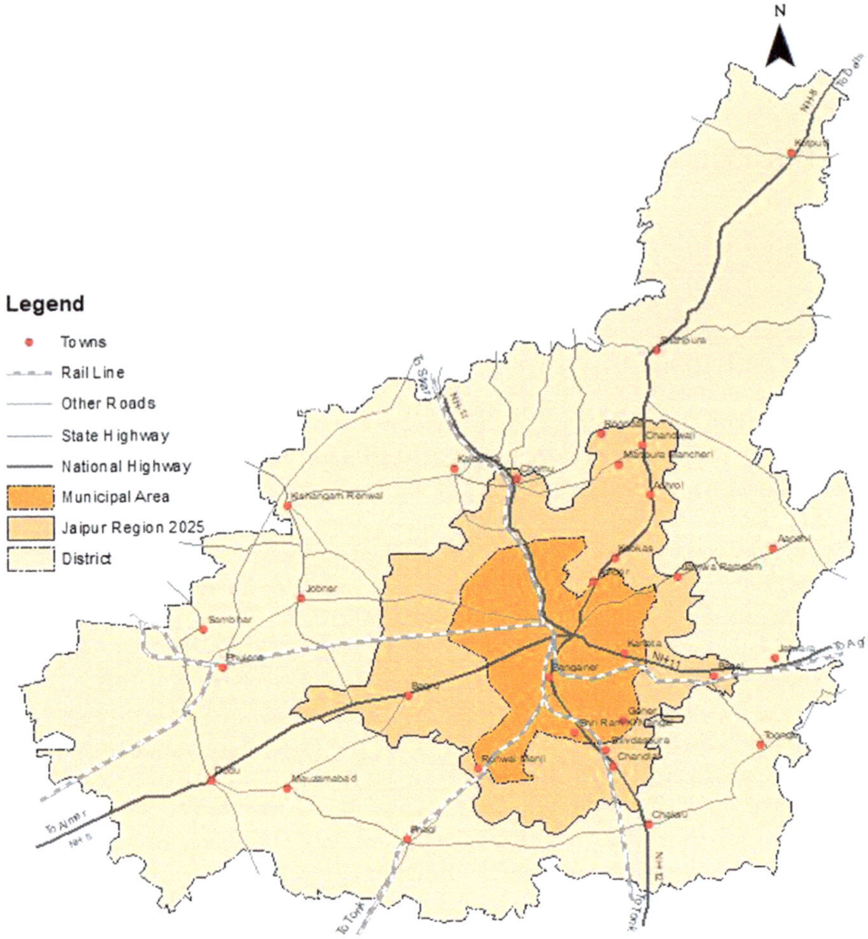

Map 9.1 Jaipur Region. *Source* Survey of India and Census of India, 2011 and Master Development Plan-2025 Jaipur Region

9.2.1 Regional Connectivity

Connectivity and linkages indicate the level of cohesiveness the urban core has with the settlements in its influence zone as well as other important settlements at the national level. Good connectivity not only helps easy movement but acts as a catalyst of growth of the region. Inter-connectivity within the region is crucial as well for the wholesome and well distributed growth of the region. Jaipur is well connected with the settlements in its region like with Chandwaji, Achrol, Kookas and Bagru through National Highway 8, Bassi, Kanota and Chomu through National Highway 11 and Shivdaspura and Chandlai through National Highway 12

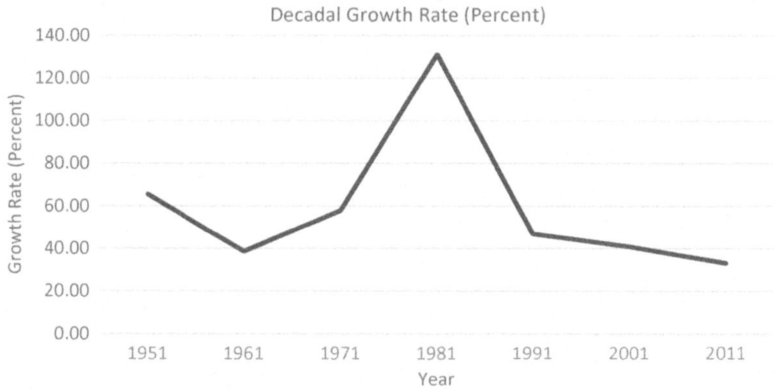

Fig. 9.1 Decadal population growth rate in Jaipur Region. *Source* Census of India, 2011, 2001, 1991 and 1981 and Master Development plan-2025–Jaipur, Jaipur Development Authority

Table 9.1 Growth of Population 1981–2011, Jaipur Region

Year	Population	Growth rate	Density (per km^2)
1981	1,472,247		501
1991	2,161,361	46.81	735
2001	3,044,946	40.88	1036
2011	4,054,451	33.15	1380

Source Census of India, 2011, 2001, 1991 and 1981 and Master Development plan-2025–Jaipur, Jaipur Development Authority

Table 9.2 Sex-Ratio Comparison-2011

Area	Sex ratio
Jaipur UA	900
Jaipur Region	902
Jaipur District	909
Rajasthan State	928
India	943

Source Census of India, 2011

(Refer Map 9.2). Being the state capital, Jaipur is connected through state highways with other towns and cities in the state. At the National level, Jaipur lies on the East West and North South corridor as well as on the Golden Quadrilateral of Delhi-Mumbai-Chennai-Kolkata [9].

As in case of the road, Jaipur lies on the major rail route connecting Delhi to Mumbai as well as other towns in the state. The air connection with rest of India also exists through its international airport at Sanganer. Due to its proximity to New Delhi, Jaipur International Airport serves as an alternate to international airport at

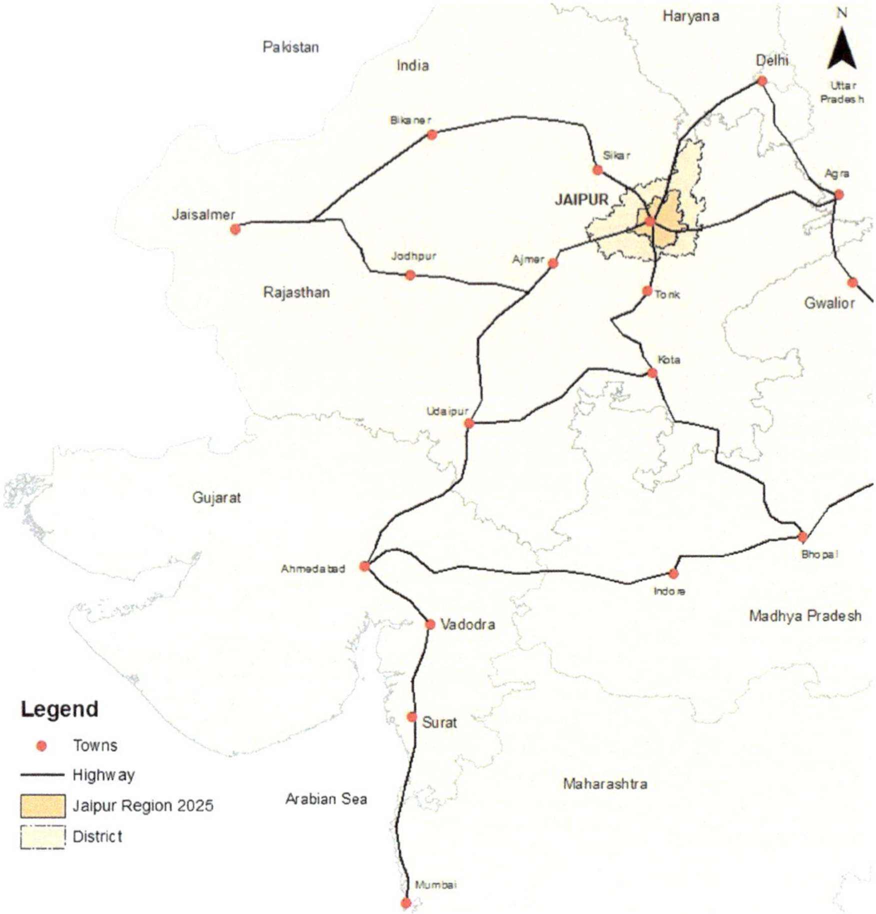

Map 9.2 Regional connectivity of Jaipur. *Source* Created by authors from Survey of India and Master Development plan-2025–Jaipur, Jaipur Development Authority

New Delhi, especially during winters, for flight diversions. The city is well connected with international cities in the Indian sub-continent and middle eastern countries.

9.2.2 *City at the Core*

Jaipur region is centrally located in the district. The Region consists of 26% of land area of the district and constitutes 61% of the district's population. The urban components concentrated in the central core in Jaipur city of the Region in 93% of the district's urban population.

9.2.2.1 Genesis of Jaipur City

Founded in 1727 CE, the settlement of Jaipur was established as the capital by Maharaja Sawai Jai Singh II, to assert a strong political statement at par with Mughal cities and as a thriving trade and commerce hub for the region. The capital was planned and developed as an administrative and trading city as opposed to the conventional agricultural base [10]. The walled city spread across 670 ha having a design capacity of 50,000 people. Since the city has grown spatially to occupy an area of 436 km^2 accommodating a population of 3,046,163 [8] and 389,987 within the walled city [11] in 2011. The city has been a major tourist attraction in the western part of the country. In 20th century due to increasing pressure resulting from population growth, modernization and urban renovations within the walled city, including the restoration work of the city walls and gates were undertaken to keep the city attractive to the increasing tourist inflow.

With attaining the status of capital of the State of Rajasthan and influx of refugee population, the city experienced a phenomenal growth rate of 65% immediately after Independence. Unplanned growth continued until 1970s. In-migration led to increased development of residential areas to cater to the growing population [12] (Refer Table 9.3).

9.2.2.2 Planning of Jaipur City

So as to control haphazard growth, in 1976, Master Plan-1991 was prepared under the Urban Improvement Act, 1959 by Town Planning Department, Rajasthan for an area of 156 km^2 [13]. The Plan was later extended up to year 1996. The Master Plan 1991 (1996) envisaged Jaipur to be the principal administrative, commercial and distribution centre of the State continuing with the earlier established character of a trade centre. Having an integrated approach, the plan was to develop Jaipur as a tourist centre.

Table 9.3 Population of Jaipur 1881–1971

Census year	Population	Percent growth
1881	142,600	–
1891	158,900	11.4
1901	160,000	0.7
1911	137,100	−14.3
1921	120,200	−12.3
1931	144,200	20.0
1941	175,800	21.9
1951	291,000	65.5
1961	403,400	38.6
1971	636,800	57.9

Source Historical Demographical Data of Urban Centers http://www.populstat.info/Asia/indiat.htm

The Plan could not be implemented as envisaged due to issues related with acquisition of the land. The lack of follow up actions to the Master Plan, like preparation of Functional Plans, Zonal Development Plans and Zoning Regulations further weakened implementation of the Plan. This led to misinterpretation of the Plan, specially the land use at the plot level [7]. Due to increasing demand and inability of Urban Improvement Trust, Jaipur (UIT) and Rajasthan Housing Board (RHB) to meet the housing demand, private developers and cooperative societies purchased agricultural land and developed housing colonies in the periphery of the then proposed planning boundary [14]. The rules for conversion of agricultural land facilitated the sub-division of the agricultural land into residential plots more easily than within the planning boundary. These developments violated not only the Master Plan 1991 (1996), but also the norms of urban planning in terms of community facilities, utilities, environmental aspects, land use and density. Jaipur urban area including these deviations, ultimately consisted of 190 km^2. In 1998, second Master Plan was approved under the Jaipur Development Authority Act, 1982 for the year 2011 [7].

The deviations to Master Plan-1991/1996 were not recorded in Master Plan 2011. Certain existing areas that had urbanised were excluded from the Master Plan 2011. The delayed notification of the Master Plan 2011 furthered unplanned growth, continuing with the trend of haphazardness. The urbanised area by 2011 was 600 km^2. as against the planned urbanizable area of 328 km^2 [17]. In 2010, Jaipur city was declared a metropolitan region [15] for which Master Plan 2025 was prepared under the Jaipur Development Authority Act, 1982 covering an area of 2940 km^2. The current region consists of 725 villages, 1 Municipal Corporation and 2 Municipal Councils [7] a population of 6.49 million. The current plan for the region recognises all the earlier deviations (Map 9.3).

9.2.2.3 Direction of Growth

Jaipur city since its inception has flourished as an administrative, economic and cultural centre. The strategic location of the town between the two hills and its culturally vibrant and resource rich hinterland sustained the town through the centuries. Patronage by the rulers during the early years of establishment led to the city growing as a dynamic and vibrant cultural centre. The richness of the region in minerals attracted population to the city leading to its spatial expansion. Figure 9.2 of Urban Sprawl of Jaipur from 1989 to 2011 [16] below depict the urban sprawl of the city clearly indicating the north-west to south direction of its expansion. The hills on the eastern side and the reserved forests area a deterrent to natural growth of the settlement. Initial growth has been along the Jaipur-Tonk road in the southern side of the region. Earlier Master Plans too had proposed the growth to be in the north-west to south direction [17]. During the last decade, proliferation has been observed along all the major highway corridors going out of the region. With higher

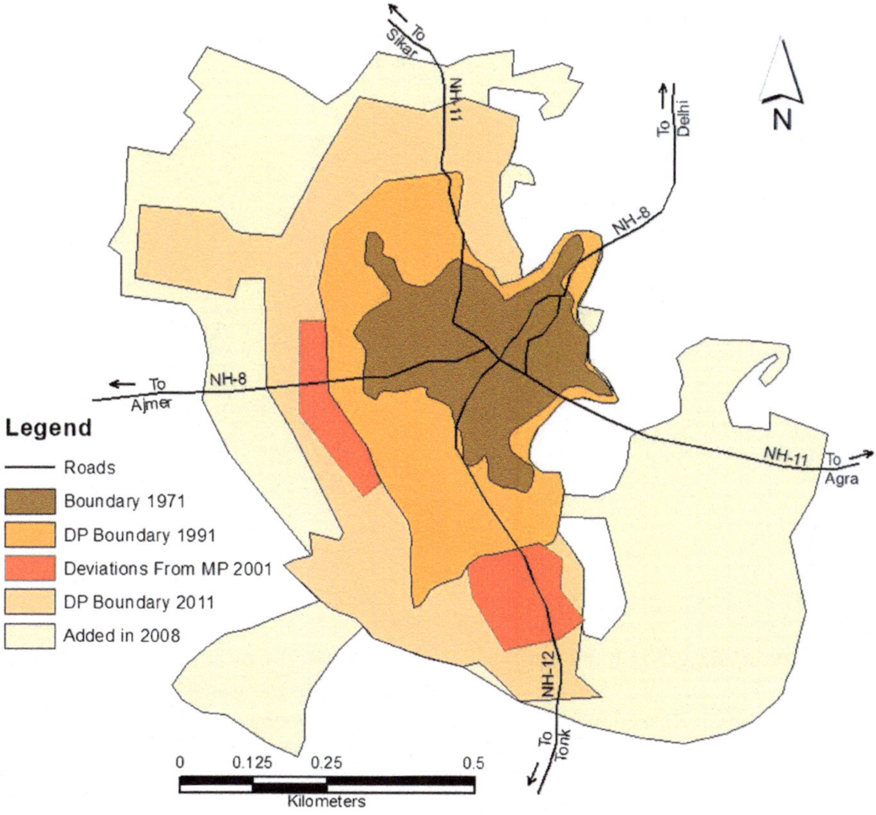

Map 9.3 Changing planning boundaries of Jaipur Region—1971 to 2025. *Source* Created by authors from Survey of India and Master Development plan-2025–Jaipur, Jaipur Development Authority

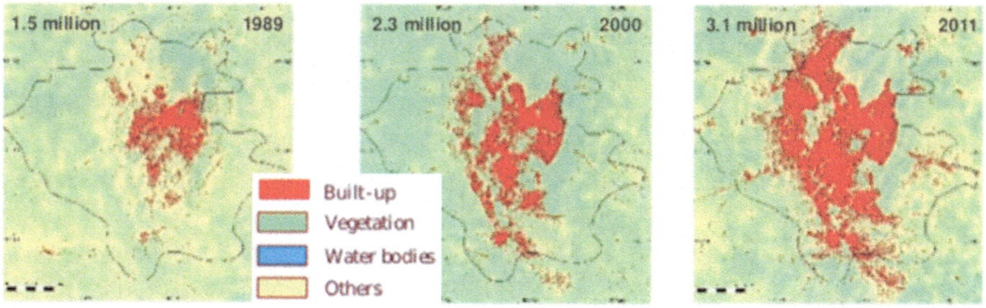

Fig. 9.2 Urban Sprawl of Jaipur 1989–2011. *Source* Revi, Aromar et al. [16], Urban India 2011: Evidences

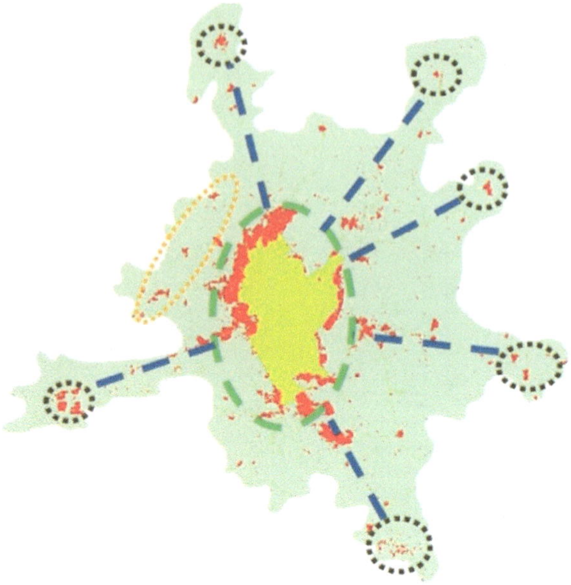

Fig. 9.3 Corridor Development in Jaipur. *Source* Baruah Neeraj G., n. d., Patterns and Processes in Spatial Modelling: Can we understand what we see?

concentration along NH-8 and the areas between the highway in the western part of the region. The current Plan envisages its growth along the transportation corridors in all directions [18] to develop the region into an amoeba shape (Refer Fig. 9.3).

9.2.2.4 Settlement Pattern of the Region

Jaipur Region comprises of 725 villages in an area of 2940 km^2 [7]. There are five urban centers as per the 2011 Census, namely Jaipur, which is the metropolitan city, Chomu with a population of 64,417 is the Class II town, Bagru is a Class II town having a population of 31,229, Akedoongar and Kanota are Census Towns[1] having population of 9062 and 11,250 respectively [4]. Amongst the villages, there is a concentration of large villages on the west and north portions of the region between National Highway (NH) 8 to Ajmer and the National Highway 8 to Delhi. Of the 5 villages with population between 10,000 and 20,000, four are in this part of the region. 15 villages with a population between 5000 and 10,000 are concentrated in this region. Another concentration of large rural settlements is in the southern part of the region, NH-12 to Tonk (Refer Map 9.4).

The settlement distribution clearly indicates that the growth in the region is happening more on the western side followed by south and rest of the region.

[1]An urban settlement that is not administered by a statutorily notified body and has population above 5000, with at least 75% of main male working population in engaged non-agricultural activities, based on which it is recognized as a town by Census of India (Censes of India, 2011).

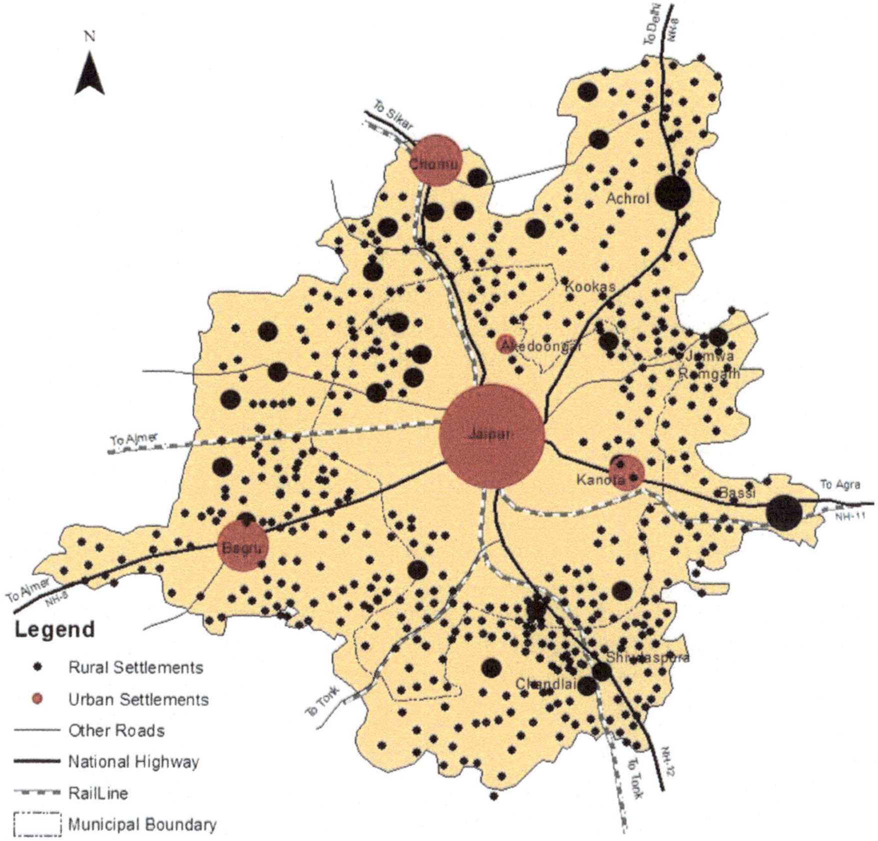

Map 9.4 Settlement distribution in Jaipur Region. *Source* Created by authors from Survey of India, Census of India, 2011 and Master Development plan-2025–Jaipur, Jaipur Development Authority

This spatial spread is attributed to the proximity to the Delhi Mumbai Industrial Corridor (DMIC) which passes through the western part of the district (Refer Map. 9.8).

9.3 Natural Systems

Jaipur, lying in the semi-arid region is surrounded in the north and east by three major hills of Aravalli range. The region is covered by a thick mantle of soil, blown and alluvium. A few isolated linear ridges and hillocks in north-east to south-west form the prominent landmark in the north-western part. The height from sea level is generally between 122 and 183 m, varying at different places. Maximum height is 515 m above mean sea level near Kalwar.

Jaipur region acts as a buffer between the arid area in the west and the humid area to the east. The soil of the region is sandy-loam Soil: This soil is semi-porous and fertile, it is generally yellowish brown with deep or light texture [19].

Jaipur experiences five seasons, having extremely hot summers and cold winters. During winters, cyclonic depressions cause precipitation. The intermediate period between summer, monsoon and winter, the area experiences spring and autumn.

The semi-arid region received annual rainfall of 527 mm between 1901 and 1971, while average annual rainfall between 1977 and 2006 was 565 mm [19]. Annual average rainfall during the period 2001–2010 reduced to 527 mm. Over 90% of total annual rainfall is received during monsoon. Total annual potential evapotranspiration is 1744.7 mm. The coefficient of variation is moderate at 32.6% indicating slightly unreliable pattern of rainfall. Though, Jaipur city has experienced floods in 1981, the district is prone to drought spells as witnessed during 1984–1989 and 1999–2002. Jaipur region acts as a buffer between the arid area in the west and the humid area to the east.

The winds are light to moderate, which strengthen during late summer and early monsoon. Dominant direction is westerly and south westerly during summers and monsoons. Westerly and north winds prevail post monsoon and during winters.

The region is mineral rich. There are occurrences of granite though confined to southern part of the district. Other than granite, marble is found in the region and beyond mainly in Bhainsalana, Raisala and Andhi. The region is rich in deposits composed of phyllites and mica schist along with granite, quartzite, feldspar, iron ore, limestone, and amethyst [19]. Existence of granite, marble and other minerals attract mining activities in the region and is a major source of these stones in western and northern India.

9.4 Water Systems

The Jaipur district is drained by small seasonal streams such as Banganga, Sota, Dhund, Bandi, Mashi, and Sabi and with their small branches. Sota and Sabi rivers in the northern part of district flow northeasterly while southwesterly flowing Banganga river passes through Shahpura, Bairath and Jamwa Ramgarh blocks and feed the famous Ramgarh lake from where it flows easterly to enter Dausa district. Mendha River in northwest portion of the district merges with famous Sambhar lake whereas Mashi river in the southwestern part flows easterly. The only natural lake in the district is the salt lake of Sambhar in Phulera tehsil, which is one of the largest sources of salt in the country [4, 19] (Refer Map 9.5).

In the early stages of Jaipur, city expansion was restricted to the foothills and thus surface water drainage patterns were not disturbed. As expansion took place, people began to fill, divert, and block the streams of Jaipur. There are as many as 518 rivulets originating from the Aravalli Hills: 398 1st order streams, 92 2nd order stream, 25 3rd order streams, and 3 4th order streams. Many natural streams began to be used for dumping garbage. Due to urban expansion, 150 streams with 113 of

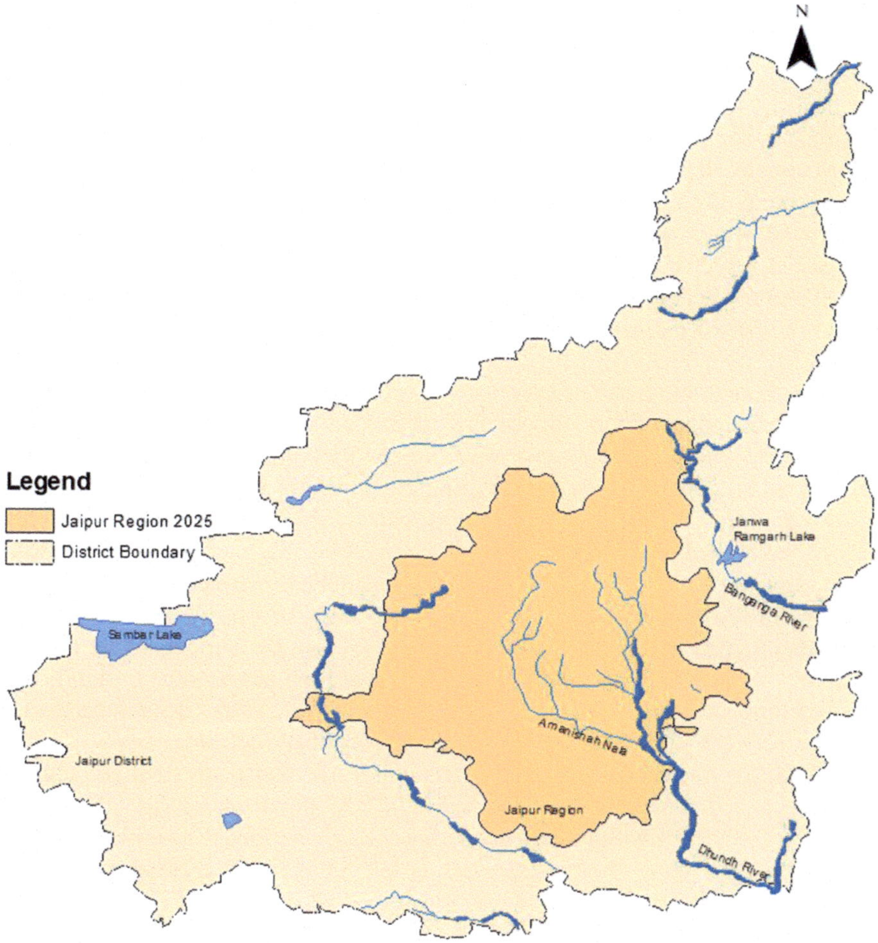

Map 9.5 Major river systems of Jaipur Region and District. *Source* Created by authors from Survey of India and Master Development plan-2025–Jaipur, Jaipur Development Authority

1st order, 37 of 2nd order, and 10 of 3rd order have been blocked or have been filled for construction purposes. This has greatly influenced the availability of clean surface water and groundwater recharge [20].

Jaipur, the first planned city of India, was originally designed with water supply in mind. Initially Jaipur depended extensively on the Ramgarh Dam as its surface water source throughout the 1900s, however due to growing water demand it became a non-viable source in the late-1980s/early-1990s leading to complete dependence on groundwater. During its rapid, pre- and post-independence urbanization, Ramgarh Lake was the main source of water for Jaipur. Ramgarh Lake only had overflows, resulting in flooding, three times during the period after its construction. Issues with Ramgarh Lake started to become evident in the 1980s, and Ramgarh Lake started to show signs of excessive drawdown and completely dried up occasionally. This transition has led to a rapidly depleting groundwater table and overexploitation of the aquifer [20].

The district gets it water supply from Ramgarh Dam on Ban Ganga river. Other important water bodies in the district are Chaparwala, Kalakh, Hingonia, Buchara and Mansagar. It needs to be highlighted that all the rivers in the district are non-perennial. Central and Rajasthan Ground Water Board together have assessed the ground water potential of the district and found that the average yearly replaceable resource is much less than consumption. Hence water availability is limited and restricts development. Main source of irrigation in the district is wells and up to 95% of the irrigation water requirements are met through ground water sources [7].

Today though, the Bisalpur Dam located south of the region contributes water to the system, the population is still primarily dependent on groundwater as its primary water source. The Bisalpur Dam is also utilized by Ajmer and the Tonk districts, and is highly dependent on yearly rainfall, which is extremely variable. The inability to meet demand and the general inadequacy of water supply is evident through the limited duration of municipal water supply, high unaccounted-for water losses, drop in supply pressure due to large quantities of water released in short durations or the depleting groundwater table affecting tube wells.

Industrial processes in and around the city have greatly and negatively affected the quality of surface waters in Jaipur. Amanishah Nala has become unsightly and foul smelling due to the discharge of industrial wastewater to its storm water drainage network [21]. There are also health risks linked to downstream usage of wastewater for domestic purposes or agricultural use. Relatively shallow aquifers are present, and thus pollutants can quickly seep down into these aquifers. In addition to industrial wastewater, hotels have also contributed to this wastewater issue as in most cases their wastewater is let out in a drain connecting to the sewerage network.

The Technical Committee of the Government of Rajasthan analyzed land use changes through remote sensing through the years around the catchment area of Ramgarh Dam. In their report the committee observed over the last few decades agricultural land area greatly increased in the region and water bodies greatly decreased from 1985 to 2012 [20]. The significant increase in agricultural activity has obstructed the flow into the reservoir and as a result the Ramgarh Dam was completely dry in 2012. Upstream of the dam, diversions for water use in agriculture have resulted in depletion of surface water sources. As well, with an increase in farmers taking double and even triple crops annually, there is more withdrawal from groundwater and depletion of the water table, thereby increasing the unsaturated thickness of the soil and absorption of the surface flow. Urbanization and industrialization have led to over overexploitation, but the lack of regulation of tube wells has also compounded the problem.

The availability, occurrence and movement of ground water is mainly controlled by the topographic features, physical characteristics and structural features present in the geological formations. Jaipur district is divided in four hydrological domains.

Table 9.4 Major land use change in Jaipur

Major land use categories	Area (km^2) 1986	Area (km^2) 1991	Area (km^2) 2003
Built–up-area	143.4	163.32	189.70
Forest area	57.45	53.52	50.01
Crop land	217.03	205.80	185.05
Waste land[a]	18.30	13.63	2.24
Total	436	436	436

Source Joshi et al. [21], Environmental Impact Study on Degradation of Natural Resources in Urban Area and Strategy for Sustainable Development
[a]The authors do not advocate the use of term waste land

Groundwater occurs at shallow depths. At deeper levels its occurs under semi-confined to confined conditions. Quality of groundwater is suitable for drinking and irrigation in major part of the region. High salinity has been observed in parts of Jamwa Ramgarh, Bassi, Phagi, Dudu, Chaksu and Sambar blocks. In most of these areas high fluoride content has been found as well.

Land use change has a great influence on groundwater recharge, and land use change in Jaipur recently shows a trend with a negative impact on groundwater. Compared to built-up land, alluvial plain (crop, waste land, and forest) is good for water percolation and thus, groundwater recharge. From 1986 to 2003 crop land area were reduced by 48 km^2 in the Jaipur region and the built-up area increased [21].

The biggest challenge before implementation of smart region Jaipur is water resource management. Jaipur region is currently experiencing growing water scarcity and diminishing drinking water sources, relying extensively on groundwater and a single surface water source, the Bisalpur Dam, which is shared with Ajmer and villages in the Tonk District and located 120 km southwest of Jaipur (Refer Table 9.4).

9.5 Cultural Systems

Rajasthan is divided into nine regions, namely Ajmer, Hadoti, Dhundhar, Gorwar, Shekhawati, Mewar, Marwar, Vagad and Mewat [22] each of which are equally rich in their heritage and culture (Refer Map 9.6). Jaipur Regions lies within Dhundhar cultural region. Over the decades since Independence, Jaipur has emerged as a melting pot of various cultures of Rajasthan, though the aspects of Dhundhar culture hold dominance.

Located on important trade routes, Jaipur became an important and vibrant center for trade and commerce. Since the beginning the City has attracted a large number of artisans, craftsmen and merchants not only from different parts of the state but from all over the country. There have been potters, utensil makers, stone carvers, building craftsmen and builders, leather workers and tanners, jewelers and

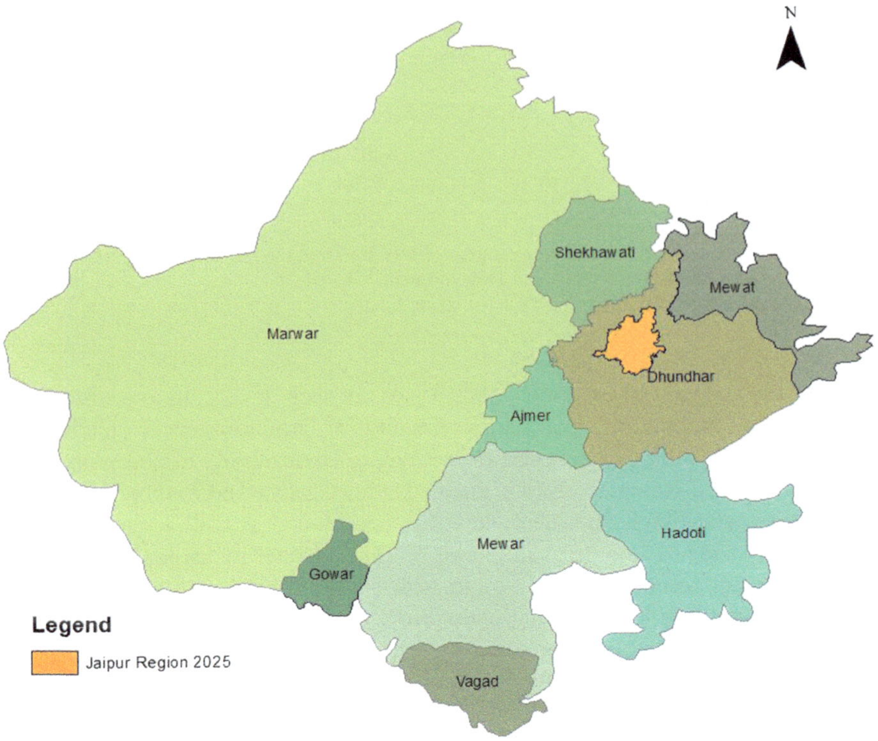

Map 9.6 Cultural Regions of Rajasthan. *Source* Created by authors from www.discoverindia.com

precious stone cutters, ivory carvers, brass ware manufacturers, enamellers, weavers, dyers and embroiderers, settled in *mohallas*[2] and organized under the traditional guild system occupied by families pursuing similar trade or craft.

When the city was planned and established by Maharaja Sawai Jai Singh, special impetus was given to commercial activities. Merchants were invited and given special incentives like prime location to settle in the city. They were also given free land and given remissions and concessions on taxes [7].

The royal patronage to various art forms like literary works, music, miniature painting and performing arts resulted in huge collections in the pothikhana (royal library) as it led to the evolution of the Dhundhar (Jaipur) style of painting, Dhundhari language, music and dance as well. The Jaipur School of painting flourished with influences from the Devgarh and Kishangarh painting styles. Later due to western influence, the traditional style deteriorated. Miniature painting continues as an art form today, drawing from Jaipur tradition style.

A system of Chhattis Kharkhane (36 departments) had been set up to revive and encourage various art forms [10]. State protection was given to learned men, poets,

[2]Community or neighborhood.

writers, painters, musicians, dancers, artists and sculptors and were further trained so as to enrich the arts and crafts traditions. Jaipur became a hub for artists after the 1857 War of Independence, when many of them who fled from Delhi found refuge in Jaipur. Besides the buildings crafts of stone carving, mirror works and inlays; an indigenous fresco technique done on araish evolved, examples of which can be seen in Bairat caravan serai, Bharmal Ki Chhatri from the period of Man Singh (influence from the Mughal Court of Akbar), and in Ganesh Pol (1639), Amber [10].

The Panchranga (five color) flag designed by Raja Man Singh continues to be the identity of the Dhundhar region which is reflected in the five coloured pagri (turban) that was worn by the rulers of Dhundhar region. The colors are an important aspect of the Dhundhari cultural heritage, with various dyeing techniques and block printing having evolved from it.

The Region has well developed folk music and dance forms, such as 'Dhudhadi', that is the Jaipur style of Galibazi (verbal art form traditionally used as medium for social awareness and reform) and 'Tamasha', a style of musical folk play introduced to the Region during the reign of Sawai Ram Singh II (1855–1880).

The tie and dye (leheriya) textile printing, textile block printing, natural dyes, metal crafts, enameling (minakari), crafting silver jewelry and artifacts, stone crafts, handmade paper and blue pottery are the most famous handicrafts today.

After Independence, Jaipur having the advantage of being an administrative capital and on the major transit routes has emerged as the melting pot of various cultural regions of the state. The colorful outfits and unique jewelery from the western parts of the state are now a part of the Region's culture which is displayed in a flamboyant way. Similarly, the famous dances of Jaipur including Ghoomar, Chari where the dancers dance on a pot with fire-lamps on their head have merged with the dance forms of surrounding regions. Traditional instruments like sarangi, ektara, and jhalar are also played while singing folk songs. Food or the local delicacies of Jaipur also reveal the culture of the city.

The culture that has developed in the Region has led to the growth of specific economic activities, peculiar to the region. Carved silver jewellery, kundan as well as meenakari jewellery, ivory carved sculptures, wood work and leather goods are an amalgamation of the art forms patronized by the rulers of not only the region but the surrounding regions as well. Known for blue pottery, miniature paintings and traditional clothes with work of bandhni, zari and zardosi, are certainly the best examples that depict rich culture of the region.

9.6 Economic Systems

The three sectors of the economy, primary, secondary and tertiary, reflect the direction of growth of a region. Jaipur's mainstay has been non-agricultural since the city was established. Tertiary sector contributed the maximum share of 48% to Gross District Domestic Product (GDDP) compared to Rajasthan's share of 57.2% to Gross State Domestic Product (GSDP) in 2009–10. The district's secondary (industrial) sector contributed little more than 1/3rd to GDDP (37.6%) while the contribution by primary sector was only 14.4% [23, 24].

Jaipur city is the economic core of the region having tourism, trade and commerce and local handicraft as its main economic activities. Organized and planned industrial areas are concentrated in six major industrial areas in the city of Jaipur. The household industry employed 5.3% of total workers as per 2011 Census in Jaipur Municipal Corporation (JMC) as against 4.7% in the Region [8].

Other than the industrially focused centers, city is a major trade and commerce hub, having a hierarchy of regional and sub-regional level mandis[3] and sub-mandis. The main mandi in Jaipur city (Surajpol Mandi) and its sub-mandis spread across the Regions account for 47% of the annual income of the district [7], which is indicative a well-developed system of grain market for collection and disposal of agricultural produce. These markets or mandis are equipped with large warehousing facilities, mainly localized in Chomu area in the northern part of the region. A strong agricultural market base reflects on the overall growth of the economy. Contributions of agricultural growth to overall growth of the economy is often invisible leading to its underestimation of importance in the overall economy of the area [25].

There has been a drastic decline in the share of the primary sector from 18.6% in 1999–2000 to 9.9% in 2008–2009, which increased to 12.4% in 2011–2012. The share of the secondary sector increased moderately from 32.8 to 37.6% during this period, while the share of the tertiary sector has stagnated around 48% though there was a 6% increase in 2008–2009 [23, 24]. Though there was a shift from agriculture to tertiary sector, the region has experienced sluggish but positive growth in secondary sector due to the policy driven industrial impetus (Refer Table 9.5).

Table 9.5 Sector wise Gross Domestic Product in Jaipur District (INR and percent)

Year/sector	Primary	Secondary	Tertiary	Total
1999–2000	191,017	336,889	499,837	1,027,744
	18.6	32.8	48.6	100.0
2002–2003	199,669	372,779	627,633	1,200,081
	16.6	31.1	52.3	100.0
2005–2006	254,932	675,828	907,456	1,838,216
	13.9	36.8	49.4	100.0
2008–2009	329,789	1,182,446	1,819,909	3,332,144
	9.9	35.5	54.6	100.0
2011–2012	697,341	1,824,279	2,329,288	4,850,908
	14.4	37.6	48.0	100.0

Source Estimates of District Domestic Product of Rajasthan, 2011–12 and Estimates of District Domestic Product of Rajasthan, 1999–00 to 2006–07

[3]Grain markets.

The majority share of income originating from the service sector in large urban areas [26] is a global phenomenon, and it has wider repercussions on the region's economy. The sluggish growth of the manufacturing sector is a cause of concern as growth of agriculture and industry is essential for sustained growth of the urban economy.

Education, health, family welfare, nutrition, sanitation, water supply, roads and social security are the basic social infrastructure that people require. The Region is lagging behind in social development due to the slow growth of the economy and low per capita income [27]. Although there has been an improvement in various social indicators, the region needs to make far more progress in its social indicators.

9.6.1 Work Participation Rate

According to the Census 2011 [8], Jaipur Region had 1.36 million main and marginal workers, who account for 33.73% of the region's total population. The work participation rate (WPR) increased from 29% in 1981 [28] to 30% in 1991 [4], to 32% in 2001 [29] and nearly 34% in 2011 [8]. In 2011, WPR was 39.26% in Rajasthan and 39.3% in India [30]. There are interesting rural-urban and male-female differentials in the Region's WPR. The WPR among males was 50.97%, as compared to a mere 15.53% among females. In rural areas, the WPR was 39.18%, while it was 33.73% in the urban areas indicating a higher dependent population. The WPR among rural males was 49.55 and 51.31% among urban males. Interestingly, the WPR among urban females was extremely low at 11.60% as compared to 31.69% among rural females, which is due to high participation of women as marginal workers in the rural areas [8] (Refer Fig. 9.4).

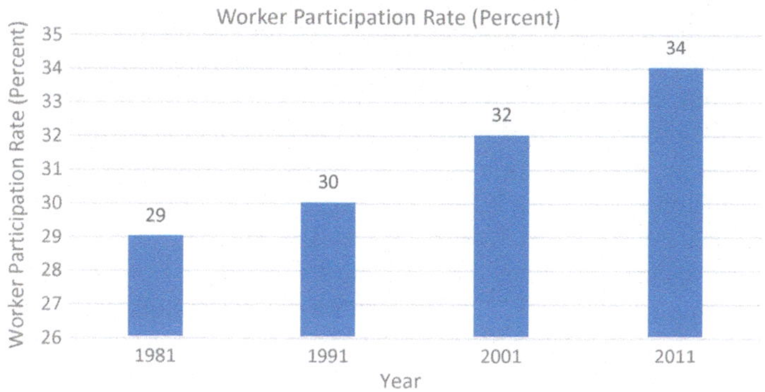

Fig. 9.4 Worker Participation Rate (%) 1981–2011. *Source* Census of India, 1981, 1991, 2001 and 2011

Table 9.6 Distribution of Workers in Jaipur Region, 2011

Area characterisation	Total workers	Agricultural workers (%)	Non-agricultural workers (%)
Jaipur Rural	316,898	53.74	46.26
Jaipur Urban	1,050,840	4.34	95.66
Jaipur Region	1,367,738	15.78	84.22

Source Census of India, 2011

9.6.2 Distribution of Workers

Table 9.6 shows the distribution of workers (main and marginal) in Jaipur Region as per Census 2011 [8]. The Region has dominance of non-agricultural activities. Only 15.78% workers were engaged in agricultural activities of which majority, that is 83% as cultivators and 17% as agricultural labourers (Refer Table 9.6 Distribution of Workers in Jaipur Region, 2011). The female WPR in the agricultural sector is much higher than the male WPR at 34 and 11% respectively. The WPR of males at 90% in non-agricultural activities is much higher than the WPR of females at 66%. Workers in the household industry constitute only 4.74% of total workers, but males have a higher share of 73%, as against 27% in the case of females. Majority of the workers are engaged in manufacturing and service sectors at 79%, wherein the percent of males is at 84% while the women are at only 16% [8].

9.6.3 Unemployment Situation

Unemployment data which is collected five yearly by the National Sample Survey Organisation (NSSO), indicates that the unemployment rates for 2011–12 was at 4.09, which was higher than the state and national unemployment rate which were at 3.1 and 3.4%. According to the estimates of Niti Aayog's (erstwhile Planning Commission) Special Group on Targeting Ten Million Employment Opportunities Per Year, the unemployment rate was higher in urban areas (4.5%) than in rural areas (2.8%) [31]. It was highest among urban male (4.7%) and lowest among rural female (1.9%). Jaipur urban area had marginally lower unemployment rate compared to the national figure for urban area (4.5%) [32].

As concluded by Fox and Dyson [33] as the city grows, its economic base expands due to increase in industrial activities and increase in trade and commerce resulting in higher employment. However, this trend is not sustainable and negative returns set in proving the theory that size of city or region and economic growth are negatively corelated. It has been noted internationally that metropolitan cities with high growth rates have low productivity levels [34]. Unemployment rate for the city was 4.09 as against the State's 3.1 and national unemployment rate of 3.4 [32].

This is indicative of the need for policy interventions in the metropolitan region of Jaipur as a lower unemployment rate would ultimately lead to smarter economy.

The mainstay of the district's economy is non-agriculture, concentrated in Jaipur region. 74% of the rural area contributed only 14% towards the district's GDP in 2011–12, whereas the secondary and tertiary activities which are concentrated in the city and its region contributed 86% towards the district's GDP [32]. The workforce participation rate of the district is 37.2%, with 65% engaged in non-agricultural activities. The corresponding numbers for the region are 34 and 86% in non-agricultural pursuits. The number of urban settlements in the district are 19 [8] of which 9 settlements are within the region accommodating 93% of urban population of the district.

For a better understanding of the economic profile of Jaipur, two of its main sectors have been studied in greater detail, namely, industrial and tourism.

9.7 Industrial Systems

The major industrial areas are at Vishwakarma Industrial Area, Jhotwara, Kanakpura, and Bindakaya in the north west, Sitapura and Sanganer towards the south and Malviya industrial area towards the south east (Refer Map 9.7). The unorganized industry is concentrated in and around Sanganer. Jaipur city has a large concentration of household industrial units, clustered within the walled city. The units engage in stone cutting and polishing, blue pottery, lac work, *gota*[4] work, sculptures and other traditional handicrafts.

The number of registered industrial Unit was 2369 in the year 2010–2011 [8]. The small-scale Industries registered by the Industries Department accounted for 1057 whereas the large and medium industries accounted for 34 in the district during 2010–2011. The industries have employed 188,680 persons in 2011. These are largely concentrated in the Jaipur Region. The industrial products manufactured in the city are ball bearings, ceramics, potter dyeing and printing, electronic appliances, brass engraving, ferrous and non-ferrous casting, gems and jewelry, marble statues, tiles and slabs, PVC and food products. The Region exports ball bearing, electronic energy meter, paper, cement, transmission line and tower, synthetic and organic color, steel/cold rolled strips, readymade garments, gems, jewelry, handicrafts, wooden furniture, leather goods, marble and granite [35].

The industrial activities of the region are concentrated in 14 industrial area which have been set up by Rajasthan Industrial Development and Investment Corporation Limited (RIICO) and Jaipur Development Authority in Bassi Industrial Area, Bassi Extension Industrial Area and Hirawala Industrial Area on Jaipur Agra highway on the eastern side of the region [7] (Refer Map 9.7). There are eight major manufacturing clusters in Jaipur region, namely rolling mills and induction furnace and

[4]Metal embroidery on cloth typical of Rajasthan.

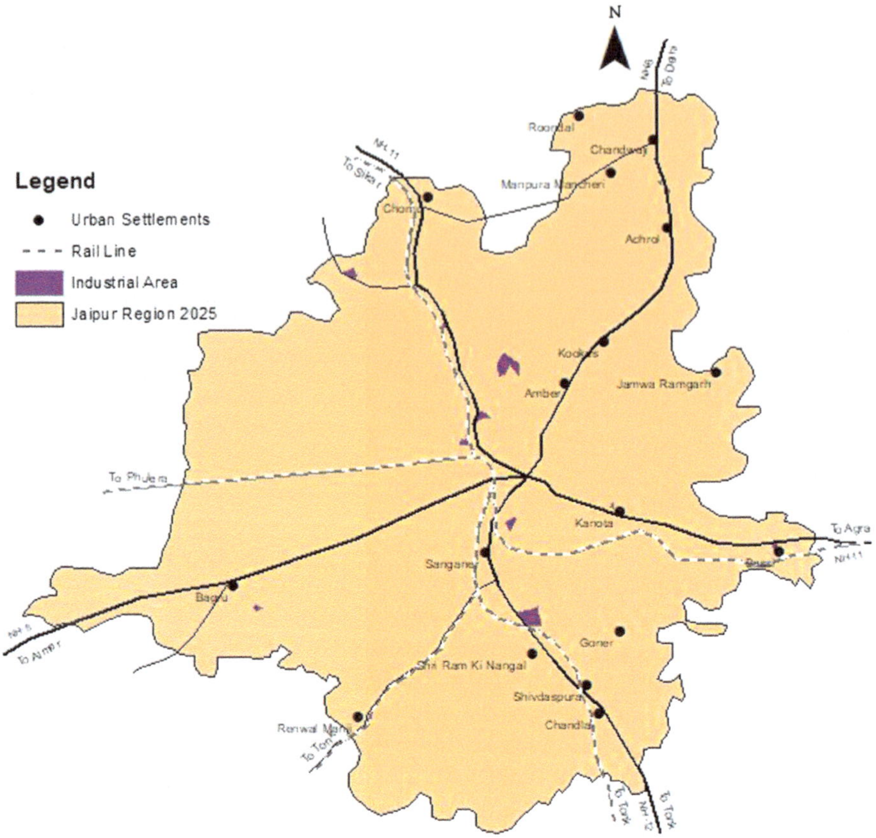

Map 9.7 Existing industrial locations in Jaipur Region, 2017. *Source* Master Development Plan-2025 Jaipur Region and 79

foundry cluster in Vishwakarma Industrial Area, Ari-Tari cluster in Naila, hand block printing cluster in Sanganer and Bagru, paper board and handmade paper and paper products in Sanganer, pipes and fittings and readymade garments spread all over Jaipur.

Sitapura Industrial Area was developed by RIICO in four phases between 1996 and 2003 covering an area of about 810 ha. The area is situated along NH-12 on Tonk Road. It is a well-developed area with quality infrastructural facilities like road network, water supply and electricity supply. The area has a concentration of gem and jewelry manufacturers, textiles and garments, electronics, pharmaceuticals and IT industry. Set up by RIICO, in Bagru, on NH-08, south-west of Jaipur city, industrial area was established a pilot project to understand its viability and suitability. Later as per Master Plan, 1991 industrial area covering about 92 ha was established. The industrial area is home to activities as stone cutting, chemical, rolling mills and warehouses. Vishvakarma Industrial Area set up RIICO in 1970 on 1400 acres of land has a concentration of gems, furniture, pottery, rubber, marbles and stone cutting, electronics, warehouse and PVC. The industrial area is located north of Jaipur city and is well facilitated with water supply, electricity and road network.

Industrial areas at Jhotwada and Bindayaka were developed by Government of Rajasthan and later transferred to RIICO. While Jhotwada Industrial Area is located in north-western part of Jaipur city has concentration of rolling mills and engineering units. The area is well developed in terms of infrastructural facilities as water supply, electricity and road network. Bindayaka Industrial Area is located west of the city between the railway line to Ajmer and Sirsi road. Developed in 1995 and spread over an area of 102 ha, the areas houses metal, rolling mills and engineering units.

To stimulate industrial growth in the Region, Special Economic Zones (SEZ) were proposed in mid 2000s in the south-western part of the Region near Bhankrota is being developed (Refer Map 9.7). SEZs are strategically located is off Jaipur Ajmer Highway (NH-8), close to the proposed Ring Road to and in proximity to DMIC. This induced growth is expected to generate 100 thousand jobs and attract INR 100 billion in investment through Information Technology (IT) industry, manufacturing, warehousing, logistics and special zones for auto and auto components and inland container depot [8].

The western flank of Dedicated Freight Corridor (DFC), a broad-gauge rail corridor connecting Delhi with Mumbai passes through Phulera in Jaipur district. The DMIC, a high-speed freight through road corridor has about 58% the state's area under its influence zone, crosses Phulera, a town in Jaipur district [36]. The district and region of Jaipur lies in the influence zone of DMIC which shall be a major boost to the industrial development in the Region. The DMIC corridor has an influence zone of 150 km on its either side resulting in all of the district falling within the influence zone. One of the Freight Logistics park along DMIC is proposed in Jaipur district at Phulera junction and industrial area is proposed between Jaipur-Dausa on NH-11 (Refer Map 9.8). Taking advantage of the central government initiative, state government is considering development of Growth Pole at Sikandra, 25 km eastward from Dausa as a multi-sectoral industrial cluster catering to stone, carpet, leather and dairy industires. Another industrial cluster proposed on DMIC and close to Jaipur metropolitan region (approximatley 75 km from Bagru) is between Ajmer-Kishangarh [8].

The industrial thrust that is being provided by the SEZ, DFC and DMIC implies development potential for various industries in areas between Sikar road and Ajmer Road, which is the western and south-western part of the Region.

The key characteristics that emerged from the study of the industrial areas is that there is in most cases negligible industrial waste management. In most cases the waste is thrown at the nearest convenient location. Moreover, there are either non-functional or absence of effluent treatment plants for treating water before disposal in drains. Polluting and non-polluting industries are found to be in same industrial areas, which indicates lack of detailed planning of the industrial site. Similar instances can be found with respect to food and construction material,

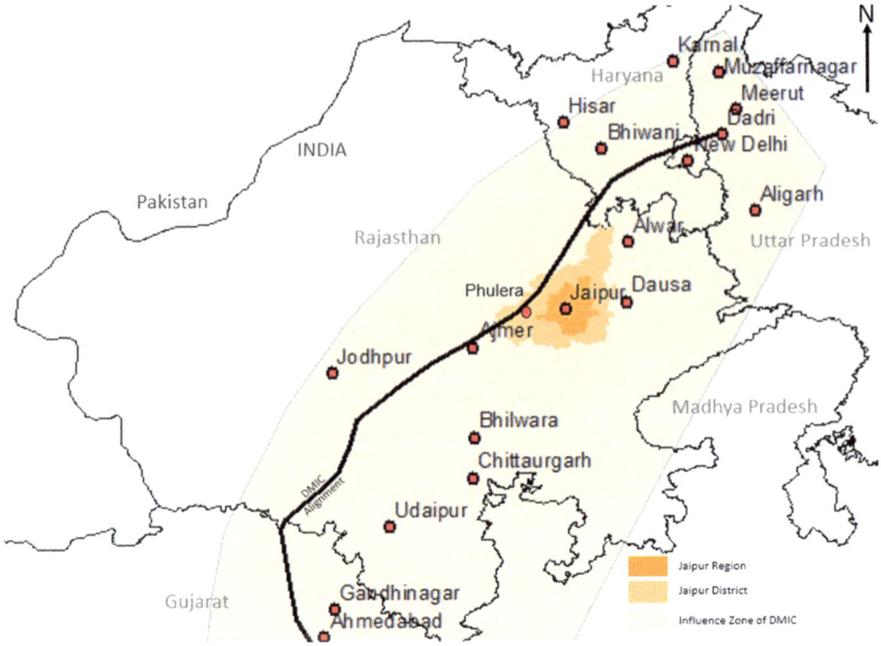

Map 9.8 Influence zone of DMIC in Jaipur Region and District. *Source* Government of India 2016, DMIC Rajasthan

labor-intensive and mechanized units. There is insufficient provision for parking spaces. The ancillary activities have not been planned for in the industrial locations, like weigh bridges, food joints, rest houses, which have come up as informal activities. The vacant plots and plan open green spaces are used as parking lots and for informal activities. Majority of the plots in the industrial locations are utilized for warehouse purpose. Other than this the by-products and waste of the industries are sold off to industries at distant locations thereby disrupting the industrial ecology of the region.

9.8 Tourism Profile

Tourism industry today has turned into a key driver of socio-economic progress through the creation of jobs, enterprises and infrastructure development globally. Past years have seen the global travel & tourism industry growing at a higher rate than significant sectors such as automotive, financial services and health care. In 2014, the travel & tourism industry generated INR 484.69 trillion and 277 million jobs for the global economy [37]. It ranked second in terms of the number of jobs generated by tourism industry in the world by supporting 36.6 million jobs in 2014.

Since after Independence of the country, tourism has experienced continued expansion and diversification, to become one of the largest and fastest-growing economic sectors. Tourism industry has also been a key driver for capital investments in the country. Capital investments in the tourism sector include spending by all sectors directly involved in the travel and tourism industry such as new visitor accommodation and passenger transport equipment, as well as restaurants and leisure facilities for specific tourism use. Such investments lead to social development of an economy as tourism infrastructure can also be utilized by the community in general.

According to the Ministry of Tourism, Rajasthan accounted for 2.7% of the tourists in the year 2013 and 7.2% of the total international tourist arrivals in India in the same year [8]. Jaipur the capital city of state of Rajasthan, is the magnificent land of numerous kingdoms, majestic forts and palaces, diverse cultures, varied landscapes and vibrant colors. Tourism has been one of the biggest revenue generators a long time. Tourism play a vital role in the economy of Jaipur Region, especially with Jaipur city being at its core. Other than the City, there are many locations in the Region that are important from tourism perspective.

Amongst the foreign tourist arriving in the State, 30% of tourists visit Jaipur. On an average the city receives 2270 tourists per day with an average stay of 3–4 days. There has been an increase of tourists from 0.50 million in 1989 to 1.79 million in 2015 [38] (Refer Table 9.7).

The peak season for tourists in Jaipur is October to March. The inflow starts declining in April and is at its low in June. The climate of the region plays a determining role in the influx of tourists. During the last four years, there has been a decline in the tourist inflow, specially the domestic, while the international tourist

Table 9.7 Tourist Arrival in Jaipur (millions)

Year	Domestic	International	Total
1989	0.35	0.15	0.50
2015	1.20	0.59	1.79

Source Rajasthan—Pragati Prativedan 2015–16, Department of Tourism and Master Development Plan 2015, Jaipur, Jaipur Development Authority

Table 9.8 Tourist Inflow in Jaipur 2012–2015

Year	Domestic	Growth rate	International	Growth rate	Total	Growth rate
2012	998,703	–	534,256	–	1,532,959	–
2013	1,104,905	10.63	566,429	6.02	1,671,334	9.03
2014	1,170,152	5.91	568,234	0.32	1,738,386	4.01
2015	1,201,152	2.65	596,756	5.02	1,797,908	3.42

Source Rajasthan—Pragati Prativedan 2015–16, Department of Tourism inflow has been fluctuating (Refer Table 9.8). Tourism activity has a direct relation with political stability in the region.

Jaipur has attracted tourists, both domestic and international, for its city planning, historical monuments, natural features, temples, and palace complexes. Jaipur's colorful and vibrant culture which is represented in its buildings, people, dance, music, food, clothes and handicrafts is a major attraction in the region. Jaipur is an important destination on the world-famous tourist circuit, Golden Triangle, comprising of New Delhi-Jaipur-Agra. However, other circuits, namely, Agra-Varanasi-Khajuraho in neighboring state of Uttar Pradesh and some circuits in the desert areas if Rajasthan have been giving tough competition to Golden Triangle connecting Jaipur with Agra and Delhi.

In Jaipur Region, most of the current tourist sites are located within the city. In order to boost tourism, the need is to develop potential sites in the Region. Proximity to National Capital Region (NCR) of Delhi, makes Jaipur an attractive weekend destination, having multiple tourist sites. Relatively lower cost of real estate and a slower pace of life compared to the hub-nub of NCR is another attraction in Jaipur.

The places of tourist attraction in the city include the Walled City consisting of City Palace, Maharaja Sawai Man Singh—II Museum, Hawa Mahal, Jantar Mantar and Bazars. The attractions outside the Walled City are mainly forts of Amber and Jaigarh, palaces like Jal Mahal and Rambagh, sacred shrines like Birla mandir, Ganesh mandir, Shila Mata temple, Galtaji and Jagat Shrimon temple, parks and gardens like Ghat ki Ghuni and Ram Niwas garden and Museums like Albert Hall.

Other than the destinations within the city, there is tourist potential in the region as well, which are not as well developed and promoted compared to the locations within the city. The areas of tourist potential can be categorized into historical, wildlife, religious, traditional medicine and natural features. There are many sites outside the city limits, within the Region at distance of a day's drive (Refer Table 9.9).

Jaipur's rich historical, cultural, architectural and natural heritage has tremendous potential of attracting foreign as well as domestic tourists. Tourist arrival in Jaipur in last two decades has been more than 3.5 times. Majority of tourist visits are related to its historic & cultural significance followed by pilgrims visiting fairs/festivals. The share of domestic tourists exceeds the international tourists.

There is a need for improving and enhancing the tourism infrastructure with respect to civic facilities, amenities and accommodation so as to maintain as well and increase the inflow of tourists. Intra-city, intra-regional and inter regional connectivity needs improvement. The increasing pressure on land for providing better infrastructure has surmounted the indirect impact on forests and wildlife.

The development of tourism infrastructure has not been in pace with the influx and growth of tourist over the decades, which has impacted the civic amenities. The region has had water supply, sewage disposal and solid waste management issues, which have aggravated over the decades.

Table 9.9 Potential Tourist Destinations in Jaipur Region

Destinations	Historical	Wildlife	Religious	Traditional medicine	Adventure sports	Natural features
Nahargarh		•				
Jamwa Ramgarh		•				•
Sambar						•
Madhogarh Tunga	•					
Bagru Fort	•					
Samode Haveli	•					
Bairath Ruins	•		•			
Chand Baori	•					
Gangaur Festival			•			
Amber Festival			•			
Elephant Village					•	
Teej Festival			•			
Chetoli					•	
Kanak Vrindavan			•			
Ayurvedic Center				•		
Sanganer	•		•			
Chomu Fort	•					
Kookas/Achrol	•					
Jaishingpur Khor	•		•			
Karauli			•			
Bichoon	•		•			
Band Buchada						•
Galta Forest						•
Jhalana Hills						•
Malayabag	•					•
Shahpura Mrig Van						•
Kadam Kund	•					

Source Collated by Authors

9.9 Institutional Systems

Jaipur being the state capital and headquarter of the district is the seat of government and houses state, district and regional level government offices. Urban areas like Jaipur, as spatial entities are administered by the ministry of urban development and Housing and its various departments. Broadly, there are three main and overlapping functions in these urban entities, namely, governance, planning and service provision and maintenance. The planning functions are under the purview of Town Planning Department and the development authority, services and maintenance is under the Urban Local Bodies (ULBs) and governance is undertaken by all the three across the hierarchical levels of these entities.

The role of the Ministry of Urban Development and Housing (MoUDH) at the state level is mainly redressal of the grievances, budget allocation, implementation of central and state level projects and schemes, extension or exclusion of municipal boundaries and overseeing the smooth running of all the departments and organizations within its purview. The Department of Local Self Governance is under the MoUDH and looks into similar matter but pertaining only to the ULBs.

Other than these, there are organization like Rajasthan Urban Infrastructure Finance & Development Corporation Limited (RUIFDCO), Rajasthan Housing Board (RHB) and other parastatal agencies that contribute towards the development and growth of the region. Rajasthan Urban Infrastructure Finance & Development Corporation Limited (RUIFDCO) is established to work as facilitator and coordinator besides serving as a link between various urban schemes operated by the central government and their implementation in the state. It is the state level nodal agency for all urban related centrally sponsored schemes. Rajasthan Housing Board (RHB), is a state government enterprise to look into the need of housing accommodation in the state. RHB through its Act is empowered to plan, develop and construct houses and schemes for housing, commercial, commercial-cum-residential schemes, industrial housing schemes,

Urban Improvement Trust (UIT) constituted under Rajasthan Urban Improvement Trust Act, 1959 looked into the acquisition of land and improving the city in a planned manner. Jaipur Development Authority (JDA) is the city level urban development authority which plays a very important role in the planning and development of the region. JDA was constituted in 1982 under the Jaipur Development Authority Act, 1982 for the purpose of planning, coordinating and supervising the proper, orderly and rapid development of the Jaipur City and areas contiguous to it. It was envisaged that the Authority would formulate and execute the plans, projects and schemes for the development of Jaipur Region so that housing, community facilities, civic amenities and other infrastructure could be properly created for the population of Jaipur Region. JDA's responsibility includes urban planning, preparation and implementation of master development plan and zonal development plans, formulation, sanction and execution of projects and schemes for development of the Jaipur regions, housing activities, preparation of master plan for traffic control & management for the region.

In the Region at the city level, the ULB in the form of Municipal Corporation exists and plays the role of providing services and maintenance of facilities. The state government has undertaken project called Rajasthan Urban Infrastructure Development Project (RUIDP) with the assistance of Asian Development Bank to provide integrated quality infrastructural facilities in selected regions. RUIDP are water supply, waste water management, roads, bridges and flyovers, slum improvement program, drainage, solid waste management, fire-fighting, emergency medical services and heritage.

Under the central government initiative between 2005 and 2014, Jawaharlal Nehru National Urban Renewal Mission (JnNURM) for improvement of the urban areas, substantial augmentation and improvement of urban infrastructure and environment was emphasized for the economic growth of the city. The vision to create a vibrant, 21st Century Renaissance City was formulated by the stakeholders and emphasized on a sustainable infrastructure base coupled with a vibrant economic base. The City Development Plan (CDP), key document of the Mission, based on which projects were detailed and implemented in Jaipur city formulated strategies on improving water supply, sewerage and sanitation, solid waste management, transportation, housing, poverty alleviation, heritage and conservation, governance and institutional development and finance. The strategies were supplemented with plans for training and capacity building, people's participation and establishment of clarity of roles and responsibilities of various agencies [39].

The state of Rajasthan is one of the most favored tourist destination in India for both domestic and International tourists. The tourism products offered by Rajasthan are in varied nature such as Heritage Tourism, Spiritual Tourism, Wild Life & Eco-Tourism, Cultural Tourism, Film Tourism, Meetings, Incentives, Conferences, Exhibitions (MICE) Tourism and Desert Tourism [8].

To promote tourism the Government of Rajasthan granted the status of industry to tourism sector in the year 1989. All the facilities and concession available to industries in the state are available to tourism units in the state. In the last few years at the state level Rajasthan is taking major initiatives for encouraging private investments in physical and social infrastructure through Public Private Partnerships so as to give the tourism industry a boost. The state government has already realized the potentials of this industry for the economic development in the state and is actively promoting tourism in the state by adopting schemes like 'Padharo Mhare Desh' meaning 'Rajasthan Invites You' [40].

The first State Tourism Policy was formulated in 2001. The main objectives of this policy was to increase employment opportunities, especially in rural areas for unemployed rural youth, optimum utilization of rich tourist resources of the state in order to attract the maximum number of domestic and international tourists, to facilitate the growth of tourism in the state and further involving the private sector in the development of tourism in the state, preservation of rich natural habitat and bio-diversity, historical, architectural and cultural heritage of Rajasthan, a special emphasis given on conservation of historical monuments in Rajasthan and to develop a ready market for the rich and varied handicrafts and cottage industries of Rajasthan [41].

Rajasthan reformulated the tourism unit policy in 2007 and updated it in the 2015. The Rajasthan Tourism Unit Policy, 2015 primarily addresses issues relating to time bound conversion of land for tourism units including new hotels and heritage hotels, time bound approval of building plans, grant of Patta[5] to heritage hotels, allotment of land for tourism units on District Level Committee rates, and smooth and speedy implementation of the provisions of related departments like Revenue, Urban Development and Housing & Local Self Government, Panchayati Raj, etc.

[5]A deed to a property.

This focus was much needed as land acquisition has been one of the biggest challenge in the past hampering not only urban but also tourism growth [41].

Another policy promoting Eco Tourism by Government was notified as Rajasthan Eco tourism Policy 2010. This policy was framed with a view to sensitize the public government departments, Non-Governmental Organizations (NGO) and others about eco-tourism and for laying down the framework of its growth in the state in a sustainable manner.

Rajasthan Investment Promotion Scheme, 2010 provides various financial incentives and subsidies for new investments. The state has also introduced a statutory Single Window System and an online electronic clearance mechanism for time bound approvals for investment proposals. The first Special economic zone dedicated for investors from Japan has been successfully established at Neemrana in the adjoining district.

Box 9.1 Tourism Policy 2015

Key features of Tourism Policy-Draft (2015)

- Investment Subsidy of 50% of Value Added Tax (VAT) and Central Sales Tax (CST) which have become due and have 19 Rajasthan Tourism Unit Policy 2015 been deposited by the enterprise for seven years
- Employment Generation Subsidy up to 10% of VAT and CST which have become due and have been deposited by the enterprise, for seven years
- Reimbursement of 25% of amount of VAT paid on purchase of plant and machinery or equipment for a period up to seven years from the date of issuance of the entitlement certificate
- Exemption from payment of 50% of Entertainment Tax for seven years
- Exemption from payment of 100% of Luxury Tax for seven years
- Land allotment in urban and rural areas at DLC rates
- 25% additional exemption from payment of stamp duty chargeable on the instrument of purchase or lease of more than 100 years old heritage property in the State, for the purpose of hotel development under the Scheme declared by the Tourism Department
- 50% additional exemption from payment of conversion charges for heritage property converted into a heritage hotel.

Source Rajasthan Tourism Unit Policy, 2015

The Tourism policy of the state has been successful in identifying the key sectors in tourism such as heritage hotels, MICE and provides fiscal benefits for developing the same. The state government has been encouraging joint ventures and contract management of private heritage properties (forts, fortresses, palaces and havelis) as well as identifying heritage government properties to be award on a lease basis.

The Rajasthan government has enacted enabling policies in spirit to promote tourism sector as it acknowledges the contribution of tourism in state's economy.

The first State Tourism Policy, 2001 does specifically mention about increasing employment opportunities in rural areas by promoting tourism. The 2001 policy too encouraged the private sector participation in developing the tourist infrastructure. The next policy in 2007 the government focused on promoting investment and develop infrastructure facilities in the tourism sector and address the shortages in accommodation. The most recent 2015 policy further defines incentives in form of subsidies and tax exemptions for and land acquisition. The policy needs to be supplemented with strong institutional framework, upgraded infrastructure and people's participation for it give boost to tourism sector as visualized [42].

Jaipur is among the first twenty identified cities under the Smart City Mission of Government of India. The vision for the city as defined by the Jaipur Smart City Limited (JSCL), a Special Purpose Vehicle is to enhance the quality of life for its citizen and to leverage its heritage and tourism through innovative and inclusive solutions. The vision has been articulated based on existing master plan, self-assessment of the city's profile and in consultation with the people of Jaipur. The city aims to achieve five goals which are aligned to the vision, namely, smart heritage and tourism, smart mobility, smart and sustainable civic infrastructure, smart multi modal mobility and smart solid waste management. The goals being project oriented have an area based development approach, wherein the walled city area has been identified for development.

9.10 Delineation of the Region

Like many metropolitan cities Jaipur has grown in many directions over the last two decades. Regional politics and spatial planning systems have an increasingly important role in the delineation of the region [43]. Authors have adopted latter as an approach to delineate the region based on population and distance between the two nodes along the transportation corridors (roads). The process of delineation began with collection of data and information related to Jaipur's existing development area. The analysis has shown that substantial development is taking place not only inside the region but outside too due to impetus provided by central government initiatives, which needs to be accounted for in the formulation of policies and development of the region. Since its notification in 2011, the region has developed in some areas which cannot be negated. Hence, the existing spatial entity that has been considered as the basis for delineating the metropolitan region is the current region of Jaipur defined in the Master Development Plan–2025.

The region is growing along the transportation corridors, especially in the south and western directions. The delineation (Refer Map 9.9) indicates the thrust of development to be largely concentrated on the stretch between NH-8, Jaipur-Ajmer to State Highway, SH-2C, Jaipur-Jobner-Phulera. The expansion of the region towards east is restricted by the hills and the reserved forest. The region may have grown southwards along Tonk road during 2001–2011. However, the analysis based on 2011 Census of India data indicates a westward direction of sprawl and linkages

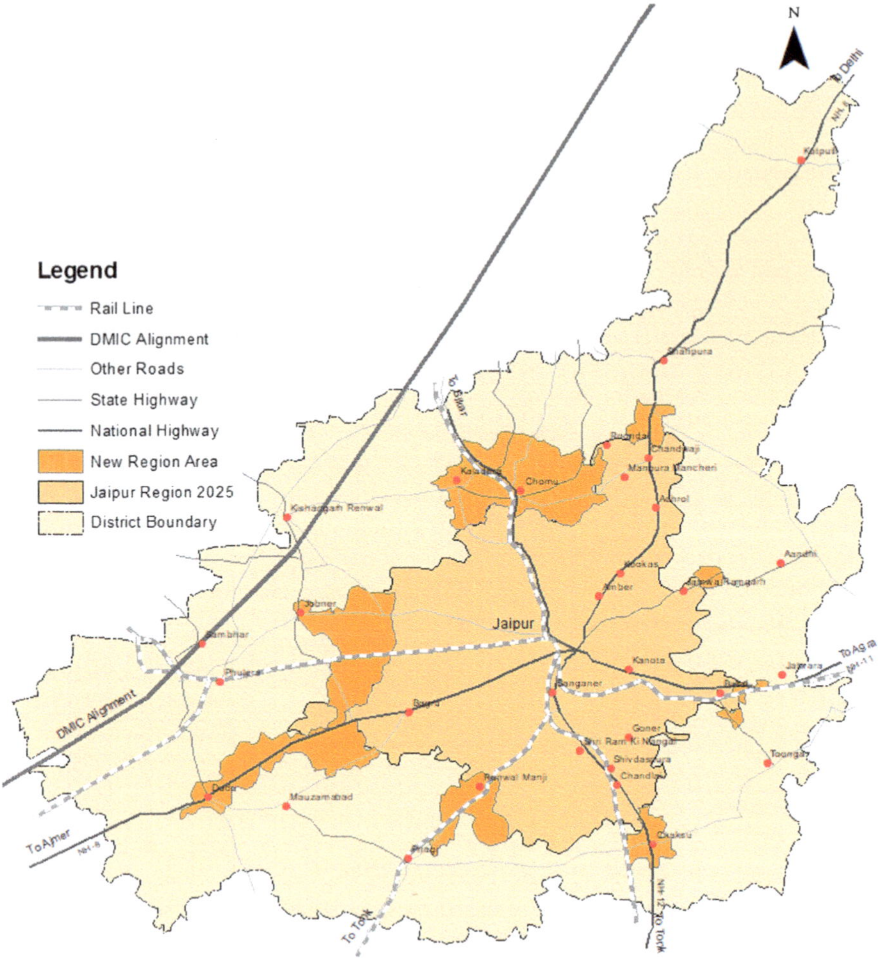

Map 9.9 Delineation of Jaipur Region. *Source* Analyzed and created by Authors based on inputs from Census of India, 2011 and Master Development Plan–2025, Jaipur Region

with the region. This change in direction of sprawl can be attributed to the DFC and DMIC, the corridors of transportation and industrial impetus, west of the region, having an important junction on DFC in the district at Phulera for the exchange of traffic between existing railway system and DFC, hence catering to the entire region.

Another growth cluster is towards the north-west, north of the region's Growth Center at Chomu, along NH-11 and SH-8B. Growth in this part of the district is primarily due to RIICO Industrial Areas as Kaldera and Manda. Moreover, the area is in the immediate influence zone of DMIC.

Other minor growth sprawls are along the highways abutting out of the region, namely up to Phagi and Chaksu in the south and Manoharpur in the north.

9.11 Issues in the Region

Jaipur Region provides exciting opportunities for growth and revitalization. However, the Region is not bereft of issues. Emerging from the analysis of the existing situation of Jaipur Region, issues have been categorized in aspects linked to various systems as natural, water, cultural, demography, economic, industrial, tourism and infrastructure.

As evident in Sect. 9.6 the region is dependent on secondary and tertiary activities. The data also indicates the region has a high dependent population and a low sex ratio. The per capita income in the city has been INR 37,984 [27] which is the highest for any city in the state. However, the WPR in the region is still lower than the state and national figures with a high disparity between the males and females. As a result, the region has a high unemployment rate too as compared to the state and national figures.

Pertaining to the infrastructure, the aspects covered in the analysis include water supply, sewerage and sanitation and solid waste management. All these aspects contribute towards strengthening the tourism activities in the region. As stated in the CDP of Jaipur [39] water supply in the city is below the permissible limits. Other than the shortage in water supply, the quality of water too has been deteriorating owing to improper disposal of sewerage and industrial waste.

Moreover, over the years the region has lost its traditional baori[6] and jhalars[7] as well, which were a source of groundwater recharge. The water table in nearly 50% of the region is at a depth more than 40 m during the post monsoon season with a seasonal variation of 2–4 m. The changed land use in the region too has impacted the groundwater recharge reducing the groundwater table. There has been a decline in the groundwater table by 4 m between 2005 and 2014 [19]. Average water level of Jaipur urban area was 28.873 m (mbgl) in 1995; that is continuously decreasing and reached at 47.488 ms (mbgl) in 2010. Over exploration of ground water, less recharging and shrinkage & dryness of surface water bodies are may be the main reasons of this depleting water level [44].

The sewerage network too does not cover 100% of the region forcing majority of population to rely on septic tanks or open defecation, particularly in rural areas and slums in urban areas. Solid waste management facilities too are lacking in the region leading to waste being dumped into drains creating environmental problems [8].

The problem of congestion and resultant vehicular pollution surmount the region particularly the core city areas and the satellite towns due to poor public transport, inadequate pedestrian and parking facilities.

The region faces problems due to multiplicity of authority with JDA, RHB, UIT and JMC being the main players. JDA and RHB exercise majority control over the JMC area. JMC does not have financial and taxation powers since it functions under

[6]A stepwell, traditional water harvesting system in Rajasthan.

[7]A spring.

the Rajasthan Municipalities Act, 1959 which is meant for municipal councils. Urban services area provided by JDA, JMC and parastatal agencies like RHB, Public Works Department, Public Health and Engineering Department and others.

Environmental issues surmount the region due to depletion of forests, water and soil erosion. Widespread deforestation in the hilly areas on the north-eastern side of the region has led to increased soil erosion. Water Pollution caused by textile industry through discharge of synthetic dyes, other chemicals and solid waste (sludge) is another challenge to be addressed (Refer Table 9.10).

Opportunities

Among all the tourist locations in India, Rajasthan is a key destination. It has emerged as a favorite destination for both domestic and foreign tourists. Data number indicates as 20 million tourists have visited in a year alone. However, the tourist sites within the city are well advertised and on the circuit. The region also comprises of many tourist sites outside the city which can be promoted.

The tourism policy of the state acknowledges the contribution of tourism industry and has tried to address the challenges to promote it further. The micro and small-scale cultural industries in the region, situated at the vicinity of Heritage Sites, have grown manifolds due to expansion of the market and inflow of the domestic and foreign travelers. Artisans and craftsmen have also seen improved access to markets, as many of them can now participate in crafts fairs across the country. Proposed DMIC corridor would increase the access to markets for the handicraft based industries.

As there is a disparity in the distribution of economic activities in the region which are mostly concentrated in the urban core of the region or in the satellite towns and industrial areas. The challenge for the policymakers is to ensure that the region continues to be attractive for the secondary and tertiary activities, which can

Table 9.10 Weaknesses and Strengths of Jaipur Region

Weakness	Strengths
• Low financial revenue and budgets • Insufficient marketing of potential tourist sites • Lack of infrastructure in terms of connectivity and other physical infrastructure • Direct international travel connection just restricted to South Asia • Limited water supply being an arid region • High temperature during specific months which makes it an unfavourable tourist destination • Balancing the development with rights of indigenous tribal communities in the region • Lack of sufficient investment funds • Lack of skilled human resources	• Presence of cultural tourism • Pre-existing industrial base in form of RIICO industrial areas and SEZs • Proximity to DMIC which can open avenues for handicraft exports • Jaipur city, a UNESCO heritage city as the core of the region • Villages have economic potential in terms of mining, small scale industries • Minerals rich zones found in the region and district—feldspar, quartz, limestone, mica, silicas and, china clay, pyrophyte • Jaipur City, an identified Smart City by central government • Pre-existing focus of state government on tourism development

be achieved not just through impetus in form of industrial incentives but by making the base strong in terms of human capital through increasing and strengthening the skilled and knowledge base of the local population.

9.12 Understanding Smart

Concept of smart in spatial context has emerged from the terms like smart city, coined by IBM and CISCO [46]; smartness, developed during some cities' debate on urban politics in early 1990s. It was later applied to smart growth which was linked to urban sprawl, sustainable development, economic environment and social equity. Over the years the term was being used in the context of policies and good practices. European Commission's Strategic Energy Technology Plan in 2009 [47] defined a smart city as 'a city that makes a conscious effort to innovatively employ Information and Communication Technologies (ICT) to support a more inclusive, diverse and sustainable urban environment'.

Smart Cities Mission is an innovative and new initiative by the Government of India to drive economic growth and improve the quality of life of people by enabling local development and harnessing technology as a means to create smart outcomes for citizens [48]. As stated by [49] Kumar, a smart city is one that is characterized by six components as smart people, smart economy, smart mobility, smart environment, smart living and smart government. Smartness in the urban areas as in the regions is to be brought about through smart solutions so that there is employment generation and the overall quality of life improves [50].

Smart is something that is Specific, Measurable, Achievable, Realistic and Timely [51]. These five parameters very aptly apply to the planning of a region. There is a need for the plans of the regions to be specific in terms of its strategies and proposals at the end of the horizon year. These achievements of the region need to be essentially measurable in tangible terms so as to understand the growth of the region. The strategies and proposals need to be achievable and realistic. Above all there needs to be timely completion of the plan in totality.

The need is not just for a smart city, as the concept of smart city is project based and is transactional at the current stage. The data created, that is Big Data, through the technological use in smart cities is aimed at transactions. At the current stage, it is incapable of generating trends [52] and the need is for smart planning.

The planning system in the country have become increasingly complex, inaccessible and inefficient, which had led to lengthy approval times. The planning which is done for the benefit of the community at large in itself becomes a barrier to the community and businesses. Planning of urban areas is largely a state subject, that urban planning falls under the purview of state governments [53]. Planning of the regions has been undertaken for metropolitan areas or special areas like National Capital Region and some area development plans like Damodar Valley. The regional plans are mostly prepared for urban nodes in complete negligence of the surrounding areas, hence lacking the regional perspective. Routra [54] had stated that there is a bias towards urban planning in the country and a lack of a regional approach to planning, which has been iterated by Mahavir in 2011 [55].

Smart Planning is understood as a system that is more effective, accessible, open and collaborative. For planning to be smart, it needs to be essentially simple and easy to understand and interpret. Planning information needs to be easily accessible to the citizens. It needs to be participatory in the true sense based on technology and create more interactivity, certainty and transparency [56]. There is an essential requirement for interconnectedness between economy, environment and social aspects within the framework of the standard outcomes of urban planning paradigm at spatial levels of area, node and network (Refer Fig. 9.5) such that there is maximum parity in the region.

Government's intent under the heritage and tourism sector is to develop the city as a world class smart heritage zone, by conserving and developing old heritage buildings along with provision of smart and sustainable infrastructure solutions and improving overall tourist experience. However, the focus is within the walled city only. The proposal is socially inclusive with benefits targeted to citizens of the lowest strata of society as well as tourist. Jaipur city proposes solutions to improve public hygiene and cleanliness; promote multi-modal mobility to improve road congestion, increase tourists' activity to grow the economy of the city and thereby improve the quality of life.

Ensuing from the combination of maximization and optimization with the integration of policies and strategies at macro and micro level would emerge a smart region of Jaipur. A smart region is a cluster of urban and rural areas which work together in order to frame a diffused smart city at the core, sharing vision, ideas, common goals and resources [57]. The premise of building smart region is the

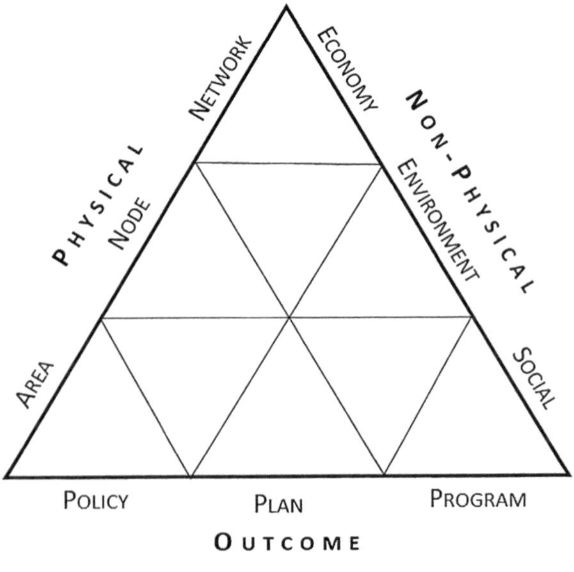

Fig. 9.5 Smart Planning Framework. *Source* Developed by the authors

capability of the various entities of the region to collaborate across administrative boundaries. A school of thought considers that the Smart Region concept aims to promote and implement innovative and intelligent technologies in the areas s of transport, energy and construction in line with sustainable development and overall life quality improvement, creating thus a live Smart Region Concept based on the Smart Cities principle [58].

The metabolism of regions is seen as an integral part of a complex network of interconnected natural, economic and social subsystems. Regions worldwide have started to look for solutions, which enable transportation linkages and high-quality urban services with long- term positive effects on the economy. For instance, high-quality and more efficient public transport that connects labor with employment is considered to be one of the key elements for regional development. Many of the new approaches related to regional services have been based on harnessing technologies, including ICT; helping to create smart regions.

Smart as per the understanding of the authors has four key elements. It should be namely attractive, efficient, digital and sustainable. In terms of a region, the place should be attractive to the citizens, tourists and entrepreneurs. It should be efficient in terms of mobility, affordability, economy and recreation. In terms of digital, it should be accessible to all so as to bring about smart thinking. Digital use needs to be intensive such that there is generation of lots of data using latest and appropriate technology. The actions in the region need to be sustainable so that the future generations are able to meet their needs. According to Colantonio [59] the concept of a smart region is far from being limited to the application of technologies.

Understanding of Jaipur region has highlighted that tourism is the industry that has to potential to develop further. Tourism is a social, cultural and economic phenomenon which entails the movement of people to countries or places outside their usual environment for personal or business purposes [60]. As in the case of other domains, there has been a natural progression towards e-tourism to smart tourism.

In most of the smart tourism projects that are born out of smart city projects, the focus is on innovation and competitiveness and development of smart end user applications that support enhanced tourism experiences [61]. In some countries the focus is on smart governance so as to transform not only the economic but also social and experiential dimensions. It has been observed that smart has become a fuzzy concept often used to derive specific political agendas and sell technological solutions.

Smart Tourism is reliant on information and communication technologies. These technologies connect the physical, information, social, and commercial infrastructure of tourism, and supplies Smart Tourism value to multiple stakeholders of a destination [45]. Smart tourism entails smart experience, smart business ecosystem and smart destination. The authors profess the smart tourism to be one that is sustainable, is technology based, is experiential to the tourists and accrues benefits to the local population.

9.13 Strategies for the Region

Smart specialization is a regional policy framework for innovation driven growth. The difference between Smart specialization and traditional industrial and innovation policies lies in the process defined as 'entrepreneurial discovery' which is an interactive process in which market forces and the private sector explore, discover and produce information about new activities and the government assesses the outcomes and empowers those actors most capable of realizing the potential [62, 63].

In the past few years, post the 2009 global economic crisis OECD has exhorted countries to go structural to make economies more competitive; to go social to address the increased inequality and lack of jobs; to go green to promote a growth path that takes due account of environmental constraints; and to go institutional to address the current confidence gap in institutions and markets [64].

Taking this as the basis, a vision suggested for Jaipur Region is 'to promote smart tourism and cultural based industries so as to promote local involvement and employment, improve quality of life yet preserve the cultural vibrancy of the region'.

The framework proposed for achieving the vision entails identifying the local strengths and overcoming the existing weaknesses so as to achieve integration of smart tourism principles of sustainability and eco-tourism based on technology. In the process, the aim would be to encourage smart industry and innovation by realigning the policies and thereby promote region's competitiveness (Refer Fig. 9.6).

Tourism needs good infrastructure to be successful. Infrastructure sector encompasses transportation, communication and utility networks. Smart tourism is especially concerned with enhancing the tourist experience at the destination enabled by the integration of ICT. As of date one of the key barriers in the case of Jaipur region for promoting smart tourism would be to integrate new forms of communication. As per analysis the region predominantly comprises of villages and class II and III town other than Jaipur city at the core (cross ref). Some parts of the region do not have access to even some basic infrastructure cross ref) which are the basis of creating a smart area. Visualizing the region as 'Smart region' which is ICT enable would mean for it to leapfrog and prioritizing the investment in technology and enhancing the promotion of tourism opportunities available in the region.

Another barrier for many regions is physical accessibility to products which enhances the exports from the region. In case of Jaipur region, there is good connectivity at not only regional and national but also international level. Moreover, the market base of the region is well developed, it being an international destination. The proposed DMIC is likely to further the accessibility of the region.

Jaipur City is already progressing towards being a Smart City under Government of India's Smart City Mission. However, the authors predicate that a Smart City cannot be smart until it is located in a smart region. As established Jaipur city shares strong linkages with is region. The Region has high potential to be developed as

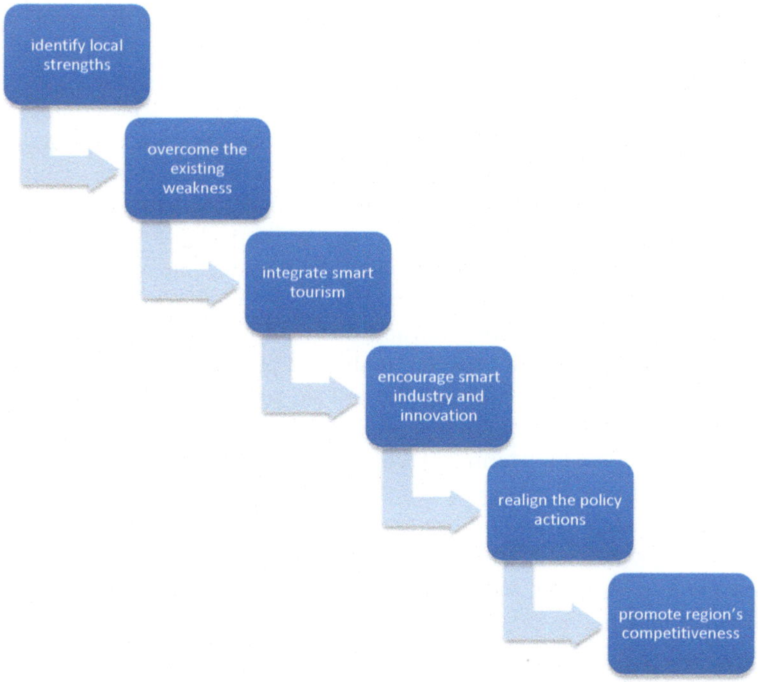

Fig. 9.6 Framework for Smart Tourism, Jaipur Region. *Source* Developed by the authors

Smart region too for which strategies are proposed to strengthen Jaipur as Smart Metropolitan Region. Four-point strategy with key emphasis on tourism in sync with the suggested vision is

- strengthening of regional competitiveness by enhancing tourism experience,
- regenerating the region's skill and knowledge base,
- smart industry and innovation, and
- strategic governance for smart specialization.

Strengthening of Regional Competitiveness by Enhancing Tourism Experience

OECD [65] states a competitive region as one that can attract and maintain successful firms and maintain or increase standards of living for the region's inhabitants. The focus in the recent years has shifted from international firms to domestic ones, hence an emphasis on the regional resources and assets. The concept of competitiveness has also been applied directly to tourism, wherein the destinations are made smart using the local resources and soft skills. As stated by Dwyer and Kim [66], Crouch and Ritchie adapted Porter's model to suggest that 'destination attractiveness' depends on four components, namely core resources and attractors as physiography, history and culture; supporting factors and resources as infrastructure; destination management which entails use of ICT and human resources and qualifying determinants like political environment.

In the context of Jaipur region, the core resources and attractors have been clearly identified and stated. The region is not well equipped in terms of infrastructure and ICT facilities. The region has a rich vibrant culture which is reflected in the people, food, attire, language and age-old customs and rituals. Culture and tourism has a mutually beneficial relationship which can strengthen the attractiveness and competitiveness of the region. Additionally, tourism provides an important means of enhancing culture and creating income which can support employment generation in the region.

So as to strengthen the interdependencies between tourism and culture, it is also important to build partnership between public and private sectors and between the region and its people [68]. Involving the private sector is essential for attracting investment and for continuous improvement of the quality that both cultural and tourism can offer. In the long term, the region will need to be increasingly smart and innovative in the way it develops, manages and markets culture and tourism. There will be a need to adapt from heritage to smart tourism (Refer Table 9.11).

In promoting smart tourism, the sustainability should not be forgotten. Three aspects according to Jucan and Jucan [69] that should be considered while promoting tourism are:

- Quality sustainable tourism to provide a valuable experience for tourist, while improving the quality of life of the host communities and protecting the environment
- Continuity—continuity of the host community should be maintained and
- Balance to be maintained between the needs of the tourism industry, environment and local community.

These may not be easy to achieve as promoting sustainability requires reconciliation of opposing interest and objectives [69], which magnifies at a regional scale. The complexity of both the tourism and cultural sectors implies that platforms must be created at smaller scale or at micro levels to support collaboration. Mechanisms must be innovated to ensure effective communication. The emphasis needs to be on policies that enhance and preserve heritage, ensure economic development of the region and employment of the local population, strengthen and/or diversify tourism products [67].

Table 9.11 Changing Form of Tourism in Jaipur

Form of tourism	Primary focus	Primary form of consumption
Heritage tourism	Past	Products
Cultural tourism	Past and present	Products and process
Smart tourism	Past, present and future	Product, process, experience and transformations

Source Adapted from OECD [68]

Regenerating the Region's Skill and Knowledge Base

The basis of any spatial entity is its people. The people of a region make its identity and develop its culture over a period of time. Smarter the people, smarter the region. Jaipur regions' biggest asset are its culturally vibrant people. Having a deep-rooted culture, the people have over the centuries developed it into their economic base that is unique to the region. These traditional practices of art, craft and resource conservation are getting lost due to the ease provided by technological innovations. There is a need to revive the lost traditions and train individuals for specific jobs or industries. Training needs to vary from soft skills to basic skills to specific skills. There needs to be a delicate balance between the use of technology to make the people and their region smart and boosting the tradition systems to enhance the quality of life of people through the strategies of smart tourism.

Smart Industry and Innovation

The Jaipur region has been dependent on mining and textile industry. However, there is a gradual decline in jobs in primary sector and stagnation in manufacturing sector. There is a strong base for the development of the cultural industries, since the region has many crafts producers, working with wood, ceramics, metals and textiles.

There is a need to foster new investment opportunities in the region that is based on cultural tourism. However, it is important to follow a balanced approach that not guided by commercial perspective alone but also factors in socio-cultural concerns.

According to Markkule and Kune [70] the smartness of a region relates to its capacity to leverage its human, structural, and relational capital, and its ability to integrate diverse actors in the region's innovation practice. In smart regions, all societal partners need to work together, and joint learning is a cornerstone of this collaboration. The educational institutes are an important instrument for codifying the lessons learned and helping other actors take the learning to the next level of practice. The educational institutes in the region can become the window on the region by disseminating the traditional practices and at the same time bringing in fresh ideas and engaging in diverse activities that build and enhance the image and reputation of the region to the wider world.

Further, responsible eco-tourism needs to be fostered in the region which would include programs that minimize the adverse effects of traditional tourism on the natural environment, and enhance the cultural integrity of local people. Hence, there is a need to formulate policies and initiatives that encourage the hospitality providers to promote recycling, energy efficiency, water reuse. In addition to planning towards protection of the environment and culture, there is a need to create economic opportunities for local communities who are an integral part of eco-tourism. Rather than focusing on the existing and potential tourist locations in the urban core of the region, rural tourism should be advocated as well, which would involve tourists witnessing or participating in activities that form the core of country life. With the decline in the primary sector being experienced in the region and people migrating to the urban core, rural tourism is a magic wand to stop rural decay.

Strategic Governance for Smart Specialization

Smart Specialization is a spatial approach characterized by the identification of strategic areas for intervention based both on the analysis of the strengths and potential of the economy [71]. The approach combines industrial, educational and innovation policies to suggest that regions identify and select a limited number of priority areas for knowledge-based investments, focusing on their strengths and comparative advantages [72].

So as to achieve this the authorities having identified tourism as the focus sector need to map and benchmark the tourism clusters, create an evidence based monitoring and evaluation system, be more effective in spending the public resources, eliminate duplicity in policies, and develop public support mechanisms for innovation in the sector and promote local culture and resource based industries [73].

The policies need to be such that the focus of growth is in areas that are already urbanized, allowing the non-urbanized parts of the region to set aside and restore more open space. Since water is a limited resource in the region any future development can only be sustainable if it is water efficient and has a reduced water footprint.

9.14 Conclusions

Jaipur region indicates an intro-regional disparity in development. Based on the region's constraints and potentials, tourism sector emerges as the most appropriate solution for steering the region's upward growth. Cultural tourism, one of the largest and fastest growing global tourism markets, boosts creative industries that is being increasingly being used to promote destinations. The analysis shows that despite the policies being promoted at the state level, there is a lack of planning for the promotion of culture and tourism at the regional scale as the focus has been of the city of Jaipur. There is need to induce balanced growth in the region by generating employment, improving the infrastructure, encouraging industry innovations and simultaneously conserving the resources so as to achieve sustainable region. However, in order to maintain balance, smart eco-tourism development needs to be ensured which is strategically planned to achieve goals taking into account the carrying capacity of the resources of the region. Since the region already has a diverse industrial structure and has further got a boost due to the proposed DMIC, smart specialization needs to be supported by the development of competitive industries in the region. Strategies are suggested for the region for enhancing the skill base of local people by promoting traditional systems to improve quality of life. This integrated approach supported by ICT can enable Jaipur Metropolitan Region to become an ecosystem that is smart and sustainable.

References

1. Pain K (2008) City-regions and economic development. In: 9th world congress of metropolis. Sydney
2. Census of India (2011) Provisional population totals urban agglomerations and cities. Census of India
3. Kulshrestha SK (2006) Dictionary of urban and regional planning. Kalpaz Publications, Delhi
4. Government of India (1991) District census handbook. Directorate of Census Operations, Jaipur, Rajasthan
5. Sivaramakrishnan KC, Kundu A, Singh BN (2005) A handbook of urbanization in india. Oxford University Press
6. Government of India (2013) Groundwater information Jaipur District Rajasthan. Central Groundwater Board, Ministry of Water Resources
7. Government of Rajasthan (2011) Master Development Plan-2025 Jaipur Region. Jaipur Development Authority
8. Government of India (2011) District census handbook. Directorate of Census Operations, Jaipur, Rajasthan
9. National Highways Development Project Golden Quadrilateral (2017) National Highways Authority of India. Available at http://www.nhai.org/gqmain_english.htm. Accessed on 15 Nov 2017
10. Jain S (2011) Walking into the microcosm of Jaipur. UNESCO, New Delhi
11. Population Foundation of India (2012) State of Inclusive WATSAN in Cities—Jaipur. Available at http://bctngo.org/themes/upload/files/460021.pdf. Accessed on 15 Nov 2017
12. Jain AK (2010) Urban planning and governance: a new paradigm. Bookwell, New Delhi
13. Government of Rajasthan (1976) Master plan for Jaipur 1971–199. Town Planning Department
14. Seetharam KS, Reddy AV (2010) State of Urban Services in India's Cities: spending and financing
15. TNN (2010) Metropolitan Planning Committee on the Anvil. Times of India
16. Aromar R et. al (2011) Urban India 2011: evidences. Autumn Worldwide Available at http://iihs.co.in/wp-content/uploads/2013/12/IUC-Book.pdf. Accessed on 17 Nov 2017
17. Government of Rajasthan (1998) Master Development Plan-2011 Jaipur Region Part 1. Jaipur Development Authority
18. Baruah NG n.d. Patterns and processes in spatial modelling: can we understand what we see? Available at https://www.geog.cam.ac.uk/research/readinggroups/gisc/patterns.pdf. Accessed on 27 Nov 2017
19. Government of India (2016) Ground Water Year Book 2014–15 Rajasthan State. Central Ground Water Board
20. Robert K, Michael R, Gray K (2013) Water Scarcity in Jaipur, Rajasthan. Jal Bhagirathi Foundation, India
21. Joshi KN, Agarwal V, Chawla B (2009) Environmental impact study on degradation of natural resources in urban area and strategy for sustainable development studies. Indian Council of Social Science Research, New Delhi
22. Singh F, Bhatia D. n.d. "VIKAS & VIRASAT" as a culturally inclusive development strategy for Rajasthan. Jaipur Virasat Foundation
23. Government of Rajasthan (2012) Estimates of district domestic product of Rajasthan 2011–12. Department of Planning
24. Government of Rajasthan (2012) Estimates of district domestic product of Rajasthan 1999–2000 to 2000–2007. Directorate of Economics & Statistics
25. Diao X (2010) Economic importance of agriculture for sustainable development and poverty reduction: findings from a case study of Ghana. OECD

26. Government of India (2016) Economic survey 2015–16: services sector remains the key Driver of Economic Growth. Available at http://pib.nic.in/newsite/PrintRelease.aspx?relid=136868. Accessed on 18 Oct 2017
27. National Institute of Urban Affairs (2015) Competitive Smart Cities in Rajasthan. CII, Rajasthan
28. Government of India (1981) District Census Handbook. Directorate of Census Operations. Jaipur, Rajasthan
29. Government of India (2001) District Census Handbook. Directorate of Census Operations. Jaipur, Rajasthan
30. Government of India n.d. Participation in economy. Available at http://www.mospi.gov.in/sites/default/files/reports_and_publication/statistical_publication/social_statistics/Chapter_4.pdf. Accessed on 30 Nov 2017
31. Government of India (2006) Rajasthan development report. Planning commission
32. Government of India. n.d. India Smart City Profile: Jaipur. Available at http://smartcities.gov.in/upload/uploadfiles/files/Rajasthan_Jaipur.pdf. Accessed on 15 Nov 2017
33. Fox S, Dyson T (2015) Is population growth good or bad for economic development?—IGC Blog. Available at: https://www.theigc.org/blog/is-population-growth-good-or-bad-for-economic-development/. Accessed 28 Sept 2017
34. Florida R (2013) The great growth disconnect: population growth does not equal economic growth. Available at https://www.citylab.com/life/2013/09/great-growth-disconnect-population-growth-does-not-equal-economic-growth/5860/. Accessed on 10 Nov 2017
35. Government of India (2016) Brief Industrial Profile of Jaipur District. MSME Development Institute, India
36. Government of India (2016) DMIC Rajasthan. Available at http://delhimumbaiindustrialcorridor.com/dmic-rajasthan.html. Accessed on 01 Nov 2017
37. FICCI (2015) Tourism infrastructure: the role states play. Available at http://ficci.in/spdocument/20610/Report-Tourism-Infrastructure.pdf. Accessed on 05 Nov 2017
38. Department of Tourism (2016) Rajasthan-Pragati Prativedan 2015–2016. Government of Rajasthan, Jaipur
39. ENVIS-SPA (2006) Jaipur City development plan: an appraisal report. Available at http://www.spaenvis.nic.in/showfile.aspx?lid=2410. Accessed on 15 Oct 2017
40. Raina CL, Raina AK (2005) Fundamentals of tourism and indian religion-principles and practices. Kanishka Publishers and Distributors, New Delhi
41. Government of Rajasthan (2017) Rajasthan Tourism. Available at http://tourism.rajasthan.gov.in/assets/pdf/Rajasthan-Tourism.pdf. Accessed on 10 Nov 2017
42. Government of Rajasthan (2015) Rajasthan Tourism. Available at http://tourism.rajasthan.gov.in/assets/pdf/rajasthan-tourism-unit-policy-2015.pdf. Accessed on 20 Oct 2017
43. Tonev P, Dvorak Z, Sasinka P, Kunc J, Chaloupkova M, Silhan Z (2017) Different approaches to defining metropolitan areas. Geographia Technica 12(1):108–120
44. Rajpoot RS, Daiman A, Kumar A n.d. Spatiotemporal characteristics of ground water level fluctuation in Jaipur urban area Rajasthan, India. Glob J Res Rev. Available at http://www.imedpub.com/articles/spatiotemporal-characteristics-of-ground-water-level-fluctuation-in-jaipur-urban-area-rajasthan-india.php?aid=10090. Accessed on 01 Oct 2017
45. Guo Y, Liu H, Chai Y (2014) The embedding convergence of Smart Cities and tourism internet of things in China: an advanced perspective. Adv Hospitality Tourism Res 2(1):54–69
46. Rosati U, Conti S (2016) What is a Smart City Project? An Urban Model or a Corporate Business Plan? Available at https://www.researchgate.net/publication/305422831_What_is_a_Smart_City_Project_An_Urban_Model_or_A_Corporate_Business_Plan/fulltext/57a1eaa708aeef8f311d8b75/305422831_What_is_a_Smart_City_Project_An_Urban_Model_or_A_Corporate_Business_Plan.pdf. Accessed on 15 Oct 2017
47. Eurocities (2009) Smart Cities. In Smart Cities Workshop. Brussels, 1 and 17 November 2009. Available at http://www.majorcities.eu/generaldocuments/pdf/eu_smart_city_initiative_workshop_report.pdf. Accessed on 15 Aug 2017

48. Government of India (2017) Smart cities mission: a step towards smart India. Available at https://www.india.gov.in/spotlight/smart-cities-mission-step-towards-smart-india. Accessed 25 Oct 2017
49. Kumar TNV (2014) Geographic information system for smart cities. Copal Publishing, New Delhi
50. Mahavir, Bedi P (2017) Integration of existing information systems as a spatial eDemocracy tool for smart cities in India. In: E-Democracy for Smart Cities. Springer
51. Knowledgehut (2017) Setting S.M.A.R.T. goals and objectives for projects successful outcomes. Available at https://www.knowledgehut.com/blog/project-management/setting-smart-goals-objectives-projects-successful-outcomes. Accessed 15 July 2017
52. Batty M (2013) Big data, smart cities and city planning. Dialogues Hum Geogr 3(3):274–278. (Sage)
53. Bedi P (2013) Spatial technological initiatives in key central government
54. Urban Development Initiatives. Spatio-economic Development Record. vol. 20 No. 1 January-February 2013, pp 16–21
55. Routray JK (1993) Urban and regional planning in India. Habitat In 17(3):55–74. Available at https://www.sciencedirect.com/science/article/pii/0197397593900166. Accessed on 01 Aug 2017
56. Mahavir (2011) A Geo-spatial approach to urban development. Available at http://mycoordinates.org/a-geo-spatial-approach-to-urban-development/all/1/. Accessed on 12 April 2017
57. State Government Victoria (2017) Smart Planning. Available at https://www.planning.vic.gov.au/__data/assets/pdf_file/0027/77463/fact-sheet-smart-planning-17072017.pdf. Accessed on 01 Nov 2017
58. Mora L (2012) The smart region. Presented at regional studies association winter conference —smart, creative, sustainable, inclusive. London, UK
59. Enviros (2017) Smart Region Concept. Available at https://en.enviros.cz/refereneces/smart-region-concept/. Accessed on 01 Nov 2017
60. Colantonio E, Cialfi D (2016) Smart regions in italy: a comparative study through self organizing maps. Euro J Bus Soc Sci 5(9):84–99
61. World Tourism Organization (2011) UNWRO Annual Report: a year of recovery 2010. UNWTO
62. Gretzel U, Sigala M, Xiang Z, Koo C (2015) Smart tourism: foundations and developments. Electron Markets 25(3):179–188
63. Hausmann R, Rodrik D (2003) Economic development as self-discovery. J Dev Econ 72(2): 603–633
64. Foray D, David PA, Hall BH (2009) Smart specialization. The concept, Knowledge Economists Policy Brief No. 9, European Commission, Brussels
65. OECD (2013) Innovation driven growth regions: the role of smart specialization. OECD
66. OECD (2017) Regional competitiveness. Available at http://www.oecd.org/cfe/regional-policy/regionalcompetitiveness.htm. Accessed on 25 Oct 2017
67. Dwyer L, Kim C n.d. Destination competitiveness: a model and determinants Available at https://pdfs.semanticscholar.org/9cd3/1126d7634a74522e57b272b6a549cca79f0c.pdf. Accessed 10 Oct 2017
68. OECD (2009) The impact of culture on tourism. OECD
69. Richards G (2001) The development of cultural tourism in Europe. In: Richards G (ed) Cultural attractions and european Tourism. CABI, Wallingford
70. Jucan CN, Jucan MS (2010) Social responsibility in tourism and sustainable development. 10(6)
71. OECD (2017) Smart Specialisation. OECD
72. Markkula M, Kune H (2015) Making smart regions smarter: smart specialization and the role of universities in regional innovation ecosystems. Technol Innov Manage Rev 5(10)
73. Law R, Leung R, Buhalis D (2009) Information technology applications in hospitality and tourism: a review of publications from 2005 to 2007. J Travel Tour Market 26:599–623

Part X
India, Kozhikode

Chapter 10
Smart Metropolitan Regional Development: Economic and Spatial Design Strategy for Kozhikode Metropolitan Region

T. M. Vinod Kumar, Namratha Radhakrishnan and Mohammed Firoz

Abstract Historically, Kozhikode was a port city and a global trading hub connecting Europe and South Asia. The subsequent political and social fabric prevalent during the post—independence period resulted in the city losing its erstwhile economic significance. Industrial and economic activity is now limited to few medium and small scale industries that are neither a major source of revenue nor employment. Economic decline, high growth of unemployment and increasing migration rates has made it important to reconsider economic strategies to transform Kozhikode. For the purpose of this study, the Kozhikode Metropolitan Region has been delineated to ensure that the area is unified by economic activities, labour markets and firms within a single administrative boundary. In this chapter we aim to study the existing economic and spatial structure of the place and propose economic and spatial strategies that can be adapted to revive the economy for a delineated metropolitan region within the district. The economic and subsequent spatial strategies for the region have been formulated based on the principles of the Third Industrial Revolution and Zero Marginal Cost Society, proposed by Jeremy Rifkin all within the framework of the sharing economy system. The fundamental principle of trust—which is the basis of sharing economy—has been integrated to further integrate the community of the metropolitan region for the implementation of the various strategies. In line with this, faith based models, which were practiced by the various communities in Kozhikode since centuries for community and social development, have also been explored for the benefit of the community.

T. M. Vinod Kumar
School of Planning and Architecture New Delhi (SPA–D), Kozhikode, India
e-mail: tmvinod@gmail.com

N. Radhakrishnan (✉) · M. Firoz
Department of Architecture and Planning, National Institute of Technology Calicut, Kozhikode, India
e-mail: namratharadhakrishnan@gmail.com

M. Firoz
e-mail: firoz@nitc.ac.in

T. M. Vinod Kumar (ed.), *Smart Metropolitan Regional Development*,
Advances in 21st Century Human Settlements,
https://doi.org/10.1007/978-981-10-8588-8_10

Keywords Smart economy · Economic transformation · Renewable energy
Micro grid · Sharing economy · Spatial code · Faith based model

10.1 Introduction

India has witnessed rapid urbanization over the past few decades. The annual growth in urban population between 2010 and 2015 was 1.1% and it is projected to add 404 million people—the largest contributor in the world—to its urban population by 2050 [1]. The number of statutory towns (all places with a municipality, corporation, cantonment board or notified town area committee as defined by the Census of India), census towns (places that satisfy three criterion, namely, (a) a minimum population of 5000; (b) at least 75% of the male main working population engaged in non—agricultural pursuits; and (c) density of population of at least 400 per km^2 as defined by the Census of India) and urban agglomerations (a continuous urban spread constituting a town and its adjoining urban outgrowths or two or more physically contiguous towns together and any adjoining urban out—growths of such towns as defined by the Census of India) has increased by 0.62, 11 and 2.15% respectively from 2001 to 2011 [2] (Fig. 10.1).

Economists and planners around the world have reached a consensus that urbanization and economic development play a dual role in creating prosperity. By 2031, approximately 75% of India's national income is expected to come from

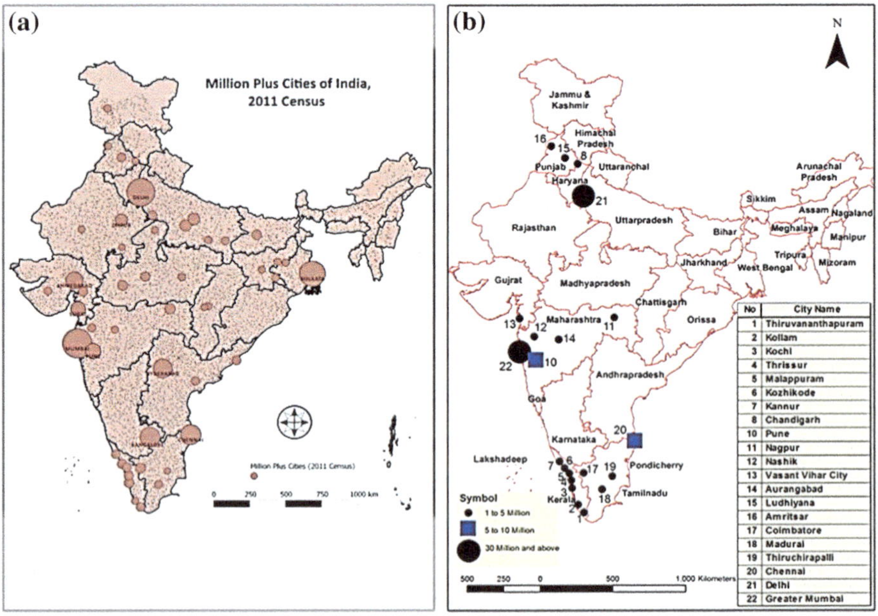

Fig. 10.1 **a** Distribution of million plus cities in India; **b** Potential million plus cities identified for smart cities upgradation; *Source* Compiled from [53]

cities [3]. As stated by François Bourguignon, emeritus professor of economics at the Paris School of Economics, these high levels of urbanization and economic development have resulted in rising inequalities and subsequent marginalization [4]. Along with economic inequalities, Indian cities have also witnessed haphazard outward growth that is outpacing their population growth [5]. As a part of the rectifying these inequalities and ensuring sustainable growth, India has focussed on adopting smart growth principles as a part of its economic agenda.

An analysis of urbanization rates of Indian states indicates that Kerala has experienced the highest level of urbanization among other states of the country. Census of 2011 indicates that the state is 47.71% urban and it has increased by 83.2% over a decade. The major contributor to this increase is the change in the workforce pattern from agriculture to other non-agriculture sectors [6]. Though the number of statutory towns in Kerala decreased by one from 60 in 2001 to 59 in 2011, the year 2014 witnessed major changes in the administration, whereby 28 new municipalities were formed, resulting in the percentage change of 45% from 2001 to 2014. Similarly, Kerala has experienced a substantial increase in the number of Census towns and urban agglomerations (UA). While urban agglomerations have only increased by approximately 12% from 2001 to 2011, there has been considerable growth in their sizes. The growth of urban agglomerations and Census towns has led to the emergence of an urban corridor that is in line with the settlement pattern of the State (Fig. 10.2).

The 73rd and 74th Constitutional Amendment Acts envisages democratic decentralization of administration and planning whereby 'urban planning including town planning' function has been assigned to the urban local bodies. However, the emergence of urban corridor in Kerala, has led to the fact that planning cannot be limited within the boundaries of the urban local bodies. The Act also mandates the constitution of the District Planning Committees (DPC) and Metropolitan Planning Committees (MPC) for the planning and development of Districts and Metropolitan areas respectively. Though, Kerala has already constituted DPCs, there are no MPCs in the State. However, the Kerala Town and Country Planning Act, 2016, recently enacted envisage plans for State, metropolitan areas, districts, local areas and micro level areas [7]. This has mandated the constitution of MPC for the planning of metropolitan areas within the State.

10.2 Metropolitan Region Defined

Scholars have defined a metropolitan region in various ways. Some have defined it as a geographic area spanning a number of local government authorities that tie together with economic connections and ultimately forming unified labour pools that link consumers and businesses with efficient infrastructure [8]. The Organization of Economic Cooperation and Development (OECD) defines it as a geographical space within which a number of economic links such as labour markets, firms, parts of supply chains and relationships between firms and local authorities [9]. The Kerala

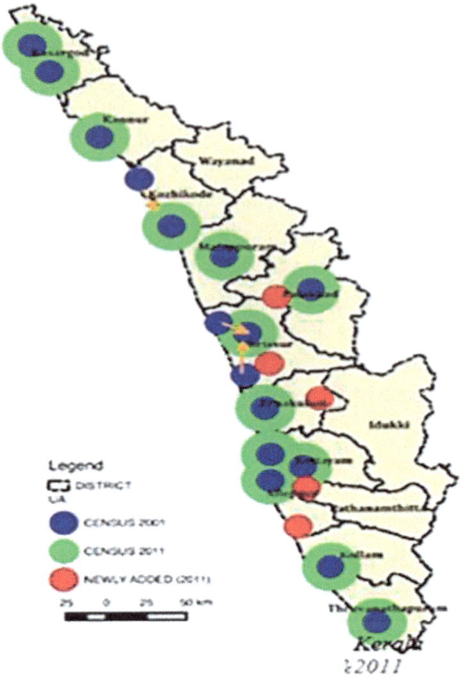

Fig. 10.2 Distribution of urban agglomerations in Kerala; *Source* Compiled from [2]

Town and Country Planning Act of 2016 defines the metropolitan area as a geographic region with population of 10 lakh or more; comprised in one or more districts and consisting of two or more panchayats, municipalities or other contiguous areas [10]. The ideal characteristics of a metropolitan region are that the boundary should not be static; the region should be decentralized and is ideal if the area is within the administrative boundary of the district.

10.2.1 Kozhikode Metropolitan Region—Boundary Delineation

In this chapter, the term "region" or "planning area" is defined as the administrative entity of the Kozhikode Metropolitan Region. A review of the definitions mentioned above have resulted in a number of parameters that need to be considered while delineating a new area such as availability of population that will act as the labour pool, existing economic activities, transport connectivity, all contained within a larger district boundary. Taking these parameters into consideration, the planning area for the Kozhikode Metropolitan Region has been delineated considering boundaries of census defined urban areas and urban agglomerations that fall within the Kozhikode district boundary. The infographic shown below indicates the delineation process for the same (Fig. 10.3).

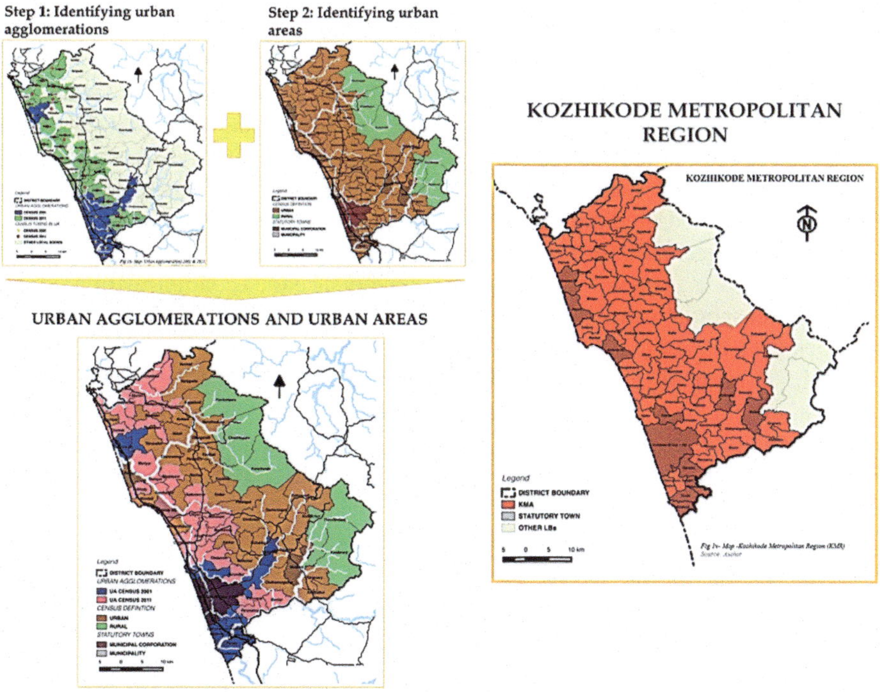

Fig. 10.3 Delineation process for Kozhikode Metropolitan Region (KMR); *Source* Compiled from [2, 27]

Box 10.1: Strategic location and regional connectivity of Kozhikode

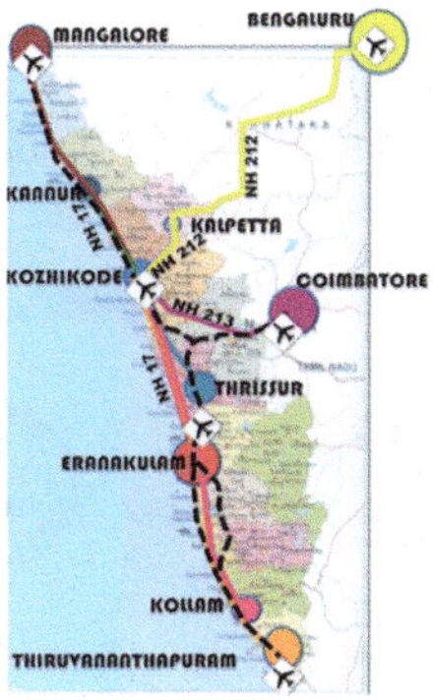

Kozhikode district is located along the south west coast of India in the Malabar region of Kerala. It is bound by the Arabian Sea on the west and the districts of Kannur, Wayanad and Malappuram to the north, east and south respectively. The district is located approximately 420 kms north of the state capital Thiruvananthapuram and 250 kms from the commercial capital namely, Ernakulam. Chennai and Bangalore, the two other major cities of south India are located within 650 kms from Kozhikode.

The district - which is well connected via air, rail and road - acts as the regional transit hub for the Malabar. The closest airport, located in Karipur, is approximately 30 kms from the city. While the Trivandrum - Mangalore - Mumbai railway line connects the district to the major coastal towns of the state and to other major cities of the country, road connectivity to all coastal cities is ensured by the National Highway (NH) 47. Two major ports namely, Kochi and Mangalore lie within 250 kms and an intermediate port namely, Beypore is located …… kms from Kozhikode - hence improving trade prospects for the district.

10.3 Study Approach and Methodology

The economic and subsequent spatial strategies for the region have been formulated based on the principles of the Third Industrial Revolution and Zero Marginal Cost Society, proposed by Jeremy Rifkin. According to Rifkin, three defining technologies emerge and then converge to create infrastructure that change power management methods and movement of economic activities across the value chains. These defining technologies are (a) New communication technology to manage economic activities; (b) New energy sources to enhance the economic activities and (c) New transportation and logistics facilities for movement of these economic activities [11].

Rifkin, in his book The Third Industrial Revolution, says that the world today is at the threshold of a new paradigm that will merge the digital communication internet, a digital renewable energy internet and a digital automated logistics and transportation internet [12]. While the first aspect is already well underway in the form of a smartphone, the second and third parameters are still in their infancy. Zero marginal cost works on the principle of efficient renewable energy technologies that will accelerate productivity such that the marginal cost of production would be near zero—making goods and services essentially free [13].

10.3.1 Rise of the Sharing Economy

Humans, as social animals are wired to share, and it has existed since ancient civilizations—barter system being the most common form of sharing [14]. The principle of sharing economy is built around sharing human and physical resources —includes the shared creation, production, distribution, trade and consumption of goods and services by people and organizations [15] (Fig. 10.4).

The benefits of sharing economy are many—increase accessibility to self-employment at house hold levels, increased business opportunities, efficient use of monetary resources, improved connectivity among different communities

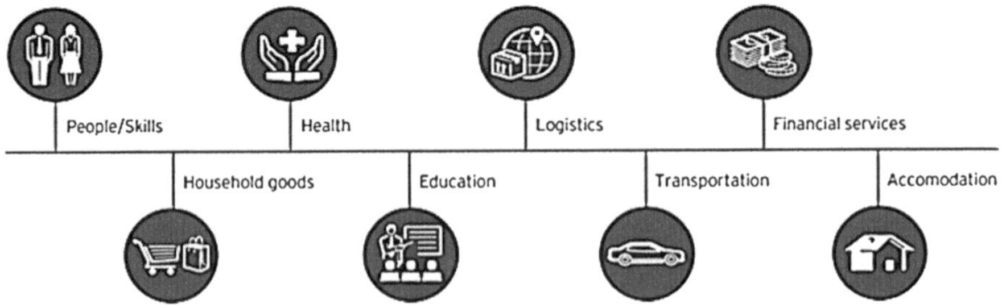

Fig. 10.4 Segments of a sharing economy; *Source* [15]

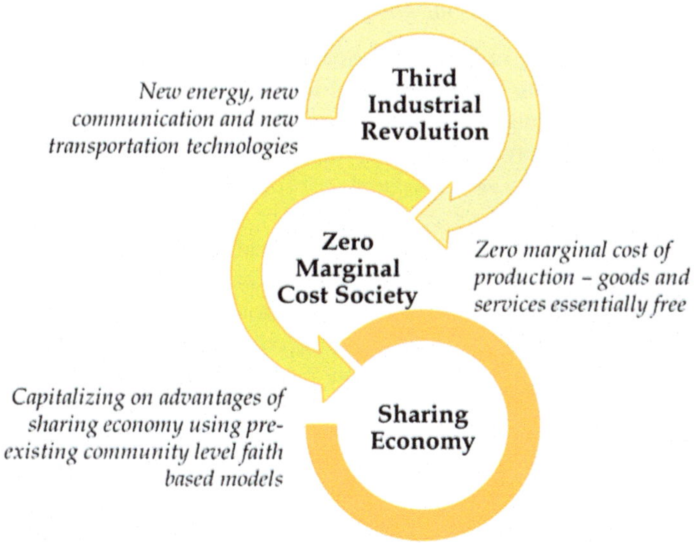

Fig. 10.5 Study approach

irrespective of region [14, 16, 17]. While sharing impacts the community in multiple ways, the foundation of the sharing economy is built on trust [18].

This fundamental principle has been adopted to integrate the community of KMR for the implementation of the various strategies. The city of Kozhikode was historically named 'The City of Truth' owing to the trustworthy nature of its people [19]. Kerala is also famous for communal harmony and Kozhikode has welcomed Arab merchants, European traders and Christian migrants among others [20, 21]. The demographic census of 2011 for Kozhikode reveals that approximately Hindu, Muslim and Christians account for 56.2, 39.2 and 4.3% in the total population respectively. Since centuries, the Hindu, Muslim and Christian communities have practised faith based charity or compulsory donation concepts namely, 'Jajmani' (Hindu), 'Zakath' (Muslims) and 'Dashamsham' (Christians). Hence, for community and social development these concepts have been explored—contributions when judiciously used can be utilized for the benefit of the community (Fig. 10.5).

10.3.2 Methodology

The three key steps in while developing the economic and spatial strategies for smart metropolitan development for the Kozhikode Metropolitan Region include, (a) Renewable energy generation and deployment; (b) Transformation of industry to a smart economy and (c) Cascading society of near zero marginal cost. These steps have been elaborated below.

(a) ***Renewable energy generation and deployment***

All of the energy used in industries, homes and vehicles still use conventional sources that entail very high environmental impacts. In lieu of climate change and strategies adopted to reduce the carbon foot print, it becomes pertinent to carry out new models of energy production. This energy can be generated by harnessing renewable sources at nearly zero marginal cost [22]. Kerala has been one of the first few states in India to focus on renewable energy generation for large scale businesses and households. This step took further strides as the Cochin International Airport Limited (CIAL) became the first high demand infrastructure project to be completely operational using solar power [23]. An interview with officials of the Kerala State Electricity Board indicates that the cost of power from solar declines at 7% per annum while the cost of grid energy increases by 3% per annum and that it will also be easier to deploy renewable energy technologies due to plunging costs of overall renewable energy generation since mass production of solar power components has brought down the unit price [24]. Energy surplus and deficit are calculated for regions which are used for energy balancing. This renewable will be generated on site at zero marginal cost by transforming businesses and homes into micro power plants. In this model, it is envisaged that the households will generate renewable energy not only for the household but also for the various economic activities in the community. A sharing economy framework built on the cultural systems of Jajmani, Dashamsham and Zakath will ensure that the production and services network will cased down to near zero in a socially responsible manner.

Internet technology and renewable energy would merge to create a powerful platform which enables millions of people to produce their own green energy in their homes, offices, and factories, and sharing it with each other in an "energy internet," just like we now create and share information online. The five pillars of renewable energy generation and deployment are:

1. Shifting to renewable energy
2. Transforming the building stock into green micro power plants
3. Deploying hydrogen and other storage technologies in every building and throughout the infrastructure
4. Using Internet technology to transform the power grid of every continent into an energy internet that acts just like the Internet
5. Transitioning the transport fleet to electric plug-in and fuel cell vehicles that can buy and sell green electricity on a smart, continental, interactive power grid [12].

Based on this, the renewable energy potential of the region has been studied first by identifying various methods of generating energy through renewable resources suitable to the region and to select the most suitable method. The energy generation capacity of the region was calculated. Then energy requirement for residential, commercial, institutional, public and semi—public land use activities has been calculated. The surplus energy has been derived based on the energy generated (supply) and the energy required (demand). The energy demand for various household industries has been estimated next followed by balancing the supply and demand regions. The micro-grids have been identified based on the balanced regions. Calculation of storage requirement and allocation of location of storage devices have been done, resulting in the spatial codes for renewable energy. The energy network has been integrated with the communication network necessary for making a 'smart' regional economy.

(b) ***Transformation of industry to a smart economy***

Smart Economy involves the creation 'Prosumers,' i.e., a person would be a consumer and producer in a smart economic environment. This would involve conversion of major economic activities to Smart economic activities based on energy and resources in a region. Rather than following a conventional system of large and medium scale industries, every individual household shall be a Prosumer and would be considered as an industry.

To create a 'Smart Economy' for the region, the existing economic activities of the region have been studied and declining activities/ industries are identified. In order to replace declining activities with smart economic activities land use conversion strategies have been formulated. Identification of the potential activities in the region has been done based on energy requirement, resources, skills etc. and on the spatial codes for renewable energy. The existing activities have been converted to smart economic activities by using ICT in every aspect of production and service systems and transforming the existing pattern of production and service by avoiding the middlemen—thus making it more profitable to the producers. The activities based on resources and energy has been spatially allocated and spatial codes for smart economy is derived. The strategies to convert smart economic activities to Zero Marginal Cost have been formulated.

(C) ***Cascading society based on Zero marginal cost***

It is based on the idea that if the marginal cost of producing each additional item falls to essentially nothing, then everything becomes free. This has become possible due to 3D printing; open-source software; the internet of things; the sharing economy; the online courses that are reshaping education; and the artificial intelligence enabling machines to replace many types of human labour.

To create a Zero Marginal Cost Society framework for the region, the resources in the region have been mapped out. Then religious and cultural aspects have been integrated and potential activities based on energy requirement, resources and skill are identified. The process and economic flow in zero marginal cost society has been worked out and spatial codes for a zero marginal cost society were prepared.

Finally, to arrive at spatial code for the metropolitan region, each spatial code for renewable energy, Smart Economy and Zero Marginal Cost Society has been consolidated followed by preparation of Land use plan and Zonal plan along with strategies for implementation (Fig. 10.6).

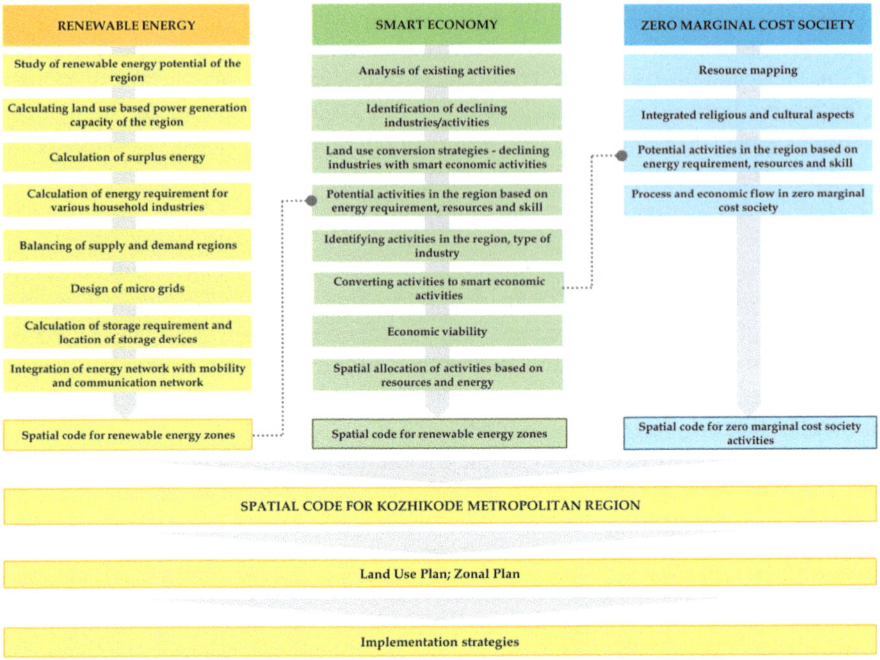

Fig. 10.6 Detailed study methodology; *Source* Compiled from [27]

10.4 Kozhikode Metropolitan Region

10.4.1 Administrative Units

Kozhikode district comprises of 1 revenue division, 3 taluks, 1 municipal corporation, 7 municipalities, 12 development blocks, 70 panchayats and 118 villages. The Kozhikode Municipal Corporation serves as the administrative headquarter for the district [25] (Table 10.1).

Table 10.1 Administrative units in Kozhikode District

Revenue Division	1	Kozhikode
Corporation	1	Kozhikode Municipal Corporation
Taluks	3	Kozhikode, Koyilandy, Vadakara
Municipalities[a]	7	Koyilandy, Vadakara, Feroke, Mukkom, Koduvally, Ramanattukara, Payyoli
Blocks	12	Balussery, Perambra, Kunnumal, Thuneri, Kunnamangalam, Thodannur, Koduvally, Meladi, Vadakara, Panthalayani, Chelannur and Kozhikode

[a]Of the total seven municipalities, Feroke, Ramanattukara, Mukkom and Koduvally were converted to municipalities in 2015 *Source* [25]

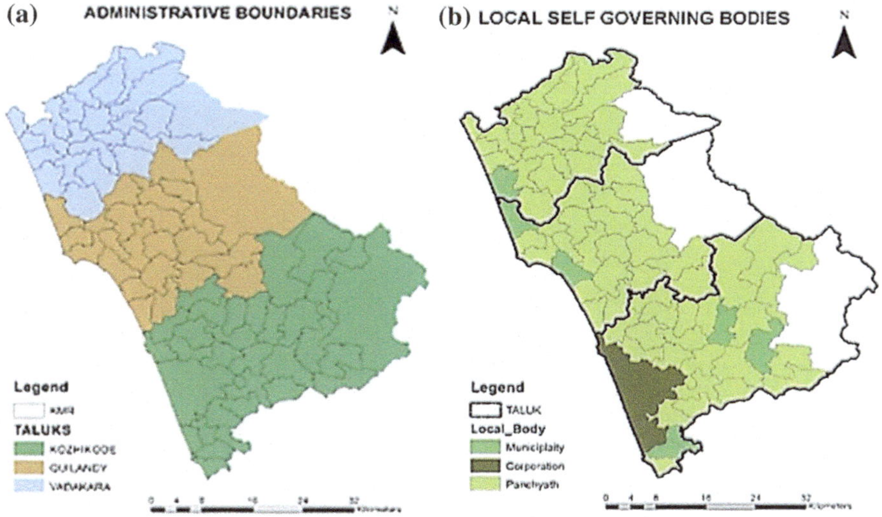

Fig. 10.7 **a** Administrative boundaries in Kozhikode District; **b** Local Self Governing Bodies in KMR; *Source* Compiled from [25, 27]

Table 10.2 Percentage share of LSGBs and population in KMR

Taluk	Kozhikode District[a]		Kozhikode Metropolitan Region (KMR)[b]			
	LSGB	Population	LSGB	% share of district	Population	% share of district
Kozhikode	32	18,93,307	30	94	18,45,809	97
Quilandy	25	3,64,359	23	92	3,40,294	93
Vadakara	23	8,16,362	22	96	7,97,150	98
Total	80	30,74,028	75	94	29,83,253	97

Source [a][26]; [b][27]

The KMR comprises of local bodies from all 3 taluks—75 of the total 80 Local Self Governing Bodies. Additionally, all 7 municipalities are also located within the planning area. The Kozhikode Metropolitan Corporation—located towards the south of the planning area—is the administrative center of the district and will also play the same role for the region (Fig. 10.7 and Table 10.2).

10.4.2 Distinct Topographical Features

The district is divided into three distinct topographical regions—363 km^2 of coastal belt; 1344 km^2 of lateritic midlands and 638 km^2 of the Western Ghat highlands—along with a coastal length of approximately 80 km [26]. While the planning area

Table 10.3 Share of topographical regions in KMR

Region	Kozhikode District[a] (km^2)	Kozhikode Metropolitan Region[b] (km^2)	% share of district
Coastal Belt	363	363	100
Midlands	1344	1344	100
Highlands	638	13	2
Total	2345	1720	–

Source [a][26]; [b][27]

of KMR is spread across all of the coastal belt and the midlands, the eastern boundary of the metropolitan region cuts through the highlands and only comprises of 2% of the highland area (Table 10.3).

10.4.3 Climate and Natural Ecosystem

Located in the tropical to sub—tropical climatic zone, the temperature varies between a maximum range of 28–32 °C and minimum range of 22–27 °C. Heavy monsoons–average rainfall of approximately 3000 mm during June to December—are a characteristic feature of this region. Hence, with the hotter months limited to approximately 3–4 months in a year, the climate of the region is pleasant for most part.

Like other coastal districts of Kerala, Kozhikode is also characterized by a river and backwater network that provides livelihoods to the coastal settlements along with cheap transport alternatives. The major rivers in the district are Kuttiadi, Korappuzha, Kallai, Chaliyar, Beypore and Kadalundi [26].

10.4.4 Demography and Spatial Implications

The KMR covers 1720 km^2 with a population of 2.9 million (computed on 2011 census figures). The density of the region is approximately 1735 persons/km^2 which places it above the state average of 860 and closer to 1508 in the capital district of Thiruvananthapuram. The population is concentrated highest in the Municipal Corporation area followed by the surrounding areas of Beypore, Feroke, Olavanna, Vadakara and Quilandy (Fig. 10.8a) . The average population growth rate of the region is 9% which is above the district and state average of 7.2 and 5% respectively.

Spatial analysis of population densities indicate that the distribution of population density in the region is determined by two factors namely, (a) topography and (b) development concentration. Changing topography has resulted in high population densities along the coastal belt which decreases progressively towards the midlands and highlands. Additionally, all major developments and subsequent employment opportunities are also concentrated in the municipal corporation and its surrounding areas. Hence within the coastal belt, the population density is highest in the Kozhikode Corporation and the surrounding areas and progressively decreases towards the north of the region (Fig. 10.8b).

The average sex ratio of the region is 1106 and is higher than the district and the state average of 1098 and 1084 respectively. Spatial analysis of the distribution of sex ratio and employment opportunities across the local bodies in the region indicates that the sex ratio is higher in the northern parts of the planning area and decreases towards the south. This is primarily attributed to lack of employment opportunities in these areas, thus leading to out migration of the male working age population in search of employment to countries in the Middle East. Migration studies state that the outflow of the working age population from Kerala to countries abroad still continues particularly because of the lack of suitable employment opportunities in the state for the increasing number of the educated youth along with the prospect of better lifestyles associated with Gulf emigration. Studies also indicate that the spatial distribution of out migrant place of origin from the state is steadily moving northward [28] (Fig. 10.8c and Table 10.4).

10.4.5 Economy, Industries and Employment

Economy

Kozhikode was once a vibrant global trading hub with an established trade route connecting Europe and South Asia. Historically, Calicut was a port city and an important center of trade and commerce especially for spices trading to several centuries around the world which thus earned its name as 'City of Spices' [21] (Fig. 10.9).

The political and social fabric prevalent during the post-independence period in the city did not support any major industrial developments. The industrial activity is presently limited to few medium scale industries related to steel processing and small scale industries manufacturing rubber products, food and dairy, textile, handloom, timber etc. The industries are neither a major source of revenue nor does it generate any significant employment opportunities.

The total Gross State Value Addition (GSVA) of the district, in 2015–2016 was approximately INR 46.7 lakhs and accounted for 9% of the state GSVA. The GSVA grew at approximately 10.8% year on year from 2014—the growth rate is comparable to the surrounding districts. Per capita income of the district was approximately INR 1 lakhs and while it is one of the highest in the region, it only ranks 8th in the state (out of 13). The economy of the district, in 2016, was primarily driven by the tertiary sector—61% of district GSVA—followed by the secondary sector—31% of district GSVA [29] (Fig. 10.10).

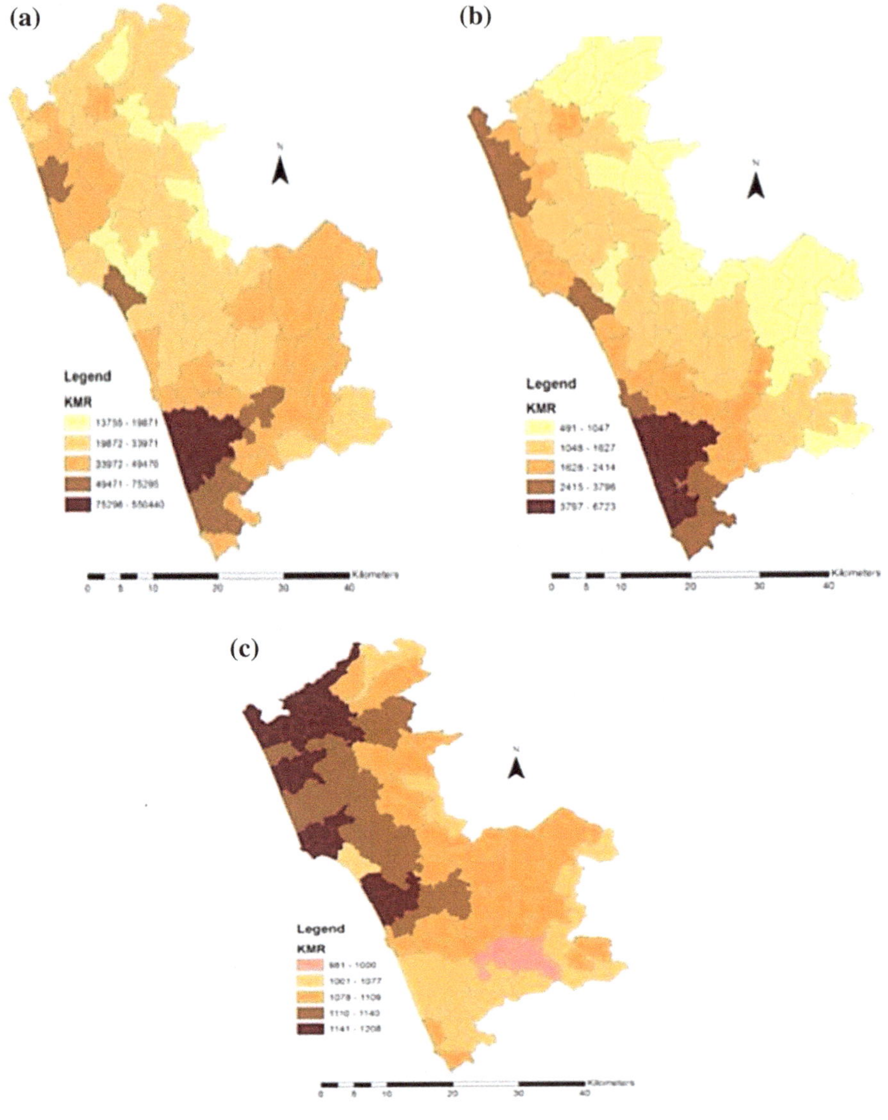

Fig. 10.8 **a** Spatial distribution of population; **b** Spatial distribution of population density; **c** Spatial distribution of sex ratio; *Source* Compiled from [2, 27]

Table 10.4 Comparison of demographic parameters between state, district and region

Parameter	Kerala[a]	Kozhikode District[a]	Kozhikode Metropolitan Region[b]
Population	33,406,061	3,086,293	2,983,253
Population growth rate (%)	5.0	7.2	9.0
Population density	860	1316	1734
Average sex ratio	1084	1098	1106

Source [a][2]; [b][27]

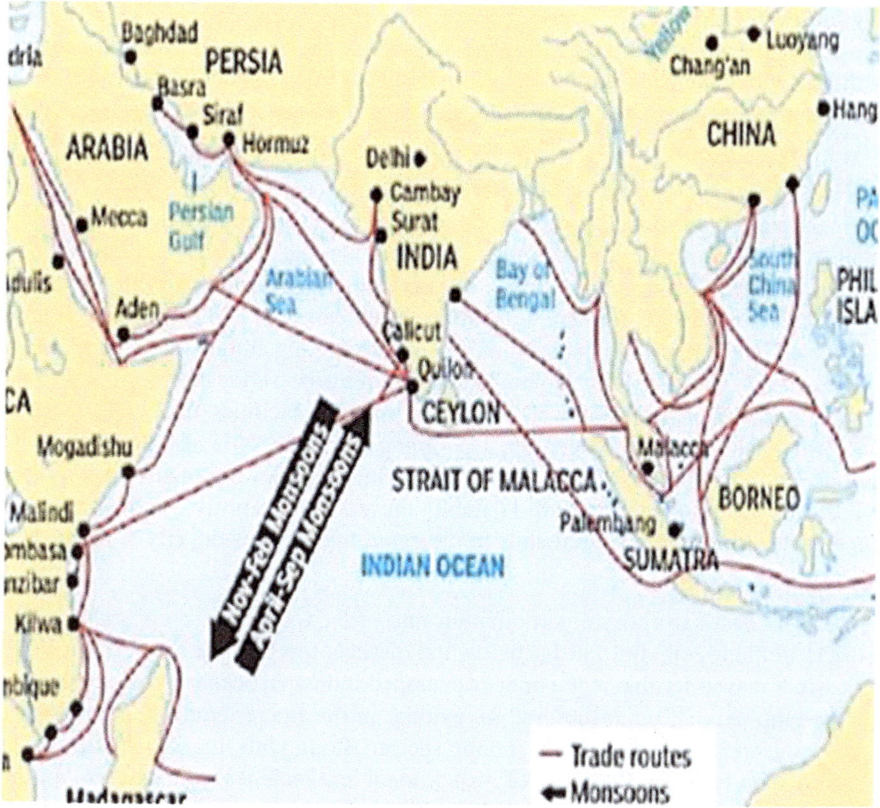

Fig. 10.9 Trade links between Calicut and rest of the world until 16th century; *Source* [21]

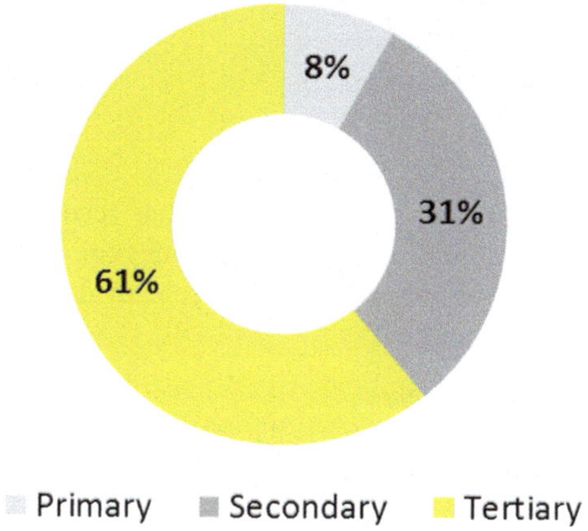

Fig. 10.10 Sectoral share of district GSVA; *Source* [29]

Key industries
The primary sector only contributed to 8% of the district economy wherein majority of the activities included agriculture cultivation of paddy, coconut, banana, tubers, spices and other tree crops along with fishing and its allied activities.

While secondary sector activities included medium scale industries with manufacturing activities focused on Food and Beverages (F&B), timber processing and tile manufacturing among others, the tertiary sector activities included tourism and retail establishments. Low profitability of other erstwhile industry and trade activities has resulted in a structural shift in the economy from manufacturing to tertiary sector activities such as tourism, education, healthcare and IT/ITeS. Lonely Planet has ranked North Kerala as the third must-visit destination among ten other places in Asia for 2017. The district is also home to some of the best education institutes in the country and the presence of well-developed medical facilities makes it the go-to place for health care in the region. Upcoming investments in the IT/ITeS sector include the development of a 50 acre cyber park. These developments are expected to position Kozhikode as the third IT hub in the state after Thiruvananthapuram and Kochi—resulting in a potential shift in the economic base of the city (Fig. 10.11).

Employment
The district has a workforce participation rate of 31% which is comparable to the districts in the region, but fall far below the districts towards the south of the state. A possible reason for this is the under developed industrial scenario and subsequent lack of employment opportunities. As evident in the figure, employees prefer private sector opportunities over the public sector. Along with this, Kozhikode ranks the third with respect to number of professional and technical job seekers—out of which majority are women (Fig. 10.12). Hence this region presents an immense opportunity with the presence of a large untapped pool of labour—both male and female—with preferences of working in the private sector enterprises. Economic decline, unemployment growth and high migration have made it important to reconsider economic strategies and transform Kozhikode.

10.4.6 Education

Kerala's experience of high human development has received international attention and this is mainly attributed to significant achievements in the education and health sectors. Education has played a central role in determining the state performance in social development.

Number of schools and management
The schools in the state are divided by stage into high school, upper primary, and lower primary and by management into government, private aided and private unaided. Kozhikode has 1269 schools which accounts for approximately 10% share of the total schools in the state. Out of the total number of schools in the district,

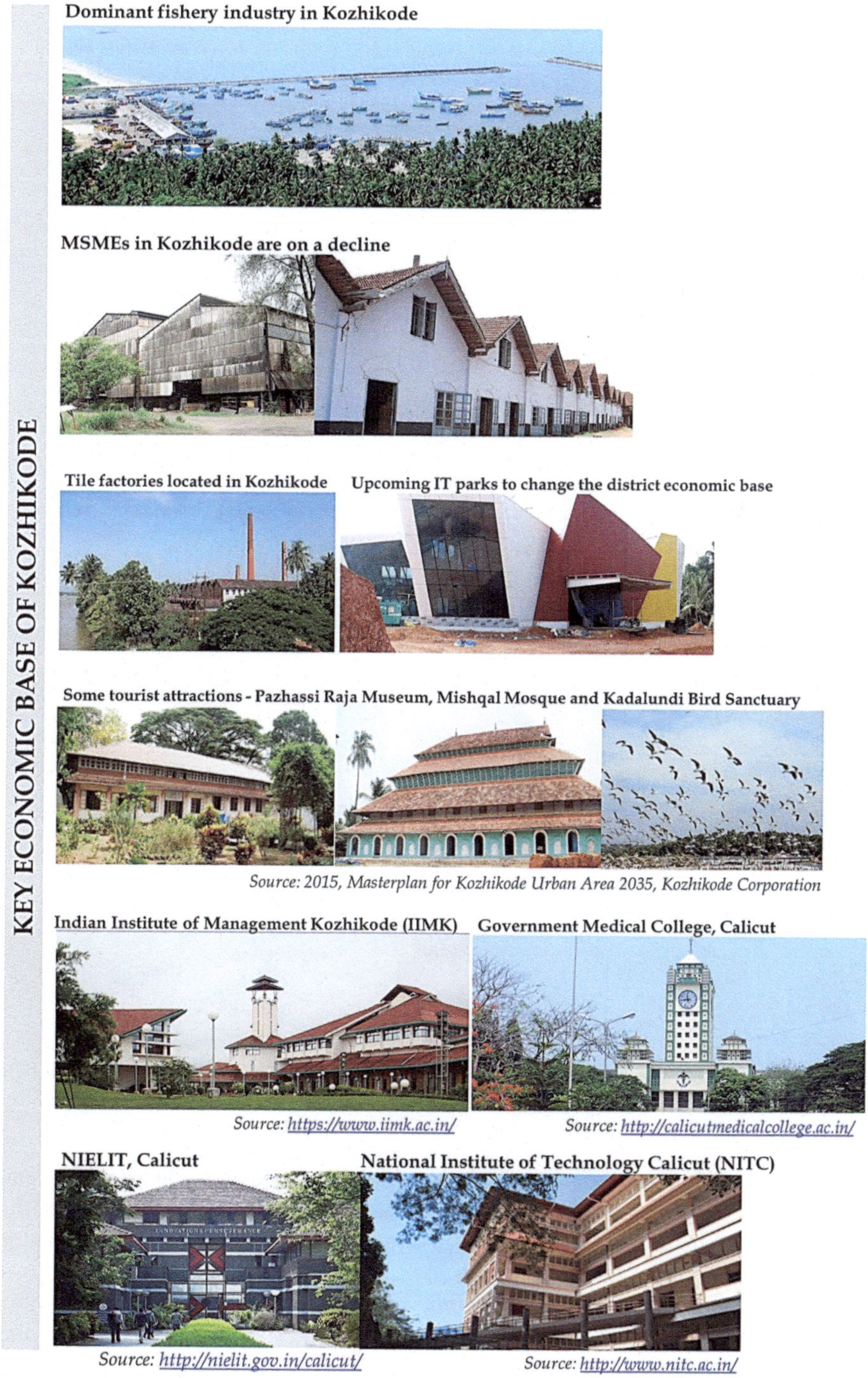

Fig. 10.11 Major economic activities in Kozhikode

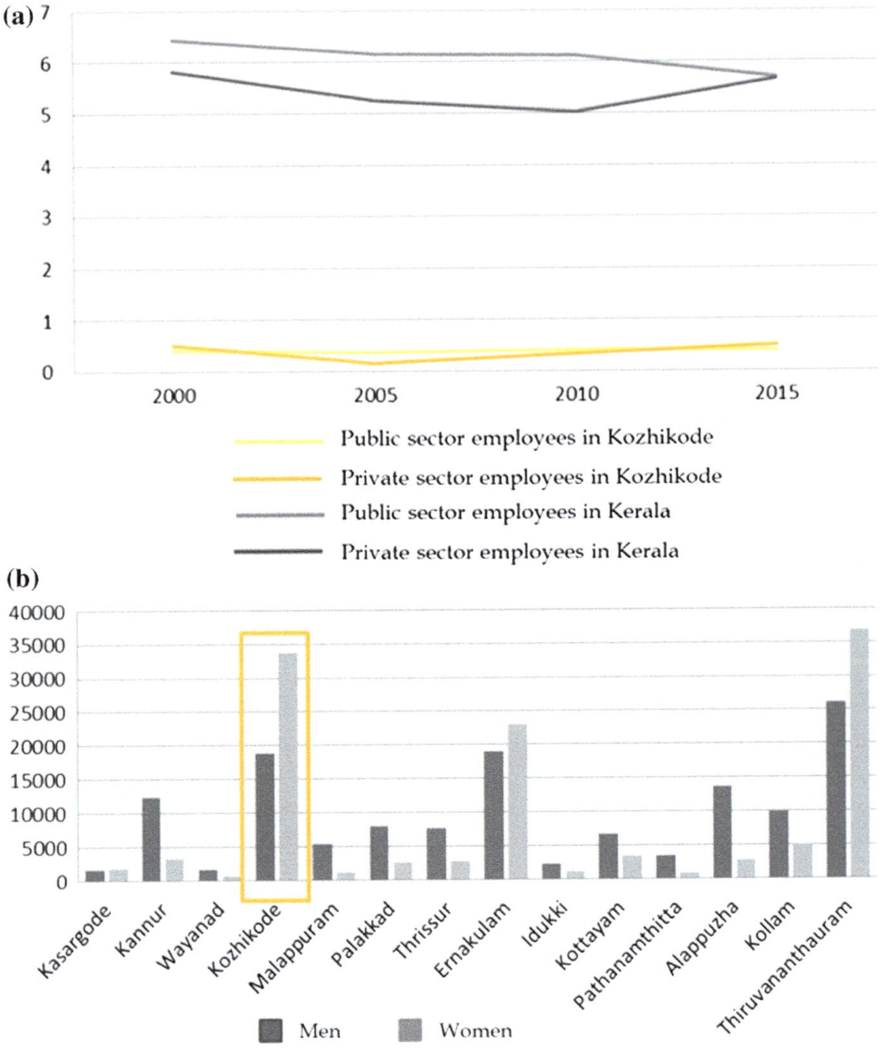

Fig. 10.12 **a** Growth rate of public and private sector employees in Kozhikode and Kerala; **b** Professional and technical job seekers by district (2016); *Source* [29]

lower primary schools account for 58% share followed by upper primary (25%) and high schools (16%) (Fig. 10.13). Approximately 50–75% of the schools in the district are private aided followed by government managed and private unaided management (Fig. 10.14).

Enrolment, dropout rate and uneconomic schools

The district has the second highest enrolment rate in the state after Malappuram and ranks seventh with respect to drop out rates. Enrolment is highest in high school and drop out is the highest in lower primary. District wise analysis indicates that

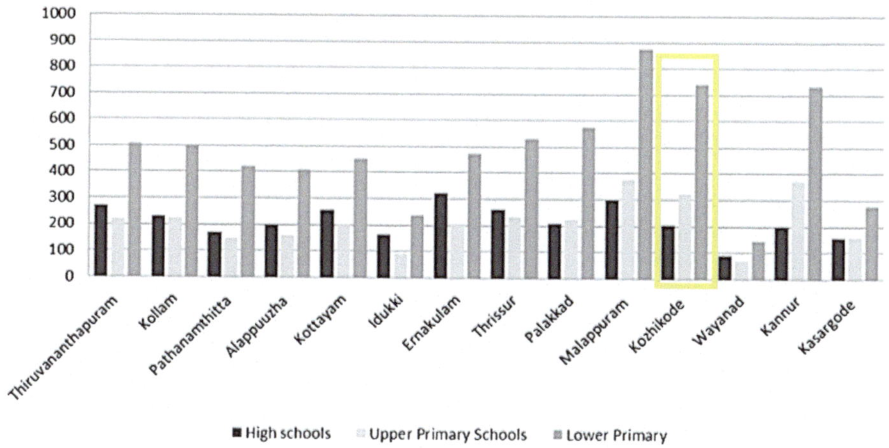

Fig. 10.13 Distribution of schools by stage across districts; *Source* [29]

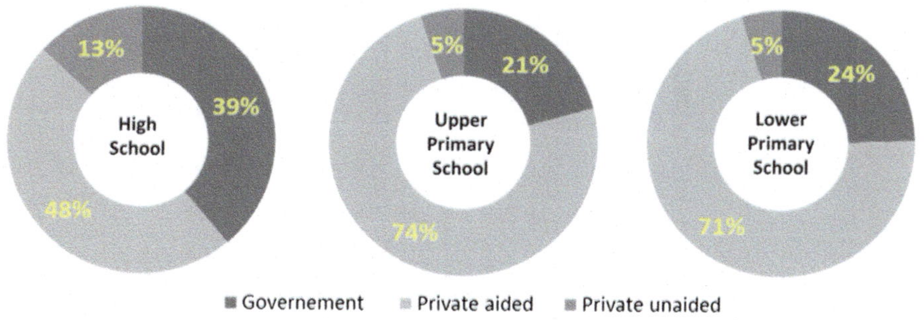

Fig. 10.14 Distribution of schools by management in Kozhikode; *Source* [29]

Kozhikode has the second highest number of uneconomic schools in the state after Kannur. Of the total number of uneconomic schools in the district, private aided schools account for approximately 72% (Fig. 10.15).

10.4.7 Health Care

The health care sector is very well developed in the region and the types of medical facilities include government hospitals, super speciality private hospitals, primary health care centers, taluk hospitals, along with ayurvedic and homeopathic hospitals and dispensaries (Fig. 10.16).

The service area of the health sector within the planning area extends to the district and adjoining districts, thus increasing its importance. The maps below indicate the spatial location of the existing allopathic, homeopathic and ayurvedic medical facilities.

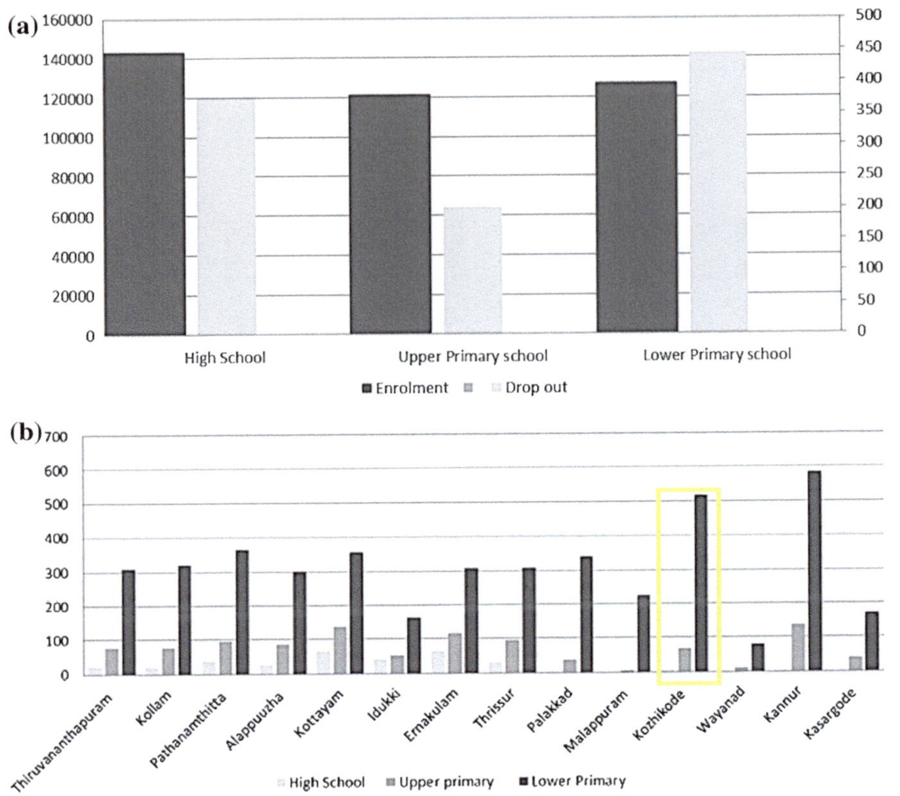

Fig. 10.15 **a** Enrolment and drop - out rates by school stage in Kozhikode; **b** Distribution of uneconomic schools by district; *Source* [29]

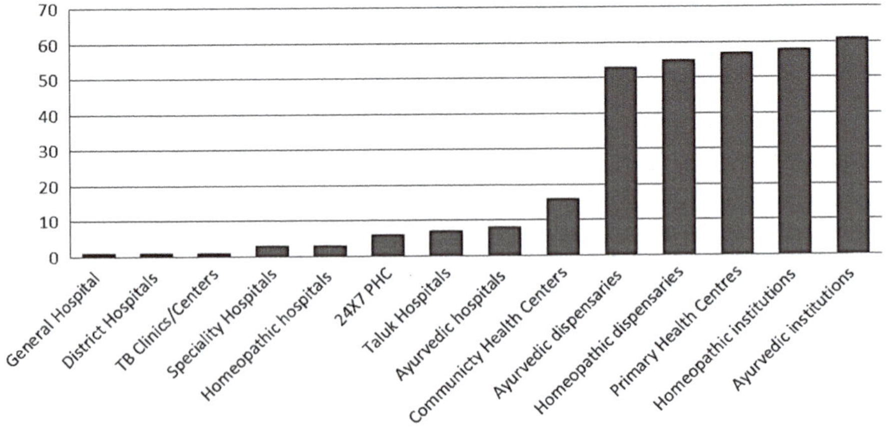

Fig. 10.16 Medical facilities in Kozhikode (2016); *Source* [29]

While the medical facilities in the planning region services areas beyond the planning region, the location of majority of the hospitals are concentrated towards the south of the planning area in the corporation area—limiting the services provided in other areas. This offers a potential for medical centers in local bodies that lack healthcare facilities (Fig. 10.17).

10.4.8 Land Use

The major land uses in the region are residential and agricultural mix and residential land use. These land uses contribute to around 64% of the land use distribution in the region [25]. Other land uses considered are mostly rural land use such as water bodies, dry agriculture, reserved forest, forest, marshy land and wet agriculture and these together accounts for 36% of the region [25] (Fig. 10.18).

10.5 Renewable Energy Generation and Deployment

The smart development within the Kozhikode Metropolitan Region is based on the fundamental concept of using renewable energy source as a means to power the economy. The first step towards this is to identify the potential source(s) of renewable energy. This has been carried out based on the renewable energy calculations carried out for solar energy, wind energy, hydro thermal energy in Kerala, by Agency for Non-conventional Energy and Rural Technology (ANERT) [30]. Based on these calculations, solar energy has been identified as a viable source of renewable energy for the region. This next step is to divide the region into renewable energy zones based on the various identified parameters. The renewable energy generated in each zone is calculated which forms the total energy supply in the region. Similarly, the energy required for each economic activity is also calculated based on infrastructure requirements and the sum of these energy requirements result in the energy demand for the region. The difference between the energy demand and energy supply indicates the energy demand gap. Energy demand gap is calculated by subtracting the energy supply from energy demand.

Step 1—Identification of solar energy potential for the region
The potential for solar, wind and hydro energy have been analysed based on the assessment carried out by ANERT), for various districts in the state. The analysis indicates that Kozhikode can capitalize on solar energy as a source of renewable energy for powering development in the region. The assessment of solar potential (via solar photovoltaic) in Kerala carried out by ANERT identifies two categories of surfaces:

(a) individual households
(b) institutional/commercial establishments.

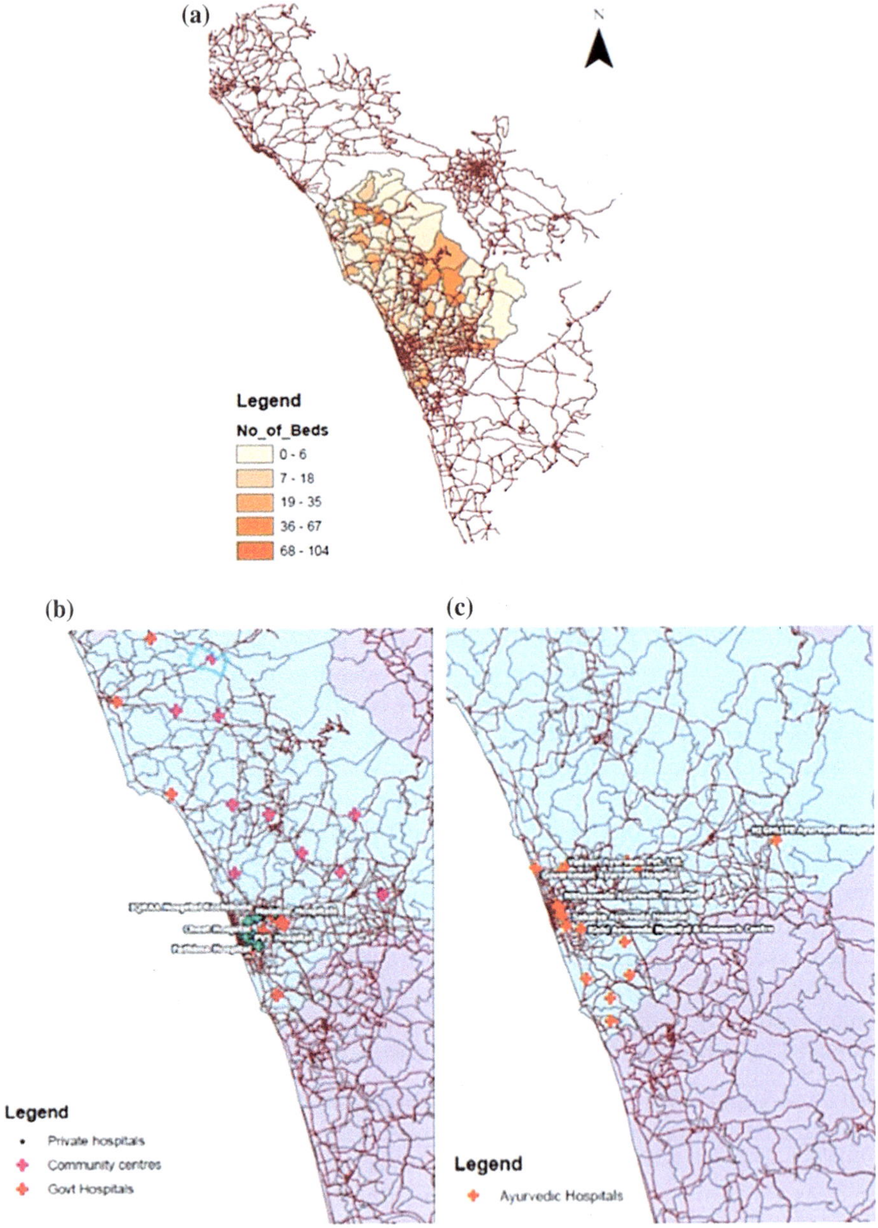

Fig. 10.17 **a** Location of public medical facilities in Kozhikode; **b** Location of private medical facilities in Kozhikode; **c** Location of ayurvedic hospitals in Kozhikode; *Source* [29, 27]

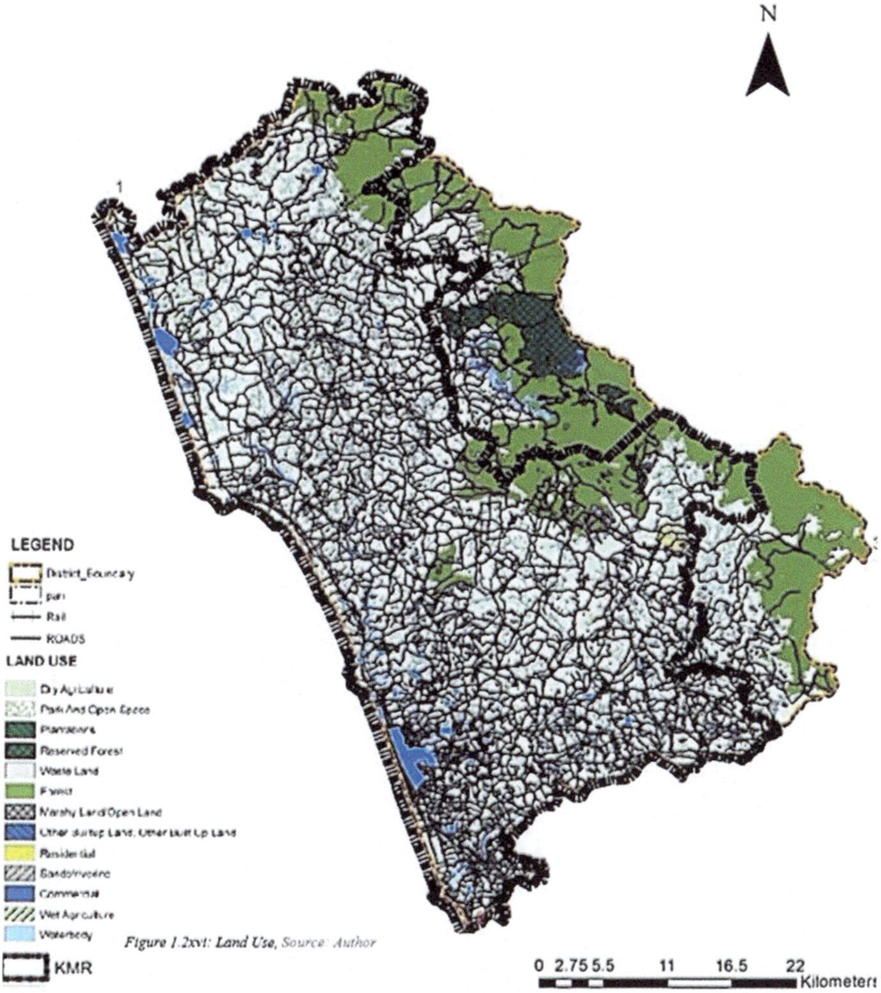

Fig. 10.18 Distribution of Land use in Kozhikode Metropolitan Region; *Source* Compiled from [25, 27]

Spot surveys of identified households with favourable rooftop systems across rural and urban households in Thiruvananthapuram, Kochi, Alleppey and Palakkad were carried out for assessing feasibility of on-site implementation. Analysis indicated that for urban households, 20% of the roof tops were partially or fully shaded, 10% were fully shaded and the remaining 70% of the households had clear shade—free roof areas—thus 90% of households can use Solar PV. As per the report, Global Horizontal Irradiance is high (above 2100 kwh/m^2) for Kozhikode and the figures for Direct Normal Irradiance fell in the moderate zone (1800–1900 kwh/m^2). Thus the region has a good potential for using solar PV for its households and establishments (Fig. 10.19).

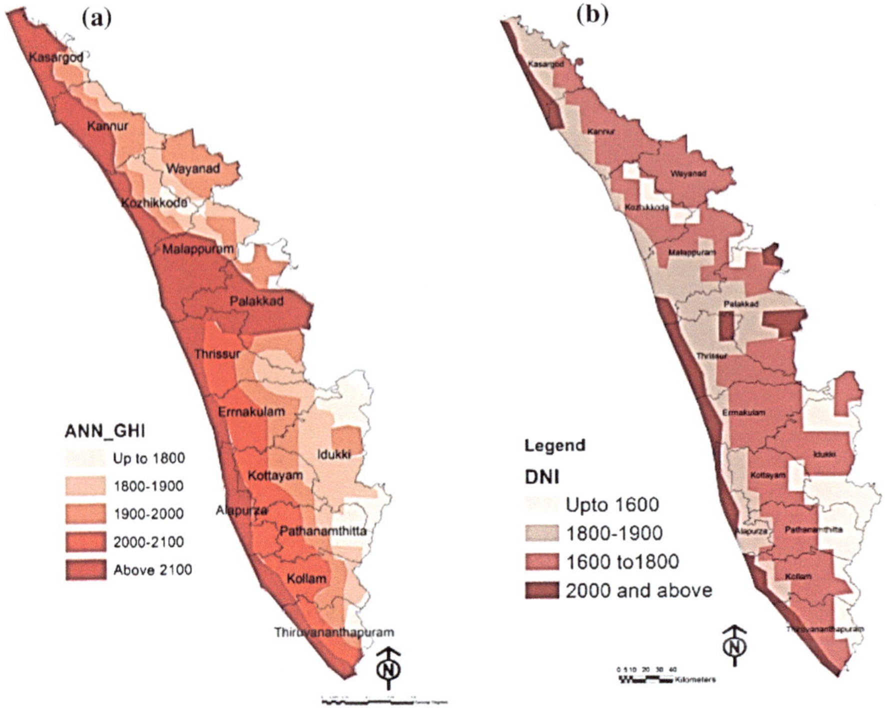

Fig. 10.19 **a** Global horizontal radiance for Kerala; **b** Direct Normal Radiance for Kerala; *Source* Compiled from [30, 27]

Step 2—Solar energy zones in KMR

Solar energy zones for the region have been identified based on the following parameters

(a) ***Solar energy potential areas***—Of the total census houses in the region, approximately 70% of the houses have potential roof top areas for solar energy generation. District level data has been interpolated at the local body level based on the population and number of households of each local body.

(b) ***Distribution and density of census households***—The figure below indicates that the municipal corporation area has the maximum density with 2000–2500 houses per km^2 followed by the city centers with 1000–2000 houses per km^2. The density gradually decreases from the coast line on the west towards the eastern boundary of the region.

(c) ***Connectivity* via *major transport corridors***—Additionally, major transport corridors and geographic features have also been taken into consideration as these become the basis for the energy distribution network. The density along the transport corridors are within the range of 500–1000 houses per square km^2 (Fig. 10.20).

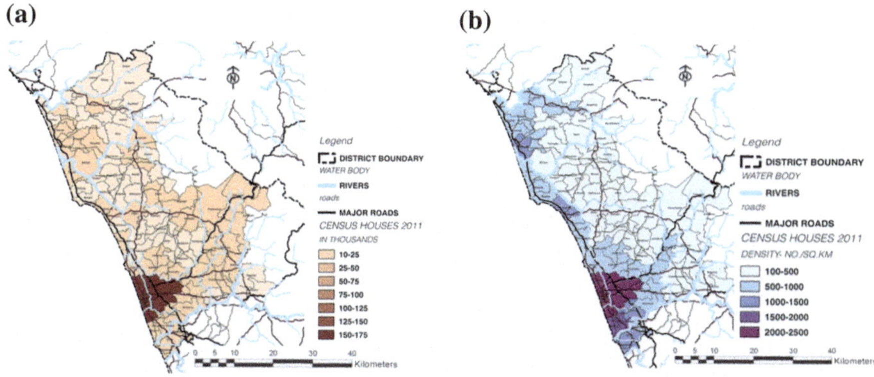

Fig. 10.20 **a** Distribution of census houses by number in KMR (2011); **b** Distribution of census houses by density in KMR (2011); *Source* Compiled from [2, 27]

On this basis, a total of 8 solar energy zones have been formed namely Zone A—Kozhikode, Zone B—Koyilandi, Zone C—Vadakara, Zone D—Kunnamangalam, Zone E—Mukkam, Zone F—Balussery, Zone G—Perambra and Zone H—Nadapuram (Fig. 10.21).

Step 3—Solar energy supply within KMR

Analysis of roof surfaces in the houses of Kerala indicate that 70% of the houses in the urban areas of Kerala are unshaded and are suitable for producing solar energy using roof top solar panels and 20% of the urban houses are partially shaded—these can also be used for solar energy generation [30]. Various factors were considered for solar energy calculation namely, (i) Total number of census houses in the region (ii) Slope (iii) Aspect (iv) Efficiency of available solar panels (v) Average roof area.

The slope analysis for the region indicates that the coastal areas, with slope less than 3%, are most suited for power generation. The results show that southeast and southwest orientations of solar panel will help in generating the maximum possible solar power. An aspect analysis was also carried out which was subsequently overlaid on the slope. Based on this, regions with slope of less than 5%, south east and south west orientations were identified (Fig. 10.22).

The Kozhikode Municipal Corporation area has the highest solar energy generation potential. Two clusters of high solar energy generation potential had emerged. One is in the southern part of the region, which includes Kozhikode Municipal Corporation and its surrounding areas—Elathur, Feroke, Beypore, Olavanna and Cheruvannur-Nallalam. The other is in the northern region of the planning area and extends from Vadakara to Payyoli. The maximum solar power that can be generated in each zone is summarized in the (Table 10.5 and Fig. 10.23)

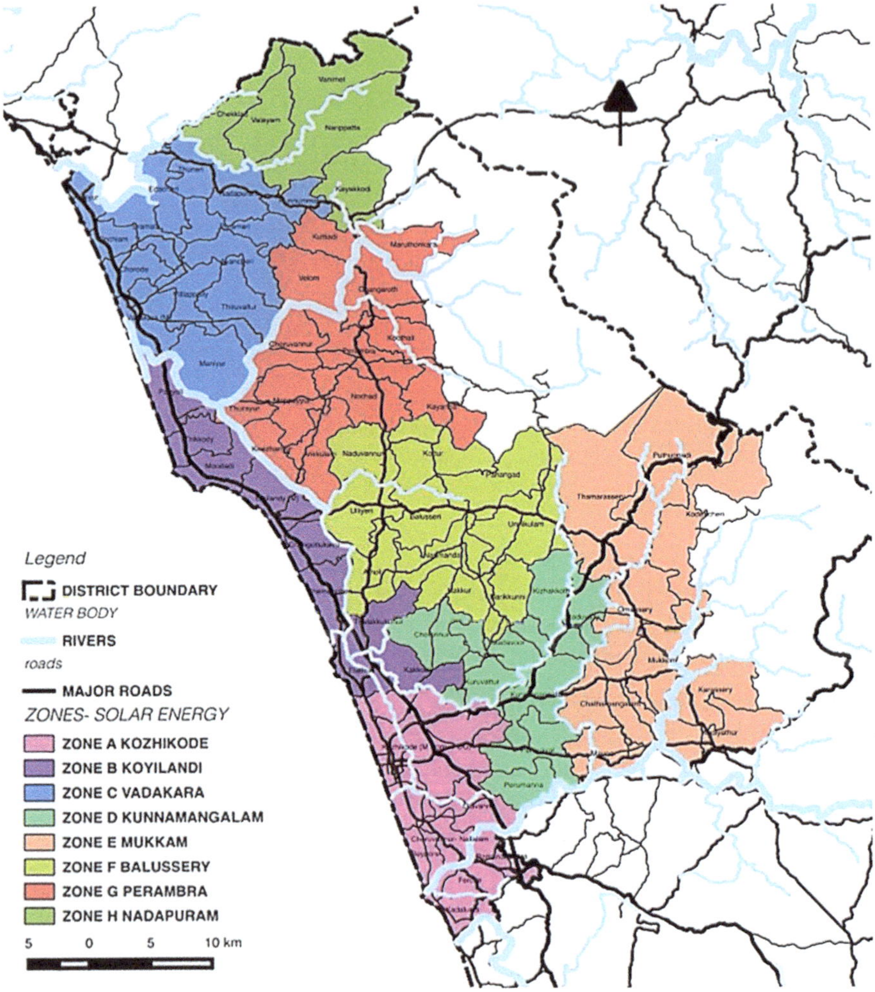

Fig. 10.21 Solar energy zones identified for KMR; *Source* [27]

10.6 Transformation of Industry to Smart Economy

This study advocates transforming existing industries to smart and ICT enabled and more economically efficient than bringing in exotic industries from outside. This will help in shaping the local talents in a smart way and reap benefits for local people without replacing them with other superior industries, capital and relate

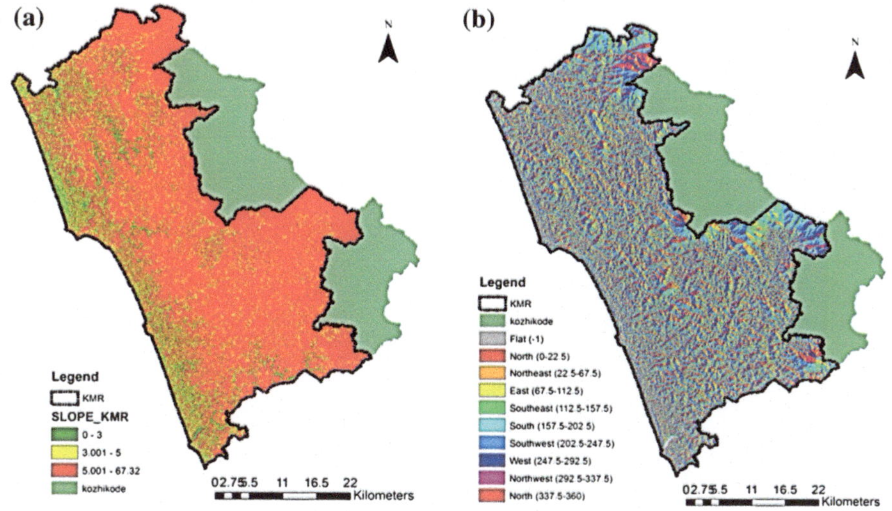

Fig. 10.22 **a** Slope analysis on KMR; **b** Aspect Analysis in KMR; *Source* Compiled from [30, 27]

Table 10.5 Solar energy generation by zone

Name of zone	Energy generated (MW/year)
Zone A—Kozhikode	65,060
Zone B—Koyilandi	28,167
Zone C—Vadakara	36,095
Zone D—Kunnumangalam	24,981
Zone E—Mukkam	25,405
Zone F—Balussery	25,308
Zone G—Perambra	23,467
Zone H—Nadapuram	9193
Total	237,676–240,000

Source [27]

manpower from outside. The major industries that have been selected as the economic base for the development of KMR have been divided into the following (Table 10.6).

The industries have been selected based on a mix of criterion namely growth rate, future investments in the industry and potential for it to be developed in the district, options for sharing economy, potential to develop in a smart economic base. Sector specific criterion have also been identified and analysed for the selected process (Table 10.7).

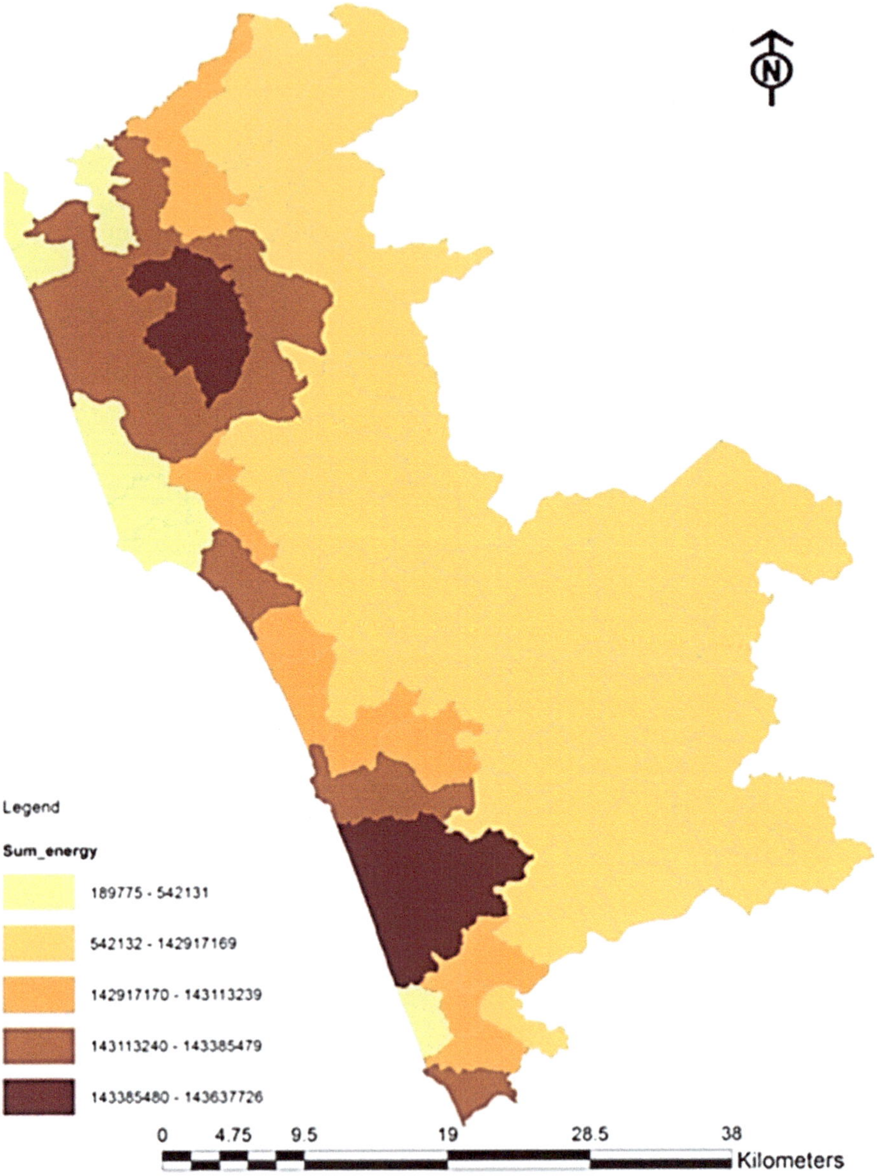

Fig. 10.23 Solar energy generation by zone; *Source* [27]

10.6.1 Revitalizing Tourism in KMR

Kozhikode is famous as a tourist destination owing to its importance as a historic place, presence of abundant natural features such as beaches, inland waterways and

Table 10.6 Industry selection for KMR

Sector	Industry/sub industry
Primary	Fishing and aquaculture
Secondary	Manufacturing *Wood based industries* *Food processing industries* *Leather and footwear industries* *Coir and agro based*
Tertiary	Trade and commerce Tourism IT/ITeS Education Healthcare

Source [27]

Table 10.7 Selection criterion for industry and sub industry for KMR

Sector	Industry/sub industry	Selection criterion
Primary	Fishing and aquaculture	• Growth rate of industry specific GSVA • Future government and private sector investments within the • Availability of raw material • Potential of the industry to develop in a smart economic base • Potential for including sharing economy options • Efficiency of existing transport network • Available workforce to be integrated into development • Computer literacy in women and children • Availability of academic and medical workforce
Secondary	Manufacturing *Wood based industries* *Food processing industries* *Leather and footwear industries* *Coir and agro based*	
Tertiary	Trade and commerce Tourism IT/ITeS Education Healthcare	

Source [27]

highlands along with other attractions such as ayurveda, kalari and Malabari cuisine. The spatial distribution of the tourist spots in the region also indicate a linear pattern with key attractions located along the coast and transport corridors in the region. Tourists in any destination are divided into leisure and business categories based on their purpose of visit and a large majority of the tourists visit Kerala for leisure purposes. Following this trend, the tourists visiting Kozhikode fall into leisure visitor category and majority of the attractions frequented include the beaches, historical sites such as museums and palaces. The department of tourism categorises these destinations into heritage, cultural and backwater tourism (Fig. 10.24).

Tourism statics from the state department of tourism indicate that the number of tourist arrivals grew by 4% over 8 years from 2007 to 2015, of which foreign tourist only account for 1–2% [31] (Fig. 10.25). Tourist arrival statistics for the year 2014 indicates that 45.5% of foreign tourists visited the Kozhikode city region followed by the emerging destination of Iringal. This emerging place is developed

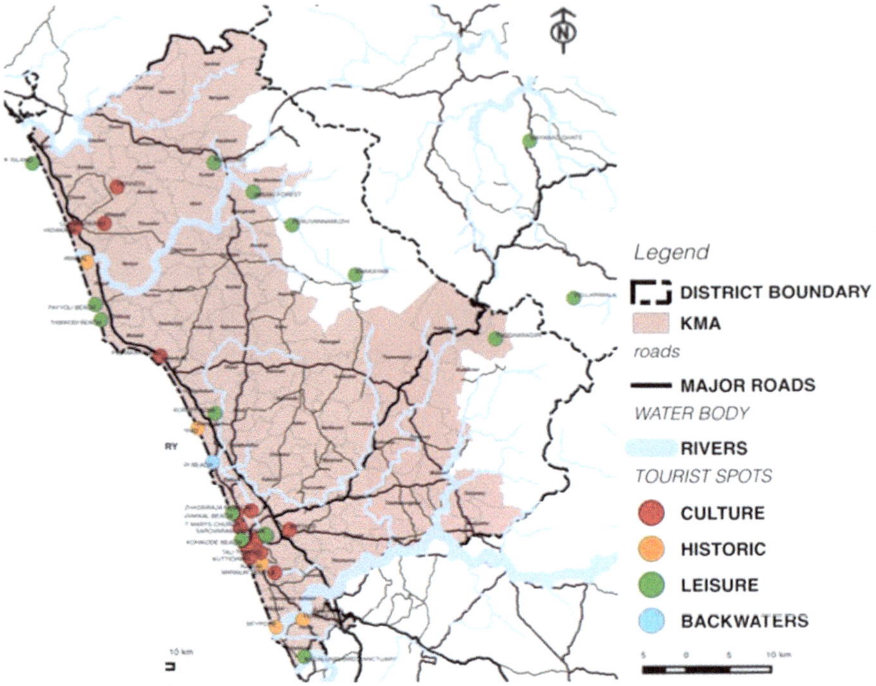

Fig. 10.24 Major tourist spots in KMR; *Source* Compiled from [31, 27]

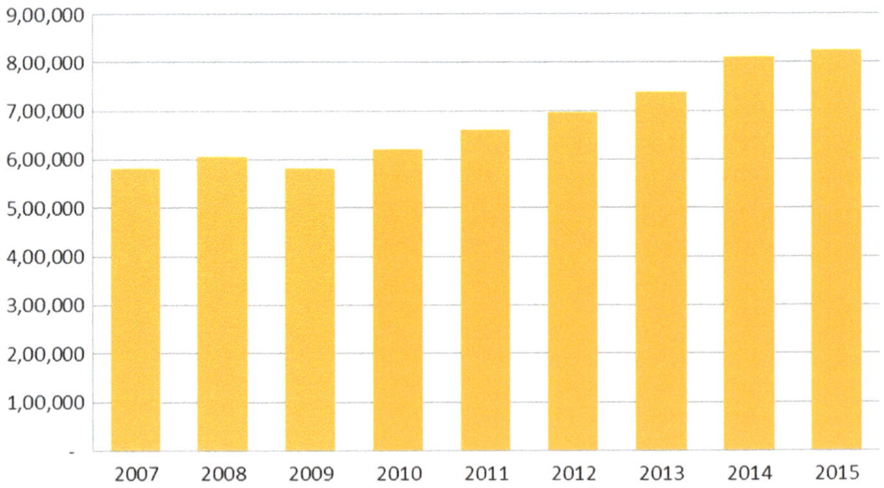

Fig. 10.25 Growth of tourist arrivals in Kozhikode; *Source* [29]

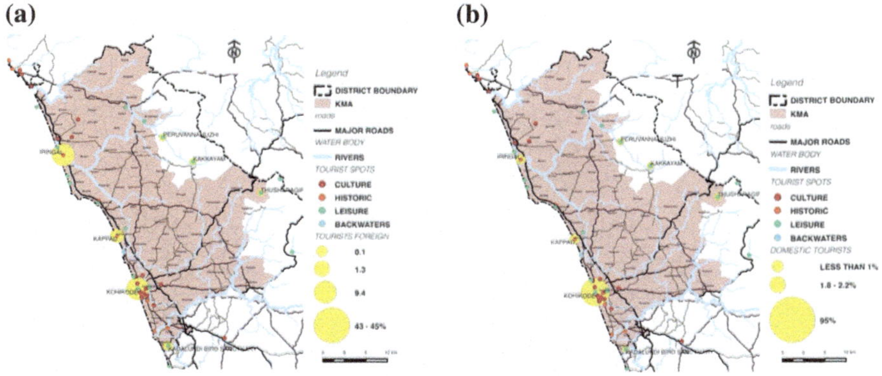

Fig. 10.26 **a** Foreign tourist arrival by destination; **b** Domestic tourist arrival by destination; *Source* Compiled from [31, 27]

as a craft village for the district attracting local artisans from all over the region. Other destinations of Kappad, Kadalundi, Peruvannamuzhi, Kakkayam and Thusharagiri—famous for natural features like beaches, lakes and waterfalls—are grossly underdeveloped and attract less than 8% of the foreign to tourists. 95% of the domestic tourists prefer to visit the Kozhikode city due to ease of access to attractions within the area. Further, surrounding areas of Iringal and Kappad attract only less than 3% of domestic visitors. Hence it is evident that the region has the potential to develop tourism circuits by integrating a variety of destinations with the existing transport network and accommodation facilities (Fig. 10.26).

Conversion to smart economy

The region follows the conventional tourism value chain and the stakeholders include transport operators such as airline carriers, railways and taxi services, tour operators and booking agents of global, national and local scale along with hotels and restaurants. They cater to the travel preferences of the customers that range from high end to the budget segment. The value chain creates linkages between various stakeholders but these linkages also very high transaction costs as every step of the service—thus increasing the total cost of the trip for the traveller (Fig. 10.27).

Except for transport from the outbound destination, most segments of the tourism value chain are handled by various companies of diverse organizational, ownership and operational business structures. The conversion of this system into a smart tourism framework will help change consumer experiences while resulting in creative tourism business models. A smart tourism value chain will work on better tourist services, innovative marketing strategies and new collaborations [32]. The smart tourism value chain for KMR will work based on upgrades made in

(a) Product(s) offered—offering more diverse and higher value tourism products
(b) Process involved—incorporating technology in the operations

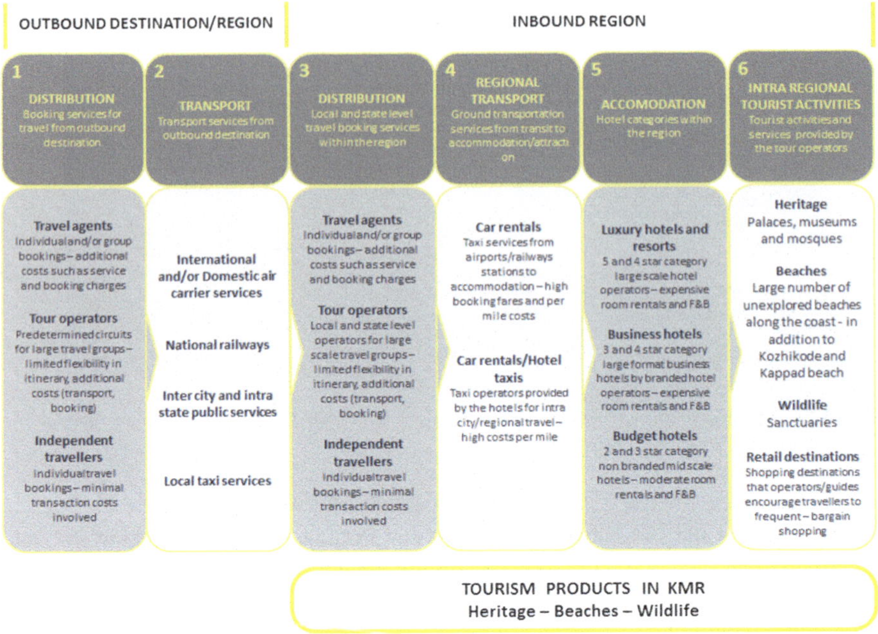

Fig. 10.27 Conventional tourism value chain; *Source* Adopted from [34]

(A) Product diversification
Kozhikode has approximately 3–4 large scale luxury hotels and resorts that attract travellers from the high end segment. These properties are primarily used for conferences, weddings and leisure stay. The budget hotels are concentrated in the corporation area of the region and they cater to primarily travellers stopping over in the city before proceeding to other destinations such as Wayanad.

Absence of hotel facilities in other locations—which may possibly lie along the off—beat path—has limited all tourist activities within the corporation area of the region. High capital costs and longer break even periods for large scale hotels and resorts have led to dwindling of investments in this segment. Hence, with limited investment in new tourism products have led to stagnation. A survey carried out by the state department of tourism in 2016 suggests that destination infrastructure needs to be further developed to revive the tourism sector [33]. Innovative facilities and business models for accommodation have taken over the hotel and restaurant segment. Studies by the Guardian Sustainable Business indicate that conventional mass tourism models need to be replaced with models that are more sustainable [33]. The region has a large number of unoccupied houses that have the potential to be developed into homestays. The homestay business has started flourishing slowly but steadily. In 2012, Kozhikode had 9 such facilities approved by the department of tourism [33]. While this has had a positive impact on small scale tourism business, the scope of these services can be further expanded to discreet location along unexplored trails in the region. Expensive restaurant food can be replaced by family and community managed eateries, thus providing employment to the local community. The homestays in these offbeat areas can be also be integrated with other prominent tourist attractions to create circuits around various themes—coastal, adventure, rural—while also providing a stable livelihood within the local communities.

(B) E-commerce platforms for streamlined processes
Both foreign and domestic tourists visiting Kozhikode choose to avoid uncertainty of travel by relying on tour operators and travel agents for providing destination related information, bundling tourism products, confirming and paying for reservations. These operations will be altered with the introduction of e-platforms on a wide scale. All booking services can be handled by online booking services such as Makemytrip, Expedia, Booking.com, Goibibo, Trivago etc. This not only provides direct access to the tourists but also facilitates in consolidating the large number of operators and agents into a few large online service providers. These systems will also help lower the costs of travel itineraries by streamlining the sales and allowing for last—minute purchases [34]. Many tour operators assure hotels of a fixed number of bookings per annum which may also include an initial down payment. This practice results in heavy price fluctuations due to competition among the operators to meet sales targets. Additionally, hospitality players are also not obliged to provide superior quality of service since the need for business is indirectly 'shared' with the tour operators Interactions with these operators indicate that a shift to ecommerce based transactions will also ensure that all facilities will maintain quality and reduce the wide disparity in prices. Another segment that can upgrade to ecommerce providers are the expensive car rentals and hotel taxis that will be replaced by Ola and Uber. These service providers provide last minute change of plans and flexibility in itineraries and standardized trip charges. Operators in Kozhikode indicate that approximately while 80% travellers use tour operators and travel agents, only 15% use online booking services with the rest 5% accounted by independent travellers and walk ins. This presents a large potential customer base that can be integrated into the system (Fig. 10.28).

10.6.2 Maximising the Fisheries Sector

The region has a coastline of 71 km and it coincides with that of the district making it the third longest coastline among all the districts of Kerala. It has 34 marine and 8 inland fishing villages. Kerala has a total of 24 fishing harbours of which 5 are

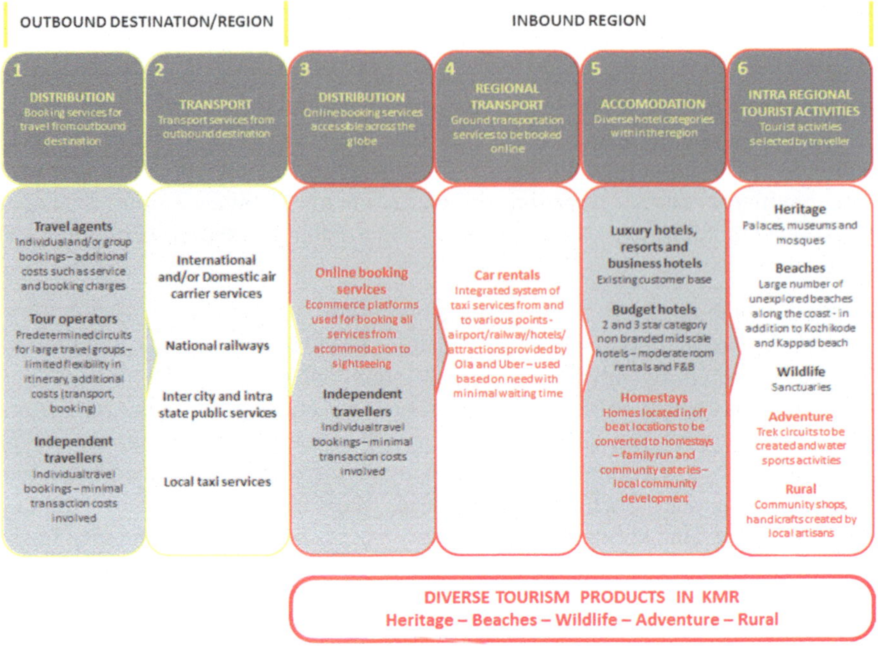

Fig. 10.28 Tourism value chain in a smart economy

located within KMR—the highest for any district. Kozhikode contributes to 12% of the total GSVA of the fisheries sector in the state. While Alapuzha and Kozhikode account for the same share of GSVA, the two top contributing districts of Kollam and Ernakulam account for 16% and 14% respectively. Active fishermen comprise of approximately 22% of the total fishing population (both marine and inland) of the district and this also accounts for 10% of the total active fishermen population of the state [35]. The district also accounted for 14% of the total fish produced in the state for 2015–2016 and ranks third after Kollam (18%) and Ernakulam (16%). 94% of the total fish produced in the district are marine fish and the district accounts for 18% of the total marine fish produced in the state (Fig. 10.29).

Conversion to smart economy

The fishery sector contributes to the coastal economy in generating income and employment to fishing communities, fish traders, processors and distributors. The conventional seafood and fishery supply chain in Kozhikode can be divided into four broad stages: capture, primary processing and collection, secondary processing and distribution and marketing. Most of the seafood is captured from the sea by boats and trawlers. Primary processing of cleaning and filleting takes place at the

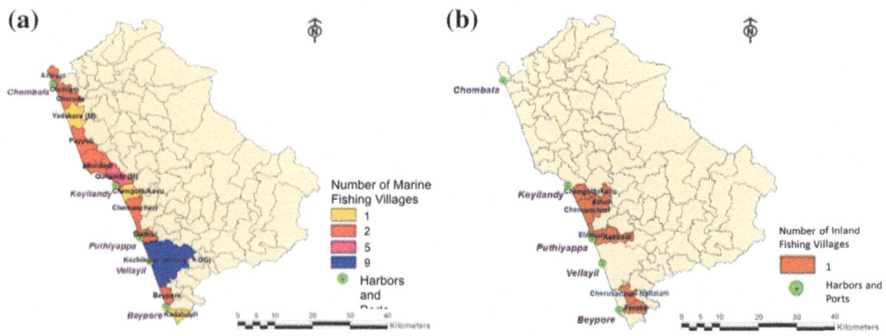

Fig. 10.29 **a** Number of marine fishing villages by local body; **b** Number of inland fishing villages by local body; *Source* Compiled from [35, 27]

fishing harbours and adjacent yards. Fresh seafood is collected at this point by large scale buyers who supply to hotels and restaurants, along with other regional consumers such as supermarkets and medium scale traders. Fresh seafood from the harbour is transported to the various processing plants for secondary processing activities which include chilling, freezing, canning and drying. Processed seafood products are exported to the overseas market via air and also sold to regional and local consumer base. The various stakeholders involved in the seafood value chain include active fishermen, handlers and assemblers at various selling points, people involved in processing activities along with transport and logistics personnel involved in distribution.

The conversion of the conventional fisheries value chain will entail intervention in two specific areas:

(a) Centralised processing and warehousing hub
(b) Introduction of online order and distribution systems

Secondary processing units in Kozhikode are spread across the district which leads to increased travel times of perishable food. The current delivery system of end products take places via various channels such as super markets, neighbourhood level markets and small scale fish mongers (door to door delivery). This entails a large number of handlers, whole sellers, and retailors which increases transaction costs at every step. A heavily decentralized food processing and storage system also requires efficient cold chain logistics to reduce food wastage. This can be rectified by (a) creating a centralized processing and warehousing hub (b) introducing online delivery system via efficient cold chain logistics (c) reducing the current seller system to two channels i.e. regional level online delivery system and small scale neighbourhood level markets (Fig. 10.30).

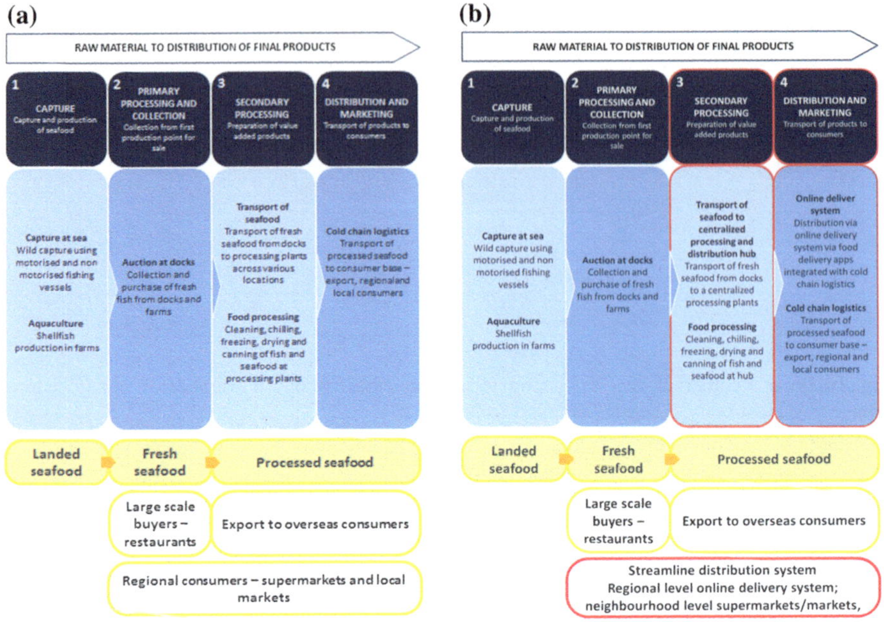

Fig. 10.30 **a** Conventional fisheries value chain; **b** Fisheries value chain in a smart economy; *Source* Adopted from [54, 55]

10.6.3 Reviving the Manufacturing Sector

The major manufacturing activities in Kozhikode include timber felling and processing, paper based products, leather products, rubber products, processed food products, coir products, granite, laterite and glass industries, ceramics and chemical and steel industries. For the purpose of this study select industries such as wood, coir, processed food and leather products have been considered for industry process transformation.

(A) ***Wood based industries***
Kozhikode Corporation has highest number of timber industries and Kallai is the most important centres for the timber business. The forests located in the south of Wayanad and Nilambur provide majority of the timber—varieties include high and mid value timber such as teak, rosewood, asan and jungle jack. The Malabar region is famous for the manufacture of furniture.

(B) ***Food processing industries***
Kozhikode is well known for processed food products such as jams, syrups, pickles, spice powders, etc. These products are mostly produced by medium, small and micro industries and these are currently located within the Municipal Corporation area. Majority of the food based industries are located in the Municipal Corporation and the surrounding areas.

(C) ***Coir industry***

Kozhikode has highest production of coconut compared to other districts with production of 948 million coconuts which accounted for approximately 16% of state production in 2014. Only about 20% of the husk is utilized for coir industry—remaining being used as fuel. Nearly 10,000 husks yield one tonne of fiber and an equal amount of pith—thus the region has the potential to manufacture around 94,800 tonne fiber. Despite this potential, the industry is faced with a number of issues, key being

(a) Non availability of labor
(b) Issues in production such as lack of adequate machinery, high power charges and intermittent power supply
(c) In adequate marketing efforts
(d) Additional charges such as transport, salesman, involvement of agent, middlemen charges, substitute goods, market rate fluctuations, inadequate storage facilities for coir and coir products

(D) ***Footwear industry***

Kozhikode accounts for the highest numbers of footwear factories are in the state. The footwear industry is an important contributor to the economy of Kozhikode and most of the footwear industries are located in Kozhikode Corporation, Feroke and Cheruvannur—Nallalam area. It has great potential for direct marketing to the customer through ecommerce. Among the total sale, flip flops, sandals and shoes account for 50, 30 and 20% of the sale, respectively [36] (Fig. 10.31).

Conversion to smart economy

The fundamental principal of converting any manufacturing activity from the conventional process flow to a smart economy process will be based on improved linkages between the producer and consumer. Based on the industry segment, the manufacturing activities may be completely centralized with manufacturing activities localized to central industrial units and sales via online platforms, maybe completely decentralized with manufacturing activity taking place at community levels and sale channel will be decided based on the catchment served—retail units for smaller catchments and online platforms for larger level catchments. Additionally, it is also proposed that some industry segments can combine both centralized and decentralized manufacturing—whereby manufacturing of larger products that are dependent on heavy engineering will take place in the centralized units and production of small scale items will be taken over by the community.

In conventional timber processing method, first level processed timber is transported from the forests to the various dealers across the region. The timber is further transported to various processing units where it undergoes further processing such as cutting and treating. The treated timber is transported to furniture manufacturing units where furniture components are manufactured on a large scale. These furniture parts are bought by sellers, both wholesale and retail sellers, located

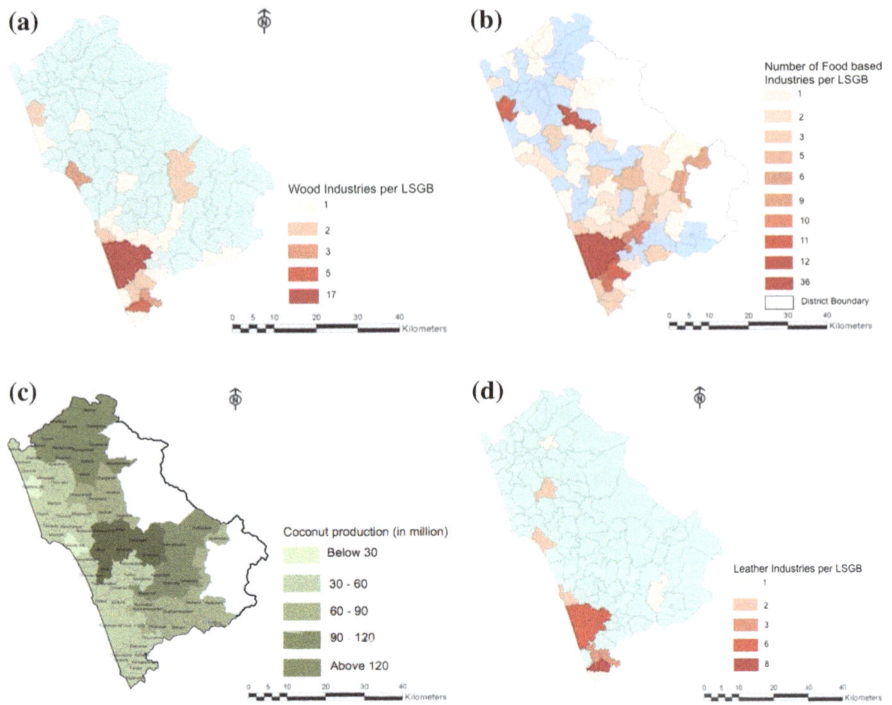

Fig. 10.31 **a** Existing location of wood industries in KMR; **b** Existing location of food based industries in KMR; **c** Existing location of coir based industries in KMR; **d** Existing location of leather industries in KMR; *Source* [27]

across various parts of the region—the parts are assembled and sold to the end consumer by order.

In a smart economy, direct transactions between the producers and consumers are enabled and the system will be managed through one or multiple e commerce portals. The consumer can place specific orders selected from an online catalogue. Based on the orders, the timber will be transported directly after primary processing to the manufacturing units. It is proposed to locate the manufacturing units and warehouses in close proximity to ensure economies of scale and efficient manpower and utility sharing. Additionally, warehouses can also have attached experience centres whereby customers can check the furniture in—person before placing the online orders. These experience centres can be located across the region and the scale can be decided based on the catchment area served.

The conventional manufacture of timber products entails large waste of timber. This timber 'waste' can converted to finished products by integrating into the household industries—this aids in (a) value addition and (b) integration of the community into an industry that was otherwise the domain of large scale manufacturers. The households can also take advantage of the ecommerce portals for sale of their products.

Food processing, in the conventional way, takes place across some key locations in the district. Majority of the processed food and beverage products are manufactured in large scales which entails heavy investment. This entails large use of preservatives and substitutes to increase shelf life of products. The 'smarter' manufacture of processed food and beverage products will be completely decentralized whereby communities can produce spice powders, pickles, oils and jams at the household level. These can be sold on online platforms to individuals and can also be transported to the number of bakeries and retails outlets located across the city.

The leather and agro based coir industry will also work in both centralized and decentralized locations. While the centralized industrial units will manufacture the bulk of the products that are primarily aimed for the export market, the local market will be served by the coir and leather products that are manufactured by artisans at the community level.

Sharing economy options for the manufacturing industry include online sales portals, online rental furniture options, shared rental of working spaces and machinery, shared storage units and warehouses among others (Figs. 10.32, 10.33, 10.34 and 10.35).

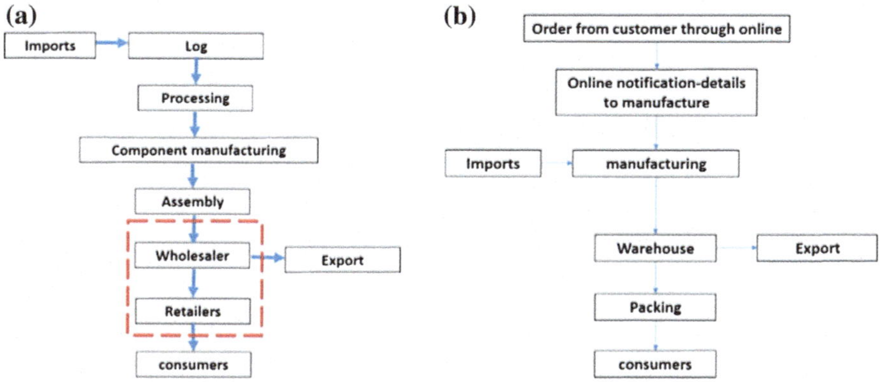

Fig. 10.32 **a** Conventional wood industry process flow; **b** Process flow of wood industry in a smart economy KMR; *Source* [27]

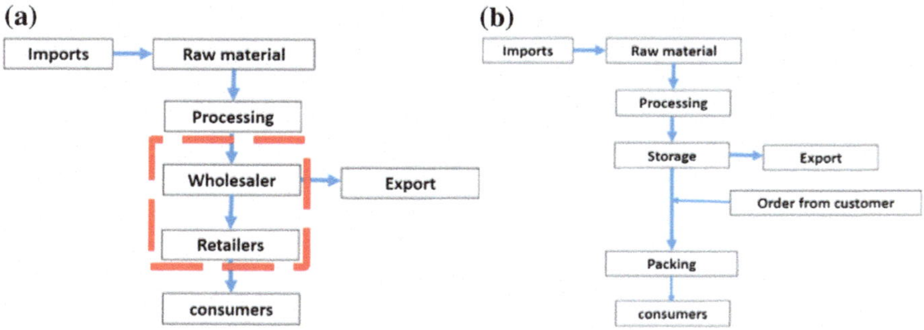

Fig. 10.33 **a** Conventional food and beverage industry process flow; **b** Process flow of food and beverage industry in a smart economy KMR; *Source* [27]

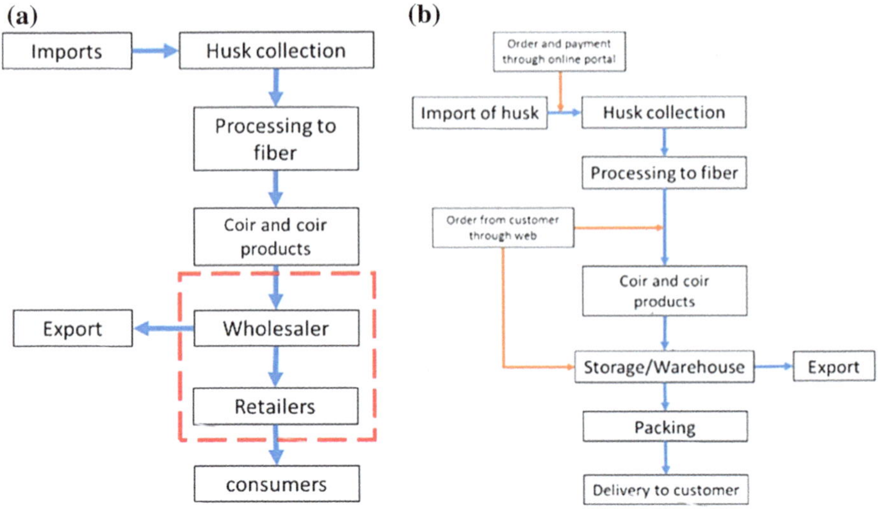

Fig. 10.34 **a** Conventional coir industry process flow; **b** Process flow of coir industry in a smart economy KMR; *Source* [27]

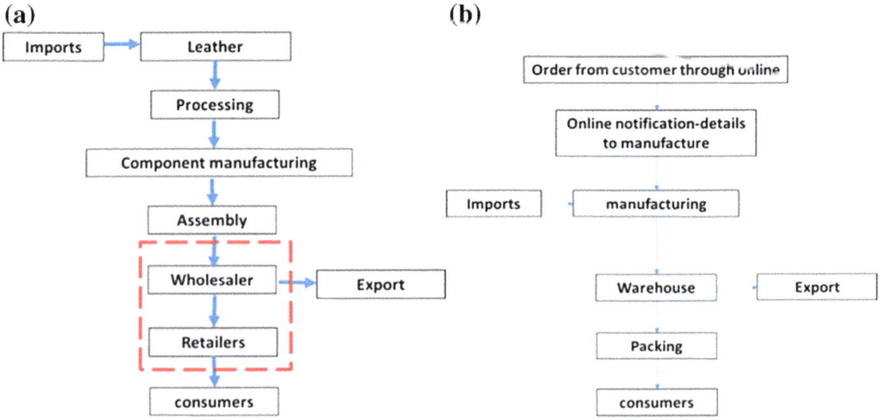

Fig. 10.35 **a** Conventional footwear industry process flow; **b** Process flow of footwear industry in a smart economy KMR; *Source* [27]

10.6.4 Robust Trade and Commerce

The trade and commerce segment in the region comprises of trade and retail establishments along with transport and allied logistics. As per the economic review of 2016, this segment contributes to approximately 22% of the total GSVA and it has also recorded a growth rate of 7% from 2013–2014 to 2015–2016 (Fig. 10.36). This segment is closely related to the growth of the small and medium

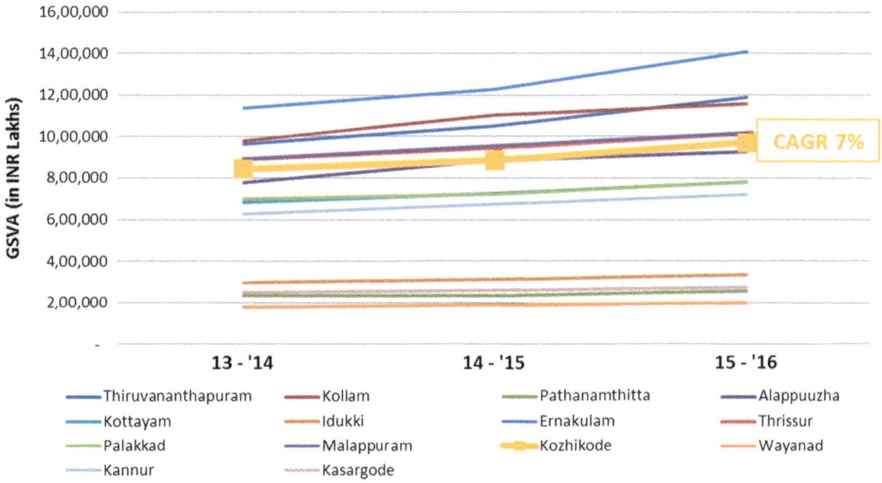

Fig. 10.36 Growth rate of trade and commerce by district (2013–2016); *Source* [29]

manufacturing industry and lack of investments in the manufacturing sector and new technology—both in the district and the state—has led to minimal growth in this segment.

The retail formats in Kozhikode have slowly changed from brick and mortar shopping to ecommerce retail. The brick and mortar retails formats include whole sale street shopping located in S M Street, and individual retail outlets and malls on Mavoor Road and Thondayad. Ecommerce shopping is still in its nascent stages and service area is primarily limited to the Corporation and its surrounding areas. These areas are catered by multi product sellers such as Amazon, Flipkart, Snapdeal etc.

The spatial distribution of the existing commercial centres is primarily governed by the distribution of transport networks and formation of subsequent nodes. Degree of development concentration has led to formation of one first order node in Kozhikode Corporation and second order nodes in Vadakara and Quilandy. Other than these, 14 other tertiary nodes are dispersed across the region. The entire population of the region is dependent on the Kozhikode Corporation for higher order services. Major retail establishments—ranging from large scale malls and multi branded showrooms to whole sale markets—are located in Mavoor Road, SM Street, Palayam, Valiyangadi, Nadakkavu and Thodayad. All these locations are easily accessible within a travel time of 60 min to 90 min from any point in the region (Fig. 10.37).

Need for additional nodes
A 60 km service area has been considered for the first order node as it was observed that people don't mind travelling up to 90 min to avail higher order services. Similarly, based on degree of services provided, service areas of 10 km and 5 km have been considered for the second and third order nodes, respectively. Using

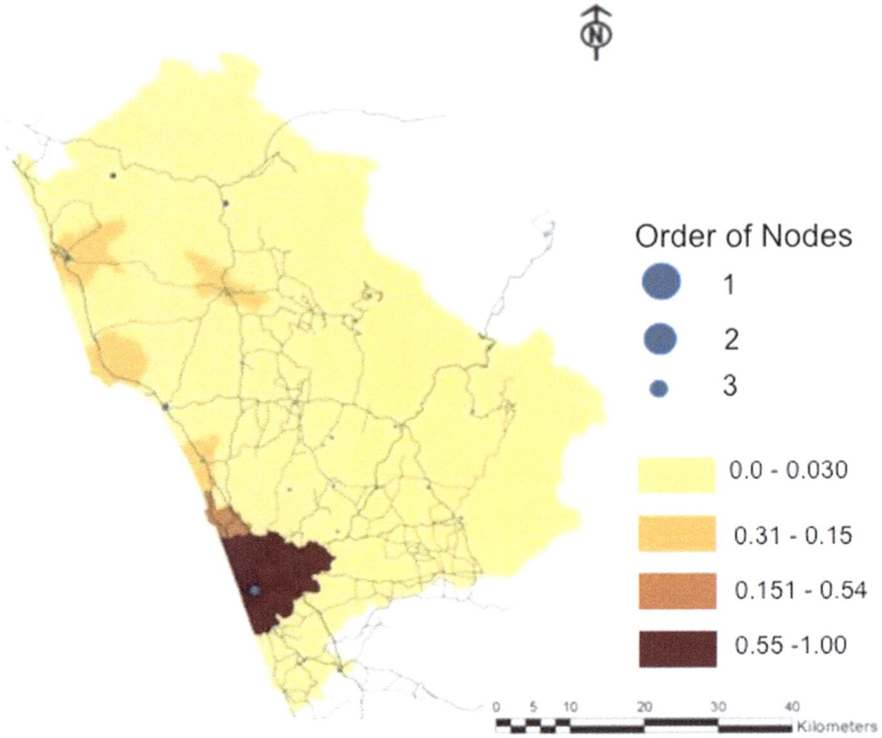

Fig. 10.37 Commercial nodes and linkages in KMR; *Source* [27]

these assumptions, it can be seen that the single first order node can service the entire region. While a combination of first and third order nodes ensure that the people in the region can avail higher order services within a travel time of 90 min, all lower order services can be availed within a travel time of 20 min. Hence, the number of existing nodes is sufficient to cater to the needs of the region and development of additional commercial nodes is unnecessary.

But nevertheless, the nearly flat temporal variation in the growth rate of the sector is a major concern. Erstwhile domestic trade segments of the region such as timber, spice, copra and agriculture, are declining and majority of the trade has been replaced by transport of commodities that cater to retail establishments in the regions. In order to maximize the operations of these commercial nodes and achieve higher growth rates, new retail formats should be explored and domestic manufacture and trade segments should be revived with the help of innovating technologies in the agriculture sector, better marketing and increased product range. Moving forward, it is envisaged that the current large scale retail spaces will function as experience centers for the various consumer labels and products that are sold via the online system (Fig. 10.38).

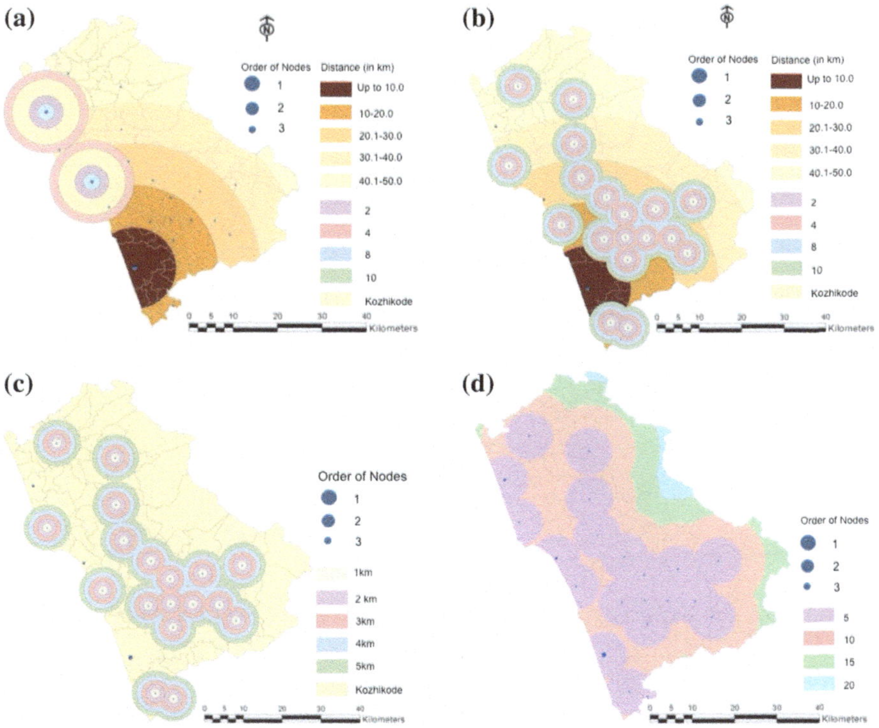

Fig. 10.38 **a** Service area for 1st and 2nd order nodes; **b** Service area for 1st and 3rd order nodes; **c** Service area for 3rd order nodes; **d** Service area for all nodes; *Source* [27]

Conversion to smart economy

Efficient trade and commercial activities is the backbone of the smart economy framework for the region and this segment will work on three pillars

(a) Well connected transportation system for the movement of goods
(b) Network of warehouses for storage space located in all commercial nodes
(c) Efficient online delivery system

The region is already serviced by a transport system of road and rail that can be utilized for well-organized system of goods movement. A network of warehouses is created by identifying unused warehouses for reuse and construction of new warehouses in the unserved areas. The Corporation area has a large number of unused warehouses. These—when retrofitted and reused—can be released as available storage space for commercial establishments. Additionally, new warehouses can also be constructed in the second and third order commercial nodes. The location of these storage spaces are governed by the service buffer and ease of its access from major transport networks. Using the above buffer analysis, each

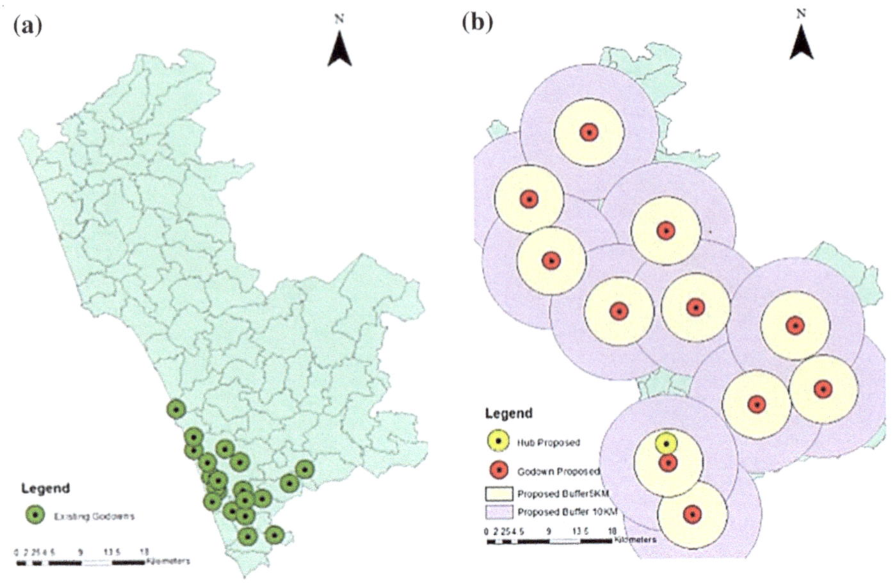

Fig. 10.39 **a** Location of existing godowns; **b** Location of proposed godowns and service buffers; *Source* [27]

warehouse can be accessed from any point in the region within a travel time of 30 min and this is aided by the efficient transport network, already in place (Fig. 10.39).

The conventional method of trade heavily relies on a number of intermediaries/middlemen ranging from assemblers, traders, distributors and shopkeepers. These intermediaries do not change the physical attribute of the product and their only function is to move the item from one point to another [37]. Each step entails added transport time and transaction costs that are ultimately loaded on the final product borne by the consumer.

In a smart economy framework, the products will be processed and sent to the nearest hub—which will then be supplied to the nearest warehouses within a travel time of 60 min. Consumers can place online orders for respective products that will reach their doorstep via smaller goods vehicles. An efficient online delivery system will reduce cost and ensure shorter delivery times by eliminating these intermediaries and restricting the transaction between the producer and end consumer. Perishable product manufacturers will also avail this system to meet the domestic and export markets. Cold food products will be transported and stored in refrigerated vehicles and warehouses and while domestic consumers will be supplied via a smaller mass transport system, exporters will use the nearest airport for overseas trade (Fig. 10.40).

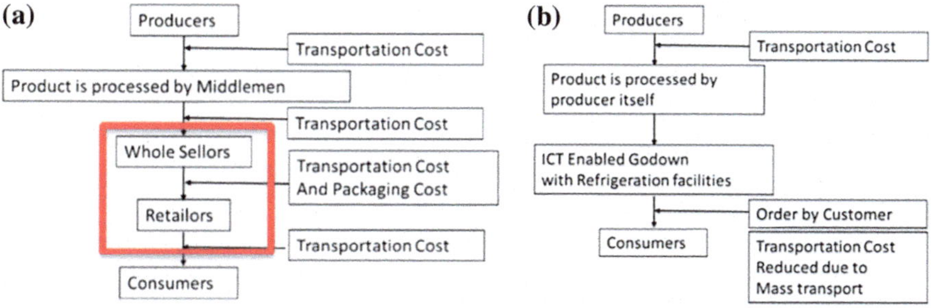

Fig. 10.40 **a** Conventional logistics process flow; **b** Process flow of logistics in a smart economy KMR; *Source* [27]

10.7 Zero Marginal Cost Society for Industry Sectors

10.7.1 Tourism

Zero marginal cost can be incorporated into the system on two specific services within the tourism industry namely, (a) transport services and (b) hotels and restaurants. The transportation of tourist can be through sharing taxi service or online booking of photo voltaic solar battery vehicles for taxi service. The power requirements of various activities in tourism can be generated from micro-grid community. The tourist can book the hotel/resort/homestay as their accommodation and can order the type of food they require so that they can cultivate as per the requirements of the tourists. The raw materials for food can be cultivated buy the micro-grid community through terrace farming and urban agriculture (Fig. 10.41).

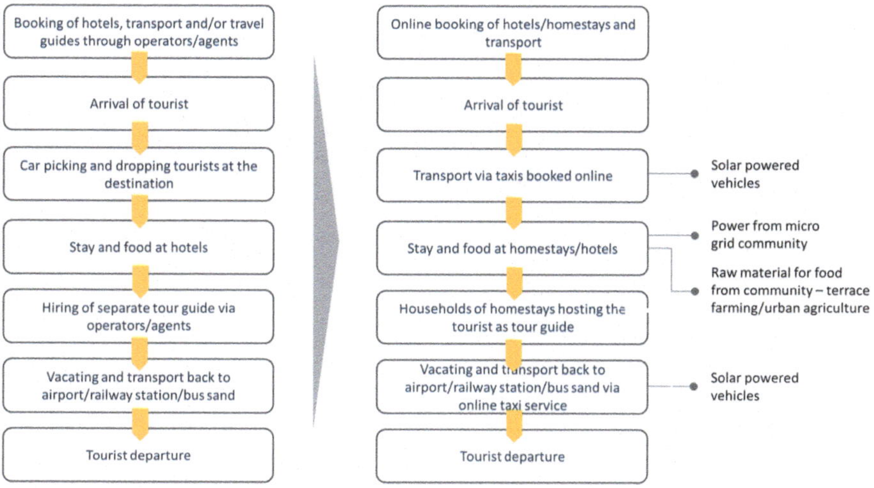

Fig. 10.41 Zero marginal cost strategies for tourism; *Source* [27]

Potential strategies to be adopted

- Homeowners sharing their apartments and houses with travellers via global online services such as Airbnb—bypassing commercial hotels.
- Providing free access to high speed internet at homestays.
- Practice urban agriculture and terrace farming for vegetable production and encourage traditional food at homes.
- People in the homestay to work as guide and/or host them for a particular tourist attraction.
- Car/Bike owners share their journey with tourists through online services like Ola, Uber etc.
- Encourage mobile restaurants which reduce use of land, infrastructure and overhead costs—thus increasing profitability.
- Linking directly to agricultural and fish traders for resources through online marketing.
- Reduce the wastage of food and make it useful for the needy.
- Encourage hotel management employment as a professional education for developing the skilled human resources.

Operation Sulaimani, a free food programme introduced to enable those who cannot afford a meal to have food with dignity and to avoid food wastage. Anyone can collect a Sulaimani Coupon from one of the authorised distribution centres like at taluk offices, selected hostels, railway station bus stand etc. which can be exchanged for a meal in one of the participating restaurants. A donation box is kept at these restaurants for the public to contribute to this programme. Any loss of revenue incurred by participating restaurants is hence compensated through the system where they match the coupons collected by each restaurants and distribution centres.

10.7.2 Fisheries

Zero marginal cost can be incorporated by reducing the large number of wholesalers and middle men from the system thus reducing the gap between the producers and consumers (Fig. 10.42).

Potential strategies to be adopted

- Online marketing and sale of fish so that customers get cleaned fish at their door step thus reducing cost of travel zero for household
- Fish sellers using their own vehicle can be the delivery persons. They can join the respective company and deliver the fish to the customers
- Delivery persons using electric powered vehicle can use power made available from solar panels connected to the warehouse or fish storage area—thus reducing dependence on petrol/diesel

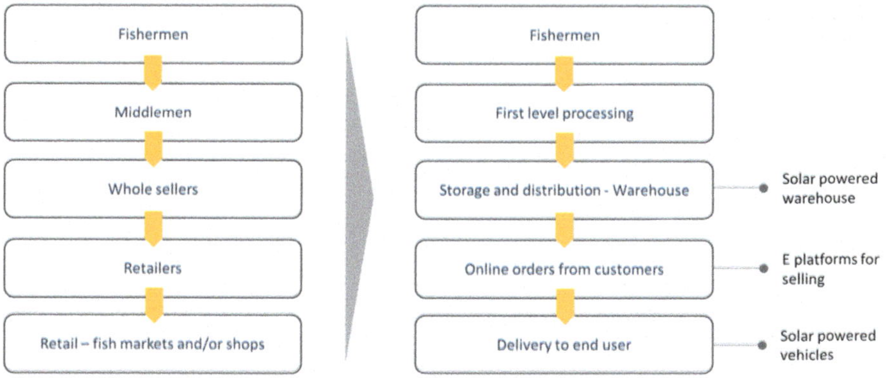

Fig. 10.42 Zero marginal cost strategies for fisheries sector; *Source* [27]

- Encouraging women in the fishing community—Collect, clean and send fish to the packing house. The quality of cleaning can be checked and the company can train them how to the cleaning to get good quality fish
- Collaboration between schools and fish markets will provide free fish for midday meals of school and in turn the energy requirements for cold storage of fish
- Boats fitted with solar panels can be used and charged batteries.

10.7.3 Manufacturing

In conventional practice the gap b/w producers and consumers is high because of that benefit for both of them are very low. Through zero marginal cost practice we can reduce this gap and can make more beneficial for both the consumers and producers. In the working system under zero marginal cost, the solar powered vehicles can be used to reduce the transportation costs and the power requirements of manufacturing units can be achieved from the micro-grid community through a renewable energy system. The marketing can be done through online services which reduce the transportation charges and the effective marketing with reasonable costs by the reduction of marginal cost to nearly zero.

The following flow chart showing the procedure in zero marginal cost society, it is clear that in this method trying to achieve a sustainable mode of process. Use of solar/Photovoltaic cells for meeting the energy requirement will help in reduce the cost production. Also, the use of online services makes it easy for producers and consumers (Fig. 10.43).

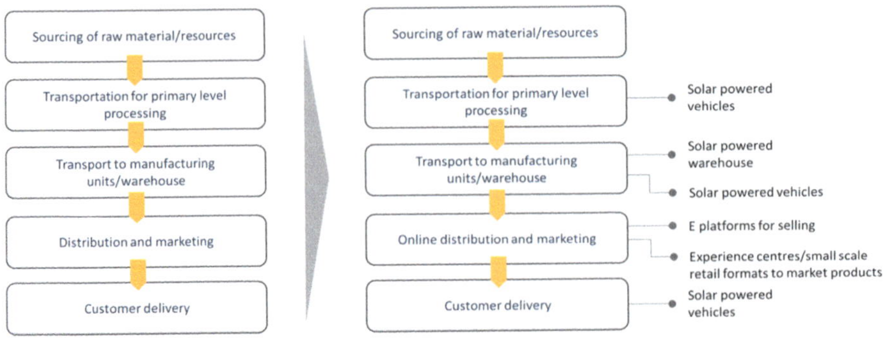

Fig. 10.43 Zero marginal cost strategies for manufacturing sector; *Source* [27]

Potential strategies to be adopted

- Utilization of renewable energy can reduce the transportation costs from raw material import to finished goods export.
- Decentralization of the activities into household level for reducing transportation costs
- Connection of all activities with the broadband backbone—results in monitoring and efficient group work
- Transportation of finished products at every stage to the existing industries for making the finished goods
- Online marketing and tie up with Flipkart, Amazon—warehouses of which are located in Bangalore and Hyderabad
- The capital cost of investment can be reduced by encouraging small scale household units as food processing units and directly linking them through online portals
- Solar powered vehicles for transportation for delivery services.
- Kudumbashree units can be linked with Micro-grid community for encouraging household food processing units by meeting the energy and power requirements through the micro-grid community.

10.7.4 Trade and Commerce

Moving forward the conventional logistics supply chain will be integrated with ICT infrastructure to result in a 'digitised trucking and logistics' system. This will transform the way freight is transported within the region and the technologies include information from the transport networks and other vehicles. These will improve utilization through remote maintenance, increase efficiency and boost safety. Ultimately, the system will be upgraded in a phase wise manner to result in a full automated and driverless trucking system (Fig. 10.44).

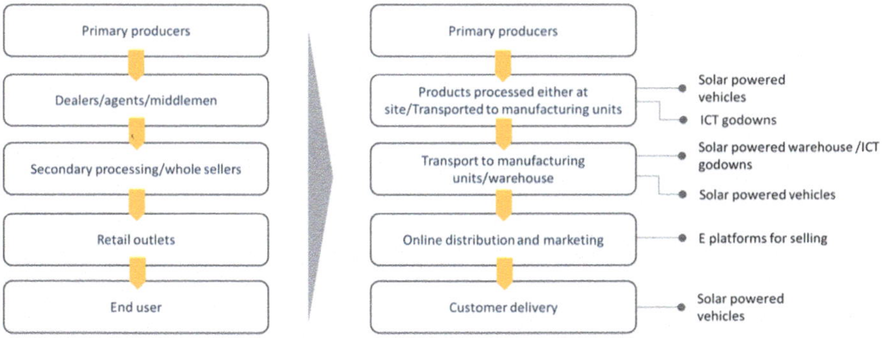

Fig. 10.44 Zero marginal cost strategies for trade and commerce sector; *Source* [27]

10.7.5 Education

The various components within the education infrastructure that entail capital costs include building, teachers and their wages, books, accommodation such as hostels and fees, commute to and from schools, and maintenance of the infrastructure. Conversion of this framework into a smart framework will include strategies as indicated in the infographic below (Fig. 10.45).

The building and infrastructure cost can be reduced by inviting voluntary organizations to contribute to the infrastructure development e.g.: PRISM (Promoting Regional Schools to International Standards through Multiple Interventions. ISRO has built the science lab while quality education and enhancement skills were developed by IIM-K. Free school transport can be provided by implementing solar powered vehicles. The services of IT companies can

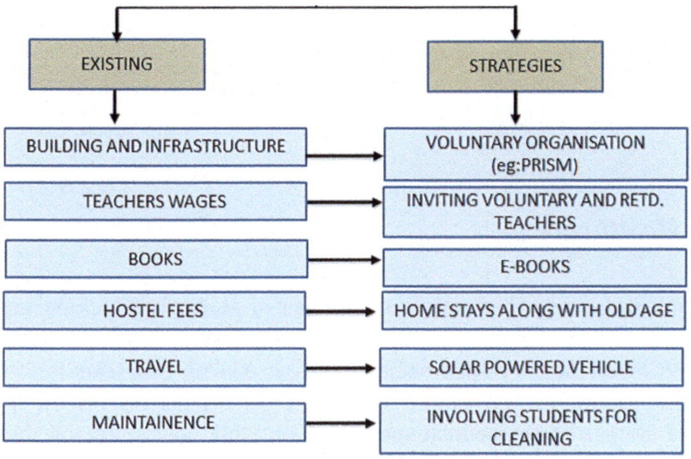

Fig. 10.45 Zero marginal cost strategies for education sector; *Source* [27]

Table 10.8 Cost savings in education after conversion to zero cost

Components	Charges*	Strategies	Revised charges
Tuition fees	INR 3500	Voluntary service from retired professionals and academicians	INR 0
Private tuition	INR 1100	E learning	INR 0
Text books, learning material	INR 600	Learning material downloadable online	INR 0
Stationary	INR 350	–	INR 350
Uniforms	INR 950	–	INR 950
Exam fee, question paper	INR 150	–	INR 150
Commute to and from schools	INR 400	Solar powered vehicles	INR 0
Subscriptions	INR 1100	E learning	INR 0
Computer education	INR 250	IT enabled services	INR 0
Study tours	INR 350	–	INR 350
Total	INR 8750		INR 1800

Source [27]

be collaborated with schools to boost the potential of educational institutions. Electronic books can be made available to the students and E-learning can be promoted. Voluntary services from professionals and retired professors can be invited so as to reduce the fees collected for salaries and for giving better coaching to the students. Providing home stays for students with old age people staying alone will reduce the hostel charges and will also promote social cohesion. By involving students for cleaning their own classes and premises, zero marginal cost environment sanitation can be attained in schools.

The various costs entailed can be reduced by adopting smart strategies into the conventional education infrastructure framework (Table 10.8).

10.7.6 Healthcare

The conventional medical practices comprises of steps that include consultation with medical practitioners, check-ups and tests, prescription of medicines and purchase of medicines by pharmacies. Zero marginal cost can be attained by implementing online health care facilities. Online health services will ensure expertise of the particular medical specialization while also reducing the expenses for the treatment. Online health account can be opened for every patient that can be

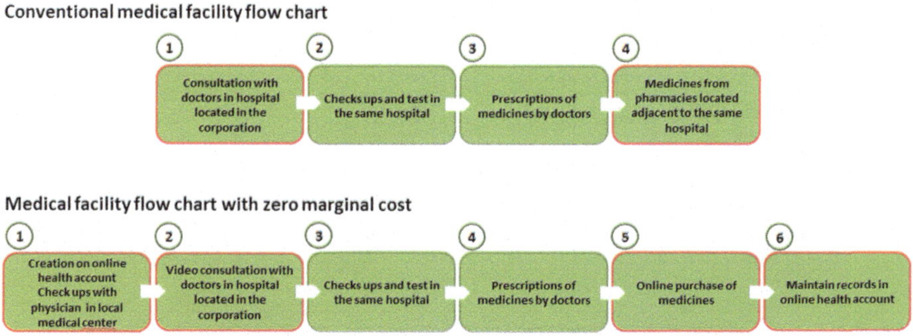

Fig. 10.46 Zero marginal cost strategies for health sector; *Source* [27]

accessed by limited number of persons (other than the patient himself). The account will keep records of health conditions, treatment and check-up history along with any parameters that need special consideration (Fig. 10.46).

10.8 Balanced Regions and Micro Grid

The ‘energy internet’ envisions ‘transforming buildings into green micro power plants’ via three steps namely (a) identification of energy surplus and energy deficit regions; (b) balancing the regions and (c) designing the micro grids.

10.8.1 Energy Surplus and Deficit Regions

Energy demand for each economic activity has been calculated using assumptions. Energy surplus or deficit regions in the KMR have been identified by subtracting the demand from the supply of the region (Fig. 10.47). A positive value indicates that the region has surplus energy and a negative value indicates that the region has energy deficit. The energy demand in the deficit regions thus identified need to be balanced either from the surplus energy available from the neighbouring local bodies or using hybrid energy systems. *The major deficit regions in the KMR are Kozhikode corporation, Beypore; Thikkodi, Moodadi, Payyoli, Onchiyam and Azhiyoor* (Fig. 10.48).

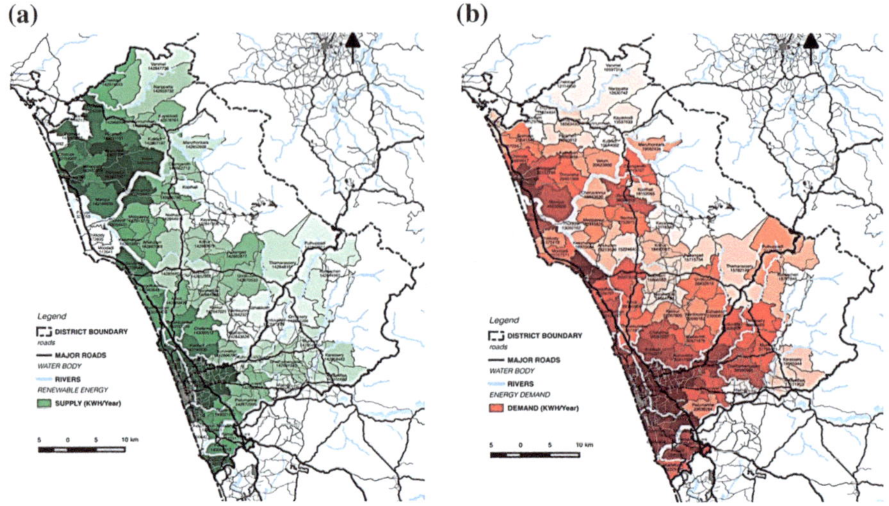

Fig. 10.47 **a** Renewable energy supply within KMR; **b** Renewable energy demand within KMR; *Source* [27]

10.8.2 Balanced Regions

From the above energy surplus and deficit local bodies in the KMR, three regions have been identified which require balancing of energy, namely, Kozhikode, Payyoli and Onchiyam. The criteria for balancing energy are presence of

(a) adjacent local bodies with surplus energy,
(b) road connectivity and
(c) existing KSEB distribution system that shall be used as the distribution network.

The Kozhikode balanced region includes the energy deficit local bodies such as Kozhikode Corporation and Beypore and the neighbouring energy surplus local bodies such as Elathur, Kakkodi, Kuruvattur, Kunnamangalam, Peruvayyal, Perumanna, Olavanna, Cheruvannur, Feroke and Kadalundi—these regions have 70–80% energy surplus. For the ease of balancing energy, Kozhikode Corporation is further divided into smaller areas based on energy demand gap such as City, Nadakkavu, Westhill, Malaparamba, Medical College and Pottamal. The energy deficits in each of the above areas are calculated. Hence, the energy deficit in these eight zones (including Beypore) shall be balanced from the ten neighbouring local bodies.

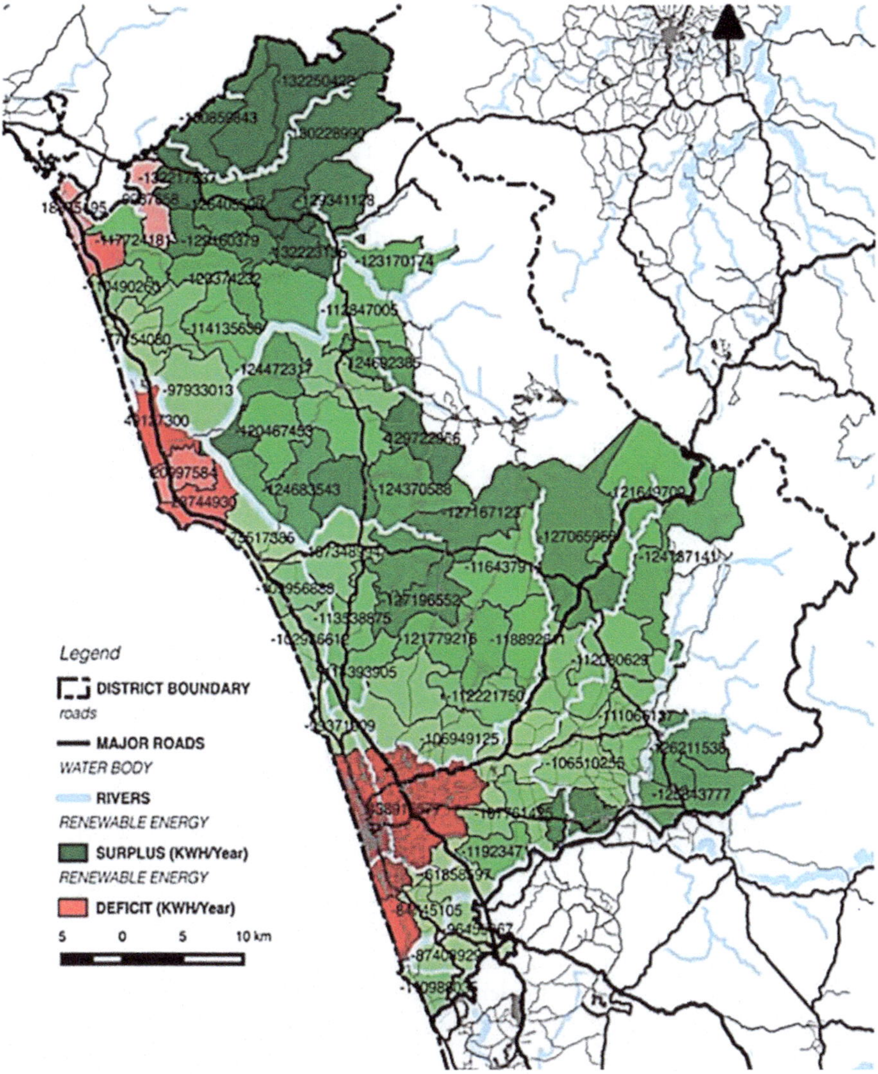

Fig. 10.48 Renewable energy surplus/deficit in KMR; *Source* [27]

The Payyoli balanced region includes the energy deficit local bodies such as Payyoli, Thikkodi and Moodadi and the neighbouring energy surplus local bodies such as Vadakara, Thurayur and Koyilandi, which has 75–95% energy surplus. Similarly the Onchiyam balanced region includes the energy deficit local bodies such as Onchiyam, Azhiyur and Edacheri and energy surplus local body Eramala that has 95% energy surplus (Fig. 10.49).

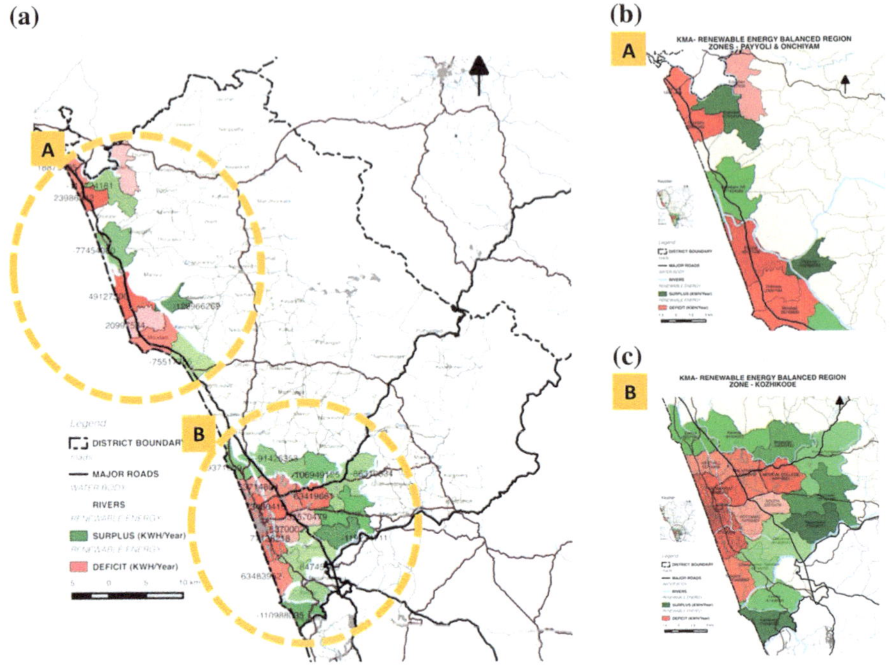

Fig. 10.49 **a** Renewable energy balanced region; **b** Renewable energy balanced regions zones in Payyoli (A) and Onchiyam (B); *Source* [27]

10.8.3 Micro Grid

The commonly used centralized power distribution system brings with it many disadvantages and the concept of a decentralized power generation and distribution system was evolved to overcome these shortfalls. This led to the development of a micro grid. A micro grid is a modern distributed power system using local sustainable power resources designed through various smart-grid initiatives. It also provides energy security for a local community as it can be operated without the presence of wider utility grid [38]. Micro grid technology generally represents three important goals of a society such as reliability, sustainability and economics [38].

A micro grid functions on four basic parameters: small scale renewable energy sources within the geographic boundary of the grid that can be connected to external power through a macro grid; various generation and consumption points in a small area that are interconnected to form a macro grid; classified as supply and demand grids with most transactions between supply and deficit grids located in close proximity to each other thus reducing T&D losses; ability to work in isolation by disconnecting from the macro grid during maintenance periods.

The major components within a typical micro grid comprises of (a) the renewable energy source; (b) grid transmission system; (c) protection system of the

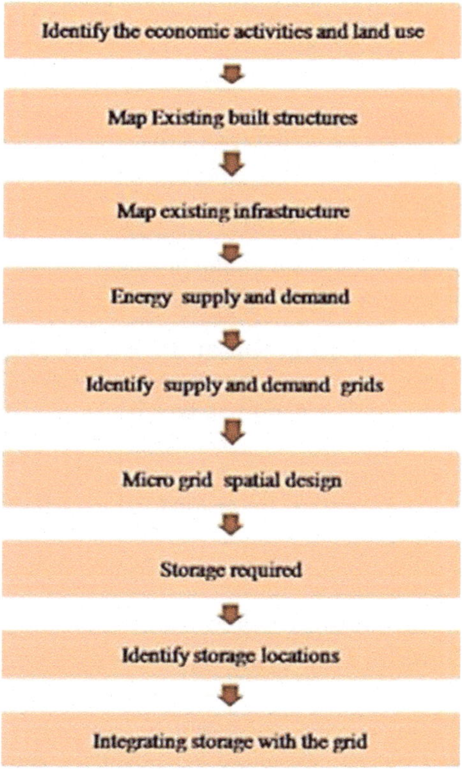

Fig. 10.50 Design methodology for micro grids; *Source* [27]

IoTs, IoEs and the IoCs integrated within the grid; (d) data monitoring system; (e) power converter; (f) end users (Fig. 10.50).

Methodology for grid design and prototype

The planning area region was initially divided into smaller zones based on natural features; roads and existing T&D infrastructure.

The economic activities and land use in each zone was identified from secondary data such as existing master plans and primary surveys. The existing built structures, infrastructure, services and utilities are mapped. The energy demand in each zone was calculated based on the activities in the zone. The energy demand calculated previously was used for this. The energy that can be generated in each zone was also calculated.

The zones were classified as supply and demand grids based on the difference between energy generated and required. The amounts of energy that can be shared between different grids were calculated. These grids were connected to each other using the existing infrastructure. The spatial design of the micro grid was done to connect the supply and demand points.

The storage requirement in the region was calculated based on the hours of energy generation in a day and the hours of energy requirement. The energy requirements for monsoon seasons were calculated. The total storage capacity required in each local body was found out. Storage locations were identified in each region based on the connectivity to multiple micro grids, existing infrastructure and the spatial structure. The storage systems were then integrated with the micro grids (Fig. 10.50).

Based on the above methodology a micro grid prototype was designed for a select location within the planning region. The broad concepts that forms the basis of the grid design include

- The energy generated in a micro grid will be consumed by the same micro grid and the excess energy will redistributed to the nearby grids
- Each macro grid will be a balanced energy area with 10 to 20 micro grids. This will reduce the span of control of each macro grid and can help in better management
- The existing infrastructure available will be used in order to minimise investment
- Roads, lanes and natural features will be used for delineation of each grid.

The design criterion for the design include

- Renewable energy zones and the balanced regions
- Area under each zone: A micro grid is best effective when the area of each grid is kept at the minimum possible value, ideally around 30–40 buildings
- Connectivity: Since the existing T and D network follows the roads and lanes, connectivity is a major factor to be considered. Utilising the existing T and D network for energy distribution within the grid can reduce the capital cost
- Calculation of potential energy generated based on number of buildings and slope
- Energy requirement calculation based on smart economic activities and other activities such as residential, institutional etc. Energy demand and supply in Kwh/year per m^2
- Calculation of excess energy and deficit energy areas within the grid as applicable—Balancing energy demand and supply at macro level considering 2 to 5micro grids as a macro grid
- Sharing of energy (Figs. 10.51, 10.52 and 10.53).

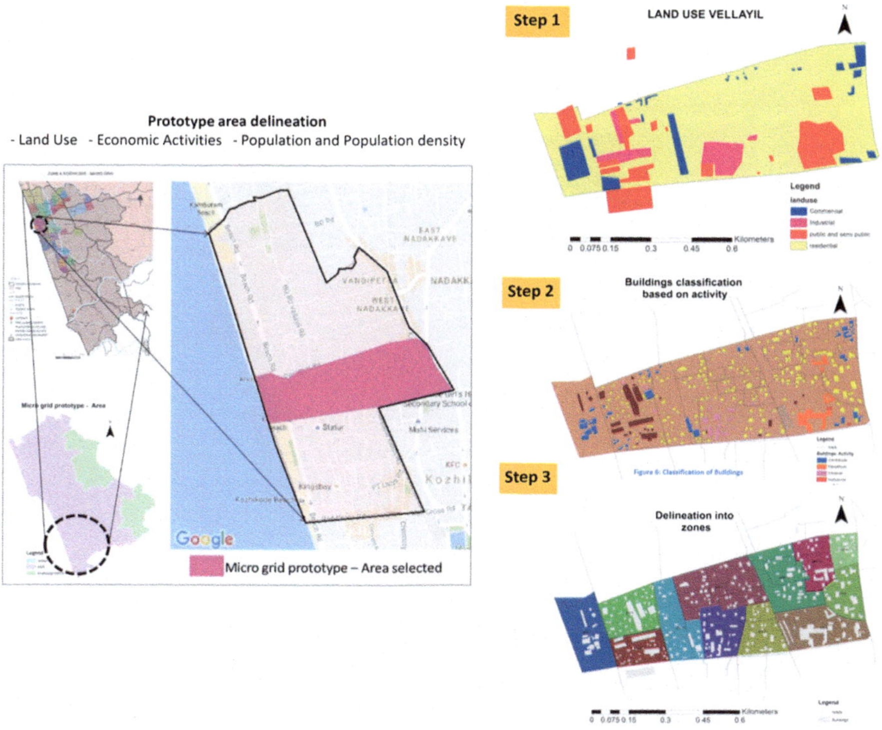

Fig. 10.51 Delineation of prototype area; *Source* [27]

10.9 Spatial Code

10.9.1 Spatial Code for Renewable Energy Zones

The spatial codes for renewable energy deployment aid the prosumers in understanding the energy scenario of the region and thereby select the area to initiate their activity or select the activity based on the area. Based on the energy demand, supply and surplus calculations, the areas are divided into developing, redeveloping and established (Table 10.9).

Based on the energy requirements of the activities and their location, the local bodies have been classified as developing, established or redeveloping. The zones have been classified as

- Mixed Activities—located in the more urbanized areas of Kozhikode Municipal Corporation, Koilandy and Vadakara
- Fisheries—located in the local bodies along the coastal areas

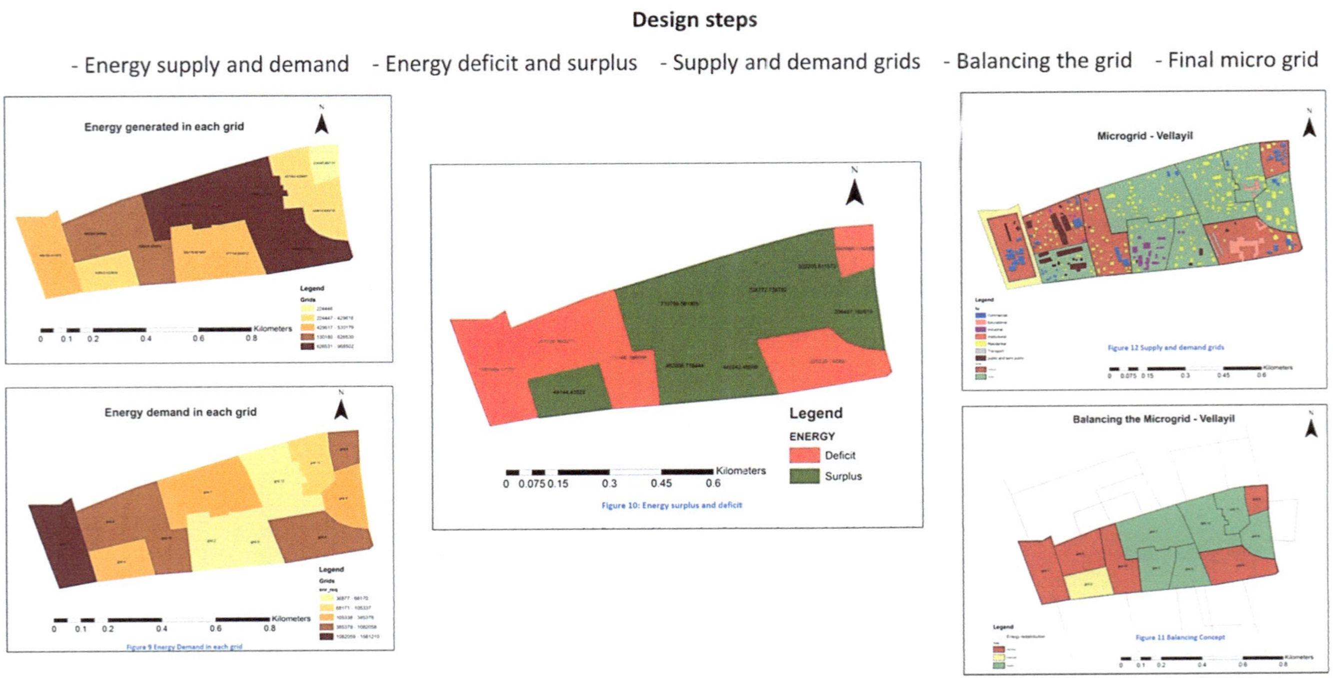

Fig. 10.52 Design steps for micro grid prototype; *Source* [27]

Fig. 10.53 Final micro design for Vellayil; *Source* [27]

Table 10.9 Classification of region within KMR by energy demand and supply

Region	Energy demand (per sq/year)	Energy supply (per sq/year)	Energy surplus (per sq/year)	Energy deficit (per sq/year)
Developing region	<2.5 kwh	>10 kwh	>1 kwh	–
Established region	2.5–5 kwh	5–10 kwh	–	<2 kwh
Redeveloping region	>5 kwh	<5kwh	–	>2 kwh

Source [27]

- Tourism—located in the some coastal zones along with the other established areas
- Household Industries—allocated in the remaining areas of the region which are envisaged to play a key role in the economic activities through the development of household industries (Fig. 10.54).

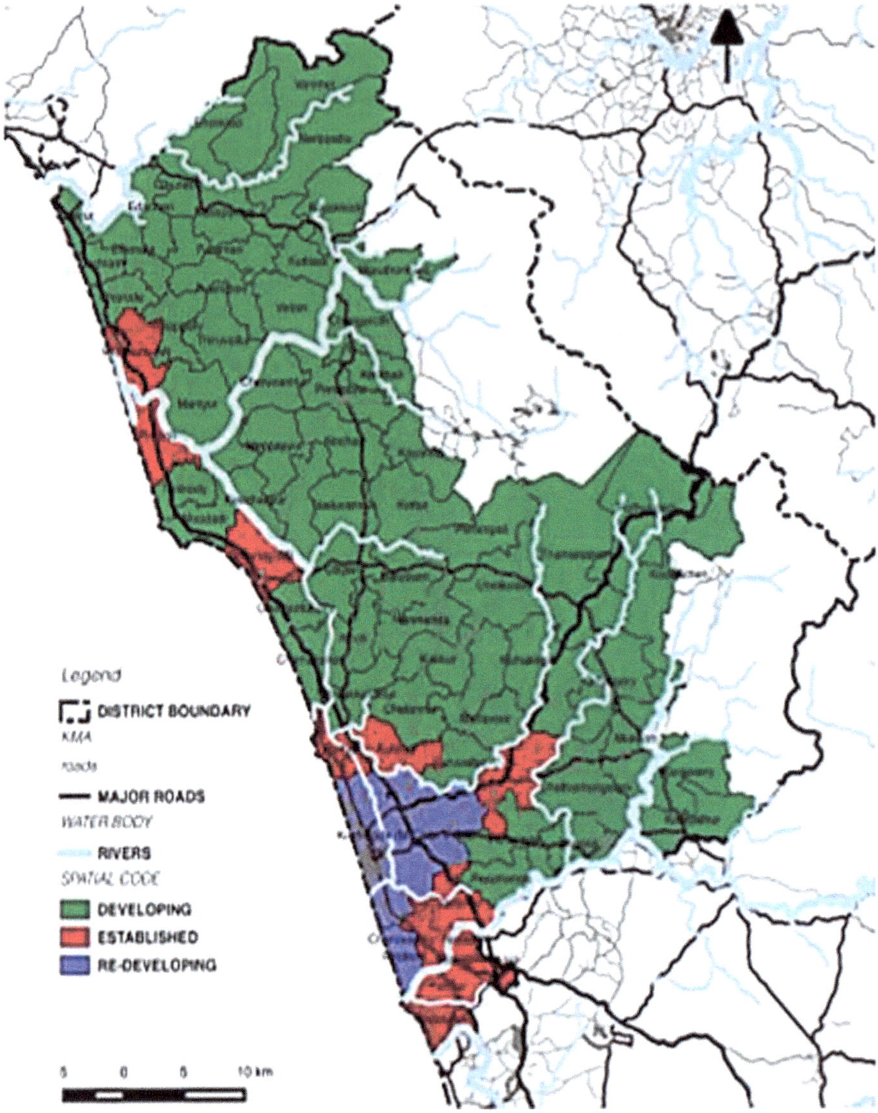

Fig. 10.54 Spatial code for Third industrial revolution; *Source* [27]

10.9.2 Spatial Code—Smart Economy

The spatial code(s) based on the smart economic activity is proposed for activities primarily within the manufacturing sector and these codes will help prosumers identify the economic potential of the locations which will in turn help them choose the likely economic activities to be established in the specific area. Potential

Table 10.10 Spatial code by region and smart industries

Name of industry	Developing	Spatial code	Established	Spatial code	Redeveloping	Spatial Code
Coir	✓	CR—D	–	–	–	–
Food processing	✓	FO—D	✓	FO—E	–	–
Fisheries	✓	FS—D	✓	FS—E	–	–
Footwear	✓	FW—D	✓	FW—E	–	–
Home Stay	✓	HS—D	–	–	–	–
Multi–functional	–	–	–	–	✓	MF—R

Source [27]

manufacturing activities such as timber, footwear, coir, food and beverages along with fisheries and the accommodation vertical of the tourism sector (homestays) have been selected for this purpose. All other service related activities are envisaged to be equally distributed across the regions. All the regions have been proposed to be mixed use zones and along with this, the Kozhikode Municipal Corporation is allocated as a multi—functional zone with the presence of all economic activities (Table 10.10 and Fig. 10.55).

10.9.3 Spatial Code—Zero Marginal Cost Society

Spatial codes for the conversion of the economic activities to a zero marginal cost society are developed based on the existing potential and resources available in the region. Agro based industries are located mainly in the eastern region outside the Kozhikode metropolitan region where the agricultural production is high. Backwater tourism is envisaged to be developed along the regions with rivers such as Korappuzha which host boat race and other activities. Fisheries are developed in the coastal region along the western coast and in close proximity to the port. Food processing activities are established based on locations where the food crops production is high. Tourism corridor is demarcated along locations that are rich in natural, cultural and historic resources. Mixed use area is demarcated considering the residential and commercial developments in the area (Table 10.11 and Fig. 10.56).

Table 10.11 Spatial code by region and industry for zero marginal cost society

Name of industry	Developing	Spatial code	Established	Spatial code	Redeveloping	Spatial code
Agro based	✓	AG—D	–	–	✓	AG—R
Backwater tourism	✓	BT—D	–	–	–	–
Coir	✓	C—D	✓	C—E	✓	C—R
Fisheries	✓	F—D	✓	F—E	–	–
Food and beverages	✓	FO—D	–	–	–	–
Industrial	–	–	–	–	✓	IR—R
Livestock	✓	LS—D	✓	LS—E	–	–
Mixed Use	✓	MU	✓	MU	✓	MU
Tourism corridor	–	–	–	–	✓	TC—R

Source [27]

10.10 Spatial Structure

The land use plan for the smart economy of the Kozhikode Metropolitan Region has been created for the activities that have been proposed as per the economic and spatial strategies.

10.10.1 Land Use Plan

The smart economy as per the third industrial revolution is based on three zones namely, (a) Supply zone; (b) Demand zone and the (c) Balanced zone. Balanced zones are expected to comprise the capacity to take up new activities based on the availability of resources. These zones will witness increased demand for more built up are and densification methods are to be adopted to meet this demand. The potential economic activities are low in the non-balanced zones. Hence six distinct zones have been identified integration smart economy and third industrial revolution

(a) ***Multifunction Zone* I**—Local bodies in this zone include Cheruvannur, Feroke, Kadalundi, Elathur and Kunnamangalam
(b) ***Multifunction Zone* II**—Local bodies in this zone include Kozhikode Corporation, Beypore, Koyilandi and Vadakara
(c) ***Mixed Zone* I *(Residential/ Manufacturing)***—Supply zone with a mix of residential and manufacturing activities such as household based coir, footwear and wood industries etc.

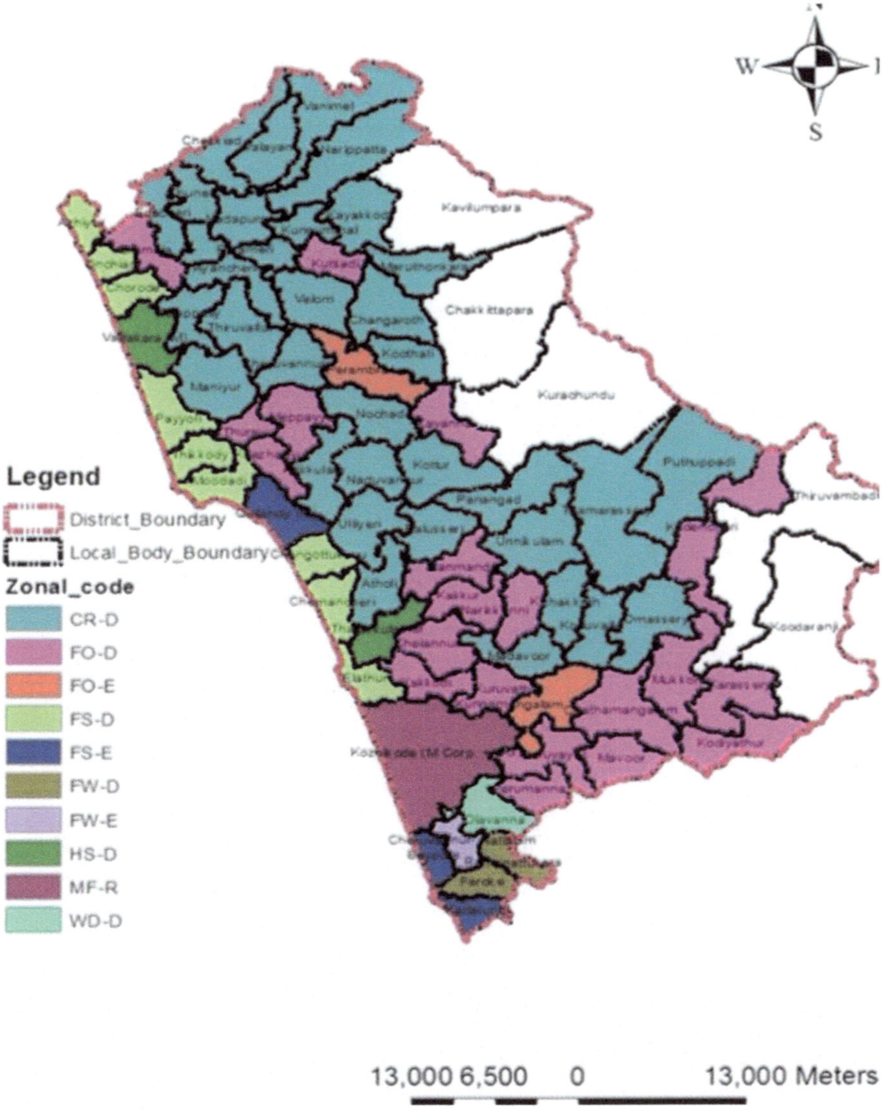

Fig. 10.55 Smart economy spatial code; *Source* [27]

(d) ***Mixed Zone* II *(Residential/Food Processing)***—Supply zone with a mixing of residential and food-processing activities such as household based food and milk products processing units etc.

(e) ***Mixed Zone* III *(Residential/Fisheries)***—Supply zone with a mixing of residential and fisheries related activities such as household based fishing activities and fish products processing units etc. The local bodies in this zone include Chengottukavu, Thikkodi, Moodadi, Azhiyur and Onchiam

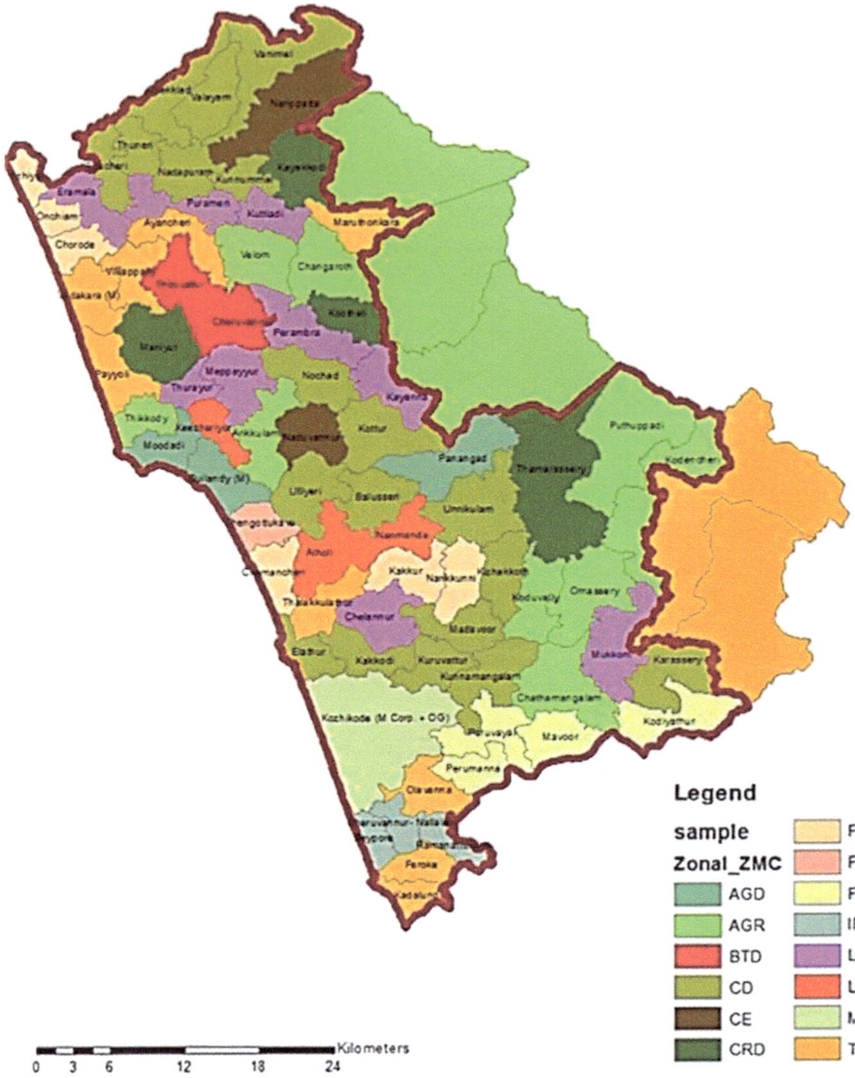

Fig. 10.56 Zero marginal cost society spatial code; *Source* [27]

(f) ***Mixed Zone* IV *(Residential/Tourism)***—Supply zone with a mixing of residential and tourism related activities such as tourist services, hotels and homestays, catering, household based coir and wood handicrafts manufacturing units, food processing units etc. The local bodies in this zone include Chemancheri, Payyoli, Kuttiyadi, Maruthonkara and Kodencheri (Fig. 10.57).

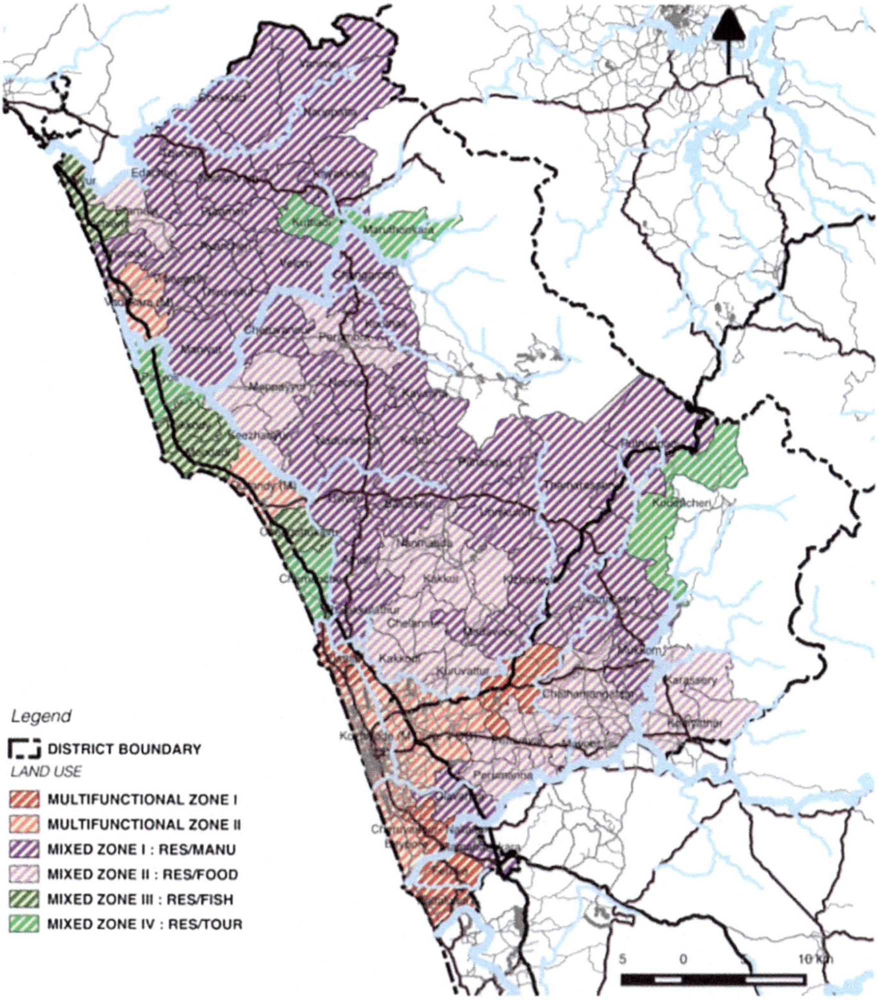

Fig. 10.57 Land use plan for Kozhikode Metropolitan Region; *Source* [27]

10.11 ICT as a Pillar for Implementation

Kozhikode district has been identified as the third IT hub in Kerala after Cochin and Thiruvananthapuram. Kozhikode district has 320 IT based enterprises and 40 IT companies in which the KINFRA Park at Kakkancherry has accommodated 16 companies and the NIT, Kozhikode has incubated about 14 companies. There are two professional organizations, Kozhikode IT initiative and Caffit, who organize IT related seminars and workshops in association with NASSCOM and KSITM. It is interesting to note that NASSCOM had maximum number of such events organized in Kozhikode, compared to the other cities in the State.

The existing IT companies in Kozhikode are mainly focussed on Product Development and IT services. The Major IT Parks are the UL Cyber Park (constructed and operated by ULCCSC, a Kozhikode based company) and Cyber Park initiative of Kerala state Government, KINFRA park at Kakkancherry and an Advanced technology cyber park at Ramanattukara in a close proximity to Kozhikode Metropolitan Region. The two IT park will create a total 100,000(100,000) direct job opportunities. Other planned projects include the Birla IT Park (at Mavoor) and Malaysian satellite city (at Kinaloor) where KINFRA has plans to set up a 400-acre (1.6 km^2) industrial park. Additionally, the expansion of the existing IT parks in the Kozhikode Municipal Corporation and Olavana panchayat has been proposed. Kozhikode has become dream destination for high quality work and life offered by the city.

Major business outsourcing centres include Cyber park, KINFRA park, IT companies etc. and other prominent knowledge outsourcing centres include NIT, NIELIT, Technical institutes, Akshaya centres and IT education centres. The availability of low cost highly skilled resources, of which a substantial portion is qualified female IT professionals, good civil infrastructure including the international airport with over 50 daily flights connecting Middle East, European and American destinations, good quality hospitals, International schools, shopping malls and entertainment facilities (Fig. 10.58).

The contribution of IT services to economic development has been widely increasing in the region. While incorporating smart technologies, major business outsourcing centers such as Cyper Park, KINFRA park and other IT companies with knowledge outsourcing centers in NIT, NIELIT, technical institutes, Akshaya centers and IT education centers. Skill development programmes provided in the existing knowledge outsourcing centers and e-learning programmes are also to be improved. Additionally, IT services will also be extended to various ICT technologies incorporated in different sectors to improve online trading and marketing, E-learning, E- Health services, Tourism services etc. (Fig. 10.59).

The expansion of IT services to different sectors will be marginalized to zero costs by introducing the Google fibre project in the Kozhikode Metropolitan Region. The primary objective of introducing this project is to improve access to broadband internet within the region. Google Fiber provides a connection with high internet speed up to 1000 megabits per second through fibre-cable network or through wifi and also enables the storing and sharing information within the region. Google will save money on its deployments in many ways, such as piggybacking on existing power line infrastructure and building its own network gear.

10.11.1 IoE, IoM and IoC for Renewable Energy

Internet of Energy (IoE)

The flow of energy generated from each household within the micro grid will be controlled, monitored and notified through the users. These activities will be carried out by mobile apps and make the functioning of the micro grid smoother.

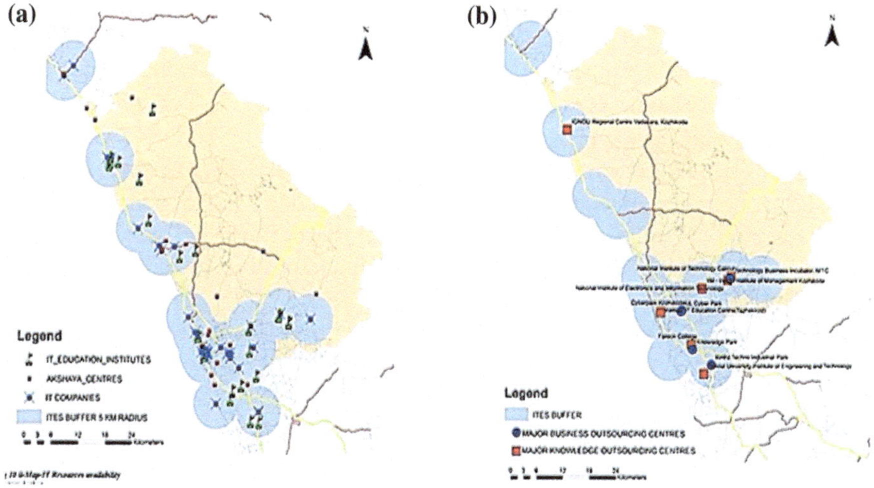

Fig. 10.58 **a** IT resources available in KMR; **b** Major and knowledge outsourcing centres in KMR; *Source* [27]

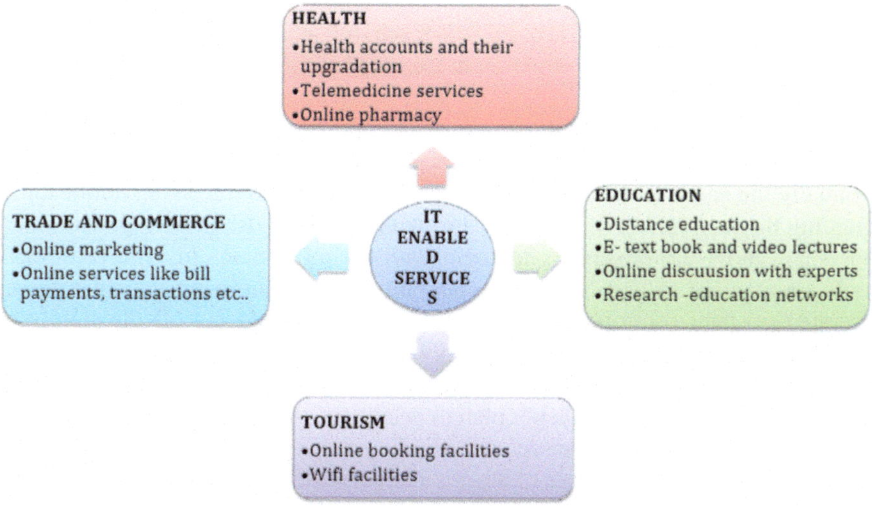

Fig. 10.59 9IT applications across various sectors; *Source* [27]

Real time monitoring and management of power flow—The micro-grids shall be connected to the cloud server and shall perform real time monitoring of energy generation and demand. Real time energy surplus and/or deficit shall be calculated, storage levels in batteries shall be monitored and energy flow from/to storage and grids shall be controlled.

Identify regular maintenance requirement—The micro grids and maintenance cell of the technical wing of energy supply shall be connected to cloud server. The efficiency of solar PV shall be calculated and analysed on hourly basis. It would identify power generation points where efficiency has reduced and send notification to the producer and maintenance cell.

Detection of unusual power usage—A system to analyse power flow between micro and macro grid and integrating storage system with macro grid is required using a smart app. The app would detect abnormal power flows and switch between grids wherever required. The app would then notify the prosumers and maintenance cell regarding the change in flow.

Ensure secure power distribution system—It is important to monitor the power flow between points and the micro-grid shall be connected to the technical wing and the cloud storage, monitor real time power consumption and power flow in the system and geographically located points on the grid to analyse parameters, detects abnormal power usage, power leakage etc. The app would notify maintenance cell and prosumers regarding the abnormalities.

Internet of Mobility (IoM)
This strategy will focus on creating a platform for a GIS based system showing charging point locations to the GPS enabled electric cars and enable the drivers to make use of the available charging points. Developing a smart app that shall notify the users regarding the time required to reach full charge, unoccupied charging points and constantly feed information to the cloud storage system would make this possible.

Internet of Communication (IoC)
This will be used to connect the prosumers, households, micro-grid community and eventually the entire KMR through the use of smart apps for receiving notifications regarding the functioning of the smart grid with respect to use of energy. The smart apps shall allow the prosumers to login with their Prosumer Id. The prosumers would be required to update in the app when they are not using the solar power generated on their rooftops. This would enable the grid to cut off the supply of energy to the house and feed the energy directly to the grid. The Micro-Grid Community Management System shall have platforms such as 'Sharing for free' where schemes, guidelines, weekly plan to decide the things that are to be shared in a week or a month and all other details regarding the venue, volunteers etc. shall be notified through the app. Other platforms include 'Making decisions' which shall enable voting to select people in charge for various activities, decisions regarding implementation of new ideas; 'Addressing Grievances' to resolve conflicts and technical issues; 'Training programs' enrolling prosumers in training programs, skill improvement as 'Sharing for free' where schemes, guidelines, weekly plan to decide the things that are to be shared in a week or a month and all other details regarding the venue, volunteers etc. shall be notified through the app. Other platforms include 'Making decisions' which shall enable voting to select people in charge for various activities, decisions regarding implementation of new ideas; 'Addressing Grievances' to resolve conflicts and technical issues; 'Training

programs' enrolling prosumers in training programs, skill improvement and tracking their progress (Fig. 10.60).

10.11.2 IoG, IoT and IoP strategies in economic transformation

IoG, IoT and IoP for tourism sector

(a) Automated comfort adjustor
This is to manage the comfort level in a room. The sensors will continuously collect data on temperature, number of people entering end existing the rooms, humidity, air flow, light, motion in the room etc. the collected data will stored in the room and the analytics will perform the analysis and automatically adjust the brightness of light, speed of fan and temperature of air conditioner. This also will automatically switch off all electric appliances when no one is in that room.

IoG, IoT and IoP for fisheries sector

(a) Automated fish catch
The fish catch can be increased by using smart technologies, specifically, the hydrowave—a patented electronic sound device that emits the natural sound of bait fish which will be activated to attract fish. The proximity sensor, motion detector are used to identify the approximate location of fishes where the hydrowave embedded nets can be stored.

IoG, IoT and IoP for manufacturing sector

(a) Automated irrigation system
The automated irrigation system will be used to regulate the irrigation in fields and lawns automatically—based on the moisture content, temperature and other parameters—thus ensuring efficient use of water. The system starts with diagnosing the moisture content and temperature using sensor, along with forecasting of weather. This data will be sent to cloud then to analytics which will subsequently forecast chances of rain. Upon receipt of rain showers, the analytics will calculate the amount of water required according to temperature and existing moisture content. The irrigation system will start automatically.
(b) Automated drying system
The automated drying system is used to adjust the drying procedure according to the temperature and weather forecasting. The weather forecasting sensor, temperature sensor, humidity sensor will collect data then send to cloud. Using analytics the temperature will measured and if external energy is required it will supplied otherwise the material will dried in natural light. The weight sensor will continuously monitor the weight, if weight decreases to a particular limit, which means the product is dried, a notification to will be sent to the owner and also the external energy will get cut off.

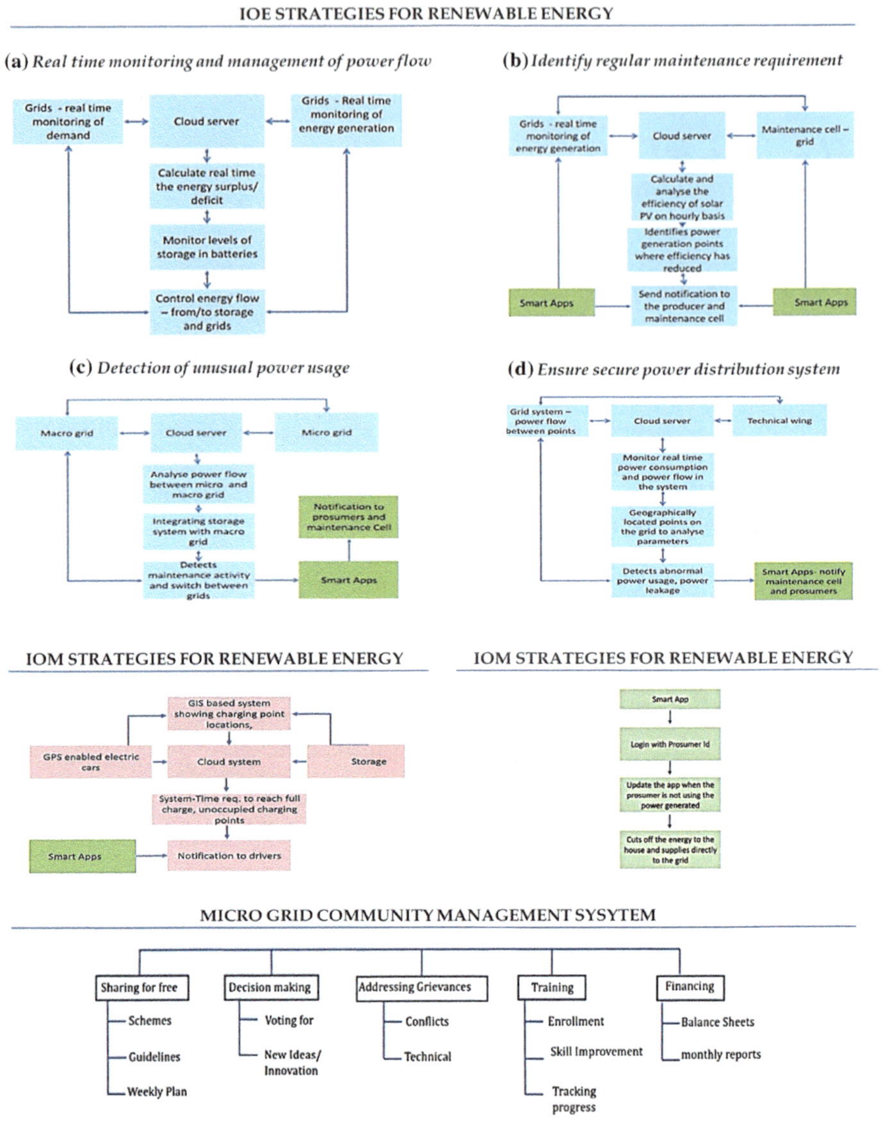

Fig. 10.60 IoE, IoM and IoC strategies for renewable energy; *Source* [27]

(c) Industrial registration

The new and interested manufacturers will be required to collect an industrial Aadhar number from the authority for the integration of manufacturers. For that the unit owner will need to register using the specific Aadhar number and building Aadhar number through the online portal. The authorities will check for the authenticity and will issue the industrial Aadhar card online upon completion of procedures.

(d) Tax collection
The industrial aadhar issued for each manufacturing unit will be linked to its owner's smart card. All the money transactions including tax collection will go through e-banking.
(e) Placing of manufacture orders
When an order is placed, the analytics will locate all the registered manufacturers under the category. Those who are not engaged in the any activity will be selected and will be ranked according to the customer feedback score. The subsequent works will be assigned to the top scorer first. If he/she is not ready to take the project it will be handed over to the succeeding worker.

IoG, IoT and IoP for trade sector

(a) *Smart mobility plan for goods movement*
The mobility plan has been prepared by considering the movement of goods between the hubs, warehouses and end users. The warehouses are located near major roads for improved connectivity. Kozhikode and Vadakara which will be the hubs and they have access to rail, water and air connectivity. The Kozhikode hub is located in close proximity to the Kozhikode railway station, 10.8 km to Beypore port and 27.8 km away from the Calicut International Airport. The Vadakara hub is located close to the railway station in Vadakara, 22.7 km from the port in Thalassery and 46.7 km away from the proposed Kannur International Airport. The goods movement between warehouses and product delivery will be done using electrically running cart rickshaws or E-Cart rickshaws and driverless electric vehicles or drones (Fig. 10.61).

Mobility of goods
The product from an industry or household will be sent to the nearest warehouse which will be transported to the warehouse nearest to the end consumer for delivery. The products for export will be sent to the nearest warehouse and then transported to nearest hub which will be sent to the seaport, closest airport or railway station depending upon the destination and vice versa for importing the products (Fig. 10.62).

(b) Automated warehouses
All products will be stored in single ware house. The items are stored on portable storage units with separate QR code on the floor for each row to identify products in that row. When an order is entered into the database, the software locates the closest automated guided vehicle (bot) to the item and directs it to retrieve it. The average size of a bot is 2 ft. by 2.5 ft. and 1 ft. height. It can move with a speed of 1.3 m/s, and it can lift 1000 lb to 2000 lb. The mobile robots navigate around the warehouse by following the QR code on the floor. Each drive unit has a sensor that prevents it from colliding with others. When the drive unit reaches the target location, it slides underneath the pod and lifts it off the ground through a corkscrew action. The robot then carries the pod to the specified packing area.

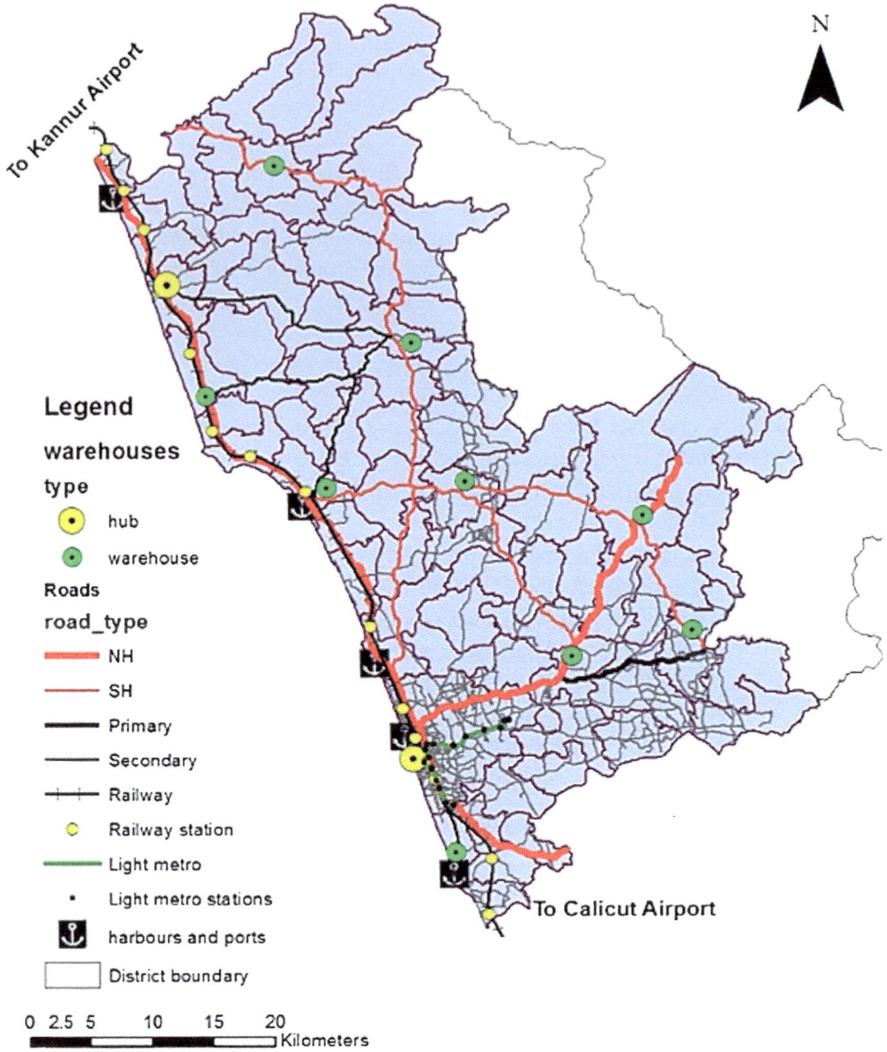

Fig. 10.61 Linkages in KMR; *Source* [27]

(c) Automated raw material quantity management
The quantity of raw material will be continuously monitored using quantity sensors. When the quantity goes below a particular limit automatic ordering will take place through sending a notification to owner.

(d) Automated cold storage
According to the quantity inside the cold storage unit and temperature inside the unit, the analytics will automatically handle the temperature in cold storage.

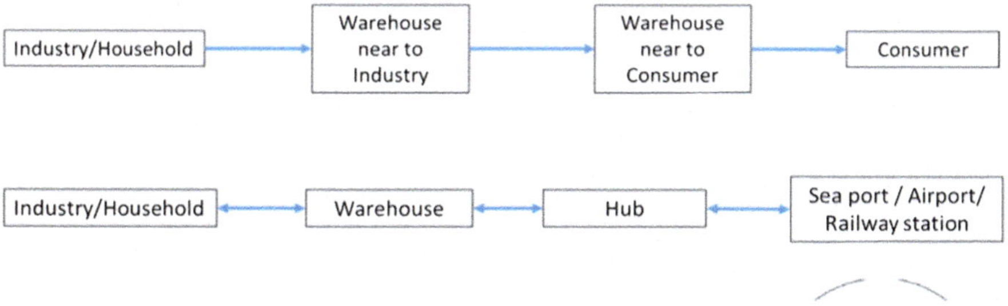

Fig. 10.62 Goods movement within and outside KMR; *Source* [27]

(e) Personalized marketing
Any data with respect to product specific consumer searches will be saved on the cloud. If there is any new product of same category or discount in the same product, analytics will remind the customer through an online notification.

IoG, IoT and IoP for health sector
A ring or a watch can be designed with heartbeat, body temperature, steps, sleep quality, blood pressure and hydration level sensors which will measure these parameters and save the data on the cloud. Analytics will continuously compare the values and if there is any medical emergency needed, the owner will get a notification regarding his health status and recommendations to be followed. There will be a notification given to the emergency medicine wing of nearby hospital which helps them to get prepared before patients reach the hospital and taking all the necessary checking.
IoG, IoT and IoP for mobility sector

(a) Real time signage
The GPS installed in all vehicles, speed detectors in the road and counting sensors in the road will calculate the average number of vehicles in the stretch and maximum speed attained at the stretch will be displayed in the starting of each section—this will avoid increase traffic congestion. This can also be used for congestion pricing.
(b) Driverless vehicles
The image recognition, speed monitor and radar sensor to access other vehicles speed will effectively manage the vehicle speed and will provide extra brake if needed.
(c) Drones
GPS is used to locate the destination, obstruction sensor to reroute and avoid collision. Image recognition will be used to land the drones. When the drones reach the nearby area customer will get notification and if he accepts only the drone will land, otherwise after some time it will move back to the ware house and the order gets automatically cancelled.

(d) Parking slot finder
Sensor to detect free parking slots and the data will be collected which will be used by the analytics to process and guide the passenger who are requesting for parking slot to the nearby parking slot (Fig. 10.63).

10.12 Institutional Framework

10.12.1 Smart Community Using Concepts of the Faith Based Model

The social importance of religion has been debated for many centuries and philosophers such as Emile Durkheim has described religion as "something eminently social" [39, 40]. Historically, religion and social institutions have been deeply interlinked wherein the primary religious institutions have always posed as the primary provider for the poor, the elderly, the orphaned, and the needy [39, 41]. Religious institutions are defined as ontological communities—communities of meaning—and these communities are social, cultural and ethnic centres of various groups that form the heart of the community [39, 42]. These communities can be used to supplement activities where primary social systems fail [42].

Faith based models were practised by communities in Kerala since many centuries. The three dominant communities in Kozhikode namely Hindu, Muslim and Christian each follow a distinct faith based model that is Jajmani, Zakath and Dashamsham, respectively. It is envisaged that the development within the Kozhikode Metropolitan Region will be mobilized by the communities entirely. The basis is that these faith based models should be and can be used for mobilizing and integrating the community as part of implementation of the development. This section will broadly indicate the application of these models in community integration and development.

Jajamani as a concept for improving linkages between the resource and service providers: The Jajmani system is the backbone of social order within Hindu societies and is social contract system of reciprocal relations [43]. Under this system, each group within a village provides a certain standardized set of services to the land owning communities with payment made in both cash and kind—while payment in earlier times were made in the form of grains and sometimes allocation of land particularly in the case of farmers, modern times are witness to currency as the most common method of payment [44, 45]. This system is characterised by the interdependence, reciprocity and cooperation between various groups. This system can be viewed as an institution or social system made up of networks of roles and norms integrated into the roles and into the system as a whole which is legitimised and supported by the strong cultural values [46].

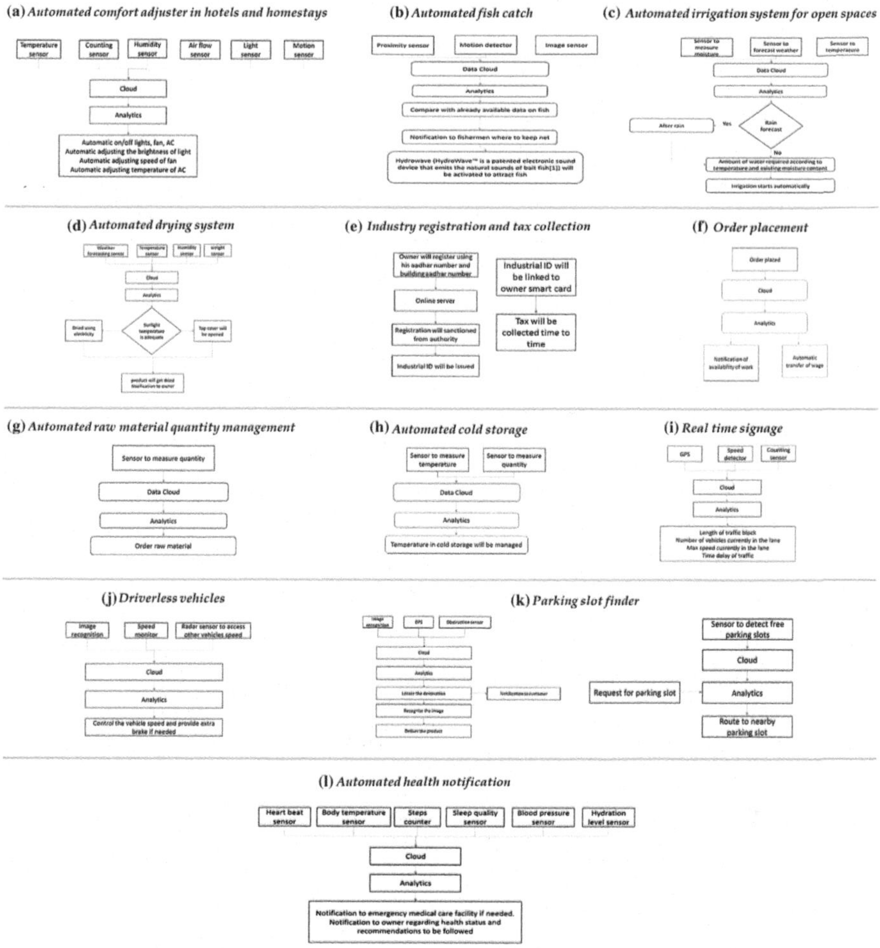

Fig. 10.63 IoG, IoT and IoP strategies for economic transformation; *Source* [27]

Zakath as a concept for mobilization of resources: 'Zakath' is one of the five pillars of faith in Islam and is to be followed by every Muslim, mandatorily. A donation of approximately 2.5% of annual household income of all Muslims, are to be distributed in the form of charity from the surplus recourses after meeting the essential requirements of the family. The Zakath system in Islam envisages equal distribution and circulation of wealth for the economic stability and prosperity of the poor and the needy. It also avoids capital formation among the rich. The magnitude of philanthropic giving in Muslim communities is estimated to total between USD 250 billion and USD 1 trillion annually. Despite having such a large philanthropic wealth base, in most countries it is being either mismanaged or misappropriated [47]. Zakath' funds, though primarily can be used for helping the poor and needy thus contributing for poverty reduction. It can also be utilized in

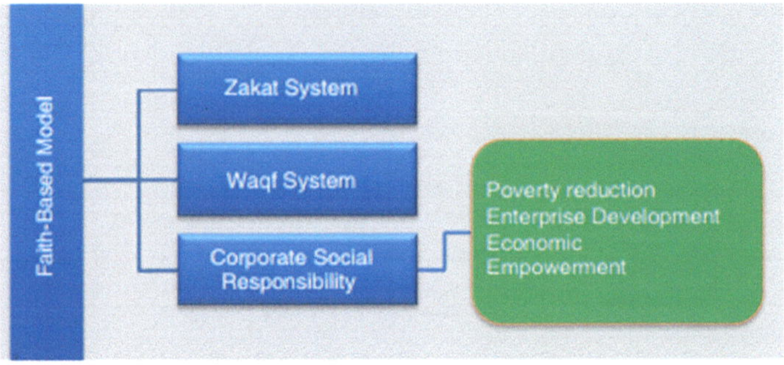

Fig. 10.64 Faith based model; *Source* [49]

building social infrastructure like school and hospitals among others, thus promoting the collective welfare of the society [48]. Therefore, there is a possibility of devising a faith based model (FBM) for social welfare community development and poverty reduction. Studies show that FBMs find more acceptability among Muslim nations because they have theoretical and theological underpinnings from the Qur'an and Hadith (a collection of quotes, sayings, daily practices etc. of Prophet Mohammed followed by Muslims apart from the religious scripture Qur'an), as opposed to conventional models that are often viewed with suspicion, sequel to Islamic revivalism [49] (Fig. 10.64).

The above mentioned model clubs the concept of 'zakath' system along with the 'waqf ' system of Muslims along with the concept of corporate social responsibility in evolving the Faith Based Model.

Dashamsham as a concept for Social and Community Development: Dashamsham is based on the fifth canonic law of the Christians (5 Commandments of the holy church) and it clearly depicts the need for helping the poor and needy by making voluntary contributions to the church. The faithful are obliged to assist with the material needs of the Church, each according to his own ability by giving a percent of income [50]. In most cases this voluntary contribution is approximately 10% of the income.

These underlying principles of the faith based models that have been practised in the communities in Kerala since many centuries will be used to help mobilise the resources which include renewable energy and funds along with the sharing of services and economic activities that make up for a vibrant and self-reliant community.

10.12.2 Triple Helix Used as a Model for Implementation

The smart metropolitan development within the Kozhikode Metropolitan Region is expected to transform the region into a knowledge based economy. This is a resultant of the full recognition of the role of knowledge and technology in

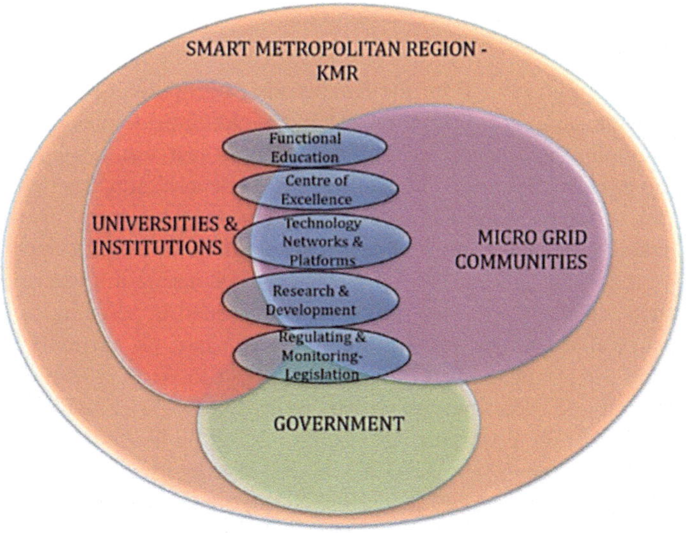

Fig. 10.65 IoG, IoT and IoP strategies for economic transformation; *Source* [27]

economic growth. While knowledge has always been central to economic development, its importance has only been growing exponentially over the last few years [51]. The three pillars of innovation are (a) product development in industry, (b) policy formulation in government and (c) creation and dissemination of knowledge within academia. These pillars which were confined to a single entity have shifted to a model where in the interaction between these three pillars as the source of new innovative organizational designs and social interactions [52]. This shift would entail various mechanisms of institutional restructuring along with rethinking the main models of innovation [52] (Fig. 10.65).

The concept of the Triple Helix of University—Industry—Government and their relationship, developed in the 1990s by Etzkowitz, has been used as the basis for the implementation framework for the smart metropolitan regional development in KMR. The triple helix model interprets the shift from a dominating industry—government dyad to a growing triadic relationship between university, industry and government in the knowledge based society. It integrates the universities and institutions, micro grid communities (industry) and Government for a community that facilitates the functioning of system which focuses on providing functional education, serves as a centre of excellence by promoting use of technology, networks and platforms that encourage interactions of the community with the various agencies to improve their knowledge and skills. While the institutions would be involved to a large extent in research and development inventing new ideas, the Government shall intrinsically play the role of a regulatory body and shall look into the aspects of monitoring and legislation.

10.13 Conclusion

Strategies for economic revival for the Kozhikode Metropolitan Region and spatial applications of the same are centred on the fundamental principles of new sources of energy generation, goods and services provided at near zero costs and a sharing economy base to help implement these strategies. Research indicates that renewable energy use for economic activities is the future for any sustainable economy. The use of renewable energy for large scale economic activities is not new in Kerala. Additionally, the existing state level policies that encourage the use of renewable energy in various aspects will also provide much needed impetus in this direction. However, renewable energy generation and utilization is currently limited to households and the application of this practice is yet to cover a community.

A holistic study of economic revival strategies for the Kozhikode Metropolitan Region indicates that the economic activities practised on household or community level can sustain on renewable energy resources. The deployment of the renewable energy using micro grids adopts the principles of near zero marginal cost and a sharing economy. Along with this, each economic activity has also been studied to explore the potential for converting to a near zero marginal cost activity. The revenue generating potential for each zone is thus maximised. Spatial codes for these zones indicate specific development potential for these zones thus creating a well-rounded spatial structure for the region.

The smart metropolitan region in Kozhikode is envisaged to transform the communities into a knowledge base society—one that will benefit from a vibrant knowledge base and collaborations between knowledge centres, communities and legislation for the effective implementation of proposed economic and spatial strategies. The practice of Faith Based Models in Kerala relies on the fundamental principles of harmony and trust—principles common to the sharing economy. This provides the potential to model these characteristics of Jajmani, Zakath and Dashamsham practiced by the Hindu, Muslim and Christian communities, to mobilize resources, for social and community development and improve linkages between services providers. These implementation frameworks will help add on to the robustness of the development within the metropolitan region of Kozhikode.

Credits This chapter is based on an academic project conducted with nine students of third semester Master in Planning in the Department of Architecture and Planning, National Institute of Technology Calicut. They were divided into three groups based on the three major steps in the methodology. We thank the students of all three groups for each of them have contributed to developing our understanding of the topic.

Team members

Anu Paul, Asha Devadas, Ashika Raoof, Fathim Rashna Kallingal, Maya K P, Sai Priya Nunna Tagore, Prabhakar Uniyal, Saratha Chandran P, Shabana Yoonus

References

1. United Nations (2014) World Urbanization Report
2. Government of India (2011) Census
3. Tewari M, Godfrey N et al (2016) Better cities, better growth : India's urban opportunity. New Clim Econ World Res Institute Indian Counc Res Int Econ. Relations. (London, Washington, DC, New Delhi)
4. Urban inequality rising far more in India, [Online]. Available: http://www.livemint.com/Politics/5JVPBWc7VGX0AMt3U3YeQM/Urban-inequality-increasing-far-more-in-India-Franois-Bour.html
5. Sprawl will cost India $1.8 Trillion Per year by 2050, [Online]. Available: https://www.citylab.com/solutions/2016/12/sprawl-will-cost-india-18-trillion-per-year-by-2050/509573/
6. Firoz CM, Banerji H, Sen J (2014) A methodology to define the typology of rural urban continuum settlements in Kerala. J Reg Dev Plan
7. Easow J, Kerala town and country planning act: a democratic spatial planning law for states in India. Spat—Econ Dev Rec
8. Cochrane SG, Mcgee M, Cochrane BYS, Mcgee M, Zandi K (2012) Global metropolitan areas : the natural global metropolitan areas : the natural. Moody's Anal
9. Competitive Cities in the Global Economy. OECD Territ Rev
10. Department of Town and Country Planning (2016) Town and Country Planning Act
11. Third Industrial Revolution Green Economy, [Online]. Available: https://www.huffingtonpost.com/jeremy-rifkin/third-industrial-revolution-green-economy_b_8286142.html
12. Rifkin J (2011) The third industrial revolution. Palgrave Macmillan, Springer Nature
13. Capitalism is making way for the age of free, [Online]. Available: https://www.theguardian.com/commentisfree/2014/mar/31/capitalism-age-of-free-internet-of-things-economic-shift
14. The rise of sharing economy and why it will work in India, [Online]. Available: https://yourstory.com/2016/02/sharing-economy/
15. The rise of the sharing economy: the Indian Landscape, 2015
16. Impact of the Sharing Economy, [Online]. Available: http://www.iicom.org/themes/item/impact-of-the-sharing-economy-2
17. 5 Ways the sharing economy is building a more social economy, [Online]. Available: https://www.huffingtonpost.com/entry/5-ways-the-sharing-economy-is-building-a-more-social_us_59500761e4b0c85b96c65a6a
18. The future of the sharing economy depends on trust, [Online]. Available: https://www.forbes.com/sites/theyec/2015/02/10/the-future-of-the-sharing-economy-depends-on-trust/#1c5476974717
19. Narayanan M (2006) Calicut: the city of truth revisited. Univ, Calicut
20. There's a place in india where religions coexist beautifully and gender equality is unmatched, [Online]. Available: https://www.huffingtonpost.com/chandran-nair/kerala-religion-gender_b_9577234.html
21. Firoz M, Kumar (2017) Transforming economy of calicut to smart economy. In: Smart economy in smart cities. Springer
22. The Third Industrial Revolution Strategy Study, 2016
23. The Rise of Renewable Energy in Kerala, [Online]. Available: http://blog.ksidc.org/the-rise-of-renewable-energy-in-kerala/
24. Kerala has potential for 10,000 MW in solar rooftop energy, [Online]. Available: http://www.thehindubusinessline.com/news/national/kerala-has-potential-for-10000-mw-in-solar-rooftop-energy-expert/article4606754.ece
25. District Urbanization Report: Kozhikode, 2011
26. Master Plan for Kozhikode Urban Area—2035, 2015
27. M in P Third Semester, "Smart metropolitan regional development: economic and spatial design strategy for Kozhikode Metropolitan Region

28. Zachariah RS (2015) Dynamics of emigration and remittances in Kerala: results from the Kerala Migration Survey 2014
29. State Economic Review 2016, 2017
30. Agency for non-conventional energy and rural technology, "The Energy Report—Kerala," 2013
31. Research and Statistics Division, "Kerala Tourism Statistics 2015," 2015
32. Gretzel KC Zhong U (2016) Application of smart tourism to cities. Int J Tour Cities
33. Kerala Tourism Trends: Trade Survey, 2016
34. Fernandez-stark K, Cggc C, Shelli R, Heil J (2011) The tourism global value chain: economic upgrading and workforce development. Cent Glob Gov Compet Glob Gov Compet DUKE Univ
35. D of F Statistical Cell, "Kerala Fisheries Statistics," 2016
36. Jisana TK (2014) Consumer Behaviour in Kerala Towards Selected Goods. Abhinav Natl Mon Ref J Res Commer Manag
37. Cooke AT (2000) "New Role fpr Middlemen. INET
38. Bayindir R, Hossain E, Kabalci E (2014) A Comprehensive Study on Microgrid Technology. Int J Renew Energy Res
39. Tirrito T, Cascio T (2003) Religious organizations in community services: a social work perspective. Springer Publishing Company, USA
40. Durkheim E (1995) The Elementary Forms of Religious Life. The Free Press, 1995
41. Simmons HC, Ethical perspectives on church and synagogue as intergenerational support systems. J Relig Gerontol
42. Bruggeemann WG (2002) The practice of macro social work. Cengage Learning
43. Weiss AM (2016) Brief notes on hinduism, caste, hierarchy, the jajmani system, and islam, [Online]. Available: http://pages.uoregon.edu/aweiss/intl442_542/Jajmani_system_notes.html
44. Lewis BV (1956) Caste and the Jajmani System in a North Indian Village. Am Assoc Adv Sci
45. Mondal P "Essay on Jajmani System in India," [Online]. Available: http://www.yourarticlelibrary.com/economics/essay-on-jajmani-system-in-india/4364
46. Kolenda PM (1963) Toward a Model of the Hindu Jajmani System. Soc Appl Anthropol
47. Alam N (2010) Islamic Venture Philanthropy: a tool for sustainable community
48. Economic Advantages and Benefits of Zakat, [Online]. Available: http://www.studylecturenotes.com/management-sciences/economics/387-economic-advantages-and-benefits-of-zakat
49. Lukman Raimi APIA (2014) Corporate social responsibility, Waqf system and Zakat system as faith-based model for poverty reduction Corporate social responsibility, Waqf system and Zakat system as faith-based model for poverty reduction. World J Entrep Manag Sustain Dev 10
50. Richert SP The Precepts of the Church and the Duties of All Catholics
51. The Knowledge Based Economy, 1996
52. Ranga H, Etzkowitz M Triple Helix systems: an analytical framework for innovation, Policy and Practice in the Knowledge Society
53. Vinod TM (2016) Presentation on Smart Metropolitan Regional Development
54. Future of Fish, "Making Sense of Wild Seafood Supply Chains," 2015
55. Supply chains are the key to change for sustainable fisheries and oceans, [Online]. Available: https://voices.nationalgeographic.org/2014/02/26/supply-chains-are-the-key-to-change-for-sustainable-fisheries-and-oceans/

Part XI
India, Surat

Chapter 11
Smarter Economic Opportunities for Surat Metropolitan Region

Bhasker Vijaykumar Bhatt, Leena Garg and Krupesh A. Chauhan

Abstract One of a considerably old civilisation settlement of Surat as a prominent urban centre is established in the western part of India and in the Southern region of the Gujarat state. The city was earmarked in the year 1992 since it recovered from a devastating calamity of plague. Since then, the city has recorded enormous progress regarding urbanisation and industrialisation and listed in the top ten fastest growing cities in India. The city has developed and retained its identity as a Diamond city and Textile city as well renowned for its unique delicacies and preparations. A metropolitan region of 1357 km^2 is considered as a study region encompassing the city dynamics of Surat. Current chapter browses through the history of the city as well discusses current scenario in the region from a development perspective. The chapter proposes certain conceptual aspects for developing the metropolitan region in a smarter way and based on the concept of zero marginal cost society to achieve economic sustainability in the region. Need for local zoning is identified and based on a quadratic bifurcation of the study area; conceptual proposals are discussed exploring various strengths and opportunities considering the scope of improvement based on current status. Surat metropolitan region has good level of natural resources and has a potential for serving the society at large. However, it calls for a proactive, systematic planning approach for smart metropolitan region planning to infuse the fundamentals of zero-marginal cost society concept. By leveraging opportunities for changes in the planning, implementation and practices

B. V. Bhatt (✉)
PG in Charge (ME-TCP) & Assistant Professor, Faculty of Civil Engineering, Sarvajanik College of Engineering & Technology, Surat, India
e-mail: er.bhasker@gmail.com

L. Garg
Faculty of Architecture, Bhagwan Mahavir College of Architecture & Planning, Surat, India
e-mail: leenagarg.1990@gmail.com

K. A. Chauhan
Associate Professor & Section Incharge, Civil Engineering Department, SVNIT, Surat, India
e-mail: kac3srb@gmail.com

T. M. Vinod Kumar (ed.), *Smart Metropolitan Regional Development*, Advances in 21st Century Human Settlements,
https://doi.org/10.1007/978-981-10-8588-8_11

followed by citizens for various activities need to be inculcated. Collaborative efforts can lead the metropolitan for a better and perspective future.

Keywords Smart Surat Metropolitan Region · SMC · SUDA · Surat

Abbreviations and Acronyms

AD	Anno Domini
BC	Before Christ
BCE	Before Common Era
BSNL	Bharat Sanchar Nigam Limited
BUDA	Bardoli Urban Development Authority, Bardoli
CE	Common Era
cm	Centimeter
DGVCL	Dakshin Gujarat Vij Company Limited, Surat
DIC	District Industries Center
ESR	Elevated Storage Reservoir (for Water)
GEB	Gujarat Electricity Board
GIDC	Gujarat Industrial Development Corporation
GIS	Geographic Information System
Govt.	Government
GPCB	Gujarat Pollution Control Board
GTPUD	The Gujarat Town Planning & Urban Development Act, 1976
GWSSB	Gujarat Water Supply, Sewage & Sanitation Board, Government of Gujarat
HR	Head Regulator
IEC	Information, Education and Communication
INR	Indian National Rupees
IoT	Internet of Things
km	Kilometer
km^2	Square kilometer (Area)
L&T	Larsen & Toubro
LNG	Liquified Natural Gas
m	Meter
MSL	Mean Sea Level
MT	Metric Tonne
NH	National Highway
NTPC	National Thermal Power Corporation
NUDA	Navsari Urban Development Authority, Navsari
NWRWSK	Narmada, Water Resources, Water Supply and Kalpasar Department, Govt. of Gujarat
PPP	Public-Private Partnership
RS	Remote Sensing
RWH	Rain Water Harvesting
SGCCI	Southern Gujarat Chamber of Commerce and Industries, Surat

SH	State Highway
SIC	Surat Irrigation Circle, Surat
SMC	Surat Municipal Corporation, Surat
SUDA	Surat Urban Development Authority, Surat
T.P.	Town Planning (as in T. P. Scheme)
UGSR	Underground Storage Reservoir (for Water)
URDPFI	Urban and Regional Development Plan Formulation and Implementation Guidelines

11.1 Introduction

The history of Surat dates to 300 B.C. The city flourished on the banks of Tapi river flowing from the East to the West that meets the Arabian Sea at about 18 km from the city centre [1]. However, it has its roots of origin beyond the era. It settled as a hamlet by then, and today it is housing more than 4.6 million citizens in its urban area [2]. Originally, it emerged as a settlement on the meandering bank of Tapi river flowing in the Southern part of Gujarat State. Over a period, the alluvial soils spread by frequent floods in Tapi river made the region fertile for agricultural activities and promoted for a larger settlement. However, the history marks the city as a Port town since the Mughal empire, and it is claimed that the port had offices of about 74 countries [1]. Figure 11.1 below shows the location of Surat district on a political map.

"It is the ninth largest metropolitan area and eighth largest city of India. Based on a study conducted by The City Mayors Foundation, Surat was globally ranked as 34th largest city by area. An annualised GDP growth rate of 11.5% was reported for Surat city over the seven fiscal years between the years 2001 and 2008. Also, the city was globally ranking 4th fastest developing cities [4]." Since then, the trading activities predominated and attracted for in-migration of people from different parts of India. Since past half-century, the port activities were diminished, and trading, as well as industrial sector, picking up as economic activity choice by the citizens. Today, the textile industries, diamond cutting-polishing, jewellery manufacturing and real estate sectors are predominating with ancillary service providers [5]. The lands have emerged a unique identity by the typical and unique culture of 'Suratis' (the Citizens of Surat), well known for a variety of delicacies, their friendly and welcoming behaviour showing acceptance to all who wish to contribute in the wealth of all. Surat is ranked Fourth among the fastest growing global cities [6]. The city has emerged as the economic capital of the Gujarat State, however, by the spatial extent and population, it is ranking second in the State following Ahmedabad. As per the records of Census of India, 2011, the urban population of Surat city was earmarked as 4.4 million with an administrative area of the ULB (Surat Municipality, established in the year 1852 AD and working as a Corporation since 1966 AD) [7] as 326.515 km^2 [8]. However, the urban region of Surat, the study area as a metropolitan region, has an extent of 1357 km^2 [9] housing a

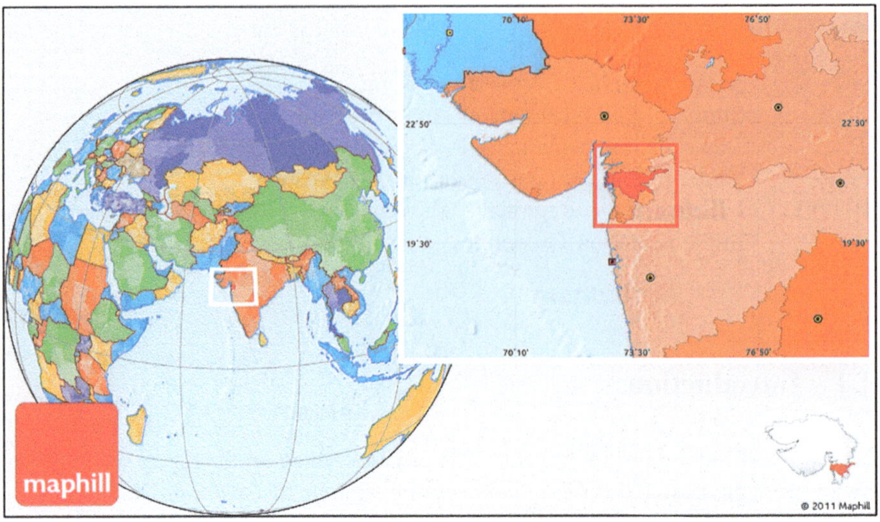

Fig. 11.1 Political location of Surat [3]

population of about 5.5 million. It forms the study area for the discussion in following sections and considered as a metropolitan region. An attempt here is made to analyse the study area delineated administratively by the Government of Gujarat as Surat Urban Development Authority (SUDA) and surrounding spaces thereof that has an inward impact. The region is located equidistant to Mumbai and Ahmedabad. The Arabian sea is at the West, the Bardoli Urban Development Authority (BUDA) boundary to the East and the Navsari Urban Development Authority (NUDA) boundary to the South of the study area. There is no development authority constituted to the North of the SUDA study region. Figure 11.2 herewith illustrates the delineated administrative boundary of the study region (SUDA) as a metropolitan region.

Administrative decentralisation of Surat city is split into seven zones (Central, North, South, East, West, South-East and, South-West zone) [10]. In addition to these zones, the study area of metropolitan region covers the inner and outer fringes of the Surat city forming a larger entity. Surat Metropolitan Region is becoming a major commercial hub of South Gujarat [5]. Surat city is home to the highest percentage of the migrant population in India, and thus a vibrant melting pot of varied cultures. As per a report, the UNESCO titled as '*Social Inclusion of Illegal Migrants in India*' [11] around 60% population of this city has its background in different states of India. Besides the region has been bestowed with a rich natural resource of a major perennial river Tapi, which drains off in the Arabian sea, a long coastline having biodiversity of flora fauna both marine, as well as terrestrial. The dang forests in the South Gujarat region (located within 60 km distance from the study region) are a home to a variety of tree species and herbs, chiefly the species of Bamboo. The South Gujarat region is also home to various tribes having indigenous

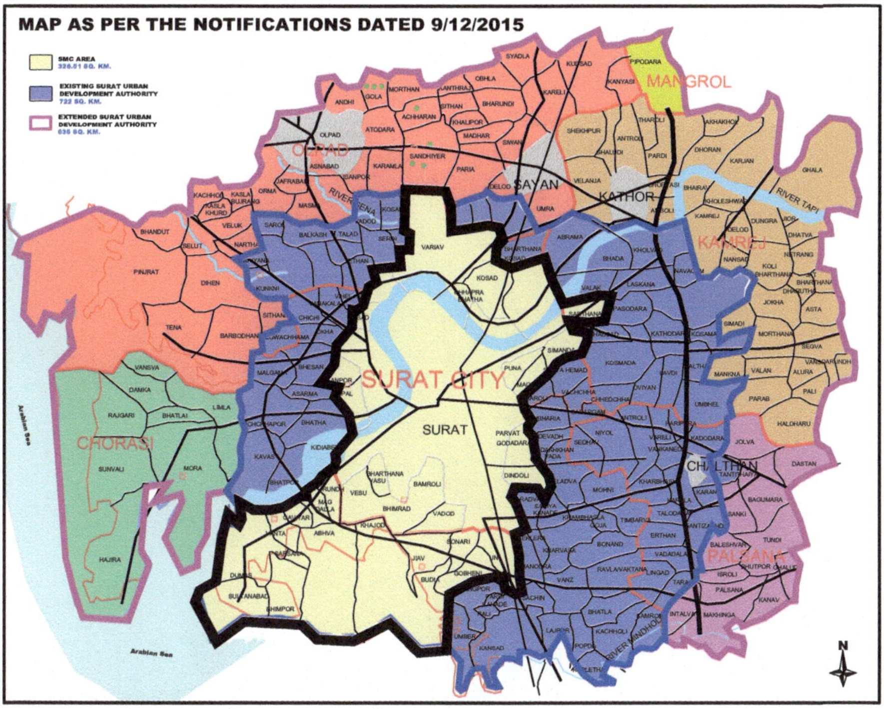

Fig. 11.2 Surat region study area (SUDA Boundary) [9]

cultural traditions that add richness to local arts and crafts giving the region a unique identity of its own. Since past three decades, the pace of development regarding several establishments and residential settlements is booming in Surat [5], at the same time certain industries are suffering due to lack of visionary and creative approach [12]. Below is a map (Fig. 11.3) showing development plan proposals of SUDA region [13] encompassing proposals for 225 village boundaries under the Authority.

A timely intervention with proactive and creative measures can save the entire region from the decay that is observed in various spheres. The concerns must focus on smarter regional development that promotes the vibrancy of varied social cultures, a healthy sustainable economy and most importantly preserving environmental resources at large leading to long-term sustainability of all the human settlements big and small within the region.

Current efforts presented here are aiming to visualise and address an ideal development based on an intervention and scope for proactive planning measures. It aims to identify existing geospatial conditions of various settlements of the study region, along with the industrial and commercial activity. Further discussions are taken up to propose strategies for smart development that integrates the smart city initiatives with the regional development.

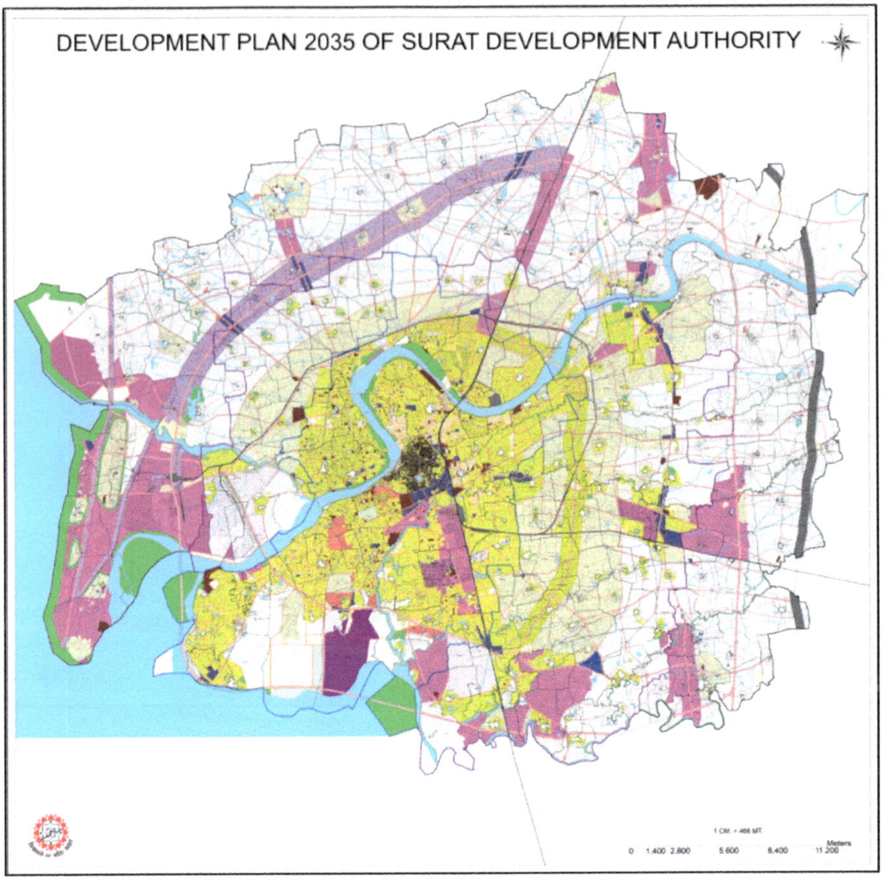

Fig. 11.3 Proposed land use plan of SUDA region [13]

11.2 Objectives of the Study

At present, the growth of local establishment is observed, that needs to be integrated with the overall vision for smart development. As the country is poised to take advantage of revolutionary economic policies of the present government, the region also awaits a total transformation in the way businesses are carried out, bringing more equity, more transparency and holistic development aided with smart technologies.

The current study is aiming at addressing not limited to but the following significant points of concern that may be considered as objectives:

1. To explore the development potential of sub-regions within the Surat metro region on smart and sustainable paradigms;
2. To identify growth extent and latent expansion possibilities as a multi-nuclei entity;

3. Identification of development goals and activities in accordance 'smart development' notion; and;
4. Forecast based guidance for the anticipated impacts of development in the entire of the metropolitan region.

11.3 Surat in Time and Space

The conventional story stretching the origin of the name of Suryapur refers to the time during the period of 1500–1520 A.D. when Surat was already a city of great trade. It seems to be possible that modern city of Surat was built on the site of old Hindu town of Suryapur. In the course of time, a significant Muslim domination changed its Hindu name "*Suryapur*" to "*Suryt*", and over a period, it was largely practised as "*Surat*" [1]. Surat became the most critical trade link between India and many other countries and was at the height of prosperity till the rise of Bombay port in the 17th and 18th centuries. Surat was also a flourishing centre for shipbuilding activities. Post rising of Bombay as an active port, the port of Surat faced a severe blow. As a result, the shipbuilding activities declined Surat witnessed considerable growth in the industrial activities (notably, textiles and allied) after India's independence. The expansion of city limits has a one of a reason as increasing industrial establishments and rising need to accommodate the workers [14].

Formerly settled on the banks of river Tapi with a fort on the eastern bank and a Custom house on the northern side of the port the city growth in the initial years was concentrated in the inner walled city area only. The wall was constructed in the year 1664 A.D., and the area within the walled city measured about 1.78 km^2. There were totally 12 gates allowing entrance to the city. The outer wall was constructed in the year 1707 enclosing an area of 7.35 km^2. The economic and infrastructural development since the 1960s, seen in the city and the region, have resulted in a spurt in an urban population in the city.

The evolution of the power loom and hand-loom sectors and diamond cutting and polishing industries in the year 1950s changed the city outlook. Parallel to the industrial expansion, Surat emerged as a major centre for trade and commerce in the region and a silent evolution has been that of the informal sector. The turnaround of the city of Surat happened after the plague in the year 1994. Within two years, Surat had been transformed from one of the filthiest cities to the second cleanest city in the country. In the centre of all this transformation process is the Surat Municipal Corporation (SMC). A systematic process to upgrade infrastructure, both quantitatively and qualitatively, has been made by the local government. Surat continues to present a 6% plus annual population growth since the year 1960s, placing Surat 8th regarding population size countrywide (in the year 2011) with a population of about 4.46 million within local government limit to the extent of 326.515 km^2 and a district population of around 6.08 million (from Census of India, 2011), a change of about 42.24% over the district population recorded in the year of 2001. Notable is

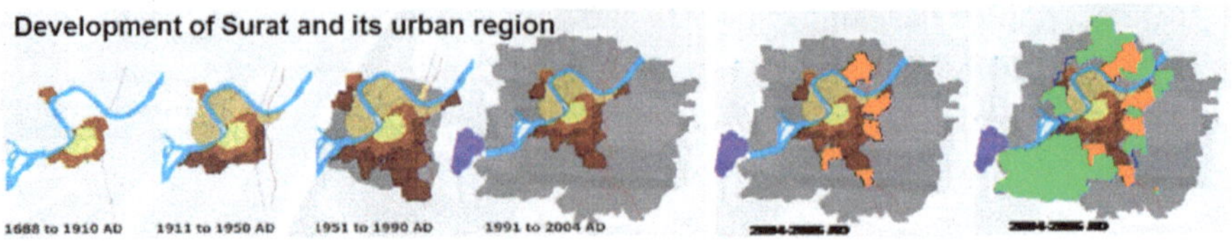

Fig. 11.4 Spatial development extent of Surat [12]

Year	SMC area			SUDA area (Except SMC)			SUDA area		
	Population	Density per sq.km	Growth Rate	Population	Density per sq.km	Growth Rate	Population	Density per sq.km	Growth Rate
1981	9,99,373	3,066	--	1,20,993	306	--	11,20,366	1,552	
1991	16,24,135	4,982	62.52%	1,55,501	393	28.52%	17,79,636	2,465	58.84%
2001	28,68,603	8,799	76.62%	2,36,521	597	52.10%	31,05,124	4,301	74.48%
2011	44,73,143	13,721	55.93%	3,31,739	838	40.26%	48,04,882	6,655	54.74%

Fig. 11.5 Population, density and growth rate for study area. *Source* Derived from [8] and [13]

also the fact that out of this about 80% population comes under urban area, and only 20% population resides in rural areas, which is far higher than national average of 32% (urban population) and 68% (rural population). While the city has successfully filled the sizeable infrastructural gap accumulated over the decades within a short span of a decade. Additionally, consistently high rate of growth in population has posed for newer demands. From time to time jurisdictional limits of SMC the local governing body has also been extended to include the outgrowth.

Following is the geospatial expanse (Fig. 11.4) and a brief regarding the hierarchy of various settlements of this region that demonstrate for SMC and SUDA region. The population extent and growth regarding SMC area, SUDA area and remaining of SUDA (Except SMC) area are illustrated (Fig. 11.5).

11.4 Demography Dynamics

Due to rapid industrialisation, with the largescale business establishments within the study region of Surat Urban Development Authority (SUDA) as KRIBHCO, L&T, ESSAR, NTPC, Reliance industries and such the population growth is increasing. Surat leveraged magnetic effects attracting the migrants with huge job potentials in the traditional textile, diamonds and construction activities Also, relatively peaceful and harmonious social environment and moderate climate have converted the migrants into permanent settlers in the city. It resulted in higher decadal growth compared to any city in the state in last two decades. This statistic of the population of the Surat city from the census year 1901–2001 is (Fig. 11.6) highlighting the centurial increase in city area by about 14 times coupled with about

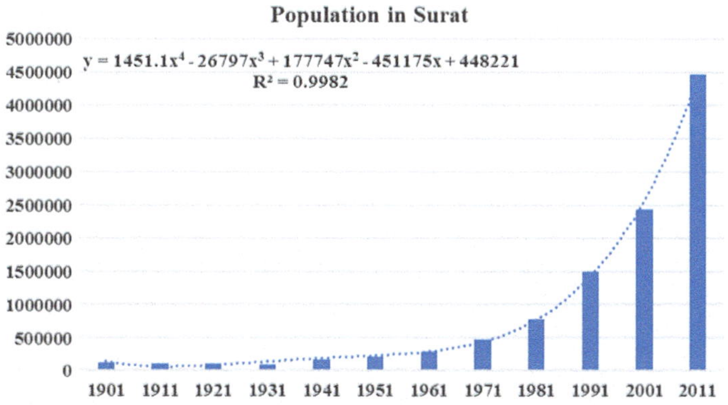

Fig. 11.6 Population of Surat [15]

20 times increase in the population. The figure depicts for the decadal population figures showing abrupt rise in population after the year 1981 highlighting effect of in-migration.

Surat is considered as a Mega City in India. In the Census year 2001, it came out with the highest growth rate among cities of Gujarat state. Moreover, Surat was the second 'Most growing City' in India as per the Census of India, 2001 with a rate of 62.38%. In the years after 1971, the population growth was recorded tremendous with a continuous rise in the population showing a polynomial trend.

Surat city recorded an exponential growth in decades of recent past. It has its significant effect in the fringe region in all directions. In the next sub-sections, the effect of population growth in the entire of the region since the decade of 1961 till 2011 is visualised. The spatial pattern of population growth depicts a possibility for the existence of correlation among city population increase and inhabitants in the surrounding areas in the study region.

Since the independence of India, an independent Census of India was first time conducted in the year 1961 after the adoption of the Constitution of India. Simultaneously, it was a period for implementation of Five-year financial plans float by the then Central Government, and in this initial duration, primary focus of the funding was for the agricultural sector. Below discussions are based on the current study area boundary with the incorporation of records of each. It shall be helpful in visualising any existence of interrelationship among various locations. However, there is always social and economic interdependencies among places in such a region.

In the map below (Fig. 11.7), it is observed that the population in different administrative constituencies based on records of the Census year 1961. The overall population in various parts were observed to be of moderate level residing in a spatially scattered manner.

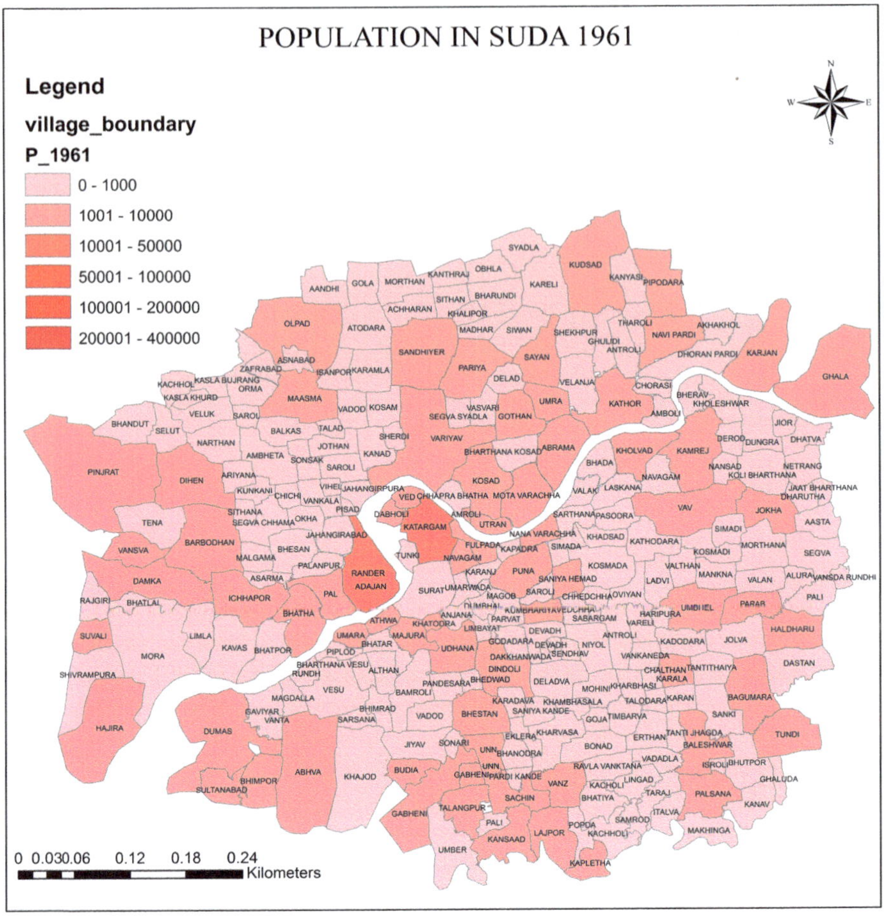

Fig. 11.7 SUDA population in the year 1961. *Source* Developed by Authors

Over decades then after, a trend of human settlement is visible and showing a concentration at the centre of the study region, known as Surat city (Figs. 11.8, 11.9 and 11.10).

The decade from the year 1991–2001 shown one of the highest degree of progression regarding population rise. The same is visualised in maps presented here with (Figs. 11.10 and 11.11) The decade recorded highest growth rate for Surat city (to the extent of 88%) in the Census of India, 2001. It brought the region to the attention at many levels.

The rise of enterprise establishments manifold is believed to be the reason behind. The sex ratio was reported to be low as 773 [8] in the year showing possibilities for substantial migrant males (workers) from different parts of the nation.

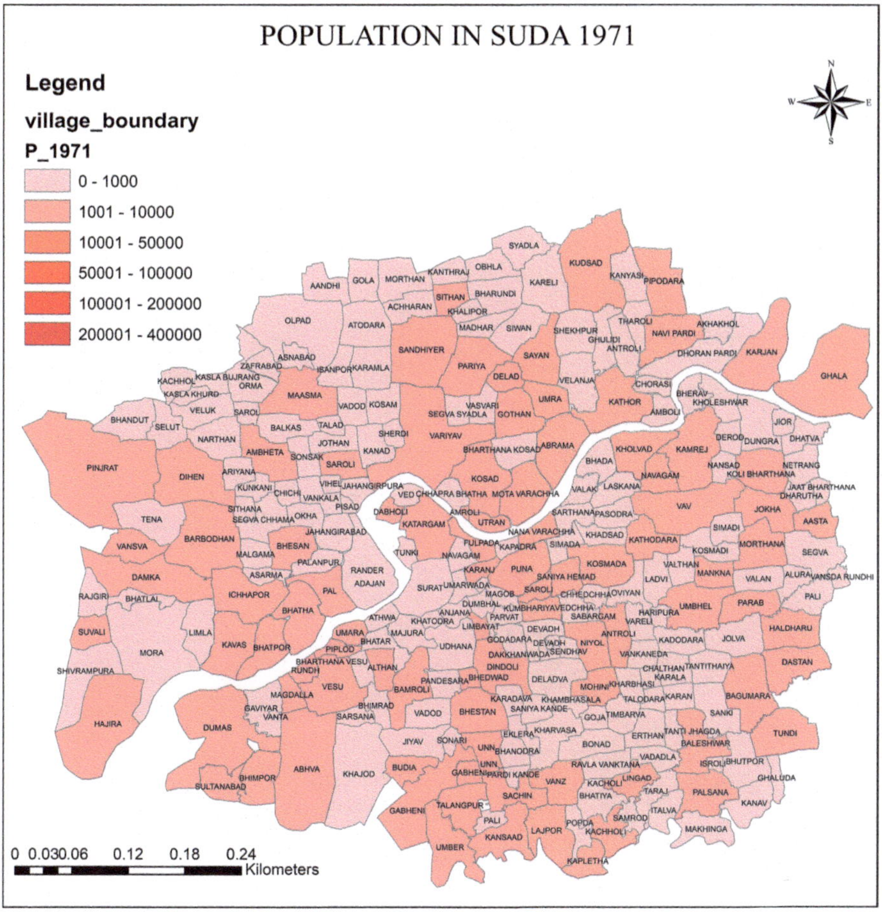

Fig. 11.8 SUDA population in the year 1971. *Source* Developed by Authors

11.4.1 Present Situation

Many highly populated localities in the region were recorded in the Census of India, 2011 for the study region. The scenario is illustrated (Fig. 11.12) where not only the Surat city but a high level of population can be seen in other constituencies also. Compared to the scenario in the year 1961 (Fig. 11.8), the population is showing a spread in the entire of the study region. The rise in the population may be because of scattering and sporadically increasing economic activities. A detailed assessment based on location for each type of enterprise establishment may reveal a realistic scenario. However, based on broad level information available, the economic activities regarding current scenario is discussed later.

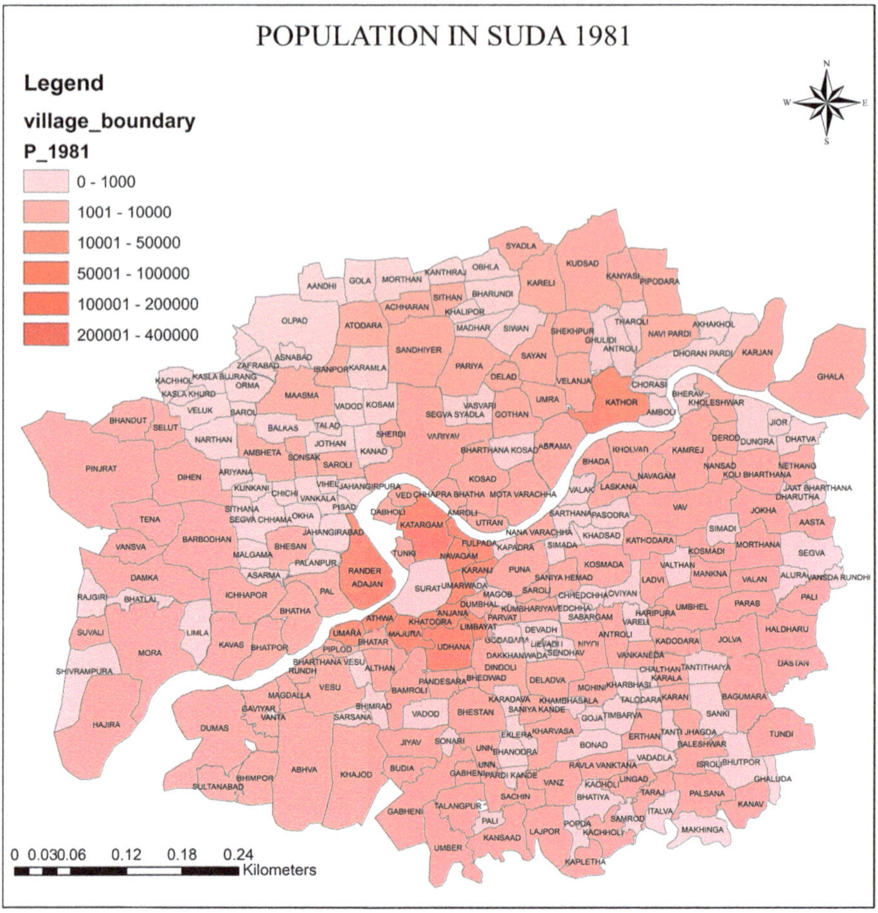

Fig. 11.9 SUDA population in the year 1981. *Source* Developed by Authors

11.5 Regional Features and Connectivity

11.5.1 Climate

The climate of Surat city can be broadly divided into four seasons viz. summer, rainy, autumn and winter. Summer for three months from March to May, Rainy from June to September, autumn from October and November and the winter season is from December to February. The climate of the city is hot and humid due to the location of the Arabian Sea nearby. The midsummers are fairly warm with temperatures extending from 37.78 to 44.44 °C. The weather is pleasing during the monsoon while autumn is moderate. The wintertime are not very cold but the temperatures in January array from 10 to 15.5 °C. The maximum humidity is about

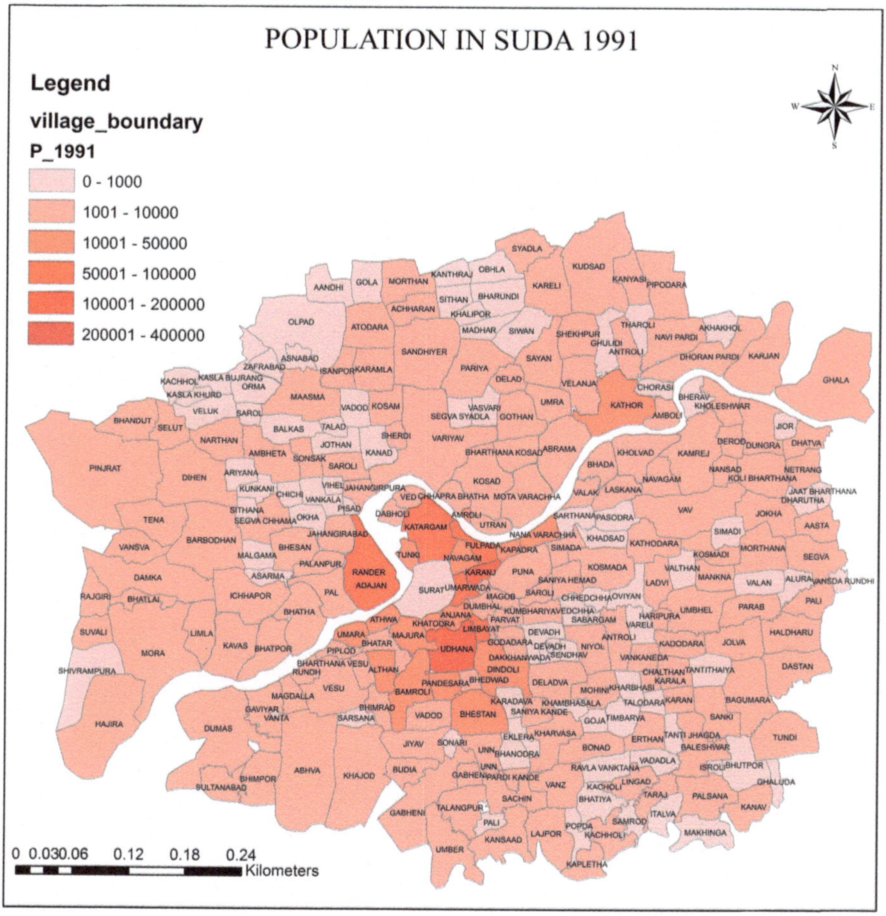

Fig. 11.10 SUDA population in the year 1991. *Source* Developed by Authors

80%. The raining term is usually five months with an average rainfall of 114 cm [1]. Surat experiences prevailing winds from the South West direction. The breeze is mostly from the sea towards the central city. Having a proximity to the Arabian Sea, the city enjoys a pleasant climate throughout the year.

11.5.2 Connectivity

The city of Surat is situated at latitude 21°12′N and longitude 72°52′E on the bank of river Tapi having a coastline of Arabian Sea on its West at a distance of about 19.4 km by boat along the Tapi stream and about 16 km by road towards Dumas. It is 13 m above the mean sea level. Surat city is located in well-developed South

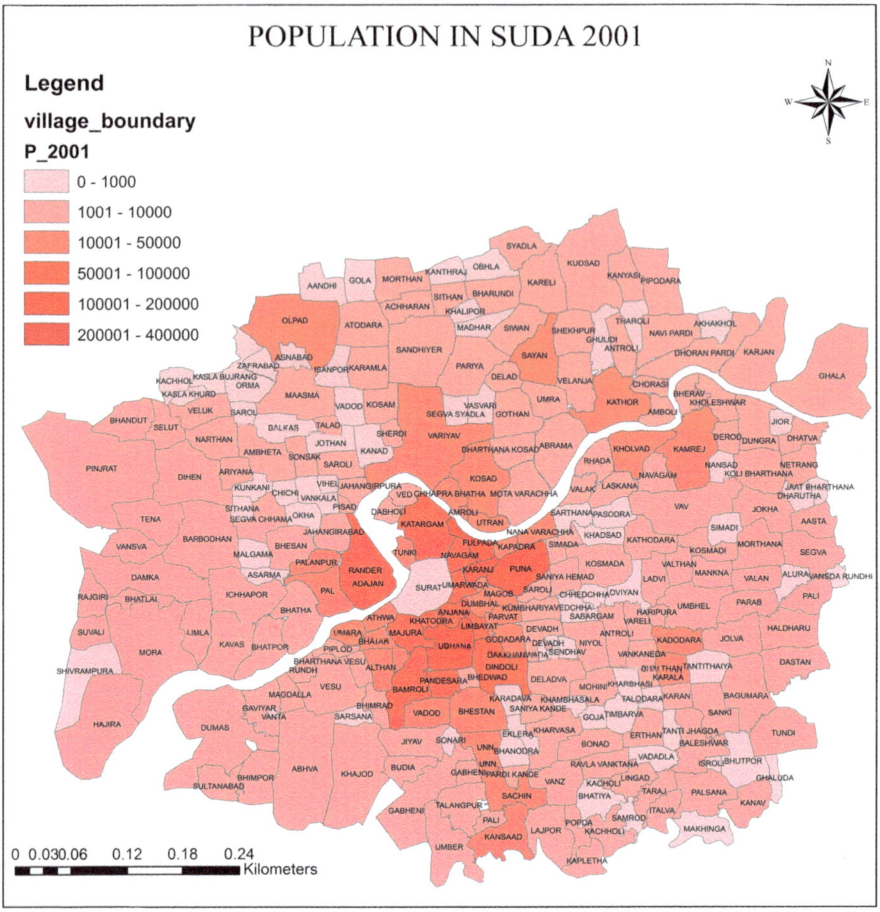

Fig. 11.11 SUDA population in the year 2001. *Source* Developed by Authors

Gujarat region. The region occupies a pivotal position on the nationally significant Ahmedabad—Mumbai corridor centrally located at 260 km North of Bombay and 224 km South of Ahmedabad. The region is setting on the 225 km long industrial belt, having direct linkages with the urban industrial centres of Vadodara, Ankleshwar, Vapi and so on. The National Highway No. 48 passes within 16 km of the SMC boundary and is one of the busiest inter-state trunk routes in the country. The city is well connected by road and rail with the major cities and towns of the state as well as states of Maharashtra and Madhya Pradesh.

The city lies at a bend of the River Tapi, where its course swerves suddenly from the south-east to south-west. With the walled city at its centre, the city forms an arc of a circle, the bends enclosed by its walls stretching for about a mile and a quarter along the bank. From the right bank of the river, the ground rises slightly towards the north, but the height above mean sea level is 13 m. The topography is controlled

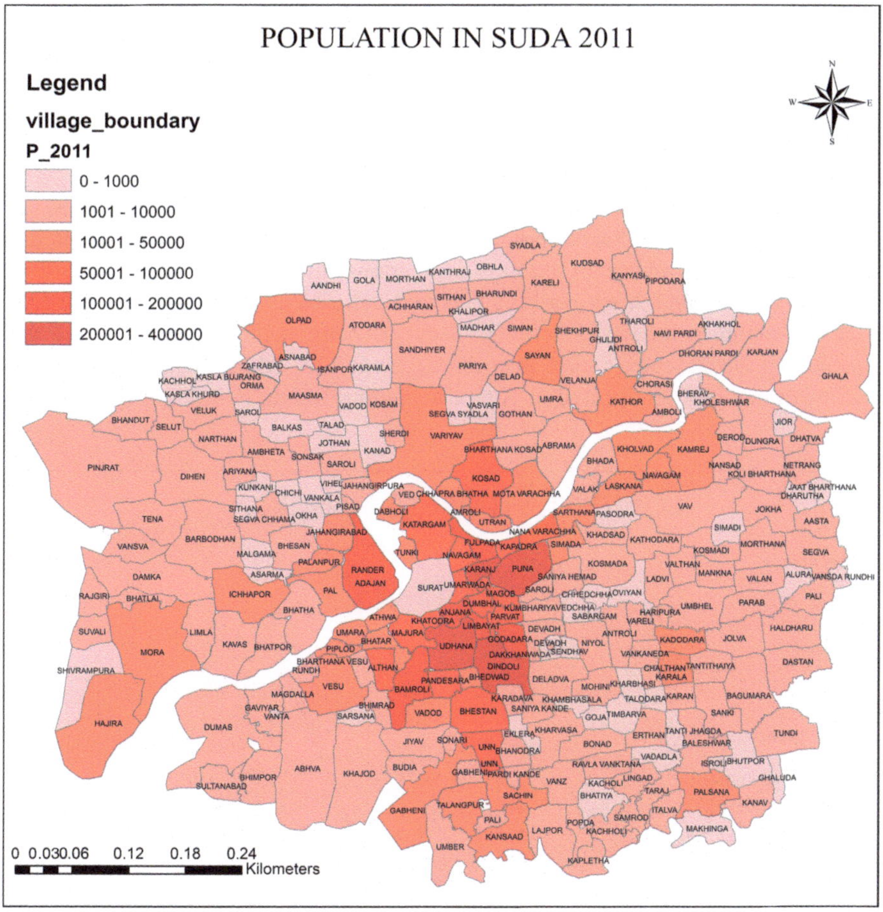

Fig. 11.12 SUDA population in the year 2011. *Source* Developed by Authors

by the river and is flat in general, and the general slope is from north-east to south-west.

Surat has a national airport providing commutation links to different metros. Daily flights are operative for Delhi, Mumbai, Jaipur and other places. The state government has also commenced an aviation master plan for Gujarat with prioritises development as well as up gradation of Surat airport. The other nearest airports are Vadodara and Ahmedabad. The upgraded Surat airport is envisaged to offer direct air connectivity with essential destinations in India and abroad. An expectation is sought for boosting up commercial activities in the city as well as in the district, leading to an upsurge in demand in the hospitality sector, which is primarily driven by the corporate tourism.

The study region also has seaports operative by private entities. One is located on the Southern bank of Tapi river at Magdalla and another on the Northern Bank in

Hazira. Both these ports are in the vicinity delta region of River Tapi. Magdalla & Hajira Ports in Surat have excellent rail and road connectivity (Fig. 11.13). Magdalla port is about 2 km away from the state highway and 15 km away from NH-48. Hajira seaport in Surat is adjacent to a high speed dual carriageway. The port is well linked with the Mumbai-Delhi Freight Corridor. The linkage is expected to become a chief strength for the industrial growth Hajira Port.

As reported by the Government of Gujarat [16], "Hajira is known as the Gateway Port to serve the hinterlands of North, West and Central India as it is situated in the midst of one of the most industrialised area in the country. It is thick water, all-weather and direct berthing port in Surat. The existing industrial portfolio of Hajira includes industrial activities such as petrochemicals, fertilisers, heavy engineering, steel, energy and port-related activities. More than 20 large and medium companies are located in Hajira, including Bharat Petroleum Corporation,

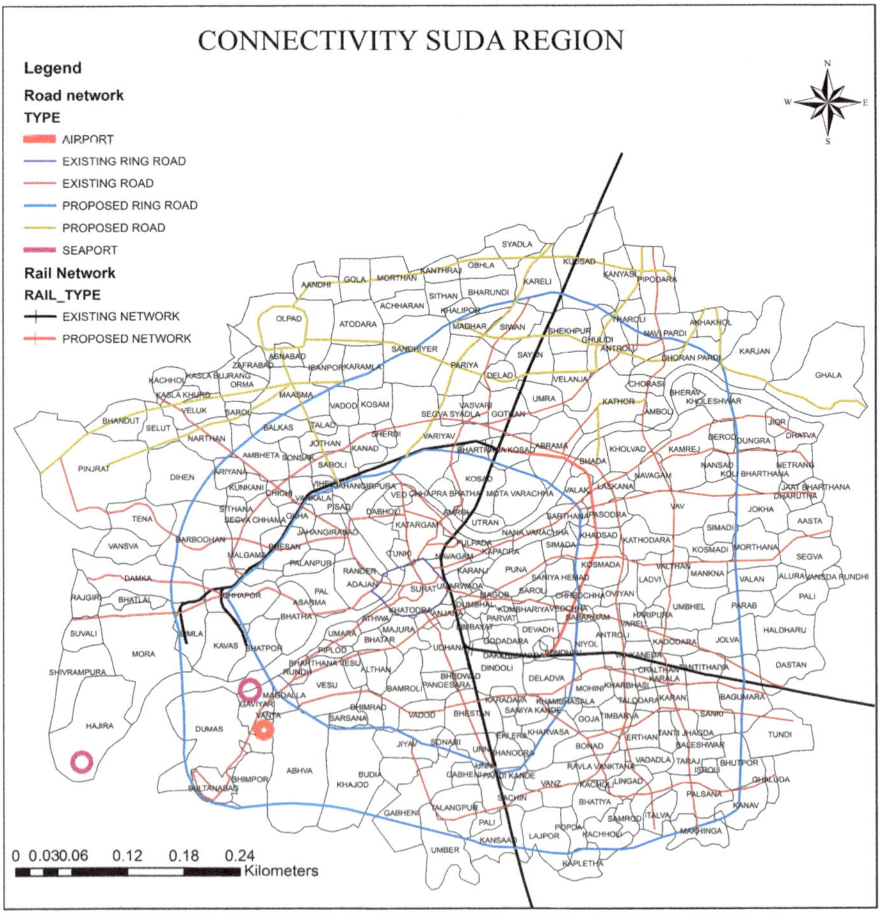

Fig. 11.13 Regional linkages in the study region. *Source* Developed by Authors

Cairn Energy, Essar Power, Hindustan Petroleum and Larsen & Toubro, Reliance, NTPC among others. Shell established an LNG Terminal at Hajira in 2004."

11.5.3 Surface Features

The terrain of the region has a mild slope towards the West from the East. If the Surat city railway station is considered as a pivot of the region, the ground level is 17 m above the MSL. It is located at a distance of about 20 km from the coast of the Arabian sea on the West.

There are more than 300 water tanks (ponds, lakes) in the region, of varying size and storage capacities. The River Tapi is flowing from the centre of the study area, and River Mindhola is located in the Southern part of the region. Both rivers traverse from the East to the West. Apart from these rivers, there exists a substantial network of streams that acts as a feeder to these rivers. Small channels are mostly rising from the end of various irrigation canals and at present catering a need for sewage and effluent disposal. With recent efforts of the SMC, the bank protection work along with buffer zone creation is in progress. Such works will ensure of non-disposal of unauthorised and unwanted wastewater disposal in these streams in future. Also, with reduced flow carrying capacity of these channels, a flooding condition takes place in the monsoon season in a few specific habitable city areas in surrounding of these streams.

The land use and land cover analysis as illustrated (Fig. 11.14) shows a high density of road network and developed land in the centre of the region. Also, there is considerable scope for planned future development with available vacant lands. Mostly, these vacant lands are utilised for the agricultural activities. Based on the observations, the cash crops (such as sugarcane, cotton, and so on) are predominating in the region [16].

The extent of available developable land in the Surat metropolitan region is better illustrated in the map below (Fig. 11.15). To the North, South and East of Surat city, mostly the land parcels are occupied with agricultural activities. The Western part of the region has forest lands, mangrove covers on the seashore and some parts with the barren uncultivable land. Based on a thought of 'The barren land parcel can be ideal for industrial activities without damaging the agricultural activities' resulted in the establishment of industrial hub of Hazira in the Western part of the study region. The South-west part of the region has much of mangrove covers in the vicinity of Mindhola River mouth. The establishment of Sachin industrial estate in the Southern part of the region is due to open barren land parcel similar to settings of the Hazira.

Built-up Spaces in the Region

A spatial extent of built-up spaces is illustrated in a map below (Fig. 11.16). It shows a highly built space in the central part of the region (the Surat city). Also, there lie built-up spaces in the Western area in Hazira pocket. Undeveloped and

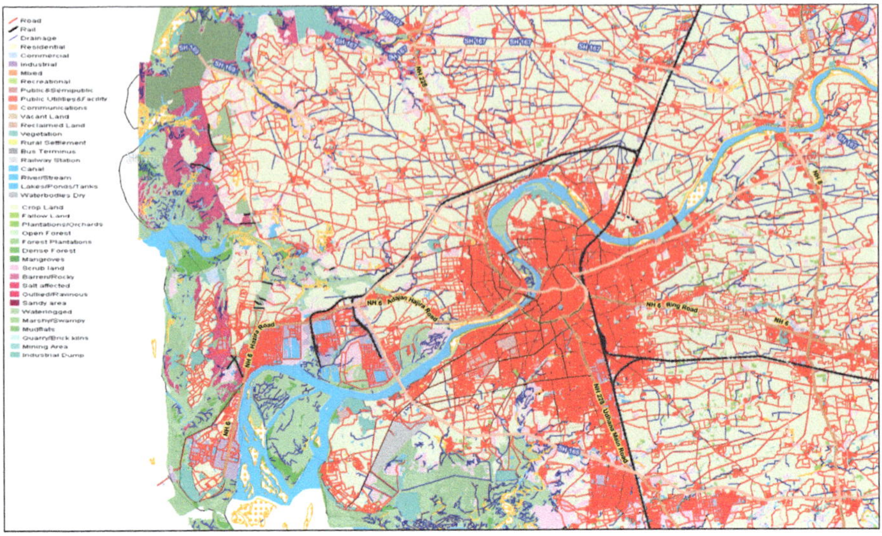

Fig. 11.14 Urban land use in the study region [16]

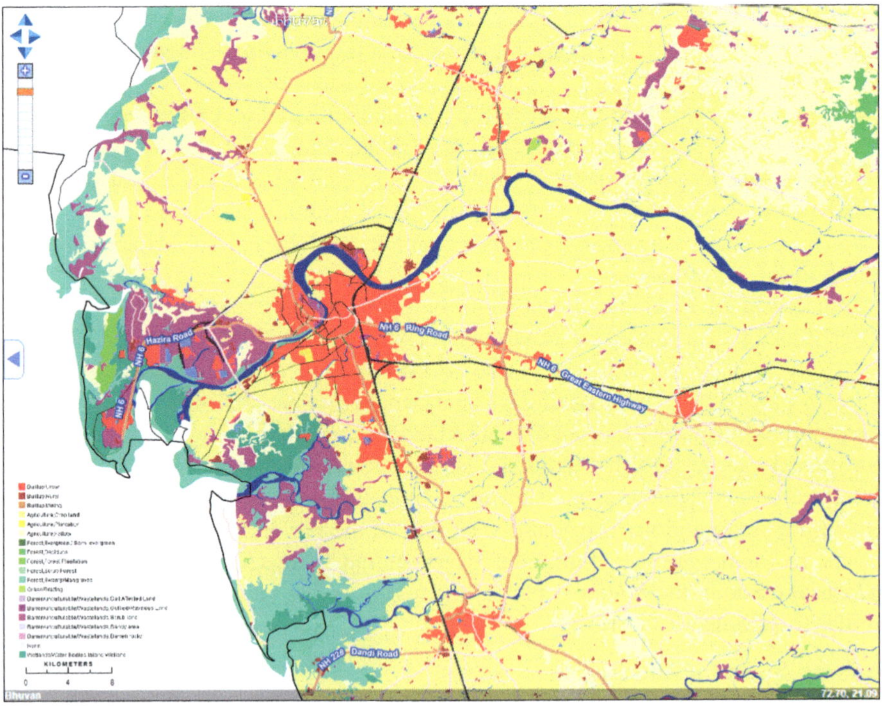

Fig. 11.15 Land use and Land cover in the study region [16]

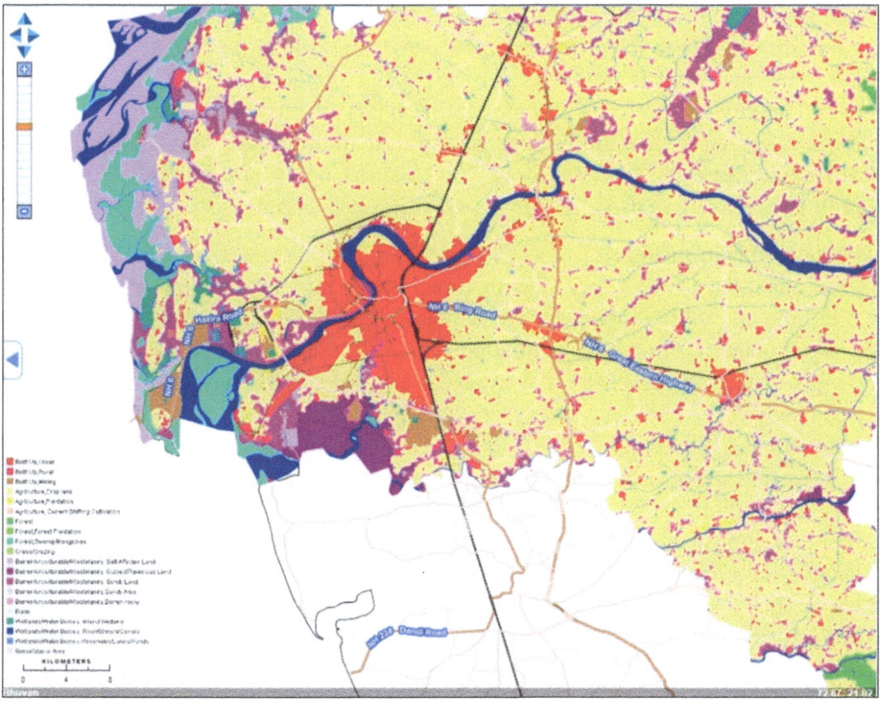

Fig. 11.16 Built land use in the study region [16]

barren lands are available in the Southern part of the study region. Apart from these significant chunks, some scattered built-up spaces can be visualised in entire of the region showing no specific pattern. However, along with the NH 48, the ribbon development is observed in some parts. In future, the same can result in an extensive and haphazardly grown area if not planned well in time.

11.5.4 Infrastructure for Services

Water Supply Services

The entire region has a natural boon regarding annual rainfall. The study region has about 1000 mm rainfall in the monsoon (for about five months) per year. Also, the soil profile allows the precipitation to get absorbed and filtered naturally. The groundwater is mostly utilised for agricultural activities. For domestic and industrial water supply, the systems are in place by the SMC or GWSSB (for villages out of Surat). Scattered places cater to their needs by using either groundwater source or by supplies through water tankers, that is seldom observed.

In the city area, the SMC has well-established facilities regarding raw water collection from the Tapi River (a perennial one) using radial wells and such mechanisms. The raw water is treated as per the quality standards imposed by the GPCB norms. The city is covered with piped radial type of water supply network, and the purpose is solved by the intermittent gravity flow through ESR and UGSR facilities established across the city. In about 500 m surrounding area of SMC limits, the facility is extended to users as per the Government directives.

Sewage and Sanitation

As the city area is well served with piped water supply network, the SMC has covered the city with a piped sewage network as well. Gravitational forces are used and the sewage is brought to a treatment facility at various locations (Fig. 11.17) serving surrounding area [18]. The treated wastewater is discharged into the natural drains nearby. The SMC has not opted for discharging this water in the Arabian sea directly. In the remaining parts of the study region, the wastewater is mostly primarily treated using a community septic tank and released in the deeper grounds using soak pits.

For the industrial units, the GIDC has facilitated for the piped network for collection of effluent and treat the same. Apart from it, as per the mandate in practice by GPCB, the industrial units consuming more water (such as colour manufacturing, chemicals, cloth dying and so on) are established with individual wastewater treatment plants within the premises. There are some scattered enterprises established outside the earmarked GIDC area, where the practice imposed by GPCB is strictly followed under the provision of acts in practice.

Fig. 11.17 Sewage treatment plants, facilities in the study region [18]

Stormwater Drainage
As the entire region is enjoying the considerably right amount of annual precipitation during monsoons each year, the natural features are already in place to support the water runoff caused by a storm.

The city area has a separate piped network in place by SMC for stormwater collection, conveyance and discharge to the River or another stream. The SMC has also constructed HR at stormwater disposal locations to avoid back-flooding from the water stream. In the urbanised area, the surface water resulting from a storm is collected in the network and safely discharged to nature without making it affect adversely in any manner. Apart from the city, remaining areas in the study region has seldom established such network facilities, and the dependency is predominated for the natural features, and roadside drains for areas in the vicinity to NH or SH.

The SMC and SUDA have laid regulations for practising rainwater harvesting to avoid surface runoff resulting in loss of large quantities of fresh water. On the other hand, in the Western part of the region, due to groundwater extraction, over a period, the salinity ingress is increasing. The act of water conservation by RWH shall reduce the effect of salinity spread in the groundwater sources.

Increasing urbanization and blockage of natural drains has resulted in flood inundation. Many such instances are recorded bringing in devastating hazards resulting in huge loss of life and properties. A map below (Fig. 11.18) shows the flood incidences in the region. The Government of Gujarat has implemented Real-time data acquisition system in the upstream of Ukai day (located at about 80 km to Surat city) that serves as a multipurpose reservoir. Ukai dam stores the water and serves for water supply related to diverse requirements in the study region. As discussed earlier, flooding in the region is observed not only due to the River Tapi but also with encroached flow path of the natural drains in the study region.

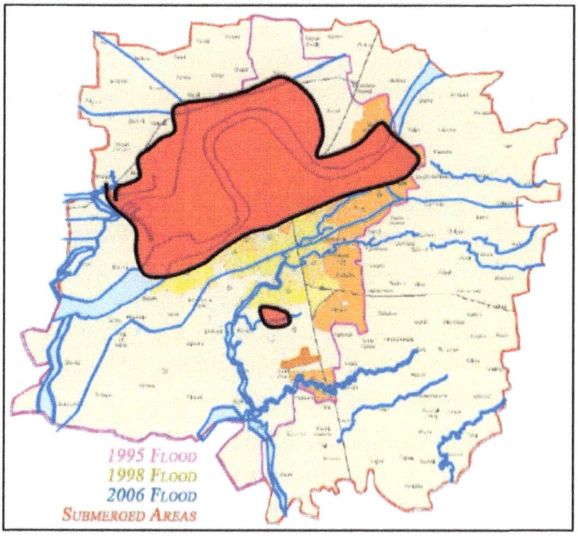

Fig. 11.18 Floods in the study region [19]

Solid Waste Management
Waste management is essential for retaining a healthy society. Within the city limits, the SMC carry out the task regularly and efficiently. The door-to-door garbage collection system is in practice for collecting domestic segregated solid waste that is transferred to stations and sent for sanitary land fill site. The SMC has incorporated for the field work, daily meetings, complaint redressal system and litter prevention system [20]. Urban streets are swept daily. The SMC uses more than 300 vehicles with more than 700 trips to haul a daily average of about 900 MT of solid waste generated. About 700 MT of solid waste is collected through the container lifting. Night scrapping activity helps in collecting about 215 MT waste on a daily basis. Additionally, about 1150 MT solid waste is collected using secondary transportation [20].

In the remaining of the study area, no regular practice is followed in absence of delegation of the responsibilities. However, open air burning activity for dumped waste is observed to be one of a local alternative in practice by the citizens. In some parts, agriculture waste recycling is also practiced locally however, it is not predominating and seems not to have any scopes for a commercial conversion.

Electricity
One of organizations of the Government of Gujarat, Dakshin Gujarat Vij Company Limited (DGVCL) has established thermal power generation plants in the study region. It operates along the Gujarat Electricity Board (GEB). The DGVCL also caters the electricity supply needs in most of the area for domestic and industrial requirements. In addition, a private entity namely, Torrent power limited has also its power generation plant in operation (Sugen) on Tapi river. The NTPC facility at Kawas on the western part of the region is a gas based power plant and generate electricity mostly for the industrial supplies. A map below shows various power generation facilities in the Surat metropolitan region [21]. So far, the power supply is made available adequately for the domestic usage however, regular cutoffs are observed on a weekly basis for the industrial establishments (Fig. 11.19).

The Government is promoting for installation for and using non-conventional power generation modes and citizens are transforming at a slower pace considering cost aspects.

Canal Network
The Surat Irrigation Circle (SIC), a setup by the Government of Gujarat under the ministry of Narmada, Water Resources, Water Supply and Kalpasar Department (NWRWSK) is actively operative in the region. Prime responsibility of the organization is to cater the needs of water for agricultural and industrial use. It has deployed a large canal network in the region and functioning since 1980s. The organization also assists in maintaining the Ukai dam and acts as a flood controlling and coordinating unit during the monsoon season.

Internet and Communications
Mobile phone and landline operators include BSNL, RJio, Airtel, Idea, Vodafone, TATA Docomo, Uninor, Telenor, and a few others. Below Fig. 11.20 illustrates the

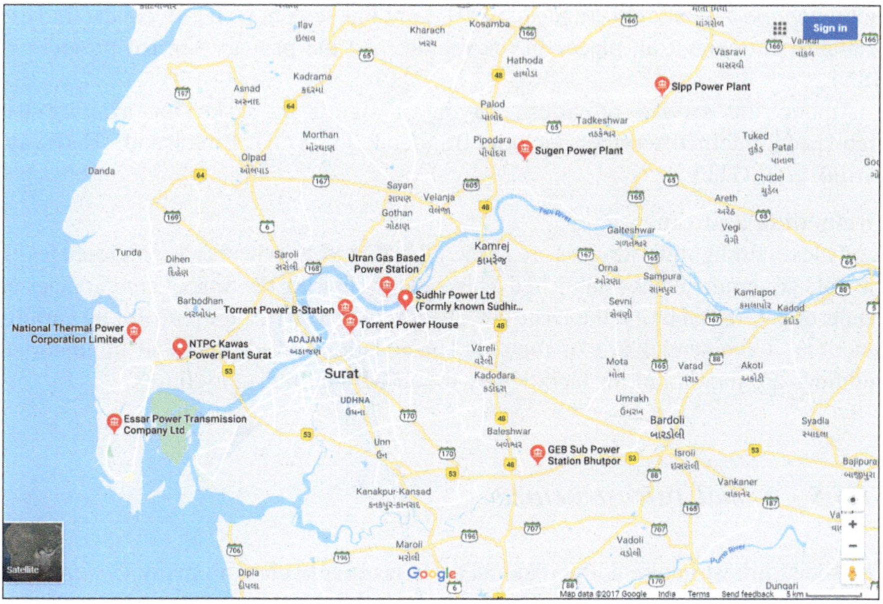

Fig. 11.19 Electricity generation plants in the study region [21]

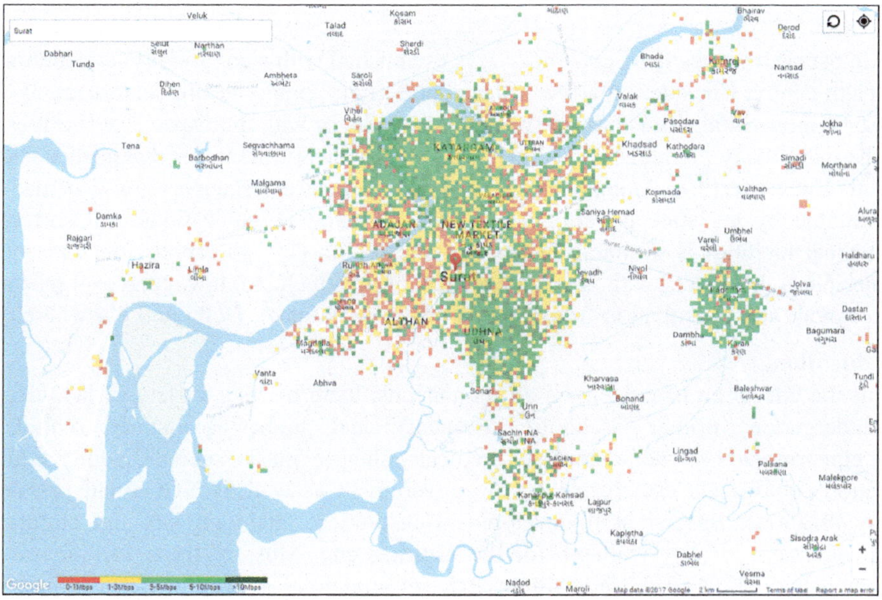

Fig. 11.20 Mobile communication network strength in the study region [22]

mobile phone network coverage area [22]. Landline telephone based connectivity is available throughout all places in the region, and the primary service provider is BSNL.

Among the service providers, having a significant stake for availing the high-speed internet services include BSNL, RJio, YOU Broadband, Hathway, Airlink and, GTPL.

Firefighting and Emergency
A city level firefighting services are rendered by the SMC whereas at a regional level, the state government has established a few centers. During the time of emergencies or events of natural calamity, the District Collector play lead role for performing various functions. Commercial use of fuels and installations of elevators in multistoried buildings are monitored for facilities by the SMC within the city limits.

11.5.5 Social Infrastructure

The social infrastructure at the regional scale is remarkable with many Government as well as private owned facilities and services. Implementation of Government programs and schemes related to health and education are observed. With advancement of technology, the level of public participation is also increasing and improving in all aspects.

Health
Almost each village level constituency is established with a community level public health centre. Considering a few villages in a group, public health centers are also established. Surat city is having two medical colleges with multispecialty facilities. There are many privately owned hospitals and medical centers for diagnostics and pathological testing. Ambulance services for medical emergencies are also made available by the state government as well as private organizations. All sort of medical facilities is available in the city and region through reputed doctors and specialists however, with specific desires of treatment, a few citizens opt for treatment and consultation from experts based in Mumbai, Delhi and elsewhere.

Education
For the education of citizens in all dimensions, there exists a variety of facilities. Kindergarten, primary schools, secondary and higher secondary schools, engineering-arts-science-commerce-medical colleges, music schools, adulty education centers, are established by the government as well as NGOs and private organizations. Special skill development institutions including art and cultural development is also available for the aspiring one. Most of these facilities are established based on the prevailing URDPFI guidelines as well as on need-based provisions for the community. However, there exist no mechanism for identifying gaps that is addressed to some extent by the authority while preparing development plan for a ten-year implementation period. For availing education, outmigration

from the region is seldom required unless ambitious and very specific. SMC has established a unique Children Traffic & Training Park that allows children to drive a cycle in a controlled system (away from actual risks) to have a perception of a road and traffic movement along with sensitization for traffic signs and signals.

Recreation and Cultural Activities
The entire region has pockets developed where citizens enjoy through a variety of recreational activities. However, the sector still has scopes for development. There are regional level gardens, lakes fronts, a recent and developing river front, amusement park, Dumas sea shore are a few to name where citizens opt for recreational happenings. Community halls, indoor and outdoor stadiums, reading halls, privately owned party-plot cum gardens are occupied with celebrations and festive activities. The stadiums are well utilized for a variety of sports. However, a forma sports education institution seems to be identical for establishment. Citizens in the region has a unique tradition for involving in celebrating and enjoying festivals; in a way people observes to have a habit of recreation in non-space specific and a different way.

A recently redeveloped lake front, Gopi Talav has been added for recreation at a regional scale. The place was built in around 1510 CE by the then merchant Malik Gopi who happened to be a Governor also [23]. While the construction of the lake was in progress, it was Moghul Era prevailing. The construction of this pond was to serve for retention of flowing water and conserve in open. Over a duration, the pond lost its importance through ignorance and changing practices of citizens. However, the SMC put efforts to recover not only the water body but also step-well in vicinity and gave a modern outlook. The facility of recreation is made available to the society by leveraging operation and maintenance on contractual arrangement on a PPP mode.

Heritage
As discussed earlier, the city history dates back and considerably old. Precincts that can be identified as a heritage are vanished over a period. There exists an absence of a dedicated heritage preservation policy as well as mechanism to implement conservation activities. In the villages surrounding the city, there exists stepwells for water conservation having heritage values and still are operative. Documentation and dissemination about these is somehow leaves a scope for improvement. Documentation carried out by the then English and French officers in form of their diary are available. Such notes reflect the practices in past and explains about certain properties if could be preserved, it would be contributing to richer heritage values.

Some special delicacies (*Ghari*[1], *Bhusu*[2], *Khaja*[3], *Khaman*[4] and so on) and preparations has a special mention in the region as a cultural heritage. Surat is very

---

[1]A traditional sweet.

[2]A mixture of fried delicacies made up mostly of gram flour.

[3]A spicy delicacy made up of fine wheat flour.

[4]A locally famous breakfast item made up by steaming of gram flour.

famous among food-lover community and attracts people for various parts of the nation. These days with technological advancement, the vendors are taking advantage of value addition to the preparations and export globally using web-based and mobile- application based order catering using available rapid and efficient goods-transportation services.

11.5.6 Administration

Following the norms set through the Constitution of India, the administration and governance is practiced in the study region. Various authorities, government departments (central and state level) are operative. The SMC is an urban local body and an exemplary local government responsible as a city manager for Surat.

The affairs at the city level administration are generally dealt by the SMC. Block level administration is in action for the remaining parts of the study region There is a considerable overlapping of geographical distribution of responsibilities. The map below (Fig. 11.21) shows the block level distribution of the study region.

The Surat district is administered by the Collector. The one is assisted by a District Development Officer as well Assistant Collector. The collector is responsible to the ministry of Revenue at the State level following the Constitutional arrangement. A district is sub-divided in blocks. Each block has a Block Development Officer in command. Each block us further sub-divided in villages where a *Talati cum secretary* is looking after the administrative and record keeping as well updating tasks. At each level, there exist a supporting rather leading public representation as Member of Legislative Assembly of State for the district or part thereof and Sarpanch at the village level. For a progressive civilization, entire system needs to operative in close coordination following a set vision for development.

The city area planning affairs are looked after by the ULB under the aegis of the regulations and a vision consultatively set in form of a Development Plan by the SUDA for the area under its authority. After its establishment in the year 1978 under the provisions of The Gujarat Town Planning & Urban Development Act, 1976; the SUDA has published a Development plan in 1986 with its subsequent revisions accepted by the Government of Gujarat during 1997, and 2004 [12]. A recent revision proposed by SUDA in the year 2016 is still pending for approval by the State Government (Fig. 11.23). However, there prevails ambiguity in terms of regulation at village level. Often it results in growth of an area where there seems no intervention in place by the SUDA. Along with SUDA, there are about 21 agencies responsible for making land parcel acquisition for various public purposes [12]. As recognized nationally, a recent development of a ring road by SUDA through a mechanism of T. P. Scheme, is an exemplary task for a government organization of developing land on no-profit no-loss bases under the provisions of the GTPUD Act.

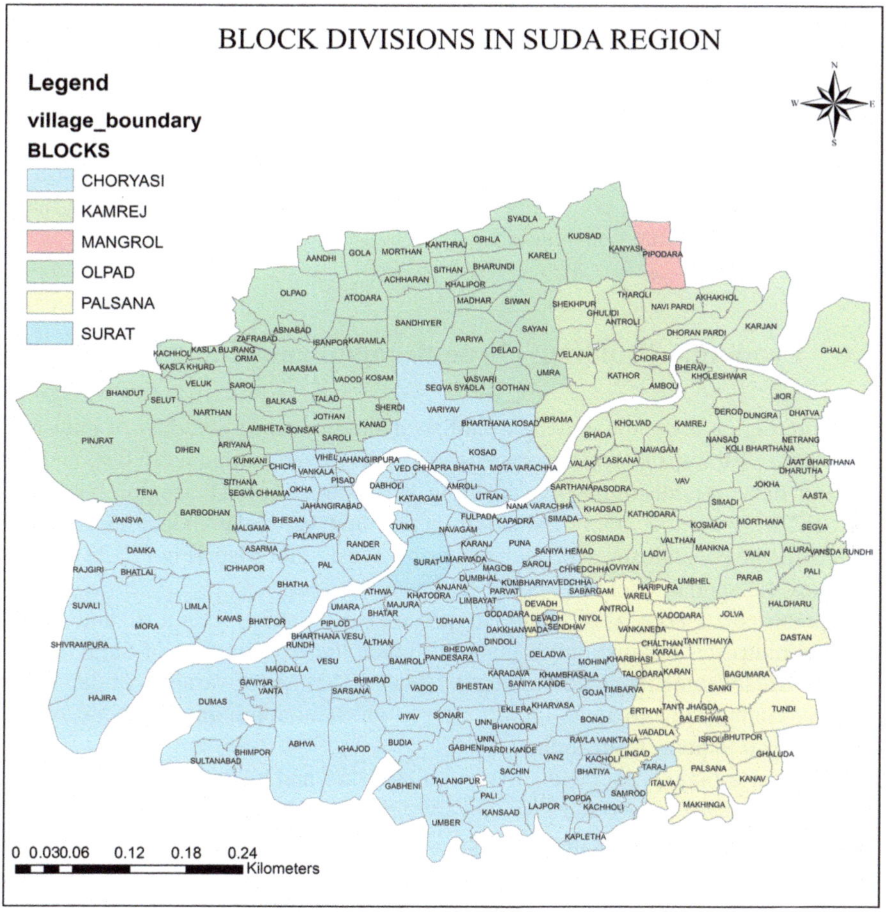

Fig. 11.21 Block level administration in the study region. *Source* Developed by Authors

So far, the SMC has bagged many awards and recognitions at the state, national and global level for initiatives towards exemplary E-governance and M-governance [24]. Since past many years, the SMC has been able to transfer almost all the record keeping through a computerized system and with recent efforts under Smart Cities mission, the systems are progressively upgraded through introduction of IoT and GIS based operations in service provision and management.

Apart from the initiatives and administration by the Government Organisations, there exists a wide scope for role of NGOs as a stakeholder. Local self-help groups, female groups, community and youth groups, national and international NGO units are operative under their set agenda and performing in support to action by the government. The SGCCI has been playing a unique role in growth of the industrial and economic activities by putting efforts for upgradation and integration among diverse sectoral activities. It has been regularly organizing skill development programs, seminars and performing various IEC tasks by forming various committees represented by members of each sector along with promotional activities and representation of views at various levels of the Government [24].

In the study area where a population greater than 0.6 million is housed and engaged in various activities, there lies a vast scope for integration among organizations and straightening up of the records unanimously. There is no Metropolitan Planning or Development committee in action at present and efforts are being made individually with partial consultation.

11.6 Economic Profiling Through Significant Industries

The entire region is by now having established and upcoming a variety of industrial units. Most of the male workers are engaged in these economic activities, and the extent of these establishments is spreading in all directions. The region has about four SEZ in operation, many industrial estates and particular industrial area earmarked for heavy industries of Hazira on the West. The Map (Fig. 11.22) below illustrates the spatial extent of spread of these activities.

A presence of sectors of diamond processing, textiles, chemical and petrochemical industries has distinguishingly attributed the development of Surat district [25]. The establishments processes about 10 out of 12 type of varieties of diamonds globally, contributing to INR 45,000 Crore, which is approximately 65% of the total diamond exports from India.

Surat is also famous as '*synthetic capital of India*' and by the year 2015, it hosts over 65,000 power looms and provides over seven lakh jobs in the district. Surat has successfully been attracting Foreign Direct Investment in various sectors like energy, oil, and petroleum. One of the largest Greenfield FDI in the country as Hazira Terminal Project is implemented through an investment of INR 3000 Crore (468 million USD).

The emergence of a petrochemical complex, gems and jewellery Park and the centrally promoted Surat SEZ are expected to further fuel the industrial and economic growth of the city. The expansion plans of Hajira port is envisaged in two phases. The first phase envisions the development of port infrastructure to handle liquefied Natural Gas (LNG) imports, and the second phase would offer port facilities for handling dry bulk and containerised cargo. The port facility would help in attracting sizeable investments in the coming future.

11.6.1 Textile Industries

The evolution of the power loom and handloom sectors led to the gradual growth of textile industries in Surat City. Today Surat is known for its textile manufacturing especially synthetic Sarees (almost 70% of Nation's synthetic Sarees are manufactured in Surat), and intricate sari works (excellent brocade work). These textiles based units are located mostly in the Central Zone and adjoining Northern, Eastern, and South-Eastern Zones of the city.

11.6.2 Diamond Industries

Another essential addition since the 1950s is the diamond cutting and polishing industry. In previous decades, especially during the 1980s, large-scale industries have come up in Surat and its peripheries, especially in the Eastern Zone of Surat City which saw a tremendous increase in population owing to diamond-cutting and Polishing Industry increase (Decadal growth of 670% for 1991–2001) in the Fringe areas of this Zone. Today Surat city accounts for 75% of nation's total rough diamonds cutting and polishing. It increased the importance of Surat in the regional context. The southern part of the city houses the industrial complexes of Gujarat Industrial Development Corporation at Sachin and Diamond Nagar.

11.6.3 Heavy Industries

The West and South-West part of the city is characterised by historical Magdalla Port and location of chemical industries and the gas-based industries at Hazira established by leading nationally significant industry houses such as ONGC, Reliance, ESSAR, Shell, KRIBHCO and GAIL.

11.6.4 Small Scale Industries

The industrial base is labour intensive. Of the total 2,78,656 small-scale units registered (2003) in the state, Ahmedabad and Surat districts lead the list with a high number of small-scale industrial units at 61,185 and 41,509 units respectively constituting 21.96 and 14.9% of total SSI units in the State. The share of Surat has increased from 12.6% in the year 1980 to 15.1% in the year 2005. Regarding factory sector, also the share of Surat is significant. In 2001, Surat district accounted for about 1900 units employing about a lakh and thirty-five thousand workers.

In the year 2000, it was reported by the Directorate of Economics and Statistics, GoG based on The Fourth Economic Census, 1998 that Ahmedabad district accounted for nearly 12% in the State total enterprises, which is the highest among all the districts. The district-wise percentage share of enterprises in State total aggregate is varying between 5 and 10%. It implies for each of the districts with Rajkot (6.09%), Mahesana (6.80%) Sabarkantha (6.49%), Kheda (6.59%), Anand (6.66%), Vadodara (5.52%) and ***Surat* (9.56%)**. Interestingly, Surat is ranking the Second position in sharing the enterprises established in the Gujarat state. Further,

Table 11.1 Enterprise establishments in the study region

Enterprise establishments	Nos
In SMC	49,437
In SUDA (Ex-SMC)	743
In Ex-SUDA Palsana	302
In Ex-SUDA Kamrej	98
In Ex-SUDA Chorasi	23
In Ex-SUDA Olpad	330
Total in SUDA (with SMC)	50,180
Total in peri-SUDA area	753

Source Compiled information from GIDC and DIC, Surat, 2014

the Quarterly Review of Gujarat Economy, 2011 discusses the District-wise details of Micro, Small and Medium Enterprises. The document reveals that during the period of 01-07-2011 to 30-09-2011 Surat district registered with 5,914 Micro, 386 small and 17-Medium enterprise with subsequent employment to 32,174, 7204 and 1286 workers respectively [16].

Based on the records from the District Industries Commissioner (DIC) and the Gujarat Industrial Development Corporation (GIDC), an analysis was performed on the overall industrial/commercial units that have been established in Surat till the year 2012. The analysis reveals that total Enterprises established in SUDA is 50,180 out of which within SMC limits are 49,437 and 743 in SUDA Ex-SMC area. Further, this is distributed among four blocks namely Palsana, Kamrej, Chorasi and Olpad in which these enterprise establishments are 302, 98, 23 and 330 respectively (Table 11.1).

11.7 Approaching Smart Development Strategies

Recently, the city was identified under the 'Smart Cities Mission' for 'Smart Development' by the Central Government of India, Ministry of Housing and Urban Affairs [26]. It has opened a path for futuristic and visionary development which can cater for inclusive as well as sustainable development. The present discussion is an extension of the goals that have been identified for smart city development, at a broader context of the entire region to which this city caters.

The smart city goals can only be met when a clear understanding of the regional activities is developed which has a bearing upon the city growth with its regional network.

For the administrative zoning purpose, the Surat city is divided into seven zones as discussed earlier, based on cardinal directions. However, there exists no such bifurcation in the remaining part of the region falling under the administration of SUDA. The government identification of block-level boundaries and village level boundaries are predominating for referring to different tasks (Fig. 11.23). A zoning like the SMC convention will need to be extended for the regional studies as well.

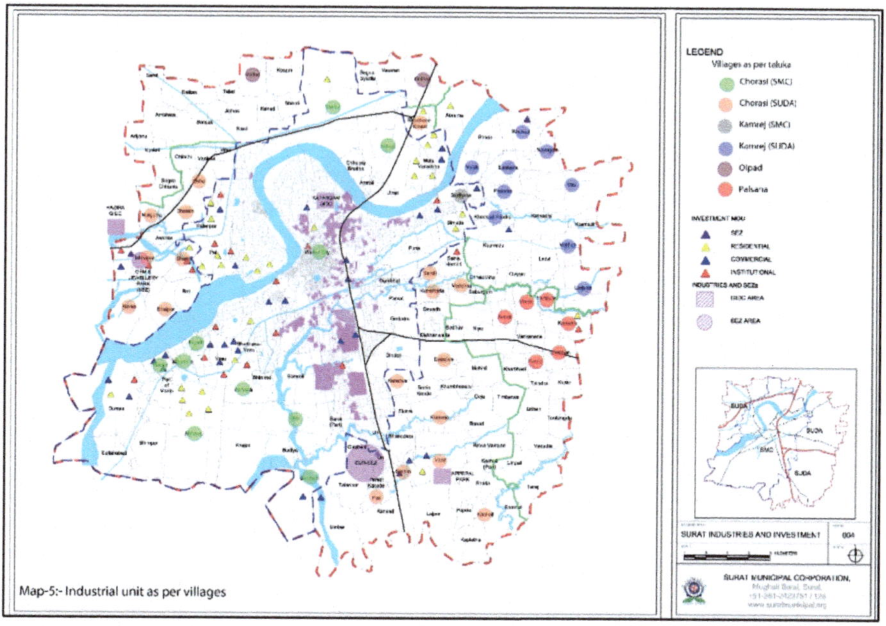

Fig. 11.22 Industrial establishments in SUDA area. *Source* Derived from SMC map and information made available by DIC-Surat and GIDC, 2016

At the juncture, a suggestive need is identified to adopt sub-division of the region by zonal delineation for formulating a smarter regional development strategy and timely implementation and active monitoring in future.

Surat, being a coastal city offers very distinctive features along its fringes. The northern and north-eastern fringe is the one with the maximum potential for further expansion and growth as it connects the city with other cities through national highway road infrastructure. Mumbai-Ahmedabad highway connects the city with these two giants of eastern India along this quadrant. No wonder that, most of the outgrowth is happening along this quadrant and eastern side of the city. The south of the city again has extension possibility; however, this zone has become saturated with small-scale industrial growth and substandard housing catering to such industries. A creek carrying most of the industrial wastewater flows in this quadrant, which needs cleaning and rejuvenation efforts. The south-western quadrant and the north-western quadrant have adjoining Arabian sea; thus, these directions have their limitation so far, the city expansion is concerned. However this area has excellent potential to be developed as eco-sensitive zones partly serving recreational purposes. Also, the western side of the city houses all mega industries of national importance, as well as the regional air terminal, thus in future, these two quadrants having western fringe boundary can be developed for low-rise low-density development. Also, the proposals for coastal highway, Mumbai-Vadodara Expressway, Damanganga-Narmada Interlinking Canal,

Dedicated freight corridor, Delhi-Mumbai Industrial Corridor, Kalpsar project, declaration of development authorities in surrounding and such mega initiatives will have their different impacts over the region.

For the study purpose, the Surat metropolitan region is broadly divided into four quadrants (Fig. 11.24), and the corresponding quadrant is discussed in detail regarding existing strengths, the opportunities offered versus the limitations.

11.8 Strategic Opportunities for Development

11.8.1 Identification of Regional Centres

Having a reference to Fig. 11.23, that proposes for development plan keeping 2035 as a horizon year by SUDA, a care is taken for proposing ring roads (not in the city) around smaller centers where urbanization is anticipated. In addition to it, all the villages are earmarked with residential zone around the Gamtal[5]. Larger part is kept as a residential zone. Areas are declared as industrial zones and some of these zones, specific industries are identified for development (for example, hazardous chemical industries). Also, a care is taken for connecting these places with residential areas by means of a proposed network of wide roads. A third of a kind, ring road is also proposed with an alignment. Mangrove forest zone is kept reserved along the coast line. Not extensively yet somewhat green buffers are also proposed. At a broad-level, the proposal seems fairly good.

The proposal seems lacking in terms of specific and result oriented efforts for specialized goods and services to have access to collaborative efforts. It means that development of industries is supposed to establish within a specific zone however, types of industries supporting each other are not specified at a micro level. Henceforth the development of pockets will be at will of the investors. It may result in farther dependencies resulting in increased transportation and logistics.

For example, a diesel engine and generator based industrial unit is dependent on mechanical part manufacturer, lubricant supplier, lifting operators (crane services), plastic parts supplier, website developer/manager and so on. Now, all these types of the of allied units will be established at various and unknown locations, maybe within reserved zone or out of the zone. In the case, a quick action of any kind will have chances for delay due to transportation of goods as well as manpower, resulting in reduced productivity and efficiency. More time spent on each task will result in more cost making the production efforts yet costlier and, contributing to the environment degradation as well. With technological advantages available, it seems not to be smarter approach. Integration of a major industry type with land provisions for its allied units (maybe derived by exploring dependencies), shall be

[5]A Gamtal is a traditionally settled pocket, in other words, it is the original hamlet.

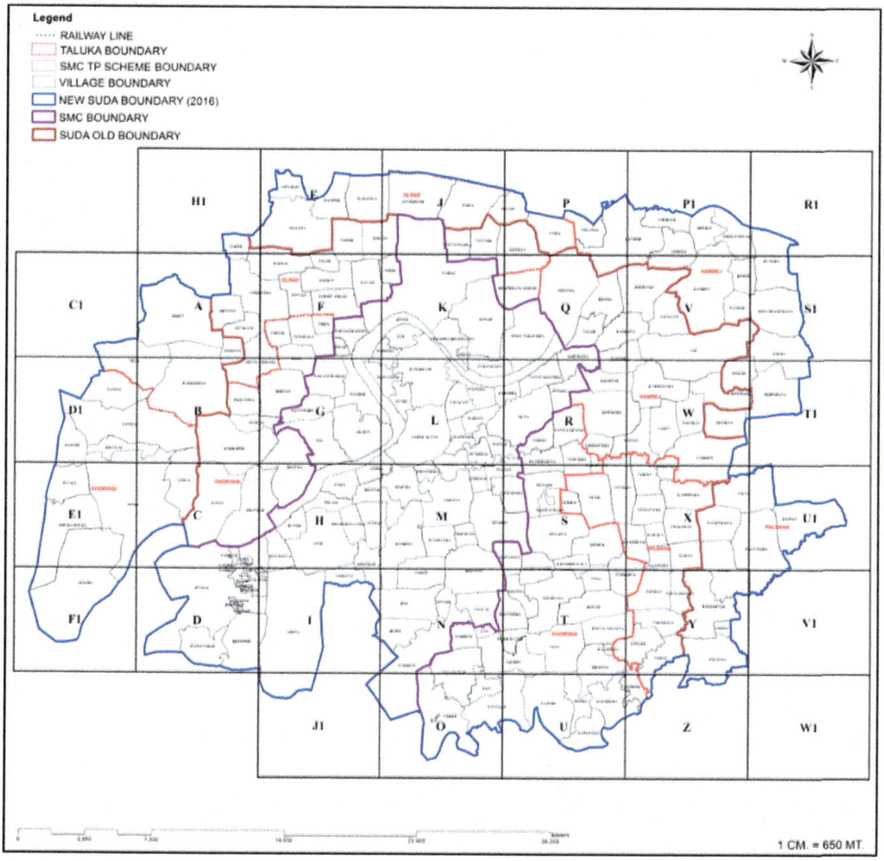

Fig. 11.23 Study region (SUDA) Administrative boundary [9]

included so that integrated approach at a regional scale is obtained. In other words, provisionally a zero-marginal cost development is proposed for implementation through self-induced evolution.

As soon the exploration of dependent industries is accomplished, next shall be the specific requirement identification for transportation needs. According to the anticipated traffic requirement, a road/rail network proposal shall be made. However, there can be an additional provision of faster mode of commutation like BRTS or MRTS corridor connecting residential colonies.

Subsections herewith discussed about integrating collaborative commons aspects in support to a zero-marginal cost development at a regional scale resulting in better opportunities, leading to a higher level prosperity.

11.8.2 Promotion of Diversified Industries Towards Scope of Economic Sustainability

Surat and surrounding region have achieved certain service level benchmarks and with rising population there will be pressure exerted on the systems in future. To avoid such circumstances, a proactive planning with a larger and inclusive purpose is very much essential. The concept of zero marginal cost society explores the aspects regarding practices over eclipse of the capitalism, cost, cooperation, IoT, and the collaborative commons [27]. In a way, the entire concept be summed up in a regional planning perspective as to have provisions for various needs that can be best utilized with collaborative efforts sustaining and supporting each other. For achieving such, a detailed exploration about dependencies as well inter-dependencies need to be worked out. Once the levels are available in relation to dependencies, a clustering approach of planning shall be taken up to make provisions for land. At the same time, an exercise is required to visualize the ancillary needs to support these clusters as transportation, logistics, service infrastructure and social infrastructure. The clustering approach at a regional level with provision of IoT based and monitored services, there shall be a considerable saving of time and resources. It also shall result in the increased productivity and efficient performance balancing economic and social engagements.

11.8.3 Transport Services

For an urban area to sustain and keep progressive, an efficient transportation service is essential for the movement of people and goods. Inter-zonal connectivity as well as intra-zonal connectivity need be addressed. There needs to be choices available to users for selecting a specific mode from accessible alternatives. The services need to be effective, efficient, safe and sustainable. A very much insightful analysis of needs shall be done based on which, planning and implementation strategies be worked out. Web-based and mobile application based movements at all identified hierarchical levels shall be in place. A detailed study for logistic needs and application for optimisation of goods movement with route optimisation shall help using mobile devices, software tools and technology hand-in-hand.

In the case of Surat, if a quadratic regional planning as suggested (Figs. 11.23 and 11.24) is visualized, a vast imbalance can be observed. Certain parts are overcrowded with facilities whereas some parts have very less connectivity. That too, the plan discusses about road provision only and not any modes of transportation.

11.8.4 Society and Social Infrastructure

In the perception here, the planning proposed for Surat region (Fig. 11.23) have earmarking of land for social infrastructure development in future. However, the approach seems to create 'cities within a city' where independence within the areas will not be obtained resulting in usual urban problem propagation. Since, zero marginal cost society concept is to be explored in relation to identify interdependence, earmarking of residential zones with social facilities shall be planned for. Such an effort shall help improving social cohesion among citizens, reduce pressure on transportation and other systems and avoid 'cities within a city.'

Apart from sustaining economic activities among each other, there shall be a need for facilities where skill development and regeneration activities be taken up. Such facilities may prove to be very effective for job seekers youth and migrants looking forward opportunities. Altogether, with a rich an integrated, updated database, a tracking and monitoring on the activities shall be done. Here, an IoT based deployment shall be useful. It becomes essential for planners and policy makers to visualize that the data, population and skills are matched with job demands in the region. Hence, there needs to be a continued and active interactive among the Government organizations, NGOs, industrial associations and other stakeholders on intervals, regularly.

11.8.5 Governance and Institutional Support

The involved organizations when are working in a loop helping all to move forward on a shared vision, sort of magnets shall be in place for individuals and organization to be part of the movement. Here, an introduction for incentives (not monetary but of kind, relaxation or subsidies in taxes and so on) shall be offered. At times, creating competitive environment helps in fulfilling a purpose to involve all. To an extent smart economy based business models shall be explored and by motivating entrepreneurs, the same shall be implemented. It calls for a committed, supportive system that can be possible only with collaborative efforts and using best skills available through in the benefit of an aspirant. Later, the one shall help the newer one to build capacities. Such collaborative approach needs to be intricate at a regional scale. In a way, there will be evolution of different and smarter business models.

11.8.6 Zero-Marginal Cost Based Metropolitan Development in Surat: A Concept Note

To achieve the state of a 'Zero-marginal Cost society', there will be an apt need to integrating various dimensions of sectoral and spatial planning, city and village

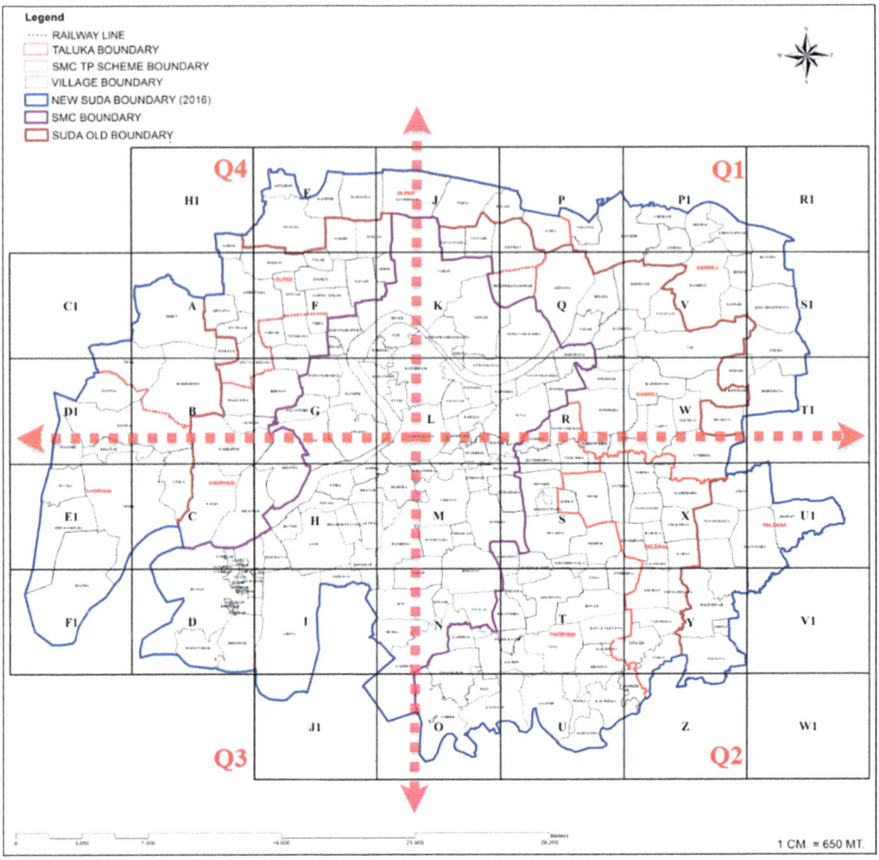

Fig. 11.24 Quadratic bifurcation of the study region. *Source* Author derivation based on [9]

level implementation, feedback from all citizens and in accordance, the institutional arrangements for monitoring and evaluation. There has been attempts made in terms of conceptualizing integrated planning at the district level by the Planning Commission of India [28]. However, it seems that the targets for integrating sectoral planning back in the year 2008 are yet not achieved with ongoing practices.

The zero-marginal cost based metropolitan development not only suggest for integrating planning for various sectors but also, there will be a need to identify sector specific and intra-sector dependencies. If done so, there shall be a considerable effect on the trips generation (for human and goods) and trip length reduction (home-based trips as well as work-based trips) for entire of the metropolitan region.

The prime step shall be the identification and demarcation of metropolitan region based on existing administrative boundaries at village level. In case of Surat, the administrative area as proposed in the current study may be considered as a metropolitan region.

A centralize data collection and database system shall be developed in the next step. As discussed in the earlier sections, Surat has a mixture of developmental aspects in concern. Demographic information, services and social infrastructure, industrial and enterprise establishment information with workforce, residential development (accommodation capacity and its utilization level) shall be analyzed on a spatial-temporal-scale. The data integration using GIS can be performed with deployed database management systems.

As soon the population, their involvement levels are identified, trip generation and trip distribution based traffic planning exercise shall be taken up. It shall help revealing movement of people and goods within the metropolitan region as well as outside of it. Based on the findings, an evaluation of existing transportation system shall be performed in terms of capacity and service level assessment.

Using the information within the database, the sector specific dependencies and level of interaction shall be worked out. Proposals for possible shifting of attraction magnets shall then be attained. Suggestions for space allocation for these shifted magnets, shall be checked for accommodation of dependent units in vicinity. For the case of Surat, a quadratic division of the entire metropolitan region shall be identified for specific economic activities.

For example, the Q1 on the North-East quadrant has more of green spaces and open for development. It has connectivity through railways, roads (SH and NH as well). It has water resources and other natural features. The land mass has not been observing floods in the past. The entire pocket seems very much suitable for the diamond (cutting, polishing and processing) and jewelry (manufacturing) based economic activities.

The Q2 (South-East) quadrant has established industries related to textiles, chemicals and processing units. It also has very good connectivity in terms of roads and railway network. There is no River however, some flooding through natural drains was observed in past. Keeping provisions for proactive measures in planning, zoning shall be worked out for dependent units.

The Q3 (South-West) sub-region is a unique land parcel in aspects that it has most of residential colonies and industrial belt of Hazira along with salinity ingresses in land. No forest lands (except mangroves on the coast line) but have much of barren and open lands available. It has an Airport of Surat, two ports (Hazira and Magdalla). The area is prone to floods in River Tapi. The ecosystem here is unique as in some parts migratory birds are landing and stay (Gaviar) during winters.

The Q4 (North-West) quadrant has the Arabian sea on its west side and Surat city on the east. The area has connectivity by roads only and no other modes. Some part of Tapi River with a few local waterbodies are featured. The area geography is flood-safe. The zone shall be developed for educational and recreational activities as there exists a large number of school establishments developed in recent past.

Apart from the quadratic zonation proposals, dependency analysis for diverse aspects shall be performed. To name a few for Surat metropolitan region are: heritage perspective (mapping of entire region for built and natural heritage), enhancement opportunities for developing green infrastructure, development of agro-tourism, adventure tourism, corporate tourism, medical tourism, eco-tourism and to some extent pilgrimage tourism are some of the potential seeking attention. One of a most important aspect here will be developing and matching skill requirements for newer opportunities along with institutional setup for the implementation.

Secondary requirements in terms of efficient transportation shall be addressed for planning in the metropolitan. Some suggestive modes include for provision of express buses, metro rails, express urban highways (non-existent present) and provision for dedicated lanes for goods transportation. Also, to make this system yet effective, a shift-based timing for various activities be introduced through regulation. Such an attempt can save enormous in terms of time and other resources by avoiding congestions on transport corridors.

By adopting and accommodating policy and regulation reforms of national agenda, some smarter practices in work patterns need to be in place. Smarter education modes of online learning need to be promoted and sincerely included in the society. The present state of the same is limited to individual interest and willingness only. Smartly avoiding physical presence of people for various tasks can result in saving the expenses in turn, can improve the prosperity in the region. There is a significant need for a pragmatic change in the planning and implementation as well as practices by citizens for smarter moves for sustaining the metropolitan region. The same can only be envisioned with micro-level yet wholistic approach by multi-disciplinary and collaborative efforts by various levels of stakeholders.

Hence, Surat needs to be attended with a proactive, systematic planning approach for smart metropolitan region planning to infuse the fundamentals of zero-marginal cost society concept. It has potential to accommodate citizens with offering of opportunities and serve the nation as efficient engine of growth. Population trends in each of the constituency shall be considered while leveraging opportunities through provisional planning.

Major limitation of the present study is unavailability of data and information in an integrated manner. All the data are not available and the information that is available is not well aligned based on which a good level spatial interface can be developed. Henceforth, the study through maps and spatial intervention was limited to population distribution and proposed development plan. However, secondary information from various references was used for discussions.

Appendix 1

See (Table A.11.1).

Table A.11.1 Major enterprise units in the study region

Sr	Particular	Textile processing	Diamond	Weaving	Chemicals
1	No of units	320	3500	25,000	70
2	No of people employee	4,00,000	Above 4,00,000	1,25,000	30,000
3	Current investment	INR 1000 Crore	INR 25,000 Crore	INR 1,27,500 Crore	INR 500 Crore
4	Current annual turnover	INR 2000 Crore	INR 2,00,000 crores	INR 54,000 Crore	INR 6000 Crore
5	Current area (million m^2)	0.74	4.5	3	0.9
6	Locations	Palsana, Pandesara, Sachin, Kadodara,	Varachha, Katargam, Mahidharpura, Puna	U. M. Rd, Sachin, Pandesara, Bamroli, Unn, Udhna udyog Nagar, Bhestan, Vasta Devdi, Kapodra Laskana, Limbayat, Parvat Patia, Ved Road, Katargam GIDC, Salabatpura, Sagrampura, Khatodra, Sayan, Kim, Pipodra, Jolva	Pandesara, Sanchin, Palsana
7	Products	Dyeing and Printing	Diamond Polishing, Cutting, Gems and Jewellery	Art silk Cloth Mfg.	Dye Intermediates, Pharmaceutical chemicals manufacturing, Textile Chemical, Agro Chemicals mfg.
8	Technology in future	Production of Colour Cotton	Jewellery Mfg Casting	Production of Colour Cotton, Waterjet, Rapier	Nano-technology, Microreactor
9	Future location	Sachin–Magdalla Road	Vesu, Piplod, Pal	Delad, Gothan	NA

Source Compilation of information from GIDC and DIC, Surat

References

1. Surat Municipal Corporation (2008) "Surat City Development Plan (2008–2014)." Surat Municipal Corporation, Surat
2. Surat Municipal Corporation (2017) "Introduction." Surat Municipal Corporation, [Online]. Available: https://www.suratmunicipal.gov.in/TheCity/Introduction. Accessed 2017
3. Maphill "Political Location Map of Surat." [Online]. Available: http://www.maphill.com/india/gujarat/surat/location-maps/political-map/. Accessed 2017
4. Surat Urban Development Authority "Welcome to Surat." Surat Urban Development Authority, [Online]. Available: http://www.sudaonline.org/. Accessed 2017
5. The Southern Gujarat Chamber of Commerce & Industry "About Surat and South Gujarat." SGCCI, Surat, [Online]. Available: https://www.sgcci.in/pages/about-surat/
6. The Times of India "Surat fourth fastest growing city in world." The Times of India, 23 July 2011, [Online]. Available: https://timesofindia.indiatimes.com/city/surat/Surat-fourth-fastest-growing-city-in-world/articleshow/9328885.cms. Accessed 2017
7. Surat Municipal Corporation "History." Surat Municipal Corporation, [Online]. Available: https://www.suratmunicipal.gov.in/TheCity/History. Accessed 2017
8. Surat Municipal Corporation "Demographics: Statistics for Surat Municipal Corporation." Surat Municipal Corporation, [Online]. Available: https://www.suratmunicipal.gov.in/TheCity/Demographics. Accessed 2017
9. Surat Urban Development Authority "Map as per the Notifications Dated 9/12/2015," [Online]. Available: http://www.sudaonline.org/wp-content/uploads/2013/08/SUDA-AUTHORITY-9_12_2015.pdf. Accessed 2017
10. Surat Municipal Corporation "Area, Population, Density, Growth Rate, Slum House Hold & Population." Surat Municipal Corporation, [Online]. Available: https://www.suratmunicipal.gov.in/TheCity/City/Stml3. Accessed 2017
11. UNESCO New Delhi "Social Inclusion of Internal Migrants in India," June 2013. [Online]. Available: http://unesdoc.unesco.org/images/0022/002237/223702e.pdf. Accessed 2017
12. Surat Urban Development Authority "A workshop on 'Initiating Preparation of Revised Development Plan of SUDA—2035'," 03 August 2013. [Online]. Available: http://www.sudaonline.org/wp-content/uploads/2013/08/RDP-2035-presentation.pdf. Accessed 2017
13. Surat Urban Development Authority "Proposed Land-Use Plan Submitted U/S 16," 2015. [Online]. Available: http://www.sudaonline.org/development-plans/proposed-land-use-plans/. Accessed 2017
14. Surat Municipal Corporation "Details Of City Limit Extension & Population." Surat Municipal Corporation, [Online]. Available: https://www.suratmunicipal.gov.in/TheCity/City/Stml1. Accessed 2017
15. Surat Municipal Corporation "Wardwise Area, Population & Density," Surat Municipal Corporation, [Online]. Available: https://www.suratmunicipal.gov.in/TheCity/City/Stml7. Accessed 2017
16. MSME-Development Institute "Brief Industrial Profile—Surat District," [Online]. Available: http://www.dcmsme.gov.in/dips/BIP%20SURAT_guj.pdf. Accessed 2017
17. National Remote Sensing Centre "Bhuvan—Indian Geo-Platform of ISRO," IIRS-ISRO, 2017. [Online]. Available: http://bhuvan.nrsc.gov.in/gis/thematic/index.php. Accessed 2017
18. Google Maps "Sewage Query," Google Inc., 2017. [Online]. Available: https://www.google.co.in/maps/search/Sewage+treatment+plants+in+Surat/@21.1610476,72.7328445,11.75z. Accessed 2017
19. O. Parishwad "Interdisciplinary approach," 03 May 2012. [Online]. Available: https://www.slideshare.net/omkarparishwad/interdisciplinary-approach. Accessed 2017
20. Surat Municipal Corporation "Solid waste management key aspects," Surat Municipal Corporation, [Online]. Available: https://www.suratmunicipal.gov.in/Departments/SolidWasteManagementKeyAspects. Accessed 2017

21. Google Maps “Search query,” Google Inc, 2017. [Online]. Available: https://www.google.co.in/maps/search/Electricity+generation+plants+in+Surat/@21.2292114,72.7695341,11z. Accessed 2017
22. Telephone Regulatory Authority India “TRAI My Speed Portal,” TRAI, 2017. [Online]. Available: http://www.myspeed.trai.gov.in/. Accessed 2017
23. Wikipedia the Free Encyclopedia “Gopi Talav,” Wikipedia, the free encyclopedia, [Online]. Available: https://en.wikipedia.org/wiki/Gopi_Talav. Accessed 2017
24. Surat Municipal Corporation “Awards,” Surat Municipal Corporation, [Online]. Available: https://suratmunicipal.gov.in/Corporation/Awards. Accessed 2017
25. Southern Gujarat Chamber Of Commerce and Industry “SGCCI at a Glance,” Southern Gujarat Chamber Of Commerce and Industry, [Online]. Available: https://www.sgcci.in/pages/sgcci-at-glance/. Accessed 2017
26. Ministry of Housing and Urban Affairs Government of India, “Cities Profile of 20 Smart Cities,” Smart Cities Mission, 28 March 2016. [Online]. Available: http://smartcities.gov.in/content/innerpage/cities-profile-of-20-smart-cities.php. Accessed 2017
27. J. Rifkin (2014) “The Zero Marginal Cost Society.” In: The Zero Marginal Cost Society: the internet of things, the collaborative commons, and the eclipse of capitalism, First ed. Palgrave Macmillan, New York, USA, p 290
28. Planning Commission of India Manual for Integrated District Planning, New Delhi, India: Planning Commission, Government of India, 2008

Part XII
Italy, Naples

Chapter 12
Spatial and Economic Smart Strategies for the 21st-Century Metropolitan City of Naples

Antonio Caperna, Eleni Tracada, Guglielmo Minervino, Elina Alatalo and Maria Cerreta

Abstract The assumption of Neoliberalism in the economy has multiplied exponentially financing speculation, and produced several "distortions" both in the social system and in the job market: the destruction of a welfare program, the attack to the right of the labor market and workers right, the powerful growing of financial institutions supported by the ICT. This means the need to identify a new epistemological approach, suggesting a conceptual framework for ecological economics based on systemic principles of life and a shift from techno-city to a human city. A model, called the homological smart city, could be a new way, based on direct citizen participation, peer-to-peer community, neuroergonomics, biophilic design, and biourban economics. The operational character of this model is explored by analyzing the most recent Italian experiences in reaction to the diffused crisis conditions. Several villages, towns and cities have seen a slow phenomenon of the revival of local communities, for the merit of grassroots' initiatives of social innovation constituted mostly of young people that, leveraging on their capabilities and a peer-to-peer network supported by the ICT, promote a novel vision for the future of their community, building a more sustainable urban system. Through a change of paradigm, the human being is put at the centre of the system and its

A. Caperna (✉)
International Society of Biourbanism, Rome, Italy
e-mail: antonio.caperna@biourbanism.org

E. Tracada
University of Derby, Derby, UK
e-mail: E.Tracada@derby.ac.uk

G. Minervino
Mediterranea University of Reggio Calabria, Reggio Calabria, Italy
e-mail: guglielmo.minervino@unirc.it

E. Alatalo
University of Tampere, Tampere, Finland
e-mail: elina.alatalo@uta.fi

M. Cerreta
University of Naples "Federico II" Naples, Naples, Italy
e-mail: maria.cerreta@unina.it

T. M. Vinod Kumar (ed.), *Smart Metropolitan Regional Development*,
Advances in 21st Century Human Settlements,
https://doi.org/10.1007/978-981-10-8588-8_12

designing, considering social innovators as the key actors of change and local assets as the key resources for the implementation of Biourbanism principles. In the above perspective, the experience of a new biourban strategy named "mushrooming", implemented in Finland, constitutes a good example of practice-oriented to consider diversification as a principle of life in a city and developed by testing with real-life conditions. The Finnish experience was started to build a network to foster interaction between small self-organized co-working communities, by taking into account spatial and economic processes that emerged due to this. These processes were able to activate connected diversification, recognized as a systemic principle of life that fits the context of urban development especially well. The principle of connected diversification drives the methodological process structured for the case study of the Metropolitan City of Naples, one of the 14 Italian metropolitan cities, with a specific attention for the 16 municipalities of the Coast Area. Starting from vulnerability and resilience concepts, the study dealt according to a multi-methodological approach, based on a GeoDesign process supported by multi-criteria analysis, multi-group analysis, and spatial analysis. The elaboration of Spatial Opportunity Maps (SOMs) is the output of a multidimensional evaluation process that leads to the identification of a biourban strategy, characterized by human smart spatial solutions, place-based and situated actions. The enhancement of the coastal area of the Metropolitan City of Naples can be considered as a prerequisite for the activation of a process-oriented to the identification of "homogeneous zones", conceived not only as areas with similar characteristics but, above all, as territories where it is possible to promote networks of opportunities between the various municipalities and their communities. Cooperation has conceived a source of mutual benefit and involves a mutual convenience, based on the constant construction of bonds and relationships and the interdependence determined by spatial proximity. Economic processes require cooperative-collaborative behaviours between the various components and become increasingly territorialized, and therefore more resilient and, at the same time, less and less associated with the production of negative environmental impact.

Keywords Metropolitan city · Homological smart city · Social innovation
Mushrooming · Connected diversification · GeoDesign · NAIADE
Naples

12.1 Economy and Capitalism in the 21st-Century City

During the last decade, the economy has been characterized by "neoliberalism" [1].

This is an expression to indicate the recent phase of global capitalism, in which the business model becomes not only the only acceptable one for organizing the economy and the finance, but also the way in which the "system" organizes the social structure and the labor market. Several authors [2–5] who use the neoliberalism concept derogatively, they describe what they distinguish as the regrettable spread of global capitalism and consumerism, as well as the equally deplorable flattening of the proactive welfare state [1, 4, 6–10].

Neoliberalism was coined at a meeting in Paris in 1938 by Ludwig von Mises and Friedrich Hayek. Neoliberalism was a reaction against social democracy policies, exemplified by Franklin Roosevelt's New Deal and the gradual development of Britain's welfare state; both were considered as indicators of Collectivism which is placed into the same spectrum as Nazism and Communism.

David Harvey provides a definition of neoliberalism as follows: "*Neoliberalism is in the first instance a theory of political economic practices that proposes that human well-being can best be advanced by liberating individual entrepreneurial freedoms and skills within an institutional framework characterized by strong private property rights, free markets and free trade. The role of the state is to create and preserve an institutional framework appropriate to such practices. The state has to guarantee, for example, the quality and integrity of money. It must also set up those military, defence, police and legal structures and functions required to secure private property rights and to guarantee, by force if need be, the proper functioning of markets. Furthermore, if markets do not exist (in areas such as land, water, education, health care, social security, or environmental pollution) then they must be created, by state action if necessary. But beyond these tasks the state should not venture. State interventions in markets (once created) must be kept to a bare minimum because, according to the theory, the state cannot possibly possess enough information to second-guess market signals (prices) and because powerful interest groups will inevitably distort and bias state interventions (particularly in democracies) for their own benefit*" [5].

Neoliberalism policy includes few key points, such as:

(i) Free market, which means liberating "free" enterprise or private enterprise from any bonds imposed by the state, no matter how much social damage this causes. Greater openness to the international trade and investment; no more price controls; total freedom of movement for capital, goods and services;
(ii) Cutting public expenditure for social services, such as education and health care;
(iii) Deregulation, or reducing government regulation of everything that could diminish profits, including protection of the environment and safety of employment;
(iv) Privatization, or selling state-owned enterprises, goods and services to private investors, such as banks, key industries, rail, electricity, schools, hospitals and even fresh water;
(v) Destruction of the concept of "community" and "the public good", replacing it with "individual responsibility".

These policies increased during the 1980s in the 20th century. The neoliberal ideological doctrine was launched in the USA and the UK by the conservative leaders, Ronald Regan and Margaret Thatcher. Therefore, in few years it turned into "the unique global doctrine" in financial policies. It was a reaction against the

1970s' energy crisis. Indeed, although "*from roughly 1950 until the early 1970s there was a period of unprecedented economic growth and egalitarian economic growth*" and "*it was also a period of some limited but real form of benefits for the population*", this changed during the mid-1970s. As emphasized by Noam Chomsky, the structural change of the financial markets initiated a profound transformation: "*Bretton Wood's restrictions on finance were dismantled, finance was freed, speculation boomed, huge amounts of capital started going into speculation against currencies and other paper manipulations, and the entire economy became financialized. The power of the economy shifted to the financial institutions, away from manufacturing*" [11].

After a couple of decades, this ideological approach to the economy has reinforced speculation tremendously [11] and a lot of "distortion", both of them acting in the social system as well as in the labor market. The destruction of the welfare program, the abuse of the rights of the labor market and the workers' rights, the growing power of the financial institutions have created a state of concealed war, in which the workers fight against the immigrants, moving out from their counties destroyed by globalization. Conversely only a minority of people in the world are the main beneficiaries of the neoliberalism policies. For the vast majority of people, this brings even more suffering than before: suffering without the small, hard-won gains of the last 60 years.

This ideology acting as a strategy has been adopted differently by the global north and south. In the north we are witnessing the reduction of the workers' rights, the marginalization of the middle class, and the reduction of the welfare state. In the global south, we are witnessing the rise of a neo-colonialist authorities in the poorest countries of the planet and momentous migration towards the north. The result of this might be considered as a social time bomb which can break definitively any form of reception.

The capitalist system has triggered this systemic crisis; there is no way to resolve it without overcoming capitalism first. The capitalist system catalyses this systemic crisis. However we should always have in mind that we cannot reduce everything to the economy. The internal contradiction of the capital has been amplified by the digital technology and generally by the role of *Technè*.

12.1.1 From Capitalism to Finance-Capitalisms

During recent years we have witnessed the era of transformation of the capitalism. According to Karl Marx, we know that the simplest formula of the circulation of commodities is C-M-C; that means the transformation of commodities into money, and the change of the money back again into commodities.

Nevertheless alongside this form we find another specifically different form: M-C-M, the transformation of money into commodities, and the change of commodities back again into money; or buying in order to sell. Money circulating in this latter way is thereby transformed into capital; it becomes capital, or, in theory, it is already capital [12].

The previous sequence is demonstrated by commercial business activities. What happens today can be exemplified by a new formula: $M\text{-}C\text{-}M_1\text{-}M_2\text{-} \ldots M_n$. The values 1, 2,…, n denote speculation; this shows that growth operates on a capital. After few changes, it becomes an abstract value with no relationship with the reality. The only aim of this financial activity is to generate profit.

When money borrowing or financial speculation occurs, M-C-M is reduced to M-M, a sequence which disrupts the capital realization process [12–14]. This speculative aspect has been emphasized by Karl Marx: "*The simple circulation of commodities—selling in order to buy—is a means of carrying out a purpose unconnected with circulation, namely, the appropriation of use-values, the satisfaction of wants. The circulation of money as capital is, on the contrary, an end in itself, for the expansion of value takes place only within this constantly renewed movement. The circulation of capital has therefore no limits*" [12].

During its race to the top, the capital needs to colonize territories and natural resources, to decrease the cost of human labor, to develop new technologies and promote new financial investment and trade rules which allow capital to get more and more profit. When this process of endless growth comes across limits (i.e. in production) and profits, then it begins to decline. Then, here comes a crisis. This occurred several times during the past two and a half centuries.

What is new now is:

(a) The disproportionate amount of speculative capital in the markets;
(b) The fact that the capital has reached and surpassed the limits of the Earth System.

Research conducted by Andrew Lawrence, a real estate analyst of Barclays Capital, shows a relation between financial speculation and the skyscraper. In his "Skyscraper Index" [15], Andrew Lawrence presents his homonymous theory by suggesting that extremely tall skyscrapers are linked to the economic cycle; the completion of the world's tallest building is either an indicator for a recession or an economic crisis. The rationale is that very tall skyscrapers are the result of an economic boom—real estate demand is up, credit is cheap and the economy is doing well. The higher the skyscraper, the higher the boom. It takes years to build a large skyscraper. Thus, skyscrapers are completed as the economic cycle tops; then, it heads down into a recession or a crisis. Examples of this link are the Singer Building and the Met Life Tower which were completed around the time of the 1907 panic, or the New York Chrysler Building, which began in 1928 and the Empire State Building, completed just as the Great Depression began. In recent years, we can find the Petronas Towers in Malaysia which were completed as the Asian financial crisis began in 1998.

Also Carol Willis's "Form Follow Finance" [16] emphasizes the importance of speculative development and the impact of real-estate phases on the forms of buildings and their spatial distribution. As a matter of fact we may say that some papers, such as "Skyscraper Height and the Business Cycle: Separating Myth from Reality" [17] by Barr, Mizrach and Mundra, Rutgers University, affirm that the drivers for skyscraper heights are not yet well understood. They suggest that skyscrapers can be used for non-economic purposes, and the tallest skyscrapers are economically "*too tall*"; these non-economic motives become obvious within predictable locations in the business cycle [17]. The conclusion is that they do not deny that psychological and egocentric motives are present in the skyscraper market; these motives do not appear to be a systematic part of it. The fact that heights rise over the business cycle indicates that height is usually a rational response to rising incomes [17].

12.1.2 Capitalism, Financial Market and Digital Technology

After three decades dominated by neoliberalism in 2008 we witnessed one of the worst economic and financial crises in the post-war economic history. After World War II, no other economic downturn has been as severe as the 2008 recession. The current crisis is the deepest, most synchronous across countries, and most global since the 1930s Great Depression. The depth and breadth of the current global financial crisis is unprecedented in the post-war financial history. It has several features in common with similar financial stress-driven crisis episodes. It was preceded by a relatively long period of rapid credit growth, low risk premiums, abundant availability of liquidity, strong leveraging, soaring asset prices and the development of bubbles in the real estate sector. Stretched leveraged positions and maturity mismatches made financial institutions very vulnerable to corrections in asset markets, deteriorating loan performance and disturbances in the wholesale funding markets. Such episodes happened before; the examples are abundant (i.e. Japan and the Nordic countries in the early 1990s, and the Asian crisis in the late1990s). However the key difference between these earlier episodes and the current crisis is its recent global dimension.

From a technical point of view, the current crisis is a product of inconsistencies of the twenty-five-year long neoliberal boom, which started in 1982 [18]. According to Joel Geier [18], the first contradiction to note is the creation of a giant debt bubble in the USA. Over the two decades preceding 2007, credit market debt roughly quadrupled from nearly $11 trillion to $48 trillion, far exceeding growth rates. Then, the deregulation of the banking system and the creation of a "*shadow banking system*" were able to keep all sorts of loans and investments, in spite the fact that these banks did not have to put up adequate capital reserves. As a result, "*they were able, through this unregulated system, to borrow thirty, forty, or fifty times above the value of their capital in order to invest in the stock market and in various new exotic debt products, such as collateralized debt obligations (CDOs), credit-default swaps (CDSs—essentially a form of insurance against debt default), and various other financial swindles, many of which were based on the packaging and repackaging of housing mortgages*" [18].

Finally, the last contradiction underlined by Geiner was the collapse of the American balance of payments that shifted from a surplus in 2000 and 2001 to a deficit in 2002. This led to a series of changes, such as, for example, United States becoming dependent on foreign capital. Foreign capital, in particular from China, Japan, and the Middle East oil exporting countries, financed the American debt and the dot-com bubble collapsed; lowered interest rates between 1 and 2% for three years generated a massive asset inflation, predominantly in housing prices.

Since the late 1980s we finally encounter the increasing sophistication of financial intermediation and the simultaneous and mutual expansion of ICT infrastructures and finance capital [19]. The result was a fast structural transformation of the economy from a "*real world production*" to a fluid financial market based on speculation and governed by digital technology. One of the new aspects that govern the financial flow is the time transformation. Adam describes ICT-generated time as instantaneous rather than durational, simultaneous rather than sequential, and globally networked rather than globally zoned [20]. It was this kind of time reckoning which underpinned the globalization of financial activity and the financialization of the capitalist economies.

The globalization of the finance capital is strictly interconnected automatically with digital technology. Ongoing advances in telecommunications extended and accelerated informational and monetary transfers between computer terminals. Specific technological advances included high-speed Internet link-ups enabling stockbrokers, institutional traders and personal investors to buy and sell shares in different stock exchanges simultaneously. This international facility became jointly owned by over 1000 banks and reliant on interbank funds transfer [21]. All these advances together created an unprecedented economic environment; multiple currency units and complex financial assets, worth trillions of dollars were traded globally in real time.

This extreme market financialization and out of control speculation generated something which was never seen in the history; the ratio between the combined global Gross Domestic Product of goods and services and the capital market was 1 to 26 in 2010 [22] compared the 1990 ratio which was 1–7. This means that, in 1990 the capital market was seven times more than the global GDP, whereas now it is 25 times bigger than the global GDP. In 20 years, the capital market has grown more than ten times larger, while the real economy only increased by three times. In 1990 the financial derivatives were only \$2 trillion. However they reached \$601 trillion in 2010, or, 30 times more!

Going back to the global economic crisis, many scholars explain that it is different from the other. Authorslike István Mészáros believe that today's crisis is evident under four main aspects that constitute a newness if compared to the previous crisis. Mészáros [23] explains that the actual crisis has a different character because:

(i) It is not restricted to a particular sphere, such as finance or commercial, or affecting this or that particular branch of production, but it has a universal character;
(ii) As opposed to the past, its scope is truly global rather than confined to a particular set of countries;

(iii) Its time scale is extended, continuous rather than limited and cyclic, as all former crises of the capital happened to be, and the 2007 speculative bubble was the iceberg peak;
(iv) Its mode of unfolding might be called creeping—in contrast with the more spectacular and dramatic eruptions and collapses of the past—while adding the provision that even the most vehement or violent convulsions cannot be excluded as far as the future is concerned; for example, when the complex machinery, now actively engaged in 'crisis-management' and in the more or less temporary 'displacement' of the growing contradictions runs out of steam.

Other authors emphasize that the crisis of global financial-led growth reflects evidence of exhaustion of the current capitalistic model of accumulation, characterized by lower growth rates and decreasing labor shares. A system which so far has only been possible by means of excessive consumerism through increasing indebtedness, accelerated depletion of resources, growing income inequalities and social exclusion and unrest. Since the end of the last century, we can find and connect root signs of a multidimensional systemic crisis, which reveals itself today beyond the destruction of millions of jobs worldwide; it increased disparities both internal as well as international, exacerbated conflicts and violence, and intensified the exploitation of natural resources. All these factors show both an economic downturn and an ecological and socio-political downturn.

According to Cairò-i-Céspeds and Castells Quintana [24]: "*we are experiencing a systemic crisis of multiple dimensions, namely an economic, human, ecological and socio-political one, reflected in global challenges of unprecedented intensity, magnitude and scope, and rooted in the very dynamics of the capitalist economic system, which has now clearly become unsustainable*". Their recent study discusses the nature of the current crisis from a multidisciplinary perspective; they analyze different dimensions of the systemic crisis by working with political, ecological and socio-economic indicators of a wide-ranging perspective. The authors' aim is to show to what extent these crises are interconnected and can be considered as different dimensions of a systemic crisis [24]. By analyzing the main dynamics behind the process of capital accumulation and economic growth, they show how the inner contradictions—social and natural—that drive the process of surplus generation lead to recurrent crisis in the capitalist economies. Finally, their analysis: "*suggests how recent dynamics indeed reflect deeper interconnected systemic disruptions, reinforcing each other and representing dimensions of a major systemic crisis of capitalism as the driving force of current socioeconomic, environmental and socio-political dysfunctions. Increasing inequalities seem to not just play a major role in the evolution of the current economic crisis, but are also behind the evolution of the other systemic dysfunctions and their interconnectivity*" [24].

12.1.3 Global Capitalism and Environmental Crisis

Strictly connected with the economic model is the environmental question. Today's economy generates unstable socio-political and environmental conditions; the gap between the rich and the poor is increasing, the natural resources are ruthlessly exploited, conflicts are growing. The loss of biodiversity and climate change are examples indicating that nature's rhythm and balance are dramatically disturbed.

The Global Risks Report 2016, 11th Edition [25] was published by the World Economic Forum within the framework of the Global Competitiveness and Risks Team. The 11th edition of the Global Risks Report has explored how global risks are becoming increasingly imminent and materializing in new and sometimes unexpected ways. From climate change to the imperative for improved water governance, from large-scale involuntary migration to reviving growth in the Fourth Industrial Revolution, global risks are affecting the lives of citizens and the functioning of institutions and economies. We now need to move beyond mitigation to adaptation and building resilience. Understanding the drivers of the global security landscape, boosting governance and strengthening policy agility are even more important. Building a better understanding of how the new security landscape and technological change will impact countries, economies and peoples' lives is, therefore, essential for building sustainable, resilient growth strategies and stable societies.

Global risks remain beyond the domain of just one actor, highlighting the need for collaborative and multi-stakeholder action; that is the key message that [25] *The Global Risks Report* series has highlighted over the past decade. Recognizing joint interests and aligning stakeholders on key priorities across the different areas of global risks is the first step to create action through collaboration. We hope that the aforementioned report will contribute to the recognition of a need for action; it will create an imperative towards greater resilience, and it will also motivate change and concrete action towards a better future for everyone. As affirmed, we do not get only climate change, but also biodiversity loss, deforestation, pollution, population growth, especially in the megalopolis of the global south, water resource vulnerability. Over-all the destruction of all ecosystems is aggravated by the current economic system.

The 2016 the Living Planet Report [26] estimates that, since 1996, the global demand for natural resources has doubled. It now takes 1.5 years to regenerate the renewable resources used in one year by humans. This means that we are eating into our natural capital, instead of living off its interest, and, therefore we are creating more ecological debt. Humanity's demands are greater than our planet's ability to sustain us.

Unfortunately it is evident that we do not have a policy, let alone adequate instruments to control a capital without nationality; corporations are much more powerful than states. The predatory attack to the natural resources has been intensified in the last few years. For example, in the energy sector we encounter an explosion of extractive projects, land and natural resource grabbing, and mega projects of infrastructure and energy which are reshaping the ecosystems. Corporations often fuel civil wars in developing countries. All this can fuel migration towards the north, generating social tension and racism.

Our vision is so fragmented and distorted that the green policies are being transformed into business. An example is the issue of Certificates of Emission Reductions (CER). CERs were created for greenhouse gas emission reductions. Well today CER have become a business product; the more a corporation buys them, the more CERs can pollute. The result is the financialization of the "*natural capital*". Note that we use the noun "*capital*" for the environment, as if it was an ordinary thing, measurable by money. This shows how wrong our approach to nature is. We are just able to think in terms of money and calculus. This mental attitude of the Western modernity has pervaded each aspect of our life. The result is that we have lost our capability to understand what beauty is, what is right or wrong, what is sacred or what a community is. We see everything as capital that we can buy, destroy, and sell as an ordinary object. This attitude of the Western thought has been discussed by Heidegger [27].

In his *Introduction to Metaphysics*, Heidegger analyses how the "*technè*" thought has transformed the world in a unique technical corpus. According to Heidegger "*technè*" means neither art nor technology, but a kind of knowledge. According to Heidegger, what is genuinely disturbing is not only that the world becomes a unique technical apparatus, but also dangerous. The fact is that we are not prepared to this radical transformation of the world. Finally Heidegger adds another grade for the *technè*, or, the fact that the most dangerous question is the lack of an alternative thought to the *technè*one. Heidegger says that we only know to "*calculate*" (="*Denkenals Rechnen*" in German), or, as Adorno and Horkheimer write: "*Anything which does not conform to the standard of calculability and utility*" is "*viewed with suspicion*"; it becomes downgraded to a mere myth or superstition [28]. This theme is elaborated in the work of the Frankfurt School, where philosophers like Theodor Adorno, Max Horkheimer, Herbert Marcuse and others viewed the culmination of Western Enlightenment in the early 20th century exactly as technique's domination upon nature, and, where the subject of history is not the humankind but the *technè*.

In Adorno and Horkheimer's words: "*What human beings seek to learn from nature* [physis] *is how to use* [techné] *to dominate wholly both it and human beings. Nothing else counts*" [28]. Because of the Western Enlightenment becoming "*totalitarian*," the world has become intelligible to the human kind when its multiple forms are calculable, hence, "*the control of internal and external nature has been made the absolute purpose of life. Now that self-preservation [of the humankind] has been finally automated, reason is dismissed*" [28]. Then again not only is reason dismissed, but also the reason itself has been subsumed into technical or instrumental causes. As Marcuse writes: "*Rationality is being transformed from a critical force into one of adjustment and compliance. Autonomy of reason loses its meaning in the same measure as the thoughts, feelings and actions of men are shaped by the technical requirements…Reason has found its resting place in the system of standardized control, production, and consumption*" [29].

The subsumption of reason under the technical attitude leads to "*the subordination of thought to pre-given external standards*," in which, thinking becomes routinized, standardized, quantifiable and predictable. According to Heidegger, this way of thinking also changes our idea about how we perceive our environment. Under conditions of *technè*, "*the earth reveals itself as [only] a coal mining district, [its] soil as a mineral deposit*" [27]. The above "*mental*" approach to the economy, society and in general towards the environment, uses new words but this does not change anything in substance. Thus, this has created also what we call "*green economy*." The ideology of totalitarian capitalism is like a fluid that creeps up on our day life in search of new markets and profit. The "*green economy*" is another invention to generate more markets for the capital.

The promise of the "*green economy*" is to decouple growth from environmental deterioration.

Bundeskoordination Internationalismus [30], a German group, has just published a new critical paper with the title "After the Failure of the Green Economy", containing ten theses as an evaluation of the green economy. The report affirms that "*Strategies of the Green Economy will not be able to outweigh the social and ecological contradictions of capitalism*" [30]; it emphasizes the fact that "*the forces responsible for the crisis are identified as the beacon of hope: states and especially markets and capital, as well as the orientation towards growth and competitiveness. With the aid of the Green Economy the driving forces for social and ecological disastrous capitalism are not called into question; on the contrary, they are to be used for a green conversion*" [30].

The "*green economy*" is not the only solution for the environmental and social problems. It works in the same epistemic environment. Unfortunately good management of the environment and the natural resources is not a vital factor for the economic system. Usually we look at it in short term without taking into consideration that, only a sound and long-time management of the natural resources can provide the basis for a really sustainable and inclusive wellbeing, food security and poverty reduction. The natural resources provide livelihoods for hundreds of millions of people. The world's ecosystems regulate the air, water and soil on which we all depend; they form a unique and cost-effective buffer against extreme weather events and climate change. Not only healthy ecosystems are essential for the long-term growth of economic sectors, such as agriculture, forestry, fisheries and tourism, but also they provide hundreds of millions of jobs at present.

Sustainability has been often studied according to the deterministic vision of the world, in which our environment is analyzed as a set of independent "*bodies*" without correlation. Today a reductionist attitude appears again in the study of sustainability. The same term "*sustainable development*" appears as an oxymoron. Serge Latouche considers this association as "*explosive and toxic*", an illusory mystification [31]. According to Latouche, it is a contradiction in terms, because sustainable means that human activity does not have to create a pollution degree which is superior to the capacity in order to regenerate the environment. This applies the principle of "*responsibility*" as stated by the German philosopher Hans Jonas: "*Operate in such a way that the effects of your actions are compatible with the permanence of an authentically human life on this earth*" [32]. On the other hand, development in daily language means the process in which someone or something grows or changes and becomes more advanced [33].

As a matter of fact we believe that this crisis is only the peak of an iceberg. It is systemic in sense of demanding answers to different questions originating from diverse social and political movements. An inclusive political and socio-economic system should support equity and social justice; it should support the innocent victims of crisis, the schools, hospitals, public transport, social housing and welfare schemes through democratic institutions. Sustainable or smart politics should reduce the emphasis on speculation and should implement democratic control over financial and economic institutions. In the same way the production system should sustain people's lives and the environment. Based on these demands, a long-term political strategy towards the transformation of existing power relations could be built. We believe in a bottom-up process rather than a process led by some elite group of economists and planners; it should be based on a different idea of the world and it should be a different epistemological view able to support a political process. The crisis can offer us an opportunity to shift paradigms, in which development means social and cultural progress, not only profit. As the crisis is systemic, our response should be also systemic. This should be the real implication of a crisis.

12.1.4 The Epistemological Reconfiguration as Necessary First Step Towards a Systemic Ecological Economy

Our main question is: are we able to develop an economy that unites opposites without coming into conflict with nature and society? What can we learn from other civilizations in term of their relationship with nature in order to allow a shift in the current way in which the economy is not only destroying our natural environment, but also our communities and social relations?

Capra and Jakobsen [34] have indicated four principles, deriving from systems' view of life for a real ecological viewpoint in economy. In their paper "A conceptual framework for ecological economics based on systemic principles of life" Capra and Jakobsen illustrate the principles as follow:

(i) "*Nature is superior to economy*";
(ii) Economy must be seen and studied as a network and the relationship is an essential attribute to take into account;
(iii) Economy must be seen as an open system. For a living economy this means that all economic processes need to be circular in three dimensions;
(iv) Economy as cognitive interactions—a sense of ethics according to the systems' view of life; all living systems interact cognitively with their environment in ways that are determined by their own internal organization [34].

The above principles provide us with a new epistemological perspective towards real sustainability. The first principle can be seen as turning upside down the values as derivatives from the industrial revolution.

The contemporary post-industrial society has destroyed the values upon which the Western civilization was built. Alasdair MacIntyre in his *After Virtue* [35] sees our moral crisis as the consequence of the historical rejection of an Aristotelian moral scheme in which the role of the virtues is dominant, reflecting a society of shared goods. MacIntyre focuses directly on the social relations and the historical character of the human life in terms of which merits are defined and understood, by tracing the profound transformation over the time, since classical Greece, the middle ages, and the modern world. The Greek approach to nature and what we call economy was completely different from the modern way. Ancient Greeks look at the world as something where nature was not generated by divine power or by man, but nature is considered as the "*eternal living fire*" which spreads and runs with proper balance [36]. In the same way, the Greek idea of economy was divided into two approaches.

Aristotle's approach to "*economy*" was coherent with the people's approach by the Greeks; he discusses economic and business questions as a subordinate subject within his treatise on Politics [37]. His predominant interest was with what rightfully should be considered economics (=*oikonomia*): the concern for morally adequate individual and public household management [37]. In other words, the very way in which Aristotle deals with economic affairs undercuts the modern separation of economics from ethics and all other concerns of life. In classical Greece, the "*oikos*" included both goods and persons—either free people or slaves—under the authority of the head of the household, what we would call the family and family wealth, a unity of persons and goods [38]. Aristotle distinguishes two different ways to organize economy: "*oikonomia*" and "*chrematistics*". According to Aristotle [37], "*oikonomia*" and "*chrematistics*" do not cover the same semantic meaning: *oikonomia* has to do with the satisfaction of the necessary and useful com-modities [39]. On the other hand, *chrematistics* has to do with what we call as art of money-making. According to Aristotle, when the world of chrematistics is under the government of the oikonomia, this is natural. On the contrary, when chrematistics work as independent body, this degenerates into an unnatural structure that transforms means to an end. So, "*economy*" is a body subordinated to natural laws, and its aim is the satisfaction of the humankind's natural wishes. This way characterized Western Civilization until Medieval era, when Thomas Aquinas in his *Summa Theologica* [40] described the fair price theory as the necessity of ethics in economics, which attempts to set standards of fairness in transactions, and the role of community sequentially considered as something more important than an individual person.

Over the last three centuries the Western approach has changed radically by forgetting the meaning of community and the values linked to it. Renaissance and Enlightenment created a new vision of the world: Hence, liberal individualism with his private desires and objectives is identified as the antithesis and the opponent of the Aristotelian community. A decisive blow arrived with the industrial revolution and a new ethical way based on individuality, profit, predominance of the economy and technology. In modern times, the meaning of the economy, as related to the family governance of persons and goods has been lost, whereas what originally constituted the "*chrematistics*" (=the acquisition of livelihoods) has been identified in practice with the "*economy*" and as such it is commonly understood today. This semantic change makes it difficult to appreciate the organic relationship that Aristotle established between ethics and economics [41], as the natural and morally permissible chrematistics is subordinated to the oikos and, as a matter of fact, the oikos is subordinated to the "*polis*" (=city) [4], in which the human ideal of social coexistence is appreciated. Strictly speaking, "the ethical" in economics is the cautious administration of the oikos within a fair polis [42].

According to Capra and Jakobsen [34], this approach is weak if we look at it in terms of a life system approach. In fact in life system approach economy is a nested system. It is part of more large sub-systems, such as society, ecosystems and the planet, and it contains the individual sub-system formed by the individual. This system works, if each sub-system works in harmony with the other sub-systems and none dominates the others. This generates equilibrium between macro and micro level characterized by a continuous building and re-building of relations between the sub-systems [34]. This way shifts the actual paradigm. Firstly, the system life is based on relationships of its sub-systems and agents, not single independent agents, but a whole structure which cannot survive as an individual one. As in ecology, analysis and study of this system asks to study all processes as part of the web of life. If economics is studied separately from social and ecological contexts, and knowledge is expressed through narrowly defined mathematical models, it becomes abstract and remote from life, according to Capra and Jakobsen [34]. The above principles suggest to understand economy as part of a whole living system and this relates to the concept of network and flows unified in the concept of metabolism [34].

The concept of a metabolic structure has a deep epistemological meaning: each ecosystem is characterized by a circular flow of energy and matter, and waste production in biological systems becomes a source of life for other sub-systems. In biological systems, all transformation processes are characterized by a low level of waste and a high level of productivity. For this reason, a real ecological economy must be characterized by a circular cycle in three dimensions, where economy, nature and culture are integrated parts within a "*living*" organism [34, 43]. In this way we get a circular economy that works as a biological structure; it is characterized by cooperation and not competition; that means cooperation between manufacturers, distributors and consumers. Cooperation with nature and culture, a real cooperation with nature, regulates input and output of natural resources in the economic value chain, and cooperation with culture is essential to develop a life-enhancing economy. "*On the one hand, knowledge exerts influence on the innovative processes in the economy and on the other hand changes in the economic sector influence cultural development*" [44]. Finally, according to Capra and Jakobsen, ecological economics, by contrast, advocates the urgently needed shift from quantitative growth to qualitative development [34].

12.1.5 From Techno (Smart?)-City to Human City

Nonetheless a real smart economy must work in a real smart city. Unfortunately, a "*smart city*" is an ambiguous expression, because it does not regulate clearly, whether it refers to community or infrastructures, nor states the purpose of its smartness, which fits mostly as a neutral quality of its technological equipment. It does not state this, because it has in fact lost its aim, which is organic to the real nature of the urban body. Many researches have underlined how much fluid is the concept of a smart city [45, 46]. It is still a pretty fuzzy label, and sometimes it is not used in any consistent way. Whilst the list of self-claiming "*smart*" neighborhoods and cities is long, so is the amount of discussion of this topic which has emerged in recent years [47–51]. Almost every scholar refers to a smart city as some kind of positive new urban economic development based on ICT innovation. This implies the use of analogous terms, such as telecommunication [52], informational [53], digital [54], or intelligent city [55]. Therefore computer and city intertwinement is the first point acknowledged by almost everybody in defining smartness. But, what do we mean when we talk about smart city?

Biourbanism teaches us to look at the city as a hyper-complex organism, a dynamic structure composed by several interconnected layers, all influencing each other in a non-linear manner [56]. This is more than a metaphor because it refers to the latest scientific developments in biology, physics, and to more general and complex dynamical systems. Biourbanism supports a human-centred urban life and goes for structural sustainability—a sustainability that overcomes mere standard ecological claims and takes into account all those dimensions that are required to make an environment really human-friendly and enhance the environment's own identity [45, 57]. A wise city should therefore hopefully copy such a model.

In "*Biourbanism as a new framework for smart cities study*" [45] authors show that in order to achieve such an aim, one should first oppose the subsumption process and refocus the urban development on human values. Now subsumption is a term introduced by Karl Marx in the draft Chap. 6 of his book, *Capital* [58], meaning the way the capital utterly activates people to work. It is a fundamental divergence with subsumption because it changes the city from within by removing five fundamental human experiences [45]:

(i) *Body awareness*, as a criterion of what is good and what is bad in the environment;
(ii) *Space as place*;
(iii) *Scales*, cutting away relationships between one dimension and another. This happens in both architecture and urban geometry as well as in social and economic structure;
(iv) *Natural patterns*, substituted by anti-patterns; and
(v) *Real needs*, like the need for rest, for meaning, for physical and emotional connection (substituted or distorted by tension, stress, and artificial desires which are induced by the need for relief).

In the context of the smart city, we know how important the role of Information Technology is. In the contemporary society, dominated by media, point (i) has to do with the medium role of ICT in support of the society of spectacle as illustrated by Guy Debord [59]. When Debord says that "*All that was once directly lived has become mere representation*", he is referring to the dominant importance of the image in contemporary society. Debord says that images have supplanted genuine human interaction [59], and digital technology has become the tool through subsumption process informing people's minds. Thus, Debord's fourth thesis explains that: "*the spectacle is not a collection of images; rather, it is a social relationship between people that is mediated by images*" [59]. In Debord, spectacle includes the transformation and control of the urban environment, such as by substituting the relics of an urban past with fantastic architecture, invading space and silence with screens, and spreading hyper-technological devices of control such as cameras, face scanners, geo-localizers [45].

But the role of digital technology is pervasive and also distorts the relationship between space and place, or the transformation of places into non-places [60]. In this context, there is a transformation of the XIX and XX centuries' physical spaces into virtual spaces. Here the physical highway becomes a flux of dematerialized data, generating a profound transformation both in the physical environment as well as in the social space and the interpersonal relationship, and the labor market. Subsumption distortion encourages gaps between social classes, different areas of the city, economic structure, and the human body and environment [45, 59] through breakdown in scales. Scales refer to the application of a natural law called scaling law; this latter not only is the expression of physical principles in the mathematical language of homogeneous functions [61], but it also refers to a law that describes the scale invariance found in many natural phenomena, such as human or animal bodies' and metabolism [62–64], or, in the way in which some characteristic of a city can evolve over the time [65, 66]. This means that, if we destroy this "*natural rule*", we may destroy part of the relationships in the natural environment and human space [45]. And finally a last element that supports subsumption is the anti-patterns [45]. Anti-patterns encourage disconnection in social and physical environment, because it changes space geometry that gets people in touch with both their built environment as well as in terms of social interrelation [45]. Disconnections created through anti-patterns can be substituted by the Constructal Law of design [45, 67, 68] and patterns of Neuroergonomics [69]. This change creates stress, confusion and disconnection in both our social structure as well as our own body.

A way to contrast subsumption process is the homological city [45]. According to this framework, a homological smart city should be built or enriched by the means of real and direct citizens' participation (what has been called p2p urbanism) [70, 71], Neuroergonomics and Biophilic design, and biourban economics, all based on the human body epistemic. Neuroergonomics becomes a way to establish an epistemological foundation of urban design which is aimed at exalting human well-being [45, 70, 71]. To conclude, homological city means rediscover and reconstruct the idea of the city and the community. These words have characterized

Western Civilisation for hundreds of years; "city" was about its human content, which was synonymous for "*community*" and "*civitas*" as the real essence of human being, and "*communitas*" (a Latin noun) commonly referring either to an unstructured community in which people are equal or to the very spirit of community. Thus, in our heritages, when we talk about "*civitas*" we refer to both the body of citizens who constitute a state, especially a city-state, a commonwealth, or the like as well as the citizenship, especially as imparting shared responsibility, a common purpose, and sense of community, where one could achieve his own human status only by acting and sharing in a society of peers.

12.2 Toward a 21st-Century Italian Socio-Economic Renaissance

Italy, since the early 90s, has found itself in a growing climate of economic stagnation as a result of the national economic model based on low labor costs to boost the economy, which has led the country into a vicious cycle of low wages, low growth in labor productivity, and lack of innovation [72]. When economic globalization [73] hit the economies of the European states, the Italian governance, especially at the local level, found itself unprepared and without any strategy to reshape its development model.

Since June 2004, the Italian unemployment rate has grown by 3.2%, reaching 11.1% in 2017, when more than 35% of people with an age of 15–24 years were unemployed. The situation in the South is even more dramatic as the unemployment rate in June 2017 was 21%, reaching 44.7% for people of 18–29 years old. For the same age group, in the Metropolitan City of Naples, the unemployment rate in 2016 was 46.4% [74]. Furthermore, a substantial portion of youths encounters considerable difficulties in finding employment. Many young people do not actively look for a job, as indicated by the inactivity rate of 26.3% for people aged 25–34 years. Although the employment slowly rose in 2017, the number of temporary employees was 2.69 million in June 2017, the highest number registered since data collection [75]. In 2016 the number of Italian's NEET has made the European record with 19.9% of youths aged 15–24 years that are not in education, employment, or training while the general EU trend dropped to 11.5% [76].

The institutional panorama is also paralyzed by a profound crisis of political delegation. The politic, at any level, is not capable of interpreting the needs of society because it is conditioned by the pressure of neoliberal lobbies, which have gained great power in the wake of financial capitalism [77]. The Italian crisis of political delegation is also due to an electoral system that gives to the parties most of the control on the candidacies, resulting in a growing separation of the governmental institutions from the civic society [78, 79].

Lastly, the whole country is in a permanent state of emergency condition, as evident in the reaction to natural calamities such as earthquakes, flooding, severe snowstorms, and water scarcity [80]. The earthquake that hit central Italy in 2016 had an estimated cost of 5 billion Euros [81]. The peninsula also has a diffused hydrogeological risk [82], intensified by a land use consumption rate that reached a peak of 8 square meters per second during the first decade of the 21st century, to slow down to 3 square meters per second in 2016 [83].

For the economic fabric, made of small and medium enterprises, the pressure from the globalized economy and the presence of national structural issues constitute a suffering climate of development that is harsh to operate in, especially in the absence of a clear and long-term national economic development strategy. Businesses need to innovate quickly or move their production abroad if not just shut down because of the impossibility of restructuring and maintaining competitiveness. Major issues for young entrepreneurs are difficulties in accessing funds, high bureaucracy, high tax wedge and complex job market regulations, obsolete technological infrastructures, labor force competences mismatching the market demand, little protection against low-cost foreign products and labor force.

12.2.1 The Revival of Local Communities

The harsh climate of crises also brings opportunities, opening the creativity of people to find solutions through innovative activities [84]. Around the world, a large number of young people have been committed to collaborate in an effort of challenging complex issues by activating innovative processes [85]. These people gather together in grassroots *initiatives of social innovation* that challenge social and environmental issues, often by empowering local communities [86]. In Italy, particularly in the South, there is an emergent phenomenon of community revival led by innovative social initiatives [87, 88].

Those grassroots' initiatives are based on social innovation, which is referred to as the participation process and relative outcomes that support social progress, leading to systemic improvements in the dynamic of the urban system [89]. Outcomes of the process are products, services, and models that simultaneously meet social needs and create new collaborative relationships [90, 91]. Solutions produced from that process are seen to be better than existing ones regarding effectiveness, efficiency, and sustainability, for which the value created impacts society as a whole [92].

In Italy, social innovation has roots in a long-time tradition of active citizenship. This term refers to the set of self-organization forms that involve the exercise of powers and responsibilities in public policy to make effective rights, protect common goods, and support vulnerable people [93]. The principle underlying active citizenship is contained in the Italian Constitution, which declares that "The state, regions, provinces, and metropolitan cities promote the autonomous initiatives of citizens, individually and associated, in performing activities of general interest, by subsidiarity" [94]. The first and most common type of formal organization of active citizenship in Italy is the *Pro Loco* associations, which have been present in most Italian municipalities since 1881 [95].

Active Italian citizens have always dealt with a broad set of issues, such as social assistance, environmental hazards, culture and heritage preservation, promotion of local resources and products, education, and sport. The dimension of active citizenship can be appreciated by looking at the third sector, as most of these citizens decide to self-organize into the form of nonprofit associations.

The last Italian census reported that in 2011 there were 301,191 active nonprofit institutions, included the associations, a growth of 28% in 10 years. The nonprofit sector represents 6.4% of all Italian economic activities with 4.7 million voluntary workers, which is 83.3% of the total of human resources involved the sector. In Italy, the third sector is the first productive reality in the industries of social assistance, cultural activities, sport, entertainment, and leisure [96].

The initiatives of social innovation considered in that chapter differentiate from other third sector institutions for their underlying goal of improving the quality of life at the level of community. They do not only want to solve one issue but instead, they aim to have a positive impact on the whole city system, economically, socially, and environmentally. The OECD recognizes precisely that point, affirming that social well-being is a goal, not a consequence of social innovation, which exists "wherever new mechanisms and norms consolidate and improve the well-being of individuals, communities, and territories regarding social inclusion, creation of employment, quality of life" [97].

Members of those initiatives use their creativity to find new solutions by relying on local assets, including natural resources [98]. Young innovators are more aware and concerned about negative externalities the society can produce on the environment. Because of this, it is not surprising that it is easy to find several initiatives that deal with ecological topics such as the reuse of waste to make recycled products, tourism and the promotion of natural resources, the recovery and reuse of the abandoned heritage, the education and sensitizing activities that focus on the recovery and promotion of their local territories.

Local Italian communities are being revived by these initiatives that self-organize with internal roles and regulations, following a bottom-up process of development that is responsive and adaptive to local circumstances. Because of the intense use of ICT, they interplay with each other without geographical limits. The result is a network of networks that operates at different scales and on an undetermined territory. The active presence of these social innovation initiatives has taken the form of a movement interacting with various sectors and industries at local, regional, and national scales [86].

12.2.2 The Italian Social Innovation Movement

12.2.2.1 Geography

The entire Italian peninsula is affected by this movement of social innovation. A web platform named *Italia che cambia* (Italy that changes) started in 2015 the mapping of the initiatives and the sharing of their basic information. So far, the web platform has mapped roughly 1800 initiatives present in all the Italian regions (Fig. 12.1). However, many others are missed as the mapping process is on a voluntary basis: every single initiative has to self-register on the platform and decide what contents to share. The map provides a representative account of the movement, geographically and regarding the area of activity. In fact, the initiatives are grouped according to the following categories: living, agriculture, art and culture, production cycles and waste, economy, education, energy, social integration, jobs and entrepreneurship, legality, mobility, health and nutrition, environmental sustainability, lifestyles, traveling [99].

The resulting picture is an eclectic movement of citizens, grouped in initiatives dealing with all the major themes regarding society and the territory. The same panorama results from a study in 2015, named *Sud Innovation*, that investigated the phenomenon in Southern Italy, from the Campania region, of which Naples is the seat, to Sicily [87].

12.2.2.2 The Habitat of Initiative of Social Innovation

Economic indicators show a difficult situation, particularly for Southern Italian regions that registered a lower GDP than the other Italian areas, making them eligible for European structural funds [100]. However, in the biennium 2015–2016, the South recorded a reversal of the trend by growing more than the North, yet, at this rate, the prediction is that the South will regain the pre-crisis level in 2028, while Italy can do so in 2019 [101].

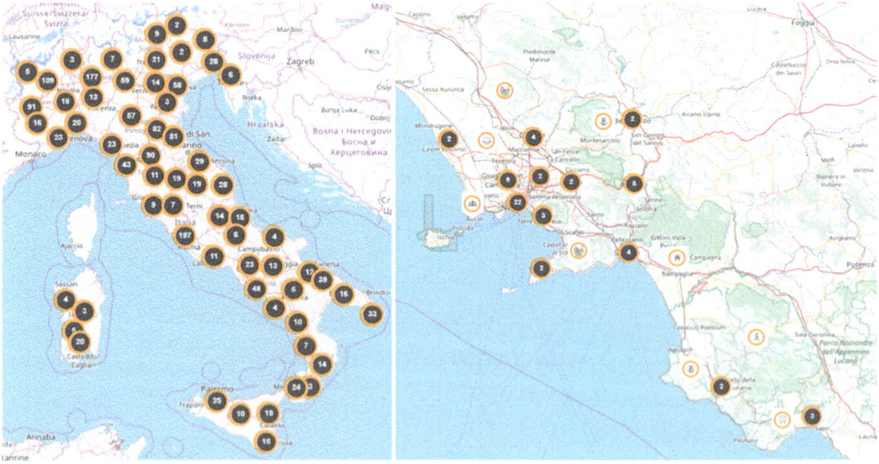

Fig. 12.1 Mapped Italian initiatives at October 2017. Italy (left), Campania Region (right). *Source* [99]

There is a vast economic and social gap between Southern and Northern Italy rooted in historical reasons, among which the out-migration phenomenon and progressive globalization have played crucial roles [102]. In the last decades, the Southern countryside has seen rural communities affected by the phenomenon of new immigrants, subjected to a low-wage, casual, and frequently irregular employment that originated in critical situations similar to those at the turn of the 19th century when the South was primarily based on an agricultural economy. However, the organization and appearance of these rural peripheries took different aspects, reflected in new economic, social, and spatial changes shaped by the influence of globalization [103].

In these less developed regions (Campania, Puglia, Basilicata, Calabria, Sicilia) there are thousands of small towns and villages, mostly located in inner areas, which have been significantly affected by economic globalization, and that present several issues, such as high unemployment rate, strong outward-migration, population ageing, abandonment and deterioration of physical assets, lack of infrastructural investments, and lack of capability of local governments in facing contemporary challenges [104]. These aspects have plummeted towns and villages on a pauperization trend that today reveals situations of distressed communities with substantial economic stagnation or decline that persists for decades [87].

That situation is particularly real for small towns and villages (up to 5000 inhabitants) which in Italy represent 69.9% of all municipalities (Table 12.1) [105]. In those territories, economic and social components show values of risk threshold or even dynamics of no return [106]. A large number of initiatives of social innovation deal with these types of settlements.

Almost all of those cities are hill towns with a medieval historical center. Although it is usually the most distressed area of the town, it represents an identity place for local inhabitants, especially for youths. They are particularly attracted to the old town, which frequently hosts the headquarters of several initiatives of social innovation. That is the case of *LongoTherapy*, a cultural association born to attempt to give to the historical center of Longobardi village (Calabria) an innovative perspective on the cultural overview of the territory by organizing cultural and

Table 12.1 Small towns and villages in the Southern Italian regions

Regions	No. of municipalities	No. of small municipalities	% of small municipalities
Abruzzo	305	249	81.6
Molise	136	125	91.9
Campania	550	335	60.9
Puglia	258	85	32.9
Basilicata	131	101	77.1
Calabria	409	323	79
Sicilia	390	205	52.6

Source [105]

leisure events that can remember all the social and cultural flow, which used to animate their community. The association has refurbished at its expense *La Casa delle Culture* (the house of cultures), a public structure unused for a long time [107]. In Succiso (Tuscany), all of the 65 inhabitants are part of a community cooperative named *Cooperativa Valle dei Cavalieri* (Cooperative valley of the knights) [108]. In 1991, after the last bar shut down, a group of youths decided to gather together, investing their money in the refurbishment of the abandoned schools. Then, they opened several business activities in the sectors of food&drink, accommodation, local products, and leisure. Today, they generate an economy of €700,000 per year having invested €1.5 million over 25 years [109].

- A large number of initiatives aim to revitalize historical centers, or at least make an impact that will benefit those heritage neighborhoods. This purpose can be found in manifestoes, statutes, or just documents of intention of a number of initiatives.

For example, the mission statement of *Belmonte in Rete* is "*we work for the sustainable development of the historical center of Belmonte Calabro*" (Calabria) [110]. In Sicily, the *Farm Cultural Park* is an independent cultural center with a strong focus on contemporary art and innovation. Located in the heart of the old town of Favara, it has acquired some of the abandoned houses and transformed them into contemporary art exhibitions, meeting spaces, open kitchens for workshops and lunches, cocktail bars, and vintage shops [87, 111].

The case of Pisticci town is emblematic. This is a historical hill town in the Basilicata region, headquarters of the *Imbianchini di Bellezza* (beauty painters), and an initiative that in 2014 in reaction to the state of abandonment of their historical center decided to paint the house façades with whitewash. The group grew fast and started to take a variety of actions. Under the topic of urban regeneration, they now organize activities directed to "*propose a model of territorial development that can combine the listening of the citizens by the public administration and the needs of the community*" [112].

From these intentions and activities, it emerges that the revival (whatever is a revitalization, regeneration, refurbishment, or recovery) of these historical urban masterpieces is a crucial point for several initiatives of social innovation.

Not only have the small historical centers hosted initiatives of social innovation. The *Borgo vecchio factory* is a social promotion project developed by the not-for-profit organizations *PUSH* and *Per Esempio Onlus* in collaboration with the street artist Ema Jons. It involves the creation of a cycle of creative painting workshops for 20 children in the neighborhood of Borgo Vecchio in Palermo (Sicily). The children of Borgo Vecchio have an active part in the project: they work with artists to create street-art—inspired by their drawings—on the walls of the neighborhood houses and buildings. The initiative has been funded by a crowdfunding campaign [113, 114].

Palermo is one of the Italian metropolitan areas that is constituted not by an interrupted city, but rather by a constellation of urban fragments, including single buildings, industrial agglomerations, entire detached districts, and medium and small municipalities with both modern and historical quarters, all scattered among agricultural fields, orchards, public parks, and woods.

From an administrative point of view, this is the effect of the Italian legislation that in 2014 aggregated a huge number of municipalities into 10 metropolitan areas [115], regardless of geographic and socio-economic differences of their components. The result is large metropolitan territories of high complexity made prevalently of scattered small towns and villages around a central city. However, the territorial coherence of such metropolitan cities can be very different from each other due to the complex morphology of the Italian peninsula. The settlements within the metropolitan city can be very close and connected, such as in the case of the Metropolitan cities of Milan and Venice, or very far and disconnected, such as for those Metropolitan cities that extend over the Apennines Mountains, such as Palermo or Reggio Calabria. The last one is mostly rural, spanning between two seas and with a huge mountain range in the middle, very different economies, and lack of infrastructure that discourages any type of synergy among its parts. Naples is a typical Italian Metropolitan city, with over three million people inhabiting 92 municipalities of which 60% is less than 10 km^2 wide, and only 12 municipalities have more than 50,000 inhabitants. It is also exposed to a high seismic risk due to the presence of the Vesuvius volcano in the middle and the Campi Flegrei volcanic complex at the north.

12.2.2.3 What Triggers an Initiative and What Drives It

In the harsh Italian climate of development, a young person faces the decision whether merely to stay inactive, emigrate if it can, or challenging itself by starting a project. Those who decide to undertake the challenge find the necessary motivation rooted in their spirit and linked to the innate sense of place toward the native territory. The motivation supports the new entrepreneur who is confident that through his efforts he will also help his native place to improve.

The innate link between a human being and a place was already identified by ancient Romans by resorting to the concept of *Genius Loci*, the protective spirit of a place. Not only places had their genius, but also every person had one; therefore, this establishes a robust innate relationship between people and places that Wilson defined as biophilia [116]. Norberg-Schultz investigated the topic within the field of architecture and pointed out how the ancients had recognized and respected the *Genius* of the place they wish to inhabit [117]. Bevilacqua suggests that it is through the sight and sentiment of astonishment that we can see the presence in a place of a *Genius Loci* [118].

That sentiment is precisely the one that young innovators have towards their native place. That deep spiritual attachment enhances the motivation and overwhelms the dissuasive power of economic indicators that would discourage any investment in that area.

In Jacurso, a less than 630 inhabitants historical village in Calabria, the initiative *Jacurso da Vivere e Imparare* (Jacurso to live and learn) seeks to record and safeguard aspects of the traditional life in reaction to the globalization effects that could erase them. They aim to restart the life of the small community through sustainable tourism based on participation, offering a full immersion through workshops, Italian language classes, guided tours with an anthropologist and contact with the local community daily life [119].

However, the effort of a single person would never be enough to change the direction of a town in economic decline. The confidence is that it is exactly around a clear project that other people will find their motivation and aggregate to the initiative. The personal motivation fed by the sense of place reveals to be the most valuable source of power for starting a grassroots initiative of social innovation that aims to increase the quality of life for the innovators and their whole community.

12.2.2.4 The Innovators

People involved in initiatives of social innovation are young, hold a degree, and are armed with the intention to improve their surrounding territories. They are aware that they have to rely mostly on their capabilities and they are willing to do so. These people are seen as protagonists of a soft revolution for which they fight using their talents such as creativity, bravery, resilience, transparency, and ability to engage people, culture, youth, ethics, visions of beauty, and humanity [120].

These young people have not experienced the welfarism of the Italian state, which until the 1990s guaranteed to their parents a safe future. Neither can they quickly aim for a permanent public job nor to a large availability of non-repayable funds to start a business, like those that in the past drugged the Italian market with the results of putting people one against the other in the run for achieving public funds [87].

Young innovators place their trust in a collective entity, the group. While the Renaissance man mastered its skills and produced masterpieces through a solitary process of development, today people gather together around projects, taking advantage of the numerous skills and knowledge shared by others through the network. The Italian social innovators manage to meet in specific events and collaborate on short- and long-term projects at different geographic scales. Interactions rely mostly on technologies such as open-access web platforms, social networks, and phone apps used to spread their media contents and reach other players and networks. For young innovators, the communication is the biggest sector of investment in terms of work-time.

12.2.2.5 Peer-to-Peer, Networking, and ICT

The phenomenon of social innovation can be described as a peer-to-peer networking system made of initiatives (nodes) and social relationships (links). ICT is used to attract, share, and spread knowledge, and to organize, promote, and grow the initiatives, also to expand the network (Fig. 12.2).

Peer-to-peer is a crucial principle for innovators of the 21st century. This form of relational dynamic, based on the equivalence of its participants, suits entirely the

Fig. 12.2 Peer-to-peer networking system of initiatives of social innovation. *Source* Authors

vision of a future society where people organize through free cooperation in the view of creating common goods, accessible to and improvable by everyone. Peer-to-peer is powered by a robust social networking activity that fosters the exchange of know-how, information, and ideas among the members.

People decide to join a network of their interest because they believe they can support it with their skills and capabilities. However, one of the implicit objectives of networking is to establish professional relationships that may turn useful for future projects, business activities, and career development.

The peer-to-peer network of innovators is also used to solve practical issues and reduce operation costs. Not only knowledge and skills are shared with other members of the network, but also physical assets, such as residences and other properties.

For example, it is a common pattern for several initiatives to organize events on specific topics where specialists are invited to make presentations and offer their point of view on the matter. This is a sharing operation as the specialist is usually asked to spend some time, maybe days, in the town. S/he is walked around like a tourist, putting them in contact with local stakeholders, allowing them to experience the context truly, and offering them free accommodation and meals. In exchange, the specialist shares his/her knowledge and opens their network with the local innovators. In this way, the network expands, and the share of know-how is not conditioned by a monetary transaction.

Trust is put in the networking system. The trusting mechanism is the most powerful weapon against the threats that would come from outside and inside the network. Members build trust by maintaining regular contact with each other, not only professional, and as much as is possible in face-to-face occasions. If a member is seen to be in contrast with the principles of the network, it is likely that s/he would be isolated, resulting in losing the support of the network, which in this way protects its members.

There is a strong interaction between the initiatives that aggregate in networks at different scales: people with people, initiatives with initiatives and groups of initiatives with other groups. This is a precise and conscious strategy that generates shared visions of systemic effects on the socio-economic fabric of the whole country, similar to what happens for reticulated business systems as in cluster productive districts [121]. The difference is that the social innovation network is established not only within a single industry sector but with whichever (single players, informal groups, associations, institutions) is interested to be part and support the network, independently of the geographic distance. This allows a holistic approach to emerge by doing networking that reflects a predisposition of the innovators in being open to new possibilities, even if they cannot foresee them yet.

The major requirement to facilitate the emergence of a peer-to-peer network is the existence of a technological infrastructure that enables and facilitates access to the network and interaction between its members. ICT facilitates and speeds the process of aggregation of people as well as the emerging and sharing of ideas. Feedback is also received immediately and discussed among members or the network, which accelerates more the development process of an initiative of social innovation than in the past.

Young innovators make extensive and sophisticated use of web platforms and phone apps for social networking, organizing activities, and spreading information. The most common platforms of interaction are Facebook (to publicize initiatives and related activities, get information from others, organize themselves through the use of the group function, etc.) and WhatsApp to keep themselves in permanent contact. Websites are instead used more like showcases or portfolios to display their best outcomes, keep a record of the activities, and provide easy access to basic information and contacts. Other platforms and software are used for online meeting places, selling and buying materials, online learning, etc.

ICT infrastructures represent, therefore, a crucial asset for young innovators that are willing to start an initiative; therefore, investments in ICT infrastructures are crucial for any city that aims to foster social and economic development.

12.2.2.6 Structure and Development Process of Initiatives of Social Innovation

Initiatives of social innovation are structurally similar to business enterprises, which differ in their aim, which is to create primarily social value [92]. Murray identified six stages of development that an initiative usually goes through: prompts, proposals, prototyping, sustaining, scaling and diffusion, and systemic change [91]. These stages have feedback loops between them and might not be sequential or even not using all the steps. The development process is very sensitive to local conditions, project ideas, and the organization model of the initiative.

Typical Italian initiatives start from a small group of people that usually know and trust each other. In most of the cases, they live in the same town or geographic area. This is especially true for those initiatives that want to address the topic of city development. In the prompting stage, they identify problems, causes, and discuss possible solutions. A group such as this is the embryonic state of a social innovation initiative. The following case is a clear example of how an initiative can be started and the relevant role the network played in that.

Belmonte in Rete is a social initiative emerging from the community of Belmonte Calabro (Calabria). Its mission is to foster sustainable development in that village. Belmonte in Rete was born with the first goal of building a shared Vision of Belmonte Calabro that images the town over the next 20 years; capable of inspiring all its inhabitants and all those who want to contribute to the development of Belmonte [39]. In 2015, one of the co-founders attended a seminar near Rome on the revitalization of historical centers. There, he met other people presenting their initiatives, some of which were based on bottom-up processes and citizen participation. Inspired by the other stories, he went home with the motivation to start something similar for his hometown, Belmonte Calabro. At the seminar, he also met the other co-founder of *Belmonte in Rete,* which as an idea, was born in Calabria, and then found the necessary partnership and support provided by the network in Rome.

A crucial factor that is frequently determinant in the emerging of an initiative is the presence of a mentor. This figure is a knowledgeable person that is attracted by the ideas of the group; thus, s/he offers help by protecting and supporting the initiative through its capabilities, experience, network, and sometimes even financing activities. A mentor is a kind of guardian angel for an initiative of social innovation [87]. Having one or more mentors is a crucial resource in the stage of generating proposals and ideas.

As part of its organization the association *Pensando Meridiano* (Calabria) recognized a group of mentors that have no responsibility nor statutory role in the association; they are there for reference, support, and accompaniment to those actions that the association identifies and plays. The mission of the association is to foster social innovation among youth generations. It also helps local institutions to develop visions for the future of their communities. Bottom-up processes and the use of ICT are at the base of their programs and initiatives [122].

Once the group is solid and the motivation high, an initiative usually develops in a more structured subject, clearly defining the aims, principles, and the way to act. Innovators have to decide which would be the most appropriate identity for the group. The most common forms are simply an informal aggregation of people that collaborate on a common agenda, a private company, or a nonprofit organization.

The last formula is the most chosen, as it communicates well the willingness to operate for making social value and not for an exclusive economic benefit. That form also provides easy acceptance by the related community, and at the same time, it guarantees openness to whoever is interested in being part of the initiative. Another reason for choosing a nonprofit formula is the simplified and cheap opening procedure as well as the access to fiscal benefits. It is easy, fast, and inexpensive to open a nonprofit association, it suits well for a broad range of industry sectors, and its structure and organization can be determined by the associates in the way it serves to better the aim of the initiative.

Other typical activities are the design of a logo, the organization of a launching event open to the whole community. Later on, many initial events and activities are planned and realized. All of the activities are finalized to develop a prototype. That would be one or more services for cases of initiatives that aim to improve a town.

Expanding the network, achieving competence, analyzing strengths and weaknesses of their territory, and gaining feedback are all activities finalized to grow and diffuse the initiative.

There are several ways the initiatives engage with the community. Workshops, photographic and video contests, education activities, conferences and festivals, guided tours of the city, sportive and leisure events are all activities ideated and implemented to work systemically for general community development. The primary intent is to bring into the local territory different points of view, ideas, and knowledge that could be used to develop proposals and projects for the city.

The *Progetto Artena* was a five-year participative strategic design project for the revival of the historical centre of Artena (Rome) started by the nonprofit organisation *International Society of Biourbanism*. After more than 80 initiatives (Fig. 12.3), during the third year of the project, a breakthrough event happened: The *Comitato Centro Storico Artena* (The neighborhood committee of the historical center of Artena) arose spontaneously by a proposal of a few citizens. Social events, recovery of neglected areas, restoration of buildings, a mediation role between citizens and the Municipality, promotion of local products, fundraising for initiatives are the kinds of initiatives the Committee promotes to improve the quality of life of the community. Today, the Committee involves around 30 residents and supporters, gathering different actors such as local business owners, makers, professionals, the Municipality, food producers, schools, and associations [123].

12.2.2.7 Relations with Local Institutions

In the Italian panorama, the public sector finds difficulties in promoting effective public policies supporting innovation and development. This is due to the lack of investment in new technologies for public offices, and unskilled human resources of the administrative apparatus, which therefore results in being slow and inefficient, even in being aware of what happens in its territory. The sole logistical-infrastructure deficit penalizes the Italian economic system for an amount of €42 billion per year. In addition, the burden of bureaucracy is on small and medium enterprises for an amount of €31 billion a year [124].

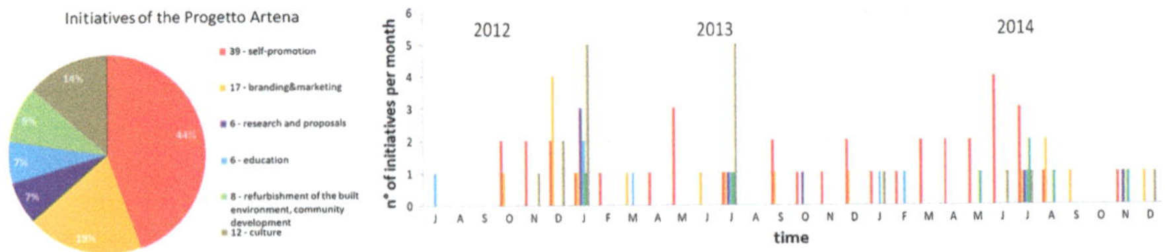

Fig. 12.3 Initiatives of the *Progetto Artena* by type and their distribution over time. *Source* Authors

All of those factors encourage a climate of light control and propensity to assist people outside the legislative framework. Simply, the most effective solutions are taken, regardless of the law and standard procedures in a climate of corruption (moral and economic) that people are keen to undergo to gain more profit by overcoming the bureaucracy and the cost of operations. A visible effect on the built environment is the presence of a massive number of illegal buildings and appropriations of public spaces.

This critical situation is produced by the overlapping of two different paradigms of governance. The first one is an informal and shared ancient model that relies on peer-to-peer social relationships to address issues. That one was proper for the Italian preindustrial municipalities, mainly self-sufficient where the preservation of the community integrity was seen as paramount to respond to outside threats. The second one is the modern hierarchical model of governance based on a powerful central authority. There are urban areas where the contemporary model of governance is weak and gives space to the remains of the older informal model to take place. However, in that situation, both models coexist, clashing with each other. Those are the situations where initiatives of social innovation find a fertile territory to propose their ideas. Their success is due to the fact that they work by using a model based on social relationships, which allows the two governance models to find points of contact. Initiatives of social innovation usually take the role of mediators between the community and institutions.

The public authority is frequently a partner. There are several forms of collaboration. Almost always there is a simple institutional patronage on singular activities that works as a sort of guarantee that the activity promoted by an initiative of social innovation is for noble and shareable purposes. Sometimes, the patronage is combined with funds or other services (a school bus to drive people, the cleaning of a place for an event, etc.) but in most cases, it goes with a concession of using a public space or a building for free. Those concessions can also stand for long periods of time, such as the case of providing a free-of-charge space as headquarters for a nonprofit organization. A more recent and growing practice is the establishment of a collaboration agreement where a public property is given for free under certain conditions, such as the maintenance and refurbishment of the property on expansion of the initiative.

Some municipalities have taken this direction by implementing sophisticated tools for the management of their unexploited assets by relying on the active involvement of third sector's subjects. For example, this is the case of Bologna, which developed an ICT platform called *Iperbole2020* [125]. To pursue the aim of cooperation, Bologna created a project called *Collaborare è Bologna* (Collaborating is Bologna), which works as a framework for several initiatives. Its aim is to promote the culture of collaboration and seeking community involvement to make information, technologies, resources, space, knowledge, and skills more accessible. The project is based on a document, "*the Regulation on collaborative forms between citizens and administration*" [126].

In Altamura (Puglia), *Iperurbano* is a network of three subjects aggregated to develop a participatory path promoted by the Municipality of Altamura to build the

Integrated Sustainable Urban Development Strategy [127]. Each of the three subjects has a story of intense activity in the promotion and development of Altamura's territory. They are *Esperimenti Architettonici* (Architectural Experiments) a research group born in Altamura in 2011 consisting of a network of students, researchers, and designers from different parts of Italy. It is set up as a collaborative platform between the web and the territory, promoting urban regeneration strategies and cultural promotion focused on the active involvement of citizens [128]; the association *Cuore di Altamura*, made of citizens residing in the historical center of Altamura, aiming to enhance the historical and cultural heritage of the ancient district [129]; the *Pro Loco Altamura*, a local association with the aim of promoting and developing the territory [130]. Saverio Massaro, President of *Esperimenti Architettonici*, is confident that this working opportunity along with the public administration will strengthen the relationships between the three organizations and pave the way for further involvement of the citizens in the governance of the whole town.

12.2.2.8 Relations with the Built Environment

The Italian built environment is largely deteriorated, especially in the peripheries and historical centers. The wild and unregulated urbanization that started during the second half of the last century has left the new generations a damaged territory with plenty of second and third property houses, infrastructures, and industrial buildings most of them illegally built and then regularized by the politics through a series of laws that threatened the environment more and more; this in contrast with the Italian Constitution, which specifically addresses safeguarding of the landscape [131]. A considerable number of empty buildings, mostly in bad condition and of poor quality, are scattered along the main arterial roads, alternated by unused plots and abandoned fields that have originated a continuous ugly urban landscape [132]. Historical centers look at those territories from the top of their hills where they are in a dreadful state, mostly abandoned, full of private properties without residents and dangerously in ruin [133].

How to use public and private unexploited assets is a long-running debate in Italy. The State's real estate has a value of almost 60 billion Euros: over 47,000 inventories, 32,691 buildings and 14,351 areas, worth 54.1 and 4.78 billion, respectively [134]. There are not enough public funds nor an effective national strategy to maintain all of them and put them into the economic system. Recent selling programs of state properties have achieved lower results than expectations [135].

This huge real estate represents an opportunity for the reintroduction of those assets into the economy of each Italian municipality. The already cited study, Sud Innovation, explores the relationship between social innovation and physical heritage. Several initiatives deal with these assets, taking care of them and bringing them into a new economic and social dynamic. The social innovation phenomenon offers a way to a model of management that is based on subsidiary and cooperation principles. At least, a part of the public and private heritage could be maintained by local citizens and put into an economically sustainable dynamic of asset management, without the necessity to recur to external funds [87]. The new approaches to enhance public assets by supporting grassroots participation have achieved success under certain conditions, such as "*the presence of a community capable of self-organized processes, the location of the buildings in urban centers, specific building characteristics in terms of type and state of preservation, the presence of a party — other than the public administration — with the ability to independently assume a managerial role in the activities promoted, and a flexible legal relationship between the administration and private parties*" [136].

The model that emerges by looking at the built environment refurbishing activities implemented by initiatives of social innovation is opposite to the one that is based on significant investments for large projects, and requiring a long time and several bureaucratic procedures. This alternative model is based on incremental developments made of micro interventions of refurbishment and reconnection to the urban system of deteriorated public spaces and buildings.

The described model of intervention has several points in common with an urban regeneration method called *urban acupuncture* [137]. Here, the underlying assumption is to cure the urban organism where it is affected by an illness or harmed in its physical part. The principle and practice of recovering the energy of a sore or fatigued spot by means of a simple touch have to do with the revitalization of this point and the surrounding area [137]. This method combines urban design with acupuncture, which comes from traditional Chinese medicine [138]. The method wants to provoke a shock that affects all the connected elements of the urban system. In fact, "*urban acupuncture is characterized by punctual interventions through the official surface of the city which aim to establish contact between the urban collective conscious and the life-providing systems of nature, including human nature*" [139].

An example is the case of the village of Artena, close to Rome. The inhabitants of the half-abandoned historical center take care of the maintenance and refurbishment of many small public spaces, they collect money to buy street furniture and to repair public fountains. They also organize open events on a range of topics related to their town's characteristics. The activities are organized by a local neighborhood committee, which was founded in 2014 with the purpose of promoting projects and initiatives aimed at improving the quality of life of the residents of the historical center of Artena [140].

12.2.2.9 Scales of Intervention

Initiatives dealing with social innovation operate on at least three scales: the local one, accounting for the largest number of initiatives, a middle scale, made of groups that focus on helping other initiatives by offering highly specialized services, and lastly, a high scale where initiatives support all the others by sharing their stories or offering a space to publicize their activities.

All of the cases presented so far, mostly operate at the local scale. An example of high-scale initiative is *Italia che cambia*. We have already seen that it offers a free mapping service on their web platform. In addition, this initiative is registered as a digital newspaper, as its aim is to document and share information about all other initiatives [99].

Italian Stories is a "*digital marketplace for immersive travel experiences focused on Italian craftsmanship.*" That initiative has the form of a business that finds and engages with local crafters, sharing their stories in a network accessible to whoever "*wants to discover the real artisan knowledge, while living unique immersive experiences.*" [141].

South Cultural Routes finds stories, places, and people, a testimony to a past that today lives and revives thanks to a new form of tourism: slow, attentive, sustainable, and green. The initiative was started by Lorenzo Scaraggi who embarked on a caravan trip across the slow cultural and tourist itineraries in Puglia, Basilicata, Calabria, Campania, and Molise [142]. Massimo Castelli defines himself as a traveler, storyteller, filmmaker, and human being. Through his *Amazing Everyday People*'s project, he video-story tells places and people he meets, aiming to inspire other people to make a positive impact on society. He largely uses ICT, drones, and other technological devices [143].

Other initiatives operate at the middle scale, networking with other initiatives and young innovators, providing support or establishing mutual help. Usually, shared projects originate from linking together two or more of these initiatives.

Founded by Andrea Paoletti and Mariella Stella in 2012, *Casa Netural*is is based in Matera (Basilicata). It is a "house" where the community can start collaborations and shared projects with social innovators from all over the world. Thanks to the Co-living project, the innovators can live and work together but, above all, meet the local population and try to imagine possible futures for the territory. In 2017, they launched an incubator project, a training course to make cultural and creative businesses sustainable thanks to the team's experience and the support of a large network of international mentors. They also are involved in the field of education, tourism, and event organization [144].

GiovanIrpinia, born 2017, is a socio-cultural association aimed at enhancing the youth, the arts and the territory. They help young people between 18 and 35 years old to become active builders of their future, collaborating in partnership with public and private bodies, associations, and institutions to make Irpinia's geographical area (Campania) more and more livable. The association wants to attract ideas and innovations that produce more work opportunities in their territory and beyond. The association is particularly close to all those young people who experience uncomfortable situations and who would like to be heard and supported in developing their own ideas and ambitions [145].

12.2.3 The Common Effort to Overcome the 21st-Century Uncertainty

Periods of profound crisis, like nowadays, produce huge cultural changes, revealing the fragilities of the in vogue society model and allow the emergence of profound spiritual doubts that permeate daily life [146]. The uncertainty about the future overwhelms the optimism that instead is predominant during more prosperous periods.

Nowadays, a general sense of discouragement and anxiety permeates the heart of young generations [147, 148]. Most of the youths are afflicted by a predominant anguish of failure, they doubt their capabilities, and they fear being alone in facing the widespread social injustices they see. In this climate of uncertainty, some of them find the courage to start to act, exactly because they are aware that neither the system nor any other person will pave them the way to the future. Therefore, they identified an effective means in concrete actions to make a change in society. They are confident that this way will create a positive feedback as the actions will become examples to be followed, bringing trust into the city community where others can be inspired and join the effort.

Making resolute actions is the same strategy that the Renaissance man relied on. It was confident in science and human capabilities to seek glory as a way to assure itself a surviving after death and therefore defeat uncertainty [149]. Like Leonardo, who retouched *Gioconda* until his last day, young innovators of the contemporary era never stop to increase and improve their baggage of knowledge through the countless experiences they get from the social networking system they are part of. It is a self-feeding pedagogic approach similar to the Socratic Method [150, 151]: as members of the network they do not have all the solutions, rather, they find them by dialogue with each other. The communities the innovators are part of the benefit from the spillover of that process of learning and know-how exchange.

The popular case of *Interazioni Creative* is a co-working initiative in Cosenza (Calabria) conceived by Deborah De Rose who decided to devote a physical and virtual space to the unexpressed talents. It is based on interactions to share ideas in a continuous synergic exchange, to innovate and allow the emerging of the community social dimension, a crucial asset to build an effective path toward the personal and community well-being [152].

12.2.4 A New Humanism at the Base of 21st-Century Urban Design Strategies

At the end of the 20th century, cities all over the world started to face a long transitional phase of profound structural change due to the effects, positive and negative, of globalization [153]. The situation was aggravated at the turn of the 21st century when a huge crisis, coming prevalently from the economic system, hit all the other society sectors and became systemic [154]. European urban communities found themselves unprepared against the increasing uncertainty recognizable in several threats, such as the collapse of markets and social welfare under policies of austerity, climate change, terrorism, and other parochialisms that bring disaggregation at national and international levels [155]. The effect of globalization has been much worst in Italy, which has experienced a long stagnation phase where almost no structural reforms have been made to readapt the economy and the society to the new challenges [156, 157]. While the national government was revealed to be unprepared and unable to set up a process of getting out from the crisis, at the local level, the Italian movement of social innovators organized to challenge the long-time unsolved issues accentuated by the crisis [88]. An emergent movement of social innovation has spread across the Italian country, taking root in many villages, towns, and cities.

Several aspects of this Italian social innovation movement suggest an analogy with the historic Italian Renaissance, of which the humanistic stance was at its basis. The Renaissance was a huge cultural revolution, born in a context of crisis that created the condition for the emergence of a humanistic view of life where its fundamental characteristic was the spiritual, moral, and civil training of the human being [158]. Similarly, the Italian social innovators revalue the human being and put trust in its capacities and virtues.

In the manifesto of the *Belmonte in Rete* initiative, it is clearly expressed that in their vision "*the human being is the main element of the urban ecosystem, [...] in opposition to the capitalist and mechanistic model that identifies persons as just passive users and consumers.*" In addition, the citizen is seen "*as an active agent, a creator of processes, physical and digital places, and networks.*" [110]

We can synthesize the Italian revival of local communities as a phenomenon that has taken place from within the city and for the merit of grassroots' initiatives of social innovation, constituted mostly of young people leveraging on their capabilities—mainly knowledge, skills, and creativity—promoting a novel vision for the future of their community.

That vision proposes to build a more sustainable urban system that can combine the economic development of the city and the personal development of those living in that system.

In the actual neoliberal society, those two aspects exclude each other as human rights are sacrificed to increase productivity [159]. At that point, the social innovation movement proposes a change of paradigm where the human being is put at the center of the system and its design. Social innovators propose that change by leveraging on local assets (physical resources, institutions, individual knowledge, creativity, skills and capabilities, entrepreneurial spirit, sense of place) to trigger a community revival process [98] able to produce positive systemic effects on the whole urban system, which would then offer a better quality of life [160]. Frequent effects are a change in the local governance towards more inclusive models, the promotion of sustainability at the ecologic level of the urban system, the healing of that system by its incremental regeneration.

In conclusion, every initiative of social innovation heals a part of the urban ecosystem producing systemic effects on the surroundings. Contemporary economic and spatial design strategies cannot disregard the fact that cities are being transformed by their citizens on the basis of a revived human-centered paradigm, an alternative to the one that dominates the modern society. The future structure of the city is difficult to foresee but new coming development strategies have to take into account those emerging phenomena.

12.3 Mushrooming Experiment Providing for New Biourban Strategies to Nourish Life-Enhancing Diversification in a City

12.3.1 Spatial Development Governed by Capitalist Economy in a City?

Looking into the relation of space and economy in city explains why the experiment of Mushrooming is so interesting. While researching contemporary spatial development, Carmona [161] has identified a battlefield of three tyrannies. They compose of creative, market-driven and regulatory modes of praxis. Carmona proposes that it is the battle of these three tyrannies to hinder spatial development, leading to compromises and thus to poor quality spaces without character. Interpreting Carmona, we can say that it is the market-driven tyranny to be the most powerful of these three. Firstly, because regulatory tyranny can be regarded mostly as an attempt to correct market failure. Secondly, because the designers are playing themselves out of the game. They are creating their own arena of fantasy, in which they can be the tyrants to value image-like aesthetic qualities over real needs of people and to neglect the constraints of economic feasibility. Market-driven tyranny is the most powerful also from a simple practical point of view. As McGlynn [162] puts it, those who have the ability to fund development, have the real power to shape our urban environment.

Building for example upon the work of Alexander [163] and Salingaros [164], Biourbanism sees it essential how built environment should support the well-being of humans as psycho-physio-social wholes. Thus, the issue of buildings as goods of market is seen troubling [165]. If the viewpoint of market dominates, buildings become instruments of consumerism, where, for example, aesthetics serve more the illusory styles that feed market than real experiences of beauty [166]. Consumerism is based on creating artificial needs, thus making people forget their real needs. Built environment that is not created to meet our real needs can never really support our well-being as humans. In addition, urban space created by the rules of market breaks human relations [45]. This breaking happens, for example, at the emergence of big shopping malls at the crossroads of motorways. Urban spaces produced by shopping malls, to which people arrive in their private cars, foster much less local human encounters than pedestrian streets filled with varied small shops.

The question arises: what to do in this market-driven reality, in order to create better cities for us? Serafini [165] points out one path: we need designers who acknowledge the actual powers and forces at play, we need designers who master economics. With the experiment of Mushrooming, the author will add to this. As designers, it surely is important for us to claim a new powerful position in the battlefield that the tyranny of market reigns. Nevertheless, there are interesting things happening outside this frontline as well. For example, in the following paragraphs the author will show how the network of self-organizing co-working places of Mushrooming mostly stays out of the real estate market, and thus, out of the frontline. Even more, it uses the left-over spaces of the market as its seedbed. For us designers, understanding alternative economics is part of the task of understanding economics. This possibility of adding understanding of alternative economics in the context of design when acting outside of the battlefield encouraged us to gather and analyze more in detail, what we could learn from the spatial and economic processes that the experiment of Mushrooming has fueled.

12.3.2 Self-organized Work as Context for the Experiment

In order to understand the experiment of open process and bottom-up urbanism in Finland we first need to get familiar with the phenomena of self-organized work. This is important also as we are interested in economic design strategies that consider livelihoods of people. Development of ICT technologies has freed many professions from the traditional places, times and communities of work. Many become self-entrepreneurs, collecting their livelihood from several projects here and there. Working life has fragmented [167, 168] meaning also that we more and more often change jobs or professions.

In most cases self-entrepreneurialism is a chosen path of self-expression, for one to choose more freely for example one's own projects or working habits. This is the largely celebrated view in literature that dangerously misses part of the picture [169]. One can also be forced to self-organize one's work, for example, in situations where bigger companies or public institutions are suddenly out-sourcing their services. In both cases there is the risk of dropping out from the everyday communities of work. Thus, people have started to form their own social anchor-points of work. They rent a space and create their own coworking communities. People work there side-by-side, sharing their everyday of work, sometimes working on same project, sometimes on their own. These self-organizing coworking communities differ from corporate-run coworking spaces firstly by being smaller and more intimate, comprising usually of 2–10 people. Secondly they differ from corporate-run coworking spaces, as they do not accentuate monetary profit in the making of livelihoods, but, they concentrate on developing wellbeing at work [170]. In this article with coworking communities we refer to the self-organizing ones, if not mentioned otherwise.

12.3.3 Mushrooming—A Network of Self-organizing Coworking Spaces

Mushrooming is a seven years old conscious experiment on bottom-up urbanism in Finland. It supports connections and interaction of self-organizing coworking communities that did not know each other before. Metaphorically said, the facilitators of this network have made mycelium (i.e. connections) grow, waiting what mushrooms (i.e. reactions) pop up and where. During the years there have been different strategies to support the growth of the network. The strategies have been a combination of conscious design and letting bottom-up activities emerge [171]. There has also been a combination of virtual platforms and interaction in real places in cities, face-to-face.

Since the beginning, Mushrooming has been developed as an open project on many scales, meaning, for example, that there has not been any fixed idea on what Mushrooming should become or what functions it should have. The network is open to anyone; its' use is free of charge and its service-page on internet is built on open source software (mushrooming.fi). Network's main activity has become to connect the individuals in search of a membership of a coworking community to the coworking communities that would need new members. There are also other things happening in the network, such as looking for help to a work project, recycling goods or organizing happenings of getting to know one-another. Mushrooming is active in the biggest cities of Finland (See on Facebook e.g. Mushrooming Helsinki) and it can grow to any country. It is used by over 7000 people and approximately by 300 coworking communities.

The understanding of Finnish coworking culture and of the Mushrooming network presented here is based on intensive on-going action research [172] that the author (Alatalo) has been doing for seven years. Together with a fluctuating group of 4–10 facilitators we initiated the network and have been experimenting with it since 2010. Conceptualizations are based on analyzing further the observations that I have marked in my field diaries. These observations have been collected in action at different situations of developing the network, such as irregular meetings of facilitators (from twice per year to periods of meeting every other week) or organized problem-solving workshops open to anyone. The author has co-organized several face-to-face happenings of Mushrooming with the intention of getting to know one-another in four Finnish cities. Many insights have come from visiting numerous collectives while co-organizing open doors-happenings in different cities at least once a year. Some of the notes come from participating to public discussions on coworking culture in Finland. Part of the research material has been gathered by following the discussions and practices on the virtual platforms of Mushrooming. This material is recorded as screen prints with notes. The author has also lived the everyday life of four coworking communities personally, of which two were established by her.

In the next paragraphs the author will not explain more in detail how Mushrooming has been facilitated [171], but she will concentrate on what interesting spatial and economic processes in the city it has launched. The aim is to add understanding of the relationship of space and alternative economies in city. The author started tracing the processes by describing and listing observations related to spatial or economic development, which were encountered in coworking communities during all the seven years of action research in Mushrooming (See Tables 12.2 and 12.3). Observations listed were often the kinds that with the help of the open networking are becoming stronger and more frequent. In other words, they are becoming processes supported and fed by networking. When tracing economic processes the author was especially interested in development of livelihoods. With spatial processes she already knew that many dealt with specialization of spaces. So, she wanted to dive deeper into that. In this phase of tracing, it was thought to be more fruitful to consider economic processes rather broadly; more as processes that are closely related to economy, in order to create a varied pool for further analyses.

12.3.4 Tracing New Economic Processes in the Network of Self-organizing Coworking Spaces

People come to self-organizing coworking spaces often during different periods of change. This means periods such as when finishing studies and entering working life, when starting a new enterprise, when changing a profession or when having

Table 12.2 Economic processes emerged by developing the network of coworking communities. On the left observation from the network, on the right the type of economic process it suggests

More possibilities to shelter over economically unsecure times	Process of continuous livelihood
Sharing economy becoming more usual	Process of growing shared resources
Gift economy becoming more usual	Processes of increasing non-monetary economies
Different professionals meeting more	Processes of innovations by skill collision
Different professionals living together more	Processes of innovations in everyday making
Lower expenses of work, culture of experimenting	Processes of prototyping
Hobbies mixed between making of livelihoods	Processes of hobbies ripening to professional skills
More diverse sets of professional skills	Birth of new professions by skill combinations
More diverse sets of professional skills	More challenges in offering the set of skills
Markets of non-professionals, micro jobs, tasks tackled by a group	New cultures of dealing and accomplishing work
More possibilities to find special peers of work	Processes of further spezialisation

Source Authors

moved to a new city. These are all vulnerable times in relation to personal economy. Coworking communities tend to keep their expenses of space and amenities low. This means that coworking communities offer shelter over unsecure economic times. It is important for self-entrepreneurs in regressive situations that are economically challenging. Mushrooming network supports the founding and survival of coworking communities and more there are these communities, more there are possibilities for people to manage over economically difficult periods. Shelter is important also for the ones taking risks to success, as you often need to cope a moment with economic insecurity before business really takes off.

Sharing economies are strong in coworking communities. Space and its' amenities are shared and maintained together. Space is often shared also for people outside the core community. In Mushrooming many coworking communities advertise if they have a special resource that other people could also use, such as old machine for working with wood or 3D-printer. Know-how is also shared, meaning that members of a community often assist each other in tasks that each is most talented to do. Sharing can build up to co-work projects, but it can be also about very simple everyday tasks. This can also be named as gift economy, where nothing specific is waited back in return of the favor done. Another signal of emerging non-monetary economies is the tendency of coworking communities to maintain a function that is not self-sufficient in regards to capital, but in which they see other values. For example, one community maintained a specific skateboarding shop, covering its shortcomings by income from other activities of the community, just to ensure that there is such a resource for the larger subculture of skateboarders in the area.

Traditionally a coworking space has been founded with old friends from school or other arenas. This means that the communities were rather homogenous. Before Mushrooming, for an individual it was difficult to find a place in coworking community outside one's own acquaintances. Nowadays half of the coworking space memberships in Helsinki are mediated through Mushrooming [170]. This means that people from varied backgrounds with different professions have started to mix. In the early phases of Mushrooming there were mostly different kinds of professionals from creative fields, such as designers, artisans, artists and knowledge workers using the network. They had had their isolated communities, but with the emergence of Mushrooming, they started to mix. During last two years also other kinds of professions have found Mushrooming, such as physicians, teachers, engineers, cooks or barbers. Now also these different professions have started to work side by side with the early adapters. This means that Mushrooming enables different talents to collide, which creates excellent ground for innovations. We can call it a process of innovations by collision of different skills.

Breeding innovations is an economic process. Innovations start new livelihoods, they are the motors of economic activities. In coworking communities the collisions of different skills is not usually programmed, but it happens by accident or little by little amongst everyday making. This kind of innovations of everyday life are often very practical, they rise from developing an answer to a question at hand. For example, an artisan might be working on jewelry. An engineer sharing the same space sees the difficulty of handling one of the phases. The engineer remembers a tool from totally different context that might help the artisan. When discussing this with the artisan, the engineer gets more knowledge about the work at hand, and suddenly understands, how some practices the artisan uses would actually help also in the current engineering project.

As stated before, the low expenses of work in coworking communities allow periods of experimenting. Experimenting happens usually in cycles of prototyping. Experiments are often fitted between work that brings the livelihood. Thus, people are not left totally without livelihoods nor is there too much pressure on prototyping to start quickly to bring income. Spaces of coworking are used also for different kinds of hobbies and rehearsing. People often mix these to the agenda of the day. This brings even more variety to the skills present at coworking spaces, which then can collide and produce new innovations. Flexibility of agendas support people to develop their skills from hobbies to the state that they can offer their talent as services, and thus, new livelihoods are born. These kinds of prototyping and rehearsing are essential early stages of innovation culture. They are the necessary phases of incubation, as you need some evidences of capability of your business before applying supportive funds or searching for companions to start-up development.

There are new professions born in coworking spaces. One way is the aforementioned ripening of a skill from hobbies to the state of a professional skill. These new professions are the kinds that do not yet have a school where you could study them and thus they are not yet even recognized as professions. Another kind of professions emerge from the multi-talentism of the people. As starting point, one has the already splintered project-structured working life of a self-entrepreneur. Then, one is encouraged to experiment with different skills and one also gets inspired by new professionals around. People develop surprising collages of skills that lead to unconventional professional capacities. One might, for example, be a priest and a cook, with skills of academic researcher and choreographer of ballet. Challenge of these multi-talents is in communicating their capacities. When asking from one person at a coworking community, what is your profession, he answered: "*It is easier, if you describe me the task for which you think you might need my skills. Then I can tell you, if and how I can help. Most probably I can also lead you to other people, who I know, who could help you in this. But it does not lead us anywhere, if I start to list the several professional degrees or other skills I have, you would never get the picture.*"

There are new cultures of dealing with work emerging. One is the way of describing the task at hand and sending it to the Mushrooming network. These tasks are usually complex and the kinds that you do not clearly know what established profession should grasp them. This process is somewhat opposite to the dominant, where you are looking for a certain professional and trusting that the professional knows how to handle the work. This kind of dealing a complex work invites groups of people to combine their skills and tackle the work. Another occurring practice is to deal with very small tasks that we call micro-jobs. Mushrooming has created channels, where coworkers have started to ask help from each other, as for example, in situations, in which one needs a photographer for couple of hours or another seeks instant hints for internet advertising. These micro-jobs are usually paid for. There has also emerged markets for non-professionals. For example, you do not always need the most talented photographer, sometimes it is enough that there just is a person to take pictures for notes. This again is a saving in expenses of the project, and thus, lowering the threshold to experiment. Markets of non-professionals offer precious situations to rehearse the perhaps ripening professional skills and add a little new source to the collection of income.

Some coworking communities have chosen specific goals in their community building, in which Mushrooming network has supported them. These goals are about specialization, which may be an important advantage when competing for sources of livelihood. Coworking communities have started to form intentional groups of professional skills, which as a compound can grasp challenging tasks. Sometimes it means a group of different professionals, who then manage a multi-purpose hub of sub-culture, for example. There are also groups who have specialized in tasks such as audio-visual production, capable of handling everything in between recording a happening to mastering television series. There are also very peculiar communities, such as a group of midwifes, who offer alternative places and ways for women to give birth. These groups have specialized little by little, usually experimenting on ways of doing that are different to dominant services in the market. Mushrooming has helped in finding people, who share the same specific interest.

12.3.5 Spatial Processes Emerged in the Network of Coworking Communities

Coworking communities take into use spaces that are left over from the dominant market. This means, for example, spaces that have lots of repairing to be done, or they are in challenging locations without good public transport. These spaces often have strange floor plans, having been in the use of small industry or built for other special functions whose spatial solutions are difficult to modify to average needs. There are also small, unused spaces here and there, such as rooms in attics, in cellars or back in the yard. Some temporary situations may be too short for stabilized businesses, when self-organizing coworkers instead see them as good opportunities. Luxurious premises at central locations are also leftovers that no single entrepreneur sees wise to house, but coworkers can take into use by cutting personal expenses low by intensive sharing.

Coworking communities prefer left-over spaces because these spaces usually have low rents, meeting the aim of keeping the expenses of work low. For coworkers there is also the wish to renovate and furnish the space by themselves, in order to experiment with what everything the working environment can be like and to realize it meeting the personal tastes and needs. This wish to renovate the space by themselves stems also from the situation, that new experiments on livelihood often require spatial features that one does not find ready-made from the normal market. Renovating is usually done little by little, both because of limited monetary resources and because of searching the best solution gradually. This is important, because it means that non-professionals are learning by making, what kind of places they actually like and how they can realize them. This applies also to professionals re-learning to forget the styles of design that they have adopted at school. When one creates an environment by really sensing, what feels good, one comes quickly to note how many visual dogmas learnt at design schools create stress. In Biourbanism, the visual dogmas, or memes in other words, which professionals and industry are keen to repeat without questioning the quality of the environment that they produce, are seen as one of the biggest challenges in contemporary architecture [164]. In a Finnish study it was found, that people are doing better in coworking communities, than in other working places in general [170]. Big portion of this well-being stems from the quality and fittingness of self-made working space as well as from the self-managed community. This leads us to the last note on why coworkers prefer left-over spaces. As the spaces do not really interest any regular buyers or renters, they provide freedom for one to do whatever experiments there are on one's mind.

The tendency of coworking communities to take left-over spaces into use and renovate them has created a phenomenon of bottom-up urban repair. This has been empowered trough the network of Mushrooming. Communities find nowadays new members more easily, thus they have more courage to renovate. Communities have started to take into use bigger premises. Some property owners have found Mushrooming and offer difficult premises with little money for coworkers. Via Mushrooming coworkers send hints of empty premises that seem promising to other communities. There is also a process of inheriting spaces, which leads to another scale of gradual renovation. When a community leaves its space for reason or another, it advertises the space in Mushrooming for other coworkers. Next coworking community steps in and renovates the space a little further. This has created paths, where the space has become renovated gradually by several communities to top standards with seemingly little efforts and resources. Backlash of this development is that rents tend to grow as space gets more renovated. Thus, Mushrooming network should assist in finding new models of contract that safeguard the rights of coworkers of not getting higher rents after having put their own workforce to renovate a space.

As described before, the left-over spaces as such are already rather special and diverse. This diversity grows even more as each coworking community repairs the space to fit the specific needs they have. This diversification of spaces of work is a welcome counter-process to for example the monotonous offices dominating the everyday of knowledge workers. Coworking communities organize plenty of open happenings, which makes their spaces momentarily public spaces and thus they diversify also the affordance of them. Through Mushrooming a kind of commons of spaces of work is created: you can find and have access to a huge diversity of spaces, from smithies to dance halls, though you need to follow the rules each community sets. Still many communities welcome visitors for short times for free and are open to new collaboration.

Urban repair described above is often dismissed as it happens slowly, scattered to spaces here and there, and hidden to privacy of the communities. Still it is a valuable process of repair in cities, working upon small scale tasks, contrary to big scale development that top-down models prefer. The diversity that this process of urban repair produces would be very difficult to produce by other traditional methods. Usually there comes the moment when coworking communities step outside from their spaces indoors and start to affect the public space in their vicinity. This comes through different neighborhood movements and shared happenings, aiming to enliven and repair the common spaces outdoors. Coworking spaces also tend to cluster. After one community has arrived to a new area, soon several other ones have found the premises as well. All these activities start to change the area, even to the degree where coworkers find themselves having assisted to the well-known process of gentrification, prices mounting to be too high for themselves to continue there.

Different kinds of coworking communities settle in different parts of the city. The ones in the commercial city center are the most familiar to us, as they are visible parts of the streetscape. They occupy the spaces of former shops with big display windows to the street. These coworking communities often consist of the laptop workers of creative fields and the choice of location is bi-fold. Firstly, they communicate about themselves and their services by being visible and easily accessible. What their space looks like and what the workers and their working looks like is in this case experienced important. Secondly, they choose the bustling city life to be in the center of services for themselves, such as cafés and restaurants. They feel that the work they are doing benefits of them sensing the zeitgeist directly from the city life around them. Also some artisan professionals enjoy the city center and use the display windows traditionally to advertise their products. Usually these artisans deal with small production such as jewelry or exclusive clothing produced in small quantities in small spaces. There are also groups of artisans that run their common shop in the center, but, they have their places of production elsewhere. Coworking communities in central locations often have meeting rooms and other special amenities that they rent out for anyone interested.

Then there are coworking communities that prefer peace and quiet, work on production and experimentation that is somewhat messy or wish to keep their expenses very low. These communities house the hidden spaces in the city center, in cellars or back in the yards. Moving a little outside the center they prefer for example the former spaces of industry. These are the spaces of high mixing of professionals. Low costs invite to practicing the hobbies as well. We have coworking communities also in the housing suburbs; there, the driver has been to bring the working as close to the living as possible. Here the coworkers are often young parents balancing through the currents of parenthood combined to working life. Often these communities activate the shared spaces of apartment buildings, but sometimes they also rent a flat and for example establish a shared visual arts studio there.

Outside the cities we have one more specific type of coworking space emerging. It is still a weak signal, but seems to be strengthening. These coworking spaces are established by people living in the countryside. They are managing generous premises with lots of spatial potentials surrounded by splendid nature. These communities have chosen the tactics of inviting interesting people with fresh ideas to come to them. This is done by offering a combination of working and living space for a limited period of time, varying from weeks to couple of months. We are talking of retreats, that artists and writers have done for centuries, but that is now introduced to any professional interested.

In Finland the growth of Mushrooming has made visible differently accentuated cultures of coworking in different cities. In the capital of Finland, where Mushrooming has existed since 2010, there are the most specialized coworking communities. For example, there is a self-organizing community of performing arts, referring to actors and performance artist. One can imagine that it is not self-evident to be able to create your livelihood out of these professions. This coworking community has established different kinds of supporting activities and services for themselves. They include workshops from developing your professional skills to open sharing of funding and project possibilities, not to forget weekly well-being meetings. Even though being practitioners of the field, they also keep themselves intertwined to the latest theoretical development by inviting inspiring persons to work with them.

The other local branches of Mushrooming are now just about two years old. During fall 2015 and spring 2016, there came wishes from coworkers in other cities of Finland to establish Mushrooming network also there. These local networks are at the moment accentuated by the coworking communities from which they started to grow. We cannot say whether the differences of Mushrooming network in each city mean that there are different coworking cultures in each of them. It can also be that only part of coworkers has yet found the network in these newcomer cities. For example, in Turku, the initiator community is focusing on design and advertising. Its motivation for establishing Mushrooming was to find new interesting people to collaborate with professionally and to keep their space full in order to manage their rent. Instead in Jyväskylä there was a group of artisans and artists, who had a need to get politically organized and to have the support of Mushrooming of the capital in order to negotiate about former industrial spaces in the center of new development. In Tampere, there were strong alternative cultures of artists and urban activists, who took Mushrooming as one more platform of their interaction. Some local users of Mushrooming have wanted to differentiate from this agenda, which has resulted even in leaving the network in the fear of becoming too strongly associated with the alternative movements towards which they are critical (Table 12.3).

12.3.6 Development by Spatial and Economic Opportunities

While collecting the economic and spatial processes that have emerged in Mushrooming, the author noticed that there are also situations where these two processes are closely intertwined. This led to one of the key findings: development

Table 12.3 Spatial processes emerged by developing the network of coworking communities. On the left observation from the network, on the right the type of spatial process it suggests

Taking into use left-over spaces	Processes of recycling spaces
Renovating and fitting the space by themselves	Relearning spatial qualities by experimenting
Courage for more challenging premises and inheriting of spaces	Process of bottom-up urban repair
Inheriting of spaces	Higher rents by upgrade
Special spaces taken into use and renovated to become even more special	Diversification of spaces of work
Spaces available via Mushrooming network	Emergence of commons of spaces of work
Urban repair and clustering of cool coworkers	Gentrification
Different types of coworking communities preferring different kinds of spaces	Growth of different coworking cultures in different parts of a city
Different contexts of birth of city based groups	Growth of different coworking cultures in different cities

Source Authors

in the network of coworking communities is sparkled by spatial and economic opportunities. For example, a coworking community is established when a spatial or economic opportunity emerges. Here one spatial opportunity can be an empty premise encountered or one economic opportunity a possibility to new livelihood by finding a new peers of work with whom to collaborate. These opportunities are encountered more and more with the help of the network. This is interesting in relation to view of the battle field of three tyrannies [161] preventing spatial development, reflected in the beginning of this part of the chapter. In the network of self-organizing coworking communities, it is not the tyranny of market that hinders the spatial development, but the opportunities to new livelihoods that sparkle the spatial development, especially in the cases such as introduced next, where economic and spatial development are closely intertwined.

The professional mixing and specialization that Mushrooming has enabled and supported during the years, has become visible also as spatial mixing and specialization. As an example of mixing, there are coworking communities that consist of knowledge workers, artists, yogis, dancers, social workers, musicians and cooks. They have developed spatially rich complexes that fit multiple uses. These multiple uses cover the spaceeither in turns or simultaneously. This means, for example, a melting pot of café-restaurant, a well-equipped gig hall, practice halls, laptop worker's rooms and studios for painting, hosting everything from discussion evenings to rare cinema projections, a flamenco school and mother-child yoga.

It is not only a mixed group of professionals coming together that produce multi-purpose spaces, the reason behind a multi-purpose space can also be about professional specialization. One of the main runners of this kind of a place mentioned, that they saw establishing a multi-purpose space as the only mean for themselves to do what they wish, which is to continue developing as professional dancers. Early in their career they understood, that it is only a fraction of dancers who get stable jobs and that kind of institutionalization is not even what they want to. The other usual try to earn the living as a dancer is to continuously apply for project-based funding from different foundations. That is very unsecure business, taking lots of effort and forcing you to concentrate on what are the wishes of the foundations. To have the freedom of expression they chose not to earn their livelihood from the dance, but to figure out something else that brings enough money to live and leaves enough time to practice dancing. Their solution was to manage a spatial compound that rents out the multiple facilities it has at low prices but with full house all the time, resulting in richly inspiring working environment for themselves as well. This meant for them to expand their professional tasks a little, from learning sound engineering to become assistants in kitchen, but not too much for dancing to get affected. It is not an easy path. One pioneering highly mixed community like this is at the moment in their third location experimenting with their third combination of actors, balancing on the edge of economic survival, but also creating a vivid and appreciated sub-cultural hub at the same time.

It can also be a special spatial opportunity that creates the community and possibilities for new livelihoods. For example, there is one interesting type of coworking communities, where specialized group of designers, artisans and other small producers grows combined to special spatial opportunities to support a specific short and vulnerable period of new enterprises. Spatially this means to have a good location with windows to the street and enough space for a shop-showroom combined to a café where possible buyers of the products can linger, and where you can organize inviting happenings. In connection to this, you need big spaces suitable for housing different kinds of workshops from clean high-tech production to rougher machinery. The prototypes developed in the workshops are presented and sold in the shop-showroom since their rough first versions. This way developers can get straight feedback from the customers throughout the whole process of prototyping. In Helsinki, one community like this housed at a time about ten starting enterprises of one to six persons in each. The production rooms were not open to everyone and many small enterprises preferred having their own room. This early incubator community worked very well and was recognized as an interesting place for trendsetters. One of the runners of this community presented a notion, which had come as a surprise to him: the turnover rate of the starting entrepreneurs was really high. Either their experiment did not succeed in half a year and they left the space, or, their experiment succeeded so well, that their needs for space grew bigger than the incubator community could provide. He saw it very important, that you have these kinds of incubator places, where failing experiments are natural part of the picture. He also saw that Mushrooming network had been precious in finding the new self-entrepreneurs as the changeability had been surprisingly high.

As Mushrooming grew to other cities, some coworking communities developed tactics in creating satellites to them. This is an example of combined economic and spatial tactic. Establishing a branch of a coworking community to a bigger city than it is originally from, presents usually an economic challenge of higher rents and higher competition of the best locations. It also presents an economic opportunity of new clients. One way to balance through this is bravely to rent a big space and directly sub-rent a good portion of it to for example a pop-up restaurant surviving with simple kitchen. Coworking community then starts in the small space left, using the hype of the new restaurant to advertise oneself for example by organizing a series of brunches with program that presents the doings of the coworking community. Restaurants often have a lifetime of 3–4 years in Finland, during which the coworking community has time to grow steady and fill the whole space if needed.

In other words, self-organizing coworkers create spatial development, that is, establish new spaces to our cities, by gripping to spatial opportunities and to opportunities to develop new livelihoods. And not only opportunities, a strong need can also be the driver, as in the case of dancers in need of developing their livelihoods. When coworking community is established based on a spatial opportunity, it usually means, that the community does not exist yet. A space is found by couple of people who then start to look for other members. Mushrooming network helps in this search. On the other hand, when the developing of a new livelihood is more the kick-starter of the coworking space, there usually is already a community that in turn is in search of a space for them. Mushrooming helps also in this case, though the adverts on empty spaces are much rarer in the network, than the adverts of memberships in a community.

Based on the activities in the network it seems that self-organizing coworking communities are established mostly by encountering a spatial opportunity. At the same we know from the field that need for spaces is bigger than availability of them. There are fitting spaces, but the channels for dealing them, such as Mushrooming, are not familiar to property owners. Thus, the central issue hindering development of self-organizing coworking culture is that the communities do not get established as much as they could, as plentiful vacant spaces and their possible users do not find each other. Mushrooming is efficient in making different professionals to find each other, but new practices should be developed to foster the establishment of coworking communities by encountering a spatial opportunity. Spatial opportunity seems really to be the key for the development of the coworking culture, as also the communities who start from the idea of developing a new livelihood, will immediately face the question of where to find a space for them. Thus, we can say that in the field of self-organizing coworking culture, it is not the market that hinders the spatial development, but the poor communication of spatial needs and offers.

12.3.7 Markets Creating the Coworking Bubble?

Since the beginning of coworking phenomena, it is observed that the orientation of coworking itself is not in the business [169]. Regardless of that, the hype around coworking is used for marketing and especially city branding to such a degree that many question [169, 173] whether it is creating a coworking bubble. There are similar kinds of expectations born of coworking, as there have been around

Florida's ideas on creative class [174] and its capacity to foster economic development. Are these expectations in vain?

One challenge that researchers have when trying to answer this question is, that most of the studies are made on corporate-run coworking places. This is because the self-organizing ones are difficult to reach, if there is no network such as Mushrooming. The world of corporate-run coworking places is different to self-organizing ones. From the perspective of corporate-run places, there is the risk of a bubble. In the arena of self-organizing coworking places, we can see promising alternative development instead. The idea of corporate-run coworking places is usually to have a rather big premise that can house, for example, a hundred coworkers. It aims to foster collaboration, but, often ends up to become more likely a drop-in office, where freelancers work alone together [173, 175]. As this happens, it breaks the coworking bubble by not fulfilling the expectations of fostering new innovations. As collaboration is not happening, there are questions, whether coworking really brings new skills or supports survival of coworkers [173]. This result may come also from the twisted and partial view of having studied mostly the corporate-run coworking places. Examples from Mushrooming show, that when self-organizing coworking places are networked, different professionals start to mix and develop new livelihoods. When costs of working are low, people are encouraged to experiments leading to innovations.

The isolation in corporate-run coworking places is fought back by hiring professional community managers, which increases costs. Corporate-run coworking places act in the markets, they compete by offering extra services to freelancers, as well as extra spatial qualities. All this increases costs even more. Here comes one thing that is creating the bubble: corporate-run coworking places become too costly to the majority of potential coworkers. It has been already noticed, that the profitability of corporate-run coworking places often is low [173]. I claim that this does not mean, that coworking culture would not be promising in fostering economic development. It just means, that the coworking culture flourishes on other economic arenas and with other logics than our dominant markets. As cases from Mushrooming presented above in this text show, the profitability in self-organizing co-working places is high; for example, in situations, where with very little resources, conditions of accommodation are created to cover economically uncertain times.

12.3.8 Mushrooming Providing for Biourban Strategies

As presented earlier in this text, in the core of the tactics of self-organizing coworking communities there is to grip to a spatial opportunity encountered in a city. Even if a community is driven firstly by economical experiments, they need the spatial opportunity to start developing. This suggests, that even though economic perspectives are strong in the self-organizing coworking culture, it is very much about spatial development, and thus, in the arena of urban designers and other city developers. We can learn for our design strategies from Mushrooming to facilitate the tactics of self-organizing coworking communities.

Biourbanism strategies are mainly considering a city as a complex living organism which cannot be strictly planned. Instead it can be rather guided and directed. New possible routes to development may be opened, new actors may be supported to act. Biourbanism suggests strategies, such as urban acupuncture [139] or peer-to-peer (P2P) urbanism [71]. Urban acupuncture finds specific elements of action in a city; these elements can launch healing and energizing processes, which may essentially be spread inside the urban fabric. These operations can activate or renovate factual urban spaces. P2P urbanism is more about building networks and enabling bottom-up development in the cities. These networks are often used for sharing resources, or, for making improvements to individual everyday lives with the help of peers. By networking with other peers, proposed small actions can be slowly transformed into city-scale phenomena.

Initiating and facilitating a network of coworking communities such as Mushrooming is a design strategy combining elements both from P2P urbanism and urban acupuncture. There is the building of a network of peers as to enable self-organization, similarly to ideas of P2P urbanism. Then, both the facilitators and the users of the network are doing acupuncturing kind of actions that launch different processes in the virtual and spatial dimensions of the network. Acupuncture in Mushrooming does not refer to everyday stabilized practices of the network, but, to new activities introduced every now and then, which start new processes and alter the functions of the network itself [171]. Emergence of new acupuncture activities is possible as the network is an open process. It has not any fixed idea of what it should be for; it is more an interaction channel for the coworking communities. This has created practices that enable the emergence of acupuncture activities from bottom up [171].

Whilst analyzing different spatial and economic processes emerged in Mushrooming, it was soon noticed that many of them are about mixing and specialization (See Tables 12.1 and 12.2). In other words, they are about diversification. In the following sentences, this diversification has been conceptualized from the point of view of the professions. The author is currently working on similar conceptualizations of diversification of spaces, but this is not yet ready for publication. The first order of the diversification in Mushrooming is about specialization by becoming other, by one isolating oneself. It refers to a situation in the beginning, when the simple act of establishing a coworking community is a diversifying thing in relation to the dominant working culture. These communities often consisted of old friends, thus, they were rather homogenous and non-surprising, and still they offered special conditions for professional development. For example, forming a group of visual artists may be very emancipating. Second order of diversification was about mixing. It happened, when these communities started to use Mushrooming for their survival, in situations, where new members were needed to share the rent; different professionals started to mix, still though being mainly creatives. The third order of diversification was about specialization, when coworking communities started consciously to use Mushrooming for finding special community members or proposing special amenities. Then, interaction was no longer about survival over unexpected situations. The fourth order of diversification was again about mixing. It happened, when other professions than creatives found Mushrooming and started to share spaces alongside them. In other words, it seems that from a larger perspective, diversification in Mushrooming takes new steps in producing diversity in alternate waves of specialization and mixing. At the same time, in the everyday life of coworking communities, specialization and mixing are often intertwined, as in the case of dancers creating themselves opportunities to develop their skills of dance by simultaneously running a multi-purpose space.

The Mushrooming experiment provides for biourban strategies, because it creates processes of diversification. Ability of a city and of human culture to have processes of diversification can be mirrored to the ability of an ecosystem to have processes of growing biodiversity. We need that diversity for situations, where circumstances change rapidly, where old dominant solutions no longer work. In evolutionary theory, when looking long perspectives of time, evolution of species seems to be dominated by long periods of stagnation that then are punctuated by sudden revolutionary change [176]. During the stagnation or balance, the genes of dominant species stay quite much the same, the mutations and alterations to genes happen mostly in the groups that, for a reason or another, become isolated from the dominant pool [177] and are small enough to be affected by alterations in the genes of even one individual. It is from these small alternate groups that new genetic development rises after dominant one fits no longer. More diverse these groups are, more possible new solutions we have.

Urbanist Shane shares the view of the importance of having the pockets of richly varied grass-roots' alternative cultures in a city [178]. He sees that these pockets play important roles in the development of a city and especially in the situations of rupture. Similarly to evolutionary theory, he sees that in situations of urban change the alternative solutions these pockets have practiced, may suddenly fit better to the new context than the practices of the dominant culture. These alternative solutions can be about new innovations but also about old practices that dominant culture remembers no more. Alternative pockets are thus important reservoirs from which fitting solutions filter every now and then to the dominant culture. What is not so often discussed, are the methods of whether and how we could support these alternative pockets, and whether supporting them could be conscious act of urban developers. Case of the Mushrooming shows that by creating open interaction processes the alternative cultures do not melt together and vanish but they actually grow even more special and more diverse. Openness of the network and transparency of self-organization are crucial in order the alternatives not to become too extreme or too other to the society.

The Mushrooming experiment provides for Biourban strategies by being a combination of P2P urbanism and urban acupuncture with the special emphasis on creating processes of mixing and specialization that then compound to processes of diversification to become life-enhancing strategies for a complex ecosystem, such as a city. Seeing diversification as life-enhancing process in a system is built also on Capra and Jakobsen's [34] systems view of life and philosophy of organism.

They propose four life-enhancing principles: nested systems, self-generating networks, open systems and cognitive interactions. Earlier in this paper the author described how coworking communities are nested systems inside a coworking network. In an earlier paper [171] the author explored how coworking network is a nested system in relation to its context. In that paper, the themes of openness and cognitive interactions were analyzed as processes of shared knowledge generation emerged in Mushrooming. At a conceptual level Mushrooming shares many features with the idea of life-enhancing ecological economics of Capra and Jakobsen; these authors talk about the principles. What new things Mushrooming can bring to this discussion, is to show what empirically happens in some case, where these principles are in function.

Stepping out from the context of coworking, we can say in more general terms that Mushrooming is about facilitating open networking of self-organizing communities. Learning from the growth of Mushrooming in new cities has shown, that, because of the local communities, it always takes local forms of appearance. It grows from the people's knowledge of their context, thus, it is a strategy that can be applied to many places. It has many scales, some of the actions considering the everyday life activities of a room, up until to phenomena becoming visible at metropolitan scale, and even among cities. It differs from many design strategies by taking years to develop; Mushrooming grows hidden in the privacy of coworking communities. It is a strategy aiming for diversification of spaces and it could be applied to other domains as well. For example, we may foresee that developing co-living by facilitating self-organizing co-living communities to interact can promote processes of diversification similar to those emerging in Mushrooming, thus, resulting in the creation of interesting new spaces of housing in connection to new ways of living together.

12.4 Naples Biourban Smart City: A Spatial Strategy for Local Innovations

12.4.1 Introduction

The European Union has dedicated relevant efforts to devising a strategy for achieving urban growth in a "smart" way for its metropolitan areas, according to other international institutions oriented to believe in a wired, ICT-driven form of development (for example, the Intelligent Community Forum, the OECD and the EUROSTAT), stressing the role of innovation and providing a suitable framework of analysis on urban innovation [179]. Indeed, the availability and quality of the ICT infrastructure is not the only definition of a smart city—taking into account the relation between ICT infrastructure and economic performance [180], the role of human capital and education in urban development—but also the interaction between human capital, urban development and economic development [181, 182].

In general terms, a smart city can be described by six key dimensions: (1) smart people, (2) smart economy, (3) smart mobility, (4) smart environment, (5) smart living, and (6) smart governance [183–186]. These dimensions are closely interlinked and describe the relations system that characterizes tangible and intangible processes of urban and territorial development. According to Vinod Kumar and Dahiya [187], among the six dimensions of a smart city—conceived as a "system"—a prominent and central role has been assumed by "smart people", characterized by excellent skills, high level of qualifications and expertise based on economy of knowledge and innovation, a high Human Development Index and a high Graduate Enrolment Ratio, supported by cultural and educational services, a lifelong learning and use of e-learning models.

At the same time, "smart people" are highly flexible and resilient to the changing circumstances, with creativity, open-mindedness, and a multicultural perspective. They are able to maintain a healthy lifestyle and are involved in sustainable and inclusive development processes. The role of "smart people" is essential for the identification of smart citizens as possessing civic intelligence—according to the conceptualization of "human smart city" [188, 189] centered on the human aspect—and the different multidimensional components able to mobilize the participation and cooperation of citizens, businesses and societal organizations, overcoming the technocentric approach. Adding "human" means to introduce a new perspective oriented to consider equitable, effective, long-term, incremental, and participatory design processes, which can integrate experimental, educational, and community mobilization, research, and policy work within a shared, coherent perspective [190, 191]. The human dimension is consistent with the Biourbanism perspective that supports a human-centered urban life, requiring a really human-friendly environment and enhancing the environment's own identity, where human well-being and human values interact and support each other [45].

According to the above considerations, decision-making processes related to territorial transformations and economic development require the elaboration of appropriate multidimensional frameworks, which can support the choice of public and private resource allocation—favoring different interests and purposes—and include the human dimension as civic intelligence. Today, we are beginning to realize that a sustainable and smart transformation of the territory, aimed at enhancing it in innovative approaches, can have positive economic effects and become an opportunity for improving the level of well-being and quality of life. Sustainable and smart territorial transformations, therefore, require a strategy to reduce its vulnerability to external forces that can compromise its integrity, sinking or undoing the value needed for the survival of a community within it. Spatial transformations can indeed determine or intensify the system of anthropic and natural pressures, thus increasing the risk of non-linear evolution of the pressures themselves [192]. The vulnerability of a territory, moreover, is not limited to its physical-environmental aspects, but it also involves the social and economic dimensions. Recognizing the resource value of the territories, which can improve the level of well-being of the local community, determines the need for new management processes, with the predisposition of complex design capabilities.

Vulnerability control becomes a condition of exposure and ability to cope with dynamic processes of territorial transformation. At the same time, attention to the resilience of a territory means to analyze spatial arrangement, spatial interactions and spatial context, and identify spatial relationships that affect the social-ecological systems. Indeed, resilience research has often emphasized the importance of spatial dimension—especially, cross-scale interactions [193–196], identifying resilience as the potential of a territorial system to remain in a particular configuration and to maintain its feedbacks and functions, involving the ability to reorganize following disturbance-driven change, and reflecting flexibility capacity to experiment and adopt innovative solutions. The identification of suitable directions and agendas, which can outline practical and human-centred implications for citizens, governments and professionals, needs a framework based on the combination and interaction of bottom-up and top-down processes, of which p2p urbanism constitutes the core [197].

Starting from the previous reflections, the paper analyzes the case study of the Metropolitan City of Naples, verifying the potential of an approach that identifies development opportunities, considering the synergistic design capacity of the various municipalities. Each municipality in the coastal belt of Naples is analyzed by selecting some indicators that describe the processes and governance actions promoted by the municipal administration (top-down) and activated from below (bottom-up) through initiatives by the citizens and local associations, expression of social innovation processes. Through the tools of spatial analysis, multi-criteria and multi-group analysis, it is possible to explain the ability to activate synergies and territorial symbiosis between the different resources that characterize coastal areas.

12.4.2 Naples Metropolitan City: Towards a Biourban Spatial Strategy

The Italian metropolitan cities are considered as territorial institutions oriented to the metropolitan area's strategic development, taking into consideration the support and management of local services, associated functions, infrastructure systems, and communication networks. Naples, in Italy's Campania Region (Fig. 12.4), is one of the 14 metropolitan cities defined "territorial entities of vast area", which have replaced the homonymous provinces—established on January 1, 2015—after the reform of Italian local authorities with the Law no. 142/1990, and then recognized by the Law no. 56/2014. In particular, the Law no. 56/2014 (also named "Legge Delrio") governs the ten metropolitan cities of regions with ordinary status, whose territories coincide with those of the pre-existing provinces: Rome, Turin, Milan, Venice, Genoa, Bologna, Florence, Bari, Reggio Calabria and Naples. To these are added the four metropolitan cities of the regions with special status: Cagliari, Catania, Messina and Palermo.

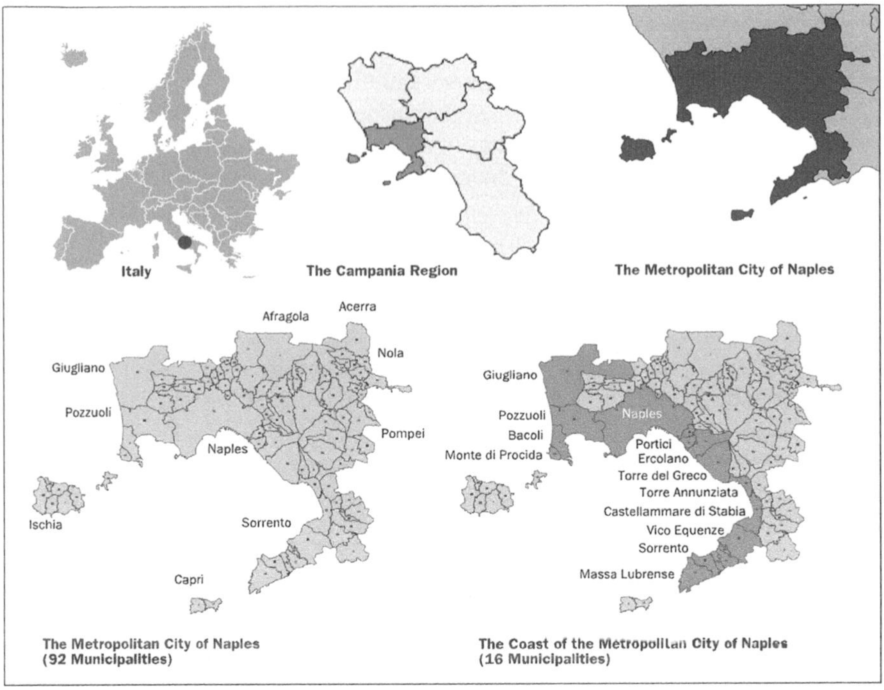

Fig. 12.4 The metropolitan city of Naples. *Source* Authors

The Metropolitan Conference Resolution no. 2, June 11, 2015 adopted the Statute of the Metropolitan City of Naples, dividing its territory into so-called "homogeneous zones" for a more suitable, balanced and functional management of local resources, which takes into account their identity characters, historical and cultural components, geomorphologic and naturalistic contexts, landscape functional interactions, and socio-economic frameworks. Indeed, the concept of "homogeneous zones" is part of the metropolitan strategic plan and is oriented to improve the territorial productivity, services to citizens' opportunities, and cooperation among the different municipalities. According to the definition of homogeneous zones, it is possible to identify common territorial interests, to activate effective shared projects, and to manage associated forms of socio-economic services and urban infrastructures [198, 199]. The need to identify some homogeneous zones, having common characteristics, has also been recognized by the European Union.

The Metropolitan City of Naples is a recent institution, and the Coordination Territorial Plan was adopted with the Resolution of Metropolitan Mayor no. 25 on January 29, 2016, published on the following February 3 and declared immediately executable. This huge asset of information, analysis and strategic lines—the result of a multi-year activity of confrontation with the territory and of constant updating—is destined to be the starting point for the future Metropolitan Territorial Plan [200].

The Coordination Territorial Plan identifies a perimeter of "areas" aimed at selecting the constituent elements of the metropolitan territory, with particular reference to the landscape and natural, cultural, environmental, geological, rural, anthropic and historic characteristics. It does not constitute a division of territory into "homogeneous territorial areas", as indicated in DM 1444/1968, being the task of the Municipalities through the local urban plan (municipal urban plan—PUC) to establish the division of communal territory into "homogeneous territorial zones" (ZTO), indicating physical and functional transformations eligible in the individual areas as well as areas not subject to processing. The homogeneous zones—to be established by a resolution of the metropolitan council—have to be constituted by aggregations of territorial contiguous municipalities, which comprise a population of not less than 150,000 inhabitants, identified on the basis of identity characters and historical reasons, geomorphologic, naturalistic and landscape contexts, functional relations and economic-social frameworks that substantiate their common membership.

The Regional Law no. 16 of 2004 and the Regional Regulation no. 5 of 2011 stipulate that the Coordination Territorial Plan contains structural and programmatic provisions. Through the structural provisions, we identify the urban planning strategies, the definition of the value and potential characteristics of the natural and anthropic systems of the territory, the determination of the areas where it is appropriate to establish protected natural areas of local interest, the definition of the infrastructure network and other works of provincial interest, the perimeter of historic centers, productive and tertiary areas, and areas of agricultural vocation.

The establishment of the Metropolitan City body is an important opportunity for the Neapolitan context, where the metropolis processes manifest themselves with clear intensity and criticality. These processes unfold on a spatial scope that crosses the provincial boundaries within which the new institution has jurisdiction: A priority for its activity is the search for agreements with neighboring municipalities on some key issues. Another crucial issue is the establishment of homogeneous areas to govern the territory, taking into account its high degree of polycentrism. At the same time, it is necessary to adopt this new partition for revision and rationalization of the current spatial spheres of the organization of services and the exercise of functions [201, 202]. The Metropolitan City has among its priority tasks to promote socio-economic development of the territory, pursuing its mission of enhancing the strategic development through six fields of intervention: strategic planning; simplification of local government; management and organization of public services of general interest; territorial marketing; support for innovation and new entrepreneurship; and the implementation of the Digital Agenda.

The present contribution proposes the structuring of a methodological process for the identification of homogeneous zones, with particular reference to coastal municipalities (Fig. 12.5), which represent a significant reality of the Metropolitan City of Naples. The methodological proposal recognizes as essential the identification of the specific characteristics of the various municipalities of the Metropolitan City and of their resources. Its aim would be the identification of homogeneous zones, which presupposes economic, social, environmental and

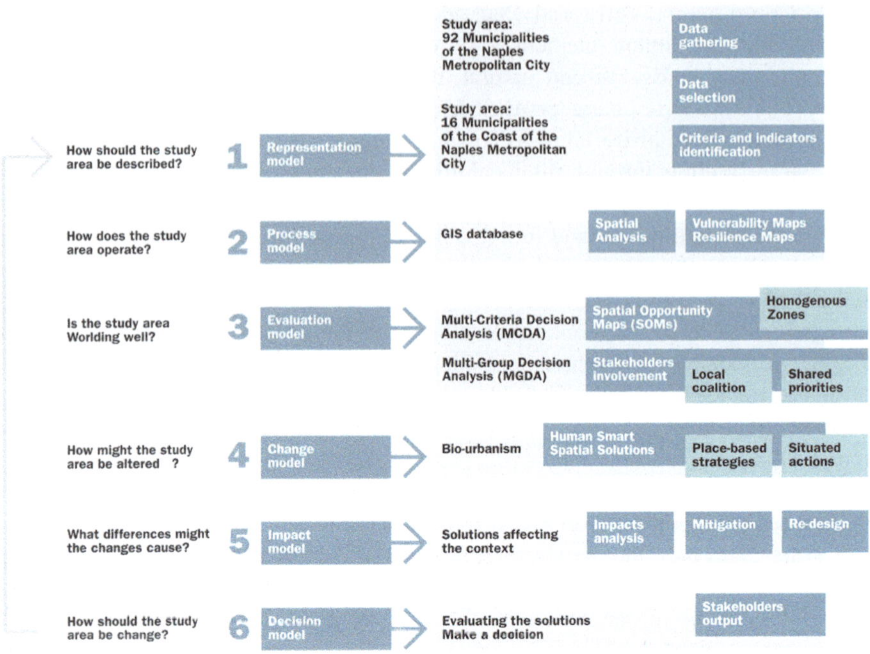

Fig. 12.5 The methodological process. *Source* Authors

cultural dynamics that can activate spatial synergies and symbiosis, and where Biourbanism principles could be implemented as the starting point of a more diffused process. The perimeter of homogeneous zones cannot be defined by a unique top-down process, but it must be characterized by top-down and bottom-up continuous feed-backs, which can activate double-view vertical processes and recognize the energy of horizontal processes developed locally, in small communities.

A smart development strategy, conceived in the sense of "human" [45, 187, 189, 197], identifies the communities and the relationship system that binds them to the territory as the essential components to activate sustainable regeneration processes.

12.4.3 Naples Metropolitan City: A Description Through Some Indicators

The Metropolitan City of Naples—with a population of more than three million people (3.107.006 inhabitants according to ISTAT, 2017)—is the third Italian metropolitan city by the number of inhabitants, and it is the first in terms of housing density. The entire metropolitan territory extends over an area of 1.178,93 km^2 and includes 92 municipalities, 12 of which count more than 50,000 inhabitants. It includes two of the three UNESCO World Heritage declared sites, and it is part of the World Heritage List for outstanding universal cultural and natural values: the historic centre of Naples (the historic city centre, the largest in Europe, has been listed by UNESCO as a World Heritage Site since 1995), the archaeological sites of Pompeii, Herculaneum, and Torre Annunziata. The volcanic Somma-Vesuvius complex is also part of the UNESCO World Network of Biosphere Reserves. Each one of the 92 municipalities consists of one historic centre of cultural relevance, where the form of human settlements identifies a broad and continuous cultural landscape with archaeological sites, significant monumental attractions, religious sanctuaries and churches.

The Naples Metropolitan City, characterized by a significant seismic and volcanic vulnerability due to the presence of Vesuvius and the Campi Flegrei, has a territory that occupies just 8.6% area of the Campania Region (13.590 km^2), where it concentrates more than half of the entire regional population. This overcrowding phenomenon has created a strong demographic and territorial imbalance with the other areas of the region, which are wider and less populated.

The territory is characterized by many heavily inhabited centers with a high population density: 60% of the municipalities of the Metropolitan City is small (less than or equal to 10 km^2), while 36% is medium in size (>10 km^2 and ≤ 25 km^2), and 11% exceeds 25 km^2, of which two municipalities (Acerra and Giugliano) are between 50 and 100 km^2, and only the municipal area of Naples exceeds 100 km^2. The population density of the territory is 2.672 ab./km^2; 12 metropolitan municipalities exceed 50,000 inhabitants; and in the list of the first 60 Italian municipalities, there are three of the Neapolitan area: Giugliano, Torre del Greco and Pozzuoli—where Portici is the first municipality in Italy with a population density of about 12,000 ab./km^2, followed shortly by Casavatore.

In the last Coordination Territorial Plan, it has been found that the urbanized area is a total of 381.12 km^2, corresponding to 32.54% of the total area of the territory. This results in an actual picture of the high demographic density that rises to over 8000 ab./km^2. In the vast hinterland, there are portions of territories where the cultivation intensity is still high, distant from the inhabited centers and located mostly in the northern parts of the area, lapping the area near the city of Caserta. Indeed, it is possible to identify some main relevant regions, recognized by common historical, geographic, social and economic characters: Naples, Nord Naples, Nord-West Naples, Nord-East Naples, Vesuvius Area, Nola Area, Campi Flegrei, and Sorrento Peninsula.

Municipalities developed along the old state roads (Via Nazionale delle Puglie, SS Sannitica, the Miglio D'Oro, and the ancient Calabrian Street) are nowadays only small and medium-sized suburbs – a kind of over-populated, decentralized neighborhoods—a mosaic of autonomous fragments, which gravitate on the city. At present, it is difficult to distinguish the boundary between the city of Naples, the adjacent areas and the hinterland, completely welded together into a seamless urban unicum, both in the vast area of the North (Giugliano, Afragola, Acerra and Nola) and that of the Campi Flegrei, the municipalities near Vesuvius, and the towns of Torre Annunziata and Castellammare di Stabia; less chaotic and out of the conurbation are the Sorrento Coast and the islands of Ischia, Capri and Procida. In the outskirts of urban and historical centers, intensive farming, abandoned industrial lands, and urban sprawl have compromised the potential environmental quality of the landscape.

The Metropolitan City of Naples is characterized by intensive, dynamic anthropogenic processes, full of potentials and also of many critical issues. The discriminatory management of urban areas has led to a series of negative impacts, which affected the physical-spatial quality and health conditions of the inhabitants. The great danger of contaminated sites has resulted in a significant ecological and social damage, in particular in "the Vesuvian Areas" (11 municipalities) and in the "Land of Fires" (32 municipalities). The excessive and speculative cementation of the Neapolitan hinterland has not created an orderly urban framework equipped with adequate structures, but has transformed much of the former province of Naples into "periphery". This transformation has not been accompanied by economic, urban and infrastructural development which can guarantee a good quality of life and services, but has increased a major commuting phenomenon towards the center of the metropolis. The historic city, however, is unable to support the over-municipal burden of a populated and over-urban hinterland that over the years has been fully integrated into its urban, economic and social fabric. Thus, from the urban point of view, Naples is one of the most compact, populous and congested metropolis in Europe. It shows relevant phenomena of increased population pressure within limited territorial areas, seen as economic, social and environmental challenges. At the same time, the area is characterized by a strategic location in the Mediterranean region that makes it an important infrastructure node for maritime and terrestrial interchanges. The Naples Port area is the main economic device of the Campania Region, one of Italy's most important passenger ports and also the headquarters of the Port System Authority of The Central Tyrrhenian Sea, which manages the ports of Naples, Salerno and Castellammare di Stabia.

The Metropolitan City of Naples is the most relevant employment centre in the region (it gives jobs for 955,800 people), and its Gross Domestic Product is 55,170 million Euro—the 4th among Italian metropolitan areas. The industrial sector is well developed and extends to a number of suburban centers. Metalworking, engineering, chemicals, petrochemicals, shipbuilding, food (pasta; food processing) and construction materials (cement) are the largest industries, but textiles, footwear, leatherworking and garment manufacturing (including relevant products such as Alfa Romeo; also shipbuilding and metal processing) are organized on a semi-craft scale. The major industries are engineering and metalworking (Pomigliano d'Arco, Casoria, Castellammare di Stabia, Naples), chemicals and petrochemicals (Naples, Pozzuoli, Torre Annunziata), construction materials (Naples), food and food processing (Torre Annunziata, S. Giovanni a Teduccio), textiles garment manufacturing and footwear. Another relevant economic sector is tourism that drives the development process. Despite all the different economic activities (tourism, agriculture with the accent on fruits, crafts such as clothing, shoes and food, as well as industries of means-of-transport), the minimum salary in the Metropolitan City is 20% lower than the Italian average, and 25% of the people remains unemployed. Poor neighborhoods have the characteristics of slums, and organized crime runs its own informal regime. Naples is unable to reduce unemployment, which is the cause of serious problems related to crime, the "black economy", and the emargination of young people. At the same time, Naples and the Metropolitan City are an important example of creativity and cultural vitality, as shown by the recent study of the European Commission [203] on "Cultural and Creative Cities Monitor"—a tool to promote mutual exchange and learning between cities to boost culture-led development processes.

Indeed, only the city of Naples counts a large museum offer, including the National Archeological Museum (one of the most important in the world for quantity and quality of its collection, especially from the Greeko-Roman period) and the Capodimonte National Museum, housing paintings from the Renaissance to the Baroque periods. Naples is also home to the biggest science museum in Italy, the Città della Scienza. Naples is considered an important laboratory and an international window for contemporary arts. The Naples Art Palace and the Donnaregina Museum are very active in this sense. Also, since 1996 and up to the second decade of the millennium, the so-called art stations have been built to combine urban transport underground with the fruition of modern art installations. The city is characterized by a long-lasting theatre tradition; Naples hosts the theatre-related events labeled Europe for Festivals, Festivals for Europe, etc. According to the main dimensions selected by the "Cultural and Creative Cities Monitor", Naples occupies a good ranking position in terms of the following criteria: "openness, tolerance and trust", "human capital and education", "creative and knowledge-based workers"—even if its total C3 Index, which assesses 29 indicators clusterized in nine dimensions, is particularly criticized. Indeed, Naples occupies the penultimate place on the list; if we consider the related sub-index, the last position is for "Cultural vibrancy", the third last position for "Creative economy", and a better position for "Enabling environment".

Taking into account the above considerations, the Metropolitan City of Naples can be described selecting some relevant indicators, which can analyze the multidimensional dynamics that characterize the area and the different municipalities.

12.4.4 Naples Metropolitan City: A Methodological Proposal for the Coast Area

In order to identify a methodological approach useful for delineating the boundaries of the "homogeneous zones" [cf. §12.4.2]—made up of municipalities activating synergic cooperation processes—the municipalities in the coastal area of the Metropolitan City of Naples are analyzed, characterized by the same geographical context and many common aspects, including the direct relationship with the sea. The municipalities of the Coast Area of Naples structure a territorial, geographical, environmental, economic, social and cultural context, where there are multiple and differentiated resources that identify complex, sometimes conflicting, characterizing features, isolated or fragmented in many cases. At the same time, they identify seaside towns that, in a different way, have activated systems of use, and relations with the coastal strip and port areas, becoming catalysts of economic, social and cultural processes, in which they compare lifestyles and development models.

In general terms, cities along the coast form cosmopolitan sites, open to many cultures—spaces of creativity and innovation for the economy, culture and society [204]. However, they are also places of significant conflicts between economy and ecology, between economy and culture, and between ecology and society. Consumption of natural resources and production of pollutants, together with the intensification of uses aimed at meeting the economic needs of coastal areas, contribute to compromise the complex ecological balance and to alter or destroy specific local characters. In many cases, the processes of environmental degradation and pollution are accompanied by the production of economic wealth, which entail not only ecological costs, but also social and cultural costs, along with a progressive reduction in the well-being and quality of life.

These cities have extraordinary value added that can play a decisive role in a strategic vision of urban development. In recent years, the concept of "water renaissance" [205, 206] has been growing in the urban sphere to define that complex process of redevelopment and revitalization of waterfront areas—which has characterized many renewal, recovery and retraining operations and valorization in many cities around the world—rediscovering the value of water in the city [207–210]. The port and coastal areas, therefore, represent sites where multiple contradictions often arise, but are also the most suitable places to reduce conflicts and turn them into synergies through innovative approaches to governance, planning and territorial management. They can become the point of entry for a sustainable development process of the wider urban system, activating synergies between port areas, where deep differences exist between cultures, lifestyles and socio-economic dynamics, but where the potential for creativity is higher and can overcome the various criticalities and possible conflicts. A key feature of a sustainable city is the ability to close the circuit of resources through circular processes, capable of reproducing/regenerating the original capital stock. In this way, cities can become key to the implementation of a new model of sustainable development based on multiple value-added systems, focused on a synergistic and circular approach, capable of breaking the current linear organization of many traditional economic systems and allowing the local economy to strengthen through integration.

12.4.4.1 The Methodological Process

In this perspective, the study of the municipalities of the coastal area of the Metropolitan City of Naples analyzes the opportunities for transformation and enhancement that the sea resource can offer to the territory, trying to identify the components that could interact to improve territorial productivity of coastal cities and make the urban and territorial regenerative city model operational.

Through the recognition of the different kinds of projects promoted, implemented and/or ongoing—following a top-down and a bottom-up approach—a knowledge framework of the transformations in place has been structured, elaborating an interpretative analysis in support of possible regeneration scenarios in a long-term vision, according to the principles of Biourbanism approach and considering the perspective of spatial vulnerability and resilience. The study is divided into six main phases (Fig. 12.5), that identify the main issues elaborated by the Steinitz framework for Geodesign [211] and advocates the use of six models to describe the overall planning process:

1. Representation Models (*How should the context be described?*);
2. Process Models (*How does the context operate?*);
3. Evaluation Models (*Is the current context working well?*);
4. Change Models (*How might the context be altered?*);
5. Impact Models (*What differences might the alterations cause?*);
6. Decision Models (*Should the context be changed?*).

The first three models comprise the assessment process, looking at existing conditions within a geographic context, considering hard data and soft data, and including stakeholders' interactions.

The second three models comprise the intervention process, looking at how that context might be changed, the potential consequences of those changes, and whether the context should be changed and by whom. In our proposal, the fourth model—the Change Model—starts from the principles of Biourbanism approach [cfr. 12.1, 12.2, 12.3 sections of this chapter] and provides the specific framework for developing and creating the proposed changes (design scenarios) that are predicated on the science- and value-based information contained in the Representation Models and assessed against that same information in the Impact Models.

The definition of Geodesign considers the term "geo", defined as geographic space or also as geo-scape, describing all types of spatial data (physical, biological, social, cultural, economic, urban, etc.) that can be geo-referenced and analyzed. Indeed, geo-scape is the planet's life zone, including everything that lies below, on, and above the surface of the earth that supports life [212]. At the same time, the word design usually refers to a process or series of activities. Design is conceived as the thought process comprising the creation of an entity [192]; it is the mental synapse that instantly sees the potential connection between the problem and possibility.

Considering the above reflections, Geodesign is the thought process comprising the creation of an entity in the planet's life zone (geo-scape); or more simply, Geodesign is designed in the geographic space (geo-scape) in order to facilitate life in the geographic space (geo-scape) [211]. The relevant link between the entity being designed and its geographic context provides the tangible basis for doing both science-based and value-based designs, providing operational linkages to a wide variety of domain-specific information and the multidisciplinary platform for doing integral/holistic design.

In our proposal, in phase 1—the Representation model—the main steps of data gathering, data selection and criteria and indicators elaboration are related to identifying the general framework of the Municipalities of the Metropolitan City of Naples and of the Cost. In particular, we identify the transformation projects of the sixteen municipalities of the coastal area: Giugliano, Pozzuoli, Bacoli, Monte di Procida, Naples, Portici, Ercolano, Torre del Greco, Torre Annunziata, Castellammare di Stabia, Vico Equense, Meta di Sorrento, Piano di Sorrento, Sant'Agnello, Sorrento, and Massa Lubrense. The transformation projects have been selected and classified by taking into account a number of significant criteria that have highlighted the type of work, location, work amount, funding type, actuators and partnerships, top-down approach and bottom-up approach, state-of-the-art implementation, and the reference period. For each municipality, the relevant "project cards" have been drafted with the identification of selected projects and specific detail information, synthesizing the transformations that have characterized the municipality territory, also highlighting the processes of governance and territorial management activated.

In the step 2—the Process model—the identification and classification of the selected projects have also been carried out with the support of the Geographic Information System (GIS), enabling information organization and their visualization by thematic maps. These maps have also been elaborated to describe the main dimensions that characterize the context, in order to understand the economic, social, and environmental dynamics related to each municipality. The various information collected were elaborated by appropriate, simple and composite spatial indicators, useful to identify the potentials and critical aspects of the areas affected by the interventions, both with reference to the coastal area and the municipal surface. Therefore, the opportunities for activating territorial regeneration processes have been highlighted by analyzing the specificities of the territories of the municipalities concerned.

By selecting some relevant indicators, the concepts of vulnerability and resilience have been expounded in three dimensions: environmental, social and economic vulnerability; and environmental, social and economic resilience. In this way, it has been possible to elaborate a multi-criteria analysis, which could explicate the territorial vulnerability and territorial resilience of the municipalities of the coastal area of the Naples Metropolitan City, and a multi-group analysis able to include the preferences of the different stakeholders.

Indeed, in step 3—the Evaluation model—a multi-criteria analysis and a multi-group analysis become useful support to identify the suitability areas, where the Change model of step 4 could be implemented. In this case, the Novel Approach to Imprecise Assessment and Decision Environments (NAIADE) [213, 214] has been implemented, starting from the results of the elaborations developed in the previous phases. Maps of territorial opportunities have been set up, where through the selection of some composite indicators and the support of spatial analysis, it has been possible to outline the construction of a territorial coalition strategy oriented to the identification of "homogenous zones", where the municipalities could enhance endogen resources and make the "Biourbanism model" operational.

In step 4—the Change Model—through the identification of Spatial Opportunity Maps (SOMs), it has been possible to verify the significance of the processes in progress and to elaborate the simulation of possible regeneration and valorization scenarios of the coastal area that could promote synergic and symbiotic relations among the various opportunities discovered, expression of connected diversifications.

The breadth of the study area, the heterogeneity of the data, and the variety of subjects involved, and the processes underway identify the coastline of the Metropolitan City of Naples as a multidimensional system characterized by complex values and expression of local specificities; it can start with the elaboration of strategic goals and actions that can generate "new values" through the identification of spatial solutions, expression of local strategies, and situated actions [215].

The step 5—the Impact model—identifies how solutions could affect the context and the related needs of mitigation and/or re-design process.

The step 6—Evaluating the solutions and Making a decision—considers the final step of the process, where the alternatives are selected, and the stakeholders choose the shared solutions. Also in this step, the support of multi-criteria analysis and multi-group analysis methods can make the process more transparent and able to consider a multidimensional perspective.

The local reality—with its spatial, economic, environmental, social and cultural characteristics—is the context where the responses to regeneration and territorial enhancement are defined. The identification of an urban regeneration, as evidenced by Audis [216], can no longer rely on traditional growth and urban transformation factors—driven only by the private propensity to invest, especially real estate—and the impacts that these investments could generate on specific territories.

In a context characterized by scarce economic resources, complex problems, and relevant conflicts, it is essential to propose quality actions in a multidimensional perspective that takes into account material and immaterial, hard and soft, objective and subjective values, use, non-use and intrinsic values, and their synergic and complementary relationships, in order to formulate "situated actions".

12.4.4.2 Implementation of the Methodological Process

This research develops and deepens the following phases of the methodological process, with the related results. In step 1—the Representation model—a collection of data and their classification has been oriented to the construction of knowledge of the transformations of the territory along the coast of the Metropolitan City of Naples. A definition of the decision problem through an interpretative analysis of processes and opportunities for the transformation of the coast is started from the analysis of potentialities and criticalities for a long-term scenario, in terms of resilience and vulnerability. The development prospects for the coastal area of the Metropolitan City of Naples have been investigated by verifying the performance of the coastal territorial system in relation to the three key aspects of sustainable development, translated into three macro-criteria: environment, economy, and society. They include the selected urban regeneration criteria for evaluating the coastal opportunities for transformation, with an identification of relevant indicators. The selected indicators differ in being simple or composite [217]. The former allows for the framing of municipalities in the environmental, social and economic systems and for assessing the transformation projects and activities analysed in each municipality; the second consists of aggregating simple indicators that are considered representative of different aspects of a multi-dimensional phenomenon. The various indicators—identified from the environmental, social and economic characteristics of the territory, and developed taking into account the information gathered for each municipality—have been structured into the GIS.

In the step 2—the Process model—the classification of indicators by means of a GIS has enabled the elaboration of a series of thematic maps that return the spatial distribution of the investigated phenomena, highlighting the potentialities and the criticalities that characterize each municipality and the coastal area, enabling the acquisition, recording, analysis, display and return of information derived from geo-referenced data. The analyzed indicators are explained according to two main criteria: territorial vulnerability and territorial resilience. The vulnerability of a system is a concept linked to the risk of an intervention and consequently, of an investment of different capital forms. In a broader perspective, the various actors (institutional, economic, social, etc.) operate by taking on the risks associated with the criticalities that depend on the characteristics of the context in which they operate. Critical aspects, in turn, exercise pressures on the territory in multidimensional terms and contribute to increasing the level of vulnerability of the system. Vulnerability measures the probability that the same criticality may result in negative impacts, and the ability of the system to withstand pressures determines the level of vulnerability [192].

A vulnerable territory is a system that has lost its ability to withstand external pressures and therefore remains exposed to the risk of adverse impacts; so, small disturbances can lead to high levels of vulnerability and consequently, put at risk the development of the territory itself. In a vulnerable system, where processing instances are difficult to handle, change cannot create development opportunities, thus reducing territorial attractiveness. Evaluating an area's attractiveness can support build development models where it is possible to establish symbiotic relations between the physical environment, social actors and the economic world, improving the productive use of the territory, promoting functional diversification, and getting the relationships system better, in order to reduce the conflict between conservation, change and transformation.

In step 3—the Evaluation model—multidimensional vulnerability assessment presupposes the selection of indicators capable of describing the complexity of the context based on specific objectives, which, in the present case, are related to the identification of smart valorization and regeneration strategies on the coastal area of Naples Metropolitan City.

The structure of vulnerability indicators on a municipal scale, identified for the present study, consists of three main dimensions (Table 12.4):

Table 12.4 The vulnerability indicators

Criteria	Indicators	M.U.	Scale	Source
Environmental dimension				
Urban context	Extension of the urbanized urban area compared to the total extension of the municipal area	%	Municipal	SIT
	Index of building degradation	%	Municipal	Istat
	Surface of degraded areas	ha	Municipal	SIT
Coastline	Extension of the artificial municipal coast compared to the total extension of the municipal coast	(ml/ml)%	Municipal	SIT
Water system	Poor quality bathing water	%	Municipal	Arpac
	Polluted outfalls	%	Municipal	Arpac
Social dimension				
Demography	Housing density	ab/kmq	Municipal	Istat
	Average age of the resident population	years	Municipal	Istat
Social inequality	Average annual income per inhabitant	€/ab	Municipal	Istat
	Unemployment rate	%	Municipal	Istat
Economic dimension				
Enterprise system	Variation in the number of local companies in 2001–2011	%	Municipal	Istat
	Variation in the number of employees in shipping and fishing companies in 2001–2011	%	Municipal	Istat
	Variation in the number of employees in manufacturing companies in 2001–2011	%	Municipal	Istat
	Variation in the number of employees in the construction companies in 2001–2011	%	Municipal	Istat
	Variation in the number of employees in the hospitality sector in 2001–2011	%	Municipal	Istat
No-profit services	Variation in the number of employees of no-profit companies in 2001–2011	%	Municipal	Istat
	Total number of social projects	no.	Municipal	Municipality
	Percentage of projects with no-profit actuators	(n/n)%	Municipal	Municipality
Real estate market	Real estate market intensity index (Ratio between NTN and Housing Stock)	%	Municipal	OMI
	Percentage variation in property prices between 2008 and 2017	%	Municipal	OMI

Source Authors

1. Environmental dimension: Physical and environmental vulnerability factors are identified, which involves the search for indicators describing the ability of a system to maintain self-regeneration capacity, or conversely, its ability to increase structural fragility [218]. In the present case, environmental vulnerability indicators have been subdivided into the following macro-categories: urban context, coastline and water system.
2. Social dimension: Indicators based on the theory of adaptive systems can be highlighted [219], underlining the multidimensional nature of the concept of vulnerability to a social system. Appropriate indicators have been identified for the following macro-categories: demography and social inequality.
3. Economic dimension: Economic vulnerability indicators identity the relevance of the economic processes and can be divided into the following macro-categories: enterprise system, no-profit services, and real estate market.

The spatial variability of vulnerability components– determined by the territorial non homogeneity of development processes and the existence of local factors—significantly influences the multidimensional level of well-being for the local community, as well as individual opportunities and enterprise development [220–222].

Each criterion is a different information layer that, if overlaid on the other layers, builds a complex picture of interrelated information. Each layer of information identifies a degree of system vulnerability. For each of the three dimensions of vulnerability, a ranking of municipalities has been elaborated in order to identify shared territorial contexts with greater or lesser vulnerabilities and analyze the most influential factors. As vulnerability can be understood as the tendency of a system to undergo the effects of external pressures with poor responsiveness, resilience identifies its ability to cope with and recover from the perturbing action effect, based on the system intrinsic capacity of adaptation and regeneration [219, 220]. In a resilient system, change, or external pressure has the potential to create new development opportunities by recovering over time an identical or similar situation and introducing variations that can make important innovations [221]. For the above reasons, it is possible to bring together the vulnerability analysis and the resilience analysis of the territorial systems.

In assessing resilience at the local scale, factors that could lead to an increase are considered. The analysis of the resilience of a territory can also be decomposed respectively into the social, economic and environmental. The resilience indicator context is the municipal one in order to compare the results of the vulnerability analysis with that of resilience analysis and produce useful information for building territorial opportunities for enhancement scenarios.

The construction of territorial resilience framework of the coast municipalities has been achieved by declining the selected dimensions with respect to some relevant indicators:

1. Environmental dimension: The elements that describe the aptitude of a system for maintaining self-regeneration capacity have been considered. Environmental resilience refers to the ability of systems to change processes, practices or structures to moderate or balance any damage or seize any opportunities arising from change. For the coastal area system of the Naples Metropolitan City, they have been considered the most likely factors to represent the ability of the territory to adapt to critical issues.
2. Social dimension: Social criteria have been identified that can describe the system's capacity to adapt to external pressures, linking each of them with some indicators. It is important to emphasize the relational dimension between demographic factors (generational regeneration capacity) and the factors that reduce social inequality.
3. Economic dimension: The ability to respond to a territory has been investigated as an ability to adapt, induced by actions from internal agents to the economic system. Therefore, some indicators that describe the current framework of the local economic system have been considered.

The complex framework of information thus generated has been elaborated using the multi-criterion and multi-group methodology of Novel Approach to Imprecise Assessment and Decision Environments (NAIADE) [213, 214], which allows for a ranking of municipalities based on their level of vulnerability and resilience. The construction of preference orders has been elaborated through the structuring of a multi-criteria decision process [223], in which the alternatives are made up of the sixteen municipalities of the Coast of the Naples Metropolitan City, and the criteria are the selected vulnerability and resilience indicators. The NAIADE method builds a preference order between municipalities, from the least vulnerable to the most vulnerable and the least resilient to the most resilient. The order of preference obtained is represented in the GIS, associating each municipality with its position in the ranking obtained from the multi-criteria analysis. Therefore, the assessment made it possible to develop a territorial vulnerability map for each of the three dimensions (environmental, social, and economic) and at the same time, elaborate vulnerability maps for each of the selected indicators (Figs. 12.6 and 12.7). Similarly, territorial resilience maps have been elaborated (Table 12.5).

The analysis of territorial vulnerability and resilience for coastal municipalities allows a comparative analysis of the critical aspects and potentials to identify smart development and valorization strategies, structuring an assessment of the opportunities for a smart spatial development. The elaboration of Spatial Opportunity Maps (SOMs) is the output of a multidimensional evaluation process that leads to the identification of specific territorial contexts with a greater propensity for some specializations relating to the economic sectors and the local transformation.

In step 3—the Evaluation model—combining the results of the NAIADE implementation and the preferences of the local stakeholders (mayors, technical experts such as engineers, architects and economists, construction companies, developers, port authorities, ship-owners, fishermen, hoteliers, restaurateurs,

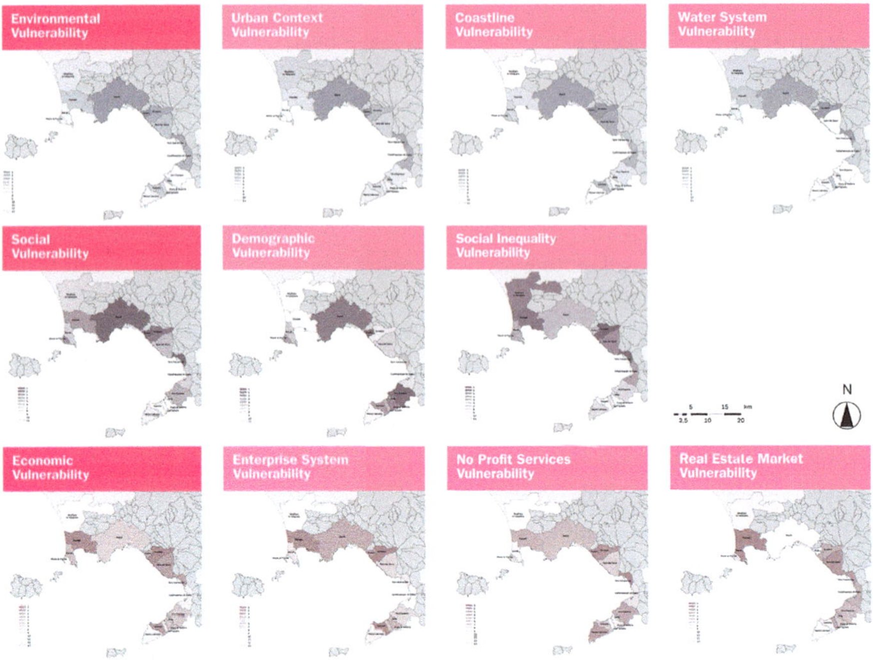

Fig. 12.6 The vulnerability maps. *Source* Authors

traders, cultural associations, environmental associations, and citizens), collected during a series of five focus groups; four areas of opportunity have been identified for the exploitation of resources of the coastal municipalities of the Metropolitan City of Naples (Fig. 12.4): Receptivity and seaside tourism; Culture, archeology and nature; Maritime economy and productivity; Buildings recovery and urban regeneration; and Local good governance.

As regards to local good governance, an index of the dynamism of public investment and local community involvement has been developed, enabling to identify the municipalities with greater propensity to manage and implement territorial transformations, combining top-down processes and bottom-up and p2p activities. Opportunity maps show—with a semaphorical scale (green for municipalities with better performance, red for municipalities with worst performances)—the ranking of municipalities for each opportunity. The evaluation through the NAIADE method returns a ranking of the Municipalities based on the selected criteria and indicators. The opportunity maps obtained allow identifying the geography of territorial resources and values that take into account the specifics of each municipality. The analysis of the results of the opportunity maps detects the existence of territorial clusters with specific peculiarities. One of the features of multidimensional evaluations should be taken into account, which depends on the correct interpretation of the results: the final ranking of the alternatives, constituted by the municipalities of the Coast, depending on the selection of the indicators

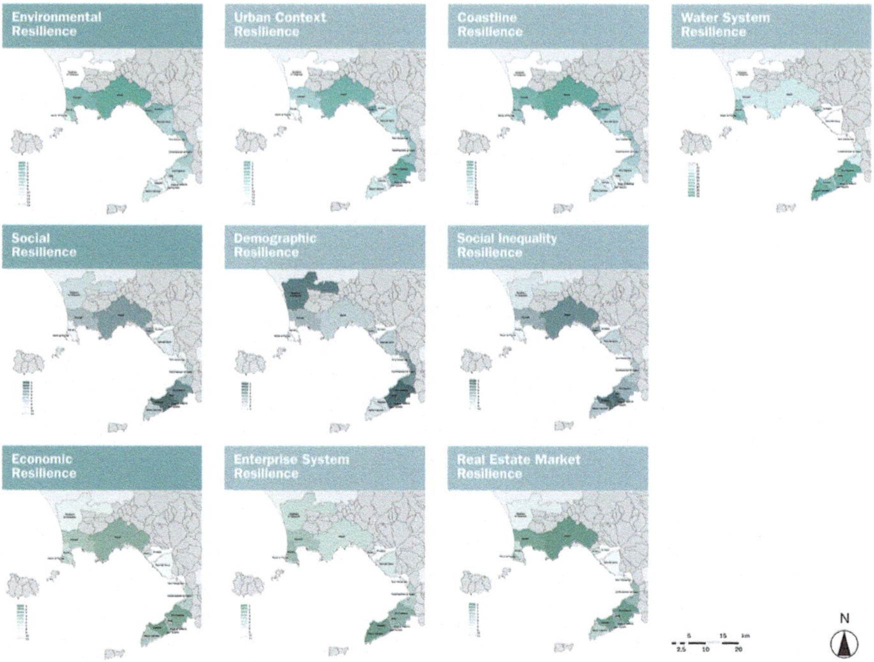

Fig. 12.7 The resilience maps. *Source* Authors

under consideration. Therefore, sorting returns an interpretation of reality, based solely on the data available, which could still not fully grasp the complex description of territorial realities, but only focus on certain aspects.

About "Receptivity and seaside tourism" opportunities, the municipalities of the Sorrento Peninsula identify a concentration of territorial opportunities where, among the most relevant factors, there is the high quality of bathing water and the absence of commercial ports, as well as the existing receptivity vocation of the area. At the same time, there is an additional aggregation of municipalities with a high degree of opportunity for receptivity: Monte di Procida, Bacoli and Pozzuoli. In this case, the absence of commercial ports and the presence of tourist ports remain among the most important factors, in addition to the remarkable extensions of the natural coastal bays, the beaches and the sandy coasts.

As regards to the opportunities related to "Culture, archaeology and nature", the municipalities of the Sorrento Peninsula play a significant role due to the absence of port areas and degraded areas of considerable size, as well as due to the morphological features of the coast, characterized by cliffs that enhance its landscaping value. In addition, the city of Giugliano di Napoli shows good performance, mainly due to the absence of port areas and the remarkable extension of the coastal strip, despite some environmental concerns.

Table 12.5 The resilience indicators

Criteria	Indicators	M.U.	Scale	Source
Environmental dimension				
Urban context	Extension of the municipal area destined to be green areas compared to the total extension of the municipal area	(mq/mq)%	Municipal	SIT
	Historical-architectural and landscape pre-existences in extra-urban areas	ha	Municipal	SIT
Coastline	Extension of the municipal coast compared to the municipal area	(ml/mq)%	Municipal	SIT
	Extension of the natural municipal coast compared to the total extension of the municipal coast	(ml/ml)%	Municipal	SIT
	Areas of historical-cultural and landscape interest along the municipal coastline	ha	Municipal	SIT
	Surface of the municipal beaches compared to the surface of the provincial beaches	(mq/mq)%	Municipal	SIT
	Accessibility	no.	Municipal	SIT
Water system	Excellent quality bathing water	%	Municipal	Arpac
Social dimension				
Demography	Youth concentration	%	Municipal	Istat
	Schooling rate	%	Municipal	Istat
Social inequality	Average annual income per taxpayer	€/taxpayer	Municipal	Istat
	Employment rate	%	Municipal	Istat
	Activity rate	%	Municipal	Istat
Economic dimension				
Enterprise system	Number of local businesses per inhabitant	%	Municipal	ATECO
	Number of employees of manufacturing companies per thousand inhabitants	%	Municipal	ATECO
	Number of employees of construction companies per thousand inhabitants	%	Municipal	ATECO
	Number of employees of companies in the hospitality sector per thousand inhabitants	%	Municipal	ATECO
	Total number of projects of public or private-public initiative	no.	Municipal	Municipality
	Total number of public-private-social projects	no.	Municipal	Municipality
	Total number of social projects	no.	Municipal	Municipality
	Total value of the interventions	1000 €/ab	Municipal	Municipality
	Percentage of projects with private actuators	(n/n)%	Municipal	Municipality
	Percentage of projects with no-profit actuators	(n/n)%	Municipal	Municipality

(continued)

Table 12.5 (continued)

Criteria	Indicators	M.U.	Scale	Source
Real estate market	Quotation price of residential properties (average municipal value)	€/mq	Municipal	OMI
	Quotation price of buildings for commercial use (average municipal value)	€/mq	Municipal	OMI
	Quotation price of buildings at production destination (average municipal value)	€/mq	Municipal	OMI
	Property market intensity index (Ratio between NTN and Housing Stock)	%	Municipal	OMI

Source Authors

About "Maritime economy and productivity" opportunities, Naples appears to be the territorial sphere with the highest level of opportunities, involving the neighboring municipality of Pozzuoli also. In addition, the towns of Torre Annunziata and Torre del Greco show a good level of opportunity, mainly due to the presence of commercial ports and a high concentration of the population which, in terms of production, translates into high availability of human capital.

As regards to "Buildings recovery and urban regeneration", the municipality of Giugliano di Napoli, on the basis of the selected indicators, presents the greatest opportunities (mainly due to the absence of the port area), together with the municipality of Portici (where, in addition to the absence of port areas, the large number of historical-architectural pre-existing buildings in the extra-urban area is relevant) and the municipalities of Vico Equense and Piano di Sorrento (where, among the most important factors, there is the limited extension of degraded areas within the municipal territory).

Finally, for the analysis of the opportunities for "Local good governance", projects for public buildings and urban spaces analyzed for each municipality are selected along with the various bottom-up interventions, promoted by associations and citizens, also in collaboration with institutions and private subjects; however, the collected data set is not complete, and some data on the municipalities of Bacoli, Monte di Procida and Sorrento are missing.

Municipalities with good performance due to the high value of per capita spending on projects on urban and urban heritage are Massa Lubrense, Torre Annunziata, Naples and Pozzuoli. In the case of Massa Lubrense, together with Torre del Greco, the number of projects activated on the territory is decisive for the purpose of defining a high level of opportunity. At the same time, the municipalities of Torre Annunziata, Naples and Pozzuoli are the only ones that have shown a certain propensity to public-private partnership to activate the processes of transformation, recovery and conservation.

The obtained opportunity maps (Fig. 12.8) become the input for an assessment of the possible strategies for valorization and regeneration of the Coast, by identifying possible synergistic coalitions between the municipalities in order to make complementary opportunities for a smart spatial network. Considering the results of Spatial Opportunity Maps, an evaluation process oriented to the identification of possible coalitions between the sixteen municipalities—taking into account territorial specificities to activate a network strategy that recognizes synergy, symbiosis and complementarities as essential components of a process of valorization and regeneration at the territorial level, and where spatial solutions are consistent with the principles of the Biourbanism approach—could be implemented.

The methodology used to analyze possible coalitions among municipalities is a multi-group analysis defined as "equity analysis", provided by the NAIADE method. This method consists of a fuzzy clustering algorithm that associates a verbal scale qualitative assessment for each pair of alternatives/groups, expressing the preference of a group with respect to the alternative and consequently, constructing the dendrogram of the coalitions—that is, the consecutive aggregation of the groups with respect to the preferences. In the present case, alternatives are made up of different opportunities for valorization, while the groups are made up of municipalities which, based on the results of the analyses obtained with the opportunity maps, express a preference over each of the specializations of the territory. It is therefore possible to structure a process of progressive formation of the coalitions, so that each municipality can recognize its role within the territorial system and can either seize the opportunities offered by the territory or build a

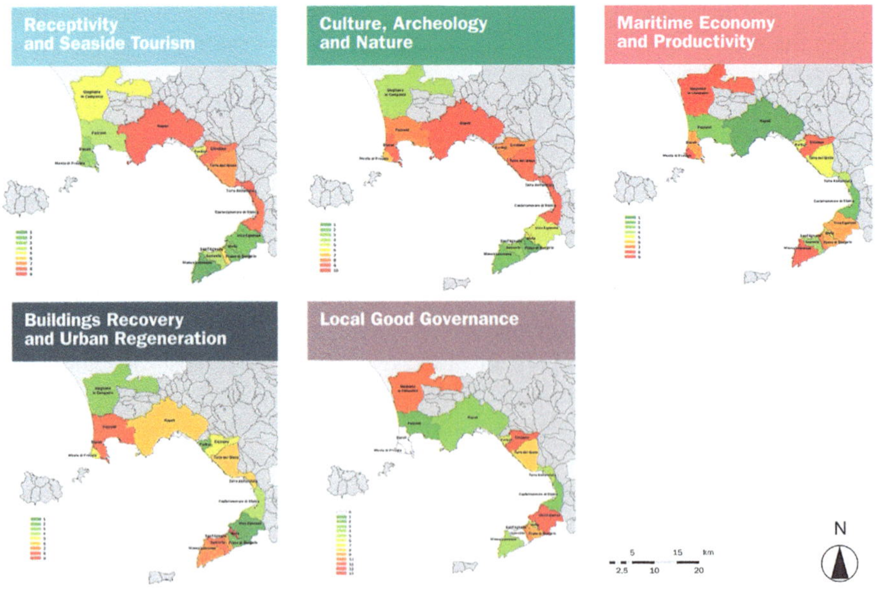

Fig. 12.8 The spatial opportunity maps (SOMs). *Source* Authors

	Bacoli	Castellammare di Stabia	Ercolano	Giugliano in Campania	Massa Lubrense	Meta	Monte di Procida	Napoli	Piano di Sorrento	Portici	Pozzuoli	Sant'Agnello	Sorrento	Torre Annunziata	Torre del Greco	Vico Equense
Receptivity and Seaside Tourism	good	very bad	very bad	moderate	perfect	good	good	extremely bad	very good	moderate	moderate	more or less bad	more or less good	extremely bad	bad	very good
Culture, Archeology and Nature	extremely bad	extremely bad	very bad	more or less good	very good	more or less bad	bad	extremely bad	perfect	very bad	very bad	very good	good	extremely bad	extremely bad	moderate
Maritime Economy and Productivity	bad	perfect	extremely bad	extremely bad	extremely bad	more or less bad	extremely bad	perfect	bad	moderate	very good	very bad	good	more or less good	moderate	bad
Buildings Recovery and Urban Regeneration	very bad	more or less good	moderate	good	bad	extremely bad	moderate	more or less bad	very good	very good	very bad	bad	very bad	more or less bad	more or less bad	perfect
Local Good Governance	moderate	perfect	very bad	bad	very good	more or less good	moderate	very good	moderate	good	perfect	more or less bad	moderate	good	more or less good	very bad

Fig. 12.9 The equity matrix. *Source* Authors

network of municipalities of the Coast with different levels of specialization and local resources. The starting point for the equity analysis is the construction of the so-called "equity matrix" (Fig. 12.9) that corresponds to each alternative/group crossing a judgment according the following verbal scale: perfect, very good, good, more or less good, moderate, more or less bad, bad, very bad, or extremely bad.

For territorial governance opportunities, there is no complete information on the municipalities of Bacoli, Monte di Procida and Sorrento, and it has been decided to give a mid-term evaluation (with its verbal "moderate" judgment). Through the fuzzy clustering algorithm, the NAIADE returns the coalition's dendrogram that allows the coalitions to be reduced to a similarity index between the groups, represented by the municipalities (Fig. 12.10).

The dendrogram permits to grasp the major or minor similarities between municipalities in terms of opportunities for coastal zone enhancement and territorial specialization. At each coalition, it is possible to calculate the value of the conflict index for each partnership alternative. This conflict index is a measure of the level of agreement between members of the coalition towards a specific alternative.

This index tends to zero if the agreement increases, but grows instead if the agreement declines. In addition, some coalitions may veto one or more alternatives, coming up with a veto on most (or even all) alternatives for cases of very low similarity index values (Fig. 12.11). The conflict index for each coalition allows to select the preferred alternatives for each coalition level and identify coalitions and alternatives with a low value of the conflict index that corresponds to the priority specializations in terms of opportunities for each municipality. The final ranking identifies the following list of priorities: "Local good governance", "Receptivity and seaside tourism", "Buildings recovery and urban regeneration", "Maritme economy and productivity", "Culture, archeology and nature".

With the existence of territorial zones with similar vocations in terms of opportunities for Coast enhancement and local coalitions, it is possible to analyze the similarity matrix between the groups, which allows estimating how similar the preferences expressed between each pair of municipalities are. In particular, it is possible to detect the presence of three main groups of municipalities with a high

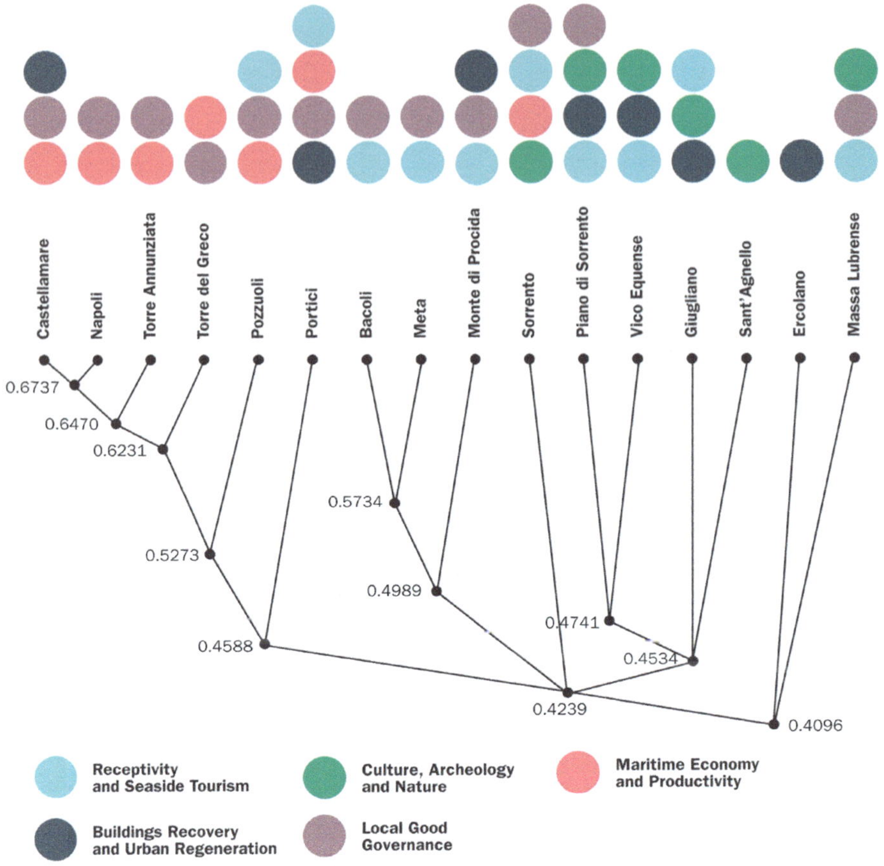

Fig. 12.10 The dendrogram of coalition. *Source* Authors

degree of similarity. The first group consists of the following municipalities: Castellammare di Stabia, Napoli, Pozzuoli; Torre Annunziata, Torre del Greco, and Portici. These municipalities show greater opportunities for the maritime economy and productivity, and for local good governance. In the second group, there are the municipalities of Bacoli, Meta, Monte di Procida, Meta di Sorrento, Piano di Sorrento, Sant'Agnello, Sorrento, Vico Equense, and Giugliano. These municipalities mainly present more opportunities for receptivity and seaside tourism, as well as for the theme of culture, archaeology and nature. In the third group, there are the municipalities of Ercolano and Massa Lubrense that consider different thematic strategies: for Ercolano buildings—recovery and urban regeneration; and for Massa Lubrense—receptivity and seaside tourism.

The evaluation results enable a strategic map to be built (Fig. 12.12), where it is possible to highlight the connected diversifications as opportunities that the identified territorial networks could be activated. Therefore, the elaboration of human

Fig. 12.11 The veto diagram: The progressive reduction of the conflict index. *Source* Authors

Fig. 12.12 The strategic map of connected diversifications. *Source* Authors

smart spatial solutions, characterized by place-based strategies and situated actions, for the enhancement of the coastal area of the Naples Metropolitan City can be considered as a prerequisite for the activation of a process oriented to the identification of "homogeneous zones", conceived not only as areas with similar characteristics, but above all as territories where it is possible to promote networks of opportunities between the various municipalities and their communities. The last two steps of the methodological process—the step 5: Impact model, and the step 6: Decision model—can be implemented after identification of the specific actions, in order to assess the related impacts, the necessary mitigations and the eventual re-design. The final decision closes the process and activates an operative interaction among the involved stakeholders.

12.4.5 Discussion and Conclusions

The evaluation process structured in the present study has made it possible to understand the specific resources of the coastal municipalities of the Naples Metropolitan City by examining the components that characterize the environmental, social and economic systems. In particular, spatial specificities have been analyzed from the identification of vulnerability and resilience characteristics through a selection of specific indicators. The vulnerability and resilience maps, elaborated by the multi-criteria approach of the NAIADE method combined with GIS, have allowed identifying the most vulnerable and resilient municipalities. Through the application of the NAIADE method, it was also possible to identify municipalities ranking in relation to the identified opportunities, by the opportunities maps elaborated in GIS, from which the main propensities of the objectives of valorization and regeneration are pursued.

From the methodological process, it emerges that at least two territorial networks can be identified along the coast of the Metropolitan City of Naples, with respect to which some opportunities are most significant. Coalitions between the municipalities, which emerge from the interpretation of the dendrogram elaborated with the multi-group approach of the NAIADE method, allow to highlight how municipalities with similar opportunities could identify a shared intervention strategy, to exploit the resources present and transform the differences into synergies and symbiosis, implementing their connected diversifications.

From the results obtained, it is possible to find new approaches and new tools for planning and managing the transformation and regeneration processes, supporting close resource flows through circular processes, capable of activating virtuous synergies between institutions, communities, and citizens. The study has been dealt according to a "multi-methodological" approach [224], with a view to addressing a complex decision-making process such as the transformation of the Coast of the Metropolitan City of Naples, characterized by multiple variables and a high level of uncertainty, in an incremental evaluation process with continuous feedback and iterations, useful to outline a process of conscious transformation and enhancement that is sensitive to the specificities that characterize the territory under consideration. A "multi-methodological" approach allows to handle complex decision-making process by simultaneously taking into account a multitude of aspects of the problem, trying to build models that can describe, interpret, and understand the complexity of the real world. A suitable evaluation path should start from the correct definition of objectives and values, highlighting their synergy and conflict relations and the relative importance assigned to them by the various stakeholders involved in the decision-making process [225]. Managing a large and complex set of quantitative and qualitative information in evaluation processes may require the use of appropriate Decision Support Systems [226] that can improve the quality and transparency of the choices.

The methodological process has been structured to identify significant opportunities that should be triggered by an urban regeneration process for the Coast of Naples Metropolitan City, recognizing the essential role of sea resource and synergies that could develop into a win-win-win–oriented territorial cohesion [227]. Here, each of the sixteen municipalities involved could play a decisive role, enhancing their environmental, social and economic resources and collaborating in building a territorial network of resources and actors.

Numerous urban regeneration practices that have activated symbiosis models allow to highlight how "co-operation" is a source of mutual benefit and involves a mutual convenience, based on the constant construction of ties and relationships and the interdependence determined by spatial proximity. Economic processes require cooperative-collaborative behaviors between the various components and become increasingly territorialized, and therefore, more resilient and at the same time, less and less associated with the production of negative environmental impacts. The economic development crisis due to a public and private resource crisis coupled with the crisis in demand for the type of product the market used to offer may become an opportunity to redesign the city and the metropolitan area on the basis of models closer to the need of citizens, where the terms "quality", "urban welfare" and "sustainability" play a leading role [216].

The smart Biourbanism model identifies a development perspective capable of stopping the ongoing degradation processes, with the purpose of rebuilding the local economy and stimulating a demand for social and cultural innovation. Indeed, the above model emphasizes that it is no longer necessary to preserve the existing forms of capital; but it has become essential to increase them, creating economic, social and environmental benefits. It is therefore based on the relationships and ties that are able to trigger synergies and symbiosis, to generate circular organization processes and show how to cooperate economically, socially and environmentally. It is therefore necessary to regenerate the relations between the inhabitants, between the inhabitants and the urban space, between the inhabitants and the rural area, between productive activities and economic activities, and between the urban space and the rural area. The model prioritises the reconstruction of links at different levels, from which the density of new development activities emerges. It assumes synergistic processes, circularization and symbiosis as tools to multiply the ties that, in turn, generate new synergies and symbiosis, triggering new circular processes. These processes shape the city and make it more attractive to locate new activities and new investments. In fact, economic productivity can be considered as a function of attractiveness, which in turn depends on the quality of the landscape, the function of the synergy and symbiosis intensity, and the processes of circularization between the different forms of capital characterizing the city-territory system. The notion of symbiosis, broadened to the urban and social sphere, is central to the regenerative city model, where the circularization of processes concerns the very organization of the city, its economy, its social system and its governance. Attention, hence, is focused on how to make the regenerative city model operational, understanding how it is possible to stimulate and multiply the links and circular relationships—that is, the "virtuous circular processes"—and thus the synergies and the symbiotic among the various resources of a territory, able to regenerate the ongoing processes and to produce virtuous processes capable of building new values.

In the age of globalization, each territory must identify its own specificities, making it different and unique, exploiting the existence of differences to become more attractive due to the productive factors it owns and from which it can derive its strategic role on a local scale or global. Territorial productivity is closely linked to the ability to activate circularization processes, combining different interests and goals, and to create new bonds, new synergies and new symbiosis projects as components of a chain that generates different, not just economic, values. They, at the same time, create new relationships shared between social actors and new knowledge, in a circuit that is self-feeding and self-supporting. This development strategy should integrate economic values, social values and environmental values into a "win-win-win approach" that will help increase the resilience of the city, understood as the city's ability to react and manage change, while integrating its urban and organizational structure, and reducing vulnerability. A development process that takes into account these assumptions cannot be based on the approaches and tools of the traditional economy, but identified in the ecological economy the bases for realizing a new urban and territorial metabolism in a long-term perspective [228, 229].

By integrating economic wealth production with ecological safeguarding and social marginalization reduction, it is possible to outline the circular processes that develop along these three main levels [230]:

1. *Economic*: Capable of activating symbiosis processes for value production, allowing synergies and stimulating circuits between society and community prosperity [231], between society and the city, among all the actors and the territory;
2. *Social*: Capable of regenerating interpersonal relationships, often debilitating in urban contexts, through the promotion of relationships and ties with the "places" recognized as identity and symbolism;
3. *Ecological*: Can support ecological processes that characterize living systems, allowing you to conserve and reproduce in time. Circular processes imitate the organization of ecological systems, which are able to self-procreate (*autopoietic*) and at the same time, "support" other (*heteropoietic*) systems. Examples of ecological circular processes emerge through re-use, recycling and regeneration, with a reduction in negative externalities. Therefore, economic, social and ecological circular processes contribute to increasing the resilience and creativity of urban systems by reducing linear metabolism and vulnerability.

Cities are conceived as vital organisms when acting as complex dynamic systems, capable of transforming and adapting to the constant pressure of external change, and capable of modifying space structure, systemic organization and typologies of functions, combining infrastructures, services and activities, and retaining their identity and constituent characters [232–236].

Activating and promoting circular processes in the territories of the port cities and the coastal strip, therefore, imply a new economic organization of the city that reduces material consumption, waste generation, energy consumption, and pollutant emissions by linking the port areas to the city-territory system through pilot experiments of circular projects, capable of making the "7 R" slider operational: Reduce, Replace, Reuse, Recycle, Recover, Refuse, Reject and Rethink [237]. It is therefore necessary to reorganize coastal cities according to processes that favor symbiotic exchanges between the productive structures of cities, between the city and the production systems, and between the city and the extra-urban territory.

12.5 Conclusions

The study team offers a deep analysis of the contemporary socio-economic situation, showing all the incongruence, inequalities, environmental problems and the general unsustainability of the actual economic system. They consider necessary a new theory and practices of smart cities and economic development, suggesting a new framework based on a Biourbanism epistemology. This framework adopts the *systemic principles of life* and suggests to shift from techno-city to human city—called homological smart city—to be contrasted with the diffused idea of the smart city. The actual idea of the smart city is an "empty box", with a vague meaning. This vagueness furnishes energy to the actual unsustainable economic system thereby exacerbating economic inequalities and environmental crisis. All this has affected contemporary city creating, especially in the global South, critical conditions.

The proposed framework offers a clear vision of what a smart city should be, and a deep concept of sustainability founded on life value. Only this epistemological revolution can permit to our society the regeneration of the socio-economic body, founded on citizens' active participation and communities' involvement, respect of nature and life, shared values and commons. Only if we shift the paradigm, we may assure a better future, able to reinforce social inclusion, support an inclusive economy and protect the physical environment.

The implementation of the homological smart city model, based on Biourbanism principles, recognizes peer-to-peer strategies and actions as the starting point for social, cultural and economic innovations, promoting a novel vision for the future of local communities. This change of paradigm considers the human being at the center of the socio-economic system and social innovators as the key-actors able to manage and implement the change. By leveraging on local assets and human resources, they trigger a community revival process able to produce positive systemic effects on the whole urban system and on local governance inclusive models, promoting incremental regeneration processes.

Diversification is considered as a principle of life in a city, conceived as a living organism that cannot be strictly planned but whose processes should rather be guided toward new futures. New strategies to function in a complex system like a city need to be developed by experimenting with real-life situations, implementing open process and bottom-up practices to build a network of interactions between small self-organized co-working communities, taking into account spatial, cultural and economic processes. These processes were able to activate connected diversification, recognized as a systemic principle of life, suitable for the urban development context. It is able to reduce territory vulnerability to external forces that can compromise its integrity, and to identify resilience as the potential of a territorial system to remain in a particular configuration and to maintain its feedbacks and functions, involving the ability to reorganize following disturbance-driven change, and reflecting flexibility capacity to experiment and adopt innovative solutions.

Recognizing the resources value of the territories, which can improve the level of well-being of the local community, determines the need for new management processes, with the predisposition of complex design capabilities, and the identification of suitable directions and agendas, which can outline practical and human-centred implications for citizens, governments and professionals, needs a framework based on the combination and interaction of bottom-up and top-down processes.

According to the above reflections, the case study of the Metropolitan City of Naples can be considered an example to test the potential of an approach that identifies development opportunities, considering the synergistic design capacity of the various municipalities of the Coast Area. Each municipality in the coastal belt of Naples is analyzed by selecting some indicators that describe the processes and governance actions promoted by the municipal administration (top-down) and activated from below (bottom-up) through initiatives by the citizens and local associations, expression of social innovation processes. Through the tools of spatial analysis, multi-criteria and multi-group analysis, has been possible to explain the ability to activate synergies and territorial symbiosis between the different resources that characterize coastal areas.

The Statute of the Metropolitan City of Naples divides its territory into so-called "homogeneous zones" for a more suitable, balanced and functional management of local resources, with the purpose to take into account their identity characters, historical and cultural components, geomorphologic and naturalistic contexts, landscapes functional interactions and socio-economic frameworks. Indeed, the concept of "homogeneous zones" is part of the metropolitan strategic plan, and is oriented to improve the territorial productivity, services to citizens' opportunities, and cooperation among the different municipalities. The authors reflect on the need to identify some homogeneous zones able to underline common identity characters and activate a strategy of connected diversification, with a specific attention for the municipalities of the Coast Area of the Metropolitan City of Naples.

Starting from vulnerability and resilience concepts, the study has been dealt according to a multi-methodological approach, based on a Geodesign process supported by multi-criteria analysis, multi-group analysis and spatial analysis.

The elaboration of Spatial Opportunity Maps (SOMs) is the output of a multi-dimensional evaluation process that leads to the identification of specific territorial contexts with a greater propensity for some specializations relating to the economic sectors and the local transformation.

The identification of a biourban strategy, characterized by human smart spatial solutions, place-based and situated actions, for the enhancement of the Coast Area can be considered as a prerequisite for the activation of a process-oriented to the identification of "homogeneous zones", conceived not only as areas with similar characteristics but above all as territories where it is possible to promote networks of opportunities between the various municipalities and their communities. "Co-operation" is conceived a source of mutual benefit and involves a mutual convenience, based on the constant construction of ties and relationships and the interdependence determined by spatial proximity. Economic processes require cooperative-collaborative behaviors between the various components and become increasingly territorialized, and therefore more resilient and at the same time less and less associated with the production of negative environmental impacts.

The methodological process has been structured to identify significant opportunities that should be triggered by an urban regeneration process for the Coast of Naples Metropolitan City, recognizing the essential role of sea resource and synergies that could develop into a win-win-win–oriented territorial cohesion. Each of the sixteen municipalities involved could play a decisive role, enhancing their environmental, social and economic resources and collaborating in building a territorial network of resources and actors.

The proposal of a smart Biourbanism model identifies a development perspective capable of stopping the ongoing degradation processes, with the purpose of rebuilding the local economy and stimulating demand for social and cultural innovation. This model is based on the relationships and ties (between the inhabitants, between the inhabitants and the urban space, between the inhabitants and the rural area, between productive activities and economic activities, and between the urban space and the rural area) able to trigger synergies and symbiosis, to generate circular organization processes and show how to cooperate economically, socially and environmentally.

The model prioritizes the reconstruction of links at different levels, from which the density of new development activities emerges. It assumes synergistic processes, circularization, and symbiosis as tools to multiply the ties that, in turn, generate new synergies and symbiosis, triggering new circular processes. These processes shape the city and make it more attractive to locate new activities and new investments, the catalyst of circularization processes between the different forms of capital characterizing the city-territory system.

References

1. Saad-Filho A, Deborah J (2005) Introduction. In: Saad-Filho A, Deborah J (eds) Neoliberalism—a critical reader. Pluto Press, London
2. Abbey R (2005) Is liberalism now an essentially contested concept? New Polit Sci 27:461–480
3. Campbell JL, Pedersen OK (eds) (2001) The rise of neoliberalism and institutional analysis. Princeton University Press, Princeton, New Jersey
4. Chomsky N (1999) Profit over people—neoliberalism and global order. Seven Stories Press, New York
5. Harvey D (2005) A brief history of neoliberalism. Oxford University Press, Oxford
6. Bourdieu P (1998) Contre-feux: Propos pour servir à la résistance contre l'invasion néo-libérale. Éditions Liber, Paris
7. Bourdieu P (1998a) L essence du néolibéralisme". Le Monde diplomatique Mars 1998. Available at: http://www.monde-diplomatique.fr/1998/03/BOURDIEU/10167
8. Bourdieu P (2001) Contre-feux 2: Pour un mouvement social européen. Éditions Raisons d Agir, Paris
9. Touraine A (2001) Beyond neoliberalism. Polity, Cambridge

10. Bernard W, Gisela N (eds) (2006) Neoliberal hegemony—a global critique. Routledge, London
11. Chomsky N (2009) Understanding the crisis. Markets, the State and Hypocrisy. In: Focus. Available at: https://chomsky.info/20090210/, 10 Feb 2009
12. Marx K (1970) Capital, vol 1. Lawrence & Wishart, London
13. Marx K (1970) Capital, vol 2. Lawrence & Wishart, London
14. Foster JB, Magdoff F (2009) The great financial crisis. Monthly Review Press, New York, pp 91–93
15. Lawrence A (1999) The skyscraper index: faulty towers. Property Report. Dresdner Kleinwort Waserstein Research
16. Carol W (1999) Form follows finance, skyscrapers and skylines in New York and Chicago. Princeton Architectural
17. Barr J, Mizrach B, Mundra K (2014) Skyscraper height and the business cycle: separating myth from reality. Rutgers University
18. Geier J (2008) Capitalism's worst crisis since the 1930s. The international socialist review 62. Available at: http://isreview.org/issue/62/capitalisms-worst-crisis-1930s
19. Hope W (2010) Time,communication, and financial collapse. Int J Commun 4:649–669
20. Adam B (2004) Time. Polity, Cambridge, UK
21. Singh K (2000) Taming global financial flows. Zed Books, London
22. Katz C (2009) The crisis of capitalism. Beyond regulation and greed. International socielist review, 65. Available at: http://isreview.org/issue/65/crisis-capitalism
23. Mészáros I (2014) Structural crisis needs structural change. Available at: https://systemicalternatives.org/2014/03/03/structural-crisis-needs-structural-change/
24. Cairó-i-Céspedes G, Castells-Quintana D (2016) Dimensions of the current systemic crisis: capitalism in short circuit? Progress in Development Studies 16(1) pp 1–23. Available at: http://journals.sagepub.com/doi/pdf/10.1177/1464993415608067
25. World Economic Forum (2017) Global risks report 2016. Available at: http://www.oliverwyman.com/content/dam/oliver-wyman/global/en/2016/jan/Global_risks_report_2016.PDF
26. WWF (2016) Living planet report 2016. WWF International, Gland, Switzerland. Available at: http://wwf.panda.org/about_our_earth/all_publications/living_planet_report/
27. Heidegger M (1990) Introduzione alla Metafisica. Mursia Publishing
28. Adorno TW, Horkheimer M (2010) Dialettica dell'Illuminismo. Einaudi, p 3
29. Marcuse E (1998) Some social implications of modern technology. In: Kellner D (ed) Technology, war and fascism: collected papers of Herbert Marcuse, vol 1. Routledge, London and New York, p 49
30. Working Group 'societal relationships with nature' (GesNat) of the Federal Coordination of Internationalismus (BundeskoordinationInternationalismus – BUKO) After the failure of the Green Economy. 10 theses of a critique of the green economy. Available at: http://social-ecology.org/wp/wp-content/uploads/2012/05/Theses-on-Green-Econ.pdf
31. Latouche S (2003) Sustainable development as a Paradox. Available at: http://rsesymposia.org/themedia/File/1151679499-Plenary2_Latouche.pdf
32. Das Prinzip Verantwortung HJ (1979) Versuch einer Ethik für die technologische Zivilisation. Frankfurt
33. Cambridge Dictionary. Dictionary. http://dictionary.cambridge.org/dictionary/english/development
34. Capra F, Jakobsen OD (2017) A conceptual framework for ecological economics based on systemic principles of life. Int J Soc Econ 44(6):831–844
35. Mac Intyre A (2007) After virtue: a study in moral theory. University of Notre Dame Press
36. Eraclitus. Fragments (2003) Penguin classics
37. Aristotle's "Politics" (2013) University of Chicago Press, 2nd edn
38. Cendejas JL (2017) Economics, chrematistics, oikos and polis in Aristotle and St. Thomas Aquinas. The Journal of Philosophical Economics. X pp 5–46

39. Leshem D (2016) What did the ancient Greeks mean by oikonomia? J Econ Perspect 30 (1):225–231
40. Tommaso d'Aquino. Summa Theologica. ESD-Edizioni Studio Domenicano
41. Encinar MI, Cendejas JL, Muñoz FF (2006) On the relationship between ethics and economics. Cuadernos de Economía 29(79):93–118
42. Dierksmeier C, Pirson M (2009) Oikonomia versus Chrematistike: learning from Aristotle about the future orientation of business management. In: Journal of Business Ethics 88(3), 15th IESE international symposium on ethics, business and society business and management: towards more humanistic models and practices, pp 417–430
43. Lovelock J (1988) The ages of Gaia. Norton, New York, NY
44. Ingebrigtsen S, Jakobsen O (2007) Circulation economics. Theory and Practice, Peter Lang, Oxford
45. Caperna A, Serafini S (2013) Biourbanism as new framework for smart cities study. In: Vinod K (ed) Geographic information system for smart cities. Copal Publishing, Ghaziabad, India
46. Hollands RG (2008) Will the real smart city please stand up? In: City: analysis of urban trends, culture, theory, policy, action, vol 12(3), pp 303–320
47. Eger J (1997) Cyberspace and cyberplace: building the smart communities of tomorrow. San Diego Union-Tribune
48. Coe A, Paquet G, Roy J (2001) E-goverance and smart communities: a social learning challenge. Soc Sci Comput Rev 19(80):80–93
49. Thorns D (2002) The transformation of cities: urban theory and urban life. Palgrave, Basingstoke
50. Caragliu A, Del Bo C, Nijkamp P (2009) Smart cities in Europe. Serie Research Memoranda 0048. University Amsterdam
51. Batty M et al (2012) Smart cities of the future. Centre for advanced spatial analysis University College London, Paper, p 188
52. Graham S, Marvin S (2001) Telecommunications and the city: electronic spaces, urban place. Routledge, London
53. Castells M (1996) The rise of the network society. Blackwell, Oxford
54. Ishido T (2002) Digital city Kyoto. Commun ACM 45(7):78–81
55. Komninos N (2002) Intelligent cities: innovation knowledge systems and digital spaces. Spon Press, London
56. Caperna et al (2011) Biourbanism manifesto. Source: http://www.biourbanism.org/biourbanism-definition/
57. Serafini S (2011) Liberazione partecipata dello spazio dall'iperreale. L'Italia come esperimento biourbanistico. Abitare l'Italia. Territori, economie, diseguaglianze, XIV Conferenza SIU, 24/25/26 marzo 2011. Planum, The journal of Urbanism
58. Marx C (1990) Results of immediate process of production. In: Capital vol 1. A critique of political economy. Penguin Classics, London pp 943–1084
59. Debord G (1970) The society of the spectacle. Black & Red Ed
60. Augé M (1995) Non-places introduction to an anthropology of supermodernity. Verso, London
61. Benjamin Widom (2009) Scholarpedia 4(10):9054 Available at: http://www.scholarpedia.org/article/Scaling_laws
62. Sornette D (2006) Critical phenomena in natural sciences: chaos, fractals, self-organization and disorder: concepts and tools. Springer Series in Synergetics, 2nd edn. Springer, Heidelberg
63. Bak P (1997) How nature works. Oxford University Press, Oxford
64. Newman MEJ (2005) Power laws, Pareto distributions and Zipf's law. Contemp Phys 46 (5):323–351
65. Gabaix X (1999) Zipf's law for cities: an explanation. Q J Econ 114(3):739–767
66. Cristelli M, Batty M, Pietronero L (2012) There is more than a power law in Zipf. Sci Rep 2:812

67. Salingaros NA (2010) Twelve lectures on architecture. Algorithmic Sustainable Design, Umbau, Solingen
68. Bejan A, Lorente S (2010) The constructal law of design and evolution in nature. Philos Trans Royal Soc B 365:1335–1347
69. Caperna A, Serafini S (2013) Biourbanistica come nuovo approccio epistemologico. In: Mortola E (ed) Partecipazione e ICT. Gangemi Editore, Rome
70. Serafini S (2011) P2P (peer to peer) Urbanism and Biourbanism. МЕЖДУНАРОДНАЯ НАУЧНАЯШКОЛА"Перспективные направления физико-химической биологии и биотехнологии" (тезисы докладов)». TomskStateUniversity, Tomsk, pp 11–15
71. Caperna A, Mehaffy M, Mehta G, Mena-Quintero F, Rizzo A, Salingaros NA, Serafini S, Strano E (2010) A definition of P2P (peer-to-peer) urbanism, AboutUsWiki, the P2P foundation, DorfWiki, peer to peer urbanism. Presented by Salingaros N at the international commons conference. Heinrich Böll Foundation, Berlin, Available at: http://wiki.p2pfoundation.net/images/P2PUrbanism_definition.pdf
72. Perri S (2013) Ascesa e caduta del modello economico italiano. Available at: http://www.economiaepolitica.it/primo-piano/ascesa-e-caduta-del-modello-economico-italiano. Accessed 12 Aug 2017, Accessed 15 Sept 2017
73. Babones S (2007) Studying globalization: methodological issues. In: Ritzer J (ed) The blackwell companion to globalization. Wiley, New York, p 146
74. ISTAT—Istituto Nazionale di Statistica (2015) I. Stat database. http://dati.istat.it. Accessed 12 Aug 2017
75. ISTAT—Istituto Nazionale di Statistica (2017) Occupati e disoccupati Giugno 2017. Available at: http://www.istat.it/it/files/2017/07/CS_Occupati-e-disoccupati_giugno_2017.pdf?title=Occupati+e+disoccupati+%28mensili%29+-+31%2Flug%2F2017+-+Testo+integrale+e+nota+mettodologica.pdf. Accessed 12 Aug 2017
76. European Commission (2017) Employment and social developments in Europe. Annual review 2017. Available at: http://ec.europa.eu/social/main.jsp?catId=738&langId=it&pubId=8030&furtherPubs=yes. Accessed 21 Aug 2017
77. Galli G (2015) Il golpe invisibile. Kaos, Milano
78. De Fiores C (2007) Rappresentanza politica e sistemi elettorali in Italia. Available at: http://www.costituzionalismo.it/articoli/253/. Accessed 18 Oct 2017
79. Schmit H (2017) La garanzia dei diritti elettorali fondamentali. Available at: https://www.academia.edu/33331644/La_garanzia_dei_diritti_elettorali_fondamentali. Accessed 1 Oct 2017, Accessed 19 Oct 2017
80. Scandone R (ed) (2015) Le catastrofi naturali in Italia. Scienze e Ricerche 10, supplemento
81. Aon Benfield (2017) 2016 Annual global climate and catastrophe report. Available at: http://thoughtleadership.aonbenfield.com/Documents/20170117-ab-if-annual-climate-catastrophe-report.pdf. Accessed 19 Oct 2017
82. ISPRA—Istituto Superiore per la Protezione e la Ricerca Ambientale (2015) Dissesto idrogeologico in Italia: pericolosità e indicatori di rischio. Available at: http://www.isprambiente.gov.it/it/pubblicazioni/rapporti/dissesto-idrogeologico-in-italia-pericolosita-e-indicatori-di-rischio-rapporto-2015. Accessed 17 Oct 2017
83. ISPRA—Istituto Superiore per la Protezione e la Ricerca Ambientale (2017) Consumo di suolo, dinamiche territoriali e servizi ecosistemici. Edizione 2017. Available at: http://www.isprambiente.gov.it/it/pubblicazioni/rapporti/consumo-di-suolo-dinamiche-territoriali-e-servizi-ecosistemici. Accessed 21 Sept 2017
84. Archibugi D, Filippetti A (2012) Innovation and economic crisis: lessons and prospects from the economic downturn. Routledge, London
85. Kania J, Kramer M (2013) Embracing Emergence: how collective impact addresses complexity. Stanford Social Innovation Review 2013: 1–7. Available at: https://ssir.org/pdf/Embracing_Emergence_PDF.pdf
86. Smith A, Fressoli M, Abrol D, Around E, Ely A (2017) Grassroots innovation movements. Routledge, New York, NY
87. Consiglio S, Riitano A (2015) Sud innovation. Franco Angeli, Milano

88. Caroli MG (ed) (2016) L'innovazione delle imprese leader per creare valore sociale. Terzo rapporto CERIIS sull'innovazione sociale. Franco Angeli Open access, Milano. Available at: http://ojs.francoangeli.it/_omp/index.php/oa/catalog/book/220
89. Moulaert F (ed) (2015) The international handbook on social innovation: collective action, social learning and transdisciplinary research. E Elgar, Cheltenham (UK)
90. European Commission (2010). This is European social innovation. Available at: http://ec.europa.eu/DocsRoom/documents/19042. Accessed 15 Oct 2017
91. Murray R, Calulier-Grice J, Mulgan G (2010) The open book of social innovation. National endowment for science, technology and the art, London
92. Phills A, Deiglmeier K, Miller D (2008) Rediscovering social innovation. Stanford Soc Innov Rev 6(4):34–43
93. Moro G (2015) La cittadinanza attiva: nascita e sviluppo di una anomalia. In: Salvati M, Sciolla L (eds) L'Italia e le sue regioni. Istituto della Enciclopedia Italiana, pp 55–77. Available at: http://www.treccani.it/enciclopedia/la-cittadinanza-attiva-nascita-e-sviluppo-di-un-anomalia_%28L%27Italia-e-le-sue-Regioni%29/
94. Costituzione della Repubblica Italiana. Parte II, Titolo V, articolo 118. Available at: https://www.senato.it/1025?sezione=136&articolo_numero_articolo=118. Accessed 15 Oct 2017
95. UNPLI—Unione Nazionale Pro Loco d'Italia. http://www.unioneproloco.it. Accessed 22 Oct 2017
96. ISTAT—Istituto Nazionale di Statistica (2012) 9° Censimento industria, servizi 2011. La rilevazione sulle istituzioni non profit: un settore in crescita. Available at: http://www.istat.it/files/2013/07/05-Scheda-Non-Profit_DEF.pdf. Accessed 23 Aug 2017
97. OECD (2011) Fostering innovation to address social challenges. Workshop proceedings. Available at: https://www.oecd.org/sti/inno/47861327.pdf. Accessed 28 Aug 2017
98. Kretzmann JP, McKnight J (1993) Building communities from the inside out: a path toward finding and mobilizing a community's assets. Center for urban affairs and policy research, neighborhood innovations network, Northwestern University, Evanston, Ill
99. Italia che Cambia. http://www.italiachecambia.org. Accessed 20 Oct 2017
100. European Commission (2014) Regional eligibility for structural funds, NUTS 2 regions, 2014–20. Available at: http://ec.europa.eu/eurostat/statistics-explained/images/8/84/Regional_eligibility_for_structural_funds%2C_by_NUTS_level_2_region%2C_2014%E2%80%9320_%28%C2%B9%29_%28%25_of_EU-27_average%29_RYB15.png. Accessed 29 Aug 2017
101. SVIMEZ—Associazione per lo sviluppo dell'industria nel Mezzogiorno (2017) Anticipazioni dei principali andamenti economici e sociali dal "Rapporto SVIMEZ 2017 sull'economia del Mezzogiorno". Available at: http://www.svimez.info/images/RAPPORTO/materiali2017/2017_07_28_anticipazioni_testo.pdf. Accessed 20 Oct 2017
102. Marfé L, Schibotto E (2012) Il divario: Globalizzazione, emigrazione e Sud. Edizioni Nuova Cultura, Roma
103. Corrado A, Colloca C (2015) La globalizzazione delle campagne: Migranti e società rurali nel Sud Italia. Franco Angeli, Milano
104. Arminio F (2012) Terracarne: Viaggio nei paesi invisibili e nei paesi giganti del Sud Italia. Mondadori, Milano
105. Anci IFEL (2015) Atlante dei piccoli comuni. Available at: http://www.fondazioneifel.it/documenti-e-pubblicazioni/item/4635-atlante-dei-piccoli-comuni-2015. Accessed 29 Aug 2017
106. Legambiente (2016) Piccolo (e fuori dal) comune. Cosa sta cambiando nell'Italia dei piccoli comuni. Available at: https://www.legambiente.it/sites/default/files/docs/dossier_piccoli_e_fuori_dal_comune_piccolicomuni2016.pdf. Accessed 11 Sept 2017
107. LongoTherapy. https://www.facebook.com/longotherapy.2014. Accessed 22 Aug 2017
108. Cooperativa Valle dei Cavalieri. http://www.valledeicavalieri.it. Accessed 25 Oct 2017
109. Grella D. (2016) Succiso, il paese-cooperativa dove ogni giorno si cambia lavoro. Available at: http://www.vita.it/it/article/2016/05/23/succiso-il-paese-cooperativa-dove-ogni-giorno-si-cambia-lavoro/139495. Accessed 25 Oct 2017

110. Belmonte in Rete. http://www.belmonteinrete.flazio.com. Accessed 22 Aug 2017
111. Farm Cultural Park. http://www.farmculturalpark.com. Accessed 22 Aug 2017
112. Inbianchini di Bellezza. http://www.imbianchinidibellezza.it. Accessed 20 Oct 2017
113. Borgo Vecchio Factory. https://www.facebook.com/borgovecchiofactory. Accessed 25 Oct 2017
114. Borgo Vecchio Factory. Indiegogo. https://www.indiegogo.com/projects/borgo-vecchio-factory-education#/. Accessed 25 Oct 2017
115. Gazzetta Ufficiale della Repubblica Italiana—Serie Generale n.81 del 7–4-2014. Legge 7 Aprile 2014, n. 56. Disposizioni sulle citta' metropolitane, sulle province, sulle unioni e fusioni di comuni
116. Wilson EO (1984) Biophilia. Harvard University Press, Cambridge
117. Norberg-Schulz C (1979) Genius loci: paesaggio, ambiente, architettura. Electa, Milano
118. Bevilacqua F (2010) Genius loci. Il dio dei luoghi perduti. Rubettino, Soveria Mannelli
119. Jacurso da Vivere e Imparare. http://jacursodavivereeimparare.it. Accessed 25 Oct 2017
120. Tucci F (2016) Soft revolution. L'Erudita, Roma
121. Levanti G (2010) Il governo dei sistemi reticolari di imprese. Franco Angeli, Milano
122. Pensando Meridiano Association. https://www.pensandomeridiano.com. Accessed 25 Oct 2017
123. Progetto Artena. http://progettoartena.com. Accessed 26 Oct 2017
124. CGI Mestre (2017) Le inefficienze dellaPAa frenano la ripresa. Available at: http://www.cgiamestre.com/articoli/23933. Accessed 23 Oct 2017
125. Iperbole2020. http://iperbole2020.comune.bologna.it. Accessed 23 Oct 2017
126. Comune di Bologna (2014) Regulation on collaboration between citizens and the city for the care and regeneration of urban commons. Available at: http://www.comune.bologna.it/media/files/bolognaregulation.pdf. Accessed 30 Aug 2017, Accessed 23 Oct 2017
127. Iperurbano. https://www.facebook.com/iperurbano. Accessed 22 Aug 2017
128. Esperimenti Architettonici. http://www.esperimentiarchitettonici.it. Accessed 22 Oct 2017
129. Il cuore di Altamura. https://www.facebook.com/Il-Cuore-Di-Altamura-523721281099589. Accessed 22 Aug 2017
130. Associazione Turistica Pro Loco Altamura. http://prolocoaltamura.it. Accessed 22 Aug 2017
131. Settis S (2010) Paesaggio Costituzione cemento: La battaglia per l'ambiente contro il degrado civile. Einaudi, Torino
132. Minervino MF (2010) Statale 18. Fandango libri, Roma
133. Lauria M (ed) (2009) Che fine hanno fatto i Centri Storici minori? Atti del seminario di studi. Centro Stampa di Ateneo, Reggio Calabria
134. Agenzia del Demanio. http://dati.agenziademanio.it. Accessed 25 Aug 2017
135. Il Fatto Quotidiano (2016) Vendita immobili pubblici, quegli incassi che cambiano di anno in anno. Ma il Tesoro nega che ci siano partite di giro. http://www.ilfattoquotidiano.it/2016/05/07/vendita-immobili-pubblici-quegli-incassi-che-cambiano-di-anno-in-anno-ma-il-tesoro-nega-che-ci-siano-partite-di-giro/2701474/. Accessed 22 Aug 2017
136. Mangialardo A, Micelli E (2017) From sources of financial value to commons: Emerging policies for enhancing public real-estate assets in Italy. Papers in Regional Science
137. Lerner J (2003) Acupuntura urbana. Record, Rio de Janeiro
138. O'Connor J, Bensky D (eds) (1981) Acupuncture: a comprehensive text. Eastland Press, Chicago, Ill
139. Casagrande M (2013) Biourban acupuncture. Treasure hill of Taipei to Artena. International Society of Biourbanism, Rome
140. Comitato di Quartiere Centro Storico Artena. https://www.facebook.com/centro.storico.artena. Accessed 24 Oct 2017
141. Italian Stories. https://www.italianstories.it. Accessed 25 Oct 2017
142. South Cultural Routes. https://www.facebook.com/SulleStradeDelSud/. Accessed 25 Oct 2017
143. Massimo Castelli. http://www.massimocastelli.eu. Accessed 25 Oct 2017
144. Casa Netural. http://www.benetural.com. Accessed 25 Oct 2017

145. GiovanIrpinia. https://www.facebook.com/GiovanIrpinia. Accessed 26 Oct 2017
146. Signorelli A (2016) La vita al tempo della crisi. Einaudi
147. Bartolini S (2013) Manifesto per la felicità. Feltrinelli, Roma
148. Gage T (2017) Fully alive: using the lessons of the Amazon to live your mission in business and life. Atria Books, New York, NY
149. Ariès P (1980) L'uomo e la morte dal Medioevo a oggi. Laterza, Bari
150. Platone (428–347 a.C). Cambiano G, Adorno F (eds) (1995–2008) Dialoghi. UTET, Torino
151. Dolci D (1996) La struttura maieutica e l'evolverci. La Nuova Italia, Firenze
152. Interazioni Creative. https://www.interazionicreative.com. Accessed 26 Oct 2017
153. Sassen S (1988) Globalization and its discontents: essays on the new mobility of people and money. The New Press, New York, NY
154. Braga JC (2009) The systemic crisis of financial capitalism and the uncertainty of change. Estudos Avançados 23(65):89–102
155. Egan MP, Nugent N, Paterson WE (eds) (2010) Research agendas in EU studies: stalking the elephant. Palgrave Macmillan, Basingstoke
156. Nesi E (2012) Storia della mia gente. Bompiani, Milano
157. Joji Sakurai—YaleGlobal Online (2014) Italy and Japan: Troubled Twins of Globalization. Available at: http://yaleglobal.yale.edu/content/italy-and-japan-troubled-twins-globalization. Accessed 24 Oct 2017
158. Garin E (1967) La cultura del Rinascimento. Laterza, Bari
159. Brera GM, Nesi E (2017) Tutto è in frantumi e danza. La nave di Teseo, Milano
160. Orsi W, Ciarrocchi RA, Lupi G (2009) Qualità della vita e innovazione sociale: Un'alleanza per uscire dalla crisi. F. Angeli, Milano
161. Carmona M (2009) Design coding and the creative. market and regulatory tyrannies of practice. Urban Stud 46(12):2643–2667
162. McGlynn S (1993) Reviewing the rhetoric. In: Hayward R, McGlynn S (eds) Making better places: urban design now. Architectural Press, Oxford
163. Alexander C, Neis HJ, Moore M (2012) The battle for the life and beauty of the earth: a struggle between two world-systems. University Press, Oxford
164. Salingaros NA (2006) A theory of architecture. Umbau-Verlag, Solingen
165. Serafini S (2016) Editor's note: the silence of design. J Biourbanism 1&2:9–23
166. Al-Sabouni M (2016) Architecture with identity crisis: the lost heritage of the middle east. J Biourbanism 1&2:81–98
167. Osnowitz D (2010) Freelancing expertise: contract professionals in the new economy. Cornell University Press
168. Vosko LF, Zukewich N, Cranford S (2003) Precarious jobs: a new typology of employment. Perspectives on Labour and Income 15(4)
169. Gandini A (2015) The rise of coworking spaces: a literature review. Ephemera 15(1): 193–205
170. Houni P, Ansio H (eds) (2015) Duuniakimpassa: YhteisöllisettyötilatHelsingissä (Working together: Co-working communities in Helsinki). Helsinginkaupungintietokeskus, Helsinki
171. Alatalo E, Jokinen A (2016) Connecting the spaces of co-work: joy of expedition in a growing trajectory. J Biourbanism 1&2:205–221
172. Kemmis S, McTaggart R, Nixon R (2014) The action research planner: doing critical participatory action research. Springer, Singapore
173. Moriset B (2014) Building new places of the creative economy. The rise of coworking spaces. In: Proceedings of the 2nd geography of innovation international conference 2014, Utrecht University, Utrecht
174. Florida R (2002) The rise of the creative class: and how it's transforming work, leisure, community and everyday life. Basic Books, Cambridge
175. Spinuzzi C (2012) Working alone, together: coworking as emergent collaborative activity. J Bus Techn Commun 26(4):399–441
176. Gould SJ, Eldredge N (1977) Punctuated equilibria: the tempo and mode of evolution reconsidered. Paleobiology 3(2):115–151

177. James M (2001) The speciation revolution. J Evol Biol 14:887–888
178. Shane DG (2006) Recombinant urbanism: conceptual modeling in architecture, urban design and city theory. Wiley, Oxford
179. Caragliu A, Del Bo C, Nijkamp P (2011) Smart cities in Europe. J Urban Technol 18(2): 65–82
180. Roller L-H, Waverman L (2001) Telecommunication infrastructure and economic development: a simultaneous approach. Am Econ Rev 91(4):909–923
181. Berry CR, Glaeser EL (2005) The divergence of human capital levels across cities. Pap Reg Sci 84(3):407–444
182. Le Galès P (2006) Le città europee. Società urbane, globalizzazione, governo locale. Il Mulino, Bologna, Italy
183. Batty M, Axhausen KW, Giannotti F, Pozdnoukhov A, Bazzani A, Wachowicz M, Ouzounis G, Portugali Y (2012) Smart cities of the future. Eur Phys J Spec Top 214:481–518
184. Dioguardi G (2015) Città-impresa per il governo della complessità urbana. Domus, Supplemento Smart City, 14–17
185. Dubey SK, Sharma D (2017) An overview of sustainable dimensions and indicators for smart city. In: Singh R, Kumar S (eds) Green technologies and environmental sustainability. Springer, Dordrecht–Heidelberg–New York
186. Lacinák M, Ristvej J (2017) Smart city, safety and security. TRANSCOM 2017. In: International scientific conference on sustainable, modern and safe transport, procedia engineering 192:522–527
187. Vinod Kumar TM, Dahiya B (2017) Smart economy in smart cities. Smart economy in smart cities. International collaborative research. Springer, Singapore, pp 3–76
188. Concilio G, Marsh J, Molinari F, Rizzo F (2016) Human smart cities: a new vision for redesigning urban community and citizen's life. In: Skulimowski AMJ, Kacprzyk J (eds) Knowledge, information and creativity support systems: recent trends, advances and solutions. Springer, Dordrecht–Heidelberg–New York, pp 269–278
189. Concilio G, Rizzo F (2016) Human smart cities: rethinking the interplay between design and planning. Springer, Dordrecht–Heidelberg–New York
190. Hart K, Laville JL, Cattani AD (eds) (2010) The human economy. Polity Press, Cambridge and Massachusetts, USA
191. Schuler D (2016) Smart cities + smart citizens = civic intelligence? In: Concilio G, Rizzo F (eds) Human smart cities. Rethinking the interplay between design and planning. Springer, Dordrecht–Heidelberg–New York, pp 41–60
192. Miller F, Osbahr H, Boyd E, Thomalla F, Bharwani S, Ziervogel G, Walker B, Birkmann J, Van der Leeuw S, Rockström J, Hinkel J, Downing T, Folke C, Nelson D (2010) Resilience and vulnerability: complementary or conflicting concepts? Ecol Soc 15(3):11
193. Walker B, Carpenter S, Anderies J, Abel N, Cumming G, Janssen M, Lebel L, Norberg J, Peterson G, Pritchard R (2002) Resilience management in social-ecological systems: a working hypothesis for a participatory approach. Conserv Ecol 6(1):14. http://www.consecol.org/vol6/iss1/art14/
194. Folke C (2003) Social-ecological resilience and behavioral responses. In: Biel A, Hansson B, Mårtensson M (eds) Individual and structural determinants of environmental practice. Ashgate, Aldershot, UK, pp 226–242
195. Resilience Alliance (2010) Assessing resilience in social-ecological systems: workbook for practitioners (Version 2.0). http://www.resalliance.org/3871.php
196. Walker B, Abel N, O'Connell D, Grigg N (2016) A resilience approach to conservation and development. Working Paper Version 1.5. CSIRO, Australia, https://www.resalliance.org/discussion-papers
197. Caperna A, Mehaffy M, Mehta G, Mena-Quintero F, Rizzo A, Salingaros NA, Serafini S, Strano E (2010) A definition of P2P (peer-to-peer) urbanism. AboutUsWiki, the P2P foundation, DorfWiki, peer to peer urbanism. Presented by Salingaros N at the international commons conference, Heinrich Böll Foundation, Berlin, 1 Nov 2010. http://www.biourbanism.org/definition-of-biourbanism/

198. Fusco Girard L, Cerreta M, De Toro P (2015) La risorsa mare per il territorio. Prospettive di sviluppo per la fascia costiera della provincia di Napoli. Giannini, Napoli, Italy
199. Fusco Girard L, Cerreta M, De Toro P (2017) Towards a local comprehensive productive development strategy: a methodological proposal for the metropolitan city of Naples. Qual Innov Prosperity/Kvalita Inovácia Prosperita 21(1):223–239
200. Vetritto G (ed) (2017) I dossier delle Città Metropolitane. Città metropolitana di Napoli. Dipartimento per gli Affari Regionali e le Autonomie Presidenza del Consiglio dei Ministri, DARA, Roma, Italy
201. De Matteis G (ed) (2011) Le grandi città italiane. Società e territori da ricomporre, Marsilio, Venezia, Italy
202. Staricco L (2016) Secondo Rapporto «Giorgio Rota» su Napoli. SRM-Centro Einaudi, Napoli, Italy
203. European Comission (2017) The cultural and creative cities monitor. Joint Research Centre, Varese, Italy
204. Sassen S (1994) Le città nell'economia globale. Il Mulino, Bologna, Italy
205. Bruttomesso R (1993) Waterfront e trasporto urbano d'acqua. Kineo 3:8–11
206. Bruttomesso R (2007) Nuovi scenari urbani per le città d'acqua. Percorsi d'acqua 2006–2007, Lecture, Milano
207. Moretti M (2008) Cities on water and waterfront regeneration: a strategic challenge for the future, Grundtvig, II meeting rivers of change-river//Cities Warsaw, Poland, 24–27 July 2008, http://www.river-cities.nazwa.pl/www/download/m.moretti_warsaw2008.pdf
208. Hou D (2009) Urban waterfront landscape planning. Master's thesis, Blekinge Institute of Technology Karlskrona, Sweden
209. Hein C (ed) (2011) Port cities: dynamic landscapes and global networks. Routledge, London, UK
210. Timur UP (2013) Urban waterfront regenerations. In: Rijeka ÖM (ed) Advances in landscape architecture. InTech, Rijeka, Croatia, pp 169–206
211. Steinitz CA (2012) Framework for geodesign: changing geography by design. ESRI Press, Redlands, CA
212. Campagna M (2016) Metaplanning: about designing the geodesign process. Landscape Urban Plann 156:118–128
213. Munda G (1995) Multicriteria evaluation in a fuzzy environment. Physica-Verlag, Heidelberg, Germany
214. Munda G (2008) Social multi-criteria evaluation for a sustainable economy. Springer, Heidelberg, New York
215. Cerreta M (2010) Thinking through complex values. In: Cerreta M, Concilio G, Monno V (eds) Making strategies in spatial planning, knowledge and values. Springer, Dordrecht–Heidelberg–New York, pp 381–404
216. Audis (Associazione Aree Urbane Dismesse) (2014) Documento di proposte per la rigenerazione urbana. http://www.reagaa.eu/images/reag/news/AUDIS_Documento_per_la_rigenerazione_urbana_16_maggio2014__02931.pdf
217. OECD (2008) Handbook on constructing composite indicators. OECD Publications, Paris, France
218. Hansen LJ, Biringer JL, Hoffman JR (eds) (2003) Buying time: a user's manual for building resistance and resilience to climate change in natural systems. WWF Climate Change Program, Berlin, Germany
219. Cutter SL, Finch C (2007) Temporal and spatial changes in social vulnerability to natural hazards. Proc Natl Acad Sci 105(7):2301–2306
220. Naudé W, McGillivray M, Stephanié R (2008) Measuring the vulnerability of subnational regions. Research paper n. 2008/54, UNU-WIDER World Development Economic Research Institute
221. Turvey R (2007) Vulnerability assessment of developing countries: the case of small-island developing states. Dev Policy Rev 25(2):243–264

222. Graziano P (2012) Rischio, vulnerabilità e resilienza territoriale. Serie Rossa, Economia—Quaderno N. 87 novembre 2012. Università Cattolica del Sacro Cuore, Piacenza, Italy
223. Roy B, Bouyssou D (1993) Aide Multicritère à la Decision: Méthodes et Cas. Economica, Paris, France
224. Cerreta M, De Toro P (2012) Integrated spatial assessment (ISA): a multi-methodological approach for planning choices. In: Burian J (ed) Advances in spatial planning. InTech, Rijeka, Croatia, pp 77–108
225. Costanza R, Folke C (1997) Valuing ecosystem services with efficiency, Fairness and Sustainability as Goals. Island Press, Washington, D.C., USA
226. Beinat E, Nijkamp P (eds) (1998) Multicriteria analysis for land-use management. Springer, Heidelberg, New York
227. de Bruijn H, Ten Heuvelhof E (2008) Management in networks: on multi-actor decision making. Routledge, London and New York
228. Fusco Girard L (2010) Sustainability, creativity, resilience: towards new development strategies of port areas through evaluation processes. Int J Sustain Dev 13:161–184
229. Fusco Girard L (2013) Creative cities: the challenge of 'humanization' in the city development. BDC 13(1):9–33
230. Ravetz J, Fusco Girard L, Bornstein L (2012) A research and policy development agenda: fostering creative equitable and sustainable port cities. BDC 12:67–69
231. Porter ME, Kramer MR (2011) Creating shared value. Harvard Business Review. https://hbr.org/2011/01/the-big-idea-creating-shared-value
232. Calafati A (2009) Economie in cerca di città. La questione urbana in Italia, Donzelli, Roma, Italy
233. Girardet H (2010) Regenerative cities, world future council and hafencity university. Hamburg, Germany
234. Micelli E (2016) Restituire alla comunità il public value delle scelte urbanistiche: Tassazione vs strumenti di partenariato. Scienze Regionali 3(3):105–113
235. Barreca A, Curto R, Rolando D (2017) Assessing social and territorial vulnerability on real estate submarkets. Buildings 7(94):1–17
236. Torre CM, Morano P, Tajani F (2017) Saving soil for sustainable land use. Sustainability 9 (350):1–35
237. Antonelli Abella T (2013) Follow the Rs: reduce, replace, reuse, recycle, recover, refuse and reject, rethink. Envirocities eMagazine 4:8–10

Part XIII
Kenya, Nairobi

Chapter 13
Achieving Regional Development Through Enhanced Connectivity in the Nairobi Metropolitan Region

Daniel Githira, Romanus Opiyo and Dennis Mwaniki

Abstract Reliable transport and ICT connectivity are frequently cited as primary factors affecting regional development patterns. This is informed by the fact that such connectivity has great potential to enhance concentration (agglomeration) of economic growth. Transport and ICT connectivity impact on the productive sector through the flow of products and labour markets in a smart manner. This chapter examines the potential of enhanced transport and ICT connectivity in driving smart development in Nairobi Metropolitan Region (NMR), as well as the operational regulatory and institutional framework that is currently facilitating (and/or creating positive prospects for) the region's smart growth. It details the historical challenges associated with connectivity in the NMR, plots the region's economic nodes and their spatial connectivity, and analyses elements of transportation and ICT connectivity in the region. The chapter concludes by highlighting key economic and governance factors which are critical in maximizing the two connectivity parameters (transport and ICT) in achieving smart NMR development.

Keywords Nairobi metropolitan region · Connectivity · Smart region

13.1 Introduction

Smart regional development focuses on infusing intelligence into the management of regions to improve their functionality while reducing costs [1]. For a region to qualify as experiencing smart development, it must have ICT-enabled governance,

D. Githira
The County Government of Taita Taveta, Nairobi, Kenya
e-mail: dangithira@gmail.com

R. Opiyo (✉)
University of Nairobi, Nairobi, Kenya
e-mail: romanop2000@gmail.com

D. Mwaniki
Freelance Consultant and Researcher, Nairobi, Kenya
e-mail: dennmwa02@gmail.com

T. M. Vinod Kumar (ed.), *Smart Metropolitan Regional Development*, Advances in 21st Century Human Settlements,
https://doi.org/10.1007/978-981-10-8588-8_13

a desired mix of land use, financial sustainability, enhanced mobility, efficient utilities, citizen involvement in governance and meaningful private-public partnerships [1, 2]. Kumar [3] defines a smart city or region as having the closely-interconnected building blocks of smart economy, mobility, environment, living, and governance. Whereas these aspects are difficult to measure in absolute terms, various direct and proxy indicators can be applied to appraise the 'smartness' of a city or region.

This chapter focuses on enhancing economic development for Nairobi Metropolitan Region (NMR) though enriching its connectivity. Accordingly, it places emphasis on smart economy and smart mobility for NMR. Based on the features of smart economy [3], the chapter explores how the NMR is interacting with innovative and creative connectivity, offering entrepreneurial leadership and opportunities, tackles economic challenges of globalization, promotes sharing economy, supports tourism, excels in productivity, offers flexibility in labour market, and utilizes its natural resources prudently. Further, it proposes connectivity strategies for addressing gaps in the realization of a smart economy for this region. This includes ensuring smart mobility and enhanced virtual connectivity.

Mobility impacts on human interaction either by impeding or facilitating circulation in the urban space [4]. Street networks, which are primary measures of spatial connectivity, enable cities to function and have a shared life [5]. Other elements of connectivity, which are components of urban form, include street layouts, transportation, densities, centricity and land use mix [6]. Each urban location, either based on size, geography, history, politics or culture has a unique set of connectivity attributes which give it a connectivity signature. Many studies have found that this signature has an apparent relationship with performance of regions [5, 7, 8]. According to Kumar [3], smart mobility is realized when a city has vibrant streets, pleasurable routes, balanced transport options, mass rapid transit systems, highly-integrated mobility systems and has regard for cycling and walking.

Indeed, better connected locations are more likely to experience prosperity than lesser connected regions. A study on the importance of spatial connectivity of centres on regional economic development in the United Kingdom established that by modelling and redesigning problematic transport linkages, the value of places can drastically be improved [9]. It is therefore unsurprising that transportation corridors are key among drivers of land use change [10]. Locations that have already established fast modes of transportation have eliminated the distance barrier, and the land-value-drop away from the business nodes as anticipated in some economic theories of land use is fast changing [11].

In assessing NMR's level of 'smartness' in terms of economy and mobility, this chapter contributes to the discussion on the influence of urban form to economic prosperity of regions. This analysis creates an opinion on economic significance of inward *vis-a-vis* outward growing regions. For spatial connectivity, it explores how improvement of linkages (roads) has impacted on growth of settlements in the NMR. As an integral part of modern economic landscape, the chapter assesses virtual connectivity in the region to propose strategies for its improvement.

13.2 The Nairobi Metropolitan Region

13.2.1 Region's Formation

Urbanization in Kenya just like in many global South countries is characterized by rapid population growth. Consequently, many adjacent urban settlements have spatially expanded forming urban agglomerations. Whereas the optimum size of a well-performing region is largely debatable, urban growth will eventually lead to agglomeration of economies, triggering the need for joint regional management [12]. Additionally, when interaction between urban regions intensifies, social-economic and political challenges extend beyond administrative boundaries, requiring establishment of special management authorities. Accordingly, the Nairobi city's sphere of influence continually expanded such that there was a felt need to create for it a metropolitan hinterland based on functionality (interaction between core and hinterland), morphology (trends of land use change) and administration (gazetted local authorities) [13].

The NMR was officially delineated in 2008 through a presidential decree and its management placed under a government ministry named Nairobi Metropolitan Development (MoNMED). At its formation, the region was zoned along local authority boundaries as (a) the core Metro, which covered the city; (b) the Northern Metro, which covered local authorities within Kiambu County; (c) the Southern Metro, which covered local authorities within Kajiado County and (d) the Eastern Metro, covering local authorities within Machakos County. The region, which is no longer identified by local authorities, covers four administrative counties[1] (Nairobi City, Kajiado, Kiambu & Machakos) (Fig. 13.1) and is approximately 32,000 km^2 with a projected population of nine million [13].

Urbanization within the NMR surpasses the average rate for Kenya (42%) with Nairobi county having rates of 100%; Kiambu, 60.8%; Machakos, 52%; and Kajiado, 41.4% [14]. Of the estimated 9 million residents in the NMR, 48% are in Nairobi county, 25% in Kiambu, 17% in Machakos, and 10% in Kajiado counties. Comparing these statistics to spatial coverages of the counties, it is apparent that population densities are highest in Nairobi, followed by Kiambu. Kajiado and Machakos have vast stretches of undeveloped land with concentration of population being majorly in urban centres.

The vision of the MoNMED was to make the NMR a world class African metropolis that is able to create sustainable wealth and offer a high quality of life to its residents, investors, people of Kenya and visitors. These were premised on the recognition that there is a close nexus between economic, social, cultural and environmental wellbeing of a region. The Ministry's mission was be to build a robust internationally competitive, dynamic and inclusive world class infrastructure to

[1]Kenya is divided into 47 independent governance structures called counties, each with its own governance structure under the leadership of a County Governor.

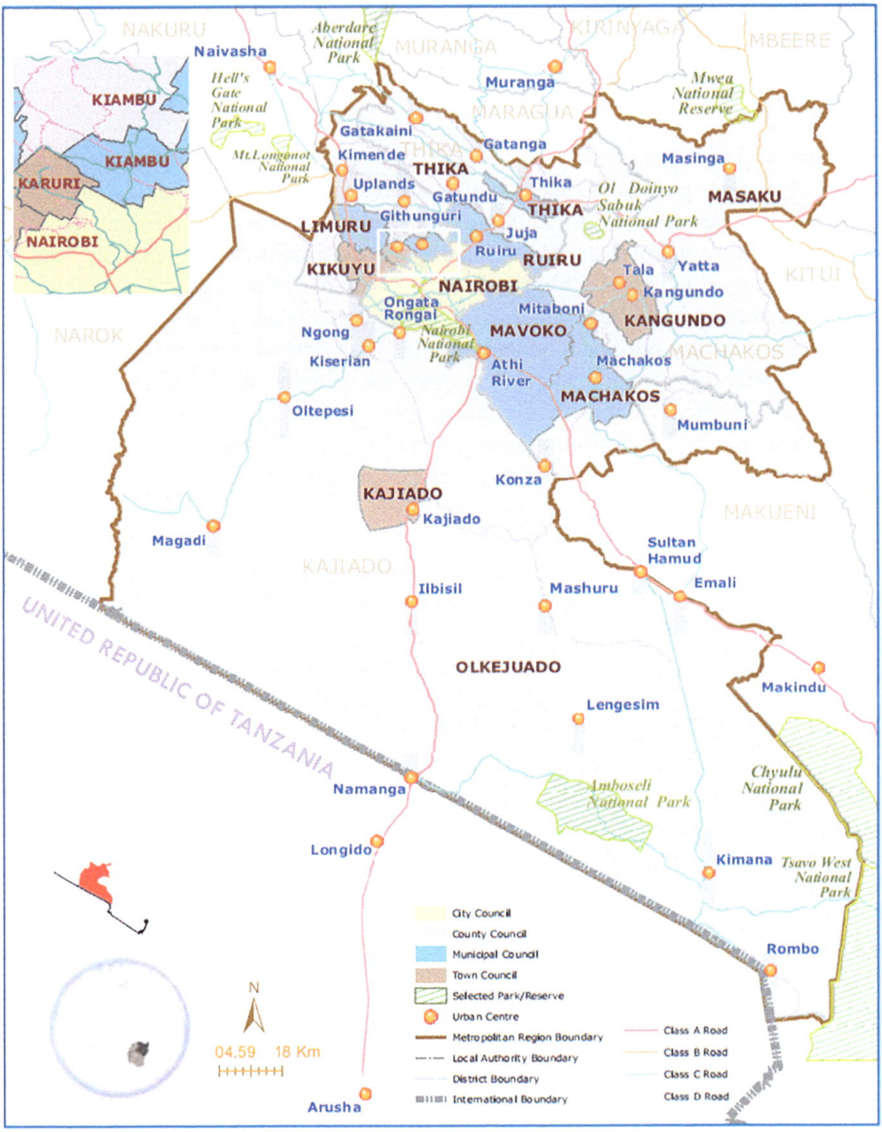

Fig. 13.1 The administrative boundaries of NMR. *Source* Government of Kenya [13]

support development, enhance linkages and accessibility to national, regional and global contexts [13].

To achieve this vision and mission, the MoNMED formulated the Nairobi Metro 2030 strategy with seven key result areas that included: building an internationally competitive and inclusive economy for the metropolitan region's prosperity; deploying world class infrastructure and utilities; enhancing the quality of life and inclusiveness; delivering a unique image and identity through effective place branding; ensuring safe and secure region; and building and sustaining inclusive and efficient metropolitan governance structures. This chapter focuses on the connectivity aspects of this strategy alongside planning opportunities that the NMR can leverage towards becoming a smart region.

13.2.2 Region's Functionality Challenges

The achievement of the Nairobi Metro 2030 strategy—and subsequently the infrastructure and connectivity goals—has been hindered by several challenges. First, the manner in which the metro region was formed has served to largely hinder its smooth operationalization and function. The top-down approach to its formation undermined ownership by some of the local authorities thus creating coordination challenges. The formation of the region brought together 15 independent local authorities, and at its inception, most of the municipalities (now defunct following the formation of county governments) affected were reluctant to actively be part of the metro region. This challenge was compounded by the shift in Kenya's governance structures in 2010. The new structure, functionalized by a new constitution, created independent county governments and empowered them to create and foresee management structures of urban settlements within their jurisdiction.

Second, the NMR seems to have a broken link between governance and institutional management. This has translated into hitches in coordination, including duplication of functions. Upon the region's establishment in 2008, MoNMD was charged with the responsibility of coordinating its activities. The governance structure however failed to acknowledge the functional, financial and political independence of the involved municipalities. At its inception, which happened under Kenya's old governance system, the metro region consisted of 15 independent local authorities, and even though these authorities partly relied on the national government for financial support, each local authority largely controlled its budget and local revenue. As a result, metro wide projects could only be funded and executed by the MoNMD, usually with limited support from affected local authorities. Further, in a bid to achieve the limited number of ministries specified by the 2010 constitution, the MoNMD ministry was transformed to a state department under the ministry of Land, Housing and Urban Development, reducing its institutional significance and budgetary allocations.

The challenges facing the region were identified in the Nairobi Metropolitan Service Improvement Program (NaMSIP) as including uncompetitive economy, inadequate infrastructure, poor transportation and mobility, poor quality of life, poor safety and security, and poor governance [15]. Under connectivity, which is of key focus in this chapter, the NaMSIP notes that the region has inadequate transport management institutions, incomplete road networks and poor intermodal connectivity. This has translated into congestion on transport networks, expensive transport services and huge economic losses.

13.2.3 Region's Economic Nodes

There are numerous economic centres in the NMR, Nairobi Central Business District being the core centre (Fig. 13.2). A formal classification of the centres within the region does not exist; however, the Nairobi Metropolitan Strategy [13] proposed the hierarch of centres to be as: (a) Regional complex, (b) Sub-regional centres, (c) Priority towns, (d) Growth centres, and (e) Market centres. Since the NMR strategy has not been fully implemented, centres such as "Priority towns"

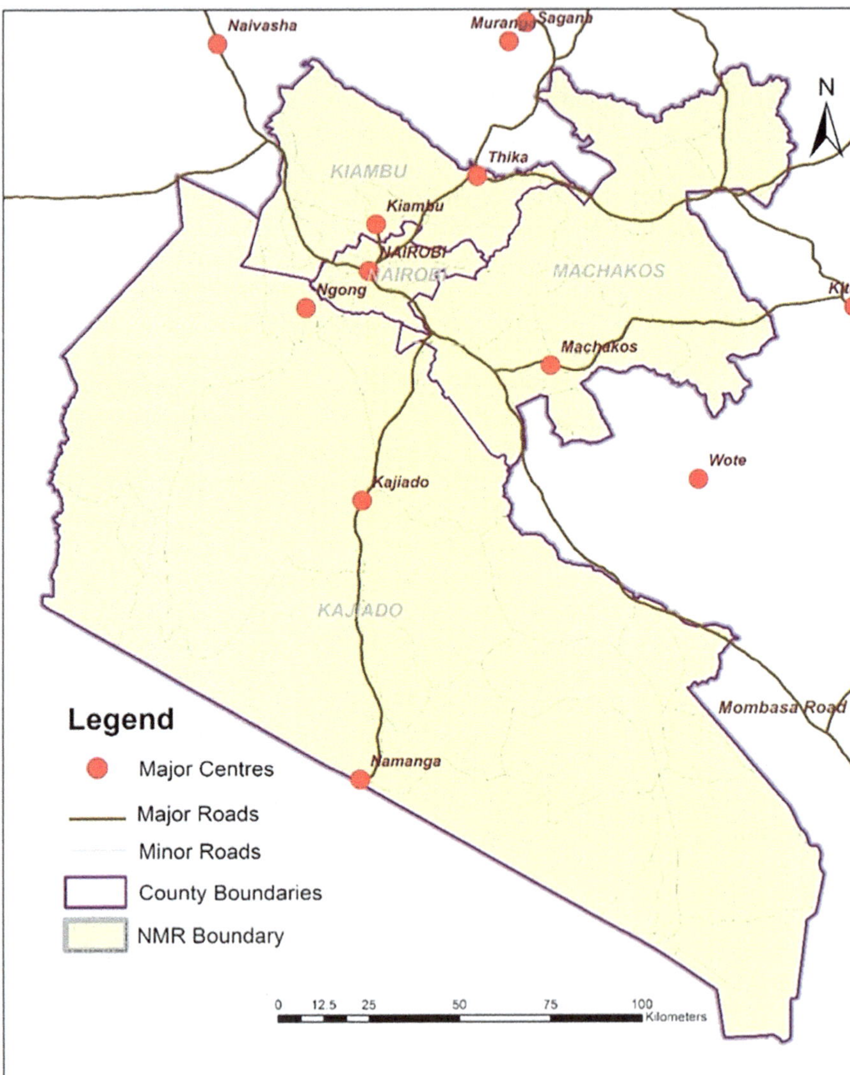

Fig. 13.2 NMR major economic centres and their connectivity

which includes Aerotropolis, Knowledge-cum-health City, Cyber City, Transport New Town, Sports City, and Amboseli New Town, are still non-existent, making this classification inoperative. By their sizes and economic importance, the centres can be classified as:

(a) **The Core Centre**

Nairobi city is the heart of the NMR region and acts as the region's organizing component. It has a Central Business District which is the Kenya's business capital. Surrounding the CBD are satellite centres that house majority of workers in the city. The city has a mixture of low, middle and high density settlements, with majority of the city dwellers residing in high density informal settlements. Other regions in the metropolitan region endeavour to have enhanced connectivity with Nairobi so as to leverage the city's economic strength.

A major challenge facing core centre is congestion—of both traffic and settlements. Prior to the establishment of the devolved governance in Kenya, the city yielded a lot of political and economic powers such that virtually everybody in the country hoped to reside in it. This resulted to rapid population growth, which in the absence of commensurate provision of housing, infrastructural services and utilities, resulted in emergence and proliferation of informal settlements [16].

Currently, the affinity to reside in Nairobi has reduced for some city resident, with most pursuing opportunities in the county regions. This has shifted the balance of power with other centres within the NMR gaining prominence, for example, Machakos, Thika, Kiambu and Kijiado. It's mobility and connectivity status is still high as compared to other parts of the metropolis.

(b) **Major Economic Hubs**

Centres that follow the Metro Core in ranking are major economic hubs, and they include Thika, Kiambu, Machakos and Kajiado. These centres (apart from Thika town) are county headquarters and therefore of high political significance. The biggest of the centres is Thika which is an industrial town linked to Nairobi by an international trunk road, class A2. Alongside Kiambu town, Thika town has a rich agricultural hinterland which makes the town a significant food supplier to Nairobi city. On the other hand, Machakos and Kajiado are in semi-arid states and only practice subsistence agriculture in maize and drought resistant crops such as sorghum. The economy of Kajiado town is driven by tourism and animal farming [17]. Whereas precise data on the growth rates of these towns are hard to obtain, spatial-temporal assessments of their developments on imagery show that these centres are top among the fast growing centres in Kenya.

(c) **Minor Centres**

Under this category are lower level centres within the NMR. They mostly connect to the Metro Core through other major centres. Such centres in Kiambu include Kikuyu, Limuru, Ruiru and Karuri. Kajiado's notable centres include Ngong (its biggest centre), Kitengela, Kiserian, Ongata Rongai, Olkejuado and Bissil; and centres in Machakos include Kangundo–Tala and Athi River. Centres within Nairobi city are considered part of the city and are therefore viewed as satellite towns. Spatial linkages of these centres to major centres is majorly by road.

13.2.4 Region's Connectivity

Regional connectivity can be spatial or virtual. While virtual connectivity focuses on how a region is connected through ICT, spatial connectivity is concerned with ease in which people, good and services are able to move from one location to another while interacting with desired spaces [4].

13.2.4.1 Transport Connectivity in the Nairobi Metropolitan Region

The previous section of this chapter has established that centres within the NMR are relatively spread out. This implies that people, goods and services are required to travel long distances between centres, underlining the need for enhancing connectivity. Although there are several airports and airstrips within the NMR, transportation within the region is majorly by road, and minimally by rail. The only reliable rail transport within the region connects Syokimau, which is at the edge of Nairobi City towards Machakos, and Nairobi CBD. Inter-county train transport is only possible by the standard gauge rail connecting Nairobi and Mombasa; the line is however on high demand for intercity traffic which limits its NMR service capacity.

Road transport therefore remains key in connecting economic centres in the NMR. Major connections include: Thika road, an A2 class road linking Nairobi and Thika road; Mombasa Road, an A 109 road that is part of the link between Nairobi and Machakos town; and Nairobi–Kajiado link which takes two routes, one through Kiserian and another through Kitengela along Mombasa road. There are other numerous roads connecting other centres, for example, Kiambu road which branches from Thika road to Kiambu town. Within Nairobi city, attempts to divert through traffic from the core metro led to the establishment of the Northern Bypass (connecting Thika Road and Limuru Road); the Southern Bypass (connecting traffic from Mombasa road to Kikuyu through Langata and Ngong); and the Eastern Bypass (connecting Ruiru and Mombasa Road). Other notable projects for improving connectivity within the core metro include construction of Outering road, and rehabilitation of several other roads.

13.2.4.2 ICT Connectivity in the Nairobi Metropolitan Region

In the case of the Nairobi metropolitan region, development of ICT infrastructure has been impressively progressing over the last decade. The region today enjoys the highest level of connectivity as a result of overall increase in the country's internet bandwidth, which grew from a mere 28 Megabits per second (Mbps) in 2004 to 193.58 Gigabytes per second (Gbps) in 2015 [18, 19]. The rolling out of the National Optic Fiber Backbone Infrastructure (NOFB) in two phases served this goal effectively, with connections in the first phase focusing on sections of the NMR. As a result, most areas are now connected with reliable internet. High investment in the mobile telecommunications industry by various players has tipped the balance on the need for physical internet hardware/cabling, as more people and offices are adopting wireless forms of internet services as opposed to landline cables. Decreasing cost of internet (data) has also made it possible for high adoption of the services. Today, the cost of accessing internet services is relatively constant between the Nairobi city core and the other outlying centres, with similar speeds, broadly implying increasing connectivity. Equally, previously unconnected centres are becoming more and more served with internet services, further contributing to enhanced connectivity. Additionally, a change of regional management to a county governance structure, where counties are developing their own web-based systems and offering online services, has been greatly contributing to enhanced connectivity within, between counties, and to the rest of the world.

13.3 An Appraisal of Smart Economy and Connectivity for NMR

13.3.1 NMR and Smart Economy

The question whether Nairobi has a smart economy or not does not have a definitive answer. Based on the conceptualization of 'smart economy' within the building blocks of 'smart cities'—as defined earlier in this chapter—analysis shows NMR as performing fairly well. The region has notable university sponsored innovations, a noteworthy entrepreneurial leadership, a flexible labour market, a climate that promotes tourism and investors and a fast growing ICT industry, among other 'smart' economy measures. The JLL [20] in their ranking of dynamic city regions placed Nairobi at position 10 out of 134 major business hubs globally. Within Africa, the city was ranked first, with notable ability to adopt technological changes and absorb population growth. Ranking criteria consisted of adventure, citizenship, cultural influence, quality of life, power, movers, open for business, heritage, entrepreneurship and open for business (see [20]). On entrepreneurship and being open for business, the city was ranked 65th and 53rd respectively.

Nairobi is recognized as the gateway to East Africa. It is a leader in global innovation, a regional tech-hub, and a frontrunner in mobile payment technologies. Employment of cutting-edge research by its expanding universities is being seen in science, industry and business. Yet, the NMR is still dealing with common urbanization challenges of the global south, effectively pulling it away from the quintessential smart region.

These challenges are related to poverty, spatial planning, infrastructural development and housing [21]. In effect, the region has apparent development neutralization forces. Examples of such situations include businesses losing profits in traffic jams; a booming housing industry with majority of city residents being unable to own a house; a highly educated populace but without job opportunities; an increasing car ownership without commensurate transport planning; and a growing population of high-income earners against a burgeoning population of low income groups. For having a capitalistically inclined economy, capital accumulation and manipulative wage labour have brought about huge income gaps, tilting the balance of the economy. The supply chain is minimally regulated by the government, and brokers and middlemen often exploit gaps in information which disadvantages primary producers. A scrutiny of these challenges show that they fall within the economic as well as other blocks of smart cities, namely mobility, people, living and governance. Accordingly, a key aspect of smart economy that needs to be addressed in the NMR is creating a balance in sustainable economic development, one that particularly considers enhanced connectivity.

13.3.2 NMR and Smart Connectivity

Similar to the case of smart economy, the NMR offers a resemblance of smart connectivity while at the same time exhibiting obvious connectivity deficiencies. This section analyses connectivity at two levels: spatial and virtual.

(a) **Spatial Connectivity**

Spatial connectivity implies mobility. In smart mobility, key indicators of smartness include availability of balanced transport options and linkages, cycling and walkable spaces, pleasurable spaces, vibrant streets, mass rapid transit, integrated high mobility systems, assured mobility for all (including people living with disabilities), and effectively managed traffic [3]. In the NMR, realization of these smart city desirables has only been met partially. In terms of mass-rapid transit, the region lacks a metro rail, a light metro, a high speed mobility 'skytrain', and even a high-capacity bus system. Public transport is majorly by low-capacity busses and vans which do not have dedicated lanes. The public vehicles share routes with private vehicles, often on very congested roads. Rail transport is limited to a few routes, and only operates in the in the morning and evening hours when people travel to and from work. Inter-modal traffic transfer plan is often lacking, and this compels train commuters to walk long distances to and from the stations.

The region has currently invested hugely on road expansion, particularly on routes connecting major business nodes [22]. Traffic has in effect been moving faster; while this is commendable, heavy traffic jams are experienced at the slightest disturbance of traffic, for example due to a road accident or weather change. Balance in transport modes is lacking in the NMR, with private transport unfavourably competing for carriageway with public transport. Walkability of streets has only received limited attention, and walkable and pleasurable streets only exist in few areas, majorly in the central business district and affluent neighbourhoods. In addition, the NMR does not have bicycle infrastructure, and only a few road users prefer this green mode of transport. This is attributed to poor road safety for cyclists. In a nutshell, it is observed that while the NMR had an evolving economy, people, environment and living, mobility is among the key areas where progress has been acutely limited.

(b) **Virtual Connectivity**

This kind of connectivity concerns linkages that do not involve physical interaction. Virtual connectivity is an integral part of smart connectivity. The NMR is a leading region in Africa in this regard, enablers of connectivity being ICT. Gustaf and Knut [21] note that among all other sectors of development, Nairobi's ICT-sector is the most developed. Application of ICT has been used in the NMR business sector to enable voice and video communication, make mobile money payment, shop and order delivery of goods, and even trade in stocks. Geo-spatial ICT applications involving GIS tools are employed by businesses for identification of business opportunities, understanding the region's landscape, monitor transportation fleets, routes planning and navigation among others [2]. The NMR's utilization of virtual connectivity in doing business is limited by poor spatial connectivity during delivery of good and services, common challenges being unpredictable traffic flows, lack of seamless modal transfers and a poorly regulated public transport industry [23]. As such, a proper foundation in 'smart' mobility is a necessity for optimal virtual connectivity. Various smart transport initiatives have been launched and tested in Nairobi, ranging from Beba pay card which was a collaborative initiative of Equity Bank and Google to come up with public transport travelling payment card system instead of cash system. Another smart transport service idea is the *magic bus* system which is an initiative by a group of university students from Earlham College in the US to enable public service vehicles commuters to book for a seat of a matatu from their phone without having the trouble of making long queues at the Bus stop. Use of NMR as test ground for these initiatives including other initiatives such as digital matatu, is an indication of primacy of NMR when it comes to absorption and adoption of innovative ideas in as far as smart connectivity is concerned.

13.4 Analysis of Economic Development, Spatial Structures and Connectivity of Regions

Studies show that there is a relationship between spatial organization or regions and their levels of economic development [24, 25]. This organization is what constitutes urban form. Patchy urban developments over a landscape, for instance, may translate into need for more spatial linkages, including motorized travel for accessing economic and social facilities such as shops, schools and parks. Such could impact on the economic success or failure a region. In this section, we include a study that assess the relationship between urban form and economic prosperity of regions; the analysis serves as input to our recommendations for creating a better connected and prosperous NMR.

Urban form is defined by urban sizes, centricity and density. In this study, we have compared city regions with sizes close to that of Nairobi and have employed spatial metrics to measure centricity of regions. These metrics analyse density within centricity measures of compactness and dispersion. In principle, regions with compact development are dense and will have high centricity; conversely, dispersed urban development will have low centricity. This index is generated from classified imagery [26].

Comparing a region's urban form and measures of economic development can give an indication of how the former impacts on the latter. However, it is to be noted that economic development is a multivariate concept with no single agreed upon definition [27]. Measures that have been used to quantify economic development have included population growth, urbanization, infrastructure, gross national product (GNP) and Gross Domestic Product (GDP) per capita, consumption per capita, labour force occupational structures, and social conditions such as life expectancy, health care, mortality, literacy rates and even calories intake among others [28]. Regions found in countries with substantially different economic potentials can therefore not be objectively compared. To overcome this limitation, this study only compares indicators of economic performance and spatial structures of metropolitan regions found in the same IEF (Index of Economic Freedom) bracket [29] with Kenya, and within the same population bracket with the NMR [30]. Based on this criteria, metropolitan regions picked for comparison are: Nairobi in Kenya, Dar Es Salam in Tanzania, Abidjan in Ivory Coast, Khartoum in Sudan, and Alexandria in Egypt.

To establish each of the 5 regions' centricity of built-up development, the study employed a landscape analysis application, Fragstats [31]. Aggregation index (AI), which is a spatial metric computed for a raster surface with a goal of establishing the extent of spread or clumping in a landscape, has been used as a measure factor for centricity. The index ranges from 0 to 100, where 100 represents a completely packed block of cells, all sharing adjacencies [32]. It is computed as given in the equation below [33].

$$\mathrm{AI} = \left[\frac{g_{ii}}{\max - g_{ii}}\right](100)$$

where g_{ii} is the joins between pixels of the classes being analysed; and max-g_{ii} is the maximum number of joins between pixels of a classes being analysed.

The study uses Global Human Settlement Layer (GHSL) [32] data as input to the Fragstats Application. The data has a spatial resolution of 38 m. All the layers for built-up areas were aggregated to a single layer (Fig. 13.3), and their aggregation indices generated.

The AI for the five regions are captured in the table alongside measure of economic wellbeing of the regions. It is to be noted that very little data on Africa is available, and this makes it challenging to rank African metropolitan regions by their per capital GDP strengths. However, Fraym [34] carried out a survey of the biggest markets in Africa, ultimately ranking 169 metropolitan regions in Africa by their estimated GDP, consumer size (which considered number of people owning assets that characterize middle-class consumers) and trade relationship and economic linkages. Rankings for the five regions under study are captured in the Table 13.1.

(a) **GDP Ranking and Aggregation**

Results show that Nairobi is the least compact of the 5 regions under analysis. Its GDP is ranked 19th in Africa behind Khartoum, Alexandria and Abidjan. It is only ahead of Dar es Salam which apparently ranks second least compact region.

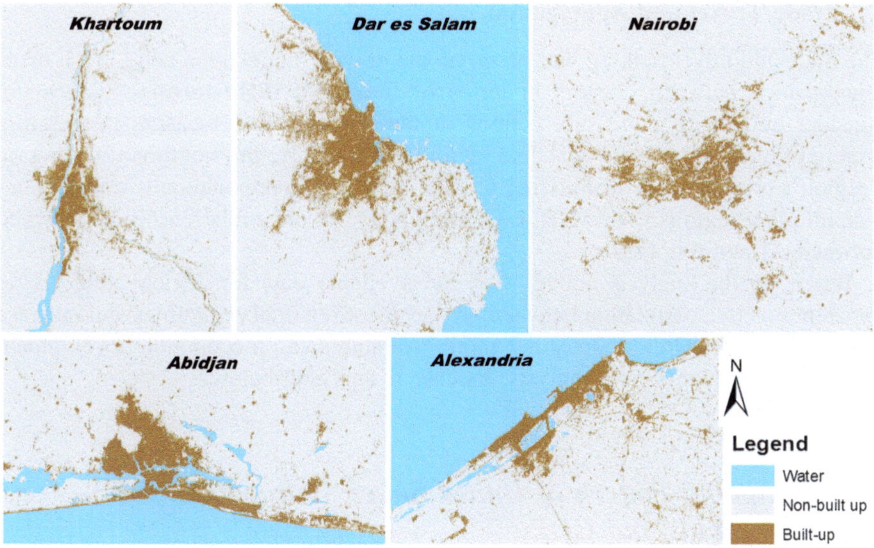

Fig. 13.3 Built up area of selected metropolitan regions

Table 13.1 AI and economic indicators of regions in African

Metro regions	GDP rank (x/169)	Consumer size (x/169)	Travel & trade (x/169)	Aggregation index of regions
Nairobi	19	13	6	58.0108
Dar es Salam	23	9	13	67.3613
Khartoum	8	7	10	69.0140
Abidjan	16	8	11	84.1308
Alexandria	6	6	16	71.2912

Whereas the five cities do not pass the threshold for a statistical analysis such as correlation and regression, it can be deduced from these findings that more aggregated regions generally perform better in GDP than less aggregated regions.

(b) **Consumer Size and Aggregation Index**

Under consumer size, Fraym considered the population within each cluster/region that owns a television, car, motorbike or a refrigerator, these being characteristics of middle-class population. Incidentally, Nairobi—which has the lowest AI—is ranked 13, lower than the rest of the cities. The rest of the regions are closely ranked with Dar es Salam following Nairobi at the lower ranks. Even though the relationship is not linear, with Abidjan being the most aggregated but ranking two regions below Alexandria, the general trend indicates reducing consumer size with reduction in AI.

(c) **Trade-Travel and Aggregation**

The trade and travel ranking was based on the gravity model approach [35]. It takes relationships between clusters in terms of trade and flight travels. It does not consider trade within the sub-sections of the clusters. This measure is therefore largely influenced by links outside a region, and therefore has not direct bearing on a region's urban form. Unsurprisingly, Nairobi—being a regional hub—ranks better than all other regions in this factor. Consequently, no clear relationship is observed between AI and this factor.

Based on the analysis above—and the available data for Africa—there is an apparent positive correlation between prosperity of regional economies and increase in centricity. The United States has more available data on economic development to allow a complementary analysis. Its case is analysed below:

13.4.1 Spatial Pattern of United States Metropolitan Regions and GDP

A complimentary study of centricity of city regions was carried out by Metropolitan Research Centre; the organization analysed spatial design of cities and generated scores for compactness or sprawl developed for each metropolitan areas based on development density, land-use mix, activity cantering and street connectivity [36]. The score of 100 represented regions that are neither compact nor sprawled, and scores above and below it represented increasing compactness and sprawl respectively. To assess how urban form impacts on economic performance of these American regions, the plot below show indices of the biggest 20 city regions in the US [37] against their gross-domestic produce per capita (Fig. 13.4).

The regression analysis shows a positive but weak relationship between increasing compactness of urban form and increasing economic development. While the relationship is not strong enough for a conclusive inference, it is notable that the relationship is positive just as it is in the case of Africa. This study, similar to the Fraym study, concludes that more compact regions are smarter and more productive.

13.5 Analysis of Spatial Connectivity and Economic Development for NMR

Nairobi has historically grown as the main political and economic hub of Kenya. At the same time, however, other small centres which have played the role of regional growth hubs around the city grew at a rapid rate, mostly stimulated by the level of productivity within the hinterlands they serve. Towns which were directly linked to Nairobi (e.g. by their virtue of supplying various services to the city), or which were

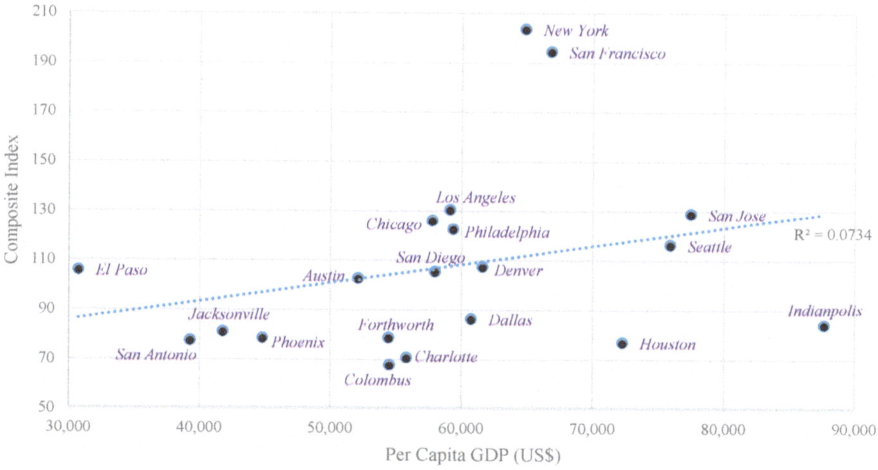

Fig. 13.4 Spatial pattern indices of American metro regions and per capita GDP

within its periphery benefitted from the broader city influence and grew faster than those in more rural hinterlands. The level of influence by the city however also relied on the nature of physical connections via road or rail systems, with towns closer to the core and where better transport networks were provided, growing faster than those with poor networks. This trend has greatly shaped the structure of the Nairobi metropolitan area, and can be associated with the existing form today, wherein majority of the larger towns are located along the main transport corridors converging at the city centre.

Investment in construction and upgrade of transport networks in the broader Nairobi metropolitan region over the past decade has greatly boosted the development of these towns, with built up areas changing by upwards of 50% in areas where such investments have taken place. This finding follows the conventional logic of development, wherein more connected and accessible areas have a greater potential for growth. While majority of the initial urban developments were residential in nature, the need for commercial services closer to the people has resulted in the development of alternative commercial centres adjacent to the main transport networks.

Using remote sensing techniques, we analysed change in built up area for locations abutting key transport infrastructure over the past ten years. We specifically compared areas where new infrastructure installations have been made over the study period against areas where no such investments have been made in order to assess the rate of development change. Figure 13.5 illustrates an analysis of the broader growth trend of the metro region.

The above figure presents low resolution automatic extraction of built up area in a select section of the Nairobi metropolitan region. As per the figure, areas close to the city, and which are along main transport corridors have experienced varying levels of growth over time, with the Nairobi–Thika connection and its connected by-passes exhibiting a high level of linear developments.

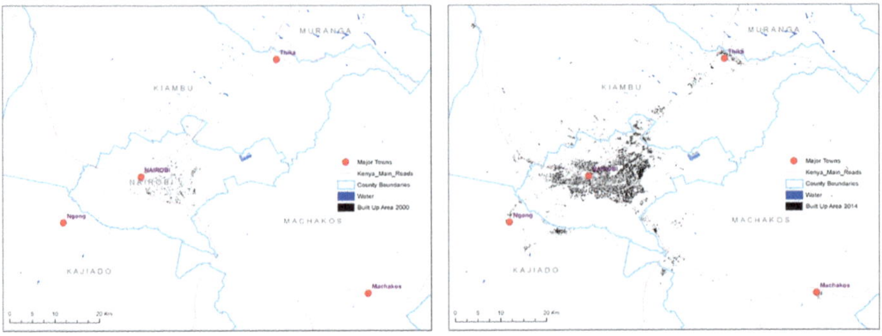

Fig. 13.5 Change in built up for select area of NMR, 2000 and 2014. *Source* Global Human Settlements Layer

The *Nairobi–Thika Connection* is the major road—53.6 km in length—connecting the industrial town of Thika and Nairobi City. Between the years 2009 and 2012, this link was upgraded from a single carriageway with mixed sections of 2 to 4 lanes to an 8-lane controlled access highway. This upgrade reduced traffic time from Nairobi to Thika from 2 hours to 40 minutes [22]. Further, it triggered a sharp appreciation of land values along the route, and since then, there have been an upsurge of investment on land and businesses along the route. Consequently, Thika town has become more connected to Nairobi, and urban zones of intermediate centres such as Ruiru, Juja, and Githurai have increased tremendously with investors and entrepreneurs relocating there for strategic access to emerging markets as well as access to space for expansion. Additionally, the construction of the road has boosted transport of daily and horticultural produce and traders to Nairobi [22].

An in-depth analysis of two single points along the highway and its connected Northern by-pass indicates that the new road developments have resulted in change in built up area by more than 40% over the period 2007–2017 (red pixels in Figs. 13.6 and 13.7).

Thome is located in between the Thika highway (south) and the Northern by-pass (north), and the area between the two roads experienced the most growth between 2007 and 2017.

Ruaka, a previously tiny village located along the Northern by-pass, has now turned into a major residential and (growing) commercial hub at the periphery of the city.

We compared the findings from the Nairobi–Thika connection with developments along another major road, the Nairobi–Namanga road, which branches off Mombasa road and links Kenya to Tanzania. In particular, we focused on the urban area comprised of Athi-River and Kitengela towns, and Kajiado town, all located along this connection. Despite these towns being located along an important road, their rate of growth (both internally and outwards) was less than 25% over the period 2007–2017 (shown by high number of green pixels in Figs. 13.8 and 13.9).

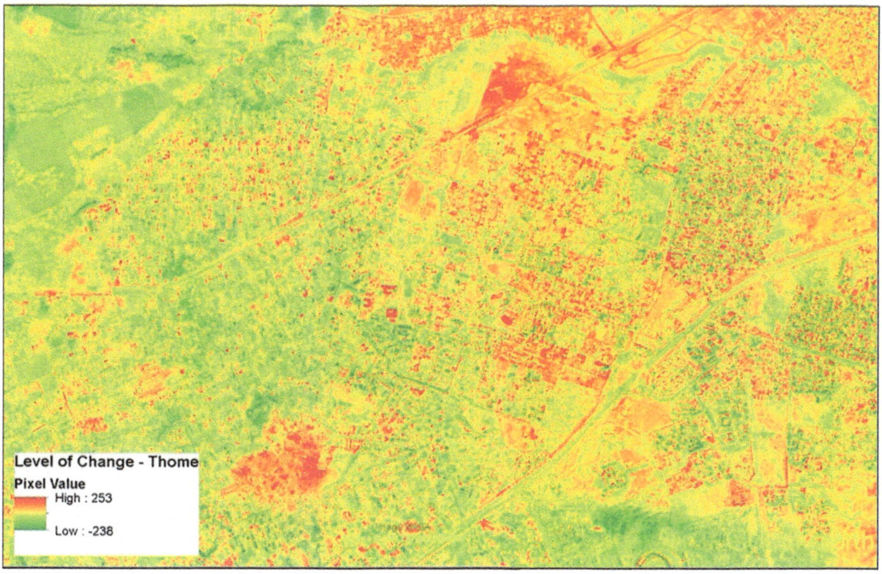

Fig. 13.6 Illustration of level of change in Thome along Thika road and the Northern by-pass

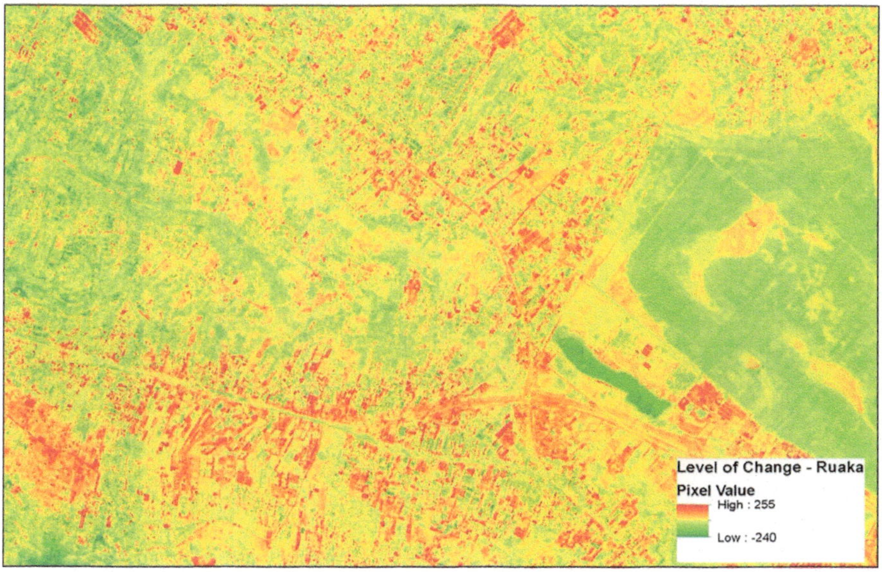

Fig. 13.7 Illustration of level of change in Ruaka along the Northern by-pass

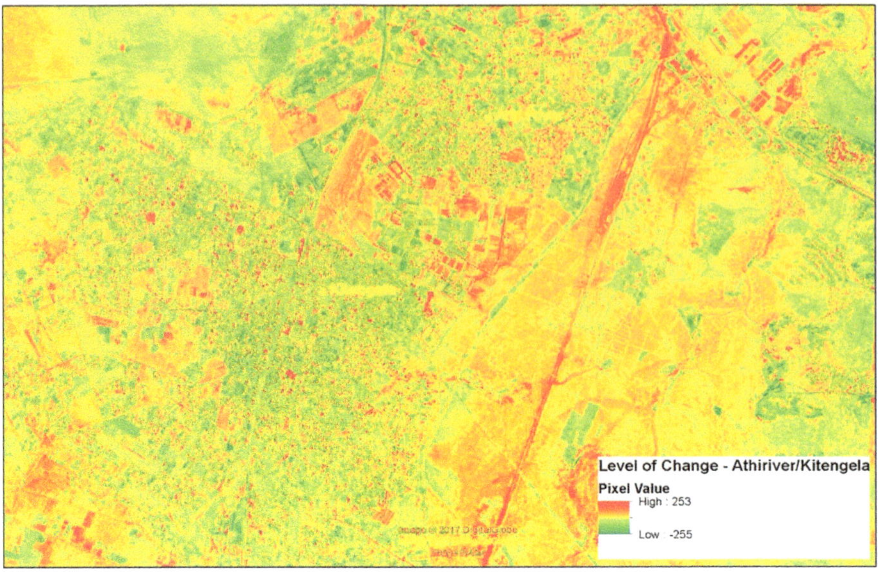

Fig. 13.8 Illustration of level of change in Athi-River—Kitengela area along Nairobi–Namanga road

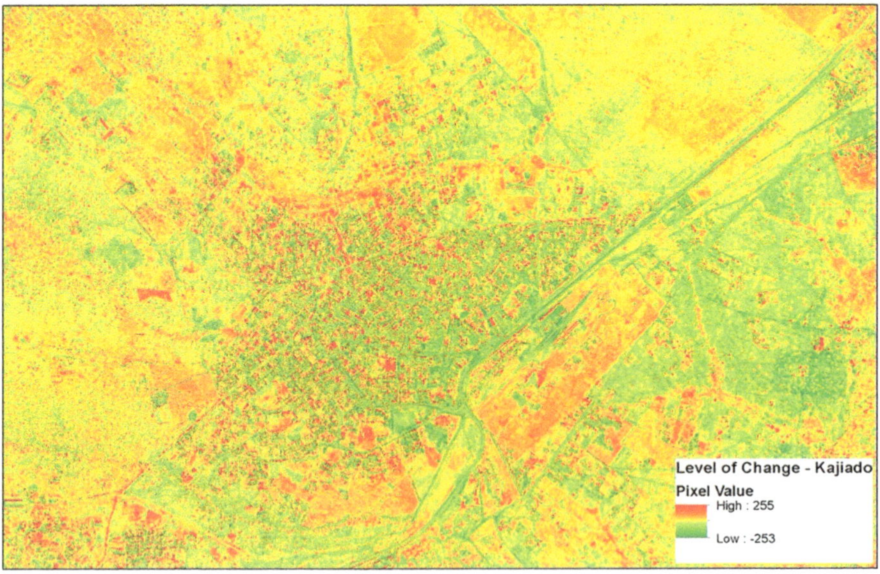

Fig. 13.9 Illustration of level of change in Kajiado town along Nairobi-Namanga road

Recent upgrades to this road stretch is attracting investments in the area, and massive changes are expected in the next five to ten years.

All locations analysed above have largely grown as residential areas, with their residents commuting on a day to day basis to the Nairobi city core. Over time however, demand for commercial services has attracted new forms of development, which constitute business premises (supermarkets, cyber-cafes, restaurants, and car sale yards), office space, ware-houses, factories and informal road-side businesses. The connectivity of the areas to internet is today also enabling many people to work from home, promoting the growth of a smart economy.

The development of commercial centres/services within the newly developed areas is a good indicator of enhanced productivity within the larger metro region. Increased connection to, and use of improving ICT structures is further reducing the need for back and forth movement to the city, while at the same time promoting economic growth. For example, with increasing prices of rental warehouse space within Nairobi's industrial area, several businesses have opted to adopt online marketing (e.g. in OLX platform) for their goods, which are stored in cheaper warehouses within the newly growing areas. These businesses only rent out small offices in strategic locations within the core city, which act as show rooms. Customers can order for the goods online and visit the show rooms, then the goods are delivered to them straight from the warehouses. This has broadly improved business efficiency and is contributing to enhanced incomes for such businesses. New jobs are also being created within these newly developing areas, and as the need for related/supporting services emerge, new business opportunities are continuously emerging. All these are largely boosting the economic development for the larger region.

13.6 Analysis of Policy, Legal and Institutional Framework Shaping NMR

In discussing legal, institutional and policy framework shaping up NMR in relation to connectivity, it will be important to borrow some of the principles raised by Ortiz [38] who noted that:

> Metropolises are not to be planned or managed as cities. The (Deoxyribonucleic acid) DNA is different. Cities, even large, are conurbations with a single administration and require a political dialogue about the future between the citizens and their administration. Metropolises are multifaceted systems with layers of different administration agencies and governance institutions. The dialogue is between those institutions. DNA is different. And you should act accordingly. [39]

This section will look at the legal, policy and institutional transport and ICT framework which can enhance smart connectivity within the Nairobi Metropolitan Region (NMR) which cumulatively lead to its development as a region.

13.6.1 Transport Legal Framework

a. **The Constitution of Kenya**

Transport connectivity matters in the constitution is mentioned thrice, with the first mention with reference to the bill of rights in Article 54(1c) *which notes that; a person with any disability is entitled to reasonable access to all places, public transport and information.* Secondly, it is mentioned twice in fourth schedule with regards to distribution of functions between the National Government and the County Government. Part 2 of the fourth schedule indicates that the functions and powers of the county in relation to county transport include:

(a) *County roads*
(b) *Street lighting*
(c) *Traffic and Parking*
(d) *Public road transport and*
(e) *Ferries and harbours excluding the regulation of international and national shipping and matters related thereto.*

Transport connectivity is given constitution prominence both at national and county level. The constitution though is not clear on metropolitan as an area but gives positive pointers which can be best operationalized at metropolitan level due to unique interaction between neighbouring counties forming a metropolitan region which can be enabled to handle connectivity mandates in an efficient and effective manner. This is the window of opportunity which the president of Kenya used to propose, the Nairobi Metropolitan Area Transport Authority. The Bill for formation of this Authority is still under consideration by the parliament.

b. **The Kenya Roads Act, No. 2 of 2007**

The act was enacted for management and provision of road Infrastructure including NMT in all classes of roads through various authorities. Under part two section 3 (1), the act establishes the Kenya National Highways Authority (KeNHA) as a body corporate. The Highways Authority is responsible for the management, development, rehabilitation and maintenance of national roads.

Some of the functions of KeNHA include:

(a) Constructing, upgrading, rehabilitating and maintaining roads under its control;
(b) Controlling national roads and road reserves and access to roadside developments;
(c) Implementing road policies in relation to national roads;
(d) Ensuring adherence to the rules and guidelines on axle load control prescribed under the Traffic Act;
(e) In collaboration with the Ministry responsible for transport and the Police. Department, overseeing the management of traffic and road safety on national roads;

(f) Collecting and collating all such data related to the use of national roads as may be necessary for efficient forward planning under this Act;
(g) Preparing the road works programmes for all national roads;
(h) Liaising and coordinating with other road authorities in planning and on operations in respect of roads.

Under part two section 9(1), the act establishes the Kenya Urban Roads Authority (KURA) as a body corporate.

The Urban Roads Authority is responsible for the management, development, rehabilitation and maintenance of all public roads in the cities and municipalities in Kenya except where those roads are national roads. Some of its functions include:

(a) Monitoring and evaluating the use of urban roads;
(b) Planning the development and maintenance of urban roads:
(c) Collecting and collating all such data related to the use of urban roads as may be necessary for efficient forward planning under this Act
(d) Preparing the road works programmes for all urban roads;
(e) Liaising and coordinating with other road authorities in planning and on operations in respect of roads.

Under section 6(1) the act establishes the Kenya Rural Roads Authority (KeRRA) a body of corporate. Under section 7(1) The Rural Roads Authority shall have responsibility for the management, development, rehabilitation and maintenance of rural roads.

This act reformed the roles of former ministry of roads and public works and gave specific mandates to different roads agencies in as far as development, management and rehabilitation of various roads is concerned. A lot has been achieved under this arrangement and duplication of roles and mandates greatly reduced which has opened up many parts of the country thus inducing development in all regions. This approach can give metropolitan authorities such as Nairobi a more visible role of coordinating connecting roads for different counties within a metropolitan area since this forms the basis at which smart connectivity routes can be developed in achieving smart mobility within a metropolis.

c. **The County Government Act**

An Act of parliament to give effect to Chapter Eleven of the Constitution; to provide for county governments' powers, functions and responsibilities to deliver services and for connected purposes. This Act gives county government powers to control development and investments including infrastructure investments in their areas of jurisdiction through various plans such as County Integrated Plans and land use plans.

This Act is important since the constitution is clear on the two levels of governments and mandates and responsibilities including connectivity and development funding is distributed through the two levels. The metropolitan area is made up of counties and therefore their operations must be within the confines of the law for optimal development driven by good transport connectivity.

d. **The Traffic Act**

This Act consolidates the law relating to traffic on all public roads. The Act also prohibits encroachment on and damage of roads including land reserved for roads. The Act ensures that there is harmony on laws guiding traffic. It is assumed that good connectivity strategies within a metropolitan area should be safe and responds to the mobility needs of the metropolitan residents and visitors. Nairobi metropolitan area is one of the region which is best endowed with good connectivity roads in the country and unfortunately it is the region which leads in terms of incidences of road related accidents leading to injuries and loss of life. It will therefore be critical to use resources within the region to reduce such negative incidences which is a threat to development and gains associated with good transport connectivity corridors.

e. **The Kenya Roads Board Act**

The act outlines the major tasks of the Kenya Roads Board (KRB) as follows:

(a) Coordinate implementation of all policies relating to the maintenance, rehabilitation and development of the network
(b) Coordinate maintenance, rehabilitation and development of the road network with a view to achieving efficiency, cost-effectiveness and safety
(c) The KRB Act provides for broad allocation of funds, with 60% going to international and national trunk roads and primary roads, 24% to secondary roads, and 16% to rural roads.

The successful implementation of the KRB Act is expected to translate into the physical improvement of the road network, improved utilization of the fuel levy funds, reduction in vehicle operating costs and travel times, open and accountable procurement of road works, and effective financial and technical auditing of road works. This has potential of fast tracking metropolitan regional development through rapid and reliable connectivity with support from KRB kitty.

f. **The Nairobi Metropolitan Area Transport Authority (NAMATA) Bill, 2017**

This is a bill whose ultimate aim is to be an ACT of Parliament to give effect to Article 189(2) of the Constitution: to establish the Nairobi Metropolitan Area Transport Authority; to provide for an integrated and sustainable Public Transport System within the Nairobi Metropolitan Area; and for connected purposes.

Article 189(2) states that:

> Government at each level, and different governments at the county level, shall co-operate in the performance of functions and exercise of powers and, for that purpose, may set up joint committees and joint authorities.

Due to mobility and connectivity challenges NAMATA bill is handy in changing the negative transport connectivity bedevilling the region. Initiatives such as the proposed five (5) Bus Rapid Transit (BRTs) routes will go along away in improving mobility and flow of goods and services within the metropolis which will definitely catalyse development for the region given that studies have shown that Nairobi Metropolitan Area controls over 50% of Kenya's Gross Domestic Product (GDP) and it is estimated that USD 1 Billion is lost annually due to congestion [40].

13.6.2 Transport Policy and Planning Framework

a. **Kenya Vision 2030**

Kenya's Vision 2030 is the current long-term development blueprint for the country. The aim of Vision 2030 is to realize "a globally competitive and prosperous country with a high quality of life by 2030." It aims at transforming Kenya into "a newly industrializing, middle income country providing a high quality of life to all its citizens in a clean and secure environment". The Vision is anchored on three key pillars: economic; social; and political governance. The economic pillar aims to achieve an economic growth rate of 10% per annum and sustain the same until 2030 in order to generate more resources to address the Sustainable Development Goals (SDGs).

The demand for transport infrastructure and services is expected to be influenced by the economic policies to be implemented under Vision 2030, population growth, urbanization, increased volume of trade and productivity both in Kenya and in the neighbouring countries. Transport sector is expected to play an even greater role than it had ever previously done in all key sectors of the economy, namely; agriculture, manufacturing, building and construction, mining and quarrying, tourism, and in the service sectors, including wholesale and retail trade. These are reliable sectors which well-connected metropolitan regions can capitalize on achieving great heights of development.

To achieve vision 2030 the Government identified the following infrastructure projects, among others, for implementation:

(1) Developing a 50-year Integrated National Transport Master Plan which is linked to the National Spatial Plan. It will ensure that the investment and location of transport infrastructure and services are consistent with other public policies. Also, it will ensure optimal transport infrastructure investment to position Kenya as the most efficient and effective transport hub of the East and Central African region and promote national aspirations for socio-economic reconstruction and development. It will also facilitate improvement and expansion of transport infrastructure in a manner that will reduce transport costs and also open new frontiers for economic development.
(2) Developing Nairobi metropolitan region Bus Rapid Transit System to cover three transport corridors
(3) Development of light rail for Nairobi and its suburbs. It is projected to serve at least 150,000 passengers daily.

The implementation of the second and third Medium Term Plan (MTP) of vision 2030 will focus on the competitiveness and rebalancing growth so as to make Kenya globally competitive. The government will increase investment in expansion, development and modernization of roads, rail, ports, ICT and telecommunications in order to make Kenya a top logistics hub which is all critical in achieving smart equity development for all Kenya metropolitan regions

b. **Integrated National Transport Policy**

The INTP guides the development of all the sub-sectors including: road, rail, aviation, maritime and inland water transport and pipeline transport system.

The Integrated National Transportation Policy (INTP) covers key challenges related to transport infrastructure planning, development and management, legal, institutional and regulatory framework for the sector, safety and security, funding, gender mainstreaming, utilization of Information and Communication Technology (ICT), and environmental considerations, among others. Its aim is to provide a policy that is conducive to the stimulation of rapid development and efficient management of a safe, widely accessible transport system that responds to modern technological advancement in a rapidly changing and globalized environment

The policy is comprehensive and anticipates key participation of various authorities as anticipated in the constitution in transportation matters through: Committing the national government to improve governance and service delivery at the local level (County) and this will call for increased responsibilities and greater managerial competence. Some of the reform initiatives at this level will involve improvement of local finances including utilization of cess.

It also gives county governments powers in collaboration with the relevant government agencies and stakeholders, to focus on development and management of transport infrastructure as appropriate, implementation of urban policy, development of local transport plans and integrating these with overall urban land use planning, environmental management, enforcement and local traffic management.

c. **County Integrated Development Plans**

This stems from Sect. 104 (1) of the County Government Act of 2012 which requires all county governments to plan for the county and no public funds shall be appropriated outside a planning framework developed by the county executive committee and approved by the county assembly. All current CIDPs are aligned to vision 2030 and are running from the year 2013 to 2017.

The CIDPs for the all 47 counties have addressed transportation matters under infrastructure and access chapter. They have listed all the ongoing and proposed transport related projects in their respective counties.

d. **Nairobi Integrated Urban Development Plan (NIUPLAN)**

This is the plan guiding development of Nairobi County after the expiry of the Nairobi Metropolitan Growth Strategy (NMGS) in the year 2000. Nairobi has been lucky to have had plans since 1920s. The first master plan was developed in 1927 to capitalize on Nairobi strategic location as the hub of the Kenya Railway line which is transport oriented. The most elaborate plan for Nairobi was the 1973 Nairobi Metropolitan Growth Strategy (NMGS).

NIUPLAN builds from the Study on Master Plan for Urban Transport in the Nairobi Metropolitan Area (NUTRANS) which was supported by JICA in 2006 in terms of addressing missing links in Nairobi transportation system. NIUPLAN is the first deliberate attempt to integrate land use in transportation planning in Nairobi. It also emphasizes on modal integration though heavy on motorized transport including heavy expenditure on public transport infrastructure such as Bus Rapid Transit (BRT).

e. **Integrated Strategic Urban Development Plans**

Various counties such as Mombasa, Kisumu, Kiambu, Nakuru, Machakos, Kitui, Embu, Nyeri have recently prepared Integrated Strategic Urban Development Plans (ISUDP) though support from World Bank, Japan International Cooperation Agency (JICA) and French Development agency (AFD), where transportation issues have been addressed exhaustively including elaborate urban transport baseline information which is good but based on the methodology used, this can be expanded to cover entire county to help the counties develop metro wide transport policies to guide provision and management of transport sectors in counties. This will help initiatives such as NAMATA for Nairobi metropolitan in linking land use and transportation planning which has been a major gap hindering connectivity in Nairobi metropolis.

13.6.3 ICT Legal Framework

At the end of the 20th century, the world witnessed a move from industrial revolution toward an information revolution. Principal to this revolution has been the rapid growth of the new technologies otherwise known as the information and communication technologies (ICTs) which include the internet, email, and mobile telephony among others [41]. Legal framework affection the NMR include:

a. **The Kenya Information and Communication Act**

An Act of Parliament to provide for the establishment of the Communications Commission of Kenya to facilitate the development of the information and communications sector (including broadcasting, multimedia, telecommunications and postal services) and electronic commerce, to provide for the transfer of the functions, powers, assets and liabilities of the Kenya Posts and Telecommunication Corporation to the Commission, the Telcom Kenya Limited and the Postal Corporation of Kenya, and for connected purposes. The Act establishes the Communication Authority of Kenya which has the mandate of regulating ICT sector in Kenya.

Other relevant regulations important in understanding the legal framework are listed as:

- The Kenya Communications Regulations 2001
- Universal Access and Services Regulations, 2010
- Interconnection and Provision of Fixed Links, Access and Facilities Regulations, 2010
- Fair Competition and Equality of Treatment Regulations, 2010.

13.6.4 ICT Policy Framework

a. National Information and Communication Technology of 2006

The policy seeks to facilitate sustained economic growth and poverty reduction; promote social justice and equity; mainstream gender in national development; empower the youth and disadvantaged groups; stimulate investment and innovation in ICT; and achieve universal access. It is based on internationally accepted standards and best practices, particularly the COMESA Model adopted by the COMESA Council of Ministers in March 2003.

The policy is based on four guiding principles: infrastructure development, human resource development, stakeholder participation and appropriate policy and regulatory framework.

13.7 Analysis of the Impacts of the County Governance on the Functionality of NMR

In synthesis, we note that devolution, which came into the fore with a change in the country's constitution in 2010, changed the governance structure of the NMR [42]. The result of this has been new growth trends that have manifested in the rapid growth of new towns, particularly those designated as county headquarters. These new towns, whose administration is vested within county-based governance systems, are growing at a fast rate and are likely to shape the country's urbanization in the next decade.

As a desired effect of this governance structure, huge investments in business are beginning to happen more in the counties. Furthermore, in their quest to be outstanding, counties have attracted huge investments in infrastructure and public facilities that migration from centres like Kajiado and Machakos to Nairobi in search of better services and facilities is drastically reducing. While this is a good thing, per se, it could weaken regional bonds between the counties in the NMR. In fact, the new setup has encouraged inter-county business competition which—though boosting development of counties within the region—has broken major links between them and Nairobi city. The primacy of Nairobi has started to feel the effects of limited capital inflow (albeit at a minimal level) as more developers eye cheaper land and growing opportunities in the counties.

In spite of their robust take-off, the newly set up county governments are still struggling to operationalize basic administration functions; additionally, they are faced with a huge challenge of administering and directing sustainable urbanization. The emerging challenges in counties are related to, among other things, limitations in staff and their capacity to perform different duties, lack of adequate laws and policies to guide actions, as well as lack of know-how in establishing workable programmes. Some basic planning structures are yet to be institutionalized in counties; for example, a survey by the Council of Governors [43] found only 4 out of 30 counties with GIS labs, a vital department in mobility planning. Challenges cites against the realization of this goal included budgetary constraints, inadequate capacity and lack of political goodwill.

Regional economic development—wise, one of the undesirable effects of the new governance structure was the introduction of new forms of taxes to meet budgetary demands. For instance, in 2013 many counties introduced produce cess charges, which is a form of tax for goods moving between counties. The implication of this is that a person transporting goods through three counties has to pay a charge to each county. This has largely impacted on the cost of doing business throughout the metro-region. For the product value of maize, for example, a 16% cess cost is paid, while in transporting vegetables and animals, the cess value of 10% paid [44].

13.8 Synthesis of Connectivity Factors Impacting Regional Economic Development

Regions with high centricity generally perform better economically than less compact regions. Characterized by high-densities and mixed-land use, compactness of a region supports in the realization of sustainable development [45]. For a region experiencing financial constraints such as the NMR, compact development will reduce per capita infrastructure demand, ease provision of public service, and create sustainability for mass rapid transit. Incidentally, the NMR has spatially spread out developments, particularly to the east and southern parts. It would have benefitted the NMR if Machakos and Kajiado were closely knit spatially. Despite availability of space for expansion, measures to limit urban extents should be prioritized if this goal is to be realized. Incidentally, spatial modelling of landscape pattern of the NMR shows that the NMR's built-up area is spatially expanding while becoming more aggregated—even though with non-linear, complex, disconnected patches [46]. This is desirable if it can be embedded in urban planning policy, and implemented comprehensively.

Further analysis on transport connectivity within the NMR shows that locations along transportation routes get an immediate boost in development when the route is upgraded. Such is the case along major transportation routes in the NMR as the analysis of Thome, Ruaka, Kitengela and Kajiado settlements have shown. This translates to mean transport connectivity is a prerequisite for regional growth. It is to be noted that, in a competitive economy, businesses anticipate improvement of transportation linkages. For example, the Nairobi's satellite towns experienced a growth of 21.4% in 2014 following the government's declaration of its plans to start up major infrastructure projects including the standard gauge railway, the western bypass and a commuter rail. Locations impacted directly by these planned projects such as Juja, Ruiru and Limuru experienced a rise in property prices by 43, 42 and 34% respectively [47]. Such was the case in Isinya town where plot values doubled in one year upon the commencement of upgrading of the Athi River–Namanga road. Today, Isinya, which was barely recognizable as settlement 5 years ago, is emerging as a leading centre in Kajiado County alongside Ngong, Kitengela, Kiseran and Ongata Rongai.

Policy analysis show that the NMR has made considerable progress in establishing the relevant policy, legal, and institutional framework. Whereas some laws were developed prior to the formation of devolved governance units—and did therefore not anticipate regional governance—the necessity for regional integration is conveyed in the constitution and recent policy documents. Implementation gaps are however manifestly wide in most of the well intentioned connectivity policies. For example, the need to develop a rapid mass transit system for the region is expressed in numerous policy strategies including the Nairobi Integrated Urban Development Plan but implementation efforts are not evident. This points to the fact that it is not enough to have policy documents and legal frameworks; the policy makers should first grasp the issues limiting implementation of existing policies and work more on implementation strategies. Such calls for the establishment of a monitoring and evaluation system able to holistically look at the existing metro-region operationalization and management setups.

13.9 Towards Enhanced Connectivity in the NMR for Economic Development

The NMR possesses huge potential in major sectors of the economy such as modernization of the service industry, industrial growth and upgrading, agricultural sector diversification and marketing. From the NMR vision and mission, for example, the Kenyan government commits to come up with various strategies including developing and pursuing a Smart strategy for Nairobi Metropolitan Region. In pursuance to this, the government has come up with various smart metropolitan enablers such as the recent formation of the Nairobi Metropolitan Area Transport Authority (NAMATA) which will provide a comprehensive and dynamic platform for addressing the challenges in the transport sector that have affected the

Metropolitan Area [48]. The body shall formulate a sustainable integrated public transport strategy that will be the basis for the orderly development of the proposed Metropolitan Area mass-transit system, which incorporates both bus rapid-transit and commuter rail. This may be a concept borrowed from Washington Metropolitan Area Transit Authority (WMATA) which can help the NMR in achieving the principle of delivering sustainable transport choices which has an impact on other principles such as promotion of clean energy and attracting investments which will lead to increase in employment and business opportunities, given that NMR has itself a big consumer market for most of its products.

Another intervention which is likely to facilitate achievement of smart NMR is the Nairobi Metropolitan Service Improvement Project (NaMSIP) which is focusing on different development projects such as sewerage, roads, boreholes, and markets. The project also pushes for capacity development of officers working in NMR counties, including purchasing of software and furniture among others. These interventions will help the NMR to achieve some aspects of interconnectedness within the NMR which is also a principle of smartness. A key component of NaMSIP that makes it promising is its partnership strategy. The project brings together various counties forming the NMR as well as the national government and the World Bank. This approach ensures that governance is strengthened in the realization of regional development goals.

Development of better roads infrastructure, particularly between centres of economic significance is an apparent intervention. Infrastructure-based or infrastructure-driven economic development school of thought which combines growth models from various world regions holds that in order to stimulate long-term economic growth and efficiency, particularly in economically lagging regions, a substantial proportion of a nation's resources must be strategically invested in long-term infrastructure assets, such as transportation, energy and social infrastructure [49]. Development of connectivity enabling infrastructure opens up areas for development, and in turn promotes economic growth and/or enhances productivity. Emphasis on 'smart' infrastructure planning must however be made. In an example, good roads comparable to Thika road, if created for major link routes such as Mombasa-Machakos, appeal to the mind as the missing link between perfect connectivity in the NMR. Yet, all facts considered along the long-term nature of planning, this would barely withstand the test of time. In the Nairobi Metro Strategy, projections show that the NMR will swell to 14.3 million by 2030 if the present conditions prevail. This translates to 58.8% growth from the present. The fact that Thika road is already experiencing occasional traffic jams particularly during peak hours show that a futuristic approach to enhancing mobility beyond road expansion is required. Perhaps much can be learnt from city regions that are bigger than Nairobi, both in developed and developed countries.

In Istanbul, arguably the most populous city region in Europe with a population of about 15 million, transport is bound to be a problem. However, the city is partially able to solve the traffic gridlock by use of public transport, which includes high capacity busses, the Metro tram and funicular for steep terrains [50]. Without these options, movement would be virtually impossible in the city region. With a population lesser than half of Istanbul, the NMR could function efficiently if smart mobility approaches were employed. At present, transport linking major NMR regions is majorly by mini-buses and 14-seater vans. For being minimally regulated, the condition of the vans are often not satisfactory for the most middle and high income earning groups, a result of which majority who own cars prefer to travel by private means.

While there are numerous proposals to establish a rail transport for the NMR, implementation of these programs has not happened majorly due to governance challenges. Only the rail link between Nairobi and Syokimau has been reliable, yet with the Nairobi-Mombasa standard gauge rail station at Syokimau, this connection is mostly utilized in feeding traffic to the intercity train. The current traffic situation in the NMR challenges the delivery of food items to the city, with most transportation only happening in very early morning hours, beyond which food perishables become stale in traffic.

13.9.1 Key Strategies for a Better Connected NMR

The NMR is potentially able to realize huge economic gains with better connectivity. From enormous losses occasioned by traffic jams, poor inter-modal linkages, gaps in land use planning, to failure to leverage advancements land use and mobility management, the NMR is considerably operating below its optimal economic potential. Business models show that cities or regions that embraced principles of smart-city management greatly reduce their cost of doing business, boost their profit margins and subsequently attract more investors [1]. To place the NMR at this strategic locus, adoption of the following strategies is desirable.

(a) **Urban Growth Management and Spatial Temporal Modelling**

The connectivity analysis in this chapter has shown that regions perform better economically when they have a compact urban form. It is to be noted that urban growth, while unchecked against sprawl, is often associated with lack of proper planning policies, speculation, and legal disputes [51]. This phenomena affect the NMR; spatial planning policies in the region, though elaborate, are not explicit in promoting compact growth. Subsequently, for a county like Kiambu where land was predominantly agricultural a decade ago, urban growth is now sprawling into agricultural regions on a thin density fabric. The fact that Kajiado and Machakos have land for expansion creates even more urgency in limiting the expanse of their urban regions. For Nairobi city, which is fully urbanized, a more realistic approach would involve protecting areas that must be preserved as non-built up, and eliminating low rise development in high-density areas. This will involve revising the city's zoning ordinance. Congestion of settlements in Nairobi will reduce if vertical growth strategies were implemented, particularly in informal settlements. These measures will lead to the realization of wider roads which would yield economic benefits.

To tackle urban sprawl, it is imperative that the NMR adopts a policy of forward planning which limits urban centres to definite boundaries and ensures strict compliance. This intervention will yield compact regions with crisp urban-rural boundaries as has been the case with many West-European towns. This policy will lead to a realization of a productive urban hinterland and aggregation of numerous fragmented market locations. More business competition will be realized as well as economies of scale.

Implementation of the aforementioned spatial policies has been made easier by the advances in spatial modelling technologies. Spatial data, complemented by social-economic data, can guide in the identification probable growth locations which can be used to direct future growth. Using spatial modelling tools, for example, the Cellular Automata, Agent Based Modelling and Logistic Regression Modelling, the NMR's planners and managers can study human behaviour in the region and model settlement locations in a manner that is desirable [52].

(b) **Smart Transportation**

With Nairobi's economy suffering over $360 million in losses from traffic jams [53], it is obvious that the NMR requires a well-thought out transport system and that fully meets the needs of the residents. Previous section of this chapter have proven that the region requires an elaborate transport plan, more so beyond intensified road expansion projects. Mass transit is often at the core of successful transport plans. In the United States, for example, adoption of high occupancy, express and carpool lanes have been used to lessen traffic congestion [54]. Accordingly, major business nodes within the NMR need to be connected by a mass transit systems, particularly rail.

The rail is favourable in that it can handle huge capacity of travellers and can adjust fast to a growing population. In this regards, a rail line taking major transportation axis connecting Nairobi to Kiambu, Thika, Machakos and Kajiado is desired. In planning, Ortiz [42] advises that connectivity networks in Nairobi should not be targeted to be orbital-radial as the city is not circular; nonetheless, prioritization of rail access should be to and from the city centre. A complementary system of bus-rapid transit will be required, especially in connecting settlements that are distant from main rail line. This arrangement will translate into reduced usage of private means of transportation and use of vans. However, it is worthwhile to note that some travellers will not use public transport even when it appears expedient for them; in this regard, a modelling approach, particularly ABM, is required to establishing travellers' behaviours, including their motivation to choose a mode of travel over another. Such a study has not been carried out for NMR. It could be well that some travellers prefer quiet vehicles or spacious seats. As such, a conclusive research may propose development of train transport with silent or family cabins among other options. Studies observe that overlooking consumer needs lead to failures even for the most appealing product [55].

Yet another key consideration of smart transportation is modal shift. In Nairobi, it is observed that there is user travellers' reluctance in using the train where a convenient mean of connecting to their final destination is not provided for. Indeed, even the time spent transitioning from one transport mode to another can contribute to favourability of a travel mode. Smart transit must therefore be ensured for all transport modes. Regions that have adopted such transportation strategies leverage the improving information and communication technologies.

ICT is an integral part of smart transportation. Current milestones in the sector allow exploitation of information technologies in traffic data management and application of mobile applications in booking and monitoring traffic. With internet usage in Kenya projected at over 67%, these ICT capabilities are primary inputs to smart regional mobility. The success being enjoyed by mobile applications (Apps) for booking taxi cab services (such as Taxify and Uber) and traffic monitoring and carpooling apps (such as Ma3Route and Waze) is a pointer to the potentials in the NMR. Advancement of these Apps will happen immediately, including through public initiatives, if a base infrastructure such a reliable, round-city metro rail can be put in place.

(c) **Regional Integration and Governance**

Managing urban growth and embracing smart transportation strategies will only work if attached to a dedicated implementer. Project implementation is a major challenge undermining plan making processes. In many regions of the world, it is not uncommon to find a costly plan making process superseding a well-intentioned plan that is not implemented. In fact, it is true to assert that the gap between plan making and implementation has been the biggest failure for planning in the NMR, and generally African cities. While, for example, the Integrated Urban Development Master Plan laid a groundwork for classification of centres, including setting out strategies that would improve Nairobi's outlook, the development projects presently being implemented do not make reference to this policy document. Thus, a plan that would be guaranteed to be implemented first requires to be articulate in defining its goals, especially at short, medium and long terms [56], and second, requires to secure commitment from the implementing authorities.

Unfortunately, the NMR is spread over four counties. This means greater efforts in respect to coordination are required in the regional planning. Indeed, regional integration is one of the principles of smart urbanism [57]. A shared planning vision for the four counties is therefore desired. This should start with harmonizing development visions of the four county authorities—as may be stated in their different planning frameworks such as their County Integrated Development Plans, County Spatial Plans and County visions. Ultimately, a regional planning authority, formed by administrators and technocrats from the four regions, should serve as the integrating body. On a continuous basis, the authority should review trade barriers, such as double taxation through county cess, and endeavour to improve connectivity of the region.

13.10 Conclusion

Nairobi Metropolitan Region (NMR) has great potential in uptake of smart and innovative ideas which can propel and sustain the region as one of the key regional development hubs in Sub-Saharan Africa. This will be achieved through completion of on-going initiatives such as Mass Rapid Transit (MRT) including Light Rail Transit (LRT) and Bus Rapid Transit (BRT). If embedded with online booking platforms such as the one currently used by the Kenya commuter train service from Mombasa to Nairobi, enhanced efficiency will be experienced in the transport sector, translating to increased metro-wide productivity.

The region is also benefitting from the national government goodwill to transform the region as shown by the commitment by the office of the President to support it to have a spatial plan, as well as a transport authority.

The chapter therefore concludes that though transport and ICT connectivity will continue to play a critical role and function in NMR development, it is not a sole panacea to fully unlock its economic and growth potential. Just like other metropolises in Africa, sustainable NMR development will require wide consultations with all relevant stakeholders in a bid to build consensus and commitments, which should take cognizance of the complex multi-level governance associated with such regions. This should be embedded on such metro region's own competitiveness and innovations which in most cases is highly boosted by their location as knowledge and economic hubs, which in itself is the main driver in speedy uptake of innovative ideas which are critical ingredients of making them sustainable regional development hubs.

Bibliography

1. Batagan L (2011) Indicators for economic and social development of future smart city. J Appl Quant Methods 6(3):27–34
2. Mwaniki D (2017) Smart city foundation, the core pillar for smart economic development in Nairobi. In: Kumar TV (ed) Smart economy in smart cities. Springer, New Delhi, p 1094
3. Vinod Kumar T (2017) Smart economy in smart cities. Springer, Singapore
4. Trova V (2012) Measures of street connectivity: spatialist_lines (MoSC). In: Hull A, Silva C, Bertolini L (eds) Accessibility instruments for planning practice. Technical University of Denmark, Copenhagen, pp 103–109
5. Peponis J, Allen D, French S, Scoppa M, Brown J (2007) Street connectivity and urban density: spatial measures and their correlation. In: Proceedings of the 6th international space syntax symposium, p 12
6. Jabareen YR (2006, Sep) Sustainable urban forms. J Plan Educ Res 26(1):38–52
7. Melo PC, Graham DJ, Brage-Ardao R (2013) The productivity of transport infrastructure investment: a meta-analysis of empirical evidence. Reg Sci Urban Econ 43(5):695–706
8. Jain V, Sharma A, Subramanian L (2012) Road traffic congestion in the developing world. In: Proceedings of the 2nd ACM symposium on computing for development—ACM DEV '12, p 1
9. Woollett N, Knight P, Redfern R (2009) The importance of transport connectivity in supporting regional economic development. In: European transport conference, 2009 association for European transport, April 2006, pp 1–20

10. Githira D (2016) Growth and eviction of informal settlements in Nairobi. University of Twente—ITC
11. El-barmelgy MM, Shalaby AM, Nassar UA, Ali SM (2014, Jan) Economic land use theory and land value in value model, vol. 2
12. Cottineau C, Finance O, Hatna E, Arcaute E, Batty M (2016) Defining urban agglomerations to detect agglomeration economies, 2004
13. Government of Kenya (2008) Nairobi metro 2030 strategy. Nairobi
14. Open Data Kenya (2017) Opendata.go.ke. [Online]. Available: http://www.opendata.go.ke/. Accessed 11 Nov 2017
15. World Bank (2011) Nairobi metropolitan service improvement project. Nairobi
16. UN-HABITAT (2006) Nairobi urban sector profile. United Nations Human Settlements Programme, Nairobi
17. J. I. C. A. JICA (2014) The project on integrated urban development master plan for the city of Nairobi in the Republic of Kenya. Nairobi
18. Export Processing Zones Authority (2005) Kenya's information and communications technology sector 2005
19. Communications Authority of Kenya (2015) First quarter sector statistics report for the financial year 2015/2016
20. Jones Lang LaSalle (JLL) (2017) City momentum index 2017. Chicago
21. Ankarcona G, Holm K (2016) The entrepreneurship ecosystem and its supports in Nairobi. Lund University
22. Road Traffic Technology (2017) Nairobi-Thika superhighway—Verdict Traffic. [Online]. Available: http://www.roadtraffic-technology.com/projects/nairobi-thika-superhighway/. Accessed 13 Nov 2017
23. Salon D, Aligula EM (2012, May) Urban travel in Nairobi, Kenya: analysis, insights, and opportunities. J Transp Geogr 22:65–76
24. Huggins R, Piers T (2017) Networks and regional economic growth: a spatial analysis of knowledge ties. Environ Plan A
25. UN-HABITAT (2014) The economics of urban form : a literature review. Nairobi
26. Kuffer M, Barros J, Sliuzas RV (2014) The development of a morphological unplanned settlement index using very-high-resolution (VHR) imagery. Comput Environ Urban Syst 48:138–152
27. Dialoke I, Edeja MS (2017) Effects of Niger delta militancy on the economic development of Nigeria (2006–2016). Int J Soc Sci Manag Res 3(3):25–36
28. Harper College Education Foundation (2017) Economic geography: measures of economic development. [Online]. Available: http://goforward.harpercollege.edu/foundation/index.php. Accessed 16 Oct 2017
29. Heritage.org (2017) Index of economic freedom. [Online]. Available: http://www.heritage.org/index/heatmap. Accessed 16 Oct 2017
30. CityPopulation.De (2017) Major agglomerations of the world—population statistics and maps. [Online]. Available: http://www.citypopulation.de/world/Agglomerations.html. Accessed 11 Nov 2017
31. UMass Landscape Ecology Lab (2017) FRAGSTATS: spatial pattern analysis program for categorical maps. [Online]. Available: https://www.umass.edu/landeco/research/fragstats/fragstats.html. Accessed 12 Nov 2017
32. He HS, DeZonia BE, Mladenoff DJ (2000) An aggregation index (AI) to quantify spatial patterns of landscapes. Landscape Ecol 15(7):591–601
33. UMassAmhrest (2017) (C116) Aggregation Index. [Online]. Available: http://www.umass.edu/landeco/research/fragstats/documents/Metrics/Contagion-InterspersionMetrics/Metrics/C116-AI.htm. Accessed 09 Nov 2017
34. Fraym (2017) Where are Africa's biggest markets? Washington, DC
35. Anderson J (2016) The gravity model of economic interaction. Boston
36. Ewing R, Hamidi S (2014) Smart growth America: making neighbourhoods great together. Houston

37. United States Census Bureau (2017) American factfinder—results. [Online]. Available: https://factfinder.census.gov/faces/tableservices/jsf/pages/productview.xhtml?src=bkmk. Accessed 16 Oct 2017
38. Ortiz P (2015) Metropolitan pentalogue: five commandments for metro management. [Online]. Available: https://www.linkedin.com/pulse/metropolitan-pentalogue-five-commandments-metro-management-ortiz/. Accessed 12 Dec 2017
39. Ortiz P (2015) Metropolitan pentalogue. Five commandments for metro management. https://www.linkedin.com/pulse/metropolitan-pentalogue-five-commandments-metro-management-ortiz
40. NaMATA (2017) Nairobi Metropolitan Area Transport Authority (NaMATA). [Online]. Available: kenas.go.ke/wp-content/uploads/NAMATA-WAD-2017-June-9th.pptx. Accessed: 09 Nov 2017
41. Kenya Human Rights Commission (2014) Internet legislative and policy environment in Kenya. Accessed from https://www.sbs.ox.ac.uk/cybersecurity-capacity/system/files/The%20ICT%20Legislative%20and%20Policy%20Ennvironment%20in%20Kenya_0.pdf
42. Ortiz P (2017) Reports for Nairobi—Pedro Ortiz online. [Online]. Available: http://www.pedrobortiz.com/display-articles/listforcity/city/36. Accessed 13 Dec 2017
43. Mutua F, Mwaniki D (2017) GIS needs assessment in Kenya. Report prepared for Kenya's Council of Governors (CoG) and United Nations Development Programme (UNDP). Nairobi
44. Bayensia Consulting Group (2016) The burden of produce cess and other market charges in Kenya. Nairobi
45. Chen H, Jia B, Lau SSY (2008) Sustainable urban form for Chinese compact cities: challenges of a rapid urbanized economy. Habitat Int 32(1):28–40
46. Mundia CN (2017) Nairobi metropolitan area. In: Murayama Y, Kamusoko C, Yamashita A, Estoque R (eds) Urban development in Asia and Africa. Springer, Singapore, pp 293–317
47. Nairobi Business Monthly (2017) Rising demand for housing spurs Kenya's real estate sector. [Online]. Available: http://www.nairobibusinessmonthly.com/rising-demand-for-housing-spurs-kenyas-real-estate-sector/. Accessed 25 Sep 2017
48. Office of the President (2017) Executive order: nairobi metropolitan area transport authority. [Online]. Available: http://www.president.go.ke/2017/02/10/executive-order-the-nairobi-metropolitan-area-transport-authority-namata/. Accessed 27 Apr 2017
49. Era AN, Budgetary OF (2011) Infrastructure investments in an age of austerity: the pension and sovereign funds perspective, pp 34–37
50. Gerçek H (2007) Transportation planning and decision-making in Istanbul: a case study in sustainable urban transport policy development, March 2006
51. Mohammed I, Alshuwaikhat H, Adenle Y (2016) An approach to assess the effectiveness of smart growth in achieving sustainable development. Sustainability
52. Arsanjani JJ, Helbich M, Kainz W, Boloorani AD (2012) Integration of logistic regression, Markov chain and cellular automata models to simulate urban expansion. Int J Appl Earth Obs Geoinf 21(1):265–275
53. Construction Business Review (2017) Nairobi mass rapid transport systems plan. [Online]. Available: http://www.constructionkenya.com/3480/nairobi-rapid-bus-transit/. Accessed 23 Sep 2017
54. Konishi H, il Mun S (2010) Carpooling and congestion pricing: HOV and HOT lanes. Reg Sci Urban Econ 40(4):173–186
55. Singh H (2006) The importance of customer satisfaction in relation to customer loyalty and retention by Harkira. Kuala Lumpur
56. UN-HABITAT (2010) Citywide strategic planning. UNON—Publishing Services Section, Nairobi
57. Spalding J, Gard A, Brown B, Hadian S, Salaris K, Toner B, Moughtin C, Shirley P, Trudeau D, Lydum D, Rahnama MR, Roshani P, Hassani A, Hossienpour SA, Vanderbeek M, Irazábal C, Knaap G, Talen E, Trudeau D, Chang H, Parandvash GH, Shandas V (2005) A typology of New Urbanism neighborhoods. Int Reg Sci Rev 28(2):267

Part XIV
Nigeria, Abuja

Chapter 14
Smart Metropolitan Regional Development of Abuja and Its Region

Femi Olokesusi, Femi Ola Aiyegbajeje and Ibitayo Modupe Arije

Abstract The heightened intolerable conditions of living and working in Lagos and the need to centralize the capital city for equal access from other parts of the country led to the creation of a new Federal Capital in Abuja in 1976. The Nigeria new capital, Abuja, was premised on some principles meant to underpin the regional development of the new capital. These are principle of equal access, equal citizenship, environmental conservation, city beautification, functional city, regional development, and national economic growth. The study assessed the metropolitan region of Abuja in view to understand and identified the usual associated urban problems facing the region, and how Information and Communication Technology (ICT) can serve as a key critical factor to the realization of smart metropolitan development in an effort to address the identified urban problems of the Abuja metropolitan region. For Abuja to achieve these goals, the use of ICT was found to be essential. Therefore, this study evaluated among others the roles and integration of ICT in the development of a Smart metropolitan Abuja region. The chapter recommends a realistic implementation of the multi-nuclei regional plan with appropriate transportation framework, stronger economic base and pro-poor land use strategies. This paradigm shift could occur if and only the Government of Nigeria exhibits robust political will in the regional development process.

Keywords Abuja smart metropolitan region · Regional development Innovation · Federal capital territory

F. Olokesusi
Afe Babalola University, Ado-Ekiti, Nigeria
e-mail: femioloke@yahoo.com

F. O. Aiyegbajeje (✉)
Department of Geography, Faculty of Social Sciences,
University of Lagos, Lagos, Nigeria
e-mail: femidavid2002@yahoo.com

I. M. Arije
UNHABITAT, Nairobi, Kenya
e-mail: arije_tayo@yahoo.com

T. M. Vinod Kumar (ed.), *Smart Metropolitan Regional Development*,
Advances in 21st Century Human Settlements,
https://doi.org/10.1007/978-981-10-8588-8_14

14.1 The Concept of Smart Metropolitan Regional Development Abuja

The Abuja smart metropolitan regional development revolves round a sustainable, inclusive and prosperous city that promotes a people-centric approach based on the integration of smart city foundation, information and communication technology (ICT), innovation, smart institutions and laws into its developmental and operation architecture. Smart metropolitan regional development enhances infrastructure development, environmental sustainability, economic and social development, ability to reduce disasters and promotes mixed neighbourhood where services are walking distances from people's residences.

Abuja is the new administrative capital city of Nigeria created in 1976. It is located within a geographical space known as the Federal Capital Territory (FCT) at the centre of the country (Fig. 14.1). The clamour for a new Federal Capital had arisen because of the compounded intolerable conditions of living and working in Lagos. The city started witnessing rapid influx of human population particularly since December 1991 when the seat of national governance moved formally from Lagos into the city.

Fig. 14.1 Abuja's location. *Source* [40]

The founding fathers of the new Federal capital wanted it to meet some of the ideals of a restructured polity [1]. Seven principles, explicit and implicit, could be identified in the philosophy that was meant to underpin the development of the new capital. These are the principle of equal access, equal citizenship, environmental conservation, city beautiful, functional city, effective regional development and rapid national economic growth [1].

Abuja, also known as Federal Capital Territory is located in the geographical centre of Nigeria. Abuja is the seat of the government of Nigeria. It has a land area of 7315 km^2. It is bounded on the north by Kaduna State, the west by Niger State, the east and southeast by Nasarawa State and the southwest by Kogi State. It falls within latitudes 9°4′ N and 7°39′ E. The Abuja's natural endowments such as; its rolling hills, isolated highlands and other endearing features make it a delight. The savannah grassland of the North and the Middle Belt, the richness of the tropical rain forests of the south and an equable climate all combined to make Abuja a soil-rich agricultural haven. Abuja is located at the centre of Nigeria designed to increase equal accessibility to every regions of the country (see Fig. 14.2).

There are six Area Councils in the FCT, each subdivided into wards headed by local councillors. Abuja Municipal Council (AMAC) otherwise known as the Federal Capital City (FCC) is the most developed and prominent of the six Area Councils. Each of the Area Councils is responsible for local administration and provision of basic social services. The Area Councils were created in 1996 when the Federal Government under the military created additional local government areas in the country. The Area councils were created to bring government close to the grassroots and are contained in Part II Section 3 of the 1999 Constitution of the Federal Republic of Nigeria.

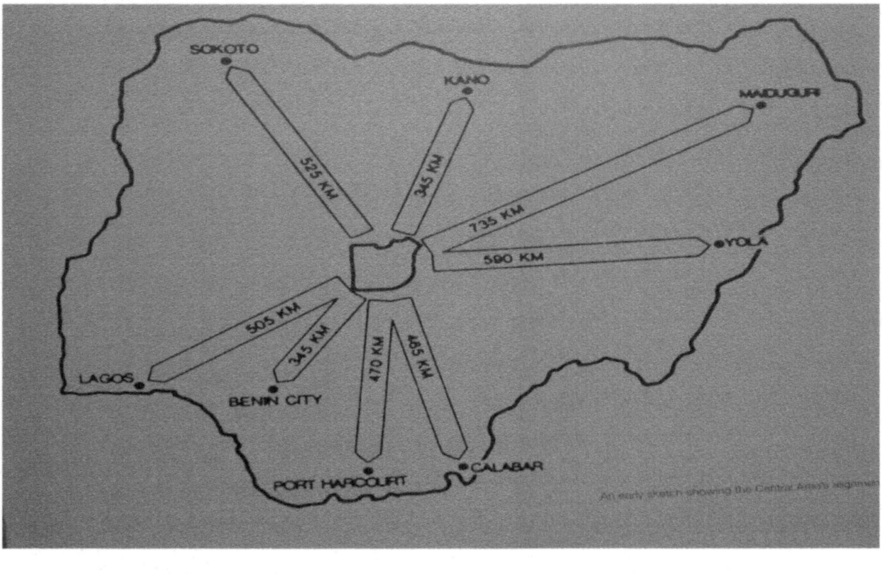

Fig. 14.2 Distance of Abuja (Federal Capital Territory) from other Nigerian cities. *Source* [41]

The Minister of the FCT is the overall administrative head and is appointed by the President of Nigeria. In 2006, the national population census stated that the FCT had a population of 778, 567 [2] and in 2016, the population of the Federal Capital Territory has risen to 3,564,126 [3]. Both Municipality and entire FCT have since 1991 undergone a huge population growth—with some areas around Abuja growing at a rate of up to 30% each year. Hitherto, the FCDA functioned as the implementing agency responsible for the provision of infrastructural facilities, management and control of all urban services such as transportation, waste management, healthcare services and regulation, water services and other essential infrastructural provision in the territory. Since about a decade ago, these mandates have been transferred to the Ministry of Federal Capital Territory.

Abuja FCT had an official population of 1.4 Million in 2006 (see Table 14.1), but given the daily influx of people into the territory, the population today could be well over 8 million people [4].

Abuja was planned as a capital where all Nigeria's ethnic groups, tribes, and religions would come together in harmony. It has avoided the violence prevalent in other parts of Nigeria, which has more than 250 ethnic groups. The variations in the population size of the area councils in Abuja as shown in Table 14.1 correspond with the functionality, livability and accessibility to the Abuja city centre. AMAC with the highest population figure serves as the city centre and Nigeria's seat of power where everybody desires to live because of the series of activities taking place. But due to the expensive accommodation, people tend to move out to other closer areas like Bwari, Gwagwalada, Kuje, Kwali and Abaji in the order of population size. The main factors determining the size of these other area councils is the livability and accessibility to Abuja capital city.

Abuja is one of the unifying factors in Nigeria because it accommodates every Nigeria of various tribes and religions, and it has been relatively peaceful except between 2010 and 2015 during which several suicide bombings traceable to Boko Haram terrorists were recorded. The city of Abuja has the potential to remain the economic growth-pole for the central region of Nigeria [5]. The city has the potential to be the cleanest and most conducive environment for livelihood if necessary policies are created and implemented.

The economic base of Abuja is yet to be as robust as it should be. Since it is the seat of Nigerian Government, blue and white collar jobs are provided by ministries,

Table 14.1 FCT Abuja and its area councils

Area councils	Headquaters	Population
Abaji	Abaji	58,642
Bwari	Bwari	229,274
Gwagwalada	Gwagwalada	158,618
Kuje	Kuje	97,233
Kwali	Kwali	86,174
Municipal	Abuja	776,298
		Total 1,406,239

Source [15]

departments and agencies of government. But, agriculture is the most prominent economic activity especially for the original land owners and unskilled migrants. Crops grown include maize, rice, vegetables, onions and yam. Fishing is another income generating activity. The region is blessed with an array of commercially viable solid minerals deposits such as Talc, Dolomite, Granite, Limestone, Feldspar, Mica, Marble, Sand and Clay. Presently, only Granite, Sand and Clay are being exploited. Industrial district was created by the International Planning Associates to host key manufacturing enterprises, however only few light manufacturing enterprises exist. Manufactured outputs include food and beverages, pharmaceuticals, paints, furniture, metal doors and windows. While FCC hosts the head offices of several formal sector businesses, informal sector small and medium enterprises dominate the region. The most prominent are commercial in nature followed by service enterprises like fashion design, printing, insurance, estate surveying, information and communication technology (ICT) products, textiles and garments and consultancy. Abuja is host to one public and five private universities plus several secondary and primary educational institutions.

Land use plans and physical development plans are context-specific and not necessarily mutually exclusive. Both land use and physical planning are guided by specific laws in the FCT. The administration of the FCT including land matters is vested in the Minister of Federal Capital Territory. The three most important Departments responsible for land management, infrastructure and physical planning in the ministry are: Department of Land Administration, Department of Engineering, and Department of Urban and Regional Planning. The Department of Urban and Regional Planning issues planning permits. The Survey Department surveys all lands and layout mapping for the purpose of land zoning and development. There is also a functional Land Use Allocation Committee. The Nigeria Urban and Regional Planning Law of 1992 provided the guide for physical planning in the FCT. Proper land use and physical planning in the FCT is confronted with some pertinent challenges, among which are the following:

- Institutions responsible for land and physical planning matters are weak due to inadequate resources including personnel, working equipment and vehicles.
- Low public awareness of the significance of physical planning laws and sustainable environmental management.
- Political interference in the physical planning process and city management.
- Inadequate involvement of the physical in the land use and physical planning process due to preference for "top-to-bottom" approach.

14.2 Methodology

As noted by [6], a considerable proportion of planning policies (rural, urban and regional) are based on facts or data, which invariably throw some light on the problem(s) at hand. The methods adopted for this study included sourcing data from secondary sources, presentation of data and proffering solutions based on available secondary data. The study utilized the United State's Environmental Protection Agency (EPA)'s model and the smart city conceptual framework to arrive at conclusion and recommendation.

14.2.1 Data Sourcing Specification

This involves data gathering from specific sources in order to have full knowledge of the circumstances surrounding the research and make generalization about the research interest. For the purpose of this study, the secondary data source was adopted.

Data Sources and Presentation A variety of existing data from various sources such as government reports and publications, periodicals, and the Internet were collected for the study. The data were consistent with the design strategy for the Abuja metropolitan region which is briefly highlighted below. Data presentations were based on the available secondary data.

14.3 Urban Problems Associated with Abuja Metropolitan Region

Abuja is a metropolitan city associated with the usual urban problems known to major cities of the world. These problems include: continuous Influx of People, constant Search for employment, traffic congestion, increased level of insecurity, poor and inadequate accommodation, high standard of living, poor electricity and water supplies.

14.3.1 Nature of Urban Sprawl in Abuja

Urban Sprawl can be defined in various ways. However, for the purpose of this work, it is define as having: segregated land uses, or a push for growth at the boundary of the metropolitan area [7]. Urban sprawl could be characterized by vigorous spatial expansion of urban areas. It involves transforming pastoral farmland into often-unattractive suburbs. Sprawl creates a disruption of a natural balance between urban and non-urban land uses, leading to a deplorable degradation of the landscape [8]. Others suggest that sprawl occurs in addition as a result of speculation over the future use of land, or as a result of physical terrain that is not suited for the continuous development [9]. Urban sprawl could also be argued to be lack of coordination of the decision to speculate which produces sprawl and not the speculation itself [9, 10].

Urban sprawl has developed in Abuja because of government's unyielding to the provision of adequate housing for the large population that has moved to the city over time. Also, the lack of transparency in the distribution of land which is dominated by elite' operators largely contributes to urban sprawl problem in Abuja. There is a need to use a more democratic process to allocate land not only to politicians, top civil servants, military officers, academics, top traditional rulers and other highly influencer individuals, but to ordinary citizens. In Abuja, urban sprawl comprises two main contrasting types of development in the same city. One type is characterized by large, peri-urban areas with informal and illegal patterns of land use and is combined with a lack of infrastructure, public facilities and basic services; this is often accompanied by little or no public transportation and by inadequate access roads [4]. The other form is the suburban sprawl, is one in which residential zones for high and middle income groups and highly valued commercial and retail complexes are well connected by individual rather than public transportation [4].

14.3.2 Continuous Influx of Migrants

The urban population in Nigeria is growing at a rapid rate. This is evident in the increase in its population of about 2.8% per annum with a 5.5% urban growth per annum [11]. The population of Nigeria was put at 170 million according to the 2006 Population Census. Over 50% of this population lives in the cities [12].

Most noticeable problem is the continuous influx of people into the FCT. One worrisome trend of the migration is that most people fleeing from many areas of conflict end up resettling in the metropolitan Abuja city simply because of the perceived secured environment. This can be deduced from persons of North-Eastern part of Nigeria that fled from the *Boko Haram* crisis and social unrests in the North, notably Bauchi, Yobe and Borno states among others, prefer to settle in Abuja due to better security apparatus that has ensured the overwhelming reduction in crime and its derivative [13]. Although, most of the migrants are located in the periphery and this is because of the lack of accommodation and/or high-cost of living in the city centre.

Also, job-seekers in search of the non-existent government jobs end up in Abuja and as they come, they also join in swelling the population of the city. The weak economic base of the territory contributes to the low employment absorptive capacity. These deficits as combination of forces have added to put a lot of pressure on facilities, amenities, infrastructure among other things being provided by FCT Administration for residents [13]. The pressure on public facilities in Abuja is noticeable in the following areas including transportation, housing, health, education security, and environmental challenges. For instance, there is acute housing deficit especially for the middle and low-income groups. Most of the schools, particularly public schools are overpopulated now because of the increase in child enrolment. The issue is the same with most health facilities, with most of the

General Hospitals now having larger number of out-patients and in-patients, especially the senior citizens and elderly person that usually have greater healthcare needs.

14.3.3 Increased Level of Insecurity

While Nigerian society was not initially highly stratified, the political developments within the country have created political elites out of retired army officers, former military and civilian heads of states and business people who have allied themselves with the army officers or political office holders (both elected and appointed). Nigerian civil servants and members of academia are also in a favored position as they are often the first to get land allocations at highly subsidized prices. These individuals are a very small percentage of the Nigerian population but are highly influential and very rich. This coupled with the increase influx of unemployed Nigerians with no means of livelihood, the level of crime and criminal activities tend to be on the increase [14].

14.3.4 Housing Inadequacies

A substantial proportion of houses that were built by householders were constructed informally and outside the official building codes and planning regulations. Production of houses for outright sale by small-scale builders and private housing firms, for sale and mortgage funded housing are very limited. The construction of housing units by companies for their workers is becoming less significant, not least since the economic down turn of the mid-1980s [15]. House hunters are therefore at the mercy of landlords and real estate agents who often demand two years rent upfront. In 2005 a modest house with three bedrooms and two bathrooms was available for rent at N500, 000 per year on the outskirts of Abuja but at the city center, the rent was between N1.5 million and N3 m. As at February 2017, it was between N2 million and N4.5 million at the city centre. In the outskirts of Abuja, the increment is very mild. There is a known reluctance on the part of corporate housing firms for providing rental housing units to low income occupants because the high return margins required are just not available from this particular group [16].

One of the major problems of housing in Abuja is largely because of non-involvement of private firms to play a major role in housing production. For instance, only few private investors are involved in housing production of the type seen in Fig. 14.3 (Goshen Villa Estate). Other examples of this type of estates include Crown Court, Peggy's Pointee, Efab Estate Lokogoma and Sunny Vale Homes. In Nigerian cities including Abuja, individuals construct a large proportion of the housing stock. A substantial proportion of houses that were built by householders were constructed informally and outside the official building codes and planning regulations. Production of houses for outright sale by small-scale builders and private housing firms, mortgage funded housing schemes and housing for sale programme is very limited in Abuja. This has resulted in subjecting prospective tenants to the mercy of landlords and real estate agents who often demand two years rent upfront. This has undoubtedly aggravated the housing problems in Abuja.

The exorbitant price of land that is indiscriminately raised by the FCT administrator (Minister) is also a contributing factor. The justification was that land was the only source of revenue in the FCT and that increasing the price would generate funds to continue to develop the area. The most unfortunate fact is that these plots of land are always sold to the land speculators thereby making it unaffordable to many unless the few rich people in the city. In a country where the minimum monthly wage is N18, 000 it is hard to see how anyone can afford a plot of land in Abuja [17]. The surrounding areas such as Kubwa, Lugbe, Bwari, and Karu etc. have become a haven for those who have to work in Abuja but cannot afford to live in the city centre, thus leading to the development of sprawl, poor housing condition and scarcity of accommodation for the teeming population [18]

14.3.5 Poor Electricity and Water Supplies

Water supply to some parts of the Abuja metropolis has always been epileptic in the recent decades. Sometimes, water supply to different parts of the metropolis may be disrupted for three days to enable technicians to carry out repairs on a damaged

Fig. 14.3 Goshen Villa Estate. *Source* [42]

water trunk main line around Gishiri in the Mabushi District of Abuja. Usually, when this happen, residents of Wuse I, Wuse II, Garki I, Garki II, Maitama, Asokoro, Wuye, Gudu, Games Village, Karu and Nyanya would experience water supply shortages. As reported in The Guardian of 5th February, 2016, residents of Karu Village in Abuja Municipal Area Council made several appeal to the FCT authorities managing Abuja city to help solve the problem of water scarcity and epileptic power supply affecting the area. Some of the residents said that the scarcity of potable water and unstable power supply was impacting negatively on their well-being. A respondent reported that the amount of money he spent daily to fuel his generating set and buy water add up to a handsome amount at the end of the month. "The major problem we have in this area is that we don't have water; no sufficient light (electricity). Sometimes we see light one day, the next three days no light. "So, this is affecting our business. We are not making impact at all because of all these social amenities that we are not getting. "Every day, I spend an average of N400 to buy fuel for my generating set. We spend a lot of money buying water from vendors. If you calculate this in a month, it amounts to big money." [19].

According to an official of Federal Capital territory Administration (FCTA), it was observed that the provision of basic amenities such as potable water and stable electricity would help to boost the standard of living of citizens. FCTA, while appealing for the intervention of government in the provision of the amenities, pledged that the residents would take ownership when provided, by protecting them from vandals. "If not for these water vendors, we would not be able to get water for use. "We have light only a few hours in a day but most times, the area is in blackout. "Electricity and water supply are essential to the well-being of the people; without electricity and water, life becomes very difficult. "If government can give us these essential amenities, we would appreciate it. If there is a water pump (tap) on this street, we (residents) can watch over it to protect it from vandals.

14.3.6 Transportation

Transportation is an essential and indispensible catalyst for activating and stimulating the pace of economic, social, political and every other human endeavours in any society. Transportation, as one of the basic infrastructures, is required for effective and efficient functioning of urban centres. This then suggest that transport infrastructure has to be developed in a logical way, so as to ensure that movement of people, goods and services are conducted quickly, economically, safely, comfortably and in an environmentally-friendly manner [20].

The city of Abuja has a fairly detailed and comprehensive Transport Master Plan originally developed to improve on the awkward experience of Lagos. The objectives of the plan [21] are to:

i. Maximize public transport mobility for residents who do not own cars (Captive Riders).
ii. Provide high quality attractive transit services to hire those with cars (Choice Riders).
iii. Minimize traffic movement passing through the various development sectors.
iv. Provide multiple highway paths between development sectors thereby avoiding network bottlenecks.
v. Achieve maximum self-containment within the outlying sectors.

The strategies to be adopted are discussed later in the chapter.

At the moment, efforts are currently on-going to meet these targets, but available evidences suggest that the city can be said to have a fairly good network of roads comprising Expressways, Arterials, Parkways and Collectors. But surprisingly most of these are concentrated on phase 1 which comprises only 95 out of the 208 districts of the FCT (see Table 14.2).

From the Table 14.2, it is obvious that for over thirty years of the existence of the Abuja, development has concentrated mostly on phase 1 area and the satellite towns, a situation which is now challenging mobility in the territory. Table 14.2 show a steady decline in the numbers of roads constructed in Abuja and revealed a lot of backlog meeting the road infrastructural needs of Abuja municipal and its environs. Also, the early relocation of the seat of government ahead of the proposed date changed the socio-economic data upon which traffic demand forecast in the plan was based relative to time and spatial distribution and thus diminishing the transportation system capacity to develop into an efficient flow [22]. Another major challenge is the non-development of the public transit ways which link all the major activity centres, markets, and offices. There is also the problem of not introducing the Light Rail Transit and the Rapid Rail Transit even as the threshold population and development levels have since been surpassed. This implied that Abuja city lacks an integrated intermodal transport system thereby making all movements within the metropolis road-based thus leading to overstretching of the existing road infrastructure.

Table 14.2 Stages of development of phase 1 Abuja city

Stages	No. of districts	Planned period of completion	Length of roads so far constructed	% of planned road lengths	Remarks
I	10	1990	263.77 km	90	With fairly adequate parking facilities
II	20	1995	77.2 km	35	Only in Wuye, Jabi, Utako and Mabushi areas
III	23	1998	53 km	25	Only in Gwarimpa I & II, Kado, Lugbe
IV	42	2006	10 km	5	Essentially link roads to the areas

Source [21]

14.3.7 Road Markings, Signs, and Traffic Signals

Road/Traffic signs are required in cites for the control, warning guidance and information of road users. They enhance traffic discipline, encourage full usage of available road space, and they add to the safety of traffic, besides improving traffic flow. These are absent in many areas of the city thereby contributing greatly to the traffic chaos experienced in the Capital territory. For traffic light controls, 85 junctions in the city have traffic lights controls, but about 25 more junctions require such facilities for effective coverage [23].

14.3.8 Traffic Congestion

In the area of transportation, the effect of the increase in population has been quite telling on the FCT because this has resulted in congestion on the roads and the highways within and leading to the city, thus making commuting a harrowing experience to and from work in Abuja a petrifying experience for most road users, The flashpoints are the entry and exit roads to the FCT, namely the outer Northern Expressway (ONEX), the Outer Southern Expressway (OSEX) and of course the Nyanya/Mararaba/AYA road (see Figs. 14.4 and 14.5). All these places are congested because even residents in far-away places like Jos, Lafia, Akwanga, Lokoja,

Fig. 14.4 A typical congested road in Abuja. *Source* [43]

Fig. 14.5 FCT Abuja Traffic. *Source* [43]

and even Minna access the Abuja city centre from these places for contracts, jobs and various other businesses. In other words they come into the city in droves very early in the morning and they begin to exit the Federal City (Abuja) by late evening, adding a lot of pressure and congestion [13]. Traffic congestion is renowned for accumulation of vehicular gaseous emissions and pollution thus contributing to global warming.

Whereas, the FCT has adopted some forms of ICT, the benefits are yet to become a reality. For instance, close circuit television technology was adopted for the city in order to improve surveillance and security, however, poor project implementation has led to its failure to impact positively on crime. Since 2004, land management in the FCT has been done with the aid of Abuja Geographic Information System (AGIS), but the problem of unapproved land use conversion and multiple allocations persist. In addition, the Federal Capital Territory Administration (FCTA) and by extension the Federal Government has devised some solutions to the Abuja's ever increasing population influx in the territory. It has formulated a policy that encompasses not only the construction of roads in the FCC but also in the Satellite Towns and Area Councils, this has set a mechanism in place to rapidly decongest the City Centre thus spreading the population of residents across the almost 8000 KM^2 of the territory, this is an unprecedented feat in the history of the FCT [13]. However, there is no discernible policy on smart city development just a there is no spatial plan for this purpose.

One of the major challenges of transportation in Abuja metropolis is the considerable stress and frustration now witnessed on the road system evident by prolonged traffic hold-ups. Traffic jams are now seen in all part of the metropolis and are very serious especially at peak hours (morning and evening). Passengers which include school children, workers and others wait endlessly daily at peak hours to board vehicles to various destinations in the metropolis. Notably, the congestion problem is most acute on the Expressways and major roads leading to the satellite towns especially during the peak periods of the day. On roads to Nyanya, Kubwa, Airport and Gwagwlada, motorists and passengers are trapped for hours daily. The heavy vehicular traffic volumes recorded on the main roads and other arterials in the metropolis as shown in Fig. 14.6 is to further illustrate the congestion problem in the metropolis.

14.3.9 Parking

Parking today is a serious challenge in Abuja. Though on—street parking facilities are provided on many routes in Abuja especially in the neighbourhoods, but increased car ownership has rendered them grossly inadequate. The parking problem is more acute on major arterials which provide access to centres of activities in the city. In an ideal situation, parking lots ought to be provided by government offices, public buildings, private offices, hospitals, schools, shopping centres among others within their premises but such is not the case in most of these places, thereby compelling visitors to park on the roads and made to pay for short-term durations. Unfortunately, off—street parking facilities are generally absent in most parts of the metropolis, and till date, parking remains a major obstruction to smooth flow of traffic in the entire Abuja metropolis.

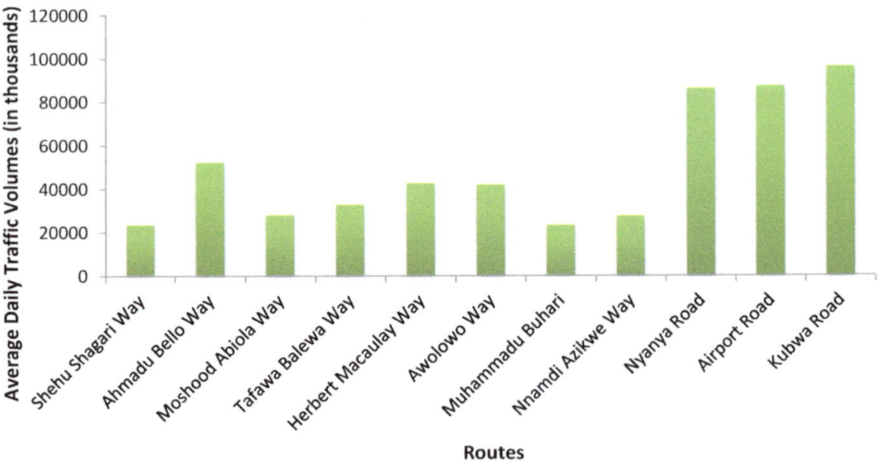

Fig. 14.6 Average daily traffic volumes on selected routes. *Source* [21]

14.3.10 Pedestrian Walkways

Good walkways are an essential complement to make public transport to be an acceptable alternative to the private car mode of transport in Abuja. The Abuja master-plan estimated that about 15% of Abuja residents will have some disabilities which may give them reduced mobility. In addition, Tourists would normally be expected to walk unhindered from their hotels to the shops and CBD for shopping or seeing. Walkways are therefore required for this growing section of urban residents [24]. But regrettably, many roads in Abuja lack such facilities. Roads to Nicon Hilton, and Sheraton hotels, NNPC building, Ahmadu Bello Way have all lost sidewalk characteristics. The construction of bridges and roads around the Stadium Complex which do not provide for walkways is strange since such a complex would generate the greatest level of non- vehicular traffic. In the CBD of Abuja, priority is given to vehicular traffic without consideration for sidewalk and provision of prams which can be used by nursing mothers. The bridges in Central Area have ramps without weave length while all the bridges linking Wuse to Central Area lack walkways as they disappear on approach to the bridges [23].

14.4 Spatial Design Strategies for the Transformation of Abuja Metropolis

Before proposing the preferred design strategies for transforming the region into a smart one, the key features of the spatial plan [25] and its implementation by the Federal Government) are presented. Figure 14.7 shows that the FCC occupies a small proportion of the Abuja region, and the population is also spatially concentrated in the FCC. Yet, there is growing population pressure on the other settlements within the region as earlier discussed.

The Federal Capital City was designed to accommodate a target population of 1.6 million people by 2000. Planners envisioned that the city ultimately would reach a population size of 3.2 million people once completely developed, after which population growth would be managed through the construction of adjacent, 'satellite' towns. The development of the city was to occur in phases, the first of which was planned for completion by 1986 and intended to accommodate up to 150,000 residents [25].

According to [26], the 1979 Master plan of Abuja was designed to be in four phases with a clearly defined target population of three million inhabitants. The city was designed as an efficient and attractive environment at each stage of its growth—from Phase 1, which was designed to accommodate 230,000 residents through Phases II and III, which were to accommodate 585,000 and 640,000 respectively, to Phase IV aimed at accommodating 1.7 Million [27]. The Master Plan is developed to provide land use pattern and development guidelines for 500 ha of government activity, 891 ha of service, 12,486 ha of residential land,

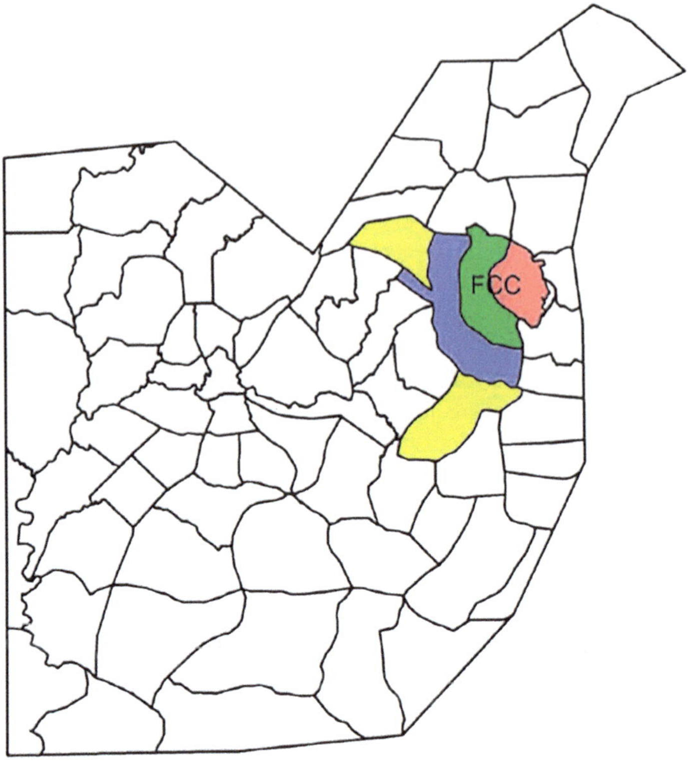

Fig. 14.7 The city of Abuja within the context of FCT. *Source* [25]

920 ha of light industry, 1840 hectares of transport infrastructure, 561 ha of commercial, and 8300 ha of open and recreational land (Table 14.1). It stipulates a maximum population of 3 million people after 25 years of development after which population growth would be accommodated in satellite towns [26]

14.4.1 The Extant Physical Development of Abuja

It is now close to 40 years that implementation of the provisions of the Master Plan and physical development of the FCC and FCT started. A lot has happened within this period. How has the development fared? What has happened to proposed land-uses? One of the indicators of smart growth is the proposed land-use within the context of a Regional Master Plan. The proposed land-use enables monitoring and control of physical development of a settlement. For the analysis of proposed and extant land-use conditions of Abuja Central Area, both the former and the latter were analysed, and the results presented in Table 14.3. The analysis revealed that <50% of the development zones of the Central Area has been developed; about

46.9% of the commercial/cultural zone has been developed, while 42.0% of the diplomatic zone has been developed.

Further analysis revealed differentiation in extant and proposed land-use for the Central Area. It revealed significant deviation between the proposed and extant land-use plan of the area (Table 14.4). The highest deviation is found in Commercial land use with 68.84% while the least been transportation/road network with 44.64%. This explains why the Central Area is not adequately developed as observed from the overall analysis above; therefore, this indicates the poor commitment of the implementations process of the Central areas development plan towards meeting the mandates of the FCC.

Infrastructural facilities are referred to as the mainstay of the city development. Owing to that, adequate provision was made for facilities in the Central Area development plan. The level of development of proposed infrastructural facilities is presented in Table 14.5. The analysis revealed that the plan is populated with high class infrastructural facilities in the Central Area, these includes proposed water supply (50,000m^3/d), electricity (1452 kV), road networks (2650 km), 4 numbers of Bus Terminus/Stations across the area, telecommunications (3 m lines), street lightening (265,000 poles) and storm water drainage (2650 km). It was discovered that the level of infrastructural development was below expectations. The highest deviation 85.61% was recorded for electricity (Fig. 14.8), water supply had 67.54%, 46.66% for telecommunications. The road networks analysis of the area

Table 14.3 Land development analysis in the central area of Abuja

Land allocation (Ha)	Administrative zone	commercial/cultural zone	Diplomatic zone
Allocated land (a)	409.12	460.26	153.42
Developed land (b)	184.10	216.32	64.44
Undeveloped land ©	225.02	243.9	88.98
% difference (d)	44.9	46.9	42.0

Note a-b = c, b/a × 100/1 = d
Source [47]

Table 14.4 Comparison of Land Use Development in the Central Area

S/No	Land-use type	Proposed (ha)	Existing (ha)	Deviations (Ha)	Percentage developed
1	Commercial	174.25	54.30	119.95	68.84
2	Public/Utilities/ Services	191.27	91.27	1000.00	52.28
3	Mixed uses	124.47	42.47	82.00	65.88
4	Open space	230.83	86.83	144.00	62.38
5	Transportation/road network	301.93	167.15	134.78	44.64
	Total	1022.8	442.02	580.78	

Source [47]

shows that the development of the road hierarchically have not been adequately carried out (Fig. 14.9), though some of the proposed roads in the development plan as at the time of the research are under constructions, the details of roads system analysis of the area and their percentage deviations are shows in Table 14.5.

Expansion and modernization of road network The most ambitious of the road projects are the Airport Road (re-named Umaru Yar-Adua Expressway)—37.5 km in length; the Kubwa/Zuba Expressway (Outer Northern Expressway) which is 39.76 km; and the Southern end-the Nnamdi Azikiwe Road. These are 10-lane super multiple carriage highways with full accompaniments of street lights, drainages, pedestrian bridges and flyovers, lays-by, culverts, telecommunication ducts and interchanges, etc. The Airport and Kubwa/Zuba Expressways have been able to reduce travel time to the City Centre from 3 h to 30 min (over 100% reduction) and billed for completion in 2014. The Southern Parkway from the Inner Southern Expressway to Ring Road completes the 10 lane ring, and completed in 2014. There is also the reconstruction of the 72 km Lower Usuma Dam—Gurara Dam—Jere Road aimed at creating an alternative route for vehicular movement out of the city (Abuja) to Kaduna through the Bwari axis.

The 3.8 km Inner Southern Expressway to Southern Parkway is now over 80% completed [28] and it's aimed at reducing traffic congestion induced by the North-eastern vehicular traffic at the AYA roundabout. Some internal roads completed are roads B6, B12 and Circle Road in the CBD with the aim of reducing travel time (Fig. 14.10).

Table 14.5 Analysis of infrastructural facilities in the FCC

S/No	Facilities	Proposed	Existing	Departure	% departure
1	Water supply	50,000 m^3/d	16,232 m^3/d	33,768 m^3/d	67.54
2	Electricity	1452 kV	209 kV	1243 kV	85.61
3	Road Networks				
	(a) Primary arterial Roads	62.15	10.24	51.91	83.52
	(b) Secondary Arterial Roads	85.08	21.49	63.59	74.74
	(c) Distributor roads	1,521.08	116.27	1,404.91	92.36
	(d) Local Access Roads	981.59	35.16	946.43	96.42
	Other roads				
	(i) Semi-tarred roads		62.21		
	(ii) Untarred roads		113.16		
4	Bus terminus station	4	–	–	–
5	Telecommunications	3 m lines	1.6 m lines	1.4 m lines	46.66
6	Street lightening	256,000 poles	80,000 poles	185,000 poles	69.81
7	Storm water drainage/foul water systems	2560 km	1015 km	1635 km	61.70

Source [47]

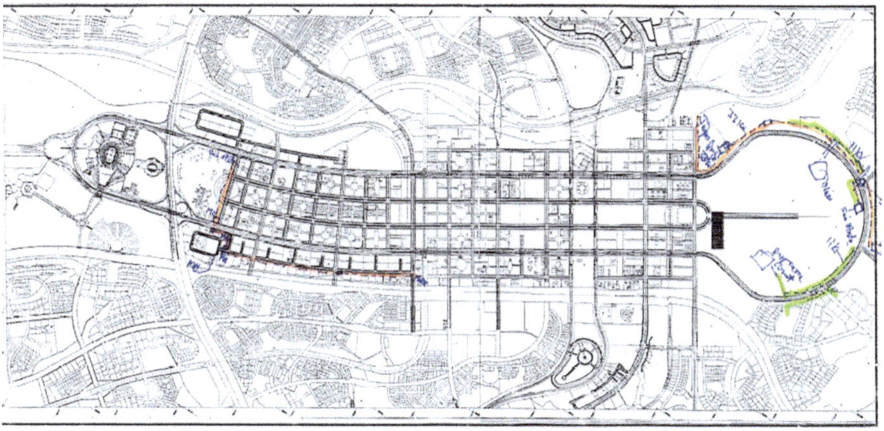

Fig. 14.8 Existing central area showing the electric cable network in Red Colour. *Source* [44]

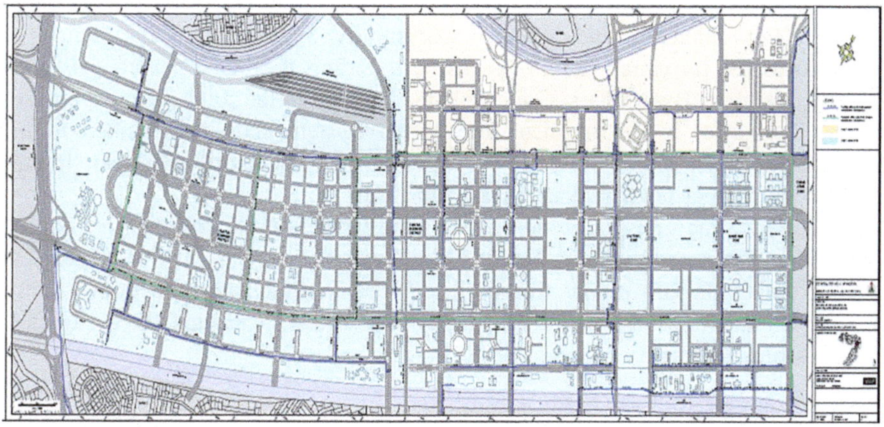

Fig. 14.9 Water supply. *Source* [45]

The Metro System The development of the Light-Rail mass transport system is one of the highlights of the Federal Capital transit initiative. Provision was made for the Abuja Metro project in the Abuja Master Plan but was not implemented until 2007. The Master Plan made provision for six light rail routes, totaling 300 km long. Construction of two of the routes (1&3) has begun and it was expected to be completed in 2015, but still on-going [28]. The remaining four routes are to be financed through the Public/Private Partnership arrangement. The rail network, when fully operational, is expected to employ 20,000 people in the FCC and to convey 700,000 passengers daily.

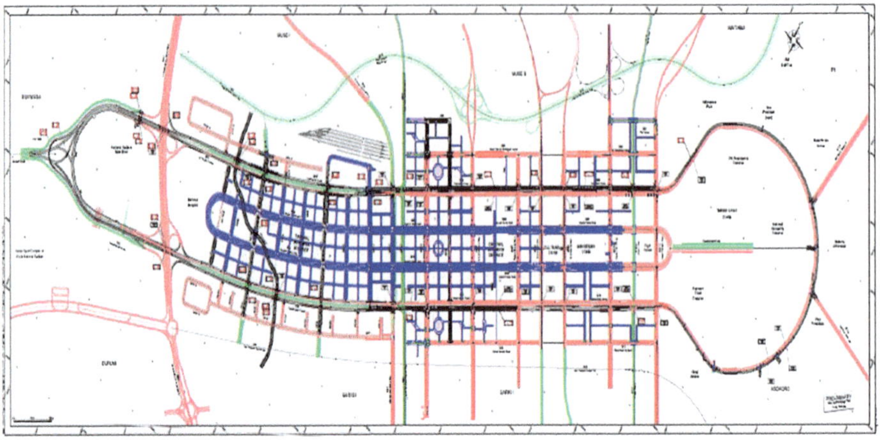

Fig. 14.10 Existing road network development in black colour. *Source* [44]

The Metro System is to be complemented by the Intra-and Inter City transport terminal. For efficiency and ease of accessibility by the poor, this should be extended to the proximal satellite towns surrounding the FCC, namely, Nyanya/Mararaba, Kubwa/Bwari, Kuje/Gwagwalada, and later to the southern-most parts of the FCT. Indeed, in the ultimate, the Metro System should be linked to the North-South West national rail line at Lokoja and the North East–South East national rail line at Gudi in Nasarawa State [29].

In the FCC, the transport zone traverses the Central Area from North to South along the axis defined by North-South-1 and Transit Way NS-2. It was intended to contain the city's key public transport services coming from and going to the Federal Capital City, the country and beyond. The zone hosted three (3) main elements plus the metro system;

(a) The National Railway main station
(b) The city Bus terminal and
(c) The Intercity Bus Terminal.

These elements if developed would have provided the city centre with vibrant transit system for its functionality and flexibility. But as at the time of this study, only the National railway station development is ongoing since the implementations of the plan, and it has been rated at 6.5% while the remaining important three envisioned facilities are yet to be implemented (Table 14.6). Furthermore, the pressure on the road by the private cars and other vehicles across the areas of central city manifesting a poor image for the city and health challenges to the residents and visitors due to hold-up induced stress and vehicular emissions. The failure in implementing the transportation zone components of the plan has strongly cropped the city image, economy, functionality and equality.

Table 14.6 Analysis of implementation of transportation facilities

S/No	Elements	Implemented (%)	Unimplemented
1	National railway main station	6.5	–
2	City bus terminal	0	–
3	Intercity	0	–
4	Transportation centre	0	–

Source [47]

14.5 Proposed Multi-nuclei Abuja Regional Land Use Plan

Smart Regional Development Plan (SRDP) refers to a strategy that builds upon existing assets, takes incremental actions to strengthen communities, and builds long-term value to attract a range of investments [30]. This development approach is a step-by-step guide to building a place-based economic and environmental development strategy. Many cities and regions around the World are adopting strategies of smart development that use less land and energy; provide safe, affordable housing options for people of all incomes and ages, and support transportation options such as walking, biking, and public transit [31].

Abuja FCT was originally designed to have more than one central business district (CBD), but presently, many other unplanned CBDs are springing up. This is largely due to the high influx of people, poor planning to cater for the inhabitants and the poor implementation of the Abuja master-plan. In the light of the foregoing, the extant land use plan is shown in Fig. 14.11 while the proposed land use plan is depicted by Fig. 14.12. The proposal in Fig. 14.12 covers the six Area councils with the headquarter of each constituting a growth center.

The Proposed Abuja Multi-Nuclei Regional Land Use Plan is conceived as one of the panaceas to the existing problems and enthronement of Smart Regional Development in Abuja. The proposed rail network links all the growth centers and AMAC. The road networks are also adequate for inter-connectivity among the growth poles. Additionally, it is proposed that the plan be revised every five years so as to avoid the problem that plagued the unrevised but now lapsed Abuja Master Plan. The following advantages are derivable from the proposed regional form:

- Reduction in trips to work and for shopping and recreation among others.
- Enhanced employment opportunities.
- Has the potential to catalyze regional development.
- Greater inter-state trade between settlements in the FCT and adjoining states.
- Population will be more evenly distributed.
- High probability that the urban prosperity and inequality gap will be reduced.
- Spatial distribution of population will be more even.
- Employment opportunities.
- Due to integration of transportation with land use in the growth centers, mobility of the urban poor will be affordable.

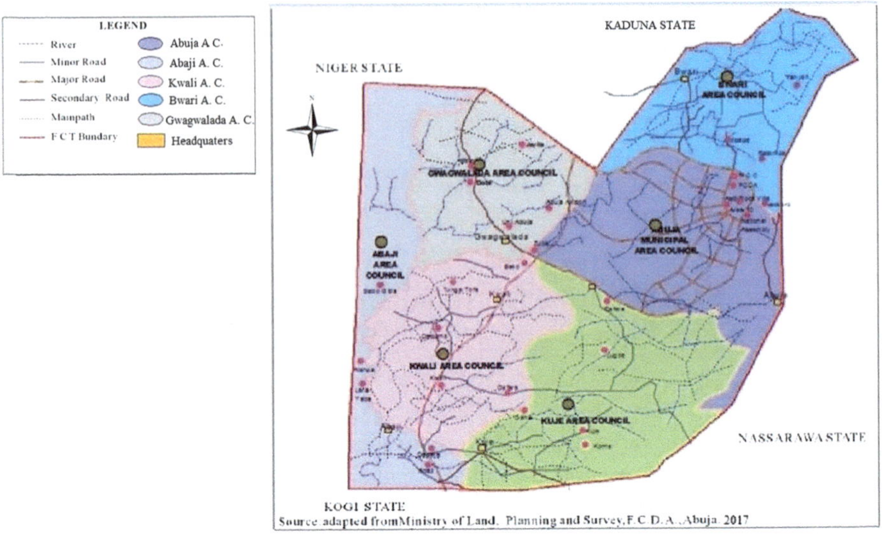

Fig. 14.11 Existing Abuja regional plan

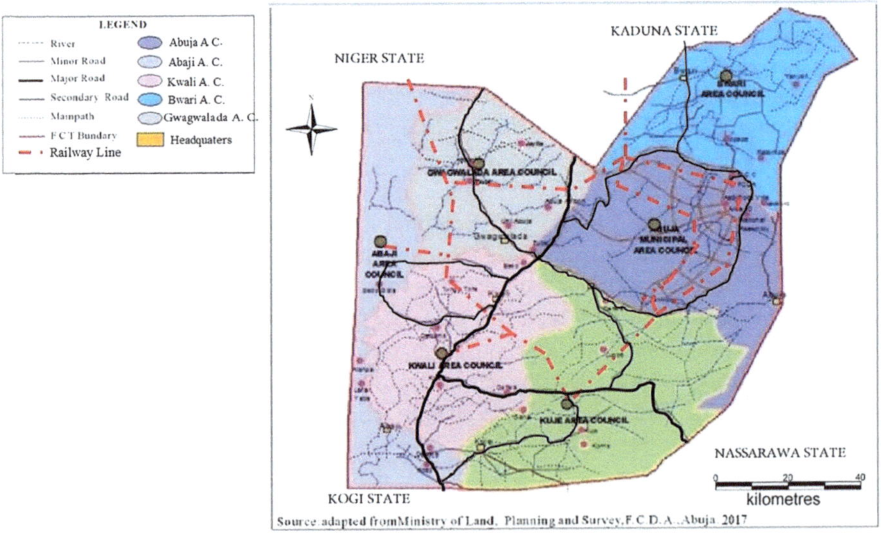

Fig. 14.12 Proposed Abuja multi-nuclei regional plan

14.5.1 The EPA's Smart Growth Programme and Its Adoption

The EPA's Smart Growth Program is a strategy adapted to enhance the smart development of the Abuja region. The EPA smart growth program is designed to inform businesses, developers, local government, and other groups about the benefits of smart growth development. This explores how investors can overcome real and perceived barriers to benefit from infill opportunities, how decisions about where to locate will impact the bottom lines of businesses, and why smart growth strategies are good fiscal policy for local governments. Thus, each of the region's main growth centers such as Abaji, Kwali, and Bwari shown in Fig. 14.12 can achieve the smart metropolitan status by first adapting the EPA's smart growth model coupled with the integration of ICT infrastructure and other regional development drivers [32]. The EPA model proposes three strategies that could transform a city. The strategies include developing a compact city by developing infrastructures, creating places that are walk-able and provision of a diverse range of choices in land uses, building types, transportation modes, housing, workplace locations and stores.

14.5.2 Compact and Infrastructural Development

In transforming Abuja into a smart region, compact development of infrastructural facilities is essential and needed. This entails using land and resources more efficiently and redeveloping old or neglected areas while retaining existing infrastructure which can create economic advantages for real estate developers, investors, businesses, and local governments. Compact development can generate more revenue per acre because it uses land more efficiently. It can reduce the costs of land and infrastructure for individual projects and the costs of providing fire and police protection, utilities, schools, and other public amenities [33]. By locating companies closer together, compact development can create a density of employment that increases economic productivity and attracts additional investment [33].

14.5.3 Walkability

Another strategy that could help in achieving the effort of making Abuja a smart metropolis is to make neighbourhoods in the city have well-connected streets and a mix of land uses near each other, making not only walking but also bicycling and transit more convenient and appealing. Projects in walkable neighborhoods command a price premium, earning real estate developers and investors a higher return on investment [33]. Improvements to streets and sidewalks to make them more appealing to pedestrians can benefit local businesses by attracting more customers. In turn, local governments benefit through additional property and sales tax revenue [33].

14.5.4 Range of Choices

Provision of varied choices for the neighbourhoods is also a way of creating a smart metropolitan region. This is because people and businesses value places that bring together a variety of activities to create vibrant environments. The demand for such places exceeds the supply. Many people in the two largest demographic cohorts, baby boomers and their children are particularly interested in lively neighborhoods with their daily needs close by. Communities with access to transit also help people reduce their transportation costs, enabling them to save money or spend more on their homes, entertainment, or other things they value. Changing demographics will likely further increase the demand for smart growth development over the coming decades; developers, investors, businesses, and local governments who respond to these market preferences could reap economic advantages [33].

14.6 ICT Integration for Smart Metropolitan Regional Development of Abuja Metropolis

Information and Communication Technology (ICT) plays a major role in all aspects of human endeavors: in politics, in economic life, as well as in social and cultural development. It is rapidly transforming our lives, the way we do business, provide and access information and services, communicate with each other. For a city to be smart, it must be inclusive at the onset of the urban planning and promote mixed neighbourhoods where social clustering is discouraged (see Fig. 14.13). The integration of ICT into the city structure is essential in solving many urban problems [34]. There are many ways of defining ICT. In 1998 Organization for Economic Co-Operation and Development (OECD) member countries agreed to define the ICT sector as a combination of manufacturing and services industries that capture, transmit and display data and information electronically. Going by this definition, it is deductive to say that transmission and display of data and information electronically is germane to smart regional development. A key element of smart urban planning is a smart street network that reduces travel time and encourages walking and social interactions [35]

Smart metropolitan regional development is not a unique initiative, but a complex set of projects and actions, different in goals, contents and technology intensity. For instance, ICT could be the pivot of a Smart regional development initiative. To achieve a smart Abuja regional development, initiative that allows for the incorporation of ICT into daily activities should be encouraged.

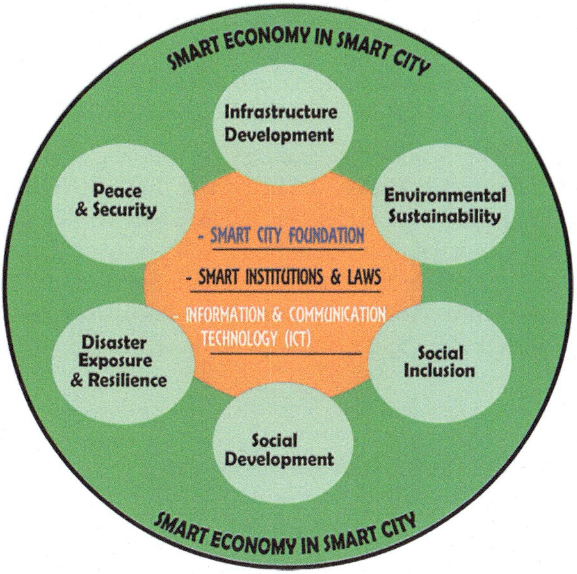

Fig. 14.13 Conceptual framework-smart economy in smart cities in the African context. *Source* [46]

14.6.1 Smart Mobility

According to [36] there are two meanings of Smart Mobility respect to the use of ICT: the first one refers to an efficient and effective mobility system and is independent from the role played by ICT, but it is rather connected to the use of appropriate technologies; while the second one relates to a mobility system characterized by a consistent and systematic use of ICT. Smart mobility system involves the use of all the paradigms composing the smart city. These paradigms include digital city, green city and knowledge city. ***Digital city***: because the traffic system could use ICT and software applications for a lot of different aims, such as optimizing traffic fluxes, support effective public transport routes, collect citizens' opinions and suggestion about urban mobility, and so on [37]. ***Green city***: because the environmental impact of transport in city is one of the main causes of city pollution [38]. ***Knowledge city***: because the smartness of transport depends also on the sharing of civic values and on the citizens' smart behaviours [39].

Smart Mobility is therefore a multifaceted topic that involves all the smart city paradigms and generating a set of heterogeneous benefits for the entire smart city stakeholders. They can act like agents of the Smart Mobility initiatives, that is, to be the movers of the actions, or gain the resulting benefits, or the both.

14.6.2 Abuja Geographic Information System

The Abuja Geographic Information System (AGIS) is a classic example of the smart initiatives of Abuja. AGIS was created in 2004 as an FCT government agency responsible for providing spatial data imagery for the administration, computerization and digitalization of spatially—related work flows in selected departments and agencies of government. It is also an independent service provider and the primary source for geo-spatial data in the FCT. AGIS now generates revenue from land related sub-heads: fresh application, ground rent, payment, sale of map products, etc. Land records and transactions have been straightened [28]. AGIS requires to be improved upon and made to be more efficient.

14.6.3 Conclusion and Recommendations

Abuja as the capital city of Nigeria offers multiple opportunities as a hub for political and administrative activities as well as a link to local, regional and global socio-political relations. Its high population density and its youthful population is a great catalyst for its economic productivity and growth. The FCT should implement the multi-nuclei regional land use plan with the strong political will it deserves so as to increase economic opportunities and livability while reducing inequality and deprivation. Regional growth management and sustainable development might not be realised in the absence of creativity and innovation. In an increasingly competitive world, coupled with the challenges of climate change and poverty, FCT would benefit tremendously from key stakeholders undertaking innovative and creative activities based on the following guiding principles:

- The city must adopt the smart city principles in the administration of the city.
- Full and immediate integration of ICT in the city's urban planning strategies.
- Commitment to a resilient, liveable, sustainable city underpinned by physical infrastructure that will drive a low-carbon economy.
- Promotion of an inclusive, employment intensive, resilient and competitive economy that harnesses the potential of inhabitants.
- Commitment to improved quality of life and development-driven resilience for all, irrespective of age, gender, religion and physical cum psychological challenges.
- Good Urban Governance, a very critical element to sustainable development of the EPA, hence its holistic application is quite germane.
- Build on the legacy of the Millennium Development Goals (MDGs) by providing the political will to successfully implement the United Nations Social Development Goals Compact, Post–2015 Development Agenda, and the Addis Ababa Action Agenda: 2015 (United Nations Social Development Goals (2015), United Nations Post-2015 Development Agenda (2015), UN-Habitat 2015) Addis Ababa.

- Promotion of mobility through provision of adequate space for streets and an efficient street network that facilitates walking and cycling.
- Co-location of residential, commercial and recreation land uses as mixed land use.
- Provision of diverse qualitative housing units that are affordable to the urban poor and non-poor.
- Link social and economic development with environmental protection and enhancement by making the most efficient trade-offs and appropriate choices, so that, the region's environmental assets are maintained at a level that meets the need of the present generation without jeopardizing the interests of future generations.

References

1. Mabogunje, A.L. 2001. The Promise, the Performance and the Prospects. The Review of Abuja Master-plan. Fountain Publication, Ibadan. pp 1–10
2. National Population Commission, 2006. National Population Commission, Abuja, Nigeria
3. Nigeria's National Population Commission and National Bureau of Statistics Estimates http://www.nigerianstat.gov.ng/report/470
4. Harvey RO, Clartk WAV (1965) Nature and economics of urban sprawl. Land Econ 1965:1–9
5. Burdett HN (2010) Urban Sprawl is now a Problem in World's Developing Countries http://potomacdigest.blogspot.com/search/label/Urban%20sprawl%20has%20a%20negative%20impact%20on%20infrastucture%20and%20sustainability%20of%20cities. Accessed 15 Mar 2017
6. Abumere SI (2004) Statistical methods in urban and regional planning. In: Agboola T (ed) Readings on urban and regional planning. Macmillan Nigeria Publishers, Ibadan, p 116
7. Johnson MP (2001) Environmental impacts of urban sprawl: a survey of the literature and proposed research agenda. Environ Plann 2001:717–735
8. Bruckner JK, Fansler DA (1983) The economics of urban sprawl: theory and evidence on the spatial sizes of cities. Rev Econ Stat 1983:479–482
9. National Association of Local Government Environmental Professionals and Smart Growth Leadership Institute (2004) Smart growth is smart business: boosting the bottom line & community prosperity. http://www.nalgep.org/publications/PublicationsDetail.cfm?LinkAdvID=52733
10. News Agency of Nigeria (2016) Residents appeal to FCT Authority over power supply and water scarcity. The Guardian 5th February, 2016. https://guardian.ng/news/residents-appeal-to-fct-authority-over-power-supply-and-water-scarcity-karu/
11. www.skyscraperscity.com
12. Imam A, Mohammed B, Wilson DC, Cheeseman CR (2008) Solid waste management in Abuja, Nigeria. Waste Manag 28(2):468–472
13. http://www.myfinancialintelligence.com/node/1574
14. www.247nigerianewsupdate.com
15. World Fact Book. CIA World Fact Book (2012) https://www.cia.gov/library/publications/the-world-factbook/geos/ni.html
16. National Population Commission (2006) Nigeria Population Commission, Abuja
17. Sumaila AG (2012) Sustainable Transportation Management for Federal Capital Territory, (FCT). Paper Presented at a One-Day Retreat for Staff of FCT Transportation Secretariat, Abuja. 4th August, 2012

18. www.thenews-chronicle.com
19. Federal Capital Development Authority (1986). The making of a new capital city for Nigeria. FCDA Official Document
20. Ogu VI, Ogbuozode JE (2001) Housing policy in Nigeria: towards enablement of private housing development. Habitat Int 2001:473–492
21. Adegboye, Kingsley (2009) Affordable housing, any hope for low wage earners? Vanguard. April 6, 2009. http://allafrica.com/stories/200904060833.html. Accessed 28 Apr 2012
22. Federal Capital Development Authority (FCDA), 2012. FCDA Document, 2012
23. Sumaila AG (2004) Road transportation in Abuja. Paper Presented at a one-day seminar on road transportation in urban centres of Nigeria: a case study of Abuja, Organized by National Economic Intelligence Committee, Abuja 2004
24. Sumaila AGF (2012) Characterization of current transportation challenges in the federal capital territory, Nigeria. J Sustain Dev 5(12)
25. International Planning Associates (1979) The master plan for Abuja the new federal capital of Nigeria, p. 27
26. Obiadi BN, Agbonome PC, Ezezue AM (2017) Abuja sustainable spatial housing design: a spatial dialectics (volumetric and unvolumetric Settlements). Mgbakoigba, J African Stud 6(2):111–129
27. Olaitan DT (2004) Sustainability and city development. A Critique of the Implementation of the Abuja Master Plan. Department of Urban and Regional Planning, University of Lagos, Akoka-Yaba, Nigeria
28. TELL Magazine, May, 2014. p. 32
29. Jiriko GK, Gwom Dung JY, Wapwera S (2015) The evolution of Abuja as a 'Smart City' a prognosis'. www.researcgate.net/publications/286458553
30. EPA (2016) Framework for creating a smart growth economic development strategy: a tool for small cities and towns. www.epa.gov/smartgrowth
31. Jonathan RC, Wallace R (2009) Smart growth guidelines for sustainable design and development. EPA, Washington, D.C
32. www.epa.gov/smartgrowth
33. Obiadi BN, Agbonome PC, Ezezue AM (2017) Abuja sustainable spatial housing design: a spatial dialectics (Volumetric and Unvolumetric settlements). Mgbakoigba. J African Stud 6 (2):111–129
34. United States Environmental Protection Agency (2012) Smart growth and economic success: benefits for real estate developers, investors, businesses, and local governments. EPA's office of sustainable communities with the assistance of Renaissance Planning Group
35. Femi Olokesusi, Femi Ola Aiyegbajeje, Gora Mboup, Dennis Mwaniki (2017) Smart city foundation for smart economy. In TM Vinod Kumar (ed) Smart economy in smart cities, Advances in 21st Century Human Settlements. Springer Nature, Singapore, pp 798–817. https://doi.org/10.1007/978-981-10-1610-3_28
36. Nam T, Pardo TA (2011) Conceptualizing smart city with dimensions of technology, people, and institutions. In: Proceedings of the 12th annual international digital government research conference: digital government innovation in challenging times. ACM (2011)
37. Olokesusi F, Aiyegbajeje FO, Mboup G, Mwaniki D (2017) Smart city foundation for smart economy. In TM Vinod Kumar (ed) Smart Economy in smart cities, Advances in 21st Century Human Settlements. Springer Nature Singapore Pte Ltd.
38. Mechant P, Stevens I, Evens T, Verdegem P (2012) E–deliberation 2.0 for smart cities: a critical assessment of two 'idea generation' cases. Int J Electron Govern 5(1):82–98
39. Zygiaris S (2013) Smart city reference model: assisting planners to conceptualize the building of smart city innovation ecosystems. J Knowl Econ 4(2):217–231
40. www.newsrescue.com
41. Julius Berger Nigeria PLC. 2006 Abuja the City. Julius Berger Nigeria PLC. p 11
42. Staricco L (2013) Smart mobility, opportunità e condizioni. J Land Use Mob Environ 3:289–354
43. https://www.tolet.com.ng/blog/5-best-residential-estates-in-abuja/

44. Engineering Department, FCDA, 2015
45. FCT Water board, 2015
46. Mboup (2015) Revised version of the conceptual framework for smart economy in Smart Cities 2014
47. Danraka IM (2013) Assessment of the implementation of the central area development plan of The Federal Capital City, Abuja. Unpublished M.Sc. Dissertation, Department of urban and Regional Planning, Faculty of Environmental Design, Ahmadu Bello University, Zaria, Nigeria

Part XV
Senegal, Dakar

Chapter 15
Towards a Smart Metropolitan Regional Development—Spatial and Economic Design Strategies: Dakar

Gora Mboup, Mame Cheikh Ngom, Cheikhou Balde and Mandiaye Ndiaye

Abstract Development of large metropolitan regions like Dakar constitutes economic and social opportunities as well as challenges depending on how the urban growth was planned designed and managed. The dynamic transformation of the Dakar Metropolitan Region from a population of 245,000 to 3.5 million inhabitants in 1950 and 2016 respectively has occurred in different paces, times and places in terms of numbers, urban form and structure (density distribution, and compactness). When a city reaches one million or more of population size, it becomes more complex for its planning, design and management compared to a small city. For instance, when a city reaches one million inhabitants, its accessibility will be largely determined by elements of its urban form and structure as well as elements of transport infrastructures and, recently, elements of Information and Communication Technologies (ICT). This accessibility determines in large the interaction between economic and spatial design strategies. After the introduction, this study is divided in six sections. **The first section** presents the "Conceptual framework of the Smart Metropolitan Regional Development" of the study; **The second section** presents "Urbanization of the Dakar Metropolitan Region in Time and Space"; **The third section** focuses on the "Urban Accessibility and Mobility"; **The fourth section** presents "Urbanization, Economic Growth and Metropolitan Regional Development" including economic growth, human development and sustainability,

---

G. Mboup (✉)
Global Observatory Linking Research to Action, New York, USA
e-mail: gmboup@gora4people.org

M. C. Ngom
University Cheikh Anta Diop of Dakar, Dakar, Senegal
e-mail: thieeko@hotmail.com

C. Balde
Municipal Development Agency, Dakar, Senegal
e-mail: debale2002@yahoo.fr

M. Ndiaye
Metropolitan Regional Development, Minister of Urban Renewal, Housing and Living Environment, Dakar, Senegal
e-mail: mnjaye@yahoo.fr

T. M. Vinod Kumar (ed.), *Smart Metropolitan Regional Development*, Advances in 21st Century Human Settlements,
https://doi.org/10.1007/978-981-10-8588-8_15

inclusion, resilience and prosperity; **The fifth section** presents "Towards a Smart Metropolitan regional Development"; **The sixth section**, based on the development of the previous sections, presents the "Spatial and Economic Design Strategies for a Smart Dakar Metropolitan Regional Development".

Keywords Smart metropolitan region · Economic & spatial design strategies Urbanization · Urban system · Urban form and structure · Streets and other public spaces · Human development · Accessibility · Sustainable · Inclusive Resilience · Prosperity

15.1 Introduction

Development of large urban agglomerations as the Dakar Metropolitan Region constitutes economic and social opportunities as well as challenges depending on how the urban growth was planned and managed. Large urban agglomerations provide opportunities for economies of scale and agglomeration, but it will also call for large investments in infrastructures to respond to the increased demand for water, sanitation, solid management, energy, streets and public spaces, and mobility, etc. They will also require efficient institutions for the management of social demand and equity such as on education and health as well as protection of people against violence and insecurity. Economies of scale and agglomeration economies are, for instance, greater in metropolitan areas where accessibility infrastructures are able to answer accessibility to services' needs with higher access to markets and resources than those where accessibility is impeded by deficient accessibility infrastructures. Efficient accessibility infrastructures will allow large-scale production of goods and services that can be distributed within the metropolitan regions and beyond with time, cost and reliability opportunities [1]. Without efficient accessibility systems, a metropolitan region losses its economic power and remains just clusters of disconnected settlements. The latter is the situation of the Dakar metropolitan region as most African metropolitan regions.

Dakar has not been able so far to respond to growth in accessibility demand and several other needs such as access to water, sanitation, management of solid waste, and streets and other public spaces, key elements of a city foundation. From a population of about 214,000 in 1950, the Dakar Metropolitan Region has a population of 3.5 million in 2016. Since 1950 Dakar has been a primate city with more than half of the national urban population living in the urban agglomeration that occupies a very tiny proportion, 0.3%, of the national territory. This has occurred through spatial as well as economics transformation and mutation. Along with the population growth, the land used has grown from 35 km^2 in 1950 to 179 square in late 1990 (with the three departments: Dakar, Pikine and Guediawaye), and now the metropolitan region including the department of Rufisque occupies an area of 551 km^2. This has been mostly done through urban inclusion as well as leapfrogging. Two main factors that characterise this spatial transformation are:

(1) human settlements with high density such as in the departments of Dakar, Pikine and Guediawaye and low density settlements in the department of Rufisque; and (2) conservation of the onset monocentrism of the department where all formal administrative and commercial activities are concentrated in Dakar, leaving the other departments as suburbs in the Dakar metropolitan region. Not only they are suburbs, but they lack most of the basic services. Finally their city foundation is weak to foster sustainable urban development.

Recognizing the unbalanced Dakar metropolitan regional development, in one hand, and the multiple problems associated to lack of basic services, on the other hand the Senegalese government has taken bold actions to transform the urban landscape of the metropolitan region from a monocentric to a polycentric metropolitan region with the creation of urban centres to decongest the Dakar urban agglomeration. This new spatial re-organization is also accompanied with economic transformation through the ambitious economic development, the Plan Senegal Emergent. This is through key main complementary programmes, reforms and plans: (1) the Plan Senegal Emergent; (2) Act III of Decentralization; (3) Metropolitan Planning; (4) Dakar Urban Master Plan 2035; and (5) Digital Senegal Strategy 2016–2025. These urban policies and programmes aim to spatially and economically transform the Dakar Metropolitan Region to be smart, sustainable, inclusive, resilient and prosperous. They aim to transform the monocentric system to a polycentric system where every urban centre equally contributes to the development of the metropolitan region. They implicitly lie on the principles of a Smart Metropolitan Regional Development which will be presented in the "conceptual framework, methodology and data" section.

The Dakar case study will consist of analysing the economic and spatial strategies undertaken by national and local authorities for a Smart Dakar Metropolitan Regional Development under the context of the historical transformation of the metropolitan region in time and space. After the introduction, this study is divided in six sections.

The first section "Conceptual framework of the Smart Metropolitan Regional Development" is based on its City Foundation (urban planning, basic infrastructures and land tenure) and Institutions as well as on the Economic Development, the Infrastructure Development, Environmental Sustainability, the Social Development, the Social Inclusion, Disaster Risk Exposure and Resilience, and Peace and Security. One element of the City Foundation, which is the spatial design, and the element of Economic Development will play the role of interface vis-à-vis the other elements of the conceptual framework. Economic and Spatial Design Strategies along cannot make a regional metropolitan region smart, the way they interplay with the other dimensions are crucial.

The second section presents "Urbanization of the Dakar Metropolitan Region in Time and Space" covering: The Formation of the Dakar Metropolitan Region from 1950 to 2015; Senegal's Urban System with Dakar being a Permanent Primate City; spatial planning and change in urban land use of the Dakar Metropolitan Region, elements of the city foundation.

The third section focuses on the Urban Accessibility and Mobility: Streets as drivers of urban accessibility; and Transport modes and infrastructures, traditional focus of urban mobility studies. Though streets constitute a fundamental element the City Foundation as determinant of the urban form and structure, we have given preference here to the prominent role they play in urban accessibility along with transport modes and infrastructures.

The fourth section presents "Urbanization, Economic Growth and Metropolitan Regional Development" including economic growth, human development and sustainability, inclusion, resilience and prosperity. Though cities are potentially engine of prosperity with their power of economies of scale and agglomeration as well as of technology innovation and diffusion of ideas, the Relationship between urbanization and development had been uneven in the Dakar Metropolitan Region. This section will assess the main reasons, including the urban form and structure, and the urban accessibility that have contributed to the relatively low economic performance of the Dakar metropolitan region compared many cities with similar population size in other developed regions.

The fifth section presents "Towards a Smart Metropolitan regional Development" including Human Development, Environmental Sustainability and City Prosperity.

The sixth section, based on the development of the previous sections, presents the "Spatial and Economic Design Strategies for a Smart Dakar Metropolitan Regional Development". It is sub-divided in their sub-sections considering the holistic approach taken by the government of Senegal as the holistic conceptual framework of the study: (a) Decentralization of the Dakar Metropolitan Region; (b) Plan Senegal Emergent (PSE): Holistic approach of economic development regional; (c) Spatial Design Strategies for a Smart Dakar metropolitan regional; (d) Creation of digitally, economic and socially served urban centres: from Monocentric to Polycentric Dakar Metropolitan Regional development; (e) Emerging Trends: Linking the Dakar metropolitan region to other regions—The Corridor Dakar-Thies-Mbour and beyond; (f) Connecting economically and spatially distinct urban centres in the Dakar metropolitan region; (g) ICT for Smart Economic and Spatial design Strategies; and (h) Means of Implementing the Economic and Spatial Design Strategies.

15.2 Conceptual Framework, Methodology and Data

15.2.1 Smart Metropolitan Regional Development: Concepts and Components

The New Urban Agenda adopted in October 2016 in Quito encourages governments, including the government of Senegal as a member of the United Nations, to commit themselves to adopting a smart-city approach that makes use of opportunities from digitalization, clean energy and technologies, as well as innovative transport technologies, thus providing options for inhabitants to make more environmentally friendly choices and boost sustainable economic growth, and enabling cities to improve their service delivery (NUA, paragraph 66) [2].

In our study, a smart metropolitan regional development is viewed as a sustainable, inclusive and prosperous metropolitan regional development that promotes a people-centric approach based on three core components—Smart Metropolitan Region Foundation, Information and Communications Technology (ICT) and Smart Institutions and Laws. Here The Metropolitan Region Foundation goes beyond the city foundation and includes elements of mobility across urban centres that form the metropolitan region. First each urban centre must have a smart city foundation and smartly connected to the other urban centres. Smart Metropolitan regional foundation, institutions and laws, and ICT are the pillars of the seven other dimensions of a smart metropolitan region: Infrastructure Development, Environmental Sustainability, Social Development, Social Inclusion, Disasters Exposure, Resilience, and Peace and Security. **Infrastructure Development** complements the basic infrastructure services under each smart metropolitan foundation and extends to actual investment and advancement of services such as transport, ICT, industrial energy, education, health, etc. **Environment Sustainability** is comprised of elements of Climate Change, Biodiversity, Waste Management, energy, transport, building and pollution. **Social Inclusion** includes aspects of participation in decision-makings as well as equal opportunities for growth and prosperity. **Social Development** encompasses elements of education, health, public space, social inclusion and social capital. **Disaster Exposure** incorporates elements of mitigation and adaptation to various disasters such as flooding, droughts, storms and earthquakes. **City Resilience** is composed of elements of city foundation, environment, social capital, and social development. **Peace & security** includes the elimination of all forms of discrimination and violence and conflicts, including domestic violence, violence in public places, crime, armed conflicts, terrorism, etc. An insecure metropolitan region limits opportunities for investment and economic growth and cannot be a smart metropolitan region (Fig. 15.1).

The opportunities for ICT to support the overall urban challenges and opportunities are enormous, and the Dakar Metropolitan Region must integrate and use ICT solutions to facilitate the greater provisioning of urban services. Following the exigency of the city of the 21st century that calls for sustainability, inclusion, resilience and prosperity, the planning and management of human settlements must take into consideration the gain in knowledge on various conditions that make cities smart, green, ecological, liveable and healthy. Through increased efficiency and innovation, ICT increases economies of scale and agglomeration, and promote diffusion of knowledge, even at the smaller human settlements. Small settlements as small firms are benefiting from ICT in making available their goods and services beyond their territories. Virtual offices, virtual networks, teleconferences are increasing large public participation and inclusion. Digital firms can start and scale up quickly with relatively little staffing or capital investment. ICTs harness the benefits of agglomeration economies in easing circulation of goods and services and encouraging polycentric urban development and allowing synergies between

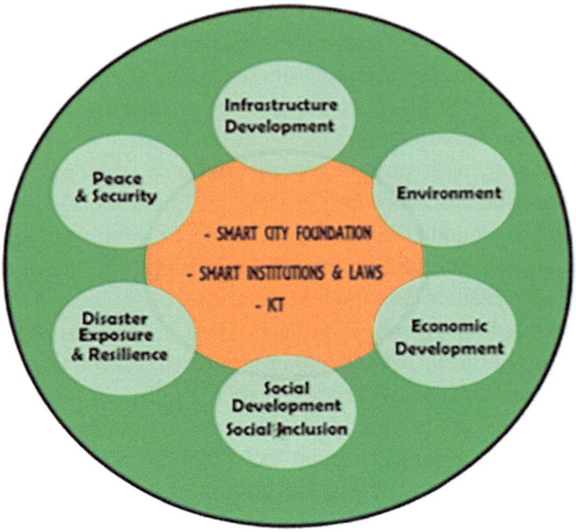

Fig. 15.1 Smart Dakar Metropolitan Regional Development Conceptual Framework. *Source* Adapted from Mboup, G. et al., 2016. Smart City Foundation—Drivers of Smart Cities. In Vinod et al., 2016. Smart Economy in Smart Cities, Springer

centres and sub-centres. They intensify urban nodes and corridors to maximize the benefits of concentration.

Numerous societal problems are explored and addressed in urban and regional planning agencies, including urban growth, unemployment and economic revitalization, transportation, environmental degradation and protection, neighbourhood decay and redevelopment, conservation of land and natural resources, provision of open space, parks and recreational facilities, etc. New urban planning instruments are becoming available with the worldwide spread of ICTs. They make it possible to adopt innovative e-planning approaches, strengthen communication between urban stakeholders, and make communication available at various stages of the planning process. Local governments can engage their citizens with real-time information to gain support for policy initiatives, identify unforeseen concerns, and recognize potential conflicts [3].

Cities are dynamic living organisms that are constantly evolving. ICT has begun to turn some places into real-time cities. This rapidly changing society makes the assessment and anticipation of future needs of city dwellers in terms of services, including transport, water, energy, employment, education and health, even more problematic. To address the complex problems of city planning it is not sufficient just to be concerned with the physical structure of the city; the interplay of intangible economic, social and environmental factors needs to be considered as well [4]. Planning procedures make use of models that show historic and present situations and communicate planned situations [5]. The introduction of ICTs allows planners and planning departments to carry out new actions or to implement conventional practices through new tools, such as GIS, virtual reality technologies, e-participation devices, including public participation GIS applications, among other tools, with the aim of improving conventional decision-making processes.

The provision of better planning and urban management services, more efficient, with lower costs and, at the same time, a more collaborative and participative, transparent and accountable planning decision-making process are some of the basic objectives usually associated with the move from conventional urban planning to e-planning [6, 7].

ICT also can increase economic productivity and competitiveness through increased smart mobility that reduces traffic congestion and commuting costs. Efficient and fast transport, in turn, can increase labour productivity by reducing commuting times, and increasing worker productivity. ICT reduces the use of motorized means and transport and promotes walking and cycling. This will improve the overall health of city residents, thereby reducing the health and economic costs of workers who are absent due to illness. Another important benefit is on consumer expenditure, as less is spent on cars and fuel. Non-motorized means of mobility support specific industries, such bicycle shops, tourism, retail activity, construction and real estate development that highlight liveability. It can also lead to a drop in the share of the household budget devoted to motorized means of transport, such as cars, which are expensive particularly in African countries.

The digital dividends are real and the Dakar metropolitan region must take the opportunity to efficiently integrate the use of ICT in all sectors of its development. From the beginning of the 21st century a digital citizenship, particularly the "Millennial" generation, has started to emerge in Dakar as in many African cities. The rapid adoption of digital technologies in the economy will have huge benefits directly as well as indirectly. However, ICT alone will produce little in the smart metropolitan regional development. 'Smart' is not an end in itself; It is the way ICT is integrated in the city development that will determine the city smartness. Maximizing the digital dividends requires better integration of ICTs with the other factors of smart cities such as: city foundation, infrastructure development, environment sustainability, social development, disaster prevention, resilience, peace and security. ICT can be seen as substitution and catalyst factors, but the other factors (or the analogue part) of the operation are crucial in making smart economy. Those factors include the city foundation, institutions and laws, infrastructure development, social development, social inclusion, environmental sustainability, disasters prevention and resilience (particularly in the context of climate change), peace and security. Most tasks have an aspect that cannot be automated by technologies and that requires human judgment, intuition, and discretion. This is where lies the crucial role of human capital associated with institutions and laws, efficient governance, management and administration. Without improvement on accountability at all levels, ICT alone cannot change the outcome of the economic productivity equation. African Countries that are able to swiftly adjust to this evolving digital economy will reap the greatest digital dividends, while the rest are likely to fall behind. This is the context where we are introducing economic and spatial design strategies for a smart metropolitan regional development to unlock the potential of Dakar metropolitan region to be smart, sustainable, inclusive, resilient and prosperous.

15.2.2 ICT Institutional Development in Senegal

Since 2001, Senegalese national authorities have elaborated the legislative and regulatory component of new technologies to create a legal environment favourable to their development [8]. The first step was the establishment by law No. 2001-15 of December 27, 2001, as amended, of the Code of Telecommunications and the Regulatory Agency for Telecommunications and Posts "Autorité de Régulation des Télécommunications et des Postes (ARTP)' responsible for providing the telecommunications sector with an effective and transparent regulatory framework, promoting fair competition to the benefit of users of telecommunications networks and services [9]. Another key step in the process of ICT development in Senegal was the creation of the State Information Technology Agency "Agence de l'informatique de l'Etat (ADIE)" which has the mandate to stimulate public action in the treatment and dissemination of information in accordance with international legal and technical standards for quality, availability, safety and performance (decree N° 2004-1038 of 23 July 2004) [10]. ADIE launched a process that led to the adoption of appropriate laws and regulations in 2005, followed by several laws on: orientation law on information society (law No. 2008-10 of January 25, 2008); electronic transactions (law No. 2008-08 of January 25, 2008); Cybercrime (law No. 2008-11 of January 25, 2008); the protection of personal data (law No. 2008-12 of January 25, 2008); Establishment of a voluntary contribution of one per cent (1%) on public procurement of goods and digital services (law No. 2008-49 of September 23, 2008); Establishment of a royalty on access or use of public telecommunications network—RUTEL (law No. 2008-46 of September 3, 2008) and; Cryptology with the creation of a National Commission of Cryptology, attached to the General Secretariat of the Presidency of the Republic, and whose permanent secretariat is provided by the Central Technical Services of Numbers and Security of Information Systems—STCC (law No. 2008-41 of August 20, 2008). The Privacy Protection Commission (CDP) is an Independent Administrative Authority (IAA) established under Law No. 2008-12 of 25 January 2008 on the protection of personal data. The year 2011 marked the adoption of a telecommunication code implementing most of the directives adopted by the Economic and Monetary Union of West Africa (UEMOA) and additional acts that the Treaty of the Economic Community of African States (ECOWAS) to create a legal environment conducive to the emergence of a regional market [8].

Under the "Plan Senegal Emergent", the Senegalese Government has particularly initiated the large digital programme called Senegal Digital Strategic Plan 2025. The slogan of the Plan is "ICT for all and for all uses by 2025 in Senegal with a dynamic and innovative private sector in an efficient ecosystem". The government is also putting in place an ambitious project of a technology park in one of the urban centres, Diamniadio, called "Diamniadio Technology Park". The park is based on the Silicon Valley model and intends to promote data revolution and higher education centres.

Dakar Digital City "Dakar Ville Numerique: Along the government programmes and policies on ICT, the city authorities of Dakar have also introduced initiatives aiming to make Dakar a digital city. The city of Dakar has launched the project "SSID: City Wi-Fi Dakar" [11]. A pilot project to install Wi-Fi in large public places was initiated by the city of Dakar in partnership with the mobile operator Tigo, a subsidiary of Millicom International Cellular Group, to make so free Internet connection to the public. The Initiative will be also extended to the public transport buses with the network 3G+. This will contribute to the expansion of Internet in a country with the where the cost of connection to Internet is still high and the speed of the network slow. City authorities will need to develop and accurate transit maps and data and train transit agencies in the use of such maps and data. Methodologies and open-source software applications are available allowing the production of transit maps at low-cost and in a short period. This will allow city authorities of Dakar to make ambitious planning and investment decisions based on accurate, comprehensive transit data [12]. Development and use of transit maps and data are becoming common in cities of developing countries as in Manila (Philippines). Mobile Surveys Applications with the introduction of the smart phone allow to automatically collect route data in an open-data platform with and open-source software system, transit staff members simply to ride the routes and allow the GPS capability of the phone to generate route coordinates that were simultaneously transmitted to the database [13].

Along the government programmes and policies on ICT, there are several other initiatives by private sectors, NGOs, Civil society, Academic that invest on the expansion of ICT. An ICT Incubator, launched in 2011 assists ICT companies, Information Technology and Communication, as well as project leaders, in their stages of creation, development and growth. It offers businesses and to project the ICT infrastructure and services necessary to ensure sustained growth and sustainable. CTIC Dakar is an example of Public Private Partnership led by the ICT Incubators Foundation of Senegal (FICTIS). Partners and sponsors include both public and private international organizations. Across the city of Dakar, CTIC are incubating companies, running an accelerator programme and a pre-incubation programme with start-ups as well as on a broad range of sectors, primarily but not exclusively commercial. HubSocial works to develop social solutions for Senegal and West Africa, particularly on the just concluded MDGs, particularly on poverty reduction, health and education. A little different is Jjiguene Tech, a female-led organisation with a mission to encourage women and girls in ICT and to keep female ICT graduates and others actively involved in the ICT business [14]. Technology hub Bantalabs have established offices in Dakar to provide open source web development, consulting and training. Facebook also invests in Senegal to launch Internet.org (a project aimed at bringing Internet access to two thirds of the world that are not connected) as well as more than a dozen free basic services within the country. According to Facebook, the services will be available to Tigo SIM cardholders. Senegal is now the sixth country in Africa, and third country worldwide, where Internet.org is available. The free basic services that will be available through Internet.org to consumers with a Tigo SIM card will include:

AccuWeather, BabyCenter & MAMA, BBC News, BING, UNICEF, Ebola Info, Facebook, UNICED Facts for Life, Girl Effect by Nike Foundation, Malaria No More, Messenger, Wattpad, Wikipedia, Wiwisport, Dakaractu.com, Senjob. The E-Riders Senegal project aims to mentor a group of young Senegalese to make it free software developers and providers of ICT services by the organizations of the civil society [15]. Jokkolabs works to support ICT communities by targeting multiple creative sectors, not just techies and open source geeks. Finally Coders4Africa is a not-for-profit with a mission to create hubs and labs across Africa where technologists and others can receive advanced training and develop ICT tools that address community needs [16]. All these projects, though not well coordinated to increase synergy, witness the era of ICT revolution in Senegal, and particularly in Dakar.

The establishment of a legal and institutional framework for the administration, implementation and use of ICT infrastructure coupled with several private and public ICT initiatives show clearly that Senegal, particularly Dakar, is moving towards a digital era. If well coordinated and articulated all these initiatives should yield to a smart metropolitan regional development. Indeed, the establishments of various laws and institutions governing the use and the development of ICT has been followed by en effective integration of the ICT in several publics as well as private sectors with the adoption of: E Governance, E commerce, E Education, E-infrastructure, etc.

15.2.3 State of ICTs in Dakar

ICT infrastructures have various components ranging from hardware categories such as mobile phone, computers and tablets to soft categories such as Internet connection, cloud storage and multiple applications. In this chapter, we present the basic components of ICT hardware, which are mobile phones and computers, as well as Internet connectivity that allows global and local inter-linkage. Whereas the other aspects are equally important, data on them is seldom available and comparable in the context of Dakar and will thus not be discussed here.

Data on mobile phone, computer and Internet had been collected in series of DHS conducted in Senegal over the past 20 years. The advantage of using the DHS surveys is that they allow for association of ICT components with economic as well as social variables such as education and literacy using the same source of information. The DHS data shows that, almost every household in Dakar has access to a mobile phone, easing potential access to Internet and multiple apps. With a level of 58% in 2005, the coverage of mobile phone is quasi universal in Dakar with a level of 99% in 2015. This spectacular boom of mobile phone makes some authors to feature Senegal as the next Silicon Valley. Recently it was demonstrated that revenues of the telecommunication companies are much higher that total costs of consumption of water and other basic goods, pointing that households in Dakar take communication via mobile top in their agenda compared to other household

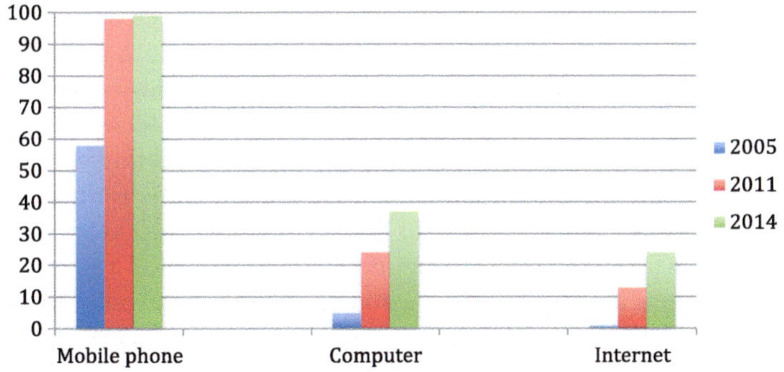

Fig. 15.2 Percentage of households with at least a mobile, a computer a connection to internet, Dakar 2005–2014

needs. Considering the fact that the mean household size in Dakar is 6, and considering the fact that each person has at least a mobile phone, we can forecast that there are at least 18 million mobile phones in the city of Dakar. This makes mobile phone business and mobile telecommunication very attractive in the city. In addition to making phone calls, the mobile phone are used to access various Internet based commercial opportunities (e.g. jobs, business opportunities through online trading etc.) as well as social communication platforms (e.g. Facebook, Viber, WhatsApp, etc.) (Fig. 15.2).

Along the boom of mobile phones, there is increase in other ICT infrastructure such as computer and accessibility to Internet, but in a lesser extent. With only 5% of level of possession in 2005, in 2015 more than one-third of people live in a household with at least a computer (37%). With similar trends, the coverage of computer is estimated at 57%, 75% and 92% in 2020, 2025 and 2030 respectively. The possession of a computer is not yet generalized in Dakar and feature in richest families and in household where some members are with high education. The association between computer possession and level of education shows that people with high education are more likely to live in a household with computer. In these households, computers may be needed for school as well as for work activities.

Access to Internet at the household level has also increased from less than 1% in 2005 to 13% in 2011 and 24% in 2014. The coverage of an Internet connection has drastically increased since 2005 where the percentage of people living in a household with an Internet connection was less than 5%. Six years later, the same percentage was multiplied by six with 24% of people living in a household with Internet in 2011. The trend is irreversible with an absolute increase of 13 points in 3 years; in 2014, more than third (37%) of people lived in a household with an Internet connection. With similar trends, this percentage of household with Internet connection will reach 62% in 2030 and 75% in 2035. However, with the desire to make Dakar a digital city, this trend can be boosted and the availability of an internet connection at the household level can be generalized way before the year 2030, and this will be in line with the increase in education and literacy observed in

Dakar during these past 20 years. Today 85% of young women and 95% of young men 15–19 years have at least a primary education level. Investing in infrastructure goes hand and hand with investing in education. Today there are 63% of young women 15-19 years and 71% of young men 15–19 years old who have an education level but live in a household where the Internet connection is not available. This represents a real missed opportunity from education. The situation is more dramatic if we consider that only 38% of young women 15–19 years old and 47% of men 15-years with secondary education or higher live in a household without accessibility to Internet. This represents a double miss of opportunity for ICT use as well as for ICT development. It is much easier to make the internet accessible to train educated people to use it and develop apps than otherwise [17]. There is also a missed opportunity with households that have computer but do not have access to Internet (13% of households). Finally the only option of these households is to use Internet via mobile phone as other households without a computer. With the boom of social media platform such as Facebook, WhatsApp and Viber, the use an Internet via mobile is almost universal in Dakar.

15.2.4 Urban Legislation, Rules and Regulation

Senegal's urban policy is institutionalized by the creation of a ministry in charge of urban issues in relation with other ministerial departments. In various forms, this ministry has always been part of the government structure since independence. Today, the Minister of Urban Renewal, Housing and Living Environment (MRUHCV), prepares and implements the policy defined by the Head of State in the areas of urban planning, restructuring and urban renewal, housing and construction [18]. The MRUHCV defines the framework for urban planning and ensures the development of cities and agglomerations, in cooperation with other ministries such as territorial planning and local communities, to the protection and enhancement of urban architectural heritage. Since independence, an arsenal of legislative and regulatory texts were prepared and adopted, including the adoption of the Urban Planning Code "Code d'Urbanisme") adopted in 2009. The Urban Planning Code defines the framework for the development of urban and peri-urban areas. It sets the rules for urban planning instruments and operations as well as other planning acts such as building permits and demolition. The provisions of this document must, however, be supplemented by other texts, in particular the law on the transfer of powers and the law on expropriation for reasons of public utility. The regulatory and legal framework of the urban sector is governed by a system related to land tenure, decentralization, planning code, etc. In the aftermath of independence, there was a diversity of tenure systems in Senegal, including the cohabitation of traditional land tenure and inherited colonization land tenure. At the time of the country's independence, the various measures taken to remedy this situation have resulted in a pluralist land system.

Along the urbanization process, national authorities have put in places planning guidelines through the Urban Master Plan of Dakar (Plan Directeur d'Urbanisme—PDU) 2035 and National Plan for Territorial Development "PNAT". The PDU 2035 aims promoting the at urban development of the region of Dakar and its surroundings by 2035. Its main objectives are: Sustainable Urban Development; Compact cities connected with a transport network; Robust and Resilient city and; Vibrant city with active interaction between information, goods and people. The PNAT proposes five development urban areas: Multifunctional urban areas; Areas for economic activities; Areas for agriculture activities; Areas for touristic activities; and Areas for conservation. The PNAT also identifies areas with high risk for habitation and any other activity. In these areas, modification of the land including by residential structure, or any cadastral operations is prohibited. This plan promotes green areas and other protected areas for environmental purposes or spaces with light recreational facilities, which, when well integrated enhance the character or the ecological value of the area. Specific focus areas of the PNAT also aim at: controlling the internal urban growth; reducing the proliferation of slums; promoting a balanced urban development; and meeting the housing demand.

The MRUHCV also develops tools, instruments and guidelines on how to design, plan and manage cities. From 1946 to 2015, there have been several urban Master Plans developed in 1946, 1961, 1967, 2000 and 2014 (revision of the PDU 2025) [19]. The PDU is a reference document that aims to plan and program the development of an agglomeration in short and medium terms (10 to 20 years), taking into account the global objectives of the regional development. It specifies the socio-economic and demographic perspectives of the region and the agglomeration and determines the means and strategies to be implemented in order to reach a harmonious and sustainable development. There are other sectorial plans, which have been also hardly implemented. The Master Plan for Urban Planning and Preservation of the Niayes and the Green Zones of Dakar (PDAS) has not played its role yet, in terms of overall urban planning. The PDAS is a specific plan for the Niayes (depressions zones) and Green Zones at the level of the Dakar Metropolitan Area, and has been validated since 2004. The Regional Land-Use Planning framework (SRAT) provides comprehensive guidelines, which are not compulsorily taken into account in the field. In fact, the SRAT is validated but the implementation does not really respect the prescriptions included in it. Finally, the actions planned in the Regional Integrated Development Plan (PRDI) finalized since 2004 are not implemented yet.

Since independence, the government of Senegal has also opted for a progressive and irreversible decentralization policy. The Act III of Decentralization adopted in December 2013 was the result of this long process. Under this Act each department of the Dakar Metropolitan Region, has been sub-divided into municipalities. Overall, the metropolitan region counts 49 municipalities. Each municipality is headed by a mayor, who is responsible to, among other things, preserve, maintain and administer the properties and assets of the municipality, ensure municipal waste management and hygiene, protection and conservation of historic sites and monuments, promotion of national and local cultures, preparation and implementation of various kind of plans (including master plans and detailed urban plans), and management of health and education programmes. The Act further gives guidelines on municipal finance, which is key in the implementation of urban programmes. This is a clear demarcation of the original urban development management where administrative and financial management of municipalities was performed at the central level; it opens the road to polycentric specialized urban centres.

Senegal has also made a special commitment at the international level to promote sustainable and resilient urbanization as promulgated in the Sustainable Development Goals (SDGs) adopted in September 2015, the New Urban Agenda adopted in October 2016, the COP 21 (Paris Declaration), the Sendai Framework on Disaster Risk Reduction, the Addis Ababa Declaration on Population and Development and the Africa Agenda 2063.

Despite the existence of comprehensive urban legislation, rules and regulations, the urbanization of Dakar has been marked par proliferation of informal settlements. Indeed, a cumbersome regulatory approval process constrains the implementation of several urban policies and plans. The intermediate causes to this cumbersome approval process include: Lack of clarity of roles and responsibilities of various administrative structures; weak coordination between the Urban line Ministry and municipalities in charge of localization urban policies. In addition to that municipalities do not have the required skills and competencies to properly plan, design and manage their cities. Awareness and capacity building of local authorities and communities are key for the success of urban policies. Local authorities do not also have sufficient resources to implement urban policies. Urban issues are not well positioned in the competition of several other interests when it comes to economic development.

15.2.5 Methodology and Data

The Smart Metropolitan Regional Development is composed of several dimensions. Put them together to assess the degree of smartness will require the development of composite indices using sophisticated multi-level statistical analysis including Principal Component Analysis. The complexity of composite index is the fact that it is strongly influenced by several factors including: the normalization, the standardization, the degree of association between variables and the number of variables that compose the index. The analysis of the Dakar Smart Metropolitan Regional Development (SMRD) will consist first of the analysis of each of its components with the production of individual indicators such as the proportion of household with access to improved water, followed by a development of a composite index, the SMRD index (SMRDI) with sub indices such as the Smart City

Foundation Index (SCFI) built upon the spatial planning, the land tenure and the basic infrastructure indicators. However, an index can also be conceptualised differently using different variables. This calls for caution when interpreting a composite index. For instance the Quality of Life is conceptualised and measured in different ways across studies. The Human Development Index includes indicators of health and education, alongside an economic component (Gross Domestic Product or GDP); the Legatum Prosperity Index considers quality of life to be multidimensional, including both wealth and wellbeing; The Economist Intelligence Unit's Quality of Life Index links life satisfaction to health, family life and community life; and the OECD's Well-being Initiative has two dimensions: 'material living conditions' and 'quality of Life' [20]. The following section will provide the different components of the SMRDI and the variables used for their development.

Putting the dimensions together to develop the Smart Metropolitan Regional Development Index (SMRDI)
The SMRDI, which is built here, is meant to serve research as well as policies. For the latter, the SMRDI will be built using existing mechanisms of the production of indicators globally as well as nationally. Building the SMRDI through existing mechanism of development of indicators such as the monitoring of the Africa Agenda 2063, the Sustainable Development Goals (SDGs), the New Urban Agenda, the Sendai Framework, the COP21, the Addis Ababa Declaration on Population and Development as well as the monitoring of national programmes and plans such as of the Plan Senegal Emergent (PSE) will ease its implementation at the local level for SMRD policies. As part of the international communities, Senegal has also endorsed all the above global agendas. Box 15.1 provides for each dimension of the SMDR variables, quantitative or qualitative, included for its measurement.

Box 15.1 Smart Metropolitan Regional Development Index (SMRDI)

Dimensions	Definitions/variables
City Foundation	*Spatial planning* Planned settlements, Streets, Other Open Public Spaces *Land tenure* Land: Documentation or perceived eviction Housing: Proportion of urban population living in slums, informal settlements or inadequate housing *Basic infrastructures* Connection to piped water, Connection to sewerage system, Connection to electricity, management of solid waste

(continued)

(continued)

Dimensions	Definitions/variables
Institutions & Laws	Urban Planning Codes, Urban Policies, Urban Plans, Unplanned settlements, Resilient building codes, standards, development permits, land use by-laws and ordinances, and planning regulations;
ICT	Connection to internet, Mobile phone, Computer, Broadband
Infrastructure development	Transport and road infrastructure, Energy, Knowledge infrastructure, Health infrastructure
Environmental sustainability	Air quality (PM10), CO_2 emissions, energy and indoor pollution
Economic development	City product, Employment Primary, Secondary and Tertiary sectors
Social development	Education: literacy, primary, secondary and tertiary enrolment Health: life expectancy, under-five mortality rates, HIV/AIDS, morbidity and nutrition: Social capital, Public space
Social inclusion	Income/consumption Inequalities: Gini coefficient, Health, Education, Access to basic services
Disaster risk reduction/resilience Strengthen resilience and adaptive capacity to climate-related hazards and natural disasters in all countries	Number of deaths, missing persons and directly affected persons attributed to disasters per 100,000 population Adoption and Implementation of national disaster risk reduction strategies in line with the Sendai Framework for Disaster Risk Reduction 2015–2030 Proportion of local governments that adopt and implement local disaster risk reduction strategies in line with national disaster risk reduction strategies
Peace & Security	Proportion of population that feel safe walking alone around the area they live; Proportion of women and girls aged 15 years and older subjected to sexual violence by persons other than an intimate partner in the previous 12 months, by age and place of occurrence; Proportion of ever-partnered women and girls aged 15 years and older subjected to physical, sexual or psychological violence by a current or former intimate partner in the

(continued)

(continued)

Dimensions	Definitions/variables
	previous 12 months, by form of violence and by age; Average share of the built-up area of cities that is open space for public use for all, by sex, age and persons with disabilities; Proportion of population that feel safe walking alone around the area they live

Source Mboup et al. 2018 Forthcoming Relevance of the Concept of Smart City in the African Context. Springer. All the variables of this table are not included in the Dakar SMRDI due to lack of data

Data sources and quality assurance

Data are generally obtained from various sources depending on the level of analysis. For the SMRDI, as an impact product, population-based data is most indicated since they provide information on people's quality of life. Since 1976, Senegal has already conducted six censuses and several household surveys such as Demographic and Health Surveys, the Multiple Indicators Cluster Surveys, the Living Standard Measurement Survey and several other household surveys. They will constitute the principal data sources of the SMRDI. In addition to these population-based data, the analysis of spatial form such as density, compactness, streets call for the use of the Geographical Information System (GIS) for the analysis of remote sensing data with the integration of population-based data. The analysis of the component of the SMRDI such as Institutions and Laws call for a Policy Analysis of urban legislations, rules and regulations in Senegal, it is more a content analysis of administrative records rather than a spatial or a population-based analysis. Finally, the development of the SMRDI will be based on quantitative as well as qualitative data: population-based, spatial and content analysis, etc.

Challenges and emerging successes in the definition of urban components for the development of the SMRDI

The development of the SMRDI will face challenges such as data comparability, the delimitation of urban boundaries, etc. Critical issues and huge challenges the development of the SMRDI will face is the choice of data sources in the situation where it is noted significant discrepancies between national sources. Lack of data comparability and non-compliance with international standard make the development of SMRDI challenging. A lack of uniform definitions across data sources poses an even greater challenge for aggregating data and comparing levels of SMRDI [21]. Urban human settlements have various definitions: urban agglomeration, metropolitan region, city proper, municipalities, etc. These definitions also vary across countries [22]. The **urban agglomeration** is defined as the built-up or densely populated area containing the city proper; suburbs, and continuously settled commuter areas. This may be smaller or larger than the metropolitan area. Other similar UN definition: Comprises a city or town proper and the suburban fringe or thickly settled territory lying outside, but adjacent to, its boundaries. A single large urban agglomeration may comprise several cities or towns and their suburban fringes. (United Nations. 1998. Principles and Recommendations for Population and Housing Censuses). This has been regularly published by UNDESA and it forms the official reference of urban population as in the Urbanization Prospects publication. The **metropolitan area** is the set of formal local government areas, which are normally taken to comprise the urban area as a whole and its primary commuter areas. The **city proper** is the single political jurisdiction, which contains the historical city centre [23].

All these three definitions of urban component have often interchangeably been used for the measurement of urban population, urban densities but referring the same name. For instance, the Dakar Metropolitan Region, administratively called region of Dakar, is sub-divided into four departments: Dakar, Pikine, Guediawaye and Rufisque. The first three departments and the urban areas of the department of Rufisque form the urban agglomeration of Dakar. The Department of Dakar forms the city of Dakar. Though they have distinct geographical boundaries, all three (the metropolitan, the agglomeration and the department of Dakar) have the same name, Dakar, making always confusion on the production and analysis of urban development statistics in the metropolitan region. For instance, in 2015 the population of the department of Dakar was 1.2 million, the population of the urban agglomeration of Dakar (excluding Rufisque) was 2.8 million, and the population of the Dakar Metropolitan Region was 3.3 million including all departments. To overcome such confusion, we will use the full name for each geographical entity.

Another important measurement, one can be interested is how the spatial growth occurs along the population growth, a crucial information for spatial planning. There is **emerging success in disaggregation of urban growth** made by UN-Habitat and partners using remote sensing and GIS to analyse changes in urban extent in different components such as infill, extension, inclusion and leapfrog. City growth within the same urban extent is considered as **city infill** that "consists of all built-up pixels added in the new period that occupy urbanized open space within the urban extent of the earlier period". City can also grow beyond its previous urban extent in a contiguous manner. This type of city growth is known as city **extension** that consists of all built-up pixels added in the new period that constitute **contiguous** urban clusters that are **attached** to the urban extent of the earlier period. The growth of city can also occur through **inclusion** that consists of all urban, rural, or suburban built-up pixels that were **outside** the urban extent in the earlier period and are now within the urban extent of the new period. There is also another trajectory of growth where city expand to **over rural open space** that were **not attached** to the urban extent of the earlier period or to new extension clusters. This is known as leapfrog [24]. All these categories of spatial growth have occurred along the urbanization of the Dakar Metropolitan Regional over a century and will be analysed here.

15.3 Dakar Metropolitan Region in Time and Space

15.3.1 History of the Formation of the Dakar Metropolitan Region

Cities form and grow in many different parts of the world for various different reasons such as rural-to-urban migration, natural population growth, economic opportunities, politics, environmental changes, social conflicts, etc. In a given country, the way these factors play are determinant for the size and the national distribution of cities that, in turn, form the urban system. The History of the Dakar Metropolitan Region is marked by several migrations dating back to the 14th century. The most prominent recorded settlements were however in the 15th and 16th centuries when the Lebou group settled in the area called "Yoff" located to the north of the city and in other areas such as Ouakam, Ngor and Camberene [25]. The second major settlement was recorded later in the 19th century when the French started to settle to the extreme south of the city, which later became the centre of the city—hosting most administrative and state buildings (*including the State house, the Parliament and the Court house*) [26]. The third major settlement was marked by huge migrations of population from rural as well as small towns to Dakar, but neither in the city centre nor in the indigenous settlements, but to the outskirts of the city forming the Pikine settlements, which are mostly unplanned and lack most basic services. These three settlement eras had varying influences on the form of Dakar metropolitan region, which has resulted into three unique spatial organization patterns—an organic pattern evident in the Yoff area, a grid system evident in the French occupied southern parts, and an irregular pattern in Pikine (Fig. 15.3).

Beside the difference in street design, the Lebou spatial design is fundamentally different from the French spatial design. The former favoured a polycentric system of governance, and promoted equality among community members, and the creation of autonomous villages, which was the opposite of the monocentric system introduced by the French. The Lebou system of governance is socially, economically and politically decentralized. Each settlement was under the leadership of the "Djaraf", assisted by the "Ndeye ji Rew" for interior and foreign affairs, and the "Saltigue" who was in charge of land, water, etc. With the influence of Islam, the function of Serigne Ndakarou was introduced in late 18th century to provide justice based on the Quran. Until today, the Lebou community conserves its socio-political structure where the Serigne Ndakarou and other local authorities still play a key role [27]. At the opposite of the indigenous' decentralized governance system, the settlement of the French brought forth a new form of monocentric system with the centre, "the Plateau", hosting large administrative buildings as well as the French residences, while the indigenous communities were moved to a newly formed settlement close to the centre, the Medina settlement [28]. The Medina settlement lacked most basic services such as piped water, sewerage systems, education facilities and health centres as well as administrative and commercial businesses. Being the only places where migrants were allowed to settle, it soon

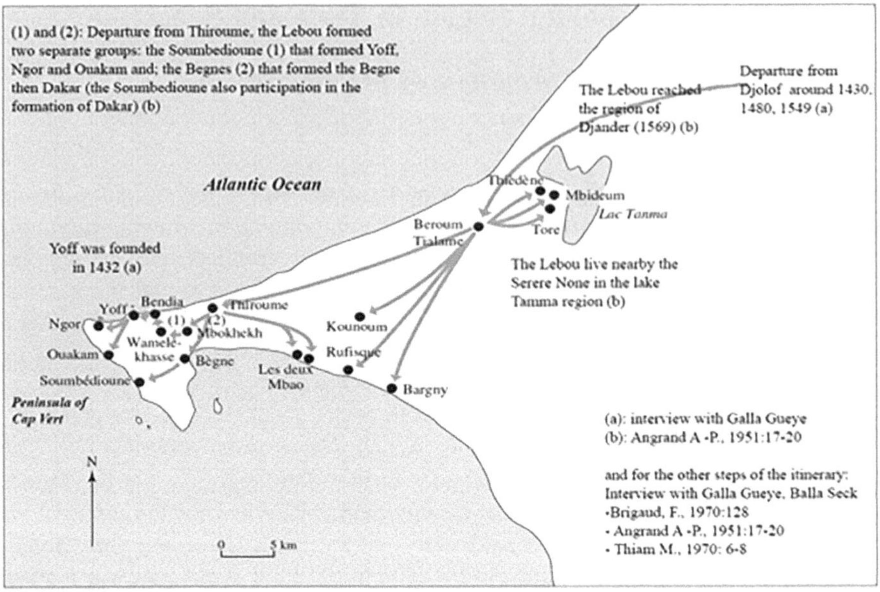

Fig. 15.3 Settlements of the Lebou group in 15th and 16th centuries in the Peninsula of Cap Vert (Dakar)

became densely populated, and since it lacked basic services, and disease outbreaks became the norm. In response, most migrants as well as indigenous communities were further relocated to the outskirts of the city, in a new settlement called Pikine. Just like Medina, the Pikine settlement was neither properly planned nor were there adequate basic service provisions. This marked the second phase of proliferation of slums that excluded the migrants as well as the indigenous communities from the comparative urban advantage a city can offer. Today, settlements of Pikine as well as Medina remain haunted by this early spatial and social segregation [29].

The Fig. 15.4 provides the historical spatial occupation of the Dakar Metropolitan Region in the 19th–21th centuries. The Dark colour illustrates settlements before 1923, with several areas of Yoff and Rufisque occupied by the Indigenous, as well as the centre of Dakar and the island of Goree mainly occupied by the French Colons. The Pink colour illustrates the development of Dakar between 1923 and 1953, particularly in the centre towards settlements of Medina and surrounding, and some dispersed settlements in the Yoff. The Statistics show that Dakar reached 214,000 of population size in 1950. Between 1954 and 1967, illustrated by the red colour, the city continued to develop with the appearance of new settlements, but concentrated near the city centre. This period also marks the development of new settlements in Pikine. In 1970, the Population of Dakar is estimated at 610,000 inhabitants. Between 1968 and 1976, the population growth was mainly in Pikine as illustrated by the dark orange colour. Between 1977 and 1980, there is further development of Medina, illustrated with the light orange

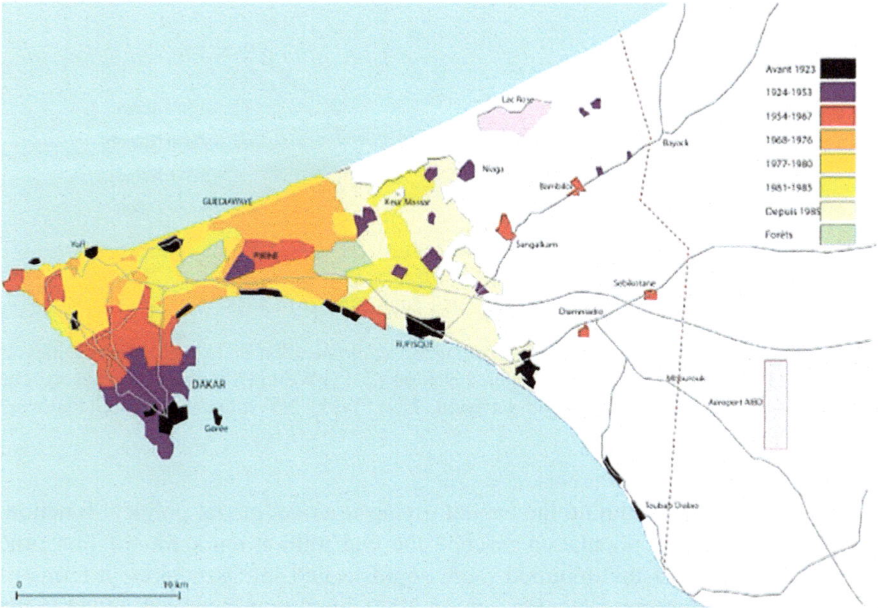

Fig. 15.4 Historical spatial development of the Dakar Metropolitan region (before 1923–since 1985). *Source* Republic of Senegal, 2014. Review of the Senegalese Urbanization

colour, with the appearance of Settlement in Gueule Tapee, Grand Dakar, Colobane, etc. In 1980, the population Dakar is estimated at 957,000 Inhabitants. Between 1981 and 1985, settlements in Pikine extended further to reach Keur Massar, illustrated by the dark yellow colour. In 1985, Dakar population reached the one million mark. Since 1985, most of the population growth of Dakar occurred in the department of Pikine towards the department of Rufisque as illustrated by the light yellow colour.

15.3.2 The Urban Numbers of Dakar Metropolitan Region, 1950–2015

As illustrated in Fig. 15.5, the dynamic transformation of the Dakar Metropolitan Region has occurred in different paces, times and places in terms of urban form and structure: density distribution, and compactness. From a population of about 214,000 in 1950, the Dakar Metropolitan Region has a population of 3.5 million in 2016. In 65 years, the population of Dakar has been multiplied by more than 16 times particularly during post-independence. During the colonial period, Dakar attracted massive migrations from rural areas, people coming to seek jobs in the newly built city centre. This flux had amplified after Senegal became independent

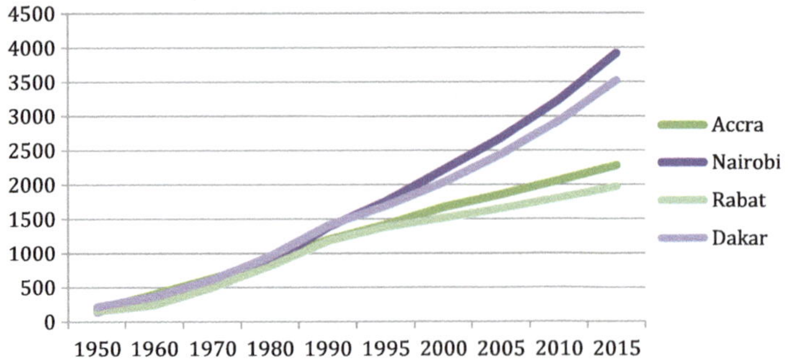

Fig. 15.5 Urban population growth of Dakar urban agglomeration and other African cities of similar sizes in 1950. *Source* Data from United Nations (2015) World Urbanization Prospects: The 2014 Revision, UNDESA, Population Division, New York, NY, http://esa.un.org/unpd/wup/index.htm

in 1960 with Dakar becoming the capital giving it an additional political function. In 1985, Dakar urban population reached the one million mark for the first time making it a large city that required more sophisticated infrastructures in terms of spatial planning, transport planning, etc. In 2000, another million was added to the Dakar urban population. In 2010, the Dakar population was nearly 3 million (2.9 million), and in 2015 it was estimated at 3.5 million. What do these figures tell us? It took 35 years for Dakar to reach the population of one million inhabitants in 1985 from a population of 214,000 inhabitants in 1950. But it took only 15 years from 1985 for another million inhabitants to be added to the Dakar population in 2000, and only 10 years for another million inhabitants to be added to the Dakar population in 2010. With this exponential population growth rate, the Dakar urban population will reach 5.6 million in 2030, and 7 million in 2040 (Fig. 15.6).

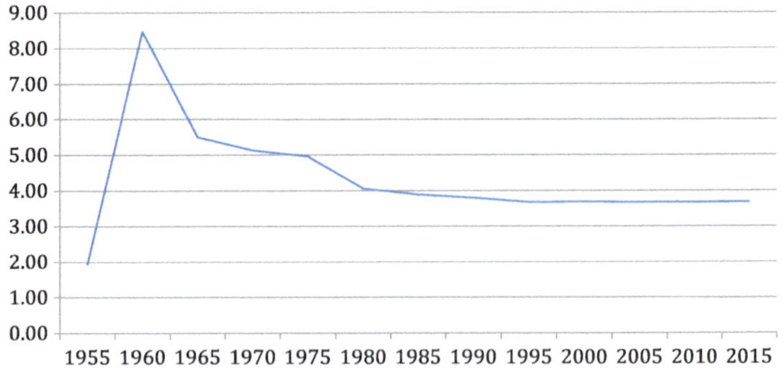

Fig. 15.6 Growth rates of the urban population of Dakar (1950–2015). *Source* Data from United Nations (2015) World Urbanization Prospects: The 2014 Revision, UNDESA, Population Division, New York, NY, http://esa.un.org/unpd/wup/index.htm

15.3.3 Urban Densities as Factor of Economies of Scale and Agglomeration

The population size is an important component of urbanization, but to address various questions in urban economics, environment and infrastructure development, information on density of the population is valuable [30]. Positive production or residential externalities, as well as negative externalities such as congestion, are function of density among others [31]. Various policy responses as well as programmes and projects such as on transport and infrastructure are subject to how population is distributed across a city. Important factors of land use such as land prices should be also considered in this exercise as part of elements that can feed the land legislation and investment in infrastructure [32]. From an economic point of view, the higher the density of an urban agglomeration is, the lower are the costs of transactions, and the more prosperous is the economy. A deficient spatial structure fragments labour and consumer markets into smaller less efficient markets; it contributes also to higher transactions costs by unnecessarily increasing distances between people and places. It increases the length of the city infrastructure network and therefore increases its capital and operating costs. Lower densities and segregated land uses increase economic costs in terms of average commuting distances, public utility provision, and energy consumption. From an environmental point of view, a deficient spatial structure can decrease the quality of life by increasing the time spent on transport, by increasing air pollution, and contributing to the unnecessary expansion of urbanized areas in natural sites. Poor environmental quality can also lower a city's productivity.

The Dakar metropolitan region displays high densities, particularly in the departments of Dakar, Pikine and Guediawaye where they are more than 15,000 inhabitants per km^2. These densities are largely sufficient to produce economies scale and agglomeration, support efficient urban transport and reduce the use of land. For the latter, with more than 50% of the national urban population, the Dakar metropolitan region occupies only 0.3% of the national land area. The first three departments, while hosting over 80% of the metropolitan region population occupies only one third of its land area (179 km^2 over 551 km^2, the total land area of the metropolitan region). However, except the economy in land use, since they are not well planned and managed, high densities have become liabilities to the city of Dakar. High densities not accompanied with provision of streets and other public spaces, and basic infrastructure often lead to high exposure to diseases and to epidemics such as cholera (Fig. 15.7).

When the densities are disaggregated across municipalities, there are huge disparities in the Dakar metropolitan area. However, these disparities do not follow the classical monocentric model of urban densities distribution, introduced by Colin Clark in 1951, that predicts a decline a density from the CDB to the suburbs. Two hypotheses are the basis of this model: (i) in all cities, excluding a business and commercial area, there are densely populated areas, which decrease when moving away from the centre, and (ii) in most of the cities, as time passes the density

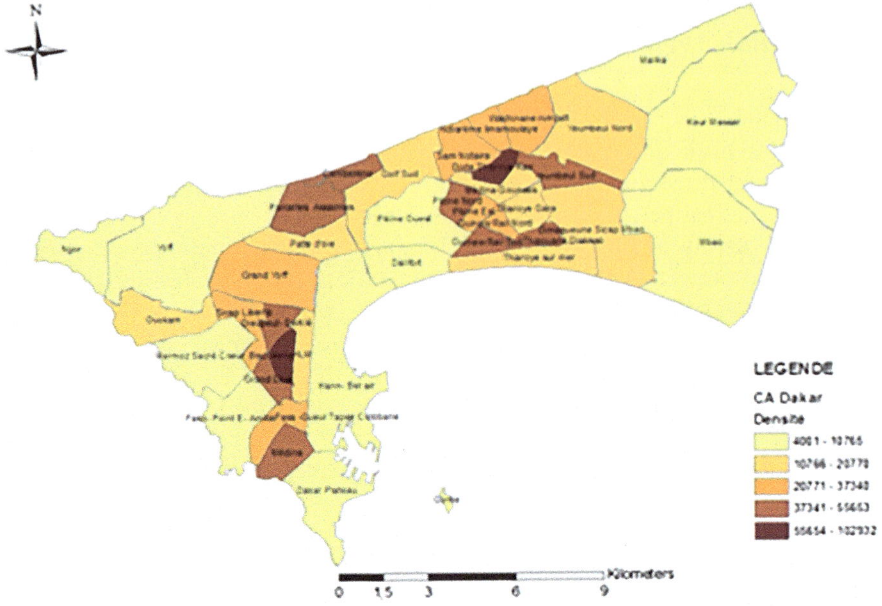

Fig. 15.7 Dakar population density map, 2015. *Source* Population density computed and geo-referenced using ARCGIS software by the authors applied to the ANSD 2015 Projection. Population, Housing, Agriculture and Livestock Census, 2013

decreases in the central areas and increases in the suburbs, thus producing a territorial expansion of the city [33]. The model proposed by Clark (1951) expresses the urban population density as negative function of the distance from the city centre as follow:

$$D = Doe^{-\gamma x},$$

D represents the population density at distance x from the centre of a city; Do is the density at the centre; e is the base of natural logarithms; γ is the distance gradient, or the rate at which density falls from the centre. The monocentric model supposes a city with constant returns Cobb-Douglas production functions for housing, consumers with identical tastes and incomes, and unit price elasticity of demand for housing [34]. Following Clark, several models have been developed that relax this strict monocentricity, including some that have two centres, some that have a centre and a beltline of employment, others that have multiple nodes or a beltline. Others related urban density to traffic planning [35], to housing market [36]. Some amenities can be of high quality of infrastructures—schools, health centres, etc.; alternatively they can be of poor quality clustering crime or pollution along with the lack of most basic services. In general, the classical theory predicts that variations in urban land cover, average density, and fragmentation among cities and countries, as well as their rates of change, can be largely explained by variations in city population, household income, buildable land, the cost of agricultural land in the urban periphery, and the cost of urban transport [37].

To verify the model monocentric density distribution in Senegal, since the city of Dakar is not in a circular form due to its peninsular form, we have opted to subdivide Dakar in three main axes: Plateau-Almadies, Plateau-Grand Dakar and Plateau Camberene. In Dakar, considering the high level of business and commercial areas in the CBD, we have opted to start the series from Medina, which is just juxtaposed to the CBD. As noted in the historical formation of Dakar, Medina was formed by early development of Dakar with its high population density. It is a walking-distance to the CBD. Based on this, all densities in the axe is lower than the density of Medina, varying from 43,519 persons per square to less than 7000 persons per km^2 in the following municipalities except the density of the Municipality of Ouakam which is 27,747 persons per km^2. The high level of the density of Ouakam witnesses the fact that the municipality of Ouakam is populated by various social economic groups, kind of middle class while the other municipalities following Medina are mostly populated by high high-income group. Overall, the density distribution from Medina to Yoff follows more or less the density model of Colin Clark. When the municipality of Ouakam is excluded, the density model from Medina to Yoff is $D = 43519e^{-432x}$. The monocentric and other forms of urban density models have major limitations in other axes from the Dakar CBD (Fig. 15.8).

In the axes of Plateau-Grand Dakar and Plateau-Camberene, instead of observing a decline of density from the centre to the outskirts of the city, we witness a combination of changes in density. Lack of specific model of densities in these axes is due to, disconnected, unplanned settlements. In most of the municipalities in these axes, households settle generally first in unplanned settlements either at the

Fig. 15.8 Population density by distance to the city centre

outskirts or anywhere in the cities where there is space to settle. In these axes, poor families that are not able to afford the expensive costs of living of the city core live in densely populated sub-urban areas. While in these two axes, in one hand we found very high densities such as in Medina, Grand Dakar, Biscuiterie and Camberene with more than 40,000 inhabitants per km^2, on the other hand, the municipalities in the axe Medina-Yoff have densities as low as 10,000 inhabitants per km^2. These figures are indicative of a segregated city and point to a need for the national and local authorities to consider a balanced spatial planning of the city of Dakar coupled with provision of basic services in poor neighbourhoods.

The situation is alarming in many municipalities in Pikine where densities exceed 50,000 habitants per km^2. Yeumbeul Sud (51,468 habitants per km^2), Djidah Thiaroye Kao (102,932 habitants per km^2) and Pikine Sud (49,665 habitants per km^2) are among the municipalities with very high population densities in Pikine. In all these very highly populated settlements, few streets are built, and they are lacking other public spaces. This can explain the permanent high prevalence of infant and child diseases in Dakar compared to other cities and the rural areas of Senegal [38].

15.3.4 The Demographic Dividends of the Dakar Metropolitan Region from 1950 to 2015

The population of Dakar metropolitan region is relatively young as shown in the age pyramids (Fig. 15.9). The median age of the population varies from 20 years in the department of Rufisque and 21 years in the department of Pikine to 23 years and 25 years in the departments of Guediawaye and Dakar respectively [39]. The age pyramid with high proportion of young people age 20–29 years old shows clearly that migrant population for both sex from other regions of the country contribute substantially to the rapid growth of the agglomeration population, particularly in the department of Dakar, Pikine and Guediawaye. The department of Rufisque displays a demographic regime of high fertility without migration effect. In fact, the migrants mostly prefer to settle in the department of Dakar followed by the department of Pikine and Guediawaye. Whatever the reference period, 66% of migrants settle first in the department of Dakar, 18% in the department of Pikine, 10% in Guediawaye and only 6% in Rufisque. The main reason is that the department of Dakar is still the only place which is economically attractive to migrant and has at the same time poor neighbourhoods where migrants can settle with low cost housing as Pikine and Guediawaye offer.

With the right empowerment, Dakar's young population can lead the vision for the city of tomorrow and work for it. The ability of the youth, as the "Millennial", to adopt emerging technologies and transfer experiences from other parts of the world must be key drivers of Dakar's smartness in the long term. Various studies show that Africa's young people will be the driving force behind economic prosperity in

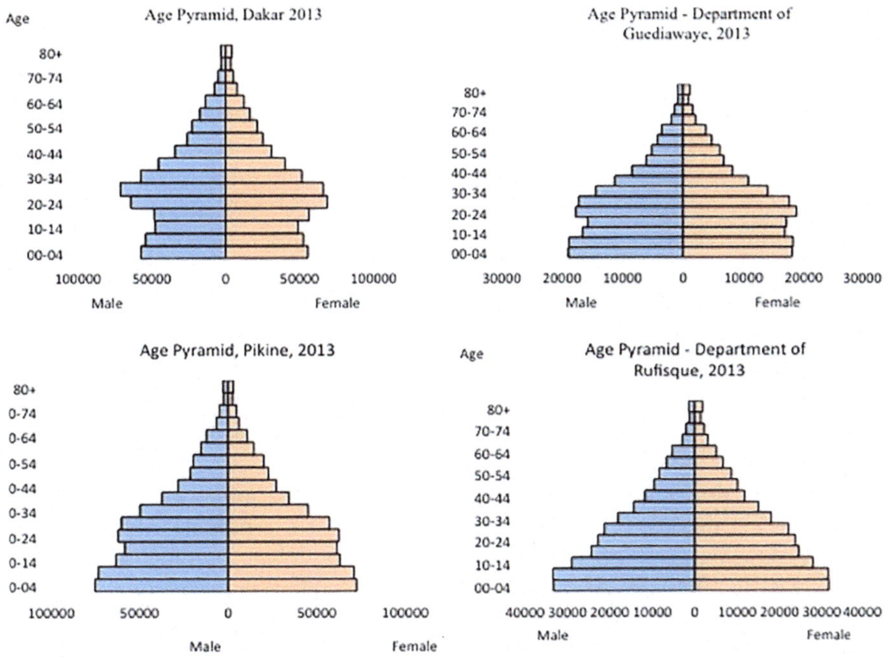

Fig. 15.9 Age pyramid Dakar Metropolitan Region. *Source* Data from ANSD, 2015. Senegal Population, Housing, Agriculture and Livestock Census, 2013

future decades, but only if policies and programmes are in place to enhance their opportunities [40]. These policies and programmes include factors such as a strong rule of law, efficient bureaucracies, government stability, lack of corruption, and a stable business environment that encourages domestic and foreign investors. In order to benefit from its youthful population, the Dakar metropolitan region must embrace smart education driven by the availability of ICT and respond to the requirement of the job market of the 21st century. It is also important to encourage young people's participation in public life, and in policies, programmes, and services. However, youthful population could also present a significant risk and threat to social cohesion and political stability if Dakar fails to create sufficient economic and employment opportunities to support decent living conditions for them [41].

15.3.5 Senegal Urban System: Dakar the Permanent Primate City

The spatial distribution of cities of different sizes throughout the national space constitutes a country's urban system. The urban system is determined by the way cities are geographically distributed in the country according to their size. In the urbanization process, some countries were able to develop a balanced system of cities with no city does predominantly hold a large share of the urban population while others concentrate their population in one city that become a primate city. The conditions under which urban primacy emerges can vary greatly and evolve with country size, population density, and the stage in its urbanization process, among other factors [42]. Changes in primacy can be affected by all of the factors affecting urbanization more generally, including natural growth, migration and reclassification of settlements from rural to urban. A country's urban system can be linked to the organization of the government at the national, regional, and local levels [43]. The share of city to the total urban population can be a good indicator of a city performance, attractiveness or a simple political preference [44]. In turn, the urban system may determine the need for specific urban development policies, consistent with the size, growth and function of each city. For instance, when all the national investments, all health, education and commercial infrastructures are vested in a particular city, such as Dakar, people do not have other choices except moving to Dakar.

The Urban primacy as defined here characterizes the urbanization process in Senegal during the last 60 years where Dakar has been a primate city with more than half of the national urban population living in the urban agglomeration. With a level of 50% in 1950, the share of Dakar in the national urban population decreased and stabilized at 48% until 1990. However, from 1990, this share continuously increased until it reaches its highest level since 2010 (54%). The permanent primacy status of Dakar is opposite from most cities in East Africa where though they were primate cities at the onset had lost their primacy along the urbanization process. For instance in Kenya, the share of Nairobi in the national population was 49% in 1960, but in 2015 Nairobi represented only 33% of the national population. In Ethiopia there was a drastic reduction of the demographic power of Addis Ababa from 47 to 17% (Figs. 15.10 and 15.11). In North Africa, expect in Egypt there is no single country that hold a share in the national urban population that reached 30%.

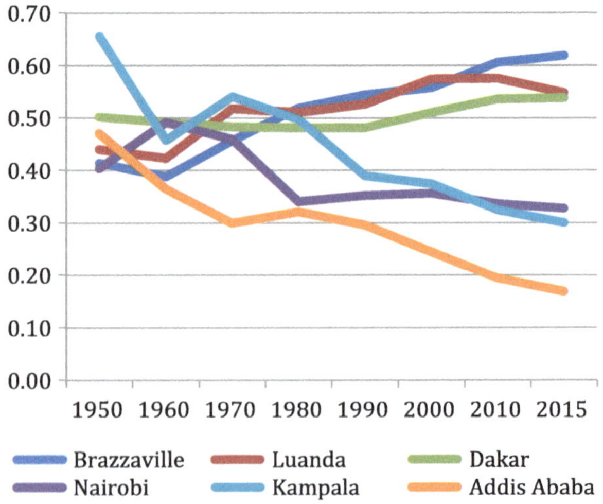

Fig. 15.10 Share of Dakar and other African cities in the national urban population (1950–2015). *Source* Computed using data from United Nations (2012) World Urbanization Prospects: The 2014 Revision, UNDESA, Population Division, New York, NY, http://esa.un.org/unpd/wup/index.htm

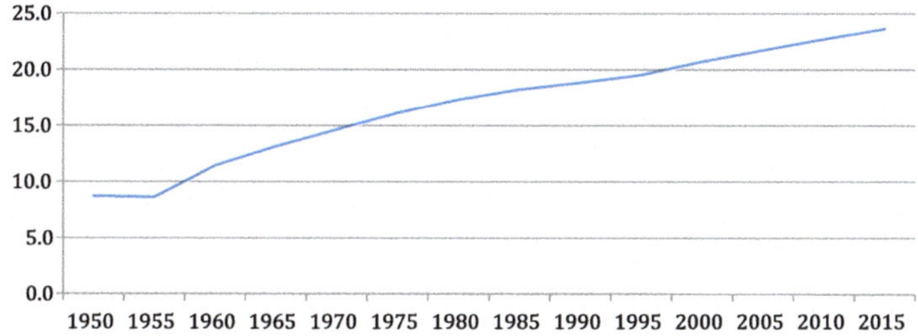

Fig. 15.11 Percentage of the total population residing in Dakar urban agglomeration (1950–2015)

The Fig. 15.12 illustrates two type of information: the size of population measured by the size of the circle and the population growth measured by the colour of the circle. The bigger the circle, the larger the size of the population with the metropolitan region of Dakar having the larger circle with its population of 3.5 million, followed by the city of Touba with its population of 817,000 inhabitants. All other cities have a population of less than 300,000 inhabitants. The highest population growth rate is observed in the city of Touba, more than 4.1%, followed by the city of Mbour with a population growth rate between 3 and 4.1% and Dakar

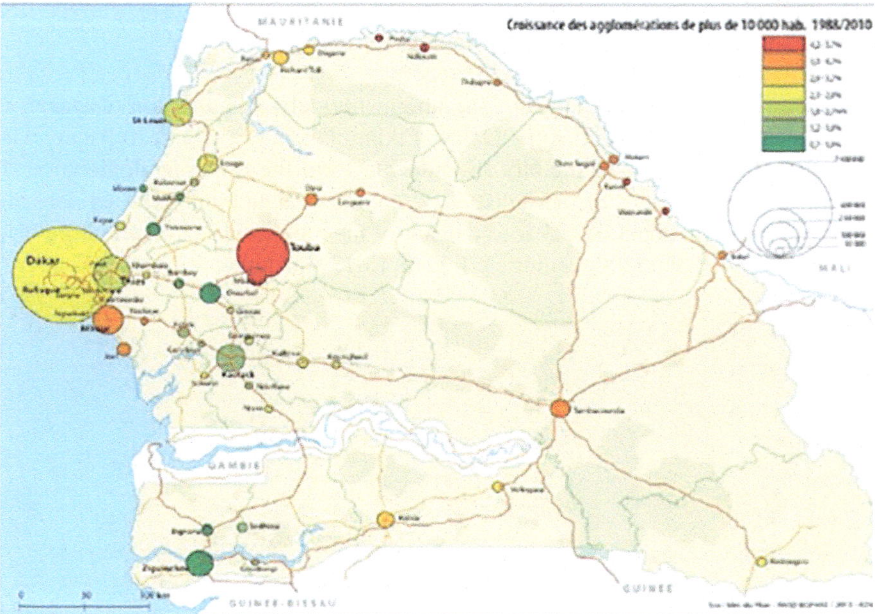

Fig. 15.12 Senegalese agglomerations with a population of 10,000 inhabitants or more in 2013 and Population growth rate between 1998 and 2010 of agglomerations with a population of 10,000 inhabitants or more. *Source* Republic of Senegal, 2014. Review of the Senegalese Urbanization

with a population growth rate of 2.5%. In the next decade, the population of Touba will exceed the million Mark and call for an adequate urban planning, design and management, a sufficient provision of basic services, a well-administered land tenure, an efficient mobility system, etc.

High primacy shares may imply a high concentration of persons in one city as observed with Dakar, enabling cities to benefit from the positive effects of high density such as economies of scale and agglomeration, diffusion of ideas and innovation. While there is evidence to support such a positive relationship between concentration of people and economic efficiency, research has also found that city primacy can create an unbalance in the urban hierarchy and bias development processes [45]. All this depends on various conditions including the country's level of economic development and size as well as urban policies and programmes [46]. The economic benefits of urban primacy should be supported by evidence that some primate cities are such because of distorted and unequal political consideration, which is the case of Dakar. Primate cities that receive preferential consideration by the state disadvantage smaller cities, are harmful to efficiency and nation-wide development, and could lead to spatial inequality [47].

15.3.6 The Dakar Metropolitan Regional Foundation

15.3.6.1 Dakar—A Monocentric Metropolitan Region

Today, the Dakar Metropolitan Region, administratively called region of Dakar, is sub-divided into four departments: Dakar, Pikine, Guediawaye and Rufisque. The first three departments and the city of Rufisque form the urban agglomeration of Dakar. However, most of the administrative, commercial and industrial services are located in the department of Dakar as a result of the monocentric system introduced by the French in 19th–20th centuries. Figure 15.13 illustrates the concentration of

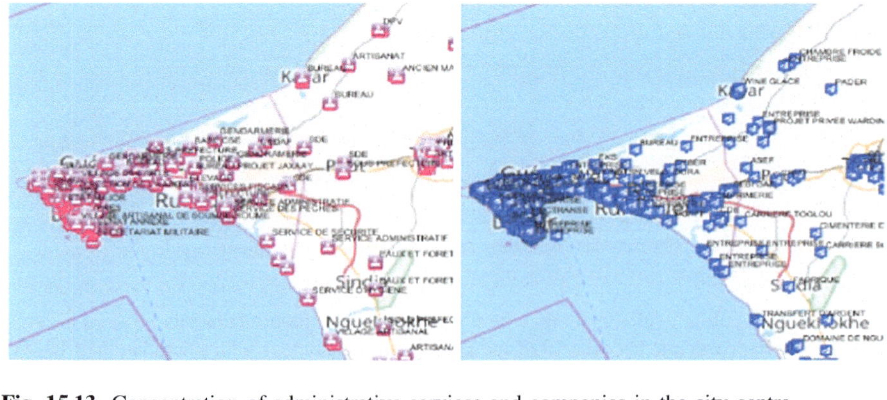

Fig. 15.13 Concentration of administrative services and companies in the city centre

administrative services in the Dakar city centre. This monocentrism of Dakar dated back the colonialism period and had been pursued by the Senegalese government after independence. Residential areas were also designed along economic class lines. This has led to social and economic fragmentation that disadvantage lower income groups in accessing basic services such as public transport, and prevent social interaction and integration. This has impacted the traffic with people having to share a main street network to and from Central Business Districts. This has led to congestion and long trips from the periphery to the centre [48] (Fig. 15.13).

Efforts have been made to decentralize the Metropolitan region with success observed with the decentralization of the financial systems.

15.3.6.2 Proliferation of Unplanned Settlements and Lack of Security of Tenure

The Urban Master Plan of Dakar has established criteria to classify houses and building on the following categories: Individual house type villa; Planned housing type real estate company; Regular spontaneous Habitat; Irregular Spontaneous Habitat; Habitat type buildings and; Village dwelling. The **Individual house type villa** consists of villa of high standing with a maximum of three levels, with very good-asphalted roads, and most sidewalks are paved. **Planned housing type real estate Company**: These are housing programs carried out by public or private bodies specialized in real estate development. This type of habitat has asphalted roads for the main streets. **Regular spontaneous Habitat**: This type of habitat, realized in self-construction on approved subdivisions, is constituted of so-called popular districts such as Medina, Gueule Tapée, Colobane, Grand Dakar and certain districts of Pikine and Guediawaye. High densities characterize this type of habitat. **Irregular Spontaneous Habitat**: This type of habitat is manifested by an anarchic and unauthorized occupation of public spaces and land, especially in the urban peripheries. It is made up of poor quality buildings, more or less precarious, sometimes even slum type, where the problems of lack of infrastructure and public facilities are acute. The streets are narrow, sandy and difficult to access for motorized vehicles. **Habitat type buildings**: buildings exceeding three floors characterize this type of housing. It is generally located in the city centre and is the administrative district of the Plateau. It has the particularity of being located in buildings with a large proportion of offices, as well as commercial premises (on the ground floor). The road is paved and the sidewalks are all paved. **Village dwelling**: These are mainly "traditional villages" in Dakar (Ngor, Yoff, Ouakam, Hann, Cambérène) and Pikine (Keur Massar, Thiaroye sur Mer, Mbao and Keur Mbaye Fall). These villages are characterized by a habitat of summary materials gradually replaced by permanent dwellings. The streets are narrow and sandy for the most part, only the primary road is paved. The shape of the lots is not regular and the houses are grouped around small squares or around a mosque. The government has also classified settlements based on their coverage in terms of infrastructures. It has classified them as: settlements with infrastructures, settlements with limited

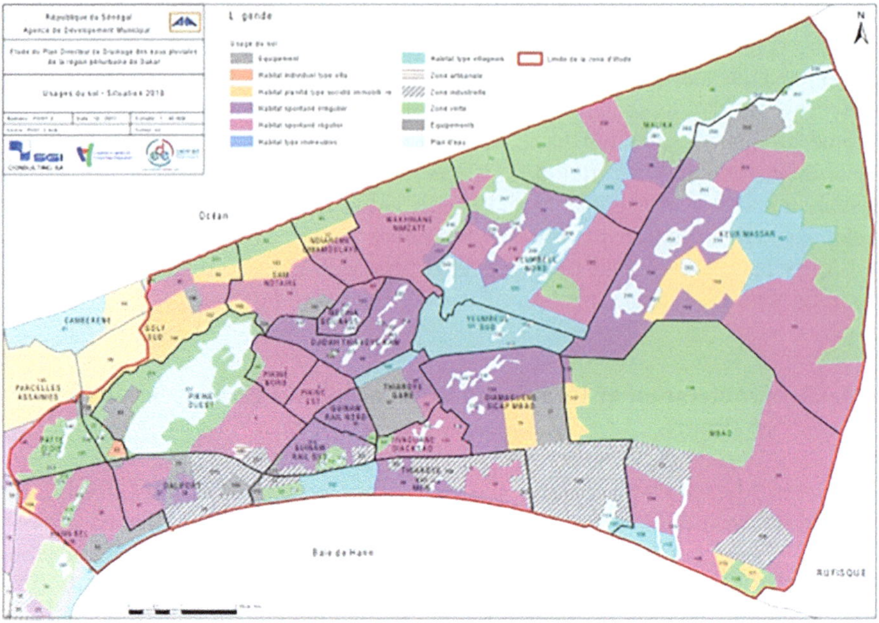

Fig. 15.14 Classification of land use in the peri-urban of Dakar

infrastructures, and irregular settlements. The 19 municipalities of the department Dakar were distributed according to these types of settlements (Fig. 15.14).

Based on the above classification of settlements, official information indicated that the percentage of population living in regular areas is 60% in Dakar, 35.8% in Pikine, and 30% in Guédiawaye. Another study "Cities of Senegal without slums" revealed that there were 49 spontaneous settlements in the city core of Dakar, covering a total area of 418 ha; and 47 large spontaneous settlements in the suburbs of Dakar covering a total area of 1856 ha. In the urban audit conducted in the city of Dakar in 2001 [49], equipped settlements were defined as settlements within a radius of 500 m that have an elementary school and a health centre in addition to be connected to water, electricity and a paved street network. This type of settlements covers an area of 1917.56 ha in 2001, i.e. 57.8% of the total built up area of the core of Dakar. Under equipped settlements, defined as settlements that do not have these infrastructures or are far away from them, cover an area of 1029.96 ha, i.e. 31.1% of the total built area of Dakar. Others are considered irregular planned settlements with traditional structures, which mostly are the first settlements of Dakar, cover an area of 367.15 ha, i.e. 11.1% of the total built up area of the core of Dakar. However, it is important to note that these settlements considered are irregular in terms of planning such as Ngor are also those hosting now wealthy people. The indigenous have sold their land to wealthy people who settled without changing the layout of the neighbourhood. Other irregular settlements are in areas such as Camberene where mainly lively some indigenous people and poor migrants that cannot afford the cost of living in the centre of Dakar. These settlements have an irregular planning and lack most of the basic infrastructures. This again supports our assumption that irregularity in Dakar does not necessary means poverty. Irregular settlements can host rich families as well as poor families. As we previously state it, informal land tenure in Dakar does not concern only the poor, but also the rich; it is across social classes. Without access to financial market households are obliged to build based on their capacity. Mortgaged houses are about 5% in Senegal.

15.3.6.3 Basic Infrastructures

In a smart city foundation, basic infrastructure such as piped water services, sewerage facilities electricity sources and solid management are considered along the city planning. They are part of the city planning prior to settlements of households. In a sustainable city foundation, use of improved water from piped water services, sewerage facilities, solid waste management, energy for lightning is quasi universal. For the city of Dakar only connections to piped water facilities and to electricity are quasi universal with levels of 92 and 95% respectively. The coverage of sewerage facilities and solid waste management is still very low, below 50% in 2015 (Figs. 15.15 and 15.16).

During these past 20 years, there is steady progress in the coverage of water connection from 64% in 1993 to 77% in 1997 and 90% in 2005. At this level, it is clear that the coverage of connection to water facilities and to electricity sources did not follow the rapid progress of unplanned settlements characterized by lack of documented land tenure. Considering that the water and electricity services are governed by public administrations until recently, their high level of coverage shows a de facto secure tenure, which was highlighted earlier. Indeed, in order to have connection to these services, the application must obtain the approval of the Ministry in charge of housing and urban development.

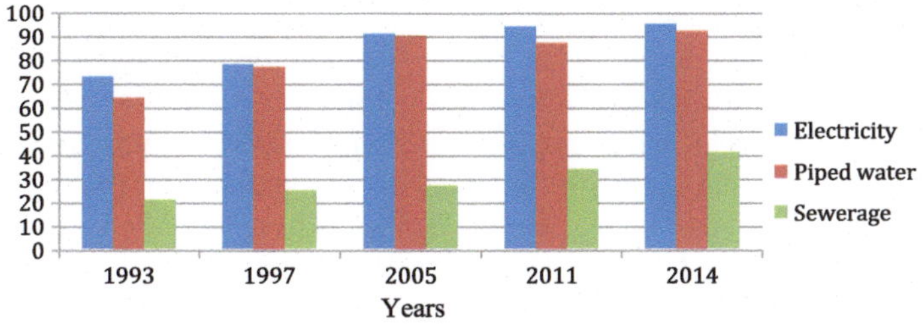

Fig. 15.15 Percentage of household connected to electricity sources, water piped and sewerage facilities, Dakar 1993–2014. *Source* Computed from Senegal demographic and health surveys (1993–2104)

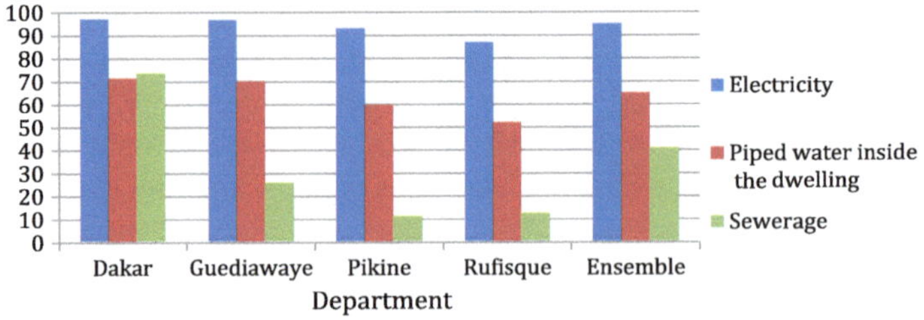

Fig. 15.16 Percentage of household connected to electricity sources, water piped inside the dwelling and sewerage facilities, Dakar 2013. *Source* ANSD, 2015. RGPHE 2013

Connections to water services and to energy sources often go together though processed by two different entities, the SDE (Sénégalaise des eaux) and the SENELEC (Société nationale d'électricité du Sénégal) respectively. Therefore it is not surprising to have similar figures for both amenities. Here the only problems that arise are during the supply, which is more, a demand–supply equation than a city foundation matter. During the foundation of the city what was important was to ensure that connection to water services and electricity are established along the planning of the city. For instance due to poor infrastructure development, many households of Dakar may stay one week without water according to the Demographic and Health Survey conducted in 2014 in Dakar [50]. Though connection to water exists, the use is not neither sufficient nor affordable [51]. Similar situation occurs for the electricity where households can stay hours and days without electricity.

Sewerage facilities are fundamental for Dakar smartness
Data on connection to sewerage facilities have been collected from the Demographic and Health Surveys held in Dakar between 1990 and 2015. During the last twenty years, there has been important progress on the connection to sewerage facilities with the proportion of people living in household connected to sewerage facilities almost doubling from 21% in 1993 to 41% in 2014. With similar progress, it is expected that by 2020, 2030 and 2035, the coverage of connection to sewerage facilities will be 49, 64 and 72% respectively. However, with an aggressive sanitation programme, this progress can be boosted and the majority of households connected to sewerage facilities before 2020, and the universal access before the year 2035. The achievement of this will however require re-planning of the city of Dakar. At the moment, the majority of the city residents rely on sanitation facilities such as septic tanks and pit latrines. Along with the lack of sewerage system, the city of Dakar lacks a properly developed drainage system—both in the rich and poor neighbourhoods. Lack of adequate investment in these two basic infrastructure components exposes the city to challenges such as management of surface water, flooding, and vulnerability of the population to water-borne diseases,

especially during the rainy seasons. In Dakar as in most cities of developing regions where there are few streets built, sewerage facilities and storm drainage are not factored during the planning of the city. The situation is worse in the suburb of Pikine where the level of connection to a sewerage facility is less than 10% in most settlements. To be a smart city, Dakar needs to take into consideration the urgent need of a sewer system.

Lack of access to sewer system in overcrowded urban areas has negative impact on health as a factor of proliferation of diseases. Waterborne diseases such as diarrhoea and respiratory infections are frequent in cities. A series of demographic and health surveys conducted in the city of Dakar during these last twenty years show consistently high levels of diarrhoea diseases and respiratory infections than in the rural areas [52]. This can be associated to low coverage of sewerage facilities, with its correlate, high frequency of floods during raining seasons. In addition to that, households are frequently crowded in tiny houses hosting kitchens without adequate ventilation.

Management of solid waste

One main characteristic of Dakar's weak city foundation is poor waste management. Due to lack of a comprehensive drainage system, wastewater flows in the streets of Dakar. Senegalese municipalities have major difficulties to cope with the waste. In Senegal, 2438 tons of solid waste is collected every day in the municipalities, or 0.5 kg per resident per day. This waste mainly comprises of organic waste (44%), paper (10%), plastic (3%), glass (3%), metals (1%) and other types of waste—textile, electronic, medical, industrial, etc. (39%) [53]. Dakar's main solid waste disposal site, the Mbeubeuss, (located in Pikine) receives 475,000 tons of garbage per year from both domestic and industrial sources [54]; yet its designation as the city's dumpsite in 1968 was not well informed by robust feasibility studies. To date, the landfill remains as an unfenced open pit where garbage is not covered with an inert material and where intense recovery activities are practiced. Proposals are however being developed to start the conversion of the waste in the dumpsite to wealth through recycling and other means. Currently, Mbeubeuss receives on average of 3500 people per day (33% of recyclers, resellers and other buyers of recovered materials) whose activities mobilize financial mass of nearly thirteen million francs (13 million FCFA) [54].

Due to lack of regular solid waste collection, household waste also finds its way to the streets. Indeed, Dakar has very limited waste disposal sites, and the few that exist are poorly managed. Poor management of waste impacts negatively on people's health and exposes children to multiple diseases. Senegalese Demographic and Health Surveys conducted in 1993, 1997 and 2005 constantly show that diarrhoea and acute respiratory infections are more frequent among children living in Dakar than those living in rural areas [55]. This high exposure to diseases in Dakar is also associated with the fact that households are overcrowded and are more affected by indoor as well as outdoor pollution than their counterparts in the rural areas and small towns.

15.3.6.4 The City Foundation: Spatial Planning, Land Tenure and Basic Infrastructures

For a city foundation to be smart, it must be inclusive at the onset of the urban planning and promotes mixed neighbourhoods where social clustering is discouraged. Having all the poor living together creates slums and fuels instability and insecurity. Inclusive urban planning eases access to basic services (water, sanitation, housing, education and health) and to decent employment for all. A key element of smart urban planning is a smart street network that reduces travel time and encourages walking and social interactions. Smart urban planning enhances infrastructure development, environmental sustainability, economic and social development; makes cities resilient and prepared to overcome natural disasters; and promotes mixed neighbourhoods where services are walking distances from people's residences. Overall the metropolitan region is not smart; disaggregated data reveal huge disparities across departments in favour of the department of Dakar (Fig. 15.17).

Using the three main components of the city foundation, we have computed a composite index that represents the overall city foundation, the City Foundation Index (CFI) with values varying from 0 to 1. A value closes to 0 means that a city lacks most of the elements of the city foundation elements, and a value closes to 1 means that a city enjoys most of the elements of the city foundation. While the CFI of the department of Dakar is estimated at 0.723, the CFIs of Pikine, Guediawaye and Rufisque are estimated at 0.402, 0.432 and 0.425 respectively. The low level of the CFI of Pikine, Guediawaye and Rufisque is due to low coverage of a sewerage system, less than 30% compared to 74% in the department of Dakar. The three departments also perform poorly in land governance and administration with a proliferation of informal settlements. While in Dakar, the majority of households enjoy security of tenure, in the three departments less than the third of households

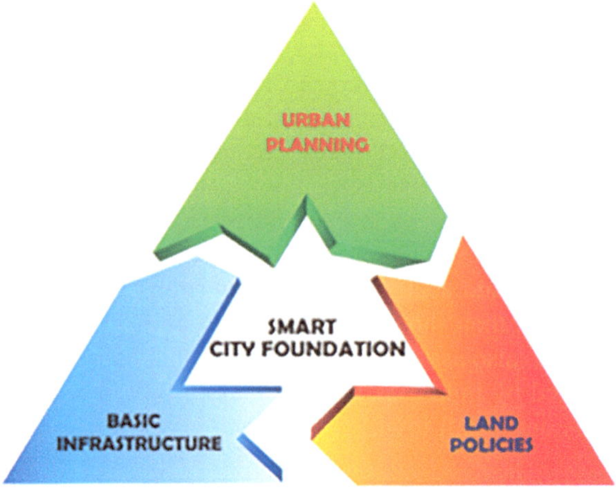

Fig. 15.17 City foundation conceptual framework. *Source* Mboup, G. et al., 2016. Smart City Foundation—Drivers of Smart Cities. In Vinod et al., 2016. Smart Economy in Smart Cities, Springer

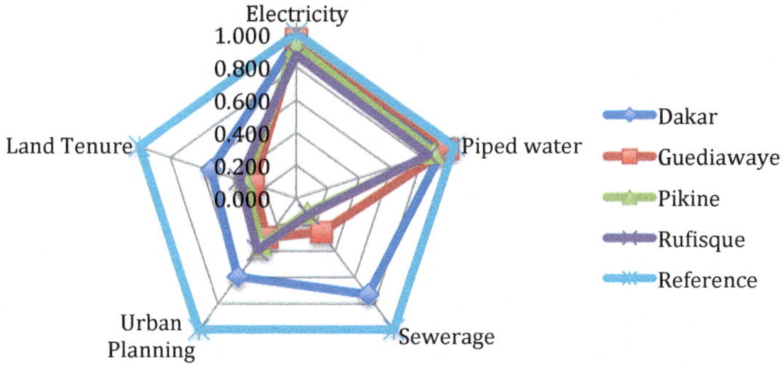

Fig. 15.18 Components of the Dakar Metropolitan Region city foundation index. *Source* Computed by the authors

have security of tenure. However, the department of Dakar is unequally served with settlements in the CBD and Fann/Point E, enjoying fully all elements that make a city foundation smart against Colobane, Grand Dakar with similar situations as observed in Pikine, Guediawaye and Rufisque. The CFI in the department of Dakar varies from 0.903 in the wealthiest municipalities to 0.207 in the poorest municipalities (Fig. 15.18).

National and local authorities must improve the foundation of the city of Dakar with smart planning, smart basic infrastructure and smart institutions and laws. Smart basic infrastructures, which are particularly of great urgency in the city, include connection to water, connection to sewerage facilities and connection to energy sources, as well as development of efficient waste management systems. Effecting these changes will enhance the economic value of land, encourage investments, reduce risks from natural hazards, increase resilience and minimize the costs of infrastructure maintenance among various other positive impacts. This calls for use of ICT and GIS in space planning, land tenure governance and administration, and management and monitoring of the provision of basic infrastructures.

15.4 Urban Accessibility and Mobility

Development of large metropolitan regions like Dakar constitutes economic and social opportunities as well as challenges depending on how the urban growth was planned and managed. It provides opportunities for economies of scale and agglomeration as well as for diffusion of ideas and innovations, but it will also call for large investments in infrastructures to respond to the increased demand for water, sanitation, solid management, energy and mobility, etc. It will particularly come with increased demand on mobility that must be satisfied with an efficient public transit accompanied with increased spaces for pedestrians and cyclists in order to safeguard the environment while creating economic growth. Economies of scale and agglomeration economies are greater in metropolitan areas where transportation infrastructures are able to answer mobility needs with higher access to markets and resources than those where people mobility is impeded by deficient transportation infrastructures. Efficient mobility will allow large-scale production of goods and services that can be distributed within the metropolitan regions and beyond with time, cost and reliability opportunities [56]. Without efficient mobility, a metropolitan region losses its economic power and remains just clusters of disconnected settlements. The latter is the situation in Dakar as in most African cities, marked by frequent traffic congestions, a major indication of the disjuncture between land-use planning and transport systems in the city. It not only exposes the limitation of a transport-oriented bias to mobility, but also reveals the inefficiency of land-use systems in Dakar.

15.4.1 Streets Key Components for Urban Accessibility

Urban accessibility is shaped by both spatial planning elements such as: density, compactness, type of city (monocentric or polycentric) and streets, and transport planning elements such as the streets infrastructures and the transport infrastructures. The first three elements of spatial planning—density, compactness, and type of city (monocentric or polycentric)—are presented in Sect. 15.3. This section will address one important element of spatial planning—streets—and elements of transport planning and system.

The Dakar metropolitan region is not benefiting from all the multiple advantages associated with well-connected streets. Dakar suffers from traffic congestion along with the exclusion of pedestrians and other non-motorized means, and further marginalization of the most vulnerable segments of society who rely the most on foot and public transports if they are affordable. The suburbs of Dakar allocate less then 5% of their land to streets compared to a level of 15% in the city core as illustrated in Fig. 15.19. In most Dakar suburbs, there is no space left for mobility except very tiny streets where people have to squeeze through. The streets are also narrow, not paved, and lack sidewalks; this result in competition for space with motorists, often exposing pedestrians to accidents. While Dakar as in many African cities has high pedestrian flows, less than 20% have pedestrian footpaths [57]. In many of these settlements, public transport services are accessible in few arterial streets where people have to walk several kilometres to reach them. During the raining season, the already limited walking spaces are flooded, forcing people to stay at home, losing many productive hours. This in turn has serious social and economic impacts on the city's economy.

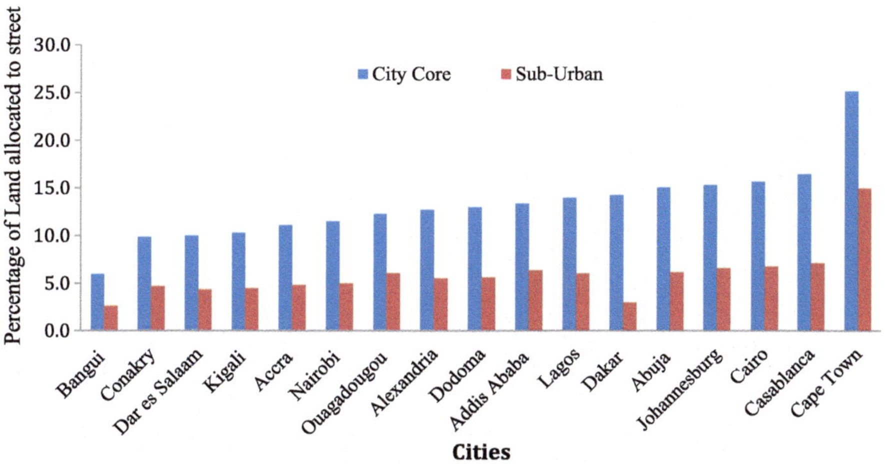

Fig. 15.19 Land allocated to street (LAS) in African cities. *Source* Figure prepared from Mboup, G. et al., 2016. Smart City Foundation—Drivers of Smart Cities. In Vinod et al., 2016. Smart Economy in Smart Cities, Springer

Composite Street Connectivity Index

Street connectivity is determined by the amount of land allocated to streets, the length of the street network and the number of intersections along the network. The proportion of land allocated to streets and the length of the street network is not sufficient to assess street connectivity. A city with wide streets within a very limited street network and low intersection density is considered a city with low street connectivity because the width of the streets is not complemented by a larger street network and higher street density. Similarly, a city with a lengthy street network and dense intersections may not qualify as a city with high connectivity if the streets are very narrow. The Composite Street Connectivity Index (CSCI) aims to assess the connectivity of a street considering its width, its length and the number of intersections, all in relation to the total land area of a city. Interestingly, some cities in this group have low levels of land allocated to streets, but higher intersection density (ID) increases the value of their CSCI. For instance, Dakar city core has a LAS index moderately low (15%), but due to its high ID, it has a CSCI similar to the other cities with higher LAS but insufficient ID. The city core of Dakar is planned in grid pattern that favours good connectivity. However, the Dakar suburbs suffer from low LAS, limited street density with few intersections Cities in this group have very poor street connectivity due to low levels of land allocated to streets, low street density and low intersection density. Their CSCI is less than half the highest level of the CSCI, which is 1. In these suburbs, which are mainly unplanned, provisions of basic services as well as means of transport remain challenges. Comprehensive city planning programmes are needed in these suburbs to improve the lives of urban dwellers.

Fig. 15.20 Composite Street Connectivity Index (CSCI)

Dakar is among the five African cities (Cairo, Alexandria, Casablanca and Johannesburg) with **a CSCI of between 0.500 and 0.650.** Their level of connectivity in the city core is sufficient to promote infrastructure development and to ease connections to basic services, such water, sanitation facilities as well as drainage systems. However, their suburban areas are very poorly connectivity, with a CSCI of below 0.300. Only one African city, Cape Town, features in the group with **Cities with a CSCI equal to or above 0.800** along with cities in developed countries. It has streets sufficiently wide to accommodate all types of users, sufficient to reach most neighbourhoods, and sufficient intersections to accommodate all users (Fig. 15.20).

Streets where the needs of all users are considered and factored (ages, gender, economic status, modal means, etc.) are urgently needed in Dakar. Mobility must be multidimensional, easy, comfortable, and safe [58]. Within the existing street network, Dakar must re-design its streets by allocating more spaces for walking, cycling and promoting the use of public spaces. On a smaller scale, Dakar is redesigning streets, particularly in new settlements to allow pedestrians and cyclists to share space with motor vehicles. Design measures that enhance the pedestrian environment include expanding sidewalks, planting trees and installing benches or other seating. In existing settlements, the master plan has introduced the redesign of certain intersections. All these initiatives have a common set of objectives: to reduce use of motorized transport modes and enhance environmental sustainability and social interaction. However, these initiatives are still not addressing problems faced by the urban poor who live in suburban areas and slums. Most suburban areas and slums are poorly served with streets; this further hinders accessibility to urban public transport [59].

Along the reduction of land allocated to streets, there is also reduction of other open public spaces as illustrated in the Fig. 15.21.

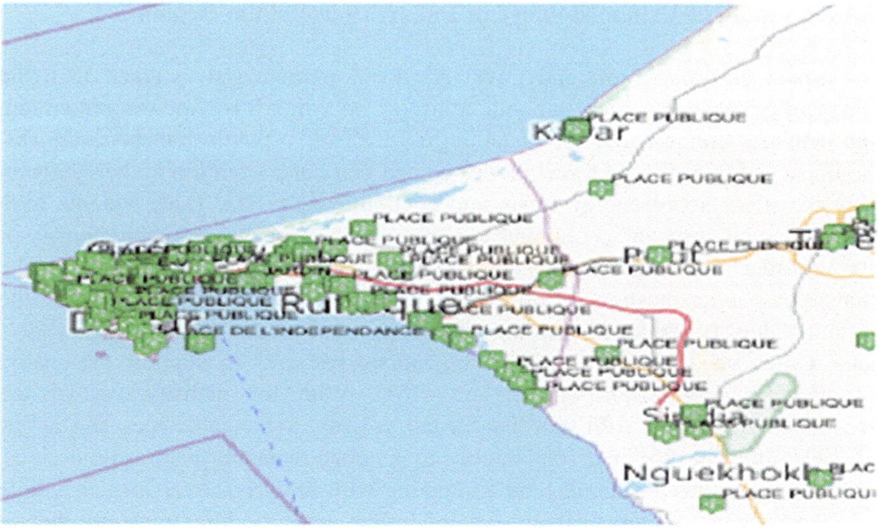

Fig. 15.21 Spatial distribution of public spaces in Dakar—A qualitative assessment

15.4.2 Transport Modes and Infrastructures Are Key Urban Accessibility

Cities are locations having a high level of **accumulation** and **concentration** of economic activities and are complex spatial structures that are supported by transport systems. The larger the city, the greater is its complexity and the potential for disruptions, particularly when this complexity is not effectively managed. The most important transport problems are often related to urban areas and take place when transport systems, for a variety of reasons, cannot satisfy the numerous requirements of urban mobility. The Sustainable Development Goals (SDGs) recognize that sustainable transport is crucial for urban economic development. With Goal 11 of the SDGs "*Make cities and human settlements inclusive, safe, resilient and sustainable*", member states have committed themselves to "by 2030, enhance inclusive and sustainable urbanization and capacity for participatory, integrated and sustainable human settlement planning and management in all countries. Member States have further committed to support positive economic, social and environmental links between urban, peri-urban and rural areas by strengthening national and regional development planning, and by 2020, substantially increase the number of cities and human settlements adopting and implementing integrated policies and plans towards inclusion, resource efficiency, mitigation and adaptation to climate change, resilience to disasters, …" [60].

15.4.2.1 State of Urban Mobility in Dakar Metropolitan Region

"No aspect of urban form and travel has been more closely studied than the influences of urban densities on public transport ridership. It is widely accepted that high densities are essential for sustaining cost effective public transport services. The impact of densities on travel modes gained particular attention in the 1990s, in the wake of a global energy crisis and economic recession. Rail, with its high up-front capital costs and economies of scale, needs to attain a threshold density of trips, in order to cost less than accommodating the same trips by car or bus. The reliance of public transport on urban densities has prompted efforts to define the minimum density thresholds required to support successful public transport services. Cities need to average 3000 inhabitants per km^2 to support reasonably cost-effective public transport services" [61]. With low density, neither the investment required for BRT, nor that for rail, is likely to be viable because stations will lack sufficient proximate populations to generate demand. In spread-out cities, public transport has a difficult time competing with the private car for the middle and upper economic or walking for the urban poor. Public transport that is cost-effective can only be achieved through high urban densities and a large share of jobs and retail activities concentrated in the urban core, or in polycentric cities with multi-directional travel patterns [61]. In Dakar metropolitan region, the population density exceeds this threshold indicating that Dakar metropolitan region transport problem is not associated to low density but to other urban form components such as a low coverage street network, urban monocentrism, etc.

Infrastructure Development is one the component of a smart metropolitan regional development. It is composed of several elements including transport, energy and ICT. Here, our focus is on the first element, which is transport. While transport data has been gathered for most cities in developed countries, information on transport is scarce in African cities, and when it exists it is poorly documented; metadata providing definition, method of measurement and sources of information is rarely available. The lack of reliable data on transport has held back the development of effective urban policies aimed to effectively tackle deficient transport systems in African cities. The most common available statistics are the distribution of trips by transport modes (large buses, minibuses, private cars, motorcycles, and walking). However, this information does not make distinction on the type of trips. Are they for going to work, to the health centres, to the school, to the market, or all together [62]. In the era of data revolution, this information must be processed, analysed in association with social, economic and demographic information and used to guide transport policies and planning in an open platform accessible to all stakeholders including the public. This information will help to overcome transportation challenges in African cities through informed planning, real-time monitoring and management that take into account multidimensional variables beyond the transport sector.

Though the statistics may be from different sources, there is a common finding which is that in most sub-Saharan African cities, people have limited mobility choice in getting to work: either they ride minibuses or they use their foot. Data

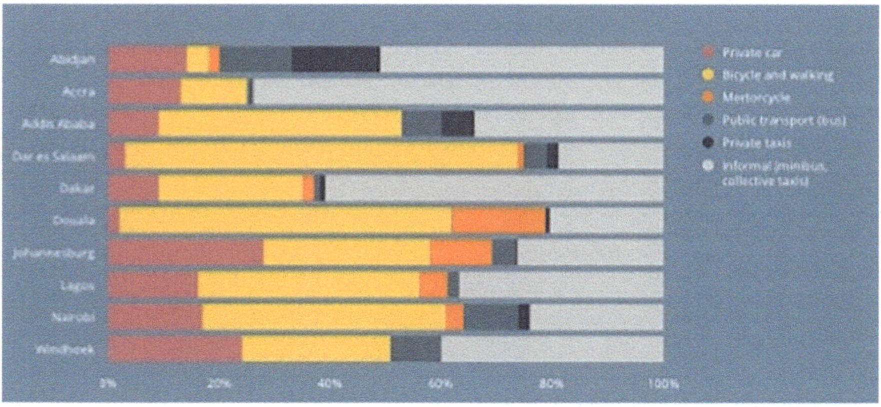

Fig. 15.22 Distribution of trips in selected African cities by modes of transport. *Source* From Lal. SV. et al. 2017 citing International Association of Public Transport 2010

collected in Dakar show that the share of minibuses in the public transport exceeds 90%. Despite various efforts taken by national and local authorities to boost the formal public transport, it remains with a very low share in Dakar, less than 5% in the total public transport, which is heavily dominated by informal means of mobility. Most of the formal public transport sector is composed with few medium-sized buses that must follow a specific itinerary, but they are barely seen in the streets submerged with the informal motorized means of transport. As illustrated in Fig. 15.22, the share of minibuses is high in Dakar, 63% in all mobility modes. Among the motorized means of mobility, this figure represents 89%.

15.4.2.2 Formal Public Transport

Despite various efforts taken by national and local authorities to boost the formal public transport, it remains with a very low share (5%) in the total public transport, which is heavily dominated by informal means of mobility. There was a subsidized public transport with the public company SOTRAC with buses in early 1970. But it was discontinued during the structural adjustment period imposed to Senegal as in many African countries by the IMF and the World Bank in mid 1980s. However, in 2001 public transport was reinforced with the introduction of newly public transport company named "Dakar Dem Dikk" (3D), with the share of three quarters from public resources and the rest from private investors [63]. The 3D is essentially composed of few medium-sized buses that must follow a specific itinerary, but they are barely seen in the streets submerged with the informal motorized means of transport. The efficiency of the public company 3D has faced, indeed, various difficulties to be sustainable in terms of supply as well as in terms of demand: lack of institutional coordination between the city council and the company 3D; structural deficit associated to the gap between the affordable fares and the operating costs leading to operating deficit; and poor operational and commercial performances associated to public service obligations without the corresponding resources. Other constraints are low commercial returns in congested roads, low productivity of employees, and attempt to serve the whole urban area without having enough buses. Other formal public transport means is the railway composed by the urban train created in 1998 "Petit Train de Banlieue" with a little share (less than 1% in 2009) to the total public transport. Efforts to modernize the urban train under the Urban Mobility Improvement Programme, "Programme d'Amélioration de la Mobilité Urbaine- PAMU" include transfer of track provision and maintenance to Transrail, a new autonomous body for operation, track improvement, and building fences along the track (compensated by new footbridges) [64]. Despite all these efforts, the share of the railway in the public transport remains little compared to the demand of passengers from Dakar to its suburbs.

15.4.2.3 Informal Public Transport

Considering the lack of capacity by the formal public transport systems to meet the mobility demand of the residents of Dakar, dominance of the informal sector operators remains irreversible. The informal transport sector, composed of small-scale economic activities with unregulated employment, is largely dominated by buses operated by private entrepreneurs It represents 95% of the public transport. It supplies small-vehicles with low investment and minimal public support, low-performance services that fill the niche between formal taxis and the 3D buses [65]. They are mainly composed of minibuses called "Car Rapide", "Ndiaga Ndiaye" and "Tata". Their operational status is legal, as the vehicles are licensed for the purpose of public transport with a specific capacity limit. Informal fleets can be a lifeline to making a living where poor municipal public transport resources have led to limited or no service to outlying residential areas and high fares. It represents the only accessible means available to the poor and provides an important service that is often well adapted to local conditions. It delivers affordable services in instances where scarce municipal resources have led to a deteriorating service and higher fares. Informal fleets reach outlying residences in sprawling areas.

However, the buses as means of public transport come with enormous externalities that authorities must not overlook in the transport planning process. In this sector, most vehicles do not fit to be in the road technically and with respect to traffic rules and regulations (technical control, license, insurance, etc.). This contributes significantly to congestion, air and noise pollution and traffic accidents. In addition, passengers, particularly women are exposed to harassment from the bus operators and other passengers. Violence against women is common in the public transport sector [66]. In a city like Dakar where the formal public transport sector is quasi absent, the role of informal transport in generating broader social and economic benefits must be assessed together with the costs entailed. Prohibiting informal transport is not a viable solution as it destroys jobs and blocks access to employment, but externalities must be assessed and reduced significantly. A combination of valuing the service and regulating them can be effective in managing related issues, such as traffic congestion, accidents and pollution.

15.4.2.4 Increased Use of Private Cars Hampers Efforts to Make Dakar a Smart Metropolitan Region

In absence of reliable public transport systems, households that can afford it, usually the upper middle class and the richest, possess private means of transportation or use taxis, which offer comfort compared to the public means of transport. The perceived advantages of convenience, privacy and status continue to make the private car an attractive means of transport in Dakar as in many African cities [67]. The growth of private motorized transport during the twentieth century had major impacts on the growth and development of the city of Dakar as many cities in developing countries. The rate of motor vehicle ownership in Dakar has significantly increased during these past 20 years, from 5% in 1995 to 13% 2014. The major factor behind the growth of car ownership in Dakar is at a cost that is becoming affordable for a growing number of people of the middle class that cannot tolerate the discomfort of the public transport with the "car rapide" and seek individual freedom. With the continuous urban growth, it is expected that Dakar will be an exclusive motorized city if the public transport is not reinforced and mixed neighbourhood promoted, considering the creation of six new urban centres as part the extension of Dakar [68].

While private cars offer the convenience of individual choice, this advantage is traded for much more land being used for road space and parking. Cars' land consumption and infrastructure costs can be a significant part of a city's budget and this costs heavily subsidized by both drivers and passengers. Emissions of pollutants in Dakar seem to be higher than the thresholds defined by international standards. Estimates of safety costs from the World Bank show that in Dakar these costs represent 2.7% of the national GDP [69]. A city submerged with cars is prone to more accidents, which generate economic and human costs. Without adequate public transport systems in place, and good planning choices that increase connectivity and proximity, congestion, pollution and energy consumption will also increase exponentially with the growth in car ownership. Unless authorities introduce rules and regulations such as road pricing, parking management and circulation policies to reduce car demand, ownership of private will continue to rise.

The monocentrism of the city of Dakar is a source of traffic congestion and an obstacle to smartness. The mobility in Dakar is heavily affected by the fact the city was conceived as a monocentric city where residential areas and workplaces are distinct, with the later concentrated in the centre of the city. Residential areas were also designed along economic class lines. This has led to social and economic fragmentation that disadvantage lower income group in accessing basic services and prevent social interaction and integration. This has impacted the traffic with every morning and evening people have to share a main street network to and from the downtown of Dakar, which is named the "Plateau". The urban structure of Dakar illustrates an unsustainable travel structure of the city, focuses on the historic centre called Plateau that hosts the main commercial and administrative services. This leads to congestion and long trips from the periphery (Pikine, Guediawaye and Rufisque) to the centre. However, this monocentrism has started to erode with the emerging on new commercial and administrative centres outside the Plateau.

Two major impacts associated with low coverage of formal public transport and increased use of private cars are traffic congestion and pollution. Traffic congestion is a major indication of the disjuncture between land-use planning and transport systems. It not only exposes the limitation of a transport-oriented bias to mobility, but also reveals the inefficiency of land-use systems in a given city. Limited road capacity, in the face of growing demand for motorized mobility, partly explains deteriorating traffic conditions. Congestion has widespread impacts on the urban quality of life, consumption of fossil fuels, air pollution and economic growth and prosperity. Congestion accounts for significant percentages of the gross domestic product in many major African cities as noted in World Bank studies from the 1990s. About 90% of the cost comes from the value of the time lost by residents, 7% from the fuel consumed and 3% from gas emissions. In addition to economic costs, congestion causes significant numbers of early deaths from respiratory illnesses, stress and physical and mental fatigue. It also degrades green areas, which, in turn, diminishes their carbon sequestration properties.

15.4.2.5 Walking—Key for Dakar Smartness

In absence of affordable, reliable public transport, various options exist depending on level of economic conditions or purchasing power. Those that can afford it will own a car or ride a taxi, while those that cannot afford it, who constitute the majority, will use their foot. Those using their foot have limited opportunities compared to others that have access to public transport. They won't accept job where they have to walk very long distance, or if they accept job far away from their residence, they will encounter several days of absenteeism, and they end up by quitting the job. Despite lack of comparative data, it seems that the share of walking to access services remains the first option in Dakar due two main factors, namely: low coverage of public transport infrastructure and unaffordability of public transport for the poor where various needs compete. While in cities of developed countries, it is assumed that a walkable street is more attractive to people for various reasons and even defines the "liveability" of a city, in the city of Dakar walking is not a choice, but a necessity due to lack of other affordable transport alternatives.

Available information shows that 60% of residents of the city of Dakar walk either to their place of work or to health centres, to schools, or to the market. However, the streets of Dakar are not designed for pedestrian use. First, the land allocated to the streets is low, only 15% in the city core and less than 5% in many sub-urban settlements. The streets are also narrow, not paved, lack sidewalks. This results in competition for space with motorists, often exposing pedestrians to accidents. It is also noted that in many African cities, 95% of the city streets also have high pedestrian flows but only 20% have pedestrian footpaths [67]. In addition, streets in the suburbs are poorly connected with less than 40 intersections per km^2. Lack of sufficient intersections make the street network of Dakar not friendly to pedestrians. It is demonstrated that for a street network to be well connected and walkable, at least 100 intersections per km^2 are required [70]. In areas such as Yeumbeul Nord, Camberene and Colobane, which are densely populated with densities of more than 50,000 inhabitants per km^2, there is no space left for mobility except very tiny streets where people have to squeeze through. In many settlements, public transport services are accessible in few arterial streets where people have to walk several kilometres to reach them. During the raining season, the already limited walking spaces are flooded, forcing people to stay at home, losing many productive hours. This in turn has serious social and economic impacts on the city's economy.

Despite challenges, walking has enormous economic, social and environmental advantages. Dakar of the 21st century must be planned as walkable with affordable means of public transport. Walking is an enabler of social cohesion and environmental sustainability with enormous social, Economic and Environmental benefits. In addition to its social and economic benefits, walking has a major advantage in reducing energy consumption, greenhouse gas emissions and pollution (air, water and noise) substantially, as it does not rely on fossil fuels unlike other modes of transport in cities. Furthermore, as walking requires significantly less road space and parking, it enables the preservation of natural habitats and open spaces. Walking also provides the daily physical activity required for a healthy lifestyle. Based on this, many motorized cities of developed countries have been changing their street planning and designing, and promoting public transport in order to reduce private motorized use and boost walking and cycling. With the 60% of people that are already walking in city, Dakar is indeed in a good position for a healthy society in a sustainable environment. However, in order to sustain this advantage, it is high time to give to pedestrians their right share in the transport infrastructure. Providing adequate infrastructures to pedestrians is cost-effective considering the enormous benefits from walking.

Integrating public transport, walking and cycling in the extension of the city of Dakar

Though Dakar is a large agglomeration, it does not benefit from the multiple advantages of high density due its unplanned urban growth and land expansion. While the advantage of a dense settlement is to ease accessibility, reduce cost of provision of basic infrastructures and other urban services, reduce erosion of natural resources, lower business costs and improve social equality, in Dakar none of these benefits are present or they are little. Considering the very high population density in many unplanned settlements of Dakar where the urban growth still high (more than 2% annual growth), it is urgent to develop and implement city extension at the fringes of the built-up area.

15.5 Urbanization, Economic Growth and Metropolitan Regional Development

Dakar as a coastal city offers multiple opportunities as a hub of economic activities as well as a link to local, regional and global economies. Today, it has an added advantage associated with its high population density and its youthful population, two important drivers of economic productivity and growth. One of the components of a Smart Metropolitan Regional Development is Economic Development, which is analysed in this section.

15.5.1 Urbanization and Economic Growth

Cities are potentially engine of prosperity due their power of economies of scale and agglomeration as well as of technology innovation and diffusion of ideas. High densities of cities also reduce transaction costs, make public spending on infrastructure and services more economically viable. Urban agglomerations with adequate urban planning and management coupled with good governance will produce wealth and sustain economic and social development [71]. The metropolitan of Dakar yet occupies a pivotal place in the national trade sector both nationally and internationally. Its autonomous ports (for coastal cities), international airports, international trade centres, touristic sites and commercial centres are assets economic growth. The Report of the Business General Census of 2016 has allowed to identify 407,882 economic units in Senegal, with the Dakar Metropolitan Region having the highest share of 40% (160,963 units), followed by the region of Thies (12%), Diourbel (9%) and Kaolack (6%); all other regions have each less 5% of national economic units. Furthermore, three-quarters of formal economic units are located in Dakar (Fig. 15.23). The metropolitan region also employed 87% of the formal sector. This shows clearly hat Dakar is the engine of the Senegalese economy. While in the Dakar metropolitan region, 52% are employed in the formal sector, in other regions, except Saint Louis and Thies (22 and 14% respectively), this figure is less than 10% [72].

The majority of economic units in Dakar are in the commercial sector (52%). The industry sector (food, textile and others) has a share of 25% in the economic units. It is important to underscore that 92% of large-size company, 88% of medium-size company and 66% of small size company are located in the Dakar Metropolitan Region (Fig. 15.24).

Industry in the Dakar Metropolitan Region is confronted with several challenges that include low production levels, inadequate competitiveness of the local market, lack of capacities of industrial firms to upgrade their production systems, geographical and structural weakness of the industrial fabric. In operational terms, initiatives concerning partnership between the State and the private sector are still minimal, particularly with regard to the promotion of entrepreneurial initiatives, the

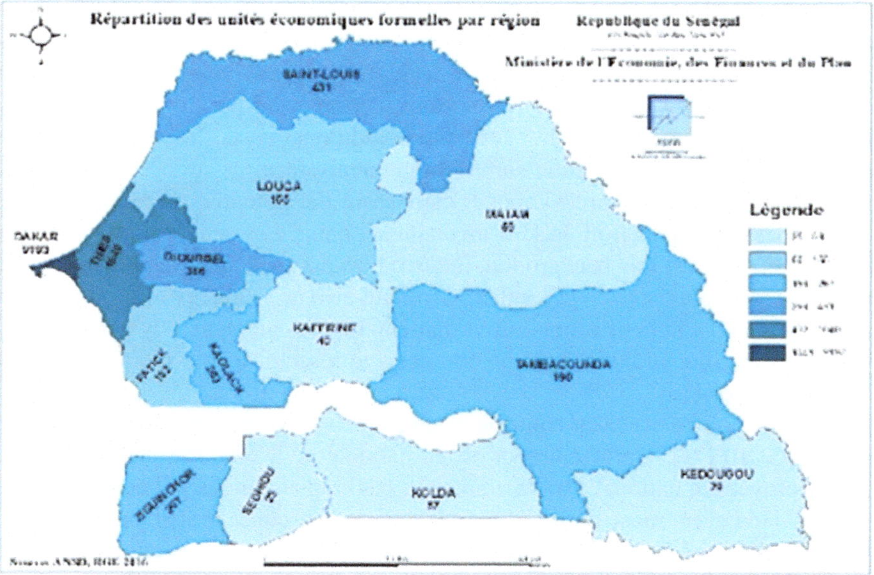

Fig. 15.23 Distribution of formal economic units by region, RGE 2016. *Source* Republic of Senegal, 2017. Recensement General des Entreprises 2016

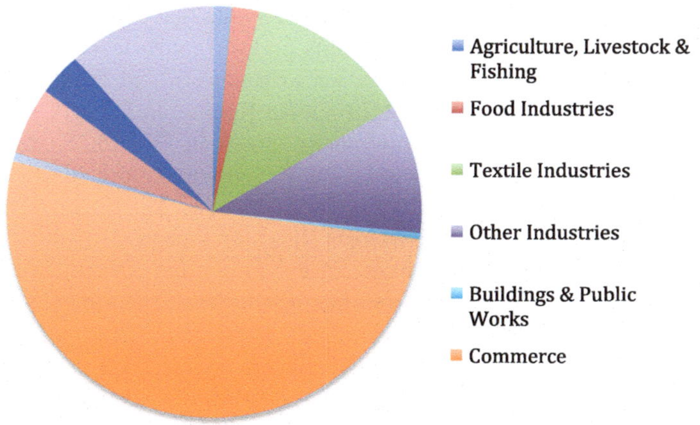

Fig. 15.24 Distribution of economic units by Sector, Dakar 2016. *Source* Data from ANSD/RGE, 2016

development of innovation through research application, the creation of integrated competitiveness poles and the training of future champions of the different sectors and the development of venture capital. The shortage and unreliability of power supply and the weak infrastructural platforms cripple the performance and competitiveness of industries, causing substantial additional costs. In addition to these drawbacks, industrial production suffers significantly from its strong concentration in the Dakar area as mentioned above, hampering the potential of provincial economic zones, as well as from lack of diversification and a system of product quality certification. Against this background, industry has not paid attention to the need to protect the environment as a general concept, but also as a strategy to improve their productivity and competitiveness. Environmental issues are generally perceived by industry as causing additional production costs to a sector that is faced with many other challenges [73].

The metropolitan of Dakar contributes up to 55% of the national Gross Domestic Product (GDP) [74]. Up to nine out of ten national civil servants, and from 50 to 90% employees in national's trade, transport, banking and industrial enterprises are in cities. However, most these urban advantages are concentrated in the city of Dakar. Though the Dakar Metropolitan Region had been administratively divided in four departments, in reality the other three departments act as suburbs of the department of Dakar, where the main commercial and administrative businesses are concentrated. The other three departments remain haunted by the early spatial and social division; they are not well planned, and they lack sufficient land allocated to streets, other public spaces, basic infrastructure and security of tenure.

Though Dakar metropolitan region is engine of national economies, association between urbanization and GDP around the world shows that the Dakar metropolitan region as many African metropolitan regions perform much less than its counterparts in other regions. Relationship between urbanization and economic development had been uneven in the Dakar Metropolitan Region. In Senegal, urbanization has not led to expected economic growth as shown in Fig. 15.25, with Senegal below the curve of association between GDP per capita and urbanization.

While countries in Asia such as China, Indonesia, Malaysia, Thailand and Vietnam and in LAC such as Mexico, Costa Rica and Argentina display an increase in GDP per capita associated to high urbanization rates, a similar association between GDP per capita and urbanization was not observed in most sub-Saharan African countries among them Senegal, Togo, Liberia and Burundi. In most of African countries, capital cities were not able to produce and distribute wealth nation-wide. Dakar has not been able to fully transform its large population in terms of economies of scale and agglomeration economies as many Europeans and American cities did more than 50 years ago, and LAC cities and Asian are recently doing. The low productivity of Dakar lies to several factors including those spatial and mobility planning presented in previous sections as well as other factors associated to urban legislation and governance and financial capacities. Unbalanced urban system, unplanned dense settlements without basic services, poor street connectivity, and poor transport infrastructures impeded Dakar economies of scale and agglomeration potential, and create scale and agglomeration diseconomies.

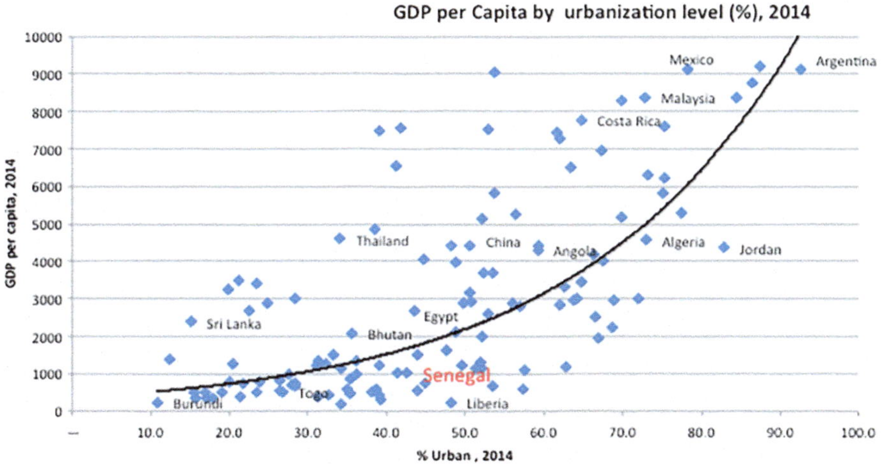

Fig. 15.25 GDP per capita by urbanization level (%), Country with GDP below US$10,000. *Source* % urban (UNDESA), GDP per Capita (World Bank)

Agglomeration diseconomies

With its permanent urban primacy status, Dakar has not been able to satisfy the employment demand, particularly from young people. The Dakar job market is, finally, predominantly informal. In addition to that, due to a deficient urban mobility, Dakar has not been able to create conditions for specialization. While urban agglomeration allows for job specialization, efficient market transactions and knowledge diffusion, if concentrated growth is not well planned—such as the integration of urban growth with efficient public transit investments, the resulting economic benefits tend to erode. Agglomeration diseconomies—i.e. the inefficiency and loss resulting from poorly planned concentrations—are expressed in the form of lost labour productivity from extreme traffic congestion, increasing air pollution and an overall decline in the quality of urban living [75]. This is the case in Dakar as in many African cities with poor connectivity leading to disconnected settlements.

Economic inequalities are also very high in the Dakar Metropolitan Region leaving municipalities with high unemployment and low access to basic services unable to afford many subsistence needs as illustrated in the Fig. 15.14. The Population of Dakar was classified according to their degree of poverty (Fig. 15.26), from to the highest living standard (blue colour) to the poorest living standard (red colour). Qualitative assessment of the Fig. 15.26 points out the fact that most of the population of Rufisque and Pikine have a very poor living standard (red colour). The orange colour that features category of families that have high living standard in Dakar, most of them are in the city core of Dakar along with families with the highest living standard. The department of Pikine, Guediawaye and Rufisque are predominantly populated with families with poor or moderate living standards.

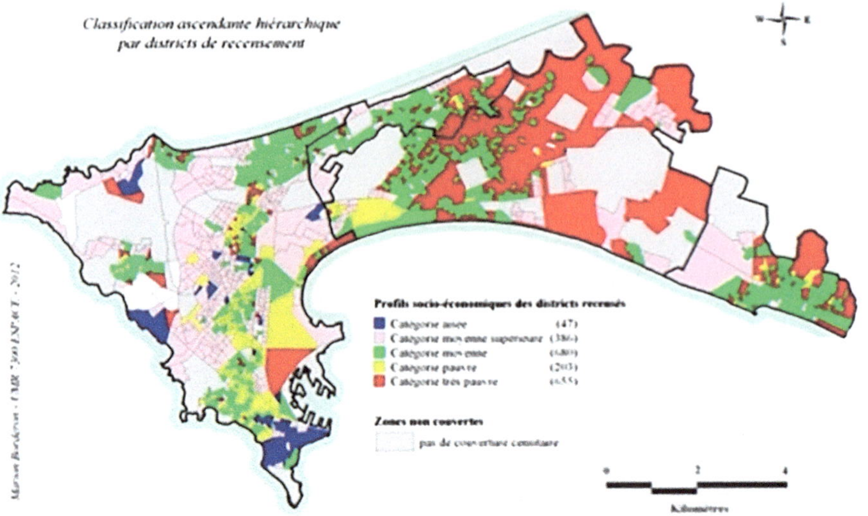

Fig. 15.26 Economic inequalities in Dakar metropolitan region. *Source* Republic of Senegal, 2014. Review of the Senegalese Urbanization

15.5.2 The Economics of Urban Land—the Hidden Assets

15.5.2.1 The State of Land Tenure

The smartness of Dakar Metropolitan Region has been long time hampered by weak institutions and laws holding back various factors of the smart city framework, making informality the norms in many sectors starting with the city foundation surrounded by informal settlements with irregular land tenure. Secure tenure goes beyond protection against eviction and includes economic and financial advantages. *Tapping in the Triangle of Economic Productivity—People, Land & Infrastructures* will create sustainable, inclusive prosperous and resilient cities. With functioning institutions and laws, land and housing assets can contribute to the planning, management and provision of services in settlements. However, in absence of functioning institutions and laws providing legal propriety rights, as observed in Dakar, most of these assets remain dead investments sheltering only people. To tap into the potential of high densities, the Dakar metropolitan must formalize its land system, which will be the driver of many other components of their foundation such as streets and public spaces, provision of basic infrastructures such as water, sanitation and energy, and waste management. It is urgent that national and local authorities recognize the wealth of their citizens and involve them in the planning, the building and the management of their city. No city can claim to be sustainable, inclusive, prosperous and resilient, when the wealth of its citizens is not fully taken into consideration in the economy. Dakar as most African Capital cities are not, however, fully benefiting from its land assets due to the fact that most of its land is considered as irregularly acquired and lacks legal ownership document such as a title deed.

Possession of ownership or tenancy document is low in Dakar where, according the Demographic and Health Surveys conducted in 2005, only 54% had a proof of legal ownership: title deed, sale certificate, power bill or other documents [76]. When disaggregated, less than 15% of the household reported having a title deed, which is considered the most secure document. However, it is interesting to note that in the city of Dakar, despite the low proportion of households with ownership or tenancy document, a large proportion feel protected against eviction (77%). Measures to reduce the risk and stress associated with lack of documents and fear of eviction are based on recognizing and respecting a plurality of tenure systems, including intermediate forms of tenure arrangements and alternative forms of land administration and land records [77]. The legal institutional framework in a given country or city plays a key role on various elements of security of tenure such as acquisition or adjudication which is the process of final and authoritative determination of the existing rights and claims of people to land.

15.5.2.2 Economic and Social Costs of Settling in Unplanned, Informal Land Areas

Due to poor land administration and governance, there is no compliance with standards of occupancy of the space leading to a high building density and irregularity of the urban fabric. Wetlands in Dakar cover an area of 40 square km of which nearly 72% (29 km^2) are hosting human settlements [78]. From 1954 to 2003, 95% of these areas have been converted into habitation [79]. The population of Dakar has been constantly exposed to flooding during raining season. Apart from habitat degradation, floods cause considerable economic losses on the various activities performed by people across various income and social connections. The impacts of floods on people and communities are enormous ranging from economic, social and health issues to environmental aspects. Asset losses degrade the quality of life of households and reduce the housing value. By Affecting social development with inaccessibility to most services, the economic development is also severely affected with significant decline of productivity of the active population. In 2009, the Post-Disaster Needs Assessment (PDNA) estimated damage and losses to total 44.5 billion FCFA nationwide (US$89 million), of which 35.5 billion FCFA (US$71 million) was for damage and loss in the Dakar region alone, with the most significant damage being on housing (61%), Transport (11%) and health (10%) [80, 81]. In terms of losses, the trade sector suffered the most losses, with 23% (mostly informal trade), followed by housing (18%), urban community infrastructure (18%), energy (17%), and transport (16%) sectors. At the household levels, an estimated 30,000 houses were affected in the Dakar region, most of which are now uninhabitable and often abandoned [82], and nearly 360,000 people representing 44% of the population of Pikine were affected. The impact of flooding related disasters remains a significant challenge to sustainable development of the city of Dakar [83]. Faced with recurring floods in most cities across the country and especially in the Dakar region, public authorities felt the urgent need to find a solution in 2009. In August 2010, the Government of Senegal decided to prepare an urban development project for rainwater management and climate change adaptation, known as PROGEP, aiming to reduce floods through an integrated and sustainable approach [84]. However, all these projects require financial supports that divert funds from other priorities and needs. For instance, the PROGEP is a five-year project (2013–2017) funded for USD 72, 9 million [85].

15.5.2.3 Provisional Estimates of Land Value of Dakar

Under its Disaster Risk Reduction (DRR) strategy as part of its Poverty Reduction Strategy (PRS) process (IMF 2007), land has been given a central place. The DRR team combined hazard and population maps, land price data, and land cover information to derive the exposure of different variables in different locations. It is estimated that the Dakar city area represents a total land value of $44 billion. This figure represents 8 times the city GDP. Considering the level of informality of 37%, we can assert that US$17.4 billion out of the US$44 billion of land value of Dakar is not convertible in the financial market to secure marketable financial transactions; it cannot also generate revenues for the development of infrastructures. In addition, due to poor urban planning and irregular land use, over $2 billion or 5% is exposed to high natural hazard potentials. In the absence of functioning land market where prices are regulated and documented, these figures must been considered as rough estimations of the exposure of economic assets to natural hazard such as flooding [86]. Information from the PROGEP corroborates the economic damages and losses associated to flooding in Dakar. At the household levels, an estimated 30,000 houses were affected in the Dakar region, most of which are now uninhabitable and often abandoned [82], and nearly 360,000 people representing 44% of the population of Pikine were affected. The impact of flooding related disasters remains a significant challenge to sustainable development of the city of Dakar [83]. These various estimates point out the importance of adequate planning and land use regulation to mitigate the city's economic vulnerability against risks such as flooding.

15.5.2.4 Financing Infrastructure Development and Economic Development Through Smart Land Tenure

At the economic and financial aspects, there are various social and economic advantages including access to the financial and economic market as illustrated by de Soto. De Soto argued that granting titles to the poor would liberate the plots they occupy and transform them into capital. This, in turn, could be used as collateral for loans to jumpstart their businesses, or improve their houses, among other gains that increase their quality of life. At the community level, regularization of land tenure will help the municipality to legally collect various taxes that can be used to improve basic infrastructures such as connection to water, sewerage facilities, energy sources and waste management facilities. This would also allow people to fully participate in the development of their communities at the policy as well as the implementation level instead of seeing proprieties as dead investments serving only for shelter [87]. Though having a title dead does not necessary lead to secure a bank loan; it may not be sufficient in itself to animate the dead capital interred in land and property, particularly in countries where banks lend only to workers with high wages and a stable job, as it seems happen in some countries covered in de Soto study. However, families with title dead may be likely to invest either to improve the quality of their homes or to increase their size. It is also important to note that land tenure goes hand and hand with urban planning [88]. Without an accurate mapping, legal title dead cannot prevent land disputes as it often occurred in African cities.

While there is argument against the de Soto theory in poor settlements, in Dakar regularization of land tenure will work for many middle and upper income households that have not been able to use their property as collateral due to lack of legal ownership document. Another important group land regularization will benefit is the Senegalese from the diaspora who usually send money to their family and build houses in Senegal. Most of these houses are built in irregular settlements. Among the Senegalese of the diaspora, certain have a high desire to return home, but they have not save enough to do so, and they cannot use their houses as collateral and start a business in Senegal. No choice, even those they are here cannot get a title dead, what about those that just come to visit their family for a month or less. These are not the poor as described in the de Soto book, they have already the value of savings as underlined by the money they sent to their family in a monthly basis. The money received from the diaspora is estimated at $1.9 billion (FCFA 971.4) by the Senegalese Ministry of Finances, much higher than the foreign aid Senegal received during the same year. It represented 12.1% of the national GDP. The Ministry of Finances considered that the diaspora are the first donors of Senegal [89]. However, the money of the diaspora is not invested in the market, it is invested in residential houses and household subsistence. Without legal recognition, these houses cannot be transformed neither into capital nor transacted through the economies of scale and agglomeration that the city of Dakar potentially offers. It may take several years before regularization, particularly when it is bought informally.

Large infrastructure projects require huge investments. However, like any large structures, they depreciate very slowly over decades or even centuries [90]. The central government transfers on which Dakar often rely will not suffice to finance all infrastructures required to take advantage of the economies of scale and agglomeration that Dakar can offer its large population of more than 3 million inhabitants with high density of over 15,000 inhabitants per km^2. Therefore national and local authorities should explore various financing options for infrastructure development. The costs of developing housing, infrastructure, and industrial premises must be coordinated with land markets and land use regulations in order to fully take advantage of the economies of scale and agglomeration of the city of Dakar. This calls for the formalization of the land tenure, which, in turn, will increase the land values that can be used to contribute to the development of basic infrastructures such connection to water, sanitation, drainage, solid waste management and streets. For instance, in Dakar due to lack of municipal finances most municipal street networks are not paved. Municipalities with formal land system have the large proportion of paved streets compared to others without formal land system [91].

Secure tenure goes beyond the legal character; it attracts investments as illustrated by the level of provision of infrastructure in municipalities considered as legal settlements comment to others considered as informal settlements. These municipalities also enjoy health centres and school facilities. They also attract the financial market because they have a legal urban plan with a sufficiently documented cadastre system. This shows the community character of land tenure that goes beyond the household and embrace infrastructure in the municipality itself. In most irregular settlements, there are few or no paved streets, and the few streets they have do not have light. Lack of documented urban plans affects the financial as well as the land market. It is noted that capacity and resource constraints are the main reason infrastructure in most municipalities have been lagging behind. In absence of formal land system, a settlement is trapped into poverty. With this, there is no doubt that wealth is associated to formal land system. Let make secure tenure work for people and communities. Dakar is expected to reach 5 million inhabitants in 2035. This will propel new demand for infrastructure such as water, sanitation, sewerage, waste management and streets among others. To meet this demand, land transactions must be eased with transparent efficient land law, administration and governance with documented land and property rights; documented guidance of land valuation and prices and; coordinated land use and urban planning. Land tenure goes hand and hand with urban planning. Urban planning supported by accurate demarcation of public and private uses are of importance [88]. Without an accurate mapping, legal title dead cannot prevent land disputes as it often occurred in African cities. In Dakar, it may take several years before regularization, particularly when it is bought informally. Formal land Registration will ease transactions and boost economies of scale and agglomeration.

Considering the weak financial revenues of the city Dakar, a land value of 44US $ must be seen as an opportunity to tap on it. It represents nearly 500 times the annual revenue of the metropolitan area estimated at US$94.8 million). Though there is a steady increase of the budget of Dakar from US$ 4 million to US$ 56 million, the city is still in the incapacity to satisfy the increased demand in most basic services such as water, sanitation, solid waste management, health and education. The budget of the city finances mainly operational against investment expenditures (59% against 41%) [92]. The budgets mainly include local taxes and levies (around 90%) for all the departments. Subsidies from central government remain minimal, less than 2%. But this situation is more due to the nature of taxes levied in each entity rather than to performance of local authorities [92].

15.6 Towards a Smart Metropolitan Regional Development

Most of the components of a Smart Metropolitan Regional Development (SMRD) are composed of several simple sub-components. Put them together to track progress on the overall objective of the SMRD will require the development of composite indices using sophisticated multi-level statistical analysis including Principal Component Analysis. The complexity of a composite index is the fact that it is strongly influenced by several factors including: the normalization, the standardization and the degree of association between variables and the number of variables that compose the index. An index can also be conceptualised differently using different variables. This calls for caution when interpreting a composite index. For instance the Quality of Life is conceptualised and measured in different ways across studies. The Human Development Index includes indicators of health and education, alongside an economic component (Gross Domestic Product or GDP); the Legatum Prosperity Index considers quality of life to be multidimensional, including both wealth and wellbeing; The Economist Intelligence Unit's Quality of Life Index links life satisfaction to health, family life and community life; and the OECD's Well-being Initiative has two dimensions: 'material living conditions' and 'quality of Life' [93].

In our study, a smart metropolitan regional development is viewed as a sustainable, inclusive and prosperous metropolitan regional development that promotes a people-centric approach based on three core components—Smart Metropolitan Region Foundation, Information and Communications Technology (ICT) and Smart Institutions and Laws, which are the pillars of the other dimensions of a smart metropolitan region: Infrastructure Development, Environmental Sustainability, Social Development, Social Inclusion, Disasters Exposure, Resilience, and Peace and Security. **Infrastructure Development** complements the basic infrastructure services under each smart metropolitan foundation and extends to actual investment and advancement of services such as transport, ICT, industrial energy, education, health, etc. **Environment Sustainability** is comprised of elements of energy, transport, building and pollution. **Social Inclusion** includes aspects of participation in decision-makings as well as equal opportunities for growth and prosperity. **Social Development** encompasses elements of education, health, public space, social inclusion and social capital. **Disaster Exposure** incorporates elements of mitigation and adaptation to various disasters such as flooding, droughts, storms and earthquakes. **City Resilience** is composed of elements of city foundation, environment, social capital, and social development. **Peace & security** included the elimination of all forms of discrimination and violence and conflicts, including domestic violence, violence in public places, crime, armed conflicts, terrorism, etc. An insecure metropolitan region limits opportunities for investment and economic growth and cannot be a smart metropolitan region.

The city of Dakar was included in the development of three indices published for the UN-Habitat and conceptualized and further developed by Mboup G. (2012,

2013 and 2014). Those indices include: the City Human Development Index, the City Prosperity Index and the urban form-based City Prosperity Index integrating the Composite Street Connectivity Index. This section will take it further for the development of the Smart Metropolitan Regional Development Index.

15.6.1 Urbanization and Human Development

The human development is made from element of economic development (GDP per capita) and elements of Social Development (Education and Health). All three elements are components of a smart metropolitan regional development.

In the last 20 years, considerable progress had been made in education as witnessed by the increase in the percentage of young people 15–19 years and 20–24 years old with at least a primary level of education. Estimated at 68% in 1993 and 72% in 1997, the percentage of women and men 15–19 years with at least a primary education is estimated at 90 and 95% in 2015 respectively (Fig. 15.27). These figures show that basic education is quasi universal in Dakar and makes the city of Dakar potentially smart. Education is, indeed, critical to meeting the challenges in the ICT era, as it also connects people to new approaches, solutions and technologies that require basic literacy.

It is important to note that similarly the literacy rate has increased from 59% in 1993 to 80% in 2014 among the young women of 15–19 years and 84% in 2005 to 86% in 2015. However, these literacy rates are 10 points lower than the proportion of people in the same age group with a primary education level, indicating that primary education does not necessary provide literacy capabilities. In Senegal, measures have been taken to improve learning quality with the introduction of pedagogical resources, especially textbooks for the core subjects of reading and mathematics with the provision one textbook to every student in a classroom; this has increased literacy scores by 5–20% [94]. By 2030, literacy will be universal in Dakar as access to education will also be universal. This will impact all sectors of the metropolitan region smartness in many levels in terms of attitudes and behaviour.

In addition to analysing the education level, we have also assessed the net primary school attended among children 6–11 years old in Dakar, which is the normal primary school age. Since 2004, the minimum age to enrol to public primary school is reduced to six years from seven years previously. The net primary school attendance rate calculated among children 6–11 years old used the series of Demographic and Health Surveys conducted in Dakar also show progress in school attendance for both boys and girls. Estimated at 60% in 2000, the net enrolment ratio at the primary education level increased to 78% for both boys and girls, and to 90% and 91% in 2015 for boys and girls respectively.

Analysis of series of DHS in Dakar shows that the percentage of young people with secondary education level has also increased during the last 20 years, from 37 to 62% for women aged 15–19 years old and from 57 to 71% for men aged 15–19 years old between 2000 and 2015; this is a notable change (Fig. 15.28).

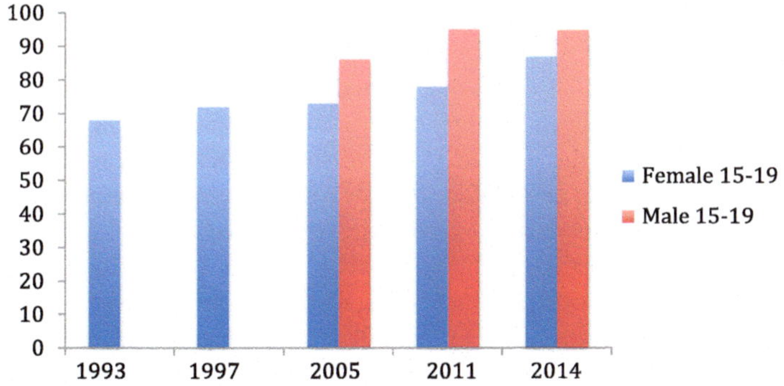

Fig. 15.27 Percentage of men and women aged 15–19 years old with at least a primary education level, Dakar

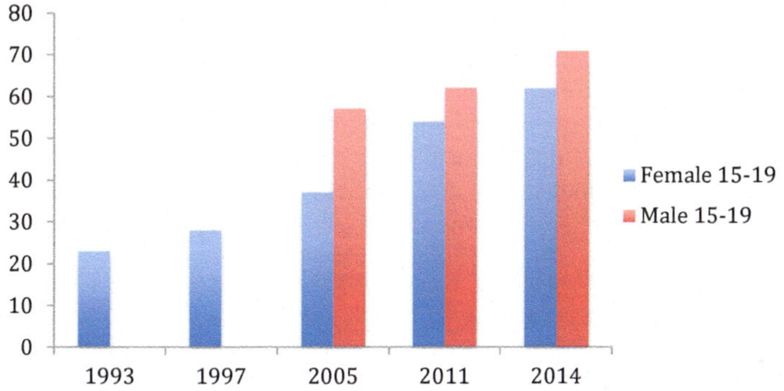

Fig. 15.28 Percentage of men and women aged 15–19 years old with at least a secondary education level, Dakar

The increase in secondary education in Dakar can be associated with two main aspects. First is the expansion of secondary education facilities in remote and poor areas taken by the government in the early 2000 allowing children to move from primary to secondary level within the same geographical area. Schooling in their neighbourhood has also mental and social advantages on children, giving them the opportunity to grow along with their parents. Second is the extension of free and compulsory education to include lower secondary, which has been possible due the political commitment of Senegal as in the MDGs and in the Dakar Framework. Senegal is, indeed, among the few African countries that devote more than 1% of GDP (1.4%) to higher education.

A healthy population is critical to realizing economic growth through increased productivity [95]. Healthy workers are more productive, bringing greater income to families and higher levels of economic growth for nations. The Demographic and Health Surveys (DHS) held in Senegal show a constant progress in access to health services in Dakar during the past 25 years. In the 1990s, access to health services was very limited and infant and child mortality rates were consequently high. With a steady improvement in health coverage, enhanced with the Millennium Development Goals (MDGs) with its Goal 4, 5 and 6, the improvement was accelerated during the 15 years of the MDGs. There is no doubt that improvement in health is significantly contributing to the Dakar smartness. Children born in Dakar have more access to health services such as antenatal (by their mothers), delivery cares and immunization than those born in other cities, town and villages of Senegal. In Dakar, access to antenatal cares is almost universal and 90% of delivery occurs in health centres compared to less than 50% in other places of Senegal (49% at national level). Over 10% of delivery in Dakar is assisted by physicians or other health care specialists compared to only 1% at the national level. Most of children in Dakar have also received all required vaccines such as BCG, polio, measles and DTP. When we consider all four types of vaccinations, the percentage children with all vaccinations is 59% compared to 42% at the national level. In Dakar coverage of child immunization is high both in non-slum and slum areas, a situation that shows that living in Dakar provide remarkably the opportunities of access to better health care. Thus, access to health care services in Dakar regardless of the economic status of families, is much better than elsewhere in Senegal.

Significant decline of infant and child mortality rates make living in Dakar smart In the context where access to health services has significantly improved, mortality has also significantly decreased, particularly among children under five years old. Infant and child mortality rates have been more than halved during the past 15 years. In particular, infant mortality has significantly decreased from 59 per 1000 in 2000 to 28 per 1000 in 2015, the child mortality from 35 per 1000 to 13 per 1000 during the same period. The under-five mortality rate, on the other hand, has decreased from 87 per 1000 to 41 per 1000. The decline of mortality rates has also been observed in other parts of Senegal, but to a lesser extent, making the figures of Dakar much better than the national figures. In 2014, the under-five mortality rate, estimated at 41 per 1000 is more than two times lower than the the rates estimated for five regions in Senegal (Diourbel, Sedhiou, Tambacounda, Kedougou and Kolda) (Fig. 15.29). Life expectancy in Dakar is estimated at 69.6 years with an advantage of 2.5 years for female compared to male, 70.9 years and 68.4 years respectively. Overall, people living in Dakar live five years longer than those of others regions; the national life expectancy is estimated at 64.8 years. It is important to note that Senegal has experienced a remarkable improvement on health with the life expectancy of 10 years higher than the level 15 years ago. In 2000, the life expectancy was estimated at 56 years.

Significant progress has also been made in the fight against malaria, which had been the first cause of deaths in Senegal. In the past fifteen years, there has been a

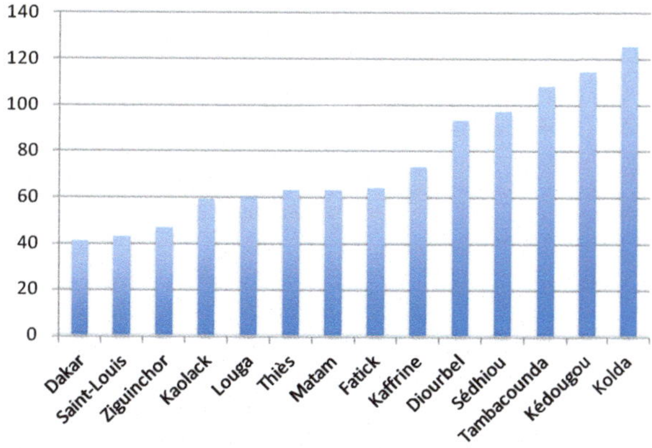

Fig. 15.29 Under-five mortality rates (per 1000) by region. *Source* ANSD, Senegaesel Demographic and Health Survey, 2014

significant decrease in malaria-related deaths from 40% in 2000 to less than 5% in 2015 contributing to the remarkable decline of mortality as observed in the same period. The prevalence of fever, among the symptoms of malaria, has also drastically decreased, estimated at over 40% to less than 20% in 2015 during the same period as indicated by Demographic and Health Surveys in Dakar. A hemoglobin level less than 8.0 g/dl is an indication of severe anaemia. In Dakar, that level is estimated at 3.2%. In Senegal, the National Programme to fight against Malaria (NMCP) has included in its strategic planning the promotion of the use of Insecticide Treated Nets (ITNs) as a major axis of intervention for reducing morbidity and mortality due to malaria. To achieve this objective, the NMCP and its partners have implemented procurement and distribution activities "Impregnated mosquito nets Insecticide to Long Term Action". This distribution is performed during routine activities through health facilities and community-based organizations. Since 2003, with the spread of chloroquine resistance, Senegal adopted sulfadoxine—pyrimethamine combination plus amodiaquine for the treatment of uncomplicated malaria. Then in 2006 according to WHO recommendations for the management of confirmed cases of uncomplicated malaria, Senegal adopted a Therapeutic Artemisinin-based combination.

Dakar is still engine of Senegal Economy as illustrated in Fig. 15.30 with the Dakar HDI much higher than national HDI; it has better economy outcomes as well as better health and education status. It occupies a pivotal place in the trade sector both nationally and internationally. Important progress has been made on education, which is critical to meeting the challenges of smart metropolitan regional development, as it connects people to new approaches, solutions and technologies that enable them to identify, clarify and tackle local and global problems. Similarly, important progress has been made on health, which is critical to realizing economic growth through increased productivity. Healthy workers are more productive, bringing greater income to families and higher levels of economic growth for nations, and then enhance smart economy.

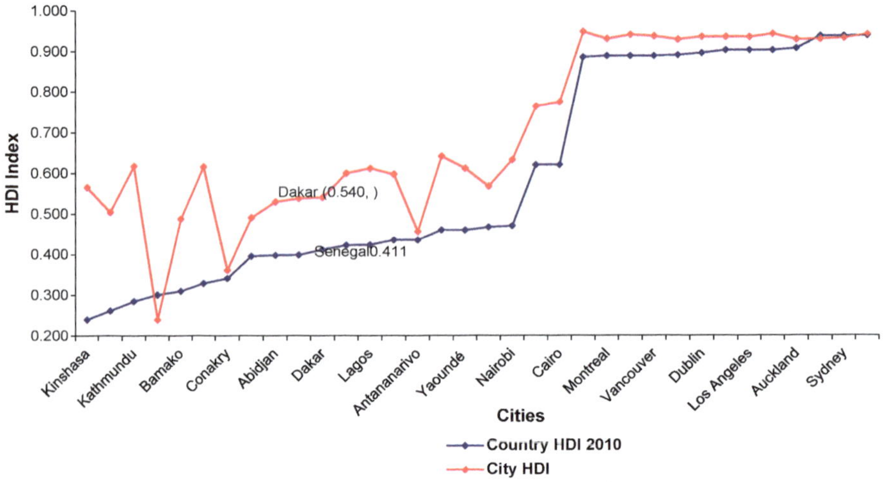

Fig. 15.30 City Human Development Index (HDI) and Country HDI, 2014. *Source* Mboup, G., 2016. Smart Social Development Key for Smart Economy. In Vinod K. et al., 2016. Smart Economy in Smart Cities, Springer

However, these advantages may also be linked to political choices that concentrate various economic infrastructures in Dakar such as: ports, international airports, the international trade and commercial centres, etc. They have also concentration of civil servants, private sectors such as banking and industrial enterprises. All this allows Dakar to contribute more than half of the national economies. However, Dakar has not been able to pay back and boost the rural economies and other Senegalese cities. Now the question is how to turn the linkages between urban and rural in the way that rural can reap the benefit of urban economies of scale and agglomeration of Dakar. This is where, ICT as a connector may play a crucial role in the inter-cooperation between rural and urban making circulation of goods and services cheaper and faster.

Though cities like Dakar are engine of economy growth, they need to be driven smoothly in an efficient administrative and governance environment. Thinking that having a large city and a high density is sufficient to produce economies and scale and agglomeration, is a simplistic view because there are many other factors that need to be present for urbanization to produce wealth and prosperity in general. In Dakar like in many African cities those ingredients are lacking leading to various negative externalities. Finally instead of enhancing economies of scale and agglomerations, Dakar as many African cities encourage diseconomies. Youth non-employment is high, and young people from impoverished urban areas can only find jobs in the informal sector with no social security coverage, paid parental leave, retirement, and unemployment benefits. The high rate of non-employment of young people in Dakar is a reason for concern; the lack of decent, sustainable jobs promotes a sense of displacement in the general youth population and often leads to crime, under-development, and a cycle of poverty. Frustrations accompanying long-term unemployment among groups of urban young men may feed political and ideological unrest and provoke violence.

15.6.2 *Urbanization and Environmental Sustainability*

Proliferation of informal sectors has led to the anarchic development of economic activities, industries and residential settlements. Growing industries located in or near the city, are key resource users as well as sources of pollution, waste, and greenhouse gases emissions. In addition to the fact that most unplanned settlements are located in flood prone areas, and wastes are poorly managed, climate change makes the situation worse [96]. Climate change contributes to sea level rise as recently noted in several coastal cities such as Dakar. Rise of the sea level accelerates coastal erosion and causes the loss of farming and habitable land. In Dakar, every year, the sea level rises by 4 mm causing a loss of 1 m of coastal land and an increase in flood vulnerability [97]. GHG emissions in the region of Dakar is also of concern with an annual estimate of 15,786,000 tons of CO_2, and 5.03 tons per capita. The main sources of GHG are industrial energy (30%), transportation (16%) and household (15%), and processing industries (12%) [98]. Indoor air pollution is a quiet and neglected killer, with lack of global awareness being one of the primary obstacles to the widespread implementation of existing, proven interventions [99]. Women usually have the added responsibility of caring for children who are then also exposed to high levels of indoor air pollution on a daily basis [100] (Fig. 15.31).

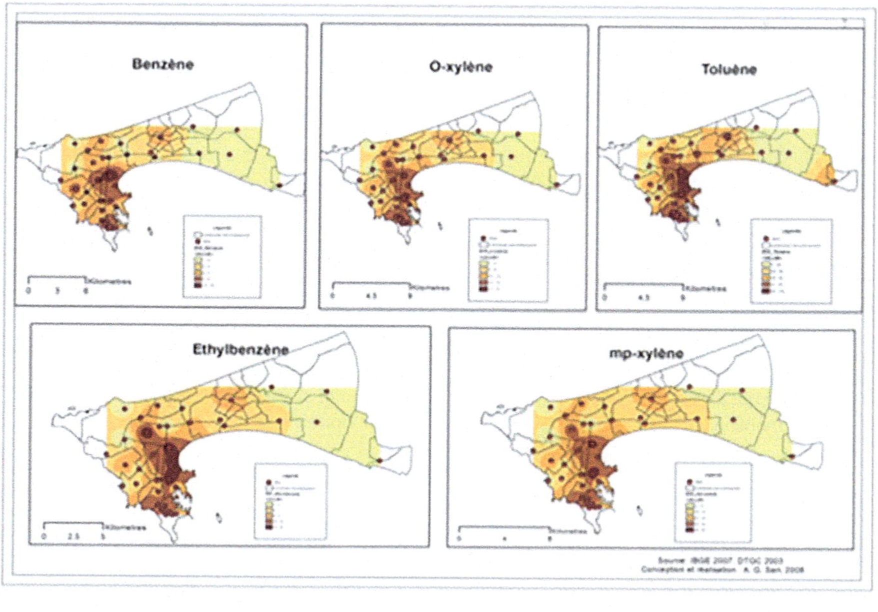

Fig. 15.31 Dispersion of pollutants in Dakar in June 2007. *Source* Gaye et al., 2009

The government of Senegal has developed an Integrated Urban Storm water Management and Climate Change Adaptation Program for peri-urban Dakar that aims at developing a national strategy on integrated urban storm water management and climate change adaptation based on the full set of investment plans and technical studies emerging from the Drainage Master Plan as well as from the Sanitation Master plan and diagnostic studies and consultations in selected cities. In addition to that, the three years Senegal's Sustainable Cities Initiative Project initiated in 2017 aims to integrate climate risks in urban planning, design and management [101].

15.6.3 Urbanization and City Prosperity

The City Prosperity Index (CPI) published by UN-Habitat in 2012 under the coordination of Mboup G includes various indices and indicators that are relevant to urban areas, and important for prosperity-oriented public policy-making. The first version of CPI published in 2012 was based on five dimensions: Productivity, Quality of Life, Infrastructure Development, Environmental Sustainability, and Equity and Social Inclusion [102]. In 2013, the CPI was revised to include the urban form dimension measured by street connectivity [103].

Dakar was classified among the Cities with weak sustainability, inclusion, resilience and prosperity index (with an index between 0.500 and 0.599). In this group, much remains to be done in terms of quality of life, infrastructure and environment. Production of goods and services is still too low, a reflection of underdevelopment. The Economic Development of Dakar is estimated at 0.347. Historic structural problems, chronic inequality of opportunities, widespread poverty, inadequate capital investment in public goods, and lack of pro-poor social programmes are critical factors behind such low degrees of prosperity. The city product of African cities in this bracket is low, as are the ratings for quality of life and infrastructure. Most of these cities perform better on the environment indicator [low emissions of fine particles (PM10)] (Fig. 15.32).

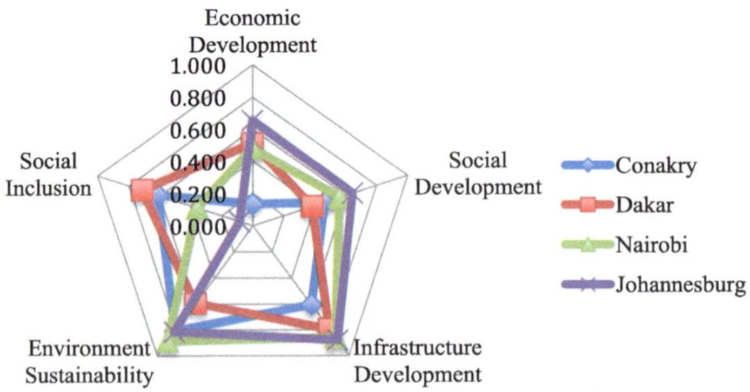

Fig. 15.32 Towards prosperity

15.6.4 Towards a Smart Metropolitan Regional Development

Among the components of the Smart Metropolitan regional Development, five have been part of the City Prosperity Index (CPI): Productivity, Infrastructure development, Environmental Sustainability, Quality of Life and Social Inclusion. The SMRD index introduces other components relevant for sustainable urban development: Disaster Risk Reduction and Resilience, Peace and Security, Institutions and Laws; It also expands productivity to Economic Development, Quality of Life to Social Development, and make the distinction between Infrastructure Development and Basic Infrastructure. It also introduced the concept of city foundation built upon three dimensions: urban planning, land policies and basic infrastructure. It also introduces ICT at the centre of the SMRD along with the City Foundation and Institutions and Law. Considering that the CFI and the CPI was already measured in previous sections, the measurement of the SMRD is incremental and built upon those indices with the inclusion of Peace & Security as well as Disaster Risk Reduction and Resilience (Fig. 15.33).

In addition to the urban challenges analysed in previous sections, the population of Dakar metropolitan region has been constantly exposed to disasters, particularly to flooding during raining seasons. Apart from habitat degradation, floods have caused considerable economic, social and environmental losses in Dakar. In 2009, the Post-Disaster Needs Assessment (PDNA) estimated damage and losses to total US$89 million nationwide, of which US$71 million was for damage and losses in the Dakar region alone, with the most significant damage being on housing (61%), transport (11%) and health (10%)[1]. Contrary, Senegal was classified among the

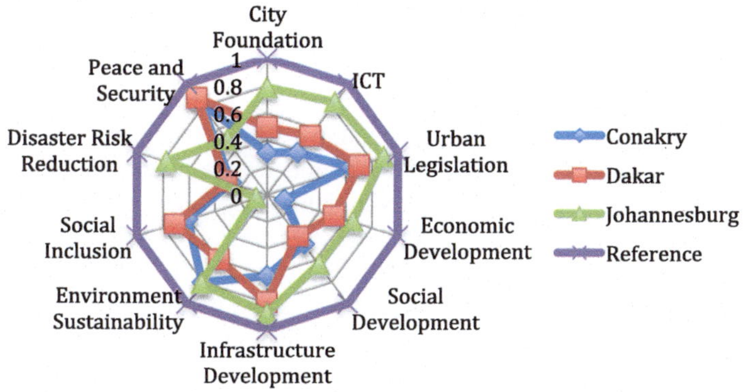

Fig. 15.33 Smart Metropolitan Regional Development Index (SMRDI). *Source* Provisional results from Mboup G. (ed.), 2018 Forthcoming Publication "Smart Economy in Smart African Cities", Springer

[1]Republic Senegal, 2010: 44-53, quoted by IAGU, 2011.

highly peaceful countries based on the Global Peace Index (GPI) ranking in 2015. The GPI scores are the aggregates of 23 qualitative and quantitative indicators across three thematic domains: the level of Societal Safety and Security; the extent of Ongoing Domestic and International Conflict; and the degree of Militarisation. Out of 163 countries, Senegal was ranked at the same level as the United Arab Emirates with a score of 1.805 compared to the lowest score, which is estimated at 1.148 for Iceland and the highest score, which is estimated at 3.645 for Syria. In Africa, only four countries—Senegal, Mauritius, Botswana and Namibia—are among the top 50 peaceful countries globally. For Guinea and South Africa, the GPI score is estimated at 2.214 and 2.376 respectively[2]. However, a high score in the GPI is not sufficient to boost the smartness of the Dakar metropolitan region, which has been suffering from multiple social, economic and environmental deprivations. With a weak City Foundation, a low Human Development Index as well as a weak City Prosperity Index, It is obvious that the Dakar Smart Metropolitan Region Index will also be weak as it is strongly determined by the three indices. Though this a preliminary assessment of the SMRDI since data are not able for some sub-indices, rapid assessment of available data point to the fact Dakar SMRDI is low due do to historic structural failure on urban development, chronic inequality of opportunities, widespread poverty, inadequate capital investment in public goods, etc.

15.7 Spatial and Economic Design Strategies for a Smart Dakar Metropolitan Regional Development

Recognizing the unbalanced Dakar metropolitan regional development, in one hand, and the multiple problems associated to lack of basic services, on another hand the Senegalese government has taken bold actions to transform the urban landscape of the metropolitan region from a monocentric to a polycentric metropolitan region with the creation of urban centres to decongest the Dakar urban agglomeration. This new spatial re-organization is also accompanied with economic transformation through the ambitious economic development, the Plan Senegal Emergent. This is through key main complementary programmes, reforms and plans: (1) the Plan Senegal Emergent; (2) Act III of Decentralization; (3) Metropolitan Planning; (4) Dakar Urban Master Plan 2035 and; (5) Digital Senegal Strategy 2016–2025. These urban policies and programmes aim to spatially and economically transform the Dakar Metropolitan Region to be smart,

[2]https://reliefweb.int/sites/reliefweb.int/files/resources/Global%20Peace%20Index%20Results%20Map_0.pdf. Download: 8 March 2018.

sustainable, inclusive, resilient and prosperous. They aim to transform the mono-centric system to a polycentric system where every urban centre equally contributes to the development of the metropolitan region.

This section is a prospective analysis of different ways economic and spatial transformations from these different programmes, reforms and plans will contribute to the smartness of the Dakar Metropolitan Regional Development. Most national and urban policies and programmes aim to spatially and economically transform the Dakar Metropolitan Region to be smart, sustainable, inclusive, resilient and prosperous. First we will analyse different spatial design strategies in the ACT III of Decentralization, the PSE, the National Territorial Plan, Master Plans, and other plans. Second we will analyse the economic transformation expected in these programmes, reforms and plans.

15.7.1 Decentralization of the Dakar Metropolitan Region

Under the Act III of Decentralization adopted in December 2013, each department of the Dakar Metropolitan Region has been sub-divided into municipalities. Overall, the metropolitan region counts 49 municipalities. Each municipality is headed by a mayor, who is responsible to, among other things, preserve, maintain and administer the properties and assets of the municipality, coordinate municipal waste management and hygiene, protection and conservation of historic sites and monuments, promotion of national and local cultures, preparation and implementation of various kinds of plans (including master plans and detailed urban plans), and management of health and education programmes. The Act further gives guidelines on municipal finance, which is key in the implementation of urban programmes. This is a clear demarcation of the original urban development management where administrative and financial management of municipalities was performed at the central level; it opens the road to polycentric specialized urban centres.

15.7.2 Plan Senegal Emergent (PSE): Holistic Approach of Economic Development

The PSE was conceived based on the Social and Economic Development National Strategy for the period 2013–2017 with the objective of: (a) 7% of economic growth by 2018 compared to a level of 4.6% in 2013; (b) Emerging Country Status by 2035. The Plan Senegal Emergent (PSE) provides further guidance on the Dakar Metropolitan Regional Development, strategically based on three axes: (1) structural economic growth and transformation; (2) human capital, social protection and sustainable development; and (3) governance, institutions, peace and security.

At the sectorial level, the PSE is glued around six main sectors: energy, infrastructure, business environment, telecommunication, human capital and finance. Flagship projects under the PSE include developing Dakar as a regional logistics hub for the production and distribution of industrial products and services in West Africa; development of multi-services and touristic hubs; development of Dakar as a referential regional campus with five world-class international schools; creation of business parks hosting international companies and institutions; and making Dakar an international medical city. The plan anticipates that these flagship projects will result in strong convergences and synergies within the region and improve the attractiveness and competitiveness of Dakar, and Senegal at large. Making Dakar a regional hub is similar to the thesis of global cities as discussed by Saskia Sassan since the 1990s. Four functions characterise metropolitan regions in the context of globalization: (1) innovation and competition; (2) decision-making and control; (3) gateway; and (4) symbol or branding [104] (Fig. 15.34).

15.7.3 *Spatial Design Strategies for a Smart Dakar Metropolitan Regional*

Along the transformation of the Dakar metropolitan regional development, national authorities have put in places planning guidelines through the Urban Master Plan of Dakar (Plan Directeur d'Urbanisme—PDU) 2035 and National Plan for Territorial Development "PNAT". The PDU 2035 aims at urban development of the region of Dakar and its surroundings by 2035. Its main objectives are: Sustainable Urban Development; Compact cities connected with a transport network; Robust and Resilient city and; Vibrant city with active interaction between information, goods

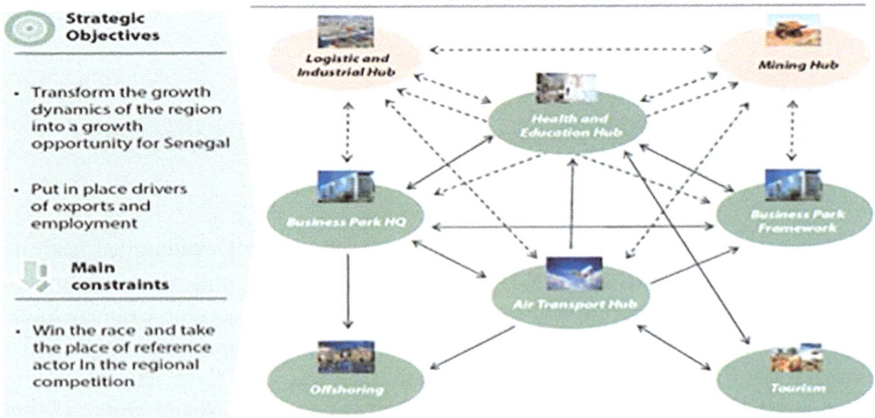

Fig. 15.34 PSE Flagships Projects. *Source* Republic of Senegal, Plan Senegal Emergent, 2014

and people. The PNAT proposes five development urban areas: Multifunctional urban areas; Areas for economic activities; Areas for agriculture activities; Areas for touristic activities; and Areas for conservation. The PNAT also identifies areas with high risk for habitation and any other activity. In these areas, modification of the land including by residential structure, or any cadastral operations is prohibited. This plan promotes green areas and other protected areas for environmental purposes or spaces with light recreational facilities, which, when well integrated enhance the character or the ecological value of the area. Specific focus areas of the PNAT also aim at: controlling the internal urban growth; reducing the proliferation of slums; promoting a balanced urban development; and meeting the housing demand.

The Department of Urbanism and Architecture (DUA) provides tools, instruments and guidelines on how to design, plan and manage cities. They have also introduced a Master Plan of Dakar in 2000 (Plan Directeur d'Urbanisme (PDU) 2025), another Master Plan of Dakar in 2014 (Plan Urbain de Dakar 2035) in 2014. The PDU is a reference document that aims to plan and program the development of an agglomeration in short and medium terms (10–20 years), taking into account the global objectives of the regional development. It specifies the socio-economic and demographic perspectives of the region and the agglomeration and determines the means and strategies to be implemented in order to reach a harmonious and sustainable development. From 1946 to 2015, there have been several urban Master Plans developed in 1946, 1961, 1967, 2000 and 2014 (revision of the PDU 2025).

However, technical and regulatory instruments, which could ease the implementation of master plans at the local level are complex and require enormous steps and times for their approval. Initiated in 2000, the PDU 2025 was approved only nine years later. The Master Plan Dakar 2025 has been ineffective in particular because of insufficient consideration of local realities, of the lack of involvement of local communities and populations in its preparation and its inability to properly treat disaster risk such as flooding. The Urban Mobility Plan in the PDU 2025 has been developed to address the critical problems of mobility in the Dakar Metropolitan Area and to reorganize the transport by promoting public transportation in Dakar, but it has not been taken into account in the construction of the major roadwork components in Dakar, which are implemented and managed by other specialized agencies. This shows a lack of coordination between the central government which is in charge in preparing and approving master plans, and the local authorities and other agencies which have the responsibilities to implement these plans at the local level.

Learning from the PDU 2025, **The Urban Master Plan of Dakar (Plan Directeur d'Urbanisme) 2035** is in this context the revision of the **Urban Master Plan of Dakar 2025**, it was initiated by the Ministry of Urban Renewal, Housing and Living Environment with support from the Japan International Cooperation Agency (JICA) in 2014. The PDU 2035 aims at urban development of the region of Dakar and its surroundings by 2035 [68]. Its main objectives are: Sustainable Urban Development; Compact cities connected with a transport network; Robust and

Resilient city and; Vibrant city with active interaction between information, goods and people.

There are other sectorial plans, which have been also hardly implemented. The Master Plan for Urban Planning and Preservation of the Niayes and the Green Zones of Dakar (PDAS) has not played its role yet, in terms of overall urban planning. The PDAS is a specific plan for the Niayes (depressions zones) and Green Zones at the level of the Dakar Metropolitan Area, and has been validated since 2004. The Regional Land-Use Planning framework (SRAT) provides comprehensive guidelines, which are not compulsorily taken into account in the field. In fact, the SRAT is validated but the implementation does not really respect the prescriptions included in it. Finally, the actions planned in the Regional Integrated Development Plan (PRDI) finalized since 2004 are not implemented yet.

15.7.4 Creation of Digitally, Economically and Socially Served Urban Centres: From Monocentric to Polycentric Dakar Metropolitan Regional Development

While developing urban policies and projects to tackle flooding and traffic congestion in the agglomeration of Dakar, national authorities have taken bold urban policies with the creation of six polycentric urban centres in the outskirts. The creation of these urban centres will decongest Dakar and will mark the transition of big cities to digitally connected towns (of less than 1 million people). Fighting the problem of the big city from outside with the formation of small cities. This represents a new form of urbanization, the digital urbanization where digitally connected towns offer urban advantages traditionally only found in big cities with high densities, such as economies of scale, agglomeration of economies, diffusion of ideas, innovation, and participation to political affairs.

In order to decongest Dakar, three categories of urban centres are being planned to take place in the Department of Rufisque, situated in the East of the Metropolitan Region:

- *Urban centres of the "Massif"* structured around national sport centres and an international exposition centre;
- *Coastal urban centres* structured around coastal touristic areas and;
- *Eco urban centres at the "Lac Rose"* characterized by the presence of microorganisms and mineral elements.

The creation of these urban centres is a combination of economic and spatial design strategies. As underscored in Chap. 1 of this book "Designing economic and spatial strategies for Smart Metropolitan Regional Development shall be conducted by the spatially identifiable economic community at micro levels from time to time to suit the ever-changing scenarios of the economic environment. However

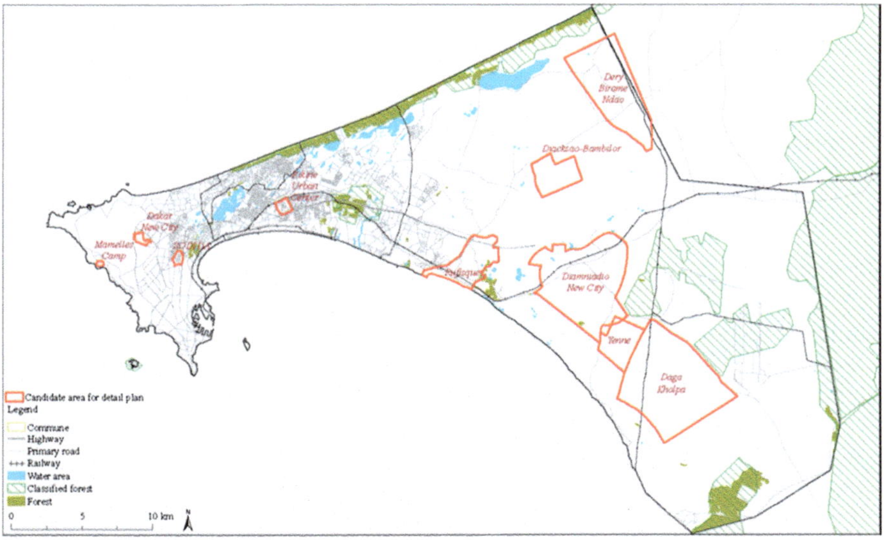

Fig. 15.35 Map of Dakar Metropolitan Region with the creation of the six urban centres. *Source* Mboup G., 2016. In Smart Economy in Smart Cities (ed. V. Kumar et al.), Springer

economic strategies may have a shorter lifespan than spatial strategies but both should fit in. Spatial strategies should be based on economic strategies each complementing the other. When this micro level communities cumulates to have a mega-community, they can make use of the vehicle of economic E-Democracy to formulate ever changing and dynamic strategies" [105].

Overall six urban centres are planned to take place in the first phase of the programme. These urban centres are created to support economies of scale and agglomeration. Among these urban centres, features the Urban Pole of Diamniadio, which is already under implementation as part of the Plan Senegal Emergent (PSE), representing the first sustainable city model with the integration of climate risk resilience. For instance, the Special Economic Zone of Diamniadio will constitute a multifunctional platform for most of the income-generating activities (industry, crafts, clothing, equipment, infrastructure, etc.). It is part of the national policy to encourage companies to relocate and diversify their activities outside the department of Dakar and attract new investors, notably with: the creation of manufacturing units with high added values; Assembly and processing industries; Construction of several logistics platforms (transit, storage) to streamline internal and regional transport flows; Creation of two zones dedicated to export services, integrated and connected (Fig. 15.35).

It is in the ICT revolution momentum that the government of Senegal has initiated the creation of urban centres in the outskirts of Dakar in order to decongest the city [106]. The government is, indeed, putting in place an ambitious project of a technology park in Urban Pole of Diamniadio, called "Diamniadio Technology Park". The park is based on the **Silicon Valley model and intends to promote**

data revolution and higher education centres. The Urban Pole of Diamniadio is also among the pilot cities of the National Sustainable Cities Initiative (SCI) as part of the Sustainable Cities Programme launched by the Global Environmental Facility (GEF) in 2016. It consists of "Planning and Managing the urban pole to decongest Dakar and to be a sustainable city model connected to Dakar and the rest of the country through walking, non-motorized, ICT means".

In the Urban Pole of Diamniadio, it is also planned to create an integrated campus of reference by 2018 with the creation of at least five world class professional schools; Business Park by 2017; Establishment of "Dakar Medical City"; Integrated tourist areas; Regional airline hub (medical, tourism, regional headquarters of companies and international institutions, education-training). The urban pole of Diamniadio is in a crossroads that connects Dakar to the rest of Senegal thanks to the RN1, the RN2, the Niayes road and the Diamniadio-AIBD highway. In addition the Diamniadio pole enjoys an opening on the sea through Bargny. The Diamniadio Special Economic Zone will be a multifunctional platform for most of the income-generating activities (industry, crafts, clothing, equipment, infrastructure, etc.). It is part of the desire of the government to encourage companies to relocate and diversify their activities outside the capital and attract new investors, in particular with: the creation of a high added value manufacturing centre; assembly and processing industries; rehabilitation of the Dakar-Bamako railway line to boost the corridor; construction of several logistics platforms (transit, storage) to streamline internal and regional transport flows; creation of two (02) zones dedicated to export services, ready to use, integrated, connected; Dakar an integrated reference campus by 2018 with the creation of at least five internationally renowned schools; Business Park by 2017 and to host by 2018, 50 seats and 2000 to 3000 high-income households receiving services (luxury housing, social services, cultural and leisure activities); implantation of "Dakar Medical City"; integrated tourist areas; regional air hub (medical, tourism, regional headquarters of companies and international institutions, education-training); AIBD's capacity to reach 6 million passengers by 2020 and 10 million by 2035.

The creation of the Diamniadio Urban Pole and other urban centres is a paradigm shift to decongest the Dakar urban agglomeration. In order to do not repeat the errors observed in the Dakar agglomeration, where Pikine and Guédiawaye are suburbs of Dakar, the Diamniadio Urban Pole is created to be a place of production, distribution and consumption of goods and services, for its population and for the rest of the country. From the perspective of urban planning, the Diamniadio urban pole is designed to promote: (1) an adequate percentage of space allocated to streets for all modes of transport; (2) efficiently designed streets that allow traffic without congestion; (3) road design that allows for shorter trip configurations with multiple choices; and (4) Mixed land use models that allow shorter travel distances.

In addition to the creation of urban pole to de-densification of Dakar city core, the Senegalese is also taking other actions outside the Dakar metropolitan region to reduce the demographic and economic power of the Dakar agglomeration. The Senegal government is also introducing the development of urban poles in the rest of the country as illustrated in Fig. 15.36.

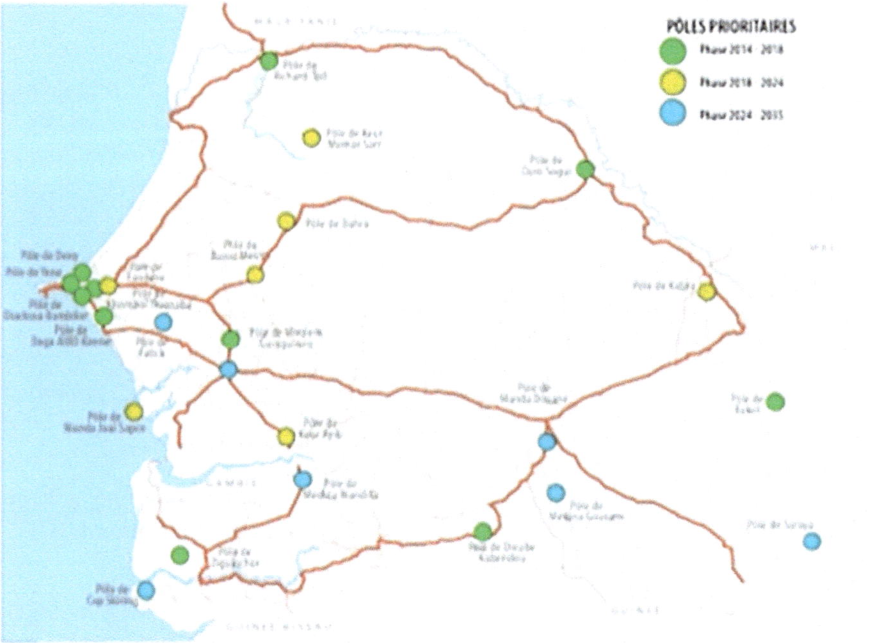

Fig. 15.36 Spatial distribution of urban poles in Senegal. *Source* Republic of Senegal, 2014. Review of Senegalese Urbanization

15.7.5 Emerging Trends: Linking the Dakar Metropolitan Region to Other Regions—The Corridor Dakar-Thies-Mbour and Beyond

Another urban development is the emergence of urban corridors, which represent a type of spatial organization with specific economic and transportation objectives. In Senegal, the government is also encouraging growth, convergence and spatial spread of geographically linked metropolitan areas and other agglomerations. These are emerging trends linking Dakar-Thies-Mbour, turning into spatial units that are territorially and functionally bound by economic, political, socio-cultural, and ecological systems. It is expected that the Dakar master plan include Thies and Mbour. Linking Senegal to Mali, Mauritania, Cape Verde, Guinea, and Guinea Bissau and beyond. This is inline with Africa intra-regional cooperation as outlined in Africa Agenda 2063. However, More holistic approach of corridor policies is needed to include economic development of areas in the corridor zone of influence and support continental cooperation (Fig. 15.37).

However, all these forms of urban growth require planning and management of the space, and efficient mobility means. Though their economic output may be enormous, their management requires considerable investment on infrastructure,

Fig. 15.37 Inter-regional urban corridors. *Source* Republic of Senegal, 2014. Review of Senegalese Urbanization

particularly to ease mobility and communication in order to constitute a large economic output, combining large markets, skilled labour and innovation [107].

15.7.6 Connecting Economically and Spatially Distinct Urban Centres in the Dakar Metropolitan Region

The Dakar metropolitan regional development with a population of more than 3 million comes with its constraints or requirements in terms of infrastructures such as road networks and transportation systems which are key for linking services such as workplaces, health centres, schools, markets, etc. Efficient mobility is required to allow localities of the metropolitan region to specialize in the production of goods and services for which they have comparative advantages and ease inter-localities cooperation. This will allow large-scale production of goods and services that can be distributed within the metropolitan region and beyond with time, cost and reliability opportunities.

The government of Senegal has quickly realized that development of the Dakar metropolitan region will be achieved only through the development of efficient mobility means. In the PSE, it is indeed, planned: the construction of road infrastructures (1170 km of paved roads, at least 4000 km of rural roads, 7 bridges, and modernization of bus stations); the construction of maritime infrastructures (restoration of inland ports of new infrastructures and port platforms); the construction of railway infrastructures (rehabilitation of 573 km of railway line, construction of a new standard gauge line) and airport facilities (rehabilitation of regional aerodromes), a tramway is also under development. The newly created international airport, located in the Urban Pole of Diamniado, is expected to have a capacity of 6 million passengers by 2020 and 10 million by 2035 (Fig. 15.38).

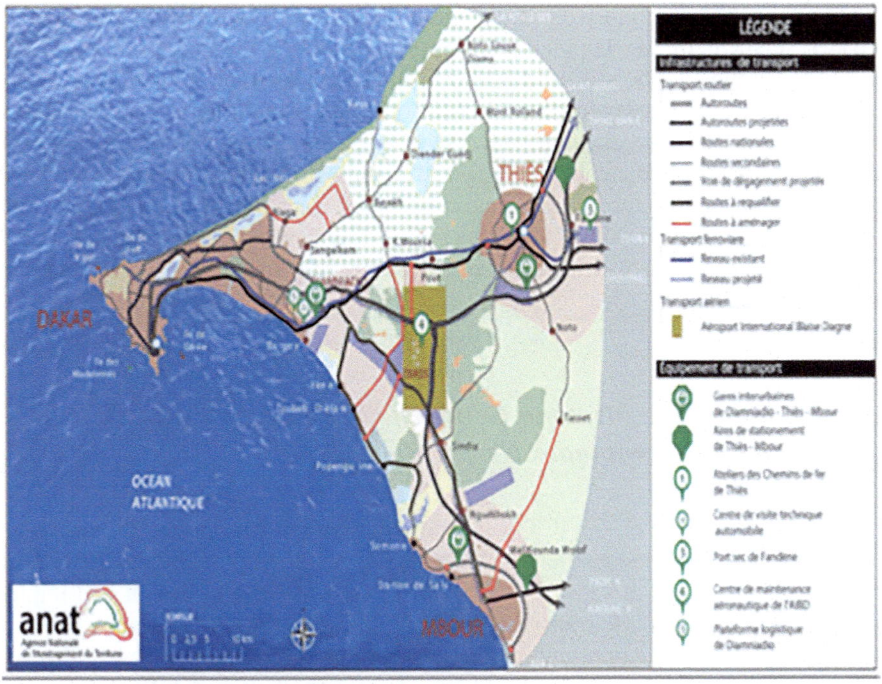

Fig. 15.38 Planning and management of transport Infrastructures in the Dakar metropolitan region (road, ail, air and Water). *Source* Republic of Senegal, 2014. Plan National d'Amenagement du Territoire

In addition to development of transport infrastructures, it is important that these newly urban centres be planed to be dense, compact, mixed land with sufficient connected streets and public spaces.

Mixed land use and polycentric urban centres for efficient urban mobility is central to the creation of the urban centres

At the beginning of the city development, services and residences are kept together as all start-up of settlements of small size; the question is to make a city planning where the population growth will be coordinated with services geographically balanced, avoiding the start-up to become the city centre of the future city, and the expansion being the periphery. There is nothing new of having people and services together, the challenges are how to keep them growing in a coordinated manner where the mixed land use is preserved [75]. It is noted that along urbanization, the city centre and the periphery become distant, and if urban growth is not well coordinated, this will end up with the formation of residential areas without services distinct from the city centre without residences as the current situation in Dakar metropolitan region. When planning city growth, it is important to integrate the mixed land use approach that has proven efficient with high economic, social and environmental returns. Mixed land use settlements have various social and economic benefits by improving accessibility to services and urban amenities for a broader segment of the population, and increasing housing options for diverse household groups. In the process of the growth of the Dakar metropolitan approach, it is important to consider the enforcement of urban planning and particularly for transport infrastructures (land in reserve and to keep inbuilt for future transport infrastructure) [108]. By combining spatial planning and transport policies, local governments would reduce people's need to travel; improve travel conditions with affordable and efficient public transport options; and manage supply and demand traffic to curb congestion, which is a major barrier to productivity and a headache for residents.

15.7.7 Modelling Multiple Connectivity Choices: Example of the Diamniadio Urban Pole

Today, the opportunities for ICT to support the overall transport challenges and opportunities are enormous. Newly cheap ICT can unlock possibilities for greater transport efficiency. Cities may be able to use real-time data to design and implement policies that increase transit accessibility, decrease travel time, substitute for expensive road construction, and reduce congestion and pollution. The ICT revolution with the rapid development and use of Internet, digital mobile communication, and "big data" analysis enable to develop less costly and more powerful "intelligent transport systems" (ITS) [109]. Big data with its wealth of information from various sources as smart phones, GPS and sensors offer unique opportunities for complex transport modelling.

Allocation, design and development of transport infrastructure and services can more and more subject to rigorous planning. Transport planning has to reflect the fundamental changes in cities and contemporary issues and problems. Transport planning is to make effective decision concerning the allocation of transport resources, including the management and regulation of existing transportation activities. Just as urban planning requires the inputs of many stakeholders and specialists, transport planning is beginning to utilize multi-disciplinary teams in order to broaden the scope of the planning process. This is now possible due to the fact that ITS allow an open platform of transport data accessible to all stakeholders and partners. It can also boost active participation of citizens in sharing their needs and concerns in terms of mobility [110, 111].

Under the Senegal Sustainable Cities Initiatives, in addition to developing transport infrastructure, the government of Senegal has introduced the model of multiple connectivity choices as presented in Fig. 15.39.

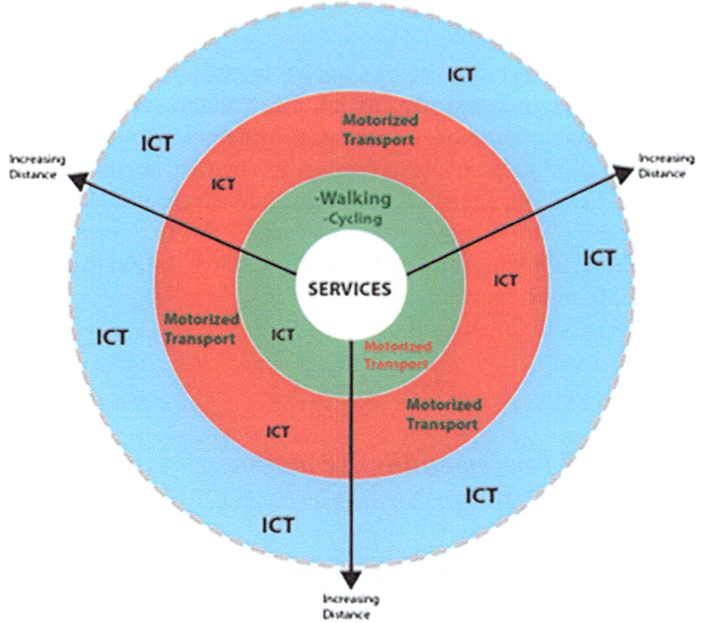

Fig. 15.39 Multiple choices of access to services. *Source* Mboup, G., 2016

Interpretation of the multiple choices of access to services: (1) the green panel illustrates a situation of mixed land used or alike where services are walking distance from residence (threshold to be determined). Here we can walk, drive and use ICT, but walking along with the use of ICT to access services is sufficient and then highly recommended; (2) the red panel illustrates the situation where services are far away from residence, in a distance making it impossible for people to effectively walk to reach most services, but are in a reasonable distance (threshold to be determined) where people can use motorized means of transport with the benefits that can offer an ITS; They can also use ICT to perform work outside their workplace or to access other services. This situation includes telecommuting as well as all other forms of ITS such as RTPI and; (3) the third scenario with the blue panel illustrates a situation where the services are very far (threshold to be determined) from residence making it difficult for people to use motorized means of transport every day to services, here the convenient option is the use of ICT to effectively reach services. This third scenario represents a complete substitution of movement of people by the use of ICT to perform work outside workplaces, such as in an offshore call centre and to access services far away from their producer and distributer location such e-banking, e-commerce, etc. This is usually the case where people reside outside the city where a company they are working with is located. ICTs have, indeed, made virtual access to services such as

employment, education and business opportunities possible, hugely changing the dynamics on physical distance limitations. In all situations, the use of ICT will ease access to services and take city towards smartness. ICT indeed offers a promising sustainable solution to access to different social and economic sectors, to perform duties or receive services. This will allow economies of scale and agglomeration, even for small firms and small settlements.
Source: Mboup, G., 2016. Creating Digital, Smart Cities for Smart Economies: From Big Cities to Digital Urban Centres. In Smart Economy in Smart Cities, Springer, September 2016

15.7.8 Means of Implementing Economic and Spatial Design Strategies

15.7.8.1 Barriers to Implementing Economic and Spatial Design Strategies

It is not the first time the government of Senegal conceives holistic urban development programmes and projects. But few had been implemented due to various barriers among them: Lack of cooperation and involvement of the population in the preparation of urban programmes and projects; Competition between urban issues and other social, economic and political issues; Regulatory, policy and institutional mechanisms; Lack of Awareness and capacity; Insufficient Financial Resources; etc.

Lack of cooperation and involvement of the population in the preparation of programmes or projects Though the need for economic and spatial design strategies is well grounded, there are various barriers to tackle in the process of findings to the problems and its root causes. Re-organization two centuries of unplanned settlements with little or no consideration of sustainability, inclusion and resilience require transformative actions such as re-planning the unplanned. In the Dakar metropolitan region, due to poor urban planning and management, land tenure in most human settlements is neither well governed nor well administered. Poor land governance is surrounded by poor land administration characterized by a poor determination, recording and dissemination of information about tenure. Without legal proof of ownership, people feel exposed to eviction and are less willing to cooperate when government authorities plan to re-organize the spatial occupation of their settlements. This constitutes an important barrier providing solution to urban development issues. Even in situation where the population is willing to cooperate, they are not involved in the urban planning and management processes. Therefore they do not feel concerned by roles and regulations in terms of spatial occupation and buildings codes established by national and local authorities.

It took over nine years before the Urban master plan of Dakar initiated in 2000 be approved. As of today, only 29 municipalities out of more than 400 municipalities have an urban master completed. This shows the complexity of urban planning and management in Senegalese municipalities.

Competition between urban issues and other social, economic and political issues National and local policy-makers seek visible and tangible outcomes for expenditure, where they can point to something concrete (literally and figuratively) that they have achieved for their constituents. In this situation, economic and spatial design strategies are often in neither in the negotiation table nor in the planning stage. Even, where they are in consideration, due to their medium or long-term nature, policy-makers will opt for other actions more visible in short term than them.

Regulatory, policy and institutional barriers The cumbersome regulatory approval process constrains the preparation of detailed economic and spatial design strategies. The intermediate causes to this cumbersome approval process include: Lack of clarity of roles and responsibilities of various administrative structures; weak coordination between the central government and municipalities in charge of preparing their own detailed economic and spatial design strategies according to the Act III of Decentralization.

Awareness and capacity barriers In most of the Dakar municipalities, local authorities and communities suffer from lack of awareness regarding sustainable urban development issues in general. In addition to that they do not have the required skills and competencies to properly integrate smartness in the planning and management of their cities. This constitutes important barriers to consider in the economic and spatial design strategies. Awareness and capacity building of local authorities and communities is key for the success of any urban project at the local level.

Financial barriers Local authorities do not have sufficient resources to plan and manage their cities. Economic and spatial design strategies are not well positioned in the competition of several other interests when it comes to development of cities. In addition to that, there are: difficulties in accessing various financial sources due to lack of appropriate financing mechanisms to assist municipalities. The availability of finance is essential for efficient implementation of urban development programmes. "Marginal role and weak capacities of municipal authorities: municipalities are marginalized and lack the required capacities and resources to plan and manage the more and more complex situations. The role of African local governments in urban development is indeed highly constrained. Many countries have sought to decentralize service delivery to local governments, but without a commensurate transfer/increase in revenue sources/financing. Moreover, most towns and cities in Africa have limited professional capacity for managing urban development. Municipalities lack legal and administrative frameworks for efficient service delivery and management of urban planning, land tenure and finance" (ECA 2014).

The absence of finance can constrain the ability of relevant authorities to implement, for instance, sustainable high-capacity public transport options. The implementation and management of metros, light rail or BRT systems require large amounts of funding. As an illustration from the UN-Habitat's Global Human Settlements Report 2013, "to live decently at the standard appropriate for middle-income countries, their average per capita income will need to be at least $5000, so the income of such a city would be around $25bn. To put this in context, the typical African country currently has a sovereign bond market worth around $2bn, and annual aid inflows are around $1bn. No African government, and no individual company, could afford to finance the cost of building an entire city. The future financing of Africa's urban structures will rely primarily upon tapping into financial markets, domestic and international, and tapping into them on such a scale is not a trivial matter....". During their conception, it is important to take into consideration key elements determinant to their cost, demand, supply, affordability, etc. Due to the huge investments that public transport such as metro, light right and BRT required and considering the limited revenues of cities, it is important to assess carefully the funding for capital investments that requires the participation of local, regional and national governments as well as the private sector. Other partners can also be of interest such as international funding from climate change financial mechanisms. It is also important to tap to non-traditional funding such land [112]. In addition to transfers from different levels of government, local authorities require innovative funding mechanisms to support implementation and operation of public transport systems beyond the fare box revenues [61].

15.7.8.2 Tapping in the Synergies Across Global, Regional and National Agendas and Programmes for Resources Mobilization

Financing the Plan Senegal Emergent and other programmes

In February 2014, the PSE Consultative Committee has estimated a budget of about US$ 6–7 billion to boost the Senegalese economy. It is expected that half of the financing of the Plan Senegal Emergent (about US$ 3.8 Billion) will be from international donors, another 25% from international investors, and another 25% from the government. These figures indicate clearly how the PSE depend heavily on external risks in time of resource scarcity globally. This represents a high risk for a timely implementation of the PSE.

Financing urban poles are also based on partnership with public and private entrepreneurs. Considering the holistic approach in creating these urban poles as presented above, various partners will work together in their implementation.

Since Senegal is among the least developed countries (LDCs) most of its programmes are donor-funded programmes. Therefore it is important to present these programmes and assess how they are linked to national programmes. Being part of the international community, the Senegal Government has also committed to implementing global and regional development agreements. Prominent among which, in the last couple of years, are those relating to the Post-2015 Development Agenda: Sustainable Development Goals adopted in September 2015; the Paris Agreement on Climate Change adopted in December 2015; The Sendai Framework on Disaster and Risk Reduction adopted in March 2015; the Addis Ababa Declaration on Population and Development earlier adopted in July 2015; as well as the latest one—the New Urban Agenda adopted in October 2016. At the regional level, Africa Agenda 2063 serves as the main interface for operationalizing and monitoring the implementation of all these global agreements in the continent.

All these agendas have a common goal, sustainable development through various inter-related channels and avenues, with the SDGs, New Urban Agenda and Africa Agenda 63 being more holistic and embrace multiple sectors compared to other agendas which are more sectorial and thematic, such Sendai Framework for Disaster Risk Reduction, Paris Declaration for Climate Change, and Addis Ababa Declaration for Population Development. The SDGs is outcome-oriented while the NUA is implementation oriented to enable processes to reach some SDGs goals, particularly the SDGs Goal 11. The Africa Agenda 63 defines the transformative actions or outcome required to transform the region to middle income region "The Africa We want". In Senegal all these agendas operate through the Plan Senegal Emergent (PSE) that defines a holistic approach for the country to transform itself and fulfil the regional aspiration of Agenda 63 towards a middle-income country. In all these programmes, the role of ICT is underscored.

Synergies across Global, Regional and National Agendas and Programmes Africa Agenda 63 SDGs, Sendai Framework, Paris Agreement, Addis Ababa Declaration, New Urban Agenda (Fig. 15.40).

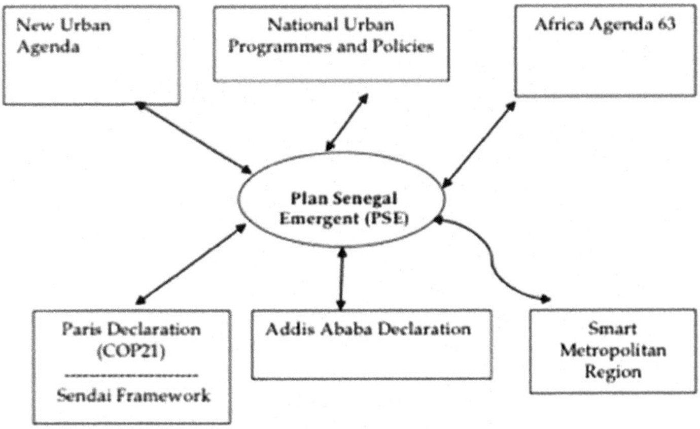

Fig. 15.40 Synergies across Global, Regional and National Agendas and Programmes. *Source* prepared by the author from different agendas

15.8 Conclusion

Development of large metropolitan regions like Dakar constitutes economic and social opportunities as well as challenges depending on how the urban growth was planned and managed. It provides opportunities for economies of scale and agglomeration as well as for diffusion of ideas and innovations, but it will also call for large investments in infrastructures to respond to the increased demand for water, sanitation, solid management, energy and mobility, etc. It will particularly come with increased demand on mobility that must be satisfied with an efficient public transit accompanied with increased spaces for pedestrians and cyclists in order to safeguard the environment while creating economic growth. The metropolitan of Dakar yet occupies a pivotal place in the national trade sector both nationally and internationally. Its autonomous ports (for coastal cities), international airports, international trade centres, touristic sites and commercial centres are assets for economic growth. The majority of economic units in Dakar are in the commercial sector (52%). The metropolitan of Dakar contributes up to 55% of the national Gross Domestic Product (GDP). Up to nine out of ten national civil servants, and from 50 to 90% employees in national's trade, transport, banking and industrial enterprises are in cities. However, most of these urban advantages are concentrated in the city of Dakar. Though the Dakar Metropolitan Region had been administratively divided in four departments, in reality the other three departments act as suburbs of the department of Dakar, where the main commercial and administrative businesses are concentrated. The other three departments remain haunted by the early spatial and social division; they are not well planned, and they lack sufficient land allocated to streets, other public spaces, basic infrastructure and security of tenure.

However, Dakar has not been able to satisfy the employment demand, particularly from young people. In addition to that, due to a deficient urban mobility, Dakar has also not been able to create conditions for specialization. While urban agglomeration allows for job specialization, efficient market transactions and knowledge diffusion, if concentrated growth is not well planned—such as the integration of urban growth with efficient public transit investments, the resulting economic benefits tend to erode. This is the case in Dakar as in many African cities with poor connectivity leading to disconnected settlements.

Recognizing the unbalanced Dakar metropolitan regional development, in one hand, and the multiple problems associated to lack of basic services, on another hand, the Senegalese government has taken bold actions to transform the urban landscape of the metropolitan region from a monocentric to a polycentric metropolitan region with the creation of urban centres to decongest the Dakar urban agglomeration. This new spatial re-organization is also accompanied with economic transformation through the ambitious economic development programme, the Plan Senegal Emergent. It is through key main complementary programmes, reforms and plans: (1) the Plan Senegal Emergent; (2) Act III of Decentralization; (3) Metropolitan Planning; (4) Dakar Urban Master Plan 2035 and; (5) Digital

Senegal Strategy 2016–2025. These urban policies and programmes aim to spatially and economically transform the Dakar Metropolitan Region to be smart, sustainable, inclusive, resilient and prosperous. They aim to transform the monocentric system to a polycentric system where every urban centre equally contributes to the development of the metropolitan region.

Another urban development is the emergence of urban corridors, which present a type of spatial organization with specific economic and transportation objectives. In Senegal, the government is also encouraging growth, convergence and spatial spread of geographically linked metropolitan areas and other agglomerations. These are emerging trends linking Dakar-Thies-Mbour, turning into spatial units that are territorially and functionally bound by economic, political, socio-cultural, and ecological systems. It is expected that the Dakar master plan include Thies and Mbour. Though their economic output may be enormous, their management requires considerable investment on infrastructure, particularly to ease mobility and communication in order to constitute a large economic output, combining large markets, skilled labour and innovation.

Along the transformation of the Dakar metropolitan regional development, national authorities have put in places planning guidelines through the Urban Master Plan of Dakar (Plan Directeur d'Urbanisme—PDU) 2035 and National Plan for Territorial Development "PNAT". The PDU 2035 aims at urban development of the region of Dakar and its surroundings by 2035. Its main objectives are: Sustainable Urban Development; Compact cities connected with a transport network; Robust and Resilient city and; Vibrant city with active interaction between information, goods and people. The PNAT proposes five development urban areas: Multifunctional urban areas; Areas for economic activities; Areas for agriculture activities; Areas for touristic activities; and Areas for conservation. The PNAT also identifies areas with high risk for habitation and any other activity. In these areas, modification of the land including by residential structure, or any cadastral operations is prohibited. This plan promotes green areas and other protected areas for environmental purposes or spaces with light recreational facilities, which, when well integrated enhance the character or the ecological value of the area. Specific focus areas of the PNAT also aim at: controlling the internal urban growth; reducing the proliferation of slums; promoting a balanced urban development; and meeting the housing demand.

It is in the ICT revolution momentum that the government of Senegal has initiated the creation of urban centres in the outskirts of Dakar in order to decongest the city [113]. The government is, indeed, putting in place an ambitious project of a technology park in the Urban Pole of Diamniadio, called "Diamniadio Technology Park". The park is based on the Silicon Valley model and intends to promote data revolution and higher education centres. The Urban Pole of Diamniadio is also among the pilot cities of the National Sustainable Cities Initiative (SCI) as part of the Sustainable Cities Programme launched by the Global Environmental Facility (GEF) in 2016. It consists of "Planning and Managing the urban pole to decongest Dakar and to be a sustainable city model connected to Dakar and the rest of the country through walking, non-motorized, ICT and motorized means". Under the

Sustainable Cities Initiatives, in addition to developing transport infrastructure, the government of Senegal has also introduced the model of multiple connectivity choices (non-motorized, motorized means of mobility and ICT).

References and Notes

1. Rodrigue et al (2013) https://people.hofstra.edu/geotrans/eng/ch1en/ch1menu.html
2. United Nations GA, October 2016. Draft outcome document of the United Nations Conference on Housing and Sustainable Urban Development (Habitat III)
3. Conroy & Berke, 2004; Conroy & Gordon, 2004; Wild & Marshall, 1999
4. Chen Y, Hamilton A, Borning A (2010) Modeling & matching and value sensitive design: two methodologies for e-planning systems development
5. Marcinkowski 2007
6. Silva CN (2010) The e-planning paradigm –theory, methods and tools: an overview. In: Silva CN (ed), Handbook of research on e-planning: ICTs for urban development and monitoring (http://www.igi-global.com/book/handbook-research-planning/41793) (http://www.igi-global.com/viewtitlesample.aspx?id=43177&ptid=41793&t=the+e-planning+paradigm+%E2%80%93+theory%2c+methods+and+tools%3a+an+overview), downloaded, 9 April 2017. The Use of Social Media for Urban Planning: Virtual Urban Landscapes Created Using Twitter. Data
7. Yigitcanlar T (2013) Planning online: a community-based interactive decision-making model (http://www.igi-global.com/viewtitlesample.aspx?id=43178&ptid=41793&t=planning+online%3a+a+community-based+interactive+decision-making+model) downloaded 9 April 2017
8. http://www.ist-africa.org/home/. Accessed 15 Jan, 2016
9. http://www.artpsenegal.net/. Accessed 15 Jan, 2016
10. http://www.adie.sn/
11. http://www.afribaba.info/ville-de-dakar-wifi-pour-une-ville-numerique-au-dakar.html
12. http://pubdocs.worldbank.org/pubdocs/publicdoc/2015/5/138414322174903l3/Transport ICT-Connections-Note02.pdf
13. http://pubdocs.worldbank.org/pubdocs/publicdoc/2015/5/337641432217490793/TransportICT-Connections-Note04.pdf Advancing Development with Mobile Phone Locational Data Improving the Effectiveness of Assistance Ryan Haddad, Tim Kelly, Teemu Leinonen, and Vesa
14. https://africanentrepreneurshipaward.com/senegal-the-west-africa-quiet-high-tech-titan
15. http://www.bokk.org/index.php?option=com_content&view=article&id=9. See also website www.xamxam-africa.org
16. http://indigotrust.org.uk/2013/10/31/hubs-incubators-and-labs-in-senegal
17. See also experiences in Africa such as http://www.sipsenegal.org/snn/ict-expansion-through-layered-vertical-relay-teaching
18. Décret n°2014-875 du 22 juillet 2014 relatif aux attributions du Ministre du Renouveau urbain, de l'Habitat et du Cadre de vie
19. Republique du Senegal, Ministere du Renouveau Urbain, de l'Habitat et du cadre de vie, 2014. Plan Directeur d'Urbanisme de Dakar et ses environs Horizon 2035
20. Mboup G (2014) Urban quality of life—concepts and measurements. http://deliver2030.org/?wp_blog=urban-quality-of-life-concepts-and-measurements. Accessed 9 Nov, 2017
21. Kiregyera, 2007; Mboup G., 2003
22. UNDESA (2014) Urbanization prospects the 2014 revision
23. UN-Habitat (2009) Urban indicators guidelines
24. Angel et al 2016 Atlas of Urban Expansion—The 2016 Edition. Volume 1: Areas and Densities. New York: New York University, Nairobi: UN-Habitat, and Cambridge, MA:

Lincoln Institute of Land Policy. Further analysis of these data has been done by Mboup G. (2017) "African Cities – Urban Form & Structure" in the forthcoming book "Smart Economy in Smart African Cities" by Springer, winter 2018

25. Ba B (1972) La société lébu. La formation d'un peuple. La naissance d'un État, Dakar, Université de Dakar. Chenal J., Op. Cit. Sylla, A., 1992. Le Peuple Lébou de la presqu'île du Cap-Vert, Dakar, Les Nouvelles Éditions africaines du Sénégal. Laborde C
26. Diop A (1995) La communauté lebu face aux développement de la politique coloniale : la question des terrains de Dakar (1887-1958), These de Doctorat, Université Cheikh Anta Diop, Dakar. Chenal J., 2009. Op. Cit
27. Dumez R, Kâ M (2000) Le territoire assiégé. Un village lébou dans la banlieue de Dakar, UNESCO (translated from French to English by the authors)
28. Chenal J 2009. Harris, 2011. Harris DA (2008) Constructing Dakar, cultural theory and colonial power relations in french african urban development cities of tomorrow.
29. ENDA Graf Sahel, (2009) Pikine:aujourd'hui et demain, Diagnostic participatif de la ville de Pikine.
30. See also Kevin A. Bryan, Brian D. Minton, and Pierre-Daniel G. Sarte
31. Kevin A. Bryan, Brian D. Minton, and Pierre-Daniel G. Sarte, The Evolution of City Population Density in the United States. Economic Quarterly—Volume 93, Number 4—Fall 2007—Pages 341–360; (Chatterjee and Carlino 2001, Lucas and Rossi-Hansberg 2002)
32. Betraud et al
33. Joan Carles Martori and Jordi Suriñach, 2001. Classical models of urban population density. The case of Barcelona Metropolitan Area. 41st Congress of the European Regional Science Association, 29 August-1 September 2001 (Zagreb)
34. Bertaud et al
35. Tanner, 1961; Smeed 1963
36. Muth, 1969
37. Much of this literature is ably surveyed by McDonald (1989). Broad reviews of the theoretical models behind this empirical work can be found in, for example, Wheaton (1979) Straszheim (1987) and Arnott et al. (1998)
38. Mboup G (2011) Cities—place of opportunities and challenges
39. ANSD (2015) Regional analysis of the RGPHE 2013
40. Ashfor S. Lori (2007) Africa's youthful population: risk or opportunity? Population Reference Bureau, Washington DC. USA
41. http://www.africaneconomicoutlook.org/en/theme/youth_employment/ downloaded on 21 October 2015. Promoting Youth Employment in Africa
42. OECD 2012, Short and Pinet-Peralta 2009
43. Kim and Law 2012
44. This definition is brought from UNDESA "Urbanization Prospects 2014"
45. Short and Pinet-Peralta 2009, UN-Habitat 2013, 2008
46. Henderson, 2000
47. Duranton 2009; Kim 2009
48. Godard Xavier, 2011. Sustainable Urban Mobility in 'Francophone' Sub-Saharan Africa. Thematic Study for the UN-Habitat's Global Report on Human Settlements 2013, Nairobi (Kenya)." "www.wds.worldbank.org/external/default/WDSContentServer/WDSP/IB/2010/10/07/000356161_20101007005655/Rendered/PDF/569540NWP0Tran10Box353751B01PUBLIC1.pdf. Access 15 Jan, 2016
49. ADM, 2001. Urban audit of Dakar
50. La pénurie d'eau à Dakar révèle une défaillance au sommet de l'état dans la sécurité intérieure et la sécurité civile. By MO in Actualités, Défense & Sécurité, Gouvernance, Logement, Urbanisation,Publications on septembre 17, 2013 http://thinktank-ipode.org
51. Mboup, 2004: Notion of water accessibility, sufficiently and affordability – The situation of Addis Ababa, 2003 Urban Inequities Survey
52. Mboup, 2012. Cities, Place of Opportunities and Challenges
53. (Bhada-Tata and Hoornweg, 2012) cited by UNIDO, 2017

54. IAGU, 2009. Décharge de Mbeubeuss : Analyse des impacts et amélioration des conditions de vie des populations de Diamalaye à Malika dans la banlieue de Dakar
55. Mboup, G., 2013. Op. Cit., UN-Habitat, 2010. State of World's Cities, 2010. Bridging the Urban Divide
56. See Rodrigue et al (2013). https://people.hofstra.edu/geotrans/eng/ch1en/ch1menu.html
57. "UN-Habitat, 2013a. Planning and Design for Sustainable Urban Mobility Global Human Settlements 2013. Earthscan from Routledge."
58. "Gehl's early work in Copenhagen suggested the need to promote non-motorized means of mobility in order to create livable streets. Based on Gehl's findings, in 1962, Copenhagen made a shift towards increased use of bicycles as an alternative to cars. Durin."
59. UN-Habitat, 2013a and 2013b
60. "United Nations, 2015b. Adoption of the Paris Agreement, Conference of the Parties, twenty first session, Paris, 13 November to 11 December 2015."
61. UN-Habitat, 2013a. Op. Cit
62. Sources of information used here are from different sources such as the Street report of UN-Habitat in 2013, the International Association of Public Transport, the Global Human Settlements Report of UN-Habitat (2013), Kumar A. and F. Barret, 2008 (Stuck in Traffic; Urban Transport in Africa) and Lall, S.V., et al. 2017, Africa's Cities: Open Door to the World, World Bank)
63. Trans Africa Consortium, 2008
64. Godard Xavier, 2011. Sustainable Urban Mobility in 'Francophone' Sub-Saharan Africa. Thematic Study for the UN-Habitat's Global Report on Human Settlements 2013, Nairobi (Kenya)
65. Godar Xavier, 2011. Op.Cit
66. www.wds.worldbank.org/external/default/WDSContentServer/WDSP/IB/2010/10/07/000356161_20101007005655/Rendered/PDF/569540NWP0Tran10Box353751B01PUBLIC1.pdf/access on 15 January 2016
67. UN-Habitat, 2013a. Planning and Design for Sustainable Urban Mobility Global Human Settlements 2013. Earthscan from Routledge
68. Republique du Senegal, Ministere du Renouveau Urbain, de l'Habitat et du cadre de vie, 2014. Pland Directeur d'Urbanisme de Dakar et ses environs Horizon 2035
69. World Bank, 2005. Sub-Saharan Africa Transport Policy Program, Working Paper 80, Washington DC (USA)
70. UN-Habitat, 2013b. Streets as Public Spaces, Drivers of Urban Prosperity. Nairobi, Kenya
71. UNDESA, 2014. UN-Habitat, 2012
72. Republic of Senegal, 2017. Rapport Global du Recensement General des Entreprises
73. UNIDO, 2016. Unpublished document
74. UN-Habitat, 2012. Prosperity of Cities. State of World's Cities, 2011/2012. Earthscan
75. Rodrigue et al, 2013. Op. Cit
76. Mboup G., 2015. Op. Cit
77. GLTN, 2008
78. http://www.dakar-info.com/les-inondations-urbaines-au-senegal-sont-de-plus-en-plus-dangereuses-et-de-moins-en-moins-gerables/, Djigo M. (2009). Les inondations urbaines au Sénégal sont de plus en plus dangereuses et de moins en moins gérables. http://www.africadt.com
79. http://iopscience.iop.org/article/10.1088/1755-1307/6/33/332025/pdf Flood risk and land occupation in Dakar outskirts. Does climate variability reveal inconsistent urban management? Cheikh Mbow, A Diop, AT Diaw University Cheikh Anta Diop, Institute of Environmental Sciences, Laboratoire d'Enseignem, Dakar, Senegal. An integrated analysis has been made using land use maps, local topography using DTM (SRTM) and field surveys to show factors and implications of flooding in the outskirts of Dakar called Yeumbeul. Physical and human driving factors are integrated in the analysis. The impacts on environment and health are analysed against policy for flood aftermath and beyond with the so-called "Diaxaay" State plan for flood prevention

80. https://www.gfdrr.org/sites/gfdrr/files/Senegal_English_August%202014.pdf. "Assessment report on post-disaster needs, urban floods in Dakar 2009", Government of the Republic of Senegal, World Bank, the United Nations System and the European Commission, June 2010.). The PNDA assessment is funded through the Global Facility for Disaster Reduction and Recovery (GFDRR)
81. Republic Senegal, 2010: 44–53, quoted by IAGU, 2011, Op.Cit
82. Republic of Senegal, AWF, EAA, Donor Roundtable aimed at financing the emergency phase of the Ten-Year Flood Program, Programs for Integrated Water Resources Management and Access to Drinking Water and Sanitation, December 2012
83. The World Conference on Disaster Reduction, held in Kobe, in Japan's Hyogo Prefecture, from 18 to 22 January, 2005
84. This project was supported by the World Bank
85. https://www.gfdrr.org/sites/gfdrr/files/Senegal_English_August%202014.pdf. The PROGEP is being implemented together with priority measures such as: The Government designated the Municipal Development Agency (ADM) with PROGEP preparation and implementation, PROGEP was officially launched in November 2012, but preparation of the master plan for storm water drainage in the outlying areas of Dakar and the preparation of technical drainage studies began in 2011. The budget of PROGEP is USD 72.9 million of which USD 55.6 million through the World Bank; USD 10.6 million from the Senegalese government, USD 4.1 million from the Nordic Development Fund and USD 2.6 million by ADM. Additional funding was being negotiated with the World Bank, IDB and AfDB
86. Hyoung Gun Wang Economist, Spatial team FEU, SDN Marisela Montoliu-Munoz Head, Spatial team FEU, SDN The Geoville Group Ndèye Fatou D. Gueye Consultant, FUE, SDN, 2009. Preparing to Manage Natural Hazards and Climate Change Risks in Dakar, Senegal. A Spatial and Institutional Approach. Pilot Study Report. See also World Bank & Republic of Senegal, 2013. Economic and Spatial Study of the Vulnerability and Adaptation to Climate Change of Coastal Areas in Senegal
87. De Soto, 2000
88. Toulmin, 2005
89. Ministry of Finance, 2017. Press Conference, Dakar, Senegal
90. Philibert 2007
91. ADM, 2001. Dakar Urban Audit
92. CDS, IAGU, 2008
93. Mboup G (2014) Urban quality of life – concepts and measurements. http://deliver2030.org/?wp_blog=urban-quality-of-life-concepts-and-measurements. Accessed 9 Nov, 2017
94. Fehrler et al., 2009
95. African Union and ECA, COM2013 (2013) Toward a demographic dividend: invest in health and education. In: Conference industrialization for an emerging Africa, Abidjan March 2013
96. Climate change has emerged as one of the defining issues of the early 21st century. (See also Fig. 1. Urban climate change vulnerability and risk assessment framework. Adapted from Rosenzweig et al. (2011); after Mehrotra et al. (2009)
97. Republic of Senegal, 2016. Sustainable Cities Initiative Project document submitted to the GEF'6's Sustainable Cities Programme prepared by Mboup, G
98. PDU 2035
99. Consequently, the United Nations Environment Programme/World Health Organization Global Environment Monitoring System (GEMS) has confirmed that the worst overall air pollution conditions and the largest indoor pollutant concentrations and exposures are found in both rural and urban areas of the developing world
100. Potential harm is very great for children under one year, since their lungs and immune system are not yet fully formed. Household use of biomass fuels has been found to significantly increase the risk of acute respiratory infections, which annually kill millions of children under age five (Smith et al., 2000)

101. Republic of Senegal (2016) Sustainable Cities Initiative Project document submitted to the GEF'6's Sustainable Cities Programme prepared by Mboup, G
102. UN-Habitat, 2012. Prosperity of Cities (Author: Mboup et al.)
103. Mboup G (2013) Streets as public spaces and drivers of Urban Prosperity, UN-Habitat, 2013
104. For details refer to the chapter 1 of this book as well as to the original publication of Saskia Sassan since the 1990s (reference [11] &[12])
105. Kumar V., chapter 1 of this book. Smart Metropolitan Development: Economic and Spatial Design Strategies
106. Republique du Senegal, 2014. Revision du Plan Directeur d'Urbanisme de Dakar et ses environs 2035.
107. See Rodriguez et al., 2013. https://people.hofstra.edu/geotrans/eng/ch1en/ch1menu.html
108. UN-Habitat (2013c) Urban planning for city leaders. Nairobi (Kenya)
109. http://www.worldbank.org/en/topic/transport/brief/connections-note-26. By Winnie Wang, Raman Krishnan, and Adam Diehl.
110. Holly Krambeck, January 2015, Mapping Manila Transit A New Approach to Solving Old Challenges "Haddad, Ryan; Kelly, Tim; Leinonen, Teemu; Saarinen, Vesa. 2014. Using Locational Data from Mobile Phones to Enhance the Science of Delivery. World Bank, Washington, DC. © World Bank. https://openknowledge.worldbank.org/handle/10986/19316 License: CC BY 3.0 IGO."
111. http://www.worldbank.org/en/topic/transport/brief/connections-note-26
112. Mboup G., 2017. Land – The Hidden Assets in African cities. World Bank Land and Poverty Conference, March 2017, Washington DC
113. "Republique du Senegal, 2014. Revision du Plan Directeur d'Urbanisme de Dakar et ses environs 2035."

Part XVI
South Africa, Johannesburg

Chapter 16
Towards a Smart Metropolitan Regional Development—Spatial and Economic Design Strategies: Johannesburg

Naledzani Mudau, Gora Mboup, Paida Mhangara and Mfaniseni Sihlongonyane

Abstract From a population of 26,303 in 1890 to 102,078 in 1896, to a population of 4.4 million in 2011, Johannesburg located in the Gauteng province, is the largest city in South Africa and is one of the 50 largest urban areas in the world. This has occurred in different paces, times and places in terms of numbers, urban form and structure (density distribution, and compactness). As a megacity, Johannesburg becomes more complex for its planning, design and management compared to a small city. For instance, elements of its urban form and structure as well as elements of transport infrastructures and, recently, elements of Information and Communication Technologies (ICT) will largely determine its accessibility. This accessibility determines to a large extent, the interaction between economic and spatial design strategies. Henceforth, this chapter is divided into six sections. **The first section** presents the "Conceptual framework of the Smart Metropolitan Regional Development" of the study; **The second section** reflects on the "Gauteng City region and Johannesburg in Time and Space"; **The third section** focuses on the "Urban Accessibility and Mobility"; **The fourth section** looks at "Smart Energy for Economic Development"; **The fifth section** presents "Urbanization, Economic Growth and Metropolitan Regional Development"; and, **The sixth section**, based on the development of the previous sections, presents the "Spatial and Economic Design Strategies for a Smart Johannesburg Metropolitan Regional Development".

N. Mudau (✉) · P. Mhangara
South African National Space Agency (SANSA), Tshwane, South Africa
e-mail: nmudau@sansa.org.za

P. Mhangara
e-mail: pmangara@sansa.org.za

G. Mboup
Global Observatory Liking Research to Action (GORA) Corp, NY, USA
e-mail: gmboup@gora4people.org

M. Sihlongonyane
School of Architecture and Planning, University of the Witwatersrand, Johannesburg, South Africa

T. M. Vinod Kumar (ed.), *Smart Metropolitan Regional Development*,
Advances in 21st Century Human Settlements,
https://doi.org/10.1007/978-981-10-8588-8_16

Keywords Smart metropolitan region · Economic & spatial design strategies Urbanization · Urban system · Urban form and structure · Streets and other public spaces · Human development · Accessibility · Sustainable · Inclusive Resilience · Prosperity · Apartheid

16.1 Introduction

The development of large urban agglomerations such as Johannesburg Metropolitan Region as well as city regions such as Gauteng City Region (GRC) provides opportunities for economies of scale and agglomeration, but it will also call for large investments in infrastructures to respond to the increased demand for water, sanitation, solid management, energy, streets and public spaces, and mobility, etc. They will require efficient institutions for the management of social demand and equity on education and health as well as protection of people against violence and insecurity. Economies of scale and agglomeration economies are, for instance, greater in metropolitan areas where accessibility infrastructures are able to provide accessibility to services' needs with higher access to markets and resources than those where accessibility is impeded by deficient accessibility infrastructures. Without efficient accessibility systems, a metropolitan region losses its economic power and remains just clusters of disconnected settlements. Therefore, efficient accessibility infrastructures will allow large-scale production of goods and services that can be distributed within the metropolitan regions and beyond with time, cost and reliability opportunities [1].

Established in 1886 following the discovery of gold on what had been a farm, the city of Johannesburg was a centre stage of the apartheid regime in South Africa that lasted from 1948 to 1990. Its foundation was historically guided by a segregationist ideology manifested through institutions and laws, urban planning and design, and access to basic services and amenities. During Apartheid, the majority of township areas designated for non-white people were situated between 25 and 30 km away from the central business district of Johannesburg, leading to multiple social and economic deprivations as well as transport challenges. Originally an acronym for "South-Western Townships", Soweto, one of the townships, originated as a collection of settlements on the outskirts of Johannesburg, populated mostly by native African residents that were forcibly by government removed from their own land.

The dismantling of apartheid in the early 1990s came with calls for inclusive development in South Africa, mostly guided by Nelson Mandela's ideal of a democratic and free society. The entire city of Johannesburg was then considered as a former apartheid city. Different policies, programmes and projects were adopted to make Johannesburg a more inclusive city. The key policies embraced participatory planning and design; development of inclusive public spaces; improvement in basic infrastructure provision and public transport particularly to the poor neighbourhoods and; development of policies that encourage inclusive human settlement and trade.

Today, Johannesburg is part of the Gauteng City Region (GCR) located in the Gauteng province which is the most densely populated province in South Africa.

The GCR is an integrated cluster of cities, towns and urban nodes that together make up the economic heartland of South Africa [2]. It has urban landscapes that reflect the spaces, identities, relationships and encounters of historical divisions, while also embedding the significant and complex generations of difference that are produced by many overlapping economic histories and functions [3]. The GCR consists of the cities of Johannesburg and Ekurhuleni, the district municipalities of Sedibeng and West Rand, and the City of Tshwane (Pretoria) (Map 16.1). Although the area of Gauteng is the smallest of the nine provinces of South Africa with only 1.4% of the country's surface area, it contributes 50% of South Africa's economic output, 42% of SA's industrial output, 34% of the country's GDP and 10% of the Africa's GDP. The size of the Gauteng economy is 1.07 Trillion Rands, based on mining, manufacturing, financial and business services, innovation or trade [4]. Johannesburg alone contributes 17% of the national GDP and approximately 47% of Gauteng's economy [5].

The Gauteng City Region had 13.4 million people living within the radius of 100 km from the Johannesburg city centre in 2011. The population of the City of Johannesburg Metropolitan municipality was 4.4 million in 2011 [6]. According to the United Nations Population Division estimates, the Population of Johannesburg urban agglomeration was 8 million in 2015 and projected to surpass the 10 million mark in 2030 [7].

For this study, the unit of analysis will be always indicated to ease interpretation and trends analysis. However, for a meaningful analysis that can be linked to policies, the Johannesburg Metropolitan Municipality under the authority of a mayor as well

Map 16.1 Gauteng City region. *Source* Gauteng City Region Observatory. www.GCRO.ac.za

as the Gauteng City Region head by a Premier are centred on and linked to Johannesburg and the other cluster of cities. For the study of the Johannesburg greater metropolitan region, it is more pragmatic to make it through the Gauteng City region for the following reasons: (1) Johannesburg is geographically and economically linked to all other cities or municipal districts of the city region; (2) the city region has an autonomous administration under the leadership of the Gauteng Premier that defines coordinated policies, programmes and projects including all municipalities of the city region; and (3) the Johannesburg Growth Development Strategy goes beyond the Metropolitan Region to embrace the entire Gauteng province as it is indicated in its the Gauteng Development Strategy outcome 4 stating, "A high performing metropolitan government that pro-actively contributes to and builds a sustainable, socially inclusive, locally integrated and globally competitive Gauteng City Region" [8]. Urban Policies and Programmes for the Greater Johannesburg Metropolitan are indeed, developed at the provincial level and trickle down to the city level.

The Johannesburg case study also consists of analysing the economic and spatial strategies undertaken by the City of Johannesburg authorities and the Gauteng City Region authorities in order to turn the city and the region into smart development in the context of the historical transformation of the metropolitan region in time and space. This is explored under six sections.

The first section "Conceptual framework of the Smart Metropolitan Regional Development" is based on its City Foundation (urban planning, basic infrastructures and land tenure) and Institutions as well as on the Economic Development, Infrastructure Development, Environmental Sustainability, Social Development, Social Inclusion, Disaster Risk Exposure and Resilience, and Peace and Security. One element of the City Foundation, is the spatial design and the element of Economic Development which plays the role of interface vis-à-vis the other elements of the conceptual framework. Economic and Spatial Design Strategies alone cannot make a regional metropolitan region smart, the way they interplay with the other dimensions are crucial.

The second section presents "Gauteng City region and Johannesburg in Time and Space".

The third section focuses on the Urban Accessibility and Mobility in the City of Johannesburg: Streets as drivers of urban accessibility; Transport modes and infrastructures, and traditional focus of urban mobility studies. Though streets constitute a fundamental issue of the City Foundation as determinant of the urban form and structure, the analysis gives preference to the prominent role they play in urban accessibility along with transport modes and infrastructures.

The fourth section presents "Smart Energy for Economic Development".

The fifth section presents "Urbanization, Economic Growth and Metropolitan Regional Development" including economic growth, human development, city prosperity and smart metropolitan regional development. Lastly,

The sixth section, based on the development of the previous sections, presents the "Spatial and Economic Design Strategies for a Smart Johannesburg Metropolitan Regional Development" as implemented through the urban policies and programmes of the city of Johannesburg and the Gauteng City Region.

16.2 Conceptual Framework, Methodology and Data

16.2.1 Smart Metropolitan Regional Development: Concepts and Components

In our study, a smart metropolitan regional development is viewed as a sustainable, inclusive and prosperous metropolitan regional development that promotes a people-centric approach based on three core components: Smart Metropolitan Region Foundation, Information and Communications Technology (ICT) and Smart Institutions and Laws. Here, the Metropolitan Region Foundation goes beyond the city foundation and includes elements of mobility across urban centres that form the metropolitan region. First each urban centre must have a smart city foundation and be smartly connected to the other urban centres. Smart Metropolitan regional foundation, institutions and laws, and ICT are the pillars of the seven other dimensions of a smart metropolitan region: Infrastructure Development, Environmental Sustainability, Social Development, Social Inclusion, Disasters Exposure, Resilience, and Peace and Security. **Infrastructure Development** complements the basic infrastructure services under each smart metropolitan foundation and extends to actual investment and advancement of services such as transport, ICT, industrial energy, education, health, etc. **Environment Sustainability** is comprised of elements of climate change, biodiversity, waste management, energy, transport, building and pollution. **Social Inclusion** includes aspects of participation in decision-makings as well as equal opportunities for growth and prosperity. **Social Development** encompasses elements of education, health, public space, social inclusion and social capital. **Disaster Exposure** incorporates elements of mitigation and adaptation to various disasters such as flooding, droughts, storms and earthquakes. **City Resilience** is composed of elements of city foundation, environment, social capital, and social development. **Peace & security** includes the elimination of all forms of discrimination violence and conflicts, such as: domestic violence, violence in public places, crime, armed conflicts, terrorism, etc. An insecure metropolitan region limits opportunities for investment and economic growth and cannot be a smart metropolitan region (Fig. 16.1).

The opportunities for ICT to support the overall urban challenges and opportunities are enormous, and the Johannesburg Metropolitan Region as well as the Gauteng City Region must integrate and use ICT solutions to facilitate the greater provisioning of urban services. Following the exigency of the city of the 21st century that calls for sustainability, inclusion, resilience and prosperity, the planning and management of human settlements must take into consideration the gain in knowledge on various conditions that make cities smart, green, ecological, liveable and healthy. Through increased efficiency and innovation, ICT increases economies of scale and agglomeration, and promote diffusion of knowledge, even at the smaller human settlements. Small settlements and small firms are benefiting from ICT in making available their goods and services beyond their territories. Virtual offices, virtual networks, teleconferences are increasing large public participation

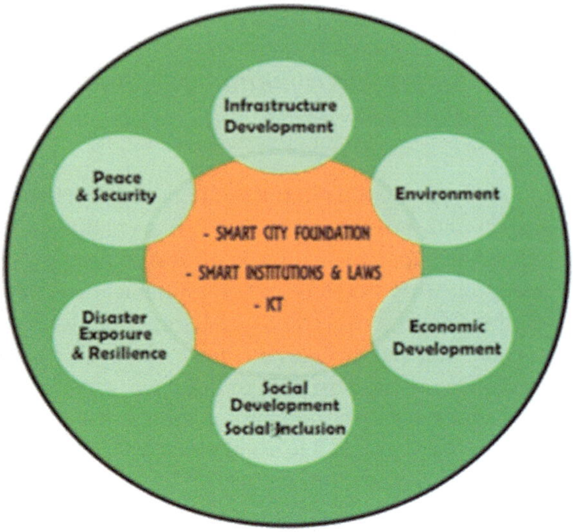

Fig. 16.1 Smart Johannesburg Metropolitan Regional Development conceptual framework. *Source* adapted from Mboup, G. et al., 2016. Smart City Foundation—Drivers of Smart Cities. In Vinod et al., 2016. Smart Economy in Smart Cities, Springer

and inclusion. Digital firms can start and scale up quickly with relatively little staffing or capital investment. ICTs harness the benefits of agglomeration economies in easing circulation of goods and services and encouraging polycentric urban development and allowing synergies between centres and sub-centres. They intensify urban nodes and corridors to maximize the benefits of concentration.

Numerous societal problems are explored and addressed in urban and regional planning agencies, including urban growth, unemployment and economic revitalization, transportation, environmental degradation and protection, neighbourhood decay and redevelopment, conservation of land and natural resources, provision of open space, parks and recreational facilities, etc. New urban planning instruments are becoming available with the worldwide spread of ICTs. They make it possible to adopt innovative e-planning approaches, strengthen communication between urban stakeholders, and make communication available at various stages of the planning process. Local governments can engage their citizens with real-time information to gain support for policy initiatives, identify unforeseen concerns, and recognize potential conflicts [9].

Cities are dynamic living organisms that are constantly evolving. ICT has begun to turn some places into real-time cities. This rapidly changing society makes the assessment and anticipation of future needs of city dwellers in terms of services, including transport, water, energy, employment, education and health, even more problematic. In addressing the complex problems of city planning, it is not sufficient just to be concerned with the physical structure of the city; the interplay of intangible economic, social and environmental factors needs to be considered as well [10]. Planning procedures make use of models that show historic and present situations and communicate planned situations [11]. The introduction of ICTs allows planners and planning departments to carry out new actions or to implement conventional practices through new tools, such as GIS, virtual reality technologies, e-participation devices, including public participation GIS applications, among other tools, with the aim of improving conventional decision-making processes. The provision of better planning and urban management services, more efficient, with lower costs and, at the same time, a more collaborative and participative, transparent and accountable planning decision-making process are some of the basic objectives usually associated with the move from conventional urban planning to e-planning [12, 13].

ICT plays an important role in the transport sector. It can increase economic productivity and competitiveness through increased smart mobility that reduces traffic congestion and commuting costs. Efficient and fast transport in turn, can increase labour productivity by reducing commuting times, and increasing worker productivity. ICT reduces the use of motorized means and transport and promotes walking and cycling. This will improve the overall health of city residents, thereby reducing economic costs of workers who are absent due to illness. Another important benefit is on consumer expenditure, as less is spent on cars and fuel. Non-motorized means of mobility support specific industries, such as bicycle shops, tourism, retail activity, construction and real estate development that highlight liveability. It can also lead to a drop in the share of the household budget devoted to motorized means of transport, such as cars which are expensive particularly in African countries.

The digital dividends are real and the Johannesburg metropolitan region as well as the Gauteng City Region must take the opportunity to efficiently integrate the use of ICT in all sectors of its development. From the beginning of the 21st century, a digital citizenship particularly the "Millennial" generation has started to emerge in Johannesburg as in many African cities. The rapid adoption of digital technologies in the economy will have huge benefits directly as well as indirectly. However, ICT alone will produce little in the smart metropolitan regional development. 'Smart' is not an end in itself; It is the way ICT is integrated in the city development that will determine the city smartness. Maximizing the digital dividends requires better integration of ICTs with the other factors of smart cities such as: city foundation, infrastructure development, environment sustainability, social development, disaster prevention, resilience, peace and security. ICT can be seen as substitution and catalyst factors, but the other factors (or the analogue part) of the operation are crucial in making a smart economy. Those factors include the city foundation, institutions and laws, infrastructure development, social development, social inclusion, environmental sustainability, disasters prevention and resilience (particularly in the context of climate change), peace and security. Most tasks have an aspect that cannot be automated by technologies and that requires human judgment, intuition, and discretion. This is where lies the crucial role of human capital associated with institutions and laws, efficient governance, management and administration. Without improvement on accountability at all levels, ICT alone cannot change the outcome of the economic productivity equation. African countries that are able to swiftly adjust to this evolving digital economy will reap greatest digital dividends, while the rest are likely to fall behind. This is the context where we are introducing economic and spatial design strategies for a smart metropolitan regional development to unlock the potential of Johannesburg metropolitan region as well as the Gauteng City Region to be smart, sustainable, inclusive, resilient and prosperous.

16.2.2 Methodology and Data

Putting the Dimensions together to Develop the Smart Metropolitan Regional Development Index (SMRDI)
The SMRDI which is built here is meant to serve research as well as policies. For the latter, the SMRDI will be built using existing mechanisms of the production of indicators globally as well as nationally. Building the SMRDI through existing mechanism of development of indicators such as the monitoring of the Africa Agenda 2063, the Sustainable Development Goals (SDGs), the New Urban Agenda, the Sendai Framework, the COP21, the Addis Ababa Declaration on Population and Development as well as the monitoring of national programmes and plans will ease its implementation at the local level for SMRDI policies. As part of the international communities, South Africa has also endorsed all the above global agendas. Box 16.1 provides for each dimension of the SMDR variables, quantitative or qualitative, included for its measurement.

Metropolitan Regional Development Index
Box 16.1 Smart Metropolitan Regional Development Index (SMRDI)

Dimensions	Definitions/variables
City foundation	*Spatial Planning* Planned settlements, Streets, Other Open Public Spaces *Land Tenure* Land: Documentation of perceived eviction Housing: Proportion of urban population living in slums, informal settlements or inadequate housing *Basic Infrastructures* Connection to piped water, Connection to sewerage system, Connection to electricity, management of solid waste
Institutions & Laws	Urban Planning Codes, Urban Policies, Urban Plans, Unplanned settlements, Resilient building codes, standards, development permits, land use by-laws and ordinances, and planning regulations
ICT	Connection to internet, Mobile phone, Computer, Broadband
Infrastructure development	Transport and road infrastructure, Energy, Knowledge infrastructure, Health infrastructure
Environmental sustainability	Air quality (PM10), CO_2 emissions, energy and indoor pollution

(continued)

(continued)

Dimensions	Definitions/variables
	Climate Change, Biodiversity, Land degradation
Economic development	City product, Employment Primary, Secondary and Tertiary sectors
Social development	Education: literacy, primary, secondary and tertiary enrolment Health: life expectancy, under-five mortality rates, HIV/AIDS, morbidity and nutrition: Social capital, Public space
Social inclusion	Income/consumption Inequalities: Gini coefficient, Health, Education, Access to basic services
Disaster Risk Reduction/Resilience Strengthen resilience and adaptive capacity to climate-related hazards and natural disasters in all countries	Number of deaths, missing persons and directly affected persons attributed to disasters per 100,000 population Adoption and Implementation of national disaster risk reduction strategies in line with the Sendai Framework for Disaster Risk Reduction 2015–2030 Proportion of local governments that adopt and implement local disaster risk reduction strategies in line with national disaster risk reduction strategies
Peace and Security	Proportion of population that feel safe walking alone around the area they live; Proportion of women and girls aged 15 years and older subjected to sexual violence by persons other than an intimate partner in the previous 12 months, by age and place of occurrence; Proportion of ever-partnered women and girls aged 15 years and older subjected to physical, sexual or psychological violence by a current or former intimate partner in the previous 12 months, by form of violence and by age; Average share of the built-up area of cities that is open space for public use for all, by sex, age and persons with disabilities; Proportion of population that feel safe walking alone around the area they live

Source Mboup G. et al. 2018 Forthcoming. "The Relevance of the Concept of Smart City in the African Context." Springer. All variables of this table are not included in the Johannesburg SMRDI due to lack of data

16.3 Gauteng City Region and Johannesburg in Time and Space

16.3.1 Gauteng City Region and Johannesburg Post Apartheid

The apartheid government reinforced the spatial segregation of South Africa by establishing laws including Group Area Act, Prohibition of Mixed Marriages, Native Settlement Act and Native Act, which controlled people' movement and place of settlement per race. The city foundation of Johannesburg has historically been guided by a segregationist ideology manifested through institutions and laws, urban planning and design, and access to basic services and amenities. During the Apartheid, the majority of township areas designated for non-white people were situated between 25 and 30 km away from the central business district, leading to significant transport challenges. Originally an acronym for "South-Western Townships", Soweto, one of the townships, originated as a collection of settlements on the outskirts of Johannesburg, populated mostly by native African workers from the gold mining industry.

The long march to freedom and social inclusion—Transforming Johannesburg to an inclusive, vibrant, sustainable and smart metropolitan region within the Gauteng City Region
With an apartheid city foundation characterized by apartheid institutions and laws, apartheid urban planning and design, and apartheid access to basic infrastructure and services, the excluded people of Johannesburg as most parts of South Africa, had to fight for their social freedom. Both peaceful and violent protests were held in the city between 1950s and 1990 when apartheid was officially abolished, each with varied outcomes. Protests held during the 1950s to push for social inclusion in aspects such as voting were generally unsuccessful both as a result of non-unified campaigns and also as a result of draconian laws aimed at suppressing any opposition to the government. Large scale anti-pass demonstrations held in the 1960s were countered with suppressive measures such as declaration of a state of emergency, banning of social movements, imprisonment and exiling of leaders, and empowerment of the police, which made it impossible to legally protest against the system. With mounting economic hardships, international pressure, and increased local protests throughout the 1980s, the government secretly began negotiations with Nelson Mandela, who had been jailed since 1964 leading tothe release of Nelson Mandela in 1990 and the subsequent repealing of segregationist laws. The first multi-racial democratic elections in 1994 won by the African National Congress and the enactment of South Africa's new constitution, which enfranchised blacks and other racial groups, marked the official end of the apartheid system [14]. Major strides have been made especially in the areas of policy and opening up of the physical space to promote social integration in the living and public settings of South Africa, particularly in the Gauteng Province and Johannesburg.

Some of the major policies developed are revitalizing South African cities through inclusive urban planning, inclusive provision of basic services, inclusive access to land.

From an Apartheid City to an Sustainable, Inclusive, Resilient and Prosperous Johannesburg City
Urban development was marked by City Improvement Districts (CIDs) and Urban Development Zones (UDZ) as well as through the creation of themed spaces that influenced the revitalisation of inner city Johannesburg. The establishment of the Urban Development Zone Tax Incentive brought leveraging influence to inner city redevelopment and attracted investment to the inner city. The Johannesburg Development Agency (JDA) facilitated the revitalising of Constitution Hill, Newtown, and Faraday Station, upgraded Main Street to make it more pedestrian-friendly, reconstructed and renovated the Drill Hall, and developed the Fashion District. At the launch of the inner city Urban Renewal Strategy in 1997, the city was positioned as the "Golden Heartbeat of Africa" [15]. Johannesburg inner also features an array of cultural venues and a wide range of creative and cultural industries as the cultural capital of the country. However the inner city also features a tale of the northern suburbs such as, Rosebank, Sandton and Midrand which have significant residential investments along with restaurant and shop-lined avenues. In contrast, the inner city environments are resulting from influx of migrants from the Gauteng province and other provinces as well as other African countries and creates a huge informal sector. Though there is increase in the number of formal dwellings in the city of Johannesburg by over 220,000 units, estimates from the World Bank indicate that, during the period 1996–2001, the informal shacks increased by 42% to reach a quarter million [16].

In 2001, the Gauteng Provincial Government (GPG) officially launched the Blue IQ initiative and embarked on an energetic drive to get an initial group of eleven mega projects off the ground. These included the development of City Deep Transport Logistics Hub, Constitution Hill, Cradle of Humankind, Dinokeng, the Gautrain, Johannesburg International Airport, the Innovation Hub, Newtown Regeneration, Wadeville–Alrode Industrial Corridor, Kliptown Regeneration, and Gauteng Automotive Quarter. Blue IQ sought to "develop world-class infrastructure, implement marketing and investment strategies, reduce bureaucratic red tape for investors and suppliers, and encourage skills training and resource building in the areas of technology." The Expanded Public Works Programme (EPWP), launched in May 2004, was central to creating employment opportunities for the poor. Nationally, the number of employed people grew from 2.7 million in 1995 to 4.9 million people in 2014. There was also a sharp rise in the ratio of household to disposable income, from approximately 54% in 2004 to almost 80% in 2007 [17]. However, a large portion of household spending was financed by credit.

The Post-apartheid was also Marked by the Suburbanization of Johannesburg
The suburbanization of Johannesburg has been recorded at several levels, notably with businesses as well as government housing programme located at the outskirts of the city. Decentralised nodes in well-established suburban areas such as Sandton,

Randburg and Midrand expanded to create what can be called 'edge cities' grew up along the post-apartheid towards polycentric cities [18]. The suburbanization was also politically motivated with the RDP housing programmes at the outskirts of cities where land is relatively affordable [19]. The RDP housing programmes might have various social and economic advantages including access to the financial and economic market as illustrated in de Soto publication (2000). De Soto argued that granting titles to the poor would liberate the plots they occupy and transform them into capital. This in turn, could be used as collateral for loans to jumpstart their businesses, or improve their houses, among other gains that increase their quality of life. At the community level, regularization of land tenure will help the municipality to legally collect various taxes that can be used to improve basic infrastructures such as connection to water, sewerage facilities, energy sources and waste management facilities. This would also allow people to fully participate in the development of their communities at the policy as well as the implementation level instead of seeing proprieties as dead investments serving only for shelter [20].

However, there is an argument that supports that having title deeds at the outskirts of cities also created some form of exclusion from jobs and amenities [21]. The RDP model delivered housing to communities far from commercial centres where work opportunities could not be found, and sometimes also far away from the existing major townships, so these locations produced inflexible and unsustainable environments, rather than vibrant places where people can access the opportunities and amenities they require to thrive. These spatial patterns have perpetuated, or at worst, increased, the large distances for most people between work and home. The costs of mobility for the low-income people are extremely high and transport constitutes a major component of the budget for them—and the levels of congestion associated with public transport are growing, as are its consequent damaging environmental impacts. Suburbanization, either through middle class and wealthy investors' decision to locate at the outskirts of cities or through the RDP housing programmes has contributed considerably to urban sprawl in the region, which has seen its morphological footprint extent southwards to areas like Evaton, Sebokeng, and Sharpeville; northwards to Ga-Rankuwa and Hammanskraal, for example; and eastwards to areas such as Benoni, Daveyton, and Springs [22].

Upgrading the Townships

Therefore, the post-apartheid was marked by inclusion policies contributing towards exclusion particularly those geographically located in townships such as Soweto. Some projects targeted townships that had been particularly affected by violence, such as Kathorus [26]. However, these projects were featured as socio-political interventions rather than urban development projects supported by official urban plans [22]. Due to lack of job opportunities in townships, most of the township residents were forced to seek employment in other areas such as Johannesburg CBD. The townships in the south and pockets in the northern parts of the province, are still marginal spaces, economically. They have large populations and little or no industrial activities, and some are isolated and burdened with informal settlements. Some continue to grow and expand the township footprint away from the core areas of the city. Distortions of the urban footprint have been exacerbated by poor state-led housing projects and weak strategic spatial planning of the public sector that has allowed the private development of space in unsustainable forms on the urban periphery. A fragmented public transport system and the shift of jobs away from mining and industry to tertiary activities have also been significant. Therefore, the urban built-up area has increased in the region, but as an unsustainable, inequitable and inefficient spatial development trend [19].

Box 16.2 is an illustration of the Urban Landscape of the Gauteng City Region by Sihlongonyane, 2018.

Box 16.2 The Urban Landscapes of the Gauteng City Region

The urban landscapes can be loosely divided into nine historical formations:

Old small towns landscape grew from the availability of agricultural land or the discovery of mineral resources to accommodate white settlers in the 1800. These landscapes have both colonial and apartheid beginnings and tend to have an old industrial plant and/or railway lines. They contributed significantly to the proliferation of dysfunctional settlements, dysfunctional space economies and, to the terrible trio of poverty, inequality and unemployment. They have declined since the slump in mineral prices and with the shift towards post-industrial production.

Metropolitan landscapes are large cities such as Johannesburg and Pretoria with large populations and social and economic relations with satellite towns of the region. These landscapes developed from old small town landscapes, of which they still have the vestiges, through trade and diversification. They contribute the highest earnings to the GDP of the province, but are characterised by remarkable economic, income, and social inequalities, and spatial fragmentation. They are responsible for unprecedented urban sprawl, new slums, and unbalanced development.

Suburban landscapes are part of metropolitan landscapes. They are places of privilege providing comfortable neighbourhoods for the middle and upper classes. They are produced largely through decentralisation urban growth away from the inner city. These are characterised by the post-industrial urban dynamics of office parks, residential estates (gated communities) and shopping complexes, which propagates fragmentation.

Gated community landscapes, developed over the last 15–20 years, are walled, residential communities or housing estates with strictly controlled access to "prevent penetration by non-residents" [23], and public spaces that are privatized.

Black township landscapes are areas created originally by the apartheid regime for 'non-white' residents—blacks (Africans), Indians and 'Coloureds', on the periphery of towns and cities, usually 20–30 km away from the CBD. These areas often have limited formal economic development apart from retail, and conflicting interests about peripheral land-use. Services such as electricity, water and heating are inadequate or unreliable.

Informal settlement landscapes are urban or peri-urban locations where houses and 'shacks' have been constructed on land that the occupants occupy illegally, without official approval. Many informal settlements are located on the outskirts of the city adjacent to black townships.

RDP (Reconstruction and Development Programme) landscapes were established through the massive government housing capital subsidy scheme. They are sprawls of formal low-cost mass housing, largely at the edge of the cities where land is cheaper, far from job opportunities. Despite the government's good intentions, they perpetuate the apartheid urban form [24].

Peri-urban Landscapes are located on the periphery of the major cities. They contain a variety of restaurants, retail shops, conference facilities, a spa, and hotels. These are strategically located to provide touristic leisure and entertainment. They are thriving economically. While not intense, they provide a lot of jobs to the lowly skilled. They include: Shell Ultra cities, cultural villages, botanical gardens, nature reserves, airports, and waterfronts. They often carry a mix of these activities. Such areas include: Muldersdrift, Walter Sisulu National Botanical Garden, Lanseria International Airport, Hartbeespoort Dam, Nature Reserves, Maropeng, and Lesedi Cultural Village.

One of the primary trends in the growth of tourism is the proliferation of African cultural villages within the Gauteng region. African cultural villages landscapes are defined as complexes that are purposely built with the help of 'cultural workers,' to aspects of cultural life for a group as it was at a specific period (or over several periods) of time [25]. The 'cultural workers' are employed to perform and demonstrate a purportedly 'traditional' way of life for visitors. These landscapes are therefore purposely-constructed tourist attractions by people recruited (sometimes as families) from the relevant 'ethno'-linguistic groups to live and work in the villages. They are tourist villages which celebrate the cultural traditions of several different people of Southern Africa.

Source Mfaniseni F. Sihlongonyane, 2018. Chapter 2—The Economic Drivers of Urban Change in the Gauteng City-Region: Past, Present and Future. In K. Cheruiyot (ed.), The Changing Space Economy of City-Regions, Springer AG 2018

Current Urban Policies and Programmes at the Gauteng Provincial Level
Under the cluster of cities of the Gauteng city region, the city of Johannesburg is the Financial Hub; the Ekurhuleni Metro is the Manufacturing Hub; the West Rand District is the agribusiness/agro-processing and green and blue economy; Sedibeng District Municipality with the New economy to be anchored on steel; the Vaal river, tourism, agriculture linked to Sasolburg; and, the City of Tshwane being the

Administrative Capital has become a key driver of knowledge, innovation and automotive industry development. The GCR has features of a polycentric city-region (one that has multiple centres), and is anchored by its three large metros: Johannesburg; Tshwane; and Ekurhuleni. Johannesburg's development is an integral part of the region in which it finds itself [8]. The Gauteng province has also developed a GCR Integ-rated Infrastructure Master Plan (GCR IIMP 2030) that promotes sustainability through the efficient use of resources and the adoption of transformative technologies [27]. It has also put in place a National Development Plan Vision 2030 to set goals for infrastructure development to enable socio-economic transformation. The ambitious GCR's programme of Transformation, Modernisation and Reindustrialisation aims to: develop economic potential in individual corridors; ensure growth and opportunities are equitably distributed across the GCR; deliver housing and economic opportunities where most needed to realise radical spatial and economic transformation; ensure infrastructure efficiency through coordinated planning, prioritisation and timing of delivery; and unlock human capital in a dynamic, innovative, competitive and connected city region (Fig. 16.2).

Current Urban Policies and Programmes at the Johannesburg Level
Recognizing the unbalanced Johannesburg metropolitan regional development, in one hand, and the multiple problems associated to lack of basic services, the Johannesburg authorities have taken bold actions to decongest the Johannesburg urban agglomeration. This new spatial re-organization is accompanied with

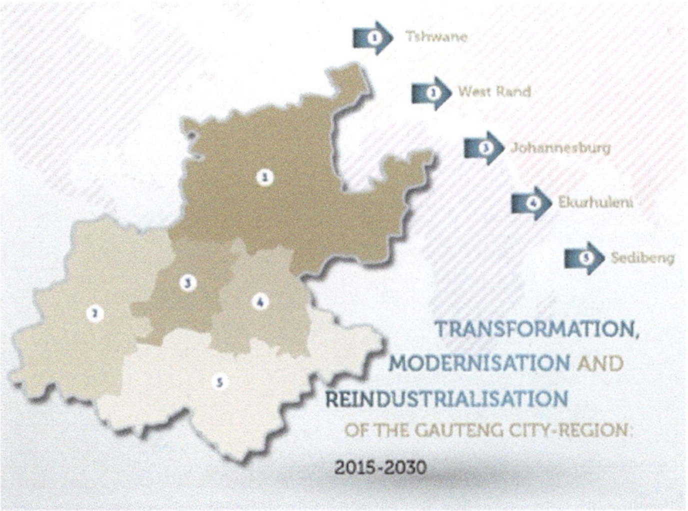

Fig. 16.2 Gauteng City Region—Cluster of Cities. *Source* Gauteng Province Government, 30 January 2015, Transformation, Modernisation and Reindustrialisation of the Gauteng City Region: 2015–2030

economic transformation through the ambitious economic development. This is under the vision:

> Johannesburg – a World Class African City of the Future – a vibrant, equitable African city, strengthened through its diversity; a city that provides real quality of life; a city that provides sustainability for all its citizens; a resilient and adaptive society. (Mayor of Johannesburg, 2017)

Today according to the Integrated Development Plan, Johannesburg has been divided in seven regions (A, B, C, D, E, E, G) as shown in Fig. 16.3.

The Johannesburg 2040 Growth Development Strategy (GDS) is an aspirational strategy that defines the type of society the city aspires to achieve by 2040. The City of Johannesburg developed its first Growth and Development Strategy (GDS) in 2006, as a long-term strategy—an articulation of Johannesburg's future development

Fig. 16.3 Regions of Johannesburg. *Source* City of Johannesburg, 2017–18 Draft Integrated Development Plan Review

path. At that time, there were numerous sectoral strategies, including amongst others: 'Joburg 2030', the Human Development Strategy (HDS), the Integrated Transport Plan and the City Safety Strategy. Each addressed a different angle of the city's development. Johannesburg has also embarked in a holistic approach of Smart City Programme.

16.3.2 The Urban Numbers of the Johannesburg and the Gauteng City Region Post Apartheid

The Gauteng City-Region
The population of Gauteng grew from 7.6 million people in 1996 to 9.2 million in 2001, and is currently estimated at 12.2 million people (GPG 2014). The population of the region is projected to reach 15.2 million in 2021 and 21.3 million in 2041. This significant increase can be attributed primarily to labour migration. Today, the region is a global city-region as an important node in the global economic system, where the linkages that bind the city have a direct and tangible effect on global affairs through socio-economic means [28]. However, the population of the GCR is not evenly distributed across cities, municipalities and districts. It is, indeed, concentrated in three cities that represent 85% of the province total population: Johannesburg (37%), Ekurhuleni (26%) and Tshwane (22%). Johannesburg remained the most urbanised municipality with 56% built-up area in 2009, followed by Ekurhuleni and Tshwane with built-up areas of 44% and 35%, respectively. Overall, the region experienced urban growth, especially in the main metropolitan centres while there is population decline among old small towns and industrial sites of larger cities, associated to decline in job opportunities. Job opportunities develop notably post-industrial centres such as shopping complexes, office parks, and science parks, etc. [22].

When the urban agglomeration is the measurement unit, statistics from the UNDESA's Urbanization Prospects 2014 Revision shows that the population of the Johannesburg urban agglomeration was nearly 10 million in 2015 (9.4 million). These statistics put Johannesburg among the four African megacities along with Cairo, Lagos and Kinshasa (Fig. 16.4).

The population growth of the Johannesburg urban agglomeration was stable until 1970 with an annual growth rate averaging 2.5%. From 1970 to 1985, it declined to reach 1.5%, and the level was maintained until 1990, a year marking the end of Apartheid. From 1990 to 2000, it increased above 4% before decreasing to reach a level, which has remained high, above 3% in 2015 (Fig. 16.5).

The same statistics show that at the national level, Johannesburg and Tshwane (Pretoria) are among the four most populated urban agglomerations in South Africa with Johannesburg leading the group with 27.1% of share on the country's total

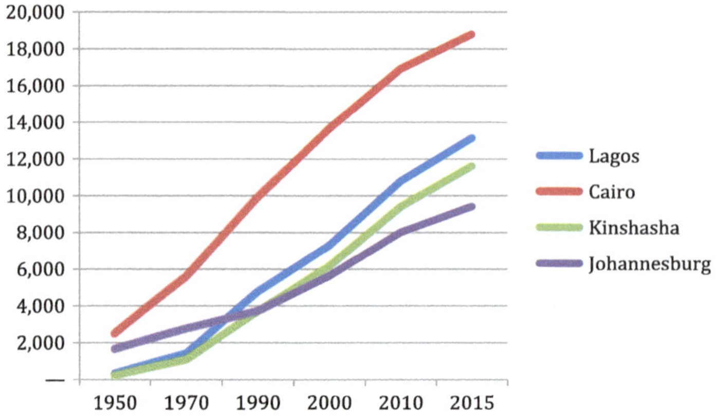

Fig. 16.4 Johannesburg Urban Agglomeration, 1950–2015. *Source* Data From United Nations (2015) World Urbanization Prospects: The 2014 Revision, UNDESA, Population Division, New York, NY, http://esa.un.org/unpd/wup/index.htm

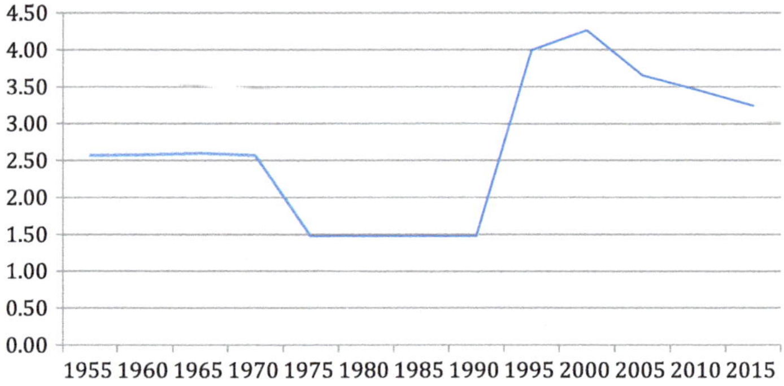

Fig. 16.5 Johannesburg Urban Agglomeration Population Growth rates (1950–2015). *Source* Data from United Nations (2015) World Urbanization Prospects: The 2014 Revision, UNDESA, Population Division, New York, NY, http://esa.un.org/unpd/wup/index.htm

urban population, followed by Cape Town (11%), Durban (8.4%) and Tshwane (6%) (Fig. 16.6).

According to 2016 Community Survey, about 37% of Gauteng's population is aged less than 25 years. The age pyramid with high proportion of young people age 20–29 years old shows clearly that migrant population for both sex from other provinces contribute substantially to the rapid growth of the Gauteng (Fig. 16.7).

Johannesburg city alone constitutes an important demographic dividend with an age-pyramid marked by a demographic regime of a decline in fertility combined with an important migration flux, particularly from young people aged from 20 to 35 years olds. The net migration into Gauteng accounted for 31% of the population increase between 1991 and 1996, and 20% between 1996 and 2001. Between 2001 and 2007, 40% of the population increase in the region was made up of net

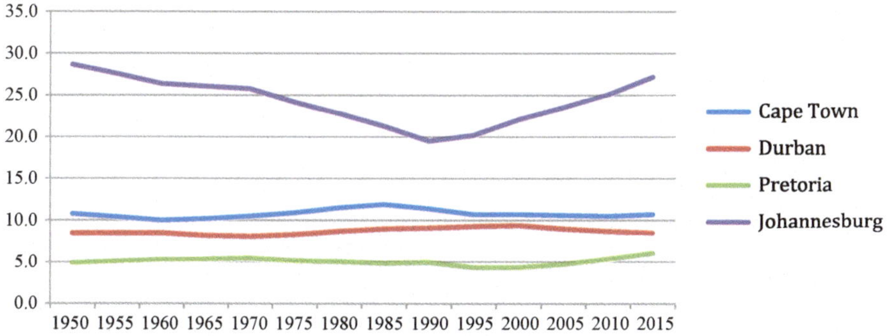

Fig. 16.6 Percentage of urban population residing in each urban agglomeration (1950–2015). *Source* Data from United Nations (2015) World Urbanization Prospects: The 2014 Revision, UNDESA, Population Division, New York, NY, http://esa.un.org/unpd/wup/index.htm

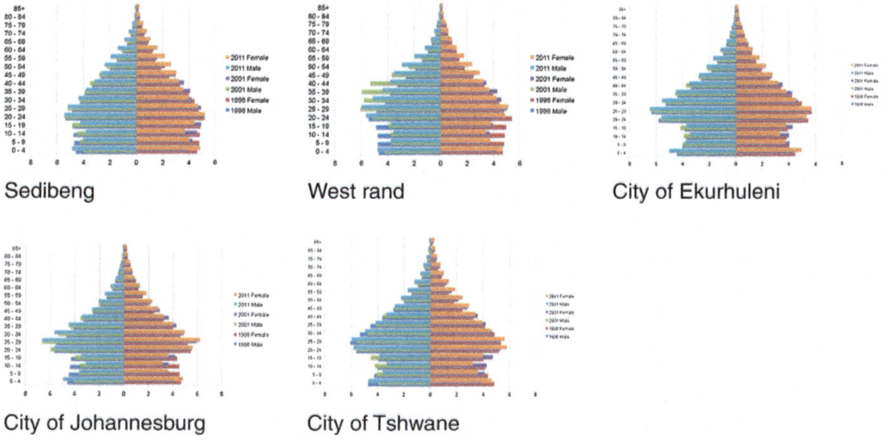

Fig. 16.7 Age Pyramid, Gauteng City Region. *Source* Census 2011 Municipal report, Gauteng, Report 03-01-52

migration. A total of 781,701 immigrants came into the region: 42,385 to Sedibeng; 18,073 to Motsweding; 43,246 to the West Rand; 177,107 to Ekurhuleni; 275,027 to Johannesburg; and 225,863 to Tshwane [29]. Johannesburg offers opportunities for employment and high quality of living conditions and remains a city of migrants. Over the last 15 years, Johannesburg has become the main destination for migrants from other South African provinces as well as from the southern African region and the African continent in general. According to 2016 community survey only 65% of people living in Gauteng were born in the province. One third was born either in other provinces (29%) or in other countries (6%).

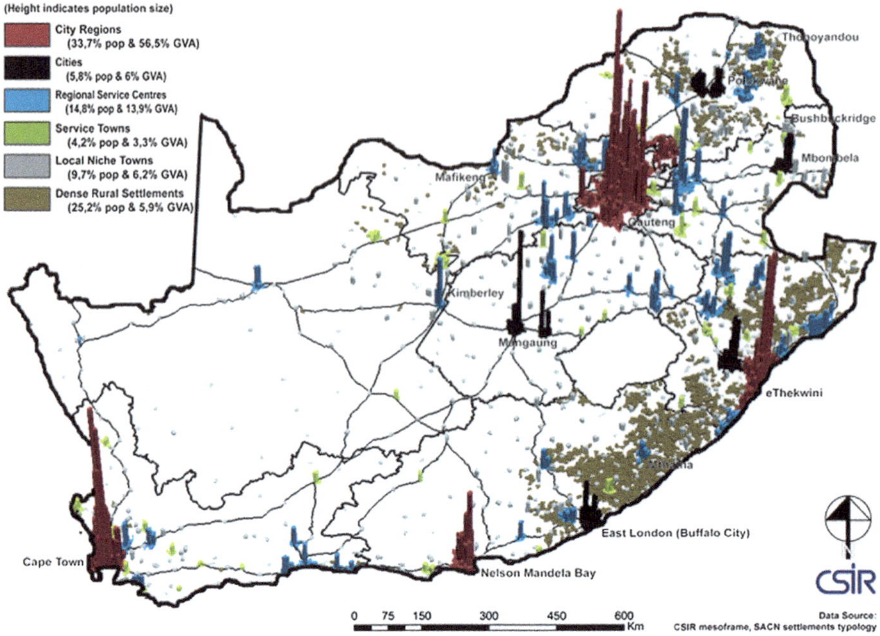

Fig. 16.8 Spatial distribution cities, towns and settlement. *Source* Spatial distribution cities, towns and settlements

The Fig. 16.8 "Spatial distribution cities, towns and settlement" provides the spatial distribution of the settlement types and the proportion of population density. Cities, towns and dense settlements play a significant role within South Africa's development settlement patterns. They can be divided in nodes and clusters on the one hand and sparsely populated areas on the other. Human Settlements are the nodes and clusters categories. Nodes are distinguished according to numbers and densities of population, levels of economic activity, as well as levels of social and economic services. They include City regions; Cities; Regional service centres (large and medium sized towns); Service towns; as well as Local & niche nodes. Clusters are largely described as densely clustered rural service centres and settlements, distinguished by high levels and densities of population distribution and relatively lower levels of economic activity.

16.3.3 Urban Densities as Factor of Economies of Scale and Agglomeration

The population size is an important component of urbanization, but to address various questions in urban economics, environment and infrastructure development, information on density of the population is valuable [30]. Positive production or residential externalities, as well as negative externalities such as congestion, are function of density among others [31]. Various policy responses as well as programmes and projects such as for transport and infrastructure are subject to how population is distributed across a city. Important factors of land use such as land prices should be considered in this exercise as part of elements that can feed the land legislation and investment in infrastructure [32]. From an economic point of view, the higher the density of an urban agglomeration is, the lower are the costs of transactions, and the more prosperous is the economy. A deficient spatial structure fragments labour and consumer markets into smaller less efficient markets. It contributes to higher transactions costs by unnecessarily increasing distances between people and places. It increases the length of the city infrastructure network and therefore increases its capital and operating costs. Lower densities and segregated land uses increase economic costs in terms of average commuting distances, public utility provision, and energy consumption. From an environmental point of view, a deficient spatial structure can decrease the quality of life by increasing the time spent on transport, by increasing air pollution, and contributing to the unnecessary expansion of urbanized areas in natural sites. Poor environmental quality can also lower a city's productivity.

In the African context where many cities are of density higher than 10,000 persons per km^2, South African cities are particularly the exception withlow density. In the case of Johannesburg, the population density is only 3300 persons per km^2 in 2016. The case of Johannesburg is common to South African cities as illustrated by the low population density also observed in Cape Town (4700 persons per km^2), Durban (3200 persons per km^2) and Pretoria (2500 persons per km^2), all being cities of 3 million or more [33].

Gauteng is the most densely populated province in South Africa and the fastest growing province with 33% population increase between 2001 and 2011. The population density is 680/km^2. The highest population density is in Johannesburg CBD, followed by Tshwane CBD. Townships and informal settlements have density of between 7476 and 22,371 per km^2 [34].

However, as illustrated in the Fig. 16.9, the low level of population density in South African cities compared to other African cities hides huge disparities in the cities such as Johannesburg where in townships and informal settlements densities are more than ten times the city average. In fact the average population density of 3050 inhabitants per km^2 of the city of Johannesburg masks densities as high as 40,000 inhabitants per km^2 observed in certain neighbourhoods of the city. These numbers are indicative of a segregated city and point despite huge efforts deployed by the city authorities since the end of the Apartheid regime to bridge the huge urban divide inherited from the cruel Apartheid regime (Fig. 16.10).

The pattern of spatial development in the Gauteng province features huge urban divides inherited from the Apartheid. While economic activities are concentrated in Johannesburg, Pretoria and the Vaal region, the population is concentrated in black townships such as Soweto, Alexandra, Tembisa, Mamelodi, Saulsville, Chris Hani, Thokoza, and Vosloorus, far away from job opportunities.

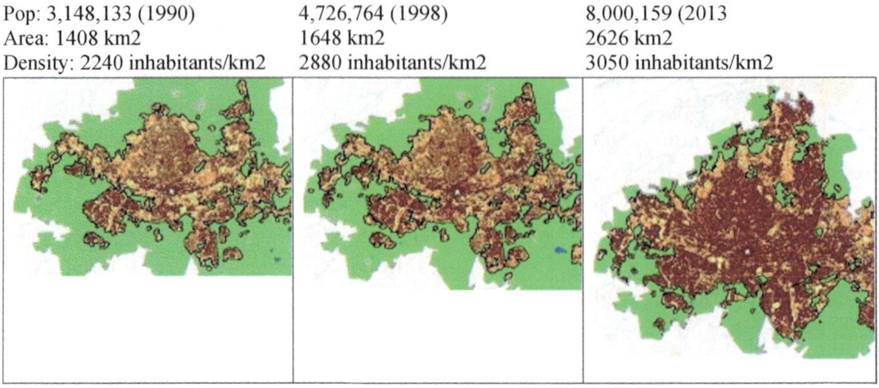

Fig. 16.9 Population, and density of Greater Johannesburg Metropolitan Areas, in 1990, 1998, and 2013. *Source* Angel et al., 2016. Atlas of Urban Expansion 2016, First Edition

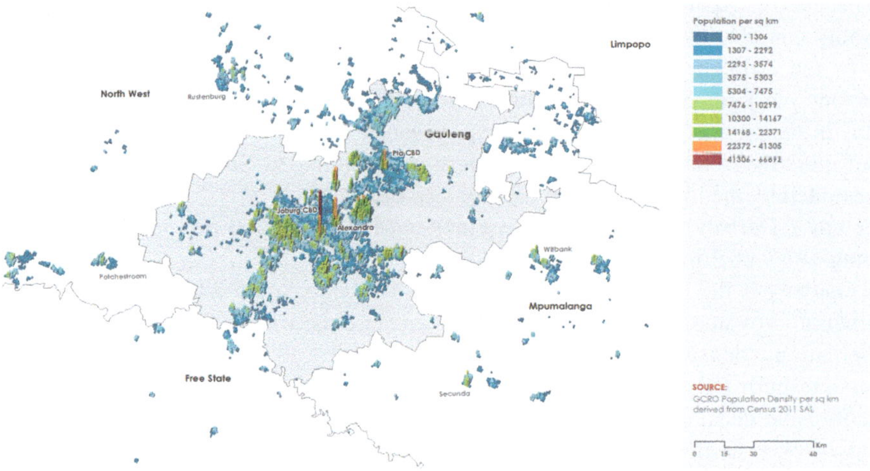

Fig. 16.10 Gauteng Population Density Map, 2015. *Source* CGRO population density per km^2 derived from census 2011

16.3.4 The Johannesburg City Foundation

16.3.4.1 Spatial Planning and Land Tenure

Access to adequate housing is one of the most important components of the South African urban development programmes. The implementation of housing development projects and other infrastructure development projects are aimed building non-racial towns and cities to fulfil the principle of a united people in non-racial residential areas. Since 1994, more than 4.3 million houses benefiting more than

10 million people had benefited from the housing programme delivered through development of low cost housing units, rental and bond subsidies. There developments were done in situ or green fields. Through Public-Private Partnership (PPP, numerous integrated projects, including residential units, business units, hospitals or social units were developed to promote social inclusion [35]. In addition, the South African government has enacted Spatial Planning and Land Use Management Act (SPLUMA) in 2013 for inclusive, equitable and efficient spatial planning at the different levels. In 2011, The Gauteng province also developed the Gauteng Spatial Development Framework (GSDF) 2030 to guide infrastructure investment and the management of activity patterns towards integration of the apartheid fragments, sustainable public transport, improved quality of life, sustainable and inclusive economic growth, protection of the natural environment, promote innovation and creativity and increased community participation in policies for development [36]. The GSDF has been aligned with the National Development Plan 2030, SPLUMA as well as with provincial and sector plans. The municipalities within Gauteng have also developed their Municipal Spatial Development Framework to address priorities within their municipalities. For instance, the city of Johannesburg Spatial Development Framework (SDF) was developed to addressing issues faced by the municipalities including inequality, urban sprawl, disconnected street networks, inefficient residential densities and land use diversity [36].

A number of programmes are in place to support the National Development Plan 2030 vision to transform settlements into sustainable spaces, where cities are self-sufficient and able to satisfy housing, economic and social needs of their communities. These future cities, through adequate planning and sufficient and timely provision of basic infrastructure such as water, sanitation, electricity, energy, roads and social amenities, will create sustainable human settlements. The concept of Mega Cities, which is embedded in these cities, "reinforces the key principles and considerations of spatial reconfiguration, township economy revitalisation and massive infrastructure investment as a means of transforming settlements and cities into productive cities that are inclusive, sustainable and founded on principles of good governance". These developmental corridors are categories into Northern, Eastern, Western, Southern and Central. Through the Mega Cities programme in the five corridors, the Gauteng Department of Human Settlements targets to build 520,000 housing units allocated into 140,000 housing units in the Central Corridor; more than 100,000 in the Eastern Corridor; more than 160,000 units each in the Northern and the Western Corridors and more than 120,000 units in the Southern Corridor by the end of the current term in 2019 [37].

Despite huge efforts taken by the Gauteng authorities to create an inclusive, prosperous society, the province is still haunted in all aspects by the Apartheid regime that excluded for over a century millions of black people from accessing basic services. Even though the province has made great progress on eradicating informal settlements the number of people living in informal settlements remains high. There are still over 1.2 million households and 4.4 million people living in informal settlements in Gauteng and the province is estimated to have the country's largest number of people living in informal settlements as illustrated in Fig. 16.11 [38].

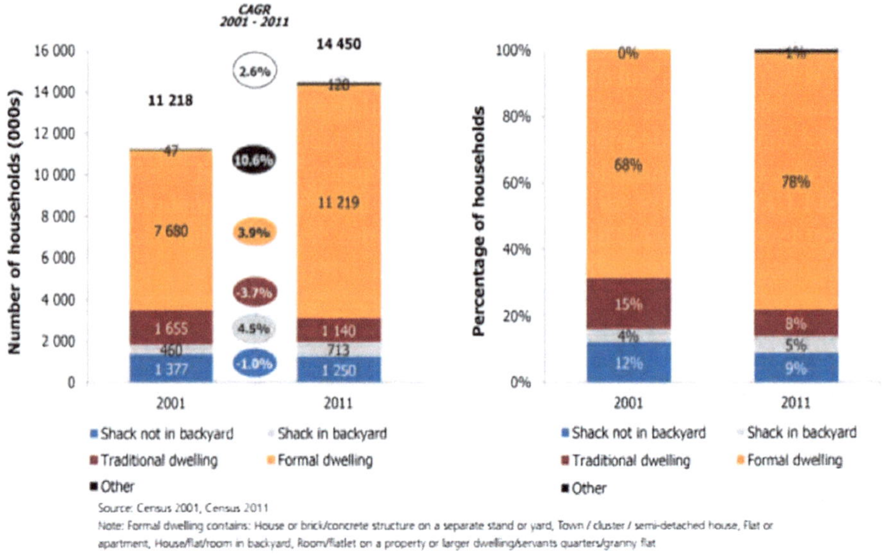

Fig. 16.11 Distribution of house in formal informal [16]

16.3.4.2 Basic Infrastructures

Water and Sanitation

The 2014 release shows that 13,2 million (85,9%) households had access to piped water in 2014 compared to 9.4 million (79.9%) in 2005. Households however, seem to grow less satisfied with the service as only 61.4% of households in 2014 indicated that they experienced 'good' quality service, compared to 76.4% of households in 2005. The percentage of households that indicated that they would pay for the water they received declined from 61.9% in 2005 to 43.7% in 2014 (Fig. 16.12).

The percentage of households in South Africa without access to an improved source of sanitation has decreased consistently over the past decade. The latest General Household Survey (GHS) found that the percentage of households with access to improved sanitation, that is flush toilets or pit toilets with ventilation pipes, has increased consistently from 62.3% in 2002 to 79.5% in 2014. The percentage of households without access to any sanitation facilities, or who were still using a bucket, at the same time declined from 12.3% in 2002 to 4.9% in 2014. The study found that households that shared sanitation facilities considered poor lighting (25.9%) and poor hygiene (23.7%) as the main problems followed by physical safety (19.5%) (Fig. 16.13).

Energy Domestic Use

Having access to energy for cooking, heating and lighting is crucial to households. The survey found that the percentage of households connected to an electricity supply from the mains has increased consistently from 77.1% in 2002 to 86% in

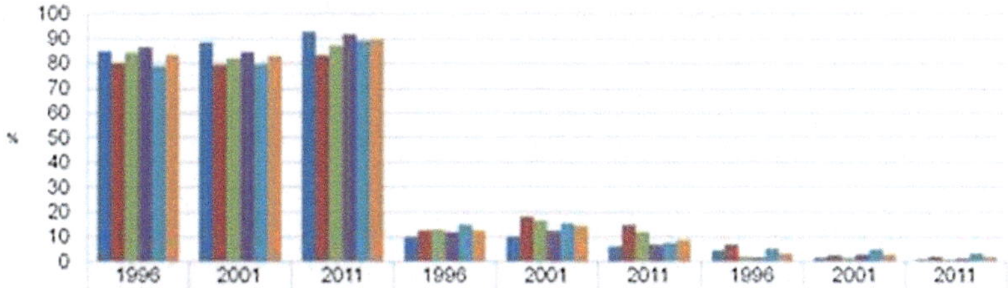

	Piped (tap) water inside the dwelling/yard			Piped (tap) water on a communal stand			No access		
	1996	2001	2011	1996	2001	2011	1996	2001	2011
Sedibeng	84,8	88,3	92,8	9,9	10,1	6,2	4,5	1,6	1,0
West Rand	79,8	79,2	83,2	12,6	18,1	14,8	6,8	2,7	2,0
Ekurhuleni	84,3	81,9	87,1	12,8	16,4	11,7	1,9	1,7	1,1
City of Johannesburg	86,5	84,5	91,6	11,7	12,6	7,0	1,5	3,0	1,4
City of Tshwane	78,9	79,7	89,2	14,9	15,5	7,4	5,2	4,8	3,4
Gauteng	83,6	82,7	89,4	12,6	14,5	8,8	3,1	2,9	1,8

Fig. 16.12 Percentage of households that have access to piped water, Census 2011

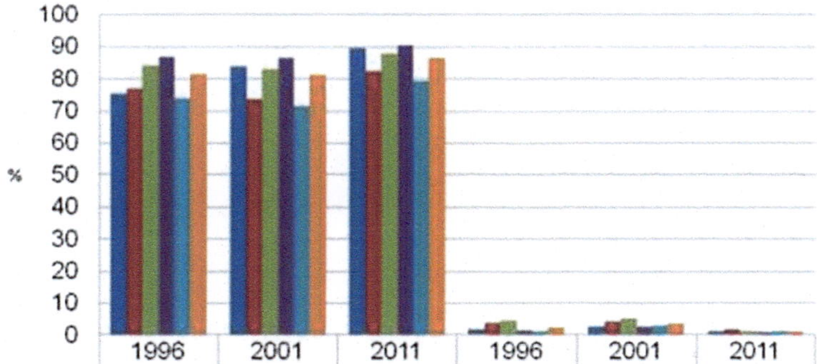

	Flush/Chemical toilet			None		
	1996	2001	2011	1996	2001	2011
Sedibeng	75,5	83,8	89,4	1,8	2,6	1,1
West Rand	76,8	73,5	82,3	3,9	4,2	1,7
Ekurhuleni	84,0	83,0	87,6	4,5	5,2	1,2
City of Johannesburg	86,9	86,5	90,5	1,5	2,8	0,8
City of Tshwane	74,0	71,5	79,4	1,3	3,1	1,3
Gauteng	81,5	81,2	86,5	2,4	3,6	1,1

Fig. 16.13 Percentage of households that have access to flush/chemical toilet, census 2011

2014. Percentage of households in Gauteng with access to electricity for ligthing increased from 85.9% in 1996 to 90.8% in 2011, while percentage of households with access to electricity increased from 80.9% to 84.2% in the same period (Fig. 16.14).

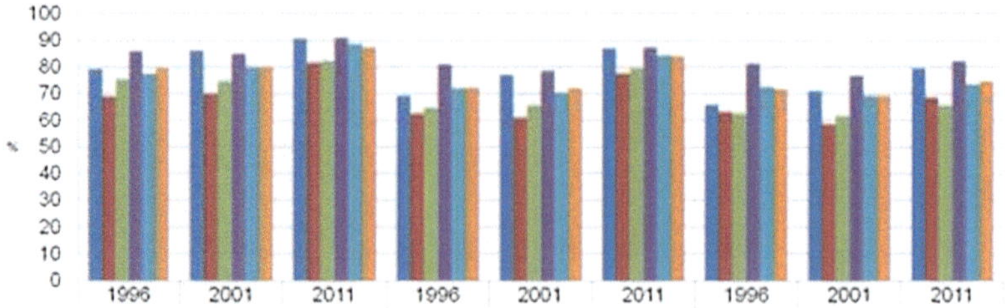

	Lighting 1996	Lighting 2001	Lighting 2011	Cooking 1996	Cooking 2001	Cooking 2011	Heating 1996	Heating 2001	Heating 2011
Sedibeng	79,2	86,1	90,6	69,3	77,1	87,0	66,0	71,2	79,4
West Rand	69,0	69,8	81,7	62,6	61,1	77,7	63,2	58,6	68,8
Ekurhuleni	75,4	74,8	82,2	64,7	65,6	79,4	62,5	61,7	65,6
City of Johannesburg	85,9	84,9	90,8	80,9	78,7	87,4	81,2	76,9	82,1
City of Tshwane	77,3	79,9	88,6	71,9	70,5	84,2	72,5	69,3	73,5
Gauteng	79,4	80,1	87,4	72,3	72,0	83,9	71,7	69,4	74,7

Fig. 16.14 Percentage of households that have access to electricity, census 2011

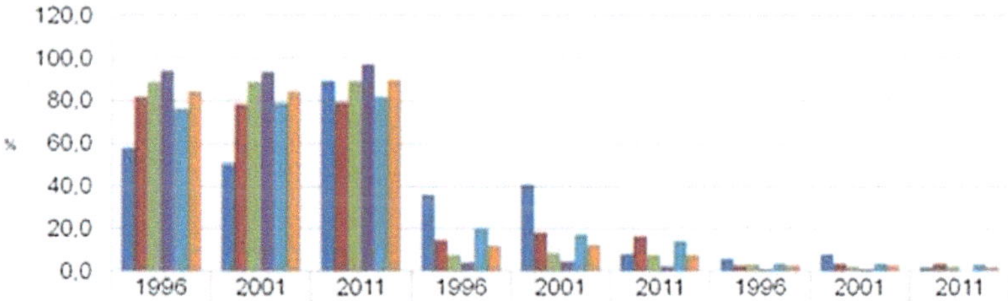

	Removed by local authority/private company 1996	Removed by local authority/private company 2001	Removed by local authority/private company 2011	Communal/Own refuse dump 1996	Communal/Own refuse dump 2001	Communal/Own refuse dump 2011	No rubbish disposal 1996	No rubbish disposal 2001	No rubbish disposal 2011
Sedibeng	58,0	50,9	89,4	36,0	40,8	8,2	5,8	8,3	1,9
West Rand	82,0	78,4	79,5	14,8	18,0	16,3	3,1	3,6	3,6
Ekurhuleni	88,8	88,8	89,4	7,8	8,4	7,8	3,3	2,7	2,5
City of Johannesburg	94,1	93,9	97,0	4,4	4,8	2,3	1,3	1,3	0,5
City of Tshwane	76,1	78,7	82,0	20,0	17,4	14,1	3,7	4,0	3,3
Gauteng	84,7	84,6	89,8	12,3	12,4	7,9	2,9	3,0	2,0

Fig. 16.15 Percentage of households that have access to waste disposal, census 2011

Management of Solid Waste

In terms of refuse removal, while the percentage of households whose refuse is removed at least once per week has increased from 56.7% in 2002 to 63.8% in 2014, large differences exist between urban and rural areas. Whereas 88.6% of households in metros, and 84.1% of households in urban areas have their trash removed once per week, 90% of rural households either have no refuse dumps or have to create their own (Fig. 16.15).

16.3.4.3 The City Foundation Index: Spatial Planning, Land Tenure and Basic Infrastructures

A smart city foundation is composed of three elements: Urban Planning and Design, Land Policies and Basic Infrastructure. For a city foundation to be smart, it must be inclusive at the onset of the urban planning and promotes mixed neighbourhoods where social clustering is discouraged. Having all the poor living together creates slums and fuels instability and insecurity. Inclusive urban planning eases access to basic services (water, sanitation, housing, education and health) and to decent employment for all. A key element of smart urban planning is a smart street network that reduces travel time and encourages walking and social interactions. Smart urban planning enhances infrastructure development, environmental sustainability, economic and social development; makes cities resilient and prepared to overcome natural disasters; and promotes mixed neighbourhoods where services are walking distances from people's residences. Overall the metropolitan region is not smart; disaggregated data reveal huge disparities across districts in favour of the city of Johannesburg (Fig. 16.16).

Using the three main components of the city foundation, we have computed a composite index that represents the overall city foundation, the City Foundation Index (CFI) with values varying from 0 to 1. A value closes to 0 means that a city lacks most of the elements of the city foundation elements, and a value closes to 1 means that a city enjoys most of the elements of the city foundation. In the Gauteng City Region, the CFI is quite high in all cities and municipalities ranging from 0.891 in Johannesburg to 0.807 in West Rand. However, due the huge urban divide inherited from the Apartheid regime, this average hides numerous social inequalities experienced particularly in the townships such as Soweto (Fig. 16.17).

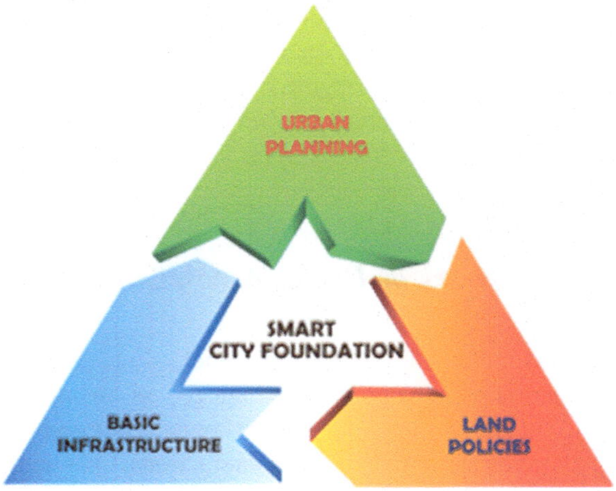

Fig. 16.16 City Foundation. *Source* Mboup, G. et al., 2016. Smart City Foundation—Drivers of Smart Cities. In Vinod et al., 2016. Smart Economy in Smart Cities, Springer

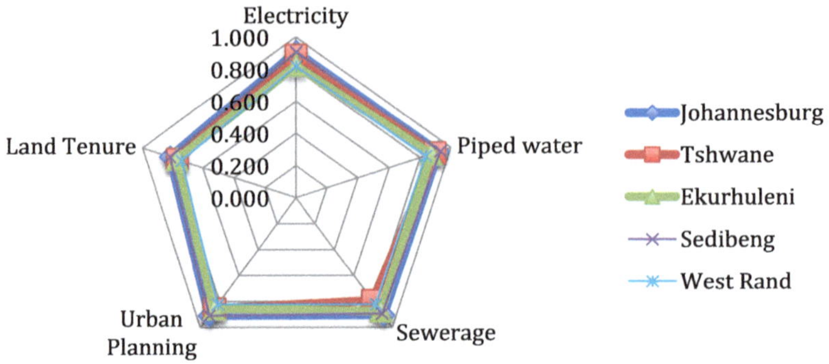

Fig. 16.17 Components of the City Foundation Index, Gauteng City Region

16.4 Urban Accessibility and Mobility

Development of large metropolitan regions like Johannesburg constitutes economic and social opportunities as well as challenges depending on how the urban growth was planned and managed. It provides opportunities for economies of scale and agglomeration as well as for diffusion of ideas and innovations, but it will also call for large investments in infrastructures to respond to the increased demand for water, sanitation, solid management, energy and mobility, etc. It will particularly come with increased demand on mobility that must be satisfied with an efficient public transit accompanied with increased spaces for pedestrians and cyclists in order to safeguard the environment while creating economic growth. Economies of scale and agglomeration economies are greater in metropolitan areas where transportation infrastructures are able to answer mobility needs with higher access to markets and resources than those where people mobility is impeded by deficient transportation infrastructures. Efficient mobility will allow large-scale production of goods and services that can be distributed within the metropolitan regions and beyond with time, cost and reliability opportunities [1]. Without efficient mobility, a metropolitan region losses its economic power and remains just clusters of disconnected settlements.

16.4.1 Streets as Drivers of Urban Accessibility

One key element of urban planning is the street, which defines the form and structure of city. In recent years streets have been recognized as an integral factor in the achievement of sustainable urban mobility. A connected street network reduces travel time and encourages walking and social interactions. One fundamental feature of sustainable streets is their connectivity in terms of planning as well as design. With regard to planning, sufficient land should be allocated to streets (at least 25% city's

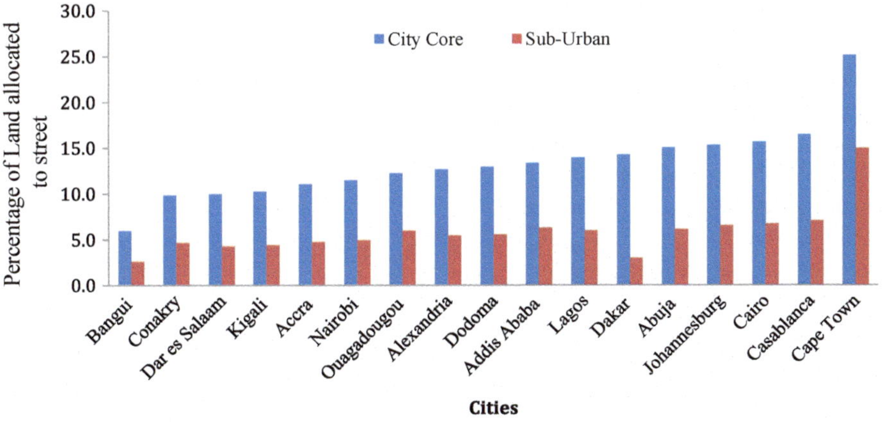

Fig. 16.18 Land allocated to street (LAS) in African cities. *Source* figure prepared from Mboup, G. et al., 2016. Smart City Foundation—Drivers of Smart Cities. In Vinod et al., 2016. Smart Economy in Smart Cities, Springer

area) and the street network should be sufficiently long to cover all areas. There must be sufficient intersections available (at 100 intersection per km^2) to facilitate shorter distances and reduce travel times, and encourage walking and social interactions [39]. A connected street network expands multimodal mobility systems with sidewalks and bicycle paths, ensures eco-efficiency of infrastructural systems, and supports density through integrated infrastructure development, thereby enhancing efficiency and access. In addition to accommodating all kinds of users (pedestrians, cyclists, motorists), sufficient land allocated to streets promotes connections to services that contribute to good health and productivity, such as clean water, sewerage facilities, drainage systems, power supply, and information and communication technologies. It enhances infrastructure development, environmental sustainability, and economic and social development. Streets that provide space only to motorists are characterized by congestion and high CO^2 emissions [40].

The Johannesburg metropolitan region is not benefiting from all the multiple advantages associated with well-connected streets. With only 10–15% of its land allocated to streets in their CBDs, and less than 5% of land in their suburbs, it needs enhancement.

Composite Street Connectivity Index

Street connectivity is determined by the amount of land allocated to streets, the length of the street network and the number of intersections along the network. The proportion of land allocated to streets and the length of the street network is not sufficient to assess street connectivity. A city with wide streets within a very limited street network and low intersection density is considered a city with low street connectivity because the width of the streets is not complemented by a larger street network and higher street density. Similarly, a city with a lengthy street network and dense intersections may not quality as a city with high connectivity if the streets are very narrow. The Composite Street Connectivity Index (CSCI) aims to assess the connectivity of a street considering its width, its length and the number of intersections, all in relation to the total land area of a city. Interestingly, some cities in this group have low levels of land allocated to streets, but higher intersection density (ID) increases the value of their CSCI. For instance, Johannesburg city core has a LAS index moderately low (15%), but due to its high ID, it has a CSCI similar to the other cities with higher LAS but insufficient ID. The city core of Johannesburg is planned in grid pattern that favours good connectivity. However, the Johannesburg suburbs suffer from low LAS, limited street density with few intersections. Cities in this group have very poor street connectivity due to low levels of land allocated to streets, low street density and low intersection density. Their CSCI is less than half the highest level of the CSCI, which is 1. In these suburbs, which are mainly unplanned, provisions of basic services as well as means of transport remain challenges. Comprehensive city planning programmes are needed in these suburbs to improve the lives of urban dwellers (Fig. 16.18).

Johannesburg is among the five African cities (Cairo, Alexandria, Casablanca and Dakar) with **a CSCI of between 0.500 and 0.650.** Their level of connectivity in the city core is sufficient to promote infrastructure development and to ease connections to basic services, such water, sanitation facilities as well as drainage systems. However, their suburban areas are very poorly connectivity, with a CSCI of below 0.300. Only one African city, Cape Town, features in the group with **Cities with a CSCI equal to or above 0.800** along with cities in developed countries. It has streets sufficiently wide to accommodate all types of users, sufficient to reach most neighbourhoods, and sufficient intersections to accommodate all users (Fig. 16.19).

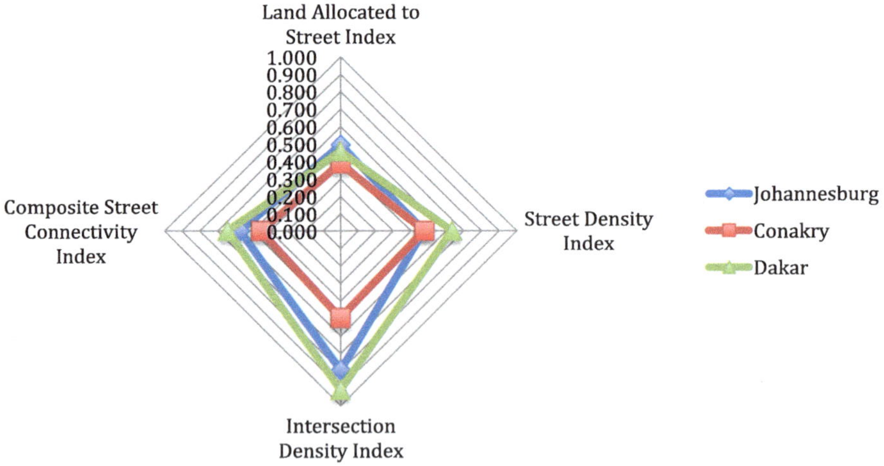

Fig. 16.19 Composite Street Connectivity Index (CSCI)

16.4.2 Transport Modes and Infrastructures Are Key for Urban Mobility

Cities are locations having a high level of **accumulation** and **concentration** of economic activities and are complex spatial structures that are supported by transport systems. The larger the city, the greater is its complexity and the potential for disruptions, particularly when this complexity is not effectively managed. The most important transport problems are often related to urban areas and take place when transport systems, for a variety of reasons, cannot satisfy the numerous requirements of urban mobility. The Sustainable Development Goals (SDGs) recognize that sustainable transport is crucial for urban economic development. With Goal 11 of the SDGs "*Make cities and human settlements inclusive, safe, resilient and sustainable*", member states have committed themselves to "enhance inclusive and sustainable urbanization and capacity for participatory, integrated and sustainable human settlement planning and management in all countries. Member States have further committed to support positive economic, social and environmental links between urban, peri-urban and rural areas by strengthening national and regional development planning, and by 2020, substantially increase the number of cities and human settlements adopting and implementing integrated policies and plans towards inclusion, resource efficiency, mitigation and adaptation to climate change, resilience to disasters, by 2030, …" [41].

16.4.2.1 State of Urban Mobility in Johannesburg Metropolitan Region

With their large population, megacities offer potential economies of scale and agglomeration. But they also call for large investments in transport infrastructures to respond to the increased demand in mobility. BRT, alone, will be insufficient to move daily millions of people. To respond to this demand, these megacities must put in place various combinations of high-capacity transport public modes, which are metro, light rail, BRT and buses. Though Johannesburg is pioneer in the development of BRT in the African continent, it is still facing huge mobility challenges associated to the diversity of their urban form as well as to the high unemployment rates. There are many settlements in the outskirts of the city that are left out from the BRTs. It may require a multimodal transport systems including, such a metro, that can accommodate a very large number of passengers. These high-capacity transport modes must also be integrated with other means of movement such walking, cycling and other private means that will serve as "feeders" [42]. These "feeders" are crucial particularly in Johannesburg where many informal townships unplanned and lack street network that can accommodate high-capacity transport modes.

Inter-city Public Transportation to Make the Gauteng City Region Works To create economies of scale and agglomeration within the Gauteng City Region, an efficient, timely transport system is needed. Without efficient public transport, economies and scale and agglomeration that can generate the GRC with its 13 million will be jeopardized. The Gauteng province introduced, indeed, the Gautrain in 2002, a Rapid Rail Link connecting Pretoria, Johannesburg, and Johannesburg International Airport (JIA, now OR Tambo International Airport) as one of eleven Spatial Development Initiatives (SDIs). Two spines were also built: the west–east spine has a commuter service linking Sandton to the East Rand (through Rhodesfield, in Kempton Park) as well as an airline passenger service between Sandton and the city's international airport. The south–north spine links the Johannesburg and Pretoria CBDs [43]. In 2007, the province launched the Gauteng Freeway Improvement Project (GFIP), the country's largest road scheme, which aimed "to upgrade and expand the provincial road network through freeway widening, building new roads, upgrading interchanges, installing traffic-management systems (cameras and electronic signs) and an automated toll system" [19]. In addition to easing economies of scale and agglomeration, these inter-city public transports were supposed to reduce traffic congestion and reduce the commuter time. Other elements such as traffic accidents and pollution associated to private cars and minibuses were also expected to reduce.

In 2007, the National Cabinet also approved the Bus Rapid Transit (BRT) scheme as a road-based component of the Public Transport Strategy (SACN2011). This was to be funded by national government, but delivered by the metros. It was designed to be more transformative and scale-able. It had modern buses running on dedicated lanes along the middle of main roads [44]. The City of Johannesburg's BRT system, named Rea Vaya, was operational from October 2013. The buses provide residents of Soweto and the southern parts of Johannesburg with direct access to the inner city and surrounding areas. The routes run from Soweto through Noordgesig, New Canada, Pennyville, Bosmont, Coronationville, Newclare, Westbury, Westdene, Melville, Auckland Park and Parktown, to the CBD. The Tshwane BRT system, named A Re Yeng, started operating in 2014 (SA Yearbook 2014). The R2.6-billion project, part of the city's revitalisation project, consists of 80 km of bus lines and about 340 buses, some of which are powered by gas, serving 62 bus stops. It runs from Mabopane through Pretoria CBD, past Menlyn, and on to Mamelodi (City of Tshwane 2014).

However due to poor spatial planning in the apartheid days, Johannesburg still faces a huge transport problem, particularly: moving people from outlying areas such as Soweto into working hubs such as Joburg CBD and Sandton City. The city's Comprehensive Integrated Transport Plan (ITP) is a very important document to guide how Johannesburg will provide, integrate and support public transport. The last plan expired in 2008 and a new plan went under review in 2011 (ITPs last five years), which considers affordable public transport, convenient and accessible transport, and a transport system with reduced crime, collisions, injuries and fatalities. Another consequence of the lack of efficient public transport systems, comes congestion, and the poorly maintained taxi ranks and train stations add to this problem. Ultimately, there are too many cars on Johannesburg's roads and a comprehensive plan, such as the ITP, is needed to ease the gridlock [45].

In South Africa, the Public Transport Strategy aims to improve public transport by establishing an integrated rapid public transport network that comprises of an integrated package of rapid rail and road corridors. Through BRT, the government aims to link different parts of a city into a network and ensure that by 2020, most city residents are no more than 500 metres away from a BRT station [42]. The BRT systems are being implemented through public–private partnerships, whereby cities build and maintain the infrastructure for the operation of the buses, stations, depots, control centres and a fare collection system. Private operators, by contrast, own and manage the buses, hire staff and provide services on a long-term contract. Johannesburg is the pioneer in BRT in the African continent. "As soon as the BRT was introduced and operational in 2010, a robust but affordable bus management system was required, as there are many financial gains that can be made with the successful implementation of such a system. Today four other South African cities —Cape Town, Durban, Tshwane and in Rustenburg have adopted the BRT and have also introduced the ITS for transport planning, management and monitoring. Eight other South African cities are planned to adopt BRT or alike for efficient pubic mobility. All buses are equipped with free Wi-Fi on the trunk route, full air conditioning, low-floor technology which supports Euro V emission levels, vehicle stability and a cashless automated electronic system that is fully monitored by camera and censors [46].

State of Mobility in the Gauteng Province

As shown in Fig. 16.20, in the three large cities of the province—Ekurhuleni, Johannesburg and Tshwane—half or more can access a public transport within 10 min walk. In All three cities, more than three-quarters of the population can access public transport within 30 min walk (87, 77 and 77% respectively). These figures indicate the efficiency of public transport in the Gauteng province compared to many African cities where formal public transport is quasi-inexistent.

Turning Informal Transport Sector Challenges into Opportunities

Despite enormous efforts made to boost the public transport, the mobility demands have not been fully satisfied, informal public motorized means as well as use of private cars for those they can afford them are still significant mobility means. While playing a critical role in the public transport sector in the Gauteng province and in cities, the informal motorized means of transport face a number of constraints, ranging from administration and managerial issues to financial obstacles. Efforts should be made to interlink them with the formal transport systems. Rather than discouraging informal operators, national and local governments can set incentives and regulations that capture the benefits they may bring while mitigating negative factors, such as road safety and pollution. Some actions already taken by some countries and cities must be reinforced. The formalization of minibuses' activities includes a renewal programme of the fleet minibuses as well as their

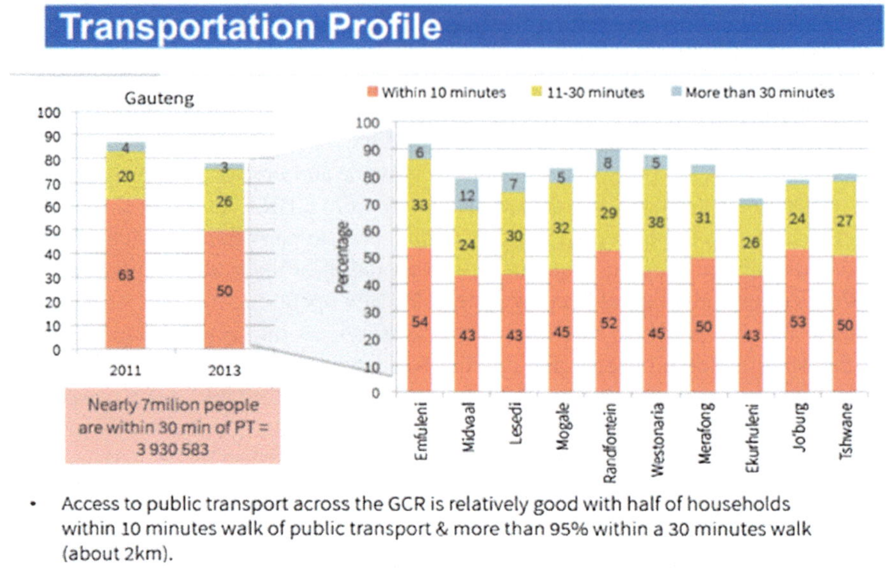

Fig. 16.20 Transport Profile, Gauteng City Region. *Source* Gauteng Province Office of the Premier

administration. This also includes assisting in creating better conditions for loans to transport operator such as creating financial structures that can make car loan possible to operators. Subsequent formalization occasionally occurs through aid-financing arrangements, for instance through trust funds guaranteeing credit lines for vehicle purchase [42].

Walkability

Over the last few years, Johannesburg has been developing 'walkable' cities such as the infrastructure in Ivory Park, consisting of new pavements, benches, landscaping, public art, designated taxi park lanes, kerbs and stormwater drainage. According to Rehana Moosajee, member of the mayoral committee for transport in Johannesburg, transport is not only about the motorised kind, but people too. The Ivory Park project focused on infrastructure that would enable walking or cycling—the most sustainable way of getting around [47]. Overall, the future transport system in Johannesburg is dependent on a shift in transport modes from motorised to non-motorised; private to public transport; and road to rail (Fig. 16.21).

The city of Johannesburg Spatial Development Framework is aimed at developing sustainable and integrated residential areas, improved public transport and connectivity to the CBD and areas of economic activities and improving service delivery and promoting green economy.

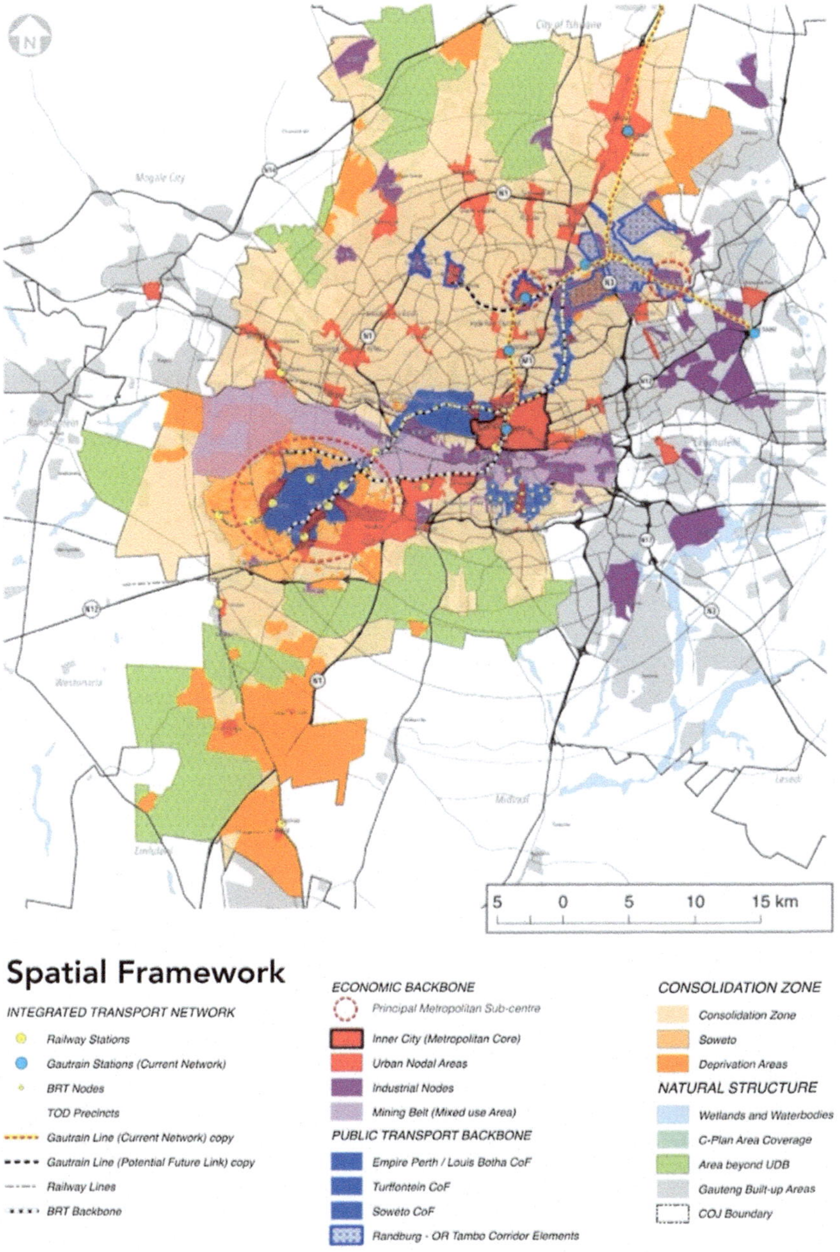

Fig. 16.21 Draft Spatial Development Framework Map of City of Johannesburg, 2016. *Source* GCRO; City of Johannesburg Draft Spatial development Framework

16.4.3 ICT and Mobility in Johannesburg

"As soon as the BRT was introduced and operational in 2010, a robust but affordable bus management system was required, as there are many financial gains that can be made with the successful implementation of such a system. The Automatic Public Transport Management System (APTMS) was developed by a private consortium to deliver an ambitious range of information and services, including dynamic passenger information. The whole system was managed by a centralized control centre with CCTV monitoring on vehicles and in stations allowing headways to be monitored and provide direct voice communication with station staff and drivers" [48].

The Gauteng Open Road Tolling Project was commissioned by the South African Roads Agency Ltd in order to finance the upgrading of a network of approximately 185 km of freeway in Gauteng. This project uses electronic and automated methods of levying and collecting toll charges using 42 Tolling Points and is estimated to generate in excess of 2 million Tolling Point vehicle passages per day. The project comprises of the following sub-systems: Road Side Systems (overhead gantries), Open Road Tolling System Back Office, Transaction Clearing House and Violations Processing Centre. This project entailed the design, supply, installation and commissioning of remote traffic monitoring systems (RTMS) and equipment at 22 sites along the N2 between the Airport and City Centre. Aspects of the remote monitoring of traffic covered include the detailed recording of traffic volumes, vehicle types and vehicle speeds during all traffic conditions. The roadside sensors are capable of recording multi-lane data up to 10 lanes separately.

16.5 Smart Energy for Economic Development

South Africa is among the world's least energy efficient countries with the country's economy very energy intensive and accounting for approximately 40% of Africa's total electricity usage. Unfortunately more than 90% of electricity generated in South Africa is based on coal. This puts South Africa as one of the highest emitter of greenhouse gasses in the world. To respond to the energy challenge, the Department of Energy has enacted a number of strategies, projects and programmes that are aimed at increased targets on the reduction in the national energy consumption, promotion of renewable energy and introduction of Independent Power Partnership (IPP) among others and development of Smart Grid. The adoption of the Constitution (1996) and Bill of Rights created the foundation for South Africa's Renewable Energy (RE) programme. The energy policies in South Africa policies include the 1998 White Paper on Energy Policy (WPEP), the 2003 White Paper on Renewable Energy (WPRE) and the 2011 National Climate Change Response White Paper Policy. The 2003 white paper seeks to ensure that renewable energy is part of the energy mix in the country. The National Development Plan (NDP) sets the overarching long- term vision for South Africa's growth and development. On the National Climate Change response 2011 white paper, South Africa commits to make a fair contribution to the global effort to stabilise greenhouse gas concentrations in the atmosphere within a time frame that is compatible with sustainable development.

The major fuels consumed in metros are electricity, which is used predominantly for lighting, heating, cooling and cooking, and petrol/diesel, which is mostly used for transport. Diesel is used for freight and passenger transport, as well as for industrial applications. Petrol is used almost exclusively for passenger transport largely for private passenger transport. Most energy is consumed by the transport sector than residential areas. This is attributed to the fact that South African cities have lower densities, which results in expanded transport network. Even though the transport sector consumes more energy, residential areas emit higher carbon emissions.

In 2012, South Africa was amongst the top investors in renewable energy. South Africa is fortunate in that, over and above its rich coal resources, it is also well endowed with non-delectable renewable energy sources, notably solar and wind. The country has an average of more than 2500 h of sunshine per year and average direct solar radiation levels range between 4.5 and 6.5 kWh/m^2 per day, placing it in the top-3 in the world. The Department of Energy has developed a Wind Atlas, which showed greater potential of wind energy along the coastal areas. Potential of small-scale hydro energy has been identified in some part of the country. With few hydro power stations, the potential of hydro energy is yet to be exploited. The country also recognised the potential of biomass resources and has developed the Biomass Action Plan, which assesses biomass resources in the country. The local municipalities are mandated to play a large role in improving energy efficiency by ensuring the existence of applicable laws, education Programme and policies. As such municipalities are expected to develop and implement energy efficiency and renewable energy strategies and projects.

Smart Grid

A smart grid is an electrical grid that uses modem technology (digital or analogue) to collect and communicate electricity related information of both the suppliers and consumers. Installation of smart grids improves the overall efficiency of the system by contributing in the efficient transmission and distribution of the electricity. Smart meters are the key component in smart cities. City power, an independent municipal entity wholly owned by City of Johannesburg, was established in 2001 to provide electricity and energy solution to CoJ. City Power contributes to smart city initiatives through development of smart grid. The metropolitan municipalities are also responsible for the implementation of smart grid, renewable energy and energy strategies.

Smart Meter

A smart meter is an electronic device that records consumption of electric energy in intervals of an hour or less and communicates that information at least daily back to the utility for monitoring and billing. Smart meters enable two-way communication between the meter and the central system Smart meters are the key component in smart cities. As part of smart grid initiative, in Johannesburg, City Power started installing smart meters (Advanced Metering Infrastructure (AMI)) in 2011 to reduce energy loss due to levels of vandalism and meter by passing to improve safety of people and energy efficiency. AMI is an integrated system of smart meters, communications networks, and data management systems that enables two-way communication between utilities and customers. The installation of smart meters resulted in increased revenue and reduced our operational costs. The use of smart meters also enables the city to communicate with the residents through sms to reduce power consumption at a particular time to reduce the demand of electricity or remind them to pay an outstanding bill. Another advantage of using smart meters is that City Power will be notified immediately when there is fault on the line. In 2016, about 75,000 smart meters were installed in Johannesburg suburbs, business areas and townships.

Renewable Energy

City of Johannesburg in line with Kyoto Protocol and the summit is determined to create a green city, in keeping with the movement to prevent climate change and improve the environment. Any of the City's development decisions are taken with the need for a green planet in mind: from greening and cleaning the environment, to reducing the energy load in buildings, to cutting carbon emissions. City of Johannesburg has a policy in place that emphasises energy efficiency and resource sustainability in urban development. Its own buildings set the trend for energy cuts, and the City has received much acclaim internationally. Most areas in South Africa average more than 2500 h of sunshine per year, and average solar-radiation levels range between 4.5 and 6.5 kWh/m^2 in one day. This makes solar the most accessible resource in the country. In 2010, the city of Johannesburg launched a solar heating programme to reduce electricity consumption which special focus to poor and low-income areas. The programme is part of the City's 2040 Vision to save energy and create employment through a low carbon infrastructure. In 2014, around 43,000 solar water heaters which generate 22.5 GW of electricity per hour were installed in the city. By-laws regarding solar water heaters were recently approved City of Johannesburg Council, and the city enforces the use of solar heaters in all newly built houses and in new building additions. Areas that have benefited from the solar heater programme include government building, townships and informal settlements. These areas include Devland, Lawley, Vlakfontein, Eldorado Park, Lenasia, Alexandra and Leha as illustrated in the figure (Fig. 16.22).

Waste to Energy Project

Johannesburg water utility made the decision to investigate biogas to electrical energy options following the realisation that continued electricity tariff increases by Eskom would have a direct effect on the operational costs of wastewater treatment in Johannesburg. In 2013, the first City of Johannesburg water utility, Johannesburg Water (JW), officially launched its pilot Biogas Project at the Northern Works Waste Water Treatment Plant. The electricity produced is for own-use within the wastewater treatment works and runs in parallel with the incoming Eskom grid. It is

Fig. 16.22 Solar geysers installation project

capable of producing approximately 10% of the plant's power requirement. This project will be incorporated into other 6 of JW's treatment works into the programme eventually.

Biogas to Energy
Biogas is produced when any type of organic matter decomposes in the absence of oxygen. The City of Johannesburg (COJ) identified that methane emissions at some of their landfill sites were in excess of the prescribed air quality licensing thresholds. They embarked on consultations with industry and identified that harvesting the landfill gas to produce electricity would solve their air quality problem. An ambitious landfill gas to electricity project has been initiated. South Africa's first independent landfill gas-to-power project has begun generation at Robinson Deep landfill in Johannesburg in 2016, providing 3 megawatts (MW) of renewable electricity, enough to supply more than 5500 homes. This project was a partnership between the city and private companies (Independent Power producers, IPP) and it will be expanded to other landfill sites.

16.6 Urbanization, Economic Growth and Metropolitan Regional Development

Towards the end of 2010, a New Growth Path (NGP) was introduced. The priority in the NGP was the identification of 'job drivers' for the creation of 'decent' jobs. Public investments were directed at infrastructure development, improving value chains in agriculture and mining, investing in a green economy, and encouraging light manufacturing sectors, tourism and other high level services. The government

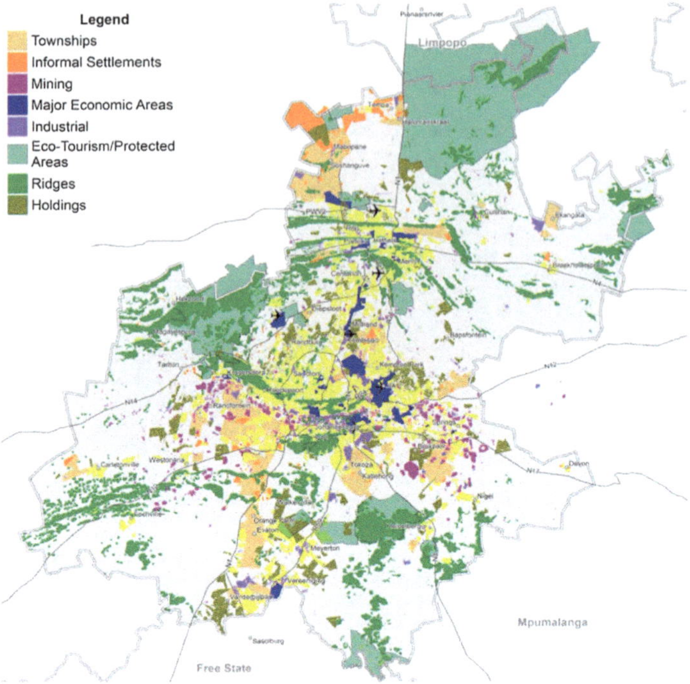

Fig. 16.23 Overall provincial spatial structure. *Source* Gauteng Spatial Development Framework 2030, http://www.gauteng.gov.za/government/departments/office-of-the-premier/Documents/GSDF%202030

planned to expand employment to the tune of five million jobs by 2020 through high GDP growth rates, and improve the employment intensity of growth, which had decline significantly over the preceding decade from 0.8 to 0.67 [49]. The South Africa government also operated a paradigm shift considering that the concept of human settlements is not just about building houses. Changing the apartheid spatial patterns, it calls for a holistic approach that allows low income households in rural or urban areas to have easy access to economic centres, social amenities and basic services such as water, electricity, recreational facilities, schools, clinics and a host of others [3].

The spatial structure of the province reveal contrasted pictures with the development of Eco-Tourism/Protected Areas (light green) in several parts of the province, but most importantly in Tshwane and Johannesburg. Major Economic Areas are mostly located in Johannesburg while the Mining activities are located in Ekurhuleni (Fig. 16.23).

16.6.1 Urbanization and Economic Growth

Contrary to most African countries, urbanization and economy growth in South Africa are closely associated above the average. The relationship between urbanization and economic development had been uneven in most African countries, and urbanization has not led to economy growth as shown in 16.24. As countries in Asia such as China, Indonesia, Malaysia, Thailand and Vietnam and in LAC such as Mexico, Costa Rica and Argentina, South Africa displays an increase in GDP per capita associated to high urbanization rates; a similar association between GDP per capita and urbanization was not observed in most sub-Saharan African countries among them Senegal, Guinea, Togo, Liberia and Burundi. In most of African countries, capital cities were not able to produce and distribute wealth nation-wide (Fig. 16.25).

The economic sectors within GRC include agriculture and forestry, mining, manufacturing, electricity, gas and water, wholesale and retail, transport and communication, finance and business services, community and social services. Estimates of income disparities in 1995 suggest that average white monthly household incomes were four and a half times larger than those of black households (Gauteng Provincial Government 1995).

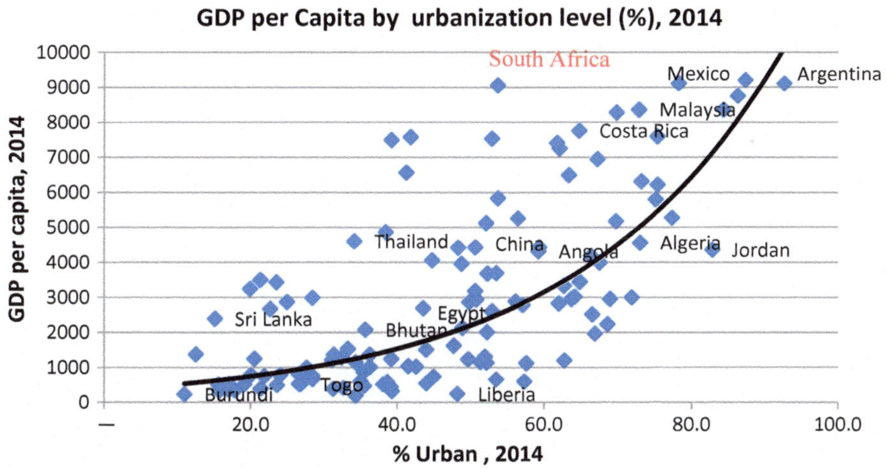

Fig. 16.24 GDP per capita by Urbanization level (%), Country with GDP below US$15,000. *Source* % urban (UNDESA), GDP per Capita (World Bank)

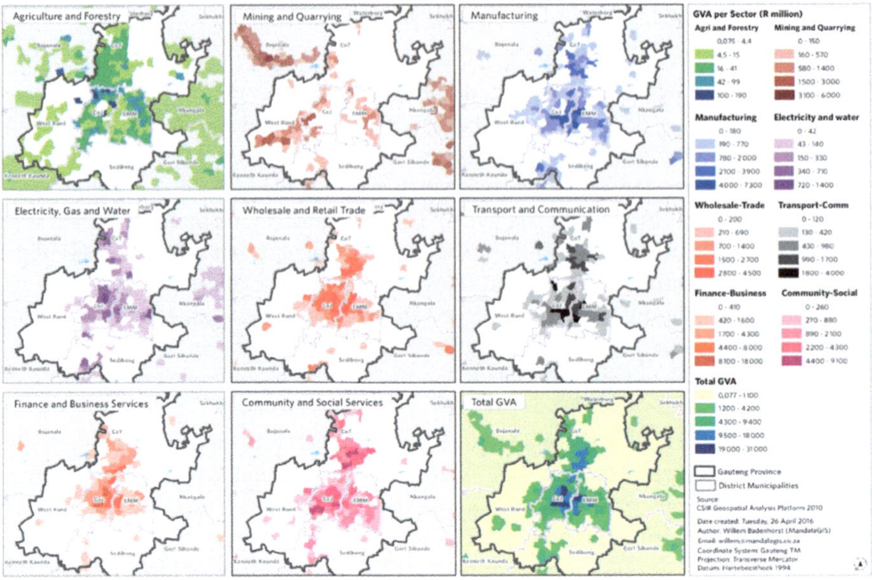

Fig. 16.25 Gauteng Economic Activities. *Source* Gauteng Spatial Development Framework 2030

16.6.2 Towards a Smart Metropolitan Regional Development

Most of the components of a Smart Metropolitan Regional Development (SMRD) are composed of several simple sub-components. To put them together to track progress on the overall objective of the SMRD will require the development of composite indices using sophisticated multi-level statistical analysis including Principal Component Analysis. The complexity of a composite index is the fact that it is strongly influenced by several factors including: the normalization, the standardization and the degree of association between variables and the number of variables that compose the index. An index can also be conceptualised differently using different variables. This calls for caution when interpreting a composite index. For instance, the Quality of Life is conceptualised and measured in different ways across studies. The Human Development Index includes indicators of health and education, alongside an economic component (Gross Domestic Product or GDP); the Legatum Prosperity Index considers quality of life to be multidimensional, including both wealth and wellbeing. The Economist Intelligence Unit's Quality of Life Index links life satisfaction to health, family life and community life; and the OECD's Well-being Initiative has two dimensions: 'material living conditions' and 'quality of Life' [50].

The city of Johannesburg was included in the development of three indices published by the UN-Habitat and conceptualized and further developed by

Mboup G. (2012, 2013 and 2014). Those indices include: the City Human Development Index, the City Prosperity Index and the urban form-based City Prosperity Index integrating the Composite Street Connectivity Index. This section will take it further for the development of the Smart Metropolitan Regional Development Index.

16.6.2.1 Urbanization and Human Development

The human development is made from element of economic development (GDP per capita) and elements of Social Development (Education and Health). All three elements are components of a smart metropolitan regional development. Johannesburg is still engine of South Africa Economy as illustrated in Fig. 16.26 with the Johannesburg HDI much higher (0.823) than national HDI (0.658); it has better economy outcomes as well as better health status than the national average. It occupies a pivotal place in the trade sector both nationally and internationally. Important progress has been made on education, which is critical to meeting the challenges of smart metropolitan regional development, as it connects people to new approaches, solutions and technologies that enable them to identify, clarify and tackle local and global problems. Similarly, important progress has been made on health, which is critical to realizing economic growth through increased productivity. Healthy workers are more productive, bringing greater income to families and higher levels of economic growth for nations, and then enhance smart economy.

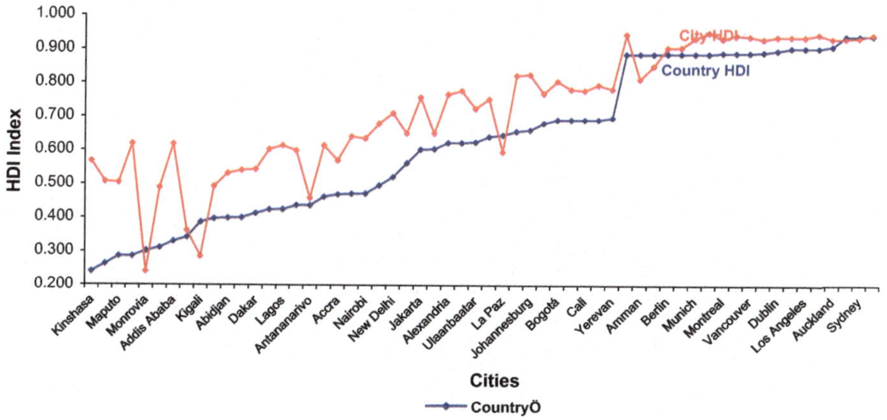

Fig. 16.26 City Human Development Index (HDI) and Country HDI, 2014. *Source* City HDI computed by the authors, and National HDI from UNDP, 2014, Human Development Report 2014

16.6.2.2 Urbanization and City Prosperity

The City Prosperity Index (CPI) published by UN-Habitat in 2012 under the coordination of Mboup G includes various indices and indicators that are relevant to urban areas, and important for prosperity-oriented public policy-making. The first version of CPI published in 2012 was based on five dimensions: Productivity, Quality of Life, Infrastructure Development, Environmental Sustainability, and Equity and Social Inclusion [51]. In 2013, the CPI was revised to include the urban form dimension measure of street connectivity [52].

Among the 20 African cities included in UN-Habitat's CPI sample, Cape Town, Johannesburg, Cairo and Casablanca are the only ones featuring solid prosperity factors (CPI of between 0.700 and 0.799). This group is heterogeneous, with some cities showing a 'less coordinated', ill-balanced development in the 'spokes'. This comes as the result of institutions, legal and regulatory frameworks and urban management practices that are being consolidated and because of this; they cannot hold together all the elements of the 'wheel' to operate with stability. Heterogeneity is also related to the stage of development of the relevant countries. However, it is important to note that inequality is inconsistent with prosperity as understood in this Report. When the equity index is included in the CPI, Cape Town and Johannesburg (which both feature very high Gini coefficients), drop from the bracket of cities with 'solid' prosperity factors and join the group with 'weak' or even 'very weak' prosperity factors, with CPI values of 0.590 and 0.479 respectively. Johannesburg like most South African cities has experienced significant economic growth, but in the past two decades life expectancy has declined substantially, affecting quality of life. Cities in emerging economies such as South Africa combine high economic growth and strong infrastructure, and are expected to move faster along the path of prosperity—but then, for the sake of balanced development, they must tackle inequalities and environmental degradation. They

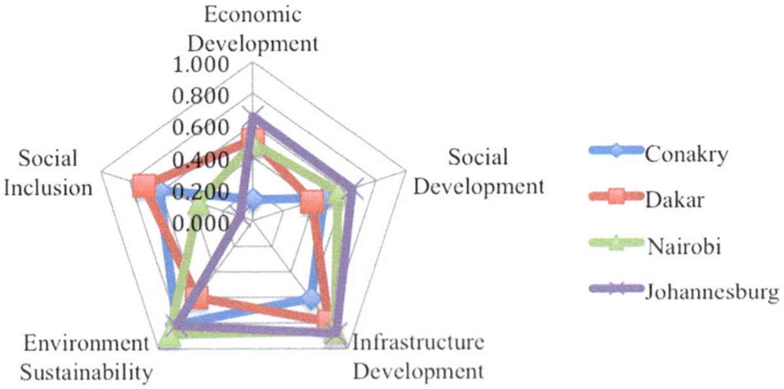

Fig. 16.27a Components of City Prosperity

must also look to improving quality of life through more ample provision of public goods (Fig. 16.27a).

16.6.2.3 Towards a Smart Metropolitan Regional Development

Among the components of the Smart Metropolitan regional Development, five have been part of the City Prosperity Index (CPI): Productivity, Infrastructure development, Environmental Sustainability, Quality of Life and Social Inclusion. The SMRD index introduces other components relevant for sustainable urban development: Disaster Risk Reduction and Resilience, Peace and Security, Institutions and Laws. It expands productivity to Economic Development, Quality of Life to Social Development, and makes the distinction between Infrastructure Development and Basic Infrastructure. It introduced the concept of city foundation built upon three dimensions: urban planning, land policies and basic infrastructure. It also introduces ICT at the centre of the SMRD along with the City Foundation and Institutions and Law. Considering that the CFI and the CPI was already measured in previous sections, the measurement of the SMRD is incremental and built upon those indices with the inclusion of Peace and Security as well as Disaster Risk Reduction and Resilience (Fig. 16.27b).

In addition to the urban challenges analysed in previous sections, the population of Johannesburg has been constantly exposed to disasters, particularly to flooding during raining seasons. Apart from habitat degradation, floods have caused considerable economic, social and environmental losses in Johannesburg. Most households in Johannesburg are built in unplanned settlements that lack most basic services, particularly in suburbs, leading to frequent flooding and several other social and economic negative externalities. In addition to disaster exposure, South Africa was classified among the less peaceful countries based on the Global Peace

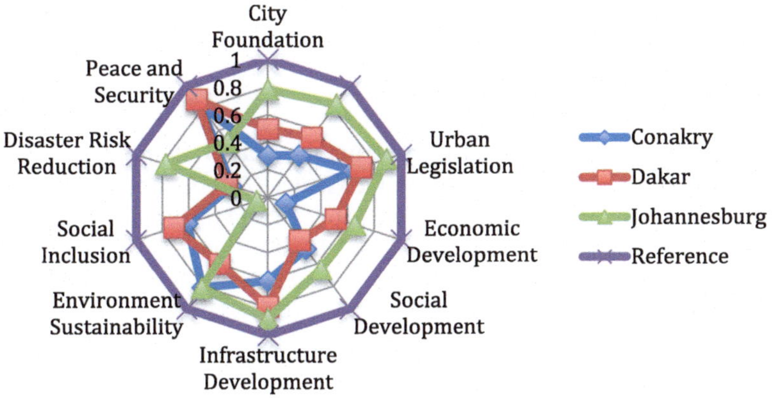

Fig. 16.27b Components of the SMRD. *Source* Provisional results from Mboup G. (ed.), 2018 Forthcoming Publication "Smart Economy in Smart African Cities", Springer

Index (GPI) ranking in 2015. The GPI scores are the aggregates of 23 qualitative and quantitative indicators across three thematic domains: the level of Societal Safety and Security; the extent of Ongoing Domestic and International Conflict; and the degree of Militarisation. Out of 163 countries, South Africa was ranked 136 with a score of 2.376 compared to the lowest score, which is estimated at 1.148 for Iceland and the highest score, which is estimated at 3.645 for Syria. In Africa, only four countries—Senegal, Mauritius, Botswana and Namibia—are among the top 50 peaceful countries globally[1]. With a high City Foundation, a high Human Development Index and a weak City Prosperity Index due to high inequalities, we have noted that high exposure to disaster coupled with high insecurity will reduce the level of Johannesburg Smart Metropolitan Regional Development Index. Though this a preliminary assessment of the SMRDI since data are not able for some sub-indices, rapid assessment of available data point to the fact Johannesburg SMRDI is low due do to historic structural failure on urban development, chronic inequality of opportunities coupled with high level of insecurity.

16.7 Spatial and Economic Design Strategies for a Smart Johannesburg Metropolitan Regional Development

This section consists of analysing urban policies and programmes put in place by national and local authorities for the transformation of Johannesburg to a Smart Metropolitan Region. The analysis will be done at two levels: (1) at the Gauteng Provincial level, and (2) at the Johannesburg city level. In addition to urban policies and programmes at the provincial level, each city, district and municipality which developed specific urban policies and programmes relevant for its territory will be looked at.

16.7.1 Urban Policies and Programmes at the Gauteng Provincial Level

Gauteng City Region Integrated Infrastructure Master Plan [53]
Towards sustainable urban development, the Gauteng has developed several urban policies and programmes including: the GCR Integ-rated Infrastructure Master Plan (GCR IIIM 2030), the National Development Plan Vision 2030, The GCR's programme of Transformation, Modernisation and Reindustrialisation, etc.
The GCR Integ-rated Infrastructure Master Plan (GCR IIMP 2030), as a comprehensive inter-sectorial plan, promotes sustainability through the efficient use of

[1]https://reliefweb.int/sites/reliefweb.int/files/resources/Global%20Peace%20Index%20Results%20Map_0.pdf. Downloaded 8 March 2018.

resources and the adoption of transformative technologies. It provides policy directions to various stakeholders including residents and investors and fosters collaboration across government and with the private sector [27]. This works in relation to broader national plans.

The National Development Plan Vision 2030 sets goals for infrastructure development to enable socio-economic transformation as follow:

- All South Africans have access to affordable, reliable and safe drinking water;
- Universal access to hygienic sanitation;
- At least 90% of South Africans have access to grid electricity, with the remainder accessing electricity from off-grid resources; and
- Improved productivity of infrastructure and increased levels of public and private investment to a combined 30% of GDP.

The GCR's programme of Transformation, Modernisation and Reindustrialisation requires infrastructure interventions to:

- Develop economic potential in individual corridors;
- Ensure growth and opportunities are equitably distributed across the GCR;
- Deliver housing and economic opportunities where most needed to realise radical spatial and economic transformation;
- Ensure infrastructure efficiency through coordinated planning, prioritisation and timing of delivery; and
- Unlock human capital in a dynamic, innovative, competitive and connected city region.

Realising an inclusive and competitive GCR requires that infrastructure development promotes equitable access, sustainability connectivity, spatial justice and mobility.

Development of Corridors Under the GCR

The GCR's space and economy are configured along five Development Corridors to drive radical economic transformation, decisive spatial transformation and re-industrialisation. Each corridor has distinct industries and comparative advantages:

- The Central Development Corridor is developed around the City of Johannesburg as the hub of finance services, ICT and pharmaceutical industries.
- The Eastern Development Corridor is developed around the economy of the Ekurhuleni Metro as the hub of manufacturing, logistics and transport industries.
- The Western Development Corridor includes the economy of the current West Rand District and the creation of new industries in support of green technology and manufacturing, new economic nodes and new cities.
- The Southern Development Corridor embraces the economy of the Sedibeng district and the creation of new industries to support and develop the agricultural sector, new economic nodes and new cities.

Fig. 16.28 Gauteng City Region's Corridor Projects. *Source* Gauteng Provincial Government, Republic of South Africa, Gauteng City Region Integrated Infrastructure Master Plan (GCRIIMP 2030)

- The Northern Development Corridor encompasses Tshwane as the nation's administrative capital city and the hub of the automotive sector, research, development, innovation and the knowledge based economy (Fig. 16.28).

The GCR IIMP identifies the following Apex projects as key to achieving the objectives of radical Transformation, Modernisation and Reindustrialisation, including: Water security; Regional sanitation schemes; New freight and logistics hubs at the international airport; New rail link; BRT; Outer freight link; and Upgrade of 19 railway stations.

It identifies smart projects such as:

- E-government: Modernising access to government services and rolling out the Gauteng Broadband Network to ensure GCR-wide access by 2019
- Improving healthcare through the construction of three district hospitals that will employ the latest technology and e-health systems
- Ensuring that all learners have access to tablet technology, broadband and smart classrooms to develop the skills required for a modern and innovative economy
- Reconfiguring the City of Johannesburg around high-density transport corridors and improving mobility through extending the Rea Vaya bus rapid transit system
- Rebuilding the western centre of Tshwane through infrastructure renewal, high density residential development and well-connected communities with access to world-class amenities
- Revitalising the economy of Sedibeng around a modern and sustainable node along the Vaal River and creating a more cohesive urban form.

The Gauteng's 25-year Integrated Master Plan also includes the transformation of Johannesburg into a sustainable, active city with accessible public spaces for all. Sustainability, active lifestyles, human interaction, and great public spaces are for improving environment and human health, advance economic development, and boost transportation systems.

Launched on 19 August 2013, the 25-year Integrated Transport Master Plan (ITMP25) is built on the principles of economic beneficiation. The ITMP25 proposes eight key interventions:

- Subsidised housing provisions;
- Land use densification in support of public transport;
- Reinforcing the passenger rail network;
- Extending the integrated rapid and road-based public transport networks;
- Strengthening freight hubs;
- Ensuring effective travel demand management;
- Mainstreaming non-motorised transport; and
- Ensuring continued mobility throughout the province;

The Gauteng programme of "Transformation, Modernisation and Reindustrialisation of the Gauteng city region, 2015–2030 includes ten pillar programmes as follow [49]:

- Radical economic transformation
- Decisive spatial transformation
- Accelerating social transformation
- Transformation of the State and governance
- Modernisation of the economy
- Modernisation of the public service
- Modernisation of human settlements and urban development
- Modernisation of public transport infrastructure
- Re-industrialise Gauteng
- Lead in Africa's new industrial revolution

Other projects include:

- **Revitalisation of township economies**

The GCR adopted a Township Economy Strategy with a detailed action plan to transform townships into vibrant economic centres of opportunity and job creation

- **Aerotropolis**

This GCR Project aims at leverage the potential of the three big airports in the GCR as economic engines for sustainable growth and job creation (to be anchored along ORTIA, Lanseria and Wonderboom Airports)

- **Gautrain expansion and extension**

This is to increase accessibility and improve mobility by expanding and extending the current network

- **Energy mix and Gauteng green agenda**

This is respond to the energy generation and distribution in the GCR. Gauteng Energy Steering Committee has been established, to work towards establishing a GCR energy mix mechanism, work with independent power producers, to generate and distribute energy, including outside of the grid.

- **ICT and Broadband Connectivity for the GCR**

This is to establish a hi-speed capacity network across Province in support of effective governance, service delivery and economic growth, and to encourage growth in ICT sector through services and infrastructure

- **Spatial Transport Transformation**
  - Metropolitan BRT Systems to increase accessibility by completing metro systems with potential extension
  - Metrorail Upgrade to implement 25 years ITMP specifically moving passenger transport from road to rail
  - Gauteng Freight and Logistics Hubs to o increase economic efficiency by improving the regional flow of goods through the expansion of freight & logistics hubs

- **Agritropolis**

This is to support the development of the Agri-sector and ensure job creation from increased agri-activity

- **Sustainable Human Settlements**

From RDP developments to the development of sustainable human settlements on well located land, with socioeconomic opportunities.

Realising an inclusive and competitive GCR requires that infrastructure development promotes equitable access, sustainability connectivity, spatial justice and mobility.

16.7.2 Urban Policies and Programmes at the Johannesburg City Level

To ease management of municipalities, the administration of Johannesburg metropolitan region is now organised into seven regions following the creation of the post-apartheid City of Johannesburg Metropolitan Municipality in 2000. "Each region is operationally responsible for the delivery of health care, housing, sports and recreation, libraries, social development, and local community-based services. Each region will develop its own Local Integrated Development Plans (LIDPs). A LIDP guides a region's future development. For this reason, the LIDP zones closely follow the boundaries of the regions. LIDPs deal with city development, management and growth over a five to 10-year period. While they deal with local issues, they take an integrated approach to issues such as transportation, housing and environmental management. An overall Metropolitan IDP looks at the bigger picture and ensures that LIDPs don't conflict or lead to wasted resources. LIDPs are revised annually so as to respond to changing conditions both locally and at a city level" [55].

The Johannesburg Smart City Programme

Despite several urban policies and programmes put in place for an inclusive sustainable urban development in Johannesburg, the metropolitan is still facing enormous challenges such as [56]:

- Slow economic growth marked by: (a) Structural weaknesses in the global economy coupled with a weak Rand; (b) Rising interest rates and inflation; (c) Income inequality and poverty remain prevalent; and, High Unemployment rate, particularly among young people; informal economy;
- Social exclusion characterised by: (a) exclusion of poor communities from access to housing and land; (b) mobility is still a challenge for the urban poor;
- Urban violence and insecurity: crime rates are still among the highest in the world;
- Climate Change: the city is facing climate change with Johannesburg being ranked 13th in the world of Greenhouse Gas (GHG) emitters and largest GHG City in South Africa;
- Corruption and lack of transparency are still on course.

Considering these multiple challenges, the city of Johannesburg has put in place a holistic smart city programme to be implemented at 85% by 2021. This programme encourages innovation and efficiency; preservation of resources for future generation; resource sustainability with, for instance, 30% diversion in waste disposed by landfill. The programme is extended to natural resources, environment pollution and carbon emissions [56]. For the year 2017/2018, the city of Johannesburg has also elaborated ten points for urban development including: (1) ensuring that the entire city embraces the environment of a new coalition government; (2) Promoting economic development and attract investment towards achieving 5% economic growth; (3) Ensuring pro-poor development that addresses inequality and provides meaningful redress; (4) Creating a culture of enhanced service delivery; (5) Creating a sense of security through improved public safety; (6) Creating an honest and transparent city that fights corruption; (7) Creating a city that responds to the needs of residents; (8) Enhancing financial sustainability; (9) Encouraging innovation and efficiency through programmes such Smart City and; (10) Preserving the city resources for future generation [56].

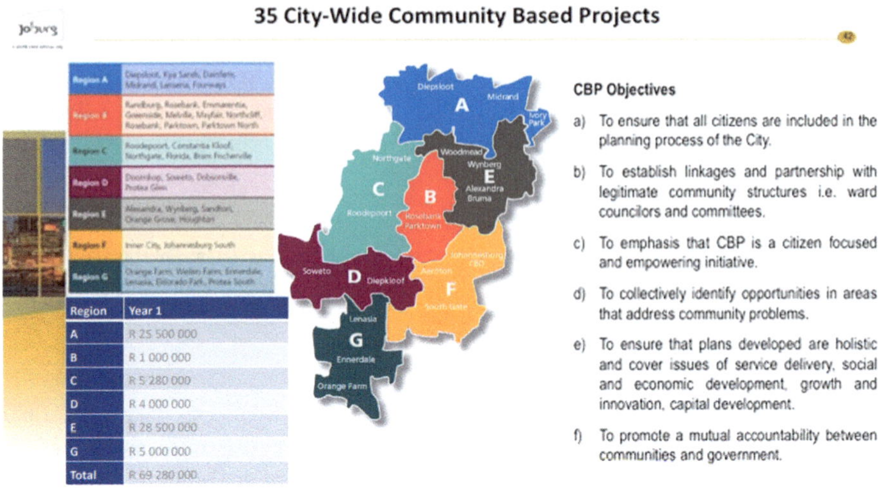

Fig. 16.29 35 Citywide Community Based Projects. *Source* City of Johannesburg, 2017–18 Draft Integrated Development Plan Review

These ten points will be delivered through the following five pillars:

- PILLAR 1: Grow the economy and create jobs
- PILLAR 2: Enhance quality of life by improving services and taking care of the environment
- PILLAR 3: Advance pro-poor development that provides meaningful redress
- PILLAR 4: Build caring, safe and secure communities
- PILLAR 5: Institute an honest, responsive and productive government.

This will also be based on city-wide community based projects as presented in Fig. 16.29 "35 City-Wide Community Based Projects.

At the spatial level, the transformation of the city of Johannesburg will consist of:

1. Compact city—combining density, diversity, proximity and accessibility, reducing distances, travel times and costs, bringing jobs and social amenities to single use, marginalised residential areas, reducing energy consumption and infrastructure costs.
2. Inclusive city—ensuring balanced service provision (hard and soft) and opportunities for all by diversifying land uses, promoting social mixing and bridging social, spatial and economic barriers.
3. Connected city—enhancing public transit and ICT infrastructure at provincial and urban scales to re-connect the city, starting from 'the Corridors of Freedom' to street and neighbourhood-level connectivity.
4. Resilient city—building a metropolitan open space system as a protection buffer, preserving valuable green infrastructure and areas of high agricultural potential,

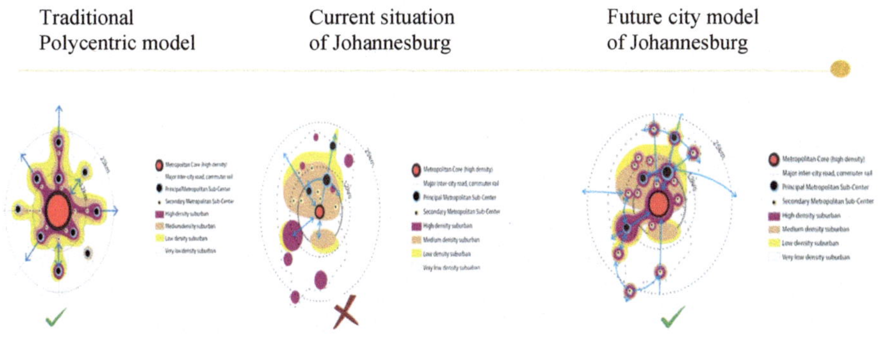

Fig. 16.30 Spatial development framework-future outlook. *Source* City of Johannesburg, 2017–18 Draft Integrated Development Plan Review

promoting sustainable energy use, reinforcing the urban development boundary and protecting biodiversity resources.

5. Generative city—focusing investment in transformation areas and nodes towards: achieving positive social, economic and environmental returns on investment; spurring economic growth and job creation and enhancing public space and promoting sustainability (social, environmental and economic) (Fig. 16.30).

In response to multiple challenges faced by its citizens, the city of Johannesburg has also developed the 2040 Growth and Development Strategy (GDS) initiated in August-September 2011. The Joburg 2040 GDS lays the foundation for multi-level, multi-scalar and integrated responses to the challenges the city faces. It aims to strengthen the adaptive capacity of the City and its citizens in order to become more resilient to potential and unpredictable futures. The GDS is closely tied to the five-yearly Integrated Development Plan (IDP). The GDS charts the long-term strategic course of the City, and makes some of the bigger, overarching decisions about what to emphasise if Joburg is to accelerate economic growth and human development. The IDP defines where we want to be after five years, and how we intend incrementally achieving long-term goals. The outcomes originate from the 2040 GDS.

Outcome 1: Improved quality of life and development-driven resilience for all
The city envisages a future that presents significantly improved human and social development realities, through targeted focus on poverty reduction, food security, development initiatives that enable self-sustainability, improved health and life expectancy, and real social inclusivity. By 2040, the city aims to achieve substantially enhanced quality of life for all, with this outcome supported by the establishment of development-driven resilience.

Outcome 2: Provide a resilient, liveable, sustainable urban environment—underpinned by infrastructure supportive of a low-carbon economy
The city plans to lead in the establishment of sustainable and eco-efficient infrastructure solutions (e.g. housing, eco-mobility, energy, water, waste, sanitation and

information and communications technology), to create a landscape that is liveable, environmentally resilient, sustainable, and supportive of low-carbon economy initiatives.

Outcome 3: An inclusive, job-intensive, resilient and competitive economy that harnesses the potential of citizens
The city of Johannesburg will focus on supporting the creation an even more competitive, 'smart' and resilient city economy, when measured in relation to national, continent and global performance. The city will promote economic growth and sustainability through the meaningful mobilisation of all who work and live here, and through collaborating with others to build job-intensive long-term growth and prosperity, from which all can benefit.

Outcome 4: A high performing metropolitan government that pro-actively contributes to and builds a sustainable, socially inclusive, locally integrated and globally competitive Gauteng City Region
The City of Johannesburg finally envisages a future where it will focus on driving a caring, responsive, efficient and progressive service delivery and developmental approach within the GCR and within its own metropolitan space, to enable both to reach their full potential as integrated and vibrant spaces.

16.8 Conclusion

Since the end of the Apartheid, Johannesburg has become the main destination for migrants from other South African provinces as well as from other African countries. One third (35%) of Gauteng residents were born in other provinces or in other countries. The Greater Johannesburg Metropolitan Area represents 14% of the national population. It is an engine of South Africa with a share of 18% of the national GDP which itself represents 24% of the African GDP. This is the reason Johannesburg is considered a global African city. Johannesburg also scored the Human Development Index in the country.

The City Foundation of Johannesburg has historically been guided by a segregationist ideology manifested through institutions and laws, urban planning and design, and access to basic services and amenities. During the Apartheid, the majority of township areas designated for non-white people were situated between 25 and 30 km away from the central business district, leading to significant transport challenges. The dismantling of apartheid in the early 1990s came with calls for inclusive development in South Africa. Different models have however been adopted to make the city more inclusive post apartheid, the key ones being on participatory planning and design; development of inclusive public spaces; improvement in basic infrastructure provision and public transport particularly to the poor neighbourhoods and; development of policies that encourage inclusive human settlement and trade.

Today Johannesburg is at the heart of the Gauteng province' cluster of economic activity.. Under the cluster of cities of the Gauteng city region, the city of Johannesburg is the Financial Hub, the Ekurhuleni Metro the Manufacturing Hub, the West Rand District the agribusiness/agroprocessing and green and blue economy, Sedibeng District Municipality with the New economy to be anchored on steel, Vaal river, tourism, agriculture linked to Sasolburg, and the City of Tshwane being the Administrative Capital has become a key driver of knowledge, innovation and automotive industry development. The region promotes spatial transport transformation through Metropolitan BRT Systems, Metrorail Upgrade and Gauteng Freight and Logistics Hubs.

In addition to the city-region projects, cities themselves are also developing their specific sustainable urban development projects and programmes. For instance, to ease management of municipalities, the administration of city Johannesburg is now decentralised into 7 regions following the creation of the post-apartheid City of Johannesburg Metropolitan Municipality in 2000. "Each region is operationally responsible for the delivery of health care, housing, sports and recreation, libraries, social development, and local community-based services. Each region will develop its own Local Integrated Development Plans (LIDPs). A LIDP guides a region's future development. While they deal with local issues, they take an integrated approach to issues such as transportation, housing and environmental management.

Despite several urban policies and programmes put in place for an inclusive sustainable urban development in Johannesburg, the metropolitan is still facing enormous challenges such as slow economic growth marked by High Unemployment rate, particularly among young people; Exclusion of poor communities from access to housing and land; Mobility is still a challenge for the urban poor; Urban violence and insecurity; and Climate change with Johannesburg being ranked 13th in the world of Greenhouse Gas (GHG) emitters and largest GHG City in South Africa.

Considering these challenges, the City of Johannesburg has put in place a holistic smart city programme to be implemented at 85% by 2021. This programme encourages innovation and efficiency; preservation of resources for future generation; resource sustainability with, for instance, 30% diversion in waste disposed by landfill. For the year 2017/2018, the city of Johannesburg have elaborated ten points for urban development including to: (1) Ensure that the entire City embraces the environment of a new coalition government; (2) Promote economic development and attract investment towards achieving 5% economic growth; (3) Ensure pro-poor development that addresses inequality and provides meaningful redress; (4) Create a culture of enhanced service delivery; (5) Create a sense of security through improved public safety; 6) Create an honest and transparent City that fights corruption; (7) Create a City that responds to the needs of residents; (8) Enhance financial sustainability; (9) Encourage innovation and efficiency through programmes such Smart City and; (10) Preserve our resources for future generation.

The future of the city is driven by many efforts of state intervention towards building a smart city and a smart province. However, the effort for a successful smart future will lie in collaborative and collective function between all stakeholder involved. The government, business and community organisations therefore should come together to build a smart urban future. A new form of networking with technology, with knowledge, action and most importantly in spirit is crucial in the making of a smart future city and a smart province.

References and Notes

1. Rodrigue et al., 2013. https://people.hofstra.edu/geotrans/eng/ch1en/ch1menu.html
2. Harrison, 2013. By 1995, the GCR consisted of "metropolitan cities, secondary cities, towns and smaller service centres, well connected by road and rail
3. Sihlongonyane MF (2018) Chapter 2—The economic drivers of urban change in the Gauteng City-Region: past, present and future. In Cheruiyot K (ed), The changing space economy of City-Regions, Springer AG, Berlin
4. Gauteng Province Government, 30 January 2015, Transformation, Modernisation and Reindustrialisation of the gauteng city-region: 2015-2030
5. City of Johannesburg: 2012/16 Integrated Development Plan: 2013/14 review
6. Statistics South Africa, Census 2011
7. United Nations, Department of Economic and Social Affairs, Population Division (2014). World Urbanization Prospects: The 2014 Revision, Highlights (ST/ESA/SER.A/352)., p. 14
8. City of Johannesburg, the 2040 GDS
9. Conroy & Berke, 2004; Conroy & Gordon, 2004; Wild & Marshall, 1999
10. Chen Y, Hamilton A, Borning A (2010) Modelling & matching and value sensitive design: two methodologies for e-planning systems development
11. Marcinkowski, 2007
12. Silva CN (2010) The E-Planning Paradigm –Theory, Methods and Tools: An Overview. In Carlos Nunes Silva, 2010. Handbook of Research on E-Planning: ICTs for Urban Development and Monitoring (http://www.igi-global.com/book/handbook-research-planning/41793) (http://www.igi-global.com/viewtitlesample.aspx?id=43177&ptid=41793&t=the+e-planning+paradigm+%E2%80%93+theory%2c+methods+and+tools%3a+an+overview), dowloaded, 9 April 2017. The Use of Social Media for Urban Planning: Virtual Urban Landscapes Created Using Twitter. Data
13. Yigitcanlar T (2013). Planning online: a community-based interactive decision-making model (http://www.igi-global.com/viewtitlesample.aspx?id=43178&ptid=41793&t=planning+online%3a+a+community-based+interactive+decision-making+model) downloaded 9 April 2017
14. Mhangara P, Mudau N, Mboup G, Mwaniki D (2016) Transforming the City of Cape Town from an Apartheid City to an inclusive smart city. In Kumar V, Diyala B (eds), Smart economy in smart cities, Springer, Berlin
15. President Thabo Mbeki promoted Johannesburg as 'an African city'
16. Estimates from the World Bank quoted by Honey 2004 cited by Mfaniseni F. Sihlongonyane, 2018
17. Hanival S, Maia J (2008) An overview of the performance of the South African economy since 1994. Pretoria: The Presidency, Government of South Africa
18. Landman 2004 cited by Sihlongonyane, 2018
19. Turok, 2012, cited by Sihlongonyane, 2018
20. See Mboup, 2017. Land the Hidden assets in African Cities, World Bank's Land and Poverty Conference 2017
21. Turok (2012) Urbanisation and development in South Africa: economic imperatives, spatial distortions and strategic responses. IIED and UNFPA, Urbanization and Emerging Population Issues, Working Paper 8
22. Sihlongonyane, 2018

23. Blakely E, Snyder M (1999) Fortress America: Gated Communities in the United States (Washington DC, Brookings Institution)
24. Horn A (2010) Telling Stories—A history of growth management in the gauteng province (South Africa). Eur Spat Res Policy 17(2):41–54
25. Van Veuren EJ (2001) Transforming cultural villages in the spatial development initiatives of South Africa. S Afr Geogr J 83(2):137–148
26. Zack, 2004 cited by Sihlongonyane, 2018
27. The Gauteng Infrastructure Coordinating Council (GICC) is an intergovernmental GCR Institution tasked with the overall coordination and delivery of infrastructure. Chaired by the Premier, the council oversees and monitors the implementation of the GCR IIMP 2030 and ensures alignment with the Presidential Infrastructure Coordinating Commission (PICC) and Strategic Infrastructure Projects (SIPs)
28. As defined by Sassen, 2003 regarding New York and London as global cities
29. Simkins C (2010) Population growth, migration, economic growth and poverty in Gauteng since 1970. A Southern Africa Labour and Development Research Unit Working Paper No. 47. Cape Town: SALDRU, University of Cape Town
30. see also Kevin A. Bryan, Brian D. Minton, and Pierre-Daniel G. Sarte
31. Kevin A. Bryan, Brian D. Minton, and Pierre-Daniel G. Sarte, The Evolution of City Population Density in the United States. Economic Quarterly—Volume 93, Number 4—Fall 2007—Pages 341–360; (Chatterjee and Carlino 2001, Lucas and Rossi-Hansberg 2002)
32. Betraud et al
33. Demographia, 2016 edition
34. http://2013.legacy.gcro.unomena.net/spaceandmobility/populationdensity
35. Gauteng Parternship Fund, Gauteng Infrastructure Funding Summit Mega Projects; 2017
36. http://www.gauteng.gov.za/government/departments/office-of-the-premier/Documents/GSDF%202030.pdf
37. Hloni Masego (2017) Plan to build 800 k houses. The New Age, May 23, 2017
38. http://www.treasury.gov.za/documents/provincial%20budget/2016/3.%20Provincial%20Appropriation%20Acts/5.%20Guide%20to%20the%20Budget/GT/Gauteng%20-%20Social-Economic%20Review%20and%20Outlook.pdf
39. Mboup G (2013) Streets as Public Spaces and Drivers of Urban Prosperity, UN-Habitat's publication, November 2013, Nairobi; Mboup G (2015). Streets as Public Spaces and Drivers of Sustainable, Inclusive and Prosperous Cities in Africa, World Bank's Land and Poverty Conference, March 2015, Washington DC; Mboup G (2015) City Foundation—Key for Sustainable, Inclusive and Prosperous Cities, World Statistics Congress, July 2015 Rio de Janeiro, Brazil
40. van Ryneveld P (2010) Assessment of public transport in SA cities. Institute for Transportation and Development Policy, New York
41. Mboup G (2016) Smart infrastructure development will make smart cities. Promoting Smart Transport and ICT in Dakar, in Smart Economy in Smart Cities, Springer, September 2016
42. United Nations (2015b) Adoption of the Paris Agreement, Conference of the Parties, twenty first session, Paris, 13 November to 11 December 2015
43. UN-Habitat, 2013a, Op. Cit
44. GDPTRW 2002
45. http://www.gauteng.gov.za/government/departments/office-of-the-premier/Documents/GCR%20IIMP%20Summary(final).pdf
46. www.itssa.org
47. GCR-IIMP
48. Heather Allen, 2012, Africa's First Full Rapid Bus System: the Rea Vaya Bus System in Johannesburg, Republic of South Africa, Case study prepared for the UN-Habitat 's Global Report on Human Settlements 2013, Nairobi, Kenya
49. Republic of South Africa, 2010
50. Mboup G (2014) Urban quality of life—concepts and measurements. http://deliver2030.org/?wp_blog=urban-quality-of-life-concepts-and-measurements. Accessed 9 November 2017

51. UN-Habitat, 2012. Prosperity of Cities (Author: Mboup et al.)
52. Mboup G (2013) Streets as public spaces and drivers of urban prosperity, UN-Habitat
53. Gauteng Provincial Government, Republic of South Africa, GCR IIMP 2030
54. Gauteng Province Government, 30 January 2015, Transformation, Modernisation and Reindustrialisation of the Gauteng city region, 2015-2030. The development of what became Gauteng province in 1994 bears unique industrial and economic features—an urban conurbation and clustering of cities and towns. This is what makes Gauteng a City Region, like the City states such as Singapore
55. City of Johannesburg, 2017–18 draft Integrated Development Plan Review
56. City of Johannesburg, Op. cit

Part XVII
USA, Pittsburgh

Chapter 17
Metropolitan Regional Scale Smart City Approaches in a Shrinking City in the American Rust Belt—Case of Pittsburgh, Pennsylvania

Sudeshna Ghosh, Sweta Byahut and Calvin Masilela

Abstract This chapter describes the smart city example from the City of Pittsburgh, Pennsylvania in the United States. Pittsburgh is a typical city in the rust belt of America that faced decline and shrinkage due to de-industrialization and globalization forces. The economic transformation of Pittsburgh into a post-industrial regime has brought its own set of unique challenges and represent a unique perspective of urban and regional planning approaches that met with varying degrees of success. The purpose of this chapter is to describe the smart city approaches in the context of a Shrinking City that have been shaped by forces of industrial decline, population and job loss, abundance of derelict industrial sites, suburban exodus and racial conflicts, along with shared notions of fighting back to comeback and succeed in the New Economy. After witnessing a prolonged state of industrial and economic decline since the 1980s, Pittsburgh had no choice but adopt planning strategies that would restructure its economy, and transform its large-scale abandoned brownfield sites to create places of vibrant economy and community. Pittsburgh's proactive policies to involve local communities in reusing old and vacant lands for economic transformation and urban greening, as well as using green infrastructure for storm water management and improving the quality of its waterways, resulted in many success stories throughout the region. Thus, our study of Smart City strategies in Pittsburgh focuses on the successes and failures of economic transformation, brownfield redevelopment and urban greening. While it

---

S. Ghosh (✉) · C. Masilela
Department of Geography and Regional Planning, Indiana University of Pennsylvania, Indiana, PA 15705, USA
e-mail: Sudeshna.Ghosh@iup.edu

C. Masilela
e-mail: Calvin.Masilela@iup.edu

S. Byahut
Department of Political Science, Auburn University, Alabama AL 36849, USA
e-mail: sweta.byahut@auburn.edu

T. M. Vinod Kumar (ed.), *Smart Metropolitan Regional Development*, Advances in 21st Century Human Settlements,
https://doi.org/10.1007/978-981-10-8588-8_17

is in the process of coming back, the region still faces several challenges, such as a large number of vacant and distressed lands, aging population and infrastructure, which needs to be addressed in the coming decades.

Keywords Deindustrialization · Economic resiliency · Brownfield redevelopment Green infrastructure · Smart city

17.1 Introduction

The concept of Smart City (SC) entails the notion of promoting better quality-of-life for residents and sustainability with increase in utilization of advanced Information and Communication Technology (ICT). In urban and regional planning context, SC strategies are increasingly becoming a popular paradigm with its focus on sustainable urban development, sustainable economic growth, environment and social sustainability, and advanced infrastructure capabilities [1]. While there is a wide range of understanding and applications of SC strategies, scholars, practitioners, policy-makers and residents commonly agree on the prevalent use of ICT in such strategies [2]. Nevertheless, the global trends in SC strategies and approaches demonstrate increasing variations that makes it difficult to identify common trends [3].

Our objective in this chapter is to study the aspects of SC approaches at a metropolitan regional scale in the city of Pittsburgh, Pennsylvania (PA) in the United States (US). The story of Pittsburgh entails a typical urban transformation story of a post-industrial city from the Global North, and thus, represents a unique perspective of urban and regional planning approaches with varying degrees of success that can be studied under the domain of SC approaches. The common trends of SC approaches in such type of cities have been shaped by forces of industrial decline, population and job loss, abundance of derelict industrial sites, suburban exodus and racial conflicts, along with shared notions of fighting back to comeback and succeed in the New Economy.

Similar to the industrial cities from Europe, the mid to late 20th century rust-belt cities in the US witnessed a trend of decline in their manufacturing base. With global shift of industrial production to the developing regions of the world and the New International Division of Labor (NIDL), the US rust-belt cities were compelled to rethink their urban and economic development strategies to compete in the 21st century. The dominance of advanced service and technology-based jobs in the New Economy resulted in adoption of economic transformation plans in the old industrial cities that would replace manufacturing jobs with advanced service sector jobs. However, there were more challenges than opportunities in the process of doing so, which set the context of urban and regional planning practices in the rust-belt cities.

Pittsburgh, one such example of a rust-belt city, was once known as the "Steel City" due to its production capacity of raw steel in the world economy (Fig. 17.1). Abundance of natural resources, such as coal, timber, iron, and limestone, and navigable waterways—the Allegheny, Monongahela and Ohio rivers, helped

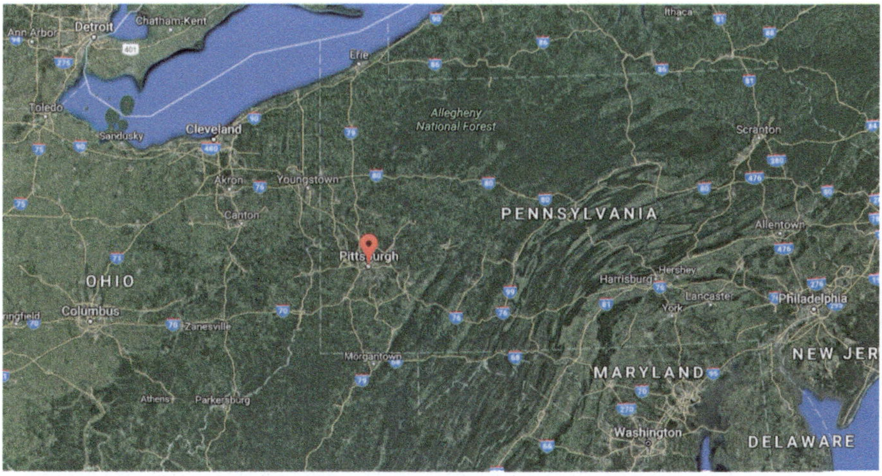

Pittsburgh is located in the southwest region of Pennsylvania; other old industrial cities in close proximity are: Cleveland, Ohio; and Detroit, Michigan, which are still struggling to comeback in the New Economy. (source: Google Maps, 2017).

Fig. 17.1 Map showing location of Pittsburgh, PA

Pittsburgh emerge as the center of steel industries in the US in between the late 19th–early 20th century (Fig. 17.2). With such comparative advantage of natural resources and navigable waterways, the city burgeoned with large-scale steel mills since the 1870s, and an economic and population base to support these activities. By 1970s, Pittsburgh produced about one-third to half of total steel production in the US [4, 5].

The era of deindustrialization witnessed a geographic shift of steel-manufacturing activities towards lesser developed countries, such as India and China. With international shift of crude steel production, the US witnessed a sharp decline in steel production. In 1973, during its peak the US produced nearly 137 million metric tons (mmt) of raw steel, which sharply declined to 67.7 mmt in 1982 [4], severely impacting the steel mills of Pittsburgh region. Pittsburgh lost its position as the Steel City, and witnessed a continuous trend of decline in its manufacturing-based economy and population since the 1980s.

Decline of its steel-based manufacturing base, led to several other consequences, such as decline in number of jobs, loss of population, brownfield sites, and vacant and abandoned properties [6]. Since the 1980s, the challenge of massive brownfield sites, property abandonment, declining tax base, and increasing poverty and crime, led Pittsburgh to be known as an example of a "Shrinking City" in the rust-belt region. The planners and policy makers had no choice but to focus on adoption of newer sets of strategies that could transform Pittsburgh and make it competitive in the New Economy. The late 20th century Pittsburgh increasingly became successful in adopting many bold policies and strategies that transformed its local economic base towards advanced service sectors: healthcare, higher education, technology, research and development, banking and finance (Fig. 17.3).

Monongahela Wharf, Pittsburgh riverfront ca. 1900-1927.
(source: Historic Pittsburgh Image Collection, Archive Service Center at University of Pittsburgh)

Golden Triangle and Old Point Bridge, Pittsburgh riverfront ca 1900-1910.
(source: Pittsburgh Historic Images, Brookline Collection, http://www.brooklineconnection.com/history/Facts/Point1900.html)

Fig. 17.2 Historic images of Pittsburgh during the Steel era

Pittsburgh Golden Triangle Skyline, 2013: Confluence of Allegheny and Monongahela rivers. (source: Dr. Kevin J. Patrick)

Fig. 17.3 Contemporary image of Pittsburgh in the new economy era

Dating back to the 1980s, Pittsburgh has been adopting various right-sizing strategies to stabilize declining neighborhoods. As a result, the city is able to provide a relatively high quality-of-life to its residents within affordable prices, despite abandonment, blighted areas, crime and poverty [7]. These strategies mostly focused on economic diversification, regeneration of brownfields, urban greening and green infrastructure practices, which form the underlying principles of sustainability in SC concepts. Since the 2000s, with high concentration of employers in the areas of medical research, such as the University of Pittsburgh Medical Center (UPMC); higher education, such as Carnegie Mellon University and University of Pittsburgh; banking and financial sectors, such as Pittsburgh National Bank (PNC); and increasing number of other Fortune 500 companies; Pittsburgh is becoming more successful in transforming its economy. A more recent example of Pittsburgh's increasing competitiveness as an innovation center in the 2010s is Uber choosing Pittsburgh as its research center for experimentation with autonomous cars or self-driving robotic vehicles.

In this chapter, we aim to explore the challenges, successes, and failures of SC strategies in the metropolitan region of Pittsburgh. While it is difficult to distinctly identify the common trends in SC approach at a global scale, it is irrefutable that smart cities are increasingly becoming popular in the US. A recent report by National League of Cities highlighted that in the US "66% of cities have invested in some sort of smart city technology" and "25% of cities that have not invested in smart city technology are exploring it" [8]. At a global scale, many SC programs boast extensive dependency on advanced ICT [2], however, majority of scholars argue that ICT is only one aspect of SC approach. Extensive ICT-based SC programs are indeed more common in the rapidly industrializing regions of the world, such as China and India. However, the strategies to attain smartness in the US rust-belt regions that experienced decades of decline are observed to be unique; emphasizing on lesser ICT dependent ideas of place making, social inclusion, civic engagement, and entrepreneurship, and integrating them with ICT applications in areas such as urban mobility and green energy.

Our broad goal is to analyze the degree to which Pittsburgh's economic and urban restructuring reflects the principles of SC strategies. We follow the six building blocks of a SC System—smart people, economy, mobility, environment, living and governance, outlined by Kumar and Dahiya [3], as they apply to Pittsburgh. Specifically, our research explores and analyzes SC strategies across the above-mentioned six themes in the Pittsburgh metropolitan area, within two distinctly identified regional planning efforts: (1) Planning for economic resiliency and redevelopment of brownfields, and (2) Planning for urban greening and green infrastructure. In addition, we explore the principles of smart growth that essentially intersect with the themes of SC strategies within these planning efforts.

17.1.1 Study Area and Its Regional Context

The study area delineated for this research includes the Metropolitan Statistical Area (MSA) of Pittsburgh, which comprises of 7 counties—Allegheny, Armstrong, Beaver, Butler, Fayette, Washington and Westmoreland (Fig. 17.4). Among these counties, Allegheny County, home to the City of Pittsburgh, serves as the major population and economic center in the region, and is one of the places where majority of the advanced services and knowledge-based jobs are located. The southern part of Butler county is recently witnessing high growth of advanced services and high-skilled jobs, partly because of its proximity to the City of Pittsburgh and Interstate connectivity. Nevertheless, many communities within Armstrong, Beaver, northern Butler, Washington, Westmoreland, and Fayette counties continue to struggle with issues of brownfields, aging and declining population (Table 17.1). There has also been a proliferation of Marcellus Shale-based natural gas extraction activities in the region since 2005, specifically in Washington County, which brought a short-term boom in the oil and gas industries although they are vulnerable to bust in the future.

The seven-county metro region also consists of 460 municipalities classified into Cities, Townships and Boroughs (Fig. 17.4) [9]. The political landscape, thus, represents a fragmented structure of local governance, and exhibit the challenges of implementing urban policies and strategies at a regional scale. Thus, planning for economic resiliency, brownfield redevelopment, and green infrastructure require overcoming barriers of local governments and jurisdictions through a co-operative model of regional governance and through stronger participation of residents and community organizations. We explore the spatial extent of Pittsburgh's successful planning efforts that essentially follows the principles of smart growth and smart city paradigms.

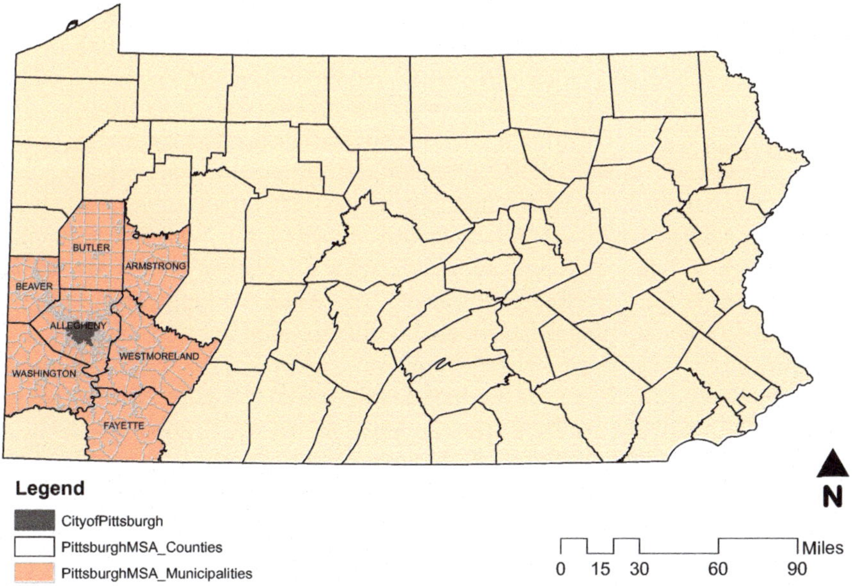

Map of Pennsylvania showing Pittsburgh Metropolitan Statistical Area (MSA) with the City of Pittsburgh at its core. (source: TIGER/Line Shapefiles, U.S. Census Bureau, 2016)

Fig. 17.4 Map of Pittsburgh Metropolitan Region in Pennsylvania

Table 17.1 Population and economic conditions in the Pittsburgh Metropolitan Region in 2015–16

	Population (2016)	Total employment (2015)	Median household income (2016)	Poverty rate (2016) (%)	Median age (2015)
Allegheny	1,225,365	701,226	$54,357	11.5	41.0
Armstrong	66,486	14,386	$45,879	13.8	45.6
Beaver	167,429	49,345	$51,887	9.7	44.8
Butler	186,847	81,472	$63,345	7.3	42.7
Fayette	132,733	37,436	$40,511	17.5	44.2
Washington	207,981	84,328	$57,534	9.7	44.2
Westmoreland	355,458	126,336	$54,142	9.8	46.8

Source U.S. Census Bureau, 2015–16, data accessed from https://datausa.oi

17.1.2 Rise and Fall of the Steel City: Historical Background of Urban Planning in Pittsburgh

Pittsburgh's recent history could easily be seen as a story of persistent loss and decline. From its humble beginnings in 1760 as a fort community of 149, Pittsburgh rose during its boom epoch of iron and steel to be the sixth largest city in the US.

By 1950, City of Pittsburgh's decennial population had peaked at 676,806. The suburban exodus that faced all industrial cities was exacerbated in Pittsburgh by the gradual collapse of the steel industry. By 2010, the population of the city had fallen 305,704, a loss of more than 50% of its peak population (Fig. 17.5) [10].

Pittsburgh can also be seen as a tale of two eras in the history of US urban planning. In the early to mid-1900s, when Pittsburgh was thriving with its steel mills and growing in employment and population, the urban environment and quality-of-life started deteriorating very rapidly. In 1944, the Wall Street Journal rated Pittsburgh a "class D" city, with little hope of recovery [11]. By 1950, Pittsburgh's polluted air, riverine environment, sprawling working class, and tangled maze of streets and bridges had singled it out as one of the most blighted and reviled cities in the US. However, Pittsburgh was not willing to resign without a fight. A private-political alliance arose in the early 1950s that gave birth to one of the earliest efforts at what we have come to know as the era of urban renewal.

In 1946, under the guidance of R. K. Mellon, David Lawrence became the first director of the Pittsburgh Urban Redevelopment Authority (URA). It is some of the more recent projects of the URA that this chapter focuses on, but their approach and substance differ substantially to the URA's early years.

The early years of the URA were years of clearance and demolition. Pittsburgh's "Renaissance I" began with the Point Park and Gateway Center project. Point Park at the confluence of the Monongahela and Allegheny rivers was an area of blight and flooding. With the financing of the Equitable Life Assurance society, a thirty-six-acre park and a twenty-three-acre redevelopment site became known as the Gateway Center. Between 1950 and 1960s seven high rise office buildings, a Hilton Hotel, a residential apartment building, and an underground garage filled the once blighted space [12]. Had the Renaissance stopped there it might only have been a story of partially filled office buildings and an enhanced city center, but it did not.

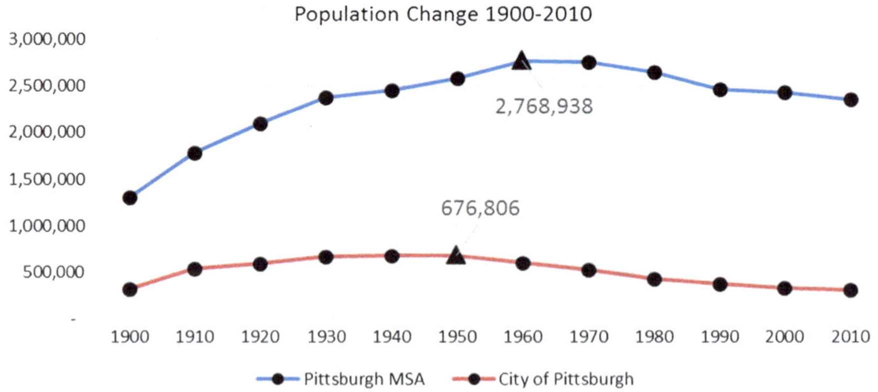

Fig. 17.5 Population change, Pittsburgh city and metropolitan region. *Source* Decennial Census 1900–2010, U.S. Census Bureau

In the Eisenhower era of demolition and highways, the URA set its sights on the Lower Hill district, a predominantly African-American neighborhood. The development of a convention center and civic arena ultimately displaced 1300 buildings, 413 businesses, and 8000 people from the Hill (Fig. 17.6). The population of the Hill fell from 17,334 in 1950 to 2459 in 1990 [11]. The loss of a stable black neighborhood and the concentration of emigres in larger black neighborhoods contributed to making Pittsburgh one of the most segregated cities in the country [13]. The reaction to this form of urban renewal was both local and national. One of the earliest proponents of preserving the legacies of a city and making them livable, Jane Jacobs, wrote in The Economy of Cities that: "so many irrelevant things have been tried here [Pittsburgh]…in immensely expensive urban-renewal and highway programs that have not helped the economy at all" [14]. And indeed, Pittsburgh's population continued to decline precipitously, stores closed, and business fled for the next 30 years.

Failure of urban renewal efforts and continuing trend of manufacturing decline deeply impacted Pittsburgh through the 1980s, 90s and 2000s. The region sharply lost more than 133,000 manufacturing jobs within only 8 years, between 1979 and 1987 [15]. This led to an era when communities after communities experienced financial difficulties and struggled to bounce back in the New Economy. In between 1987 and 2004, five municipalities in the Pittsburgh metropolitan region including the City of Pittsburgh were identified and listed as distressed communities experiencing "severe" financial difficulties under Act 47 of 1987, the Municipalities Financial Recovery Act of Pennsylvania; placing them under State assistance for recovery strategies. These municipalities were: City of Aliquippa (Beaver), Borough of Braddock (Allegheny), Borough of Rankin (Allegheny), City of Duquesne (Allegheny), and City of Pittsburgh (Allegheny), arranged in time-series [16]. Since then, several planning initiatives have been undertaken from local to regional level with assistance from the local, state and federal governments for economic and urban restructuring.

17.1.3 Rise in the New Economy: Present Conditions and Urban Planning in Pittsburgh

The process of economic restructuring was challenging along with the issues of reclaiming, and redeveloping major brownfield sites, where the huge steel mills once stood [6]. Levels of environmental contamination often varied from site to site. While federal level programs provided funding and strategy-framework to clean up contaminated sites, stigma associated with real or perceived levels of contamination often posed difficulty in successful redevelopment of brownfield sites [17]. Nevertheless, many successful examples can be found where new office spaces, research centers, and mixed-used developments were developed over time.

The 1990s and 2000s also brought a change in the planning and a change in the URA's approach to renewal. The early efforts at renewal still left large areas of

Aerial photograph of Lower Hill District in 1956 superimposed with urban renewal plan and the proposed Civic Arena. (source: Fullilove, 2016, https://www.nap.edu/read/23576/chapter/4)

Lower Hill District with the Civic Arena post urban renewal in 1961. (source: Fullilove, 2016, https://www.nap.edu/read/23576/chapter/4)

Fig. 17.6 Urban renewal of Lower Hill District in the 1950s–60s abandoned factories and decaying neighborhoods. But in planning's new paradigm, brownfields are also treated as an "opportunity". Local governments today to restore the vitality of urban life have turned to a notion of redevelopment that includes infill, mixed use, private-public partnership, and walkability; a framework of smart growth. Smart growth is also smart development. Blakely and Bradshaw tell us that "industry and business regard livability as an important locational factor," and that governments need to "identify their quality-of-life attributes, build on them and effectively promote them to the business community" [18].

Malik Banson, Director of the Kingsley Association, is quoted on the home page of the URA as saying: "Today's URA is not our fathers' URA. Today's URA is not only interested in deals, but most importantly, in where and how the deal develops. They partner with us to support the growth of ideas at the neighborhood level in unique and special ways. They have put 'community' back into community development" [19].

Thus, at the turn of the 21st century, Pittsburgh witnessed a significant change in its planning approaches, shifting towards bottom-up approach, and proliferation of neighborhood level community and non-profit organizations. This was coupled up with remarkable progress with its economic restructuring process, heavily relying on growth of hospitals and healthcare services, high-technology industries, centers of research and higher education (Fig. 17.7). Interestingly, a majority of these sectors are non-profit sectors and are property tax-exempt, and thus not contributing much to ease out the financial challenges for the local governments, particularly for the jurisdiction of the city of Pittsburgh [20]. Nevertheless, revenue generated from income taxes of, and the expanding housing market needed by population base employed in these economic sectors are beneficial for the local governments.

Monongahela Wharf, Pittsburgh riverfront in 2009. (source: Dr. Kevin J Patrick)

Fig. 17.7 Brownfield redevelopment to create the Downtown Innovation District

17.2 Analyzing Smart City and Smart Growth Strategies

We analyze the degree of successes, failures and challenges in Pittsburgh's economic and urban restructuring based on the six themes of Smart City System by Kumar and Dahiya [3]: (i) smart people, (ii) smart economy, (iii) smart mobility, (iv) smart environment, (v) smart living and (vi) smart governance. We also explore the principles of smart growth, outlined by International Economic Development Council (IEDC), to understand to what extent they essentially intersect with the themes of Smart City strategies, based on their relevance to our case study of Pittsburgh.

17.2.1 Smart City System Strategies and Applications

The literature on Smart Cities stress on the need of city-wide or regional scale planning, monitoring and control of advanced infrastructural capacities that efficiently depends on ICT systems. The ICT systems act as the "digital nervous systems" obtaining and processing real-time data on infrastructure (e.g. water supply, electricity, traffic signals and others) and optimizing their functions [2]. As a result, SC strategies with such heavy dependence on ICT systems are expected to "sense and act" [21]. The current trends of investments in cities around the globe with goals to become "smart" widely range across several areas. Some of the most common investments are in the areas of electric supply (e.g. smart energy grid systems), public lighting, water supply management, waste management, natural resource management, transportation and mobility, residential and office buildings, and others. Neirotti et al. [2] argue these as the "hard domain" of SC strategies where placement of ICT systems is crucial for optimal functioning of these systems and to promote overall goals of sustainability. However, there are other areas of implementation for SC strategies, such as public welfare, social inclusion, innovation and entrepreneurship, place making, education, citizenship and civic engagement, asset-based planning, cultural heritage and others, where ICT systems have limited role in successful outcomes (Fig. 17.8). Moreover, scholars argue that ICT systems cannot solely transform cities to "smart cities" without reliance on its residents, and thus ICT systems are only complimentary to "human and organizational capital" [2].

Nevertheless, implementation of SC strategies or smart solutions to city's needs can happen through smart communications, and there have been significant innovations to improve communications across various actors in a city and these have applied a wide array of tools, such as smart city meters for utility management, intelligent traffic signals, e-governance, wifi-kiosks, RFID sensors in pavements, to other measures. The recent survey by National League of Cities reported about 66% of mid-sized US cities have installed tools that come under the "smart" category [8].

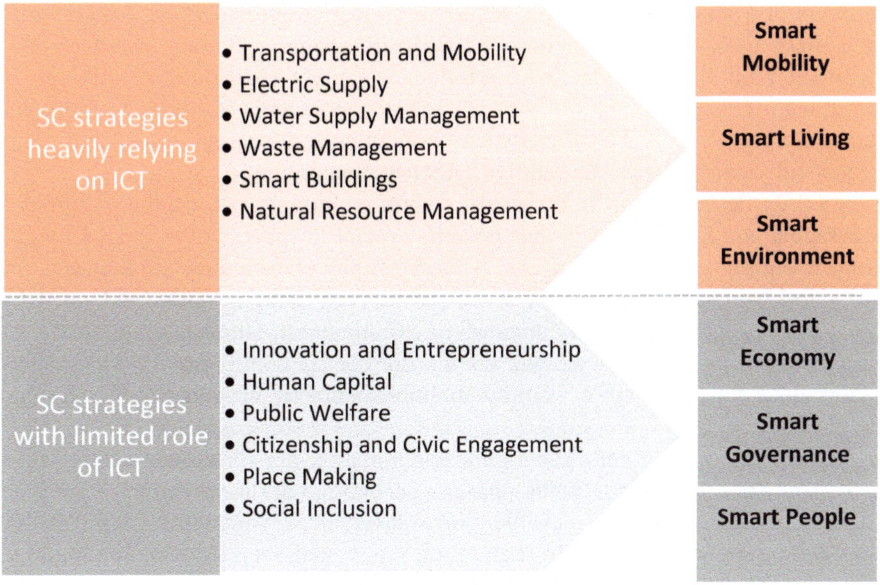

Fig. 17.8 Common areas of implementation for smart city strategies

Moreover, in December 2015, the US Department of Transportation (USDOT) launched "Smart City Challenge" to encourage mid-sized US cities to transform their SC visions into implementable solutions. The common SC applications identified among 78 mid-sized US cities were to address the issues of inadequate transit access to minority and underserved communities, carbon emissions from vehicles, traffic congestion in freeways and arterial streets, and inefficiencies in parking systems. To address these issues, US cities are witnessing a wide range of strategies for implementation of SC visions, such as bringing in autonomous vehicles to improve mobility, electrification of city fleets to reduce carbon footprints, installing vehicle-to-vehicle Dedicated Short Range Communications (DSRC) technology for efficient communication, deploying mobility marketplaces to allow residents to plan for multimodal trips, expanding bikeshare and rideshare options, installing smart street signals that prioritize bus systems and ensure safety of pedestrians and cyclists, and improving real-time tracking of transit systems. In addition, one of the core themes of SC strategies is to emphasize on data collection, processing, analyzing and sharing, where immense amount of data is collected from DSRC infrastructure, crowdsourced data from smartphone users, and installed sensors throughout a city [22]. In 2016, Pittsburgh was selected among the seven finalists for the Smart City Challenge by USDOT, mainly because of its innovative strategies in the areas of infrastructure to support electric and autonomous vehicles, and smart LED street lights with sensors to monitor air quality [23, 24].

It is well understood that adoption of SC strategies will vary from city to city, and few scholars went further to analyze what types of factors determine a city's adoption of a set of SC strategies. While SC strategies are widely adopted in cities to address their urban problems and find sustainable solutions, scholars have identified a number of factors that can influence a city's adoption of specific sets of SC strategies. Most common of these factors are: economic trends and conditions, demographics trends and characteristics, geographic and location factors, and urban trends and characteristics. In addition, efficient functioning of SC strategies, whether heavily dependent on ICT systems or not, requires human capital. Economic, geographic and human factors thus become crucial to success of a Smart City [2]. From economic perspective, a city witnessing higher economic growth is more likely to channelize its financial resources derived from its GDP and tax base towards investments in "hard domains" of SC strategies. Human capital instigates the city's capacity to innovate and attracts further economic growth. Geographic factors determine the city's viability to implement city-wide networks of smart infrastructure.

Based on these economic, geographic and human factors driving implementation of SC strategies, cities undergoing phases of economic decline and thus, scarcity of financial resources may find it challenging to invest in SC strategies, specifically the ones within the hard domain. In such cases, cities, without a significant amount of tax base and access to financial resources have no other choice but to emphasize on strategies of economic development first to attract more jobs and more population. Thus, fostering an environment of innovation and entrepreneurship, investing in its human capital, along with strategies of social inclusion and civic engagement, which are not necessarily the hard domains of SC, are argued to result in more sustainable forms of economic development and serves as foundations for Smart Cities [2].

17.2.2 Smart Growth Strategies

While the notion of Smart City is a relatively new paradigm in planning, the concept of Smart Growth has remained in the planning arena since the late 1990s [25]. Specifically, in the US planning context, applications and implementations of Smart Growth strategies became more popular and common in practice in the 2000s compared to implementations of Smart City strategies. Essentially, "smart growth is based on mixing land uses, using land and infrastructure efficiently, creating walkable neighborhoods that are attractive and distinctive, providing transportation and housing choices, and encouraging community and stakeholder collaboration in development decisions" [26].

The concept of smart growth emerged as a reaction to prevalence of (1) subdivision type zoning, promoting large single family residential lots and urban sprawl, (2) separation of places of work from places of home, leading to more auto-dependency, and (3) unsustainable pattern of land consumption entailing gigantic shopping malls, office parks and commercial strips, in the planning practices of the 20th century. Such practices were increasingly being critiqued for promoting social, racial, and economic segregation over space, increasing the costs of infrastructure expansions, and encouraging automobile dependencies and carbon footprints in American cities.

The concept of Smart Growth, as outlined by U.S. Department of Environmental Protection Agency, places emphasis on 10 key principles:

1. Mix land uses
2. Take advantage of Compact Building Design
3. Create a range of Housing Opportunities and Choices
4. Create Walkable Neighborhoods
5. Foster Distinctive, Attractive Communities with a Strong Sense of Place
6. Preserve Open Space, Farmland, Natural Beauty, and Critical Environmental Areas
7. Strengthen and Redirect Development towards existing communities
8. Provide a variety of Transportation Choices
9. Make Development Decisions Predictable, Fair and Cost Effective
10. Encourage Community and Stakeholder Collaboration in Development Decisions.

While the concepts of Smart Growth do not emphasize on dependency of ICT systems, the essential softer domains of SC concepts, based on environment, equity and economy, are visibly present in these aspects. More specifically, smart growth principles of "Compact Building Design," fostering "Strong Sense of Place," providing "Transportation Choices," and encouraging "Community and Stakeholder Collaboration" are areas where Smart Growth principles intersect with the SC strategies. With increase in smart energy grids, DSRC technology, technological platforms to share crowdsourced data from smartphone users, smart streetlights, mobility marketplaces and ridesharing applications, technological platforms to encourage civic engagements, and other applications of SC strategies, Smart Growth principles can also be efficiently promoted and implemented.

17.3 Smart Initiatives in Pittsburgh

As discussed earlier, the approach to urban planning in the City of Pittsburgh and its surrounding region went through many phases of evolution since the early 20th century. The first phase of revitalization post World War era, or Renaissance I, focused on demolition of vacant industrial sites and decaying neighborhoods to create space for office towers, parks and highway systems, but the City had to bear the consequences of social isolation, racial conflicts, suburbanization, and spatial mismatch. Renaissance II, or the second phase of revitalization since the 1980s, shifted the planning focus from demolition and renewal to finding ways to build a strong and diversified economic base for the city, and resulted in growth of

Fig. 17.9 Historic timeline of urban and regional planning phases in Pittsburgh

high-tech industries, education, healthcare, culture and tourism. The more recent phase, or Big Splash, continues to focus on reviving high-end retail and hotels, and offices, housing and amenities [27] (Fig. 17.9). However, Pittsburgh still remained a segregated city at the turn of 21st century, and there is a renewed and stronger vision to promote sustainability with advanced use of technology and SC strategies.

The key challenges identified by the City of Pittsburgh in the 2010s are transportation and economic challenges, which are also tied up with the issues of social disintegration and isolation of low-income and minority neighborhoods that were essentially the consequences of urban renewal and demolition-based planning approaches of the early 20th century. The City of Pittsburgh also identified these areas as potential areas of SC policies. In 2016, City of Pittsburgh recognized the need of collective action to respond to the city's transportation and energy needs efficiently and equitably. SmartPGH consortium was established and envisioned to promote an integrated approach that enables use of new technology to provide economic opportunity, efficient and smart mobility, and better environment for the residents. SmartPGH is also designed as a platform for real-time data sharing by smart phone users. Last year, this program proposed development of a few "smart spine corridors," which will be dominated by autonomous and connected vehicles, and powered by clean energy [23, 24]. While these smart initiatives are very recent in the Pittsburgh region, such initiatives would not have been possible without the planning visions since the late 1980s that increasingly recognized the importance of economic restructuring and urban greening in a declining city.

In this section, we illustrate Pittsburgh's SC strategies and applications through a number of planning initiatives at a local and regional scale. We focused on two important planning efforts that play crucial roles in planning for shrinking cities: (1) *Planning for Economic Resiliency and Brownfield Redevelopment*, and (2) *Planning for Urban Greening and Green Infrastructure*. While these aspects of planning are not heavily dependent on ICT systems, smart technology and communication techniques are increasingly being applied in the Pittsburgh region to foster innovation and entrepreneurship, promote social inclusion and community engagement, and build and attract human capital. Pittsburgh's story of comeback in the New Economy is set in this backdrop, which is now leading towards further implementation of ICT intensive SC strategies in the areas of mobility, energy, and carbon emission.

The broader metropolitan region of Pittsburgh has already been witnessing a phase of economic restructuring with shift from manufacturing to advanced service based economy since the late 1980s, however the uneven regional pattern of economic and urban restructuring compels us to rethink about their success from the perspective of economic resiliency. We analyze the extent of success and challenges of this economic transformation and the degree of its economic resiliency to be considered as SC strategy. We specifically discuss planning examples of the Bakery Square, Summerset at Frick Park, Southside Works, and Oakland District, where efforts to plan for economic resiliency and redevelop brownfield sites, and integrating them with green infrastructure planning led to creation of a mixed-use neighborhoods, green spaces and diverse economic opportunities that overall contributed to improved quality-of-life of the residents and environmental sustainability.

Practice of Green Infrastructure (GI) in old industrial cities is different from traditional GI planning that focuses on preservation of open spaces and natural habitats in the urban peripheries. In a post-industrial context, urban greening has the potential for returning surplus and derelict lands to productive uses, reduce surplus lands, and stabilize real estate markets. Many municipalities in the Pittsburgh metropolitan regions are increasingly mobilizing their citizens and local organizations, and becoming more pro-active in re-using abandoned properties and transforming them into green spaces and/or integrating them in a network of GI. Community gardens, landscaped spaces and other strategies are practiced at the local levels to enhance quality-of-life and foster socio-economic and environmental sustainability, which are integral part of SC approach. We investigate the effectiveness of GI strategies focusing on regeneration of vacant properties as parks, restored habitats, vegetation, or for storm water management.

17.3.1 Planning for Economic Resiliency and Brownfield Redevelopment

Planning for economic resiliency is emerging as a popular approach in the developed regions of the world, partly as a reaction to economic decline, disasters and shocks. The mechanisms of economic resiliency planning can be challenging due to unpredictability of local and regional economy in the globalized era of 21st century, but essentially focuses on capacities of local economy to withstand vulnerabilities and cope with consequences of crises [28]. Historically, the Pittsburgh region has experienced cycles of growth and decline spanning for centuries, thus, the need to embrace the concept of resiliency is unequivocally supported by politicians, policy makers and community residents.

Pittsburgh is a typical example of a post-industrial city from the western world in the recent times. For decades, the city struggled with challenges of loss of its economic base which led to further urban problems rooted in racial conflicts, flight of white population from the city's core, attraction of suburbs and problems of brownfield sites and vacant lots. The urban future of such a declining city did not look optimistic in the 1980s–90s, and city was labelled a shrinking city with aging and poor population. To change its future, Pittsburgh had to adopt economic resiliency plans since the mid-1980s. In between, 1985 and 1995, the city established a region-wide economic agenda. The plan focused on strategies to diversify the city's economy and transform its economic base from manufacturing to advanced technology, international marketing, and communication systems. The premise of such efforts was deeply rooted in identifying and capitalizing on the region's assets, taking advantage of the emerging economic trends, and fostering growth, innovation and entrepreneurship in these growing and competitive sectors of the regional economy [29].

One of the core elements of Pittsburgh's economic resiliency plans since the 1980s was a strong focus on clean-up and redevelopment of old brownfield sites (Fig. 17.10). Successful redevelopment of brownfield sites, which were idled and abandoned with closures of steel mills in the region, is considered one of the key factors that contributed to successful post-industrial economic restructuring in Pittsburgh (Fig. 17.11). Large-scale brownfield sites, which mostly occupied prime lands along the riverfront, had the potential to offer land for high-tech and knowledge-based companies and institutions. However, strategies to transform former industrial sites into viable uses are often complex due to real as well as perceived risks associated with contamination [30]. Hence, brownfield redevelopment required large-scale funding in the initial stages, which led local and regional government agencies to collaborate with State and federal government and private institutions. The visions of long-term economic and urban benefits of brownfield redevelopment encouraged multiple stakeholders collaborate in numerous projects, and eventually it emerged as an appropriate tool for efficient land management in the region.

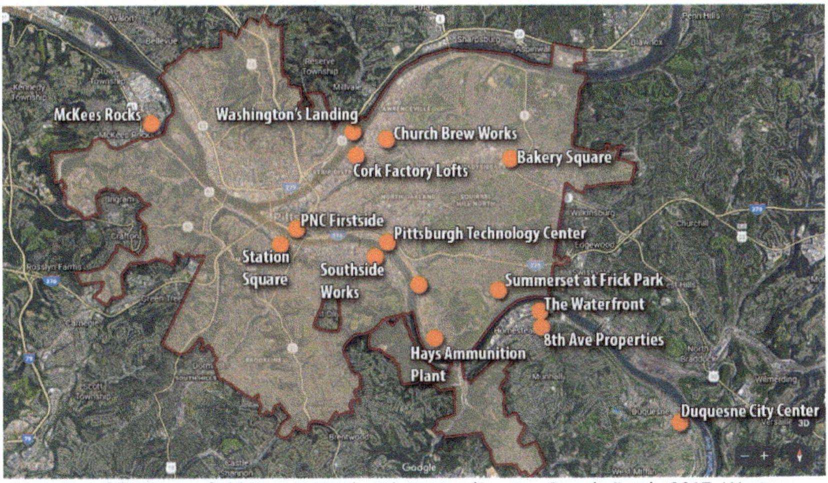

Mapping of Old Brownfield Sites in Pittsburgh Region (Source: Google Earth, 2017; Western Pennsylvania Brownfields Center, 2017)

Fig. 17.10 Map of brownfield sites in Pittsburgh

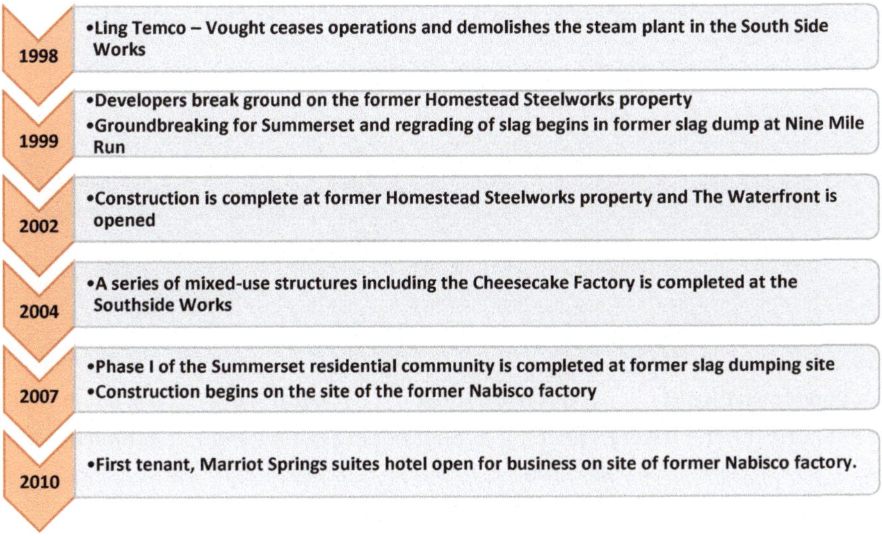

Fig. 17.11 Historic timeline of brownfield redevelopments in Pittsburgh

The Pittsburgh metropolitan region offers opportunities to explore degrees of successes and challenges in many of the successful cases of brownfield redevelopment along the Allegheny, Monongahela and Ohio riverfronts, where former industrial sites were transformed into research institutes, office spaces, mixed use developments (Fig. 17.12). We explore the effective plans and policies that led to successful examples of brownfield redevelopment at a regional scale. However, there are still vast amounts of brownfield sites in the broader region that are in abandoned and derelict state, which needs to be integrated with the changing urban fabric of Pittsburgh.

Along with economic transformation and brownfield redevelopment, Pittsburgh's planning approach in the 1980s–90s was also witnessing a radical change from the early 20th century. While the post-World War planning approaches in Pittsburgh witnessed a dominant top-down approach that led to social and economic disintegration, the planning practices since the 1980s witnessed an increasing trend of bottom-up approach with many volunteer groups and non-governmental organizations beginning to pop up throughout the city. The energy created by mobilization and engagement of community interest groups, and the low cost of housing, rental market, and living also encouraged young adults, referred to as "boomerangers," to start new businesses and create more jobs that stimulated the local economy. Over time, Pittsburgh was gradually being referred to as a "comeback" city in the context of planning and development. Thus, Pittsburgh's transformation from an industrial city to a shrinking city to a comeback

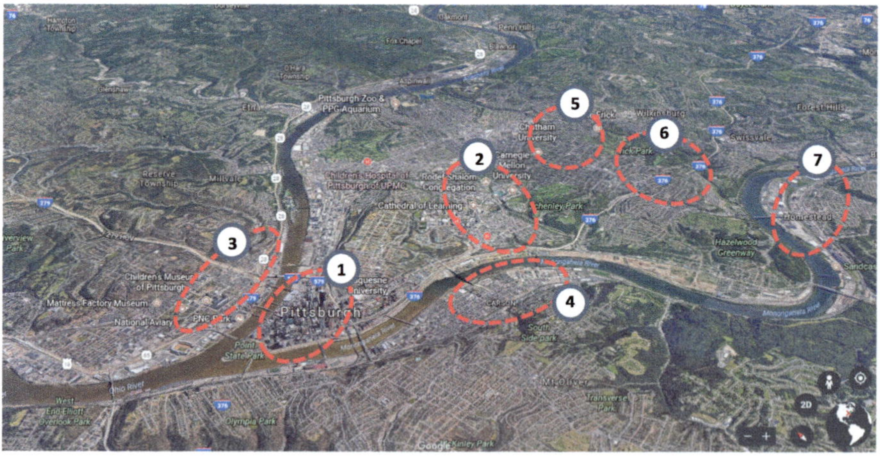

1. Downtown District **2. Oakland District** **3. North Shore** **4. South Side**
5. East Liberty/ Bakery Square **6. Summerset at Frick Park** **7. Homestead**

Pittsburgh's Innovation Districts and Revitalized Neighborhoods located in Old Brownfield Sites (Source: Google Earth, 2017)

Fig. 17.12 Spatial distribution of successful economic and brownfield redevelopments city resulted from realistic assessment of the regions assets and challenges, innovative and bold long-range economic planning, successful partnerships among various private and public stakeholders, and effective participation of local community groups and residents [29]. Since the 2000s, Pittsburgh started witnessing emergence of high-tech and research jobs in Science, Technology, Engineering and Medical (STEM) areas as the driving economic base of the region. The Locational Quotient of these jobs in Pittsburgh, as reported by the Brooking report of September 2017, are almost 200 to 400% of the national average (Fig. 17.13) [31].

In the following part, we discuss the specific examples of brownfield redevelopment and the strategies adopted to ensure economic transformation and environmental cleaning of the sites.

17.3.1.1 Example of South Side: East Carson Street Commercial Corridor Revitalization

East Carson Street is the commercial backbone of a once flourishing residential neighborhood of steel workers. With the decline of the steel industry during the 1970s, homes and storefronts were abandoned leaving a deserted and dilapidated neighborhood (Fig. 17.14). In 1982, the South Side Local Development Company (SSLDC) was created to encourage investment in the corridor. When the company was founded, property values in the neighborhood were 66.66% of the city's median value. In 2008–09, those values rose up to 170% of the city's median value. About 60% of the retail space along East Carson Street was vacant in 1982, and that rate went down to about 4.5% by 2010 [32]. Over 23 years, the SSLDC has leveraged $16 million in funding for businesses and housing redevelopment. By 2010, 225 older Victorian structures were rehabilitated, largely between 10th and 24th street using matching state funds. Feeling it had accomplished its mission, the SSLDC phased out in 2012.

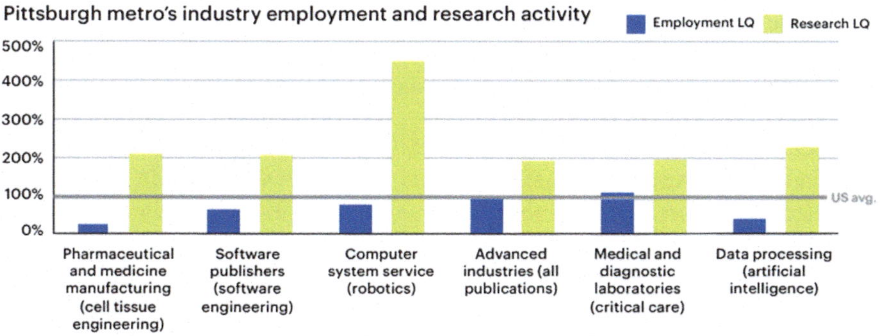

Pittsburgh metropolitan area's Industry, employment and research activity. (source: Andes et. al., 2017).

Fig. 17.13 Research location quotient in Pittsburgh Metro Region

Southside Works, Former Brownfield Site. (source: Rivers of Steel National Heritage Area)

Fig. 17.14 Former brownfield sites at south side

Partnering with the Pittsburgh URA and using state funds, CBDG grants, and mortgage revenue bonds over 100 market rate housing units were developed between 1995 and 2010. Private developers added 330 lofts and condominiums. In all the public invested $119 million since 1985, generating $368 million in private investment. The SSLDC estimates 3500 jobs have been created.

Just to the east of the commercial district was the 123-acre LTV Steel South Sides Works. In 1993, the URA purchased this parcel of land and began a redevelopment effort. Located in the South Side Fats and running from 25th to 33rd Streets, the project has invested $128 million in public funds to generate a total investment of $450 million. The once moribund site is now home to residential, retail, commercial/office, light industrial, hotel, public parks, sports training facilities; specific uses, among many others, include Cheesecake Factory, REI, Hofbrauhaus, national, McCormick and Schmick's restaurants, national corporate headquarters for American Eagle Outfitters, Maya Design, UPMC Orthopedic Clinic led by Dr. Freddie Fu, football training facility for Pittsburgh Steelers and University of Pittsburgh. Recreational amenities open to the public include a trail and public park along the Monongahela River, a public park over a railroad tunnel and numerous plazas featuring art and other public amenities. Up to 352 residential rental units have been constructed and 5400 jobs created. Once generating zero property tax for the city, the URA estimates that the area now generates $6.03 million a year. Property values increased between 160% and 250% between 2000 and 2007 [33].

Southside Works, Redeveloped Site. (source: Google Earth, 2017).

Fig. 17.15 Contemporary image of south side

Redevelopment and livability are two cornerstones of smart growth. Here the two have combined to revitalize a decaying neighborhood and an abandoned industrial zone (Fig. 17.15). As Beth Marcello, a former SSLDC director, puts it: "Carson Street was one of those neighborhoods where there was no reason to be there…but now it has a wonderful walkable business district with almost everything you could want, a high rate of ownership, and a lot of pride" [26] (Table 17.2).

17.3.1.2 Example of Oakland District: Pittsburgh Technology Center and Establishment of Eco Innovation District

Across the Monongahela from the South Side and connected by the Hot Metal Bridges, is the Pittsburgh Technology Center. Built on the site of the former Jones & Laughlin Hot Mill this forty-eight-acre site has been converted into riverfront office park and regional research center. J&L was, by far, the major competitor to the Carnegie Steel, the top steel producer at the time. At its peak it produced almost 3.4 million tons of pig iron, steel and other products, while employing almost 22,000 people [34].

Purchased by the URA in 1983, ground breaking occurred in 1993 with project completion in 2001 (Fig. 17.16). This site was the first use of Tax Increment Financing (TIF) in Pennsylvania to fund the completion of the $104 million development. Because of its almost immediate success, the $7.5 million taken from TIF was repaid 12 years ahead of schedule. Total public investment has been $54,200,000. The public investment has largely been spent on environmental cleanup (tar pits, waste oil, and ferrous cyanide), site remediation, and open space. The URA has created a suburban-like environment with tree covered walkways and a greensward along the river. Two miles from the city center, the location was chosen because of its proximity to the University of Pittsburgh and Carnegie

Table 17.2 Economic indicators of Downtown, Oakland, East Liberty, Southside, and city of Pittsburgh

	Total Population			% Employment in Professional, Scientific, Educational, and Health Services		
	1990	2000	2010	1990	2000	2010
Downtown District (CBD)	3,785	5,222	5,325	35.1	43.1	43.7
Oakland District	21,548	20,417	19,328	52.7	50.9	50.7
East Liberty District	7,973	6,871	6,088	39.4	40.9	46.4
Southside District	11,849	10,733	11,421	30.7	36.5	34.5
City of Pittsburgh	369,879	334,563	308,003	36.0	41.0	44.5
	Average Household Income*			Per Capita Income*		
	1990	2000	2010	1990	2000	2010
Downtown District (CBD)	$76,720	$90,243	$81,870	$33,391	$33,259	$30,153
Oakland District	$47,326	$47,584	$46,442	$19,825	$19,714	$18,584
East Liberty District	$35,198	$42,507	$36,263	$18,713	$23,021	$19,575
Southside District	$41,773	$51,939	$63,564	$20,176	$26,898	$34,113
City of Pittsburgh	$55,316	$60,888	$59,963	$23,519	$27,113	$27,346

*All dollar values have been adjusted for inflation to 2016 values

Source U.S. Census Bureau, 1990, 2000, 2010, data accessed from https://www.socialexplorer.com

Location of Pittsburgh Technology Center, Former Brownfield Site. (source: Rivers of Steel National Heritage Area)

Fig. 17.16 Former brownfield site at the South Oakland District

Pittsburgh Technology Center, Redeveloped Site. (source: Google Earth, 2017)

Fig. 17.17 Pittsburgh technology center at the South Oakland District

Mellon University's Oakland campuses, the research component has been a cooperative effort between two nationally ranked universities and the business community. Other tenants include Union Switch and Signal, Aristech, and the Oakland Consortium (Fig. 17.17) [19, 34]. Private investment has been $140 million. The Center generates $1 million in property tax revenue and employs 1000 people in high tech jobs. The URA is considering the development of an additional 1 million square feet because of the success of the site.

Another initiative focusing on supporting clean technology development and infrastructure projects includes setting up the Uptown Eco Innovation District which focuses on identifying ways in which redevelopment can improve the environment, support the needs of current residents and expand entrepreneurship and job growth and positively transform communities. The Eco Innovation district is an urban plan that intends to revitalize Uptown Pittsburgh and transform it into an innovative urban ecosystem where it will enhance equitable land use, attract and guide new investment, expand the local economy, reduce the city's environmental footprint, and ensure equity and access to local opportunities. This plan is being developed by a series of local and regional stakeholders. This is with collaboration among several agencies including the Uptown Partners of Pittsburgh, Oakland Planning and Development Corporation, City of Pittsburgh, Sustainable Pittsburgh, Urban Redevelopment Authority of Pittsburgh, Port Authority of Allegheny County, and Allegheny County Economic Development, neighborhood residents and groups, universities, and other partners (Fig. 17.18) [31, 34].

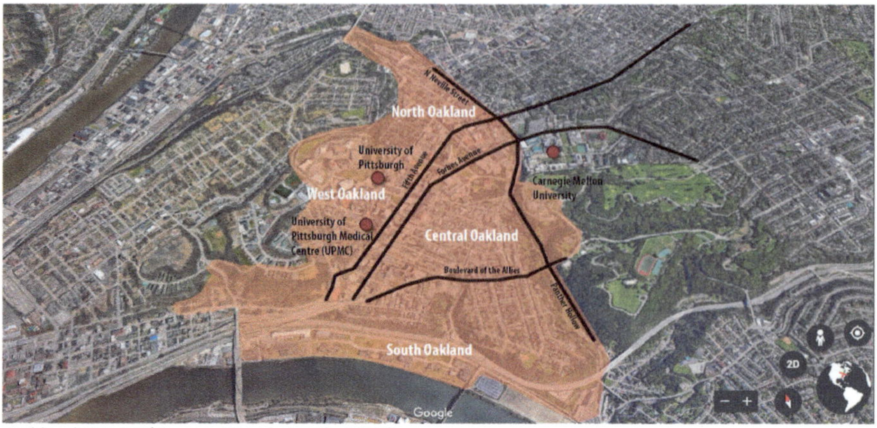

Oakland District (Source: Google Earth, 2017)

Fig. 17.18 Oakland Innovation District

17.3.1.3 Example of East Liberty/Bakery Square

East Liberty is perhaps one of the clearest examples of the failure of the early urban renewal efforts and the smart growth strategies of today. Boasting movie houses, department stores, a roller skating rink, and many retail shops, home to the National Biscuit Company (Nabisco), East Liberty was a booming regional business district until 1958. At that time faced with the first commercial vacancies in decades, congestion, poor parking, and flight to the suburbs, East Liberties business leaders began to call for change [35]. Turning to Pittsburgh's Urban Redevelopment Authority for help, East Liberty began a process of massive urban renewal. The URA proposed creating an outdoor pedestrian mall on Penn Avenue, to be surrounded by parking lots that could mimic automobile driven suburban shopping malls. The plan required the demolition of roughly half of the communities on 254 acres. Ultimately the URA demolished 1200 homes, reduced the size of the shopping district by 1 million square feet and closed the middle of East Liberty to traffic. The neighborhood lost hundreds of small businesses and in the 40 years that followed [11]. In 1959, East Liberty had 575 businesses, by 1970 it was 292, and by 1979 only 98. During the four decades that followed East Liberty lost 4500 people.

At the same time, the Pittsburgh housing authority planners noted that the African-American families displaced by the URA's earlier decision to demolish the lower hill district to make way for the Civic Arena were crowded into Homewood. Their solution was to build three 20-storey housing projects along the new Penn Circle Roads. Viewing the East Mall Apartments, a 17 storey, low income housing project that the Post-Gazette describes as "a host to drug dealers, crime and vandalism," The director of the URA in the 1950s and 1960s Bob Pease, is quoted as saying: "Right or wrong, I don't know" [11].

The story of East Liberty today is very different. In 1979, the East Liberty Quarter Chamber of Commerce formed the non-profit East Liberty Development Inc. (ELDI). Focusing on reopening Penn Avenue, Highland Avenue, and Broad Street to traffic, the non-profit also worked to restore and redevelop property along Penn Avenue [35]. The efforts faltered making slow progress, but under new leadership between 1996 and 2006, ELDI and the Urban Redevelopment Authority (URA) worked to attract new "big box" retailers to East Liberty and to remove the 20-storey housing projects that surrounded the neighborhood. First, ELDI and the URA used tax increment financing (TIF) to lure two national retailers to the neighborhood: Home Depot and Whole Foods. Both of these stores thrived, and their success convinced small local merchants and other national retailers to invest in the neighborhood. Second, after a complex and time-consuming set of transactions, two of the three housing projects that visually barricaded the neighborhood were demolished in 2005, and the third was demolished in May 2009. The 1400 high-rise public housing units were replaced by 450 mixed income units [36].

Working with the ELDI and URA, Mosites Company brought Whole Foods to East Liberty in 2002 with a $7.6 million along a stretch of Center Avenue. Eastside II a $32.5 million project houses Borders and Walgreens. The projects have used Federal tax credits, a $10 million Housing and Urban Development (HUD) loan and $2 million grants, and tax increment financing (TIF) [36].

In 2007, Walnut Capital purchased the old Nabisco plant and began the Bakery Square project for $5.4 million. The site received $1 million Environmental Protection Agency (EPA) grant for remediation to clean up asbestos and PCBs. The total cost of the project was $135 million. The rerouting of Penn Circle to two-way traffic was done at a total cost of $5.6 million. Of that $2.59 million was financed with TIF. An additional $7.41 million in TIF was used for traffic easing and property tax abatement. Bakery Square offers 932 public parking spaces, the 110 room Marriot Spring Hill Suites. The ground floor has 121,060 square feet of retail space hosting an Anthropologie, Coffee Tree Roasters, Jimmy Johns, Panera Bread, and 41,550 square feet of fitness center. The biggest client is Google which ultimately occupied 115,000 square feet and employs 500 people. The expectation is that Bakery Square will create 1600 jobs [34].

In 2012, East Liberty got another boost, a $15 million U.S. Department of Transportation TIGER IV grant which supplement the $34 million construction of a new transit center. The center will be a hub for 1000 bus arrivals and departures each day and have a pedestrian link to the Shadyside neighborhood. Mosites Co. is planning on Eastside III and IV development additions on surrounding land [37].

With apartments renting from $1300 for a 500-square foot studio to $4000 for a 1500 square foot two-bedroom apartment, East Liberty is returning to its glory days of the 1950s. A vibrant upscale shopping neighborhood with businesses and apartments, directly connected to public transit, it serves as an example of the difference between urban renewal and smart city growth (Fig. 17.19).

Old image of Brownfield Site at Bakery Square
(Source: Pittsburgh City Photographer Collection, University of Pittsburgh)

Historic shot of the Nabisco Factory, Bakery Square, Current Google Office.
(Source: Historical Society of Western Pennsylvania)

Fig. 17.19 East liberty/bakery square

Redeveloped Site at East Liberty/ Bakery Square (Source: Google Earth, 2017)

Fig. 17.19 (continued)

17.3.1.4 Example of Summerset at Frick Park

Summerset at Frick Park is one of the largest residential redevelopment projects since WWII. A traditional neighborhood development project, the site will ultimately comprise 710 new residential units and 105 acres deeded as an extension of Frick Park [38, 39].

Located approximately 5.9 miles from downtown Pittsburgh and adjacent to Squirrel Hill, Summerset lies along Nine Mile Run, one of the city's largest streams entering the Monongahela. In 1910, Frederick Law Olmstead saw it as one of the best opportunities for a large urban park, however its proximity to the industrial Pittsburgh lent it a different fate. In 1922, it was purchased by the Duquesne Slag Company who for 50 years used it to dump slag. By 1972, there was 17 million cubic yards of slag along the river piled as high as 120 feet [34]. In 1996, the 238-acre site was purchased by Pittsburgh Urban Redevelopment Authority for $38 million. Initial environmental assessment found two major problems. High levels of chromium remained in the slag and Nine Mile Run was seriously polluted.

State and Federal funds of $7.7 million were put into stream remediation, stemming direct sewage contamination as well as several non-point source problems. The slag heap itself was not removed, rather, it was graded and covered with three feet of topsoil. Roads and other infrastructure were also publicly funded bringing total local, state, and federal dollars to $89,554,000.

The residential development was done by a private partnership: Summerset Land Development Associates. The private investment so far has been $151,900,000. Property taxes are estimated at $5.7 million a year on full build out. Home values in adjacent neighborhoods have risen between 44 and 132% during the period 2000 and 2007 compared to 18% for the city [40] (Fig. 17.20).

Summerset at Frick Park, Nine Mile run Browns Road, October 31,1927 (Source: Pittsburgh City Photographer Collection. University of Pittsburgh)

Redeveloped Site, Summerset at Frick Park (Source: Google Earth, 2017).

Fig. 17.20 Summerset at frick park

17.3.2 *Planning for Urban Greening and Green Infrastructure*

Practice of Green Infrastructure (GI) in old industrial cities is different from traditional GI planning that focuses on preservation of open spaces and natural habitats in the urban peripheries. Randolph [41] defines GI as "an interconnected network of green space that conserves natural ecosystem values and functions and provides associated benefits to human populations." While this is a popular approach in GI planning, industrial cities rarely possess open spaces in natural conditions, rather they boast a legacy of abandoned and derelict properties. In this section, we describe some of the GI strategies focusing on regeneration of vacant properties as parks and gardens, vegetation, or for storm water management [41].

17.3.2.1 Vacant Land Development Through Adopt-a-Lot Program

Like many rust-belt cities, Pittsburgh has considerably large number of vacant, distressed, or undeveloped properties. In a post-industrial context, urban greening has the potential for returning surplus and derelict lands to productive uses, reduce surplus lands, and stabilize real estate markets [42]. These lots are a legacy of Pittsburgh's economic shift which resulted in a large amount of vacant and distressed lots which create several issues such as a decreased tax-base, public health issues, social isolation, environmental hazards and overall a lower quality of life for the surrounding neighborhood. Consequently, 12% of all properties in Pittsburgh were vacant in 2000, 36% of which were abandoned or blighted [43]. These distressed sites include parcels that are currently vacant, condemned, or tax-delinquent. These properties that have become public responsibility places an enormous burden on the City resources and do not contribute taxes to pay for public services. In 2015, it was estimated that there were over 28,000 vacant lots could have cost the city about $20 million to maintain, of which the city-owns about 19% by area. An Adopt-A-Lot program is being implemented since last year which aims to convert these lots from blight to asset [44] (Fig. 17.21).

Most of the city-owned vacant lots have liens against them which could take several years to clear for sale and enable redevelopment. Ignoring these vacant lots would result in accumulation of trash and overgrowth, degrading public safety and giving the impression of blight. Many municipalities in the Pittsburgh metropolitan region are increasingly mobilizing their citizens and local organizations, and becoming more pro-active in re-using abandoned properties and transforming them into green spaces and/or integrating them in a network of GI. Pittsburgh's *Open Space Plan* analyzed suitable uses for these vacant lots, and determined the feasibility of turning them into formal parks and open spaces. Concurrently, the Land Recycling Task Force also evaluates vacant lands. Community gardens, landscaped spaces and other strategies are practiced at the local levels to enhance quality of life

Fig. 17.21 Map showing vacant lots in Pittsburgh. *Source* Pittsburgh Vacant Lot Toolkit Policy Guide, City of Pittsburgh 2015

and foster socio-economic and environmental sustainability, which are integral part of SC approach.

Adopt-A-Lot is a low-cost smart city initiative to fight blight, since maintaining each one of those vacant lots would cost the city almost $600/year, amounting to over $5.6 million for maintenance of only the city-owned lots. As an interim measure, the city has enabled residents to start gardens in the 7286 vacant lots it directly owns (and other government owned lots, approximately 9500 lots) through the Adopt-A-Lot program which allows them to plant flower, edible, or rain gardens on vacant lots. The Vacant Lot Toolkit policy and resource guides help residents and community groups to convert city-owned lots into community assets and clarify city processes [44, 45]. It identifies specific vacant land management strategies such as: developing a unified and coordinated vision with city departments, clarifying and developing processes for short-, medium-, and long-term reuse of vacant lots, including creating a single point of contact, an Adopt-A-Lot license for short-term temporary projects, a longer Adopt-A-Lot lease for 1-year leases (renewable for 3 years), and permission for commercial farm or tree farm or other uses [46]. Financial sustainability can be achieved thought the "market garden" model, where community organizations operating small scale community gardens or urban farms can sell unprocessed produce from growing sites on vacant lots in a farmstand-style manner. However, in Pittsburgh, there is a prohibition on selling items produced on city-owned vacant lots due to concerns about lease fees, liability, and the difference between these community uses, in contrast with truly "commercial" uses that should pay higher fees. Commercial uses such as urban farms, commercial farmers markets, larger scale agricultural and nursery enterprises, are some of the most intensive uses that can generate employment, but need to be properly scrutinized. Another short-term proposal includes creating a vacant lot public arts program where artwork can be selected and displayed. Longer term plans for city-owned vacant land redevelopment include a "highest and best use" analysis, and implementation of green infrastructure particularly in floodplains and wetlands, and on properties adjacent to waterways. Lots along retail corridors in core development areas are likely to be more suitable for redevelopment. This year, the city created an online database of over 3000 city properties, most of them vacant lots, which provides potential buyers with information on properties available for sale [47].

17.3.2.2 Planning for County-Wide Stormwater Management

The city of Pittsburgh and other cities in the Pittsburgh metropolitan region face multiple challenges as their aging infrastructure is unable to meet the current water-quality requirements. While a majority of the rain events in Pittsburgh are not extreme, due to impermeable surfaces and clay soil there is not much absorption, and because of the steep topography in many areas (Fig. 17.22), water tends to flow fast and accumulate in low-lying and valley areas.

Like many other cities, Pittsburg's stormwater and sewage infrastructure built before 1940s relies on a Combined Sewer System (both sewage and stormwater is designed to be carried in the same pipes), so when stormwater exceeds the capacity of the pipes, untreated sewage mixed with stormwater overwhelms the system and overflows at several points before it reaches treatment plant. This results in increased flooding, water quality degradation, stream erosion, reduced groundwater recharge, and loss of aquatic life. Therefore, there is a need for more resilient infrastructure to handle the environmental problems due to untreated sewage and stormwater. Older infrastructure with combined sewer and stormwater system is not only inadequate to handle current needs, but even more ill-equipped to meet future demand due to changes in the climate, land use, and rainfall patterns. Green infrastructure (GI) techniques such as rain gardens, riparian buffers, and porous pavements have been proven to be effective for reducing the volume, the rate, as well as pollutants of stormwater runoff.

In response to EPA's citation for sewage overflows which are in violation of the Clean Water Act, a local non-profit environmental organization, *3 Rivers Wet Weather* (3RWW) was created in 1998 in a unique partnership by Allegheny County Sanitary Authority (ALCOSAN) and the Allegheny County Health Department (ACHD) to support 82 municipalities and the City of Pittsburgh to deal with the regional wet weather overflow problem. In 2008 ALCOSAN entered a consent decree with EPA that required a reduction of 85% overflow, and each of the

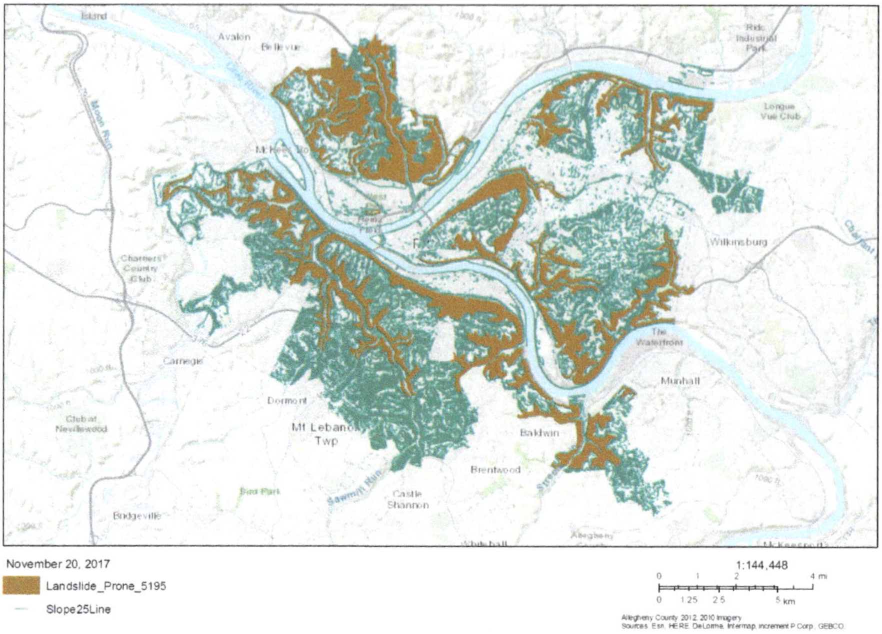

Fig. 17.22 Map showing areas with steep slopes and landslide prone areas. *Source* City of Pittsburgh GIS, http://gis.pittsburghpa.gov/zoning/

municipalities within the service area had a similar consent decree. In 2013, ALCOSAN came up with their *Wet Weather Plan,* expected to cost $3.6 Billion, which would eliminate sanitary sewer overflows and reduce overflow from combined sewers by 2026 [48]. However, ALCOSAN plan focused exclusively on massive capital-intensive gray infrastructure projects such as expanding treatment facilities and a new treatment plant, massive underground storage tunnels, and new and larger capacity drains [49]. The plan ignored GI solutions such as rain gardens, bioswales, green roofs, and tree plantation, all of which have the potential to reduce overflows in combined systems by reduction at source [50]. In any case, EPA deemed ALCOSAN's plan deficient, and incapable of complying with the consent decree [51]. Pittsburgh and Alleghany County debated benefits, feasibility and cost-effectiveness of GI solutions such as rain gardens, porous pavement and other environmentally friendly methods. In 2012, Allegheny County Sanitary Authority conducted a feasibility study for using GI as a cost-effective, long-term sustainable solution for Pittsburgh's wet weather problems with the help of 3RWW for which they used an EPA tool to identify 'best management practices' and suitable locations for different types of low-impact development/GI techniques, such as bioretention—rain garden; constructed wetlands—wet pond; grassed swale—bioswale; infiltration basin—trench; porous pavement; and vegetated filter strip—grass buffer [52]. In 2015, EPA granted a 1.5 year extension to produce a sewer plan that would include GI such as green roofs, rain barrels, permeable pavement, bioswales and stormwater planters.

At the end of 2016, the City of Pittsburgh and PWSA introduced their *Green First* approach, based on analysis of 13,700 acres and proposed runoff management of 1835 acres over the next 20 years which will reduce CSOs through stormwater management. This plan prioritizes source reduction and the implementation of enhanced green best management practices over large scale traditional conveyance and storage structural alternatives. It has established a set of guiding principles for developing stormwater management plans for selected six sewer-sheds. Pittsburgh adopted the *Clean and Green Plan* and a storm water management plan to address problems of poor water quality, CSOs/SSOs, illegal discharge of sewage in storm sewers, local surface flooding, basement sewage flooding, and problems with 80–100 years old sewers [53, 54]. Over the next 20 years ALCOSAN plans to spend $2 billion to implement the plan, mixing both gray and green infrastructures (such as porus pavement, bioswales, green alleyways, etc.), and aims to reduce about 9 billion gallons of sewage flowing into regional waterways during heavy rainfalls. To demonstrate effectiveness of GI techniques, demonstration projects are underway in the city of Pittsburgh city in three locations—one of them is redesign of two city-owned Centre/Herron lots that flood frequently with stone pavers, bioswales, water storage under soil and pea gravel, which will reduce CSO by 750,000 gallons a year; and another is an extensive stormwater park on two vacant lots that will reduce CSO by 800,000 gallons and reduce local flooding [55–57]. One rain garden in the East Liberty neighborhood was able to capture all stormwater runoff in 2016, and ably handled a "once-in-167-year-storm," absorbing three inches of rain in two hours [58].

It is interesting to note that Pittsburg has moved to implementing smarter, cost-effective, GI techniques to reduce the overflow of sewage into its rivers during wet weather events, moving away from the traditional, strictly gray infrastructure approach. This is expected to keep one billion gallons out of the combined sewer system annually, dealing with approximately 10% of the stormwater problem in Allegheny County under current conditions. Other benefits of GI include cost-effective public realm investment, re-establish riverfront connections; complete streets design approach fostering healthy, walkable communities and; creation of resilient infrastructure. But as the climate changes, larger storms and increased rainfall are expected in the area, there is a need for a more dynamic plan with resilient infrastructure [59].

17.3.2.3 Planning for Climate Protection and Resiliency

Pittsburgh has demonstrated a commitment to renewable energy and wants to use cleaner sources of energy because it is vital to creating a more sustainable city. Each year the City of Pittsburgh purchases 25% of its energy from renewable sources, which is enough to power 3500 homes per year. Renewable energy is generated from resources that are indefinitely replenished naturally: sunlight, wind, water, hydrogen, biomass, and geothermal heat. In 2007, Pittsburgh signed the U.S. Mayors Climate Protection Agreement, committing to implementing local climate protection solutions which will result in reduced taxpayer dollars and energy use. Pittsburgh's first greenhouse gas inventory was undertaken to measures the amount of GHG emitted from various sources, which is useful to target actions having the most impact. Pittsburgh is already experiencing climate change effects with colder winters, and Pennsylvania can expect longer and hotter summers, decreased winter snowpack and increased rainfall. Pittsburgh will face local climate threats such as increased severe weather events and flooding, higher prices and shortage of basic goods, increased rate of illnesses and other heat-related health problems. Currently, Pittsburgh is in the process of developing its third climate action plan (PCAP 3.0) to create policies and projects to reduce greenhouse gas emissions within city limits to reduce the severity of regional impacts and move towards a low carbon economy [60]. Further, the city of Pittsburgh, affiliated agencies, and the Green Building Alliance have been working through their Green Garage Initiative to retrofit lighting to LED technology in city owned buildings. After retrofitting the city garages buildings in January, city garage buildings saw a reduction of approx. 60% cut in energy consumption and costs. Also, the city and partners are working towards developing a lighting infrastructure fund for other municipal and privately-owned garages. The city also encourages developers to use clean technology in their building designs through workshops and trainings code enhancements and design standards [61]. The draft of Pittsburgh's Climate Action Plan Version 3.0 was created to align with the Mayor's climate goal commitments in line with the Paris Accords, pledging to meet the 1.5 °C target [62]. This plan builds synergies and overlaps across the 6 key areas: energy generation & distribution, buildings & end use efficiency, transportation & land use, waste & resource recovery, food & agriculture, and urban ecosystems. Many actions relate to energy sources and usage are included in the first 3 key areas listed above, focusing on improved energy efficiency and increased fuel shift. Other 3 key areas focus more on waste reduction and proper resource management.

17.3.2.4 Citizen's Participation and E-governance

The city witnessed proliferation of volunteer groups and non-profit organizations throughout the region, representing strengthening of bottom up approaches in planning. A nonprofit business named Growth through Energy and Community Health (GTECH) spun off from Carnegie Mellon University, is planting sunflowers, switchgrass, and other vegetation on vacant lots throughout Pittsburgh for biofuel production [43].

17.4 Existing Challenges

Several challenges can be outlined in the Pittsburgh's efforts of economic transformation and brownfield redevelopment. One of them is related to the large-scale brownfield sites that still dominate the urban landscape of the region, and requires an enormous amount of financial resources for economic transformation. A city's tax base can never to sufficient to address these issues in a short span of time, and Pittsburgh needs to continuously re-invent and re-innovate it's strategies for brownfield redevelopment at the metropolitan scale. A majority of success stories are concentrated within the Center City that are in close proximity to the Universities, Research Centers, and Interstates; while the locations in the broader region farther away from the center-city are still struggling to comeback. Municipalities, such as Braddock, which are still struggling with issues of brownfields and vacant lots along with aging population and infrastructure, are adopting more right-sizing strategies and planning for shrinkage by converting abandoned properties into open and green spaces that can reduce the cost of maintenance and create a better living environment. However, this will remain a challenge for these communities to attract jobs and human capital in the short-term.

There are several challenges that Pittsburgh faces in urban greening and redevelopment of vacant land. These include funding, maintenance costs, accessibility, and poor design and site programming. In Pittsburgh, a greening forum identified legal issues such as liability, liens, and zoning, as well as a confusing system for the acquisition and reuse of vacant lots, as challenges to urban greening activities [43]. Most of the city-owned vacant lots have liens against them which could take several years to clear for sale and enable redevelopment of these sites. Therefore, Pittsburgh needs to rely more on interim measures such as Adopt-A-Lot to maintain and put them to productive use. Constraints in redeveloping these vacant properties include the dispersed nature of these properties, their variation in size, their title status, and that some of them have historic resource value or a historic designation. These sites also present a problem in land use planning and determining their future land use, but present opportunities and potential of shaping Pittsburgh's future urban form. Another barrier specific to Adopt-A-Lot program implementation is that since the program is often on derelict sites, the applicant must check for possible soil contamination, which requires a soil test. In case if lead content is higher than 1000 PPM, a new site must be selected by the applicant. Pittsburgh faces multiple challenges and constraints in implementing GI practices such as sites with steep slopes, clay soil, high-intensity rainfall, and space constraints. EPA has developed fact sheets and white papers to address these concerns and provide guidance on constrained sites: (1) *Challenges regarding space constraints*: Significant portion of future development in Pittsburgh is expected to be on previously developed sites (redevelopment) or within urban areas (infill), which presents opportunity to incorporate GI into urban areas [63]. However, incorporating green infrastructure into these areas on limited space may be challenging due to existence of buried utilities, mature trees, basements, buildings, and roads pose obstacles;

(2) *Challenges regarding steep slope*: The Pittsburgh area has a dramatic landscape dominated by steep hills and valleys [64]. Since many GI practices enhance infiltration of water into the soil, care must be taken when designing green infrastructure for the Pittsburgh area. Development is restricted on steep slopes, and according to most ordinances, lands with slope greater than 25% should be left undisturbed and roads are typically built with slopes of less than 5%. Many strategies are adopted to manage storm water at its source for slopes of up to 25%; (3) *Challenges with abundant and frequent rainfall*: Pittsburgh area receives 37–45 inches of rainfall per year, and combined with humid climate and frequent rain events, GI is sometimes considered inappropriate for the region. However, it receives most of its annual precipitation as small rain events of one inch or less, which implies that GI works very well with Pittsburgh's climate and rainfall pattern [65]; and (4) *Challenges regarding clay soils*: The region has clay soil which poses a challenge to GI practices, since clay allows little to no infiltration of water to the groundwater table [66]. This becomes even more challenging is when soil has been disturbed and compacted by construction as compacted soil results in very little infiltration and results in ponding [63–66]. However, in spite of these physical constraints, appropriate GI practices are working well in the Pittsburgh region.

In summary, the big challenges still confronted by many municipalities in Pittsburgh region are:

- Abandoned industrial sites
- Declining and aging population base
- Lack of skills and diversity among population base
- Lack of economic diversity
- Lack of advanced service sector jobs
- Urban dereliction and abandonment.

A few municipalities are witnessing economic and population growth with success in attracting Information Communications and Technology (ICT) firms, and other advanced service sector jobs, such as Cranberry, Wexford and others. However, the challenges are to distribute such growth at a regional scale, and to recognize the need for a region-wide planning effort that calls for co-operation and resource-sharing among the fragmented landscape of local governance within the region.

17.5 Future Visions and Conclusions

Rust belt cities such as Pittsburgh have particularly high rates of vacant industrial and residential lots, and thus need effective strategies to repurpose urban spaces and address the insidious nature of blight and abandonment [43]. Such strategies define Pittsburgh's smart city initiatives. Effective application of green infrastructure, and economic redevelopment strategies for vacant lands and brownfields demonstrates a "right-sizing" model that is suitable for other similar shrinking cities in the American rust belt. However, there is a lack of strong regional-scale comprehensive strategies for smart management and monitoring of blight and abandoned, both for residential and industrial sites. While addressing challenges of small-scale abandoned and vacant lots require less financial and human resources, dealing with large-scale abandoned and vacant sites require stimulation of co-operation and planning efforts at the regional, state or even federal level.

The smart city strategy in Pittsburgh constitutes replacing vacant and abandoned properties both at large and small scales. The large-scale efforts involve public funding and multiple government agencies working with community organizations to transform brownfield sites into places of innovation district. The small-scale efforts involve strategies of urban greening and green infrastructure planning to convert surplus, blighted land into green space that costs less for the city to maintain. A green infrastructure network for shrinking cities involves the regeneration of vacant properties for new parks, community gardens, restored habitat, flood mitigation and storm water treatment sites, and urban agriculture plots linked with existing green spaces; and strongly emphasizes on community group participation and a bottom up approach of planning. Green infrastructure can be deployed to address a major economic and social consequences of a shrinking city, which is the decay and blight caused by vacant and abandoned properties. Pittsburgh is demonstrating that green infrastructure has the potential to strengthen the resiliency of the built environment while it transforms to adapt to the modern post-industrial context.

Nevertheless, Pittsburgh still needs to continue such efforts of right-sizing as a strategy to become smart and competitive in the New Economy. In doing so, the region should also place emphasis on the differential effects of smart city benefits across various racial and ethnic groups, gender and age groups, and educational levels of its population. Thus, the regional scale metropolitan planning in Pittsburgh needs to focus on the aspects of smart governance, which should be able to harness the collective efforts of public agencies, private organizations as well as local community groups. This transformation towards smart governance can ensure a common vision across municipalities, foster leadership in local communities, and lead towards more equitable access to resources and a sustainable future.

Acknowledgements We wish to thank Dr. Robert Begg, Professor Emeritus, Indiana University of Pennsylvania, for his helpful comments and feedback, and Dr. Kevin J. Patrick, Professor, Indiana University of Pennsylvania, for sharing photographs of Pittsburgh for this study.

References

1. Nam T, Pardo TA (2011) Conceptualizing smart city with dimensions of technology, people, and institutions. In: The proceedings of the 12th annual international conference on digital government research, College Park, MD, USA, 12–15 June. Available at: https://inta-aivn.org/images/cc/Urbanism/background%20documents/dgo_2011_smartcity.pdf. Accessed 15 Feb 2017
2. Neirotti P, De Marco A, Cagliano AC, Mangano G, Scorrano F (2014) Current trends in smart city initiatives: some stylized facts. Cities 38:25–36
3. Kumar TMV, Dahiya B (2017) Smart economy in smart cities. Chapter 1. In: Kumar TMV (ed) Smart economy in smart cities. Springer, Singapore
4. Haller W (2005) Industrial restructuring and urban change in the pittsburgh region: developmental, ecological, and socioeconomic tradeoffs. Ecol Soc 10(1):13–42
5. Deitrick S (1999) The postindustrial revitalization of Pittsburgh: myths and evidences. Commun Dev J 34(1):4–12
6. Perkins N (2007) A tale of two brownfield sites in Western Pennsylvania: making the best of times from the worst of times in Western Pennsylvania's steel valley. Urban Aff 34(3):503–532
7. Hollander JB, Pallagst K, Schwarz T, Popper F (2009) Planning shrinking cities. Progress Plann 72(4):223–232
8. National League of Cities (2017) The future of equity in cities. Available at: http://nlc.org/sites/default/files/2017-11/Future%20of%20Equity%20in%20Cities%202017.pdf. Accessed 17 Nov 2017
9. U. S. Census Bureau (2013) State-based metropolitan and micropolitan statistical areas maps. Available at: https://www.census.gov/geo/maps-data/maps/statecbsa.html. Accessed 25 Feb 2017
10. U. S. Census Bureau (1900–2010) Decennial census. Available at: http://www.socialexplorer.com/. Accessed 25 Feb 2017
11. Fitzpatrick, D (2000) The story of urban renewal. In: Pittsburgh Post-Gazette, 21 May
12. Muller EK (2006) Renaissance and renewal. Pittsburgh and the Appalachians: cultural and natural resources in a postindustrial age, 7–20. Available at: https://www.upress.pitt.edu/htmlSourceFiles/pdfs/9780822942825exr.pdf. Accessed 15 Oct 2017
13. Fullilove M (2016) Urban renewal and production of inequalities. Chapter 3. In: Framing the dialogue on race and ethnicity to advance health equity: proceedings of a workshop. National Academies Press
14. Jacobs J (1969) The economy of cities. Vintage
15. Pittsburgh Post-Gazette (2012) In desperate 1983, there was nowhere for Pittsburgh's economy to go but up. Available at: http://www.post-gazette.com/business/businessnews/2012/12/23/In-desperate-1983-there-was-nowhere-for-Pittsburgh-s-economy-to-go-but-up/stories/201212230258. Accessed 15 Oct 2017
16. Pennsylvania Department of Community and Economic Development (2017) Act 47 Financial distress: a timeline of financially distressed municipalities and recover plans. Available at: http://dced.pa.gov/local-government/act-47-financial-distress/. Accessed 25 Feb 2017
17. Lange D, McNeil S (2004) Clean it and they will come? Defining successful brownfield development. J Urban Plann Develop 130(2):101–108
18. Blakely E, Bradshaw T (2002) Planning local economic development: theory and practice. Sage Publication, Thousand Oaks, p 81
19. See http://www.ura.org/about/about_us.php
20. Deitrick S, Briem C (2007) The impact of nonprofit, large landowners on public finance in a fiscally distressed municipality: a case study of Pittsburgh, Pennsylvania. Lincoln Institute of Land Policy Working Paper WP07SD1. Available at: http://ucsur.pitt.edu/wp-content/uploads/2014/11/Impact-of-Large-Nonprofit-Landowners-in-Pittsburgh.pdf. Accessed 5 Nov 2017

21. Hall RE (2000) The vision of a smart city. In: 2nd international life extension technology workshop, 28 Sep, Paris, France
22. U.S. Department of Transportation (2017) Smart city challenge: lessons for building cities of the future. Available at: https://www.transportation.gov/policy-initiatives/smartcity/smart-city-challenge-lessons-building-cities-future. Accessed 5 Nov 2017
23. City of Pittsburgh (2016) Beyond traffic: the smart city challenge. Available at: https://www.eenews.net/assets/2016/03/31/document_pm_04.pdf. Accessed 5 Nov 2017
24. City of Pittsburgh (2016) USDOT smart city challenge: city of pittsburgh vision narrative. Available at: https://www.transportation.gov/policy-initiatives/smartcity/smart-city-challenge-pittsburgh-pa-final-application. Accessed 5 Nov 2017
25. Daniels T (2001) Smart growth: a new American approach to regional planning. Plann Pract Res 16(3–4):271–279
26. International Economic Development Council (2006) Economic development and smart growth. Available at: https://smartgrowthamerica.org/economic-development-and-smart-growth/. Accessed 15 Oct 2017
27. Pallagst K (2009) Shrinking cities in the United States of America: three cases, three planning stories. In: The future of shrinking cities: problems, patterns and strategies of urban transformation in a global context, 81–88. Available at: https://s3.amazonaws.com/academia.edu.documents/3304566/11Pallagst_PA_final.pdf. Accessed 15 Oct 2017
28. Hill E, Wial H, Wolman H (2008) Exploring regional economic resilience (No. 2008, 04). Working Paper, Institute of Urban and Regional Development. Available at: https://iurd.berkeley.edu/wp/2008-04.pdf. Accessed 15 May 2017
29. Carter D (2012) Hope for the future of the postindustrial city. Chapter 3. In: Kapp PH, Armstrong PJ (eds) SynergiCity: reinventing the postindustrial city. University of Illinois Press
30. U.S. Environmental Protection Agency (1997) Brownfields overview. https://www.epa.gov/brownfields/brownfield-overview-and-definition. Accessed 30 Aug 2016
31. Andes S, Horowitz M, Helwig R, Katz B (2017) Capturing the next economy: Pittsburgh's rise as a global innovation city. Available at: http://www.heinz.org/UserFiles/Library/Capturing_the_next_economy-Brookings_Sept_2017.pdf. Accessed 30 Oct 2017
32. Jones DN (2010) South side real estate board says mission accomplished. In: Pittsburgh Post-Gazette, November 1. Available at: http://www.post-gazette.com/local/city/2010/11/01/South-Side-real-estate-board-says-mission-accomplished/stories/201011010306. Accessed 20 Feb 2017
33. Urban Redevelopment Authority (2011) South city works: brownfield site redevelopment. Available at: http://www.ura.org/working_with_us/brownfieldProjects/SouthSideWorks_BrownfieldBrief.pdf. Accessed 20 Aug 2017
34. Western Pennsylvania Brownfields Center (2017) Case studies of brownfield redevelopment. Available at: https://www.cmu.edu/steinbrenner/brownfields/Case%20Studies/pdf/pittsburgh%20technology%20center%20-%20LTV.pdf. Accessed 20 Aug 2017
35. See http://www.eastliberty.org/about-east-liberty/
36. O'toole C (2010) Slumbering Pittsburgh neighborhood reawakens. New York Times, March 2. Available at: http://www.nytimes.com/2010/03/03/realestate/03pittsburgh.html. Accessed 10 Oct 2017
37. Belko M (2012) Grant to boost development in East Liberty. In: Pittsburgh Post-Gazette, June 20. Available at: http://www.post-gazette.com/local/city/2012/06/20/Grant-to-boost-development-in-East-Liberty/stories/201206200181. Accessed 10 Oct 2017
38. Li X, Yang H, Li W, Chen Z (2016) Public-private partnership in residential brownfield redevelopment: case studies of Pittsburgh. Procedia Eng 145:1534–1540
39. Li X, Yang H (2016) Turning brownfield redevelopment-the case of summerset at frick park neighborhood. In: Advances in mathematics and computer science and their applications, 78–83. Available at: http://www.wseas.us/e-library/conferences/2016/venice/MAMUA/MAMUA-12.pdf. Accessed 10 Oct 2017

40. Urban Redevelopment Authority (2011) Summerset at frick park: brownfield site redevelopment. Available at: http://www.ura.org/working_with_us/brownfieldProjects/Summerset_BrownfieldBrief.pdf. Accessed 10 Oct 2017
41. Randolph J (2004) Environmental land use planning and management. Island Press, Washington, DC
42. Schwarz T (2011) Re-thinking the places in between: stabilization, regeneration, and reuse. Reinventing America's legacy cities: Strategies for cities losing population. In: The 110th American Assembly, Detroit, MI, 167–184. Available at: http://americanassembly.org/sites/default/files/download/publications/chapter_6_and_case_study.pdf. Accessed 10 Feb 2017
43. Schilling J, Logan J (2008) Greening the rust belt: a green infrastructure model for right sizing America's shrinking cities. J Am Plann Assoc 74(4):451–466
44. City of Pittsburgh (2017) Background of Adopt-A-Lot. Available at: http://pittsburghpa.gov/dcp/adoptalot/background. Accessed 3 March 2017
45. See http://wesa.fm/post/9500-vacant-lots-pittsburgh-could-soon-be-ready-adoption#stream/0
46. Pittsburgh City Planning (2015) Pittsburgh vacant lot toolkit policy guide. Available at: http://planningpa.org/wp-content/uploads/3.-Pittsburgh-Vacant-Lot-Toolkit.pdf. Accessed 3 Mar 2017
47. See https://public-pgh.epropertyplus.com/landmgmtpub/app/base/landing
48. ALCOSAN (2010) Wet weather plan executive summary. Available at: http://alcosan.org/Portals/0/Wet%20Weather%20Plan/Executive%20Summary.pdf. Accessed 3 Mar 2017
49. Hopey D (2012) Alcosan plan to fix overflowing sewer system focuses on 'gray' construction. In: Pittsburgh Post-Gazette, July 29. Available at: http://www.post-gazette.com/local/city/2012/07/29/Alcosan-plan-to-fix-overflowing-sewer-system-focuses-on-gray-construction/stories/201207290184. Accessed 3 May 2017
50. Jones DN (2013). The cost of clean water act is an unequal burden. In: Pittsburgh Post-Gazette, December 29. Available at: http://www.post-gazette.com/news/environment/2013/12/29/The-cost-of-Clean-Water-Act-an-unequal-burden/stories/201312290058. Accessed 3 May 2017
51. See http://www.cleanriverscampaign.org/pittsburghs-wet-weather-problem/
52. See https://www.epa.gov/water-research/epa-science-matters-newsletter-helping-pittsburgh-and-other-cities-expand-green
53. Pittsburgh Water and Sewer Authority (2011) Draft of the green first plan available at: http://apps.pittsburghpa.gov/pwsa/PWSA_City_Wide_Report_11-10-16_DRAFT.pdf. Accessed 3 May 2017
54. See http://apps.pittsburghpa.gov/mayorpeduto/9.26.16_Stormwater_Post-Agenda_Presentation.pdf
55. See http://www.govtech.com/fs/Allegheny-County-Pa-Emphasizes-Green-Infrastructure-in-2-Billion-Stormwater-Control-Effort.html
56. See http://wesa.fm/post/pwsa-looks-prove-dep-green-infrastructure-works#stream/0
57. See https://www.pghcitypaper.com/pittsburgh/pwsa-ramps-up-its-green-infrastructure-program/Content?oid=4776698
58. See https://www.alleghenyfront.org/can-green-infrastructure-really-solve-pittsburghs-stormwater-problems/
59. See http://apps.pittsburghpa.gov/pwsa/PWSA-City-Wide-GI-Assessment-public-summary.pdf
60. See http://pittsburghpa.gov/dcp/sustainability&resilience.html
61. See http://pittsburghpa.gov/innovation-performance/innovationroadmap/citys-internal-capacity.html
62. City of Pittsburgh. (2017). Climate action plan version 3.0. Available at: http://apps.pittsburghpa.gov/redtail/images/606_PCAP_3_0_Draft-_9-26-17.pdf
63. See https://www.epa.gov/sites/production/files/2015-10/documents/pittsburgh-united-space-constraints-508.pdf

64. See https://www.epa.gov/sites/production/files/2015-10/documents/pittsburgh-united-steep-slopes-508.pdf
65. See https://www.epa.gov/sites/production/files/2015-10/documents/pittsburgh-united-abundant-rainfall-508.pdf
66. See https://www.epa.gov/sites/production/files/2015-10/documents/pittsburgh-united-clay-soils-508.pdf

Part XVIII
Conclusion

Chapter 18
International Collaborative Research: Smart Metropolitan Regional Development: Economic and Spatial Design Strategies and Conclusions of Cities Case Studies

T. M. Vinod Kumar

Abstract This chapter has two parts. In the first part, the organizational details of the international research collaborative project "Smart Metropolitan Regional Development: Economic and Spatial Design Strategies" is discussed, in the second part are presented in consultation with the team leaders of the city study, and their general conclusions of the study.

Keywords Organization study · Results

18.1 Smart Metropolitan Regional Development: Economic and Spatial Design Strategies

The Smart Metropolitan Regional Development energises, reorganises and transform the legacy economy to smart economy embracing ICT, creating sustainable zero marginal cost society, triggering sharing economy, designing the five pillars of the third industrial revolution by smart communities and so on. The state of the art of Smart Economy in Smart Cities has been documented by Springer, (edited by T. M. Vinod Kumar) in 2016 using in depth 13 city case in Ottawa, St. Louis, Stuttgart, Bologna, Cape Town, Nairobi, Dakar, Lagos, New Delhi, Varanasi, Vijayawada, Kozhikode and Hong Kong studies from ten countries namely Canada, China, Germany, India, Italy, Kenya, Nigeria, Senegal, South Africa and USA by about 50 authors. This book applies the experiences of these cities for Smart Metropolitan Development using the design of economic and spatial strategies. This

T. M. Vinod Kumar (✉)
School of Planning and Architecture, New Delhi, India
e-mail: tmvinod@gmail.com

T. M. Vinod Kumar (ed.), *Smart Metropolitan Regional Development*,
Advances in 21st Century Human Settlements,
https://doi.org/10.1007/978-981-10-8588-8_18

is undertaken by a series of case studies leading to finally drawing conclusions about this concept. This task is attempted largely by universities located in the US, Europe, Africa and Asia.

These 16 city studies conducted in this book start with a state of the art study of the growth of metropolitan cities around the world which includes meta cities of 20 million population and above, mega cities of 10 million and above and metro cities of one million and above of urban agglomerations. This is followed by an appraisal of economic and spatial design strategies used by the official plans of 17 cities both from the east and the west. Finally, the major conclusion drawn from the book is summarised in this concluding chapter after discussing the organisation of this international research program.

Smart Metropolitan Regional Development can be conceived and developed emphasizing some total of integration of internet of enabling ICT technologies for smart economic development, embracing an economy of production and distribution in a carbon neutral environment monitored in real time transportation system that performs intelligent and smart mobility based on real-time information drawn from big data for the Smart Economy. Some can think of re-embracing capitalism that can create. Smart Metropolitan Regional Development, or others can think of going after socialism or communism with a command economy and an ever-growing bureaucracy to implement socialism or communism. The reality, however, points towards a different direction. The authors of this book have adopted only one simple and dependable way of looking at Smart metropolitan city economic development and governance through smart activation of Smart People. Smart People need not embrace Capitalism since they have found that without much capital, not owning a brick and motor store, or a mall, they can be part of Amazon or eBay seller which is the biggest market place in the world sharing the ICT-enabled marketing and logistic system at low marginal cost and price. In the same way, they can also be a part of the largest taxi services like Uber with only one taxi at his disposal while Uber does not own even one taxi defeating the central concept of Capitalism. Since Uber just provides location based computer and smart phone platform it does not have a huge and very expensive (with very limited benefit) bureaucracy that dictates movements unlike Socialism and Communism. Since real-time information and big data guide all these economic activities, socialism and its consequent ills of proliferating and expensive bureaucracy for tax payers with no value addition to the urban economy are no use for the Smart metropolitan city economic development. During the colonialism era, everything about life was centred around the bureaucrats of the colonialist. Even when many countries became free from colonialism this dependency on bureaucracy was evident as if bureaucrat regulator can give leadership to economic development which by the very job of the bureaucrat cannot when the democratic republic is run by people. At best, he can implement policies of the elected government enshrined in legislations in close cooperation with people. This does not mean any regulation or no law and order in Smart

Economy. Like tax compliance by electronic filing of income and wealth tax, Smart Governance is there to replace age old and dysfunctional bureaucracy enshrined in a brick and motor building in a prominent place in the city, a legacy of the sixteenth century in many colonial countries. We consider Smart People and smart community and their E-democracy as all powerful and capable building blocks of Smart Cities for Smart Economic development replacing capitalism and socialism at one go. For Smart People, there should be opportunities in Smart Cities for continuing training Smart People to make them smarter today compared to yesterday. Smart People can be everyone in a city irrespective of their wealth, educational qualification and social background and, therefore, an inclusive concept since all of them have a constructive role to play in Smart Cities. They can be below the poverty level or above, which does not matter but all of them should have the wish to be Smart People and can be part of never-ending learning mode to be smart. Smart People through their E-Democracy and E-Governance plan, design and govern the Smart Cities. Smart People are the creators, governors, regulators, managers and maintainers of Smart Metropolitan Regional Development. The required Smart Cities technologies which are ICT-enabled can easily be designed by Smart People's creativity, and prototypes are made in Fabrication Laboratory (Fab Lab) if located in Smart Cities for mass production and use. Being the creators of the Smart metropolitan city technologies, Smart People can maintain, repair, innovate and evolve the existing technologies to more cost-effective and functionally superior, next generation technologies which can be shared profitably with other smart cities. No one is running away from Smart People's creation and redevelopment of next generation Smart metropolitan city technologies.

18.2 Inter Related International Collaborative Research Projects on Smart Cities

This book is the fifth in a series Professor T. M. Vinod Kumar conceived, coordinated, implemented and edited about articulating the various roles of Smart People in Smart Cities.

The first book entitled "Geographic Information System for Smart Cities" edited by him [1] was aimed at creating a comprehensive self-awareness of city functioning every second and every day in real time which is the foundation of Smart metropolitan city. Geospatial technologies, sensors and analytics can be used to reach the awareness and use it in real time for various types of use by Smart People. How it can be used for a variety of urban issues commonly observed globally is what that book is all about.

These Smart People thereby progress towards their self-directed goals, such as they demand Smart Living and Smart Economic Development. They aspire to the highest level of quality of life in a Smart metropolitan city environment which they can very well afford and can expand many folds the economic development opportunities to satisfy higher income and employment needs to sustain Smart People. No smart person in a city is an island or elite, but they share a common destiny and common urban space, urban realm, and social and physical infrastructure. Government as the regulator is required that none of the Smart People is denied of all city provides for irrespective of their income level and social status or they are above or below the poverty level. Hierarchy of government exists in a city, but their governance needs to be for a Smart metropolitan city that is fully aware of itself every second and as against Government who comes to know about the issue when a case is filed in the court which takes many decades to get a final judgement. The existing governance systems are obsolete being a product of sixteenth century or earlier designed for colonial rulers, built on the model of East India Company's administration in India for a sixteenth century which cannot be used or Smart metropolitan city economic development. However, those who aspire to live in Smart Cities are in the twenty-first century and no more part of an exploitative empire under the iron hand of a colonial administrator. Therefore, the twenty-first century Smart Cities require Smart metropolitan city e-governance system that was the subject matter of the second book entitled "E-Governance for Smart Cities" [2].

The third book in this series is, "Smart Economy in Smart Cities" [3]. This book explores possibilities for rapid change in the income level and employment opportunities of those Smart People below or above the poverty level in a Smart metropolitan city, and to make the NDP growth rate to a desired higher level consistently for the next many decades. Then, the current trend of urban local economic development is required to be converted to Smart metropolitan city Economic Development. For example, 10% NDP growth rate envisaged for next three decades in India and many other countries can only be realized through Smart metropolitan city Economic Development. Smart Cities and the related conceptualization boasts of the Smart Economy but not much has been systematically researched or documented about it so far. This calls for a study of many cities across the world to document what constitutes a Smart Economy. There are two groups of cities being studied in this book. Some of them have been designated as Smart Cities by learned societies, but others are not but aspire to be Smart Cities. These call for different approaches to research design and studies. It was seen from case studies both these cases in differing countries emphasize different approaches, establishing that there are no cook book solutions. The cities being studied in this book are spread in several major continents and regions, including North America, Europe, Africa, Indian subcontinent and East Asia. They are Ottawa in Canada; Stuttgart in Germany; Bologna in Italy; Dakar in Senegal; Lagos in Nigeria; Nairobi in Kenya; Cape Town in South Africa; New Delhi, Varanasi, Vijayawada, and Kozhikode in India; Hong Kong in China, Cape Town, Dakar, Nairobi and Lagos in Africa.

The fourth book in this series is "E-Democracy for Smart Cities" [4]. The world over, participatory democracy is worshipped and preached but what is practised is representative democracy at the city level and beyond. It is believed that in meta cities, megacities and metropolitan cities, only representative democracy with elected representatives will work. However, democracy practised in small cities like Athens in Greece, Licchavi in India in ancient times and many parts of the world documents face-to-face democracy in practice. In these cities, everyone in a city sat together and jointly decided on all aspects of the city. Citizens not only participated in decision making but acted together as one government and even as administrator and regulator. With the advent of ICTs in Smart Cities of the twenty-first century, it is possible to go back to the face-to-face democracy that, by any measure, is much superior to representative democracy. The fourth book is all about E-Democracy in Smart Cities in action. It is divided into three parts, State of the Art Surveys, Domain Studies and Tools and Issue of E-Democracy in Smart Cities.

The fifth book in the series is this book" Smart Metropolitan Regional Development: Economic and Spatial Design Strategies" edited by T. M. Vinod Kumar and published by Springer-Nature. Here metropolis also includes meta cities with 20 million and above population, mega cities with 10 million and above population and metro cities with one million and above. Here these cities however large these cities may be, need to be converted to smart metropolis using the specific design of economic and spatial strategies. The city studies for the "Smart Metropolitan Regional Development" result in many insights on many smart spatial and economic strategies using Internet of Things, Internet of Democracy and Internet of Governance oriented to the specific issue of a town and its potential; taking into consideration that the Smart metropolitan city is an integrated six systems in which Smart Economy is an integral part. It can relate to Smart Mobility, Smart Environment or Smart Living. Vinod Kumar has discussed the Smart metropolitan city System in the first chapters of four books already published [1–4] and dealt with it further in Chap. 1 of this book. Based on the elaboration of Smart metropolitan city System, if one must develop city-, region or country-specific economic and spatial design strategies for a Smart metropolitan city, it must be designed based on local ecological and cultural system of the city and not a type universal design. Location-specific and culturally acceptable economic and spatial strategies can be locally evolved, governed and managed. This is the only way local culture will find expression in Smart metropolitan city economic and spatial strategies by utilizing local, creative talents of smart people in many institutions in Smart Cities. There are 16 cities (Fig. 18.1) being studied in this project namely Pittsburgh in USA, Stuttgart in Germany and Naples in Italy, Dakar in Senegal, Conakry in Guinea, Abuja in Nigeria, Johannesburg in South Africa and Nairobi in Kenya, Ahmedabad-Gandhi Nagar, Bangalore, Chandigarh, Jaipur, Kozhikode, New Delhi, Surat in India and Hong Kong and greater Pearl River Delta Region from China.

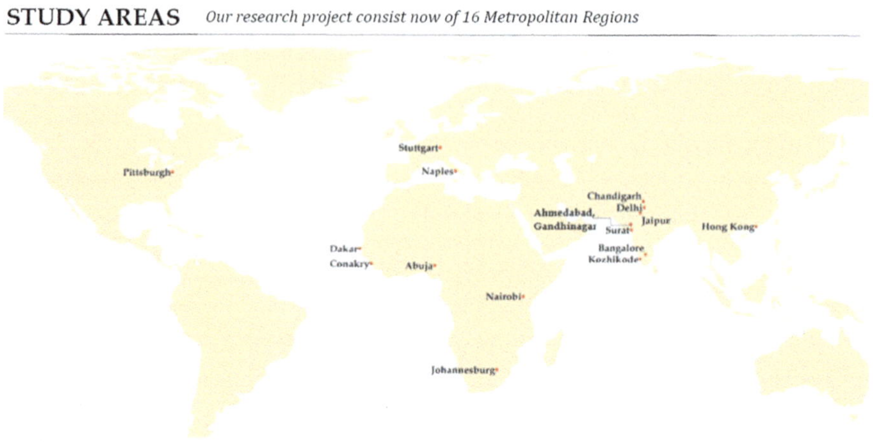

Fig. 18.1 Participating metropolitan areas for "smart metropolitan regional development: economic and spatial design strategies"

The sixth book in this series edited by T. M. Vinod Kumar will be published in 2019 by Springer-Nature is entitled "Smart Environment for Smart Cities" as a product of international collaborative research. This book, is aimed at developing the Design and Protocol and Practice of Smart Environmental Resources Management for Smart Cities. Environment Resources are common proprieties where an active role of Government and People are required and hence its management is a joint and synchronous effort of E-Democracy, E-Governance and IOT system in a 24 h 7-day framework on any environment resource in any smart cities. The smart environmental resources management is a practice that uses information and communication technologies, Internet of Things, Internet of Governance (E-Governance) and Internet of People (E-Democracy) along with conventional resource management tools to realise the coordinated, effective and efficient management, development, and conservation that improves ecological and economic welfare in an equitable manner without compromising the sustainability of development ecosystems and stakeholders. This book will present many city case studies (Nova Scotia in Canada, Hog Kong in China, Ahmedabad, Gandhi Nagar, Chandigarh, Kozhikode, New Delhi, Patna, Surat, Gurgaon from India, Rome in Italy, Yokohama in Japan, Nairobi in Kenya, Ibadan in Nigeria, Bangkok in Thailand, and Dubai in UAE), is centred on one or all environmental resource each in a city.

18.3 The Fifth Book on Smart Metropolitan Development

The editor and coordinator of the book series T. M. Vinod Kumar and many authors who participated in the earlier four books felt that there is a gap in knowledge about Smart Metropolitan Regional Development as far as large metropolises, mega cities and meta cities are concerned. Instead of converting a part of the city to smart metropolitan city, is it feasible to transform a 20 million plus population meta city by design and implementation of economic and spatial strategies? Yokohama in Japan started in a piece meal manner like any smart metropolitan city but now it is a total smart metropolitan city project on their own. This is at a juncture when the smart metropolitan city is developed for a very small part of large meta city like New Delhi Municipal Committee area as if it is an urban design project. Indeed, it is the duty of each country to fill in this gap at near zero marginal cost so that research is sustainable. Universities have a responsibility to conduct the research to fill in the gap. This can only be done with a year-long and in-depth study of selected cities around the world. No attempt was made to make a stratified sample of such cities the world over, and these city studies do not represent the universe of such large "smart" cities. We believe that each city is unique but the experience of dealing with Smart metropolitan city regional development in many cities world over can be of use to aspiring Smart Cities to a certain extent anywhere. We also believe that no city can copy the experience of another city or clone another Smart metropolitan city regional development practice since every city is a unique sociocultural, ecological and economic system.

Funding for such collaborative research project was another issue. Universities and research centres dominated in collaborating these six-smart metropolitan city research projects. We also found that along with Universities, some not-for-profit national and international networks and institutions, city governments and regional governments in certain countries also came forward to participate in this collaborative research programme. The editor and coordinator of the project again felt that this international project shall not seek any external funding other than the internal resource mobilization from within the participating universities. He could convince all participating institution on this. For this book National Institute of Technology, Calicut (NIT-C) came forward to help in an international Research Consultative meeting on Smart Metropolitan Regional development: Economic and Spatial Design Strategies. This activity was co coordinated by Firoz Mohammed and Bimal of the Department of Architecture and Planning, NIT-Calicut. The total expenditure of the International Consultative Committee meeting was a share between NIT C and participant researchers. While NIT-C supported all expenditure within Calicut City, the other expenditure was met by researchers/authors. Those who could not visit NIT Calicut were provided with a facility for video conferencing thereby saving on travel and time cost.

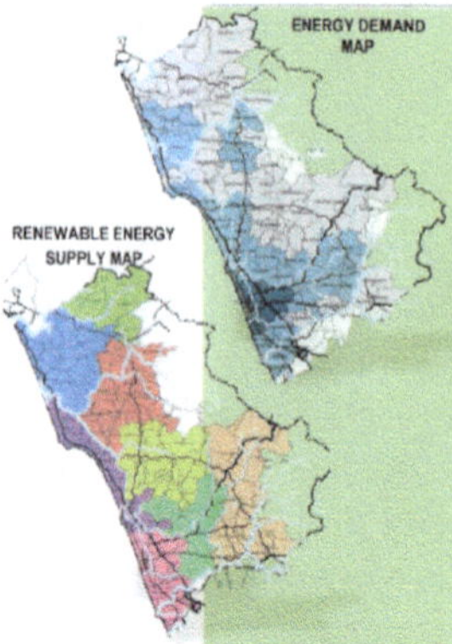

International Consultative Meeting on "Smart Metropolitan Regional Development: Economic and Spatial Design Strategies"

17th & 18th December 2016

Department of Architecture and Planning
National Institute of Technology Calicut

"Smart Metropolitan Regional Development: Economic and Spatial Design Strategies"

Preamble

Around **50** Authors from **10** countries worked in **13** cities for **2** years to produce a book. "**Smart Economy in Smart Cities**": An International Collaborative Research: Ottawa, St.Louis, Stuttgart, Bologna, Cape Town, Nairobi, Dakar, Lagos, New Delhi, Varanasi, Vijayawada, Kozhikode, Hong Kong" (2016: Springer)

A few participants felt the need to share the practices of **Smart Economic and Spatial development** of these 13 cities to other cities by preparing **Metropolitan Regional Development Plans** as Models using the experiences of each city as relevant to the city in consideration.

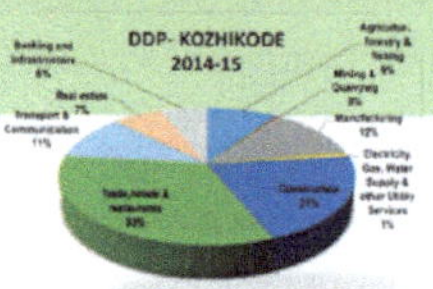

Contact Personnel:

For general queries:
Dr.Mohammed Firoz. C
firoz@nitc.ac.in
Mob: +919895254353

For queries related Video Conferencing:
Ar. Bimal.P
bimalp@nitc.ac.in
Mob: +91889100580

Organizing Team

Dr.Naseer M.A.
Head of the Department
naseer@nitc.ac.in

Prof. T.M. Vinod Kumar
Visiting Professor & Mentor of the postgraduate programme in Planning
tmvinod@gmail.com

Postgraduate Students of Urban planning at NIT Calicut.
ar.fathim@gmail.com

Dr. Mohammed Firoz C.
Assistant Professor
firoz@nitc.ac.in

Ar. Bimal P.
Assistant Professor
bimalp@nitc.ac.in

Dr. PP Anil Kumar
Associate Professor
ppa@nitc.ac.in

Ar.Chitra.K.
Assistant Professor
chithrak@nitc.as.in

Department of Architecture and Planning
National Institute of Technology Calicut
NIT Campus P.O. CALICUT, 673601.Kerala, India

Objectives

The objective of this consultative meeting is a follow-up of ideas expressed in the Preamble. Here the status of any Metropolitan region shall be presented and a methodology to work out a study shall be discussed in the consultative meeting.

International Consultative Meeting on "Smart Metropolitan Regional Development: Economic and Spatial Design Strategies"

17th & 18th December 2016

Department of Architecture and Planning
National Institute of Technology Calicut

Smart Economy
"Sharing Economy"

Invitees:

Invitees are the study teams across the world who agreed to work on the theme 'Smart Metropolitan Regional Development: Economic and Spatial Design Strategies' with representations from different continents like Asia (8), Africa (6), Europe (3) and North America (1). This constitutes 19 universities and 2 research organisations.

Duration of the Research:

Two Semesters: One in the field and another to document by teachers of the teamwork that will eventually be published as a book.

Participation in the meeting:

It is expected that the entire team shall be participating in this two days' deliberation. Video conferencing facilities shall be made available to those participants who are not able to visit National Institute of Technology Calicut for the deliberation and they may participate live from their respective cities/countries.

Fund Sharing:

Smart Economy is a '**Sharing Economy**' and this Consultative Meet shall share all the expenses. National Institute of Technology Calicut shall meet all the expenses within the campus (such as local conveyance and food and accommodation for the participants) and the participants shall bear off-site expenses.

Venue:

Department of Architecture
National Institute of Technology Calicut

Date:

17th & 18th December 2016

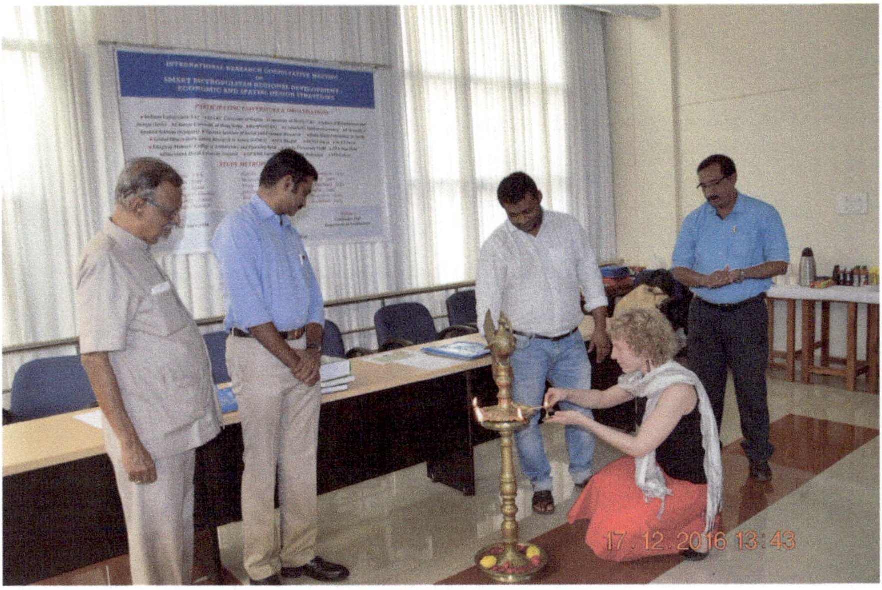
17.12.2016 13:43

17.12.2016 12:27

Time Slots for International Research Consultative Meeing

Day	UTC	Slots	Presenting Team (Please Book your slot here)	Mode of presentation	Hong kong/China +8	Thailand +7	India +5:30	Kenya - Naerobi +3	Nigeria +1	Germany +1	Italy +1	Spain +1	Guinea +0	Senegal - Dakar +0	Liberia - Monrovia +0	Florida -5	Canada - Toronto -5	Columbia - South Africa -5
Day1	07:30	1	Inaguration & Introduction	Physical	15:30	14:30	13:00	10:30	08:30	08:30	08:30	08:30	07:30	07:30	07:30	02:30	02:30	02:30
	08:15				16:15	15:15	13:45	11:15	09:15	09:15	09:15	09:15	08:15	08:15	08:15	03:15	03:15	03:15
	08:15	2	Calicut - India	Physical	16:15	15:15	13:45	11:15	09:15	09:15	09:15	09:15	08:15	08:15	08:15	03:15	03:15	03:15
	09:00				17:00	16:00	14:30	12:00	10:00	10:00	10:00	10:00	09:00	09:00	09:00	04:00	04:00	04:00
	09:00	3	Calicut - India	Physical	17:00	16:00	14:30	12:00	10:00	10:00	10:00	10:00	09:00	09:00	09:00	04:00	04:00	04:00
	09:45				17:45	16:45	15:15	12:45	10:45	10:45	10:45	10:45	09:45	09:45	09:45	04:45	04:45	04:45
	09:45	4	Surat - Gujrat - India	Video	17:45	16:45	15:15	12:45	10:45	10:45	10:45	10:45	09:45	09:45	09:45	04:45	04:45	04:45
	10:30				18:30	17:30	16:00	13:30	11:30	11:30	11:30	11:30	10:30	10:30	10:30	05:30	05:30	05:30
	10:30	5	Refreshment Break	Refreshment Break	18:30	17:30	16:00	13:30	11:30	11:30	11:30	11:30	10:30	10:30	10:30	05:30	05:30	05:30
	11:15				19:15	18:15	16:45	14:15	12:15	12:15	12:15	12:15	11:15	11:15	11:15	06:15	06:15	06:15
	11:15	6	SPA Bhopal - India	Video	19:15	18:15	16:45	14:15	12:15	12:15	12:15	12:15	11:15	11:15	11:15	06:15	06:15	06:15
	12:00				20:00	19:00	17:30	15:00	13:00	13:00	13:00	13:00	12:00	12:00	12:00	07:00	07:00	07:00
	12:00	7	Buffer	Buffer	20:00	19:00	17:30	15:00	13:00	13:00	13:00	13:00	12:00	12:00	12:00	07:00	07:00	07:00
	12:45				20:45	19:45	18:15	15:45	13:45	13:45	13:45	13:45	12:45	12:45	12:45	07:45	07:45	07:45
	12:45	8	Naples - Italy	Physical	20:45	19:45	18:15	15:45	13:45	13:45	13:45	13:45	12:45	12:45	12:45	07:45	07:45	07:45
	13:30				21:30	20:30	19:00	16:30	14:30	14:30	14:30	14:30	13:30	13:30	13:30	08:30	08:30	08:30
	13:30	9	Concluding Discussions	Video/Physical	21:30	20:30	19:00	16:30	14:30	14:30	14:30	14:30	13:30	13:30	13:30	08:30	08:30	08:30
	14:15				22:15	21:15	19:45	17:15	15:15	15:15	15:15	15:15	14:15	14:15	14:15	09:15	09:15	09:15
		BREAK						DINNER BREAK										
	14:45	10			22:45	21:45	20:15	17:45	15:45	15:45	15:45	15:45	14:45	14:45	14:45	09:45	09:45	09:45
	15:30				23:30	22:30	21:00	18:30	16:30	16:30	16:30	16:30	15:30	15:30	15:30	10:30	10:30	10:30
	15:30	11			23:30	22:30	21:00	18:30	16:30	16:30	16:30	16:30	15:30	15:30	15:30	10:30	10:30	10:30
	16:15				00:15	23:15	21:45	19:15	17:15	17:15	17:15	17:15	16:15	16:15	16:15	11:15	11:15	11:15
	16:15	12			00:15	23:15	21:45	19:15	17:15	17:15	17:15	17:15	16:15	16:15	16:15	11:15	11:15	11:15
	17:00				01:00	00:00	22:30	20:00	18:00	18:00	18:00	18:00	17:00	17:00	17:00	12:00	12:00	12:00
	17:00	13			01:00	00:00	22:30	20:00	18:00	18:00	18:00	18:00	17:00	17:00	17:00	12:00	12:00	12:00
	17:45				01:45	00:45	23:15	20:45	18:45	18:45	18:45	18:45	17:45	17:45	17:45	12:45	12:45	12:45
Day 2	07:30	14	*Hong kong - China	Physical	15:30	14:30	13:00	10:30	08:30	08:30	08:30	08:30	07:30	07:30	07:30	02:30	02:30	02:30
	08:15				16:15	15:15	13:45	11:15	09:15	09:15	09:15	09:15	08:15	08:15	08:15	03:15	03:15	03:15
	08:15	15	* Dakar -	Physical	16:15	15:15	13:45	11:15	09:15	09:15	09:15	09:15	08:15	08:15	08:15	03:15	03:15	03:15
	09:00				17:00	16:00	14:30	12:00	10:00	10:00	10:00	10:00	09:00	09:00	09:00	04:00	04:00	04:00
	09:00	16	Buffer	Buffer	17:00	16:00	14:30	12:00	10:00	10:00	10:00	10:00	09:00	09:00	09:00	04:00	04:00	04:00
	09:45				17:45	16:45	15:15	12:45	10:45	10:45	10:45	10:45	09:45	09:45	09:45	04:45	04:45	04:45
	09:45	17	*Johanasburg	Physical	17:45	16:45	15:15	12:45	10:45	10:45	10:45	10:45	09:45	09:45	09:45	04:45	04:45	04:45
	10:30				18:30	17:30	16:00	13:30	11:30	11:30	11:30	11:30	10:30	10:30	10:30	05:30	05:30	05:30
	10:30	18	Ahmedabad - India	Video	18:30	17:30	16:00	13:30	11:30	11:30	11:30	11:30	10:30	10:30	10:30	05:30	05:30	05:30
	11:15				19:15	18:15	16:45	14:15	12:15	12:15	12:15	12:15	11:15	11:15	11:15	06:15	06:15	06:15
	11:15	19	Refreshment Break	Refreshment Break	19:15	18:15	16:45	14:15	12:15	12:15	12:15	12:15	11:15	11:15	11:15	06:15	06:15	06:15
	12:00				20:00	19:00	17:30	15:00	13:00	13:00	13:00	13:00	12:00	12:00	12:00	07:00	07:00	07:00
	12:00	20	Abuja	Physical	20:00	19:00	17:30	15:00	13:00	13:00	13:00	13:00	12:00	12:00	12:00	07:00	07:00	07:00
	12:45				20:45	19:45	18:15	15:45	13:45	13:45	13:45	13:45	12:45	12:45	12:45	07:45	07:45	07:45
	12:45	21	*Nairobi	Physical	20:45	19:45	18:15	15:45	13:45	13:45	13:45	13:45	12:45	12:45	12:45	07:45	07:45	07:45
	13:30				21:30	20:30	19:00	16:30	14:30	14:30	14:30	14:30	13:30	13:30	13:30	08:30	08:30	08:30
	13:30	22			21:30	20:30	19:00	16:30	14:30	14:30	14:30	14:30	13:30	13:30	13:30	08:30	08:30	08:30
	14:15				22:15	21:15	19:45	17:15	15:15	15:15	15:15	15:15	14:15	14:15	14:15	09:15	09:15	09:15
		BREAK						DINNER BREAK										
	14:45	23	* Conakry	Physical	22:45	21:45	20:15	17:45	15:45	15:45	15:45	15:45	14:45	14:45	14:45	09:45	09:45	09:45
	15:30				23:30	22:30	21:00	18:30	16:30	16:30	16:30	16:30	15:30	15:30	15:30	10:30	10:30	10:30
	15:30	24	Buffer	Buffer	23:30	22:30	21:00	18:30	16:30	16:30	16:30	16:30	15:30	15:30	15:30	10:30	10:30	10:30
	16:15				00:15	23:15	21:45	19:15	17:15	17:15	17:15	17:15	16:15	16:15	16:15	11:15	11:15	11:15
	16:15	25	Jaipur - India	Video	00:15	23:15	21:45	19:15	17:15	17:15	17:15	17:15	16:15	16:15	16:15	11:15	11:15	11:15
	17:00				01:00	00:00	22:30	20:00	18:00	18:00	18:00	18:00	17:00	17:00	17:00	12:00	12:00	12:00
	17:00	26	Concluding Discussion	Video/Physical	01:00	00:00	22:30	20:00	18:00	18:00	18:00	18:00	17:00	17:00	17:00	12:00	12:00	12:00
	17:45				01:45	00:45	23:15	20:45	18:45	18:45	18:45	18:45	17:45	17:45	17:45	12:45	12:45	12:45

* No request from team - Assigned by organizers

\# Alternate slot proposed by organizers

18.4 Design of the Collaborative Research Programme

Research Collaborations worked out is purely voluntary and no financial support that binds a project together. Since collaborators are universities, Government, research institutions, professional networks and not-for-profit associations from Asia, Europe, Africa and America, complete independence for pursuing the research was there, free of the baggage of ideologies of granting organisation. They need not accept existing smart cities policies of study cities in their research. Coordinator Editor of the project has no financial or administrative control over any institution participating in the project since he was not in receipt of any grants and did not distribute it. Typologies of the institutions involved in this international project are given in Fig. 18.2. All these autonomous institutions are guided by the highest standard of scholarship and timely completion of research and publication.

18.5 Research Questions on Smart Metropolitan Regional Development

The kind of collaboration in this international research project requires that all participating institutions shall formulate their own research questions and research the methodology which is of use to the country where these study cities are located. Depending upon the type of city some of which are leading Smart Cities, and some are not Smart Cities, the approaches must differ?

However, the paucity of empirical evidence on Smart Metropolitan Regional Development using economic and spatial strategies opens a new area of research: What strategy intervention brings about Smart Metropolitan Regional Development? This is the central focus of the book.

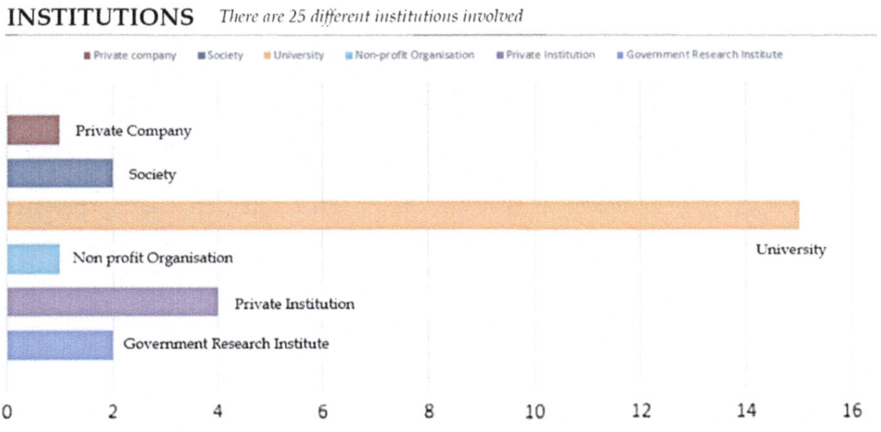

Fig. 18.2 Typologies of research institutions participating in this book

Do cities awaken social, cultural development and ecological (environmental) management through smart metropolitan development? This question lies at the heart of the proposed international collaborative research programme, and unpacking it gives us four interrelated research questions, as follows:

I. What constitutes Smart Metropolitan Development? This will need identification of the key ingredients and their role in making Smart Economic and Spatial changes in Smart Cities.
II. What changes Smart Metropolitan Regional Development brings to social development, cultural preservation, heritage conservation and ecological management? This calls for understanding the inter-linkages between Smart Economy in Smart Cities on one hand, with social development, cultural preservation, heritage conservation and ecological management on the other.
III. How and what processes facilitate the changes to the smart metropolitan region? Do cities bring to social development, cultural preservation, heritage conservation and ecological management? These may include:

    (a) innovation–diffusion (by ICTs and other modes),
    (b) spatial planning,
    (c) sectoral planning (including economic, social development, cultural preservation and ecological management),
    (d) heritage conservation and management plan and
    (e) institutional and governance processes, among others.

IV. How and what changes can be brought to improve the processes to achieve improved/optimal results? These changes related to the various processes as mentioned in Research Question iii.

A deeper understanding of changes in the social, cultural and ecological system of the Smart metropolitan city with the advent of Smart Economy and Smart People for smart metropolitan development is the focus of study. This research programme and the institutions selected for this purpose as academic collaborators are an effort to address this research gap.

18.6 Scope of Research

The following outlines the areas that may be covered when conducting research under the "Smart Metropolitan Regional Development" programme. This is an indication only, and it is left to the team to decide what is appropriate.

I. A time series study of changes in the urban economy by sectors and identifying distinct features of evolving to Smart Economy.
II. Study of theories of local economic development and smart economic development at metropolitan world city level and modelling for study city.

III. Study of the multi nuclei spatial organization of evolving city in comparison with envisaged spatial organization given by the Master Plan.
IV. The concept of accessibility in the Master Plan and its changes with respect to increasing use of ICT in Smart Cities.
V. Changing needs for spatial access in a Smart Metropolitan Economy for goods and services with an increase in the use of ICT and consequent changing needs.
VI. Changing the role of the hierarchy of service areas in a Smart Metropolitan Economy in a Smart metropolitan city as influenced by increasing use of ICT.
VII. Evolving structure of metropolitan urban agglomeration and changes required in a Smart metropolitan city.
VIII. Evolving structure of cities in urban agglomeration and changes required in view of the increase in the use of ICT.
IX. Change of spatial standards in a Smart metropolitan city.
X. Changes required in zonal policies and plans.
XI. Study of legislation of Metropolitan Planning Committee (for instance, in India) and suggest changes as per special requirement of Smart metropolitan city.
XII. Change of role of community-based organizations (for example, Residential Associations in India) in a Smart metropolitan city with an increase in the use of ICT.
XIII. Change of role of Ward Committee in a Smart metropolitan city with an increase in the use of ICT.
XIV. Change of role of Municipal Council in a Smart metropolitan city with an increase in the use of ICT.
XV. Change of role of Metropolitan Planning Committee in a Smart metropolitan city with an increase in the use of ICT.

Note: The scope of research can be further elaborated by the collaborating institutions but need not be uniform for all study cities. Each department of university participating in this research programme shall incorporate relevant Smart Metropolitan Development features appropriate to the goals for each department. The coordinator of this project does not intend to dictate the direction of the research and have a diverse group of collaborating universities, and they should orient their study strictly based on academic goals of their department.

18.7 Study Metropolitan Regions

The Metropolis will be selected as study area by each of the collaborating universities independently, which will be the place the one-year and two-semester combined effort to conduct this research. Universities participating in this programme adopted different types of collaboration. Some universities used, their doctorate and post-doctorate students, while others used students at masters and first professional degree level. A post-doctoral student in the department can work on a narrow subject area in the study as individual work. While graduate and undergraduate students can work on design solutions for the Smart Metropolitan Regional Development, Research institutions can charter their own strategic areas of research.

18.8 Project Details

The Smart Metropolitan City will be selected as study area by each of the collaborating universities independently, which will be the place the one-year and two-semester combined effort to conduct this research. Universities participating in this programme adopted different types of collaboration.

The project details of the study city are given in Table 18.1.

18.9 Way of Working the Programme

18.9.1 Integrating Smart Metropolitan Regional Development Research with Academic Programmes

This international collaborative research programme, with the participation of 10 countries and 16 study cities as tabulated above was conducted by many diverse university departments, research institutions and others as shown in the graph and table above. In Smart Metropolitan Regional Development, the basis of smart metropolitan development, work, place of work, nature of economic activities in work place, carrying capacity of economic activities, livelihood and income are not variable that can be studied only by Department of Economics alone. Purposefully, we did not enlist department of economics in this study. Authors feel that GDP rate will be considerably higher in Smart Cities, and it will be computed periodically by respective cities and/or nations using the country's economic and statistics infrastructure. Therefore, there is no need for a computational study for Smart metropolitan city economic variables and modelling by this research programme. For the conventional economist, the place does not matter but for land economists, regional planners, urban planners, landscape architects and applied scientists, success and failure of economic activities depend on upon the location, local resources, place and people. For example, heritage tourism depends on the character of heritage and economic activities that can be woven around it and for ecotourism it is the ecology of the area and tourism that is within the carrying capacity of the ecosystem. The conventional macroeconomist lacks such perspective and capacity to develop such areas. In fact, there is no department of economics participating in

Table 18.1 Project details

S No.	Study metropolitan regions	Countries	Authors	Institutions
1.	Dakar	Senegal	Dr. Gora Mboup	Global Observatory Linking Research to Action (GORA), New York, USA
			Dr. Mame Cheikh Ngom	University Cheikh Anta Diop (UCAD), Dakar, Senegal
			Cheikhou Balde	Municipal Development Agency (ADM), Dakar, Senegal
			Mandiaye Ndiaye	Ministry of Urban Renewal, Housing and Living Environment (Ministere du Renouveau Urbain, de l'Habitat et du Cadre de Vie), Dakar, Senegal
2.	Conakry	Guinea	Dr. Gora Mboup	GORA Corp, New York, USA
			Ibrahima Camara, Nene Mariama Balde	City and Territorial Management Ministry (Ministere de la Ville et de l'Amenagement du Territoire), Guinea, Conakry
			Khalil Fofana	GORA Corp, Toronto, Canada
			Mustafa Sangare	GORA Corp, Washington DC (USA)
3.	Abuja	Nigeria	Prof. Femi Olokesusi Dr. Femi Aiyegbgajeje Ibitayo Modupe Arije	Afe Babalola University, Ado-Ekiti, Nigeria University of Ibadan, Nigeria UN-Habitat, Nigeria
4.	Nairobi	Kenya	Dennis Mwaniki Romanius Opiyo Daniel Githira Keziah Mwang'a	GORA Corp, Nairobi, Kenya University of Nairobi, Kenya Consultant, Nairobi, Kenya Gran Sasso Scientific Institute, Italy

(continued)

Table 18.1 (continued)

S No.	Study metropolitan regions	Countries	Authors	Institutions
5.	Johannesburg	South Africa	Gora Mboup	GORA Corp, New York, USA
			Paida Mhangara, Naledzani Madau	South African National Space Agency (SANSA), Tshwane, South Africa
6.	Stuttgart	Germany	Michael Hertwig, Manfred Dangelmaier, Joachim Lentes, Satyendra Singh, Nikolas Zimmermann	Fraunhofer Institute of Industrial Engineering University of Stuttgart, Institute of Human Factors and Technology Management University of Applied Sciences Stuttgart
7.	Naples	Italy	Dr. Antonio Caperna, Cerreta Maria, Eleni Trscada, Guglielmo Minervino, Roberta Mele, Giuliano Poli Elina Alatalo	Department of Architecture (DiARC), University of Naples "Federico II", Italy College of Engineering and Technology, University of Derby (UK) International Society of Biourbanism, School of Biourbanism and Design, Italy Università Mediterranea, Italy Tampere University, Finland
8.	Pittsburgh	USA	Dr. Sudeshna Ghosh, Dr. Calvin Maselala, Dr. Sweta Byahut	Indiana University of Pennsylvania, Aurburn University
9.	Hong Kong	China	Dr. Sujata Govada, Tim Rogers	School of Architecture, Chinese University of Hong Kong, Institute for Sustainable Urbanization
10.	Ahmedabad-Gandhinagar	India	Prof. Jignesh G. Bhatt Dr. Omkar K. Jani	Dharmsinh Desai University (DDU), Nadiad—387001, Gujarat, India Gujarat Energy Research and Management Institute (GERMI), Gandhinagar—382007, Gujarat, India

(continued)

Table 18.1 (continued)

S No.	Study metropolitan regions	Countries	Authors	Institutions
11.	Bangalore	India	Amit Chatterjee, Binayak Choudhury, Premjeet Dasgupta, Gaurav Vaidya	School of Planning and Architecture, Bhopal
12.	Jaipur	India	Prof. H. B. Singh Dr. Prabh Bedi Dr. Neha Gupta	School of Planning and Architecture, New Delhi
13.	Delhi	India	Prof. Ashok Kumar et al.	School of Planning and Architecture, New Delhi
14.	Kozhikode	India	Prof. T. M. Vinod Kumar, Dr. C Mohammed Firoz, Nmrata Radhakrishnan	National Institute of Technology, Calicut
15.	Surat	India	Bhasker Vijaykumar Bhatt, Dr. Leena Garg, Dr. Krupesh A. Chauhan	Sarvajanik College of Engineering and Technology, Surat Bhagwan Mahavir College of Architecture and Planning, Surat SV National Institute of Technology, Surat
16.	Chandigarh	India	Asfa Siddiqui KK Kakkar Suvankar Halder Pramod Kumar	Indian Institute of Remote Sensing, Dehra Dun

this project which in no way doubting their specific capacity to GDP and employment computation. Since this exercise involves the vast accumulation of empirical knowledge and analysis in the study city, a two-semester work of all students and faculty in collaborating universities is envisaged as part of their ongoing academic programme. Participating universities decided to have one year of the Smart metropolitan regional programme focussing on this topic in selected cities in several of their ongoing academic subjects and programmes. We found that exploring boundary of knowledge is within the terms of reference of any university. This can be conducted as design studio programmes or small projects as parts of theory classes. It can also be independent post-doctoral work. Within each partner university, the collaborating faculty may design this research programme, and thereafter faculty compiles the research findings for publication based on students and faculty work. In most of the partnering universities, there are studio programmes with 9–15 h per week programmes where a project is executed by a group of students under close supervision and guidance of faculty with higher faculty–student ratio, and there are theory papers where a student must work independently on a topic under the guidance of generally one r two faculty. This work can be a mini-project, which she/he completes within one semester or a term paper. The department integrates the research programme with their academic programme by deliberately having studio programmes and term papers centred on this area of study.

18.9.2 Role of Students

This international collaborative research programme is essentially meant for students being part of an internal academic programme of the university. We consider they are the main actor and shall be given important role in this programme. Perhaps many that age group will live in the Smart Cities than their older faculty. Under the direction of faculty new concepts were introduced in the studio and empirical studies were conducted around these concepts.

18.9.3 Role of Faculty

The faculty is the designer of the program within the framework of existing curricula in design studios and theory courses of each participating university.

I. The project duration is one academic year or two semesters.
II. They guide and monitor student work as usual as part of the academic programme.
III. They monitor students' input to the monthly progress report.
IV. They rewrite the output of the project for a book to be published by an international publisher.

18.9.4 Co-design and Co-production of Knowledge

This international collaborative research programme is founded on the principles of co-design and co-production of knowledge. In today's interconnected world, such collaboration is physically and intellectually possible—thanks to the Internet and ICTs. The collaborative aspect of the research programme will be actualized in the form of:

I. Co-design the programme with the partner academic institutions.
II. Co-production of knowledge through an interactive process of sharing, reviewing and finalizing research findings.

III. Within each partnering institution, co-design and co-production of knowledge can be implemented through design studio/laboratory work between faculty and students.

18.9.5 Research Output

The key output of the "Smart metropolitan Regional Development" research programme will be a book edited by the coordinator Professor T. M. Vinod Kumar, to be published by Springer-Nature, an internationally reputed publisher in 2018.

18.10 Bulletin

During the conduct of research about 12 months through in 2017, three Bulletin has been used to communicate with the international teams of researchers. These Bulletin highlights study metropolitan city profiles selected by the various study teams independently, and introduces to the research network the research methodologies adopted, and the profiles of authors of the research output for the book, "Smart Metropolitan Regional Development: Economic and Spatial Design Strategies". The Bulletin is jointly edited by a Bulletin team, Elina Alatalo, Sudeshna Ghosh, Gora Mboup and Asfa Siddiqui.

The list of Bulletin produced with respective case studies is shown in Table 18.2.

Table 18.2 List of bulletins

No.	Country	City	Bulletin number/date
1	Germany	Stuttgart	1/15 March 2017
2	Italy	Naples	1/15 March 2017
3	The USA	Pittsburgh	1/15 March 2017
4	Dakar	Senegal	2/1 June 2017
5	Guinea	Conakry	2/1 June 2017
6	Nigeria	Abuja	2/1 June 2017
7	South Africa	Johannesburg	2/1 June 2017
8	Kenya	Nairobi	2/1 June 2017
9	India	Ahmedabad-Gandhinagar	3/1 September 2017
10	India	Bangalore	3/1 September 2017
11	India	Chandigarh	3/1 September 2017
12	India	Jaipur	3/1 September 2017
13	China	Hongkong	3/1 September 2017
14	India	Kozhikode	3/1 September 2017
15	India	New Delhi	3/1 September 2017
16	India	Surat	3/1 September 2017

A cover page of the Bulletin 3 and one sample page of the Bulletin is given below.

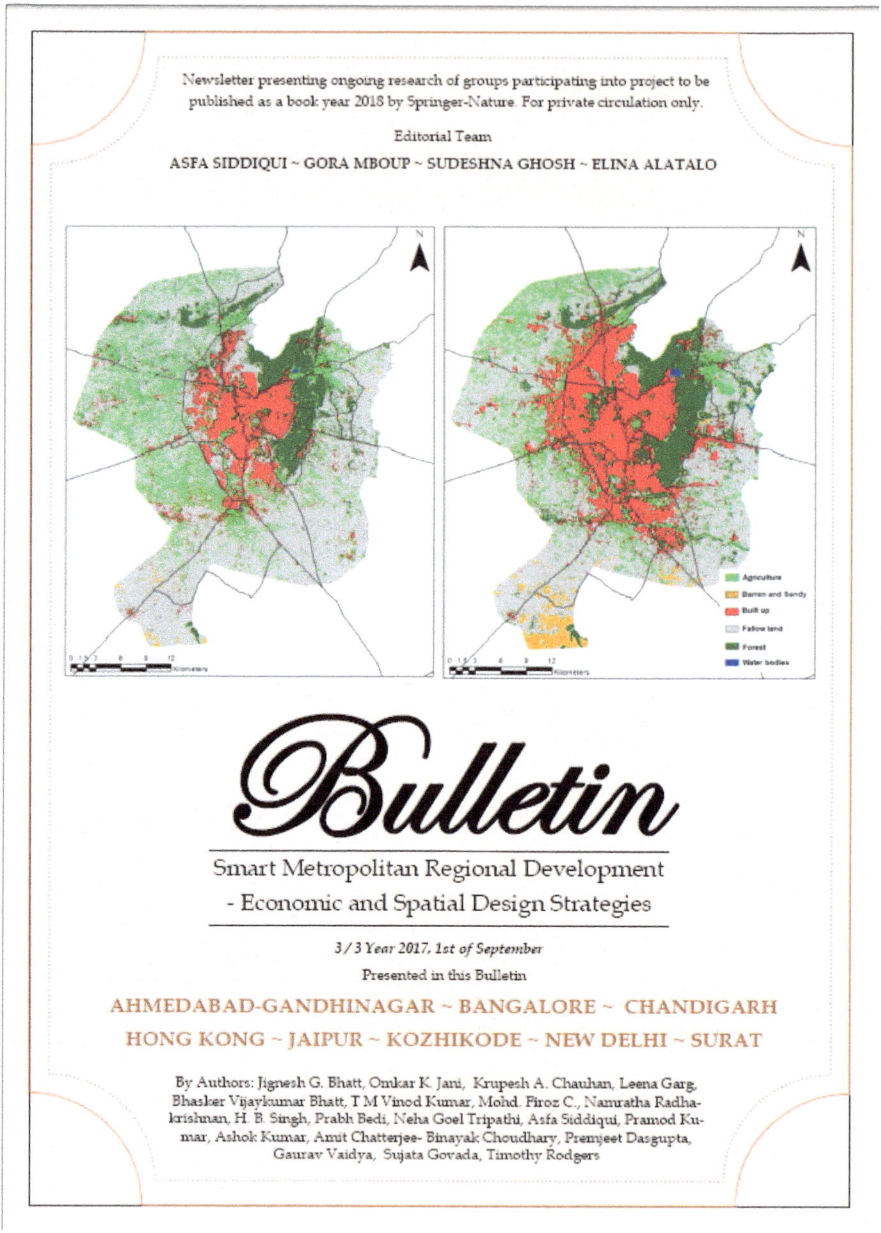
Newsletter presenting ongoing research of groups participating into project to be published as a book year 2018 by Springer-Nature. For private circulation only.

Editorial Team

ASFA SIDDIQUI ~ GORA MBOUP ~ SUDESHNA GHOSH ~ ELINA ALATALO

Bulletin

Smart Metropolitan Regional Development - Economic and Spatial Design Strategies

3/3 Year 2017, 1st of September

Presented in this Bulletin

AHMEDABAD-GANDHINAGAR ~ BANGALORE ~ CHANDIGARH HONG KONG ~ JAIPUR ~ KOZHIKODE ~ NEW DELHI ~ SURAT

By Authors: Jignesh G. Bhatt, Omkar K. Jani, Krupesh A. Chauhan, Leena Garg, Bhasker Vijaykumar Bhatt, T M Vinod Kumar, Mohd. Firoz C., Namratha Radhakrishnan, H. B. Singh, Prabh Bedi, Neha Goel Tripathi, Asfa Siddiqui, Pramod Kumar, Ashok Kumar, Amit Chatterjee- Binayak Choudhary, Premjeet Dasgupta, Gaurav Vaidya, Sujata Govada, Timothy Rodgers

A page of Bulletin is as shown

PITTSBURGH

PENNSYLVANIA, UNITED STATES

Sudeshna Ghosh - Sweta Byahut - Calvin Masilela

Introduction

The mid to late 20th century rust-belt cities in the United States (US) witnessed a trend of decline in its manufacturing base, compelling them to rethink urban and economic development strategies to compete in the 21st century. Pittsburgh, one such example of a rust-belt city, was once known as the "Steel City" due to its production capacity of raw steel in the world economy. With comparative advantage of natural resources and navigable waterways, the city burgeoned with large-scale steel mills since the 1870s and an economic and population base to support these activities (Detrick, 1999). The geographic shift of manufacturing activities towards lesser developed countries since the late 20th century triggered a process of decline in the old manufacturing cities of the developed world. Pittsburgh lost its position as the Steel City, and witnessed a continuous trend of decline in its manufacturing-based economy and population in the past three decades.

The challenge of economic decline coupled with massive brownfield sites, population loss, property abandonment, poverty and high crime, instigated planners and policy makers to adopt new strategies that could transform Pittsburgh and make it competitive in the "New Economy." Since the 1980s, Pittsburgh was successful in adopting many bold policies and strategies that transformed its local economic base towards advanced service sectors: healthcare, higher education, technology, research and development, banking and finance.

Dating back to the 1980s, Pittsburgh has been adopting various right-sizing strategies to stabilize declining neighborhoods. As a result, the city is able to provide a relatively high quality-of-life to its residents within affordable prices, despite abandonment, blighted areas, crime and poverty (Hollander et. al., 2009). These strategies mostly focused on economic diversification, regeneration of brownfields, urban greening and green infrastructure practices, which form the underlying principles of sustainability in Smart City concepts .

With high concentration of employers in the areas of medical research, such as University of Pittsburgh Medical Center (UPMC); higher education, such as Carnegie Mellon University and University of Pittsburgh; banking and financial sectors, such as Pittsburgh National Bank (PNC); and other Fortune 500 companies; Pittsburgh is becoming more successful in transforming its economy. Recently, Uber chose Pittsburgh as its research center for experimentation with driver-less smart cab services.

Pittsburgh Skyline, 2013: Confluence of Allegheny and Monongahela rivers. (source: Dr. Kevin J Patrick)

3

18.11 Summary of Conclusions and City Case Studies

Team leaders of authors were requested to prepare a summary of conclusions of their city studies and the Editor compiled it and is given below.

18.11.1 Smart Metropolitan Regional Development: Economic and Spatial Design Strategies

An overview view of smart metropolitan regional development is presented as a backdrop for research studies of this book in the first chapter. First, metropolitan region is defined followed by its form and functions. Then, smart metropolitan regional development is defined. The chapter studies the global metropolitan cities development and shows how metro cities, megacities and meta-cities are emerging across the world by regions. This process had accelerated in the last decade. Metropolitan cities are continuously exposed to external economic changes and require intermediate range strategies to face it which is not there in a 20-year Master Plan. These periodic economic strategies of cities call for differing spatial strategies to intervene in emerging global situation. To reshape these emerging metropolitan challenges, there is a need to design economic and spatial strategies. Therefore, this chapter concludes with a critical analysis of 17 metropolises in their official plans with their economic and spatial design strategies attempted there.

This chapter introduces the international collaborative research studies of 16 smart metropolitan regional development of the book drawn from 10 countries from USA, Europe, Africa and Asia. The emerging scenario of metropolises, mega cities and meta-cities temporally and spatially across the world based on UNHABITAT statistics and projection is discussed. GDP creation of nations is taking place very actively mostly in these three types of cities across the world which are projected for 2025. Spatially and economically, higher population and GDP growth rate in these cities is now evident and happening mostly in the eastern and southern hemisphere than in the western and Northern. This book concentrates on more number of studies from the east and south than the west and north. City society through their legal framework of constitution brings about the creativity of design of an economic and spatial strategy to take these cities to the next level facing all local challenges through their official Master Plans. The design of economic and spatial strategies is presented for 17 cities based on official plans. Not all these cities give equal importance to economic strategies in contrast to spatial strategies. Often there are plans with less attention to strategies in general and economic strategies. Some of them do not mention economic strategies at all. Metropolitan Planning and implementation should move from geographical space to community and then to prosumer households both for planning and implementation, but this is happening only in a few cities presented in this chapter in a very limited way. These plans shall be considered as a design for societal change and should not be just for bringing out the infrastructure provision required for the bureaucracy to implement. What is happening is also not satisfactory. There is no possibility for a generalised design of spatial and economic strategies for metropolises since each city is unique and mobilising communities for the design is the only way through E-Democracy with intensive use of ICT and IOT by society and continuing training for smart community based metropolitan development by every household.

1. This first chapter explores large cities of one million and above with special references to the design of economic and spatial strategies of the smart metropolitan region. With such concentration of people in the limited area these cities have an important role in economic development of the nation. Further bigger the size of the city in population and area the spatial issues are more complex and challenging. Although Master Plan is expected to sort out these issues, the author would like to explore how far it is achieved in a study of 17 metropolises around the world.
2. Metropolitan cities are continuously exposed to external economic changes and require medium term strategies say 5 years to face it. These economic strategies of cities call for differing spatial strategies to intervene in emerging global situation.
3. There has been an attempt to convert many metropolitan cities, largely using selective investment strategies and procedural administrative rules arbitrarily not sanctioned in the constitution in a limited part of cities with no economic and spatial rationale to smart cities world over. In many smart metropolitan city projects, only a part of the city has been taken up for the smart metropolitan city program and in many cases, it is status and elitist area of cities like New Delhi Municipals Committee area in Delhi and not where low-income people live under 100 smart cities project of India. What is required is to take the whole metropolis to convert it to smart metropolis using smart economic and spatial strategies in many steps and in many years. Designing economic and spatial strategies for Smart Metropolitan Regional Development shall be conducted by the spatially identifiable economic community at micro levels from time to time to suit the ever-changing scenarios of the economic environment.
4. If a city also performs an important commercial, cultural and political function for its region or even the whole country can be called a metropolis. Such a major importance can usually be assumed for cities boasting some 500,000 inhabitants in Europe and one million plus in India. On the other hand, Global cities are those selected few metropolises whose political, cultural and commercial influence extends across the entire globe (e.g. New York City, Tokyo or London).
5. As social spaces, metropolitan areas or regions can be characterised by the following four dimensions: Metropolitan regions are defined as an accumulation of specialised metropolitan facilities including public and private services. In terms of actors and actions, metropolitan regions constitute an arena for key regional stakeholders to exchange knowledge on joint regional

objectives, strategies and projects, as well as on the necessary organisational structures. In the context of spatial development, metropolitan regions are a normative and guiding concept intended to contribute to innovation, creativity and economic growth. About the symbolic dimension of urban and regional development, metropolitan regions are the medium of symbols, norms and values which convey aspects associated with the specific brand image of metropolis and urbanity.

6. Authors of the book "Geographic Information System for Smart Cities" Edited by T. M. Vinod Kumar examined several definitions of smart cities and were not satisfied and came out with their own definition. "The smart metropolitan (area/region) city is a knowledge-based city that develops extraordinary capabilities to be self-aware, functions 24 h and 7 days a week, communicate, selectively, knowledge in real time to citizen end users for a satisfactory way of life, with easy public delivery of services, comfortable mobility, conservation of energy, environment and other natural resources, and creates energetic face to face communities and a vibrant urban economy even at time of national economic downturns". All six components of smart cities such as Smart Economy, Smart People, Smart Governance, Smart Mobility, Smart Environment and Smart Living is implied in this definition has been presented in detail in the four books of this series by the author.
7. The concept global economy with partnership and division of labour of global network of cities is not new, but those who talk about a globalised economy insist that there have been distinct changes in its structure and modes of production. Whereas earlier economic activities crossed national boundaries, globalisation includes a deeper integration, where transnational corporations orchestrate production from various locations. Global places indicate a rising of networked society globally which can work effectively in a smart metropolis with high endowments of ICT and IOT.
8. The Global Urbanisation has five basic elements namely, new innovative technology, the centrality of information made possible by instant communication, an increasing trend toward the standardisation of economic and social products, growing cross-national integration, and mutual vulnerability stemming from greater interdependence.
9. UNHABITAT defined the metro cities as 1 million plus agglomeration 'mega-city' as with 10 million plus population and 'Meta city' which describes 'massive conurbations of more than 20 million people or above. Since ICT drives the mega city and meta city formation, these cities are smart cities in making.
10. In 2016, there were 512 cities with at least 1 million inhabitants (metropolitan region/cities) globally. By 2030, a projected 662 cities will have at least 1 million residents. As hubs of trade, culture, information and industry, they will be vested with such power that at many levels they will act as city states that are independent of national and regional mediation. Today mega cities are home to less than 10% of the global urban population.

In 2016, there were 31 megacities globally and their number is projected to rise to 41 by 2030.

11. There were eight meta cities above 20 million in 2016 in the universe, which is likely to be twice about 15–16 in number or little less in 2030 which may be considered as the accelerated graduation of mega cities to meta cities. However, the number of mega cities increase during this period which includes meta cities were from 31 to 41 which is not as impressive as that of the meta-cities.
12. Meta cities and Megacities are a key to globalisation, a state of interconnectedness around the globe that transcends and largely ignores national boundaries. Global urban economies rely on advanced producer services such as finance, banking, insurance, law, management consultancy, advertising and other services. The technology revolution has made it possible for business enterprises to hire these services anywhere in the world.
13. If mega cities including meta cities represent the economic hub of the future it is shifting towards Asia from America and Europe. Integrated spatial and economic strategies can help this transition. As per the current trend GDP growth rate in Asia is much higher than Europe and Americas and it is likely to continue for few decades. It looks like an Asian and African era is emerging. Growth rates of many mega cities in Asia and some in Africa are growing at higher rate than elsewhere
14. When compared to China, India and the US; the largest urban population is in China followed by India and US. The total urban population in India is higher than the US. The percent of the urban population in mega cities in India are more than US and China in that order. China has all its metro and mega cities in the East facing the sea leaving a vast stretch in the western area without mega, meta and metro cities. China is fast moving towards a one billion urban population. Among the urban population, the largest percent people in India live in megacities which are more than 10 million population. In Kerala state, India 76% of the urban population lives in million plus cities, followed by Maharashtra state (59%), and Gujarat (55%). The lowest percentage of people in India living in one million cities and above are Haryana (16%) and Bihar (17%). The largest number of million plus cities are in Kerala (7), and Uttar Pradesh (7) followed by Maharashtra (6) and Gujarat (4). The lowest number of one million plus cities are in Haryana and Bihar with one each.
15. There are already networks and corridors cutting across the boundary of nations which has given rise to the formation of urban settlements in 2016 in a globalised world.
16. Often the share of GDP produced in this mega and meta-city are considerable in comparison to total National GDP.
17. The projected GDP 2025 of top 30 Urban Agglomeration was used with projected population computed by UNHABITAT to derive average real GDP/population growth rate projected during 2008–2025. This shows that

GDP/population growth rates of largest 30 urban economies in the universe are generally low but there are few exceptions in the south and eastern urban agglomeration.

18. Tabulating under descending order of high GDP growth rate of first 30 urban agglomerations by GDP growth rate 2008–25 and then cities are organised by countries and found that these urban agglomerations are all in Asia and Africa. It can be seen the higher GDP performance cities are in large number in India followed by China and other countries. This justifies a maximum number of case studies in this book from Asia and Africa.
19. There is a division of labour across city regions transcending country boundaries sharing the economic responsibilities as per capabilities. Global urban economies rely on advanced and standardised producer services such as finance, banking, insurance, logistics, law, management consultancy, advertising and other services. The technology revolution and uniformity of standard practices have made it possible for business enterprises to hire these services anywhere in the world. The national hierarchy of cities and the spatial division of labour within the economy is superimposed by a global division of labour. Cities and metropolitan regions become part of an emerging international hierarchy based on a competitive division of labour at the global level by international connections that affect financial flows and the knowledge-intensive service sector.
20. Large cities are interconnected and influence global and reinforce spatial interdependent functional structure with appropriate linkages. The connection is twofold within its city region and outside the city region transcending other national boundaries. The city is connected to hinterland and outside world simultaneously in a metropolitan region.
21. World cities are characterised by a sum of political power (both national and international) and organisations related to government; national and international trade, whereby cities function as gateway for their own and sometimes also neighbouring countries; providing superior banking, insurance and related financial services; advanced professional activities of all kinds; information gathering and diffusion. The form and extent of a city's integration with the world economy and the functions assigned to the city in the new spatial division of labour will be decisive for any structural changes occurring within it.
22. The world cities exhibit the following;

    i. The nature of a city's integration with the world economy is decisive for any structural changes occurring within it.
    ii. Key cities throughout the world are used by global capital as 'basing points' for the organisation of production and markets.
    iii. The global control functions of world cities are manifest in the structure of their industrial structure and job markets.
    iv. World cities are major sites for the concentration and accumulation of capital.

v. World cities are destinations for large numbers of migrants—both domestic and international.
vi. World city formation exposes the major contradictions of industrial capitalism, particularly spatial class polarisation.

23. There are two trends of thoughts about Meta cities, Megacities, and metro cities. One tries to attain a position of the global city by deliberately executing spatial and economic strategies to achieve that goal following what is discussed in the earlier paragraphs. This is in addition to solving many city specific issues of metropolitan development. The second approach is how to make a metropolitan area a smart metropolis by appropriate spatial and economic strategies. Smart is ICT implementation in all activities including economic activities. Economic strategies may involve converting the region with all its economic activities to the smart economy as discussed in the book in this book series "Smart Economy in Smart Cities".
24. Since this book focuses on the area of design of economic and spatial strategies to achieve the overall goals of smart metropolitan development. A survey, of such designs of several official plans is undertaken here. Cities are selected randomly with no sampling plans. The following discusses only the appraisal of these economic and spatial strategies in several metropolises.

    i. **Delhi NCR Metropolitan Region**. Although Delhi will be the second largest meta city in the world in 2030 after Tokyo and therefore is a global city. There is no such consideration in the NCR Plan. Also, there is no deliberate promotion of the city to a smart metropolitan city with appropriate ICT-IOT design and related spatial strategies. The economic strategy adopted by the NCR planning board is not efficient enough for a city such as Delhi. The stated Delhi strategy of making trade and commerce barrier free which is the aim of Goods and Service Act 2017 nationwide within NCR will not suffice for Delhi. Other specific issues mainly administrative issues are not addressed for further easing to do business and giving impetus for the development of industries. Interstate agreement on unified policies can create chaos and may reach a long time to reach consensus. This might be a hindrance to further development of NCR region. Increased ease of doing business, transparent laws and regulations for the same have not been given due consideration. Also, focus on people and skill development for economic development is not visible. Smart mobility based spatial strategy is not stated.
    ii. **Vancouver Metropolitan Region**. In the case of Vancouver, there is a strong economic strategy in place. Local businesses and talents are nurtured which results in a booming diverse economy. Brain—drain is prevented by attracting foreign Canadians and immigrants with attractive and affordable housing and environment for families. This is a strategy to attract a skilled workforce. Integration of universities with business centres for productive R&D yet another important step in

economic growth. They have thoroughly invested in the clustered growth of industries with an increased focus on green jobs to remain sustainable in the long run.

iii. **Melbourne 2030**. Planning for Sustainable Growth. Plan Melbourne does not identify how the government will respond to the impacts of climate change. The Metropolitan Planning Authority must work in close partnership with Local Government, in the future planning for Urban Renewal Areas. Initiatives to 'streamline' the planning system which limits community involvement in the planning process, or reduces Local Government's decision-making responsibilities, are not supported.

iv. **Shanghai Metropolitan Plan**. Shanghai is now facing great opportunities in turning itself into an international economic, financial, and trade centre. China's continuous economic growth provides a solid base for Shanghai to move toward this goal. Shanghai has set its long-term strategic objectives for social and economic development. By 2010, Shanghai is planned to become one of the international economic, financial and trade centres of the world. Shanghai aims to form the economic scale and comprehensive strength of a world metropolis; optimise urban spatial distribution, modernise the city's physical infrastructure, participate in international labour division and the circular flow of the international economy, introduce the operational mechanism of a socialist market economy, and pursue the balanced social, economic and environmental development. Economic Strategies plan is to optimise and upgrade the industrial structure, trying to improve the pivotal status of Shanghai in global city network and process to accelerate the technological innovation.

v. **Mumbai Metropolitan Regional Development**. The Economic strategy given by the metropolitan plan is very broad. Mumbai Metropolitan Regional Plan 2016–36 mainly addressed issues of growing urbanisation, uneven distribution of jobs increasing commutes, lack of affordable housing and basic infrastructure in the region, environmental degradation and inadequate governance. The existing situation of Mumbai Metropolitan Region is analysed sectoral and cross-sectoral to arrive at issues that need to be addressed in the Regional Plan 2016–36.

vi. **New York**. The economic strategy adopted clearly touches every aspect of the economy. The plan also emphasised on strengthening the fundamentals such as workforce which is very important in achieving economic development. The New York Metropolitan Region has a separate spatial plan, transportation plan and economic development plan. The strategies adopted clearly ensure sustainable and long-term growth in jobs and income to contribute to the resurgence of the broader economy of New York State. Also, they have a detailed transportation development plan which has a shared vision for sustainability.

vii. **Calgary Metropolitan Region**. The Calgary Regional Partnership (CRP) and member municipalities will work together to ensure a diversified and globally competitive region that continues to enjoy a high quality of life and is able to attract and retain a viable and adequate regional workforce and member municipalities will endeavour to achieve a distribution of jobs creation and economic activities throughout the region consistent with transit and complete mobility policies that encourage the location of jobs close to where people live. Recognising the connections and relationships that exist between communities, the Calgary Metropolitan Plan (CMP) acknowledges and respects the vital and historic importance of rural lands, industry and culture in the region. What are the industries and where those can be established to boost the economy and employment is not stated? What are the land management techniques suitable in this region are not mentioned? There is a lot of potentials to develop tourism which is not at all described. They mentioned that various studies need to be done to analyse the economic boost but what are those not stated. The Calgary Metropolitan Plan is the blueprint for accommodating growth in future. The plan giving the various strategies to make Calgary as a healthy environment in enriched communities, with sustainable infrastructure and a prosperous economy.

viii. **Kuala Lumpur Metropolitan Regional Development Plan**. The vision and goals for Kuala Lumpur have been formulated with the aim of creating a sustainable city. City Hall Kuala Lumpur (CHKL) shall ensure that the planning of the City shall strike a balance between physical, economic, social and environmental development. Local Agenda 21 shall be adopted to encourage citizen participation towards creating a sustainable society. This is in line with government policies of implementing sustainable development strategies as stipulated in the Habitat Agenda of the Rio Declaration.

ix. **London Metropolis**. The London Plan sets out a new approach for planning in London. It emphasises growing inward and upward so that it can reduce the costs of growth, create walkable communities, revitalise our urban neighbourhoods and business areas, protect our farmlands, and reduce greenhouse gases and energy consumption. The plan sets out to conserve our cultural heritage and protect our environmental areas, hazard lands, and natural resources. Through the London Plan, the community is planning for vibrant, healthy, safe and fulfilling neighbourhoods, attractive and viable mobility alternatives and affordable housing that is accessible to those who need it. At the root of The London Plan is the goal of building a city that will be attractive as a place to live and invest in a highly competitive world and one that will offer the opportunity of prosperity to everyone—one their own terms and in their own way.

x. **Berlin Plan 2035**. The Berlin Strategy provides an inter-agency model for the long-term, sustainable development of the capital. With one-third of the city comprising of open spaces, a compact polycentric development, highly tolerant society, the development plan further proposes to enhance these key selling points of Berlin. The development plan builds its foundation on the strong points of Berlin and proposes strategies to tackle the challenges. Provisions of affordable housing, further increasing the short distances to amenities, enabling a start-up friendly environment are some of the key proposals of the plan. Cultural diversity and tolerance in the society are used as a selling point for the attraction of workforce and tourists. Community participation is given utmost importance. The economy focuses on the educational institutions, research centres, attracts skilled labour from all over the world. Using a range of strategies and goals, it sets out the areas and directions in which this growing city should develop and highlights the areas that will form the focus of its future development.

xi. **Dhaka Structure Plan 2016–2035**. The Dhaka structure plan covers the basic issues such as effective land use management, transport for efficient connectivity, enhancing employment and productivity, public Facilities, and protecting the natural and healthy environment. The spatial concept of dividing the planning area into two broad categories i.e. Urban Promotion Area (UPA) and Urban Control Area (UCA) is an efficient way to assure uniform distribution of development. Here, the basic requirements are made available at each functional region reducing the dependency on the core city. At the same time under the UCA flood flow zones, Water Retention Area, National and Regional parks, Forest Area, and large-scale Heritage sites are preserved. Instead of focussing on individual plot based development, the concept of planned unit development is introduced. The plan talks about increasing the productivity of the informal sector, which is a good step towards mitigating urban poverty. The garment sector, leather manufacturing which boosts women employment is also planned to make it more efficient. While the report covers all the broad aspects required for the development of a region, the involvement of the public in the planning process would provide a more achievable and efficient plan.

xii. **Master Plan for Patna Metropolitan Region**. The Metropolitan plan for Patna, though it takes into consideration, the Economic aspect of the city, doesn't provide a sufficient Economic Strategy for development. The plan has a concept for the spatial development of the city. The spatial strategy is fairly achieved. The Implementation plan simply explains a case study for TP schemes from Ahmedabad. There is no strategy adopted. Also, it doesn't consider each project in detail. The Economic approach in the Master Plan is only by addressing the land requirement for various employments generated. There is no mention about Economic Strategy. Patna being the only important city in the state has acquired the strong position in regional trade and business. It is necessary to have an economic strategy while planning for Patna. The Master plan lacks this. Instead, an overall development which also leads to economic development is attempted. Implementation plan doesn't consider every project in detail.

xiii. **Montreal Metropolitan Development Plan 2020**. Economic strengths are utilised and creative market and the smart economy are aimed to achieve through these strategies. A clear and comprehensive economic strategy is provided in the Master plan. The implementation strategy to carry out each strategy is also explained. The idea of Smart Metropolis is carefully dealt here which is considered as the most innovative and sustainable development concept in the current scenario. Montreal is a large city with a diversified economy needed a spatial and economic strategy which is formulated and explained well in the Master plan. Issues and potentials were identified and addressed carefully.

xiv. **Helsinki**. The spatial structure for the economic/business development is mentioned. But the kind of economic activities is not clear even though the zones are marked. The spatial zoning of economic activities is done such a way that it is around the city centre and large-scale industries are placed away from the centre which has easy access from nearby places. Retail activities are evenly distributed in all area.

xv. **Istanbul 2014–2023**. The economic strategy discusses all the sectors to make a globally decisive, high value added, innovative and creative economy. The focus has been given to entrepreneurship to encourage the work participation thereby eliminating unemployment that would lead to a better vibrant economy. The main idea of entrepreneurship is to be appreciated as the city will sustain its own economy rather than depending on external revenue as in trade. Even though the strategies and objectives are discussed elaborately the implementation is not detailed which makes it difficult for the region to put these concepts into reality. The Metropolitan Plan discusses a vision for Istanbul, spatial and economic strategy. The very important part was the public participatory approach so that the people of the place make the place. All the strategical domains are made with the opinions of the citizens and thereby solve their issues and each project is for the development of the city and the citizens. The economic domains include the transformation of the industry, ascending R&D, encouraging entrepreneurship, and increasing employment. All these directly or indirectly stand for the citizens. The spatial domains are increasing spatial quality, holistic urban transformation, protecting the culture and history, effective disaster management, sustainable transport.

xvi. **Limerick**. The Limerick city is a place where the economic activity is degrading due to various factors like recession, improper management of assets like tourist spots educational institutions etc. The people of the place are facing unemployment also. The investments from other countries decreased due to the recession. The strategies are made to overcome all these difficulties and to make the Limerick as a thriving economy better than any other surrounding Irish economies. The economic strategy they prepared had three divisions based on their weak points as well as potential, but the spatial strategy is not a well prepared one. It does not take care of the start-ups and business people. It concentrated on the infrastructure like road and buildings with potential, also the quality of standard of living of people, the tourism and recreation will increase the economic activities but for long-term economic growth the facilities for the investors also to be taken care of.

xvii. **Tokyo**. The smart metropolitan city strategy in Tokyo, the "Smart Energy City", was launched in May 2012 and builds on the emergency measures of energy savings which were enforced following the Fukushima accident in March 2011. Tokyo has further developed as a smart metropolitan city through the promotion of renewable energy sources and smart meters in the central wards and the larger metropolitan area, along with several smart community initiatives. The Tokyo Metropolitan Government's (TMG) Bureau of Environment is very active in promoting energy efficiency and has achieved a 15% cut in CO_2 emissions in the industrial sector for instance. Despite such measures though, energy efficiency and carbon dioxide emission targets may be difficult to achieve by 2020. The TMG is cooperating with the Tokyo Organising Committee of the Olympic and Paralympic Games (Tokyo 2020 Committee) to build on the existing green policies of the TMG and elaborate a concrete strategy for sustainable Games preparation and operation. Several obstacles can slow down enterprises' market access in Japan. The complexity of Japanese regulations and the lack of transparency of business practices can be challenging in some sectors for new companies which are unfamiliar with the language and the business culture. While 2020 is a catalyst and deadline for projects that are approved by the central and local government, it does not slow down the development of other smart community projects elsewhere in Japan.

The above presented constitute the conclusions of Chap. 1. What follows is the presentation of conclusions from Chap. 2 onwards of this book.

18.11.2 Hong Kong Study

Background: The Greater Bay Area (GBA), originally known as the Pearl River Delta (PRD), comprises 9 cities in Guangdong Province of Mainland China (Dongguan, Foshan, Guangzhou, Huizhou, Jiangmen, Shenzhen, Zhaoqing,

Zhongshan, and Zhuhai), and the two cities of Hong Kong and Macao in their respective Special Administrative Regions (SAR). The GBA has and continues to rapidly develop into a significant and influential mega region in China, Southeast Asia and the world. With the new initiative from the Central Government towards the development of the Guangdong-Hong Kong-Macao Bay Area (GBA), along with the Belt and Road initiative and adoption by Hong Kong and other cities in the region, the GBA is set to take its development to the next level with smart regional metropolitan development. As such, the GBA provides a great opportunity for Hong Kong to reposition itself within the region, and hence has been considered more seriously in recent years and is referred to in the 2017 Policy Address.

Within less than 50 years, the economic reforms of China and the Open-Door Policy in 1978 transformed the primarily rural agricultural region into the highly populated, sprawling and urbanized high-tech value-added manufacturing region it is today. During this period, Hong Kong played a significant role in the growth and development of the GBA, with its more mature economy providing crucial support through its financial institutions, legal and professional services, and developed international trade and logistics hub, allowing businesses in the GBA to scale and reach global markets for their products. This in turn established economic, social, and political connections, collaboration and cooperation between Hong Kong and cities in the GBA such as Shenzhen, Guangzhou, Zhuhai and Macao among others, as well as improvements in infrastructure and transportation networks that have enhanced regional accessibility and integration. Going forward however, Hong Kong's role and competitive advantage in the region is being challenged as other cities in the GBA continue to grow larger than Hong Kong both demographically and economically. As such, Hong Kong is at a time where it must both reposition itself within the region and safeguard its global role, especially as Hong Kong becomes more and more integrated with the rest of the GBA in Mainland China.

Assessing Smart City Smart Region: Over the years, the Hong Kong and GBA governments have developed and implemented numerous initiatives, both hard and soft, which have brought the two places together through business collaborations, social interactions, as well as environmental cooperation and political agreements among other new initiatives. The design of the spatial and economic strategies of Hong Kong and the GBA has therefore reflected these developments and initiatives over time.

Predominately, Hong Kong has played a vital role in the growth and development of the GBA and its industries by undertaking the role of the logistic, transport and economic hub, and global connector of the region. As a result, partnerships and collaborations in the form of economic agreements to encourage investment and ease regulations, setting up of professional institutes and Science and Technology parks to facilitate R&D and innovation, and increased business development between Hong Kong and Mainland companies, have helped spur the economic growth of the region. Facilitating this growth is the provision of new major infrastructure projects and transportation links such as the HZMB and XRL, which have unfortunately suffered from delays and cost overruns, taking longer than anticipated to implement. This is in addition to extensive existing multi-modal transport networks and routes, with a growing number of border crossing points that facilitate the flows of people, goods and services between Hong Kong and cities in the GBA. Moreover, it is great to see that there is a push towards more sustainable transportation throughout the GBA with the promotion of electric vehicles and dock less bicycle sharing systems.

Hong Kong and the GBA have also been cooperating on numerous utility and energy infrastructure projects within the region through partnerships and investments in new facilities that allow the joint production and sharing of energy, as well as regional environmental policy and target plans to transition to clean energy and reduce harmful emissions in both road and marine transportation. Moreover, collaboration on improving the living environment and ease of cross-border commuting for both Hong Kong and Mainland citizens has also been occurring. This has led to the proliferation of the number of cross-border commuters for work, leisure, education and healthcare services, as well as increasing investments and partnerships between higher academic institutions in Hong Kong and the GBA, as many universities have set up campuses or partner institutions in Guangdong province to capitalize on the talent and resources in the GBA. This has established significant spatial and social connections between Hong Kong and the GBA, which will only grow more prominent as closer integration occurs.

In planning for the future, recent spatial master plans and policy directives of Hong Kong and the GBA reflect a more collaborative and integrated regional approach to future and studies regarding the development of the GBA. This includes high-level national support from top levels of Chinese Government, in the "Outline of the 13th Five Year Plan for the National Economic and Social Development of the People's Republic of China" led by the National Development and Reform Commission, which addresses the development plan for a city cluster in the Guangdong-Hong Kong-Macau Bay Area, i.e., the GBA, and emphasizes Hong Kong's unique functional positioning and strengths. As well as the development of regional joint-jurisdiction initiatives, plans and studies among the GBA governments of Hong Kong, Macau and Guangdong to cooperate and collaborate on regional issues and decisions have materialized, such as the 2009 Planning Study on Coordinated Development of the Greater Pearl River Delta Township; 2012 Regional Cooperation Plan on Building a Quality Living Area; and Hong Kong's initiative as part of its 2017 Policy Address to set up the Guangdong-Hong Kong-Macao Bay Area Development Office to more effectively coordinate regional GBA issues by engaging and connecting with businesses and key stakeholders in the region, among others. In Hong Kong, particularly within the public sphere, there is a growing consideration and discussion regarding the need to plan for a future within its regional context. A recent workshop on "Smart City Smart Region" was held on 25th October 2017, in which discussions focused on how Hong Kong can transition towards a smarter city that is more integrated with the GBA. And on December 16th 2017 at a symposium titled "Future Cities, City Futures: Hong Kong and the World", the Hong Kong Government has recognised the importance of the GBA in future planning by explicitly talking about the Hong Kong 2030+ Territorial Development Strategy in the context of the GBA.

Towards Smarter Regional Development: In general, a clearer definition of the meaning 'smart' is needed, and as a society we need to go beyond thinking just about technology to develop smart cities, with more thought put towards focusing on people, place and planet that will result in a more liveable city that is smarter, sustainable and more resilient. Much of Hong Kong's opportunities will result by considering the city within the context of the GBA and will come from a smart and sustainable development approach that capitalizes on Hong Kong's inherent strengths and unique resources vis-a-vis other areas of the GBA that are complementary to the region, such as its strategic location, international positioning, established economy and industries, and educated talent pool. Hong Kong should also build on its advantage as a gateway city, in order to further its dominant role as a hub in the region and embrace opportunities beyond the GBA looking towards Southeast Asia and the Belt and Road Initiative. Given its strengths, Hong Kong must not become complacent, and should continue to strive to diversify its economy by embracing and supporting the creative, innovative, information and technology industries given its abundant resources in academia, talent, including its research and development capabilities.

At a social level, it is believed that attitude and mind-set change is needed, especially among the younger generations, to accept and embrace the regional opportunities that more cooperation and integration with the GBA will bring. It was recognized that more education, awareness and exposure for Hong Kong people, including the younger generations, with more frequent visits to the region and beyond, with more knowledge sharing among peers is the best way forward to achieve this. Improvements to living quality could come in the form of higher quality public spaces to encourage more diverse and engaging social interactions, and greater recreational interaction with the GBA cities. At the technology level, people in Hong Kong should be encouraged to be more forthcoming to share their data and information, and contribute 'smart' ideas with academia, the public and private sectors for more big data to be used to optimise and create solutions that directly benefit the specific needs and aspirations of the society.

Most importantly however, is the need for strong leadership and a strategic vision, guidance and direction of Hong Kong's Government and key decision-makers. At a political level, a bolder, and less risk-averse proactive Government with the ability to adopt more translucent and flexible processes is considered necessary to effectively and efficiently enact change to make Hong Kong more liveable, and transition to a truly smart, sustainable and resilient city that will effectively be able to collaborate and better integrate with the GBA region. Additionally, greater ability to turn public criticism into positive change through communication and understanding of what the younger generation needs and aspire for the future of Hong Kong is critical in uniting its society for enhanced local and regional integration.

Going forward, all future spatial, economic, social strategies and developments of the government, businesses and citizens of Hong Kong should be carefully considered and planned within the context of the GBA region, and not just locally. Since the 1970s, thc design and collaboration of spatial and economic strategies between Hong Kong and the GBA have been an integral component of the success and growth of the region's economy and urban development. The strategies have enabled Hong Kong to become one of the world's global cities and economies, as well as cities such as Shenzhen and Guangzhou to rapidly grow and develop their manufacturing and industrial economies to become the factory of the world. Now the GBA has grown and swelled into the mega-city region it is today, the next chapter for Hong Kong and the GBA will be one of closer regional integration and collaboration, and a push towards high-tech innovation and knowledge based advanced manufacturing. These spatial and economic strategies will need to be updated and improved upon to support and encourage the complementary growth of Hong Kong and cities in the GBA as one united mega-city region; through reductions in travel times, increased accessibility and cross-border commuting, improved urban liveability and quality affordable housing, enhanced academic and professional knowledge exchange, data and information sharing, and construction of collaborative smart developments along the border of Hong Kong and Shenzhen to promote greater integration and collaboration in society, technology and urban development.

Finally, it is best if Hong Kong as a city and Hong Kong citizens begin to understand and capitalize on the opportunities available, and embrace integration with the GBA as the way forward for Hong Kong to re-position itself and remain competitive in the new era of mega-city regions. This will require bolder, more risk-adverse decision-making authorities, greater community stakeholder engagement, and a need to look at Hong Kong as part of the GBA, which may mean establishing new spatial nodes within urban areas such as to potentially develop Victoria Harbour with the several development nodes and create new opportunities for leisure and recreations within the city and beyond. Hong Kong needs to be well prepared for future integration and growth of both new and existing industries, demand for housing, and the advancement of innovations in technology. As the global economy and spatial structure changes over time, Hong Kong needs to act swiftly yet smartly to move forward and retain the city's strategic global role and regional positioning.

18.11.3 Stuttgart Study

The Stuttgart region is a very prosperous area. The economical backbone of the region is the manufacturing of industrial goods. The globalization is a trigger for changing economical dependencies. Industrial processes are changing because of the new challenges. Digitalization is a current topic which has an influence on economical procedures. The main topic is the area "Industry 4.0". Next to the impact generated by the new thinking of manufacturing the digitalization open up new approaches.

The Stuttgart Metropolitan Region is situated in the centre of Baden-Wuerttemberg. This federal state is economically the most powerful area in

Europe. It creates 3% of the economic value in the European Union. The economic potential is mainly based on producing goods for industrial purposes, as machines, process technology and automotive parts. Based on the generated turnover, companies in these fields are contributing around 80% of the total economic value in the production sector. The companies in Baden-Wuerttemberg are mainly small to medium sized enterprises additional a high number is still family-run.

Coming to Stuttgart Metropolitan Region the focus lies in automotive products either cars or supply parts and products. In Stuttgart city and its surrounding which can be called Stuttgart Metropolitan Region over 2.7 million people live. The high density of people plus the widely spread industrial areas of different sizes generates a high number of commuters. The economic power with its potential for wealth and jobs let increase all living areas in Stuttgart Metropolitan Region. Additionally, Stuttgart was rated number one in the culture of all Metropolitan Regions in Germany. The Stuttgart Metropolitan Region is also seen as High-Technology location. The reason is the high amount of headquarters or big production and research location of multi-national companies with technical products. Well-known are Daimler, Bosch, Porsche, IBM, HPE and Hewlett Packard. Characteristically is the mixture of global companies and highly innovative medium-. sized companies. Innovation and research are well integrated into the Metropolitan Region. With a high number of universities and university of applied science and world-known research institutions (5 Fraunhofer-institutes, 2 Max-Planck-institutes and 6 institutes of DLR, plus others) the region has a high potential for innovative research and development in technology, products and services.

Location of industrial production and living have shorter distances. Additional it is common sense that sustainability is increasing in importance. Companies in Germany have already integrated the thinking on sustainability in their strategies because legal regulation forces them. However, the potential can be even extended when all entities in an industrial estate collaborate and use optimal technology. In this chapter it is presented how a framework for collaborative long-term development could be designed. It is presented how an implementation could be done on the different level of detail. This framework is currently on the way of implementation in the different industrial area in the Stuttgart Metropolitan Region and Baden-Wuertemberg. As stated the distance between living and industrial manufacturing zones is decreasing which requires new ways of processing.

Current discussions on sustainability are influencing the economic thinking. Obvious is the importance of sustainable development with respect to environment and climate. The climate conference in Paris was finished with a consensus of all present parties. All important economic nations created a common sense of climate targets. Reasons for that are limited resources and a huge amount of emissions.

Nevertheless, the society will only accept limitations, if not a reduction of living is a result. Based on that, it is necessary to implement a new way of doing existing structures. Additionally, it is required to implement changes without reduced economic potentials. The approach of symbiosis can support these developments. However, technology is a required extension to reach the target of sustainability.

In this chapter, different views are discussed to sustain long-term development. The first view is holistic. The idea of creating a symbiosis between enterprises in an industrial estate is easy to realize. A way is presented how industrial estates can be transformed towards sustainability. It is presented which methods can be used to support the transformation process under respect of strategical guidance. Based on the size and capacities of the industrial estate approaches of different detail and complexity. For the transformation companies change and also partners in the network. The approach shall help to identify suitable partners and matching cooperation. The long-term orientated creation of networks is supporting sustainable links with a high potential to reduce negative effects without reduced economic power.

Next to sustainable development of industrial estates the companies need a strategy for their development. A development path towards urban manufacturing is helping the management of companies to create a strategy. A concrete step-by-step approach is a base for an enterprise-internal transformation. Starting as conventional production enterprise a way for sustainable manufacturing is presented whereby different areas must change. The development path presented is supporting the strategic change process. With simple steps in different fields the companies are supposed to reach the target stage easier than without the development path. The target stage is depending on the size and power of the company independent of all conditions and can be defined freely.

The additional technological driver is analysed in its possibility of supporting the sustainability. Additive Technology is a new technology which supports a total change in production processes. Furthermore, the new technology supports sustainability by design because the potential currently is seen is a reduction of used material, reduce of waste, energy efficient processes, less effort in the adaption of processes even with a big variety of products. Nevertheless, the new technology requires a change in thinking and an adaption in processes. It is presented what areas in the value creation process are influenced by changes. An insight on use-cases is completing the analysis of the additive manufacturing technology.

18.11.4 Conakry Study

Considering its geographical location as a peninsula in the Atlantic Ocean, Conakry has the potential to be an agro-industrial, green city in addition to its huge potential of fish production. Wetlands areas are sources of income from agriculture, livestock, crafts, among other activities. They play an important role in maintaining the water quality and the prevention of natural hazards. These are also ideal places for reception and reproduction of waterfowl which are indispensable elements for the ecological balance of aquatic environments and key links in the food chain, hence the importance of the preservation of wetlands. If well planned and designed, Conakry can be a smart metropolitan region where citizens enjoy a high quality of life. In addition to its geographical advantages, the Conakry Peninsula has a population of nearly of 2 million that, as large urban agglomerations, constitutes economic and social opportunities as well as challenges depending on how the urban growth is planned and managed. However, Conakry as most African cities has not been able so far to respond to growth in accessibility demand and several other needs such as access to water, sanitation, management of solid waste, and streets and other public spaces key elements of a city foundation.

Considering the proliferation of informal settlements along the rapid urbanization of the Conakry Peninsula and surroundings, the Government of the Republic of Guinea is developing a holistic approach, the "Grand Conakry Vision 2040", for smart metropolitan regional development through economic as well as spatial design strategies. The Grand Conakry Vision 2040 considers three spatial design strategies: (1) at the metropolitan level to develop and strengthen urban polarities; (2) at the agglomeration level to channel and structure urban extensions and; (3) at the peninsula level to develop urban renewal approaches. The main objective of the Grand Conakry Vision 2040 is, indeed, to "improve the living conditions of the residents of Conakry and adapt land-use planning and policies to the rapid urbanization of the metropolitan region". This is built on seven ambitious goals for a sustainable Conakry for all: (1) Strengthening territorial balances to create a network of solidarity-based cities in a preserved environment; (2) Controlling the city limits to prevent urban sprawl and preserve urbanization; (3) Optimizing the port system to decongest the city and protect the populations; (4) Restructuring the centrality to rebuild an efficient urban system on the peninsula; (5) Restoring the landscapes to build a healthier, safer and more enjoyable city; (6) Making the city sustainable to provide housing for everyone in mixed and lively neighbourhoods; and (7) Thinking waste as a resource to protect people, the environment and generate wealth.

An important aspect noted in the Grand Conakry Vision 2040 is that the spatial design of the metropolitan region is aligned with specialization: (1) Kindia has a strong potential for agricultural and agro-industrial development as well as for tourism development; Fria is an old mining industrial city (including aluminium); Tanéné, as a crossroads city, will ease the circulation of good in the metropolitan region and beyond; Boffa has a diversified economy (fishing, agriculture, solar salt, tourism, mining, etc.); Maférinya has developed agricultural production (particularly in pineapple); and Forécariah has various agricultural activities (banana, pineapple, mango, oil palm, etc.), particularly for export.

However, this specialization calls for efficient transport infrastructures that must be assessed, planned and implemented. Efficient mobility allows localities of urban agglomerations to specialize in the production of goods and services for which they have comparative advantages and ease inter-localities cooperation. This will also allow large-scale production of goods and services that can be distributed within the metropolitan region and beyond with time, cost and reliability opportunities. The spatial design of the Grand Conakry Vision 2040 is, indeed, aligned with the needs for transport with most of the urban centres of the metropolitan region connected to the main road of the city (PK36). In addition to that, transport strategies for the Grand Conakry Vision 2040 include: (a) Establishment of traffic restriction;

(b) Structuring a road network at the scale of the Greater Conakry metropolitan area; (c) Reinforcing the railway; (d) Organize maritime connections; (e) Develop a BRT system on its own site and improve the existing urban network; (f) Implement strategies for stationing; and (g) Gather typologies of structures and modes of maintenance of the road network. At long term (2020–2040), transport strategies will include: (a) deployment of a network of peninsular-scale secondary roadside stations (associated with the markets) and urban polarities; (b) the completion of the metropolitan road network and the organization of interurban public transport; (c) the continuation of the improvement of accessibility in neighbourhoods in relation to urban renewal actions; and (d) Structuring of the passenger rail mode.

18.11.5 Ahmedabad-Gandhi Nagar Study

With India transforming as matured democracy, the government is focusing on improving quality of life of citizens by urban renewal and infrastructure development vide ambitious smart cities project. Energy, the electrical power, has been the most crucial and the resource always in scarcity in India and proving itself as a major bottleneck. Therefore, India has been transforming legacy conventional non-smart non-intelligent unidirectional electrical power grids into modern smart grids which are bidirectional and intelligent in nature by leveraging ICT, IoTs, e-Governance and e-Democracy. Smart grids are likely to serve as energy backbones of smart cities and involve the high interactive participation of citizens in energy management, based on humanitarian and customer centric approach. Different types of Prosumers (Producers + consumers), their different energy requirements at different timings, different types of energy resources and their switching feasibilities considering different aspects have been integrated.

The Ahmedabad-Gandhinagar twin city metropolitan region is situated in between Ahmedabad and Gandhinagar cities and Naroda is a region of interest and study. This location is one of the most enterprising, industrial and commercial one in the state of Gujarat, India. The economic potential of the region is found mainly focused upon industrial, commercial as well as large scale residential citizens. The region is surrounded by reputed industrial, commercial, educational-research institutes and heritage monuments and demonstrates extremely encouraging potential for creative research, technological developments, as well as interesting possibilities of meaningful cultural exchanges.

The historic city of “Ahmedabad” or “Amdavad” or “Ahmadabad”, the largest city of Gujarat, has been recently declared as India’s first UNESCO World Heritage City in July, 2017. In 2010, Forbes magazine rated Ahmedabad as one of the fastest-growing city in India and the world. In 2011, Ahmedabad was rated India’s best megacity to live in by leading market research firm IMRB. Again in 2012, The Times of India chose Ahmedabad as India’s best city to live in. With a population of more than 6.3 million and an extended population of 7.8 million, Ahmedabad has been the fifth most populous, sixth largest and seventh most populous urban agglomeration metropolitan area in India. Ahmedabad is located on the banks of the Sabarmati River, 30 km (19 mi) from the state capital Gandhinagar, which is twin city of Ahmedabad forming Ahmedabad-Gandhinagar twin city metropolitan region. Gandhinagar, the capital of the state of Gujarat in Western India, located approximately 23 km north of Ahmedabad, on the west central point of the Industrial corridor between Delhi and Mumbai. Built with parks, extensive plantation and recreational areas along the river, Gandhinagar has a green garden-city atmosphere. Naroda is a fast-growing area on the northeast side in Ahmedabad city, situated on the emerging Gandhinagar-Ahmedabad-Vadodara (GAV) corridor. With the establishment of the Naroda Industrial Area in the 1980s, it progressed well as separate town and later incorporated into Ahmedabad in 1996. Being eight km from Ahmedabad International Airport and located on the SP Ring Road, Naroda has over the last two years transformed from a neglected industrial area to desirable location for homes. The Naroda GIDC industrial park hosts national and multinational corporations. Several major township projects are being developed along the Ahmedabad-Vadodara expressway with Naroda being at the centre of the development.

This chapter presents an approach for metropolitan region development by effective energy management, active citizen participation and e-governance by making a proposal recommending deployments of smart grid and smart buildings with the integration of renewables, ICT and IoTs. In this work, ensuring 24 × 7 electricity supply along with limiting carbon footprint has been taken up as the major challenge and connectivity has been identified as a major bottleneck. Main objectives of this work have been to study existing economic and spatial strategies and recommend suggestions for the smart development of the metropolitan region. A five-step methodology has been suggested for the implementation of recommended approach:

(1) Integration of renewables

This has been covered by authors in "E-Governance for Smart Cities" Book [5] in this book series with a case of Gandhinagar Solar Photo Voltaic Rooftop Program in by including case-study of application of hybrid communication technologies deployed to serve need based data along with Development of Remote Energy Parameter Monitoring System. The proposed and implemented system has distinct features such as an affordable cost, scalability and anytime-anywhere monitoring, to encourage the inclusion of more sensors for enhanced data acquisition, improved spatial resolution for more fine-grained measurements and better monitoring of critical regions.

(2) Smart Energy (Deployment of Smart Grid)

Authors had also presented [6] democratic and citizen-centric approach of design-implementation of architectural details along with presenting a useful framework to make Smart Grid more inclusive, effective and comprehensive. Descriptions of communication technologies in form of instrumentation telemetry deployed to timely serve need based bidirectional information between utility and end users have also been included.

(3) Smart People (Prosumers) and their participation

In E Democracy for Smart Cities book of this book series authors presented, smart grid as an energy backbone of Smart City is immensely vital and serving at the core of Smart City realization. Evolving e-Democracy, the smart grid includes the highly interactive participation of citizens in energy consumption domain, based on humanitarian and customer centric approach. Different types of customers, their different energy requirements at different timings, different types of energy resources and their switching feasibilities considering different aspects have been integrated. Critical smart grid subsystems (such as BAS, HAN, AMI, DR, etc.), ICT integration and GUIs have been identified as some of the major design considerations. 'Transformation of Conventional Consumer into Smart Prosumer' has been a major outcome of the work presented, since the customer has been now enabled as producer of electricity, thereby contributing to the grid and getting credits, which is adjustable against consumption.

(4) Smart Buildings with energy efficiency

Bhatt and Verma [7, 8], indicated the transformation of legacy stand-alone security systems into intelligent computerized-network based building automation systems, and presented design-development of IoT based working models for security and HAN-BAN. Development of a web-based virtual instrument to a run-time couple local/remote monitoring and control of the building has been the major outcome. The presented proof-of-concepts of IoT systems could be employed with suitable ICT for converting existing buildings into smart buildings with improved energy efficiency.

(5) Implementation with minimal financial implication

Massive national level awareness campaigns should be organized for citizens' awareness and encouragement for enthusiastic participation. Funding could be arranged from The World Bank, IMF, United Nations and developed countries to the government of India, which could be sent to state governments via Special Purpose Vehicles (SPV) such as RAPDRP. State governments should encourage the Prosumers via various state level programs in form of subsidies and other techno commercial support mechanisms. Under 'Make in India', indigenous manufacturers (SME and entrepreneurs in particular) should be encouraged for local manufacturing of devices, systems, software and engineering integration-maintenance supports. The entire mechanism should be operated as a single integrated system with single window clearance and e-governance.

18.11.6 Bangalore Metropolitan Region (BMR) Study

BMR with an area of 8005 km^2, and a population of 11.69 million is divided into 3 districts namely—Bangalore Urban, Bangalore Rural and Ramanagra. BMR has only one Corporation namely the Bruhat Bangalore Mahanagar Palika i.e. BBMP and 10 Urban Local Bodies (ULBs). Population influx in Bangalore is an obvious result of the continuous flow of migrants to Bangalore from surrounding areas and other regions. The current regime of the urban economy is that of agglomeration of an export oriented Information Technology sector which was initiated by developments in the early 1980s. Research Institutions such as ISRO, DRDO, IISc, IIM together with the Public-Sector Enterprises such as BEL, HAL etc. had established a base in the city for innovation oriented high technology production mainly in the areas of Electronics, Telecommunication and Defence. The largest IT firms in India such as Infosys, Wipro, Satyam and TCS were all small start-ups in the city and became large global conglomerates. Bangalore city is a key growth centre in the whole BMR with 81% of the working population residing in the city. The diversification of economic activity is quite extensive in Bangalore and a majority of the population is engaged in non-agricultural economic activities. The city development plan for Bangalore, 2006 reported that less than 1% of the workforce was engaged in agriculture.

The pattern of growth, with the city core becoming increasingly saturated, and new urbanization centres developing at the periphery, can be traced to multiple factors. In periphery areas of Bangalore, loss of agricultural land, water body, and green spaces are faster, and it is imperative that urban planning efforts concentrate of city periphery where unplanned growth is taking place at an extremely rapid rate. Primacy in BMR always hints the existence of regional disparity and Bangalore city always face problems of high immigration, high demand for urban infrastructure and deteriorating quality of life.

In the case of Bangalore there is a strong need to think beyond the city of Bangalore. Peeling away the layers of spatial governance jurisdiction we can see that it is the outermost layer, which is the BMR, which is of utmost importance. If the smart region vision must be given a spatial context, then it must be BMR. BMR is an amalgamation of the urban and rural, with the overwhelming presence of Bangalore City at its core. However, beyond it there are many smaller towns which are in no way less important. Some of these towns would in fact qualify as regional cities, given the tremendous rise in their population over the last decade or so. All these towns have their own local governments (Town Municipal Councils and City Municipal Councils). If a smart region is to evolve, the need of the hour is to promote cooperation between the entities and between them and BMC. The cooperation should be directly based on digital platform and smart networking should be explored.

In this context, the role of industry assumes great importance. The legacy businesses must transform into digitally oriented enterprises. Bangalore is already the technology capital of India, driven largely by IT. While the IT industry in India is likely to emerge from the current downturn as much stronger riding on the back of technology transformation, it is the need of the hour to support them, and other industries, by providing facilities for digital innovation and incubation. The role of industry bodies will be of prime importance here, but again, the need is to connect them with the smart network of local self-governments, so that a synergy is achieved in terms of spatial decision-support and parameter-based prioritization without requiring complex interfacing among industry and multiple layers of government agencies. To achieve this, each of the regional cities around BMC must be developed as smart cities first. Recently Bangalore city has broken into the smart-city club. But it needs to be emphasized that merely promoting Bangalore City as a smart city will be a self-defeating exercise as the irreversible process and the tremendous rate of urbanization means that the region itself is transforming beyond limits of imagination and must be integrated into the context of a ‘smart region’. After all, a smart Bangalore city which gets promoted in isolation will only distort the market further and lead to lopsided urbanization, costing society dear in terms of crowding, quality of public spaces, travel times, etc. If we look at the underlying concepts of smart city, we see themes like smart energy, smart health, smart living, smart mobility, and smart economy. Thus, the transformation of BMR into a smart region would mean more than location specific actions within the area. While smart energy interventions could be a thrust in the BMC area, piloting of such interventions could be easier in the smaller City Municipal Councils or Town Municipal Councils in the region. In fact, the population explosion happening in these regional centres outside BMC demands that they should also be considered for smart energy solutions. Smart health also needs to be scaled up to regional level through linkage between health institutions spread across the region, thus benefiting the rural settlements also. The first steps toward smart living at a regional scale would involve promotion of a dispersed settlement strategy dovetailed with transit oriented development and backed by incentives for affordable housing development and specialized economy oriented growth centres. Smart mobility would entail smart demand-responsive integrated regional connectivity catering to diverse commuter segments. Smart mobility strategies, with fast and high-capacity regional networks supported by efficient intermodal options, would complement smart living strategies as discussed. For spatial and economic development to convert BMR to smart, the definite tools like smart growth, transit-oriented development, mixed land use, industries dispersal and specialized zones etc. adopted. The ultimate spatial structure also recommended for the smart BMR.

18.11.7 Chandigarh Study

The “City Beautiful” Chandigarh was the outcome of modernist thinking foreseen by Pandit Jawaharlal Nehru in 1951. The dream remodelled into reality in the form of a planned city was the first of its kind designed by the celebrated architect planner Trio. With a total design capacity of 5 Lakhs (the population expected to be met in 3 phases of development), sooner started falling short when compared to the demand within the insufficient space of 114 km^2. Also, the greenbelt estimated within the 16 km Periphery Control (earlier 8 km) were compromised to the dearth of urbanisation, principally leading to unplanned growth. The convenience of livelihood supported by unparalleled architecture and picturesque purlieu made Chandigarh the most idolised location for settlement. Being landlocked from all sides, Chandigarh found expansion in the new settlements Panchkula and Mohali and the three were together referred as the "Tri-City".

With slowing pace of growth rate in Chandigarh and the need of settlement in search of job prospects, Mohali and Panchkula were magnets of growth. This was the time when the conceived dream of "no construction in the Periphery" was overlooked against development demand. It is essential to mention that no legal document talked about a regional development agenda, driving coordinated planning in the Chandigarh region (16 km Periphery area). Efforts for regional planning like Chandigarh Urban Complex (CUC)-1977, Interstate Regional Plan for Chandigarh Region (ISCR)-2001 and Chandigarh Interstate Metropolitan Regional Plan (CISMeR) 2021 had no legal enablement. The lack of support from adjoining states made it have no hold and no powerful bearing on the region. In short, the Chandigarh Region had no plans that could rephrase and analyse its social, economic, spatial, environment or governance status.

For the purpose of this study, the 16 km Periphery marked initially in 1962 and bearing a legal sanctity was undertaken referred hereinafter as "Chandigarh Region" covering important growing settlements like Panchkula, Mohali, Kharar, Zirakpur, Banur, Dera Bassi, Mullanpur (New Chandigarh) and other smaller settlements. These settlements had their own vision of growth, where some had residential status (Panchkula, part of Mohali) and others had industrial status (all other major towns like Mohali, Zirakpur and Kharar). The demographic analysis of the region revealed that whereas the population growth in Chandigarh saw a declining decadal growth rate trend from nearly 114% in 1961–71 to 17.20% in 2001–11, settlements like Mohali, Panchkula, Zirakpur and Dera Bassi showed an increasing trend in growth. The various sub-regions within the regions were governed by different master plans of the settlements, primarily clubbed into three major documents—Chandigarh Master Plan-2031, GMADA 2008-38 and Panchkula Master Plan.

The spatial growth assessment for Chandigarh region suggests that nearly 400 km^2 of the area will be urbanised in 2048 (as per the MLP Neural Network based modelling growth scenarios using Land Cover databases of 2000, 2006 and 2012) as compared to 292.90 km^2. in 2012, 238.27 km^2 in 2006 and 224 km^2. in 2000. If we analyse the region's area outside the Chandigarh UT and within the 1 km Periphery, nearly 10.1% is the built-up area in 2000 (151.52 km^2), nearly 11.0% in 2006 (164.47 km^2) and 14.4% in 2002 (215.67 km^2). The growth trend in terms of spatial growth suggests an increase of 31.13% in the built-up area from 2006 to 2012.

The geospatial data and models have proven to be smart tools grossly leading the process of decision making for Master Plan preparation. The trio-technique of

Remote Sensing, Geographic Information System and Global Navigation Satellite System helped in various stages of plan making starting from database inventory preparation to plan to monitor for simulating future growth. With the help of modelling techniques, a future scenario of 2048 and 2024 was predicted, helping planners for an integrated sustainable planning.

Monitoring the demographic and spatial development scenario in Chandigarh Region, that the important drivers of growth for the region were proximity indices (proximity to the Chandigarh UT, roads, existing settlement, railways, CBD) and the gains and losses due to changes between 2000 and 2012. The highest impact factor (assessed using Cramer's V) was the change during 2000 and 2012 (0.4569), proximity to Chandigarh UT (0.4050) and proximity to CBD (0.3711). It was observed that out of the four layer outside the UT covering the Chandigarh Region, the layer just outside the UT showed the highest potential for growth. The growth scenarios also raised questions on the conservation of environment of the Region. The chapter correlated the concept of layers (layer 1–7) to that proposed in CISR-2001 (Planning Zones 1–4). It is felt that the Regional Plan should ponder on the issues relevant in the current growth scenarios like the preservation of agriculture land, protection and demarcation of forest land and its buffer for No Construction, conservation of water bodies in and around Chandigarh, propose afforestation and guided urbanization (proposed as per spatial growth projections for the future 2024 and 2048). It is further felt that top down and bottom-up approach to planning should be followed at various levels like micro, meso and macro level for a holistic planning. It was felt that the idealistic planning for Chandigarh had lesser attention on planning for informal sector and planning for the poor (very much relevant in Indian social structure). Hence, the Regional plan shall focus on inclusive planning where space is designated for the different strata of the society without abandoning any one.

To transform the existing scenario into a Smart Regional Development scenario, it is essential to have a legal status to the plan abiding the current and future scenarios withstanding the drivers of growth along with holistic planning, smart tools and techniques like geospatial technology can be of greater use in the process of transformation. Open source information portals like Bhuvan-NUIS developed by Indian Space Research Organization is one such platform guiding the process of planning forward. National Urban Information System published a manual in this regard at 1:10,000 scale for thematic mapping using satellite imagery, ground data and other collateral data from various partner institutions as a source. Planning techniques like smart connectivity (using Transit Oriented Development and expressways), Smart Decision making (through holistic, inclusive and holistic planning), Smart techniques (spatial techniques for assessing energy potential of the city like Solar Energy transforming the city to a sustainable and self-sufficient city) and smart development can be implemented for transforming Chandigarh into a Smart Metropolitan Regional Development Plan.

18.11.8 Delhi Study

National Capital Region centred on Delhi is a massive but complex region. With area exceeding 45,000 km^2, the NCR is counted among the largest regions of the world. A multiplicity of plans (hierarchy plans with implications for the built environment), planned (citizens), and planners (planning organizations), the NCR is also one of the most complex regions in the country. Complexity increases, and transparency is reduced because of less use of ICT and IoT in the governance of the NCR. This chapter makes five proposals for making NCR a smart region. Governance of the NCR is taken up as the enduring challenge. It is proposed that elected regional body for the NCR is better than the existing proposed NCRPB. In this hierarchy, sub-regional planning authorities should be created by state governments for all the four states under local government acts. These will also be elected bodies. Lastly, district planning and development authorities should be set up by making a constitutional amendment to the existing DPC framework as provided in the 74th amendment act. In this line of thinking, organizations like Delhi Development Authority will have no place and must be abolished and replaced with an elected sub-regional body of Delhi Government with all powers of DDA handed over to state government. Smart spatial planning strategy is that the NCR should implement regionally significant proposals and leave local projects to appropriate agencies. Whosoever implements these projects, all projects must emanate from the hierarchy of plans, more particularly the regional plan. Third, the smart transport strategy should consider the faster implementation of RRTS and road based systems by the end of the plan period by 2021 as implementing agency is already set up. Fourth, environment and economy are entangled, but first the environmental strategy. The environmental strategy proposed, involves preparation of new energy plan based on wind and solar policy for meeting enhanced future energy needs. Lastly the economic strategy is taken up. The significance of ICT and IoT is highlighted for regional planning and development. The first thing the NCR should consider when making a smart strategy for economic growth is that it should include plans about how it could use advanced technologies for regional development. No doubt automotive safety applications, smart fleet management, and intelligent transportation systems are some of the well-known applications. There are other areas where these technologies should be considered for use such as easily available and affordable broadband internet networks are hugely important for producers as well as consumers of various services and goods. Internet of things could offer better trade opportunities if affordable broadband internet networks become available all times at all places. Most importantly, Internet of Things should be embedded in the development of physical infrastructure. The next NCR plan should consider making governance and economic policies for the development of ITC or IoT through massive public private partnerships and alliances.

18.11.9 Jaipur Study

Jaipur gained the status of a Million Plus City in 1991 [9] and was recognized as a metropolitan region in 2010. The Plans have been prepared for the metropolitan city of Jaipur on the premise of the core city and its hinterland since 1998, with the approach of four tier development system of core city, outer and inner ring of satellite towns and the intermediate rural settlements and areas. The current Master Development Plan for the horizon year 2025 too is based on the multi-tier spatial approach of the district, region, city and satellite towns. Within the regions, there are three distinct sub-regions, the urban area, agricultural area and the eco-sensitive area.

The region is covered by a thick mantle of soil and alluvium and extends towards the north and the east by hill ranges and many isolated peaks of Aravalli range surrounded by plains. Non-perennial rivers like Banganga, Dhundh, Amani Shah Nala and Bandi drain the region. Climatically the region experiences extremes of cold and heat and except for the monsoons season the weather is dry. Physiologically as well as climatically, the region is suitable for all year cultivation with some portion being double cropped. Agricultural activity is dependent on irrigation through tube wells as rainfall is scanty. Fruits and vegetables are grown in abundance. The region is rich in minerals, which is an input to many industries in the region.

After Independence, Jaipur, having the advantage of being an administrative capital and being on the major transit routes has emerged as the melting pot of various cultural regions of the state, though there is dominance Dhundhar culture. It is the cultural aspects represented in art and craft on stones, wood, leather, jewellery, clothes, paintings, pottery, monuments, food, music and dance that has helped the region emerge as a tourist centre at an international level. These cultural aspects have been patronised by the rulers of the region and have got embedded in the lives and work of the people.

The three sectors of the economy, primary, secondary and tertiary, reflect the direction of growth of a region. Jaipur's mainstay has been non-agricultural since the city was established. The city had been established as a trade and commerce centre which continues to be so till date. Over the years the city and the region have gained importance not only as a commercial hub for local handicraft but for agricultural and industrial produce.

Tourism has played a vital role in the economy of Jaipur Region, especially with Jaipur city being at its core. Other than the City, there are many locations in the Region that are important from a tourism perspective. Jaipur has attracted tourists, both domestic and international, for its city planning, historical monuments, natural features, temples, and palace complexes. Jaipur's colourful and vibrant culture which is represented in its buildings, people, dance, music, food, clothes and handicrafts is a major attraction in the region. Jaipur is an important destination on the world-famous tourist circuit, Golden Triangle, comprising of New Delhi-Jaipur-Agra. Other than the destinations within the city, there is tourist potential in the region too, which are not as well developed and promoted compared to the locations within the city. The areas of tourist potential can be categorized into historical, wildlife, religious, traditional medicine and natural features. During the last four years, there has been a decline in the growth rate of tourist inflow, specially the domestic, while the international tourist inflow has been fluctuating [10].

In terms of the GDP, though the tertiary sector contributes maximum towards the economy, the growth of the secondary sector has been steady but slow. It is important to note that the success of the tertiary sector in the region is dependent a lot on the secondary sector as the traditional handicrafts which showcase the vibrant culture of the region is a big pull factor for the tertiary sector's tourism component.

Industrial impetus has been given in the region through planned intervention in the 1970s when the master plan was prepared to control the haphazard growth as well has direct the development of the city. The industrial activities of the region are concentrated in 14 industrial area [11] which have been set up by Rajasthan Industrial Development and Investment Corporation Limited (RIICO) and Jaipur Development Authority and unorganized industry is concentrated in and around Sanganer. Jaipur city has a large concentration of household industrial units, clustered within the walled city. The units engaged in stone cutting and polishing, blue pottery, lac work, gota work, sculptures and other traditional handicrafts. Other than this, industrial thrust is being provided by the SEZ, DFC and DMIC which implies development potential for various industries in areas between Sikar road and Ajmer Road, which is the western and south-western part of the Region.

The industrial areas in most instances have negligible industrial waste management practices as in most cases the waste is thrown at the nearest convenient location. Moreover, there are either non-functional or absence of effluent treatment plants for treating water before disposal in drains. Polluting and non-polluting industries are found to be in same industrial areas, which indicates lack of detailed planning of the industrial site. Moreover, there is total lack of planning and intervention in improving the quality of life of the traditional artists who mostly work in household industries in core city areas.

The region experiences a lower Work Participation Rate (WPR) (34%) as compared to the state (39.26%) and national (39.3%) figures, though the WPR has been increasing in the region over the past few decades [12]. Interestingly, the rural parts of the region have a lower dependent population when compared to its urban counterpart. As the city grows [13], its economic base expands due to increase in industrial activities and increase in trade and commerce resulting in higher employment. However, this trend is not sustainable and negative returns set in proving the theory that size of city or region and economic growth are negatively corelated. It has been noted internationally that metropolitan cities with high growth rates have low productivity levels.

The intriguing fact about Jaipur and its region is that the since beginning of the post-independence planning era, there have been marked deviations in its growth and development, be it related to the spatial spread of the urban area or the growth of population. The reasons have been related to a land acquisition within the planning boundary, lack of follow up actions to the master plan, like functional plans, zonal development plans and zoning regulations and later delayed notification of the subsequent master plans. The current master plan envisaged the development of the settlements along the transportation corridors and eventually merging of the intervening areas giving rise to an amoeba shaped region with a higher concentration towards the north east, east and south. However, the settlement pattern of the last census clearly indicates the concentration of growth on the western part of the region, attributed to the external factors like Dedicated Freight Corridor (DFC) and Delhi Mumbai Industrial Corridor (DMIC). The alignment of both these corridors is at approximately 40 km from the periphery of the region.

Keeping the fact in mind that historically the city and the region has not developed as per the master plans, there is a need to take corrective actions by providing the appropriate impetus for the growth of the region.

Jaipur Region is experiencing environmental imbalance due to depletion of its natural resources and overarching influence man-made activities. Widespread deforestation mainly in the hilly areas has resulted in an increase in the rate of soil erosion due to the combined impact of wind and water erosion, a higher rate of soil erosion has accelerated the process of silting in the river channels and water reservoirs. It has further disturbed the natural habitat of wildlife and biotic process in the region. The resultant impact of deforestation, erosion, siltation, salination processes has created environmental challenges in the Region affecting the groundwater recharge resulting in lowering of the groundwater table. It has affected the micro-climate of the region as well in terms of temperature and humidity.

The influx of tourists has brought pressures on the urban fabric and infrastructure. The region has been experiencing a shortage of water supply, which is the key infrastructure requirement for any tourist activity. There is a need to assess the carrying capacity of in terms of the region is a tourist destination with a view to regulating tourist traffic.

Key issues in the region are seen with respect to its harsh climate which becomes a discouraging factor for tourists from April to August, impacting the livelihood of the population who are dependent on the inflow of the tourists. Limited water availability due to the region being in a semi-arid zone enhances the issues related to tourism and industry.

Existing industrial and commercial base in the region is one of its strongest points. To further enhance the two sectors, secondary and tertiary, there is need to introduce appropriate policy measures. The strengths of the region are its diverse minerals base, tourist potential which is concentrated not only in the city but is well distributed across the region. There is need to take advantage of the DFC and DMIC which are in very close proximity to the region.

Jaipur has been identified to be developed as a Smart City under the Government of India's Smart City Mission [14]. However, Smart City cannot be smart until it is in a smart region or else it would emerge as a semi-smart entity having a piecemeal or sectoral emphasis on its road to smartness. As established Jaipur region's core, that is its metropolis shares a strong economic and socio-cultural linkage with is region. The Jaipur region in its entirety has high potential to be developed as Smart Region. Strategies are proposed to strengthen Jaipur as Smart Metropolitan Region.

Census of India defines as Urban Agglomeration (UA) a continuous urban spread constituting a city and its adjoining outgrowths, or two or more physically contiguous towns together with or without outgrowths of such towns [15]. The UAs having a population above one million are called Million Plus Cities. Internationally, these are considered synonymous with the term metropolis. The planners understand the metropolis to be, a city governed by one or more municipal bodies, having a population above one million, and functionally serving its influence region as the dominating centre of trade, commerce, art, culture, health care, recreation, education, research, administration and political activity. A metropolitan region is the area under the influence of the development impulse of a metropolis and comprising the metropolitan core and the metropolitan periphery [16].

Cities have grown to be metropolitan regions mainly due to the concentration of industries attracting population from rural areas and other urban centres not only in the immediate influence zone of the city but from other parts of the country as well. The high rate of migration to these regions has led to overcrowding and environmental deterioration. The growth and distribution of population within the metropolitan regions are not uniform. There is a trend of corridor development in these regions along transport networks [17].

The problems that have come to be associated with these very large regions need to be addressed with the use of technology and innovation. The answers to these may lie in the smart context which has been explored in the context of Jaipur Region.

Adapting from the project management practices, the need of the hour is to make the region S. M. A. R. T., that is Specific, Measurable, Achievable, Realistic and Timely [18]. The vision envisaged for the region is smart tourism based development through appropriate industrial impetus to promote local involvement and employment to improve the quality of life and yet preserve the cultural vibrancy of the region.

Smart as per the understanding of the authors has four key elements. It should be namely attractive, efficient, digital and sustainable. In terms of a region, the place should be attractive to the citizens, tourists and entrepreneurs. It should be efficient in terms of mobility, affordability, economy and recreation. In terms of digital, it should be accessible to all to bring about smart thinking. Digital use needs to be intensive such that there is a generation of lots of data using latest and appropriate technology. The actions in the region need to be sustainable so that the future generations can meet their needs.

To achieve the vision four-point strategy has been proposed that is strengthening of regional competitiveness by enhancing tourism experience, regenerating the region's skill and knowledge base, smart industry and innovation, strategic governance for smart specialization, the key emphasis being tourism.

Tourism brings economic development and the creation of direct and indirect jobs in hotels, travel agencies, transport companies, restaurants, guides, show and entertainment business, monuments, parks and other related sectors like insurance, health, housing, human resources and training institutions. In the case of Jaipur Region, it will lead to the growth and development of industries as well as the emphasis here is on smart cultural tourism.

Smart Tourism is understood to be reliant on four core information and communication technologies: IoT, mobile communication, cloud computing, and artificial intelligent technology. These technologies connect the physical, information, social, and commercial infrastructure of tourism, and supplies Smart Tourism value to multiple stakeholders.

The authors emphasize that Smart Cultural Tourism which is technology based needs to be embedded within the framework of sustainable Tourism, Responsible Eco-Tourism and Rural Tourism for the well-integrated development of the region. Smart Tourism needs to be sustainable, that meets the needs of the present tourist and host regions while protecting and enhancing opportunity for the future by minimizing the adverse impacts of traditional tourism on the natural environment and at the same time enhance the cultural integrity of the local people and offer opportunities to the tourists for experiencing in activities that form the core of country life.

18.11.10 Kozhikode Study

Kozhikode district is located along the south west coast of India in the Malabar region of Kerala. It is bound by the Arabian Sea on the west and the districts of Kannur, Wayanad and Malappuram to the north, east and south respectively. The district is located approximately 420 km north of the state capital Thiruvananthapuram and 250 km from the commercial capital namely, Ernakulam. Chennai and Bangalore, the two other major cities of south India are located within 650 km from Kozhikode. The district—which is well connected via air, rail and road—acts as the regional transit hub for the Malabar. The two major ports Kochi and Mangalore lie within 250 km and an intermediate port namely, Beypore is in close proximity to Kozhikode—hence increasing trade prospects for the district.

Historically, Calicut was a port city and a global trading hub with an established trade route connecting Europe and South Asia. The city gradually lost its economic significance during the post-independence period due to the political and social fabric prevalent during this time. The industrial activity is presently limited to few medium scale industries related to steel processing and small-scale industries manufacturing rubber products, food and dairy, textile, handloom, timber etc. The industries are neither a major source of revenue nor does it generate any employment opportunities. Statistics from the Economic Review of 2016 prepared by the State Planning Board indicate that the economy of the district, in 2016, was primarily driven by the tertiary sector—61% of district GSVA—followed by the secondary sector—31% of district GSVA. The primary sector only contributed to 8% of the district economy wherein majority of the activities included agriculture cultivation of paddy, coconut, banana, tubers, spices and other tree crops along with fishing and its allied activities. While secondary sector activities included medium scale industries with manufacturing activities primarily focused on F&B, timber processing and tile manufacturing among others, the tertiary sector activities included tourism and retail establishments. Additionally, many cyber parks have also been proposed for the district that is in various stages of development. With Kozhikode being positioned as the next IT destination of the state after Kochi, a potential shift in the economic base of the city is envisaged.

Economic decline, unemployment growth and high migration have made it important to reconsider economic strategies and transform Kozhikode. Low profitability of other erstwhile industry and trade activities has resulted in a structural shift in the economy from manufacturing to tertiary sector activities such as tourism, education, healthcare and IT/ITeS. The district is also home to some of the best education institutes in the country and the presence of well-developed medical facilities makes it the go-to place for health care in the region. Upcoming investments in the IT/ITeS sector include the development of a 50-acre cyber park. These developments are expected to position Kozhikode as the third IT hub in the state.

As per the results of the Census of 2011, urbanization in the district indicates that approximately 67% of the total population is defined 'urban' and the number of census areas has increased by 173% from 2001 to 2011. The growth of the urban agglomeration in Kozhikode has resulted in a continuous, vast stretch along the coast—like the urbanization pattern observed in the state. This poses a great potential for development on a metropolitan regional level. Previous planning efforts indicated in the masterplan prepared by the Kozhikode Corporation include the Interim Development Plan (1967–1981), Development Plan for Calicut Urban Area (1981–2001), Perspective Plan of 2003, City Development Plan (2006) and various town planning schemes. The various proposals in these plans range from overall district level interventions to small scale schemes, all limited to the Kozhikode Corporation area.

The metropolitan region has been defined, by both local and international organizations, as a geographic region spanning a number of local government authorities unified by economic activities, labour markets, firms and their inter relationships. It is also considered ideal if the area is within a single administrative boundary. The Kozhikode Metropolitan Region has been delineated taking into consideration these various parameters and it covers an area of 1720 km^2 with a total population of approximately 2.9 million.

In this chapter we aim to study the existing economic and spatial structure of the place and propose economic and spatial strategies that can be adapted to revive the economy of a delineated metropolitan region within the district. The economic and subsequent spatial strategies for the region have been formulated based on the principles of the Third Industrial Revolution and Zero Marginal Cost Society, proposed by Jeremy Rifkin all within the framework of the sharing economy system. The principle of sharing economy is built around sharing human and physical resources—includes the shared creation, production, distribution, trade and consumption of goods and services by people and organizations. While sharing impacts the community in multiple ways, the foundation of the sharing economy is built on trust. This fundamental principle has been adopted to integrate the community of KMR for the implementation of the various strategies. The demographic census of 2011 for Kozhikode reveals that approximately Hindu, Muslim and

Christians account for 56.2, 39.2 and 4.3% in the total population respectively. Since centuries, the Hindu, Muslim and Christian communities have practised faith based charity or compulsory donation concepts namely, 'Jajmani' (Hindu), 'Zakath' (Muslims) and 'Dashamsham' (Christians). Hence, for the community and social development these concepts have been explored—contributions when judiciously used can be utilized for the benefit of the community.

The methodology adopted for economic revival and the spatial strategies include three key steps. The first includes renewable energy generation and its deployment. All of the energy used in industries, homes and vehicles still use conventional sources that entail very high environmental impacts. In lieu of climate change and strategies adopted to reduce the carbon foot print, it becomes pertinent to carry out new models of energy production. This energy can be generated by harnessing renewable sources at nearly zero marginal cost. Internet technology and renewable energy would merge to create a powerful platform which enables millions of people to produce their own green energy in their homes, offices, and factories, and share it with each other in an "energy internet," just like we now create and share information online.

The second entails conversion of all conventional industrial activities to a smart economy i.e. a smarter means of production and consumption of products. Smart Economy involves the creation 'Prosumers,' i.e., a person would be a consumer and producer in a smart economic environment. This would involve the conversion of major economic activities to smart economic activities based on energy and resources in a region. Rather than following a conventional system of large and medium scale industries, every individual household shall be a Prosumer and would be considered as an industry. To create a 'Smart Economy' for the region, the existing economic activities of the region have been studied and declining activities/industries are identified. To replace declining activities with smart economic activities land use conversion strategies have been formulated. Identification of the potential activities in the region has been done based on energy requirement, resources, skills etc. and on the spatial codes for renewable energy. The existing activities have been converted to smart economic activities and the economic viability is assessed. The existing pattern of production services which includes 'middlemen' at every stage is transformed where in the profits to the producers will be increased by removing these 'middlemen' and increasing accessibility of the producers to the consumers—this conversion is aided by maximizing the use of ICT in every stage of the production and service system. The activities based on resources and energy has been spatially allocated and spatial codes for the smart economy is derived. The strategies to convert smart economic activities to Zero Marginal Cost have been formulated.

Thirdly, approaches to convert these smart economic strategies to near zero marginal cost have also been formulated. It is based on the idea that if the marginal cost of producing each additional item falls to essentially nothing, then everything becomes free. To create a near zero marginal cost society framework for the region, the resources in the region have been mapped out. Along with this, the religious and cultural aspects of Jajmani, Zakath and Dashamsham have also been integrated and potential activities based on energy requirement, resources and skill are identified. The process and economic flow in zero marginal cost society have been worked out and spatial codes for a zero-marginal cost society were prepared. Finally, to arrive at the spatial code for the metropolitan region, each spatial code for renewable energy, smart economy and near zero marginal cost society has been consolidated followed by preparation of Land use plan and Zonal plan along with strategies for implementation. A knowledge based society of Kozhikode Metropolitan Region will use the implementation methods formulated based on the concepts of the Triple Helix Model for the implementation of the economic and spatial strategies.

18.11.11 Surat Study

Surat metropolitan region has many unique aspects. The region is located in the western part of India and centrally on a very important industrial corridor between Mumbai and Ahmedabad. It has an almost flat terrain and suitable land available for development. In past three decades, the urbanization has observed pace creating manifold pressure on the Government systems and natural resources in the region. There seems an apt need for a proactive, systematic planning approach for smart metropolitan region planning to infuse the fundamentals of zero-marginal cost society concept. By leveraging opportunities for changes in the planning, implementation and practices followed by citizens for various activities need to be inculcated. Collaborative efforts can lead the metropolitan for a better and perspective future

The Surat metropolitan region (considered with a geographical area of about 1500 km^2) is surrounded by the Arabian Sea on the west, the NUDA at the south and the BUDA at the east. The northern part is lacking for a declaration of any developmental agenda and is open for expansion in future. The entire area has a variety of good transportation and connectivity. Land covers show a high density of road network and developed land in the centre of the region mostly concentrated in the Suart city and its close vicinity. There are more than 300 water tanks (ponds, lakes) in the region, of varying size and storage capacities. The River Tapi is flowing from the centre of the study area, and River Mindhola is in the Southern part of the region. Both rivers traverse from the East to the West. Apart from these rivers, there exists a substantial network of streams that acts as a feeder to these rivers. The extent of available developable land is large in the Surat metropolitan region.

Following national guidelines, the region has sufficient service and social infrastructure in place considering current requirements. Electricity and communication networks are available to the citizens. Canal network is sufficient enough to cater the needs for agricultural activities. Some parts of the region are prone to flooding in the monsoon. However, administration of the region has much of overlapping by various functionaries.

The entire region is by now having established and upcoming a variety of industrial units. Most of the male workers are engaged in these economic activities, and the extent of these establishments is spreading in all directions. The region has about four SEZ in operation, many industrial estates and particular industrial area ear-marked for heavy industries of Hazira on the West.

In the recently proposed development plan keeping 2035 as a horizon year by SUDA, a care is taken for proposing ring roads (not in the city) around smaller centers where urbanization is anticipated. In addition to it, all the villages are earmarked with residential zone around the Gamtal. The larger part is kept as a residential zone. Areas are declared as industrial zones and some of these zones, specific industries are identified for development (for example, hazardous chemical industries). Also, a care is taken for connecting these places with residential areas by means of a proposed network of wide roads. A third of a kind, ring road is also proposed with an alignment. Mangrove forest zone is kept reserved along the coast line. Not extensively yet somewhat green buffers are also proposed. At a broad-level, the proposal seems good. The proposal seems lacking in terms of specific and result oriented efforts for specialized goods and services to have access to collaborative efforts. It means that development of industries is supposed to establish within a specific zone however, types of industries supporting each other are not specified at a micro level. Hence-forth the development of pockets will be at the will of the investors. It may result in farther dependencies resulting in increased transportation and logistics.

Surat and surrounding region have mostly achieved certain service level benchmarks and with rising population there will be pressure exerted on the systems in future. To avoid such circumstances, a proactive planning with a larger and inclusive purpose is very much essential. The concept of zero marginal cost society explores the aspects regarding practices over eclipse of the capitalism, cost, cooperation, IoT, and the collaborative commons. In a way, the entire concept is summed up in a regional planning perspective as to have provisions for various needs that can be best utilized with collaborative efforts sustaining and supporting each other. For achieving such, a detailed exploration of dependencies as well inter-dependencies need to be worked out. Once the levels are available in relation to dependencies, a clustering approach to planning shall be taken up to make provisions for land. At the same time, an exercise is required to visualize the ancillary needs to support these clusters as transportation, logistics, service infrastructure and social infra-structure. The clustering approach at a regional level with the provision of IoT based and monitored services, there shall be a considerable saving of time and resources. It also shall result in the increased productivity and efficient performance balancing economic and social engagements.

To achieve the state of a 'Zero-marginal Cost society', there will be an apt need to integrating various dimensions of sectoral and spatial planning, city and village level implementation, feedback from all citizens and in accordance, the institutional arrangements for monitoring and evaluation. The zero-marginal cost based metropolitan development not only suggest for integrating planning for various sectors but also, there will be a need to identify sector specific and intra-sector dependencies. If done so, there shall be considerable. A quadratic sub-division of the entire region is proposed and discussed for strategic development with details. By adopting and accommodating policy and regulation reforms of national agenda, some smarter practices in work patterns need to be in place. Smarter education modes of online learning need to be promoted and sincerely included in the society. The present state of the same is limited to individual interest and willing-ness only. Smartly avoiding the physical presence of people for various tasks can result in saving the expenses in turn, can improve the prosperity in the region. There is a significant need for a pragmatic change in the planning and implementation as well as practices by citizens for smarter moves for sustaining the metropolitan region. The same can only be envisioned with the micro-level yet wholistic approach by multi-disciplinary and collaborative efforts by various levels of stakeholders.

Hence, Surat needs to be attended with a proactive, systematic planning approach for smart metropolitan region planning to infuse the fundamentals of zero-marginal cost society concept. It has potential to accommodate citizens with the offering of opportunities and serve the nation as an efficient engine of growth.

18.11.12 Naples Study

Chapter 12 "Spatial and economic smart strategies for the 21st-century Metropolitan City of Naples" is articulated in four main parts. The first part introduces the concept of economy and capitalism in XXI Century's city, starting from the assumption of Neoliberalism. Authors analyse how this ideological approach to the economy has multiplied exponentially speculation and many "distortion", both in the social system, as well as workers market: the destruction of a welfare programme, the attack to the right of the labour market and workers right, the growing powerful of financial institutions supported by the ICT powerful. Author's proposal is to find a new epistemological approach, suggesting a conceptual framework for ecological economics based on systemic principles of life and a shift from techno-city to a human city. They propose a model called homological smart city as a new way, based on direct citizen participation, peer-to-peer community, and neuroergonomics, biophilic design, and Biourban economics, all based on the human body epistemic.

The second part considers the perspective of a 21st-century Italian socio-economic renaissance to overcome a growing climate of economic stagnation. Indeed, in reaction to the crisis, several villages, towns and cities have seen a slow phenomenon of the revival of local communities. It has taken place from within the city and for the merit of grassroots initiatives of social innovation constituted mostly of young people that, leveraging on their capabilities and a peer-to-peer network supported by the ICT, promote a novel vision for the future of their community, building a more sustainable urban system able to combine the economic development of the city and its inhabitants. Through a change of paradigm, the human being is put at the centre of the system and its designing, considering social innovators as the key-actors of change. By leveraging on local assets, they trigger a community revival process able to produce positive systemic effects overall urban system and on local governance inclusive models, promoting incremental regeneration processes.

The third part is dedicated to exploring a new Biourban strategy, named "mushrooming", oriented to consider diversification as a principle of life in a city. Biourbanism considers the city as a living organism that cannot be strictly planned but whose processes should rather be guided toward new futures. New strategies to function in a complex system like a city need to be developed by experimenting with real life situations. An experiment of open process and bottom-up urbanism was started in Finland about building a network to foster interaction between small self-organised co-working communities, considering spatial and economic processes that emerged due to it. These processes were able to activate connected diversification, recognized as a systemic principle of life that fits the context of urban development especially well.

The fourth part analyses the case of the Metropolitan City of Naples, one of the 14 Italian metropolitan cities. The Statute of the Metropolitan City of Naples divides its territory into so-called "homogeneous zones" for a more suitable, balanced and functional management of local resources, with the purpose to consider their identity characters, historical and cultural components, geomorphological and naturalistic contexts, landscapes functional interactions and socio-economic frameworks. Indeed, the concept of "homogeneous zones" is part of the metropolitan strategic plan, and is oriented to improve the territorial productivity, services to citizens opportunities, and cooperation among the different municipalities. The paper reflects on the need to identify some homogeneous zones able to underline common identity characters and activate a strategy of connected diversification, with a specific attention for the municipalities of the Coast Area of the Metropolitan City of Naples. Starting from vulnerability and resilience concepts, the study was dealt according to a multi-methodological approach, based on a GeoDesign process supported by multi-criteria analysis, multi-group analysis and spatial analysis. The elaboration of Spatial Opportunity Maps (SOMs) is the output of a multidimensional evaluation process that leads to the identification of specific territorial contexts with a greater propensity for some specializations relating to the economic sectors and the local transformation. The identification of a Biourban strategy, characterized by human smart spatial solutions, place-based and situated actions, for the enhancement of the coastal area of the Naples Metropolitan City can be considered as a prerequisite for the activation of a process oriented to the identification of "homogeneous zones", conceived not only as areas with similar characteristics but above all as territories where it is possible to promote networks of opportunities between the various municipalities and their communities. "Co-operation" is conceived a source of mutual benefit and involves a mutual convenience, based on the constant construction of ties and relationships and the interdependence determined by spatial proximity. Economic processes require cooperative-collaborative behaviours between the various components and become increasingly territorialized, and therefore more resilient and at the same time less and less associated with the production of negative environmental impacts.

The study team offer a deep analysis of the socio-economic situation, showing all the incongruence, inequalities, environmental problems and the general unsustainability of the actual economic system. For they consider necessary a new theory and practices of smart cities and economic development. They suggest a new framework based on a Biourbanism epistemology, adopting the systemic principles of life and suggesting shifting from techno-city to human city (homological smart city) to be contrasted with the diffused idea of the smart city. Their framework offers a clear vision of what a smart city should be, and a deep concept of sustainability founded on life value. Only this epistemological revolution can permit to our society the regeneration of the socio-economic body, founded on citizens active participation and communities' involvement, respect for the nature and life. Only if we shift the paradigm, we may assure a better future, able to reinforce social inclusion, support an inclusive economy and protect the physical environment.

18.11.13 Nairobi Study

Transport and ICT connectivity will continue to play a critical role and function in metropolitan regions development, but it is not a sole panacea to unlock all their economic and growth potential. Sustainable metropolitan development will require wide consultations with all relevant stakeholders in a bid to build consensus and commitments, which should take cognizance of the complex multi-level governance associated with such regions. This should be embedded on such metro region's own competitiveness and innovations which in most cases is highly boosted by their location as knowledge and economic hubs, which is the main driver in the faster uptake of innovative ideas.

18.11.14 Abuja Study

Abuja is both the administrative and political headquarters of Nigeria. The city has an original master-plan to guide its growth, but unfortunately, the master-plan is yet to be fully implemented. Abuja being the administrative head-quarter, and centrally located too, it has been able to attract large numbers of people into the city. Abuja is a metropolitan city associated with the usual urban problems known to major cities of the world. These problems include: continuous Influx of People, in search of employment, traffic congestion, increased level of insecurity, poor and inadequate accommodation, high standard of living, poor electricity and water supplies. Many these problems were however assessed and found to be dominant especially in every part of the city. In traffic, it is a predominant problem in every part of the metropolis especially the city centre areas because of the high activities that take place during the working days. The problem of urban sprawl whereby many unattractive suburbs keeps springing up due to poor management of the Abuja master-plan is a common problem in Abuja region. Urban sprawl has developed in Abuja because of government's unyielding to the provision of adequate housing for the large population that has moved to the city over time.

Another major problem facing Abuja city is how to control and cater for the continuous influx of migrants from every part of the country. This is evident in the increase in its population of about 2.8% per annum with a 5.5% urban growth per annum. One worrisome trend of the migration is that most people fleeing from many areas of conflict end up resettling in the metropolitan Abuja city simply because of the perceived secured environment. This can be deduced from persons of North-Eastern part of Nigeria that fled from the Boko Haram crisis and social unrests in the North, notably Bauchi, Yobe and Borno states among others, prefer to settle in Abuja due to better security apparatus that has ensured the overwhelming reduction in crime and its derivative. Although, most of the migrants are in the periphery and this is because of the lack of accommodation and/or high-cost of living in the city Centre. Also, job-seekers in search of the non-existent government jobs end up in Abuja and as they come, they also join in swelling the population of the city. The resultant effect of this uncontrollable influx of migrants could threaten the security of the residents. It has also mounted a serious pressure on public facilities in Abuja is noticeable in the following areas including transportation, housing, health, education security, and environmental challenges.

The uncontrollable migrant inflow has also led to the problem of housing in Abuja. This is because the available houses cannot meet the unending demand of houses. This has resulted in high accommodation charges especially at the city centre. The implication of this ugly trend is the spread of unattractive suburbs in the region. The problem of water supply is not left out. Water supply to some parts of the Abuja metropolis has always been epileptic in the recent decades. Sometimes, water supply to different parts of the metropolis may be disrupted for three days to enable technicians to carry out repairs on a damaged water trunk main line around Gishiri in the Mabushi District of Abuja. Usually, when this happens, residents of Wuse I, Wuse II, Garki I, Garki II, Maitama, Asokoro, Wuye, Gudu, Games Village, Karu and Nyanya would experience water supply shortages. Transportation is an essential and indispensable catalyst for activating and stimulating the pace of economic, social, political and every other human endeavour in any society. Transportation, as one of the basic infrastructures, is required for the effective and efficient functioning of urban centres. Although, despite all the efforts of providing basic transportation infrastructure, it is still inadequate because it cannot meet the growing population and Abuja depends largely on road transportation only. This therefore suggests that there is need to introduce another mode that is very effective in mass transit to ease the incessant transport problems such heavy traffic congestion, the high cost of transport, commuters being stranded at the bus stop thereby spending longer hours before getting to their various destinations. Abuja is very notorious for heavy traffic congestion which is second to Lagos in this regard.

For Abuja to really achieve the status of a smart city these problems discussed above needs to be vigorously tackled. The study suggests that one of the ways for Abuja to achieve an all-inclusive city, it must adopt the smart city concept where a smart city is viewed as a sustainable, inclusive and prosperous city that promotes a people-centric approach based on three core components—Smart City Foundation, Information and Communications Technology (ICT) and Smart Institutions and Laws. A smart city foundation is composed of three elements: urban planning and design, land policies and basic infrastructure, all of which integrate ICT into their developmental and operational architecture. The study further suggests that the city must adopt the smart city principles in the administration of the city; full and immediate integration of ICT in the city's urban planning strategies; commitment to a resilient, liveable, sustainable city underpinned by physical infrastructure that will drive a low-carbon economy; promotion of an inclusive, employment intensive, resilient and competitive economy that harnesses the potential of inhabitants; commitment to improved quality of life and development-driven resilience for all, irrespective of age, gender, religion and physical cum psychological challenges; good urban governance, a very critical element to sustainable development hence its holistic application is quite germane; build on the legacy of the Millennium Development Goals (MDGs) by providing the political will to successfully implement the United Nations Social Development Goals Compact, Post-2015 Development Agenda, and the Addis Ababa Action Agenda: 2015 (United Nations Social Development Goals (2015), United Nations Post-2015 Development Agenda (2015), UN-Habitat 2015) Addis Ababa; promotion of mobility through provision of adequate space for streets and an efficient street network that facilitates walking and cycling; co-location of residential, commercial and recreation land uses as mixed land use; provision of diverse qualitative housing units that are affordable to the urban poor and non-poor and link social and economic development with environmental protection and enhancement by making the most efficient trade-offs and appropriate choices, so that, the region's environmental assets are maintained at a level that meets the need of the present generation without jeopardizing the interests of future generations.

18.11.15 Dakar Study

From a population of about 214,000 in 1950, the Dakar Metropolitan Region has a population of 3.5 million in 2016. In 65 years, the population of Dakar has been multiplied by more than 150 times particularly during post-independence. During the colonial period, Dakar attracted massive migrations from rural areas, people coming to seek jobs in the newly built city centre. This flux had amplified after Senegal became independent in 1960 with Dakar becoming the capital giving it an additional political function. In 1985, Dakar urban population reached the one million marks for the first time making it a large city that required more sophisticated infrastructures in terms of spatial planning, transport planning, etc. It took

35 years for Dakar to reach the population of one million inhabitants in 1985 from a population of 214,000 inhabitants in 1950. But it took only 15 years from 1985 for another million inhabitants to be added to the Dakar population in 2000, and only 10 years for another million inhabitants to be added to the Dakar population in 2010. With this exponential population growth rate, the Dakar urban population will reach 5.6 million in 2030, and 7 million in 2040.

Development of large metropolitan regions like Dakar constitutes economic and social opportunities as well as challenges depending on how the urban growth was planned and managed. It provides opportunities for economies of scale and agglomeration as well as for diffusion of ideas and innovations, but it will also call for large investments in infrastructures to respond to the increased demand for water, sanitation, solid management, energy and mobility, etc. It will particularly come with increased demand for mobility that must be satisfied with an efficient public transit accompanied with increased spaces for pedestrians and cyclists to safeguard the environment while creating economic growth. The metropolitan of Dakar yet occupies a pivotal place in the national trade sector both nationally and internationally. Its autonomous ports (for coastal cities), international airports, international trade centres, touristic sites and commercial centres are assets economic growth. Most economic units in Dakar are in the commercial sector (52%). The metropolitan of Dakar contributes up to 55% of the national Gross Domestic Product (GDP). Up to nine out of ten national civil servants, and from 50 to 90% employees in national's trade, transport, banking and industrial enterprises are in cities. However, most these urban advantages are concentrated in the city of Dakar. Though the Dakar Metropolitan Region had been administratively divided into four departments, the other three departments act as suburbs of the department of Dakar, where the main commercial and administrative businesses are concentrated. The other three departments remain haunted by the early spatial and social division; they are not well planned, and they lack sufficient land allocated to streets, other public spaces, basic infrastructure and security of tenure.

However, Dakar has not been able to satisfy the employment demand, particularly from young people. In addition to that, due to a deficient urban mobility, Dakar has also not been able to create conditions for specialization. While urban agglomeration allows for job specialization, efficient market transactions and knowledge diffusion, if concentrated growth is not well planned—such as the integration of urban growth with efficient public transit investments, the resulting economic benefits tend to erode. This is the case in Dakar as in many African cities with poor connectivity leading to disconnected settlements.

Recognizing the unbalanced Dakar metropolitan regional development, in one hand, and the multiple problems associated to lack of basic services, the Senegalese government has taken bold actions to transform the urban landscape of the metropolitan region from a monocentric to a polycentric metropolitan region with the creation of urban centres to decongest the Dakar urban agglomeration. This new spatial re-organization is also accompanied by economic transformation through the ambitious economic development, the Plan Senegal Emergent. It is through key main complementary programmes, reforms and plans: (1) the Plan Senegal

Emergent; (2) Act III of Decentralization; (3) Metropolitan Planning; (4) Dakar Urban Master Plan 2035 and; (5) Digital Senegal Strategy 2016–2025. These urban policies and programmes aim to spatially and economically transform the Dakar Metropolitan Region to be smart, sustainable, inclusive, resilient and prosperous. They aim to transform the monocentric system to a polycentric system where every urban centre equally contributes to the development of the metropolitan region.

Another urban development is the emergence of urban corridors, which present a type of spatial organization with specific economic and transportation objectives. In Senegal, the government is also encouraging growth, convergence and spatial spread of geographically linked metropolitan areas and other agglomerations. These are emerging trends linking Dakar-Thies-Mbour, turning into spatial units that are territorially and functionally bound by economic, political, socio-cultural, and ecological systems. It expected that the Dakar master includes Thies and Mbour. Though their economic output may be enormous, their management requires considerable investment in infrastructure, particularly to ease mobility and communication to constitute a large economic output, combining large markets, skilled labour and innovation.

Along the transformation of the Dakar metropolitan regional development, national authorities have put in places planning guidelines through the Urban Master Plan of Dakar (Plan Directeur d'Urbanisme—PDU) 2035 and National Plan for Territorial Development "PNAT". The PDU 2035 aims at the urban development of the region of Dakar and its surroundings by 2035. Its main objectives are: Sustainable Urban Development; Compact cities connected with a transport network; Robust and Resilient city and; a Vibrant city with active interaction between information, goods and people. The PNAT proposes five development urban areas: Multifunctional urban areas; Areas for economic activities; Areas for agriculture activities; Areas for touristic activities; and Areas for conservation. The PNAT also identifies areas with high risk for habitation and any other activity. In these areas, modification of the land including by residential structure, or any cadastral operations is prohibited. This plan promotes green areas and other protected areas for environmental purposes or spaces with light recreational facilities, which, when well-integrated enhance the character or the ecological value of the area. Specific focus areas of the PNAT also aim at: controlling the internal urban growth; reducing the proliferation of slums; promoting a balanced urban development; and meeting the housing demand.

It is in the ICT revolution momentum that the government of Senegal has initiated the creation of urban centres on the outskirts of Dakar in order to decongest the city. The government is, indeed, putting in place an ambitious project of a technology park in Urban Pole of Diamniadio, called "Diamniadio Technology Park". The park is based on the Silicon Valley model and intends to promote data revolution and higher education centres. The Urban Pole of Diamniadio is also among the pilot cities of the National Sustainable Cities Initiative (SCI) as part of the Sustainable Cities Programme launched by the Global Environmental Facility (GEF) in 2016. It consists of "Planning and Managing the urban pole to decongest Dakar and to be a sustainable city model connected to Dakar and the rest of the country through walking, non-motorized, ICT means". Under the Sustainable Cities Initiatives, in addition to developing transport infrastructure, the government of Senegal has also introduced the model of multiple connectivity choices (non-motorized, motorized means of mobility and ICT).

18.11.16 Johannesburg Study

Johannesburg metropolitan region is part of the Gauteng Province, The Gauteng Province has developed a Gauteng City Region that covers the Johannesburg metropolitan region and the city of Pretoria which is spatially and economically linked to Johannesburg but is not part of the Johannesburg Metropolitan Region. The city of Pretoria is the administrative capital of South Africa and is part of the Gauteng Province, Urban Policies and Programmes are made under the Gauteng City Region including the metropolitan and the administrative capital.

From a population of 26,303 in 1890 to 102,078 in 1896, Johannesburg located in the Gauteng province, is the largest city in South Africa and is one of the 50 largest urban areas in the world. The population of the Greater Johannesburg Metropolitan Area including suburban regions such as Ekurhuleni, the West Rand, Soweto and Lenasia is over 10 million making it a megacity. Since the end of the Apartheid, Johannesburg has become the main destination for migrants from other South African provinces as well as from other African countries. One third (35%) of Gauteng residents were born in other provinces or in other countries. The Greater Johannesburg Metropolitan Area represents 14% of the national population. It is an engine of South Africa with a share of 18% of the national GDP which itself represents 24% of the African GDP. This is the reason Johannesburg is considered a global African city. Johannesburg also scored the Human Development Index in the country.

Johannesburg was the centre of the apartheid regime in South Africa that lasted from 1948 to 1990 and has been named as being among the unequal cities in the world. The City Foundation of Johannesburg has historically been guided by a segregationist ideology manifested through institutions and laws, urban planning and design, and access to basic services and amenities. During the Apartheid, most township areas designated for non-white people were situated between 25 and 30 km away from the central business district, leading to significant transport challenges. The dismantling of apartheid in the early 1990s came with calls for inclusive development in South Africa. Different models have however been adopted to make the city more inclusive post-apartheid, the key ones being on participatory planning and design; development of inclusive public spaces; improvement in basic infrastructure provision and public transport particularly to the poor neighbourhoods and; development of policies that encourage inclusive human settlement and trade.

Today Johannesburg is in the Gauteng province. The province itself is developing City Region projects that cover: the city of Johannesburg, Ekurhuleni Metro,

West Rand District, Sedibeng District Municipality and the City of Tshwane. Under the cluster of cities of the Gauteng city region, the city of Johannesburg is the Financial Hub, the Ekurhuleni Metro the Manufacturing Hub, the West Rand District the agribusiness/agroprocessing and green and blue economy, Sedibeng District Municipality with the New economy to be anchored on steel, Vaal river, tourism, agriculture linked to Sasolburg, and the City of Tshwane being the Administrative Capital has become a key driver of knowledge, innovation and automotive industry development. The Gauteng city region promotes spatial transport transformation through Metropolitan BRT Systems, Metrorail Upgrade and Gauteng Freight and Logistics Hubs.

In addition to the city-region projects, cities themselves are also developing their specific sustainable urban development projects and programmes. For instance, to ease management of municipalities, the administration of city Johannesburg is now decentralised into 7 regions following the creation of the post-apartheid City of Johannesburg Metropolitan Municipality in 2000. "Each region is operationally responsible for the delivery of health care, housing, sports and recreation, libraries, social development, and local community-based services. Each region will develop its own Local Integrated Development Plans (LIDPs). A LIDP guides a region's future development. While they deal with local issues, they take an integrated approach to issues such as transportation, housing and environmental management".

Despite several urban policies and programmes put in place for an inclusive sustainable urban development in Johannesburg, the metropolitan is still facing enormous challenges such as slow economic growth marked by High Unemployment rate, particularly among young people; Exclusion of poor communities from access to housing and land; Mobility is still a challenge for the urban poor; Urban violence and insecurity; and Climate change with Johannesburg being ranked 13th in the world of Greenhouse Gas (GHG) emitters and largest GHG City in South Africa.

Considering these challenges, the City of Johannesburg has put in place a holistic smart city programme to be implemented at 85% by 2021. This programme encourages innovation and efficiency; preservation of resources for future generation; resource sustainability with, for instance, 30% diversion in waste disposed of by landfill. For the year 2017/2018, the city of Johannesburg has elaborated ten points for urban development including to: (1) ensure that the entire City embraces the environment of a new coalition government; (2) Promote economic development and attract investment towards achieving 5% economic growth; (3) Ensure pro-poor development that addresses inequality and provides meaningful redress; (4) Create a culture of enhanced service delivery; (5) Create a sense of security through improved public safety; (6) Create an honest and transparent City that fights corruption; (7) Create a City that responds to the needs of residents; (8) Enhance financial sustainability; (9) Encourage innovation and efficiency through programmes such Smart City and; (10) Preserve our resources for future generation.

Five pillars are the bedrock on which these ten points will be delivered:

PILLAR 1: Grow the economy and create jobs
PILLAR 2: Enhance quality of life by improving services and taking care of the environment
PILLAR 3: Advance pro-poor development that provides meaningful redress
PILLAR 4: Build caring, safe and secure communities
PILLAR 5: Institute an honest, responsive and productive government.

At the spatial level, the transformation of the city of Johannesburg will consist of:

1. Compact city—combining density, diversity, proximity and accessibility, reducing distances, travel times and costs, bringing jobs and social amenities to single use, marginalised residential areas, reducing energy consumption and infrastructure costs.
2. Inclusive city—ensuring balanced service provision (hard and soft) and opportunities for all by diversifying land uses, promoting social mixing and bridging social, spatial and economic barriers.
3. Connected city—enhancing public transit and ICT infrastructure at provincial and urban scales to re-connect the city, starting from 'the Corridors of Freedom' to the street and neighbourhood-level connectivity.
4. Resilient city—building a metropolitan open space system as a protection buffer, preserving valuable green infrastructure and areas of high agricultural potential, promoting sustainable energy use, reinforcing the urban development boundary and protecting biodiversity resources.
5. Generative city—focusing investment in transformation areas and nodes towards: achieving positive social, economic and environmental returns on investment; spurring economic growth and job creation and enhancing public space and promoting sustainability (social, environmental and economic).

18.11.17 Pittsburgh Study

The mid to late 20th century rust-belt cities in the US witnessed a trend of decline in its manufacturing base with global shift of industrial production to the developing parts of the world and the New International Division of Labour (NIDL). As a result, the rust belt cities were compelled to rethink their urban and economic development strategies to compete in the New Economy of the 21st century. An increasing trend of adopting economic transformation plans was witnessed among the rust-belt cities with the varying trend of success. The common theme of these plans was to replace the manufacturing jobs with advanced service and technology-based jobs. However, there were more challenges than opportunities in the process of doing so, which set the context of urban and regional planning practices in the rust-belt cities.

Pittsburgh, one such example of a rust-belt city, was once known as the "Steel City" due to its production capacity of raw steel in the world economy. The abundance of natural resources, such as coal, timber, iron, and limestone, and navigable waterways—the Allegheny and Monongahela rivers, helped Pittsburgh emerge as the centre of steel industries in the US in between the late 19th- early 20th century. With such comparative advantage of natural resources and navigable waterways, the city burgeoned with large-scale steel mills since the 1870s that produced one-third to half of the total steel production in the US by 1970s, and grew additionally with an economic and population base to support steel-production activities. Nevertheless, the city's steel production declined with an international shift of steel-manufacturing activities to India and China in the 1980s, and led to several consequences, such as economic and population decline, massive brownfield sites, and vacant and abandoned properties. The planners and policy makers had no choice but to focus on adopting newer sets of strategies that could transform Pittsburgh from a shrinking city and make it competitive. The late 20th century Pittsburgh increasingly became successful in adopting many bold policies and strategies that transformed its local economic base towards advanced service sectors: healthcare, higher education, technology, research and development, banking and finance.

Pittsburgh is a tail of two eras in the history of American urban planning. In 1950, Pittsburgh's polluted air, riverine environment, sprawling working class, and tangled maze of streets and bridges had singled it out as one of the most blighted and reviled cities in Americas. A private-political alliance arose in the early 1950s that gave birth to one of the earliest efforts at what we have come to know as the era of urban renewal. The early years of the Pittsburgh's Urban Renewal Authority were years of clearance and demolition. Pittsburgh's "Renaissance I" began with the Point Park and Gateway Centre project. Point Park at the confluence of the Mon and Allegheny rivers was an area of blight and flooding. With the financing of the Equitable Life Assurance society, a thirty-six-acre park and a twenty-three-acre redevelopment site became known as the Gateway Centre. Between 1950 and the 1960s seven high rise office buildings, a Hilton Hotel, a residential apartment building, and an underground garage filled the once blighted space. In the Eisenhower era of demolition and highways, the URA set its sights on the Lower Hill district, a predominantly African-American neighbourhood. The development of a convention centre and arena ultimately displaced 1300 buildings, 413 businesses, and 8000 people from the Hill. The population of the Hill fell from 17,334 in 1950 to 2459 in 1990. The loss of a stable black neighbourhood and the concentration of emigres in larger black neighbourhoods contributed to making Pittsburgh one of the most segregated cities in the country.

Failure of urban renewable efforts and continuing trend of manufacturing decline deeply impacted Pittsburgh through the 1980s, 90s and 2000s. The region sharply lost more than 133,000 manufacturing jobs within only 8 years, in between 1979 and 1987. This led to an era when communities after communities experienced financial difficulties and struggled to bounce back in the New Economy. In between 1987 and 2004, 5 municipalities in the Pittsburgh metropolitan region including the City of Pittsburgh were identified and listed as distressed communities experiencing "severe" financial difficulties under Act 47 of 1987, the Municipalities Financial

Recovery Act of Pennsylvania; placing them under State assistance for recovery strategies.

The 1990s and 2000s also brought a change in the planning and a change the URA's approach to renewal. The early efforts at renewal still left large areas of abandoned factories and decaying neighbourhoods. Although federal level programs provided funding and strategy-framework to clean up contaminated sites, the stigma associated with real or perceived levels of contamination often posed difficulty in successful redevelopment of brownfield sites. But in planning's new paradigm, brownfields are also treated as an "opportunity". Local governments today to restore the vitality of urban life have turned to a notion of redevelopment that includes infill, mixed use, a private-public partnership, walkability; a framework of smart growth. Thus, at the turn of the 21st century, Pittsburgh witnessed a significant change in its planning approaches, shifting towards a bottom-up approach, and proliferation of neighbourhood level community and non-profit organizations. This was coupled up with remarkable progress with its economic restructuring process, heavily relying on the growth of hospitals and healthcare services, high-technology industries, canters of research and higher education. Interestingly, most of these sectors are non-profit sectors and are property tax-exempt, contributing to financial challenges for the local governments.

The 2010s planning approach brought further changes by aligning smart growth ideas with smart city practices. While the notion of Smart City is a relatively new paradigm in planning, the concept of Smart Growth has remained in the planning arena since the late 1990s. Essentially, smart growth is based on mixing land uses, using land and infrastructure efficiently, creating walkable neighbourhoods that are attractive and distinctive, providing transportation and housing choices, and encouraging community and stakeholder collaboration in development decisions. While the concepts of Smart Growth do not emphasize on the dependency of ICT systems, the essential softer domains of Smart City concepts, based on environment, equity and economy, are visibly present in these aspects. Recently, smart city approach has gained much popularity in the US, and Pittsburgh is fast emerging as one of the leading cities in the US in this area.

We specifically illustrate Pittsburgh's strategies of Smart City through many planning initiatives at a local and regional scale, where efforts to (1) plan for economic resiliency and redevelop brownfield sites, and (2) green infrastructure led to creation of a mix-use neighbourhoods, green spaces and diverse economic opportunities that overall contributes to improved quality-of-life of the residents.

Pittsburgh's economic resiliency plans since the 1980s, strongly focused on clean-up and redevelopment of old brownfield sites. Successful redevelopment of brownfield sites, which were idled and abandoned with closures of steel mills in the region, is considered one of the key factors of post-industrial economic restructuring in Pittsburgh. Most of these cases of brownfield sites happened with public-private partnerships among URA, Port Authority, and other public agencies and private developers. Case studies of Brownfield redevelopment sites demonstrate the use of financial tools such as Tax Increment Financing (TIF), and others. The public investments are largely spent on environmental clean-up (tar pits, waste oil, and ferrous cyanide), site remediation, and open space creation, mainly because of the reluctance of private companies to take up such responsibilities. Many of the successful brownfield site redevelopment is also close to the University of Pittsburgh and Carnegie Mellon University campuses; which enabled collaborative cooperative research efforts between the top ranked national universities and the business community spurring the growth of high-tech jobs in the region. An example of the Eco Innovation district, an economic resiliency and brownfield redevelopment plan that intends to revitalize Uptown Pittsburgh and transform it into an innovative urban ecosystem where it will enhance equitable land use, attract and guide new investment, expand the local economy, reduce the city's environmental footprint, and ensure equity and access to local opportunities. This plan is being developed by a series of local and regional stakeholders. This is with collaboration among several agencies including the Uptown Partners of Pittsburgh, Oakland Planning and Development Corporation, City of Pittsburgh, Sustainable Pittsburgh, Urban Redevelopment Authority of Pittsburgh, Port Authority of Allegheny County, and Allegheny County Economic Development, neighbourhood residents and groups, universities, and other partners. Such strategies of transforming large-scale brownfield sites into employment generating areas, specifically high-tech and innovation oriented jobs, have helped change Pittsburgh economic base from steel-manufacturing to higher education, medical research, banking, finance and high-tech sectors.

Like many rust-belt cities, Pittsburgh has a considerably large number of vacant, distressed, or undeveloped properties. In a post-industrial context, urban greening has the potential for returning surplus and derelict lands to productive uses, reduce surplus lands, and stabilize real estate markets. These lots are a legacy of Pittsburgh's economic shift which resulted in a large amount of vacant and distressed lots which create several issues such as a decreased tax-base, public health issues, social isolation, environmental hazards and overall a lower quality of life for the surrounding neighbourhood. 12% of all properties in Pittsburgh were vacant in 2000, 36% of which were abandoned or blighted. These distressed sites include parcels that are currently vacant, condemned, or tax-delinquent. These properties that have become public responsibility places an enormous burden on the City resources and do not contribute taxes to pay for public services. In 2015, it was estimated that there were over 28,000 vacant lots could have cost the city about $20 million to maintain, of which the city-owns about 19% by area. Community gardens, landscaped spaces and other strategies are practiced at the local levels to enhance the quality of life and foster socio-economic and environmental sustainability, which are an integral part of smart city approach. An Adopt-A-Lot program is being implemented since last year which aims to convert these lots from blight to asset. Adopt-A-Lot is a low-cost smart city initiative to fight blight, since maintaining each one of those vacant lots would cost the city almost $600/year, amounting to over $5.6 million for maintenance of only the city-owned lots.

Pittsburgh's storm water and sewage infrastructure built before the 1940s relies on a Combined Sewer System (both sewage and storm water is designed to be carried in the same pipes), so when storm water exceeds the capacity of the pipes, untreated sewage mixed with storm water overwhelms the system and overflows at several points before it reaches treatment plant. This results in increased flooding, water quality degradation, stream erosion, reduced groundwater recharge, and loss of aquatic life. Therefore, there is a need for more resilient infrastructure to handle the environmental problems due to untreated sewage and storm water. Older infrastructure with combined sewer and storm water system is not only inadequate to handle current needs, but even more ill-equipped to meet future demand due to changes in the climate, land use, and rainfall patterns. Green infrastructure (GI) techniques such as rain gardens, riparian buffers, and porous pavements have been proven to be effective for reducing the volume, the rate, as well as pollutants of storm water runoff. It is interesting to note that Pittsburgh has moved to implementing smarter, cost-effective, GI techniques to reduce the overflow of sewage into its rivers during wet weather events, moving away from the traditional, strictly gray infrastructure approach. This is expected to keep one billion gallons out of the combined sewer system annually, dealing with approximately 10% of the storm water problem in Allegheny County under current conditions. Other benefits of GI include cost-effective public realm investment, re-establish riverfront connections; complete streets design approach fostering healthy, walkable communities and; creation of resilient infrastructure. But as the climate changes, larger storms and increased rainfall are expected in the area, there is a need for a more dynamic plan with resilient infrastructure.

Pittsburgh has demonstrated a commitment to renewable energy and wants to use cleaner sources of energy because it is vital to creating a more sustainable city. Each year the City of Pittsburgh purchases 25% of its energy from renewable sources, which is enough to power 3500 homes per year. Renewable energy is generated from resources that are indefinitely replenished naturally: sunlight, wind, water, hydrogen, biomass, and geothermal heat. In 2007, Pittsburgh signed the U.S. Mayors Climate Protection Agreement, committing to implementing local climate protection solutions which will result in reduced taxpayer dollars and energy use. Pittsburgh's first greenhouse gas inventory was undertaken to measures the amount of GHG emitted from various sources, which is useful to target actions having the most impact. Pittsburgh is already experiencing climate change effects with colder winters, and Pennsylvania can expect longer and hotter summers, decreased winter snowpack and increased rainfall. Pittsburgh will face local climate threats such as increased severe weather events and flooding, higher prices and a shortage of basic goods, increased rate of illnesses and other heat-related health problems. Currently, Pittsburgh is in the process of developing its third climate action plan (PCAP 3.0) to create policies and projects to reduce greenhouse gas emissions within city limits to reduce the severity of regional impacts and move towards a low carbon economy. Further, the city of Pittsburgh, affiliated agencies, and the Green Building Alliance have been working on their Green Garage Initiative to retrofit lighting to LED technology in city owned buildings. After retrofitting the city garages buildings in January, city garage buildings saw a reduction of approx. 60% cut in energy consumption and costs. Also, the city and partners are working towards developing a lighting infrastructure fund for other municipal and privately-owned garages.

In summary, several challenges can be outlined in the Pittsburgh's efforts of economic transformation and brownfield redevelopment. One of them is related to the large-scale brownfield sites that still dominate the urban landscape of the region, and requires an enormous amount of financial resources for economic transformation. A city's tax base can never to sufficient to address these issues in a short span of time, and Pittsburgh needs to continuously re-invent and re-innovate its strategies for brownfield redevelopment at the metropolitan scale. Most of the success stories are concentrated within the Centre City that is near the Universities, Research Centres, and Interstates; while the locations in the broader region farther away from the centre-city are still struggling to comeback. Municipalities, such as Braddock, which are still struggling with brownfield issues along with aging population and infrastructure, are adopting more right-sizing strategies and planning for shrinkage by converting abandoned properties into open and green spaces that can reduce the cost of maintenance and create a better living environment. However, this will remain a challenge for these communities to attract jobs and human capital in the short-term.

References

1. Vinod Kumar TM, Associates (eds) (2015) Geographic information systems for smart cities. Copal Publishing Group, New Delhi
2. Vinod Kumar TM (ed) (2016) E governance for smart cities. Springer, Singapore
3. Vinod Kumar TM (ed) (2017) "Smart economy in smart cities" international collaborative research: Ottawa, St Louis, Stuttgart, Bologna, Cape Town, Nairobi, Dakar, Lagos, New Delhi, Varanasi, Vijayawada, Kozhikode, Hong Kong. Springer, Singapore
4. Vinod Kumar TM (ed) (2018) E-democracy for smart cities. Springer, Singapore
5. Bhatt J, Jani O (2015) E-governance for photo voltaic powergrid: solar city Gandhinagar, Gujarat, India. In: Kumar TMV (ed) E-governance of smart cities. Springer, pp 177–230
6. Bhatt J, Jani O (2017) Smart grid: energy backbone of smart city and e-democracy. In: Kumar TMV (ed) E-democracy for smart cities. Springer, pp 319–366
7. Bhatt J, Verma H (2015) Design and development of wired building automation systems. Energy Build. https://doi.org/10.1016/j.enbuild.2015.02.054
8. Bhatt J, Verma H (2010) RS-485/MODBUS based intelligent building automation system using LabVIEW. In: 4th international conference on computer applications in electrical engineering-recent advances (CERA-09). IIT Roorkee, Roorkee, India, p 5
9. Government of India (1991) District census handbook, Jaipur. Rajasthan: Directorate of Census Operations
10. Department of Tourism (2016) Rajasthan-Pragati Prativedan 2015–2016. Government of Rajasthan, Jaipur
11. Jaipur Development Authority (2011) Master development Plan-2025 Jaipur region, vol 1. Government of Rajasthan, Jaipur
12. Census of India (2011) District census handbook, Jaipur. Rajasthan: Directorate of Census Operations
13. Fox S, Dyson T (2015). Is population growth good or bad for economic development? IGC Blog. Available at: https://www.theigc.org/blog/is-population-growth-good-or-bad-for-economic-development/. Accessed 28 Sept. 2017

14. Government of India (2016) Cities profile of 20 smart cities. Available at http://smartcities.gov.in/content/innerpage/cities-profile-of-20-smart-cities.php. Accessed 25 Sept 2017
15. Census of India. Census of India—Census Terms. Available at http://censusindia.gov.in/2011-prov-results/paper2/data_files/India2/1.%20Data%20Highlight.pdf. Accessed 20 July 2017
16. Kulshrestha SK (2006) Dictionary of urban and regional planning. Kalpaz Publications, Delhi
17. Sivaramakrishnan KC, Kundu A, Singh BN (2005) A handbook of urbanization in India. Oxford University Press
18. Knowledgehut (2017) Setting S.M.A.R.T. Goals and objectives for projects successful outcomes available at https://www.knowledgehut.com/blog/project-management/setting-smart-goals-objectives-projects-successful-outcomes. Accessed 15 July 2017

Printed by Printforce, the Netherlands